W9-BKL-942

Biological Effects of Electromagnetic Radiation

OTHER IEEE PRESS BOOKS

Engineering Contributions to Biophysical Electrocardiography, *Edited by Theo C. Pilkington and Robert Plonsey*
The World of Large Scale Systems, *Edited by James D. Palmer and Richard Saeks*
Electronic Switching: Digital Central Office Systems of the World, *Edited by Amos E. Joel, Jr.*
A Guide for Writing Better Technical Papers, *Edited by C. Harkins and D. L. Plung*
Low-Noise Microwave Transistors and Amplifiers, *Edited by H. Fukui*
Digital MOS Integrated Circuits, *Edited by M. I. Elmasry*
Geometric Theory of Diffraction, *Edited by R. C. Hansen*
Modern Active Filter Design, *Edited by R. Schaumann, M. A. Soderstrand, and K. R. Laker*
Adjustable Speed AC Drive Systems, *Edited by B. K. Bose*
Optical Fiber Technology, II, *Edited by C. K. Kao*
Protective Relaying for Power Systems, *Edited by S. H. Horowitz*
Analog MOS Integrated Circuits, *Edited by P. R. Gray, D. A. Hodges, and R. W. Brodersen*
Interference Analysis of Communication Systems, *Edited by P. Stavroulakis*
Integrated Injection Logic, *Edited by J. E. Smith*
Sensory Aids for the Hearing Impaired, *Edited by H. Levitt, J. M. Pickett, and R. A. Houde*
Data Conversion Integrated Circuits, *Edited by Daniel J. Dooley*
Semiconductor Injection Lasers, *Edited by J. K. Butler*
Satellite Communications, *Edited by H. L. Van Trees*
Frequency-Response Methods in Control Systems, *Edited by A.G.J. MacFarlane*
Programs for Digital Signal Processing, *Edited by the Digital Signal Processing Committee*
Automatic Speech & Speaker Recognition, *Edited by N. R. Dixon and T. B. Martin*
Speech Analysis, *Edited by R. W. Schafer and J. D. Markel*
The Engineer in Transition to Management, *I. Gray*
Multidimensional Systems: Theory & Applications, *Edited by N. K. Bose*
Analog Integrated Circuits, *Edited by A. B. Grebene*
Integrated-Circuit Operational Amplifiers, *Edited by R. G. Meyer*
Modern Spectrum Analysis, *Edited by D. G. Childers*
Digital Image Processing for Remote Sensing, *Edited by R. Bernstein*
Reflector Antennas, *Edited by A. W. Love*
Phase-Locked Loops & Their Application, *Edited by W. C. Lindsey and M. K. Simon*
Digital Signal Computers and Processors, *Edited by A. C. Salazar*
Systems Engineering: Methodology and Applications, *Edited by A. P. Sage*
Modern Crystal and Mechanical Filters, *Edited by D. F. Sheahan and R. A. Johnson*
Electrical Noise: Fundamentals and Sources, *Edited by M. S. Gupta*
Computer Methods in Image Analysis, *Edited by J. K. Aggarwal, R. O. Duda, and A. Rosenfeld*
Microprocessors: Fundamentals and Applications, *Edited by W. C. Lin*
Machine Recognition of Patterns, *Edited by A. K. Agrawala*
Turning Points in American Electrical History, *Edited by J. E. Brittain*
Charge-Coupled Devices: Technology and Applications, *Edited by R. Melen and D. Buss*
Spread Spectrum Techniques, *Edited by R. C. Dixon*
Electronic Switching: Central Office Systems of the World, *Edited by A. E. Joel, Jr.*
Electromagnetic Horn Antennas, *Edited by A. W. Love*
Waveform Quantization and Coding, *Edited by N. S. Jayant*
Communication Satellite Systems: An Overview of the Technology, *Edited by R. G. Gould and Y. F. Lum*
Literature Survey of Communication Satellite Systems and Technology, *Edited by J. H. W. Unger*
Solar Cells, *Edited by C. E. Backus*
Computer Networking, *Edited by R. P. Blanc and I. W. Cotton*
Communications Channels: Characterization and Behavior, *Edited by B. Goldberg*
Large-Scale Networks: Theory and Design, *Edited by F. T. Boesch*
Optical Fiber Technology, *Edited by D. Gloge*
Selected Papers in Digital Signal Processing, II, *Edited by the Digital Signal Processing Committee, IEEE*
A Guide for Better Technical Presentations, *Edited by R. M. Woelfle*
Career Management: A Guide to Combating Obsolescence, *Edited by H. G. Kaufman*
Energy and Man: Technical and Social Aspects of Energy, *Edited by M. G. Morgan*
Magnetic Bubble Technology: Integrated-Circuit Magnetics for Digital Storage and Processing, *Edited by H. Chang*
Frequency Synthesis: Techniques and Applications, *Edited by J. Gorski-Popiel*

Biological Effects of Electromagnetic Radiation

Edited by

John M. Osepchuk

Consulting Scientist
Raytheon Research Division

Associate Editors

James W. Frazer, Ross University Medical School, Dominica, West Indies
Om P. Gandhi, Departments of Electrical Engineering and Bioengineering, University of Utah
Arthur W. Guy, School of Medicine, University of Washington
Don R. Justesen, Veterans Administration Medical Center, Kansas City
Sol M. Michaelson, School of Medicine and Dentistry, University of Rochester
John C. Mitchell, USAF School of Aerospace Medicine, Brooks Air Force Base

A volume in the IEEE PRESS Selected Reprint Series, prepared under the sponsorship of the IEEE Committee on Man and Radiation.

The Institute of Electrical and Electronics Engineers, Inc., New York

PRINTED IN THE UNITED STATES OF AMERICA

Sole Worldwide Distributor (Exclusive of the IEEE)

JOHN WILEY & SONS, INC.
605 Third Ave.
New York, NY 10158

Wiley Order Number: 471-88787-0
IEEE Order Number: PC01594

Library of Congress Cataloging in Publication Data
Main entry under title:

Biological effects of electromagnetic radiation.

(IEEE Press selected reprint series)
Includes bibliographies and indexes.
1. Electromagnetism—Physiological effect—Addresses, essays, lectures. I. Osepchuk, John M.
QP82.2.E43B554 1983 599'.019151 82-23380
ISBN 0-87942-165-7

Contents

Foreword

This reprint volume first gleamed in the eye of Mark Grove, founding Chairman of the IEEE's Committee on Man and Radiation, in 1972. The aim of publishing a compendium of benchmark papers on the biological response to electromagnetic radiation at shortwave and microwave frequencies was approached with two objectives in mind: to provide the reader easy access to works that form the roots of Hertzian radiobiology, and to pay homage thereby to the engineers, physical scientists, and biologists whose creative works have shaped the investigative enterprise. The decade that elapsed between inspiration and fruition of the book were occasioned by several disappointments and delays. Reprints of some early works were difficult to obtain. The initial log of papers nominated for reprinting was large enough to fill several compendia. Also, with the passage of time, additional nominations were made. The need to reduce the page count to an acceptable number forced exclusion of many excellent papers, which posed a dilemma of choice often resolved by selection of the shorter of two works of equal merit.

The dilemma was resolved in 1979 by the strategem of appointing chapter editors and charging them with writing concise reviews in their areas of expertness in which benchmark papers could be cited, whether reprinted or not. Although this strategem gives less than their due to many deserving authors, it does provide them recognition and it does provide the reader indirect address and access to their works.

Some papers of high merit doubtless have not been included or cited in this collection; the later the year of the original paper's publication, the more likely its absence, which reflects the intent to illuminate the roots of the subject matter. I hope in years to come that the Committee on Man and Radiation will address, in subsequent reprint volumes, the trunk and foliage of this fascinating tree of technical and scientific knowledge. At such time as one of these books may appear, it should contain English translations of the original works of Western and Eastern European scientists, many of whom, like D'Arsonval of France, Schwan of Germany, and Presman of the Soviet Union, actually sowed the seeds of Hertzian radiobilogy. Their neglect in the present volume is a sorrowful testament to the semipermeable barrier of language: Most of those who superintended and most of those who will read this volume are not facile in French, German, and Russian. My European colleagues who read the material that follows—and *they* are facile readers of English—must accept an apology born of Yankee illiteracy, not Yankee hubris, for the absence of many important papers.

Many individuals have contributed to this volume. I thank the authors of chapters for their hard labors unrecompensed save for love of the subject matter. Peter Barber, Joan Breslin, Reed Crone, Allen Ecker, Richard Emberson, Om Gandhi, Edward Hunt, Richard Phillips, and Leo Young helped much in divers ways, as did past and present members of the Committee on Man and Radiation, from whose ranks chapter authors were recruited. I give especial thanks to John Osepchuk, the Editor, whose patience over the years was matched only by his dogged insistence on progress as he skillfully moved Mark Grove's brainchild from conception to completion.

Don. R. Justesen
Chairman, 1979, 1980
The Committee on Man and Radiation
Kansas City, MO
February 21, 1982

Preface

This book is the culmination of efforts begun in 1973 by the Committee on Man and Radiation (COMAR) to generate a collection of reprints useful to a broad audience interested in the scientific and technical literature on biological effects of non-ionizing radiation, or more popularly "microwave" and/or RF radiation (or electromagnetic radiation (EMR)). The idea was to present key reprints which would help the reader understand the bottom-line results and also some of the reports which have greatly influenced this field of research and its practical relevance. A collection of reprints cannot give the reader a coherent integrated history and overview of the field. What it can do is permit the reader to review for himself original literature and thereby permit more conviction in his own assessment of what the field is all about. The editors of the various chapters have tried to select "*key*" reprints to help the reader achieve the stated objective. Sometimes papers are chosen for their historical importance and not necessarily for scientific accuracy or validity as presently viewed. Some of the papers are chosen as representative of different schools of thinking, for example, the Eastern European bloc. Other papers are chosen because of their efficient summation of blocs of research.

This is a complicated and in some ways "murky" field with considerable controversy and growing relevance to practical political and regulatory questions. The chapter editors have found it nigh impossible to fulfill all the above objectives in one volume. The collective judgment of the irreducible minimum of *key* reprints amounted to 1,200 pages. After being forced to reduce this by more than a factor of two, the editors and I finally had to make some arbitrary choices and exclusions. These will not satisfy the expected critics. In our defense, we maintain that the final selection represents only *one* and not *the* collection of key reprints that will be useful. Furthermore, we point out that not only reprints are included but also quite extensive and updated lists of key references. Finally, the updated chapter introductions by the various editors help explain the bases and nature of the key reprint selections.

The reprints are presented in seven separate parts with associated references and editorial commentary. The apportionment of space to the various subjects is debatable, but the use of separate parts is of importance to the reader. The name of the game in this field has evolved into this three-task structure:

1) For a given biological body immersed in a given electromagnetic environment, determine the internal field distribution and resultant energy absorption and their most useful measures like "specific absorption rate" (SAR).
2) For a given internal field distribution or average indicators (average SAR) in animals or humans, determine the biological or medical effects.
3) Given the information from (2), evolve safety standards on exposure of people, susceptibility standards for electronic devices to be implanted into people and beneficial medical applications of EMR.

The first part, ably presented by an engineering pioneer of the field, Professor A. W. Guy, reflects the centrality of task (1) which falls within the professional scope of the IEEE—namely developing models for biological bodies, calculating internal distributions, absorption and scattering, and confirming or extending its knowledge by experiment. We have the privilege of including in Part 1 an original paper by E. L. Hunt *et al.* on the key technique of calorimetry which permits "wet-bench" scientists to determine averaged SAR values in experimental animals.

Parts 3 and 4, edited by D. R. Justesen and S. M. Michaelson, respectively, represent a necessarily abbreviated overview of the main bulk of the literature on "microwave bioeffects"—namely the interdisciplinary research on effects in animals and clinical or epidemiologic studies with man. We have devoted a separate part to effects on the central nervous system, as there is the general perception, especially in Eastern Europe, that these are the most sensitive and practically pertinent effects, whether this is right or wrong. Thus, these two parts represent task (2) in our above trilogy for the field.

Parts 5, 6, and 7, edited by O. P. Gandhi, J. C. Mitchell, and myself, respectively, reflect task (3), that is, practical applications to safety standards, whether on exposure, emission, or susceptibility, and medicine in diathermy, hyperthermia, and diagnostic techniques. This task is only now developing so that viewpoints are not necessarily definitive, particularly for medical applications.

Part 2 was left out of the review relative to tasks (1-3). Whereas the latter represent the areas of engineering, phenomenological or applied sciences, and practical arts, Part 2 represents the basic science efforts in this field. In sum, the most common working hypothesis on "mechanisms" of biological effects involves the role of localized energy absorption (that is, heat) in inducing biophysical or biochemical events triggering the chain of complicated events in the living creature. Even here, the true scientific model of biophysical interactions is not complete. The review by J. W. Frazer is an attempt to present some of the leading past and current views on mechanisms of interaction at the molecular and cellular level. This is a very controversial area with some excitment. It remains to be seen whether more accurate scientific models of interaction merely confirm the commonly accepted pattern of bioeffect data or suggest new effects, whether hazardous or beneficial.

Knowledge is never static and it is the privilege of COMAR to contribute to the professional and technical accomplishments of the IEEE by presenting the most up-to-date knowledge in this field in a form useful to not only scientific workers, but also others for whom this subject has practical relevance. As editors, we have tried to be scholarly and fair to the many contributors in this field. No doubt we have offended some, but we rest on our statement of noble intentions.

The reader may well question why a project like this begun in 1973 comes to fruition eight years later. I am afraid that we, COMAR, must confess the weakness of our operation (we are only a part-time voluntary committee). The task apparently is either too time consuming or too awesome. In any case, over the years various chapter editors suffered prodding or replacement; all the while, material needed updating. Personally, I am gratified at the completion of this project and, despite my hints of editorial agony, I am proud of the contributors to this volume and the resultant product.

I would like to express gratitude to Leo Young, Past-President of IEEE, who helped found COMAR and first suggested this project. We must acknowledge the important efforts of past COMAR chairman, Mark Grove; and Allen Ecker, Peter Polson, and E. L. Hunt; as well as the important roles of the present and immediate-past chairmen, Professor Gandhi and Dr. Justesen. Lastly, I must acknowledge the patience of W. R. Crone of the IEEE Press who persevered over the years in gently prodding us to completion.

John M. Osepchuk
Waltham, MA
June 4, 1981

Biological Effects of Electromagnetic Radiation

Part I
Quantitation of Electromagnetic Fields in Biological Systems

The classical problem in studies concerning the biological effects of microwave radiation is the establishment of the relationship between physical characteristics of the fields and the magnitude of the effect. The first step in solving this problem is the quantitation of the relationship between the properties of the exposure fields and the absorbed energy (rate of energy absorption) or fields in the exposed tissues.

The need to solve this problem was not recognized until approximately a half century after d'Arsonval established the use of electromagnetics in medicine in 1890 [1]. As electromagnetics became popular in physical medicine in the form of short wave diathermy, the use of poorly conceived dosimetric techniques led to many improper assumptions and conclusions concerning field effects on tissues. The dose or dose rate available from various generators operating at different frequencies and power outputs during treatment of patients was quantified only in terms of output current to the electrode or coil applicators. It was implied in advertisements that the heating of deeper tissues would be enhanced with greater machine output. Since the tissue heating seemed to vary considerably with frequency, even with the same apparent output of various machines, many researchers jumped to the conclusion that there were selective and specific properties of the various wavelengths investigated. Finally, in 1941, an interdisciplinary research team of engineers and physicians, E. Mittleman, S. L. Osborne, and J. S. Coulter [2], conceived and developed the first true RF dosimetry by quantifying the temperature rise in exposed tissues in terms of the volume-normalized rate of absorbed energy expressed in units of watts per 1000 cc, surprisingly close to the now-accepted mass-normalized specific absorption rate (SAR) expressed in units of watts per kilogram.

These investigators clearly demonstrated that the degree of tissue heating and therapeutic reaction was dependent solely on the rate of energy absorption and not on wavelength. However, subsequent research on clinical use of shortwave diathermy made little use of the techniques advanced by Mittleman *et al.*, and the microwave therapy and bio-effect research in the following two decades also lacked the application of volume or mass-normalized dosimetric concepts. Virtually all of the reported biological effects were related to incident power density so that it was difficult, if not impossible, to correlate data from different animal or human experiments so necessary for establishing human safe-exposure guides, or diathermy treatment doses.

Schwan [3–7] and Cook [8, 9] began to set the foundation in the early 1950's for later analytical work by characterizing the dielectric properties of biological tissues. This allowed microwave fields and their associated patterns of heating in exposed tissues to be analyzed later by Schwan through the use of simple models which consisted of plain layers of simulated muscle, fat, and skin [7, 10–13]. The existence of standing-waves within the tissues due to reflection at transitions of markedly different water content such as the fat-muscle or bone-muscle interface were predicted. The predictions were confirmed in the early 1960's by Lehmann *et al.* [14, 15]. It was also theoretically demonstrated by Saito and Schwan in the early 1960's that field-force effects can be elicited [16, 17]. These effects are the so-called *pearl chain* formations in which small particles of biological material, when suspended in a liquid of differing dielectric constant, are polarized by an electric field, which results in mutual attraction among the particles. Analysis indicated, however, that the field strengths required to produce pearl-chain formations are far in excess of those which would normally denature protein and thereby destroy tissue by excessive elevation of temperature.

During the period of the tri-services supported research in the late 1950's and 1960, preliminary work on experimental approaches to microwave dosimetry was done. Schwan and his colleagues described methods which could be used to fabricate tissue-equivalent phantom models of biological bodies and portions of the human anatomy [18] and to experimentally measure the energy absorption during exposure to microwave radiation [19]. In 1960, Mermagen reported his use of tissue equivalent phantoms to study energy absorption characteristics as a function of animal position in a near field [20]. He recognized the need for better microwave dosimetry and stated,

> "Since our group is interested in the absorption of microwave power rather than the incident flux, it should be more realistic to employ a unit of measurement commensurate with such power absorption. Therefore, if we measure W/cc in an absorber whose dielectric constant is similar to that of tissue and whose volume would represent a finite attenuation of microwave beams again similar to that of tissue, then a comparison might be achieved between experimental results from different investigators' laboratories."

In the early 1960's, the first analyses on biological models that revealed frequency dependencies in the coupling of electromagnetic fields to an intact electrically small body were performed by V. A. Franke in the Soviet Union [21] and discussed by Presman [22]. The models used in the theoretical analyses varied from circular cylinders to prolate spheroids of homogeneous dielectric with the same properties as muscle. Volume-normalized energy absorption rate and heat produced in prolate spheroid models of man was determined at lower

Manuscript received June 15, 1981.

RF frequencies in these studies through the use of quasi-static mathematical solutions. In other theoretical studies, spherical models were used by Schwan's group to determine relative absorptive cross sections as a function of frequency of the incident RF field [23]. It was found that the absorptive cross section—the surface area of the model's electrical "silhouette"—varies widely with frequency, reaching maximal values at the model's resonant frequency. From the mid through late 1960's, more realistic tissue-simulating models were developed and used for experimental measurements of field coupling by Guy and his colleagues [24, 25].

Finally, in the late 1960's, a true mass-normalized RF dosimetry using phantom models was introduced to actual practice in laboratory bio-effects research by Justesen [26]. This allowed the use of multi-mode cavities for exposing freely moving animals so that feeding, signaling, and stimulation for operant and classical conditioning experiments were possible.

During the same time period, distributive dosimetry—measures of the anatomical distribution of focal SAR's in biological models and bodies of small mammals—was developed by Guy through the technique of thermography [25, 27]. Finally, the first in a succession of analytical and empirical studies of the electrical and geometrical constraints on SAR was initiated by Gandhi [28-30]. The orientation of an animal with respect to the vectors of an incident plane wave was found to be a powerful controlling influence on the quantity of energy absorbed in an RF field. The importance of considering environmental factors, such as temperature and humidity, rather than focusing only on dosimetric quantities was pointed out by Mumford [31].

Paralleling the development of dosimetric concepts were the introduction in the early 1970's of twin-cell calorimetry by Hunt and Phillips—the "platinum rod" of whole-body dosimetry—[32, 33] and the development of continuous automatic integration of momentary energy absorption rates via differentiation of transmitted and reflected power-meter values in a special, environmentally controlled waveguide system developed by Ho *et al.* [34]. Ho's system provides an electrical, if indirect, means of determining in real time the whole-body SAR in a living unrestrained animal. Later, pharmacologically-induced ectothermia was introduced by Justesen's group (via steroids) [35] and Gandhi's group (via Na-Pentobarbital) [36]; this permits the whole-body SAR to be calculated directly from increments of body temperature in living animals.

Later, Guy was to deploy circularly polarized, guided waves in specially equipped cylindrical waveguides which enable precise control of SAR's and virtually continuous irradiation of small animals that are fed, watered, and de-excremented with minimal disturbance of the field [37, 38].

Theoretical work and the development of instrumentation for quantifying electromagnetic-field interactions with biological materials increased substantially in the 1970's with many important developments. Interaction of plane-wave sources with layered tissues were studied [39-42]. New and novel instruments were developed for better characterization of exposure fields [43-47]. More sophisticated mathematical and physical models of biological tissues and anatomical structures generated a much better understanding of absorptive characteristics in tissues of differing composition and geometry. Theoretical models of the human head that consisted of a brain and spherical shells for simulation of the skull and scalp led to the observation by Shapiro that electrical *hot spots* (localized regions of intensified energy absorption) could occur deep within the brain at frequencies associated with resonance [48]. Absorption of RF energy at the surface of the head was considerably less due to focusing of the field, which results from the high dielectric constant and the spherical shape of the head. The theoretical results were expanded and verified experimentally to predict energy absorption by animal and human bodies of a wide range of sizes exposed at various frequencies [49-55]. Better and more detailed analyses were developed for various geometric objects which simulate bodies of man and animals, that is, cylinders [56-60], prolate spheroids [61-63], and ellipsoids [64-66]. Theoretical analyses were verified experimentally via thermography [49, 67, 68], newly-developed calorimetric techniques [69-71], and special temperature-sensing probes composed of microwave-transparent materials such as fiberoptics guides [72-75] and high-resistance leads [76, 77]. Cetas has reviewed and discussed the relative merits of these different techniques [78]. Probes were also developed for direct measurement of electric fields within exposed tissues [47, 79, 80]. Paralleling the use of microwave transparent materials for new dosimetry instrumentation was the adoption of these materials for use in recording physiological signals from live laboratory animals under microwave exposure [49, 81-83].

Finite-difference techniques and other numerical studies, in conjunction with high-speed computers, were developed along with sophisticated programs for calculating the EM fields and associated heating patterns in arbitrarily shaped bodies [84-91]. Mathematical models were also developed to include the effects of convective cooling via blood flow in calculating steady-state temperatures for the various parts of the body, including critical organs such as the eyes and brain [92-94]. The continued development and use of phantom models has contributed significantly to our understanding of energy absorption in biological bodies [27, 67, 68, 83, 95-101].

The simultaneous applications of these new developments of the 1970's in various laboratories was the stimulus needed to allow the microwave auditory effect to be quantified and understood as a complex thermal acoustic phenomenon which was quantified for nonbiological materials by the physicists more than a decade earlier [81, 102-105].

The collective advances in analytical solutions and experimental studies, in improved devices for measuring field strengths of plane-wave fields and for determining SAR's in exposed animals, and in non-perturbing thermal and electrical sensors culminated in publication of the first *Dosimetry Handbook* for RF radiation by Curtis Johnson and colleagues, of the University of Utah and the Air Force School of Aerospace Medicine [106, 107]. These collective results have been empirically cast in a succinct form by Durney *et al.* [108].

I have mentioned major analytical, experimental, and instrumental advances that have taken place during three decades of work on the biological response to RF waves. While much work remains to be done, especially in distributive dosimetry of experimental animals and man, a solid base for further advancement has been constructed.

My selections for this volume reflect some of the more important developments; because of limitations on space, I have selected works that for the most part are most informative from both a historical and a technical standpoint to the investigators of the 1980's who, it is hoped, will continue to advance the field.

Arthur W. Guy, Ph.D.
Associate Editor

REFERENCES

[1] S. Licht, Ed., *Therapeutic Heat and Cold*, New Haven, CT: Licht Publisher, 1965.

[2] E. Mittlemann, S. L. Osborne, and J. S. Coulter, "Short wave diathermy power absorption and deep tissue temperature," *Arch. Physical Therapy*, vol. 22, pp. 133–139, 1941.

[3] H. Schwan, "Temperature dependence of the dielectric constant of blood at low frequencies" (in German), *Zeitschrift fur Naturforschung (Tubingen)*, vol. 3B, pp. 361–367, 1948.

[4] H. Schwan, "Electrical properties of blood ultrahigh frequencies," *Amer. J. Physical Med.*, vol. 32, p. 144, 1953.

[5] H. Schwan, "The electrical characteristics of muscle tissue at low frequencies" (in German), *Zeitschrift fur Naturforschung (Tubingen)*, vol. 9B, pp. 245–251, 1954.

[6] H. P. Schwan and K. Li, "Capacity and conductivity of body tissues at ultrahigh frequencies," *Proc. Inst. of Radio Engr.*, vol. 41, pp. 1735–1740, 1957.

[*7] H. P. Schwan and G. M. Piersol, "Absorption of electromagnetic energy in body tissue: review and critical analysis," *Arch. Physical Med.*, vol. 33, p. 34, 1953.

[8] H. F. Cook, "The dielectric behavior of some types of human tissues at microwave frequencies," *Brit. J. Appl. Phys.*, vol. 2, pp. 292–300, 1951.

[9] H. F. Cook, "A comparison of dielectric behavior of pure water and human blood at microwave frequencies," *Brit. J. Appl. Phys.*, vol. 3, p. 249–255, 1952.

[10] H. P. Schwan and K. Li, "The mechanism of absorption of ultrahigh frequency electromagnetic energy in tissues, as related to the problem of tolerance dosage," *IRE Trans. Med. Electron.*, vol. PGME-4, pp. 45–49, 1956.

[11] H. P. Schwan, "Absorption and energy transfer of microwaves and ultrasound tissues: characteristics," in *Med. Physics*, vol. 3, O. Glaser, Ed., Chicago, IL: Year Book Publishers, pp. 1–7, 1960.

[12] H. P. Schwan, chapter 3 in *Therapeutic Heat, Physical Medicine Library*, vol. 2, S. H. Licht, Ed., New Haven, CT: Licht Publisher, pp. 55–115, 1958.

[13] H. Schwan, "The biophysical basis of physical medicine," *J. Amer. Med. Assoc.*, vol. 160, pp. 191–197, 1956.

[14] J. F. Lehmann, A. W. Guy, V. C. Johnson, G. D. Brunner, and J. W. Bell, "Comparison of relative heating patterns produced in tissues by exposure to microwave energy at frequencies of 2456 and 900 megacycles," *Arch. Physical Med.*, vol. 43, pp. 69–76, 1962.

[15] J. F. Lehmann, V. C. Johnson, J. A. McMillan, D. R. Silverman, G. D. Brunner, and L. A. Rathbun, "Comparison of deep heating by microwaves at frequencies 2456 and 900 megacycles," *Arch. Physical Med.*, vol. 46, pp. 307–314, 1965.

[16] M. Saito, H. P. Schwan, and G. Schwarz, "Response of nonspherical biological particles to alternating electric fields," *Biophys. J.*, vol. 6, p. 313, 1966.

[17] M. Saito and H. P. Schwan, "The time constants of pearl chain formation," in *Biological Effects of Microwave Radiation*, vol. 1, New York, NY: Plenum, pp. 85–97, 1961.

[18] H. P. Schwan, "Theoretical considerations pertaining to thermal dose meters," *Proc. Third Annual Tri-Service Conf. on Biological Effects of Microwave Radiating Equipments*, Univ. Calif., Berkeley, CA, pp. 94–106, August 25–27, 1959.

[19] O. M. Salati and H. P. Schwan, "A technique for relative absorption cross-section determination," *Proc. Third Annual Tri-Service Conf. on Biological Effects of Microwave Radiating Equipments*, Univ. Calif., Berkeley, CA, pp. 107–112, August 25–27, 1959.

[20] H. Mermagen, "Phantom experiments with microwaves at the University of Rochester," *Proc. of Fourth Annual Tri-Service Conference on the Biological Effects of Microwave Radiation*, vol. 1, New York University Medical Center, New York, NY, pp. 143–152, August 16–18, 1960.

[21] V. A. Franke, "Calculations of the absorption of energy from an electromagnetic field by means of semi-conductor models resembling the human body" (in Russian), in *Collection of Scientific Papers of the VCSPS Inst. of Industrial Safety*, Leningrad, vol. 3, pp. 36–45, 1961.

[22] A. S. Presman, *Electromagnetic Fields and Life*, New York, NY: Plenum, pp. 49–51, 1970.

[23] A. Anne, M. Saito, O. M. Salati, and H. P. Schwan, "Relative microwave absorption cross sections of biological significance," in *Biological Effects of Microwave Radiation*, vol. 1, New York, NY: Plenum, pp. 153–176, 1960.

[24] A. W. Guy and J. F. Lehmann, "On the determination of an optimum microwave diathermy frequency for a direct contact applicator," *IEEE Trans. Bio-Med. Eng.*, vol. BME-13, pp. 76–87, 1966.

[25] A. W. Guy, J. F. Lehmann, J. A. McDougall, and C. C. Sorensen, "Studies on therapeutic heating by electromagnetic energy," in *Thermal Problems in Biotechnology*, New York, NY: ASME, pp. 26–45, 1968.

[26] D. R. Justesen and N. W. King, "Behavioral effects of low level microwave irradiation in the closed space situation," *Proc. Biological Effects and Health Implications of Microwave Radiation Symp.*, Richmond, VA, (BRH/DBE 70-2) (PB 193 898), Sept. 17–19, 1969.

[*27] A. W. Guy, "Analyses of electromagnetic fields induced in biological tissues by thermographic studies on equivalent phantom models," *IEEE Trans. Microwave Theory Tech.*, vol. MTT-19, pp. 205–214, 1971.

[28] O. P. Gandhi, "Frequency and orientation effect on whole animal absorption of electromagnetic waves," *IEEE Trans. Bio-Med. Eng.*, vol. BME-22, pp. 536–542, 1975.

[29] O. P. Gandhi and M. J. Hagmann, "Some recent results on deposition of electromagnetic energy in animals and models of man," *Proc. of Workshop on the Physical Basis of Electromagnetic Interactions with Biological Systems*, Univ. Maryland, College Park, MD, pp. 243–260, June 15–17, 1977.

[30] O. P. Gandhi, E. L. Hunt, and J. A. D'Andrea, "Deposition of electromagnetic energy in animals and in models of man with and without grounding and reflector effects," *Radio Sci.*, vol. 12, pp. 39–48, 1977.

[31] W. W. Mumford, "Heat stress due to RF radiation," *Proc. IEEE*, vol. 57, pp. 171–178, 1969.

[*32] E. L. Hunt, R. D. Phillips, D. M. Fleming, and R. D. Castro, "Dosimetry for whole-animal microwave irradiation," in this volume.

[33] R. D. Phillips, E. L. Hunt, and N. W. King, "Field measurements, absorbed dose, and biologic dosimetry of microwaves," *Ann. N.Y. Acad. Sci.*, vol. 247, pp. 499–509, 1975.

[34] H. S. Ho, E. I. Ginns, and C. L. Christman, "Environmentally controlled waveguide irradiation facility," *IEEE Trans. Microwave Theory Tech. Symp. Issue*, vol. MTT-21, pp. 837–840, Dec., 1973.

[35] D. L. Putthoff, D. R. Justesen, L. B. Ward, and D. M. Levinson, "Drug-induced ectothermia in small mammals: the quest for a biological microwave dosimeter," *Radio Sci.*, vol. 12, pp. 73–81, 1977.

[36] O. P. Gandhi, M. J. Hagmann, and J. A. D'Andrea, "Part-body and multi-body effects on absorption of radio-frequency elec-

*Reprinted in this volume.

tromagnetic energy by animals and by models of man," *Radio Sci.*, vol. 14, pp. 15-22, 1979.
[37] A. W. Guy and C. K. Chou, "A system for quantitative chronic exposure of a population of rodents to UHF fields," *Biological Effects of Electromagnetic Waves, Selected Papers of USNC/URSI Annual Meeting*, Boulder, CO, pp. 389-410, Oct. 20-23, 1975.
[38] A. W. Guy, J. Wallace, and J. A. McDougall, "Circularly polarized 2450-MHz waveguide system for chronic exposure of small animals to microwaves," *Special Issue, Bio. Effects of Electromagnetic Waves, Radio Sci.*, vol. 14, no. 6S, pp. 63-74, 1979.
[39] A. W. Guy, "Electromagnetic fields and relative heating patterns due to a rectangular aperture source in direct contact with bilayered biological tissue," *IEEE Trans. Microwave Theory Tech.*, vol. MTT-19, pp. 214-233, 1971.
[40] H. S. Ho, A. W. Guy, R. A. Sigelmann, and J. F. Lehmann, "Microwave heating of simulated human limbs by aperture sources," *IEEE Trans. Microwave Theory Tech.*, vol. MTT-19, pp. 224-231, 1971.
[41] H. S. Ho, "Comparison of calculated absorbed dose rate distributions in phantom heads exposed to 2450 MHz and 915 MHz plane wave and slot sources," *Proc. Symp. Biological Effects and Measurement of Radio Frequency/Microwaves*, Rockville, MD, pp. 191-200, Feb. 16-18, 1977.
[42] H. S. Ho, "Contrast of dose distribution in phantom heads due to aperture and plane wave sources," *Ann. N.Y. Acad. Sci.*, vol. 247, pp. 454-472, 1975.
[43] E. Aslan, "Electromagnetic leakage survey meter," *J. Microwave Power*, vol. 6, pp. 169-177, 1971.
[44] R. R. Bowman, "Two isotropic electric-field probes for complicated fields," *Proc. Microwave Power Symp.*, pp. 29-30, May 1972.
[45] S. Hopfer, "The design of broadband resistive (microwave) radiation probes," *IEEE Trans. Instrum. Meas.*, vol. IM-21, no. 4, pp. 416-424, 1972.
[46] H. I. Bassen and W. A. Herman, "Precise calibration of plane-wave microwave power density using power equation techniques," *IEEE Trans. Microwave Theory Tech.*, vol. MTT-25, pp. 701-706, 1977.
[47] H. Bassen, "Internal dosimetry and external microwave field measurements using miniature electric field probes," *Proc. Symp. Biological Effects and Measurement of Radio Frequency/Microwaves*, Rockville, MD, pp. 71-80, Feb. 16-18, 1977.
[*48] A. R. Shapiro, R. F. Lutomirski, and H. T. Yura, "Induced fields and heating within a cranial structure irradiated by an electromagnetic plane wave," *IEEE Trans. Microwave Theory Tech.*, vol. MTT-19, pp. 187-196, 1971.
[*49] C. C. Johnson and A. W. Guy, "Non-ionizing electromagnetic wave effects in biological materials and systems," *Proc. IEEE*, vol. 60, pp. 692-718, 1972.
[50] H. S. Ho and A. W. Guy, "Development of dosimetry for RF and microwave radiation—II—Calculations of absorbed dose distributions in two sizes of muscle-equivalent spheres," *Health Physics*, vol. 29, pp. 317-324, 1975.
[51] H. S. Ho and H. D. Youmans, "Development of dosimetry for RF and microwave radiation—III—Dose rate distribution in tissue spheres due to measured spectra of electromagnetic plane wave," *Health Physics*, vol. 29, pp. 325-329, 1975.
[52] H. N. Kritikos and H. P. Schwan, "The distribution of heating potential inside lossy spheres," *IEEE Trans. Bio-Med. Eng.*, vol. BME-22, pp. 457-463, 1975.
[53] J. C. Lin, A. W. Guy, and C. C. Johnson, "Power deposition in a spherical model of man exposed to 1-20 MHz EM fields," *IEEE Trans. Microwave Theory Tech.*, vol. MTT-21, pp. 791-797, 1973.
[54] C. M. Weil, "Absorption characteristics of multilayered sphere models exposed to UHF/microwave radiation," *IEEE Trans. Bio-Med. Eng.*, vol. BME-22, pp. 468-472, 1975.
[55] W. T. Joines and R. J. Spiegel, "Resonance absorption of microwaves by the human skull," *IEEE Trans. Bio-Med. Eng.*, vol. BME-21, pp. 46-47, 1974.
[56] H. S. Ho, "Energy absorption patterns in circular triple-layered tissue cylinders exposed to plane wave sources [calculated for sources of 433, 750, 918, and 2450 MHz]," *Health Physics*, vol. 31, pp. 97-108, 1976.

*Reprinted in this volume.

[57] H. S. Ho, A. W. Guy, R. A. Sigelmann, and J. F. Lehmann, "Microwave heating of simulated human limbs by aperture sources," *IEEE Trans. Microwave Theory Tech.*, vol. MTT-19, pp. 224-231, 1971.
[58] H. S. Ho, "Dose rate distribution in triple-layered dielectric cylinder with irregular cross section irradiated by plane wave sources," *J. Microwave Power*, vol. 10, pp. 421-432, 1975.
[59] H. Massoudi, C. H. Durney, and C. C. Johnson, "Geometrical-optics and exact solutions for internal fields and SAR's in a cylindrical model of man as irradiated by an electromagnetic plane wave," *1977 International Symp. Biological Effects of Electromagnetic Waves*, Airlie, VA, p. 49, Oct. 30-Nov. 4, 1977.
[60] T. K. Wu and L. L. Tsai, "Electromagnetic fields induced inside arbitrary cylinders of biological tissue," *IEEE Trans. Microwave Theory Tech.*, vol. MTT-25, pp. 61-65, 1977.
[61] C. H. Durney, C. C. Johnson, and H. Massoudi, "Long-wavelength analysis of plane wave irradiation of a prolate spheroid model of man," *IEEE Trans. Microwave Theory Tech.*, vol. MTT-23, pp. 246-253, 1975.
[62] C. C. Johnson, C. H. Durney, and H. Massoudi, "Long-wavelength electromagnetic power absorption in prolate spheroidal models of man and animals," *IEEE Trans. Microwave Theory Tech.*, vol. MTT-23, pp. 739-747, 1975.
[63] P. W. Barber, "Electromagnetic power deposition in prolate spheroid models of man and animals at resonance," *IEEE Trans. Bio-Med. Eng.*, vol. BME-24, pp. 513-521, 1977.
[*64] H. Massoudi, C. H. Durney, and C. C. Johnson, "Long-wavelength analysis of plane wave irradiation of an ellipsoidal model of man," *IEEE Trans. Microwave Theory Tech.*, vol. MTT-25, pp. 41-46, 1977.
[65] H. Massoudi, C. H. Durney, and C. C. Johnson, "Long-wavelength electromagnetic power absorption in ellipsoidal models of man and animals," *IEEE Trans. Microwave Theory Tech.*, vol. MTT-25, pp. 47-52, 1977.
[66] H. Massoudi, C. H. Durney, and C. C. Johnson, "Comparison of the average specific absorption rate in the ellipsoidal conductor and dielectric models of humans and monkeys at radio frequencies," *Radio Sci.*, vol. 12, pp. 65-72, 1977.
[67] A. W. Guy, J. F. Lehmann, and J. B. Stonebridge, "Therapeutic applications of electromagnetic power," *Proc. IEEE*, vol. 62, pp. 55-75, 1974.
[68] A. W. Guy, M. D. Webb, and C. C. Sorensen, "Determination of power absorption in man exposed to high frequency electromagnetic fields by thermographic measurements of scale models," *IEEE Trans. Bio-Med. Eng.*, vol. BME-23, pp. 361-371, 1976.
[69] D. I. McRee, "Determination of energy absorption of microwave radiation using the cooling curve technique," *J. Microwave Power*, vol. 9, pp. 263-270, 1974.
[70] J. W. Allis, C. F. Blackman, M. L. Fromme, and S. G. Benane, "Measurement of microwave radiation absorbed by biological systems, 1, Analysis of heating and cooling data," *Radio Sci.*, vol. 12, pp. 1-8, 1977.
[71] C. F. Blackman and J. A. Black, "Measurement of microwave radiation absorbed by biological systems, 2, Analysis by Dewar-flask calorimetry," *Radio Sci.*, vol. 12, pp. 9-14, 1977.
[72] T. C. Rozzell, C. C. Johnson, C. H. Durney, J. L. Lords, and R. G. Olsen, "A nonperturbing temperature sensor for measurements in electromagnetic fields," *J. Microwave Power*, vol. 9, pp. 241-249, 1974.
[73] C. C. Johnson, O. P. Gandhi, and T. C. Rozzell, "A prototype liquid crystal fiberoptic probe for temperature and power measurements in RF fields," *Microwave J.*, vol. 18, pp. 55-59, 1975.
[74] C. C. Johnson and T. C. Rozzell, "Liquid crystal fiberoptic RF probes, Part I: Temperature probe for microwave fields," *Microwave J.*, vol. 18, no. 8, 1975.
[75] T. C. Cetas, "A birefringent crystal optical thermometer for measurements of electromagnetically-induced heating," presented at USNC/URSI Annual Meeting, 1975.
[*76] R. R. Bowman, "A probe for measuring temperature in radio-frequency-heated material," *IEEE Trans. Microwave Theory Tech.*, vol. MTT-24, pp. 43-45, 1976.
[77] L. E. Larsen, R. A. Moore, J. H. Jacobi, F. A. Halgas, and P. V. Brown, "A microwave compatible MIC temperature electrode for use in biological dielectrics," *IEEE Trans. Microwave Theory Tech.*, vol. MTT-27, pp. 673-679, 1979.
[78] T. C. Cetas, "Thermometry in strong electromagnetic fields,"

Proc. Workshop on Physical Basis of Electromagnetic Interactions with Biological Systems, Univ. Maryland, College Park, MD, pp. 261–282, June 15–17, 1977.

[79] H. Bassen, P. Herchenroeder, A. Cheung, and S. Neuder, "Evaluation of an implantable electric-field probe with infinite simulated tissues," *Radio Sci.*, vol. 12, pp. 15–22, 1977.

[80] A. Cheung, "Electric field measurements within biological media," *Proc. Workshop on Physical Basis of Electromagnetic Interactions with Biological Systems*, Univ. Maryland, College Park, MD, pp. 217–242, June 15–17, 1977.

[81] C. K. Chou, R. Galambos, A. W. Guy, and R. H. Lovely, "Cochlea microphonics generated by microwave pulses," *J. Microwave Power*, vol. 10, pp. 361–367, 1975.

[82] V. V. Tyazhelov, R. E. Tigranian, and E. P. Khizhniak, "New artifact-free electrodes for recording of biological potentials in strong electromagnetic fields," *Radio Sci.*, vol. 12, pp. 121–129, 1977.

[83] C. K. Chou and A. W. Guy, "Quantitation of microwave biological effects," *Proc. Symp. on Biological Effects and Measurement of Radio Frequency/Microwaves*, Rockville, MD, pp. 81–103, Feb. 16–18, 1977.

[84] K. M. Chen and B. S. Guru, "Focal hyperthermia as induced by RF radiation of simulacra with embedded tumors and as induced by EM fields in a model of a human body," *Radio Sci.*, vol. 12, pp. 27–38, 1977.

[85] K. M. Chen and B. S. Guru, "Induced EM fields inside human bodies irradiated by EM waves of up to 500 MHz," *J. Microwave Power*, vol. 12, pp. 173–183, 1977.

[86] K. M. Chen and B. S. Guru, "Internal EM field and absorbed power density in human torsos induced by 1-500-MHz EM waves," *IEEE Trans. Microwave Theory Tech.*, vol. MTT-26, pp. 746–755, 1977.

[87] S. M. Neuder, "A finite element technique for calculating induced internal fields and power deposition in biological media of complex irregular geometry exposed to plane wave electromagnetic radiation," *Proc. Symp. on Biological Effects and Measurement of Radio Frequency/Microwaves*, Rockville, MD, pp. 170–190, Feb. 16–18, 1977.

[88] M. J. Hagmann, O. P. Gandhi, and C. H. Durney, "Improvement of convergence in moment-method solutions by the use of interpolants," *IEEE Trans. Microwave Theory Tech.*, vol. MTT-26, pp. 904–908, 1979.

[89] M. J. Hagmann, O. P. Gandhi, and C. H. Durney, "Numerical calculations of electromagnetic energy deposition in a realistic model of man," *1977 Int. Symp. on Biological Effects of Electromagnetic Waves*, (abstract), Airlie, VA, p. 55, Oct. 30–Nov. 4, 1977.

[90] M. J. Hagmann, O. P. Gandhi, and C. H. Durney, "Upper bound on cell size for moment-method solutions," *IEEE Trans. Microwave Theory Tech.*, vol. MTT-25, pp. 831–832, 1977.

[91] O. P. Gandhi and M. J. Hagmann, "Some recent results on deposition of electromagnetic energy in animals and models of man," *Proc. Workshop on Physical Basis of Electromagnetic Interactions with Biological Systems*, Univ. Maryland, College Park, MD, pp. 243–260, June 15–17, 1977.

[92] A. K. Chan, R. A. Sigelmann, A. W. Guy, and J. F. Lehmann, "Calculation by the method of finite differences of the temperature distribution in layered tissues," *IEEE Trans. Bio-Med. Eng.*, vol. BME-20, pp. 86–90, 1973.

[93] A. F. Emery, P. Kramar, A. W. Guy, and J. C. Lin, "Microwave induced temperature rises in rabbit eyes in cataract research," *J. Heat Trans.*, pp. 123–128, ASME, Feb. 1975.

[94] A. F. Emery, R. E. Short, A. W. Guy, K. K. Kraning, and J. C. Lin, "The numerical thermal simulation of the human body when undergoing exercise or nonionizing electromagnetic irradiation," *J. Heat Trans.*, vol. 98, pp. 284–291, ASME, 1976.

[95] A. Y. Cheung and D. W. Koopman, "Experimental development of simulated bio-materials for dosimetry studies of hazardous microwave radiation," *IEEE Trans. Microwave Theory Tech.*, vol. MTT-24, pp. 669–673, 1976.

[96] C. K. Chou and A. W. Guy, "Microwave and RF dosimetry," *Proc. Workshop on Physical Basis of Electromagnetic Interactions with Biological Systems*, Univ. Maryland, College Park, MD, pp. 165–216, June 15–17, 1977.

[97] Q. Balzano, O. Garay, and F. R. Steel, "Heating of biological tissue in the induction field of VHF portable radio transmitters," *IEEE Trans. Veh. Technol.*, vol. VT-27, pp. 51–56, 1978.

[98] Q. Balzano, O. Garay, and F. R. Steel, "Energy deposition in simulated human operators of 800-MHz portable transmitters," *IEEE Trans. Veh. Technol.*, vol. VT-27, pp. 174–181, 1978.

[99] S. J. Allen, "Measurements of power absorption by human phantoms immersed in radio-frequency fields," *Ann. N.Y. Acad. Sci.*, vol. 247, pp. 494–498, 1975.

[100] S. J. Allen, C. H. Durney, C. C. Johnson, and H. Massoudi, *Comparison of Theoretical and Experimental Absorption of Radiofrequency Power, Report no. SAM-TR-75-52*, prepared by Univ. Utah for USAF School of Aerospace Medicine, Brooks AFB, TX, 1975.

[101] S. J. Allen, W. D. Hurt, J. H. Krupp, J. A. Ratliff, C. H. Durney, and C. C. Johnson, *Measurement of Radiofrequency Power Absorption in Monkeys, Monkey Phantoms, and Human Phantoms Exposed to 10-50 MHz Fields, Report no. SAM-TR-76-5*, USAF School of Aerospace Medicine, Brooks AFB, TX, 1976.

[102] A. W. Guy, C. K. Chou, J. C. Lin, and D. Christensen, "Microwave induced acoustic effects in mammalian auditory systems and physical materials," *Ann. N.Y. Acad. Sci.*, vol. 247, pp. 194–218, 1975.

[103] K. R. Foster and E. D. Finch, "Microwave hearing: evidence for thermoacoustic auditory stimulation by pulsed microwaves," *Science*, vol. 185, pp. 256–258, 1974.

[104] R. M. White, "Generation of elastic waves by transient surface heating," *J. Appl. Phys.*, vol. 34, pp. 3559–3567, 1963.

[105] L. S. Gournay, "Conversion of electromagnetic to acoustic energy by surface heating," *J. Acoust. Soc. Amer.*, vol. 40, p. 1322, 1966.

[106] C. C. Johnson, C. H. Durney, P. W. Barber, H. Massoudi, S. J. Allen, and J. C. Mitchell, *Radiofrequency Radiation Dosimetry Handbook, Interim Report #SAM-TR-76-35*, prepared for the USAF School of Aerospace Med., Brooks AFB, TX, by Univ. Utah, 125 pp., September 1976.

[107] C. H. Durney, C. C. Johnson, P. W. Barber, H. Massoudi, M. F. Iskander, J. L. Lords, D. K. Ryser, S. F. Allen, and J. C. Mitchell, *Radiofrequency Radiation Dosimetry Handbook*, second ed., *Report no. SAM-TR-78-22*, prepared by Univ. Utah for USAF School of Aerospace Med., Brooks AFB, TX, 1978.

[108] C. H. Durney, M. F. Iskander, H. Massoudi, and C. C. Johnson, "An empirical formula for broad-band SAR calculations of prolate spheroidal models of humans and animals," *IEEE Trans. Microwave Theory Tech.*, vol. MTT-27, pp. 758–763, 1979.

THE ABSORPTION OF ELECTROMAGNETIC ENERGY IN BODY TISSUES[1]

A Review and Critical Analysis

Herman P. Schwan, Ph.D., and Geo. Morris Piersol, M.D.

PART I

BIOPHYSICAL ASPECTS

In recent years, considerable interest has been shown with regard to the frequencies at which ultrahigh frequency electromagnetic waves produce the most effective therapeutic results (1–5). Furthermore, concern has developed over the effects of powerful electromagnetic radiation sources on personnel (6–9). The purpose of this review is to evaluate the existing literature, to discuss the biophysical mechanisms by which electromagnetic radiation is absorbed in tissue, and to describe its medical applications.

The discussion is divided into: A) Classification of various effects of high frequency currents and radiation. B) Brief outline of the limitations of ultrashort wave diathermy. C) Summary of the physical laws which characterize the absorption and propagation of electromagnetic waves in tissue. D) Consideration of electrical properties of various body tissues. E) Analysis of their influence on the propagation of the radiation in tissue. F) Studies of physiological character concerning the effects of electromagnetic radiation on tissue material. G) Hazards associated with the absorption of electromagnetic radiation. H) Clinical applications. Sections F–H will appear in Part II of this review.

CLASSIFICATION OF VARIOUS EFFECTS OF HIGH FREQUENCY CURRENTS AND RADIATION

Effects on biological material can be placed in three categories (3): 1) thermal, 2) specific thermal, and 3) nonthermal. Volume heating is the general heating which any type of conductor or semiconductor, such as tissue, may receive under the influence of electrical currents or waves. Specific thermal effects (structural heating) exist when boundaries between different types of tissues or particles on a microscopic scale, such as small cell complexes or even bacteria, etc., can be selectively heated without substantial heating of the surrounding material. Those effects, which cannot be explained on a thermal basis, are classified as nonthermal effects. We are not attempting to discuss the literature dealing with the existence of nonthermal effects. To the best of our knowledge, none of the statements about these can withstand criticism. (See for example, the summaries of Krusen (10) and Osborne and Holmquest (11)). Therefore, we will confine our consideration to thermal effects of high frequency currents and radiation.

It has been shown that specific thermal effects, such as the selective heating of bacteria, are not possible. The selective temperature rise which a particle may experience by developing internal heating due to absorption of electrical energy is inversely proportional to the square of the particle size. Even under favorable conditions for selective heating in particles of microscopic size and colloidal dimensions, the temperature rise is less than 1/1000° C. (12–15). The analytical investigations upon which this statement is based are substantiated by experiments with water droplets in oil (14), yeast cells (16), fruit flies (17), and paramecia (18). Selective heating may occur only when the particle size is at least 1 mm. in diameter. On this basis, it may be possible to heat groups of cells of sufficiently large size. This is especially true in the transient period, i.e., while the particle or cell complex approaches the final temperature. Then the ratio of particle temperature rise to environmental temperature rise is especially great. Since this fact has not been recognized generally, it may be illustrated by figure 1. If a water particle, 2 mm. in diameter, is immersed in oil, it is heated selectively to 100° C. within 3 seconds by a field strength of about 10 kilovolts per cm. Equal heating occurs at first in 24 seconds when about 6 kilovolts per cm. are applied, and it requires 2 minutes with 5 kilovolts per cm. However, during the 3-second pulse the oil surrounding the water droplet has heated up only 3°, while during the 24-second pulse about 23° C. are reached and in the case of permanent application of the electrical field, 50° C. are obtained[2]. Similar examples can be given for larger objects, i.e. selective heating can be achieved especially if it is possible to apply short pulses of high power in objects of macroscopic size. It must be emphasized that a water in oil preparation is especially suitable for selective heating and that in the biological field such possibility of favorable circumstances does not exist. More favorable conditions might be created by overlapping the beams of two different sources of radiation, thereby creating a volume in which more heat is developed than in the surrounding medium. In conclusion, it may be said that although there is a future possibility of improving techniques for selective heating of cell complexes of at least cubic cm. volume, no possibility exists on physical grounds for selective heating of submacroscopic particles. This permits us to restrict our following discussions solely to volume heating.

There are two possible approaches to the study of the effects of radiation. In animal experiments, conclusions are based on the study of temperature changes in the animal in the radiation field. A more basic approach lies in the study of the fundamental tissue properties responsible for the conversion of the "primary"

Received for publication April 9, 1954.

[1] From the Department of Physical Medicine and Rehabilitation, School of Medicine, University of Pennsylvania, Philadelphia 4, Penna.

These studies were aided by a contract between the Office of Naval Research, Department of the Navy, and the University of Pennsylvania, NR119-289.

[2] These values hold if we choose for convenience the temperatures in oil at a distance $2R$ from the center of the sphere, as representative for the heating of the oil. For a more detailed discussion, see reference 14.

radiation energy into heat. Furthermore, if one understands how such data may be interpreted in terms of the molecular components of tissues, it is possible to extrapolate this information to experimentally unexplored frequency ranges, thereby predicting more suitable techniques.

In the chain of events which lead from the application of the radiant energy to its therapeutic effect, it is possible to distinguish the following steps:

Absorption of radiant energy. The absorption of radiant energy is dependent upon the various electrical constants of the individual tissues. Energy absorption leads to the development of heat. This we define as "primary heat" developed in the irradiated body. Primary heating produces temperature differences between various tissues and even differences within homogeneous material. This happens

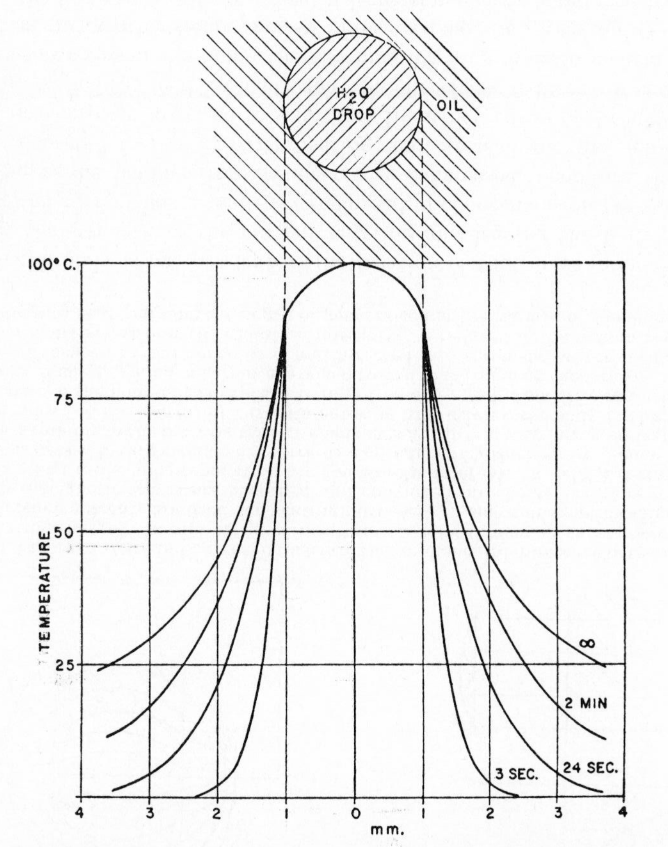

FIG. 1. Temperature rise in water droplets suspended in oil, plotted inside and outside the water. The curves hold for the case in which a high frequency field is applied for 3 seconds with a field strength of 11 Kilovolts/cm. in oil, for 24 seconds with 6 Kilovolts/cm., for 2 minutes with 5 Kilovolts/cm. and for unlimited time (∞) with 4.5 Kilovolts/cm. They illustrate how it is possible to obtain highly selective heating with short pulses of high intensity.

because primary heat development is most pronounced where the radiant energy is strongest, i.e. near the surface to which the radiation is applied. *The next mechanism of importance is the flow of heat from warmer to cooler areas.* This depends on the ability of the matter under consideration to conduct heat, i.e. its heat conductivity. It is also affected greatly by the blood flow, which may carry heat from one body area to another. It is important to realize that heat flow will diminish temperature differences. This means that final temperature distributions are characterized by curves which are flatter than those of primary heat development. It is the final heat distribution which produces the reaction observed by the clinician.

It is the task of the biophysicist to establish the basic laws governing primary effects. He and the physiologist work together and study actual heat distribution and associated phenomenon, such as blood flow. The clinician records the final results. All three groups cooperate to obtain a complete, clear picture of the effects of radiation. From a practical viewpoint, the actual temperature curve, i.e. the combined results of primary heat development and heat flow are more important than the mechanism of absorption. However, in the majority of cases, we can ascertain a complete knowledge of actual temperature distributions only in animal experiments. Extrapolation of this information to the human body is difficult because the amount of heat which can be absorbed by any body depends on the volume and surface of the irradiated body[3]. Here knowledge of the fundamentals of the absorption mechanism is most helpful. It permits us to predict how much radiant energy will be absorbed in the body; how far it penetrates; and the kind of tissue which will experience special heat development. We may then estimate the difference between primary heat absorption curves and actual heat distribution curves. If these conclusions agree with those derived from animal experiments, we arrive at well founded results to be applied to mankind.

It should be pointed out that the mechanism by which heating is brought about by microwave radiation is fundamentally different from that which takes place when so-called short wave diathermy is applied. Only in the latter instance the patient becomes a part of either a high frequency condensor or an induction field. The principal difference between the application of short wave diathermy and radiation diathermy is illustrated in figure 2. In short wave diathermy[4] ultrahigh frequency currents are passed through the body. Kirchoff's law of current conduction states that the sum of all currents ($i_1, i_2 \cdots i_n$) at any given cross section

[3] Under normal conditions the human body transmits on its own about 0.01 watt/cm.^{2} energy in the form of heat through radiation, sweating, and heat convection to the outside. Under extreme conditions, this amount may be increased as much as 10 times. If similar figures hold for other animals, an animal with one-tenth the body surface of the human can absorb only one-tenth as much radiant energy as a human, before continuously raising the total body temperature as a result of absorption of thermal energy. Heat dissipation in this case would be less than heat absorption.

[4] The term "ultrashort wave diathermy" is misleading. Short wave diathermy has nothing to do with short waves. It utilizes high frequency currents produced by equipment which is potentially able to emit ultrashort waves, if equipped with proper radiation devices.

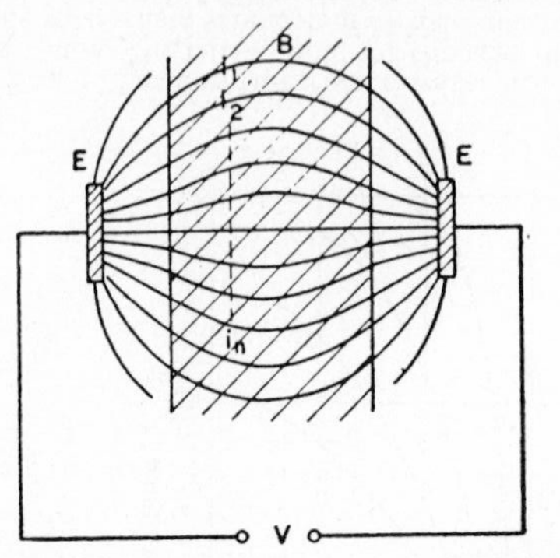

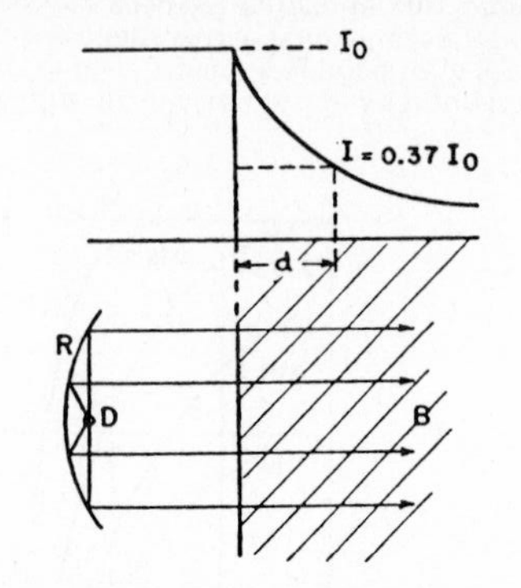

FIG. 2. Figures a (*left*) and b (*right*) illustrate the fundamental difference between current and radiation therapy. In high frequency current therapy (known as short wave therapy) high frequency currents are used for heating purposes. In electromagnetic radiation therapy, electromagnetic waves are absorbed and converted into heat. The laws for current conduction and radiation propagation are completely different. *a, left:* A segment of the body (B) is passed by high frequency currents as produced by the high frequency voltage V and made available to the body by the two electrodes E. Current heating necessitates always two electrodes and the sum of all currents entering and leaving the body is the same. *b, right:* A part of the body (B) is exposed to high frequency radiation. The radiation is generated by a dipole D, which in turn is energized from a high frequency generator. The dipole radiation is focused by one reflector R and directed into the body, where it is absorbed. The upper figure shows how the radiation intensity I is reduced in the human body due to conversion into heat and how the depth of penetration of the radiation D is defined.

through the body is independent of the location of that cross section[5]. Consequently, the strength of the current entering and leaving the body is equal. Because the current has a tendency to spread, its concentration per square cm. diminishes as the distance from the electrodes is increased. Therefore, the surface may be expected to have greater heating than areas inside the body. If two condensor electrodes are used to send high frequency currents through a body, concentration of the field intensity may be effectively reduced by spacing the electrodes away from the body, thereby creating a more uniform field (19–22). Radiation, on the other hand, consists of electromagnetic waves and is transmitted into the body by a reflector (fig. 2). Radiation energy is reduced by absorption. Its intensity decreases according to an exponential law when applied to a homogeneous material. The term "depth of penetration" can be applied correctly only in the case of radiation. Figure 2B shows how the intensity is reduced due to absorption. It is always optimal on the surface.

LIMITATIONS OF ULTRASHORT WAVE DIATHERMY

Esau was the first one abroad to recognize the value of UHF currents in effectively increasing the penetration of the heat into the human body (23), as did Schereschewsky in this country (24). Early claims of selective heating with

[5] This statement is correct only if all the currents are ohmic ones, i.e. due to ionic conduction. Where significant capacitive currents exist, as they do in the ultrahigh frequency field, this statement must be amended to read that the vector sum of all currents is constant.

REVIEW OF PHYSICAL MEDICINE AND REHABILITATION

this form of diathermy were based on investigations by McLennan and Burton (25) and by Paetzold (26). They found that a high frequency current of a given value will produce optimal heating of either an electrolyte or any other conductor, when the conductivity is altered to fulfill the relationship

$$60\lambda\,\kappa = \epsilon \qquad (1)$$

(λ = wavelength of oscillator producing high frequency current, κ = specific conductance, and ϵ = relative dielectric constant of electrolyte). Some investigators concluded erroneously from this that by proper choice of a wavelength which fulfills equation 1 for any given type of tissue a maximum amount of heat in this body tissue could be developed. However, Schaefer (27) and, more recently, Higasi (28) showed that this is not the case. The practical value of Paetzold's principle of selective heating in a series combination of various tissues (skin, fat, muscle, body organs) depends on the variation of the electrical properties of different types of tissue. Determinations of the electrical properties of tissues with high water content prove that their data are very similar to each other. Hence, the distribution of heat in all tissues with high water content is fairly uniform, provided that equal current density is assumed. Schaefer gave a detailed analysis of this problem which substantiates the impossibility of selective heating of any type of muscular or organic tissue (27, 29–31).

However, fat and bone have very different dielectric properties. Analysis shows, therefore, that the heat which is developed per volume unit of subcutaneous fat is much greater than that produced in deep tissues which have a higher water content (31, 32). Furthermore, it was recognized that the ratio of the total amount of heat developed in fatty tissue per unit volume to the heat development in muscle decreases as the frequency is increased. This statement, which was originally based on an analytical argument utilizing knowledge of the dielectric data of fat and muscle, has been corroborated in experimental studies by actual temperature measurement. It may be concluded, therefore, that even at the highest frequencies which are practical for current therapy, i.e. about 100 Mc., fatty tissue is heated more than muscular tissue. In figure 3A and 3B the ratio of heat developed in fat to that in muscle is shown as calculated from dielectric data and as determined from temperature measurements.

Concurrently with the use of ultrahigh frequency currents applied to the human body for therapeutic purposes by means of two current electrodes (capacitive heating), another method was proposed for the production of deep heating with ultrahigh frequency currents. It was suggested that an induction field produced by coils energized from an ultrahigh frequency generator be employed for this purpose (inductive heating (23, 33, 34)). Electric currents (Eddy currents) are set up by the rapidly oscillating magnetic field surrounding the windings of the coil. This is a direct consequence of one of Maxwell's equations, which states that the electrical potential created by an alternating magnetic field is proportional to the product of magnetic field strength and frequency. It was hoped that it would be possible to obtain much more uniform deep heating than with the capacitive heating technique. This was anticipated on the basis that the magnetic

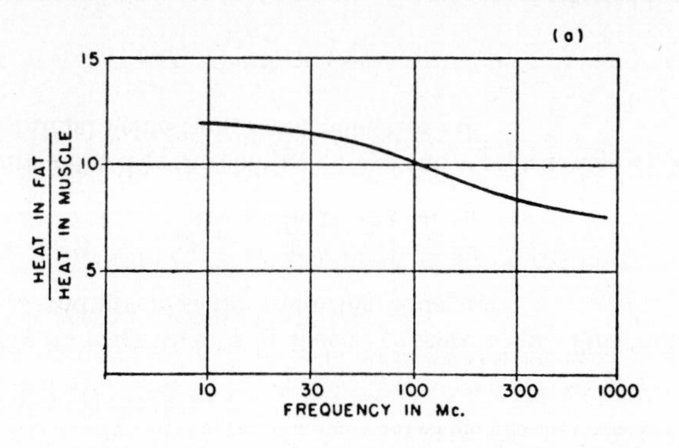

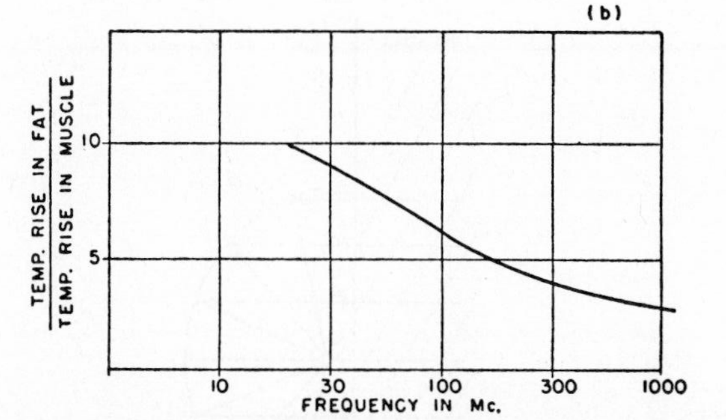

FIG. 3. Illustrates the strong tendency of high frequency current therapy to heat selectively subcutaneous fatty tissue. *a, upper:* Heat development in fat compared to that in muscle as calculated from dielectric data at various frequencies. *b, lower:* Temperature rise in fat compared to that in muscle at various frequencies (taken from Paetzold, ref. 1). The ratios of heat development and temperature rise in fat and muscle are in both cases much greater than one. They decrease with increasing frequencies. Differences in actual values between calculated and observed results may be partially due to heat conduction and in part due to variation in fat impedance data.

field is influenced only by the magnetic properties and these do not vary within the human body.[6] However, in view of the proportionality of the electric current with the frequency, either extremely powerful and expensive equipment or very high frequencies are necessary to produce sufficiently great currents useful for heating purposes. The use of very high frequencies is the only feasible way to proceed, but this in turn leads to another predicament (35–37). If coils of low inductance are used, nonuniform magnetic fields result with high magnetic field strength in the immediate neighborhood of the windings and consequent preference for surface heating. On the other hand, capacities which exist between the windings of longer coils produce capacitive currents which are conducted by the

[6] The magnetic permeability of all biological material is equal to that of vacuum within a fraction of 1 per cent and the magnetic losses of biological substances may be neglected likewise in our discussion.

surface of the body and cause consequent surface heating. In summary, it may be stated that the situation with the induction field is more complicated than with the capacitive electrode arrangement. However, it has been possible to work out a compromise which gives somewhat improved deep heating at frequencies in the neighborhood of about 20 to 30 Mc. The presently used frequency of 27 Mc. was chosen largely on this account. But it must be stated that even with this compromise, fat heating is still substantially stronger than the heating of deep tissues. So far, no data exist which permit conclusive comparative evaluation of the effectiveness of currents applied with electrodes (capacitive heating), and currents generated by coils (inductive heating). However, it is safe to state that while short wave diathermy, employing high frequency currents for heating purposes, definitely provides better deep heating than is obtainable with surface heating (hot pack, infrared) it certainly does not yet provide a good tool capable of bringing the major part of the available energy into the deep tissues below the subcutaneous fat.

RADIATION DIATHERMY

Attempts to reduce the undesirable fat heating by employing sufficiently high frequencies with the condensor field method are limited. This is due to the fact that the highest possible frequency at which the patient circuit can resonate, without excessive radiation losses, is in the neighborhood of 300 Mc.[7] (Paetzold (38) and Osswald (39)).

Successful employment of higher frequencies requires the use of the radiation field which was first proposed by Esau (23) and later in greater detail by Hollman (40–42) and Paetzold (43, 44). The first radiation experiments with small power were carried out by Bruenner-Ornstein and Randa (45) and by Denier (46). It was hoped that the low conductivity and consequent high penetration of radiation in fat would permit significant heating of the deep body tissues. Using a plane fat layer in series with muscle Hollman presented a mathematical analysis of the problem. The results of his analysis are very encouraging with regard to the value of the ultrahigh frequency electromagnetic radiation field for therapeutic purposes. However, Hollman's evaluation is incorrect because he assumed incorrect data for the dielectric properties of tissues. He obtained his dielectric data by extrapolation from material which had been obtained by others (47–50) at lower frequencies. Also, he did not take into account the marked changes in dielectric properties which occur at ultrahigh frequencies. These occur at frequencies above 300 Mc. due to polarity of water (see details in following section).

Hollman was the first to point out the advantages of impedance matching device or material between radiation source and body surface. When radiation strikes the surface of the body, part of it is absorbed and the remainder is reflected. If a coupling material with electrical properties identical to those of the body is placed between the source of energy and the body surface, no energy loss due to reflection will occur. Unfortunately, however, under such circumstances the coupling medium will absorb energy and the body will not receive the total

[7] The "patient circuit" is the lumped circuit which is used to energize the patient.

available amount. The best solution of the problem is obviously a coupling medium which is nonabsorbent and yet reduces losses due to reflection at the body surface to a minimum. Its properties may be determined as follows: The coefficient of reflection of the body interface (fig. 4) is given in the following equation

$$p = \frac{Z_2 - Z_1}{Z_2 + Z_1} \tag{2}$$

Z_1 and Z_2 are the characteristic impedances of the two media 1 and 2, for example the coupling medium and muscle. This reflection coefficient is in general a complex figure. Only its magnitude is of special concern to us. This value is derived from equation 2 and given in the following equation.[8]

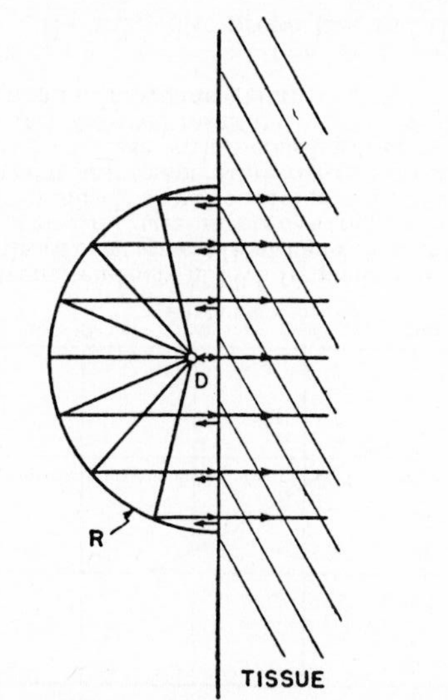

FIG. 4. Illustrates reflection of electromagnetic energy on the tissue surface. The radiation from a dipole D is focused by a reflector R and thrown against the tissue surface. The boundary between reflector and tissue surface permits only part of the radiation to enter the body while another part is reflected. The amount of reflection depends on the electrical properties of both the body and the coupling material which fills the space between reflector and body as explained in the text.

$$r^2 = \frac{a^2 + b^2 + 1 - 2a}{a^2 + b^2 + 1 + 2a} \tag{3}$$

where a and b are defined as the components of the complex ratio $Z_1/Z_2 = a + jb$. It reaches an optimal value (minimum reflection) of

$$r^2 = \frac{1 - a}{1 + a} \tag{4}$$

[8] The parameters r and p refer to field strength. The amount of reflected energy is characterized by r^2.

when $a^2 + b^2 = 1$. The condition $a^2 + b^2 = 1$ is identical with the statement that the amplitude of the ratio Z_1/Z_2 is equal to unity. The ratio Z_1/Z_2 is identical with $\sqrt{\epsilon_2^+/\epsilon_1^+}$ where ϵ_1^+ and ϵ_2^+ are the complex dielectric constants of the two media 1 and 2[9]. From this, it is obvious that minimal reflection losses occur when the magnitude of the two dielectric constants ϵ_1^+ and ϵ_2^+ are made equal.

Consider for example the situation where the frequency is about 1000 Mc., the dielectric constant ϵ_2 about 50 and the loss tangent of the medium 2 about equal to 0.5. This characterizes the case when electromagnetic waves are directed into muscular tissue. Here the calculation predicts that the coupling medium should have a dielectric constant of about 65 to minimize reflection losses. Under such circumstances, the amount of energy lost by reflection will be only about 7 per cent. This means an increase in the absorption of available energy by a factor of nearly 2.5 compared to the case where the coupling material is air. It illustrates the value of a coupling medium.

Another possibility for reducing reflection from the body surface is the insertion of a properly designed slab of dielectric material between radiation source and body surface (51, 52). Antennas of radiation sources are usually designed for free space radiation, i.e. for the characteristic impedance of vacuum (377 Ohms)[10]. If the input impedance of the dielectric slab-tissue configuration is made to equal 377 Ohms, perfect impedance match is achieved, i.e. no reflection will occur. It is well known by electrical engineers that this requires a dielectric slab which has an "electrical" length of one-quarter wavelength and a characteristic impedance Z which will transform the input impedance of the body surface Z_b into a value $Z_1 = Z^2/Z_b = 377$.

Consider for example a wavelength of 10 cm. in air and in muscular tissue as representative of the body tissue. The characteristic impedance for electromagnetic waves in muscular tissue is about $Z_b = 50$ Ohms. Its transformation into 377 Ohms obviously requires a characteristic impedance of the dielectric slab of $Z = \sqrt{377 \times 50} = 137$ Ohms. Such an impedance is found by plane electromagnetic waves in a material with a dielectric constant of $(377/137)^2 = 7.5$. This dielectric constant will cause the electrical wavelength, which is assumed to be 10 cm. in air, to have in the slab a value of only $10/\sqrt{7.5} = 3.6$ cm. One quarter of this, i.e. 0.9 cm. in thickness establishes, therefore, a slab which has an equivalent electrical length of one-quarter of a wavelength. This shows that it takes only rather thin slabs in front of the body to achieve perfect impedance match. It will be seen later that fatty tissue has dielectric constants which are near 7.5. Fatty tissue, in itself, may, therefore, establish more or less perfect impedance match to the microwave radiation, if its thickness is either 9 mm. or an uneven multiple of it. On the other hand, if the fatty layer is any multiple of 18 mm. thick (elec-

[9] The complex dielectric constant is defined by $\epsilon^+ = \epsilon\,(1 - j \tan \delta)$ where ϵ dielectric constant, $\tan \delta$ loss factor, j imaginary unit vector.

[10] The characteristic impedance of vacuum or air is defined as that resistance which can replace the antenna load (antenna assumed to radiate into free space), without any change in high frequency generator performance due to change in load.

trical length = half wavelength), it has no or only a small effect on the absorption of electromagnetic energy by the muscular tissues. It follows from this discussion that the use of dielectric slabs as proposed by Gersten, Wakim, and Krusen (51) and by Martin, Rae, and Krusen (52) for purposes of reducing reflection is of questionable value.

Actually, the addition of such slabs can make matters worse, for example when dielectric slab and subcutaneous fat layer add to a $\lambda/2$-layer. Enormous variations in actually absorbed energy with presently used microwaves are, therefore, obvious, depending on the thickness of the subcutaneous fat layer. The disagreement with regard to the actual heating value of microwaves in clinical practice may in large part be due to the fact that the physician does not realize that the indication of the Watt meter on his instrument has nothing whatsoever to do with the energy actually absorbed by the patient.[11]

Other important contributions in the late 30's and 40's were the investigations of the radiation field of UHF generators in media with properties similar to those of tissue. Experimental investigations (43, 44, 53) were corroborated by mathematical analysis (54) and a satisfactory agreement was obtained in many instances (54). It was recognized that the radiation field in front of a reflector, which is energized by a dipole oscillator, shows two marked regions. One is near the source and shows maxima and minima due to the superposition of signals arriving from different locations in the source. This region of oscillating field strength may be called, therefore, the "interference field". It was found to extend from the source to a distance of $R^2/\lambda - \lambda/4$, where R is the radius of the source reflector. Beyond this, the "distance field" extends. In it, energy reduces monotonously. There is a considerable reduction of the wavelength in water and tissue, due to the high dielectric constant of electrolytes. This reduction amounts to nearly 90 per cent because the effective wavelength changes inversely with the square root of the dielectric constant and the dielectric constant of electrolytes is about 80. This tends to increase the size of the interference field estimated by the formula R^2/λ. However, it can be shown readily that the interference field is nevertheless restricted in size. Table 1 demonstrates this. It gives the "critical distance" R which separates interference and distance field at various frequencies for the two cases of air and a material with a dielectric constant comparable to tissue in front of the radiation source.

These investigations clarified to what extent the simple assumption of plane wave propagating from source into the body requires modification and the difficulty of concentrating the waves into one beam of high field strength. A number of experiments were performed with water and electrolytes as media in

[11] A discussion of the dependence of the amount of absorbed radiant energy on frequency, skin, and fatty layer thickness has been given by Foelsche. (T. Foelsche: *The Energy Distribution in Multilayer Arrangements Caused by Decimeter—and Centimeter Irradiation and Its Relationship to Diathermy*, in print and personal communication.) Foelsche points out that the amount of absorbed energy may fluctuate under unfortunate conditions between 10 and 100 per cent, depending on how dry the skin is and assuming a wavelength in air of 10 cm.

Table 1. *Extent of interference field as function of frequency for air (ϵ = 1) and muscular tissue (ϵ = 64) in front of the radiation source*

Frequency	R = 10 cm.		R = 3.2 cm.	
	$\epsilon = 1$	$\epsilon = 64$	$\epsilon = 1$	$\epsilon = 64$
Mc.	*cm*	*cm.*	*cm.*	*cm.*
6000	20	2.5	2	0.25
3000	10	1.25	1	0.13
1500	5	0.63	0.5	0.06
600	2	0.25	0.2	0.03
300	1	0.13	0.1	0.01
150	0.5	0.06	0.05	0.01

Data are given for two different reflector sizes (radius R = 10 cm. and 3.2 cm.). The data illustrate that the interference field is in most cases very small.

front of the source, since properly selected electrolytes resemble to some degree tissues with high water content, so far as their dielectric properties are concerned (53).

The absorption of electromagnetic energy in electrolytes is rather strong. Therefore, the intensity decreases rapidly. Figure 5 gives typical curves of this effect obtained by Paetzold and pertaining to a case where the total radiation pattern exists in electrolytes (53). The situation is somewhat improved when the reflector itself is filled with a material which has high dielectric constant, but low dielectric losses and only the space outside the reflector is filled with material such as body tissue of high water content or electrolytes. Electromagnetic waves of about 1 m. wavelength, as utilized by Paetzold and Osswald, cannot be concentrated in the same manner as light waves, since they do not follow the laws of geometric optics. This is true because the ratio of the wavelength of radiation to the size of the reflector is not small enough. It cannot be made so without producing either a substantial decrease in the primary depth of penetration in tissue due to short wavelength (see fig. 11) or utilization of reflectors of impractically large size as proposed by Spiller (55). However, Paetzold's data already indicate a concentration of energy along the long axis of the reflector, especially in the distance field. Using a smaller wavelength than investigated by Paetzold and others, it is possible to increase this concentration. The relatively high absorption of energy in material similar to tissue (electrolytic conductivity about 0.01 Ohm^{-1} $cm.^{-1}$) does not prove that the radiation technique is unsuitable for deep heating. One of the major goals of any heat therapy should be to deliver the energy, which is transformed into heat, on the deep tissue side of the subcutaneous fat layer. If this is achieved, the relatively low thermal conductivity of fat and its poor supply with blood will be sufficient to guarantee deep heating, even when the energy is absorbed rather rapidly in the deep tissue. Fatty tissue has a much lower electrical conductivity than muscular tissue, as is shown later. Therefore, with wavelengths of approximately 30 cm. to 1 meter, fat may be penetrated with a somewhat concentrated beam and the total energy of radiation, delivered into the deep tissue, will suffer little energy loss in fat, because the depth of penetration

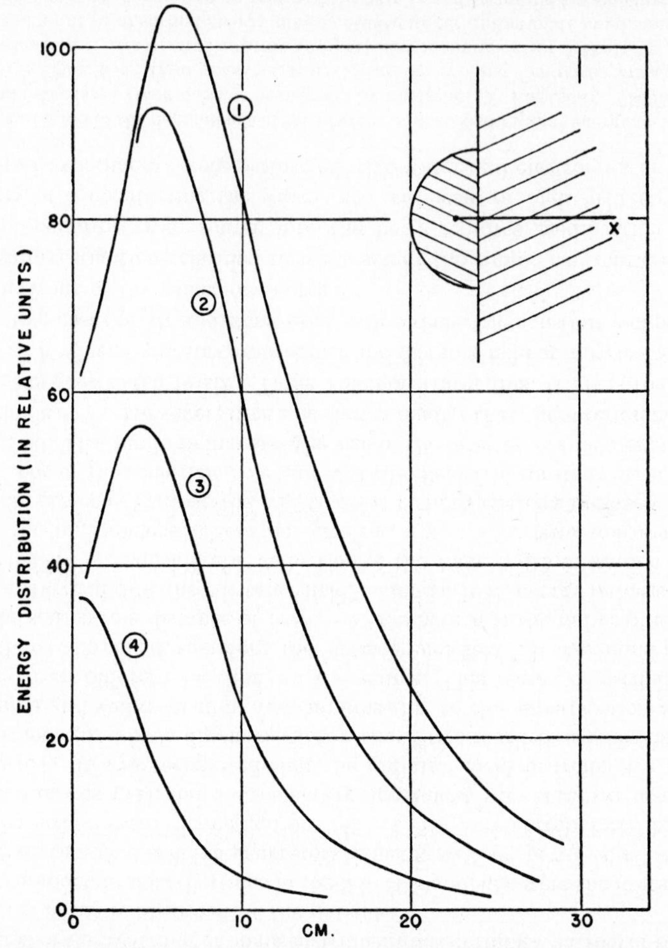

FIG. 5. Energy distribution of electromagnetic radiation along the axis of a reflector as given by Paetzold and Osswald (53). Reflector radius 10 cm., frequency about 300 Mc. The curves hold for specific resistance values of electrolytes as follows: *1*, 5700 Ohm cm., *2*, 1500 Ohm cm., *3*, 600 Ohm cm., *4*, 160 Ohm cm. *Curve 1* illustrates pattern in distilled water, *2* in fatty tissue, and *4* in deep tissue. It is seen that a strong interference maximum, which is created in water and fatty tissue about 10 cm. from the radiation source, is suppressed by heavy conductive losses in muscular material. The curves are, therefore, indicative of the ability of electromagnetic radiation to pass fat without substantial losses and be absorbed rather readily in the deep tissues.

in fat is high. However, reflection of radiant energy from fat-muscle interfaces complicates the situation as shown below, when the wavelength chosen is too small.

ELECTRICAL PROPERTIES OF TISSUE

Electrical properties of material are characterized by two constants. They are the dielectric constant, or permittivity as it is often called, and the specific

resistance. The dielectric constant of a material is equivalent to the capacitance of a 1-cm. cube, corrected by a constant factor which depends on the units of measurement of the capacitance ($1/4\pi$ if capacitance is measured in cm. and $1/36\pi \cdot 10^{11}$ if measured in Farad). The specific resistance, measured in Ohm cm., is the resistance of the cm. cube. The conductivity of the material is the inverse of the specific resistance. Dielectric constant and conductivity are measures for the two types of current which pass through matter when a unit potential is applied. A capacitive current does not cause any heating; the resistive current does according to Joule's law.

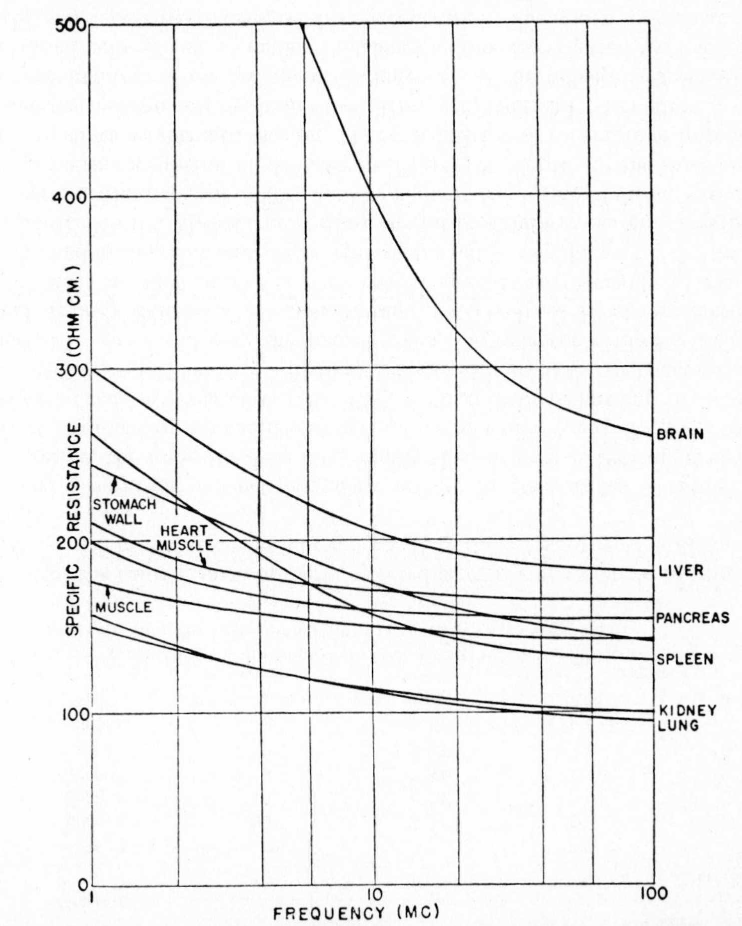

FIG. 6. Frequency dependence of the specific resistance of various body tissues in the frequency range from 1 to 100 Mc. Brain tissue shows exceptionally high resistance values due to its higher lipoid content. The curves are typical examples. Reproducibility is 10 to 20 per cent due to individual variation from sample to sample. All curves demonstrate a typical decrease in resistance as the frequency increases. Temperature 23° C., human autopsy material. Data taken from Rajewsky *et al.* (56).

The rate of heat development (dQ) in a given volume (dV) of body tissue is expressed by $dQ = E^2\kappa dV$, where κ is the electrical conductivity of the tissue and E is the field strength. The field E, in turn, at any point in the same tissue depends upon: 1) the potential difference applied to the body, 2) the geometric configuration, i.e. the electrode arrangement and the complexity of the body, and 3) the dielectric constant ϵ and specific resistance ρ of the various body tissues.

Heat development in the body may, therefore, be expressed in terms of simple physical quantities. To do so, one must know the dielectric constant and specific resistance of the various body tissues.

Detailed investigations of the resistivity and dielectric constant of body tissues in the total frequency range of diathermy up to 100 Mc. have been carried out by Osswald (50) and Rajewsky and his co-workers Osken, Schaefer, Schwan (48, 49, 56). Figure 6 gives the specific resistance measurements by Rajewsky and co-workers (56), which are characteristic for the frequency dependence. The curves represent typical examples. The individual variation from sample to sample was found to be about 10 or 20 per cent. Hence, it is possible to state that in this series of measurements no significant difference between the specific resistance values of various tissues has been noticed. Exceptions are liver and especially brain, which shows unusually high resistance, possibly due to its high lipoid content. Figure 7 shows dielectric constant values as measured by Osswald at 20° C. over a somewhat more restricted frequency range. Here again, it is seen that the dielectric constant decreases in all cases, as the frequency increases. All curves in figures 6 and 7 approach constant values at high frequencies. However, it will be shown later that a mechanism different from that responsible for the frequency behavior below 300 Mc. starts to affect the electrical constants very strongly above 300 Mc. This fact was not known to Hollman and explains

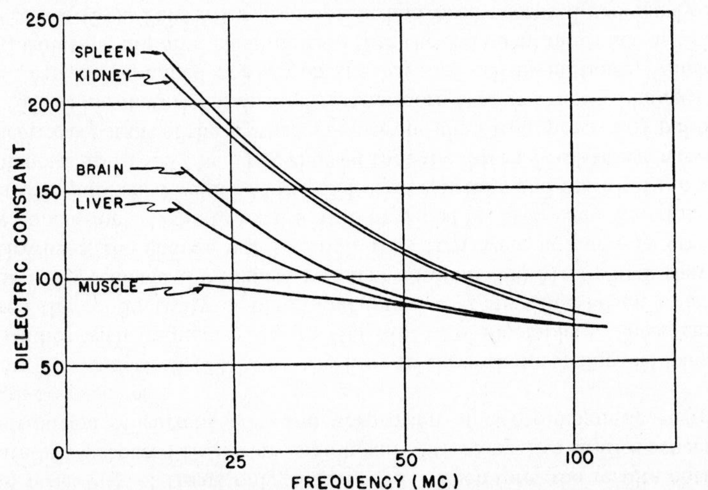

FIG. 7. Frequency dependence of the dielectric constant of various animal body tissues in the frequency range from 25 to 100 Mc. Temperature 20° C. The dielectric constant decreases in all instances with increasing frequency. Data taken from Osswald (50).

TABLE 2. *Specific resistance and dielectric constant for fat and tissue with high water content*

	Dielectric Constant	Specific Resistance
Fat	12	2200–4300
Muscle and organ tissue	85–120	125–290

The table demonstrates the difference in electrical data of fat and muscular tissue of high water content. Frequency about 50 Mc. Temperature 20° C. Data from Osswald (50).

his prediction of erroneous absorption data of electromagnetic radiation of ultrahigh frequencies as discussed before. Table 2 summarizes some fat tissue data as obtained by Osswald (50) near 50 Mc. and compares them with average data for the body tissues with high water content. Both electrical conductivity and dielectric constant are about tenfold lower for fat than for tissues with high water content, which explains the completely different heat development in both types of tissue as discussed before in the case of heating with high frequency currents.

These data are all the result of measurements on excised tissue. Therefore, they have been the subject of criticism, since it is sometimes assumed that the properties of living tissues differ from those of excised tissue. However, the theoretical and experimental contributions of Cole (57, 58), Daenzer (59, 60), Fricke (61), Rajewsky *et al.* (56), Schaefer (62), and Schwan (63, 64) show that resistivity and dielectric constant of biological material in the frequency range between 1 Kc and 30 Mc. are determined by its cell membranes and its electrolyte content. The electrical properties of cell membranes are not altered for a period of 1 to 2 days after death (65, 66). Cell membranes act as thin layers of high capacity and resistance. These membranes are short circuited when the frequency is much greater than 1 Mc. and, therefore, no longer restrict the current flow. Their high resistance hinders the current flow appreciably only at low frequencies. This explains the decrease in effective resistance and capacity shown by tissues when the frequency is increased. Therefore, it is obvious and in complete agreement with the analytical and experimental work that changes caused by the breakdown of cell membranes affect the low frequency electrical data primarily and have no effect on its characteristics at diathermy frequencies near 100 Mc. (65, 66).

Figure 8 shows a simplified equivalent electrical circuit corresponding to the electrical behaviour of tissues with high water content. The cell membranes have a high capacity of 1 μ Farad per $cm.^2$ surface and a resistance of more than 100 Ohms per $cm.^2$ (67) and are characterized by the capacitors M. The material inside the cell has a specific resistance value of about 100 Ohm cm. (68–70) and is reached by electrical current only after passing the cell membranes M. The external tissue fluid has a resistance of similar values, and appears shunted across the cell impedance. Only physical properties of the various parts of micro- and macrostructure of the tissue determine its electrical properties. Physiological reactions which determine metabolism and permeability of cell membranes affect the electrical properties of membranes only at relatively low frequencies and the resistivity of tissue only slightly (71). Therefore, the results obtained with carefully prepared tissues are significant with respect to studies concerning the distribution of current flow and absorption of electromagnetic energy at ultrahigh frequencies.

Malov (72) and Gsell (73) were the first to show that conclusions based on tissue studies with frequencies up to 100 Mc. were incomplete. These authors measured the conductivity of blood and found that its value began to change again as it approached the highest frequency (about 500 Mc.) which they had used. However, the change was so slight that they were not able to arrive at definite conclusions. Definite results were obtained in 1948 when Rajewsky and Schwan (68) published values of the dielectric constant and resistivity of blood at frequencies up to 820 Mc. They showed that the internal resistance, as well as the dielectric constant of the erythrocyte interior differ from previously published values (67). This material indicated furthermore that frequency changes near 100 Mc. can be explained neither on the basis of cell membranes, which are perfect insulators, nor on the assumption that the cell membranes act as polarizable elements (74). This latter assumption had been accepted previously for a number of biological interfaces (67).

Recently, more blood data from different sources has become available (Herrick, Jelatis, and Lee, 75; England, 76; Cook, 69, 77; Schwan and Li, 70). Figure 9 presents all these data giving complete coverage of the frequency spectrum from 50 to 30,000 Mc. The change in dielectric constant below 300 Mc. was analyzed by Schwan (74) and attributed to structural parameters, e.g. cell membranes. The change in dielectric constant and conductivity occurring above 300 Mc. is in agreement with the prediction based upon the polar properties of water. The electric polarity of water molecules makes them rotate with the electrical field at frequencies lower than a certain critical value (characteristic

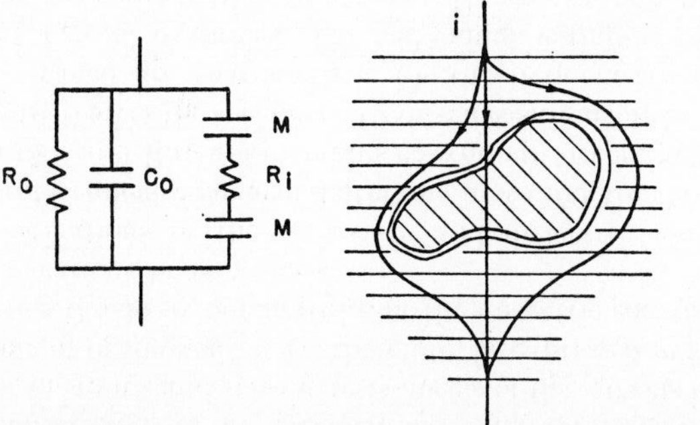

FIG. 8. Equivalent electric circuit, simulating the electrical properties of biological cells. An electric current i, approaching a biological cell, will divide into two parts. One part of it will bypass the cell by means of its surrounding fluid as characterized in the circuit by the elements R_0 and C_0. The other part will penetrate through the cell membranes, characterized in the circuit by capacitors M, and the cell interior (R_i). The resistance of the capacitors to the electrical current (reactance) changes strongly with frequency, thereby causing a frequency dependence of the ratio of the two currents entering and bypassing the cell. This causes a frequency dependence of capacity (dielectric constant) and specific resistance of biological material.

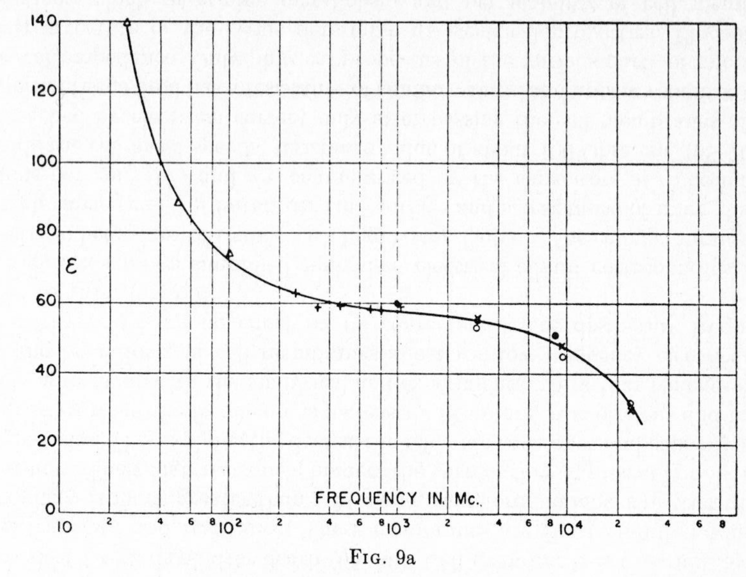

FIG. 9a

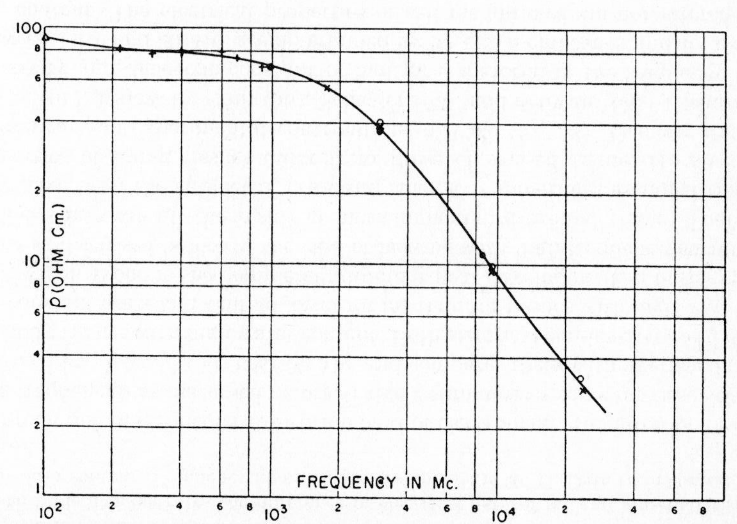

FIG. 9b. Dielectric constant and specific resistance of whole blood as function of frequency. Temperature 37° C. Figure 9a shows the dielectric constant and 9b the specific resistance. The increase, which occurs as the frequency decreases below 1000 Mc., is due to the structure of blood (presence of cell membranes with high resistance and capacity) and is especially pronounced in the frequency dependence of the dielectric constant (structural frequency dependence). The decrease, which occurs in both dielectric constant and specific resistance as the frequency increases above 1000 Mc., is due to the polar qualities of water molecules (polar frequency dependence). The two regions of frequency dependence are separated by a small frequency range where the electrical properties are relatively frequency independent. Results are given by Osswald ∆ Schwan and Li + England ○ Herrick et al. ● and Cook ×.

frequency), where the inertia of the molecules becomes too big to permit it to follow the rapid variation in field intensity. This critical frequency for water is in the neighborhood of about 20,000 Mc. and causes a rapid drop in both resistance and dielectric constant in accordance with the theory of polar media as developed originally by Debye (78). Using a dielectric constant of 54 in the region on the curve between the two frequency changes above and below 1000 Mc., Rajewsky and Schwan (68) predicted the frequency behaviour of blood up to 30,000 Mc. The evaluation was based on the equations which hold for media with polar quantities:

$$\epsilon = \epsilon_\infty + \frac{\epsilon_0 - \epsilon_\infty}{1 + (f_0/f)^2} \tag{4}$$

$$\kappa = \kappa_0 + (\kappa_\infty - \kappa_0)\,\frac{(f_0/f)^2}{1 + (f_0/f)^2} \tag{5}$$

where the indices ∞ and o characterize conductivity values $\kappa = 1/\rho$ and dielectric constant values at frequencies far above and below the "critical frequency" f_0 and

$$\epsilon_0 - \epsilon_\infty = \frac{\kappa_\infty - \kappa_0}{f_0}\,18 \cdot 10^{11} \tag{6}$$

The influence of protein on the dielectric properties of blood has been discussed by Cook (69, 77) and is responsible for the difference between the dielectric constant of 76, which one would anticipate in the flat region near 300 Mc. on the basis of the water content of blood, and the experimental value of 54. Protein molecules may be considered as ellipsoids of revolution, composed of material of low conductivity and a dielectric constant as low as 2 or 3. Protein molecules have polar properties, i.e. they carry electrical charges which make them rotate in alternating electrical fields. This polarity can be disregarded in our present discussion, because detailed investigations have shown that the characteristic frequency of various types of proteins is near 1 Mc. (Oncley, 79). In other words, the protein stops rotating with the field already at frequencies which are much lower than those under discussion at present. Protein matter, therefore, may be considered as dielectric holes in water, which is a medium of comparatively high dielectric constant. This approach leads to a reasonable analytical understanding of the dielectric constant value of 54 found at 300 Mc.

Because blood data provide the prototype for all mechanisms involved in tissues with high water content, it has been discussed here in detail. Measurements of tissues have been carried out by Osswald (50), Rajewsky, Schaefer, Osken, and Schwan (56), Cook (80), Herrick *et al.* (75), England (76), Schwan, Carstensen, and Li (4, 70) Hartmuth (81), and Kebbel (82). All of this material fits together well if one disregards Kebbel's values, which are obviously in error. It is summarized in table 3. Typical curves in figure 9 show the frequency dependence, which is determined by cellular structure, electrolytes, and protein. This table shows also that other tissues, such as subcutaneous fat and bone have different values. A complete analysis of data in tissues with high water content

REVIEW OF PHYSICAL MEDICINE AND REHABILITATION

TABLE 3. *Dielectric constant and specific resistance of various body tissues at 37° C.*

	Frequency								
	25 Mc.	50 Mc.	100 Mc.	200 Mc.	400 Mc.	700 Mc.	1000 Mc.	3000 Mc.	8500 Mc.
				Dielectric constant ϵ					
Muscle	103–115	85–97	71–76	56	52–54	52–53	49–52	45–48	40–42
Heart muscle				59–63	52–56	50–55			
Liver	136–138	88–93	76–79	50–56	44–51	42–51	46–47	42–43	34–38
Spleen	>200	135–140	100–101						
Kidney	>200	119–132	87–92	62	53–55	50–53			
Lung				35	35	34			
Brain	>160	110–114	81–83						
Fat		11–13		4.5–7.5	4–7		5.3–7.5	3.9–7.2	3.5–4.5
Bone marrow		6.8–7.7					4.3–7.3	4.2–5.8	4.4–5.4
				Specific Resistance ρ					
Muscle		113–147		95–105	85–90	73–79	75–79	43–46	12
Heart muscle				95–115	85–100	78–95			
Liver	185–210	173–195	154–179	110–150	105–130	85–115	98–106	49–50	15–17
Spleen		128–151							
Kidney		90–145		90	85	76–77			
Lung		260–450		160	140	130			
Brain	220	190–210	180–195						
Fat		1700–2500		1050–3500	900–2800		670–1200	440–900	240–370
Bone marrow		2800–5000					1000–2300	445–860	210–600

Values 25, 50, and 100 Mc. given by Osswald (50) at 200, 400, and 700 Mc. by Schwan and Li (70), at 1000, 3000, and 8500 Mc. by Herrick *et al.* (75). The values of Schwan and Li have been measured at 27° C. and are corrected for 37° C. in accordance with known temperature coefficients (see table 3). Data from other authors are omitted to avoid the necessity of introducing additional frequency parameters, but support this material.

has been given by Schwan and Li, but data of fat and bone are not as well understood (70).

Temperature coefficients between 25 and 100 Mc. have been determined by Osswald (50); between 200 and 900 Mc. by Schwan and Li (70). Table 4 summarizes the results. The dielectric constant has a positive temperature coefficient at the lower frequencies and a negative one at higher frequencies, with a transition period in between. Tissues with a high water content always have negative temperature coefficients of the specific resistance, with higher values at lower frequencies. The data of fatty tissue show similar trends, even though actual values are different. The temperature coefficients have been analyzed by Schwan and Li (70) for high water content tissues and related to the temperature coefficients of electrolytes. Their frequency dependence has been attributed to the changes occurring in the frequency dependence of the dielectric constant below 300 Mc. when the temperature varies.

ABSORPTION AND REFLECTION OF ELECTROMAGNETIC WAVES IN TISSUE

From dielectric data, we can determine the behaviour of electromagnetic waves in tissue as a function of frequency. The absorption coefficient μ for plane electromagnetic waves in a single homogeneous substance, is defined by the relationship $E = E_0 \exp(-\mu d)$ where E is the electrical field at a distance d measured from the

TABLE 4. *Temperature coefficient of specific resistance and dielectric constant of various body tissues (in per cent per degree Centrigrade)*

	Frequency				
	50 Mc.	200 Mc.	400 Mc.	900 Mc.	
Muscle	0.3		−0.2	−0.2	$100\frac{\Delta\epsilon}{\epsilon}$/°C.
Liver	0.3	0.2	−0.2	−0.4	
Spleen	1.0				
Kidney	0.5	0.2	−0.2	−0.4	
Brain	1.1				
Blood	0.3				
Serum and 0.9% saline	−0.4	−0.4	−0.4	−0.4	
Fat		1.3		1.1	
Muscle	−2.5	−1.5	−1.3	−1.0	$100\frac{\Delta\rho}{\rho}$/°C.
Liver	−2	−1.8	−1.8	−1.4	
Spleen	−2.3				
Kidney	−1.6	−2.0	−2.0	−1.3	
Brain	−1.4				
Blood	−2.7				
Serum and 0.9% saline	−2.0	−1.7	−1.6	−1.3	
Fat	−1.7 − 4.3	−4.9		−4.2	

Values at 50 Mc. from Osswald, those at 200, 400, and 900 Mc. from Schwan and Li. The temperature coefficient of the specific resistance is for all body tissues with high water content comparable to that of saline solution while the coefficient of the dielectric constant approaches only at very high frequencies the value for saline solution. It changes with frequency strongly and becomes positive at lower frequencies. The temperature coefficients of fatty material are completely different from those of tissues with high water content.

point where E is equal to E_0. The relationship of coefficient μ to the dielectric constant ϵ and the specific resistance ρ may be expressed by the equation

$$\mu^2 = \left(\frac{2\pi}{\lambda}\right)^2 \frac{\epsilon}{2}\left[\sqrt{1+\left(\frac{60\lambda}{\epsilon\rho}\right)^2} - 1\right] \tag{7}$$

where $\lambda = 3.10^{10}$/frequency is the wavelength of the radiation in air and expressed in cm. The reflection coefficient r^2 gives the relative amount of the total energy which will be reflected at each air-tissue interface. It is arrived at by the equation

$$r^2 = \frac{(n-1)^2 + \left(\frac{\mu\lambda}{2\pi}\right)^2}{(n+1)^2 + \left(\frac{\mu\lambda}{2\pi}\right)^2} \tag{8}$$

The index of refraction n of the tissue in turn is related to its dielectric properties as given by the equation

$$n^2 = \frac{\epsilon}{2}\left[\sqrt{1+\left(\frac{60\lambda}{\epsilon\rho}\right)^2} + 1\right] \tag{9}$$

Figures 11 and 12 give $\frac{1}{2}\,\mu$ and $1 - r^2$ for different frequencies and apply to high water content tissue, such as muscle, heart, kidneys, etc. $1/\mu$ is characteristic for the depth of penetration since it is the distance inside the tissue at which the electrical field strength is equal to $1/e$ (e natural logarithmic base = 2.7) of its surface value. If divided by two, it is identical with the depth of penetration, defined as the depth where energy is reduced to $1/e$. $1 - r^2$ is obviously identical with the amount of air-borne energy which is utilized for the production of heat in tissue and not reflected from the body surface. The curves hold for a temperature range of 25° to 30° C. They are deduced from the dielectric data as given previously and extrapolated to higher frequencies on the basis of the above discussed mechanisms, which determine the electrical properties of tissue (equations 4 and 5). The ratio $60\lambda/\epsilon\rho$ becomes smaller than one for all tissues, when the wavelength is less than 1 meter (frequencies above 300 Mc.). The formulas 7, 8 and 9 may then be approximated by:

$$\mu = \frac{60\pi}{\rho\sqrt{\epsilon}} \tag{10}$$

$$r^2 = \frac{\left[\sqrt{\epsilon}\left(1+\frac{1}{8}\left(\frac{60\lambda}{\epsilon\rho}\right)^2\right)^2 - 1\right]^2 + \frac{(30\lambda)^2}{\rho^2\epsilon}}{\left[\sqrt{\epsilon}\left(1+\frac{1}{8}\left(\frac{60\lambda}{\epsilon\rho}\right)^2\right)^2 + 1\right]^2 + \frac{(30\lambda)^2}{\rho^2\epsilon}} \tag{11}$$

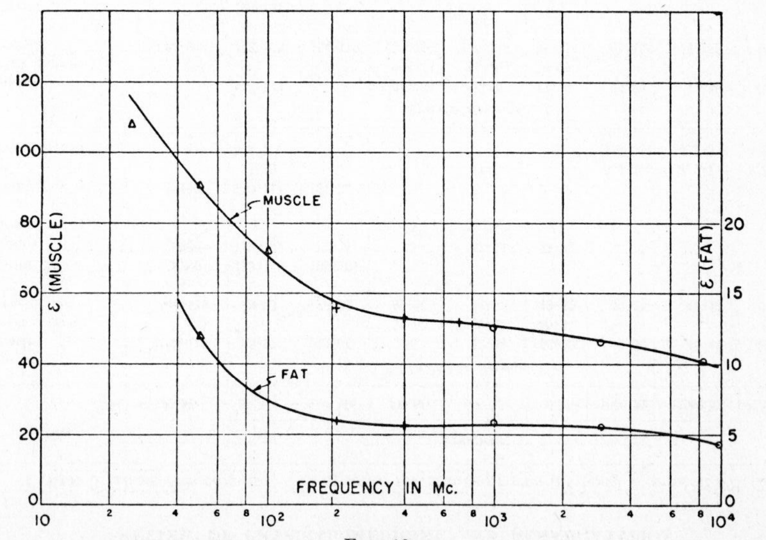

FIG. 10a

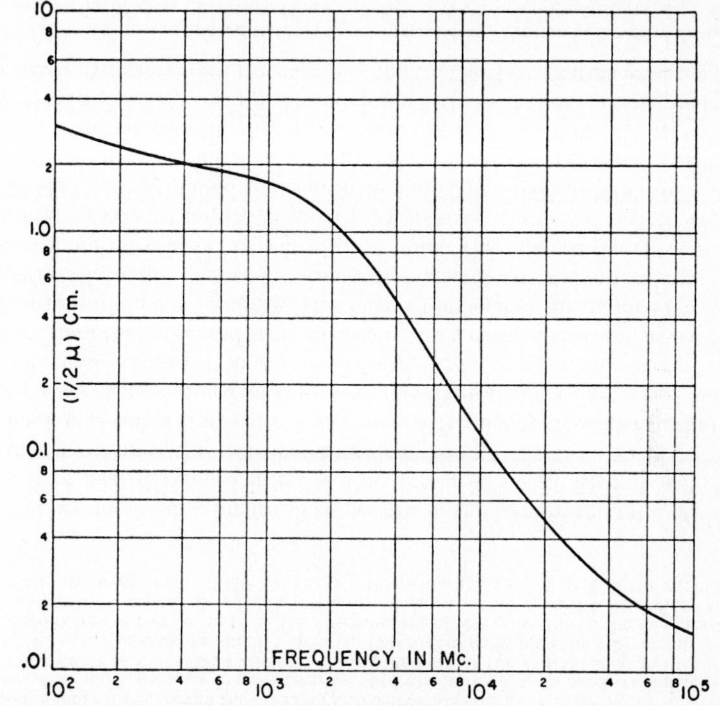

FIG. 10b. Dielectric constant and specific resistance of muscular and fatty tissue are presented as function of frequency in figures 10a and 10b. Temperature 37° C. Muscular tissue is representative for all body tissues with high water content and blood. Its dielectric constant values are much higher and its resistance values much lower than those of fatty tissue. The data for muscular tissue are reproducible within a few per cent while especially the resistance values for fatty material vary from sample to sample very much. This is indicated by highest and lowest values as given by the various authors and establishes a range of variation as indicated in the figure b by an area limited by curves of highest and lowest values. The number of measurements which have been carried out with fatty material is very limited. The area of variation can be, therefore, only suggestive of the true range of variation until more data have been published. Values taken from Osswald Δ Schwan and Li + and Herrick *et al.* ○. Data from other authors support this evaluation.

Equation 10 shows the depth of penetration $\frac{1}{2}\mu$ to be proportional to the specific resistance ρ. The latter decreases about 25 per cent when the temperature is raised from 27° to 37° C., i.e. to body temperature. Equation 11 demonstrates that the reflection coefficient is less dependent on ρ and consequently upon temperature changes than is the absorption coefficient.

The curves in figures 11 and 12 reflect the two regions where electrical data are dependent upon frequency as given in figure 10. The factor, characterizing the total energy absorption $(1 - r^2)$ changes over a range of one decade only, while the depth of penetration changes three decades. One centimeter depth of penetration is predicted for a 12-cm. wavelength (2500 Mc.) which is presently used in microwave diathermy, and only 0.1 cm. penetration for a 3-cm. wavelength (10,000 Mc.). When the wavelength is less than 1 mm., the depth of penetration approaches a constant value of the same size as the penetration of infrared radia-

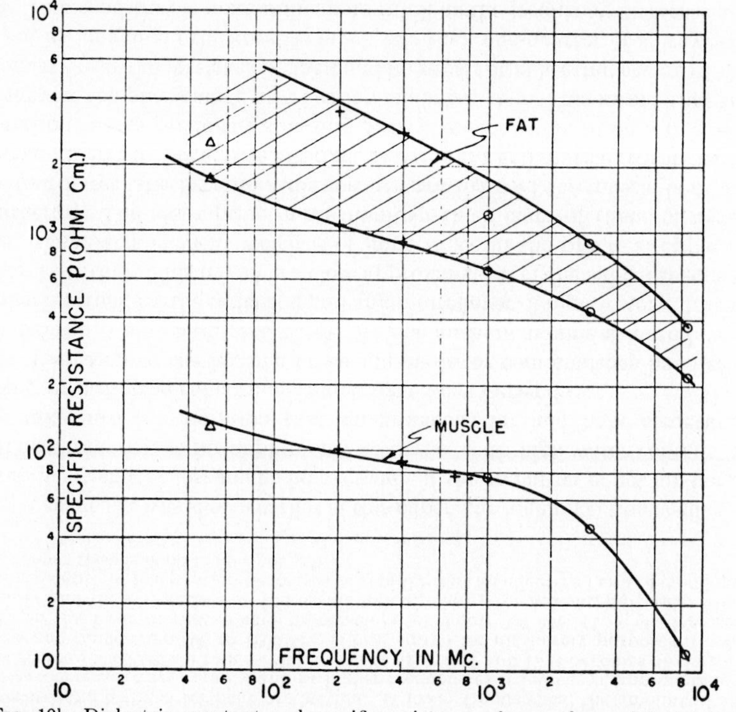

FIG. 11. Depth of penetration $\frac{1}{2}\mu$ of electromagnetic radiation in deep tissues as function of frequency. Temperature 25°–30° C. The depth of penetration decreases above 1000 Mc. with increasing frequency rapidly and amounts to only 9 mm. for presently used 2450 Mc. The depth of penetration values are even lower at body temperatures.

tion. Hardy and Muschenheim (83) have demonstrated the latter to be smaller than 0.6 mm. in human skin. These depths of penetration are much smaller than has been generally believed. It must be understood that actual temperature measurements in living tissue give greater values because of the heat flow occurring from warmer to cooler tissues. Physiological temperature investigations are of limited value, therefore, in demonstrating fundamental differences in efficiency between various forms of heat therapy. All of this supports the fact that diathermy equipment using wavelengths less than 10 cm. is but little superior to infrared radiation for purposes of deep heating (3, 5, 84).

The $1 - r^2$ curve in figure 12 shows that 90 per cent of the incident air-borne radiation enters the tissue at very short wavelengths approaching infrared. One of the chief advantages of infrared is this efficient utilization. However, the energy is rapidly absorbed once it enters the tissue, and is of little value in deep heating. The reflection coefficient r^2 is much greater at practical diathermy frequencies. At wavelengths above 10 cm. (frequency below 3000 Mc.) less than 40 per cent of the incident air-borne energy enters the tissues, as seen in the

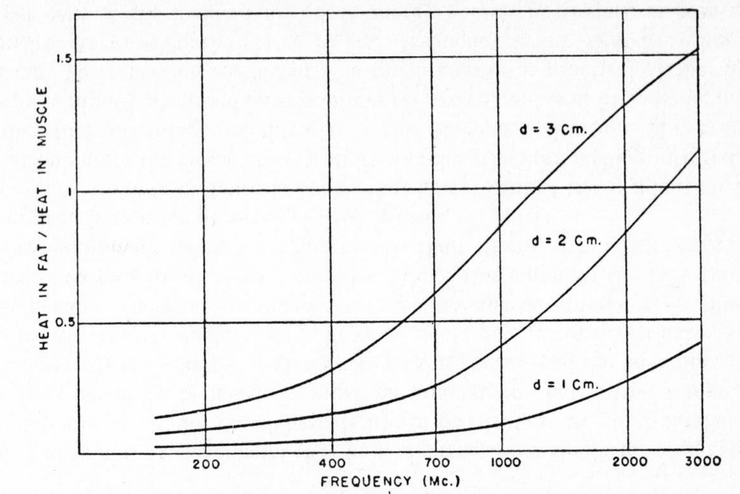

FIG. 12. Percentage of airborne electromagnetic radiation ($1 - r^2$) which is absorbed by tissue with high water content, such as muscle and other deep tissues, as function of frequency. The data hold for 25–30° C. and are relatively temperature independent (see text). Throughout the frequency range from 1000 to 10,000 Mc. about 40 per cent of the energy is absorbed and utilized for heat production. At lower frequencies, the amount of absorbed energy decreases very rapidly and at higher frequencies it increases, approaching 100 per cent above 100,000 Mc. From a comparison of figures 11 and 12 it becomes obvious that the frequency range from 500 to 1000 Mc. is best suited for diathermy purposes if one judges only on the basis of tissues such as muscle, etc. Below 500 Mc. ($1 - r^2$) decreases very rapidly and thereby causes an increasing loss of energy for heating purposes. Above 1000 Mc. the depth of penetration decreases so rapidly that an effective heating development in the deeper tissues is no longer possible.

$1 - r^2$ curve. As was shown earlier in this paper, the high reflection coefficient and consequent energy loss might be overcome if the radiation is not air-borne.

This discussion so far applies only to tissues with high water content. It does not take into consideration that subcutaneous fat and bone structures may affect seriously the conclusions which have been derived.

A discussion of the heating of fat-muscle layer combinations has been given by Schwan, Carstensen, and Li (4, 5). The authors assume a finite layer of fat which is struck by the radiation and an infinite layer of high water content tissue in series with it. This model is a close approximation of the subcutaneous fat and deep tissue arrangement which is of importance in the practical application of diathermy. The assumption of an infinitely thick layer of tissue of high water content is realistic because radiation penetrating into the muscle is rapidly absorbed as shown above. The interface between fat and muscle reflects part of the radiation energy producing standing wave phenomenon in front of this interface. However, the phase angle of reflection is almost 180°. This means that the field intensity near the interface is small and no substantial heating occurs in the fatty layer, as long as its thickness is much less than one-quarter of a wavelength in fat. This condition is more difficult to fulfill as the frequency increases and substantial heating occurs. Figure 13 shows this in the form of the ratio of the total amount of heat developed in fat to that produced in the deep tissue[12]. As the frequency increases, this ratio increases also.

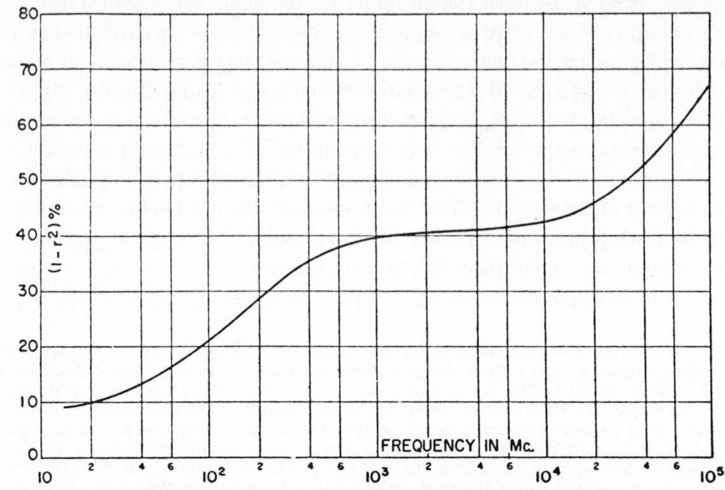

FIG. 13. The ratio of total heat developed by electromagnetic radiation in subcutaneous fat to the total heat developed in muscular tissue is given as function of frequency and thickness of fat layer. It is seen that the heat developed in fat increases with fat layer thickness and frequency. The curves characterize the ability of electromagnetic radiation to pass without energy loss through the fat layer and deliver its energy in form of heat into the deep tissues. Lower frequencies than presently used are necessary to make radiation efficient from this point of view and to deliver the major part of the radiant energy into the deep tissue (ratio factor must be smaller than 1). The curves show also the very strong influence of the thickness of the fat layer on the effectiveness of the radiation in the deep tissues and explain why it is impossible to make reliable statements with regard to the effectiveness of presently used equipment (2450 Mc.) without consideration of the size of the subcutaneous fat layer.

This law is important, since it indicates a frequency dependence that is the opposite to that obtained in the case of current therapy (see page 376). The ratio increases rapidly with the thickness of the fatty layer. However, it is much smaller than that observed with ultrashort wave current therapy, as seen from a comparison of figures 3 and 13. This means that one of the major advantages of radiation diathermy lies in its ability to transmit the radiant energy to deep tissues more effectively. It will also be noted that the efficiency of radiation diathermy could be improved greatly by increasing the wavelength of 12 cm. (2500 Mc.) presently used to between 30 and 50 cm. (600 to 1000 Mc.). If this is done, the fat heating will be substantially decreased, and the depth of penetration in deep tissues improved with only a small energy loss, due to slightly increased reflection at the body surface. The adoption of 2500 Mc. for therapeutic purposes was based, therefore, on a premature evaluation of the radiation situation.

[12] The curves are calculated for average electrical data of fat and change with the dielectric properties of fatty material. However, the curves are representative, since the same trend always is observed with frequency and thickness.

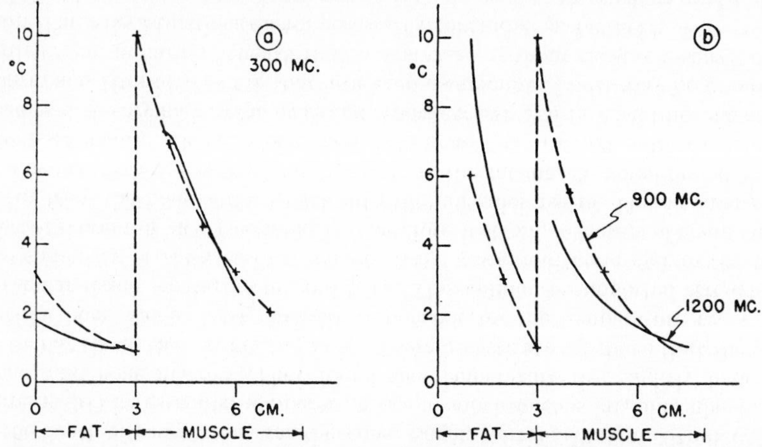

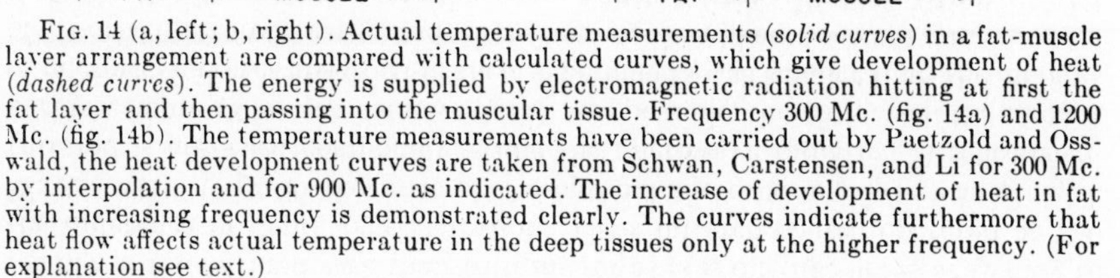

FIG. 14 (a, left; b, right). Actual temperature measurements (*solid curves*) in a fat-muscle layer arrangement are compared with calculated curves, which give development of heat (*dashed curves*). The energy is supplied by electromagnetic radiation hitting at first the fat layer and then passing into the muscular tissue. Frequency 300 Mc. (fig. 14a) and 1200 Mc. (fig. 14b). The temperature measurements have been carried out by Paetzold and Osswald, the heat development curves are taken from Schwan, Carstensen, and Li for 300 Mc. by interpolation and for 900 Mc. as indicated. The increase of development of heat in fat with increasing frequency is demonstrated clearly. The curves indicate furthermore that heat flow affects actual temperature in the deep tissues only at the higher frequency. (For explanation see text.)

The dielectric data of bone and bone marrow indicate that we may anticipate in front of bone similar reflection phenomena, as those noted with a fat muscle combination. The skin on the other hand does not present any major problem, at least for wavelength values greater than 10 cm. The thickness of skin is so small in comparison to the wavelength of the radiation that it may be disregarded in the present discussion. This statement is based on the fundamental law that the transmission of an electrical signal through any tissue layer is not affected by the layer if a) its thickness is less than one-quarter of the wavelength employed and b) the layer is so thin that its absorbs no appreciable part of the total radiation energy.

Experimental Studies

All the previous discussions have been based on analytical arguments which utilized known electrical data derived from a study of various tissues in order to arrive at figures for the "primary" heat development. Of greater interest to the physician, however, are actual temperature values. It is, therefore, valuable to compare the analytical results, above described, with experimental investigations. Experimentally, temperature variations with depth may be less pronounced than those predicted from the curves obtained from the determination of primary heat development. This is due to heat flow. On the other hand, the significance of selective fat heating is emphasized if we transform heat development into actual temperature rise. The amount of temperature rise that results from heat

development may be calculated by using the following equation: Heat development = temperature rise × specific heat × density. Both specific heat and density are lower in fat than in muscular tissue.

Ahrens, Esau, Osswald, and Paetzold (44, 85, 86) have carried out temperature measurements in fat-muscle phantoms at frequencies of 300 and 1200 Mc. Figure 14 summarizes their results. This figure not only shows the temperature rise in the fat-muscle combination for 25 cm. and 100 cm. wavelength (1200 and 300 Mc.), but also compares these findings with the heat development curves calculated by Schwan, Carstensen, and Li (4) for 300 and 900 Mc. (*dashed curves*). The comparison reveals marked similarity in the curves. In both cases, temperature rise, as well as heat development, confirms the enormous increase in selective heat development and consequent temperature rise in fat as the frequency increases. Further, their results coincide as to the limited depth of heat penetration into muscular tissue. However, the variation due to frequency in depth of penetration in the muscular material is not as great as predicted from the heat development curves. This indicates that heat flow plays an increasingly important role as the frequency increases. At 100 Mc. good agreement still exists between the calculated heat development curve and temperature curve in the muscular tissue, which means that heat flow at that frequency does not affect appreciably the temperature distribution. On the other hand, at 1200 Mc. depth of penetration of the radiation (indicated by the heat development curve) is significantly smaller than the "effective" depth of penetration shown by the temperature curve.

Measurements which support this conclusion have been carried out by Murphy, Paul, and Hines (87), on the hind limb of dogs. Their results also show a decrease in the effective depth of penetration with increasing frequency. However, their depth of penetration values are, at higher frequencies, significantly higher than those given in figure 11. The explanation is that the values in figure 11 represent the depth of penetration of the radiation while those obtained by Murphy, Paul, and Hines include the effects of heat flow. This is proven by the latter investigators, who also measured effective depth of penetration values for infrared penetration. Their values for infrared are of the order of 1 to 1.5 cm. and represent solely the effect of heat flow, since the total absorption of infrared radiant energy occurs within a fraction of a millimeter (83). Practically the same values are obtained for 3 cm. radiation (10,000 Mc.) indicating that the primary depth of penetration with radiation at that wavelength is very small as compared to 1 cm. and is not able, therefore, to increase effectively the value found for infrared due to heat flow alone. This statement is supported also by the primary depth of penetration which is given in figure 11 for the 3 cm. radiation with only 0.1 cm. and is extremely small in comparison to the effective depth of penetration due to heat flow. It may be stated that the primary depth of radiation must be greater than the depth of penetration due only to heat flow, in order to justify the choice of the more expensive and difficult electromagnetic radiation over the use of the simpler infrared. In accordance with figure 11, any electromagnetic radiation device which is to give significantly increased deep heating in comparison t infrared must operate at wavelengths above 20 cm. Murphy, Paul, and Hines (87) have included in their study the use of "irradiation with 1600 cm. wavelength". Under the latter condition, their curves indicate a smaller depth of penetration than that obtained at a lower wavelength (higher frequency). This seems to contradict our previous statement that depth of penetration of electromagnetic radiation decreases as the frequency increases. However, the description of their experimental technique shows that in reality high frequency current diathermy was used instead of irradiation. The current was supplied by a generator, which potentially would be able to produce 1600 cm. wavelength radiation, if hooked up with a reflector device (see the fundamental difference between high frequency current therapy and radiation therapy, page 374). Their results, therefore, merely prove that radiation techniques of proper wavelength provide better effective depth of penetration than does a high frequency current operating at 19 Mc. under the conditions outlined by these authors.

Rae, Herrick, Wakim, and Krusen (88) carried out temperature measurements in skin, subcutaneous tissue, and muscular tissue at 1.5 cm. and 3 cm. depth. They report that the temperature rise at a depth of 1.5 cm. in muscular tissue is slightly higher than in fat for a frequency of 2450 Mc. Krusen (89) also reports a greater temperature rise in the muscular tissue than in subcutaneous fat. Essman and Wise (90) found that 2450 Mc. radiation has a greater depth of penetration than infrared. However, their statement that surface injury is obtained more readily in the infrared case than with microwaves is due not necessarily only to the difference in depth of penetration. Their infrared source operated at 250 Watts, while only 80 per cent of the optimal microtherm power was utilized. Infrared is absorbed completely, while the microwave radiation is reflected in part (see figure 12 for quantitative data). It is possible, therefore, that the amount of microwave power absorbed in their experiments was smaller than the amount of absorbed infrared energy and, therefore, was less likely to cause damage even if the depth of penetration was the same. Engel, Herrick, Wakim, Grindlay, and Krusen (91) investigated temperature rise in subcutaneous fat, muscular tissue, bone, and bone marrow. They found the temperature rise slightly higher in muscular tissue than in fat and bone. These investigators believe that the presence of bone was at least partially responsible for the temperature rise in the muscular tissue separating fat and bone. This opinion is supported strongly by the similarity of the electrical data of fatty tissue and bone marrow from an electrical point of view (see table 3). Therefore, a strong reflection of energy must occur from the bone surface back into the muscular tissue. This is completely analogous to the reflection which occurs at the fat-muscle boundary above as discussed before. Boyle, Cook, and Buchanan (92) carried out experimental studies with very high frequencies and report excessive surface heating. Salisbury, Clark, and Hines (93) state that a dangerous amount of heat may be generated with 12 cm. radiation beneath the body surface, without causing a sensation of pain. These investigations do not necessarily contradict each other.

In many instances heat development is at its peak at the surface of the body. This is true either when there is practically no subcutaneous fat, or when the fat

layer is excessive as outlined by the analytical study by Schwan, Carstensen, and Li, summarized in figure 13. However, the actual temperature curves also are affected by the efficiency of the cooling of the body surface. Therefore, a maximum temperature is created somewhere below the surface. The exact location depends on surface cooling, the amount of subcutaneous fat, the blood flow, and frequency. It is, therefore, difficult to predict. Boyle, Cook, and Buchanan applied a waveguide applicator to the surface of the body and thereby reduced the possibility of effective surface cooling. This automatically places the maximum of temperature nearer to the surface, especially when radiation of such high frequency as applied by them, is used. Clark's (94) statement that maximum heating occurs in the range from 8 to 12 cm., if the wavelength is varied, may be explained by the interplay of the variations in the reflection and absorption coefficients. Figure 12 shows that the amount of absorbed energy decreases rapidly with increasing wavelengths above 50 cm. Therefore, temperature rise will be less pronounced at longer wavelengths.[13] Depth of penetration, on the other hand, decreases rapidly at wavelengths below 10 cm. as the frequency increases. This makes possible more efficient surface cooling, since the total amount of radiation energy is transformed into heat, in an increasingly small volume beneath the body surface. Again decrease in temperature rise will result. Both effects combined will produce a maximum temperature rise in homogeneous tissue in the frequency range between 300 and 3000 Mc. (wave-length 10 to 100 cm.). If it is assumed that subcutaneous fat overlies muscular tissue, the development of heat with a decrease of wavelength shifts its location even more rapidly nearer the body surface, as indicated in our discussion of the relative heating of fat and deep tissues (fig. 13). This will shift the range of optimal heating to somewhat smaller wavelengths and may explain Clark's statements.

Osborne and Federick (95), Rae, Wakim, Herrick, and Krusen (96), and Horvath, Miller, and Hutt (97) claim that 12 cm. irradiation heats superficial tissue better than the deeper tissues. This seems contradictory to the results of others as reported above (88, 89, 91), but is not surprising in view of the complexity of the dependence of heat development upon the relative amount of fat, as discussed in detail above, and the dependence of effective depth of penetration upon the degree of surface cooling (fig. 13). Michaelis (98) reports about Sequin's measurements, which again support our statement that higher frequencies (shorter wavelengths) are absorbed more rapidly than smaller ones. These measurements have been carried out at 3 cm. and 21 cm. wavelengths.

An attempt to evaluate analytically the effects of temperature flow has been made by Cook (99). In order to obtain equations, which can be handled mathematically, linear heat flow had to be assumed. Under conditions which approximately fulfill this assumption, agreement between calculated and experimental temperature curves was satisfactory. However, when capillary dilatation, due to heating, becomes pronounced, discrepancies between theory and experiments appear. These studies, carried out at about 10 cm. are important, since they

[13] It is assumed that the power output of the high frequency radiation source is kept constant as frequency is varied.

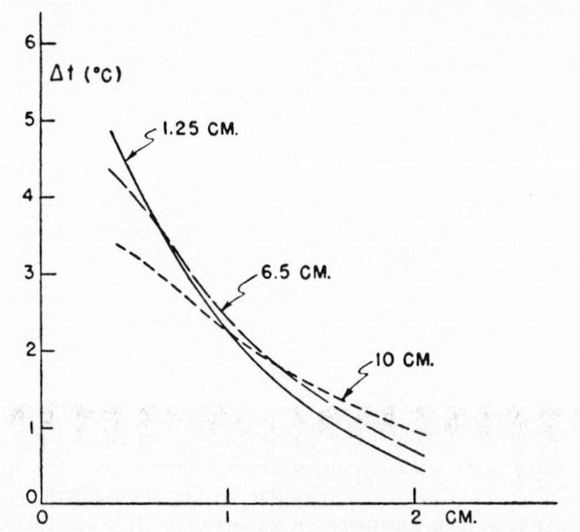

FIG. 15. Temperature rise Δt in centigrade in muscular tissue as function of depth for various wavelengths, as calculated by Cook. A total absorption of 0.42 W/cm.² was assumed over a period of 2 minutes and an initial linear increase of temperature from 30° C. at the surface of the tissue to 34° C. in 2 cm. depth. Only linear heat flow in the direction of propagation of the electromagnetic radiation is assumed in the calculation. The curves show that "effective" depth of penetration is much less frequency dependent for such small wavelength values than "primary" depth of penetration of electromagnetic radiation (see for comparison fig. 11). Radiation of less than 10 cm. wavelength is, therefore, not superior to infrared for purposes of deep heating, if only homogeneous tissues similar to muscle are considered.

present in a particular case a detailed outline of the interplay of radiation absorption, heat flow, and effective temperature rise. Cook concluded from his study that the temperature is highest at the surface only when the surface is in contact with a poor thermal conductor. His calculations of temperature rise as a function of depth and wavelength are summarized in figure 15, which is taken from his article. This figure shows that, due to thermal flow, no significant dependence of effective depth of penetration on wavelength exists if the wavelength is smaller than 10 cm. (3000 Mc.).[14] This is in complete agreement with the analysis which we presented above on the basis of Paetzold, Osswald and Murphy *et al.*'s data and *proves that electromagnetic radiation of a wavelength smaller than* 10 *cm. is no longer superior to infrared for purposes of deep heating.*

In conclusion, it can be stated that the efficiency of presently used microwave apparatus operating at 2450 Mc. is unpredictable, from a practical point of view. It will not provide a significantly greater temperature increase in deep tissue than infrared. On the other hand, it is capable of overbridging fatty subcutaneous material and transforming the major part of its energy into heat beneath the fat layer, if the latter is less than one-half inch thick. If the fat layer is thicker,

[14] More pronounced temperature differences within the first 5 mm. of tissue from the surface were reported by Cook. However, differences will be reduced greatly due to heat loss from the surface. Such cooling effects are neglected in Cook's calculation, and his values up to 5 mm. depth are omitted, therefore.

excessive surface heating will again result. Whether surface or deep tissues are heated depends primarily on certain factors, such as surface cooling, the thickness of a fatty layer, and heat exchange within the tissues. This explains contradictory reports as to the heating efficiency of the 12 cm. microwaves.

REFERENCES

1. PAETZOLD, J. High Frequency Techniques in Medicine, Chapter in *Advancements in High Frequency Techniques*, Vol. 2, pp. 791-799. F. VILBIG AND Z. ZENNECK. Akademische Verlagsges., Leipzig, 1943.
2. MURPHY, A. J., PAUL, W. D., AND HINES, H. M. Arch. Phys. Med,. **31**: 151, 1950.
3. SCHWAN, H. P., CARSTENSEN, E. L., AND LI, K. J.A.M.A., **149**: 121, 1952.
4. SCHWAN, H. P., CARSTENSEN, E. L., AND LI, K. Trans. A.I.E.E., Communications & Electronics, Sept., 1953, p. 438.
5. SCHWAN, H. P., CARSTENSEN, E. L., AND LI, K. Arch Phys. Med. & Rehab., **35**: 13, 1954.
6. HINES, H. M., AND RANDALL, J. E. Elec. Eng., **71**: 879, 1952.
7. CLARK, J. W. Proc. I.R.E., **38**: 1028, 1950.
8. HERRICK, J. F., AND KRUSEN, F. H. Elec. Eng., **72**: 239, 1953.
9. RICHARDSON, A. W., DUANE, T. D., AND HINES, H. M. Arch. Ophth., **45**: 382, 1951.
10. KRUSEN, F. H. *Physical Medicine*, pp. 393-99. W. B. Saunders Co., Philadelphia, 1941.
11. OSBORNE, S. L., AND HOLMQUEST, H. J. *Technic of Electrotherapy*, pp. 573–608. Charles C Thomas Co., Springfield, Ill., 1944.
12. KRASNY-ERGEN, W. Annalen der Physik, (5) **23**: 277, 1935.
13. KRASNY-ERGEN, W. Annalen der Physik, (5) **23**: 304, 1935.
14. SCHAEFER, H., AND SCHWAN, H. P. Annalen der Physik, (5) **43**: 99, 1943.
15. SCHAEFER, H., AND SCHWAN, H. P. Strahlentherapie, **77**: 123, 1947.
16. KULKA, D. Wiener med. Wchnschr., **I**: 210, 1935.
17. MALOV, N. N. Strahlentherapie, **53**: 326, 1935.
18. WENK, P. Strahlentherapie, **65**: 657, 1939.
19. PAETZOLD, J., AND BETZ, P. Z. ges. exp. Med., **94**: 696, 1934.
20. GEBBERT, A. Klin. Wchnschr., **13**: 905, 1934.
21. PAETZOLD, J., AND SCHAEFER, H. *Natural Sciences and Medicine in Germany, 1934-49*, Vol. 22, pp. 17–19. Biophysics II. Dieterich'sche Verlagsbuchhandlung, Wiesbaden, Germany, 1948.
22. SCHAEFER, H., AND STACHOWIACK, R. Z. Techn. Physik, **21**: 367, 1940.
23. ESAU, A. French Patent 666264, Jan. 21, 1928.
24. SCHERESCHEWSKY, J. W. Pub. Health Rep. **41**: 1939, 1926; **43**: 927, 1928.
25. MCLENNAN, J. C., AND BURTON, A. C. Canad. J. Research, **5**: 550, 1931.
26. PAETZOLD, J. Z. f. techn. Physik, **13**: 212, 1932.
27. SCHAEFER, H. Z. f. exp. Med., **98**: 257, 1936.
28. HIGASI, K. *Theory and Application of High Frequency Phenomena*, Chap. 2. Edited by Y. ASAMI. Monograph Series of the Res. Inst. Appl. Electricity, No. 1, Hokkaido, Univ., Sapparo, Japan, 1950.
29. SCHAEFER, H. Z. f. exp. Med., **100**: 706, 1937.
30. SCHAEFER, H. Deutsche med. Wchnschr., **27**: 955, 1938.
31. SCHAEFER, H. *Ultrashortwaves, Results of Biophysical Research*, Chap. I, Sec. B. Edited by B. RAJEWSKY. Georg Thieme, Leipzig, Germany, 1938.
32. ESAU, A., PAETZOLD, J., AND AHRENS, E. Naturwissenschaften, **24**: 520, 1936.
33. MERRIMAN, J. R., HOLMQUIST, H. J., AND OSBORNE, S. L. Am. J. M. Sc., **187**: 677, 1934.
34. KOWARSCHIK, J. Klin. Wchnschr., 42, 1934.
35. PAETZOLD, J., AND WENK, P. Strahlentherapie, **55**: 692, 1936.
36. PAETZOLD, J., AND WENK, P. Fortschritte a. d. Gebiet d. Roentgenstr., **54**: 83, 1934.
37. PAETZOLD, J. High frequency techniques in medicine, Chapter in *Advancements in High*

Frequency Techniques Vol. 2, pp. 788–790. F. Vilbig and Z. Zenneck. Akademische Verlagsges. Leipzig, 1943.
38. Paetzold, J. Radiologica, **1:** 530, 1937.
39. Osswald, K. Strahlentherapie, **64:** 530, 1939.
40. Hollman, H. E. Hochfrequenztechn. u. Elektroak., **50:** 81, 1937.
41. Hollman, H. E. Strahlentherapie, **64:** 691, 1939.
42. Hollman, H. E. *Ultrashortwaves—Results of Biophysical Research*, Vol. 1, Chap. 4, pp. 232–49. Georg Thieme, Leipzig, Germany, 1938.
43. Paetzold, J. Fortschr. a. d. Gebeit d. Roentgenstrahlen., **58:** 69, 1938.
44. Paetzold, J. Wiss. Veroeffentl. Siemens-Werken, **19:** 135, 1940.
45. Bruenner-Ornstein, M., and Randa, K. Strahlentherapie, **59:** 267, 1937.
46. Denier, A. Wien. med. Wchnschr., **87:** 748, 1937.
47. Rajewsky, B., and Schaefer, H. Deutsche med. Wchnschr., **63:** 1065, 1937.
48. Schaefer, H. Fortschr. a. d. Gebiet Roentgenstrahlen., **54:** 59, 1936.
49. Rajewsky, B., Osken, H., and Schaefer, H. Naturwissenschaften, **25:** 24, 1937.
50. Osswald, K. Hochfrequenztechn. u. Elektroak., **49:** 40, 1937.
51. Gersten, J. W., Wakim, K. G., and Krusen, F. H. Arch. Phys. Med., **31:** 281, 1950.
52. Martin, G. M., Rae, J. W., and Krusen, F. H. South. M. J., **43:** 519, 1950.
53. Paetzold, J., and Osswald, K. Strahlentherapie, **66:** 303, 1939.
54. Born, H. Hochfrequenztechn. u. Elektroak., **62:** 20, 1943.
55. Spiller, K. D. Elektrotechn. ZS., **71:** 27, 1950.
56. Rajewsky, B. *Ultrashortwaves—Results of Biophysical Research*, Vol. 1, Chap. 2. Georg Thieme, Leipzig, Germany, 1938.
57. Cole, K. S. J. Gen. Physiol., **12:** 29, 1928.
58. Cole, K. S. J. Gen. Physiol., **15:** 641, 1932.
59. Daenzer, H. Annalen d. Physik, **20:** 463, 1934.
60. Daenzer, H. Annalen d. Physik, **21:** 783, 1934/35.
61. Fricke, H. Physics, **1:** 106, 1931.
62. Schaefer. Ztschr. f. d. ges. exp. Med., **92:** 341, 1933.
63. Schwan, H. Annalen d. Physik, **40:** 509, 1941.
64. Schwan, H. Ztschr. f. Naturforschung, **3b:** 361, 1948.
65. Rajewsky, B., and Osken, H. Deutsche med. Wchnschr., **20:** 780, 1937.
66. Rajewsky, B., Inouye, K., and Osken, H. Deutsche med. Wchnschr., **63:** 1221, 1937.
67. Cole, K. S. Tabulae Biologicae, **19:** 24, 1942.
68. Rajewsky, B., and Schwan, H. Naturwissenschaften, **10:** 315, 1948.
69. Cook, H. F. Brit. J. Applied Physics, **3:** 249, 1952.

REVIEW OF PHYSICAL MEDICINE AND REHABILITATION

70. Schwan, H., and Li, K. Proc. I. R. E., **41:** 1735, 1953.
71. Cole, K. S., and Curtis, H. J. From Otto Glasser's *Medical Physics*, Vol. II, p. 82. Chicago, Year Book Publishers, 1950.
72. Malov, N. N. C. R. (Doklady) Acad. Sc. U.R.S.S. (N.S.) **24:** 437, 1939.
73. Gsell, G. Z. Physik, **43:** 101, 1942.
74. Schwan, H. P. Am. J. Phys. Med., **32:** 144, 1953.
75. Herrick, J. F., Jelatis, D. G., and Lee, G. M. Federation Proc., **9:** 60, 1950, and personal communication.
76. England, T. S. Nature, **166:**, 480 1950.
77. Cook, H. F. Nature, **168:** 247, 1951.
78. Debye, P. *Polar Molecules*. Chemical Catalogue Co., Inc., New York, 1929.
79. Oncley, J. L. Chem. Rev., **30:** 433, 1942.
80. Cook, H. F. Brit. J. Applied Physics, **2:** 295, 1951.
81. Hartmuth, L. Dielectric Properties of Biological Substances in the Range from $\lambda = 9$ to $\lambda = 180$ cm. Ph.D. thesis work, Univ. of Frankfurt, Germany.
82. Kebbel, W. Ztschr. Hochfrequenztechn. u. Elektroakustik, **53:** 81, 1939.
83. Hardy, J. D., and Muschenheim, C. J. Clin. Investigation **15** (**1**): 1, 1936.
84. Schwan, H. P., Carstensen, E. L., and Li, K. Electronics, **27:** 172, 1954.
85. Esau, A., Paetzold, J., and Ahrens, E. Naturwissenschaften, **26:** 477, 1938.
86. Paetzold, J., and Osswald, K. Naturwissenschaften, **26:** 478, 1938.
87. Murphy, A. J., Paul, W. D., and Hines, H. M. Arch. Phys. Med., **31:** 151, 1950.
88. Rae, J. W., Herrick, J. F., and Krusen, F. H. Arch. Phys. Med., **30:** 199, 1949.
89. Krusen, F. H. Proc. Roy. Soc. Med., **43:** 641, 1950.
90. Essman, L., and Wise, C. S. Arch. Phys. Med., **31:** 502, 1950.
91. Engel, J. P., Herrick, J. F., Wakim, K. G., Grindlay, J. H., and Krusen, F. H. Arch. Phys. Med., **31:** 453, 1950.
92. Boyle, A. C., Cook, H. F., and Buchanan, T. J. Brit. J. Phys. Med., **13:** 2, 1950.
93. Salisbury, W. W., Clark, J. W., and Hines, H. M. Electronics, **22:** 66, 1949.
94. Clark, J. W. Proc. I. R. E., **38:** 1028, 1950.
95. Osborne, S. L., and Frederick, J. N. J.A.M.A., **137:** 1036, 1948.
96. Rae, J. W., Herrick, J. F., Wakim, K. G., and Krusen, F. H. Arch. Phys. Med., **30:** 199, 1949.
97. Horvath, S. M., Miller, R. N., and Hutt, B. K. Am. J. Med. Sc., **216:** 430, 1948.
98. Michaelis, M. Brit. J. Phys. Med., **12:** 38, 1949.
99. Cook, H. F. Brit. J. Applied Physics, **3:** 1, 1952.

Analyses of Electromagnetic Fields Induced in Biological Tissues by Thermographic Studies on Equivalent Phantom Models

ARTHUR W. GUY, MEMBER, IEEE

Abstract—One of the most vexing problems in studies involving the interaction of electromagnetic fields and living biological systems and tissues is the quantification of the fields induced in the tissues by nearby sources. This paper describes a method for rapid evaluation of these fields in tissues of arbitrary shape and characteristics when they are exposed to various sources including plane wave, aperture, slot, and dipole sources. The method, valid for both far- and near-zone fields, involves the use of a thermograph camera for recording temperature distributions produced by energy absorption in phantom models of the tissue structures. The magnitude of the electric field may then be obtained anywhere on the model as a function of the square root of the magnitude of the calculated heating pattern. The phantoms are composed of materials with dielectric and geometric properties identical to the tissue structures which they represent. The validity of the technique is verified by comparing the results of the experimental approach with the theoretical results obtained for the case of plane layers of tissue exposed to a rectangular-aperture source and cylindrical layers of tissue exposed to a plane-wave source. This technique has been used successfully by the author for improving microwave applicators.

Introduction

LIVING biological tissues may be intentionally exposed to electromagnetic energy for therapeutic or diagnostic purposes, or in connection with studies on the behavior of living systems under the influence of electromagnetic fields. Tissues may also be accidentally exposed to electromagnetic sources, such as radars, microwave ovens, industrial microwave equipment, and diathermy apparatus.

In many of these cases both the tissue geometries and the electromagnetic source configuration are complex, and any theoretical calculation of the fields existing in or near the tissues is difficult, if not impossible.

The theoretical analyses of electromagnetic fields in biological tissues are complicated chiefly by the complex geometries and are limited to special simplified cases, such as plane, cylindrical, or spherical layers of tissues exposed to plane wave, dipole, and aperture sources.

A man's body is covered by a thin layer of skin next to a thicker layer of subcutaneous fat over muscle or other tissue of high water content. The absorption and reflective properties of a man exposed to near-zone or radiation fields in the low-frequency portion of the microwave band is related chiefly to the geometry and dielectric properties of these three layers of tissue. Consequently, most theoretical treatments of induced field distributions in tissues have been based on the highly simplified models composed of one or more of these tissues exposed to the fundamental sources depicted in Fig. 1(a)–(e). The more realistic case, involving irregular boundaries and an arbitrary source, shown in Fig. 1(f), is not amenable to an exact theoretical solution, so one must analyze it by means of an experimental technique such as the one discussed in this paper. The fundamental problem of plane layers of tissues exposed to plane waves, shown in Fig. 1(a), has been treated by Schwan *et al.* [1]–[3] and Lehmann *et al.* [4], [5]. This model has been used to illustrate many of the major frequency-dependent deficiencies in microwave diathermy apparatus. These deficiencies include 1) excessive dielectric heating in the subcutaneous fat layer; 2) insufficient penetration in the deep muscle layers where dielectric heating is desired; and 3) very little control or knowledge of the actual energy absorbed by the tissues, varying with the physical characteristics of the patient. Since this model is based on a plane-wave source, however, it does not show the relationship between the source size and the induced fields. Theoretical treatment of triple tissue layers exposed to finite source distributions are complex, and the cost for numerical evaluation of the expressions can be prohibitive. This is especially true when entire field distributions are needed as functions of frequency, thickness of tissue layers, source size, and source geometry. Consequently,

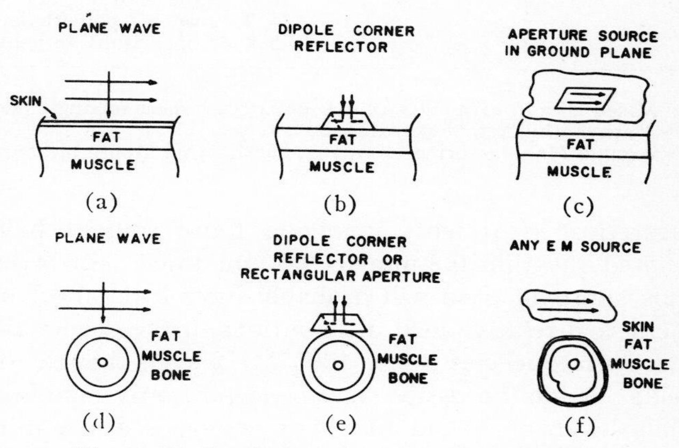

Fig. 1. Various human tissue geometries exposed to electromagnetic sources.

Manuscript received May 6, 1970; revised June 25, 1970. This work was supported by Social Rehabilitation Service Grant RT-3.

The author is with the Department of Physical Medicine and Rehabilitation, University of Washington School of Medicine, Seattle, Wash. 98105.

Reprinted from *IEEE Trans. Microwave Theory Tech.*, vol. MTT-19, pp. 205–214, Feb. 1971.

TABLE I

COMPOSITION AND PROPERTIES OF PHANTOM MODELING MATERIALS FOR HUMAN TISSUES

Modeling Material	Composition[a] (percentages by weight)	Dielectric Constant (see Fig. 2)	Loss Tangent (see Fig. 2)	Specific Heat	Density
Fat and bone (dry solid plastic)	84.81 laminac polyester resin 0.45 catalyst[b] 0.24 acetylene black 14.5 aluminum powder	4.6–6.2	0.17–0.55	0.24–0.30	1.30
Muscle (moist jellied plastic)	76.5 saline solution (12-g salt/liter) 15.2 powdered polyethylene 8.4 "Super Stuff" (a jelling agent)[c]	49–58	0.33–1.7	0.86	1.0

[a] Mixed 20 min with 600 r/min 4-in-diameter shear mixing blade.
[b] Methyl ethyl ketone peroxide, 60 percent.
[c] Whamo Manufacturing Company, 835 East El Monte St., San Gabriel, Calif.

theoretical treatments involving finite sources have ignored the thin 0.2-mm to 0.4-mm thick skin layer. This approximation will probably have little effect on the actual relative field distributions in the deeper tissues at frequencies below 1000 MHz even though the fields outside the tissues may be significantly modified. Plane layers of fat and muscle tissues exposed to a finite source consisting of a dielectric-loaded dipole and corner reflector, shown in Fig. 1(b), was examined by Guy *et al.* [6], [7] by using this approximation. The simplification of ignoring the skin was justified in this study since the aim was to optimize the applicator configuration for maximum energy penetration into muscle with minimal electric field intensity in the fat. The finite-aperture source and plane-tissue model in Fig. 1(c) has also been treated theoretically by the author [8] in a compansion paper and is used as one of the theoretical models for verifying the techniques in this paper. Cylindrical models exposed to both plane waves and finite sources, as depicted in Fig. 1(d) and (e) have been treated theoretically by Ho *et al.* [9], [10]. Absorption in spherical models has been analyzed by Anne *et al.* [11] in terms of total absorption characteristics. The electromagnetic-induced temperature patterns in the irregular geometry exposed to an arbitrary source [Fig. 1(f)] have been analyzed experimentally through the use of thermistors and thermocouples implanted in living fat and muscle tissue by Lehmann *et al.* [12], [13]. These measurements are difficult to perform without disturbing the electromagnetic fields, however, and are time-consuming and expensive. It is also difficult to relate the temperature distributions to electromagnetic fields in this case, since the former is a complicated function of blood flow. The use of the thermographic and phantom modeling techniques described in the following sections appears to be a more quantitative and reliable solution to the problem of defining the fields.

PHANTOM MODELS

Phantom modeling materials, with the composition and properties given in Table I, have been developed to simulate human fat, muscle, and bone. These materials

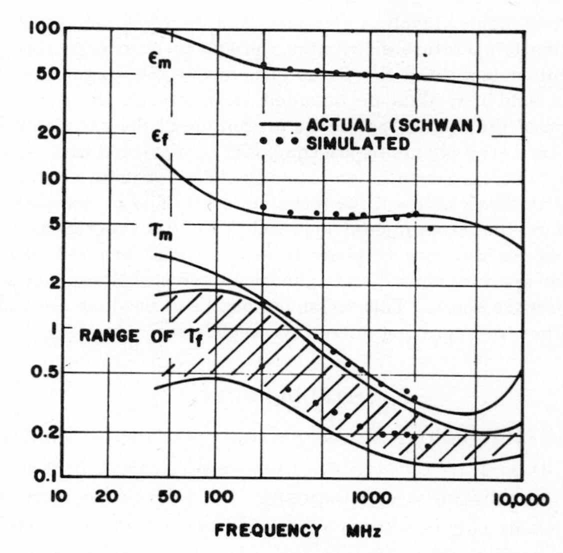

Fig. 2. Comparison between dielectric properties of actual and simulated human fat and muscle. ($\epsilon_{f,m}$ are dielectric constants; $\tau_{f,m}$ are loss tangents of fat and muscle, respectively.)

have complex dielectric properties, illustrated in Fig. 2, that closely resemble the properties of human tissues reported by Schwan *et al.* [1], [2] where the parameters ϵ_f and ϵ_m are the dielectric constants and τ_f and τ_m are the loss tangents of fat and muscle, respectively. The dielectric properties were measured by a standard method described by Von Hippel [14]. The method relates the dielectric properties of a known thickness of dielectric in a coaxial transmission line to the input and termination impedance. The calculations were performed on a laboratory digital computer. The specific heat of each material was measured by the method of mixtures [15]. A known mass of heated material was placed in a known mass of water in a calorimeter. The specific heat was calculated from the initial and final temperature of each substance. Since the dielectric properties of bone nearly approximate those for fat at the frequencies considered in this study, the modeling material for fat may also be used for bone. The synthetic muscle can also be used to simulate other tissues with high water content. The dielectric constant can be varied over a wide range by varying the percentage

Fig. 3. Various phantom tissue models and electromagnetic diathermy sources.

Fig. 4. Planar phantom tissue models and aperture sources.

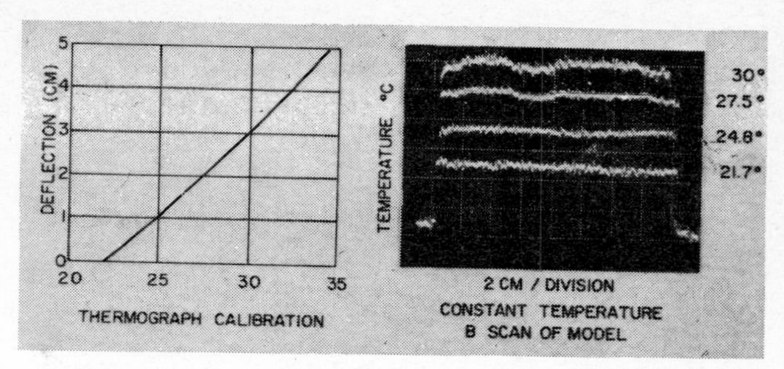

Fig. 5. Response and calibration of thermograph camera B scan for the cross-sectional surface temperature of phantom thigh model.

of polyethylene powder, and the conductivity can be controlled by the salinity of the material. The properties of the synthetic fat can be varied over a wide range to simulate other tissues of low water content by varying the amounts of aluminum power to control the dielectric constant and of acetylene black to control the conductivity. A simulated tissue structure composed of these modeling materials will have the same internal field distribution and relative heating functions in the presence of an electromagnetic source as the actual tissue structure. Phantom models of the tissue geometries depicted in Fig. 1 are fabricated as shown in Figs. 3 and 4. They include stratified layers of muscle and fat of various thickness and both circular and irregular cylindrical structures consisting of simulated fat, muscle, and bone to simulate various parts of the anatomy. These models were fabricated and used for a diathermy optimization study by the author [7]. The additional layer of skin was not required in the study since it was concerned mainly with electromagnetic heating in the fat and muscle. The experimental models were not modified for the work reported here since the theoretical models used for verifying the method also excluded the skin layer. The models are designed to separate along planes perpendicular to the tissue interfaces so that cross-sectional relative heating patterns can be measured with a thermograph. A thin 0.0025-mm-thick polyethylene film is placed over the precut surface on each half of the model to prevent evaporation of the wet synthetic muscle.

The technique for using the model is the following. The model is first exposed to the same source that will be used to expose actual tissue. The power used on the model will be considerably greater, however, in order to heat it in the shortest possible time. After a short exposure the model is quickly disassembled and the temperature pattern over the surface of separation is observed and recorded by means of a thermograph. A modified Sierra Philco camera is used for these experiments and the results are recorded on an analog magnetic tape for later transfer to photographic film. The exposure is applied over a 5- to 60-s time interval depending on the source. After a 3- to 5-s delay for separating the two halves of the model, the recording is done within a 5-s time interval or less. Since the thermal conductivity of the model is low, the difference in measured temperature distribution before and after heating will closely approximate the heating distribution over the flat surface except in regions of high-temperature gradient where errors may occur due to appreciable diffusion of heat. Since the thermograph responds to infrared radiation from the heated model, it is sensitive to changes in both temperature and emissivity of the surface of the model. The response of the thermograph is illustrated in Fig. 5. Single B scans were made along a line across one of the precut surfaces of the same phantom thigh model shown in the right of Fig. 3. Prior to each scan the model was heated in a chamber so that the temperature was constant throughout the model. Each scan was made for a different model temperature, and the infrared radiation from the model was sensed along a line passing through the center of the model face across the synthetic fat, muscle, and bone. The deflection of the thermograph oscilloscope trace is calibrated in terms of temperature on the chart on the right side of the figure. The trace appears to deflect nearly the same for all tissues scanned by the camera at a given temperature. There does appear to be a very slight decrease in emissivity for the simulated fat and bone tissues as observed from the traces for 27.5° and 30°C, however. If this slight variation is neglected, it appears that the thermograph response can be considered linear and independent of tissue type for a temperature range extending from room temperature to at least 6° above room temperature. The electric field strength in the simulated tissue can be related directly to the magnitude of the heating pattern. This is illustrated by considering the example of a plane synthetic fat and muscle geometry with a plane wave at normal incidence, as illustrated in Fig. 1(a). The internal temperature dis-

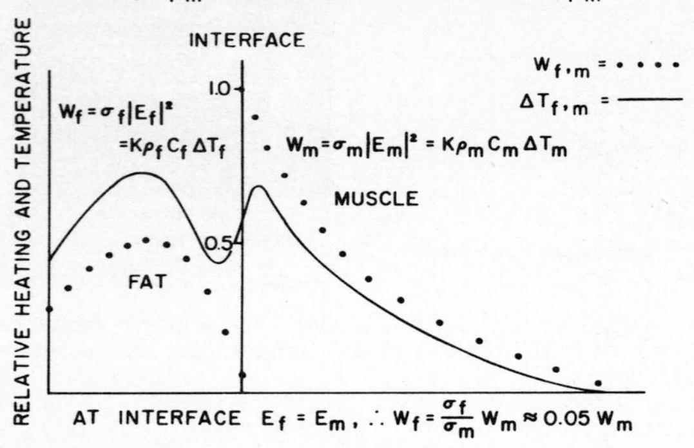

Fig. 6. Comparison of recorded temperature scan and relative heating in a tissue model exposed to a plane wave at normal incidence.

tribution along a line perpendicular to the tissue interfaces is shown in Fig. 6. The short term temperature rise $\Delta T_{f,m}$ in the fat and muscle materials is denoted by the solid curve. The heating distribution $W_{f,m}$ (W/m³) in the fat and muscle given by the dotted lines is related to the temperature changes by the relations

$$W_{f,m} \approx K\rho_{f,m}C_{f,m}\Delta T_{f,m} \tag{1}$$

where $C_{f,m}$ and $\rho_{f,m}$ are the respective specific heats and densities of the synthetic fat and muscle, and K is the factor for converting $W_{f,m}$ to W/m³, taking account of unit conversion and exposure time. The magnitudes of the electric field strengths $E_{f,m}$ in the fat and muscle are

$$E_{f,m} = \sqrt{\frac{W_{f,m}}{\sigma_{f,m}}} \tag{2}$$

where $\sigma_{f,m} = \omega\epsilon_0\epsilon_{f,m}\tau_{f,m}$ are the respective electrical conductivities, ω is the angular frequency, and ϵ_0 is the permittivity of free space. The accuracy of the estimated relative heating patterns and fields at the interface can be further improved for this simple case by correcting the error due to the heat flow across the interface between the high-temperature muscle (due to high electrical conductivity) and the lower temperature fat (due to low electrical conductivity). This can be done by noting that the electric fields on each side of the interface must be equal by the boundary conditions. Therefore, there should be a discontinuity in the relative heating curve given by

$$W_f = \frac{\sigma_f}{\sigma_m} W_m \approx 0.05 W_m.$$

We can use this relation to provide a correction for heat diffusion in the vicinity of tissue interface. This is accomplished by extrapolating the accurate portions of the curve several millimeters to this known interface discontinuity. Note that this holds true only where the electric fields are tangential to the fat–muscle interface. The validity of this thermographic technique is verified in the following sections by comparing recorded results with calculations derived from theoretical studies on models of regular shape. We will consider plane tissue layers exposed to a rectangular-aperture source and cylindrical tissue layers exposed to a plane-wave source. In addition to the results obtained for the regular shaped models, some experimental data are presented for an irregular model corresponding to a section of human thigh tissue. For convenience all of the following theoretical and experimental data discussed in this paper will be presented in terms of relative heating distribution normalized to the maximum heating in the model. Absolute values of electric field intensity can easily be derived from (1) and (2) using the information from Table I and Fig. 2.

Plane Tissue Layers Exposed to Rectangular-Aperture Source

The experimental data were taken by first exposing the center of an assembled model of fat and muscle to a direct-contact-aperture source, as illustrated in Fig. 1(c) and discussed in the companion paper [8]. The source illustrated in the top left of Fig. 4 consisted of a 12- by 16-cm aperture excited by a TE_{10} waveguide mode. The aperture was excited with the electric field polarized along the wider portion of the aperture, similar to the theoretical apertures discussed in the companion paper [8] except that no ground plane was used for these experimental models. The waveguide source was carefully designed with mode filters to allow a perfect transition from the coaxial feed to a single TE_{10} mode distribution across the aperture. The waveguide was loaded with an aluminum oxide sand to lower the waveguide cutoff frequency so that 915-MHz power could be transmitted through the guide. The phantom models used for this portion of the study, also shown in Fig. 4, were assembled by first constructing a 30- by 30- by 14-cm box with $\frac{1}{4}$-in-thick plexiglass sides. The top and bottom surfaces consisted of solid synthetic fat of uniform thickness. The box was cut so that it could be separated into two 30- by 15- by 14-cm halves, each filled with synthetic muscle. The exposed cut surfaces were covered with 1-mm-thick polyethylene film to prevent loss of moisture. The models were constructed with fat thicknesses of 0.97, 2.0, and 2.96 cm, and a muscle thickness greater than 10 cm which was sufficiently thick to allow nearly total absorption of the transmitted energy. The experimental data were taken by first exposing the center of the assembled model to the waveguide source so that the polarization of the electric field was parallel to the plane of separation for the model. A coordinate system corresponding to that used in the companion paper [8] was used with the origin at the center of the aperture, the x–y plane parallel to the aperture, the x axis parallel to the electric fields, and the z axis oriented toward the center of the model perpendicular to the fat–muscle interface. The waveguide was energized with sufficient power over a duration of 5–60 s, so that the internal temperature rise of the model was enough to

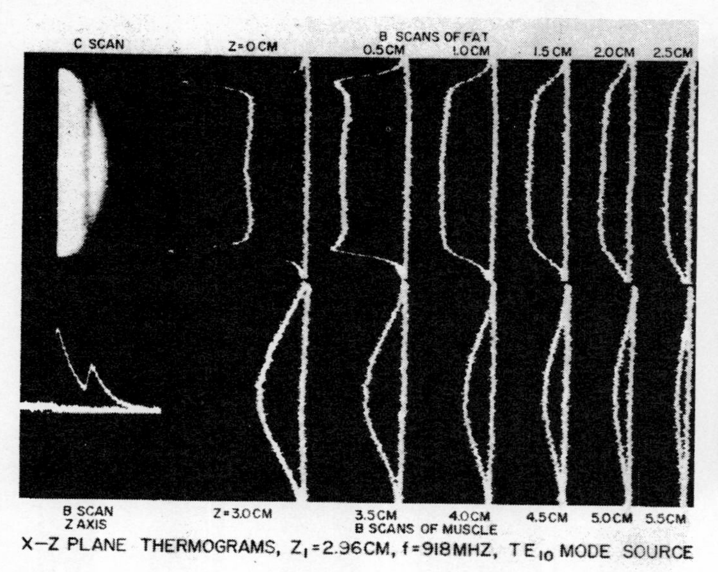

Fig. 7. Thermograms of plane layered tissue model with fat thickness $z_1=2.96$ cm exposed to 12- by 16-cm waveguide aperture. (Power: 650 W for 15 s.)

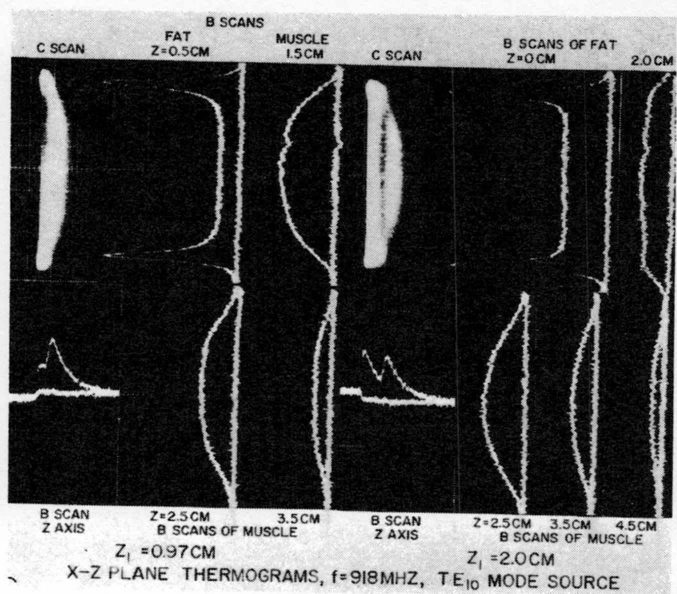

Fig. 8. Thermograms of plane layered tissue models with fat thicknesses $z_1=0.97$ and 2.0 cm exposed to 12- by 16-cm waveguide aperture. (Power: 650 W for 15 s.)

obtain a well-defined thermographic photograph of the plane of separation. The thermograph camera was set to obtain a C scan, displaying a two-dimensional picture of the entire area heated (intensity proportional to temperature) in the x–z plane as shown in the upper left portion of Fig. 7. The results are for a fat thickness $z_1=2.96$ cm. The scale on the oscilloscope indicator was set so that one large division equals 2 cm. The horizontal midline with the small subdivisions on the photograph corresponds to the z axis through the geometric center of the aperture and perpendicular to the flat interfaces of the phantom tissues. The vertical midline with the small subdivisions corresponds to the fat and muscle interface. Photographs of B scans shown in the upper right of the figure were also taken in the x–z plane corresponding to various depths $z=0$, 0.5, 1.0, 1.5, 2.0, and 2.5 cm in the synthetic fat. The photographs which are double exposures taken both before and after irradiation of the model are oriented so that the deflection to the left is proportional to temperature as a function of x (vertical direction on photograph). The temperature difference ΔT between the superimposed B scans (with the same vertical x scale as the C scans) is approximately proportional to the relative heating distributions and square of the electrical field over the region scanned, as described in the previous section. The temperature scale corresponds to 2.5°C/div. The family of B scans in the lower right of the figure were recorded for the larger values of depth z, corresponding to the muscle region. The B scan at the lower left of the figure is a scan taken along the z axis of the applicator. Note the discontinuity due to the difference in electrical conductivities of the two media. Fig. 8 illustrates the thermograms taken for the models with 0.97- and 2.00-cm-thick layers of fat exposed to the same source. Note the "hot" spots corresponding to the rapidly diverging fields near the edge of the aperture at z near zero and $x=\pm 8$ cm. The values of $\Delta T_{f,m}$ for these examples were converted to relative heating functions $W_{f,m}$, as described in the previous section, and are compared to theoretical values as derived by the method discussed in the companion paper [8]. The results are compared in Figs. 9–11, where the solid lines denote the theoretical values and the dotted lines denote the experimentally derived values of relative heating. Since the source distributions and related patterns are symmetrical with respect to the y–z plane, patterns covering only one-half of the x–z plane ($0 \leq x \leq 10$ cm) are illustrated in the figures. The relative heating pattern for the fat is illustrated on the right and the patterns for the muscle are illustrated on the left on each figure. The sharp spikes in the relative heating patterns near the surface of the model, due to the edge of the applicator, theoretically approach infinity at $z=0$ and $x=\pm 8$ cm. The differences between the experimental and theoretical magnitudes of the spikes can be attributed to heat diffusion produced by the large thermal gradient. The fact that the relative heating due to the edge is very localized as predicted by theory can be confirmed by observing the sharp reduction in the magnitude of the heating spikes with the distance between the aperture and model as illustrated in Fig. 12. Note also that the experimentally derived heating at the surface of the model is as much as 25 percent higher than that predicted by theory. This could be attributed to the absence of the infinite ground plane or modification of the aperture distribution by the tissues for the experimental case. With the exception of these differences, the curves agree reasonably well and illustrate that the thermographic method can be relied upon to quantify field dis-

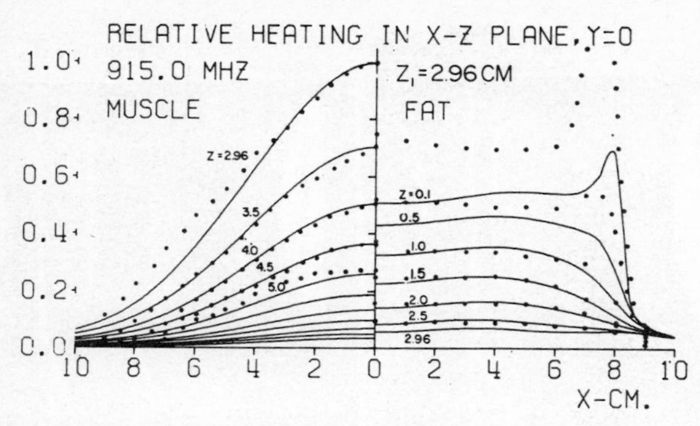

Fig. 9. Comparison between theoretical and measured relative heating patterns for plane phantom tissue model with fat thickness $z_1=2.96$ cm.

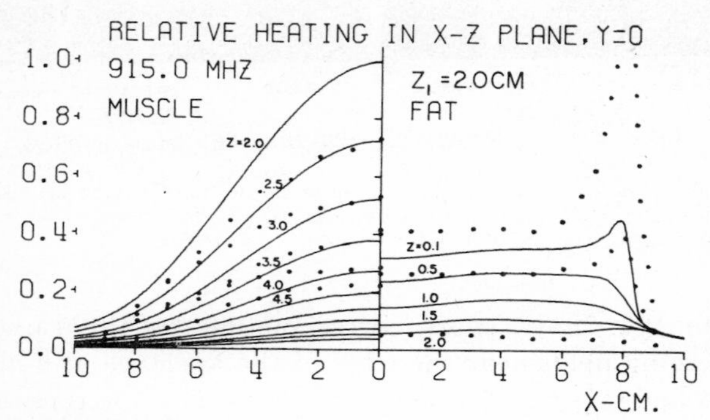

Fig. 10. Comparison between theoretical and measured relative heating patterns for plane phantom tissue model with fat thickness $z_1=2.00$ cm.

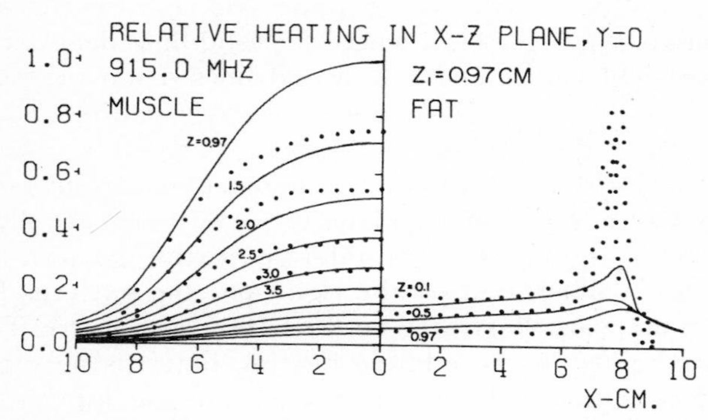

Fig. 11. Comparison between theoretical and measured relative heating patterns for plane phantom tissue model with fat thickness $z_1=0.97$ cm.

tributions in tissues when the geometry and dielectric properties are known. This experimental approach applied to rectangular models was used as an invaluable aid in improving microwave diathermy applicators. The improvements needed in diathermy are increased penetration of fields and heating in muscle tissue with decreased heating in the subcutaneous fat layer. Also, it is desirable to develop an applicator for application very close to, or in direct contact with, the area to be treated in order to provide more control of the energy in the tissues requiring therapy with less radiation to vulnerable tissues, such as the eyes and testes,

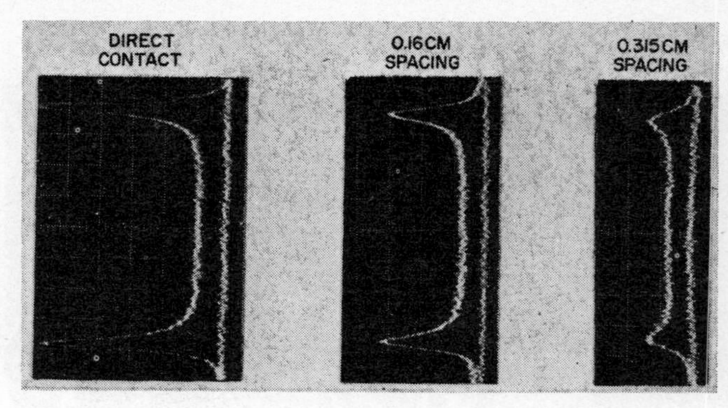

Fig. 12. Effect of spacing of TE_{10} mode aperture on relative heating at surface of 0.97-cm-thick fat.

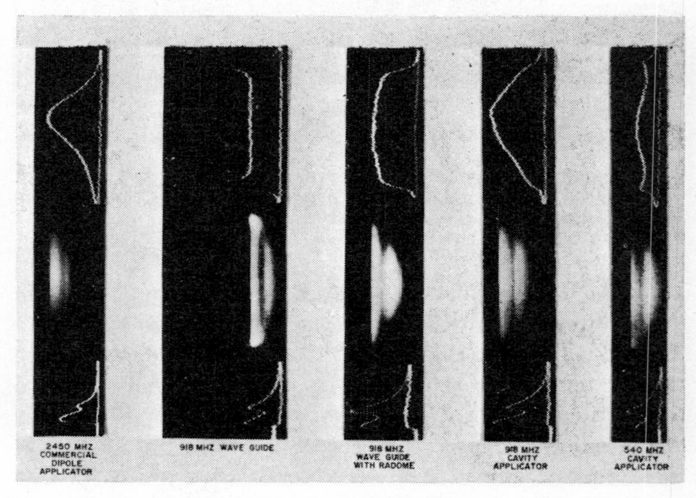

Fig. 13. Thermograms for plane phantom tissue model exposed to experimental diathermy applicators.

Fig. 13 shows a comparison of thermographic data taken for a plane phantom model with 2 cm of fat exposed to various experimental diathermy applicators. Intensity cross-sectional C scans are illustrated in the middle of the figure; B scans taken across the top of the phantom model in the x direction are illustrated at the top of the figure; and B scans taken along the z axis for each applicator are illustrated at the bottom of the figure. At the far left are the results for the standard 2450-MHz diathermy applicator pictured in the middle of Fig. 3. The heating patterns due to this applicator placed 5 cm from the model are undesirable due to excessive fat-to-muscle heating ratio and poor penetration into the muscle. The next scans are for the direct contact waveguide applicator pictured at the left in Fig. 4. They show a marked improvement in heating patterns everywhere except for the undesirable edge heating effects.

The third set of thermograms from the left, however, show that these heating spikes can be eliminated with further improvement in the heating patterns by the addition of a 0.5-cm-thick dielectric plate to the waveguide applicator as suggested from a theoretical study [8]. This type of dielectric loaded waveguide applicator is

Fig. 14. Cylindrical phantom tissue models.

Fig. 15. Large-diameter cylindrical phantom tissue model exposed to 2450-MHz approximate plane-wave source in anechoic shielded room.

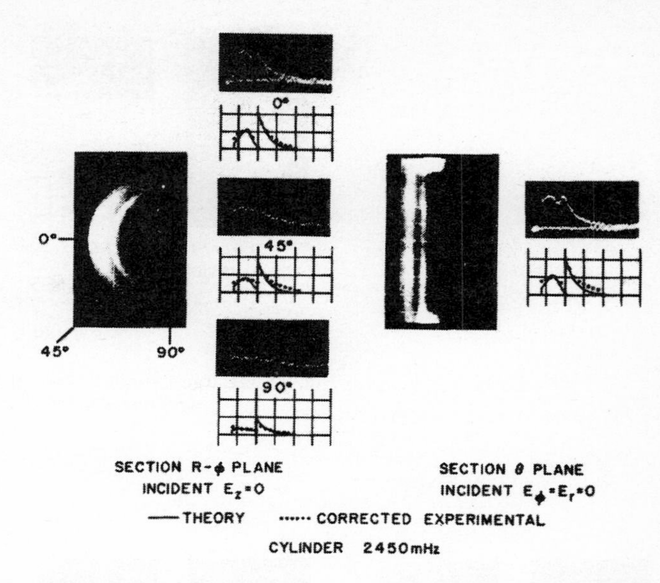

Fig. 16. Thermograms for large-diameter cylindrical phantom tissue model exposed to 2450-MHz approximate plane-wave source. (Power: 2500 W for 30 s.)

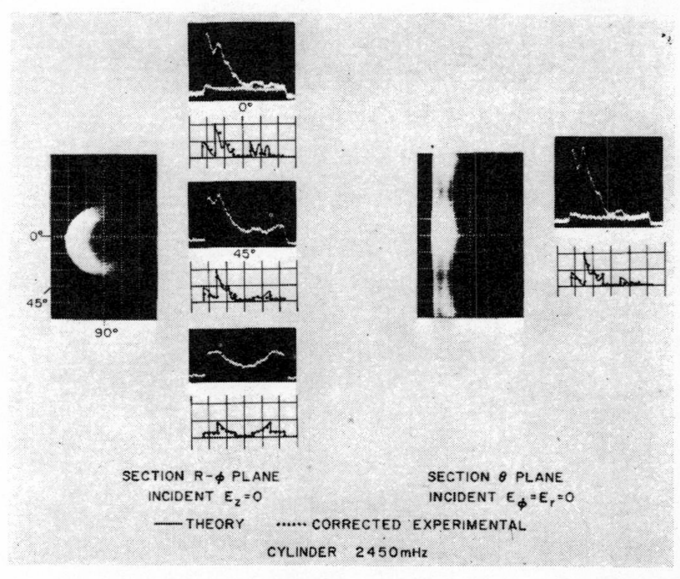

Fig. 17. Thermograms for small-diameter cylindrical phantom tissue model exposed to 2450-MHz approximate plane-wave source. (Power: 2500 W for 30 s.)

very unwieldy and heavy, however, so a smaller cavity applicator, illustrated on the right of Fig. 4, was fabricated. A less desirable heating pattern is obtained for this case as shown by the second group of thermograms from the end. This is due to a more complex aperture distribution resulting from higher order modes excited by the adrupt transitions between the coaxial feed and the 12- by 16-cm aperture. The aluminum oxide dielectric filler used in this cavity applicator was replaced with silicon carbide filler having a higher dielectric constant and the cavity was excited at the lower frequency of 540 MHz. The patterns for this applicator, shown on the far right of Fig. 13, show a marked improvement in muscle-to-fat heating ratio at the lower frequency, as predicted by the theoretical study. We may conclude from these results that the aperture excited with a pure TE_{10} mode and separated a slight distance from the tissues by a dielectric plate provides the more desirable field patterns for therapeutic purposes. We may also conclude that continued improvement is obtained in the patterns as the frequency is lowered.

Circular Cylindrical Tissues Exposed to Various Sources

Triple-layered circular cylindrical tissue models roughly simulating portions of human thigh and arms, shown in Fig. 14, were exposed to a number of sources. The large cylinder consisted of simulated bone of outside radius 1.9 cm, muscle of outside radius 6.3 cm, and fat of outside radius 8.9 cm. The smaller cylinder, composed of the same materials, had respective interface radii of 0.95, 3.18, and 4.45 cm. Thermographs were taken of the models after they were exposed to a 2450-MHz approximate plane-wave source consisting of the far-zone field of a horn antenna in an anechoic chamber, illustrated in Fig. 15. Figs. 16 and 17 illustrate the results in terms of a standard cylindrical coordinate system corresponding to the cylinder geometry. The data on the left side of each figure were taken from an $R-\phi$ plane surface of the cylinder with the incident magnetic field parallel and the electric field perpendicular to the z axis of the cylinder at $\phi=0°$. The data on the right side

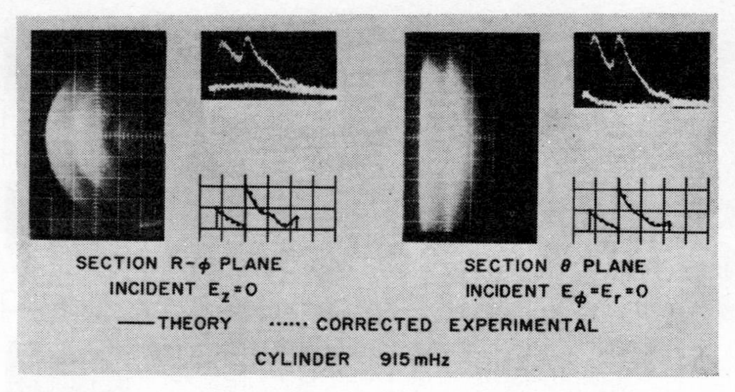

Fig. 18. Large cylinder exposed to 915-MHz 12- by 16-cm cavity-aperture source. (Power: 180 W for 40 s.)

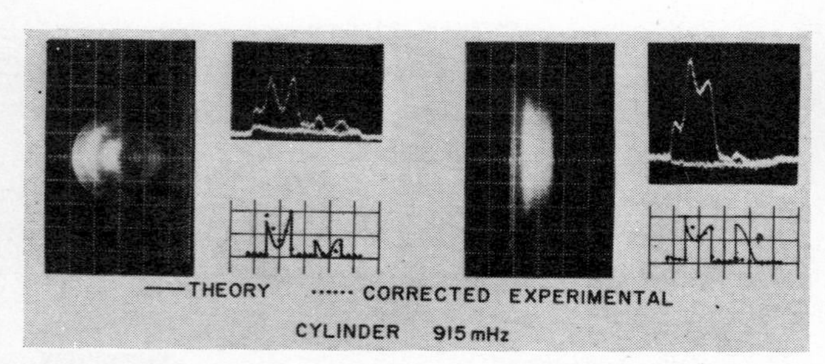

Fig. 19. Small cylinder exposed to 915-MHz 12- by 16-cm cavity-aperture source. (Power: 180 W for 40 s.)

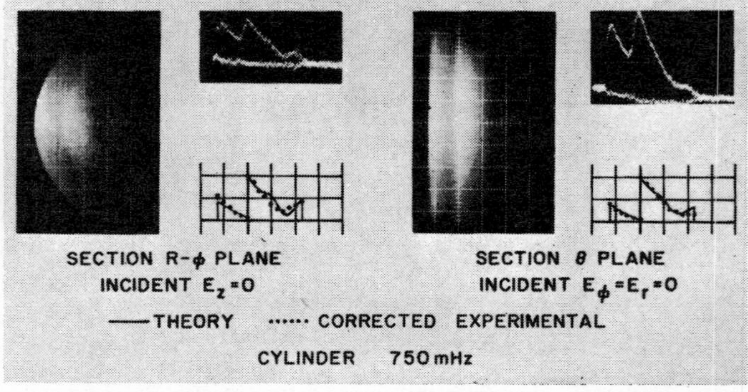

Fig. 20. Small cylinder exposed to 750-MHz 12- by 16-cm cavity-aperture source. (Power: 190 W for 40 s.)

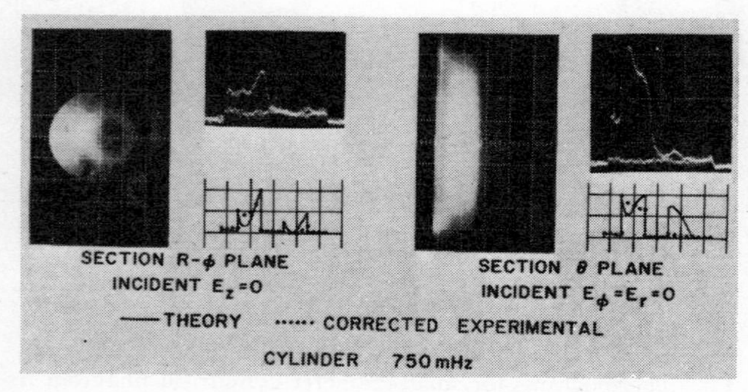

Fig. 21. Small cylinder exposed to 750-MHz 12-by 16-cm cavity-aperture source. (Power: 190 W for 40 s.)

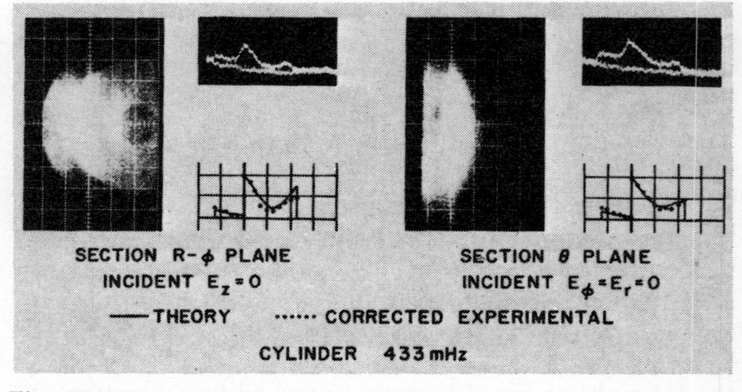

Fig. 22. Large cylinder exposed to 433-MHz dipole diathermy applicator. (Power: 210 W for 40 s.)

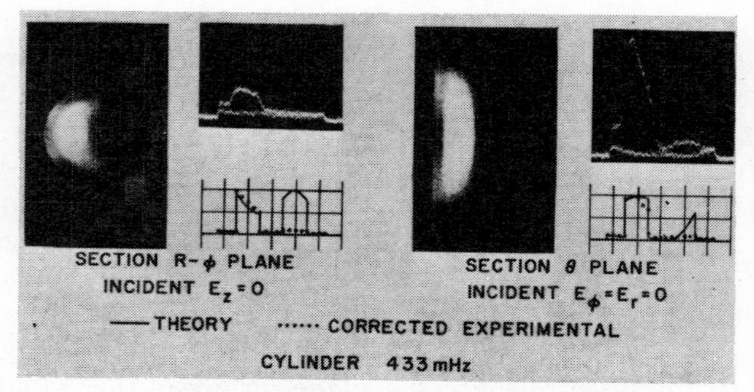

Fig. 23. Small cylinder exposed to 433-MHz dipole diathermy applicator. (Power: 210 W for 40 s.)

of the figure were taken from the $R-z$ plane surface of the cylinder with the incident electric field parallel and the magnetic field perpendicular to the z axis of the cylinder at $\phi=0°$. The B scans for each model were taken in the $R-\phi$ plane along lines corresponding to $\phi=0°$, 45°, and 90°, as marked on the C scans of each left-hand figure. The single B scan for the right-hand figure was taken along the R axis in the $\phi=0$ plane. The temperature information was converted into relative heating patterns expressed by dotted lines and compared to theory as expressed by solid lines in the figures below the B scans. The spatial scales are 2 cm/div and the temperature scales are 2.5° C/div. The theoretical results due to Ho *et al.* [9] show good agreement with the measured relative heating curves. The models were also exposed to other type sources including the direct-contact cavity applicator illustrated on the right of Fig. 4, operating at 750 and 915 MHz, and a commercial European 433-MHz (12-cm-long capacity-loaded dipole) diathermy applicator shown at the left of Fig. 3. The thermographic results from exposing the models to these linearly polarized sources are illustrated in Figs. 18–23. The format for each figure is the same as for Figs. 16 and 17, except all B scans are limited to the $\phi=0$ axis. Theoretical curves due to a plane-wave source are compared to the experimental results for these cases also. Although the actual sources used were finite in size, the agreement in the results for the exposed left sides of the cylinders are surprisingly close. The patterns clearly show the increased penetration of the fields into the muscle and the decreased field amplitude in the subcutaneous fat as the source frequency is lowered. Reflections from the fat–muscle interface are clear at all frequencies, while reflections from the bone are apparent only at the lower frequencies where penetration is sufficiently deep to produce the visible heating effects.

Irregular Cylindrical Tissues Exposed to Various Sources

The greatest benefits gained from the technique discussed in this paper are achieved through its application to irregular tissue structures that cannot be analyzed theoretically or practically by other methods. We can illustrate this by applying the various sources discussed above to an irregular shape phantom model of the human thigh pictured on the right in Figs. 3 and 14. The results are illustrated in Fig. 24. The B scans were taken along the line through the center of each source perpendicular to the tissue interfaces on the flat side of the model. The relative heating patterns were evaluated and plotted below each B scan, as shown, by accounting for the specific heat and weight of each modeling material and correcting for heat flow in the regions of high-temperature gradient at the interfaces. Note the excessive heating in the subcutaneous fat at 2450 MHz which decreases markedly with increasing frequency. Also, there is a definite increase in penetration in the muscle as the source frequency is lowered. This penetration is sufficient at lower frequencies to produce reflection of energy at the surface of the bone.

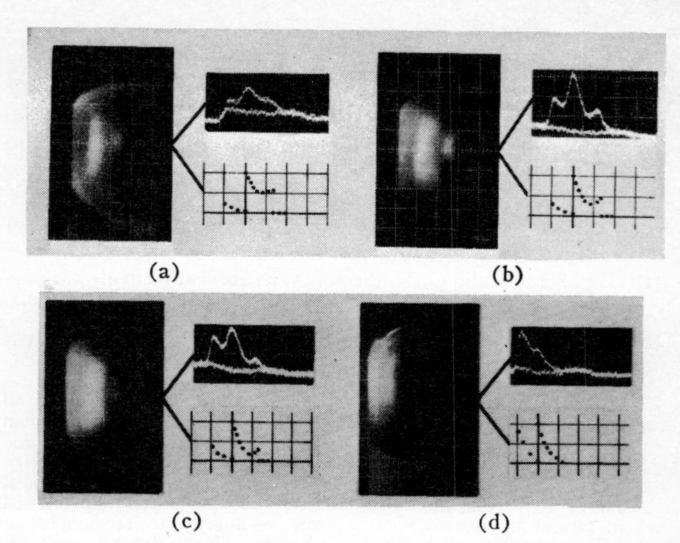

Fig. 24. Thermograms and relative heating patterns of phantom thigh models exposed to various electromagnetic sources. Frequency: (a) 433 MHz; (b) 750 MHz; (c) 915 MHz; (d) 2450 MHz. Dotted line indicates corrected experimental.

Conclusions

The experimental techniques discussed in this paper provide a method for rapidly evaluating electric fields and the associated heating patterns in any tissue geometry due to an arbitrary electromagnetic source when the tissue geometry and dielectric properties are known. The technique has been verified by direct comparison with known theoretical results. The results have provided a means for verifying the theoretical expressions for diathermy heating patterns in models that are amenable to theoretical analyses and also have provided quantitative data on models too complex to analyze by theoretical methods. This technique can also be applied with equal ease to the problem of assessing fields induced in body tissues by accidental exposure to radars, microwave ovens, or other microwave power equipment. Although the models used in this study were fabricated for the specific purpose of optimizing therapeutic heating of musculature, it would not be difficult to synthesize phantom models of other regions of the body where information on electromagnetic field distributions are desired. For example, the skull, brain matter, eyes, spinal cord, and spine could all be modeled in a complex phantom structure of the human head and neck, or the entire body complete with various cavities could be modeled. These models could be scaled to any size by proper adjustment of frequency and electrical conductivity. The technique would be useful for analyzing fields in other lossy media such as food and wood, or other substances exposed to microwave heating equipment.

Acknowledgment

The author would like to thank C. Sorensen and J. A. McDougall who synthesized the phantom models, constructed some of the applicators, and collected the data for this paper.

References

[1] H. P. Schwan and G. M. Piersol, "The absorption of electromagnetic energy in body tissues. Part 1. Biophysical aspects," *Amer. J. Phys. Med.*, vol. 33, Dec. 1954, pp. 370–404.

[2] H. P. Schwan, E. Carstenson, and K. Li, "The biophysical basis of physical medicine," *J. Amer. Med. Ass.*, 160:3: Jan. 21, 1956, pp. 191–197.

[3] H. P. Schwan, "Radiation biology, medical applications, and radiation hazards," in *Microwave Power Engineering* (Electrical Science Series, vol. 2), E. C. Okress, Ed. New York: Academic Press, pp. 215–232.

[4] J. F. Lehmann, A. W. Guy, V. C. Johnston, G. D. Brunner, and J. W. Bell, "Comparison of relative heating patterns produced in tissues by exposure to microwave energy at frequencies of 2450 and 900 megacycles," *Arch. Phys. Med. Rehabil.*, vol. 43, Feb. 1962, pp. 69–76.

[5] J. F. Lehmann, G. D. Brunner, J. A. McMillan, and A. W. Guy, "A comparative evaluation of temperature distributions produced by microwaves at 2456 and 900 megacycles in geometrically complex specimens," *Arch. Phys. Med. and Rehabil.*, vol. 43, no. 10, 1962, pp. 502–507.

[6] A. W. Guy and J. F. Lehmann, "On the determination of an optimum microwave diathermy frequency for a direct contact applicator," *IEEE Trans. Biomed. Eng.*, vol. BME-13, Apr. 1966, pp. 76–87.

[7] A. W. Guy, J. F. Lehmann, John A. McDougall, and C. C. Sorensen, "Studies on therapeutic heating by electromagnetic energy," in *Thermal Problems in Biotechnology* (ASME Winter Annual Meeting, New York, Dec. 3, 1968), pp. 26–45.

[8] A. W. Guy, "Electromagnetic fields and relative heating patterns due to a rectangular aperture source in direct contact with bilayered biological tissue," this issue, pp. 214–223.

[9] H. S. Ho, A. W. Guy, R. A. Sigelmann, and J. F. Lehmann, "Electromagnetic heating patterns in circular cylindrical models of human tissue," in *Proc. 8th Annu. Conf. Medical and Biological Engineering* (Chicago, Ill., July 1969), p. 27-4.

[10] ——, "Microwave heating of simulated human limbs by aperture sources," this issue, pp. 224–231.

[11] A. Anne, M. Saito, O. M. Salati, and H. P. Schwan, "Relative microwave absorption cross sections of biological significance," in *Biological Effects of Microwave Radiation*, vol. 1. New York: Plenum Press, 1960, pp. 153–176.

[12] J. F. Lehmann, A. W. Guy, B. J. DeLateur, J. B. Stonebridge, and C. G. Warren, "Heating patterns produced by short-wave diathermy using helical induction coil applicators," *Arch. Phys. Med. Rehabil.*, vol. 49, Apr. 1968, pp. 193–198.

[13] J. F. Lehmann, B. J. DeLateur, and J. B. Stonebridge, "Selective muscle heating by shortwave diathermy with a helical coil," *Arch. Phys. Med. and Rehabil.*, vol. 40, Mar. 1969, pp. 117–123.

[14] A. Von Hippel, Ed., *Dielectric materials and Applications*. Cambridge, Mass.: M.I.T. Press, pp. 65–66.

[15] R. C. Weast, Ed., *Handbook of Chemistry and Biophysics, 47th ed.* 1966–1967, p. F-85.

Dosimetry for Whole-Animal Microwave Irradiation

EDWARD L. HUNT, RICHARD D. PHILLIPS, DALE M. FLEMING, AND RICHARD D. CASTRO

***Abstract*—A whole-animal calorimeter for measuring the differential heat content of two rat carcasses was used to measure the additional energy absorbed by one during its exposure to 2.45 GHz microwaves in a multimodal resonating cavity. With this system, absolute dose measurements for whole-animal exposures are made available for the first time. Simplified physical models of the rat made of a gel mixture and distilled water produced overestimates of the energy absorbed by actual rats of comparable size.**

THE AMOUNT of energy absorbed by the animal during its exposure to microwave radiation is typically estimated from measurements of the heat induced in a simplified physical model of the animal. Such models usually simulate the weight or volume of the animal, are shaped simply as spheres or cylinders, and are composed of water or a homogeneous mixture of materials which simulates the average dielectric properties of the whole animal or of particular tissues [1]. Estimates of absorbed energy based on models are of uncertain accuracy since animals are heterogeneously structured of variously layered, complexly shaped tissues which vary in composition and dielectric properties (lossiness).

A physical dosimetry system was developed based on absolute calorimetry of the whole animal in order to provide a standard for calibrating a cavity exposure system for irradiation of rats. The dosimetry system was also used to determine the accuracy with which simplified physical models of the animal could be used for such calibrations. The dosimetry system measures the heat generated by brief irradiation in a fresh carcass using a differential (twin-well) calorimeter in which a nonirradiated carcass serves as the reference heat source for the thermopile. The general procedure consists of placing a pair of carcasses in the calorimeter and determining the differential body heat content for the pair. After both carcasses are temporarily removed to irradiate one and sham irradiate the other, the additional heat produced by irradiation is measured. Use of the dead animal eliminates the problems of physiological heat production and loss and requires only the assumption that death does not immediately alter the lossiness of animal tissues.

A parallel dosimetry procedure was used to measure the energy absorbed in two physical models of the rat, one composed of distilled water and the other of a gel mixture used to simulate the lossiness of muscle tissue.

Manuscript received June 15, 1981.

The authors are with Battelle, Pacific Northwest Laboratories, Richland, WA 99352.

This research was supported by the Office of Naval Research Contracts N00014-70-C-0197 and N00014-70-C-0332 with funds provided by the Naval Bureau of Medicine and Surgery.

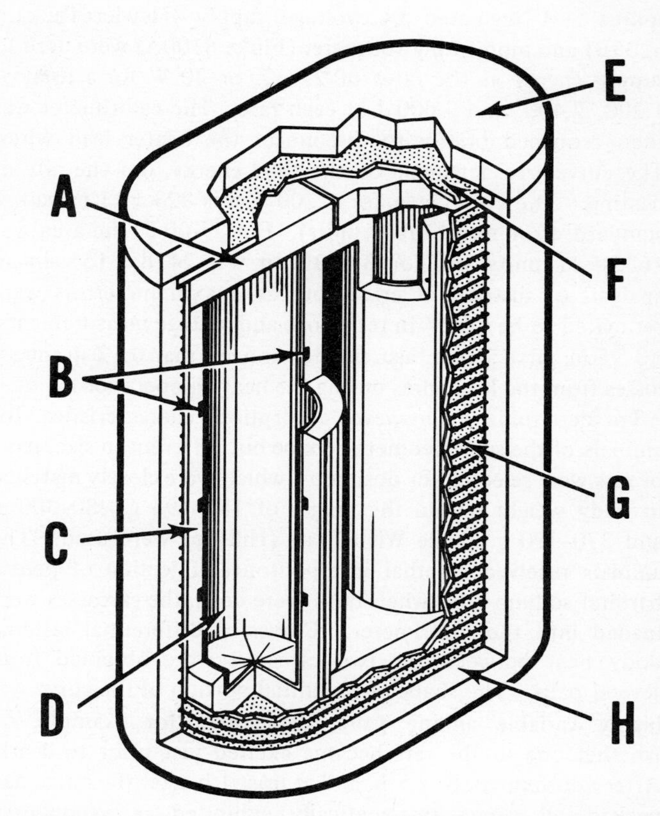

Fig. 1. Schematic of twin-well calorimeter, partially sectioned, showing: A–polystyrene cover of one well (removable), B–thermocouples positioned on outside surface of well, C–polystyrene insulation layer around well, D–thin-walled aluminum well (6.35 cm diam, 25.4 cm deep), E–polystyrene outside cover of calorimeter (removable), F–cover plate of copper jacket (removable), G–heater wire wrapped around copper jacket, and H–polystyrene insulation around the calorimeter.

The twin-well calorimeter designed and used at this laboratory is shown schematically in Fig. 1. It consists of two thermally-insulated thin-walled aluminum wells which act as isothermal reference planes for the thermopile [2]. The thermopile consists of 24 iron constantan junctions connected in series and positioned in alternating sequence at the outside surfaces of the two wells. The thermopile voltage produced by differential well temperatures is amplified (Keithley 150B) and recorded on a strip chart (Hewlett-Packard 7100B). A constant ambient temperature level for the calorimeter is provided by a copper jacket surrounding the insulated wells. Controlled heating maintains this jacket at 29.60 ± 0.05°C with a coil made of constant resistance wire (magnanin alloy) wrapped around the jacket. A thermistor bead attached to the jacket is used in a null-seeking bridge circuit which operates a voltage controller for the heater.

The calorimeter was calibrated by adding known amounts of heat to one of the wells and measuring the area of the resultant voltage-time curve. A 380 g Lucite rod, to which heat could be added electrically, was loaded into one well [3]. After the calorimeter returned to thermal equilibrium, the calorimeter was opened briefly to connect the heater wires to the power source. A regulated DC voltage supply (Hewlett-Packard 6206B) and monitoring voltmeter (Fluke 8100A) were used to supply energy at the rates of 2, 6.7, or 20 W for a total of 1,200, 2,400, and 3,600 J at each rate. The calorimeter was then reopened briefly to disconnect the heater lead wires. The curve area depended on the total energy, not the rate of heating. The curve area for 1,200 J was 823 ± 11 (mean, ± standard error in arbitrary units). For 2,400 J, the area was 1635 ± 11 units and for 3,600 J, it was 2428 ± 15. In an analysis of instrument error sources, maximum errors were estimated to be ±0.2% in resistance and voltage measurements, ±0.5% in curve area measurements, and less than 0.1% in energy losses from the lead wires during the heating operation.

To determine microwave absorption characteristics for animals of the same geometric shape but different in size, pairs of rats were selected for dosimetry which were closely matched in body weight within the ranges of 185–195 g, 280–300 g, and 370–390 g. Male Wistar rats (Hilltop) were used. The animals received a lethal intraperitoneal injection of pentobarbital sodium and, when both were dead, the carcasses were loaded into the calorimeter. Curves of differential residual body heat between the two carcasses were obtained from several pairs. The shape of the initial portion of the curve was highly variable among pairs, depending, for example, on whether one of the rats became excited just prior to death. After approximately 2.5 h, and at least 1 h after the curve had peaked, all curves systematically exhibited an exponential decay toward thermal equilibrium. The time constant was related to carcass mass and averaged 2.6, 3.2, and 3.6 h, respectively, for the smallest, middle-sized, and largest carcasses.

To ascertain whether the physical calibration values were valid with rat carcasses, additional pairs of rats in the 370–390 g weight range were used for adding known amounts of heat to the warmer carcass of the pair. A rubber-coated electrical heater, similar to that in the Lucite rod [3], was inserted into the colon of the animal at the time of death. Approximately 1 h after the exponential decay of the differential heat curve was clearly evident, the same procedures used for adding heat in the physical calibration were employed. A low rate of heating, 2 W, was used to limit the temperature rise of tissues adjacent to the heater. Two determinations were made for each of the energy levels; 1,200, 2,400, and 3,600 J. Figure 2 shows the result of this procedure. The curve area after the introduction of additional heat at time 0 corresponds to the sum of (a) the residual part of the initial heat difference between rats and (b) the heat added electrically to the warmer carcass. Although the added heat could be estimated by subtracting the area calculated to be the residual part of the initial heat difference curve from the summation curve, a simpler equivalent procedure was used. Since the terminal portion of the summation curve (terminal clear area in Fig. 2) decays at the same characteristic rate as the residual part of the initial heat differential curve, the terminal portion can be deleted from the measurement of the summation curve and the resultant area (hatched area in Fig. 2) represents the added energy. For 1,200 J, the average curve area was 820 ± 13, for 2,400 J it was 1658 ± 14, and for 3,600 J it was 2449 ± 13. These values differed from those obtained in the physical calibrations by only −0.3%, +1.4%, and +0.9%, respectively, for the three energy levels. This method for measuring added energy was used in the dosimetry with the temporary removal of both carcasses from the calorimeter and irradiation of the warmer carcass substituted for the electrical heating procedure [4].

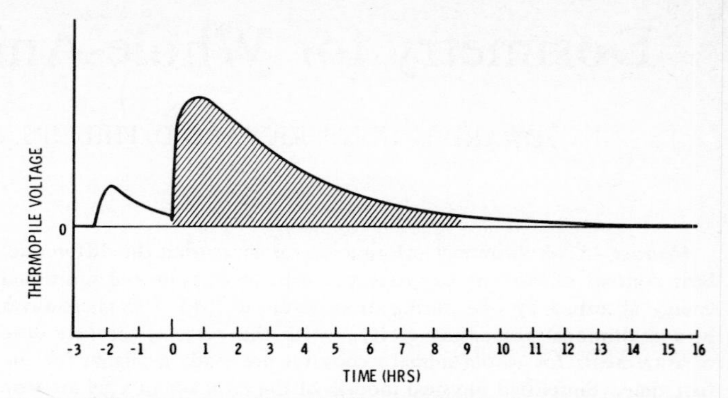

Fig. 2. Strip-chart recording of differential body heat curve for two rat carcasses in twin-well calorimeter with 2,400 J of energy added by electrical heating to the warmer carcass at 0 h. Voltage scale and area are in arbitrary units. The residual portion of the initial body heat difference is represented by the tail of the curve (clear area); the area measured with the tail deleted (hatched area) represents the energy added electrically.

The holder used to contain each carcass during dosimetry irradiations was of the Bollman type [5]. It was modified to provide efficient thermal insulation by using expanded-bead polystyrene, which also produces minimum distortion of, and energy absorption from, the incident microwave field [6].

Microwave irradiation was made using the multimodal resonating cavity method [7], modified to provide a measurement of net power delivered to the cavity. The cavity (Varian TCS-2.5A), equipped with a mode stirrer, was powered through waveguide by a magnetron source (Varian PPS-2.5A) operated at 2.45 GHz (12.24 cm wavelength) with a repetition rate of 120 pulses/s and pulse duration 2.5 ms. A 3-port circulator was located in the waveguide to protect the magnetron and minimize moding. A bidirectional coupler (Varian EW3-DMP3S) located in the cavity end of the waveguide provided the means for sampling and monitoring forward and reflected power. Net power delivered to the cavity was calculated as the difference between the forward and reflected power values read on calibrated meters. In this arrangement, the experimental animal serves as the working load in the cavity and, when positioned in a standard location, the animal absorbs energy stored in the cavity in proportion to the net power delivered to the cavity.

For irradiation, both carcasses were removed from the calorimeter, encased in their holders, and the warmer one placed in a standard position within the cavity. All dosimetry exposures were for 1 min with the forward power set at 105 W. The reflected power value was read during irradiation and the

TABLE I

ABSORBED ENERGY CHARACTERISTICS IN PHYSICAL DOSIMETRY OF WHOLE-ANIMAL MICROWAVE IRRADIATION USING 1-min EXPOSURES OF FRESH RAT CARCASSES TO A PULSED 2.45 GHz FIELD IN A MULTIMODAL RESONATING CAVITY. N = 5/GROUP. RESULTS EXPRESSED AS THE MEAN ± STANDARD ERROR OF THE MEAN.

Body-Size Groups (grams)	Net Cavity Energy[a] (joules)	Total Absorbed Energy (joules)	Absorption Coefficient[b] (millijoules/joule)	Dose Density[c] (millijoules/joule/gram)
189.6 ± 1.1	5,562 ± 90.6	2,412 ± 54.0	434 ± 9	2.29 ± 0.05
288.6 ± 3.0	5,946 ± 27.0	2,862 ± 24.0	482 ± 8	1.67 ± 0.03
374.8 ± 2.8	6,000 ± 27.0	3,108 ± 54.0	518 ± 8	1.38 ± 0.01

[a]Net cavity energy (1 min) = 60 × (forward power−reflected power), where forward power was 105 W.

[b]Absorption coefficient = $\frac{\text{total absorbed energy (millijoules)}}{\text{net cavity energy (joules)}}$

[c]Dose density = $\frac{\text{absorption coefficient (millijoules/joule)}}{\text{body weight (grams)}}$

net energy calculated. During the exposure, the other carcass remained in its holder outside the cavity. Immediately following the exposure, the irradiated preparation was removed from the cavity and both carcasses promptly returned to the calorimeter in the same order of their removal. The time between the end of irradiation to the return of both carcasses to the calorimeter was approximately 50 s.

Table I shows the dose values obtained from the irradiation of five rat bodies in each of the three weight groups. The efficiency of energy coupling was found to increase with an increase in animal size. This improved coupling was evident in both the net energy delivered to the cavity and, as shown by the increased absorption coefficient, the coupling of the field in the cavity to the animal load. Although the total amount of energy absorbed by the whole animal (shown by the absorption coefficient) increased with body size, the amount of energy absorbed per unit of body mass decreased.

The calorimeter provides an absolute measurement of the added heat still present in the exposed carcass when both bodies are returned to the calorimeter at about 50 s postexposure. The portion of heat produced by irradiation that has been lost from the carcass to its environment during the irradiation procedure is not measured calorimetrically; since the exposed carcass is the warmer one, it loses heat faster. The total additional heat in the irradiated body at the end of exposure was estimated to be about 2% greater than that measured calorimetrically. This estimate was empirically confirmed using carcasses in the 370–380 g weight range by the imposition of postirradiation delays to resumption of calorimetry of 4, 2, and <1 min, with five pairs per group. A curvilinear extrapolation back to the end of irradiation produced a dose estimate 1.75% greater than the value measured with the standard delay of 50 s. Table I is based on values uncorrected for this error but only slightly underestimates the absolute energy values which would be appropriate for whole-animal dosimetry. Corrected energy values are used routinely in the use of this dosimetry system to calibrate the cavity for irradiation of live rats.

The accuracy with which simplified physical models of rats can be used to estimate energy absorbed from the cavity field was determined in a second series of calorimetric measurements. The rat models were constructed to fit into the twin-well calorimeter and irradiation holder and were made in two sizes, containing 190 g and 290 g of either distilled water or gel mixture [8]. The gel was designed to simulate the absorption characteristics of muscle tissue [9]. To produce temperature levels and differential heat contents similar to those of the rat bodies, pairs of models were heated overnight in a laboratory drying oven set at 34°C. To produce the differential heat content, one was removed several minutes before both were transferred to the calorimeter. The procedures used for irradiation and for calorimetry were identical to those used for the rat bodies.

The values obtained for total absorbed energy and average dose density with the two modeling materials paralleled those obtained with rat carcasses of comparable weight. The average dose densities for the water models weighing 190 and 290 g, respectively, were 2.43 ± 0.03 and 1.74 ± 0.02 mJ/g. These values were 6.1% and 3.1% higher than those obtained for the 190 and 290 g rat carcasses, respectively. Corresponding values for the gel models were 2.53 ± 0.08 and 1.88 ± 0.08 mJ/g, exceeding the values obtained for rat carcasses by 10.5% and 12.6%, respectively.

The twin-well calorimetry method provides a physical dose measurement which is, for the first time, of sufficient precision to use as a standard in investigating biological effects of whole-animal microwave irradiation. Before now, dose scaling was accomplished using measurements from physical models to estimate the dose or, alternatively, using reliable biological endpoints as dosimeters. In much the same manner that the skin erythema dose unit was useful in early investigations of X-ray effects, measures such as the incidence of lens cataracts, local temperature increases, and (recently, at this laboratory) the latency for thermal induction of seizures [10] have been used for scaling absorption of microwave energy.

Physical models have provided useful information about the transmission and penetration of microwave energy, and the dependence of absorption on wavelength and animal size [11]; however, dose estimates based on physical model measurements entail assumptions of the model's similarity to the animal when, in fact, they differ in size, shape, structure, and dielectric properties. The dose estimates obtained from the models used in this investigation were consistently higher than doses measured with the animal bodies [12]. With both animals and models however, energy coupling was more efficient with larger loads. Also, for both, the average dose densities decreased with an increase in size. The latter effect is related to the limited penetration of 2.45 GHz microwaves and the lower surface area-to-volume ratio of the larger loads. Consequently, increasingly less energy per gram would be absorbed in an increasingly larger central mass. For the animal, this might be verified by making localized measurements of incident or absorbed energy in deep tissues.

In providing a practical laboratory standard for measuring absorbed microwave energy, the calorimetric dose measurement is independent of, and makes no assumptions about, the geometry, frequency, or modulation characteristics of the incident energy field. Accordingly, this measurement provides a basis for investigating the relative biological effectiveness of various types of microwave fields.

References and Notes

[1] F. G. Hirsch, *IRE Trans. Med. Electron.*, vol. PGME-4, p. 22, 1956; H. Mermagen in *Biological Effects of Microwave Radiation*, M. F. Peyton, Ed., pp. 143–152, New York, NY: Plenum, 1961; R. L. Carpenter and C. A. Van Ummersen, *J. Microwave Power*, vol. 3, no. 3, 1968; A. W. Guy, *IEEE Trans. Microwave Theory Tech.*, vol. 19, no. 2, p. 205, 1971; H. S. Ho, A. W. Guy, R. A. Sigelmann, and J. F. Lehman, *IEEE Trans. Microwave Theory Tech.*, vol. 19, no. 2, p. 224, 1971; D. R. Justesen, D. M. Levinson, R. L. Clarke, and N. W. King, *J. Microwave Power*, vol. 6, p. 237, 1971; N. W. King, D. R. Justesen, and R. L. Clarke, *Science*, vol. 172, p. 398, 1971; J. C. Sharp and C. J. Paperiello, *Radiat. Res.*, vol. 45, p. 434, 1971; A. W. Guy and S. F. Korbel, *J. Microwave Power*, vol. 7, p. 287, 1972.

[2] The wells were protected from contamination with animal matter by enclosing each animal in a 1-mil polyethylene bag.

[3] The Lucite rod–6cm diam, 15.4 cm long. Imbedded inside it was a 33-Ω heater coil of magnanim alloy constant-resistance wire wrapped around a 6.35 mm diam Teflon rod.

[4] The effect of temporarily removing both carcasses from the calorimeter on the decay portion of the differential heat curve was tested using carcasses weighing approximately 375 g. When the decay portion of the curve was clearly evident, both carcasses were removed in rapid sequence, placed in thermally-insulated holders for periods up to 10 min, and then returned to the calorimeter. Less than 15 s and 5 s, respectively, were required for removal from and return to the calorimeter. This maneuver produced only a transient perturbation in the decay curve.

[5] J. L. Bollman, *J. Lab. Clin. Med.*, vol. 33, p. 1348, 1948.

[6] The holder consisted of two end blocks of polyfoam (15.3 × 15.3 cm, 5.8 and 1.9 cm thick, respectively) joined by kiln-dried birch rods (5.5 mm diam, 30 cm long) pressed through the end blocks to form a slightly oval-shaped space (5–6 cm minimum diam, 18–20 cm long) which would accommodate rats of various sizes. Spacing between rods was approximately 1 cm. Complete thermal insulation was provided by two formed blocks of polyfoam that could be positioned rapidly to completely surround the holder between the end blocks. The rat carcass was placed in the holder with the tail bent under and parallel to the body axis.

[7] D. R. Justesen, D. M. Levinson, R. L. Clarke, and N. W. King, *J. Microwave Power*, vol. 6, p. 237, 1971.

[8] The models were constructed of cylindrical polyfoam tubes (0.5 cm wall thickness). With end plugs in place, the inside dimensions were 3.8 cm diam × 16.7 cm long for the 190 g size, and 4.5 cm diam × 18.7 cm long for the 290 g size. The inside surface was thinly coated with silicone rubber to prevent leakage or evaporation.

[9] A. W. Guy, personal communication, 1971. The mixture used in this study was one of several developed by Guy to simulate the lossiness of various tissues. The mixture corresponding to muscle consisted of a slurry of 76.4 parts of 1.2% saline and 15.2 parts fine polyethelene powder, mixed with 8.4 parts "Superstuff" (Whamo) powdered gel.

[10] T. S. Ely, D. E. Goldman, and J. Z. Hearon, *IEEE Trans. Bio-Med. Eng.*, vol. 11, p. 123, 1964; R. L. Carpenter and C. A. Van Ummersen, *J. Microwave Power*, vol. 3, p. 3, 1968; R. D. Phillips, E. L. Hunt, and R. D. Castro, *J. Microwave Power*, vol. 6, p. 90, 1971; R. L. Carpenter, E. S. Ferri and G. J. Hagan, *J. Microwave Power*, vol. 7, p. 285, 1972.

[11] H. P. Schwan, *IEEE Trans. Bio-Med. Eng.*, vol. 19, p. 304, 1972; C. C. Johnson and A. W. Guy, *Proc. IEEE*, vol. 60, p. 692, 1972.

[12] Other materials might be used in physical models which would provide more accurate estimates of the absorbed dose for the animal.

Induced Fields and Heating Within a Cranial Structure Irradiated by an Electromagnetic Plane Wave

ALAN R. SHAPIRO, RICHARD F. LUTOMIRSKI, AND HAROLD T. YURA

Abstract—**The induced fields and the static heating patterns within a multilayered spherical model that approximates the primate cranial structure irradiated by plane waves in the microwave spectrum are calculated. The relation of the model to the biological structure and the sensitivity of the results to the uncertainties in the dimensions and electrical properties of biological material are investigated. A method of solution for both the scattered and the interior fields for a sphere with an arbitrary number of electrically different concentric layers is developed in a form readily amenable to machine computation. It is shown that the semi-infinite slab model is inappropriate for calculating the microwave radiation dosage for the human head and similar structures.**

I. Introduction

THE literature on the interaction of microwave radiation with biological specimens has for the most part been concerned with the average value of the induced fields and the consequent heating of the specimen averaged over its volume [1]. The treatment of nonuniform heating has generally been restricted to infinite slab models with multiple layers [1], [2]. The effect of the curvature of the surface of the specimen on the total absorbed energy has also been treated by Schwan [1] but not the internal distribution of that energy. He showed that the absorption cross section for a lossy sphere with a concentric outer layer having different electrical properties differs appreciably from that for a homogeneous sphere and is sensitive to the thickness of this outer layer. Furthermore, while the absorption cross section for homogeneous spheres varies considerably if the ratio of the diameter to the wavelength is between 0.1 and 1, i.e., Mie scattering, for the bilayer sphere these variations extend to both lower and higher values of this ratio.

The variations in absorption cross section with the ratio of size to wavelength are due to the rapid changes in the standing-wave pattern that is excited within the irradiated object. Schwan's calculations and measurements suggested that the Q of such a cavity is sufficient to support significant field gradients despite the lossy nature of biological tissue. At microwave frequencies and particularly at those frequencies allocated for medical diathermy the ratios of the major dimensions

Manuscript received October 5, 1970; revised November 16, 1970.
Any views expressed in this paper are those of the authors. They should not be interpreted as reflecting the views of The Rand Corporation or the official opinion or policy of any of its governmental or private research sponsors.
The authors are with The Rand Corporation, 1700 Main St., Santa Monica, Calif. 90406.

Reprinted from *IEEE Trans. Microwave Theory Tech.*, vol. MTT-19, pp. 187–196, Feb. 1971.

of the human head to the wavelength fall in this regime where nonuniform heating is to be expected. Furthermore, for this case the infinite slab model can give large errors for the internal fields. In the present paper, therefore, we report a calculation of the induced fields and the static heating patterns for a multilayered spherical model that approximates the primate cranial structure irradiated by a plane wave at 3000 MHz.

The simplest problem of this type is that of a plane wave falling upon a homogeneous sphere. This problem was first solved by Mie [3] and has been concisely presented by Stratton [4]. A solution for the scattering coefficients from a sphere with one concentric shell has been given by Aden and Kerker [5] and the scattering from an inhomogeneous spherically symmetric object has been calculated by Wyatt [6]. The present paper gives solutions in a form readily amenable to machine computation for both the scattered and the interior fields for a sphere with an arbitrary number of concentric layers.

II. The Model

To facilitate an experimental check of these calculations, we chose to model the rhesus macaque monkey, a primate commonly employed in such experiments. Based upon available cephalometric and stereotaxic data [7], [8], one finds that the portion of the head enclosing the brain is closely approximated by an ellipsoid of revolution whose midpoint is located in the midsagittal plane about one centimeter anterior to the origin of the Horsley–Clarke coordinates.[1] The cross sections in coronal planes are almost circular with an eccentricity of about 0.2, while those in the sagittal and horizontal planes are more elliptic with an eccentricity of about 0.6. The semiaxes are about 35 and 45 mm in the average fully grown female, with a range of about ± 10 percent, and somewhat larger in the male. This ellipsoid includes the brain, eyes, and olfactories, and its lowest point is in the region of the first vertebrae.

The primary features of the internal structure (Fig. 1) are a boney shell (cranium) which completely encompasses the brain and has apertures for such things as the optic nerves and the brain stem; the cerebra with their sharply curved frontal lobes; the cerebellum; and the cerebrospinal fluid (CSF) which fills the subarachnoid[2] spaces and the ventricles of the brain. The cerebral hemispheres can be fairly well represented by hemiellipsoids with their common midpoint at the origin of the Horsley–Clarke coordinates and having an eccentricity of about 0.9. The major axis of each cerebrum

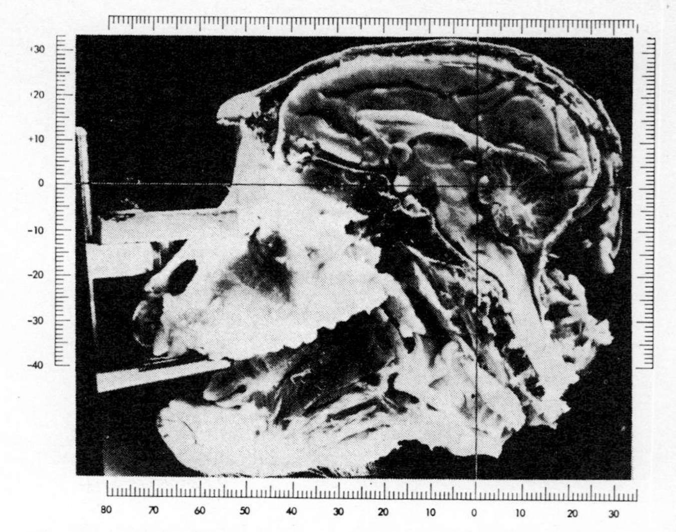

Fig. 1. *Macaca nemestrina* (midsagittal view). Male; age ~3 years; weight ~4.3 kg; scale: Horsley–Clarke coordinates, mm.

is defined by a line connecting the poles of the occipital and frontal lobes in the midsagittal plane and makes a small angle with the Horsley–Clarke horizontal plane. The thickness of the skin, subcutaneous fat, bone, dura, and subarachnoid CSF are difficult to measure; they are not uniform over the brain in one individual and vary between individuals. Estimates of these dimensions were made from a study of stereotaxic atlases of the rhesus [8] and the sensitivity of the calculated fields to a twofold variation was tested. The total volume of CSF in man is given as 140 ml and of this only 23 ml is in the ventricles. The ventricles are small cavities within the cerebra and the midbrain, the latter being in intimate contact with the hypothalamus and other important regulatory elements.

The various tissues and fluids within the primate head differ in their electrical properties. The published data on their chemical composition [10], [11] and the dielectric constant and electrical conductivity of biological materials [1], [12] have been used to construct the model. As Schwan has pointed out, the major differences in the electrical properties of tissues is a function of their water content. The head consists of low water-content material such as fat and bone, high water-content skin, dura and brain, and fluids such as the CSF and blood, which have still higher water content. In fact the water content of CSF (99 percent) is even greater than that of blood plasma (94 percent) and considerably higher than whole blood [11]. The chemical composition and the electrical properties exhibit a range of values in the normal animal. *In vitro* measurements [12] suggest that the range for the dielectric constant can be as great as ± 30 percent of the mean and for the electrical conductivity ± 40 percent. The sensitivity of the analysis to the uncertainty in the values for these properties is investigated below. The electrical properties of subcutaneous fat and of the

[1] The Horsley–Clarke coordinates [9] are the standard rectangular coordinates for stereotaxic atlases. The origin lies 10 mm above the auditory meatus in the midsagittal plane and is measured perpendicular to the basihorizontal (Frankfurt) plane.

[2] Between the cranium and the brain there are three membranous coverings, which are, starting from without, the dura, the arachnoid, and the pia. The subarachnoid space is that between the arachnoid and the pia.

TABLE I
THE MODEL
($f = 3$ GHz)

Region (p)	Tissue Modeled	Thickness (mm)	r_p Radius of Surface Bounding $p, p+1$ (mm)	κ_p Dielectric Constant $T=37°C$	σ_p Conductivity $T=37°C$ (ohm·m)$^{-1}$	$(\sigma/\omega\epsilon)^2$	$(2\beta)^{-1}$ Intensity Attenuation Length (mm)	n Index of Refraction
1	brain	sphere	26.8±1	42±13	2±0.8	0.08	8.6	6.6
2	CSF	2±1	28.8±0.25	77±23	1.9±0.8	0.02	12.3	8.8
3	dura	0.5±0.25	29.3±1	45±14	2.5±1	0.11	7.2	6.8
4	bone	2±1	31.3±0.3	5±1.5	0.2±0.1	0.06	29.8	2.3
5	fat	0.7±0.3	32±0.5	5±1.5	0.2±0.1	0.06	29.8	2.3
6	skin	1±0.5	33	45±14	2.5±1	0.11	7.2	6.6

cranial bone are taken to be the same. The skin and the dura are also assumed to be similar while the brain is assumed to be slightly less conductive based upon low-frequency data. Finally, the cerebrospinal fluid is assumed to have the electrical properties of physiological saline solution.

The mathematical difficulties encountered in the solution of inhomogeneous boundary-value problems are greatly increased when the dimensions of the surfaces of discontinuity are of the order of the wavelength and only spherical and ellipsoidal bodies have been treated [4]. Numerical techniques can in principle be employed for structures of arbitrary shape but they tend to be expensive and time consuming. Before embarking upon such an ambitious project it was decided to investigate the fields in a vastly simplified spherical model of the monkey head but one which retains some of the features that should lead to nonuniform internal distributions. These are a body bounded by a closed surface, of dimensions on the order of a wavelength, composed of thin layers of electrically different materials.

The spherical model chosen for calculation has its center at the origin of the Horsley–Clarke system with its outer radius approximately equal to the semiminor axis of the ellipsoidal model described above and represents a dorsal view of the head. It consists of five concentric layers representing skin, subcutaneous fat, bone, dura and cerebrospinal fluid and an inner sphere representing the brain. The resulting dimensions and electrical parameters are shown in Table I.

III. ANALYSIS

Fig. 2 shows the orientation of the plane wave with respect to a rectangular and spherical coordinate system having its origin at the center of the sphere. The wave is assumed to propagate in the positive z direction and the electric field is linearly polarized in the x direction. The inner sphere and each succeeding concentric shell will be denoted by the indices $p = 1, 2, \cdots, N$, respectively, where the outermost region $p = N$ denotes the surrounding medium. Each region is assumed to

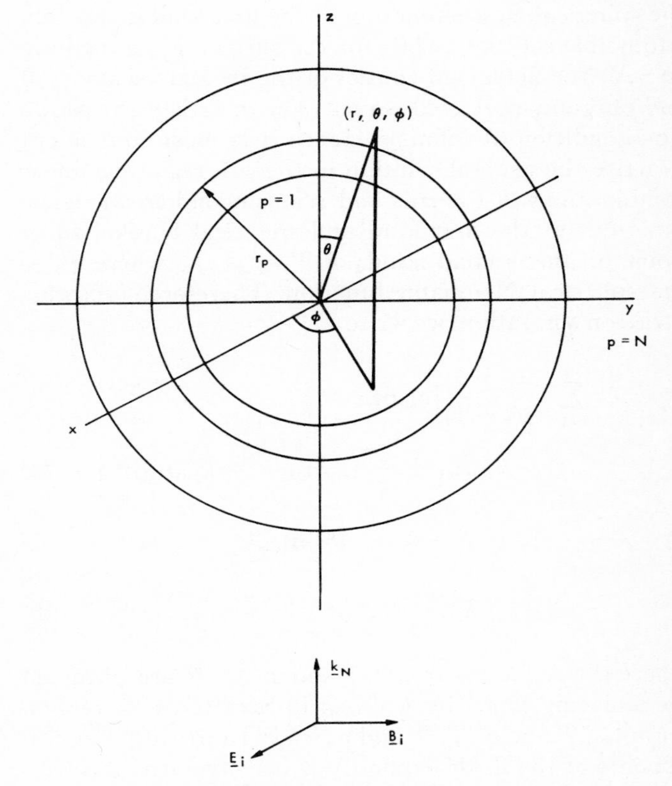

Fig. 2. Plane wave incident upon a sphere with $N-2$ concentric spherical shells.

have a different conductivity σ_p and relative dielectric constant κ_p; the containing medium is taken to be vacuum, whence $\sigma_N = 0$, $\kappa_N = 1$. The magnetic permeabilities are all taken to be unity, and a time dependence of $e^{-i\omega t}$ is assumed and suppressed. The complex propagation constant for the pth region is $k_p = \mathrm{Re}(k_p) + i\,\mathrm{Im}(k_p)$ where [4, p. 276]

$$\left.\begin{matrix}\mathrm{Re}\\ \mathrm{Im}\end{matrix}(k_p)\right\} = \frac{\omega}{c}\left[\frac{\kappa_p}{2}\left(\sqrt{1 + \frac{1}{(\epsilon_0\omega)^2}\left(\frac{\sigma_p}{\kappa_p}\right)^2} \pm 1\right)\right]^{1/2} \qquad (1)$$

and where ω is the angular frequency, ϵ_0 is the free space permitivity $= 8.85 \times 10^{-12}$ F/m, and c is the velocity of light *in vacuo*. The crux of the analysis lies in expanding the incident and secondary fields in vector spherical harmonics appropriate to each region and matching the tangential components of the fields at each boundary to determine the expansion coefficients.

The expansion of the incident plane wave is given in [4, p. 564], and an identical notation will be used. This incident wave will induce fields in all of the regions. The general solution of the vector wave equation for the transverse fields in any shell can be written as a linear combination of the $\boldsymbol{m}$ and $\boldsymbol{n}$ functions used by Stratton [4]. In general, the radial dependence is an arbitrary solution to the spherical Bessel equation, i.e., any linear combination of the spherical Bessel functions of the first and second kinds. For the inner sphere ($p=1$) the spherical Bessel function of the first kind is the only admissible solution, while for the surrrounding medium ($p=N$) the fields will consist of the incident wave (j's) and outgoing scattered waves (h's) to satisfy the radiation condition at infinity. Hence it is most convenient to write the general solution in the pth region as linear combinations of the $\boldsymbol{m}$'s and $\boldsymbol{n}$'s using spherical Bessel functions of the first kind and spherical Hankel functions of the second kind; $h_n^{(1)}=j_n+iy_n$ where y_n is the spherical Neumann function. Therefore, following Stratton's notation, we write

$$\boldsymbol{E}_p = E_0 \sum_{l=1}^{\infty} i^l \frac{2l+1}{l(l+1)} \left[a_{lp}\boldsymbol{m}_{ol l}^{(1)} - ib_{lp}\boldsymbol{n}_{el l}^{(1)} + \alpha_{lp}\boldsymbol{m}_{ol l}^{(3)} - i\beta_{lp}\boldsymbol{n}_{el l}^{(3)}\right] \quad (2a)$$

$$\boldsymbol{B}_p = -\frac{k_p}{\omega} E_0 \sum_{l=1}^{\infty} i^l \frac{2l+1}{l(l+1)} \left[b_{lp}\boldsymbol{m}_{el l}^{(1)} + ia_{lp}\boldsymbol{n}_{ol l}^{(1)} + \beta_{lp}\boldsymbol{m}_{el l}^{(3)} + i\alpha_{lp}\boldsymbol{n}_{ol l}^{(3)}\right] \quad (2b)$$

where the functions $\boldsymbol{m}_{(e,o)1l}^{(3)}$ and $\boldsymbol{n}_{(e,o)1l}^{(3)}$ are obtained by replacing $j_n(\rho_p)$ by $h_n^{(1)}(\rho_p)$ in Stratton's expression for $\boldsymbol{m}_{(e,o)1l}^{(1)}$ and $\boldsymbol{n}_{(e,o)1l}^{(1)}$ and $\rho_p=k_p r$. In writing the components of the fields explicitly it is convenient to introduce the following notation. Let

$$j_{lp} = j_l(\rho_p), \qquad h_{lp} = h_l(\rho_p)$$

$$\eta_{lp} = \frac{1}{\rho_p}[\rho_p j_{lp}]', \qquad \xi_{lp} = \frac{1}{\rho_p}[\rho_p h_{lp}]'$$

$$P_l^1 = P_l^1(\cos\theta)$$

where primes denote differentiation with respect to ρ_p, and the superscript on the Hankel function has been dropped. For machine calculations it is preferable to eliminate expressions containing derivatives and represent η_{lp} and ξ_{lp} by

$$\eta_{lp} = \frac{1}{2l+1}\left[(l+1)j_{l-1,p} - lj_{l+1,p}\right]$$

and

$$\xi_{lp} = \frac{1}{2l+1}\left[(l+1)h_{l-1,p} - lh_{l+1,p}\right].$$

Similarly, using the recursion relations for the associated Legendre polynomials, it can be shown that

$$\frac{2l+1}{l(l+1)} \frac{\partial P_l^1(\cos\theta)}{\partial\theta} = \frac{l}{l+1} \frac{P_{l+1}^1(\cos\theta)}{\sin\theta} - \frac{l+1}{l} \frac{P_{l-1}^1(\cos\theta)}{\sin\theta}$$

where

$$\lim_{\theta\to 0} \frac{P_l^1(\cos\theta)}{\sin\theta} = \frac{l(l+1)}{2}$$

$$\lim_{\theta\to \pi} \frac{P_l^1(\cos\theta)}{\sin\theta} = \frac{(-1)^l l(l+1)}{2}.$$

A. Solution for the Expansion Coefficients

The fields can then be determined from (2a) and (2b) once the a_{lp}, α_{lp}, b_{lp}, β_{lp} are known. The boundary conditions at the interface between regions p and $p+1$ are the continuity of the tangential components of the fields; i.e.,

$$\begin{aligned} (\boldsymbol{E}_p)_\theta &= (\boldsymbol{E}_{p+1})_\theta, & (\boldsymbol{E}_p)_\phi &= (\boldsymbol{E}_{p+1})_\phi, \\ (\boldsymbol{B}_p)_\theta &= (\boldsymbol{B}_{p+1})_\theta, & (\boldsymbol{B}_p)_\phi &= (\boldsymbol{B}_{p+1})_\phi. \end{aligned} \quad (3)$$

Substituting the $\boldsymbol{m}$ and $\boldsymbol{n}$ functions into (2) and applying (3) shows that simultaneous equations of the following form must be satisfied:

$$\sum_l \left(A_{l,p}\frac{P_l^1}{\sin\theta} + B_{l,p}\frac{\partial P_l^1}{\partial\theta}\right) = \sum_l \left(A_{l,p+1}\frac{P_l^1}{\sin\theta} + B_{l,p+1}\frac{\partial P_l^1}{\partial\theta}\right) \quad (4a)$$

$$\sum_l \left(A_{l,p}\frac{\partial P_l^1}{\partial\theta} + B_{l,p}\frac{P_l^1}{\sin\theta}\right) = \sum_l \left(A_{l,p+1}\frac{\partial P_l^1}{\partial\theta} + B_{l,p+1}\frac{P_l^1}{\sin\theta}\right) \quad (4b)$$

or, with $S_{l,p}=A_{l,p+1}-A$ and $T_{l,p}=B_{l,p}-B_{l,p+1}$,

$$\sum_l \left(S_{l,p}\frac{P_l^1}{\sin\theta} + T_{l,p}\frac{\partial P_l^1}{\partial\theta}\right) = 0 \quad (5a)$$

$$\sum_l \left(S_{l,p}\frac{\partial P_l^1}{\partial\theta} + T_{l,p}\frac{P_l^1}{\sin\theta}\right) = 0. \quad (5b)$$

Equations (5a) and (5b) can be manipulated in the following manner to show that these modes are orthogonal and thus the coefficients for each mode can be independently determined. Multiplying (5a) by $P_{l'}^1$, (5b) by $(\partial P_{l'}^1/\partial\theta)\sin\theta$, and adding yields

$$\sum_l\left\{S_{l,p}\left[\frac{P_{l'}^1P_l^1}{\sin^2\theta}+\frac{\partial P_{l'}^1}{\partial\theta}\frac{\partial P_l^1}{\partial\theta}\right]\sin\theta + T_{l,p}\left[P_{l'}^1\frac{\partial P_l^1}{\partial\theta}+\frac{\partial P_{l'}^1}{\partial\theta}P_l^1\right]\right\}=0. \quad (6)$$

Integrating over θ from 0 to π makes the coefficient of $T_{l,p}$ vanish because $P_l^1(\pm1)=0$. The integral of the coefficient of $S_{l,p}$ is [4, p. 417]

$$\frac{2[l(l+1)]^2}{2l+1}\delta_{l',l}$$

where $\delta_{l',l}$ is the Kronecker delta. Hence $S_{l',p}=0$, and a similar calculation shows that $T_{l',p}=0$. The continuity of the fields therefore implies that two sets of simultaneous equations are satisfied, which can be written in matrix form:

$$\begin{pmatrix}a_{l,p}\\ \alpha_{l,p}\end{pmatrix}=\left(Q_{l,p+1}{}^{ij}\right)\begin{pmatrix}a_{l,p+1}\\ \alpha_{l,p+1}\end{pmatrix} \quad (7a)$$

$$\begin{pmatrix}b_{l,p}\\ \beta_{l,p}\end{pmatrix}=\left(R_{l,p+1}{}^{ij}\right)\begin{pmatrix}b_{l,p+1}\\ \beta_{l,p+1}\end{pmatrix} \quad (7b)$$

where the elements of Q and R are

$$Q_{lp}{}^{11}=(\Delta_{lp})^{-1}\left[\xi_{l,p}j_{l,p+1}-\frac{k_{p+1}}{k_p}h_{l,p}\eta_{l,p+1}\right]$$

$$Q_{lp}{}^{12}=(\Delta_{lp})^{-1}\left[\xi_{l,p}h_{l,p+1}-\frac{k_{p+1}}{k_p}h_{l,p}\xi_{l,p+1}\right]$$

$$Q_{lp}{}^{21}=(\Delta_{lp})^{-1}\left[\frac{k_{p+1}}{k_p}j_{l,p}\eta_{l,p+1}-\eta_{l,p}j_{l,p+1}\right]$$

$$Q_{lp}{}^{22}=(\Delta_{lp})^{-1}\left[\frac{k_{p+1}}{k_p}j_{l,p}\xi_{l,p+1}-\eta_{l,p}h_{l,p+1}\right]$$

$$R_{lp}{}^{11}=(\Delta_{lp})^{-1}\left[\frac{k_{p+1}}{k_p}\xi_{l,p}j_{l,p+1}-h_{l,p}\eta_{l,p+1}\right]$$

$$R_{lp}{}^{12}=(\Delta_{lp})^{-1}\left[\frac{k_{p+1}}{k_p}\xi_{l,p}h_{l,p+1}-h_{l,p}\xi_{l,p+1}\right]$$

$$R_{lp}{}^{21}=(\Delta_{lp})^{-1}\left[j_{l,p}\eta_{l,p+1}-\frac{k_{p+1}}{k_p}\eta_{l,p}j_{l,p+1}\right]$$

$$R_{lp}{}^{22}=(\Delta_{lp})^{-1}\left[j_{l,p}\xi_{l,p+1}-\frac{k_{p+1}}{k_p}\eta_{l,p}h_{l,p+1}\right]$$

and

$$\Delta_{lp}=j_{l,p}\xi_{l,p}-h_{l,p}\eta_{l,p}.$$

All of the Bessel functions are now understood to be evaluated at the values of ρ_p and ρ_{p+1} corresponding to the interface between the two regions.

We now use the fact that $\alpha_{l,1}=\beta_{l,1}=0$ (i.e., the Hankel function, which is divergent at the origin at the innermost sphere, must be absent for this region), and $a_{l,N}=b_{l,N}=1$. Forming the products of the matrices and denoting them by

$$(Q_{lT}{}^{ij})=\prod_{p=1}^{N}(Q_{lp}{}^{ij});\qquad (R_{lT}{}^{ij})=\prod_{p=1}^{N}(R_{lp}{}^{ij})$$

we obtain

$$\begin{pmatrix}a_{l,1}\\ 0\end{pmatrix}=\left(Q_{lT}{}^{ij}\right)\begin{pmatrix}1\\ \alpha_{l,N}\end{pmatrix} \quad (8)$$

and

$$\begin{pmatrix}b_{l,1}\\ 0\end{pmatrix}=\left(R_{lT}{}^{ij}\right)\begin{pmatrix}1\\ \beta_{l,N}\end{pmatrix}. \quad (9)$$

Equations (8) and (9) each represent a pair of simple simultaneous equations from which $a_{l,1}$, $b_{l,1}$, $\alpha_{l,N}$, and $\beta_{l,N}$ can be determined. Application of the Q and R matrices given by (7) then generates the remainder of the expansion coefficients.

B. Mean Intensity, Joule Heating, and Scattering

For problems where a knowledge of the local variation in mean values of the components of the electromagnetic stress tensor are desired, one can readily compute terms of the form

$$\overline{\mathrm{Re}\,(\boldsymbol{E}_p)_i\cdot\mathrm{Re}\,(\boldsymbol{E}_p)_j}=\tfrac{1}{2}\mathrm{Re}\,[(\boldsymbol{E}_p)_i(\boldsymbol{E}_p)_j{}^*] \quad (10)$$

from (2), with a similar expression for the magnetic terms. Similarly, one can compute the mean intensity (with $i=j$) and the mean energy dissipated in heat per unit volume per second in the pth region:

$$w_a(r,\theta,\phi)=\tfrac{1}{2}\sigma_p(\boldsymbol{E}_p\cdot\boldsymbol{E}_p{}^*). \quad (11)$$

The average density of the power absorbed throughout the model, $\langle w_a\rangle$, can be obtained from the scattering and total cross sections, Q_s and Q_t, respectively. The absorption cross section, Q_a, is simply Q_t-Q_s [4, p. 569], where

$$Q_t=\frac{2\pi}{k_N{}^2}\mathrm{Re}\sum_{l=1}^{\infty}(2l+1)(\alpha_{l,N}+\beta_{l,N}) \quad (12)$$

and

$$Q_s=\frac{2\pi}{k_N{}^2}\sum_{l=1}^{\infty}(2l+1)(|\alpha_{l,N}|^2+|\beta_{l,N}|^2). \quad (13)$$

Finally

$$\langle w_a\rangle=\frac{3}{8\pi}\sqrt{\frac{\epsilon_0}{\mu_0}}\frac{E_0{}^2Q_a}{r_{N-1}{}^3}. \quad (14)$$

IV. Results and Discussion

The mean square field and the mean dissipation were calculated for the interior of the model described in Table I irradiated by a plane wave at a frequency of 3 GHz. In this case $2r_6/\lambda_0=0.66$ and $k_Nr_6=2.07$, the regime of interest. The results are shown in Figs. 3–8 in

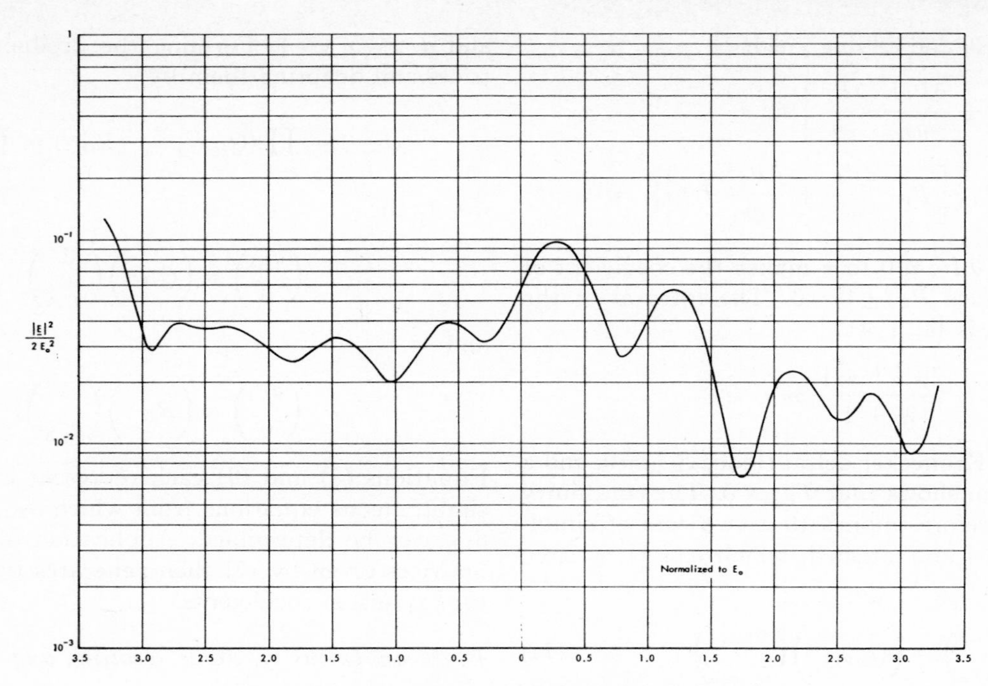

Fig. 3. Mean square electric field along the polar axis.

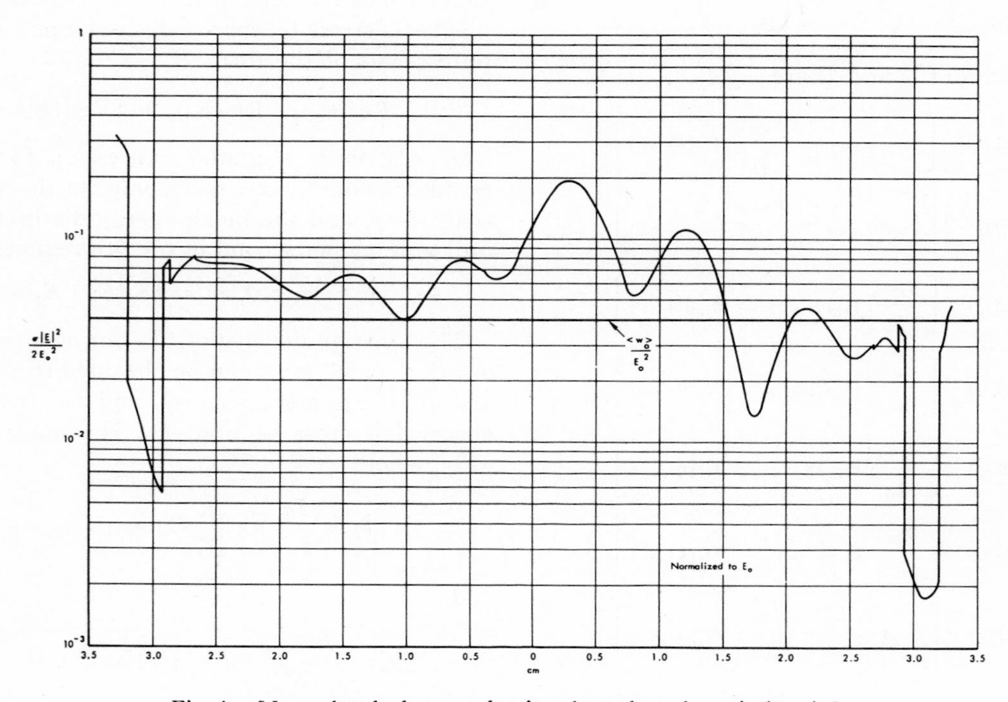

Fig. 4. Mean absorbed power density along the polar axis $(\Omega \cdot \mathrm{m})^{-1}$.

which the ordinates are normalized to the amplitude of the unperturbed incident field E_0. With reference to Fig. 2, the internal distribution is shown in Figs. 3 and 4 along the direction of propagation (the z axis); in the equatorial plane of the model along the direction of electric polarization of the incident wave (the x axis) in Figs. 5 and 6; and in the direction of incident magnetic polarization in Figs. 7 and 8.

The most striking feature of these results is the magnitude of the peaks in the center of the model. The mean square field and dissipation are approximately the same as and in one case (Figs. 7 and 8) greater than at the surface despite the fact that the intensity attenuation length[3] for the inner sphere is only about one-third of its radius. Thus a calculation which assumed the body to be semiinfinite in extent would have yielded intensities more than an order of magnitude lower than

[3] The attenuation length is taken as one-half the skin depth δ ($\delta = 1/\beta$, where β is usual attenuation factor) and shown in Table I.

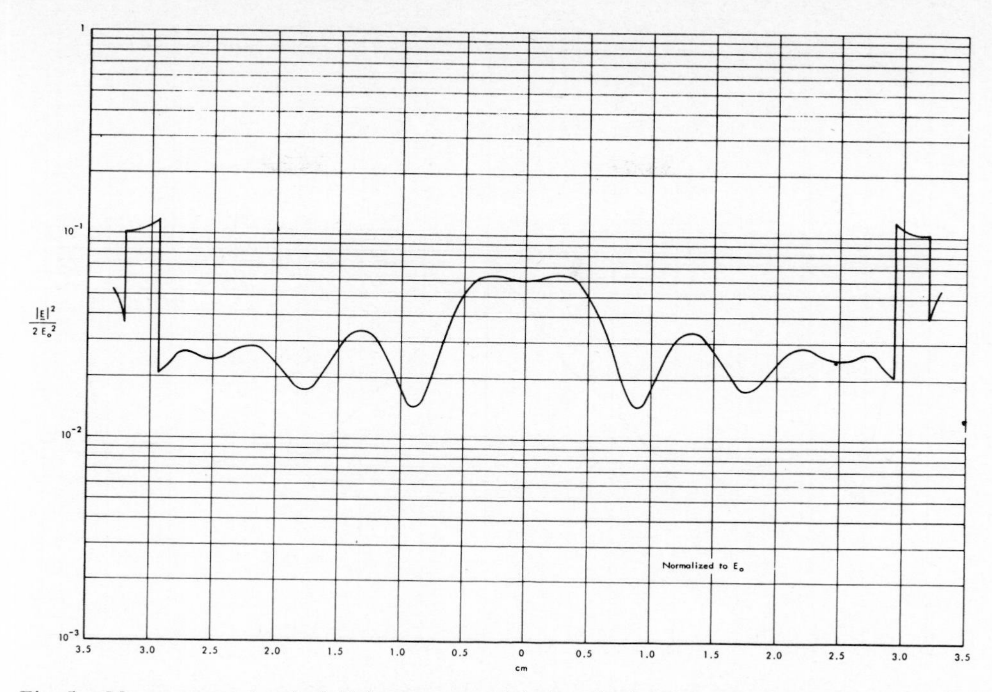

Fig. 5. Mean square electric field in the equatorial plane in the direction of electric polarization.

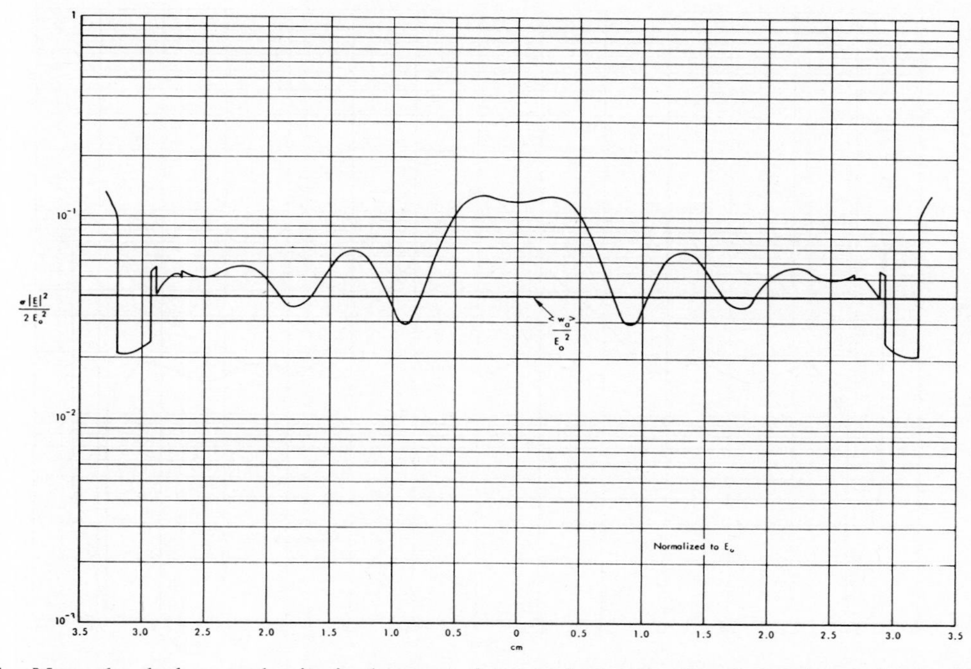

Fig. 6. Mean absorbed power density in the equatorial plane in the direction of electric polarization $(\Omega\cdot m)^{-1}$.

those given here. The general dependence of the interior fields on $k_N a$, where a is the radius of a homogeneous lossy sphere, can be seen in Fig. 9. For a homogeneous sphere having the electrical properties of the inner sphere of our model, the induced field is seen to go from a uniform distribution at $k_N a = 0.157$ (Fig. 9(a)) through one with a null at the center to distributions (Fig. 9(b)) that are peaked at the center ($k_N a = 1.26$). For a homogeneous sphere with the conductivity of biological tissue, the skin effect phenomenon will become dominant for $k_N a \gtrsim \pi$; however, in multilayered spheres the internal mode structure can still be significant at larger values of $k_N r_{N-1}$.

The average density of absorbed power throughout the model $\langle w_a \rangle$ is given by (14) as approximately 0.04 E_0^2 W/m³ for this case ($Q_a \simeq 4.4 \times 10^{-3}$ m²). The ordinates in Figs. 4, 6, and 8 are directly comparable to $\langle w_a \rangle / E_0^2$. Examination of these figures shows that the peak heating is as much as five times the average value. Gradients in the local heating as great as a factor of two

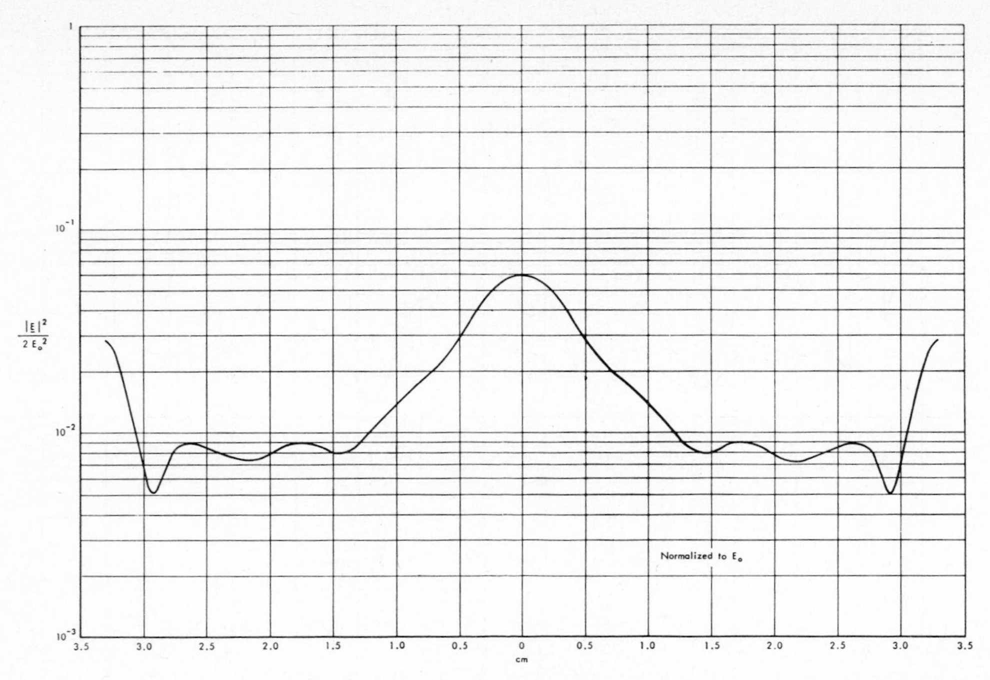

Fig. 7. Mean square electric field in the equatorial plane in the direction of magnetic polarization.

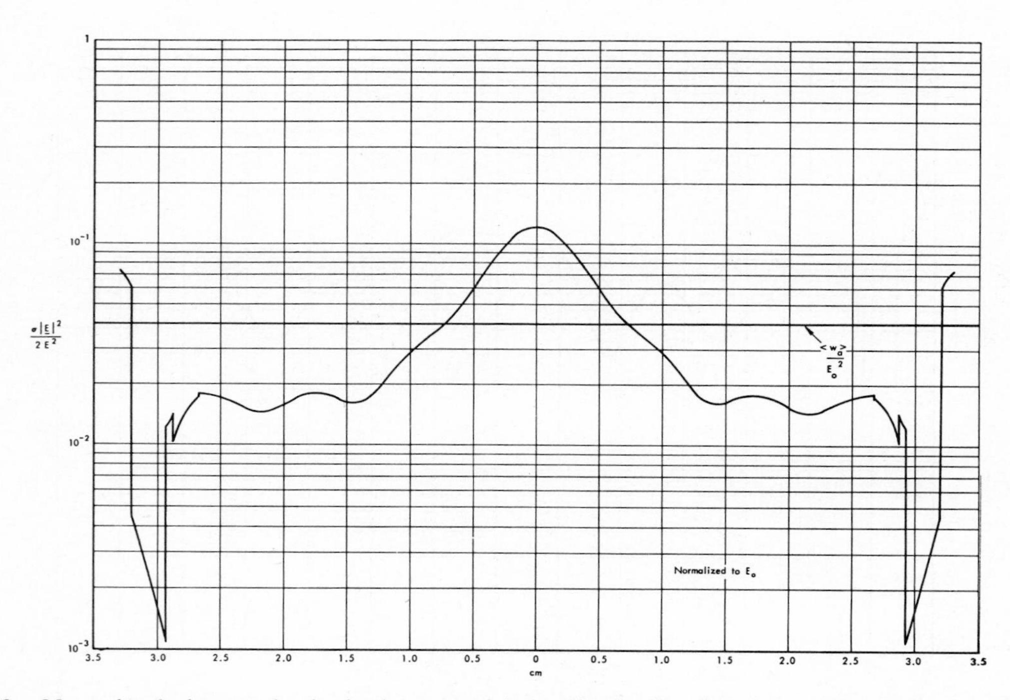

Fig. 8. Mean absorbed power density in the equatorial plane in the direction of magnetic polarization $(\Omega \cdot m)^{-1}$.

over a displacement of one millimeter are predicted within the inner sphere. Such variations imply temperature gradients induced in the body, at least for short times before diffusion becomes important.

There are appreciable differences in the spatial distribution of heating in the three directions plotted. The largest gradients occur in the direction of propagation (Fig. 4), whereas the most peaked distribution is in the equatorial plane in the direction of the incident magnetic polarization (Fig. 8). Thus the dosage in medical diathermy of the head may be a function of polarization and aspect.

For the range of parameters used in this investigation, it proved necessary to use very accurate representations of the functions in (2a) and (2b) in order to obtain convergence of these series. This was accomplished by expanding the spherical Bessel and Hankel functions in terms of trigonometric functions and by using double precision arithmetic (16 decimal digits) on the real and imaginary parts.

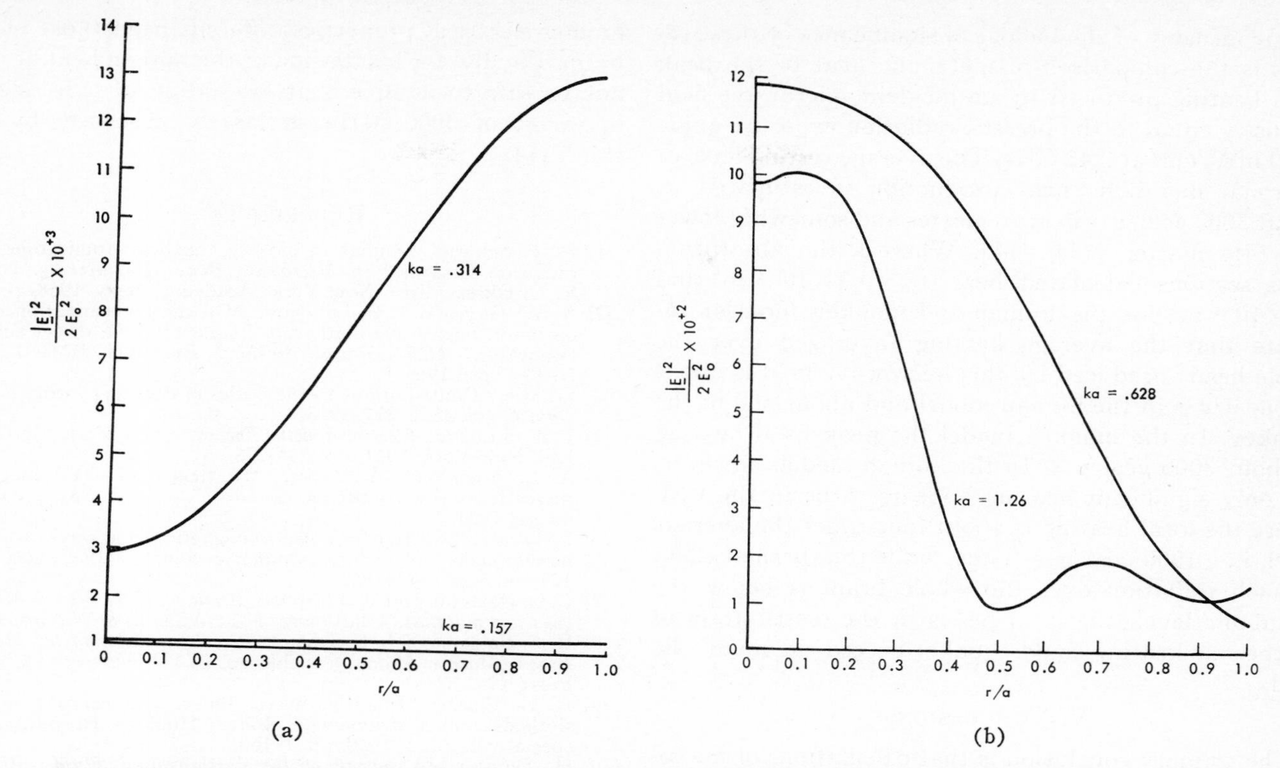

Fig. 9. Mean square electric field versus distance from center of sphere. $\theta=\pi/2$; $\phi=0$; $f=3\times10^9$Hz; $\sigma=2$ mho/m; $\kappa=42$. (a) For spheres with $k_N a=0.05\pi$, 0.1π. (b) For spheres with $k_N a=0.2\pi$, 0.4π.

The sensitivity of these results to the uncertainty in both the thickness of the layers of tissue and their electrical properties was examined. Over the range of thickness variations shown in Table I the maximum deviations from the nominal value of the absorption cross section Q_a were +10 percent, −29 percent. Both extremes occurred for changes in the simulated fat–bone layer. The test for sensitivity to κ_p and σ_p showed extremes in Q_a of +16 percent and −10 percent. In this case the extremes occurred for changes in the simulated brain–CSF regions. The details of the distributions were not examined for these variations. However, it is noted that these variations in the absorption cross section are less than the ratio of peak to average heating in the nominal configuration.

A brief investigation was conducted of the change in the distribution and the average heating with the frequency of the incident field. For the nominal values in Table I the maximum absorption cross section occurred at about 3.8 GHz ($Q_a \sim 5.3\times10^{-3}$ m²) and a neighboring minimum occurred at about 1.6 GHz ($Q_a \sim 3.7\times10^{-3}$ m²). This constitutes a range in average heating of about ±18 percent around the midvalue. Over this same range of frequencies the internal distributions did not vary appreciably in the transverse plane but the pronounced dip in the direction of propagation was greatly accentuated at 1.6 GHz. Of course this is a narrow frequency range within which $k_N r_{N-1}$ lies between 1.1 and 2.6. For $k_N r_{N-1}<0.1$ the average heating drops off rapidly. At $k_N r_{N-1}=4.6$ ($f=8.2$ GHz) the average heating was down to about 20 percent of that at 3.8 GHz but the peak in the internal distribution was about ten times the average value.

It is of interest to compare the heating that might be produced in the head of a macaque monkey at the medical diathermy frequency of 2.45 GHz with that calculated for a similar model having the overall dimensions of the human head (~18-cm diameter). The thickness of the skin, dura and CSF in the human was assumed to be the same as in the monkey but the fat–bone region was increased to 5.5 mm. The results are that although the absorption cross section for the model of the human head is larger than for the monkey ($Q_a=3.3\times10^{-2}$ m² and 3.8×10^{-3} m², respectively) the average heating of the human model is only about 42 percent of that in the monkey. The overall characteristics of the internal distribution in the human model are different from those in Figs. 3–8. The peaks in the region corresponding to the midbrain are much less pronounced in the human model and are only one-tenth the average heating. However, the heating at the center is still about one hundred times greater than would be calculated by assuming the body to be semi-infinite in extent. Furthermore, only about one-quarter of the heating is deposited in the layer corresponding to skin, whereas about 10 percent is deposited in those layers representing the dura and CSF. It is of interest to note that the internal distribution is very different for the monkey model at a frequency such that $k_N r_{N-1}$ is equal to that for the human model at 2.45 GHz (i.e., 4.6).

One measure of the biological significance of these results is the comparison of metabolic heat to the peak local heating produced by an incident microwave field intensity equal to the present radiation exposure guide of 10 mW/cm² at 2.45 GHz. The average resting level of neuronal metabolic heat production is estimated at about 3000 μcal/g·s in gray matter and somewhat lower in white matter [13], [14]. Whereas the absorption cross sections calculated here ($Q_a = 3.3 \times 10^{-2}$ m² and 3.8×10^{-3} m² for the human and monkey models) indicate that the average heating (averaged over the whole head) produced by this microwave field is about 250 μcal/g·s in the human model and about 600 in the monkey. In the monkey model the peak local heating is about 3000 μcal/g·s. In the human model, however, the only significant heating appears to be in the CSF where the local heating is about four times the average level, i.e., 1000 μcal/g·s. Thus, while the thermal effect of such radiation over the whole brain is below the metabolic level, the local peaks in the distribution of diathermic heating could equal or exceed the metabolic load.

V. Conclusions

The primary conclusion is that calculations of microwave "dosage" using semi-infinite slab models is not appropriate for those portions of the animal subject where the ratio between the local radii of curvature and the wavelength lies between about 0.05 and 5. A corollary is that inferences drawn from animal experiments should take these geometric effects into account. Particular attention should be paid to radiation of the head where the brain with some fairly sharp curvatures (e.g., the frontal lobes) is encased in a fat–bone layer having electrical properties different from those of the brain. Finally, for irradiation of the human head it may not be safe to assume that "radiation of a frequency in excess of 3000 MHz is largely absorbed by the skin" [1].

References

[1] H. P. Schwan, "Radiation biology, medical applicatons, and radiation hazards," in *Microwave Power Engineering*, vol. 2, E. C. Okress, Ed. New York: Academic Press, 1968, p. 215.

[2] A. W. Guy and J. F. Lehmann, "On the determination of an optimum microwave diathermy frequency for a direct contact applicator," *IEEE Trans. Bio-Med. Eng.*, vol. BME-13, pp. 76–87, April 1966.

[3] G. Mie, "Contributions to the optics of diffusing media," *Ann. Physik*, vol. 25, p. 377, 1908.

[4] J. A. Stratton, *Electromagnetic Theory.* New York: McGraw-Hill, New York, 1941, pp. 563–573.

[5] A. L. Aden and M. Kerker, "Scattering of electromagnetic waves from two concentric spheres," *J. Appl. Phys.*, vol. 22, no. 10, 1951.

[6] P. Wyatt, "Scattering of electromagnetic plane waves from inhomogeneous spherically symmetric objects," *Phys. Rev.*, vol. 127, no. 5, p. 1837, 1962.

[7] C. G. Hartman and W. L. Straus, *Anatomy of the Rhesus Monkey* (*Macaca mulatta*). Baltimore, Md.: Williams & Wilkins, 1933.

[8] R. S. Snider and J. C. Lee, *A Stereotaxic Atlas of the Monkey Brain* (*Macaca Mulatta*). Chicago, Ill.: University of Chicago Press, 1961.

[9] R. H. Clarke, "Investigation of the central nervous system. Methods and instruments," Johns Hopkins Hospital Rep., special vol., pt. 1, 1920, pp. 1–160.

[10] H. Davson, *Physiology of the Cerebrospinal Fluid.* Boston, Mass.: Little, Brown and Company, 1967.

[11] *Blood and Other Body Fluids*, Biological Handbooks, Federation of American Scientists for Experimental Biology, Washington, D. C., 1961, p. 326.

[12] H. P. Schwan, "Electrical properties of tissue and cell suspensions," in *Advances in Biological and Medical Physics*, vol. 5. New York: Academic Press, 1957.

[13] J. G. McElligott and R. Melzack, "Localized thermal changes evoked in the brain by visual and auditory stimulation," *Exp. Neurol.*, vol. 17, pp. 293–312, 1967.

[14] Louis Sokoloff, "Metabolism of the central nervous system in vivo," in *Handbook of Physiology*, sec. 1, vol. 3. Washington, D. C.: American Physiological Society, 1960, p. 1843.

Nonionizing Electromagnetic Wave Effects in Biological Materials and Systems

CURTIS C. JOHNSON, SENIOR MEMBER, IEEE, AND ARTHUR W. GUY, MEMBER, IEEE

Invited Paper

***Abstract*—Electromagnetic waves from the lower radio frequencies up through the optical spectrum can generate a myriad of effects and responses in biological specimens. Some of these effects can be harmful to man at high radiation intensities, producing burns, cataracts, chemical changes, etc. Biological effects have been reported at lower radiation intensities, but it is not now known if low-level effects are harmful. Even behavioral changes have been reported. Most of the effects are not harmful under controlled conditions, and can thereby be used for therapeutic purposes and to make useful diagnostic measurements. The problem of microwave penetration into the body with resultant internal power absorption is approached from both the theoretical and the experimental viewpoints. The results are discussed in terms of therapeutic warming of tissues and possible hazards caused by internal "hot spots." The absorption and scattering effects of light in biological tissues are reviewed. Molecular absorption peaks in the optical spectrum are useful for making molecular concentration measurements by spectroscopy. Much of the related work in the literature is summarized, some new results are presented, and several useful applications of wave energy and medical instruments are discussed.**

I. Introduction

HEALTH OFFICIALS and medical personnel have recently been confronted with several problems associated with wave propagation effects in tissues and living systems such as man. They have turned to the engineering profession for help, but have found engineers generally unable to provide the needed guidance. Physiologists and biologists have attempted to fill this void, with results that clearly demand an informed engineering critique. It seems curious indeed that although some of the early fundamental experiments with light and electromagnetic waves were performed over 100 years ago, the application of these energy forms to man himself is pitifully inadequate. This paper is presented in an attempt to define the overall problem in some small measure, and illustrate the new applications methodology and techniques. It quite frankly represents an attempt to stimulate the engineering profession to respond to this urgent medical and social need.

The paper consists of two basic parts. Section II describes electromagnetic radiation from, say, 1 MHz to 300 GHz, where the wavelength is large compared to cell sizes. There is little scattering, thus most of the wave reflection and transmission line concepts are applicable. Section III describes radiation effects from the far infrared through the ultraviolet spectrum. Here there is significant wave scattering and molecular absorption effects, which require different theoretical and experimental approaches.

II. Radio Frequency and Microwave Effects

A. Introduction

Electromagnetic fields in the spectrum between 1 MHz and 100 GHz have special biological significance since they can readily be transmitted through, absorbed by, and reflected at biological tissue boundaries in varying degrees, depending on body size, tissue properties, and frequency. There is very little scattering by tissues in this frequency range. These characteristics can result in either medically beneficial effects or biological damage or harm, depending on the circumstances. The frequency range receiving the most attention in terms of biological interaction is in the microwave spectrum of 300 to 10 000 MHz. This is due to the widespread use of high-energy densities in highly populated areas and to the better absorption characteristics in the tissues of man in this frequency range.

The effect of microwaves on the biological system may be categorized into two major areas, one involving medically beneficial effects and the other involving harmful effects. The effects may be further classified as either thermal effects resulting from high-level microwave power or low-level effects that may or may not be a thermal effect. To the uninitiated, a glance at the present state of knowledge in this area will provide a confusing picture indeed! One will see microwave power densities up to 590 mW/cm^2 being used in clinics for routine diathermy treatment of many areas of the human body [1], [2], and units of refrigerated human whole blood at 4°C brought up to body temperature within a minute, for immediate transfusion, by microwave heating [3]. One will also see research on the application of high-power microwave energy for treatment of cancer [4] and for quickly eliminating hypothermia after open heart surgery. Yet at the same time we note that the maximum recommended safe power density for long-term human exposure varies from 10 mW/cm^2 in the United States to as low as 0.01 mW/cm^2 in the USSR. We hear strong arguments between the microwave oven industry, the military, and the Public Health Service, and also between reputable scientists, on where realistic safety levels for microwave exposure should lie. The most recent and most intense public concern stems from both the passage of the Radiation Control for Health and Safety Act of 1968 [5], [6], and the increasing sales of microwave ovens [7]. This coupled with conflicting results from the past research done in the United States and the USSR, and inadequate quantitative data on biological effects

Manuscript received February 24, 1972; revised March 20, 1972. *This invited paper is one of a series planned on topics of general interest—The Editor.* Facilities and administrative support for the research described were provided by Social Rehabilitation Service Research and Training Grant 16-P-56818/0-09; Bureau of Radiological Health Grant 8-RO1-RL00528-02; National Institutes of Health Grants GM-16436 and GM-16000; and the Department of Rehabilitation Medicine, Department of Electrical Engineering, and Center for Bioengineering at the University of Washington, Seattle.

The authors are with the Department of Electrical Engineering, Department of Rehabilitation Medicine and Center for Bioengineering, University of Washington, Seattle, Wash. 98195.

Reprinted from *Proc. IEEE*, vol. 60, pp. 692–718, June 1972.

of microwaves, has raised many questions that remain unanswered. Fortunately, with the current public concern with radiation safety, an increasing number of electromagnetic engineers are becoming interested in studies on both biological effects and safe medical applications of electromagnetic energy. This new interest should result in a better quantitative understanding of the mechanisms of interaction.

B. Observed Effects in Tissue

1) Thermal Effects: The most investigated and documented effect of RF power on biological tissues is the transformation of energy entering the tissues into increased kinetic energy of the absorbing molecules, thereby producing a general heating in the medium. The heating results from both ionic conduction and vibration of the dipole molecules of water and proteins. The power absorbed by the tissues will produce a temperature rise that is dependent on the cooling mechanisms of the tissue. The patterns of the fields producing the heating are complex functions of the frequency, source configuration, tissue geometry, and dielectric properties of the tissues. The temperature patterns are further modified by the thermal properties of the tissues and neurocirculatory mechanisms. When the thermoregulatory capability of the system or parts of the system is exceeded, tissue damage and death can result. This occurs at absorbed power levels far above the metabolic power output of the body. Death usually results from the diffusion of heat from the irradiated portion of the body to the rest of the body by the vascular system. As the absorbed energy steadily increases, the protective mechanisms for heat control break down, resulting in an uncontrolled rise in body temperature. Michaelson [8]–[10] has demonstrated these effects in dogs and rats.

The absorption or heating patterns induced by radiation of the biological system will be nonuniform and dependent on the dielectric properties of the tissues. The absorption is high and the depth of penetration low in tissues of high water content such as muscle, brain tissue, internal organs, and skin, while the absorption is an order of magnitude lower in tissues of low water content such as fat and bone. Reflections between interfaces separating tissues of high and low water content can produce severe standing waves accompanied by "hot spots" that can be maximum in either tissue, regardless of dielectric constant or conductivity. Skin burns over the rib cage of test animals exposed to microwave power are examples of this. Local lesions of the skin and underlying tissues due to thermal effects from microwave exposure have been observed. These microwave burns tend to be deep, like fourth-degree burns, due to the deep penetration of the energy. Tissues with poor blood circulation or temperature regulation such as the lens of the eye, the gall bladder, and parts of the gastorintestinal tract are vulnerable. Experiments have shown that severe and injurious selective temperature increases can occur in these tissues with only slight increases in rectal and oral temperature. Injurious effects may occur at different temperatures depending on the tissue. For instance, it is well known that the testes undergo degenerative changes when maintained at normal body temperature over a long period of time. Microwave radiation has produced damage of the testes by increasing the temperature of the glands to as little as 35°C [11].

The heating characteristic of RF power has been used for nearly three quarters of a century by the medical profession to provide deep heating for therapeutic purposes. The therapeutic temperature range of 43°C to 45°C is very close to the temperature range where destructive changes can occur. This therapeutic technique called diathermy is discussed in great detail in the literature by Schwan [2], Scott [12], Moor [1], Krusen *et al.* [13], and Lehmann [14]. Local effects of diathermy include increases in blood flow owing to arteriolar and capillary dilation, increases in filtration and diffusion across biologic membranes, and possible greater capillary membrane permeability with resulting escape of plasma proteins. Vigorous heating can result in cellular responses associated with an inflammatory reaction. Enzyme reactions can take place due to changes in metabolic rate and proteins may be denatured with resulting products such as polypeptides and histamine-like substances becoming biologically effective. Diathermy has altered the physical properties of fibrous tissues in tendons, joint capsules, and scars to allow them to yield much more to stress. Other effects are relaxation of muscle spasms and increases in pain threshold of nerves.

2) Nonthermal Effects: The term "nonthermal effect" generally relates to an effect that is not associated with an increase in temperature. One such effect, observed by Herrick [15], Heller [16], and Heller *et al.* [17], [18], as well as Wildervank *et al.* [19], is due to forces acting on particles and is called the pearl chain effect. This effect is seen when suspended particles of charcoal, starch, milk, erthrocytes, or leucocytes (blood cells) are placed in a continuous or pulsed RF field in the range of 1–100 MHz. The particles form into chains parallel to the electric lines of force. For each particle type there is a frequency range where the effect occurs at minimum field strength. The chain formation, quantified by Satio *et al.* [20], [21], and Furedi *et al.* [22], [23], is due to the attraction between particles in which dipole charges are induced by the RF fields. The effect is discussed in detail by Schwan [2] and Presman [24]. Schwan has indicated that the effect occurs in biologic tissue at field levels where damaging thermal effects will occur.

Another nonthermal effect is the dielectric saturation occurring in solutions of proteins and other biological macromolecules due to intense microwave fields. It is suggested by Schwan [25] that such fields can cause polarized side chains of the macromolecules to line up with the direction of the electric field, leading to a possible breakage of hydrogen bonds and to alterations of the hydration zone. Such effects can cause denaturation or coagulation of molecules which was confirmed experimentally by Fleming *et al.* [26]. Resonance absorption in living cells is considered to be such an effect [27]–[29], but it is doubted by some investigators that it can occur at nonthermal power levels. Paramagnetic resonance analysis is used to determine dipole movement of protein molecules to elucidate the structure of crystalline proteins [30]. Field effects in water and protonic semiconductors in biological materials are reported [31]. Organisms such as bacteria, planaria, and snails react to weak electric and magnetic fields [32], [33]. Neuromuscular responses occur in birds in electromagnetic fields [34]. A very interesting phenomenon is the hearing of sounds corresponding to the frequency of modulation by people exposed to radar beams of low average power [35]–[37]. Direct and indirect effects on the central nervous systems (CNS) have been reported by the Soviets at levels below 10 mW/cm^2 [38]–[40]. They state that the CNS is the most sensitive of all body systems to microwaves at intensities below thermal thresholds. They

TABLE I

PROPERTIES OF ELECTROMAGNETIC WAVES IN BIOLOGICAL MEDIA

		Muscle, Skin, and Tissues with High Water Content							
						Reflection Coefficient			
						Air–Muscle Interface		Muscle–Fat Interface	
Frequency (MHz)	Wavelength in Air (cm)	Dielectric Constant ϵ_H	Conductivity σ_H (mho/m)	Wavelength λ_H (cm)	Depth of Penetration (cm)	r	ϕ	r	ϕ
1	30000	2000	0.400	436	91.3	0.982	+179		
10	3000	160	0.625	118	21.6	0.956	+178		
27.12	1106	113	0.612	68.1	14.3	0.925	+177	0.651	−11.13
40.68	738	97.3	0.693	51.3	11.2	0.913	+176	0.652	−10.21
100	300	71.7	0.889	27	6.66	0.881	+175	0.650	−7.96
200	150	56.5	1.28	16.6	4.79	0.844	+175	0.612	−8.06
300	100	54	1.37	11.9	3.89	0.825	+175	0.592	−8.14
433	69.3	53	1.43	8.76	3.57	0.803	+175	0.562	−7.06
750	40	52	1.54	5.34	3.18	0.779	+176	0.532	−5.69
915	32.8	51	1.60	4.46	3.04	0.772	+177	0.519	−4.32
1500	20	49	1.77	2.81	2.42	0.761	+177	0.506	−3.66
2450	12.2	47	2.21	1.76	1.70	0.754	+177	0.500	−3.88
3000	10	46	2.26	1.45	1.61	0.751	+178	0.495	−3.20
5000	6	44	3.92	0.89	0.788	0.749	+177	0.502	−4.95
5800	5.17	43.3	4.73	0.775	0.720	0.746	+177	0.502	−4.29
8000	3.75	40	7.65	0.578	0.413	0.744	+176	0.513	−6.65
10000	3	39.9	10.3	0.464	0.343	0.743	+176	0.518	−5.95

further state that there is a reaction of the nervous system to microwave energy on skin receptors or brain cells. The Soviets also claim nonthermal effects on the cardiovascular system, including decreased arterial pressure and heart rate [30]. As a result of these interactions, they have set their level of safe exposure 1000 times lower than that of the United States. Some of the Soviet work appears to be corroborated by investigators in the United States. For example, Frey [41], [42] has observed changes in CNS activity in animals and auditory effects in humans due to very low incident power levels. Korbel [43], [44] also reports behavioral effects in white rats exposed to low levels of microwave power. These CNS and behavioral effects are the source of a great deal of controversy at the present time.

C. *Theoretical Description*

In order to understand some of the characteristics of radio-frequency and microwave interactions with biological materials, a list of some important wave parameters is given in Tables I and II. The first column lists selected frequencies between 1 MHz and 10 GHz. The frequencies of 27.12, 40.68, 433, 915, 2450, and 5800 are significant since they are used for industrial, scientific, and medical heating processes. The frequencies of 27.12, 915, and 2450 are used for diathermy purposes in the United States, whereas only 433 MHz is authorized in Europe for these purposes. The second column tabulates the corresponding wavelength λ in air, and the remaining columns pertain to the wave properties of a tissue group. Table I gives data for muscle, skin, or tissues of high water content, while Table II is for fat, bone, and tissues of low water content. Other tissues containing intermediate amounts of water such as brain, lung, bone marrow, etc., will have properties that lie between the tabulated values for the two listed groups. The tables list the dielectric properties, the depth of penetration, and the reflection characteristics of various tissues exposed to electromagnetic waves as a function of frequency.

1) Dielectric Properties: The dielectric behavior of the two groups of biological tissues tabulated in Tables I and II has been evaluated most thoroughly by Schwan and his associates [45]–[47] and by other researchers including Cook [48]–[50] and Cole [51]. The interaction of electromagnetic wave fields with biological tissues is related to these dielectric characteristics. The tissues are composed of cells encapsulated by thin membranes containing an intracellular fluid composed of of various salt ions, polar protein molecules, and polar water molecules. The extracellular fluid has similar concentrations of ions and polar molecules, though some of the elements are different.

The action of electromagnetic fields on the tissues produces two types of effects that control the dielectric behavior. One is the oscillation of the free charges or ions and the other the rotation of dipole molecules at the frequency of the applied electromagnetic energy. The first gives rise to conduction currents with an associated energy loss due to electrical resistance of the medium, and the other affects the displacement current through the medium with an associated dielectric loss due to viscosity. These effects control the behavior of the complex dielectric constant $\epsilon^*/\epsilon_0=(\epsilon'-j\epsilon'')$, where ϵ_0 is the permittivity of free space, ϵ^* is the complex permittivity, ϵ' is the dielectric constant, and ϵ'' is the loss factor of the medium. The effective conductivity σ (due to both conduction currents and dielectric losses) of the medium is related to ϵ'' by $\epsilon''=\sigma/\omega\epsilon_0$ and the loss tangent is given by $\tan\delta=\epsilon''/\epsilon'=\sigma/\omega\epsilon'\epsilon_0$. The quantity ϵ^* will be dispersive due to the various relaxation processes associated with polarization phenomena. These are indicated by the dielectric properties given in Tables I and II. The decrease in dielectric constant ϵ_H and increase in conductivity σ_H for tissues of high water content with increasing frequency is due to interfacial polarization across the cell membranes. The cell membranes, with a capacity of approximately 1 $\mu F/cm^2$, act as insulating layers at low frequencies so that currents flow only in the extracellular medium, accounting for the low conductivity of the tissues. At sufficiently low frequencies, the charging time constant is small enough to completely charge and discharge the membrane during a single cycle, resulting in a high tissue capacitance and therefore a high dielectric constant. When frequency is increased, the capacitive reactance of the cell decreases, resulting in increasing currents in the intracellular

TABLE II

PROPERTIES OF ELECTROMAGNETIC WAVES IN BIOLOGICAL MEDIA

		Fat, Bone, and Tissues with Low Water Content							
						Reflection Coefficient			
						Air–Fat Interface		Fat–Muscle Interface	
Frequency (MHz)	Wavelength in Air (cm)	Dielectric Constant ϵ_L	Conductivity σ_L (mmho/m)	Wavelength λ_L (cm)	Depth of Penetration (cm)	r	ϕ	r	ϕ
1	30000								
10	3000								
27.12	1106	20	10.9–43.2	241	159	0.660	+174	0.651	+169
40.68	738	14.6	12.6–52.8	187	118	0.617	+173	0.652	+170
100	300	7.45	19.1–75.9	106	60.4	0.511	+168	0.650	+172
200	150	5.95	25.8–94.2	59.7	39.2	0.458	+168	0.612	+172
300	100	5.7	31.6–107	41	32.1	0.438	+169	0.592	+172
433	69.3	5.6	37.9–118	28.8	26.2	0.427	+170	0.562	+173
750	40	5.6	49.8–138	16.8	23	0.415	+173	0.532	+174
915	32.8	5.6	55.6–147	13.7	17.7	0.417	+173	0.519	+176
1500	20	5.6	70.8–171	8.41	13.9	0.412	+174	0.506	+176
2450	12.2	5.5	96.4–213	5.21	11.2	0.406	+176	0.500	+176
3000	10	5.5	110–234	4.25	9.74	0.406	+176	0.495	+177
5000	6	5.5	162–309	2.63	6.67	0.393	+176	0.502	+175
5900	5.17	5.05	186–338	2.29	5.24	0.388	+176	0.502	+176
8000	3.75	4.7	255–431	1.73	4.61	0.371	+176	0.513	+173
10000	3	4.5	324–549	1.41	3.39	0.363	+175	0.518	+174

medium with a resulting increase in total conductivity of the tissue. The increase in frequency will also prevent the cell walls from becoming totally charged during a complete cycle, resulting in a decrease in ϵ_H. At a frequency of approximately 100 MHz and above, the cell membrane capacitive reactance becomes sufficiently low that the cells can be assumed to be short-circuited. In the frequency range of 100 MHz to 1 GHz, the ion content of the electrolyte medium has no effect on the dispersion of the dielectric constant so the values of ϵ_H and σ_H are relatively independent of frequency. Schwan [45], [52] has suggested, however, that suspended protein molecules with a lower value of dielectric constant act as "dielectric cavities" in the electrolyte, thereby lowering the dielectric constant of the tissue. He attributes the slight dispersion of ϵ_H to the variation of the effective dielectric constant of the protein molecules with frequency. The final decline of ϵ_H and increase of σ_H at frequencies above 1 GHz can be attributed to the polar properties of water molecules which have a relaxation frequency near 22 GHz.

The dielectric behavior of tissues with low water content is quantitatively similar to tissues with high water content, but the values of dielectric constant ϵ_L and conductivity σ_L are an order of magnitude lower and are not quantitatively understood as well. This is due to the fact that the ratio of free to various types of bound water is not known. There is also a large variation in tissues of low water content. Since water has a high dielectric constant and conductivity compared to fat, the net tissue dielectric constant and conductivity will change significantly with small changes in water content.

The values of ϵ and σ will also vary with temperature. In the microwave region, where dispersion is small, the variation is given by

$$\frac{\Delta\sigma}{\sigma} = 2\%/°C$$

and

$$\frac{\Delta\epsilon}{\epsilon} = -0.5\%/°C.$$

The dielectric properties of the tissues play an important part in determining the reflected and transmitted power at interfaces between different tissue media. They also determine the amount of total power a given biological specimen will absorb when placed in an electromagnetic field.

2) Propagation and Absorption Characteristics of Waves

a) Plane tissue layers exposed to plane waves: Plane wave propagation characteristics in plane layered biological tissues may be examined to show how radiation is absorbed when the radius of curvature of the tissue surface is large compared to a wavelength. The propagation constant $k_{H,L}$ for power transmission through biological tissues can be written in terms of the complex dielectric constants $\epsilon_{H,L}{}^*$ and free space propagation constant k_0 in the standard form

$$k_{H,L} = k_0(\epsilon_{H,L}{}^*/\epsilon_0)^{1/2} = \beta_{H,L} - j\alpha_{H,L} \quad (1)$$

where the wavelengths $\lambda_{H,L} = 2\pi/\beta_{H,L}$ are significantly reduced in the tissues due to the high dielectric constants. Tables I and II indicate that the factors of reduction are quite large, between 6.5 and 8.5, for tissues of high water content, and between 2 and 2.5 for tissues with low water content. In addition to the large reduction in wavelength, there will be a large absorption of energy in the tissue which can result in heating. The absorbed power density $P_{H,L}$ resulting from both ionic conduction and vibration of dipole molecules in the tissues is given by

$$P_{H,L} = \frac{\sigma_{H,L}}{2} \mid E \mid^2 \quad (2)$$

where E is the magnitude of the electric field. One may note from the conductivities listed in Tables I and II that absorption in tissue of higher water content may be as high as 60 times greater than in that of low water content for the same electric fields. The absorption of microwave power will result in a progressive reduction of wave power density as the waves penetrate into the tissues. We can quantify this by defining a depth of penetration $d = 1/\alpha$ or a distance that the propagating wave will travel before the power density decreases by a factor of e^{-2}. We can see from Tables I and II that the depth of penetration for tissues of low water content is as much as 10 times greater than the same parameter for tissues of high water content.

Since each tissue in a complex biological system such as man has different complex permittivity, there will in general be reflections of energy between the various tissue interfaces during exposure to microwaves. The complex reflection coefficient due to a wave transmitted from a medium of complex permittivity ϵ_1^* to a medium of permittivity ϵ_2^* and thickness greater than a depth of penetration is given by

$$\rho = re^{j\phi} = \frac{\sqrt{\epsilon_1^*} - \sqrt{\epsilon_2^*}}{\sqrt{\epsilon_1^*} + \sqrt{\epsilon_2^*}}. \tag{3}$$

The values r and ϕ for various interfaces are tabulated in Tables I and II. Note the large reflection coefficient for an air–muscle or a fat–muscle interface. When a wave in a tissue of low water content is incident on an interface with a tissue of high water content of sufficient thickness (greater than the depth of penetration), the reflected wave is nearly 180° out of phase with the incident wave, thereby producing a standing wave with an intensity minimum near the interface. If the wave is propagating in a tissue of high water content and is incident on a tissue of low water content, the amplitude of the reflected component is in phase with the incident wave, thereby producing a standing wave with an intensity maximum near the interface. If there are several layers of different tissue media with thicknesses less than the depth of penetration for each medium, the reflected energy and standing wave pattern are influenced by the thickness of each layer and the various wave impedances. These effects may be obtained from the standard transmission line equations. The distribution of electric field strength E in a given layer is

$$E = E_0[e^{-jkz} + \rho e^{jkz}] \tag{4}$$

where E_0 is the peak magnitude of the field and ρ is the reflection coefficient. From (3), the equation for absorbed power density in the tissue layer, we obtain

$$P = \frac{\sigma E_0^2}{2}[e^{-2\alpha z} + r^2 e^{2\alpha z} + 2r\cos(2\beta z + \phi)]. \tag{5}$$

Schwan [2], [45] has made extensive calculations of these absorption distributions in various tissues. Typical distributions are shown in Fig. 1 for a wave transmitted through a subcutaneous fat medium into a muscle medium. The absorption is normalized to unity in the muscle at the fat–muscle interface. The relative absorption curves shown are not changed for smaller fat thicknesses. The severe discontinuity between the absorbed power in the muscle and that in the fat is quite apparent. Also, it can be seen that the standing wave peaks become larger in the fat and the wave penetration into the muscle becomes less with increasing frequency. This illustrates clearly the desirability for using frequencies lower than the 2450-MHz allocation for diathermy. Subcutaneous fat may vary from less than a centimeter in thickness to as much as 2.5 cm in thickness for different individuals. Deep heating for diathermy requires the transmission of energy through this subcutaneous fat layer to the muscle layer. Optimum results are attained with maximum heating in the muscle. The absolute values of absorbed power density in the tissue layers are dependent on incident power density, skin thickness, and fat thickness. Fig. 2 illustrates the absorbed power density at the muscle interface and the peak absorbed power density in a skin layer 2 mm thick as a function of fat thickness for an incident power density of 1 mW/cm². These values may be used to determine the absorbed power at other locations in the muscle and fat by relating them to the curves in Fig. 1. The peak absorbed power density is always maximum in the skin layer for the plane layered model. This is significant since the thermal receptors of the nervous system are located there and will indicate pain when the incident power density reaches levels that could thermally damage the tissue. With surface cooling of the skin, however, by natural environmental conditions or by controlled clinical procedures, the temperature increases may be higher in the fat or muscle. The peak absorption in the various tissues may vary over a wide range with fat thickness and frequency. It is apparent that frequencies below 918 MHz can penetrate more deeply into the tissues. The implications of this in terms of both radiation hazards and therapeutic applications are apparent. The first two figures clearly indicate the advantages of lower frequencies for diathermy, including 1) increased penetration into the muscle tissue, 2) less severe standing waves and resulting "hot spots" in the fat, and 3) better control and knowledge of the ab-

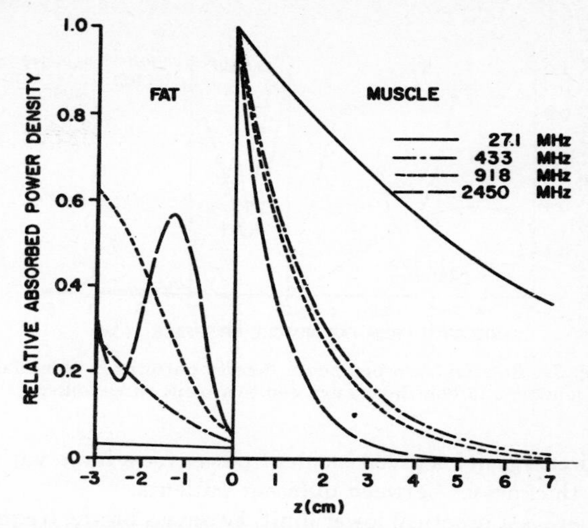

Fig. 1. Relative absorbed power density patterns in plane fat and muscle layers exposed to a plane wave source.

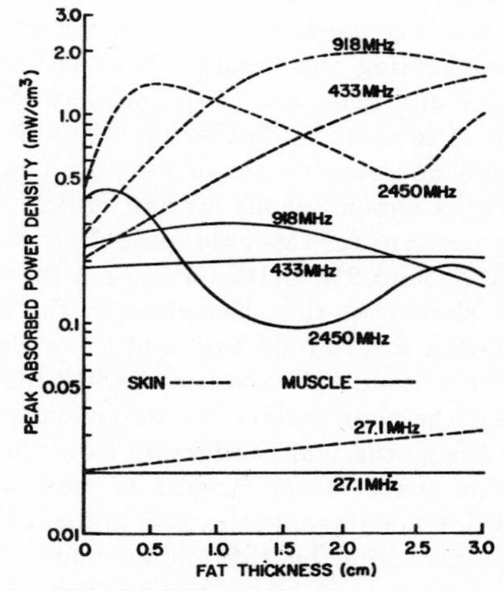

Fig. 2. Peak absorbed power density in plane skin and muscle layers as a function of fat thickness.

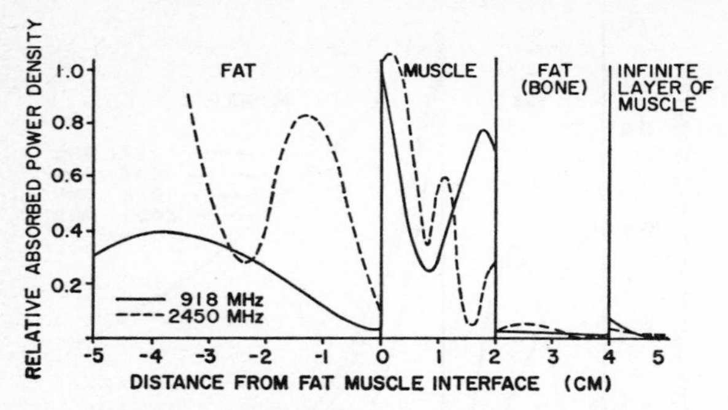

Fig. 3. Relative absorbed power density patterns in plane fat, muscle, and bone layers exposed to a plane wave source.

sorbed energy for a given incident power for a large variation of fat thicknesses between different patients.

There is a practical lower limit, however, on the frequency that can be used. As the frequency is decreased, the applicator needed becomes increasingly large until it is no longer possible to obtain desired selective heating patterns. If the applicator is not increased in size as frequency is lowered, only superficial heating will result. This has been discussed in detail by Guy and Lehmann [53], and Guy [54].

A problem of interest in diathermy is the determination of how effective microwaves are in heating a layer of bone beneath a layer of subcutaneous fat and muscle. Fig. 3 illustrates heating patterns for this case using diathermy frequencies of 2450 MHz and 918 MHz for a 2-cm-thick bone. The results clearly show that the absorption in the bone is very poor due to both a severe reflection and a low electrical conductivity. Since a standing wave peak at 918 MHz occurs in the muscle near the bone surface, we would expect significant bone heating due to thermal conduction from the muscle.

b) Spherical tissue layers exposed to plane waves: Both outside and inside body geometries also influence the amount of microwave absorption by the human body. If the entire body, or a body member such as the head or arm, is illuminated by a microwave beam of large diameter, the amounts of energy absorbed by the tissues are functions of not only the tissue layer thicknesses and cross-sectional area exposed but also the size of the body or member compared to a free space wavelength and the body surface curvatures. We can illustrate the effects of body size and curvature on absorption characteristics by considering a spherical shape body composed of tissue with a high water content.

The electric fields induced in a sphere or spherical layer (shell) of tissue by an incident plane wave field can be calculated from the general vector spherical wave solutions of the wave equation

$$\boldsymbol{E} = E_0 e^{j\omega t} \sum_{n=1}^{\infty} (j)^n \frac{2n+1}{n(n+1)} (a_n \boldsymbol{m}_{01n} - j b_n \boldsymbol{n}_{01n}) \quad (6)$$

where the functions $\boldsymbol{m}_{01n}$ and $\boldsymbol{n}_{01n}$ are defined and the coefficients a_n and b_n are obtained as described by Stratton [55, pp. 563–567]. The equations may be evaluated on a digital computer as described by Anne *et al.* [56]–[58], and Shapiro [59]. Fig. 4 illustrates the relative absorbed power density patterns (called relative heating) for a simplified homogeneous spherical model of a cat or monkey size brain and a human size brain exposed to a 1 mW/cm² plane wave source. The origin of the rectangular coordinate system used in the figures is located at the center of the sphere with wave propa-

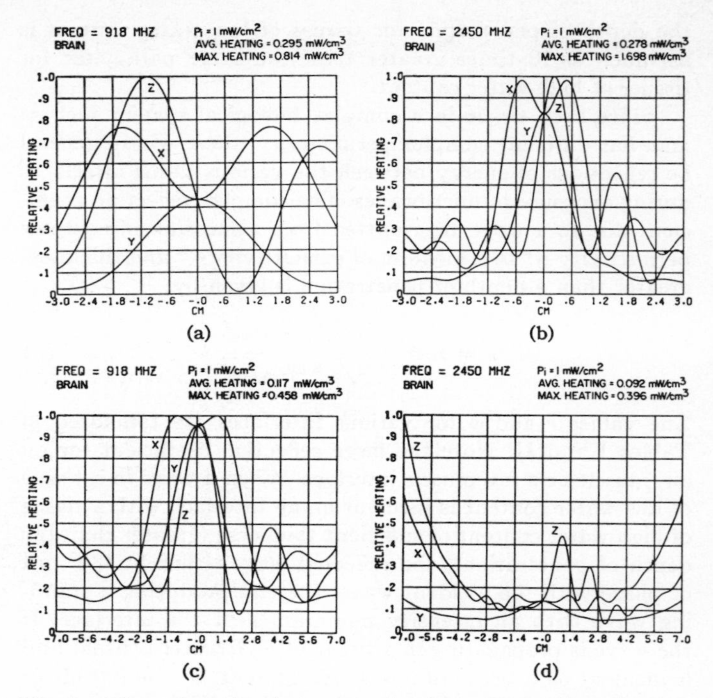

Fig. 4. Theoretical absorbed power density patterns along the x, y, and z axes models of brain tissue exposed to a plane wave source. (Incident power density 1 mW/cm², propagation along the z axis, and electrical field polarized along the x axis with origin at center of sphere.)

gation along the z axis and the E field polarized along the x axis. The maximum absorption (at the peak of the most severe standing wave) and the average absorption along the x, y, and z axes are illustrated on each plot. The dielectric properties for brain tissue used were based on values reported by Schwan [60] ($\epsilon = 35$ and 30.9, and $\sigma = 7$ and 11 mmho/cm for 918 and 2450 MHz, respectively). The figures clearly illustrate the intense fields and associated absorbed power density directly in the center of the human head and 1.2 cm off center in the animal head for 918-MHz exposure. One may note that the maximum absorption in the animal brain is larger by a factor of two than in the human brain. With 2450-MHz exposure, on the other hand, there is maximum absorption in the anterior (front) portion of the simulated human brain while there is maximum absorption in the center portion of the simulated animal brain. The animal brain receives a maximum absorption four times that of the human brain. Compared to Figs. 1–3, the plots clearly illustrate that plane tissue models cannot be used to evaluate absorbed power densities for situations where body size or radius of curvature are not large compared to a wavelength. It is significant to note that for human brain exposed to 918-MHz power, the absorption at a depth of 2.3 times the depth of penetration (depth of penetration = 3.2 cm) is twice the absorption at the surface. This corresponds to a factor greater than 200 times that expected based on the plane tissue model. At 2450 MHz, the absorbed power density at a depth approximately equal to 4.7 times the depth of penetration is 0.43 times that at the surface corresponding to a factor greater than 5000 times that expected from the plane tissue model. The regions of intense absorbed power density are due to a combination of the high refractive index and the radius of curvature of the model which produces a strong focusing of power toward the interior of the sphere which

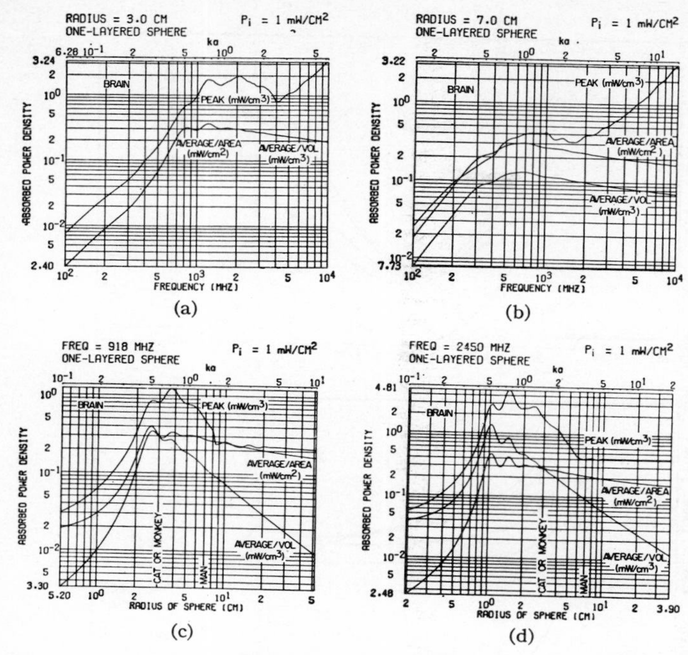

Fig. 5. Absorbed power density characteristics for spherical models of brain tissue exposed to 1 mW/cm² plane wave source.

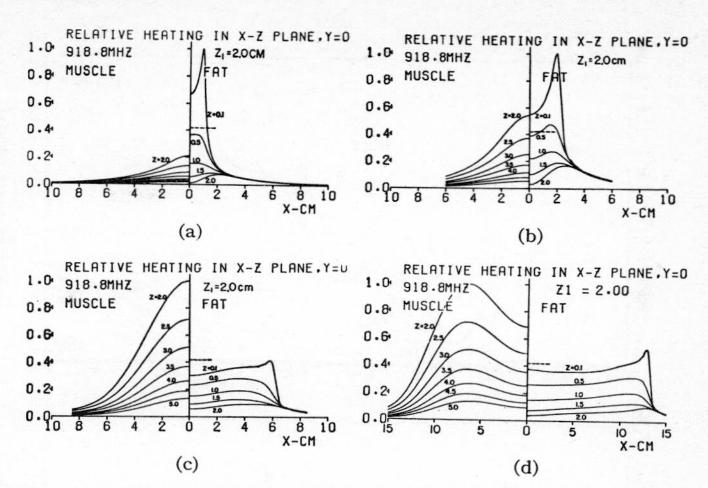

Fig. 6. Relative absorbed power density patterns in plane layers of fat and muscle exposed to TE_{10} mode waveguide aperture source with $a=12$ cm, $f=918.8$ MHz, and $z_1=2$ cm for various aperture heights. For (a)–(d) the values of b are 2, 4, 12, 26 cm, respectively.

more than compensates for transmission losses through the tissue. These results are significant in view of the many reported CNS effects at frequencies near 1 GHz. Fig. 5 illustrates the calculated peak absorbed power density per unit volume, average total absorbed power density per unit volume, and average total absorbed power density per unit area for various size spheres of brain tissue as a function of frequency. An incident power of 1 mW/cm² is assumed. Note the wide variation in absorbed power characteristics with different sphere sizes and frequencies of exposure. The graphs at the top of the figure indicate that there are sharp rises in peak absorption with increasing frequency followed by several "peaks" in the absorption curves. These peaks are related to the occurrence of "hot spots" or maxima in the internal absorption or heating patterns similar to those indicated in Fig. 4. As the frequency is further increased beyond the values where the peaks in absorption occur, the "hot spots" disappear and maximum absorption occurs at the exposed surface of the spheres. This is due to the decreasing depth of penetration with frequency. At frequencies beyond this point the peak absorption at the surface increases with frequency since the constant incident power is absorbed in a decreasingly smaller volume. The peak internal heating for the human head size sphere is maximum in the UHF frequency range centered near 915 MHz. This again is significant in terms of the large number of reported CNS effects for human exposure in the UHF frequency range. The phenomenon of hearing radar pulses is also reported in this frequency range [35], [36]. The graph indicates a further increase in peak absorption at frequencies above the UHF range, but this again is due to decreasing depth of penetration resulting in increased surface absorption. The curves at the bottom of Fig. 5 also indicate some interesting phenomena. They show that peak absorption can vary over an order of magnitude depending on brain size. The curves clearly indicate some important points to consider when one relates results between different size animals, insects, or humans exposed to RF radiation. Similar calculations have been made for muscle spheres to obtain clues to internal heating effects due to whole body radiation. Results are very similar to the curves for brain tissue given in Fig. 5. Peak absorbed power density inside the body is important since it is related to the location where localized effects or damage may occur as a function of frequency and animal size. The average absorbed power per unit volume is important because it is related to the time it takes for an exposed animal's thermoregulatory system to become overloaded or to the time that a steady-state thermal condition is reached. There is more than an order-of-magnitude variation in this value for a sphere representing man and for those representing small animals such as mice or rats. The curves depicting average power per unit area are important since they are related to the steady-state power absorption that an animal can experience without overloading its thermoregulatory system. Although the spheres are only rough approximations to actual animal bodies, they give rough approximations on how power levels for microwave effects can be extrapolated from one animal to another or to man as a function of frequency.

c) Cylindrical tissue layers exposed to plane waves: Power absorption density patterns may also be calculated for other simple tissue geometries representing portions of the anatomy. We can roughly approximate human limbs by concentric cylindrical layers of bone, muscle, fat, and skin and express the fields in each layer by an infinite series of Bessel functions of the first and second kind as discussed by Stratton [55, pp. 349–374]. For example, the electric field parallel to the z axis of the cylinder is expressed as

$$E_z = \sum_{n=0}^{\infty} 0[A_n J_n(kr) + B_n Y_n(kr)] e^{jn\theta} \quad (7)$$

where k is the wavenumber in the medium and the coefficients A_n and B_n are obtained by expanding the plane wave source expression into a series of Bessel functions and applying boundary conditions. Similar equations may be written for the wave polarized with the magnetic field parallel to the z axis. Ho *et al.* [61] has evaluated the equations and determined the fields and absorbed power densities for cylinders corresponding to human arms and legs exposed to plane waves. The results show the same increase in muscle-to-fat

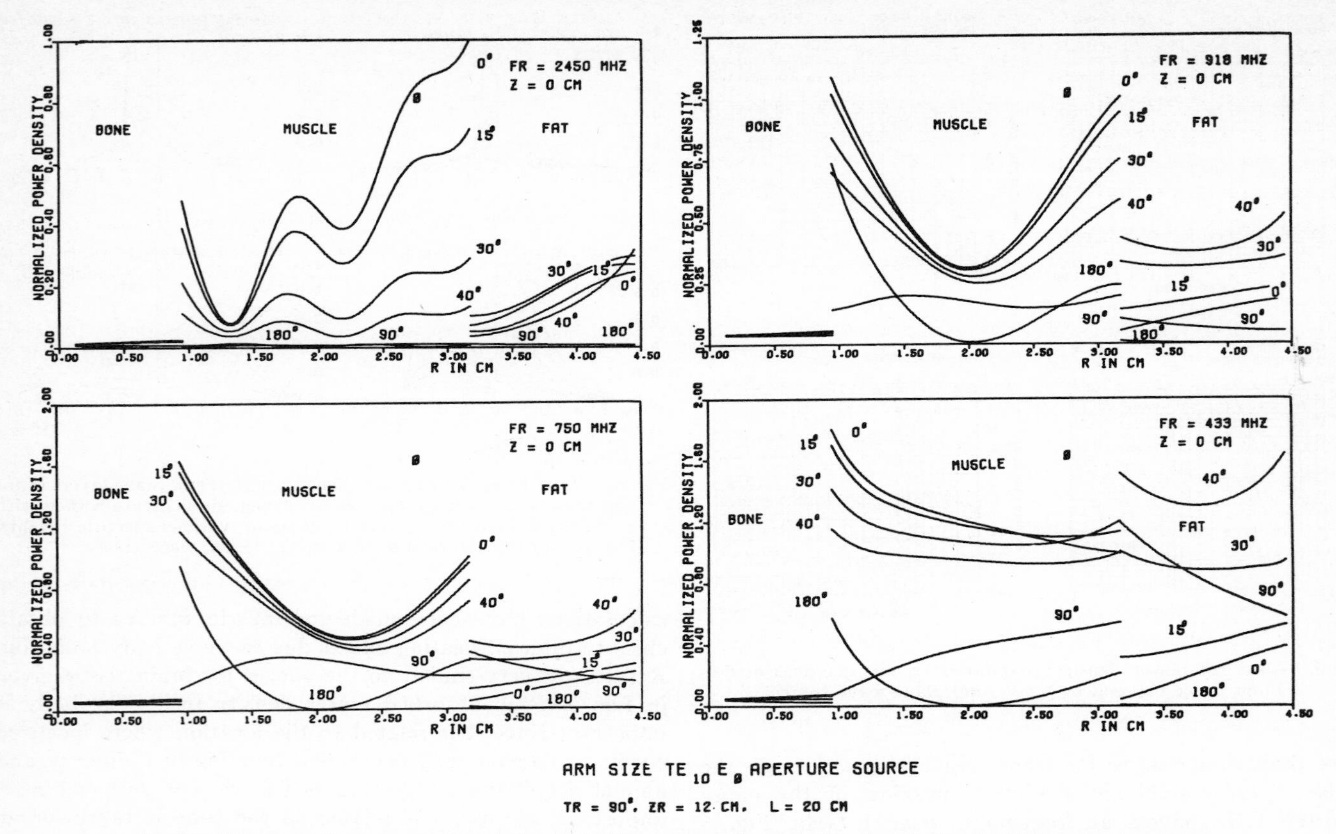

Fig. 7. Heating patterns in a cylindrical model of human arm due to a direct contact cylindrical aperture source.

absorbed power density ratio as observed for planar models exposed to lower frequency. The results also show a greater depth of energy penetration into the muscle due to the curvature of the tissues.

d) Tissues exposed to near-zone fields: If other than a plane wave source is used to expose biological tissues, the absorbed power density patterns are also very dependent on source size and distribution. Many applications of microwave power in medicine and studies on the biological effects of microwave power require an understanding of the absorbed power patterns due to tissues exposed to aperture and waveguide sources. Guy [54] has analyzed the case where a bilayered fat and muscle tissue layer is exposed to a direct contact aperture source of width a and height b. A fat tissue layer of thickness z_1 and dielectric constant ϵ_f^* in contact with a semi-infinite muscle tissue layer with a dielectric constant ϵ_m^* is assumed. The electric fields $\boldsymbol{E}_{f,m}$ in the fat and muscle tissue may be expressed as Fourier integrals

$$\boldsymbol{E}_{f,m}(x, y, z) = \frac{1}{(2\pi)^2} \int_{-\infty}^{\infty} \int_{-\infty}^{\infty} \boldsymbol{T}_{f,m}(u, v, z) e^{j(ux+vy)} du dv \quad (8)$$

where $\boldsymbol{T}_{f,m}$ are the Fourier transforms of the electric fields at the fat and muscle boundaries, derived from the boundary conditions at $z=0$ and $z=z_1$ in terms of the Fourier transform of the aperture

$$\boldsymbol{T}_a(u, v) = \int_{-\infty}^{\infty} \int_{-\infty}^{\infty} \boldsymbol{a}_z \times \boldsymbol{E}_f(x, y, 0) e^{-j(ux+vy)} dx dy. \quad (9)$$

The aperture field is denoted as $\boldsymbol{E}_f$ $(x, y, 0)$ and $\boldsymbol{a}_z$ is a unit vector along the z axis. The expressions may be evaluated numerically and the absorption patterns plotted by means of a digital computer. As an example we may consider a waveguide aperture source and evaluate it as a diathermy applicator for use at 918 MHz. Fig. 6 illustrates the complete heating curves in the x–z plane for $a=12$ cm and $b=2$, 4, 12, and 26 cm. Heating at the fat surface for a plane wave exposure is denoted by the dashed line on the figures. The results show that the relative heating varies from intense superficial heating in excess of that produced by a plane wave to deep heating greater than produced by a plane wave as aperture size is increased.

The absorbed power density patterns in multilayered cylindrical tissues exposed to an aperture source can also be determined by using a summation of three-dimensional cylindrical waves, expressing the aperture field as a two-dimensional Fourier series and matching the boundary conditions. Ho *et al.* [62], [63] has calculated the absorbed patterns for a number of different aperture and cylinder sizes. Typical results are shown in Fig. 7 for a human arm-size cylinder exposed to a surface aperture source 12 cm long in the direction of the axis. The patterns are plotted as a function of radial distance from the center of the cylinder for various circumferential angles ϕ from the center of the aperture. The patterns are normalized to the values at $\phi=0°$ at the muscle interface. The difference between the patterns in the cylindrical tissues and those illustrated for the plane layers indicates the importance of tissue curvature when assessing the effectiveness and safety of devices designed for medical application of microwave energy.

All of the theoretical results discussed in this section strongly point to the ineffectiveness of the 2450-MHz frequency as a diathermy frequency as pointed out in earlier reports by Schwan [45], [46], Lehmann [14], [65], and Guy

[54], [65]. Although the lower frequencies of 915 MHz authorized in the United States or 433 MHz authorized in Europe appear to be better choices, it appears from the theoretical data that 750 MHz would be the best choice. By their nature the frequencies that provide the best therapeutic heating would also be frequencies that could be most hazardous to man in an uncontrolled situation.

D. Measurements in Biological Tissues

The only way that realistic tradeoffs can be made between risks of biological damage and health benefits in the use of electromagnetic energy is to have available a quantitative description of what these effects are. Without proper instrumentation or a clear understanding of how electromagnetic fields interact with the tissue on a microscopic and macroscopic scale, or on the entire body structure, it is difficult indeed to say whether an observable effect may be thermal or nonthermal in origin or is merely an artifact due to the nature of the experimental approach. It is important that investigations of these effects be conducted in such a way that all aspects of the research are quantified, including the fields induced both inside and outside the tissues; the type and degree of effect; whether the effect is harmful, harmless, or merely an artifact; whether it is a thermal or a nonthermal effect; and how it relates to the results obtained by other investigators. Body size of the experimental animal must be taken into account, along with accurate *in vivo* dosimetry, so that results from one investigator obtained from rats can be related to those from another investigator using cats, monkeys, dogs, frogs, or a tissue sample in a test tube. Since body absorption cross sections and internal heating patterns can differ widely, as evidenced in the discussion in the last section, an investigator may think he is observing a low level or a nonthermal effect in one animal because the incident power is low while in actuality the animal may be exposed to as much absorbed power in a specific region of the body as another larger animal is with much higher incident powers.

Fig. 8 illustrates some of the problems encountered in measuring the fields and related effects in biological systems. Here we consider a controlled experiment in which an animal is exposed to a laboratory radiation source and we wish to see what the implication of observable biological effects is to a human exposed to some arbitrary source. For the human it is customary and desirable to define the hazard in terms of an incident power flux density or an electric field amplitude (denoted by I) as measured with a survey meter without the presence of the subject. When the subject enters the environment, however, many complicating factors arise. These include an unknown amount of scattered power (denoted by S) and transmitted power (denoted by T) and internal reflections (denoted by R). The survey meter will not give an accurate indication of the incident field I or the absorbed fields under these circumstances. Standing waves will occur on the outside of the man which may give a wide range of different readings on the survey meter, depending on its position. The greatest complicating factors are the result of interference patterns within the tissues, resulting in regions of intensification of absorbed power or hot spots and also regions of low absorbed power. These absorbed power patterns will vary depending on the source, frequency, body size, and geometry and the environment around the subject as described in the previous section.

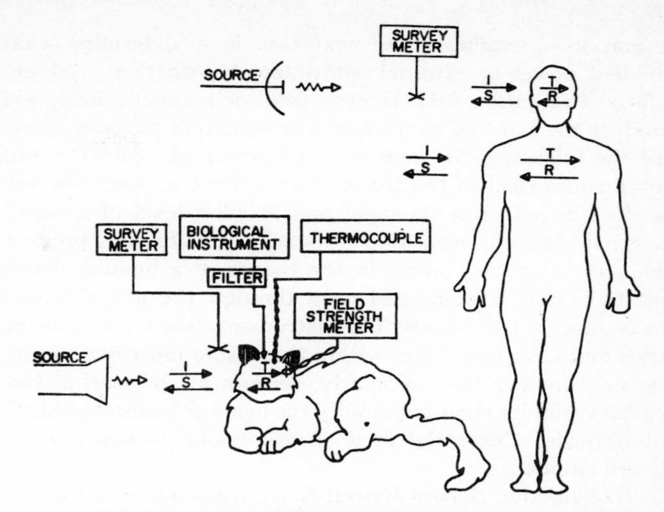

Fig. 8. Instrumentation associated with man and test animals exposed to electromagnetic radiation.

The only practical way to quantify biological damage accurately in terms of incident power levels for establishing safety standards for humans is through animal experimentation or irradiation of biological specimens *in vitro*. The question is how do we relate the biological effects to the fields and extrapolate the results to the establishment of safety standards for humans. In the controlled animal or biological tissue experiment we have the option to set up any field configuration we desire, where an effective power density can be determined with a survey meter. We cannot assume, however, that if a certain power density level produces quantifiable effects of damage in the animal or specimen, similar effects can take place in the human with the same effective incident power. With the animal or specimen there is scattering, absorption, and internal reflections uniquely associated with the animal's body and tissue characteristics or the specimen geometry which result in an absorbed power density pattern different from that for the case of human exposure. The approach that makes most sense is to quantify the actual fields, current density, or absorbed energy density in the tissues or specimen and relate this to the biological effects or damage that may occur. There has been some question among investigators concerning the most meaningful parameters to measure and how to relate them between the experimental animal and man. Bowman [66] has discussed the problem in great detail and suggests the measurement and use of the square of the magnitude of the electric field $|E|^2$ or energy density $U_E = \frac{1}{4}\epsilon'|E|^2$ as the most useful parameters to quantify hazards. He suggests the measurement of $|E|^2$ with implanted dipole antennas. Schwan [67], on the other hand, proposes that current density in the tissues is the most useful parameter to relate to the hazards. In this paper we propose the use of the absorbed power density as the most useful parameter since it is related directly to the well-established thermal effects or damage that may occur. It really makes no difference which of the above parameters are chosen, since they are all directly proportional to the magnitude or the square of the magnitude of the electric field in the tissues. One may measure these parameters directly by implanted dipoles and thermocouples, or indirectly by thermography, and relate them to electrophysiological phenomena measured while the test animal is exposed to radiation. Once this information is available, the next task is to determine what incident power or external field intensity, whether predominantly a radiation field, electric field, or magnetic field, will produce the same effect in man. The essentials to know, then, are the following: 1) what level of power per gram or unit volume absorbed in the tissue of an animal or specimen will produce an effect or damage; and 2) what level of incident power or fields as measured by a survey meter will produce the same absorbed power in the tissues of a human. These questions can be answered only through the use of sound measurement techniques. The instrumentation for measuring fields and absorbed power within the tissues must not modify the field in any way. Similarly, sensors for measuring biological variables must not modify the fields or be susceptible to interference. Currently used measurement techniques are described next.

1) Radiation Survey Meters: Since the Radiation Control for Health and Safety Act was passed in 1968 [5], there has been considerable improvement in radiation survey meters for measuring radiation power densities in air. Typical designs are illustrated in Fig. 9. The meter usually consists of a sensor consisting of two orthogonal electric dipole elements, each terminated in a thermocouple or microwave diode element and coupled via small-diameter high-resistance wires to a voltmeter calibrated to record power density directly in mW/cm². A thermocouple model described by Aslan [68] consists of a pair of thin-film vacuum-evaporated electrothermic elements that function as both antenna and detector. The sensor materials are antimony and bismuth deposited on a plastic or mica substrate, all secured to a rigid dielectric material for support. The length of the dipoles is small compared to a wavelength to allow the unit to monitor power with minimum perturbation on the RF field. The dc output of the sensor is directly proportional to the RF power heating the element. The hot and cold junctions of the electrothermic element, separated by 0.75 mm, are in the same ambient environment, thereby providing an output relatively independent of ambient temperatures. With the thin-film elements oriented at 90° to each other and connected in series, the total dc output is independent or orientation and field polarization about the axis of the probe and is proportional to the square of the electric field vector. If the proportionality constant relating E to H is known or remains constant, such as 377 Ω in the far field, the output can be calibrated in terms of power density. In the near field, the meter will read an effective power density or simply the square of the electric field divided by 377. Lead wires carrying the dc output of the thermocouples are shielded with ferrite materials and maintained perpendicular to the plane of the antennas. They will therefore be invisible to the propagating wave when the antenna is placed parallel to the phase front. The dc output is connected to an electric voltmeter calibrated to read field density directly in mW/cm². The meter has an appropriate time constant to read average power when the meter is used to measure modulated RF power density. The second configuration illustrated in Fig. 9 and described in the literature by Rudge [69] is similar, except that it uses a pair of matched diodes as sensing elements. Care must be exercised with this design such that the diodes are operated in the square low range.

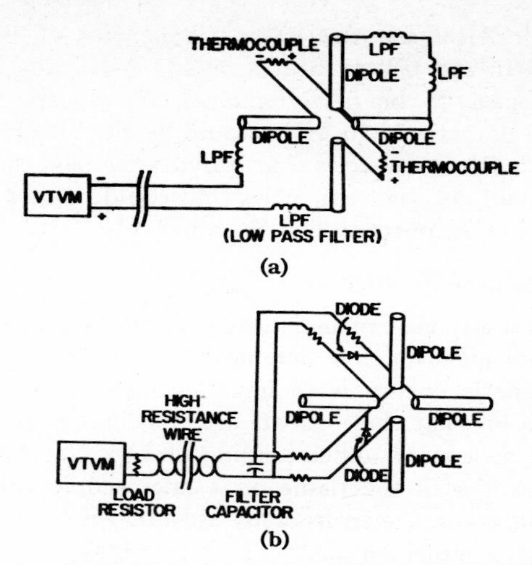

Fig. 9. Simplified sketches of various field survey meter designs. (a) Thermocouple method. (b) Diode method.

Survey meters of this type can be used meaningfully only to measure power density in a radiation-type field or the square of the electric field intensity in a near zone field. This information is not enough, however, to indicate what is happening in the tissues of an exposed subject.

2) Measurements with Implanted Probes: One of the most vexing problems in studies involving the interaction of electromagnetic fields and living biological systems and tissues is the quantification of the fields induced in the tissues by nearby sources. Electromagnetic fields or quantities related to the fields can be measured both *in situ* and *in vivo* in test animals or even humans by means of implanted microwave diodes thermocouples or thermistors. Thermocouples and thermistors were used extensively in the past for measuring the temperature rise in tissues exposed to radiation. Osborne and Frederick [70] used thermocouples inserted into the eyes of dogs to measure temperature increases associated with microwave heating of the eyes. The thermocouples were inserted before and after the microwave exposure. Richardson [71] also used thermocouples in the eyes of rabbits to measure temperature increases associated with opacity formation by exposing the eyes to microwave radiation. Lehmann and associates [64], [72] used thermistors for measuring temperature increases in the thigh of live pigs exposed to shortwave and microwave diathermy equipment. Similar experiments were also carried out with volunteers [73], [74]. There are several problems associated with the use of thermocouples or thermistors to ascertain absorbed power: 1) the element senses only the temperature of the tissue which is also a function of other mechanisms such as thermal diffusion, blood flow, and the thermoregulatory characteristics of the animal; 2) if the sensor is left in the tissue during irradiation, it can be directly heated by the RF fields or it can significantly modify the fields and the associated temperature rises; and 3) the sensor is relatively insensitive to low-power densities.

All of these problems can be eliminated through a technique that utilizes a small diameter plastic or glass tube sealed at one end and implanted at the location where a measurement of the absorbed power is desired. The tube, illustrated in Fig. 10, is long enough so that the open end, fitted with a plastic guide, protrudes from the tissue. A very small diameter thermocouple is inserted into the tube with the sensor located at the probe tip and an initial temperature is recorded. The thermocouple is quickly withdrawn from the tube and the

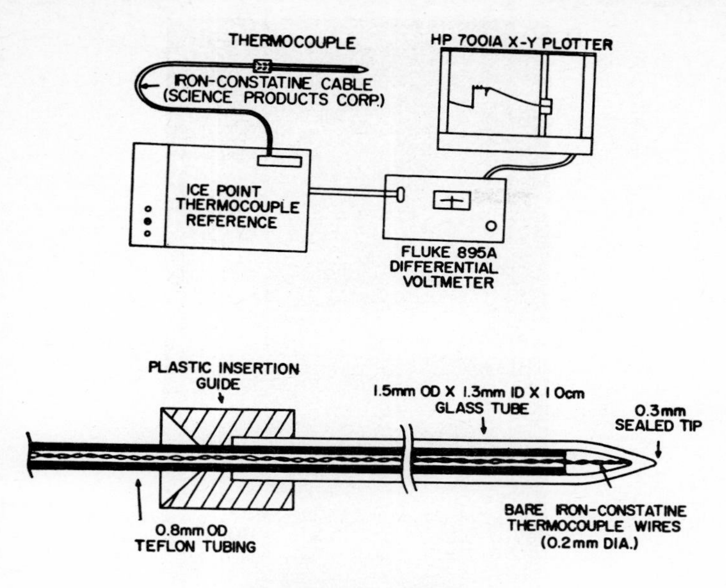

Fig. 10. Thermocouple method for power absorption density measurements *in vivo*.

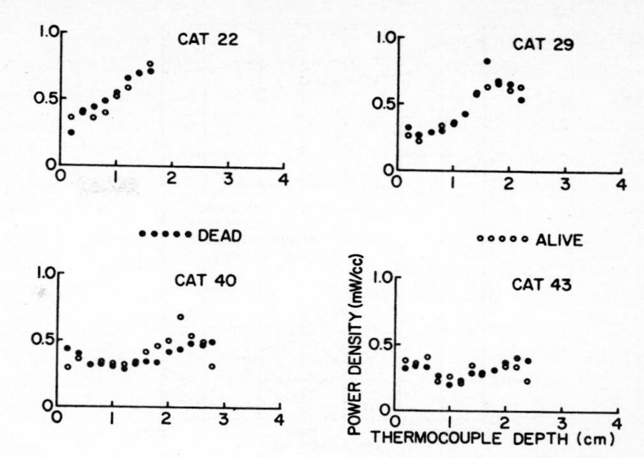

Fig. 11. Power absorption density in brain of cat exposed to 915-MHz microwave power calculated from thermocouple measurements. (Values are based on an incident power density of 2.5 mW/cm².)

animal is exposed under the normal conditions of the experimental protocol with the following exceptions. Instead of using the power level normally chosen for a given experiment, a very high power burst of radiation of duration sufficient to produce a rapid but safe temperature rise in the tissue is applied to the animal. The thermocouple is then rapidly returned to its original position and the new temperature is recorded for several minutes. The temperature versus time curve then is extrapolated back in time to the period when the power was applied and, based on the density and specific heat of the tissue, the absorbed power is calculated from the difference between initial and final extrapolated temperatures. The short exposure period insures that there is no loss of heat due to cooling or diffusion, so the expression

$$P = \frac{4.186\rho c \Delta T}{t} \qquad (10)$$

may be used to calculate the absorbed power density P in W/cm³, where ρ is the tissue density in g/cm³, c the specific heat of the tissue, ΔT the temperature change in degrees Celsius, and t the time of exposure in seconds. The measured absorbed power can then be used to relate the input power of the source to the absorbed power in the tissue under normal lower power exposure conditions. Fig. 11 illustrates power absorption profiles measured in the brains of various cats exposed to radiation. The scale corresponds to the peak absorbed power density per 2.5 mW/cm² applied to the head. Since the results are the same for both live and dead brain tissue, it is apparent that the time of exposure is sufficiently short that blood cooling effects are negligible.

The same techniques involving microwave diodes and dipoles that are used for direct measurement of the fields in air can also be used in tissues. There are difficulties, however, since the ratio of dipole length to feedline separation must be kept large to maintain accuracy while at the same time the dipole must be sufficiently short to implant with a probe. Bowman [75] proposes an implantable probe using three orthogonal dipole and diode combinations of this type with small carbon filaments as lead wires. Guy [76] has used a microwave diode, with the pigtail leads cut to $\frac{1}{2}$ cm as a dipole antenna similar to that shown in Fig. 9, to make field measurements at the brain surface of a cat. The major problem with this type of sensor is that it must be calibrated for each tissue that it is placed in to account for changes in dipole source impedance. The results of using the thermocouple and diode sensors imbedded in tissues are described further in a later section. These techniques are slow and cumbersome to implement, however, and it is not always clear where to implant the probes since the regions of maximum absorbed power in the animals are generally unknown. More complete dosimetry information can be obtained by an indirect thermograph technique.

3) Measurement of Absorbed Power Density by Thermography: Guy [77] has described a method for rapid evaluation of absorbed power density in tissues of arbitrary shape and characteristics when they are exposed to various sources, including plane wave, aperture, slot, and dipoles. The method, valid for both far- and near-zone fields, involves the use of a thermograph camera for recording temperature distributions produced by energy absorption in phantom models of the tissue structures. The absorbed power or magnitude of the electric field may then be obtained anywhere on the model as a function of the square root of the magnitude of the calculated heating pattern. The phantoms are composed of materials with dielectric and geometric properties identical to the tissue structures they represent. Phantom materials have been developed which simulate human fat, muscle, brain, and bone. These materials have complex dielectric properties, illustrated in Fig. 12, that closely resemble the properties of human tissues reported by Schwan [60], where the parameters ϵ_f, ϵ_m, and ϵ_b are the dielectric constants and τ_f, τ_m, and τ_b are the loss tangents of fat, muscle, and brain tissue, respectively. The modeling material for fat may also be used for bone. The synthetic muscle can also be used to simulate other tissues with high water content. The dielectric constant can be varied over a wide range by varying the percentage of polyethylene powder which simulates protein molecules and the conductivity can be controlled by the salinity of the material. The properties of the synthetic fat can be varied over a wide range to simulate other tissues of low water content by varying the amounts of aluminum powder to control the dielectric constant and of acetylene black to control the conductivity. A simulated tissue structure composed of these modeling materials will have the same

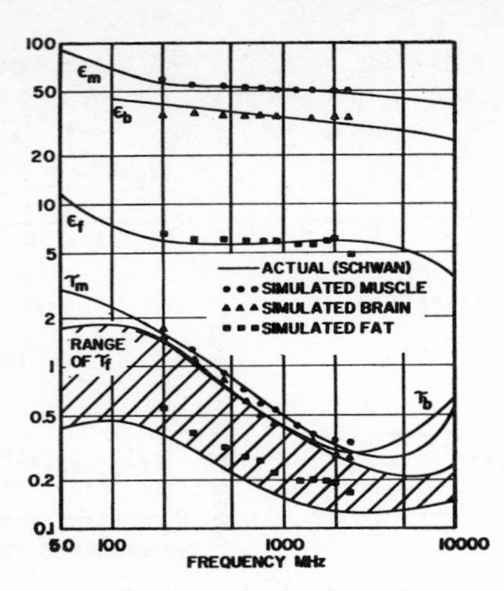

Fig. 12. Dielectric properties of actual and simulated human tissues (fat, muscle, and brain).

Fig. 13. Phantom tissue models.

internal field distribution and relative heating functions in the presence of an electromagnetic source as the actual tissue structure. Phantom models of various tissue geometries can be fabricated as shown in Fig. 13. They include stratified layers of muscle and fat of various thickness, circular and irregular cylindrical structures consisting of synthetic fat, muscle, and bone, and spheres of synthetic brain to simulate various parts of the anatomy. The models are designed to separate along planes perpendicular to the tissue interfaces so that cross-sectional relative heating patterns can be measured with a thermograph. A thin 0.0025-cm-thick polyethylene film is placed over the precut surface on each half of the model to prevent evaporation of the wet synthetic tissue.

The technique for using the phantom model is now described. The model is first exposed to the same source that will be used to expose actual tissue. The power used on the model will be considerably greater, however, in order to heat it in the shortest possible time. After a short exposure, the model is quickly disassembled and the temperature pattern over the surface of separation is observed and recorded by means of a thermograph. The exposure is applied over a 5- to 60-s time interval depending on the source. After a 3- to 5-s delay for separating the two halves of the model, the recording is done within a 5-s time interval or less. Since the thermal conductivity of the model is low, the difference in measured temperature distribution before and after heating will closely approximate the heating distribution over the flat surface, except in regions of high-temperature gradient where errors may occur due to appreciable diffusion of heat.

Fig. 14 illustrates the results of applying the method to the simulated spherical brain structures described previously (refer back to Fig. 4). The thermograms at the left of Fig. 14 are C scans taken over the surface of the separated hemispheres where brightness is proportional to absorbed power and each division is equivalent to 2 cm. The thermograms in the middle are B scans taken before and after exposure to the microwave sources where vertical deflection is proportional to absorbed power along the z axis of the sphere. The thermograms at the right are also B scans taken along the x axis of the sphere. The graphs below the B scans are comparisons between the theoretical and the measured absorbed power. The results agree well, with the exception of the deviation between the theoretical and the experimental values of the large sphere exposed to 918-MHz power. This is due to the

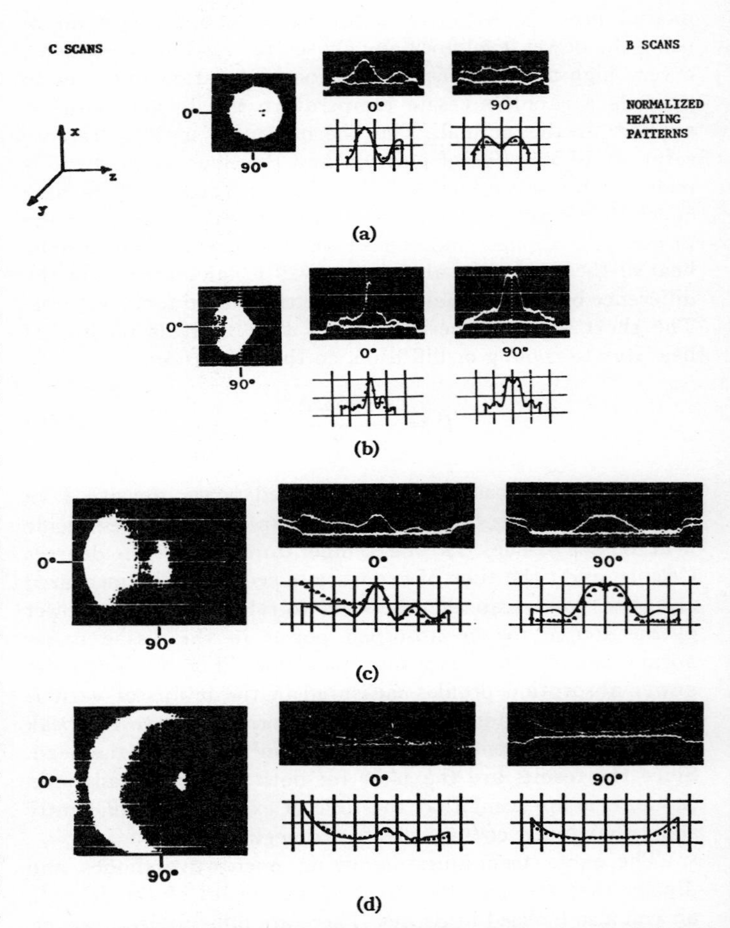

Fig. 14. Thermograms of phantom brain tissue. Scale: C scans, 1 div = 2 cm; B scans, 1 horizontal div = 2 cm, 1 vertical div = 2.5°C; and normalized patterns, 1 horizontal div = 2 cm. (Propagation in z direction with E field polarized along x axis of indicated coordinates.) (a) 6-cm diam, 918 MHz. (b) 6-cm diam, 2450 MHz. (c) 14-cm diam, 918 MHz. (d) 14-cm diam, 2450 MHz. ———, plane wave theory; · · · · ·, experimental plane wave; ▲▲▲▲▲, experimental aperture source.

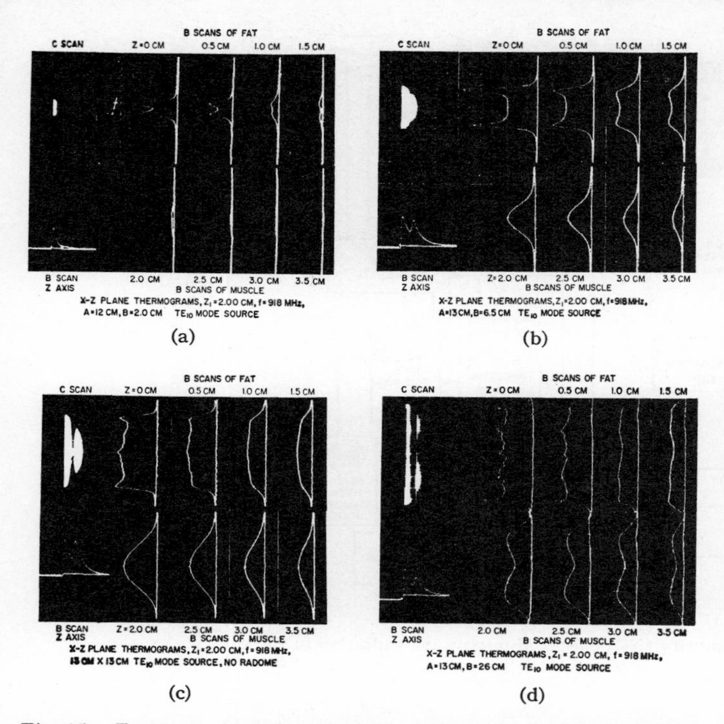

Fig. 15. Temperature distribution patterns obtained by thermography in plane layers of simulated fat and muscle exposed to a waveguide source of electromagnetic fields.

diverging fields of the finite aperture source that was used to irradiate the phantom model at this frequency.

Fig. 15 illustrates the thermograms obtained from a plane bilayered simulated fat and muscle model exposed to a waveguide source of varying height h. Since the specific heat and density of the fat is a factor 0.35 to 0.45 smaller than that of the muscle, the temperature curves in the fat $(0 \leq z_1 < 2.0)$ must be reduced by this factor to be representative of absorbed power. When this reduction is made there is close correspondence between these experimental results and the theoretical results in Fig. 6. The thermograms clearly show that the fat-to-muscle heating ratio is minimized for $b = 13$ cm ($b =$ one wavelength) and becomes excessive for aperture heights less than one-half wavelength.

The thermograph technique described for use with phantom models can also be used on test animals. The animal under test or a different animal of the same species, size, and characteristics must be sacrificed, however. The terminated animal is frozen with dry ice in the same position used for exposure conditions. It is then cast in a block of polyfoam and bisected in a plane parallel to the applied source of radiation used during the experiment. Each half of the animal is then covered with a plastic film and the bisected body is returned to room temperature. The same procedure used on the phantom model is then used with the reassembled animal to obtain absorbed power patterns over the two-dimensional internal surface of the bisected animal. Typical results are described later.

It is important to note that no metal electrodes or leads should be used to record physiological data from a region in the animal while it is being exposed to radiation. The metal leads can seriously perturb the fields in the tissues and large increases in absorbed power can occur in the tissue near the termination of the probe, as discussed in the following section. High-resistance wire leads and recording electrodes must be transparent to the fields. Also, it is necessary to filter all leads carrying physiological signals and shield the recording instrumentation to prevent the introduction of artifacts into the recording equipment. The next section describes the results of animal experiments using the techniques described in this section.

4) Absorbed Power Measurements in a Cat Brain Exposed to 918-MHz Radiation: An experimental protocol was developed to quantify the absorbed power and associated effects on the CNS of a cat exposed to 918-MHz microwave radiation. According to Fig. 4, similarities exist in absorbed power patterns in the CNS of test animals and humans exposed to 918-MHz radiation. Therefore, this frequency appears desirable to use for experimental testing. The experimental exposure technique and associated instrumentation are illustrated schematically in Fig. 16. Cats averaging about 2.3 kg were anesthetized with alpha-chloralose, supported in a stereotaxic instrument frame, paralyzed with Flaxedil®, and placed on artificial respiration. The radiation was directed with maximum intensity in and around the thalamic region of the brain by a controlled continuous-wave microwave power source and cavity radiator.

The response of the thalamic somatosensory area of the cat's brain to electrical stimulation of the skin on the contralateral forepaw was recorded both with and without the presence of microwave radiation. The gross thalamic electrical response was detected by means of a saline-filled glass electrode with a 6-μ-diameter tip. The electrode was placed in the thalamus and adjusted for optimum thalamic response. The electrode's position was verified in several animals by histological studies.

Metal electrodes were avoided to prevent perturbation of the microwave fields in the brain tissue. The electrode and associated reference electrodes were coupled to low-pass microwave filters by a pair of 1-mil-diameter high-resistance wires. The filter was designed to provide more than 150-dB attenuation to the microwave with no more than 20 pF of shunt capacitance to the input of a pair of field-effect transistors. The output of the transistors was fed through a processing differential amplifier to an analog tape recorder and computer of average transients. The body temperature of the cat was held constant by a heating pad connected to a rectal temperature control unit. The brain temperature was recorded during the time that the microwaves were off by placing a thermocouple in a glass pipette with a sealed tip at the homologous point in the opposite thalamus (assuming the brain was symmetrical) at the same depth as the recording electrode.

The thermocouple was removed during the exposure times to prevent any fringing field effects. The electrical response in the thalamus was recorded continuously with microwaves alternatively on and off in 15-min intervals over a total period of 8 to 12 h. The dosimetry was based on the following methods of calibration.

1) The power density in the main beam of the applicator was measured as a function of distance with a Narda model 8110 electromagnetic radiation monitor. The measurements were made for the applicator located in free space and also in the operating position on the stereotaxic support.

2) The actual field at the surface of the cat's brain was measured by means of a calibrated microwave diode coupled with high-resistance wire to a digital voltmeter shunted by a 10-kΩ resistor.

3) The temperature was measured with the thermocouple

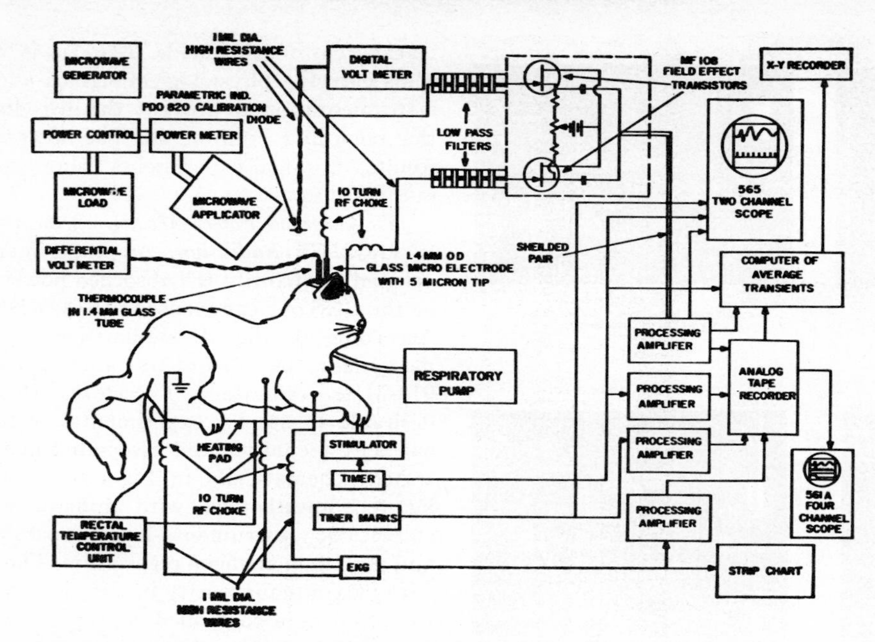

Fig. 16. Block diagram of instrumentation used to quantify CNS effects of microwave radiation in the cat.

TABLE III

CALIBRATION OF 918-MHz MICROWAVE APPLICATOR (8 CM AWAY, 1 W IN) FOR TYPICAL CAT

Region	Absorbed or Incident Power	Method
Free space	$2.6\ mW/cm^2$ (8 cm *away*) $1.7\ mW/cm^2$ (11 cm *away*)	Narda 8110 monitor
In stereotaxic support without cat	$2.48\ mW/cm^2$	Narda 8110 monitor
Maximum heating area of thalamus of cat	$2.0\ mW/cm^3$	thermographic
Maximum heating area in phantom sphere	$1.5\ mW/cm^3$ $2.1\ mW/cm^3$ $1.4\ mW/cm^2$	thermographic (8 cm from aperture source) $2.6\ mW/cm^2$ (theoretical $1.7\ mW/cm^2$ plane wave)
Surface of cat brain	0.4 to 0.8 (mW/cm^3) $0.54\ mW/cm^3$	microwave diode thermographic
Maximum heating area of thalamus of cat	$1.88\ mW/cm^3$	thermocouple (live brain)
Homologous point in opposite thalamus of electrode position	$1.23\ mW/cm^3$	thermocouple (live brain)

before and after a short-term exposure to high-power microwaves as a function of position in the cat's brain and the changes were converted to absorbed power information.

4) The cat's body was frozen in the stereotaxic support, cast in a Styrofoam block, bisected, returned to room temperature, and rejoined back into the stereotaxic support for a short-term high-intensity exposure. Immediately after the exposure, the thermograph recordings of the induced internal temperature patterns in a half-section of the cat's head were made and converted to power absorption patterns. Typical calibration results are tabulated in Table III.

A photograph of the bisected cat is shown in Fig. 17. Thermograms were taken of the cat's head and also of an equivalent phantom sphere for several different conditions. One set of thermograms was taken without the presence of any recording electrodes, the second set was done with the presence of a saline-filled glass electrode, and the final set was taken with the presence of a standard coaxial metal electrode. No differences were found between the thermograms taken for the first two cases, but a significant change was noted for the metal electrode. Figs. 18 and 19 show the results for the phantom model and cat's head with and without the metal electrodes. The results clearly indicate that a standard metal coaxial electrode normally used for neurophysiological experiments can produce very serious modifications and intensifications of the peak absorbed power density by as much as two orders of magnitude. The widespread use of such electrodes, both in the Soviet Union and in this country, in assessing the biological effects of microwave radiation on the CNS could be resulting in the interpretation of highly localized thermal effects as nonthermal or low-level effects.

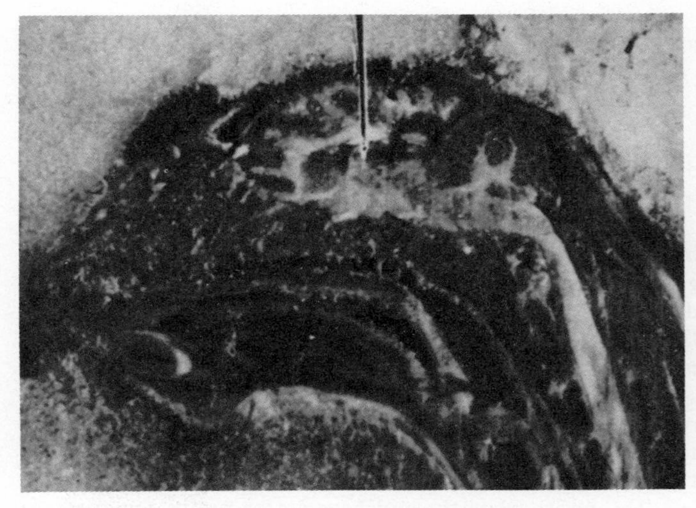

Fig. 17. Photograph of half-section of cat's head with electrode position shown.

The EKG of the cat was recorded and monitored on a strip chart recorder. The EKG was observed to increase with microwave power density and temperature at much higher levels than that required to produce thalamic changes.

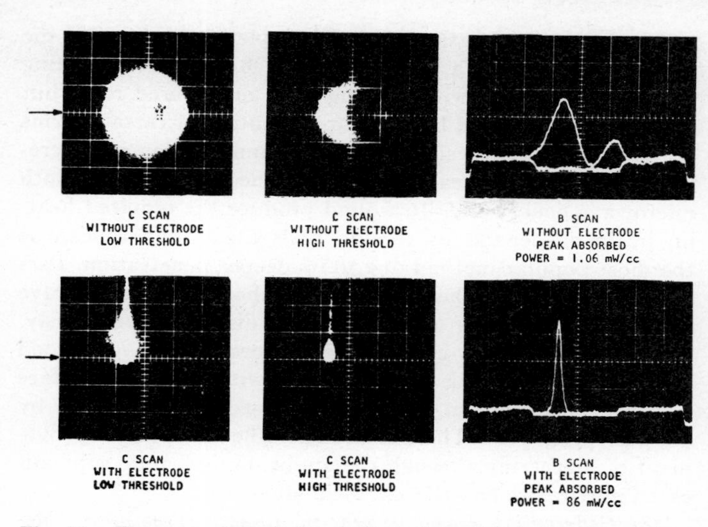

Fig. 18. Thermographic study of effect of coaxial electrode on microwave absorption pattern of spherical phantom of cat brain (6-cm diam). Scale: C scans, 1 div = 2 cm; B scans, 1 horizontal div = 2 cm, 1 vertical div = 2.5°C. (Incident power density = 2.5 mW/cm².)

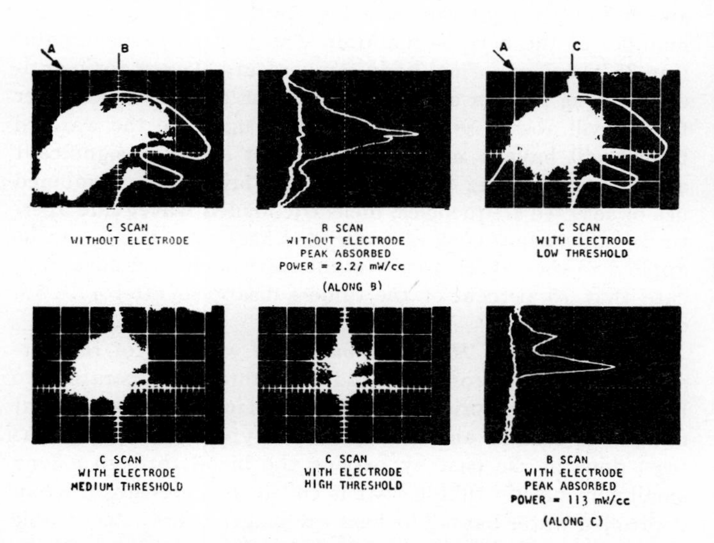

Fig. 19. Thermographic study of effect of coaxial electrode on microwave absorption pattern in brain of cat 49 exposed to 918-MHz aperture source. Scale: C scans, 1 div = 2 cm; B scans, 1 horizontal div = 2 cm, 1 vertical div = 2.5°C. Subject exposed from top with horizontal E vector in plane of paper. (A = direction of incident power B = vertical center, and C = vertical probe line, 1 mm to left of center.) (Incident power density = 2.5 mW/cm².)

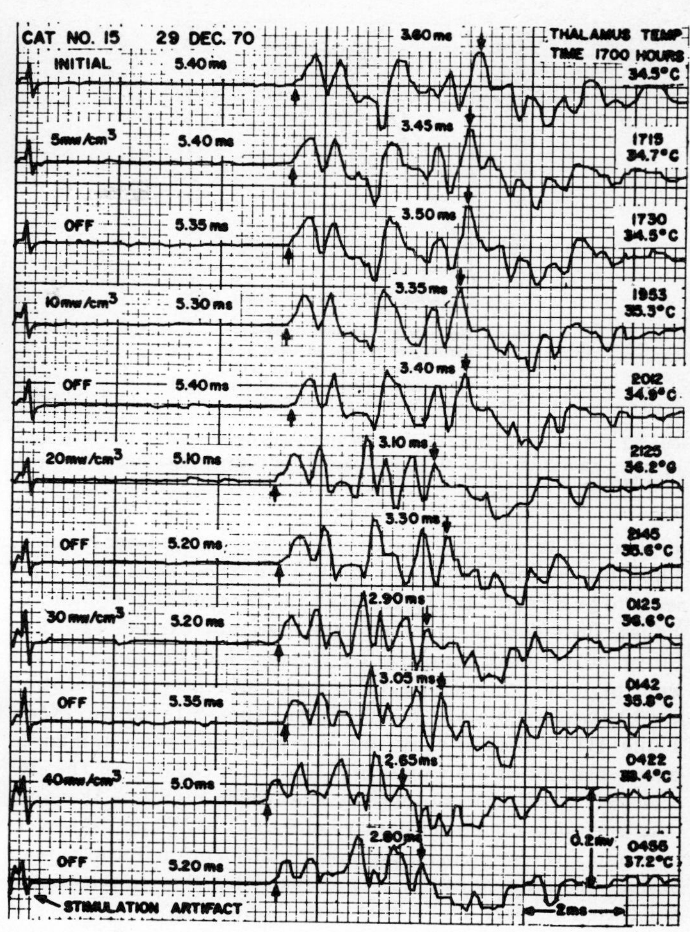

Fig. 20. Effects of microwave radiation on thalamic response of cat to stimulation of contralateral forepaw.

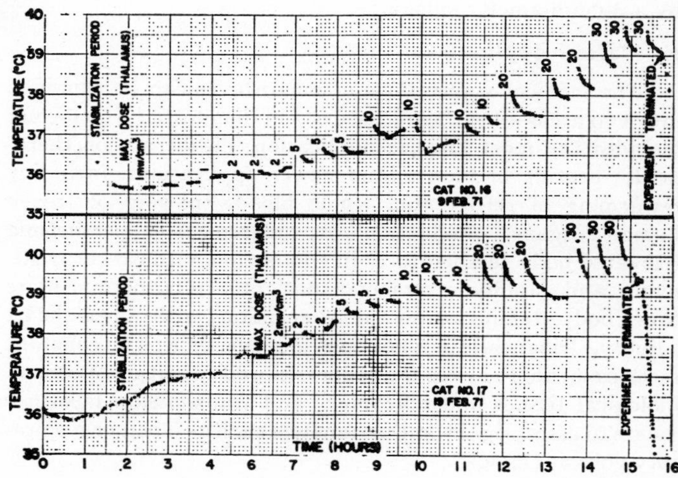

Fig. 21. Thalamus temperature of cats exposed to 918-MHz microwave radiation for 15-min intervals (power off during temperature recordings).

The thalamus response characteristics were processed for improved signal-to-noise ratio, both on-line and off-line, by a computer of average transients. The results were plotted by means of an x–y recorder with typical results as shown in Fig. 20. Typical thalamus temperature curves are illustrated in Fig. 21, and the measured power absorption patterns are given in Fig. 22. The measurable effects of the microwaves appear to be an induced temperature rise in the thalamus with an associated decrease in latency time of neural responses within the exposed area. Each curve in Fig. 20 is the averaged thalamus response based on 50 stimuli obtained either at the end of each exposure or at the end of each period with no radiation. The peak microwave power absorption density and temperature in the thalamus area are noted on each curve. The latency times between the stimulus and the initial thalamus response (denoted by an upward arrow) and between the initial thalamus response and a distinguishable later event (denoted by a downward arrow) are also noted. The latency between the stimulus artifact and the first arrow represents a neural pathway between the cat's paw and brain, whereas the latency between the arrows is probably more closely associated with pathways within the brain. Experimental results also indicate that both latencies decrease with increasing body temperature of the cat when produced by an increase in temperature of the hot pad, whereas microwaves

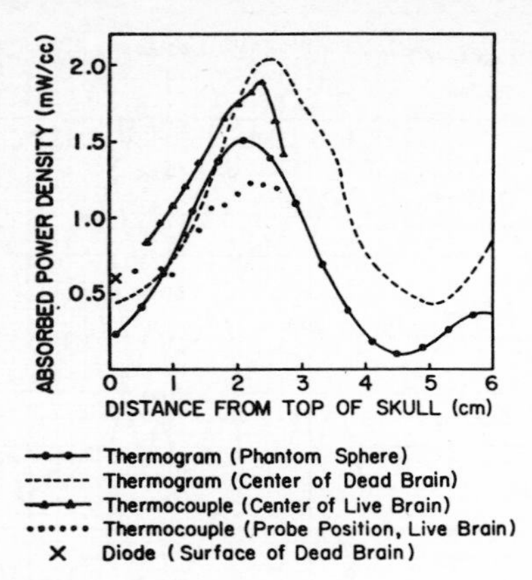

Fig. 22. Measured absorbed power patterns in cat brain due to microwave radiation from 918-MHz aperture source (spacing 8 cm with 1-W input power).

applied to the head have more of an effect on the later latency. The changes are reversible and have time constants that seem to be directly associated with the thermal effect of microwaves, provided the temperature does not exceed 42°. The threshold for both temperature changes and latency changes was found to correspond to a maximum power absorption level between 2.5 and 5.0 mW/cm³ at the center of the brain. Table III indicates that this corresponds to an indicated power density of 3.0 to 6.0 mW/cm² as measured in the unoccupied stereotaxic support by the Narda monitor. It would take a plane wave power density of 5 to 11 mW/cm², however, to produce the same maximum power absorption in the human brain, assuming it can be represented by the simple model illustrated in Fig. 4. On the other hand, Fig. 5 indicates it would take an incident power density of only 0.5 to 1.0 mW/cm² at a frequency of 2450 MHz to produce the same maximum power absorption in the animal head represented by a 3-cm-diameter sphere.

E. Medical Applications

The medical applications of microwaves can be classified into two areas—heating of tissues and diagnostic. The former includes the most historic application, diathermy or therapeutic heating of tissues, and newer applications, including rewarming of refrigerated whole blood, thawing of frozen human organs, production of differential hyperthermia (elevated body temperature) in connection with cancer treatment, and rapidly reversing a patient's hypothermic state (low body temperature) in connection with open heart surgery. Diagnostic applications include dielectric constant measurements to assess the properties and condition of certain biological tissues and reflectance and transmission measurements to assess significant parameters such as blood or respiratory volume changes.

1) Diathermy: The oldest application of microwaves to medicine is "diathermy," a clinical technique used to achieve "deep heating," i.e., heat induced in tissue beneath the skin and subcutaneous fatty layers. Sufficient deep heating can elevate the temperature to the point where therapeutic benefits are achieved through local increases in metabolic activity and in blood flow by dilation of the blood vessels. The beneficial results are believed to arise from the stimulation of healing and defense reactions of the human body. Diathermy has been used successfully in the treatment of musculoskeletal diseases, such as arthritic and rheumatic conditions, fibrositis, myositis, pain, sprains and strains, and many other ailments too numerous to mention here. A complete treatment of the subject is covered in a work by Licht [78]. The deep heating cannot be obtained with heating pads or infrared rays, but must be accomplished by the transformation of certain forms of physical energy, such as ultrasound, radio, or microwave, into heat beneath the subcutaneous fat layer. Both microwave energy and ultrasound produce the required heating in the deeper tissues, though lately ultrasound appears as the most popular method due to its deeper penetration. Part of this stems from the historic poor choice of a microwave diathermy frequency of 2450 MHz discussed previously. Clearly, the presently used frequency does not provide a good method for achieving deep heating with minimal surface heating. This problem has been discussed in great detail by Schwan [45], [46]. There is considerable room for improvement of diathermy through the application of more sophisticated microwave and clinical techniques.

2) Differential Hypothermia in Cancer Treatment: Research on a new application of microwave heating of body tissues in the treatment of cancer is now being conducted [4]. The technique involves the use of microwaves to selectively and uniformly heat the cancer or tumor area while the remainder of the body is maintained in a hypothermic condition 25°C below normal body temperature. A very toxic anticancer drug is then administered to the subject. The cooler tissues will absorb very little of the drug while the warmed tumor will have a metabolic rate that allows a significant amount of the drug to be absorbed. Through the combined use of selected frequencies, dielectric-loaded waveguide apertures, and surface cooling, controlled heating patterns can be applied to the cancer area. Recent experiments on mice indicate that 75 percent of the tumors disappear after 4 to 5 h of treatment.

3) Warming of Human Blood: The warming of refrigerated bank blood from its 4° to 6°C storage temperature to body temperature prior to transfusion is important to prevent dangerous cardiac and general body hypothermia. This has been done in the past by passing the blood through a long small-core plastic tubing that is coiled in a thermostatically controlled water bath. The heat exchanger offers considerable resistance to the blood, slowing down the rate of transfusion which presents problems when rapid blood replacement is necessary. Restall *et al.* [3] has developed a microwave blood warmer that will heat a unit of blood (approximately 500 ml) in its original plastic container from 4°–6°C to 35°C in 1 min. This warmed blood can then be rapidly administered to the patient since both the viscosity has been lowered and the warming coil eliminated. The unit is warmed by rotating it in a microwave cavity driven by a 2450-MHz 1000-W magnetron. Extensive laboratory tests indicate no deleterious effects in the blood. Transfusions to 37 volunteer patients also indicate no abnormal effects.

4) Rapid Elimination of Hypothermia: A standard technique used in open heart surgery is to reduce the body temperature in order to induce a hypothermic state prior to surgery. Hypothermia slows down the metabolic rate of the body cells so that nutrients and oxygen requirements are reduced sufficiently to allow the heart to be stopped for surgery. Surface *cooling* is desired prior to surgery since the peripheral body cells are cooled *before* the body core temperature is reduced to the point where the blood cannot provide nutrients and oxygen. During the rewarming stage after surgery, however, *core heating* is desired so that blood temperature is sufficient to allow proper metabolism prior to the

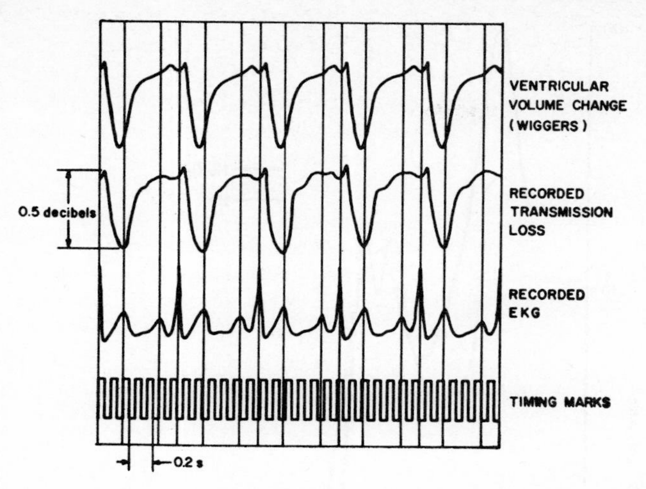

Fig. 23. Comparison of cardiac ventricular volume change with recorded microwave transmission loss through chest cavity.

rewarming of peripheral cells. If surface heating is used, peripheral cells will require oxygen and nutrients at a level that the blood cannot provide due to the lower body core temperature. Core heating can be provided for adults and large children by pumping the blood through heat exchangers. This is a time-consuming process, however, that restricts the total time allocated for surgery. This type of core heating cannot be used for infants below a certain size due to physical limitations of the apparatus. Microwaves do offer a method for rapidly achieving the core heating for any size patient by selectively heating certain portions of the body. Experiments have just begun on the development of such a device by Guy.

5) Rapid Thawing of Frozen Tissues: Rapid thawing of human organs or other biomaterials such as semen which are cryopreserved can be accomplished with microwaves [79], [80]. Many studies have shown that an increase in the survival of biological materials is associated with rapid thawing rates. With proper developmet of microwave applicators, thawing rates up to 10 times faster can be achieved over conventional methods.

6) Diagnostic Studies: Moskalenko [81], [82] has formulated methods for assessing the changes in microwave or shortwave reflectance and transmission which are caused by significant parameters such as blood or respiratory volume changes. Guy has demonstrated the feasibility of Moskalenko's method by measuring the transmission loss of a 915-MHz microwave beam as it passes through the human chest. Fig. 23 illustrates that the variation in microwave loss is proportional to the ventricular volume changes in the heart. The potentialities of this approach appear very favorable. Lonngren [83] has been able to determine the degree of pulmonary emphysema in postmorten analysis of a lung by measuring the dielectric properties of the dried diseased portions. This technique can also be applied to study the composition of many biomaterials whether in small samples or in a living subject. A great deal can be learned about the composition of biomaterials and biological systems through measurements of their electrical properties at microwave frequencies. An example is the strong dependence of the refraction index and absorption characteristics of various tissues on their water content. Thus it is theoretically possible to assess the water content of subcutaneous fat in a living subject without penetrating the skin. Another possibility is the analysis of biological membrane systems with respect to the role of structured or bound water in their function by measuring the refractive index and absorption characteristics [84]. Other applications are the large-scale modeling of certain molecular structures of macro-molecules so that some of the dynamic reactions observed at optical frequencies can be more easily studied at microwave frequencies. The optical activity of certain biological helical molecules prompted several studies involving model systems of copper helices exposed to electromagnetic radiation [85], [86]. Certain aspects of the thermoregulatory system can be studied by using microwaves as a controlled source of heat [87]–[90].

III. Optical Effects

A. Introduction

Optical radiation plays a very significant role in mankind. On the whole, man is very well adapted to the sun's optical and infrared radiation and depends on it for warmth, the processing of food through photosynthesis, and as the principal source of sensory information through the eyes. Optical hazards to man are few, and mostly come from artificial sources such as lasers. The ultraviolet portion of the spectrum can cause superficial cell damage (sunburn).

Despite the important role of optical energy in the animal and plant world, its use in medical and biological applications has been limited. Examples include quantitative biochemical analysis in the clinical chemistry laboratory, optical and electron microscopy, endoscopy (including the use of coherent fiberoptic bundles for probing in the body), spectroscopy, and holography. New principles, techniques, and devices, such as lasers, light-emitting diodes, fiberoptics, optical data processing, holography, integrated optical circuits, phototransistors, etc., are now available which are expanding the use of light in biological applications.

Of particular interest in this paper are the optical propagation characteristics in biological materials. We will not generally be concerned with cell damage and hazardous effects that are confined to the UV spectrum or high intensities. These are described elsewhere [10]. Optical propagation in biological materials is dominated by scattering, because the cellular structure creates medium inhomogeneities of the order of an optical wavelength. There are many optical absorption peaks caused by a variety of biochemical molecules which create a variable optical absorption as a function of wavelength. These combined scattering and absorption properties will be illustrated for some selected biomaterials, and an analytical foundation will be laid for describing these effects.

B. Biological Materials

It is important to know some of the characteristics of cells and tissues in the human body in order to appreciate the optical wave interactive processes that can occur. Very little will be said about anatomical configurations; the interest here is primarily in tissue and cellular structure.

1) Cells: Cells come in all sizes and shapes, and are commonly several microns in diameter. Muscle cells may be a few millimeters long, and nerve cells over a meter long. The gross characteristics of a cell include a thin membrane that holds the cell together, cytoplasm which is a gel-like material within the membrane, and usually a nucleus. Within the cytoplasm are several types of smaller structures called organelles which perform specific metabolic functions. Vesicles partition the cell interior so that materials may be separated and compartmentalized for specific chemical reactions. Organelle sizes vary from fractions of a micron up to a micron, and are thus close to optical wavelengths. Cell membranes are approximately 75 Å thick.

2) Tissues: Cells are grouped together and combined with other materials to form several characteristic types of materials called tissues. There are four basic tissues types—epithelial, connective, muscular, and nervous.

Epithelial tissue consists of cells in single or multilayered membranes that cover or line a surface. Epithelial tissues perform the functions of protection and regulation of secretion and absorption of materials. Simple squamous tissue is a single layer of flat cells, commonly used in blood vessels where a very thin lining is necessary to allow rapid diffusion of electrolytes. Cuboidal and columnar tissues are constructed from cells of cube and column shapes, and are found in airways, the digestive tract, and the bladder.

Connective tissue consists of cells and nonliving materials such as fibers and gelatinous substances which support and connect cellular tissue to the skeleton. Connective tissue comprises much of the intercellular substances that perform the important function of transporting materials between cells. Specialized examples of connective tissue are bone and cartilage. Subdermal connective tissue contains collagen and elastic fibers which give the skin its mechanical properties of toughness and elasticity.

Muscle tissue consists of cells that are 1–40 mm in length and up to 40 μ in diameter. Muscles contain an extensive blood supply, and are thus filled with blood vessels and capillaries with their attendant connective tissue. A large group of muscle fibers are commonly bound together in a sheath. Skeletal muscle has a regular internal striated fine structure due to an ordered array of protein filaments of the order of 100 Å in diameter and 1–2 μ long. This structure has been used as an optical diffraction grating to measure the filament lengths, spacings, and motion during muscle contraction.

Nervous tissue is used to sense, control, and govern body activity. It consists of nerve cells called neurons. Neurons have long projections called axons, which are very analogous to transmission lines. Neurons are located in every portion of the body, sending information to the CNS from a variety of information receptors, and from the CNS to muscles, organs, glands, etc.

3) Erythrocytes: The normal shape of the erythrocyte (red blood cell) is a biconcave disk. There are no internal organelles or supporting structures in the mature erythrocyte, thus the biconcave shape is due to a variety of other factors such as molecular and chemical constitution, membrane properties, and osmotic pressures. The erythrocyte broad diameter is about 7 μ and its thickness varies from 1 μ in the center to 2 μ near the edges. The cell consists of hemoglobin molecules packaged in a thin membrane. A principal function of the erythrocyte is to absorb and release oxygen and carbon dioxide at the cell surface, and the biconcave shape is an ideal one for this purpose. The erythrocyte contains no nucleus, and may not divide to form new cells. Erythrocyte formation occurs in the bone marrow. Young erythrocytes are continually being fed into the blood stream and old ones absorbed and reprocessed in the liver. Under normal conditions, there are approximately 5×10^6 erythrocytes per mm^3. About 40 percent of the volume of whole blood consists of erythrocytes; the remaining 60 percent is a nearly transparent solution of water and salt with a variety of electrolytes and protein molecules. The volume percentage of erythrocytes in whole blood is called hematocrit H.

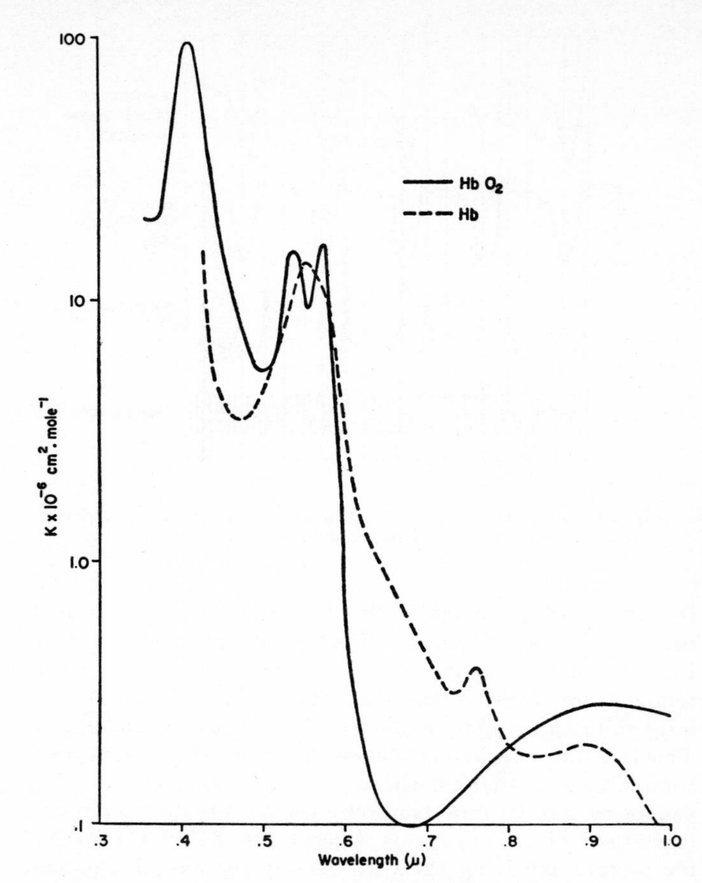

Fig. 24. Optical absorption spectrum of hemoglobin and oxyhemoglobin plotted in terms of specific absorption coefficient κ versus wavelength. The difference between the two curves causes the change in color of blood as its oxygen content changes and makes possible optical spectroscopic techniques for measuring oxygen saturation.

C. Optical Properties of Biological Materials

Very little is known about the optical properties of most biological materials. There is considerable data on hemoglobin solutions and whole blood, some data on skin, and some interest in cellular effects [91]–[95]. Ultraviolet irradiation has been a subject of interest for many years due to the hazards involved [96], [97].

1) Blood: Light has been used for many years to determine the oxygen content in blood. The normal healthy pink appearance of a person compared to a grey color or "blue baby" is a common diagnostic tool based on optical spectroscopic effects in blood. Erythrocytes contain hemoglobin molecules, Hb, which are easily oxygenated to oxyhemoglobin molecules HbO_2. An important parameter is oxygen saturation OS, defined as the ratio of HbO_2 to total hemoglobin. There is an index of refraction discontinuity between the erythrocyte and its surrounding plasma medium (about 1.40 to 1.35) [98], thus scattering occurs which complicates the optical measurement of hemoglobin contained inside the erythrocyte. Thus in order to obtain the absorption spectrum of the hemoglobin molecules, it is useful to rupture the erythrocyte membrane and release the hemoglobin into solution. When this is done, the medium is called hemolyzed blood. The specific absorption coefficients (defined later) of Hb and HbO_2 is hemolyzed blood are illustrated in Fig. 24, obtained from composite data by

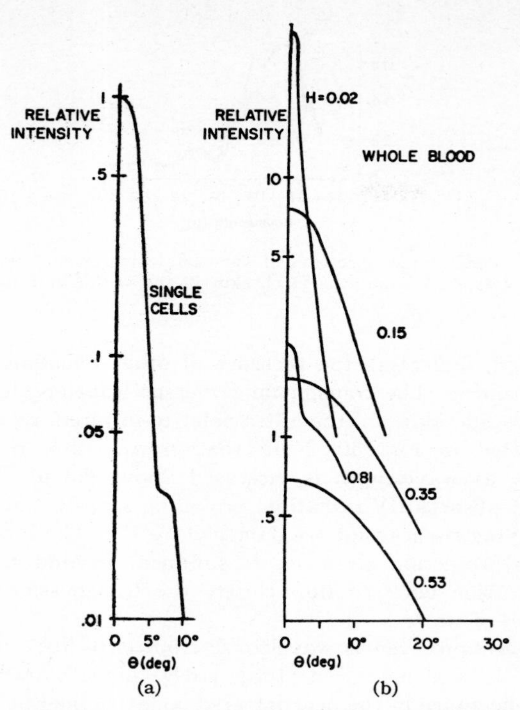

Fig. 25. Forward scattered intensity plotted versus scattering angle θ for erythrocytes. (a) Isolated erythrocytes. (b) Whole blood, sample thickness $d=75$ μ, $\lambda=0.8$ μ, for different hematocrit values.

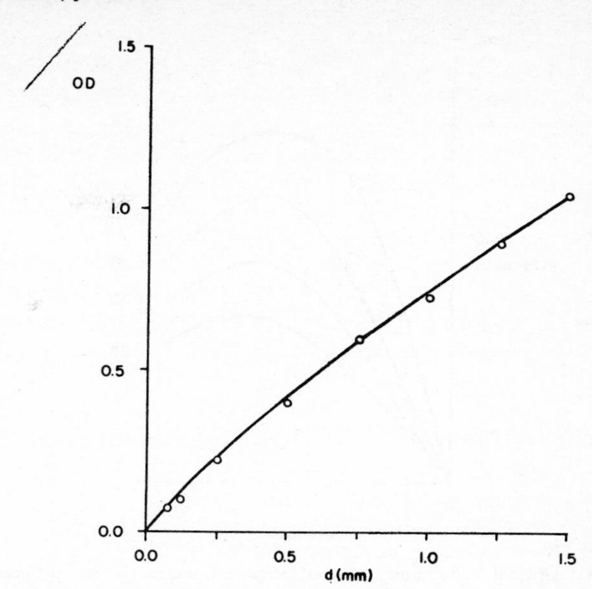

Fig. 26. Optical density plotted versus blood layer thickness d. Experimental data points are shown for $\lambda=0.80$ μ, $H=0.35$.

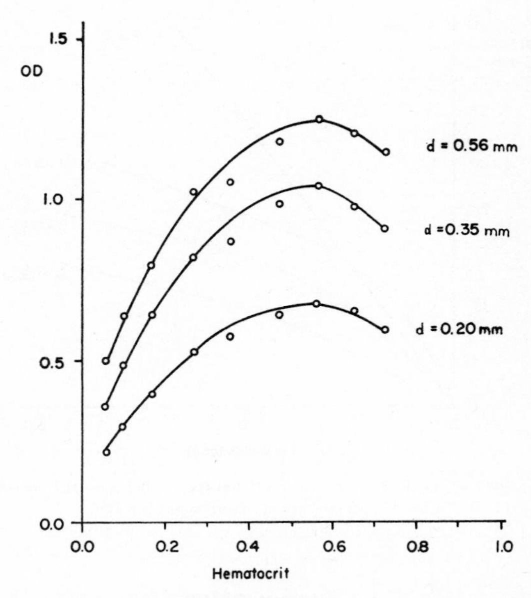

Fig. 27. Optical density plotted versus hematocrit for different blood sample thicknesses d, for $\lambda=0.80$ μ.

Barer [99] and Horecker [101]. A strong absorption band is centered at 414.5 nm, called the Soret band, with minor absorption peaks in the 550-nm region. Oxyhemoglobin HbO_2 has low absorption in the red portion of the spectrum compared to Hb. Thus blood "looks" red when oxyhemoglobin molecules are predominant. At $\lambda=548$, 568, 587, and 805 nm, the absorption values are equal, and these wavelengths are called *isosbestic* points.

Observers have noted that there is a greater net absorption when hemoglobin is packaged in erythrocytes (whole blood) as compared to being in solution (hemolyzed blood). The increased absorption is due to erythrocyte scattering effects. The relative intensity of forward-scattered light of isolated erythrocytes as a function of scattering angle θ has been obtained experimentally [101]–[103] and is shown in Fig. 25(a). The shoulder in the curve near 8° has been attributed to a secondary diffraction peak effect. Scattering data for whole blood have also been obtained [104], as shown in Fig. 25(b). Note that for low hematocrit H a very narrow beam of transmitted light is obtained, indicating that much of the transmitted light is not scattered. As hematocrit increases, the transmitted beam broadens, indicative of increased scattering. The $\theta=0°$ intensity is lowered with increased H due to increased absorption and scattering. Note the striking effect at high hematocrit $H=0.81$ of increased transmittance and the reappearance of an unscattered narrow beam on top of a broader scattered beam. The increased transmittance is due to a decreased amount of scattering, as the absorption must necessarily increase. Scattering for low hematocrit values principally occurs at the erythrocyte sites. Increased hematocrit means an increased number of erythrocytes and increased scattering. For higher hematocrit values $H>0.5$, the erythrocytes pack together to form a homogenous mass of absorbing hemoglobin material, and the scattering occurs at the plasma cavities located between masses of red blood cells. The plasma cavities decrease with increased hematocrit, thus decreased scattering is observed with increased hematocrit at larger H values.

Measurements have been obtained for transmittance T and reflectance R of thin whole blood samples. Fig. 26 shows density $OD=\log_{10}(T^{-1})$ versus sample thickness d at $\lambda=0.80$ μ. Note that a straight line Lambert–Beer law result is not obtained. The curvature is due to scattering effects, as described later. Fig. 27 shows an increase and subsequent decrease in OD versus H, for different d values [104]. Similar results for reflectance are shown in Fig. 28 for different sample depths [105]. The increase and subsequent decrease in OD and reflectance as a function of hematocrit is caused by changes in scattering from the erythrocytes.

2) Skin: Optical (particularly ultraviolet) effects in skin

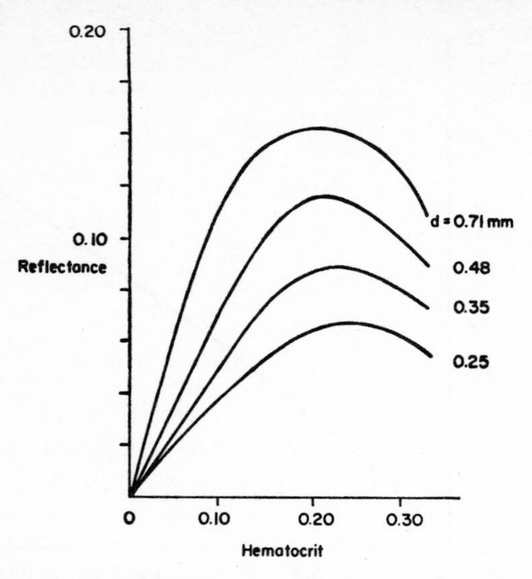

Fig. 28. Optical reflectance plotted versus hematocrit for different blood sample thicknesses d, for fully oxygenated blood at $\lambda = 0.63\ \mu$.

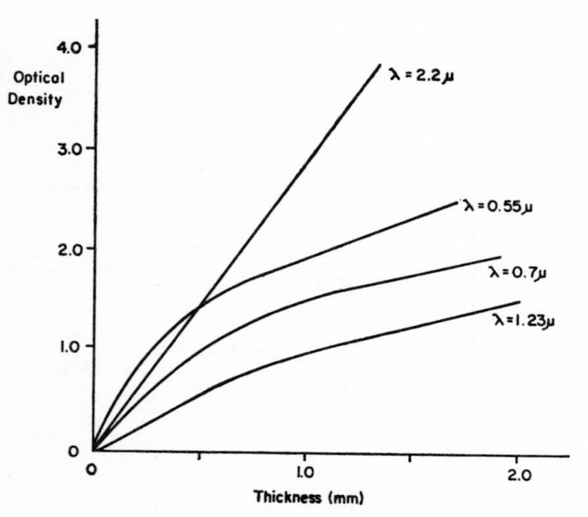

Fig. 29. Optical density plotted versus skin layer thickness of white human skin at four wavelengths.

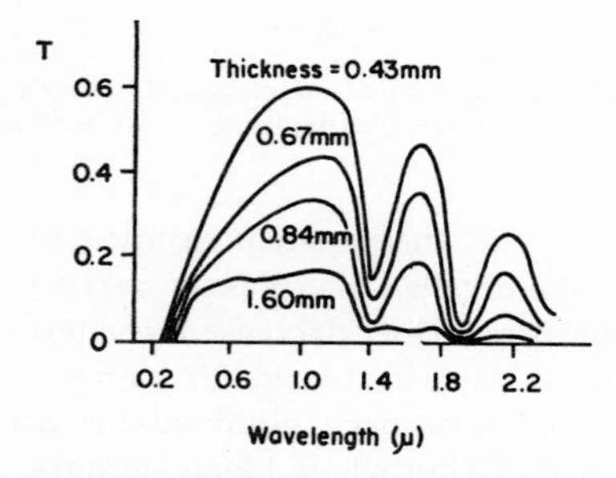

Fig. 30. Spectral transmittance of excised white human skin in the visible and infrared spectrum.

have been studied extensively [106]. A comprehensive paper by Hardy *et al.* [107] establishes the scattering and absorbing properties of skin. Fig. 29 shows OD versus skin layer thickness for several wavelengths [107]. The 2.2-μ curve is a straight line indicating that at this longer wavelength, skin transmission is largely governed by the Lambert–Beer law. At shorter wavelengths, considerable bending of the curves is

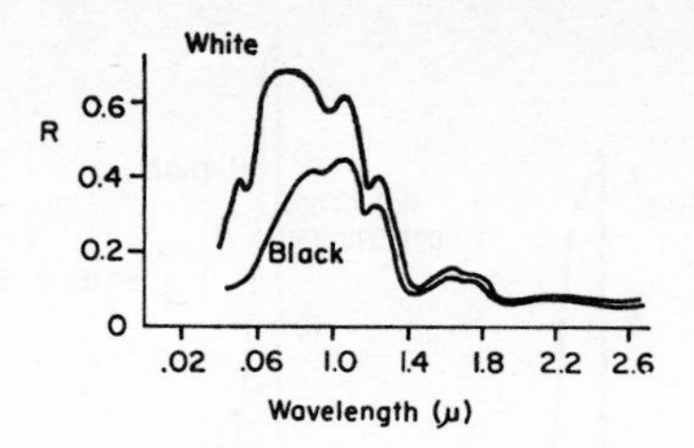

Fig. 31. Reflectance spectrum of very fair complexioned white skin, and very dark American black skin in the visible and near infrared.

observed, indicating the presence of other phenomena such as scattering. The transmittance versus wavelength characteristics for skin in the ultraviolet-to-infrared regions are illustrated in Fig. 30. Note that transmission rises very rapidly as wavelength is increased above 0.3 μ. The skin heavily absorbs UV radiation, providing a protection for the underlying tissues that are damaged by UV. The inefficiency in this protection gives rise to sunburn. Beyond 1.4 μ, the transmission curves follow closely the transmission characteristics of water.

Reflectance versus wavelength from both light and dark skins has been measured [108], [109], and is shown in Fig. 31. Reflectance in the near infrared is determined by scattering sites in skin and the transmission characteristics of water.

D. Scattering and Absorption

A homogeneous absorbing medium with no scattering obeys the simple one-dimensional exponential law of absorption called the Lambert–Beer law. The absorption per centimeter α causes a reduction in wave intensity I^+ traveling in the $+z$ direction according to $dI^+ = -\alpha I^+ dz$. This is integrated to obtain the Lambert–Beer law

$$I^+(z) = I^+(0)e^{-\alpha z} \tag{11}$$

where $I^+(0)$ is the intensity at $z = 0$. Transmittance T becomes $T = e^{-\alpha z}$. The Lambert–Beer law is often applied to a solution of molecules that absorb optical energy, for example, various protein molecules. Clearly, α is dependent on the molecular concentration C as well as the specific absorption properties of the molecular species given by the parameter κ:

$$\alpha = C\kappa.$$

The concentration C is expressed in $\text{moles}\cdot\text{cm}^{-3}$, and κ, defined as the *specific absorption coefficient* for the material, has the dimensions $\text{cm}^2\cdot\text{mole}^{-1}$. Values of κ for Hb and HbO_2 were given previously in Fig. 24. For a sample thickness d, optical density becomes

$$OD = 0.4343 C\kappa d. \tag{12}$$

Note that for a purely absorbing material described by the Lambert–Beer law, a plot of OD versus d results in a straight line, the slope of which is proportional to the specific absorption coefficient for the absorbing medium and the concentration. OD measurements can be used to determine the concentration of molecular species, knowing κ and d.

When the radiation wavelength approaches the size of an object or inhomogeneity in the medium, scattering occurs. substantial deviations from the Lambert–Beer law due to scattering are present in whole blood, and in skin for the shorter wavelengths. The historical analytical approach to this problem originated with the work of Schuster [110] in his attempt to describe optical scattering through atmospheres of distant stars. This work was subsequently modified [111]–[113], and popularized by Kubelka and Munk [114], and later by Kubelka [115]. Subsequent modifications have been extensive [116]–[119]. We are particularly interested in the application of Schuster's work to the understanding of optical absorption and scattering in biological material, such as blood and skin [120]–[122].

Schuster's two-flux theory is one-dimensional and applies only to diffuse flux. A $+z$ traveling diffuse intensity I^+ generates a $-z$ propagating diffuse intensity I^- due to scattering sites in the absorbing material. The scatterers in the material are characterized by an absorption cross section σ_a and a backscattering cross section σ_s^-. The differential intensity dI^- generated within dz for a scatterer volume density ρ is $dI^- = \rho\sigma_s^- I^+ dz$. By analogy, the amount of I^+ generated by the presence of I^- is $dI^+ = \rho\sigma_s^- I^- dz$. General differential relations describing I^+ and I^- in the material may now be written. Each wave loses intensity due to absorption and backscattering, and gains intensity from backscattering of the other wave.

$$\frac{dI^+}{dz} = -\rho(\sigma_a + \sigma_s^-)I^+ + \rho\sigma_s^- I^- \tag{13}$$

$$\frac{dI^-}{dz} = \rho(\sigma_a + \sigma_s^-)I^- - \rho\sigma_s^- I^+. \tag{14}$$

Note that when $\sigma_s^- = 0$ in (13), then (11) is obtained with $\alpha = \rho\sigma_a$. Since ρ is proportional to C, the specific absorption coefficient κ is proportional to σ_a.

We desire solutions to (13) and (14) for a semi-infinite absorbing and scattering medium occupying the space $z > 0$. It is assumed that at $z = 0$, an I^+ wave is impressed on the medium of intensity $I^+(0)$.

Exponential solutions are appropriate.

$$I^+(z) = I^+(0)e^{-\rho\sigma z} \tag{15}$$

$$I^-(z) = I^-(0)e^{-\rho\sigma z} \tag{16}$$

where $I^-(0)$ and σ are

$$I^-(0) = \frac{\sigma_a + \sigma_s^- - \sigma}{\sigma_s^-} I^+(0) \tag{17}$$

$$\sigma = \sigma_a \sqrt{1 + 2\frac{\sigma_s^-}{\sigma_a}}. \tag{18}$$

Of great interest in practical biomedical instrumentation is reflectance R, defined as $I^-(0)/I^+(0)$:

$$R = 1 + \frac{\sigma_a}{\sigma_s^-}\left(1 - \sqrt{1 + 2\frac{\sigma_s^-}{\sigma_a}}\right). \tag{19}$$

Solutions of (13) and (14) for a slab medium occupying the space $0 < z < d$ are now sought. In this case, an exponential solution of the form

$$I^+(z) = C_1 e^{\rho\sigma z} + C_2 e^{-\rho\sigma z} \tag{20}$$

is obtained, and from (13) and (14) the corresponding solution for I^- is

$$I^-(z) = \frac{\sigma_a + \sigma_s^- + \sigma}{\sigma_s^-} C_1 e^{\rho\sigma z} + \frac{\sigma_a + \sigma_s^- - \sigma}{\sigma_s^-} C_2 e^{-\rho\sigma z}. \tag{21}$$

It is again assumed that $I^+ = I^+(0)$ at $z = 0$. At $z = d$ we set $I^-(d) = 0$. Equations (20) and (21) may be solved in accordance with these boundary conditions to obtain the results

$$\frac{I^+(z)}{I^+(0)} = \frac{e^{\rho\sigma z} - Ae^{\rho\sigma(2d-z)}}{1 - Ae^{2\rho\sigma d}} \tag{22}$$

$$\frac{I^-(z)}{I^+(0)} = \frac{\sigma_a + \sigma_s^- + \sigma}{\sigma_s^-}\left(\frac{e^{\rho\sigma z} - Ae^{\rho\sigma(2d-z)}}{1 - Ae^{2\rho\sigma d}}\right) \tag{23}$$

where σ is given (18), and

$$A = \frac{\sigma_a + \sigma_s^- + \sigma}{\sigma_a + \sigma_s^- - \sigma}. \tag{24}$$

Using (22) and (23), expressions for reflectance and transmittance for a slab medium may be obtained.

$$R = \frac{\sigma_a + \sigma_s + \sigma}{\sigma_s^-}\left(\frac{1 - e^{2\rho\sigma d}}{1 - Ae^{2\rho\sigma d}}\right) \tag{25}$$

$$T = \left(\frac{1 - A}{1 - Ae^{2\rho\sigma d}}\right)e^{\rho\sigma d}. \tag{26}$$

For application of these relations to whole blood, one must write the dependence of scattering on hematocrit H. The absorption cross section σ_a depends on the relative proportion of hemoglobin and oxyhemoglobin in the erythrocyte, and is believed not to vary significantly with H. The backscattering parameter σ_s^- clearly depends on H. An isolated erythrocyte has a backscattering cross section denoted as $\hat{\sigma}_s^-$. As hematocrit increases, the scattering due to erythrocytes decreases significantly, as explained previously. Hematocrit is related to scatterer density ρ and erythrocyte volume ν by the relation $H = \rho\nu$. The scatterer parameters become

$$\rho\sigma_a = \frac{H\sigma_a}{\nu} \tag{27}$$

$$\rho\sigma_s^- = \frac{\hat{\sigma}_s^- H(1 - H)}{\nu} \tag{28}$$

where the simple multiplying factor $(1 - H)$ describes the variation of σ_s^- from $\hat{\sigma}_s^-$ at $H = 0$ to zero at $H = 1$ [123].

Although the two-flux theory has been very popular and has helped to explain qualitatively some optical effects in biological materials, quantitative results have not been obtained since the theory is restricted to a one-dimensional geometry and to diffuse flux. These restrictions greatly limit the applicability to experimental apparatus. Nonetheless, application of (25) and (26) describes very well the qualitative behavior of Figs. 26–28. In addition, the equations describe observed changes in OD and R with oxygen saturation which affects σ_a. We can conclude that the parabolic-shaped OD and R versus H curves in Figs. 27 and 28, as well as the nonlinear OD versus d curves in Figs. 26 and 29, are due to scattering effects in the absorbing material.

E. Optical Diffusion Theory

More sophisticated attempts to describe wave scattering and absorption effects can be categorized into either a "phenomenological" approach based on Schuster's work and progressing into neutron transport theory and radiative transport theory [124], [125], or an "analytic" approach based on Maxwell's field equations and dealing with averaged field quantities [126]. An optical diffusion theory that assumes isotropic scattering has emerged from the phenomenological approach as a means of describing some of the observed phenomena in biological materials with a promising degree of quantitive precision and with analytical simplicity [122], [127], [128]. A simple visual observation of light propagation in blood is highly suggestive of a diffusion process, despite the fact that erythrocyte scattering is known to be not isotropic (see Fig. 25). Quantitative measurements have since shown the partial validity of a photon diffusion theory for blood [129]–[131].

The basic integral equation from radiative transfer theory and other approaches is [132]

$$\rho(r) = \rho_0(r) + \int_{V'} \frac{\omega\rho(r')e^{-R/\lambda}}{4\pi\lambda R^2}\, dV' \tag{29}$$

where

- r field point position (x, y, z),
- r' and V' source point position (x', y', z') or source volume in the integral,
- ρ total photon density,
- ρ_0 density of unscattered photons (from an optical source),
- ω probability that a photon collision results in scattering (not absorption),
- λ photon mean free path,
- R $=|r-r'|$.

This integral equation may be obtained from photon conservation considerations taking into account absorption and isotropic scattering [133]. It is assumed that the probability for a photon to have a free flight over a distance R is $\exp(-R/\lambda)$.

The $\exp(-R/\lambda)$ term in the integral of (29) causes the integrand to receive its major contributions close to $R=0$, where $|r'| \approx |r|$. This suggests an approximation for $\rho(r')$ in the integrand consisting of the first term in a Taylor's series expansion.

$$\begin{aligned}\rho(r') &\cong \rho(r) + [(r' - r)\cdot\nabla]\rho(r)\\ &= \rho(r) - \mathbf{R}\cdot\nabla\rho(r).\end{aligned}$$

Using this approximation in (29) gives the standard diffusion differential equation

$$\left(\nabla^2 - \frac{3(1-\omega)}{\omega\lambda^2}\right)\rho = -\frac{3\rho_0}{\omega\lambda^2} \tag{30}$$

which describes the spatial distribution of total photon density ρ in terms of the unscattered source photon density ρ_0. The optical intensity or photon flux, denoted as $\mathbf{\Gamma}_s$, is desired. $\mathbf{\Gamma}_s$ is related to ρ by the expression $\mathbf{\Gamma}_s = -D\nabla\rho$, where D is the diffusion coefficient. In terms of the parameters used here, $D=\omega v\lambda/3$, where v is the photon velocity.

By the nature of diffusion theory, the basic differential relation (30) does not apply near a boundary. A simple approximate boundary condition commonly used is $\rho=0$. More rigorously, one must use the integral form (29) in order to obtain a relation between the parameters that hold at the boundary. The planar boundary between a diffusing medium and a nondiffusing medium requires an evaluation of the integral over a half-space, assuming $\rho(r')=0$ in the other half-space. This results in a relation between ρ, $\nabla\rho$, and ρ_0 at the boundary [133]:

$$\rho(1-\omega/2) + \frac{\omega\lambda}{4}\nabla\rho\cdot\hat{n} = \rho_0 \tag{31}$$

where $\hat{n}$ is a unit vector pointing normally away from the diffusing medium. Note that for $\rho_0=0$, $\omega=1$, (31) gives

$$\rho = 0.5\lambda\frac{\partial\rho}{\partial n} \tag{32}$$

whereas the accepted value from radiative transfer theory is

$$\rho_s = -0.71\lambda\frac{\partial\rho_s}{\partial n} \tag{33}$$

where $\rho_s=\rho-\rho_0$. These two boundary relations are for different photon parameters and derived under different approximations, so it is difficult to say which is more correct for this application. The form of the results is identical.

It has been found that use of the differential equation with the boundary condition gives surprisingly good results. In light of the usual difficulty encountered in trying to solve the integral equation, the diffusion theory approximation is extremely important and valuable. The diffusion equation can be used to find ρ, where ρ is required to satisfy a boundary condition. Photon flux $\mathbf{\Gamma}_s$ is then obtained from ρ.

Solutions to diffusion equation (30) can be obtained using Green's function techniques. For example, for an infinite scattering medium, one can use the infinite medium Green's function to write the solution for ρ:

$$\rho(r) = \frac{3}{\omega\lambda^2}\int_{V'}\frac{e^{-R/\lambda_d}}{4\pi R}\rho_0(r')dV' \tag{34}$$

where $\lambda_d=\lambda\sqrt{\omega/3(1-\omega)}$ is a diffusion "penetration depth." Simple solutions are obtained in the far-field region where $|r| \gg |r'|$ at positions far from the region of source photons ρ_0:

$$\rho(r) \cong \frac{3}{4\pi\omega\lambda^2}\frac{e^{-r/\lambda_d}}{r}\int_{V'}\rho_0(r')dV'.$$

For a slab medium in the region $0<z<d$, a Green's function solution has been obtained using the simple $\rho=0$ boundary condition at $z=0$ and d [131]. Assuming no ϕ variations in a cylindrical r, ϕ, z coordinate system gives

$$\rho = \frac{6\sin\left(\frac{n\pi z}{d}\right)}{d\omega\lambda^2} \cdot \sum_{n=1}^{\infty}\int_0^{\infty}\int_0^{d}\left[\sin\left(\frac{n\pi z'}{d}\right)\rho_0(z', r')dz'\right] \cdot \begin{bmatrix} I_0(\lambda_n r) & K_0(\lambda_n r') \\ I_0(\lambda_n r') & K_0(\lambda_n r)\end{bmatrix} r'dr' \tag{35}$$

where

$$\lambda_n^2 = \left(\frac{1}{\lambda_d}\right)^2 + \left(\frac{n\pi}{d}\right)^2.$$

The upper terms in the square brackets are used for $|r'| > |r|$, and the lower terms are used for $|r'| < |r|$. This result has been applied to the case of a cylindrical beam of incident

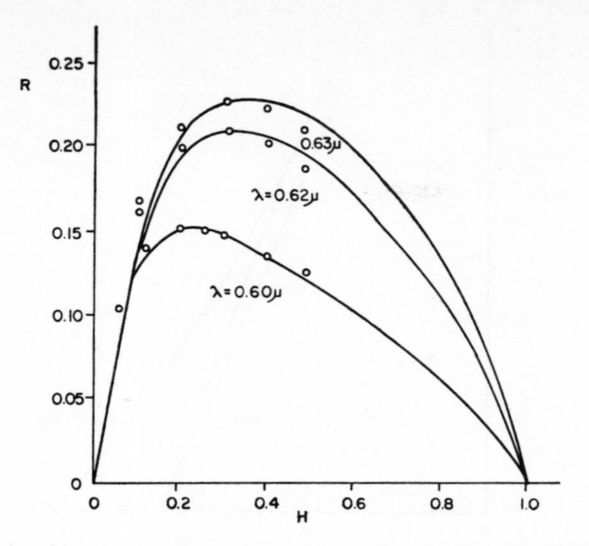

Fig. 32. Reflectance plotted versus hematocrit of whole human blood at three red wavelengths. Experimental data are shown by the circles, and theoretical diffuse reflectance curves from optical diffusion theory are shown by the solid lines, normalized at peak reflectance for the $\lambda=0.60$ μ curve only.

photons ρ_0 of radius b. The result for diffuse reflectance R is

$$R = \frac{2\omega}{d} \sum_{n=1}^{\infty} A_n[1 - (-1)^n e^{-d/\lambda}](Y - Z) \qquad (36)$$

where

$$Y = \lambda_n b[K_0(\lambda_n b)I_1(\lambda_n b) + I_0(\lambda_n b)K_1(\lambda_n b)]$$

$$Z = \frac{2a}{\lambda_n b^2} I_1(\lambda_n b)K_1(\lambda_n a)$$

$$A_n = \frac{\lambda}{1 + \left(\frac{\lambda}{\lambda_d}\right)^2 + \left(\frac{n\pi\lambda}{d}\right)^2 + \left(\frac{d}{n\pi\lambda_d}\right)^2}$$

and a is the radius of the receiving aperture.

We are now in a position to make quantitative comparison of the optical photon diffusion theory with experimental results on whole blood. Assuming absorption changes with wavelength proportional to the specific absorption coefficient κ, and scattering changes due to changing hematocrit by (28), a comparison of (36) with experimental results [105] is shown in Fig. 32. Peak reflectance is normalized to the experimental peak reflectance for the $\lambda=0.60$ data only. The experimental data points compare encouragingly well with the diffusion theory. This approach is now being used to help understand a variety of optical propagation effects in biological materials and to help design optical bioinstrumentation.

F. Optical Bioinstrumentation

The absorbing of scattering properties of biological materials can be utilized to make important material property measurements. For example, knowing the specific absorption coefficient of a molecular species, the concentration can be obtained by an optical absorption measurement. In a clinical chemistry laboratory, various reagents are combined with specific biologically important ions and molecules to generate new colored compounds especially suited to optical measurements. A vast and automated technology has been built up based on this procedure.

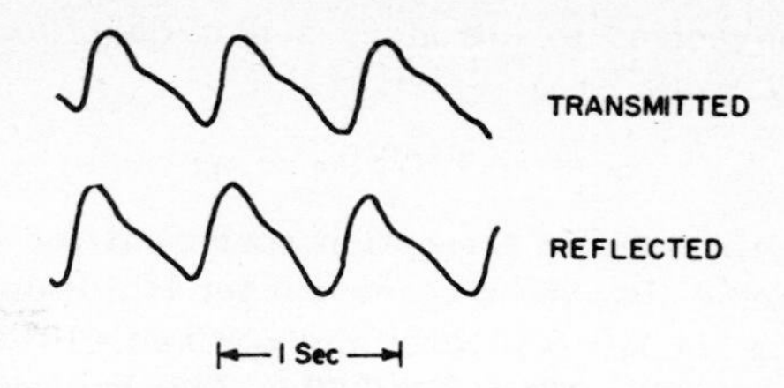

Fig. 33. Finger transmittance and reflectance pulsations at the heart rate.

Optics technology has also been applied to making important measurements on natural biological tissue. Two important applications will be discussed; the propagation of light through the skin and underlying tissue to measure blood volume (photoplethysmography) or anatomy (transilluminaltion), and optical spectrophotometric techniques to measure oxygen content in blood (oximetry).

1) Transillumination and Photoplethysmography: Cellular tissue is relatively transparent in the near infrared compared to the visible, transmitting from 2 to 3 percent in a 1-cm-thick sample [134]. However, blood is about 100 times more absorbing [135]. Obviously, air-filled structures, such as sinus cavities and the oral cavity, and clear fluid-filled structures, such as certain types of cysts and cerebrospinal fluid in the brain, absorb very little optical radiation. These gross discrepancies in optical absorption can be used to make qualitative judgments about the amount of a specific tissue type in an illuminated area. For example, by illuminating the oral cavity and sinuses, it can be determined by transmitted light through the face whether the sinuses are clear or congested. Transillumination of cysts in the testes has proved useful in determining whether the cysts are solid or accumulations of water.

Hydrocephalis (water on the brain) in infants is caused when the circulation of cerebrospinal fluid in the brain and spinal column is obstructed. When this occurs, the head is enlarged by pockets of clear fluid. The location and extent of this fluid can be observed by the use of a flashlight or chielded lamp in a darkened room [136]. A pulsed infrared transillumination device has been developed which can quantitatively measure head optical density [137].

The very large difference in absorption between blood and tissue in the near infrared means that transilluminated tissue total absorption may show periodic changes due to changes in tissue blood volume. This effect has been used to obtain indications of peripheral blood volume and pulse volume changes. This technique is termed photoplethysmography [138] and has been used to look at changes in the distribution of blood flow due to a variety of physiological causes. Reflected and transmitted light patterns give very similar results, as expected from the diffusion-like optical propagation processes involved [139]. Typical transmittance and reflectance variations with time through a finger are shown in Fig. 33.

2) Oximetry: One of the principal applications of optical propagation in blood is for the determination of blood oxygen saturation OS. Spectrophotometric techniques for measuring OS using transmitted light have been developed. The blood is hemolyzed, releasing HbO_2 of concentration C_0 and Hb of concentrtaion C_r into solution.

It is convenient to introduce a new definition for κ in terms of OS.

$$\kappa = \kappa_r + OS(\kappa_0 - \kappa_r) \tag{37}$$

where κ_0 is the specific absorption coefficient for HbO_2 and κ_r is the specific absorption coefficient for Hb. From the data shown in Fig. 24, $\kappa_0 = 0.10 \times 10^6$ and $\kappa_r = 0.80 \times 10^6$ at $\lambda = 0.66$ μ, and $\kappa_0 = \kappa_r = 0.20 \times 10^6$ at $\lambda = 0.80$ μ. This expression states that the average κ of the two-species medium consists of a simple average of the two coefficients in proportion to their relative concentrations. With this expression introduced into (12), one obtains

$$OD = 0.4343Cd[\kappa_r + OS(\kappa_0 - \kappa_r)]$$

where $C = C_0 + C_r$ is the total hemoglobin concentration. Measurements at two wavelengths are made, commonly near 0.66 μ where a large difference exists between the specific absorption coefficients for the two species of hemoglobin, and at 0.80 μ where the two specific absorption coefficients are equal. The ratio of the two optical densities yields a result for OS:

$$OS = \frac{\kappa_{r1}}{\kappa_{r1} - \kappa_{01}} - \frac{\kappa_{r2}}{\kappa_{r1} - \kappa_{01}} \frac{OD\ (\text{red})}{OD\ (\text{infrared})} \tag{38}$$

where subscript 1 refers to the red wavelength and subscript 2 refers to $\lambda = 0.80$ μ. In practice, one uses $OS = A - B\ OD(\text{red})/OD(\text{infrared})$, where A and B are constants obtained from a calibration procedure. An oximeter instrument measures the two OD values, performs the indicated algebraic operations in (38), and presents the OS value. A popular instrument of this type is the Instrumentation Laboratories model 182 CO-oximeter [140].

Oxygen saturation determinations can also be made using light reflected from whole blood. Reflection oximetry is convenient because it is unnecessary to hemolyze the blood and pass it through special apparatus. One commercially available instrument requires the deposition of a small amount of whole blood into a glass cylinder [141]. Light is reflected from the base of the cylinder and monitored to make the determination of OS. Another technique for measuring OS continuously in the living body involves the use of a fiberoptic catheter [142]–[148] in which fiberoptics are used to carry light to and from a measuring site. The reflectance equation for a semi-infinite medium, given in (19), provides the basis for this instrument. In order to determine OS, it is convenient to define σ_a in a way analogous to (37):

$$\sigma_a = \sigma_{ar} + OS(\sigma_{a0} - \sigma_{ar}) \tag{39}$$

where σ_{ar} is the average absorption cross section of an erythrocyte containing only Hb, and σ_{a0} is the average absorption cross section of an erythrocyte containing only HbO_2. The backscattering cross section σ_s^- is assumed not to vary with OS or wavelength over a limited range. The ratio of backscattered light intensity at infrared and red wavelengths has been shown from empirical studies to be related to OS by

$$OS = A - B \frac{R\ (\text{infrared})}{R\ (\text{red})} \tag{40}$$

where A and B are constants that depend on blood parameters. This result can be approximately obtained from (19). Moaveni [149] has determined approximate values for average erythrocyte cross sections at $\lambda = 0.80$ μ and $H = 0.42$ of $\sigma_s^- = 0.47$ μ^2 and $\sigma_a = 0.12$ μ^2. Using these cross-sectional values and inferring changes in σ_a with wavelength proportional to the specific absorption coefficient for hemoglobin, as shown in Fig. 24, (19) and (39) can be used to plot theoretical OS versus infrared to red reflectance ratio curves. For $\lambda_r = 0.66$ μ, and infrared wavelengths of 0.80, 0.92, and 1.00 μ, theoretical results in Fig. 34 show nearly linear curves for $\lambda_{ir} = 0.80$ μ and 0.92 μ, agreeing with the empirical result of (40).

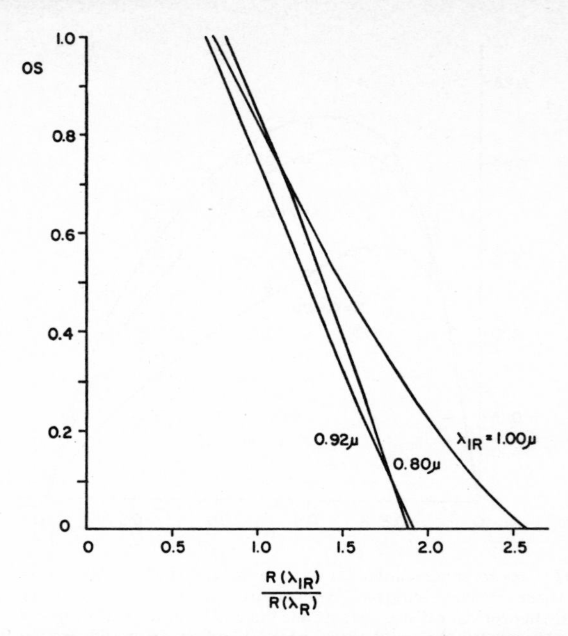

Fig. 34. Theoretical plot of oxygen saturation versus infrared-to-red reflectance ratio for three different infrared wavelengths. Red wavelength is $\lambda = 0.66$ μ.

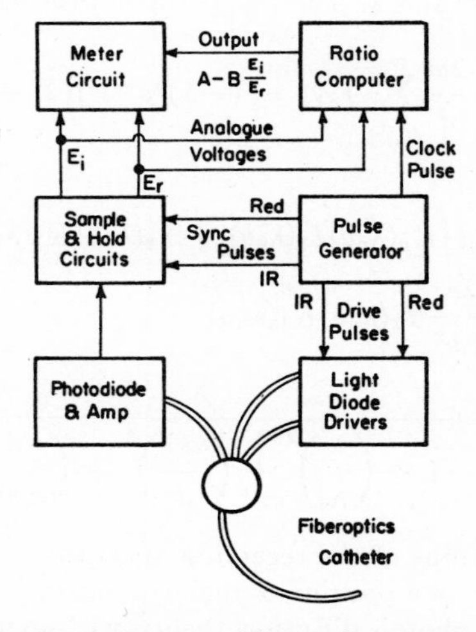

Fig. 35. Block diagram of fiberoptic catheter oximeter using pulsed light-emitting diodes.

Fortunately, gallium arsenide phosphide and gallium arsenide light-emitting diodes are available at $\lambda \cong 0.65$ μ and $\lambda \cong 0.92$ μ so that practical instruments can be developed with linear output [148]. Such an instrument is illustrated in Fig. 35 in block-diagram form. Three sets of intermeshed fiberoptic strands are located in a hollow tube called a catheter. Two sets of fiberoptic strands transmit red and infrared light to the tip of the catheter, while a third set of fibers transmit light reflected from blood at the tip of the catheter back to a

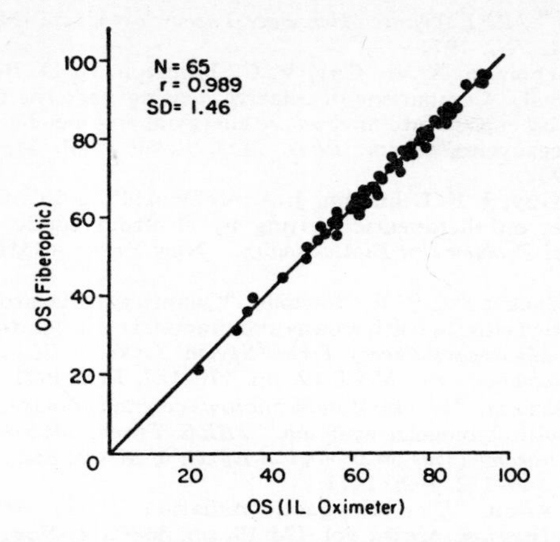

Fig. 36. Calibration plot showing correlation between oxygen saturation determined by the fiberoptic catheter oximeter and simultaneously drawn samples analyzed by the CO-oximeter. *In vivo* results were obtained from the adult catheterization laboratory and the coronary care unit.

photodetector. A pulse generator develops a 200-Hz square-wave pulse train which triggers the driving circuits for the light-emitting diodes. The diodes are pulsed alternately, producing two separate channels of red and infrared wavelengths, respectively. The spectral width of the light-emitting diodes is approximately 40 nm. The pulse generator also sends synchronous pulses to sample-and-hold circuits so that when the red light-emitting diode is turned on, a channel to the red circuit is opened; when the infrared light-emitting diode is turned on, a channel to the infrared circuit is open. Reflected red and infrared light returning from the blood is detected, amplified, and sent to sample-and-hold circuits. The output of the sample-and-hold circuits are two analog voltages E_r and E_i corresponding to the levels of red and infrared reflectance, respectively. The ratio of these two reflectance levels is proportional to the *OS* as given in (40). A direct conversion to *OS* can be made by introducing the *A* and *B* constants determined from calibration experiments. *In vivo* results, given in Fig. 36, show good correlation between *OS* determined by the Instrumentation Laboratories CO-oximeter and the fiberoptic catheter oximeter.

Acknowledgment

The information compiled in this paper is due to the contributions of many people. The authors specifically wish to thank the following for their assistance: Dr. J. Lehmann; Dr. H. Ho; C. Sorenson; J. McDougall; Dr. F. Harris; Dr. C. H. Durney; P. Cheung; J. Cole, M.D.; P. Hayden, M.D.; L. Reynolds; Dr. M. Moaveni; Dr. J. Shaken; and Dr. A. Ishimaru.

References

[1] F. B. Moor, "Microwave diathermy," in *Therapeutic Heat and Cold*, S. Licht, Ed. New Haven, Conn.: Licht, 1965, sec. 12, pp. 310–320.

[2] H. P. Schwan, "Biophysics of diathermy," in *Therapeutic Heat and Cold*, S. Licht, Ed. New Haven, Conn.: Licht, 1965, sec. 3, pp. 63–125.

[3] C. J. Restall, P. F. Leonard, H. F. Taswell, and R. E. Holaday. "Warming of human blood by use of microwaves," in *Summ. 4th IMPI Symp.* (Univ. of Alberta, Edmonton, Canada, May 21–23, 1969), pp. 96–99.

[4] R. P. Zimmer, H. A. Ecker, and V. P. Popovic, "Selective electromagnetic heating of tumors in animals in deep hypothermia," *IEEE Trans. Microwave Theory Tech.* (*Special Issue on Biological Effects of Microwaves*), vol. MTT-19, pp. 238–245, Feb. 1971.

[5] *An Act.* Public Law 90–602, 90th Cong., H. R. 10790, Oct. 18, 1968.

[6] W. Moore, Jr., *Biological Aspects of Microwave Radiation—A Review of Hazards.* U. S. Dep. HEW, PHS, Rep. TSB-68-4.

[7] M. Rosenstein, W. A. Brill, and C. K. Showalter, *Radiation Exposure Overview—Microwave Ovens and the Public.* U. S. Dep. HEW, PHS. Rep. OCS 69-1, July 1969.

[8] S. M. Michaelson, R. A. E. Thomson, and J. W. Howland, *Biologic Effects of Microwave Exposure.* Radiation Control for Health and Safety Act of 1967, Hearings before the Committee on Commerce, U. S. Senate, 90th Cong., 2nd Session, S.2067, S.3211, and H.F.10790, pp. 1443–1551.

[9] S. M. Michaelson, "The tri-service program—A tribute to George M. Knauf, USAF (MC)," *IEEE Trans. Microwave Theory Tech.* (*Special Issue on Biological Effects of Microwaves*), vol. MTT-19, pp. 131–146, Feb. 1971.

[10] ——, "Human exposure to nonionizing radiant energy—Potential hazards and safety standards," *Proc. IEEE*, vol. 60, pp. 389–421, Apr. 1972.

[11] C. J. Imig, J. D. Thomson, and H. M. Hines. "Testicular degeneration as a result of microwave irradiation," *Proc. Soc. Exper. Biol. and Med.*, vol. 69, pp. 382–386, Nov. 1948.

[12] B. O. Scott, "Short wave diathermy," in *Therapeutic Heat and Cold*, S. Licht, Ed. Baltimore, Md.: Waverly, 1958, pp. 255–283.

[13] F. H. Krusen, J. F. Herrick, U. Leden, and K. G. Wakim, "Microkymatotherapy; Preliminary report of experimental studies of heating effects of microwaves (radar) in living tissues," in *Proc. Staff Meet. Mayo Clin.*, vol. 22, p. 209, 1947.

[14] J. F. Lehmann, "Diathermy," in *Handbook of Physical Medicine and Rehabilitation*, Krusen, Kottke, Ellwood, Eds. Philadelphia, Pa.: Saunders, 1976.

[15] J. F. Herrick, "Peral-chain formation," in *Proc. 2nd Tri-Service Conf. Biol. Effects Microwave Energy*, University of Va., Charlottesville, Tech. Rep. ARDC-1 TR-58-54, ASTIA Doc. 131 477, pp. 83–93, 1958.

[16] J. Heller, "Effect of high-frequency electromagnetic fields on microorganisms," *Radio Electron.*, vol. 6, 1959.

[17] J. Heller, and A. A. Teixeira-Pinto, "A new physical method of creating chromosomal aberration," *Nature*, vol. 183, p. 905, 1959.

[18] J. Heller, and G. Mickey. "Non-thermal effects of radio frequency in biological systems," in *Dig. Int. Conf. Electronics*, vol. 21, p. 2, 1961.

[19] A. Wildervank *et al.*, "Certain experimental observations on a pulsed diathermy machine," *Arch. Phys. Med.*, vol. 40, p. 45, 1959.

[20] M. Satio *et al.*, "R- F- field-induced forces on microscopic particles," in *Dig. Int. Conf. Med. Electronics*, vol. 21, p. 3, 1961.

[21] M. Satio *et al.*, "The time constants of pearl chaim formation," in *Biological Effects of Microwave Radiation*, vol. 1. New York: Plenum, 1961, p. 85.

[22] A. Furedi and R. Valentine, "Factors involved in the orientation of microscopic particles in suspensions influenced by radio-frequency fields," *Biochim. Biophys. Acta*, vol. 56, p. 33, 1962.

[23] A. Furedi and I. Ohad, "Effects of high-frequency electric fields on the living cell," *Biochim. Biophys. Acta*, vol. 79, p. 1, 1964.

[24] A. S. Presman, *Electromagnetic Fields and Life.* New York: Plenum, 1970, pp. 59–61.

[25] H. Schwan, "Molecular response characteristics to ultra-high frequency fields," in *Proc. 2nd Tri-Serv. Conf. on Biol. Effects of Microwave Energy* (Rome, N. Y.), p. 33, 1958.

[26] J. Fleming *et al.*, "Microwave radiation in relation to biological systems and neural activity," in *Biological Effects of Microwave Radiation.* New York: Plenum, 1961, p. 239.

[27] D. A. Frank-Kamenetskii, "Plasma effects in semiconductors and biological effect of radiowaves," *Dokl. Akad. Nauk.* (USSR), vol. 136, pp. 476–478, 1961.

[28] A. S. Presman, "The physical basis for the biological action of centimeter waves," *Achiev. Mod. Biol.* (USSR), vol. 11. pp. 40–54, Libr. Cong., Washington, D. C., ATD P 65-17, 1956.

[29] J. E. Roberts, and H. F. Cook. "Microwaves in medical and biological research," *Brit. J. Appl. Physiol.*, vol. 3, pp. 33–39, 1952.

[30] A. S. Presman, I. Yu. Kamensky, and A. N. A. Levitina. "The biological effect of microwaves," *Achiev. Mod. Biol.* (USSR), vol. 51, pp. 84–103, Libr. Cong., Washington, D. C., ATD P 65–68, 1961.

[31] A. A. Teixeira-Pinto, L. L. Nejelski, J. L. Cutter, and J. H. Heller, "The behavior of unicellular organisms in an electromagnetic field," *Exp. Cell. Res.*, vol. 20, pp. 548–564, 1960.

[32] F. A. Brown, Jr., "Response to pervasive geophysical factors and the biological clock problem," *Symp. Quant. Biol.*, vol. 25, pp. 57–70, 1960.

[33] ——, "Extrinsic rhythmicality: A reference frame for biological rhythms under so-called constant conditions," in *Rhythmic Functions in the Living System.* New York: New York Acad. Sci., 1961; also in *Ann. N. Y. Acad. Sci.*, vol. 98, pp. 775–778.

[34] J. A. Tanner, "Effect of microwaves on birds," *Nature*, vol. 210, p. 636, 1966.

[35] A. H. Frey, "Auditory system response to RF energy," *Aerospace Med.*, vol. 32, pp. 1140–1142, 1961.

[36] ——, "Human auditory system response to modulated electromag-

netic energy," *J. Appl. Physiol.*, vol. 17, pp. 689–692, 1962.

[37] H. C. Sommer, and H. E. von Gierke, "Hearing sensations in electric fields," *Aerospace Med.*, vol. 35, p. 834, 1964.

[38] Yu. A. Kholodov, "The effect of an electromagnetic field on the central nervous system," *Priroda* (USSR), vol. 4, pp. 104–105, Libr. Cong., Washington, D. C., ATD P 65-68, FTD-TT 62-1107., ASTIA Doc. 284 123, 1962.

[39] ——, "Effect of a UHF electromagnetic field on the electrical activity of a neuronally isolated region of the cerebral cortex," *Bull. Exp. Biol. Med.* (USSR), vol. 57, pp. 98–104, Libr. Cong., Washington, D. C., ATF P 65-68, 1964.

[40] ——, "The effect of electromagnetic and magnetic fields on the central nervous system," NASA Technical Translation, TTF-465, Washington, D. C. 1967.

[41] A. H. Frey, "Brain stem evoked responses associated with low intensity pulsed UHF energy," *J. Appl. Physiol.*, vol. 23, pp. 984–988, Dec. 1967.

[42] ——, "Biological function as influenced by low-power modulated RF energy," *IEEE Trans. Microwave Theory Tech.* (*Special Issue on Biological Effects of Microwaves*), vol. MTT-19, pp. 153–164, Feb. 1971.

[43] S. F. Korbel, "Behavioral effects of low intensity UHF radiation," in *Biological Effects and Health Implications of Microwave Radiation* (Symp. Proc. Med. College of Virginia, Richmond, Va., Rep. RRH/DBE 70-2), pp. 180–184, Sept. 1969.

[44] S. F. Korbel and W. D. Thompson, "Behavior effects of stimulation by UHF radio fields," *Psychological Reps.*, vol. 17, pp. 595–602, 1965.

[45] H. P. Schwan and G. M. Piersol, "The absorption of electromagnetic energy in body tissues, pt. I," *Amer. J. Phys. Med.*, vol. 33, 371–404, , 1954.

[46] ——, "The absorption of electromagnetic energy in body tissues, pt. II," *Amer. J. Phys. Med.*, vol. 34, pp. 425–448, 1955.

[47] H. P. Schwan, "Electrical properties of tissues and cells," *Advan. Biol. Med. Phys.*, vol. 5, pp. 147–209, 1957.

[48] H. Cook, "The dielectric behavior of some types of human tissues at microwave frequencies," *Brit. J. Appl. Phys.*, vol. 2, p. 295, 1951.

[49] ——, "Dielectric behavior of human blood at microwave frequencies," *Nature*, vol. 168, p. 247, 1951.

[50] ——, "A comparison of the dielectric behavior of pure water and human blood at microwave frequencies," *Brit. J. Appl. Phys.*, vol. 3, p. 249, 1952.

[51] K. Cole and R. Cole, "Dispersion and absorption in dielectrics," *J. Chem. Phys.*, vol. 9, p. 34, 1941.

[52] H. Schwan, "Alternating current spectroscopy of biological substances," *Proc. IRE*, vol. 47, p. 1841–1855, Nov. 1959.

[53] A. W. Guy, and J. F. Lehmann, "On the determination of an optimum microwave diathermy frequency for a direct contact applicator," *IEEE Trans. Biomed. Eng.*, vol. BME-13, pp. 76–87, Apr. 1966.

[54] A. W. Guy, "Electromagnetic fields and relative heating patterns due to a rectangular aperture source in direct contact with bilayered biological tissue," *IEEE Trans. Microwave Theory Tech.* (*Special Issue on Biological Effects of Microwaves*), vol. MTT-19, pp. 214–223, Feb. 1971.

[55] J. A. Stratton, *Electromagnetic Theory*. New York: McGraw-Hill, 1941.

[56] A. Anne, "Scattering and absorption of microwaves by dissipative dielectric objects: The biological significance and hazard to mankind," Ph.D. dissertation, Univ. of Pennsylvania, Philadelphia, Pa., 106 p., Cont. NONR 551505, ASTIA Doc. 408 997, 1963.

[57] A. Anne, M. Satio, O. M. Salati, and H. P. Schwan, "Penetration and thermal dissipation of microwaves in tissues," Univ. of Pennsylvania, Philadelphia, Pa., Tech. Rep. RADC-TDR-62-244. Cont. AF 3-(602)-2344, ASTIA Doc. 284 981, 1962.

[58] ——, "Relative microwave absorption cross sections of biological significance," in *Biological Effects of Microwave Radiation*, vol. 1. New York: Plenum, 1960, pp. 153–176.

[59] A. R. Shapiro, R. F. Lutomirski, and H. T. Yura, "Induced fields and heating within a cranial structure irradiated by an electromagnetic plane wave," *IEEE Trans. Microwave Theory Tech.* (*Special Issue on Biological Effects of Microwaves*), vol. MTT-19, pp. 187–196, Feb. 1971.

[60] H. P. Schwan, "Survey of microwave absorption characteristics of body tissues," in *Proc. 2nd Tri-Serv. Conf. on Biol. Effects of Microwave Energy*, pp. 126–145, 1958.

[61] H. S. Ho, A. W. Guy, R. A. Sigelmann, and J. F. Lehmann, "Electromagnetic heating patterns in circular cylindrical models of human tissue," in *Proc. 8th Int. Conf. on Med. and Biol. Eng.* (Chicago, Ill., Sess. 27.4,), 1969.

[62] H. S. Ho, R. A. Sigelmann, A. W. Guy, and J. F. Lehmann, "Electromagnetic heating of simulated human limbs by aperture sources," in *Proc. 23rd Ann. Conf. on Eng. in Med. and Biol.* (Washington, D. C., 1970), p. 159.

[63] H. S. Ho, A. W. Guy, R. A. Sigelmann, and J. F. Lehmann, "Microwave heating of simulated human limbs by aperture sources," *IEEE Trans. Microwave Theory Tech.*, vol. MTT-19, pp. 224–231, Feb. 1971

[64] J. F. Lehmann, A. W. Guy, V. C. Johnston, G. D. Brunner, and J. W. Bell, "Comparison of relative heating patterns produced in tissues by exposure to microwave energy at frequencies of 2450 and 900 megacycles," *Arch. Phys. Med. Rehab.*, vol. 43, pp. 69–76, Feb. 1962.

[65] A. W. Guy, J. F. Lehmann, J. A. McDougall, and C. C. Sorensen, "Studies on therapeutic heating by electromagnetic energy," in *Thermal Problems in Biotechnology*. New York: ASME, 1968, pp. 26–45.

[66] P. F. Wacker and R. R. Bowman, "Quantifying hazardous electromagnetic fields: Scientific basis and practical considerations," *IEEE Trans. Microwave Theory Tech.* (*Special Issue on Biological Effects of Microwaves*), vol. MTT-19, pp. 178–187, Feb. 1971.

[67] H. P. Schwan, "Interaction of microwave and radio frequency radiation with biological systems," *IEEE Trans. Microwave Theory Tech.* (*Special Issue on Biological Effects of Microwaves*), vol. MTT-19, pp. 146–152, Feb. 1971.

[68] E. E. Aslan, "Electromagnetic radiation survey meter," *IEEE Trans. Instrum. Meas.*, vol. IM-19, pp. 368–372, Nov. 1970.

[69] A. W. Rudge, "An electromagnetic radiation probe for near-field measurements at microwave frequencies," *J. Microwave Power*, vol. 5, pp. 155–174, Nov. 1970.

[70] S. L. Osborne and J. N. Frederick, "Microwave radiations. Heating of human and animal tissues by means of high frequency current with wavelength of twelve centimeters (the microtherm)," *J. Am. Med. Assoc.*, vol 137, pp. 1036–1040.

[71] A. W. Richardson, T. D. Duane, and H. M. Hines, "Experimental lenticular opacities produced by microwave irradiations," *Arch. Phys. Med.*, vol. 29, pp. 765–769, Dec. 1948.

[72] J. F. Lehmann, G. D. Brunner, J. McMillan, D. R. Silverman, and V. C. Johnston, "Modification of heating patterns produced by microwaves at the frequencies of 2456 and 900 Mc. by physiologic factors in the human," *Arch. Phys. Med.*, vol. 45, pp. 555–563, Nov. 1964.

[73] J. F. Lehmann, V. C. Johnston, J. McMillan, D. R. Silverman, G. D. Brunner, and L. A. Rathbun, "Comparison of deep heating by microwaves at frequencies 2456 and 900 megacycles," *Arch. Phys. Med.*, vol. 46, pp. 307–314, Apr. 1965.

[74] J. F. Lehmann, D. R. Silverman, B. A. Baum, N. L. Kirk, and V. C. Johnston, "Temperature distributions in the human thigh, produced by infra-red, hot pack and microwave applications," *Arch. Phys. Med.*, vol. 47, pp. 291–299, May 1966. Abstract, in *Modern Med.*, Nov. 21, 1966.

[75] Private correspondence.

[76] A. W. Guy, F. A. Harris, and H. S. Ho, "Quantitation of the effects of microwave radiation on central nervous system function," in *Proc. 6th Ann. Int. Microwave Power Symp.* (Monterey, Calif., May 1971).

[77] A. W. Guy, "Analyses of electromagnetic fields induced in biological tissues by thermographic studies on equivalent phantom models," *IEEE Trans. Microwave Theory Tech.* (*Special Issue on Biological Effects of Microwaves*), vol. MTT-19, pp. 205–214, Feb. 1971.

[78] *Therapeutic Heat and Cold.* S. Licht, Ed. Baltimore, Md.: Waverly, 1958.

[79] A. A. Beisang, C. H. Mayo, N. Pace, R. C. Lillehei, and E. H. Graham, "Rapid thawing of extended semen with microwaves," in *Summ. 4th IMPI Symp.* (Univ. of Alberta, Edmonton, Canada, May 21–23), p. 89, 1969.

[80] A. A. Beisang, R. H. Deitzman, G. J. Motsay, E. H. Graham, and R. C. Lillehei, "Microwave absorption in frozen physiological solutions containing glycerine," in *Summ. 4th IMPI Symp.* (Univ. of Alberta, Edmonton, Canada, May 21–23), p. 94, 1969.

[81] Y. E. Moskalenko, "Utilization of superhigh frequencies in biological investigations," *Biophysics* (USSR) (English transl.), vol. 3, pp. 619–626, 1958.

[82] ——, "On the application of centimeter radio waves for electrodeless recordings of volume changes of biological objects," *Biophysics* (USSR) (English transl.), vol. 5, pp. 225–228, 1960.

[83] K. E. Lonngren, "An application of microwaves to medical research," in *Summ. 4th Ann. IMPI Symp.* (Univ. of Alberta, Edmonton, Canada, May 21–23), p. 91, 1969.

[84] H. P. Schwan, and P. O. Vogelhut, "Scientific uses: Microwave studies of biological systems. Microwave properties of bound water and macromolecules," in *Microwave Power Engineering*, vol. 2. New York: Academic Press, 1968, pp. 235–244.

[85] K. F. Lindmann, "Über eine durch ein Isotopen System von Spiral-Formigen Resonatoren erzeugte Rotation–Polarisation der elektromagnetischen Wellen," *Ann. Physik.*, vol. 63, pp. 621–644, 1920.

[86] M. H. Winkler, "An experimental investigation of some models for optical activity," *J. Phys. Chem.*, vol. 60, pp. 1656–1659, 1956.

[87] A. J. H. Vendrik and J. J. Vos, "Comparison of the stimulation of the warmth sense organ by microwave and infrared," *J. Appl. Physiol.*, vol. 13, pp. 435–449, 1958.

[88] E. Hendler, R. Crosbie, and J. D. Hardy, "Measurement of heating

of the skin during exposure to infrared radiation," *J. Appl. Physiol.*, vol. 12, pp. 177–185, 1958.

[89] E. Hendler and J. D. Hardy, "Infrared and microwave effects on skin heating and temperature sensation," *IRE Trans. Med. Electron.*, vol. ME-7, pp. 143–152, July 1960.

[90] E. Hendler, J. D. Hardy, and D. Murgatroyd, "Skin heating and temperature sensation produced by infrared and microwave irradiation," in *Temperature: Its Measurement and Control in Science and Industry*, vol. 3. New York: Reinhold, 1963, pp. 211–230.

[91] P. J. Wyatt, "Differential light scattering: A physical method for identifying living bacterial cells," *Appl. Opt.*, vol. 7, pp. 1879–1896, Oct. 1968.

[92] ——, "Identification of bacteria by differential light scattering," *Nature*, vol. 221, p. 1257, 1969.

[93] P. T. Phillips, "Evolution of an instrument, the differential I," *Bioscience*, vol. 21, p. 865, May 1971.

[94] R. J. Fiel, E. H. Mark, and B. R. Munson, "Small angle light scattering of bioparticles," *Arch. Biophys. Biochem.*, vol. 141, pp. 547–551, 1970.

[95] V. Twersky, "Absorption and multiple scattering by biological suspensions," *J. Opt. Soc. Amer.*, vol. 60, pp. 1084–1093, Aug. 1970.

[96] F. Daniels, Jr., J. C. Van der Leun, and B. E. Johnson, "Sunburn," *Sci. Amer.*, pp. 38–46, July 1968.

[97] G. Moreno, M. Lutz, and M. Bessis, "Partial cell irradiation by ultraviolet and visible light: Conventional and laser sources," *Int. Rev. Exp. Path.*, vol. 7, pp. 99–137, 1969.

[98] R. A. MacRae, J. A. McClure, and P. Latimer, "Spectral transmission and scattering properties of red blood cells," *J. Opt. Soc. Amer.*, vol. 51, pp. 1356–1372, 1961.

[99] R. Barer, "Spectrophotometry of clarified cell suspension," *Science*, vol. 121, pp. 709–715, May 1955.

[100] B. L. Horecker, "Absorption spectra of hemoglobin and its derivatives in the visible and near infrared regions," *J. Biol. Chem.*, vol. 148, pp. 173–183, 1943.

[101] T. J. Livesey, and F. W. Billmeyer, Jr., *J. Colloid Interface Sci.*, vol. 30, p. 447, 1969.

[102] R. F. Fiel, and H. M. Scheintaur, "Small angle light scattering by erythrocytes," *J. Colloid Interface Sci.*, vol. 37, pp. 249–250, Sept. 1971.

[103] D. K. Kreid, M. R. Kannin, R. J. Goldstein, "Measurements of light scattering characteristics of red cells, red cell 'ghosts' and polystyrene spheres," in *Proc. 24th Ann. Conf. on Eng. in Med. and Biol.* (Las Vegas, Nev., Oct.–Nov. 1971), p. 143.

[104] C. C. Johnson, "Near infrared propagation in blood," *J. Assc. Adv. Med. Instr.*, vol. 4, pp. 22–27, Jan.–Feb. 1970.

[105] N. M. Anderson and P. Sekelj, "Reflection and transmission of light by thin films of non-hemolysed blood," *Phys. Med. Biol.*, vol. 12, pp. 185–192, 1967.

[106] F. Urbach, Ed., *The Biologic Effects of Ultra-Violet Radiation.* New York: Pergamon, 1969.

[107] J. D. Hardy, H. T. Hammel, and D. Murgatroyd, "Spectral transmittance and reflectance of excised human skin," *J. Appl. Physiol.*, vol. 9, pp. 257–264, Sept. 1956.

[108] J. A. Jacquez and H. F. Kuppenheim, "Spectral reflectance of human skin in the region 235–1000mμ," *J. Appl. Physiol.*, vol. 7, pp. 523–528, 1955.

[109] J. A. Jacquez, J. Huss, W. McKeehan, J. M. Dimitroff, and H. F. Kuppenheim, "Spectral reflectance of human skin in the region 0.7–2.6μ," *J. Appl. Physiol.*, vol. 8, pp. 297–299, 1955.

[110] A. Schuster, "Radiation through a foggy atmosphere," *Astrophys. J.*, vol. 21, pp. 1–22, Jan. 1905.

[111] L. Silberstein, "The transparency of turbid media," *Phil. Mag.*, vol. 4, p. 1291, 1927.

[112] J. W. Ryde, "The scattering of light by turbid media—Part I," *Proc. Roy. Soc. London, Ser. A.*, vol. 131, p. 451, 1931.

[113] J. W. Ryde, and B. S. Cooper, "The scattering of light by turbid media—Part II," *Proc. Roy. Soc. London, Ser. A*, vol. 131, p. 464, 1931.

[114] P. Kubelka and F. Munk, "Ein Beitrag zur Optik der Farbanstriche," *Z. Tech. Physik*, vol. 12, p. 593, 1931.

[115] P. Kubelka, "New contributions to the optics of intensely light-scattering materials—Part I," *J. Opt. Soc. Amer.*, vol. 38, pp. 448–457, May 1948.

[116] S. Q. Duntley, "The optical properties of diffusing materials," *J. Opt. Soc. Amer.*, vol. 32, p. 61, 1942.

[117] R. G. Giovanelli, "A note on the coefficient of reflection for internally incident diffuse light," *Optica Acta*, vol. 3, p. 127, 1956.

[118] W. L. Butler, "Absorption of light by turbid materials," *J. Opt. Soc. Amer.*, vol. 52, p. 292, 1962.

[119] J. K. Beasley, J. T. Atkins, and F. W. Billmeyer, "Scattering and absorption of light in turbid media," in *Proc. 2nd Interdisc. Conf. on Elect. Scattering.* R. S. Stein and R. L. Rowell, Eds. New York: Gordon and Breach, 1967.

[120] L. Amy, "Sur la couleur des corps par reflexion," *Rev. Optique*, vol. 16, p. 81, 1937.

[121] F. A. Rodrigo, "The determination of the oxygenation of blood in vitro by using reflected light," *Am. Heart J.*, vol. 45, pp. 809–822, 1953.

[122] R. L. Longini and R. Zdrojkowski, "A note on the theory of backscattering of light by living tissue," *IEEE Trans. Biomed. Eng.*, vol. BME-15, pp. 4–10, Jan. 1958.

[123] V. Twersky, "Interface effects in multiple scattering by large, low-refracting, absorbing particles," *J. Opt. Soc. Amer.*, vol. 60, pp. 908–914, July 1970.

[124] S. Chandrasekhar, *Radiative Transfer.* New York: Dover, 1963.

[125] V. Kourganoff, *Basic Methods in Transfer Problems.* New York: Dover, 1963.

[126] V. Twersky, "On scattering of waves by random distribution," in *Proc. Am. Math. Soc. Symp. on Stocastic Processes in Math. Phys. and Eng.*, vol. 16. New York: McGraw Hill, 1964, pp. 84–166.

[127] R. J. Zdrojkowski and R. L. Longini, "Optical transmission through whole blood illuminated with highly collimated light," *J. Opt. Soc. Amer.*, vol. 59, pp. 893–903, Aug. 1969.

[128] C. C. Johnson, "Opital diffusion in blood," *IEEE Trans. Biomed. Eng.*, vol. BME-17, pp. 129–133, Apr. 1970.

[129] R. J. Zdrojkowski and N. Pisharoty, "Optical transmission and reflection by blood," *IEEE Trans. Biomed. Eng.*, vol. BME-17, pp. 122–128, Apr. 1970.

[130] A. Cohen and R. L. Longini, "Theoretical determination of the blood's relative oxygen saturation in vivo," *Med. Biol. Eng.*, vol. 9, pp. 61–69, 1971.

[131] L. O. Reynolds, "Three dimensional reflection and transmission equations for optical diffusion in blood," M. S. thesis, Elec. Eng. Dep., Univ. of Washington, Seattle, 1970.

[132] P. I. Richards, "Multiple isotropic scattering," *Phys. Rev.*, vol. 100, pp. 517–522, Oct. 1955.

[133] C. H. Durney, private communication.

[134] C. M. Cartwright, "Spectrotransmission of the human body," in *Medical Physics*, N. O. Glasser, Ed. Chicago, Ill.: Yearbook Publishers, 1944.

[135] K. Kramer, J. O. Elan, G. A. Saxton, and W. N. Elan, "Influence of oxygen saturation, erythrocyte concentration and optical depth on the red and near-infrared regions," *Am. J. Physiol.*, vol. 165, pp. 229–246, 1951.

[136] D. B. Shurtleff, E. L. Foltz, and D. Fry, "Clinical use of transillumination," *Arch. Dis. Childh.*, vol. 41, pp. 183–187, Apr. 1966.

[137] C. C. Johnson, B. Henshaw, and D. Watkins, "A pulsed transilluminator," in *Proc. 8th ICMBE* (Chicago, Ill., July, 1969), pp. 13–19.

[138] J. Weinman, "Photoplethysmography," ch. 6 in *A Manual of Psychophysiological Methods.* Amsterdam, The Netherlands: North-Holland, 1967.

[139] G. Uretzky and Y. Palti, "A method for comparing transmitted and reflected light photoelectric plethysmography," *J. Appl. Physiol.*, vol. 31, pp. 132–135, July 1971.

[140] A. H. J. Maas, M. L. Hamelink, and R. J. M. De Leeuw, "An evaluation of the spectrophotometric determination of HbO_2, HbCO, and Hb in blood with the CO-oximeter IL-182," *Clin. Chem. Acta*, vol. 29, pp. 303–309, 1970.

[141] M. L. Polanyi and R. M. Hehir, "New reflection oximeter," *Rev. Sci. Instrum.*, vol. 31, pp. 401–403, Apr. 1960.

[142] M. L. Polanyi and R. M. Hehir, "In vivo oximeter with fast dynamic response," *Rev. Sci. Instrum.*, vol. 33, pp. 1050–1054, Oct. 1962.

[143] Y. Enson, W. A. Briscoe, N. L. Polanyi, and A. Cournand, "In vivo studies with an intra-vascular and intracardiac reflection oximeter," *J. Appl. Phys.*, vol. 17, pp. 552–558, May 1962.

[144] P. L. Frommer, J. Ross, Jr., D. T. Mason, J. H. Gault, and E. Brunwald, "Clinical applications of an improved, rapidly responding fiberoptic catheter," *Am. J. Card.*, vol. 15, pp. 672–678, May 1965.

[145] W. J. Gamble, P. G. Hugenholtz, R. G. Monroe, M. Polanyi, and A. S. Nadas, "The use of fiberoptics in clinical cardiac catheterization, 1. Intracardiac oximetry," *Circulation*, vol. 31, Mar. 1965.

[146] D. C. Harrison, N. S. Kapany, H. A. Miller, N. Silbertrust, W. L. Henry, and R. P. Drake, "Fiberoptics for continuous in vivo monitoring of oxygen saturation," *Am. Heart J.*, vol. 71, pp. 766–774, June 1966.

[147] G. A. Mook, P. Osypka, R. E. Sturm, and E. H. Wood, "Fiberoptic reflection photometry on blood," *Cardiovas. Res.*, vol. 24, pp. 199–209, Apr. 1968.

[148] C. C. Johnson, R. D. Palm, D. C. Stewart, and W. E. Martin, "A solid-state fiberoptics oximeter," *J. Assn. Adv. Med. Instr.*, vol. 5, pp. 77–83, Mar.–Apr. 1971.

[149] M. K. Moaveni, "A multiple scattering field theory applied to whole blood," Ph.D. dissertation, Elec. Eng. Dep., Univ. of Washington, Seattle, 1970.

Long-Wavelength Analysis of Plane Wave Irradiation of an Ellipsoidal Model of Man

HABIB MASSOUDI, MEMBER, IEEE, CARL H. DURNEY, MEMBER, IEEE, AND
CURTIS C. JOHNSON, SENIOR MEMBER, IEEE

Abstract—**Expressions are derived for the induced electric fields in an ellipsoidal model of man, and experimental animals irradiated by an electromagnetic (EM) plane wave when the wavelength is long compared to the dimensions of the ellipsoid. Calculations of the power absorbed by an ellipsoidal model of man are given for six different orientations of the ellipsoid with respect to the incident plane wave field vectors. The results show that the induced fields and the absorbed power in the ellipsoid are strong functions of frequency, size, and orientation with respect to the incident EM field vectors. The results for the ellipsoidal model of man are also compared with those of the prolate spheroidal model.**

I. Introduction

A LONG-WAVELENGTH analysis of electromagnetic (EM) plane wave (perturbation technique) has recently been developed and applied to prolate spheroidal models of man and experimental animals [1], [2]. The results of power absorbed calculations in the prolate spheroid models of man and some experimental animals show that orientation of the body with respect to the incident plane wave vectors is an extremely important variable which can make an order-of-magnitude difference in EM power absorption. This strong dependence of EM power absorption upon orientation has also been observed experimentally [3]. Experiments have also been conducted at the School of Aerospace Medicine, Brooks Air Force Base, to measure the EM power absorption in a 70-kg saline-filled prolate spheroidal human phantom, twenty 3.5-kg saline-filled prolate spheroidal monkey phantoms, and twenty live rhesus monkeys [4]. Results of the calculations for spheroidal models have been compared with measurements of power absorbed by saline-filled spheroidal phantoms, and good agreement between calculations and measurements has been found. However, agreement between theory and measurements for live monkeys was not as good as that for prolate spheroidal phantoms [4]. It was found that a significant difference in power absorption occurred when the monkeys were rotated 90° about their long axis. The prolate spheroidal model did not predict this because a prolate spheroid has circular cross sections normal to its long axis, whereas for a monkey, or in general for a primate, cross sections taken normal to the long axis appear more elliptical than circular. A principal conclusion from these comparisons is that an ellipsoidal model will be a superior representation of primates (man, monkey, and others), and the prolate spheroidal model seems to be adequate only for rodents (mice, rats, and others) since for these species, cross sections taken normal to the long axis appear approximately circular.

In this paper, the perturbation technique, described in [1], is applied to analyze the internal fields in an ellipsoid irradiated by a plane wave for each of the six major orientations of the incident fields with respect to the ellipsoid. Expressions for average absorbed power and power distribution inside the ellipsoid are given. The results of power absorbed calculations in ellipsoidal models of man are compared with those of prolate spheroidal models. Curves of power absorption versus frequency show that the absorbed power is a strong function of size and orientation of the ellipsoid in the incident fields.

Manuscript received January 7, 1976; revised June 2, 1976. This work was supported by the U.S. Air Force School of Aerospace Medicine, Brooks Air Force Base, TX 78235.

H. Massoudi is with the Department of Electrical Engineering, University of Utah, Salt Lake City, UT 84112.
C. H. Durney is with the Department of Electrical Engineering and Bioengineering, University of Utah, Salt Lake City, UT 84112.
C. C. Johnson is with the Department of Bioengineering, University of Utah, Salt Lake City, UT 84112.

II. First-Order Internal Fields for the Ellipsoid Irradiated by an EM Plane Wave

In this section the perturbation technique is applied to find the solution of the zeroth- and first-order equations for a plane wave incident on a tissue ellipsoid. The equation of ellipsoid in the rectangular coordinate system is

$$\frac{x^2}{a^2} + \frac{y^2}{b^2} + \frac{z^2}{c^2} = 1$$

where a, b, and c are the semiprincipal axes of the ellipsoid with $a > b > c$.

Expressions for the internal electric fields and the absorbed power are given for each of the six primary polarizations. For convenience in referring to these polarizations, the following definition of polarization is made. The polarization is defined in terms of which of the vectors $\boldsymbol{E}^i$, $\boldsymbol{H}^i$, and $\boldsymbol{K}$ are parallel with the three axes of the ellipsoid. ($\boldsymbol{E}^i$ is the incident electric field vector, $\boldsymbol{H}^i$ the magnetic field vector, and $\boldsymbol{K}$ the propagation vector.) The vector parallel to the longest axis is listed first, the one parallel to the next longest axis is listed second, and the one parallel to the shortest axis is listed last. Thus *EHK* polarization is the one in which the incident electric field vector is parallel to the longest axis (length a), the incident magnetic field vector is parallel to the next longest axis (length b), and the propagation vector is parallel to the shortest axis (length c).

A. Derivations for EKH Polarization

The first polarization considered is *EKH* polarization. For this polarization $\boldsymbol{E}^i \parallel \hat{x}$ and $\boldsymbol{H}^i \parallel -\hat{z}$. Since the per-

Reprinted from *IEEE Trans. Microwave Theory Tech.*, vol. MTT-25, pp. 41–46, Jan. 1977.

turbation technique has been described in [1], only the outline of the procedure and the results are given here.

1) Each set of incident, internal, and scattered electric and magnetic fields is expanded in a power series of $(-jk)$, where $j = (-1)^{1/2}$ and k is the free-space propagation constant.

2) Equations for the nth-order field terms are obtained by requiring the series expansions of the incident, scattered, and internal fields to satisfy both Maxwell's equations and the boundary conditions.

The results for the internal fields are as follows:

$$\nabla \times \boldsymbol{E}_0 = 0 \tag{1}$$

$$\nabla \times \boldsymbol{E}_n = \eta_0 \boldsymbol{H}_{n-1}, \qquad n \geq 1 \tag{2}$$

$$\nabla \times \boldsymbol{H}_0 = \sigma \boldsymbol{E}_0 \tag{3}$$

$$\nabla \times \boldsymbol{H}_n = \sigma \boldsymbol{E}_n - \frac{\varepsilon_r}{\eta_0} \boldsymbol{E}_{n-1}, \qquad n \geq 1 \tag{4}$$

$$\nabla \cdot \boldsymbol{E}_0 = 0 \tag{5}$$

$$\nabla \cdot \left(\sigma \boldsymbol{E}_n - \frac{\varepsilon_r}{\eta_0} \boldsymbol{E}_{n-1} \right) = 0, \qquad n \geq 1 \tag{6}$$

$$\nabla \cdot \boldsymbol{H}_n = 0 \tag{7}$$

where $k = \omega\sqrt{\mu_0 \varepsilon_0}$, $\varepsilon_r = \varepsilon/\varepsilon_0$ (real), σ is the conductivity of the ellipsoid, and an $e^{j\omega t}$ time variation has been assumed. The equation for curl and divergence of the incident and scattered fields can be obtained from (1) to (7) by setting $\varepsilon_r = 1$ and $\sigma = 0$.

The relations between the nth-order internal and external fields at the boundary are

$$\boldsymbol{n} \cdot \boldsymbol{E}_0 = 0 \tag{8}$$

$$\hat{\boldsymbol{n}} \cdot \left(\sigma \boldsymbol{E}_n - \frac{\varepsilon_r}{\eta_0} \boldsymbol{E}_{n-1} \right) = -\frac{\hat{\boldsymbol{n}}}{\eta_0} \cdot (\boldsymbol{E}_{n-1}^i + \boldsymbol{E}_{n-1}^s), \qquad n \geq 1 \tag{9}$$

$$\hat{\boldsymbol{n}} \times \boldsymbol{E}_n = \hat{\boldsymbol{n}} \times (\boldsymbol{E}_n{}^i + \boldsymbol{E}_n{}^s) \tag{10}$$

$$\boldsymbol{H}_n = \boldsymbol{H}_n{}^i + \boldsymbol{H}_n{}^s \tag{11}$$

where $\hat{\boldsymbol{n}}$ is the outer unit normal vector at the boundary.

The zeroth-order field $\boldsymbol{E}_0$ must satisfy (1), (5), and (8), which are equivalent to the equations for the field inside a conducting ellipsoid in a uniform static electric field. The solution is $\boldsymbol{E}_0 = 0$. The scattered electric field $\boldsymbol{E}_0{}^s$ is the same as the field induced by a conducting ellipsoid in a uniform static electric field. The solution for $\boldsymbol{E}_0{}^s$ can be found by using the ellipsoidal coordinates (ξ,η,ζ). $\boldsymbol{E}_0{}^s$ is given by Stratton [5] as

$$\boldsymbol{E}_0{}^s = C_1 \,\mathrm{grad} \left\{ [(\xi + a^2)(\eta + a^2) \cdot (\zeta + a^2)]^{1/2} \int_\xi^\infty \frac{d\xi}{(\xi + a^2)R_\xi} \right\} \tag{12}$$

where

$$C_1 = -\left[\sqrt{(b^2 - a^2)(c^2 - a^2)} \cdot \int_0^\infty \frac{d\xi}{(\xi + a^2)R_\xi} \right]^{-1} \tag{13}$$

and

$$R_\xi = [(\xi + a^2)(\xi + b^2)(\xi + c^2)]^{1/2}. \tag{14}$$

The equations for the zeroth-order scattered magnetic field $\boldsymbol{H}_0{}^s$ and internal magnetic field $\boldsymbol{H}_0$ are equivalent to the equations for a conducting ellipsoid in a uniform magnetic field, and since the ellipsoid is nonmagnetic, the solutions are $\boldsymbol{H}_0{}^s = 0$ and $\boldsymbol{H}_0 = -\hat{\boldsymbol{z}}/\eta_0$.

As described in the prolate spheroid derivation [1], the first-order electric field, $\boldsymbol{E}_1$ will be written as the sum of two terms, $\boldsymbol{E}_1 = \boldsymbol{E}_1' + \boldsymbol{E}_1''$, where

$$\nabla \times \boldsymbol{E}_1' = 0 \tag{15}$$

$$\nabla \cdot \boldsymbol{E}_1' = 0 \tag{16}$$

$$\hat{\boldsymbol{n}} \cdot \boldsymbol{E}_1' = \left(-\frac{1}{\sigma \eta_0} \right) \hat{\boldsymbol{n}} \cdot (\boldsymbol{E}_0{}^i + \boldsymbol{E}_0{}^s) \text{ on the surface} \tag{17}$$

$$\nabla \times \boldsymbol{E}_1'' = \eta_0 \boldsymbol{H}_0 \tag{18}$$

$$\nabla \cdot \boldsymbol{E}_1'' = 0 \tag{19}$$

$$\hat{\boldsymbol{n}} \cdot \boldsymbol{E}_1'' = 0 \text{ on the surface.} \tag{20}$$

Since the curl and divergence of $\boldsymbol{E}_1'$ are zero, $\boldsymbol{E}_1'$ can be found from $\boldsymbol{E}_1' = \nabla \phi_1'$ where ϕ_1' satisfies Laplace's equation. In ellipsoidal coordinates, Laplace's equation is given by Stratton [5].

The properties of the ellipsoidal harmonics that satisfy Laplace's equation can be found in the literature [6], [7]. The potential ϕ_1' is an ellipsoidal harmonic which is to be found. First we will write $\boldsymbol{E}_0{}^i$ and $\boldsymbol{E}_0{}^s$ in terms of scalar potentials. Since $\boldsymbol{E}_0{}^i = \hat{\boldsymbol{x}}$, we set

$$\boldsymbol{E}_0{}^i \equiv \nabla \phi_0{}^i = \nabla(x) = \nabla \left[\frac{(\xi + a^2)(\eta + a^2)(\zeta + a^2)}{(b^2 - a^2)(c^2 - a^2)} \right]^{1/2} \tag{21}$$

or

$$\phi_0{}^i = C_2 f_1(\xi) f_2(\eta) f_3(\zeta) \tag{22}$$

with $f_i(\alpha) = (\alpha + a^2)^{1/2}$, ($i = 1,2,3$ and $\alpha = \xi,\eta,\zeta$), and $C_2 = [(b^2 - a^2)(c^2 - a^2)]^{1/2}$. Also, we write

$$\boldsymbol{E}_0{}^s = \nabla \phi_0{}^s. \tag{23}$$

From (12) and (23),

$$\phi_0{}^s = C_1 f_1(\xi) f_2(\eta) f_3(\zeta) \int_\xi^\infty \frac{d\xi}{f_1{}^2(\xi) R_\xi}. \tag{24}$$

From (22) and (24),

$$(\phi_0{}^i + \phi_0{}^s) = C_2 f_1(\xi) f_2(\eta) f_3(\zeta) \left[1 - \frac{\int_\xi^\infty \frac{d\xi}{f_1{}^2(\xi) R_\xi}}{A_0} \right] \tag{25}$$

where

$$A_0 = \int_0^\infty \frac{d\xi}{(\xi + a^2) R_\xi}. \tag{26}$$

The boundary condition given in (17) can be written as

$$\hat{\xi}\cdot(\nabla\phi_1') = -\frac{1}{\sigma\eta_0}[\hat{\xi}\cdot\nabla(\phi_0{}^i + \phi_0{}^s)], \qquad \text{at } \xi = 0$$

or

$$\left.\frac{1}{h_\xi}\frac{\partial\phi_1'}{\partial\xi}\right|_{\xi=0} = \left(-\frac{1}{\sigma\eta_0}\right)\frac{1}{h_\xi}\frac{\partial}{\partial\xi}(\phi_0{}^i + \phi_0{}^s)\Big|_{\xi=0} \tag{27}$$

with

$$h_\xi = \frac{1}{2}\left[\frac{(\xi-\eta)(\xi-\zeta)}{(\xi+a^2)(\xi+b^2)(\xi+c^2)}\right]^{1/2}.$$

From (27), it can be seen that ϕ_1' must have the same η and ζ variation as $(\phi_0{}^i + \phi_0{}^s)$. We presume, therefore, that ϕ_1' is a function of the form

$$\phi_1' = C_3 g(\xi) f_2(\eta) f_3(\zeta). \tag{28}$$

Substitution of (28) into the Laplace's equation results in

$$R_\xi\frac{d}{d\xi}\left(R_\xi\frac{dg(\xi)}{d_\xi}\right) - \left(\frac{b^2+c^2}{4} + \frac{\xi}{2}\right)g(\xi) = 0 \tag{29}$$

with R_ξ given as in (14). The solutions of the preceding second-order differential equation are

$$g_1(\xi) = (\xi + a^2)^{1/2} \tag{30}$$

and

$$g_2(\xi) = (\xi + a^2)^{1/2}\int\frac{d\xi}{(\xi + a^2)R_\xi}. \tag{31}$$

Only $g_1(\xi)$ is an admissible solution for the internal potential ϕ_1' because $g_2(\xi)$ is infinite at $\xi = -c^2$ whereas $g_1(\xi)$ is finite at all points within the surface $\xi = 0$. Therefore,

$$\phi_1' = C_3[(\xi + a^2)(\eta + a^2)(\zeta + a^2)]^{1/2}. \tag{32}$$

Substituting (32) and (25) into (27) gives

$$C_3 = -\frac{1}{\sigma\eta_0}\frac{C_2}{A_1} \tag{33}$$

with

$$A_1 = \frac{abc}{2}\int_0^\infty \frac{d\xi}{(\xi + a^2)R_\xi}. \tag{34}$$

The expression for ϕ_1' in rectangular coordinates, after substituting (33) into (32), can be written as $\phi_1' = -1/(\sigma\eta_0 A_1)x$, and then $\boldsymbol{E}_1'$ can be written as

$$\boldsymbol{E}_1' = -\frac{1}{\sigma\eta_0 A_1}\hat{x} \tag{35}$$

where A_1 is given by (34).

The expression for $\boldsymbol{E}_1''$ is found by solving (18)–(20). Since $\eta_0\boldsymbol{H}_0 = -\hat{z}$,

$$\nabla\times\boldsymbol{E}_1'' = -\hat{z}. \tag{36}$$

For convenience, we set

$$\boldsymbol{E}_1'' = \boldsymbol{F}_1 + \nabla\psi. \tag{37}$$

According to (19), then, $\nabla\cdot\boldsymbol{F}_1 = 0$ and $\nabla^2\psi = 0$. A solution of $\boldsymbol{F}_1$, by inspection, is $\boldsymbol{F}_1 = \frac{1}{2}(y\hat{x} - x\hat{y})$. Substitution of (37) into the boundary condition given in (20) results in

$$\hat{\xi}\cdot\boldsymbol{F}_1 + \hat{\xi}\cdot(\nabla\psi) = 0, \qquad \text{at } \xi = 0. \tag{38}$$

The scalar potential ψ is an ellipsoidal harmonic. In addition, (38) shows that ψ must have the same η and ζ variation as $\boldsymbol{F}_1$. The solution for ψ in this case will have, in rectangular coordinates, the form

$$\psi = B_1 xy \tag{39}$$

which is an ellipsoidal harmonic of the second kind [7]. To find the constant B_1, we substitute the expressions for $\boldsymbol{F}_1$ and ψ into (38):

$$\tfrac{1}{2}\hat{\xi}\cdot(y\hat{x} - x\hat{y}) + B_1\hat{\xi}\cdot(x\hat{y} + y\hat{x}) = 0, \qquad \text{at } \xi = 0. \tag{40}$$

Now the unit vector $\hat{\xi}$ in ellipsoidal coordinates must be related to the unit vectors $\hat{x}$, $\hat{y}$, and $\hat{z}$ in rectangular coordinates. This may be done by writing [8]

$$\hat{\xi} = \frac{1}{h_\xi}\frac{\partial r}{\partial\xi} = \frac{1}{h_\xi}\left[\frac{\partial x}{\partial\xi}\hat{x} + \frac{\partial y}{\partial\xi}\hat{y} + \frac{\partial z}{\partial\xi}\hat{z}\right]. \tag{41}$$

Substitution of (41) into (40), after a few algebraic steps, gives $B_1 = (a^2 - b^2)/2(a^2 + b^2)$. Therefore, the final expression for $\boldsymbol{E}_1''$ is

$$\boldsymbol{E}_1'' = \frac{a^2}{a^2 + b^2}y\hat{x} - \frac{b^2}{a^2 + b^2}x\hat{y}. \tag{42}$$

Using the definition that $\boldsymbol{E}_1 = \boldsymbol{E}_1' + \boldsymbol{E}_1''$ and the expansion series for the internal electric field, the electric field to first order inside the ellipsoid for *EKH* polarization is

$$\begin{aligned}\boldsymbol{E} &= -jk(\boldsymbol{E}_1' + \boldsymbol{E}_1'')\\ &= -jk[(A_x + C_z y)\hat{x} + B_z x\hat{y}]\end{aligned} \tag{43}$$

where $A_x = -1/\sigma\eta_0 A_1$, $B_z = -b^2/(a^2 + b^2)$, $C_z = a^2/(a^2 + b^2)$, and A_1 is given in (34). The expression in (43) will be used for absorbed power calculation in the next section.

Expressions for $\boldsymbol{E}_1'$,$\boldsymbol{E}_1''$ and the internal electric field inside the ellipsoid to first order for the *EHK*, *KHE*, *KEH*, *HEK*, and *HKE* polarizations are derived by following the same procedure as described previously for *EKH* polarization. The final results for each of the aforementioned polarizations are given as follows.

B. Results for the Other Polarizations

EHK Polarization: The incident fields for this polarization are chosen to be $\boldsymbol{E}^i \parallel \hat{x}$, $\boldsymbol{H}^i \parallel \hat{y}$. The internal electric field to first order is

$$\boldsymbol{E} = -jk(\boldsymbol{E}_1' + \boldsymbol{E}_1'') = -jk[(A_x + B_y z)\hat{x} + C_y x\hat{z}] \tag{44}$$

where $B_y = a^2/(a^2 + c^2)$ and $C_y = -c^2/(a^2 + c^2)$.

KEH Polarization: The incident fields are $\boldsymbol{E}^i \parallel y$, $\boldsymbol{H}^i \parallel \hat{z}$. The internal electric field to first order is

$$\boldsymbol{E} = jk(\boldsymbol{E}_1' + \boldsymbol{E}_1'') = jk[(A_y + B_z x)\hat{y} + C_z y\hat{x}] \tag{45}$$

where $A_y = -1/\sigma\eta_0 A_2$, with

$$A_2 = \frac{abc}{2}\int_0^\infty \frac{d\xi}{(\xi + b^2)R_\xi}. \quad (46)$$

B_z, C_z, and R_ξ are given previously.

KHE Polarization: The incident fields are $\boldsymbol{E}^i \parallel \hat{z}$, $\boldsymbol{H}^i \parallel \hat{y}$. The internal electric field to first order is

$$\boldsymbol{E} = jk(\boldsymbol{E}_1' + \boldsymbol{E}_1'') = jk[(A_z + C_y x)\hat{z} + B_y z\hat{x}] \quad (47)$$

where $A_z = -1/\sigma\eta_0 A_3$, with

$$A_3 = \frac{abc}{2}\int_0^\infty \frac{d\xi}{(\xi + c^2)R_\xi}. \quad (48)$$

B_y, C_y, and R_ξ are given previously.

HEK Polarization: The incident fields are $\boldsymbol{E}^i \parallel \hat{y}$, $\boldsymbol{H}^i \parallel \hat{x}$. The internal electric field to first order is

$$\boldsymbol{E} = jk(\boldsymbol{E}_1' + \boldsymbol{E}_1'') = jk[(A_y + C_x z)\hat{y} + B_x y\hat{z}] \quad (49)$$

where $B_x = c^2/(b^2 + c^2)$, and $C_x = -b^2/(b^2 + c^2)$.

HKE Polarization: The incident fields are $\boldsymbol{E}^i \parallel \hat{z}$, $\boldsymbol{H}^i \parallel \hat{x}$. The internal electric field to first order is

$$\boldsymbol{E} = -jk(\boldsymbol{E}_1' + \boldsymbol{E}_2'') = -jk[(A_z + B_x y)\hat{z} + C_x z\hat{y}]. \quad (50)$$

The expressions for the first-order electric fields inside the ellipsoid for each of the six polarizations will be used in the next section to calculate the power absorbed by the ellipsoid.

III. Absorbed Power Calculations

For biological applications, it is very important to know the space density of absorbed energy rate or absorbed power, expressed in terms of watts per kilogram assuming a tissue density of 1 g/cm³.

Expressions for the first-order time-averaged specific absorbed power inside the ellipsoid is found by using the first-order internal fields given in the previous section. It should be noted here that these expressions are valid only if $\varepsilon_2 \gg \varepsilon_1$, which is the case of typical biological tissue at lower frequencies, ε_1 and ε_2 being the real and imaginary parts of the complex relative permittivity, respectively. The time-averaged specific absorbed power inside the ellipsoid is given by

$$P(x,y,z) = \tfrac{1}{2}\sigma \boldsymbol{E}\cdot\boldsymbol{E}^* \text{ W/m}^3 \quad (51)$$

and the space-averaged specific absorbed power is given by the volume integral

$$P_{av} = \frac{1}{V}\int_{z=-c}^{c}\int_{x=-a}^{a}\int_{-f(x,z)}^{f(x,z)} P(x,y,z)\, dx\, dy\, dz \quad (52)$$

where

$$f(x,z) = b\left(1 - \frac{x^2}{a^2} - \frac{z^2}{c^2}\right)^{1/2}$$

and $V = 4\pi abc/3$ is the volume of the ellipsoid.

Using the expressions for the first-order internal electric fields in (51) and (52) gives the following expressions for the time-averaged specific absorbed power and the space-averaged specific absorbed power for the six polarizations:

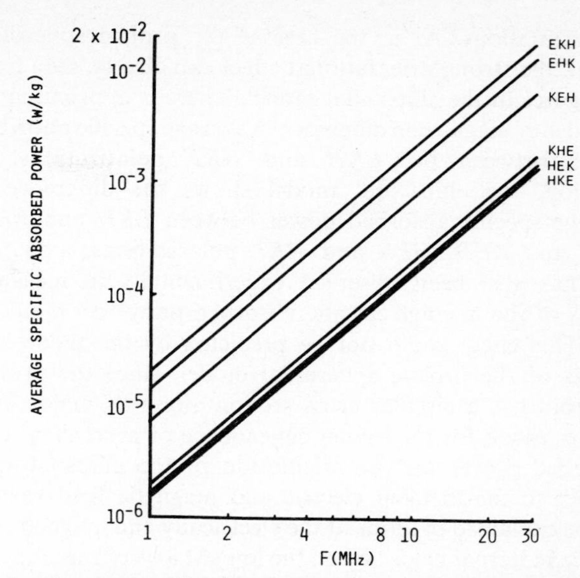

Fig. 1. Average specific absorbed power in an ellipsoidal model of man for the six standard polarizations. $a = 0.875$ m, volume = 0.07 m³, $b/c = 2.0$, $\sigma = 0.6$ mho/m a constant. Incident power density is 1 mW/cm².

EKH:

$$P(x,y,z) = D[(A_x + C_z y)^2 + B_z^2 x^2] \quad (53)$$

$$P_{av} = D\left(A_x^2 - \frac{a^2 B_z}{5}\right) \quad (54)$$

EHK:

$$P(x,y,z) = D[(A_x + B_y z)^2 + C_y^2 x^2] \quad (55)$$

$$P_{av} = D\left(A_x^2 + \frac{c^2 B_y}{5}\right) \quad (56)$$

KEH:

$$P(x,y,z) = D[(A_y + B_z x)^2 + C_z^2 y^2] \quad (57)$$

$$P_{av} = D\left(A_y^2 - \frac{a^2}{5} B_z\right) \quad (58)$$

KHE:

$$P(x,y,z) = D[(A_z + C_y x)^2 + B_y^2 z^2] \quad (59)$$

$$P_{av} = D\left(A_z^2 + \frac{c^2}{5} B_y\right) \quad (60)$$

HEK:

$$P(x,y,z) = D[(A_y + C_x z)^2 + B_x^2 y^2] \quad (61)$$

$$P_{av} = D\left(A_y^2 + \frac{b^2}{5} B_x\right) \quad (62)$$

HKE:

$$P(x,y,z) = D[(A_z + B_x y)^2 + C_x^2 z^2] \quad (63)$$

$$P_{av} = D\left(A_z^2 + \frac{b^2 B_x}{5}\right) \quad (64)$$

where $D = \frac{1}{2}\sigma E_0^2 k^2$, E_0 is the peak value of the incident electric field, and all the other parameters appearing in (53)–(64) are given in the previous section.

Some of the power absorption characteristics are shown in the curves in Figs. 1 and 2.

Fig. 1 shows the average specific absorbed power in an ellipsoidal model of man as a function of frequency for the six orientations. As in the case of the prolate spheroidal model, the strong orientational effect can also be seen from the figures in the ellipsoidal model. There is approximately an order of magnitude difference in average specific absorbed power between the *EKH* and *HKE* polarizations. In addition, the ellipsoidal model shows the difference in average specific absorbed power between *EKH* and *EHK*, *KHE* and *KEH*, *HEK* and *HKE* polarizations, a feature that has also been observed experimentally in measurements of the average specific absorbed power on monkeys [4]. This effect could not be predicted by the theoretical results of the prolate spheroidal model, since the prolate spheroid has a circular cross section along its major axis.

The reason for the strong dependence of average specific absorbed power on the orientation of the ellipsoid with respect to the incident electric and magnetic field vectors can be explained in terms of the electrically and magnetically induced internal fields. When the longest axis of the ellipsoid is aligned with the incident electric field, *EKH* and *EHK* polarizations, the magnitude of the internal electric field approaches that of the incident electric field because the fields are "mostly" tangential to the boundary and the boundary conditions require the tangential fields to be continuous. For the case when the incident electric field is along the shortest axis of the ellipsoid, *KHE* and *HKE* polarizations, the electric field coupling is much weaker as again expected from the field boundary conditions because, in this case, the fields are "mostly" normal to the boundary and the normal boundary conditions require the internal fields to be weaker. Therefore, the electrically induced fields can be classified as strong, intermediate, and weak for *EKH* and *EHK*, *KEH* and *HEK*, *KHE* and *HKE* polarizations, respectively. The internal electric field induced by the incident magnetic field forms loops about the incident magnetic field vector corresponding to eddy currents, and the strength of this magnetically induced electric field appears to be in some sense proportional to the ellipsoidal cross-sectional area perpendicular to the incident magnetic field. Thus strong magnetic coupling occurs when the incident magnetic field vector is along the shortest axis of the ellipsoid. *KEH* and *EKH* polarizations. The magnetically induced fields can also be classified as strong, intermediate, and weak for *EKH* and *KEH*, *EHK* and *KHE*, *HEK* and *HKE* polarizations, respectively.

From the foregoing discussion, one would expect the average specific absorbed power to be the greatest for the *EKH* polarization because both electric and magnetic field coupling are strong, and the average specific absorbed power for *HKE* polarization to be the smallest because both electric and magnetic coupling are weak. This is indeed the case as shown in Figs. 1 and 2. The intermediate steps in terms of absorbed power, *EHK*, *KEH*, *KHE*, and *HEK* polarizations, are the result of the various combinations of electric and magnetic coupling.

Fig. 2 illustrates the normalized average specific absorbed power in an ellipsoidal model of man for each polarization, as a function of b/c. It is interesting to note that the average specific absorbed power, for the *EKH* and *KEH* polarizations, increases as the b/c ratio increases. This occurs because the increase in b/c ratio has two effects. The first is that the cross-sectional area normal to the incident magnetic field increases, for the two polarizations, causing an increase in the strength of the magnetically induced $\boldsymbol{E}$ field. The second effect is that an increase in the b/c ratio thins out the ellipsoid and makes it less shielded from the incident $\boldsymbol{E}$ field and therefore produces a strengthened electrically induced $\boldsymbol{E}$ field, specially for the *EKH* polarization. For $b/c = 1$, the ellipsoid takes the shape of a prolate spheroid, and for this value of b/c the results of the calculations of absorbed power in ellipsoids are the same as those previously obtained for a prolate spheroid having the same volume and the same height as the ellipsoid [1].

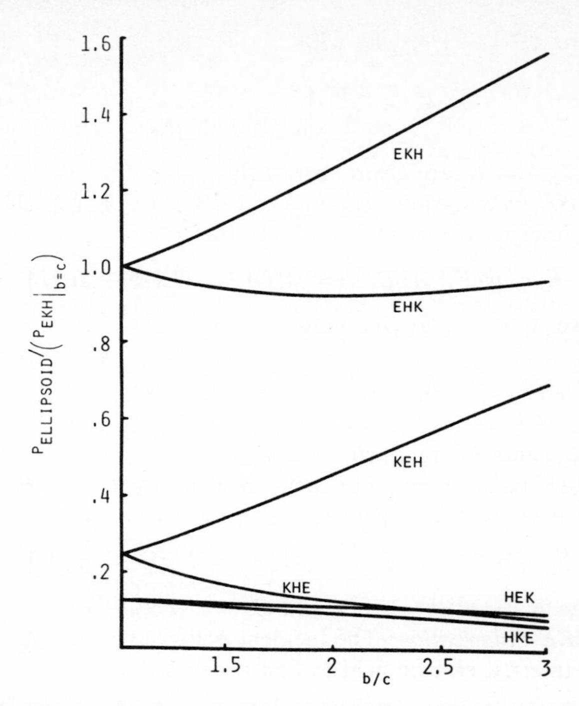

Fig. 2. Relative absorbed power in ellipsoids as a function of polarization and b/c. $a = 0.875$ m, volume $= 0.07$ m^3, frequency $=$ 10 MHz, $\sigma = 0.6$ mho/m.

Theoretical results have also been obtained for the ellipsoidal model of a sitting monkey, with good qualitative agreement with measurements of power absorbed by live monkeys made at Brooks Air Force Base [4]. The relatively good qualitative agreement between the theoretical and experimental data indicates that the ellipsoid is a better model for some experimental animals and man than the prolate spheroid.

A better quantitative agreement between the theory and experiment might be achieved by choosing the optimum combination of the dimensions of the ellipsoidal model, *a*, *b*, and *c*, and the electrical properties of the model to best fit the animal data.

IV. Summary and Conclusions

The first-order internal electric fields and specific absorbed power in ellipsoidal models of man and experimental animals irradiated by an EM plane wave have been obtained for the case when the wavelength is long compared to dimensions of the ellipsoid $a/\lambda < 0.1$. The expressions for the internal fields and specific absorbed power, and the curves of specific absorbed power versus frequency, show that the internal electric fields and the specific absorbed power depend on the body's dielectric properties and geometry, as well as the frequency and polarization of the incident wave.

Comparison of the theoretical specific absorbed power in prolate spheroidal and ellipsoidal models with the corresponding experimental data on live monkeys indicates that the prolate spheroidal models are not adequate for primates, although they may be for rodents since their cross sections taken normal to the long axis appear to be approximately circular. However, for primates, with approximately elliptical cross sections, the ellipsoidal model would obviously be a superior representation.

The expressions for the internal electric fields, specific absorbed power, and space-average specific absorbed power should prove to be very valuable in studies of radiation hazards to man for the long-wavelength case. A very important application of this analysis will be in the extrapolation to man of the results of animal experiments involving biological effects due to EM radiation. Since the results of this analysis show marked differences in EM absorption characteristics for man compared to that of animals at the same frequency and same incident field level, in extrapolating animal effects to man it will be necessary to relate the biological effects to the internal fields or power absorption and then relate the internal fields or power absorption to the incident fields.

In a future communication, we will apply this analysis to obtain data showing the internal specific absorbed power distribution and average specific absorbed power in different test animals and different human body types.

Appendix

The constants A_1, A_2, and A_3 occurring in (34), (46), and (48) are related to semiprincipal axes of the ellipsoid a, b, and c, and to the incomplete elliptic integrals of the first and second kinds [9]. These relations are as follows:

$$A_1 = abc\left(\frac{[F(\phi,k) - E(\phi,k)]}{[(a^2 - b^2)(a^2 - c^2)^{1/2}]}\right) \tag{A1}$$

$$A_2 = abc(a^2 - c^2)^{1/2} \cdot \left[\frac{[E(\phi,k) - (b^2 - c^2)F(\phi,k)/(a^2 - c^2) - ak^2 \sin\phi \cos\phi/b]}{[(a^2 - b^2)(b^2 - c^2)]}\right] \tag{A2}$$

$$A_3 = abc\left[\frac{[b \tan\phi/a - E(\phi,k)]}{[(b^2 - c^2)(a^2 - c^2)^{1/2}]}\right] \tag{A3}$$

with

$$F(\phi,k) = \int_0^\phi (1 - k^2 \sin^2\theta)^{-1/2}\, d\theta \tag{A4}$$

$$E(\phi,k) = \int_0^\phi (1 - k^2 \sin^2\theta)^{1/2}\, d\theta \tag{A5}$$

and

$$k = \left(\frac{a^2 - b^2}{a^2 - c^2}\right)^{1/2} \tag{A6}$$

$$\phi = \sin^{-1}\left(\frac{a^2 - c^2}{a^2}\right)^{1/2} \tag{A7}$$

where $F(\phi,k)$ and $E(\phi,k)$ are the incomplete elliptic integrals of the first and the second kinds, respectively. In this Appendix we made use of the customary symbol k for modulus of these elliptic integrals, and it should not be confused with the parameter k used for the free-space propagation constant in the main body of this paper. It can be shown that the order of the relative magnitude of the constants A_1, A_2, and A_3 is the inverse of the order of the three parameters a, b, and c. That is, if $a > b > c$, then $A_1 < A_2 < A_3$. Furthermore, one finds that $A_1 + A_2 + A_3 = 1$.

References

[1] C. H. Durney, C. C. Johnson, and H. Massoudi, "Long-wavelength analysis of plane wave irradiation of a prolate spheroid model of man," *IEEE Trans. Microwave Theory Tech.*, vol. MTT-23, pp. 246–253, Feb. 1975.

[2] C. C. Johnson, C. H. Durney, and H. Massoudi, "Long-wavelength electromagnetic power absorption in prolate spheroidal models of man and animals," *IEEE Trans. Microwave Theory Tech.*, vol. MTT-23, pp. 739–747, Sept. 1975.

[3] O. P. Gandhi, "Frequency and orientation effects on whole animal absorption of electromagnetic waves," *IEEE Trans. Biomed. Eng.* (Commun.), vol. BME-22, pp. 536–542, Nov. 1975.

[4] S. J. Allen, W. D. Hurt, J. H. Krupp, J. A. Ratliff, C. H. Durney, and C. C. Johnson, "Measurement of radio frequency power absorption in monkeys, monkey phantoms and human phantoms exposed to 10–50 MHz fields," *Proc. 1975 Annual USNC-URSI Meeting*, Boulder, CO, p. 200, Oct. 1975.

[5] J. A. Stratton, *Electromagnetic Theory.* New York: McGraw-Hill, 1941.

[6] E. T. Wittaker and G. N. Watson, *A Course of Modern Analysis*, New York: Macmillan, 1946.

[7] P. M. Morse and H. Feshbach, *Methods of Theoretical Physics*, Part II. New York: McGraw-Hill, 1953.

[8] M. R. Spiegel, *Mathematical Handbook of Formulas and Tables* (Schaum's Outline Series). New York: McGraw-Hill, 1968, p. 124.

[9] P. F. Byrd and M. D. Friedman, *Handbook of Elliptic Integrals for Engineers and Scientists*, 2nd Edition, Revised. New York, Heidelberg, Berlin: Springer-Verlag, 1971.

A Probe for Measuring Temperature in Radio-Frequency-Heated Material

RONALD R. BOWMAN

***Abstract*—Measuring temperature in material being heated by radio-frequency (RF) fields is difficult because of field perturbations and direct heating caused by any conventional leads connected to the temperature sensor. A temperature probe consisting simply of a thermistor and plastic high-resistance leads appears to practically eliminate these problems. The design goals are described, and the performance of an initial test model of this type of probe is discussed.**

INTRODUCTION

For bioeffects research and the control of potentially hazardous electromagnetic fields, a need exists to measure temperature in subjects and models during exposure to intense fields [1]–[7]. This problem would be trivial except for the fact that conventional thermocouples and thermistors use leads that grossly distort the internal field structure and also produce intense heating directly due to the induced radio-frequency (RF) currents [1]–[7]. One solution to this problem utilizes fiber optics coupled to a liquid-crystal temperature transducer [6]. Another approach uses a prepositioned, electrically nonconductive well that allows rapid insertion of the temperature sensor after the field source is turned off [2], [4]. Others have developed a probe consisting of a Wheatstone bridge circuit, extremely fine electrodes connected to a thermistor, and high-resistance plastic leads [5]. The temperature probe described here also uses a thermistor but is simpler in design (see the next section) and should produce considerably less heating (due to the use of higher resistance leads). As of this writing, parts are being obtained and fabricated to construct a probe with a 1-mm-OD tube, a thermistor with dimensions less than 0.5 mm, and high-resistance plastic leads with resistances of about 160 kΩ/cm.[1] This short paper describes the test results for a probe that was made from parts that were immediately available. Since this initial model of this type of probe has much greater sensitivity, stability, and dynamic range than the probe described in [6], it is believed that these early test results are of interest.

Manuscript received May 16, 1975; revised July 14, 1975. This work is a contribution of the National Bureau of Standards, not subject to copyright.

The author is with the Electromagnetics Division, U. S. Department of Commerce, National Bureau of Standards, Boulder, CO 80302.

[1] *Note Added During Review:* This probe has been fabricated. It has a response-time constant of less than 0.2 s, short-term stability better than 0.01°C, and a high-resistance-line heating error (see section on experimental tests) of less than 0.005°C for a heating rate of 1°C/min.

PROBE DESIGN AND CONSTRUCTION

As shown in Fig. 1, the probe consists simply of two pairs of very-high-resistance leads connected to a small high-resistance thermistor (about 750 kΩ at 25°C and a coefficient $\simeq -0.04/°C$). The thermistor resistance is sensed by injecting a constant current through one pair of leads and measuring the voltage developed across the thermistor by means of a high-impedance amplifier connected to the other pair. If the current generator and amplifier have high impedances compared to the leads, the thermistor can be measured accurately despite the large and unstable lead resistances. This technique is commonly used when the lead resistance is significant, but in the present application the lead resistances will typically be 10 MΩ rather than the usual lead resistances that are of the order of 10 mΩ. The main difficulties in realizing good probes of this type are fabricating high-resistance lines with lineal resistances of 100 or more kilohms per centimeter and attaching these leads reliably to the thermistor.

Leads with the required high resistance can be made by either thick- or thin-film processes, but it may be difficult to make long leads using these processes. The present probe design uses plastic high-resistance leads developed earlier for use with electromagnetic hazard meters [8], [9]. For the initial test model, the cross section of the leads is about 0.25 by 0.25 mm and their lineal resistance is about 40 kΩ/cm. The leads are bonded to the thermistor with silver-loaded epoxy.

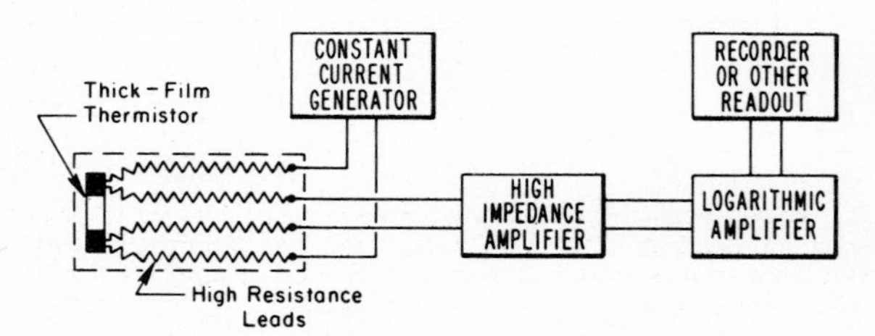

Fig. 1. Schematic of probe and associated electronics.

Reprinted from *IEEE Trans. Microwave Theory Tech.*, vol. MTT-24, pp. 43–45, Jan. 1976.

EXPERIMENTAL TESTS

For the following tests, the current generator was set to provide about 0.35 μA, which generates less than 0.1 μW of heating in the thermistor and about 0.2 μW of heating along the high-resistance leads. At 25°C this current causes about a 0.25-V drop across the thermistor. The signal from the voltage amplifier was passed through a logarithmic amplifier to provide a signal that is nearly linear with temperature change. After calibration, the probe was placed in a constant-temperature bath. The noise and drift on the signal to the recorder corresponded to less than ±0.01°C. When dipped into a beaker of water, the probe displayed a 90-percent response time of about 10 s. This rather slow response is due in part to the use of an unneccessarily large (3-mm-OD) probe-tube and in part to the fact that the thermistor was not potted into the tube.

The probe was placed as shown in Fig. 2 and exposed to 2.0-GHz fields. The high-resistance lines were parallel to the incident electric field to maximize the RF heating of these lines. With the water flow off, the RF power was adjusted to give approximately 1°C/min initial heating rate in the water at the probe tip when the RF power was turned on. (This heating rate requires about 70-mW/cm^3 power absorption and corresponds to a relatively high internal-field exposure for electromagnetic bioeffects experiments.) As will be seen, some thermistor heating due to heat generation in the high-resistance lines is apparent in the thermistor heating curve following a turn-on of the RF power.[2]

An approximate expression for the response of the thermistor without the high-resistance leads can be determined as follows. The heating rate in the water should be nearly constant until the water temperature rises enough to establish strong convection currents. Assuming a constant heating rate S in the water and that the thermistor heating rate is proportional to the difference between the thermistor temperature and the water temperature, it is easy to show that the thermistor temperature change T is given by

$$T = St - S\tau[1 - \exp(-t/\tau)] \qquad (1)$$

where the water heating begins at time $t = 0$ and τ is the time constant of the probe for still water. Since τ is about 4.3 s for this probe, a 1°C/min heating rate in the water will correspond to an eventual lag of about $S\tau = 0.07$°C in the thermistor response.

Fig. 3 shows the thermistor response as predicted by (1) for 1 min of 1°C/min water heating. The actual response of the thermistor is shown by the "water-flow-off" curve of Fig. 4. Apparently, the inherent lag of the thermistor temperature is more than compensated by the RF-generated heat from the high-resistance lines. The excess temperature can be shown by establishing a rapid water flow rate through the tube. With the same RF power level and using a water velocity of more than 30 cm/s, the temperature rise of the flowing water is less than 0.01°C. Then the thermistor heating above the constant water temperature is shown by the water-flow-on curve of Fig. 4 and is about 0.12°C. This error is probably not important, and it will be much less for the fully developed probes.

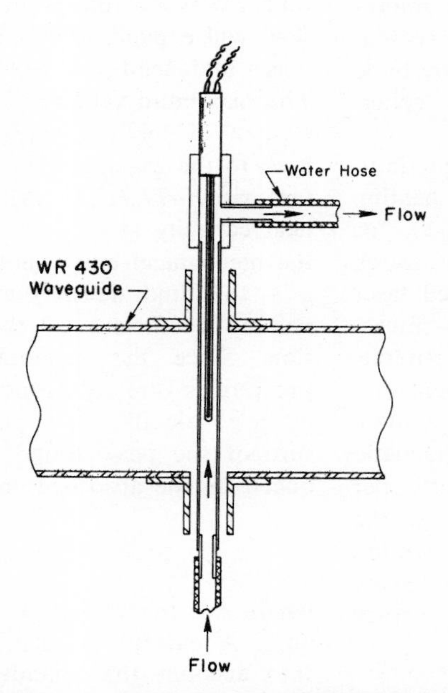

Fig. 2 Experimental arrangement for measuring the temperature error due to RF heating of the high-resistance leads. The probe is inserted into a thin-walled plastic tube filled with water and aligned parallel to the electric vector in a WR 430 waveguide.

PERTURBATION OF THE INTERNAL FIELD

In addition to the direct heating of the thermistor by the leads, significant errors can occur because the field structure in the material of interest will be different with the probe in place [2], [4]–[6]. Metallic leads are particularly troublesome because of their very high conductivity. Even Nichrome has a conductivity of about 10^6 S/m. For comparison, muscle tissue has a conductivity of about 1 S/m. An effective solution to the field-perturbation problem, as well as the direct-heating problem, is to use leads with a conductivity comparable to that of the subject or model material of interest; e.g., saline-filled glass tubes [2, p. 705]. The probes described here use leads with a conductivity of only 4 S/m,[3] and most of the bulk

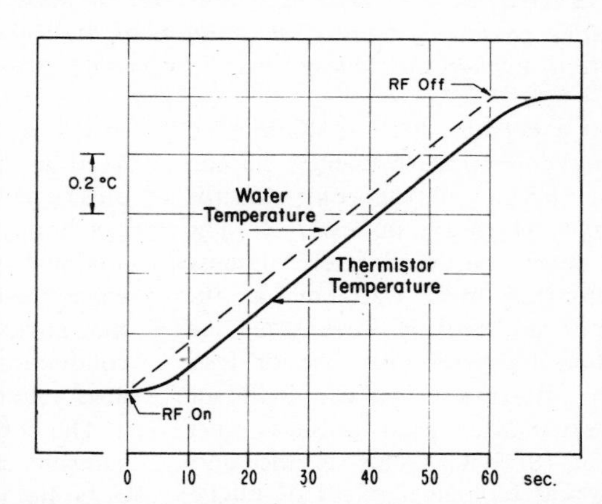

Fig. 3. Theoretical thermistor heating curve for the case of no RF heating of the high-resistance leads.

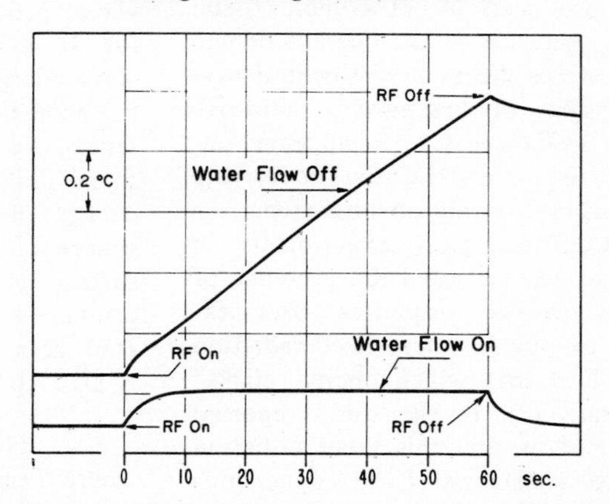

Fig. 4. Measured probe response with water flow off and on, with the RF power level constant. With the water flow on, the probe response of about 0.12°C is due to RF heating of the high-resistance lines.

[2] The induced RF currents in the thermistor will also cause the thermistor to heat; but, because the thermistor material has much lower electrical conductivity than the lead material, the heat generated directly in the thermistor is believed to be negligible.

[3] Carbon-loaded material with less conductivity can no doubt be made, but does not appear to be readily available.

of the probe consists of nonconducting material. The field-perturbation effects of the initial model were not determined; however, thermographic [2, p. 705] tests for these effects will be performed for the next model to quantify any errors associated with the leads or the small amounts of highly conducting material used to make contact with the thermistor.

REFERENCES

[1] C. C. Johnson, "Research needs for establishing a radio frequency electromagnetic radiation safety standard," *J. Microwave Power*, vol. 8, pp. 367–388, Nov. 1973.
[2] C. C. Johnson and A. W. Guy, "Nonionizing electromagnetic wave effects in biological materials and systems," *Proc. IEEE*, vol. 60, pp. 692–718, June 1972.
[3] W. H. Vogelman, "Microwave instrumentation for the measurement of biological effects," in *Biological Effects of Microwave Radiation*, vol. I, M. R. Peyton, Ed. New York: Plenum, 1961, pp. 29–31.
[4] A. W. Guy, F. A. Harris, and H. S. Ho, "Quantification of the effects of microwave radiation on central nervous system function," in *Proc. 6th Annu. Int. Microwave Power Symp.* (Monterey, Calif., May 1971).
[5] L. E. Larsen, R. A. Moore, and J. Acevedo, "A microwave decoupled brain-temperature transducer," *IEEE Trans. Microwave Theory Tech.*, vol. MTT-22, pp. 438–444, Apr. 1974.
[6] T. C. Rozzell, C. C. Johnson, C. H. Durney, J. L. Lords, and R. G. Olson, "A nonperturbing temperature sensor for measurements in electromagnetic fields," *J. Microwave Power*, vol. 9, Sept. 1974.
[7] R. D. McAfee, L. L. Cazenavette, and H. A. Shubert, "Thermistor probe error in an *X*-band microwave field," *J. Microwave Power*, vol. 9, Sept. 1974.
[8] F. M. Greene, "NBS field-strength standards and measurements (30 Hz to 1000 MHz)," *Proc. IEEE (Special Issue on Radio Measurement Methods and Standards)*, vol. 55, pp. 970–981, June 1967.
[9] R. R. Bowman, "Some recent developments in the characterization and measurement of hazardous electromagnetic fields," in *Proc. Int. Symp. Biologic Effects and Health Hazards of Microwave Radiation* (Warsaw, Poland), Oct. 15–18, 1973, pp. 217–227.

Microwave Hearing: Evidence for Thermoacoustic Auditory Stimulation by Pulsed Microwaves

Abstract. *Acoustic transients can be thermally generated in water by pulsed microwave energy. The peak pressure level of these transients, measured within the audible frequency band as a function of the microwave pulse parameters, is adequate to explain the "clicks" heard by people exposed to microwave radiation.*

When a person's head is illuminated with pulsed microwave energy, he can perceive "clicks" in synchrony with the individual microwave pulses (*1–3*). The pulses must be moderately intense (typically 0.5 to 5 watt/cm^2 at the surface of the head). However, they can be sufficiently brief (50 μsec or less) that the maximum increase in tissue temperature after each pulse is very small ($< 10^{-5}$ °C). This is the only unequivocal biological effect of microwave radiation that is not accompanied by or produced by observable tissue heating. Because of the current debate over possible effects on the central nervous system of low-power, radio-frequency radiation (*2*), it appears important to understand the underlying mechanisms for this phenomenon.

Electrophysiological experiments in cats have demonstrated the presence of auditory evoked responses after exposure to pulsed microwave radiation identical to that which elicits "clicks" in humans (*3*). In the study reported here we show that this same radiation generates substantial (>10 dyne/cm^2) sound transients in water, the major constituent of soft tissue. The peak pressure level of these transients, measured within the audible frequency band as a function of the microwave pulse parameters, is consistent with psychophysical observations of the loudness and threshold of this microwave "hearing" effect. We believe that the "clicks" are the perception, by bone conduction, of these thermally generated sound transients.

The conversion of electromagnetic to acoustic energy by the surface heating of a liquid is well known (*4, 5*), for example, in connection with shock waves produced by a *Q*-switched laser (*6*). It is apparent that pulsed microwave energy is also capable of generating acoustic transients in absorbent materials. To illustrate this effect, assume that a uniform beam of electromagnetic energy of intensity I_0 (in watts per square centimeter) is directed at the surface of a fluid with an absorption coefficient α. The radiation intensity $I(x)$ at a distance x from the surface is given by

$$I(x) = I_0 T \exp(-\alpha x)$$

where T is the fraction of the incident energy transmitted into the fluid, the remaining energy being reflected. For 2450-Mhz microwave energy incident upon physiological saline (or upon soft tissue) at 37°C, T and $1/\alpha$ are approximately 0.4 and 1 cm, respectively (*7*). Assume that the beam is turned on at time $t = 0$ and off at time $t = \tau$. The fluid will expand as it is heated by this pulse and send out a pressure wave. The maximum velocity, u, of any small element of the fluid will be proportional both to the maximum rate of temperature rise ($\alpha T I_0 / C_p J \rho$, where C_p is the heat capacity at constant pressure, J is the mechanical equivalent of heat, and ρ is the fluid density) and to β, the volume coefficient of thermal expansion. Since the instantaneous sound pressure is directly proportional to the particle velocity, an approximate measure of the peak sound pressure produced by the absorbed energy is (*5*)

$$P_0 = \frac{c \beta I_0 T}{C_p J} \quad (1)$$

where c is the velocity of sound in the fluid. A careful calculation must take into account the boundary conditions upon the fluid. It can be shown (*5*) that a pulse of electromagnetic energy of duration τ directed at a free (that is, unconstrained) fluid surface should produce both a positive and a negative

Reprinted with permission from *Science*, vol. 185, no. 4147, pp. 256–258, July 19, 1974.

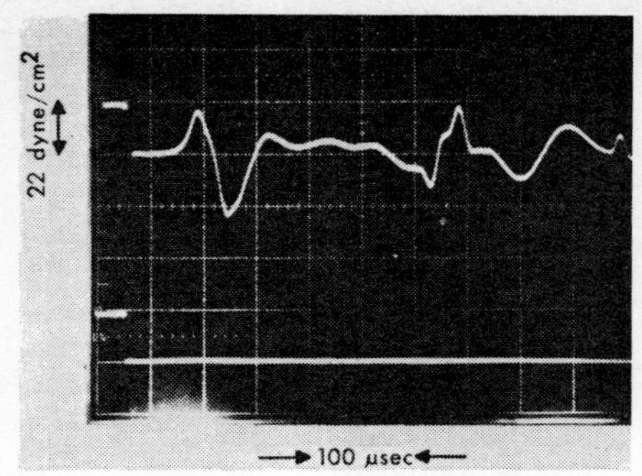

Fig. 1. Oscilloscope traces showing the acoustic transient (upper trace) produced by a 27-μsec, 2450-Mhz microwave pulse of intensity 5.3 watt/cm² (lower trace) incident upon 0.15*N* KCl solution at 25°C. The solution was contained in a cubic Lucite tank, 30 cm on a side. The latency between the microwave pulse and the first recorded transient is equal to the propagation time of sound from the front surface of the tank to the hydrophone; a reflection from the rear surface of the tank is also shown. This microwave pulse would elicit a "click" in most subjects.

pressure transient, of peak amplitude

$$P_{max} = \pm (P_0/2)[1 - \exp(-\alpha c \tau)] \quad (2)$$

corresponding to the leading and trailing edges of the pulse, respectively. For incident energy of intensity 1 watt/cm² totally absorbed by water at 37°C, $(P_0/2)$ is approximately 6.5 dyne/cm².

We have verified these predictions, using a sensitive, electrically well-shielded hydrophone (an experimental model on loan from Chesapeake Instrument Corporation) and containers of various geometries filled with 0.15*N* KCl solution. A microwave generator (Applied Microwave Laboratories model PH-40) coupled to a standard gain horn (Waveline model 299) was used to produce pulses of 2450-Mhz radiation with a maximum peak intensity of approximately 10 watt/cm² and widths of 2 to 27 μsec. The microwave power density was measured with a probe (Narda model 8300). All measurements were carried out in an anechoic chamber lined with microwave-absorbent material.

A typical result showing positive and negative pressure transients with P_{max} = 20 dyne/cm² is given in Fig. 1. The solution, at 25°C, was contained in a cubic Lucite tank, 30 cm on a side. The 27-μsec pulse had a peak intensity of approximately 5.3 watt/cm² at the fluid surface. Most persons exposed to this radiation would hear a distinct "click." If 60 percent of the incident microwave power is assumed to be reflected by the water surface, Eq. 2 predicts a peak sound pressure of 10 dyne/cm²; considering the relatively large errors inherent in microwave power measurements and the nonuniform surface heating produced by the diverging beam, the agreement between Fig. 1 and the theory is satisfactory. The shape of the transient resembles that predicted in (*5*) but is distorted by the presence of the Lucite wall. That the signal is due to thermal expansion of the water was confirmed by using a sample of distilled water and cooling the solution. Between 0° and 4°C the signal was inverted and at 4°C the signal vanished, in agreement with the temperature dependence of the thermal expansion coefficient of water. We have also observed in vitro acoustic transients produced by microwave pulses in such tissues as blood, muscle, and brain; similar transients have been observed in vitro in tissue after irradiation by a *Q*-switched laser (*8*).

The first reflection from the rear surface of the tank is shown in Fig. 1. Superficially, the initial transient and the following series of reflections (the "flutter echo") resemble a burst of white noise, which is exponentially damped with a time constant of about 3 msec as the sound energy is transferred to the walls of the container. The frequency spectrum of the burst depends upon the microwave pulse length and upon the geometry of the container; shortening the microwave pulse similarly shortens the initial transient but does not alter the decay time of the reflected pulses.

Using a variable band-pass filter (Kron-Hite), we have measured the band pressure levels of the acoustic transients that follow microwave pulses of different widths (Fig. 2). The solution at 25°C was contained in a large expanded polystyrene tank. The pressure levels were increased by 2.9 db to simulate measurements at 37°C. The incident total energy density per pulse, $I_0\tau$, was kept constant at 80 μj/cm². This is the approximate average threshold for microwave "hearing" in human subjects with normal hearing exposed to 2450-Mhz radiation, for pulses shorter than about 25 μsec repeated at a rate of three per second (*3*). For relatively long pulses, Fig. 2 shows that the P_{max} of the burst is directly proportional to I_0, consistent with Eq. 2. For shorter pulses, P_{max} depends upon $I_0\tau$ and upon the filter bandwidth. This is due to the exponential factor in Eq. 2 and to the broadened spectral distribution of the transients. Within the frequency band audible by bone conduction (200 hertz to 20 khz), the transition occurs at pulse widths of 20 to 25 μsec. In a psychophysical study Frey and Messenger showed, for pulsed 1245-Mhz radiation, that the loudness of the microwave "hearing" sensation depends only upon I_0 for pulse widths greater than about 30 μsec; for shorter pulses, their data show that the loudness is a function of the product, $I_0\tau$

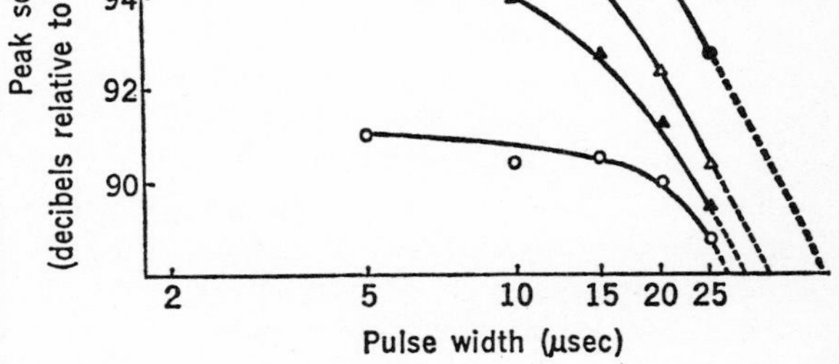

Peak sound pressure (decibels relative to 0.0002 dyne/cm² in H_2O)

Pulse width (μsec)

Fig. 2. The peak sound pressure of the microwave-generated acoustic transient, as a function of the microwave pulse width and filter bandwidth. Filter bandwidth: ●, 200 hertz to 60 khz; △, 200 hertz to 40 khz; ▲, 200 hertz to 30 khz; ○, 200 hertz to 20 khz. The 0.15*N* KCl solution was in a large rectangular expanded polystyrene tank, at 25°C. The sound pressures were increased by 2.9 db to simulate measurements at 37°C. The dotted lines indicate the function $P_0 \propto I_0$ (Eq. 1). This figure shows that, for short pulses, $P_0 \propto I_0\tau$, in agreement with psychophysical observations of the microwave "hearing" effect. The microwave intensity was adjusted so that the incident energy density per pulse was 80 μj/cm², the threshold for microwave "hearing" in an average subject.

(*1*). These results are consistent with our observations.

We observe that in water a microwave pulse, at the threshold for microwave "hearing" in humans, produces pressure transients of approximately 90 db relative to 0.0002-dyne/cm^2 peak amplitude within the frequency band 200 hertz to 20 khz. If occurring within the head, this stimulus could elicit a "click" when some of the sound energy is coupled into the skull. Neither the duration of the flutter echo produced inside the head by a microwave pulse nor the threshold for perception of this unusual stimulus is known. However, 90 db is above the expected threshold (~ 80 db) for perception, by bone conduction, of millisecond bursts of white noise incident upon a subject's skull in water (*9*). It is therefore reasonable to believe that these thermally induced transients elicit the microwave "hearing" sensations in humans.

KENNETH R. FOSTER
EDWARD D. FINCH
Naval Medical Research Institute, National Naval Medical Center, Bethesda. Maryland 20014

References and Notes

1. A. H. Frey and R. Messenger, Jr., *Science* **181**, 356 (1973); A. H. Frey, *J. Appl. Physiol.* **17**, 689 (1962).
2. A. H. Frey, *Inst. Elec. Electron. Eng. Trans. Microwave Theory Tech.* **MTT-19**, 153 (1971).
3. A. W. Guy, E. M. Taylor, B. Ashleman, J. C. Lin, *Inst. Elec. Electron. Eng. Int. Symp. Digest Tech. Pap. G-MTT* (1973), p. 321; A. W. Guy, C. K. Chou, J. C. Lin, D. Christensen, *Proc. N.Y. Acad. Sci.*, in press.
4. R. M. White, *J. Appl. Phys.* **34**, 3559 (1963).
5. L. S. Gournay, *J. Acoust. Soc. Am.* **40**, 1322 (1966). The quantity β is the volume coefficient of thermal expansion, not the linear coefficient as stated by Gournay. This can be demonstrated by an independent derivation of the fundamental differential equation for thermally driven pressure waves in fluids, equation 9 of this reference, from the first-order acoustic equations for a fluid [see F. A. Firestone, in the *American Institute of Physics Handbook*, D. E. Gray, Ed. (McGraw-Hill, New York, 1957), p. 3-33.
6. C. L. Hu, *J. Acoust. Soc. Am.* **46**, 728 (1969).
7. H. N. Kritikos and H. P. Schwan, *Inst. Elec. Electron. Eng. Trans. Biomed. Eng.* **BME-19**, 53 (1972).
8. S. F. Cleary and P. E. Hamrick, *J. Acoust. Soc. Am.* **46**, 1037 (1969).
9. W. R. Thurlow and R. Bowman, *ibid.* **29**, 281 (1957); P. M. Hamilton, *ibid.*, p. 792.
10. We thank the Chesapeake Instrument Corporation of Shadyside, Maryland, for the generous loan of the hydrophone used in this work. Supported by the Bureau of Medicine and Surgery Work Unit No. MF51.524.-015.0018BE7X. The opinions or assertions contained herein are the private ones of the authors and are not to be construed as official or reflecting the views of the Navy Department or the naval service at large.

12 March 1974

An Empirical Formula for Broad-Band SAR Calculations of Prolate Spheroidal Models of Humans and Animals

CARL H. DURNEY, MEMBER, IEEE, MAGDY F. ISKANDER, HABIB MASSOUDI, MEMBER, IEEE, AND C. C. JOHNSON

Abstract—**An empirical relation for calculating approximate values of the average specific absorption rate (SAR) over a broad-frequency range for any prolate spheroidal model is derived for *E*-polarized incident plane waves. This formula provides a simple and inexpensive method for calculating the SAR for human and animal models, which otherwise requires complicated and expensive methods of calculation. The formula satifies the f^2 SAR behavior at lower frequencies, the resonance characteristic at intermediate frequencies, the $1/f$ behavior past resonance, and the dependence on the dielectric constant at the geometrical optics limits. An expression for the resonance frequency f_0 in terms of the dimensions of the model is also derived. The unknown expansion coefficients were determined by curve-fitting all the data available in the second edition of the *Radiofrequency Radiation Dosimetry Handbook*. Numerical results obtained from the empirical relations are generally in good agreement with those calculated by other methods. Limitations of the formula and suggestions for its improvement are also discussed.**

I. Introduction

THEORETICAL STUDIES of the specific absorption rate (SAR) of electromagnetic (EM) energy by biological models have been of increasing interest in recent years because of the continuing need to evaluate the hazardous levels of the EM waves, and to refine the presently available safety standards. Of particular interest is the analysis of the prolate spheroidal models of man and animal that have been shown to give results that correlate well with those of more realistic models, as well as with the experimental results. Such calculations, however, are expensive and complex, particularly at frequencies near resonance. They also involve several theoretical methods, each one valid in a limited frequency range. Consequently, the *Radiofrequency Radiation Dosimetry Handbook* was published to provide average SAR values over a very wide frequency range for many animal and human models [1]. In many laboratory experiments, however, it often happens that an animal of different type or size than those specific cases included in the Handbook is used. In this case the researcher is left with the choice of either extrapolating the data available in the Handbook or calculating the SAR for the specific case of interest. The former is an inconvenient approximation of limited accuracy, while the latter is not only expensive and time consuming, but also beyond the interests or capabilities of many research organizations.

It is, therefore, desirable to have a simple method of calculating the average SAR over a broad range of frequencies. Such a method would be very valuable even if it were only to give results within 10 or 15 percent of the values calculated by more sophisticated techniques, since even the most accurate methods of calculation are based on models of humans and animals that are quite approximate. In the sequel, an empirical formula for calculating the average SAR over a very broad frequency range, for prolate spheroidal models of any human or animal, is developed using a combination of antenna theory, circuit theory, and curve fitting. The formula is derived only for the *E*-polarized incident plane waves, since it has been shown [1] that the highest SAR occurs for *E* polarization (incident electric field parallel to the major axis of the spheroid), and the case of greatest SAR is most important for evaluation of possible hazards.

Manuscript received August 28, 1978; revised April 9, 1979. This work was supported by the USAF School of Aerospace Medicine, Brooks Air Force Base, under Contract F41609-76-C-0025/P00004.

C. H. Durney, M. F. Iskander, and H. Massoudi are with the Department of Electrical Engineering and the Department of Bioengineering, University of Utah, Salt Lake City, UT 84112.

C. C. Johnson, deceased, was with the Department of Electrical Engineering and the Department of Bioengineering, University of Utah, Salt Lake City, UT 84112.

II. Formulation of the Equation

Consider a prolate spheroidal model of a semimajor axis a and a semiminor axis b. In deriving a simple empirical formula that characterizes the average SAR as a function of frequency, it is important to take into account the following characteristics that are found to be common among the SAR's for free-space irradiation by an *E*-polarized incident plane wave.

1) For $a<\lambda/10$, where λ is the free-space wavelength, the SAR is approximately proportional to f^2. This f^2 behavior is exact for constant conductivity and can be derived from the long-wavelength approximation [2].
2) Each model has a resonant frequency at which the maximum absorption of the incident RF power occurs. The resonant frequency depends basically on the a and b of the model [3].
3) The SAR increases faster than f^2 just below resonance. Beyond resonance it is found from the experimental data that the average SAR decreases ap-

Reprinted from *IEEE Trans. Microwave Theory Tech.*, vol. MTT-27, pp. 758–763, Aug. 1979.

proximately as $1/f$ [4]. The latter behavior is valid up to a high-frequency limit that varies with the dimensions of the model. For a spheroidal model of man size, for example, the $1/f$ behavior is expected to be valid for frequencies up to 6.7 f_0, when f_0 is the resonant frequency [5].

4) At very high frequencies, where the wavelength is much smaller than the size of the irradiated object, the geometrical optics approximation is valid. In this case, the SAR does not depend on the frequency, but varies only with the permittivity ϵ.

A simple formula that satifies all of the above requirements is given by

$$\mathrm{SAR}=\frac{A_1 f^2/f_0^2\left[1+A_3(f/f_0)u(f-f_{01})+A_4A_5(f^2/f_0^2)u(f-f_{02})\right]}{(f^2/f_0^2)+A_2\left[f^2/f_0^2-1\right]^2}\ \mathrm{W/kg} \tag{1}$$

where the SAR is in watts per kilogram, $f_0<f_{01}<f_{02}$, A_1, A_2, A_3, and A_4 are functions of a and b, and A_5 is a function of ϵ. Also, $u(f-f_{0i})$ is a unit step function defined by

$$u(f-f_{0i})=\begin{cases}0, & f<f_{0i}\\ 1, & f>f_{0i}\end{cases}$$

where $i=1$ or 2. The step functions with f_{01} and f_{02} are used to provide the characteristic f^2 behavior at low frequencies, the $1/f$ behavior above resonance, and the frequency independent behavior at very high frequencies, as shown.

From (1) it is clear that if $f<f_{01}$, the SAR expression reduces to

$$\mathrm{SAR}=\frac{A_1 f^2/f_0^2}{(f^2/f_0^2)+A_2\left[f^2/f_0^2-1\right]^2}\ \mathrm{W/kg} \tag{2}$$

which is the same as that for a series *RLC* resonance circuit [6][1]. Also, if $f^2\ll f_0^2$, it is easy to show that the SAR in (2) is proportional to f^2.

For $f_{01}<f<f_{02}$, the first unit step function will be nonzero, and hence (1) will reduce to

$$\mathrm{SAR}=\frac{A_1(f^2/f_0^2)\left[1+A_3 f/f_0\right]}{(f^2/f_0^2)+A_2\left[f^2/f_0^2-1\right]^2}. \tag{3}$$

For $A_3 f/f_0\gg 1$ and $f^2/f_0^2\gg 1$, (3) reduces to

$$\mathrm{SAR}=\frac{A_1A_3 f^3/f_0^3}{A_2 f^4/f_0^4}=\frac{A_1A_3 f_0}{A_2}\frac{1}{f} \tag{4}$$

which is the $1/f$ behavior described by Gandhi [5].

In the geometrical optics limit $f>f_{02}>f_{01}$, and hence all the terms in (1) will be included. For $f\gg f_0$, however, it can be easily shown that

$$\mathrm{SAR}=\frac{A_1A_4A_5}{A_2} \tag{5}$$

[1]The second term in the denominator of the equation on p. 413 of [6] should be squared. Also the resonant frequency for the given circuit parameters is negative.

which is obviously frequency independent and depends only on ϵ as described by the function A_5.

The coefficients A_1, $A_2,\cdots,A_5$, f_{01}, and f_{02} were determined by least-square fitting all the data available in the *Radiofrequency Radiation Dosimetry Handbook* [1], as described below.

III. Numerical Procedures and Results

Since (1) is a nonlinear function of the parameters, the method of differential corrections together with Newton's iterative method was used [7]. The method of solution involves approximating (1) with a linear form that is convenient to solve iteratively. By estimating approximate values of the unknown coefficients $A_1^{(0)}$, $A_2^{(0)},\cdots$, and of $f_{02}^{(0)}$ and expanding (1) in a Taylor's series with only the first-order terms retained, we obtain

$$\mathrm{SAR}\approx\mathrm{SAR}^{(0)}+\Delta A_1\left(\frac{\partial\mathrm{SAR}}{\partial A_1}\right)^{(0)}+\cdots+\Delta A_5\left(\frac{\partial\mathrm{SAR}}{\partial A_5}\right)^{(0)}+\cdots \tag{6}$$

where the superscript 0 is used to indicate values obtained after substituting the first guess $(A_1^{(0)},A_2^{(0)},\cdots,A_5^{(0)},f_{01}^{(0)},f_{02}^{(0)})$ for values of the unknown parameters into (1). Equation (6) is obviously a linear function of the correction terms ΔA_1, $\Delta A_2,\cdots,\Delta A_5$, and hence the least-square curve-fitting method can be used directly to determine these correction terms. The correction terms, when added to the first guess, give an improved approximation of the unknown coefficients; i.e., $A_1^{(1)}=A_1^{(0)}+\Delta A_1$, $A_2^{(1)}=A_2^{(0)}+\Delta A_2$, etc. When the improved estimates $A_1^{(1)}$, $A_2^{(1)}$, etc., are subsequently substituted as new estimates of the unknown coefficients, the Taylor's series reduces to

$$\mathrm{SAR}\approx\mathrm{SAR}^{(1)}+\Delta A_1\left(\frac{\partial\mathrm{SAR}}{\partial A_1}\right)^{(1)}+\cdots+\Delta A_5\left(\frac{\partial\mathrm{SAR}}{\partial A_5}\right)^{(1)}+\cdots$$

where $\mathrm{SAR}^{(1)}$ and its derivatives are obtained by substituting the values of $A_1^{(1)}$, $A_2^{(1)},\cdots$, etc., into (1). Again, the correction terms ΔA_1, ΔA_2, etc., are determined using the least-square curve fitting method. The procedure is continued until the solution converges to within a specified accuracy. Numerical values of the coefficients obtained after curve-fitting 18 specific models available in the *Radiofrequency Radiation Dosimetry Handbook* are given in Table I.

The second step in the numerical procedure involves expressing the values of the expansion coefficients so

TABLE I
NUMERICAL VALUES OF THE COEFFICIENTS EMPLOYED IN THE EMPIRICAL FORMULA (1)

Model	A_1	A_2	A_3	A_4	f_{o1}/f_o	f_{o2}/f_o
Average man	0.231	2.433	0.340	0.300	1.157	1.430
Skinny man	0.382	2.997	0.680	0.227	1.490	3.256
Fat man	0.1156	1.781	0.750	0.250	2.050	3.973
Average woman	0.239	2.325	0.559	0.242	1.370	2.620
Small woman	0.275	2.332	0.582	0.241	1.236	2.700
Large woman	0.188	2.190	1.000	0.350	1.740	5.500
10-year-old child	0.363	2.485	0.389	0.244	1.490	2.880
5-year-old child	0.385	2.330	0.376	0.228	1.040	2.690
1-year-old child	0.323	1.677	0.607	0.190	1.610	4.080
Rhesus monkey	0.282	1.000	0.592	0.160	2.340	4.687
Squirrel monkey	0.319	0.807	0.020	0.178	1.470	2.730
German shepherd	0.142	1.340	0.630	0.230	1.135	3.260
Brittany spaniel	0.174	1.250	0.600	0.205	1.300	3.510
Beagle	0.144	1.315	0.471	0.251	1.480	2.857
Guinea pig	0.513	0.707	0.212	0.117	1.800	3.333
Large rat	0.653	1.464	0.212	0.260	1.470	2.830
Medium rat	0.750	1.255	0.083	0.242	1.540	2.310
Small rat	1.050	1.000	0.040	0.180	0.950	1.630

obtained in terms of the a and b values of the 18 models used. Although each of the A's is expected to be a nonlinear function of a and b, the functions are chosen to be linear in the expansion coefficients. This allows a straightforward least-square curve fitting. The following are the expressions obtained by this procedure:

$$A_1 = -0.000994 - 0.01069a + 0.000172a/b + 0.000739(1/a) + 0.00566a/b^2 \quad (7)$$

$$A_2 = -0.00091 + 0.0414a + 0.39917a/b - 0.0012(1/a) - 0.00214a/b^2 \quad (8)$$

$$A_3 = 4.822a - 0.0835a/b - 8.733a^2 + 0.001575(a/b)^2 + 5.3688a^3 \quad (9)$$

$$A_4 = 0.3353a + 0.0753a/b - 0.804a^2 - 0.0075(a/b)^2 + 0.64a^3 \quad (10)$$

$$f_{01}/f_0 = -0.421a + 1.239a/b + 1.09a^2 - 0.2945(a/b)^2 + 0.0195(a/b)^3 \quad (11)$$

$$f_{02}/f_0 = 21.8a + 0.502a/b - 50.81a^2 - 0.068(a/b)^2 + 34.12a^3 \quad (12)$$

and

$$A_5 = |\epsilon/\epsilon_{20}|^{-1/4} \quad (13)$$

where a and b are in meters and ϵ_{20} is the dielectric constant of material having a permittivity equal to 2/3 that of muscle tissue at 20 GHz [1].

The resonance frequency f_0 can be estimated by assuming that the resonance is some combination of two conditions.

1) The length of the prolate spheroid is equal to $\lambda/2$ [1].

2) The circumference of the spheroid is equal to λ.

With these two assumptions, the following empirical formula for f_0 was obtained by curve-fitting data in the *Radiofrequency Radiation Dosimetry Handbook*:

$$f_0 = 2.75 \times 10^8 [8a^2 + \lambda^2(a^2 + b^2)]^{-1/2} \text{ Hz} \quad (14)$$

where a and b are in meters.

IV. RESULTS AND CONCLUSIONS

A comparison between values of f_0 obtained from (14) and those obtained using the extended boundary condition method are given in Table II. It is clear that (14) provides a quick and easy method of calculating the resonant frequency with approximately a 5-percent error at most.

Numerical results obtained from (1) are shown in Figs. 1 and 2, for the cases when the expansion coefficients are obtained from Table I, and from (7) to (13). As expected, the results are better with the coefficients in Table I, but the results from (7) and (13) give a very useful approximation. The poor agreement between the results given by the empirical relation and the Handbook values at the high-frequency end of the curve occurs because the frequency dependent values of ϵ were not used in A_5 in the empirical formula, but were included in the Handbook calculations. Much closer agreement would be obtained if the frequency dependence of ϵ were included in A_5.

One weakness in the empirical relation is that it often gives poor results at frequencies near f_{01}, which is caused by the abruptness of $u(f-f_{01})$. This limitation is not serious, however, since good results can be obtained by smoothing in the curve near f_{01}.

Gandhi [5] has given empirical relations for the resonant frequency and the SAR at resonance. A comparison with his results and some Handbook results are shown in Table III. Blank entries in the Handbook column indicate that calculations by the methods used for the Handbook are not yet possible for these particular cases, which illustrates another benefit of the empirical relation. That is, it can be used to obtain approximations for cases not included in the Handbook calculations. Although the error in these cases is difficult to estimate because more sophisticated calculations are not available, the results given by the two empirical methods show reasonable agreement. Examples of such results are shown in Figs. 3 and 4.

The empirical relation given here is useful for calculating the SAR for spheroidal sizes between a man and a rat, corresponding to the range of data used in the curve-fitting. However, it appears that the formula is also useful for some models smaller than a rat. For example, some calculations were made for models of mice and good results were obtained over a broad frequency band [1]. However, for some other spheroids smaller than rat-sized

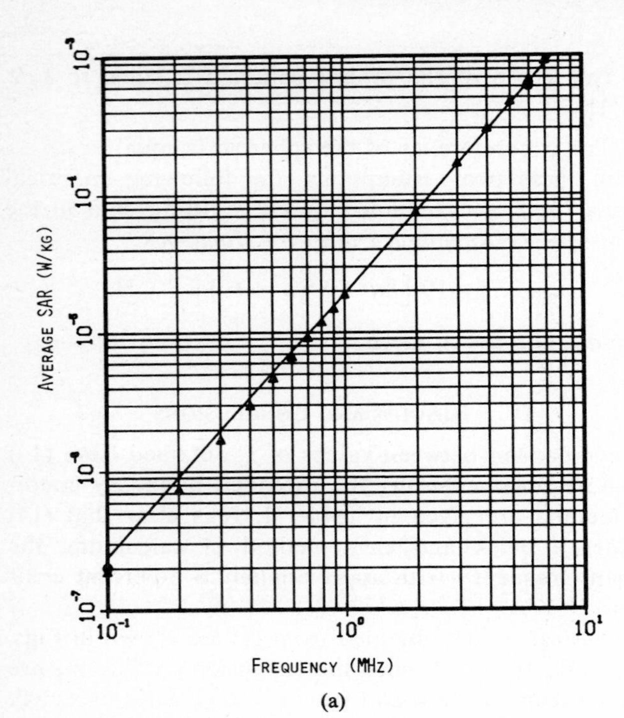

(a)

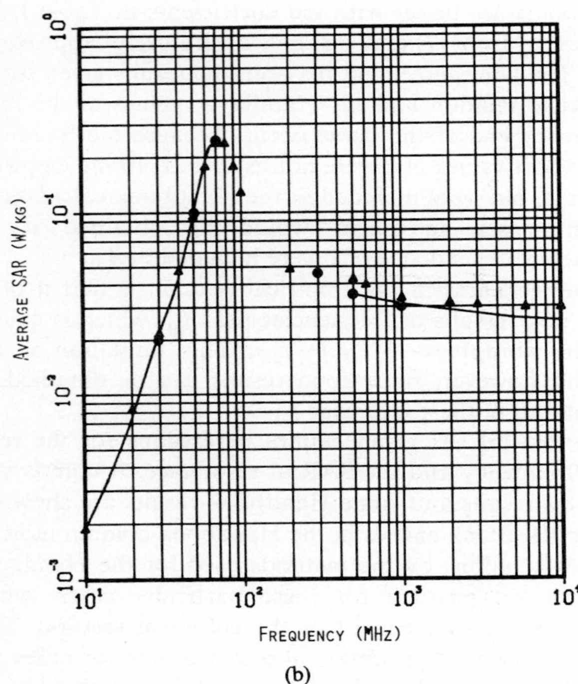

(b)

Fig. 1. Comparison of average SAR calculated by the empirical formula with the curve obtained by other calculations for a 70-kg average man. For the prolate spheroidal model, $a = 0.875$ m and $b = 0.138$ m. The incident plane wave is E-polarized with a density of 1 mW/cm^2. —— Calculated values [1]; ▲▲ empirical formula with the coefficients obtained from (7)–(14); and ●● empirical formula with the coefficients obtained from Table I.

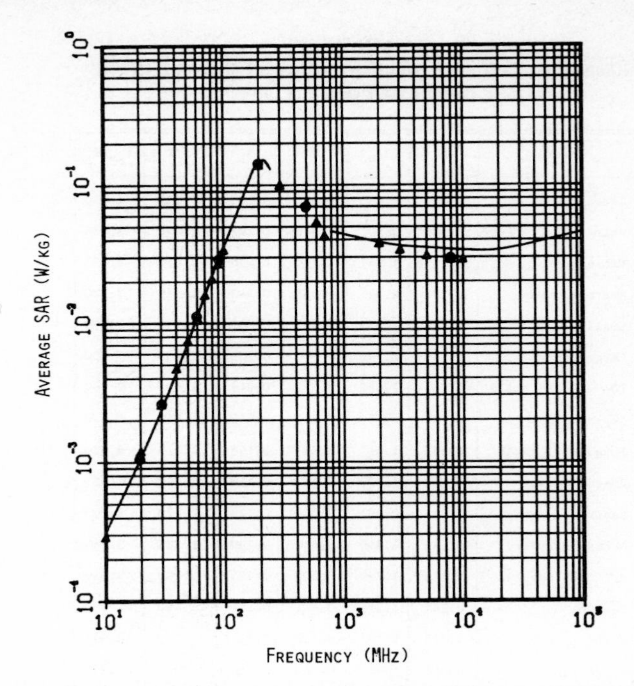

Fig. 2. Comparison of average SAR calculated by the empirical formula with the curve obtained by other calculations for a 13.5-kg beagle. For the prolate spheroid model, $a = 28.5$ cm and $b = 10.63$ cm. The incident plane wave is E-polarized with a power density of 1 mW/cm^2. —— Calculated values [1]; ▲▲ empirical formula with the coefficients obtained from (7)–(14); and ●● empirical formula with the coefficients obtained from Table I.

TABLE II
COMPARISON BETWEEN THE VALUES OF THE RESONANCE FREQUENCY f_0 OBTAINED USING THE EBCM [1] AND ESTIMATED FROM THE EMPIRICAL RELATION (14)

Model	a (m)	b (m)	f_o (MHz) From the RF Handbook [1]	f_o (MHz) From (14)	Percentage Error in Frequency
Average man	0.875	0.138	70	73.8	-5.6
Sitting rhesus monkey	0.200	0.0646	320	316.3	1.2
Squirrel monkey	0.115	0.0478	550	540.5	1.7
Beagle	0.285	0.1063	210	220.0	-4.8
Guinea pig	0.11	0.0355	600	575.6	4.1
Small rat	0.07	0.0194	950	910.2	4.2
Medium rat	0.10	0.0276	650	637.3	2.0
Large rat	0.120	0.0322	530	531.6	0.3

spheroids, the results were not as good. The accuracy of the empirical relation for the spheroids smaller than rat sized seems to depend strongly on the value of a/b.

For some spheroidal models, the SAR given by (1) at low frequencies does not fit as well as for other spheroidal models because of changes in ϵ with frequency. Since the coefficients in (1) were obtained by curve fitting without any attempt to include frequency dependence of the permittivity explicitly, fluctuations in SAR caused by a strong frequency dependence of the permittivity are not accounted for very well by (1). An example of this is shown in Fig. 5. Methods of including the frequency dependence of the permittivity explicitly in an empirical formula are being considered.

Although the empirical relation derived in this paper does have some limitations, as described above, it provides a very simple method for quickly calculating the approximate SAR as a function of frequency for prolate spheroidal models irradiated by plane waves with E-polarization, and as such should be very useful to those

TABLE III
COMPARISON BETWEEN THE SAR AND f_0 VALUES OBTAINED FROM [1], USING THE EMPIRICAL RELATION BY GANDHI [5] AND FROM (1)

Model	Empirical Relation by Gandhi [5]		From RF Handbook [1]		This Empirical Formula	
	SAR (W/kg)	F(MHz)	SAR (W/kg)	F(MHz)	SAR (W/kg)	F(MHz)
Average man	0.23	65.03	0.24	70	0.252	73.8
Average ectomorphic (skinny) man	0.34	64.68	—	—	0.382	73.6
Average endomorphic (fat) man	0.12	64.56	—	—	0.122	72.9
Average woman	0.22	70.97	—	—	0.242	80.2
Small woman	0.25	79.18	—	—	0.278	89.1
Large woman	0.18	65.78	—	—	0.193	74.5
10-year-old child	0.31	82.94	—	—	0.341	93.7
5-year-old child	0.34	101.56	—	—	0.378	115.3
1-year-old child	0.29	153.39	—	—	0.325	173.6
Rhesus monkey	0.38	285.	0.29	310	0.272	316.3
Squirrel monkey	0.40	496.	0.32	550	0.289	540.5
Baboon (Hamadryas)	0.22	166.4	—	—	0.153	184.3
German shepherd	0.21	126.73	—	—	0.146	141.3
Brittany spaniel	0.25	165.70	—	—	0.175	184.4
Beagle	0.20	200.01	0.15	200	0.142	220.0
Rabbit	1.33	284.72	—	—	0.953	322.6
Guinea pig	0.69	518.72	0.55	600	0.499	575.6
Small rat	1.48	815.96	1.1	950	1.06	910.2
Medium rat	1.04	569.44	0.78	750	0.749	637.3
Large rat	0.92	475.75	0.7	550	0.660	531.6
Small mouse	1.62	2109.0	—	—	1.18	2297.1
Medium mouse	2.03	1632.1	—	—	1.47	1803.9
Large mouse	1.92	1496.8	—	—	1.4	1663.0

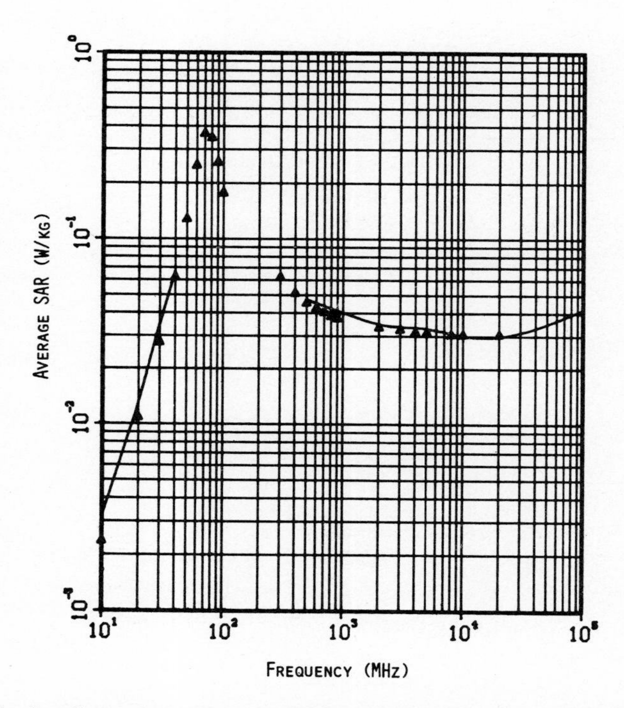

Fig. 3. Comparison of average SAR calculated by the empirical formula with the curve obtained by other calculations for a 47.18-kg skinny man. For the prolate spheroid model, $a = 0.88$ m and $b = 0.113$ m. The incident plane wave is E-polarized with a power density of 1 mW/cm^2. —— Calculated values [1], and ▲▲ are values obtained using the empirical formula.

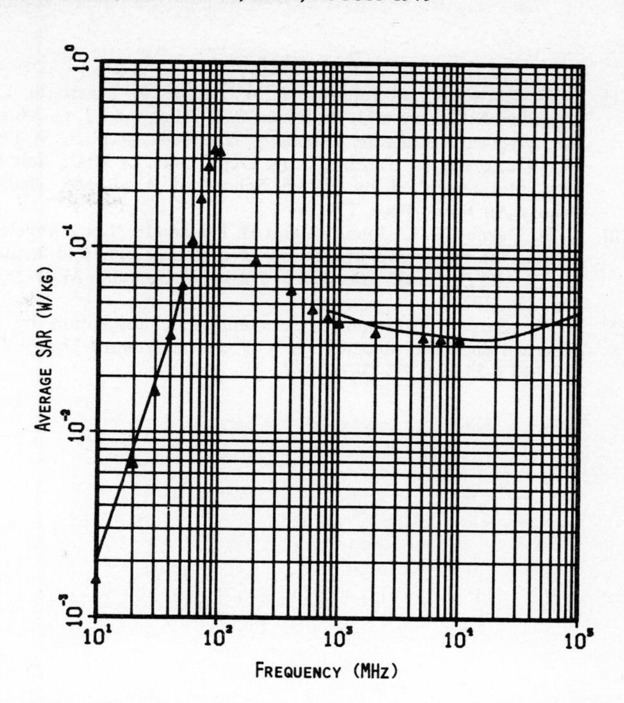

Fig. 4. Comparison of average SAR calculated by the empirical formula with the curve obtained by other calculations for a 32.2-kg 10-year-old child. For the prolate spheroid model, $a = 0.69$ m and $b = 0.106$ m. The incident plane wave is E-polarized with a power density of 1 mW/cm^2. —— Calculated values [1], and ▲▲ are values obtained using the empirical formula.

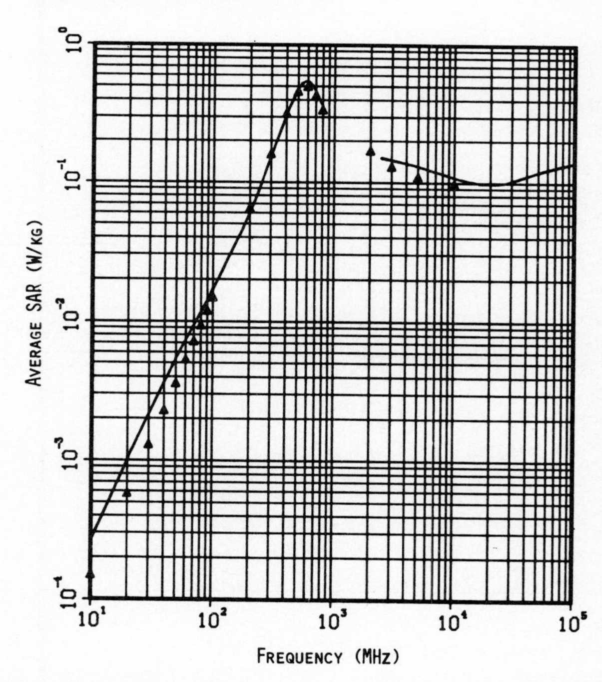

Fig. 5. Comparison of average SAR calculated by the empirical formula with the curve obtained by other calculations for a 0.58-kg guinea pig. For the prolate spheroid model, $a = 11$ cm and $b = 3.55$ cm. The incident plane wave is E-polarized with a power density of 1 mW/cm^2. —— Calculated values [1], and ▲▲ are values obtained using the empirical formula.

involved in microwave-biological research, as well as others interested in power absorption in lossy dielectric spheroids.

REFERENCES

[1] C. H. Durney, C. C. Johnson, P. W. Barber, H. Massoudi, M. F. Iskander, J. L. Lords, D. K. Ryser, S. J. Allen, and J. C. Mitchell, *Radiofrequency Radiation Dosimetry Handbook*, Rep. SAM-TR-78-22, 1978, prepared by Elect. Eng. Dep., Univ. of Utah, Salt Lake City, and published by USAF School of Aerospace Medicine, Brooks Air Force Base, TX.

[2] C. H. Durney, C. C. Johnson, and H. Massoudi, "Long-wavelength analysis of planewave irradiation of a prolate spheroid model of man," *IEEE Trans. Microwave Theory Tech.*, vol. MTT-23, pp. 246–254, 1975.

[3] P. W. Barber, "Resonance electromagnetic absorption by non-spherical dielectric objects," *IEEE Trans. Microwave Theory Tech.*, vol. MTT-25, pp. 373–381, 1977.

[4] O. P. Gandhi, E. L. Hunt, and J. A. D'Andrea, "Deposition of electromagnetic energy in animals and in models of man with and without grounding and reflector effects," *Radio Sci.*, vol. 12, pp. 39–48, 1977.

[5] O. P. Gandhi and M. J. Hagmann, "Some recent results on deposition of electromagnetic energy in animals and models of man," presented at the Int. Symp. on biological effects of electromagnetic waves, Airlie, VA, Oct. 30–Nov. 4, 1977. See also, *Radio Sci.*, vol. 12, no. 6(S), pp. 39–48, 1977.

[6] H. R. Kucia, "Accuracy limitation in measurements of HF field intensities for protection against radiation hazards," *IEEE Trans. Instrum. Meas.*, vol. IM-21, pp. 412–415, 1972.

[7] L. H. Lafara, *Computer Methods for Science and Engineering*. New York: Hayden, 1973, ch. 3.

State of the Knowledge for Electromagnetic Absorbed Dose in Man and Animals

OM P. GANDHI, FELLOW, IEEE

Abstract—**The paper gives the EM absorbed dose for man and animals at various frequencies for the plane wave irradiation condition for different orientations of the body relative to incident fields. Also included are the results for the whole-body absorption for conditions of electrical contact with ground and in the presence of reflecting surfaces of high conductivity and multiple animals. The data are given for the distribution of power deposition in man models for the resonance conditions of highest whole-body electromagnetic absorption. The highlights of the results obtained with proportionately scaled saline- and biological-phantom-filled models of man have been confirmed by experiments with small laboratory animals, from 25-g mice to 2250-g rabbits.**

I. Introduction

THE EXPANDING usage of electromagnetic (EM) radiation has necessitated an understanding of its interaction with humans. Such knowledge is vital in evaluating and establishing radiation safety standards, determining definitive hazard levels, and a physical understanding of the effects which have been reported in the literature. The need for research in this area is clearly exemplified by the widely disparate safety levels that are used worldwide at the present time. In the United States the safety level is 10 mW/cm^2 for long-term human exposure at any frequency and regardless of physical environment, whereas the recommended maximum safe-power density in Eastern Europe and the USSR is 1000 times less at a level of 0.01 mW/cm^2. Several countries, including Canada [1] and Sweden [2], are in the process of revising their safety standards downward from the previously accepted levels of 10 mW/cm^2.

Biological studies of the effects of EM radiation have used laboratory animals such as rats, rabbits, etc., for the study of behavioral and/or biochemical changes. For these experiments to have any projected meanings for humans, it is necessary to be able to quantify the whole-body power absorption and its distribution for the irradiation conditions. It is furthermore necessary that dosimetric information be known for humans subjected to irradiation at different frequencies and for realistic exposure conditions.

Unlike the field of ionizing radiation, where the absorption cross section of a biological target is directly related to its geometrical cross section, the whole-body EM energy absorption has been shown [3]–[13] to be strongly dependent on polarization (orientation of electric field $\vec{E}$ of the incident waves), frequency, and physical environments such as a conducting ground and other reflecting surfaces. A prescribed power density of, say, 10 mW/cm^2 tells almost nothing about the absorbed dose except perhaps at very high frequencies where the wavelength of irradiation is an order of magnitude or more smaller than the dimensions of the animal. This is best illustrated by examples from Schrot and Hawkins' work [14] on times of lethality of rats and mice at several frequencies and for different polarizations of incident waves. For a free-space irradiation power density of 150 mW/cm^2 at 985 MHz, mice oriented along the electric field (E-orientation) convulsed in an average time of 9 min, while similar animals oriented along the microwave magnetic field (H-orientation) lived through an experimental observation time of 60 min without significant stress. Also, identical power densities at several frequencies resulted in substantially different times to convulsion. For mice irradiated with an incident power density of 150 mW/cm^2 in the E-orientation, mean times to convulsion of 3260 and 160 s were observed for 710 and 1700 MHz, respectively.

Manuscript received April 5, 1979; revised August 15, 1979.
The author is with the Department of Electrical Engineering, University of Utah, Salt Lake City, UT 84112.

Reprinted from *Proc. IEEE*, vol. 68, pp. 24–32, Jan. 1980.

In this paper we will present the highlights of the current knowledge on EM absorption by man and animals. Most of the work to date has concentrated on the mathematically and experimentally simpler plane wave, far-field irradiation conditions, but recent research has also included conditions such as the presence of ground and reflecting surfaces of high conductivity and multiple animals.

II. Techniques

1) Carefully proportioned, reduced scale models of man have been used to determine the mass-normalized rates of EM energy absorption (specific absorption rates or SAR's) at different frequencies and for different conditions of irradiation. This work is detailed in [7]. The highlights of these results have been checked by experimentation with small laboratory animals [10], [13] from 25-g mice to 2245-g rabbits. The SAR's are determined by measuring the colonic temperature elevation of anesthetized animals or by calorimetric determination of the absorbed dose by freshly euthanized animals.

2) Prolate spheroidal [4], [9], [15], and ellipsoidal [16] models have been used for the theoretical calculations for man and animals at frequencies up to and slightly beyond the resonant region. For very high frequencies, a geometrical optics method has been developed to estimate the power absorption in prolate spheroidal [17] and cylindrical models [18] of man. It is shown that the dependence of whole-body-averaged SAR on both frequency and polarization of the incident fields may be estimated using prolate spheroidal or ellipsoidal models, but the distribution of energy deposition through the body cannot be determined with such crude models.

3) Moment-method solutions [6] for an improved block model of man [11] have given good correlation with experimental data. These calculations have also led to the identification of the frequency regions for peak absorption (resonance) in arms and the head.

The highlights of the theoretical and numerical techniques are summarized in a companion paper[1] by C. H. Durney.

4) The effects of layering on energy deposition have recently been studied for a multilayered [19] model of man. The layering information required for the multilayered model was obtained from published anatomical cross sections [20], [21]. Specific tissue thicknesses were used for 79 horizontal cross sections of man and a layering resonance (interpreted as being due to impedance matching provided by the thicknesses of the outer layers) calculated for each of the individual cross sections based on a planar model. A highlight of these calculations is to show a broad layering resonance frequency of 1800 MHz for an adult human being.

III. Current Knowledge for Electromagnetic Absorbed Dose in Man and Animals

A. Free-Space Irradiation Condition

The condition that has been studied the most extensively [3]-[11], [15]-[19] is that of free-space irradiation of single animals. The whole-body absorption of EM waves by biological bodies is strongly dependent [3], [4] on the orientation of the electric field ($\vec{E}$) relative to the longest dimension (L) of the body. The highest rate [3], [5], [6], [9] of energy deposition occurs for $\vec{E} \parallel \hat{L}$ (E-orientation) for frequencies such that the major length is approximately 0.36 to 0.4 times the free-space wavelength (λ) of radiation. Peaks of whole-body absorption for the other two configurations (major length oriented along the direction ($\vec{k}$) of propagation, $\vec{k} \parallel \hat{L}$ or k-orientation, or along the vector $\vec{H}$ of the magnetic field, $\vec{H} \parallel \hat{L}$ or H-orientation) have also been reported [3], [10] for $\lambda/2$ on the order of the weighted averaged circumference of the animals.

For each of the orientations of the major length along the E-, k-, and H-, respectively, two distinct exposure conditions are possible. These are as follows.

For E-orientation ($\vec{E} \parallel \hat{L}$):

1) power propagating from front to back;
2)* power propagating from arm to arm.

For k-orientation ($\vec{k} \parallel \hat{L}$), power propagating from head to toe:

3) E from front to back;
4)* E from arm to arm.

For H-orientation ($\vec{H} \parallel \hat{L}$):

5) E from front to back;
6)* E from arm to arm.

A 5-15 percent larger whole-body absorption is found for cases 2 and 6 for E- and H-orientations, respectively. The most difference in absorption [7] is found for k-orientation, where a 50 percent increase in the overall absorption is measured for electric field from arm to arm as compared to the case where the electric field is from front to back of the body. These observations have since been confirmed by the below-resonance calculations [16] based on an ellipsoidal model of man.

Curves for whole-body absorption (fitted to the experimental data [7], [10]) for models of man exposed to radiation in free space are given in Fig. 1. For each of the orientations, configurations corresponding to higher power deposition (identified above) were used.

For the most absorbing E-orientation, the whole-body absorption curve A may be discussed in terms of five regions:

Region I—Frequencies well below resonance ($L/\lambda < 0.1$-0.2). An f^2 type dependence derived theoretically and checked experimentally by Durney *et al.* [22].

Region II—Subresonant region ($0.2 < L/\lambda < 0.36$). An $f^{2.75}$ to f^3 dependence of total power deposition has been experimentally observed for this region.

Region III—Resonant region ($L/\lambda \simeq 0.36$-0.4). A relative absorption cross section, [7] defined by electromagnetic absorption cross section[2]/physical cross section, S_{res} on the order of $0.665\ L/2b$ (derivable also from antenna theory) has been measured for this region, where L is the major length of the body and $2\pi b$ is its weighted average circumference. For a 70 kg, 1.75 m tall adult human being, $L/2b \simeq 6.3$ and S_{res}, therefore, is 4.2. This gives an SAR $\simeq 2.16 \times 1.75/L_m$ for an incident power density of 10 mW/cm^2, where L_m is height of the individual in meters. The resonance frequency f_r in MHz is on the order of $(62$-$68) \times 1.75/L_m$.

Region IV—Supraresonant region to frequencies on the order of 1.6 S_{res} times the resonance frequency (for human beings,

[1] C. H. Durney, "Electromagnetic Dosimetry for Models of Humans and Animals: A Review of Theoretical and Numerical Techniques," this issue, pp. 33-40.

*Configurations corresponding to higher power deposition.

[2] This is defined by the rate of energy deposition divided by the incident power density.

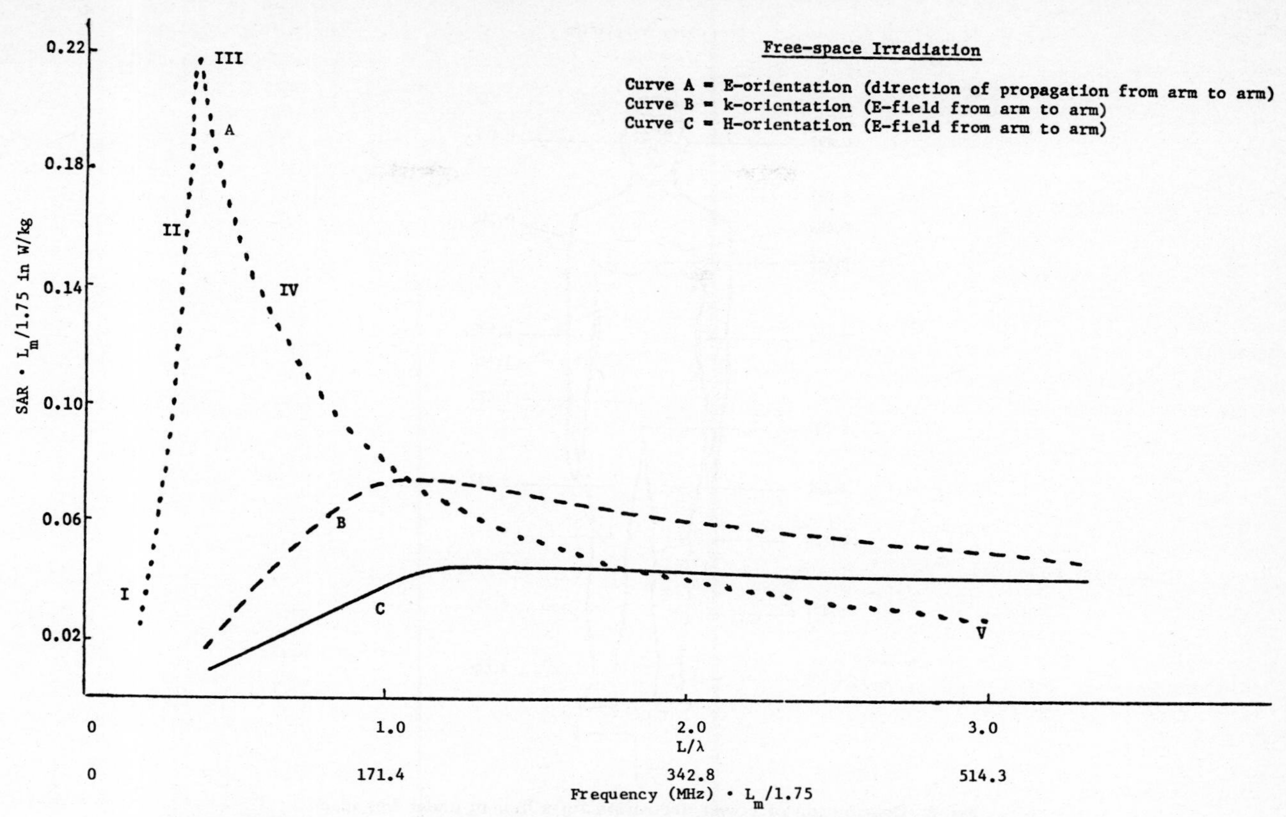

Fig. 1. Whole-body-averaged SAR for models of human beings for incident fields of 1 mW/cm² [9].

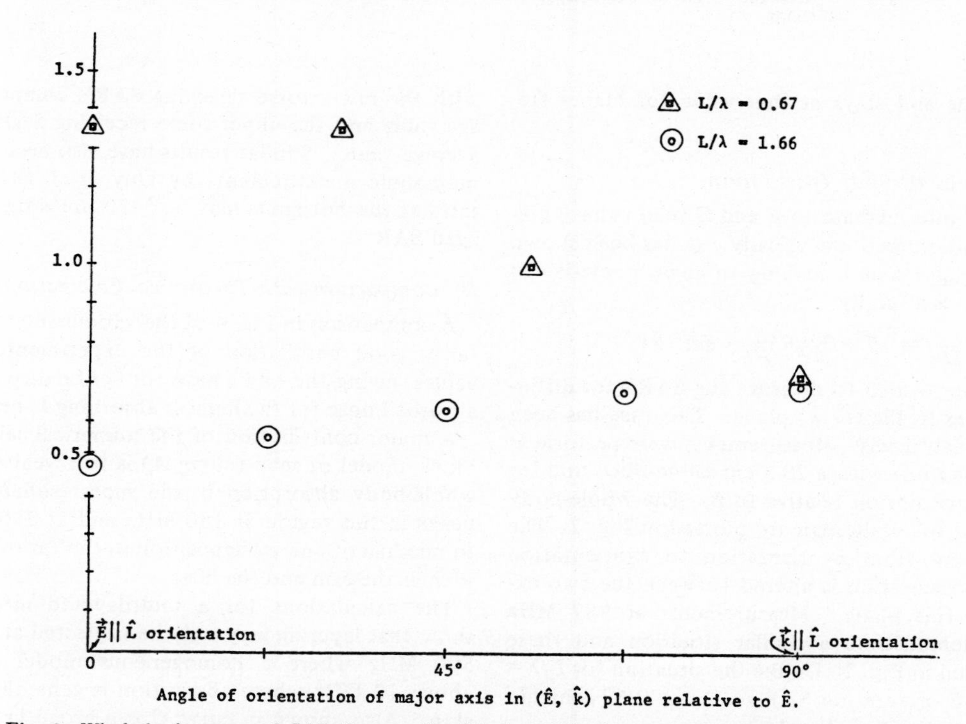

Fig. 2. Whole-body-averaged SAR for a saline-filled model of man for irradiation in the free field. The data points are for angles intermediate to $\vec{E}\|\hat{L}$ and $\vec{k}\|\hat{L}$ orientations.

this covers the region $f_r < f < 7f_r$). A whole-body absorption reducing as $(f/f_r)^{-1}$ from the resonance value has been observed.

Region V–$f \gg f_r$ region. The EM absorption cross section should asymptotically approach the "optical" value which is (1 – power reflection coefficient) or about one half of the physical cross section.

In comparing the absorptions for various orientations (Fig. 1) it is noted that the resonances for k- and H-orientations are not very sharp. In fact, for H-orientation, the value gradually

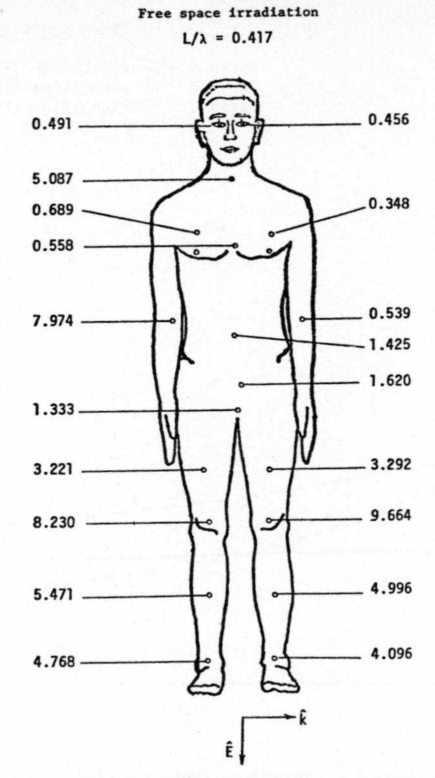

Fig. 3. Distribution of power deposition for a human under free-space irradiation. The numbers indicated are relative to whole-body-averaged SAR of $(1.75/L_m) \cdot 1.88$ W/kg for 10 mW/cm² incident fields.

reaches a peak value and stays at that value for higher frequencies.

B. Intermediate Angles of Body Orientation

Body orientations intermediate to E and H vectors have previously [15] been considered analytically. It has been shown that the SAR for major axis $\hat{L}$ making an angle θ relative to $\vec{E}$ in the $(\hat{E}, \hat{H})$ plane is given by

$$[(\text{SAR})_{\vec{E}\parallel\hat{L}} \cos^2\theta + (\text{SAR})_{\vec{H}\parallel\hat{L}} \sin^2\theta].$$

Experiments were performed to measure the SAR's for different body orientations in the $(\hat{E}, \hat{k})$ plane. This case has been difficult to handle analytically. Measurements were performed at 2450 MHz ($L/\lambda = 1.66$) with a 20.3 cm saline-filled doll for different angles of orientation relative to $\vec{E}$. The whole-body SAR's calculated for full-scale man are plotted in Fig. 2. The SAR varies smoothly from E-orientation to k-orientation values as the body orientation is altered between the two extreme positions in this plane. Measurements at 987 MHz ($L/\lambda = 0.67$) also demonstrate a similar situation and these values too are plotted in Fig. 2. Unlike the situation for $L/\lambda = 1.66$, however, the E-orientation SAR for $L/\lambda = 0.67$ is higher than that for k-orientation (see Fig. 1).

C. Distribution of Power Deposition under Resonance Conditions

The measured pattern of energy deposition for conditions of highest absorption is shown in Fig. 3. For free-space irradiation SAR's considerably higher than the whole-body average are observed for the neck, the legs, and the front elbow region, with the lower torso receiving SAR's comparable to the average value and the upper torso receiving SAR's lower than the average value. Similar results have also been obtained by thermographic measurements by Guy *et al.* [8]. The deposition rates at the hot spots may be 5–10 times the whole-body averaged SAR.

D. Comparison with Theoretical Calculations

A comparison in Fig. 4 of the various models of man shows a fairly good correlation of the experimental and theoretical values, giving thereby a basis for confidence in the whole-body absorbed dose for the highest absorbing E-orientation.

A major contribution of the numerical calculations with the block model of man (curve A) is to reveal a fine structure to whole-body absorption in the supraresonance region. Minor peaks in this region at 150 MHz and at 350 MHz are ascribed to maxima of energy deposition in the various body parts [11] such as the arm and the head.

The calculations for a multilayered model [19] of man show that layering may only be neglected at frequencies below 500 MHz where a homogeneous model is appropriate, or above 10 GHz where absorption is generally restricted to the skin. Also shown in curve C is a broad layering-caused peak in whole-body absorption at a frequency of 1800 MHz, where a deposition 34 percent larger than that for the homogeneous model (curve B) is obtained.

E. Empirical Equations for SAR for E-Orientation

The subresonant frequency dependence and the observed $1/f$ dependence in the supraresonant region have been used to de-

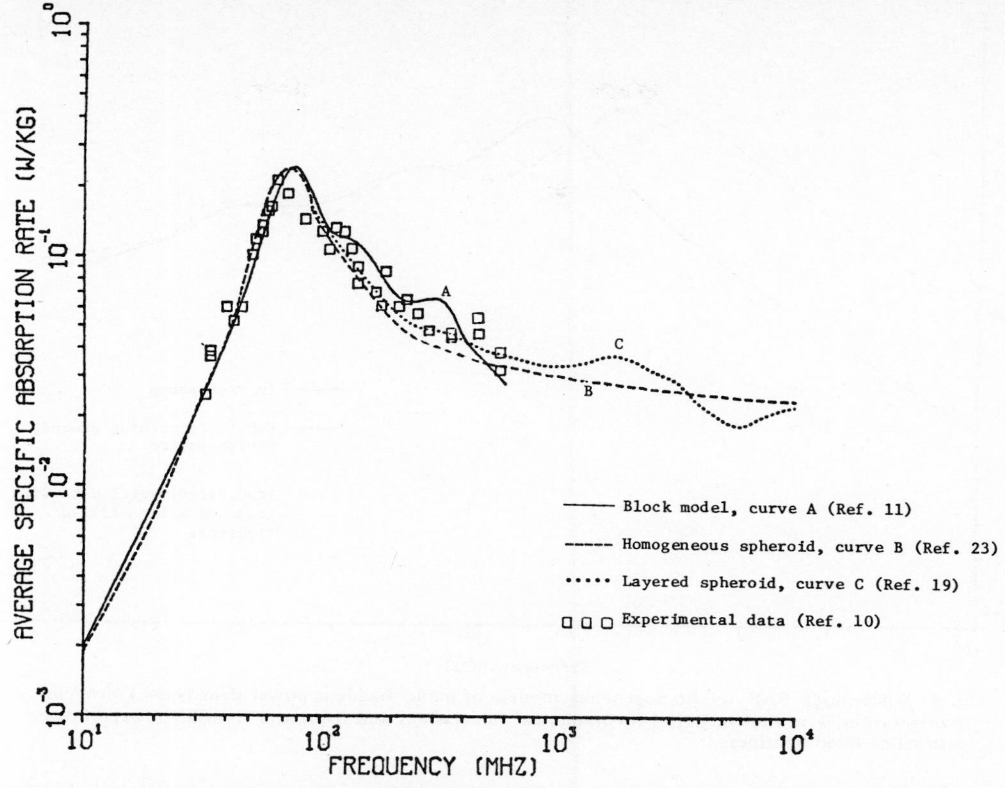

Fig. 4. Whole-body averaged SAR for homogeneous and multilayered models of man for E-orientation. Incident power density of 1 mW/cm^2; $\vec{k}$ ventral to dorsal for numerical calculations, curve A, and from arm to arm for experiments with saline-filled figurines.

TABLE I

EMPIRICAL EQUATIONS FOR WHOLE-BODY-AVERAGED SAR FOR MAN MODELS FOR CONDITIONS OF FREE-SPACE IRRADIATION

$\vec{E} \| L$ polarization

$$\text{Resonant frequency } f_r = 11\,400/L_{cm} \text{ MHz} \quad (1)$$

For subresonant region — $0.5 f_r < f < f_r$:

$$\text{SAR in mW/g for 1 mW/cm}^2 \text{ incident plane wave field} = \frac{0.52\, L_{cm}^2}{\text{mass in g}} \left(\frac{f}{f_r}\right)^{2.75} \quad (2)$$

For supraresonant region -- $f_r < f < 1.6\, S_{res}\, f_r$:

$$\text{SAR in mW/g for 1 mW/cm}^2 \text{ incident plane wave field} = \frac{5950}{f_{MHz}} \frac{L_{cm}}{\text{mass in g}} \quad (3)$$

where L_{cm} is the long dimension of the body in centimeters, and

$$S_{res} = 0.48 \sqrt{\frac{L_{cm}^3}{\text{mass in g}}} \quad (4)$$

velop the empirical equations [10] in Table I for whole-body-averaged SAR for man models for E-orientation. Since human subjects cannot be used for experimentation, the empirical equations of Table I have been by experiments with six animal species from 25-g mice to 2250-g rabbits [13] and found to be fairly accurate. For reasons not as yet understood, the measured SAR values for experimental animals are approximately 59 percent higher than those given by (2) and (3), which were derived from experiments with model figurines.

IV. Electromagnetic Absorption for Man and Animals in the Presence of Nearby Ground and Reflecting Surfaces

A. Ground Effects

Only highly conducting (metallic sheet) ground of infinite extent has been considered to date. All measurements and calculations of grounding effects have assumed that a man is standing on or above such a ground. Fig. 5 shows the calculated [12] values of SAR for the block model of man with

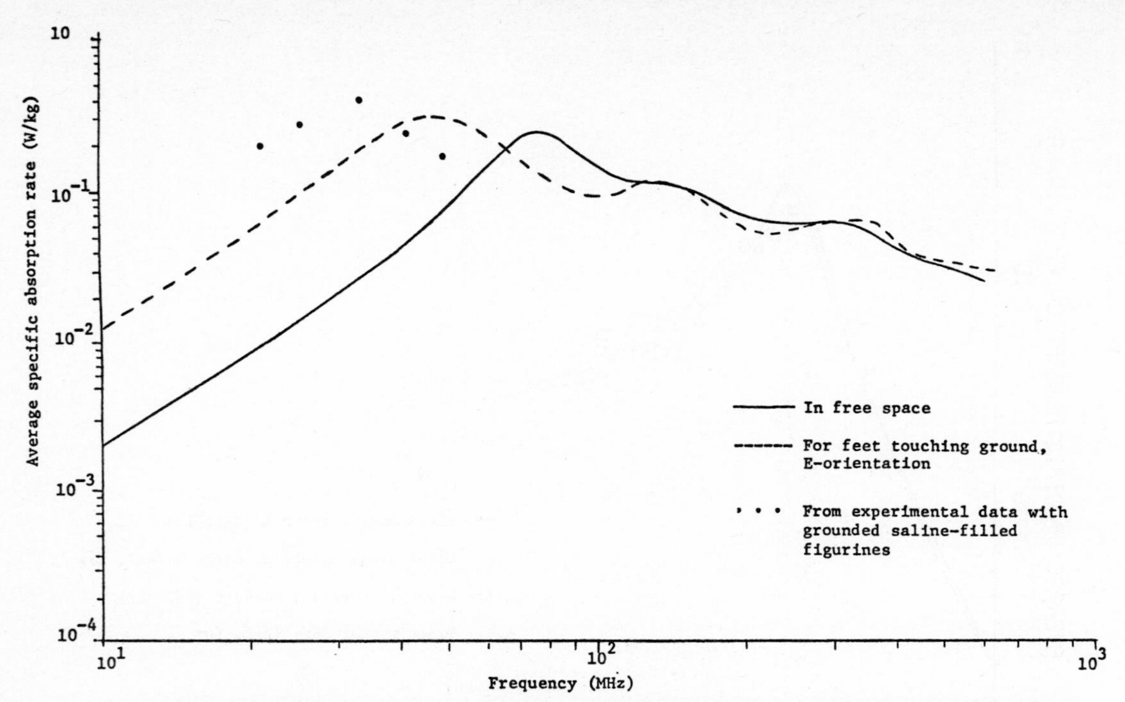

Fig. 5. Whole-body SAR for homogeneous models of man. Incident power density of 1 mW/cm², E-orientation, $\vec{k}$ ventral to dorsal for numerical calculations and from arm to arm for experiments with saline-filled figurines.

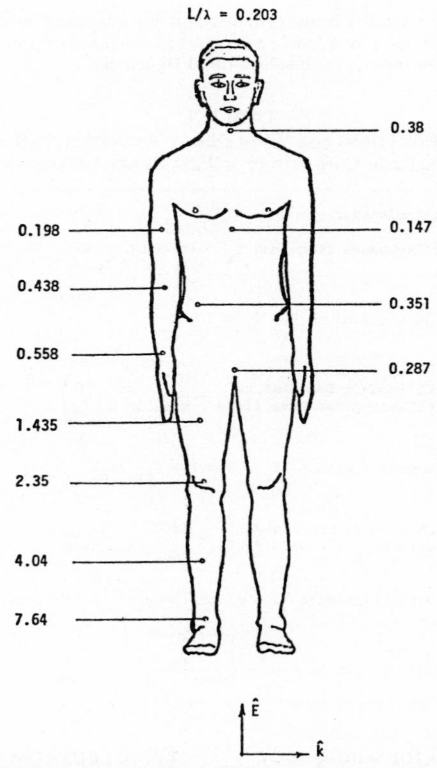

Fig. 6. Distribution of power deposition for a human being with feet in electrical contact with the ground. The numbers are relative to whole-body-averaged SAR values of $(1.75/L_m) \cdot 4.0$ W/kg for 10 mW/cm².

feet in conductive contact with ground and compares the same with the values calculated for free-space irradiation. Also shown in the same figure are the values measured [7], [10] with saline-filled figurines that were exposed to irradiation in the E-orientation at different frequencies in the monopole-above-ground [5] radiation chamber. Even though the peak SAR's of calculated data and measured values are in reasonable agreement, the resonant frequencies for the two models are

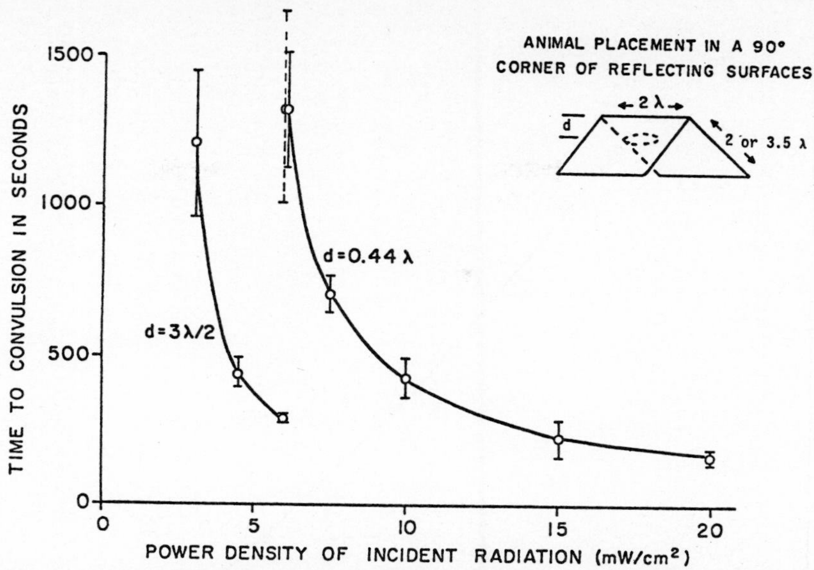

Fig. 7. Times to convulsion ($\pm SD$) of 100-g rats at the resonance frequency of 987 MHz for two distances of the animal from the corner of the reflector. (The dashed line and the data for $d = 3\,\lambda/2$ were obtained with a corner reflector of dimensions $3.5\lambda \times 2.0\lambda$.)

different. The calculated resonant frequency of man in conductive contact with ground is 47 MHz as compared to 77 MHz for man in free space. The corresponding frequencies calculated from saline-filled figurines are 34.5 and 68 MHz, respectively. The reason for this discrepancy is not clear.

The measured pattern of energy deposition for grounded resonance condition is shown in Fig. 6. The highest SAR's in this case are observed for the ankles and the legs. Like the case of free-space irradiation, the deposition rates at the hot spots are, once again, a factor of 5-10 times larger than the whole-body averaged SAR under these conditions.

The nature of the ground effects on SAR (for E-orientation) is such that even a small separation [12], [24] from ground (to break conductive contact) is sufficient to eliminate much of the ground effect. For separations from ground more than 7-10 cm, the total energy deposition and its distribution are identical to those for free-space irradiation conditions. Even for a man model in conductive contact with a perfect ground, the energy deposition in the supraresonance region ($f > 2\text{-}3f_r$) is comparable [12] (see Fig. 5) to that for conditions of free-space irradiation.

B. Reflector Effects

Here, too, only highly conducting reflecting surfaces [10], [12] have been considered to date and most of the work has concentrated on frequencies close to the resonance region and for E-orientation. Highly enhanced SAR's by factors as large as 5 and 27, respectively, have been measured using saline-filled figurines in the presence of flat and 90° corner type reflectors of aluminum of plate dimensions no bigger than a few wavelengths. For resonant targets, the measured enhancements (over free-space values) of EM absorption in human models are in conformity with the gains [25], [26] of a half-wave dipole antenna in the presence of such reflectors. Indeed, for incident plane waves for E-orientation, most of the observed results are as though the irradiated target acted like a pick-up half-wave dipole in the presence of reflecting surfaces. The times-to-convulsion of ~100-g rats at incident power densities of 3-20 mW/cm² have confirmed [10] the highlights of these results. Some of the results are illustrated in Fig. 7.

V. Head Resonance

As aforementioned in Sections III. A. and IV., a frequency region for the highest rate of energy deposition in the head has been identified. The head resonance [13], [27] occurs at frequencies such that the head diameter is approximately one quarter of the free-space wavelength. For the intact (adult) human head, the resonance frequency is estimated to be on the order of 350-400 MHz. At head resonance the absorption cross section is approximately three times the physical cross section with a volume-averaged SAR that is about 3.3 times the whole-body-averaged SAR. Both values greatly exceed numbers reported earlier for spherical models of the isolated human head [28], [29]. For reasons not as yet understood, the head resonance is observed for E- and k-orientations but not for H-orientation.

Numerical calculations [27] using 144 cubical cells of various sizes to fit the shape of the human head (340 cells for the whole body) give local SAR's at "hot spots" (above the palate area and the upper part of the back of the neck) that are about five times the average values for the head.

Enhanced absorption in the head region at head-resonant frequencies may be important in studies of the behavioral effects, blood-brain barrier permeability, cataractogenesis, etc., and needs to be examined at length.

VI. Multianimal Effects [13]

It has been shown that for resonant biological bodies close to one another, antenna theory may be used to predict the modification in SAR relative to free-space values. For two resonant targets separated by 0.65 to 0.7 λ, the highest SAR, 105 percent of free-space value, can result for man and animals for E-orientation for plane waves incident broadside to the line joining the centers of the targets. For three animals in a row with an interanimal spacing of 0.65 λ, the central animal SAR would be roughly two times, while the two end animals

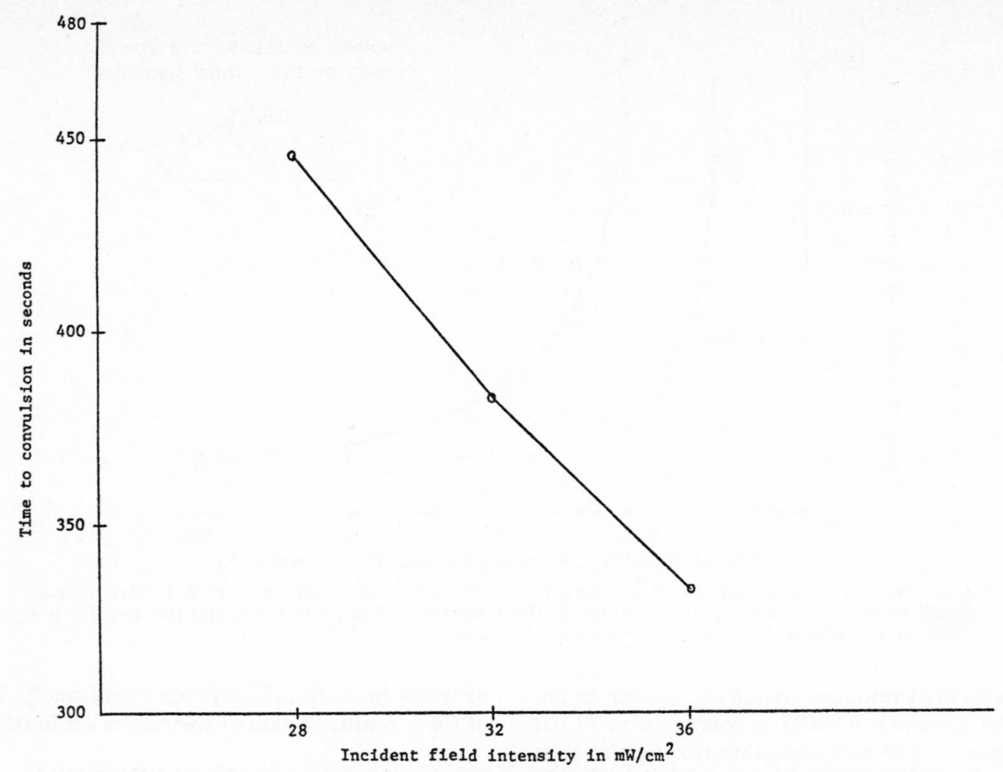

Fig. 8. Mean times to convulsion of 100-g rats at 2450 MHz for a distance $d = 3\lambda/2$ from a 90° corner of reflecting plates.

will receive an SAR that is approximately 1.5 times that for an isolated animal.

Full implications of the multibody effects on SAR are not completely understood, even though block model calculations and pilot experimental data [13] with anesthetized rats at 2450 MHz show that similar enhancements may also occur for subresonance and supraresonance regions.

VII. Relative Doses of Electromagnetic Absorption at Different Frequencies

To demonstrate the great disparity of EM absorption at different frequencies for a given animal size, equivalent power levels for a given time-to-convulsion were measured for 100-g rats placed $3\lambda/2$ in front of a 90° corner reflector at 987 and 2450 MHz. Another objective of these experiments also was to see if whole-body integrated dose rather than its distribution was the important parameter in determining the time to convulsion.

Substantially different distributions [7] have previously been reported for the two frequencies with reduced scale models of man, and a similar situation is expected for 100-g rats. On account of the $1/f$ reduction in the absorbed energy for the supraresonant region, a free-space whole-body deposition rate of 2.45/0.987 or 2.48 times is expected at 987 MHz as compared to that at 2450 MHz. Furthermore, a reflector-caused enhancement in the SAR is considerably larger for near-resonance size targets than for supraresonant conditions. A ratio of 2.96 is expected on this account for the reflector-caused enhancements in the SAR's between 987 and 2450 MHz [10] ($L/\lambda \simeq 0.42$ and 1.04 at the respective frequencies). A total SAR of $1/(2.48 \times 2.96) = 1/7.34$ times the value at 987 MHz is therefore projected for irradiation of 100-g animals at 2450 MHz.

To validate the hypothesis, that whole-body integrated dose is the all-important parameter, comparable times to convulsion should be observed for 5 mW/cm² at 987 MHz (Fig. 7) and for 36.7 mW/cm² incident fields at 2450 MHz. The selection of 5 mW/cm² for $d = 3\lambda/2$, 987 MHz is made to avoid the asymptotic regions and to be approximately on the knee of the time-to-convulsion curve (Fig. 7).

The measured times to convulsion for 100-g[3] rats at 2450 MHz for three incident power levels are plotted in Fig. 8. Time-to-convulsion of 365 seconds comparable to 5 mW/cm² incident fields at 987 MHz is observed for 33.2 rather than 36.7 mW/cm² projected from measurements with saline-filled man models of comparable L/λ. Thus there is a departure of only about 10 percent between the projected and observed values. Indeed, the difference may have been even smaller if heavier animals comparable in weight to those for Fig. 7 were used. This is because higher SAR's result for smaller animals, and this may have contributed to somewhat lower values of field intensities needed at 2450 MHz for comparable times to convulsion.

These experiments demonstrate that because of a rapid hemodynamic dispersion of heat, the whole-body integrated dose and dose rates are important parameters in the study of living animals.

[3] Somewhat lighter animals with weights of 81.1 ± 5.6 were used for these experiments. For the experiments of Fig. 7, for $d = 3\lambda/2$, the animals weights [10] were 104 ± 8.3 g.

VIII. Conclusions

The state of knowledge on electromagnetic absorption for man and animals has been presented in this paper. It is demonstrated that the electromagnetic densitometry is not of primary importance and for a given power density, the absorbed dose can vary by orders of magnitude depending upon the frequency, animal size and its orientation, physical environments, etc.

Even though a great deal of progress has been made in the field of RF dosimetry, information is sorely lacking for realistic conditions of exposure such as the effects of finite size, finite conductivity, ground and reflectors, and for the important problem of near-field and/or partial body exposures such as are encountered by workers involved in the operation of EM radiation equipment for communications, radar, and for industrial and biomedical applications. Work has recently been started [31], [32] in the important area of near-field exposures and the results are beginning to be presented at technical meetings. It will be a few years before a sufficient amount of data are available to draw the broad conclusions such as are now possible for far-field plane wave exposure conditions for humans.

References

[1] M. H. Repacholi, "Proposed exposure limits for microwave and radio frequency radiations in Canada," *J. of Microwave Power*, vol. 13, pp. 199-211, 1978.

[2] K. Hansson-Mild, personal communication.

[3] O. P. Gandhi, "Polarization and frequency effects on whole animal absorption of RF energy," *Proc. IEEE*, vol. 62, pp. 1166-1168, 1974.

[4] C. H. Durney, C. C. Johnson, and H. Massoudi, "Long wavelength analysis of plane wave irradiation of a prolate spheroid model of man," *IEEE Trans. Microwave Theory Tech.*, vol. MTT-23, pp. 246-253, 1975.

[5] O. P. Gandhi, "Conditions of strongest electromagnetic power deposition in man and animals," *IEEE Trans. Microwave Theory Tech.*, vol. MTT-23, pp. 1021-1029, 1975.

[6] K. M. Chen, B. S. Guru, and D. P. Nyquist, "Quantification and measurement of induced fields inside finite biological bodies," *Biological Effects of Electromagnetic Waves* (Selected Papers of the USNC/URSI Annual Meeting, Boulder, CO), Oct. 20-23, 1975, vol. II, pp. 19-43 [HEW Publication (FDA) 77-8011, U.S. Government Printing Office, Washington, D.C. 20402].

[7] O. P. Gandhi, K. Sedigh, G. S. Beck, and E. L. Hunt, "Distribution of electromagnetic energy deposition in models of man with frequencies near resonance," in *Biological Effects of Electromagnetic Waves* (Selected Papers of the USNC/URSI Annual Meeting, Boulder, CO), vol. II, Oct. 20-23, 1975, pp. 44-67. [HEW Publication (FDA) 77-8011, U.S. Government Printing Office, Washington, DC 20402.]

[8] A. W. Guy, M. D. Webb, and C. C. Sorenson, "Determination of power absorption in man exposed to high frequency electromagnetic fields by thermographic measurements on scale models," *IEEE Trans. on Biomed. Eng.*, vol. 23, pp. 361-371,1976.

[9] P. W. Barber, "Electromagnetic power absorption in prolate spheroidal models of man and animals at resonance," *IEEE Trans. Biomed. Eng.*, vol. BME-24, pp. 513-521, 1977.

[10] O. P. Gandhi, E. L. Hunt, and J. A. D'Andrea, "Deposition of electromagnetic energy in animals and in models of man with and without grounding and reflector effects," *Radio Sci.*, vol. 12, no. 6(S), Nov.-Dec. 1977, pp. 39-47.

[11] M. J. Hagmann, O. P. Gandhi, and C. H. Durney, "Numerical calculation of electromagnetic energy deposition for a realistic model of man," *IEEE Trans. Microwave Theory Tech.*, vol. MTT-27, Sept. 1979.

[12] M. J. Hagmann and O. P. Gandhi, "Numerical calculation of electromagnetic energy deposition in man with ground and reflector effects," *Radio Sci.*, to be published.

[13] O. P. Gandhi, M. J. Hagmann, and J. A. D'Andrea, "Part-body and multibody effects on absorption of radio frequency electromagnetic energy by animals and by models of man," *Radio Sci.*, to be published.

[14] J. Schrot and T. D. Hawkins, "Interaction of microwave frequency and polarization with animal size," in *Biological Effects of Electromagnetic Waves* (Selected Papers of the USNC/URSI Annual Meeting. Boulder, CO), vol. II, 1975, pp. 184-192. [HEW Publication (FDA) 77-8011, U.S. Government Printing Office, Washington, DC 20402.]

[15] C. C. Johnson, C. H. Durney, and H. Massoudi, "Long-wavelength electromagnetic power absorption in prolate spheroidal models of man and animals," *IEEE Trans. Microwave Theory Tech.*, vol. MTT-23, pp. 739-747, 1975.

[16] H. Massoudi, C. H. Durney, and C. C. Johnson, "Long-wavelength analysis of plane wave irradiation of an ellipsoidal model of man," *IEEE Trans. Microwave Theory Tech.*, vol. MTT-25, pp. 41-46, 1977.

[17] G. I. Rowlandson and P. W. Barber, "RF energy absorption by biological models: calculations based on geometrical optics," in *Abstracts 1977 Int. Symp. Biological Effects of Electromagnetic Waves* (Airlie, VA), p. 50, 1977.

[18] H. Massoudi, C. H. Durney, and C. C. Johnson, "The geometrical optics solution and the exact solution for internal fields and SAR in a cylindrical model of man irradiated by an electromagnetic plane wave," in *Abstracts 1977 Int. Symp. Biological Effects of Electromagnetic Waves* (Airlie, VA), p. 49, 1977.

[19] P. W. Barber, O. P. Gandhi, M. J. Hagmann, and I. Chatterjee, "Electromagnetic absorption in a multilayered model of man," *IEEE Trans. Biomed. Eng.*, vol. BME-26, pp. 400-405, 1979.

[20] A. C. Eycleshymer and D. M. Schoemaker, *A Cross Section Anatomy*. New York: Appleton, 1911.

[21] D. J. Morton, *Manual of Human Cross Section Anatomy*. Baltimore, MD: Williams and Wilkins, 1944.

[22] H. Massoudi, C. H. Durney, C. C. Johnson, and S. Allen, "Theoretical calculations of power absorbed by monkey and human prolate spheroidal phantoms in an irradiation chamber," in *Biological Effects of Electromagnetic Waves* (Selected Papers of USNC/URSI Annual Meeting, Boulder, CO), vol. II, 1975, pp. 135-157. [HEW Publication (FDA) 77-8011, U.S. Government Printing Office, Washington, DC 20402.]

[23] C. H. Durney, C. C. Johnson, P. W. Barber, H. Massoudi, M. Iskander, S. J. Allen, and J. C. Mitchell, *Radiofrequency Radiation Dosimetry Handbook: Second Edition.* Dep. Bioengineering, Univ. Utah, Salt Lake City, 1978.

[24] O. P. Gandhi, J. A. D'Andrea, B. K. Jenkins, J. L. Lords, J. R. Mijanovich, and K. Sedigh, "Behavioral and biological effects of resonant electromagnetic absorption in rats," Tech. Rep. UTEC MD-75-174, Microwave Device and Physical Electronics Lab., Univ. Utah, Salt Lake City, 1975.

[25] H. Jasik, Ed., *Antenna Engineering Handbook*. New York: McGraw-Hill, 1961.

[26] H. V. Cottony and A. C. Wilson, "Gains of finite-size corner reflector antennas," *IRE Trans. Antennas Propagat.*, vol. AP-6, pp. 366-369, 1958.

[27] M. J. Hagmann, O. P. Gandhi, J. A. D'Andrea, and I. Chatterjee, "Head resonance: numerical solutions and experimental results," *IEEE Trans. Microwave Theory Tech.*, vol. MTT-27, Sept. 1979.

[28] W. T. Joines and R. J. Spiegel. "Resonance absorption of microwaves by the human skull,"*IEEE Trans. Biomed. Engin.*, vol. BME-21, pp. 46-48, 1974.

[29] C. M. Weil, "Absorption characteristics of multilayered sphere models exposed to UHF/microwave radiation," *IEEE Trans. Biomed. Engin.*, vol. BME-22, 1975, pp. 468-476.

[30] M. F. Iskander, P. W. Barber, C. H. Durney, and H. Massoudi, "Near-field irradiation of prolate spheroidal models of humans," presented at the Open Symp. Biological Effects of Electromagnetic Waves, 19th General Ass. of URSI, Helsinki, Finland, August 1-8, 1978 (to be published).

[31] I. Chatterjee, O. P. Gandhi, M. J. Hagmann, and A. Riazi, "A method of calculating electromagnetic absorption under near-field exposure conditions," presented at the Bioelectromagnetics Symp., Univ. of Washington, Seattle, WA, June 18-22, 1979 (to be published).

Part II
Biophysical Chemical Basis of RF Field Interactions

The varied approaches to biophysical interactions meant to predict or measure RF field perturbations of systems ranging from water to organized lamellar systems have led us to give capsule sketches of selected papers and a reasonably extensive reference list in order that the interested reader may select his own area of interest. We have not attempted to be exhaustive, but only to select those papers best illustrating a particular point. Many worthy papers have not been included.

"Life is, indeed, 'a thing of watery salt', enmeshed in a macromolecular framework. Its rich play of behavior emerges from the inhomogeneities of its simplest components as these become magnified through interaction and combination. . . . If the stuff of life is built from water, simple ions, molecules and macromolecules, then particularities which seem trivial at the molecular level must magnify into singularities of great moment at higher levels of organization."

(R. W. Gerard, introduction to *A Physical Theory of the Living State*, G. H. Ling, New York, NY: Blaisdell Pub. Co., 1962.)

The interaction of microwaves with biological systems has historically been considered to be predominantly with water and dissolved ions, producing heat. The extent of interaction is dependent on frequency and the induction of rotational modes in water and dissolved ions [75]. Schwan and his colleagues [21, 33, 64] have measured the dielectric properties of several biological systems over a period of some 30 years and conclude that the major properties are those of an ionic solution, with some correction, a small fraction of the total water present, for the water of hydration of several proteins and other macromolecules.

Biophysical Consequences of General Hyperthermia

Because temperature elevation is quantitatively an important aspect of field interactions with biosystems, some understanding of the nature of bioresponse to hyperthermia is required to evaluate the effects of RF interaction. The kinetics of thermal results on tissue processes, enzyme rates, thermal denaturation, and diffusivity are all parts of classical cellular physiology [28] while processes such as electrokinesis and other transport processes have been known to have thermal coefficients for many years [24].

Recently, the development of conceptions of cellular membranes as lipid bilayers, with molecular groupings extending outside and inside of this bilayer, and with the bilayer including enzymatically active proteins with processes extending outside and inside the membrane, has emphasized the importance of the state functions of the membrane; these state functions have temperature coefficients which could lead to entropy transitions over a fairly narrow temperature range [65, 66, 3]. Lelchior and Steim [39] have measured such transitions in a variety of bacteria and in some mammallian cultured cells. Lords [44], Reed *et al.* [60], and McArthurs *et al.* [48] have exposed several tissues to 5 mW cm^{-2} fields, and have noted the appearance of neurotransmitter dependent effects after a short period of radiation. These sets of experiments have attempted to control thermal excursions, but microthermal events at interfascial planes in the tissue could hardly have been avoided at fields of this amplitude. While one legitimately speculates that such effects could have been produced by the types of lipid entropy transitions measured by Lelchior and Steim, it is also possible that other types of reactions could occur, such as complement activation with resultant neurotransmitter release [36].

It is usually considered that mammals exist in a restricted temperature range below 40°C, excluding muscles during heavy exercise and hyperpyrexic responses. Other living organisms, however, indicate that adaptation to differing temperatures may be under chromosomal control. Thus, adaptation of Thermophilus bacteria to high temperatures was reviewed by Amelunxen and Murdock [5], who showed that some species require an adaptation period before growth at high temperatures can occur, and that this adaptation is accompanied by an alteration of composition of membrane lipids, several proteins, and some of the RNA's. The thermal stability of these gene products is increased considerably from those which exist at lower temperatures. Similarly, Wickner [77] has reviewed the thermally unstable F factors in high frequency recombinant mutants of *E. coli*, conferring temperature sensitivity on several mutant lines.

Justesen [35] has used some of the Thermophilus species reviewed by Amelunxen and Murdock and radiated them with a "method of thermal equivalency," in which he maintained high growth temperatures by application of microwave fields, and used as controls other cultures maintained at nearly the same temperature by means of ordinary heating. He found that several membrane blebs occured in the radiated species which did not occur in thermally maintained cultures. There was a later initiation of log phase growth in the culture maintained with microwave radiation. The high temperature growth cutoff by this species is quite sharp, so that small thermal excursions may have contributed to some of the results; the appearance of the blebs is interesting in view of Lelchior and Steim's work on thermally induced entropy transitions in membranes.

Further evidence for genome intervention in adapting to different thermal environments has been obtained from drosophila exposed to a one-hour period of higher temperature followed by intensive analysis of chromosomes, mRNA, and proteins synthesized subsequent to that event. New chromosomal "Puff" regions associated with active transcription have been located, as well as alterations in at least six mRNA's and six new proteins [50, 61]. Johnson *et al.* [34] noted that

Manuscript received July 15, 1981.

humans exposed to therapeutic hyperthermia were quite apt to develop thermally resistant invasive tumors, an effort duplicated in cultured cells, though little other definition of this alteration appears to have been done.

Drosophila was exposed to 16.5 GHz fields by Braver [11], who was able to find no genetic alterations associated with such exposure. Unfortunately, he did not examine his specimens for the type of alteration that would have been expected from hyperthermia. On the other hand, Conover *et al.* [14] were able to obtain numerous teratological changes in rats exposed to high intensity fields sufficient to produce hyperthermia of 4–5°C over normal rectal temperatures. Again, this work demonstrates some form of genome alteration in developing cells, due to hyperthermia.

The thermal sensitivity of DNA repair processes was reviewed by Drake and Baltz [16]. Background ionizing radiation is sufficient to cause a probability of mutation of 10^{-8}/base pair/day, or 100 mutations per day per liver if DNA repair processes were inactivated. The peculiar thermal sensitivity of these repair processes in at least some tissues forms part of the rationale for the use of hyperthermia as an adjunct to radiation therapy of malignancies.

Pyle *et al.* [59] have shown that short, high-intensity pulses of fields at 2450 mHz were sufficient to cause lethality in zebra fish embryos at fields and temperatures not expected to produce such lethality on the basis of thermal controls. Electron micrographs of some of the embryos seem to indicate the initiation of cell membrane disruptions in such processes. In a more recent publication where adequate thermal controls were used, this group shows that survival curves for thermal insult and field application are completely separate from each other. Similarly, Modak *et al.* [51] exposed mice to radiation fields designed and demonstrated to radiate the brain selectively, and found that pulses calculated and measured to produce a temperature change of 2–3°C in 15–25 ms was sufficient to cause complete, flaccid paralysis followed at times by a convulsive episode and depressed activity for hours after the initial insult. Brain acetylcholine content measured with a rapid inactivation technique was still declining at the end of the observation period. Such efforts as these have previously led to the speculation that the rate of temperature change might be an important predictive variable in attempting to predict the effects of transient hyperthermia.

Less severe hyperthermia more protracted in time has repeatedly been shown to produce adrenal hormone release in response to pituitary hormones. Professor Michaelson [49] has felt for many years that many, if not most, of the effects of RF fields are explicable on the basis of such a hormonal response; indeed, his group has demonstrated the alterations in ACTH, corticosterone [45], and other endocrine responses which would be expected on the basis of hyperthermia. Liburdy [42, 43] has also shown that some of the alterations of the immune system are consistent with endocrine-induced alterations following the classical "Alarm" response of Selye, in agreement with Michaelson's contention of many years. The fact that these are non-trivial effects in fields adequate to produce hyperthermia of 1–2°C can be appreciated by reviewing the anti-inflammatory and immunosuppressive effects of the adrenal steroids [55], and appreciating the probable alteration of response to bacteriological or viral challenge this could cause. Unfortunately, the various types of mammals do not have the same sensitivity to steroids, so that extrapolating an effect from the steroid sensitive rodents to the relatively less sensitive primates introduces modeling levels not yet commonly attached to RF modeling conceptions.

Levels of hyperthermia producing lethal responses were once thought to be on the order of 45–56°C. However, Boddie *et al.* [10] found that liver perfusion at carefully defined temperatures of 42.5–43°C could be lethal, particularly when the perfusions were continued for 20 minutes or longer. This is consistent with the types of changes at similar temperatures found by Rupp *et al.* in rats exposed to HF band radiation.

"Simple" generalized hyperthermia, then, is capable of inducing alterations of genetic expression in somatic cells, state changes and permeability changes in membrane systems, and triggering endocrine responses appropriate to thermal stress with wide ranging effects on the immune system and many other biochemical processes.

Regional Hyperthermia

Hyperthermia induced by RF fields can be quite sharply localized by the nature of current carrying transmembrane structures or by design of field applicators. An example of field localization by applicator design has been described by Stavinoha *et al.* [70], who have used rapid microwave-induced thermalization of the brain to study high-turnover intermediates in brain metabolism and circulation of the blood in brain regions. Here we use this technique as an extreme example of the fact that when very high fields deposit energy in small regions, biological inactivation occurs without remarkable excursion in the colonic temperature. Thus, regional distributions demonstrated by Guy's thermographic techniques for many years become important in predicting details of regional hyperthermal response to RF fields.

On a smaller scale, Tyasolev [73, 74] and his coworkers have used model lipid membrane systems quite similar to those currently used for immunologic processes [36] to study "Microthermal" effects on antibiotic ion carriers. In general, Tyasolev interprets his results as being due to change in average collision temperature of the structure of the charge carrier itself or the "Pore" in which it is located, though the precise microwave effect produced depends on the nature of the charge carrier and the presence of calcium ion in the bathing medium. Under conditions producing striking changes in conductivity when the field is applied, repeated measurements of bulk temperature just external to the membrane reveal very small, if any, temperature changes on field application. The characteristics of "black lipid membrane" systems and a summary of the types of compounds forming ion/charge carriers in this system has been extensively reviewed by Hladky [27]. The arrangement of Tyasolev's membranes assures that a transmembrane RF potential is introduced, and there is a linear dependence on the amplitude of the microwave effect and the SAR. Schwan [64] has done extensive theoretical and experimental work indicating that transmembrane potentials of the magnitude used by Tyasolev are unlikely to be found in most cellular systems. However, in view of the lateral mobility of known receptor sites in membrane systems, as for instance in electrophoretic experiments with concanavilin A sites, as shown by Moo Ming Poo and Robinson [58] or Jaffe [30], and the importance of such groups in cell surface recognition processes as described by Hood *et al.* [29], or in immunogenetic properties as reviewed by Old and Stockert [53(a)], it may be wise to await the results of RF studies specifically designed to apply the concepts developed by Tyasolev to lateral mobility processes before making decisions as to the biological import of microthermal pore related processes.

One aspect of Tyasolev's work deserves very careful consideration. He found that use of an ion-selective charge carrier could make the microwave induced alteration in conductivity quite specific. Biologically relevant carrier mechanisms are already known to be ion selective, so it is quite possible that maintenance of an ionic gradient in the presence of such a charge carrier could indeed produce a microthermal event of biological relevance. In an experiment reported by Olcerst [53], a temperature-permissive microwave effect on efflux of sodium and rubidium from red cells was found, but the fields employed did not themselves generate hyperthermia in bulk fluids sufficient to produce the effects seen. While effects on ion fluxes themselves might seem easily reversible and unlikely to have permanent biological effect, a recent conference on growth regulation by ion fluxes demonstrates the large number of proliferative processes which could be affected by such a concept [38].

Field-Dependent Effects Not Directly Associated with Thermal Environments—"Non-thermal" Interactions

The term "athermal effect" has probably generated more heated debate than any other term in bioelectromagnetics and related literature for the past 30 years. The term has been objected to on the basis of physically impossible epistemology, but has already outlived several of its detractors and seems firmly embedded in the language. It has, however, several different usages, only some of which indicate a bioeffect of sufficient prominence to be measurable but not be associated with an extensive temperature change in any body region.

One use of the term, fairly common in some of the older literature, was simply any bioeffect produced without an alteration of the colonic temperature. We have already indicated that knowledge of field distributions, as in the case of brain inactivators, should make such use less frequent. Microthermal environments, as used by Tyasolev, deserve much more intensive study, and could very well be of importance in several situations not presently accounted for.

The origin of the term, however, was to describe effects produced at cell surfaces and in colloidal solutions which could not directly be associated with the temperature of the solution (see Turner [75] for early history). This usage is against a biological background which indicates that low frequency fields up to a few hundred Hz need be associated with amplitudes of 10^{-8} – 10^{-9} V/cm to produce demonstrable effects on neural electrical activity [1], a fact not known when the original argument began, as several other processes were not, such as have become grouped together under headings of Functional Linkages, including cooperative, transductive, or allosteric types of molecular interactions [63(a)].

Zimm [79] has given a most cogent explanation of the nature of cooperativity based on an Ising model. Briefly, the term describes the shift in affinity constants which are likely to occur on aggregation of subunits of several different kinds of molecules, of which hemoglobin was one of the first examples [17]. The shape of the hgb-oxygen association curve is sigmoidal; most mass-action hypothesis would have it hyperboloid. A low constant for first association, followed by cooperative interaction between subunits increasing affinity and resulting in proton ejection, results in the characteristic sigmoid shaped curve of a cooperative process. Alternatively, allosterism refers to a nonsubstrate interaction with an enzymatic protein controlling the rate of enzyme reaction. Such allosterism can itself initiate further cooperative processes and conformational changes in other protein subunits, conferring another characteristic of cooperative processes—a very high system gain—so that a small perturbation in a chemical sense confers very large changes in reactivity. Transductive alterations utilize either or both of these principles to alter a physical pathway which is unrelated but embedded in the same structure, as in membrane-associated complexes. These very general principles have been applied to helix-coil transitions in nucleic acids and proteins [79], behavior of membrane lipids [54], and ion-association phenomena, especially those involving calcium.

Most of these processes use much older concepts of organic chemistry: Ingold's inductomeric effect and direct field effect, quoted by Ling [43(a)], Streitweiser's solvolysis concepts and leaving group stabilities, also quoted by Ling. Molecular polarizabilities in these contexts simply show the possibility of small electron displacements allowing considerable freedom in bond rotation or altering the probability of association in metal-ligand complexes.

More recently, Gelen and Karplus [23] and Levy and Karplus [41] have shown that an investment of 5 kcal in bond angle changes in pancreatic trypsin inhibitor can alter the energy required for tyrosil group rotation by nearly 200 kcal/mole, and that several other chemical potentials in this molecule are similarly altered. Remembering that, at frequencies above 300 mHz, rotational responses of most molecules is usual, part of the polarizability constant for this molecule includes a work function for such a transition. Thus, a small amount of invested energy has resulted in a group availability not at all expected on usual thermodynamic considerations.

That such considerations apply to macroscopically observable events is dramatically shown by collagen membranes when ionic solutions containing sodium are exchanged for lithium, followed by an energetic contraction of a substance usually thought of as a structural tendon material. Under slightly different conditions, a low-frequency pulsatile flow was shown to induce a potential across collagen membranes by Grodzinsky and Melcher [26]. This potential depends on ionic concentration, pH, and is probably related to effective pore radius.

Frequencies at which these kinds of results can be obtained are quite low, on the order of a few (1–8)Hz. The effect is also reversible; application of a potential can induce a mechanical response. Thus a fairly small influence on bond angles (a fairly small proportional change in helix content) produces a grossly observable response. This is a system which was worked out as a problem in equilibrium mechanochemistry by Rubin *et al.* [62].

One would expect such changes in mechanical properties to be associated with alterations in acoustic properties. This problem has been a consistent line of research by Prohofsky's group for several years [18, 19, 20, 46]. They have found that there is an optical/acoustic mode crossover for some DNA homopolymers at 9 GHz. The methodology merits consideration since it has successfully been used to predict much of the infrared absorption vibrational frequencies. Several of the modes calculated depend on a "cooperative" interaction between dimers. Higher-order polymers are not yet calculated.

Somewhat similar calculations which fit experimentally derived spectra have been accomplished for carefully defined proteins by Peticolas [57] who calculated longitudinal vibrational modes at 5/cm for a 100-peptide model protein. In view of work done by Maret *et al.* [47] with Brillouin scattering analysis of oriented DNA fibers, experimental validation of field driven acoustic modes appears not too far from realization.

Albanese and Bell [2] have treated cellular contents as two types of molecular species and show that, if the Einstein promotion coefficient is disproportionate between them, the proportion of molecules in an excited energy state at any time is dependent on mechanisms of decay. Fields which could perform such a promotion are quite low in the type of Ising system (not specifically denoted such by the authors) examined. The properties of rigid colloidal suspensions exposed to fields of varying frequency were examined as a polarization phenomenon which could be matched by flow birefringence by Thurston and Dowling [72]. As expected from Debye theory, larger and larger phase delays occurred above a frequency of 1–2 mHz. A discussion of polarized fluorescence in dc and low-frequency fields by Sokerov and Weil [68] indicates dielectric saturation phenomena at about 6 kV/cm in sonicated DNA particles labeled with acridine orange. On the other hand, Raman spectra by Klainer and Frazer [37] of samples exposed to 100 mHz fields showed peptide vibrational alterations in protein chemotrypsin and disruption of G-C hydrogen bonding in *E. coli* t-RNA. In an extensive series of infrared spectra of calf-thymus DNA, Cody *et al.* [12] were unable to demonstrate significant alterations of DNA under the same exposure conditions. Auer *et al.* [6], using dc pulses, showed that high amplitude fields were able to cause viral and macromolecular trapping in red cells. Quite obviously, there is a strong frequency component in producing such effects, as well as a strong dependency on the magnitude of intramolecular association.

Dielectric dispersion of DNA solutions and the various dielectric theories which could be used to treat the data were reviewed by Cole [13], who showed that estimates of permanent dipoles were related to the assumed molecular geometry and that there could be a large discrepancy in interpretation, depending on the use of Onsager, Kirkwood, Toupin and Lax, or Wertheim approximations. In a somewhat similar vein, Adey [1] mentions the very large low-frequency dielectric of brain as possibly due to large numbers of polarizable molecules in the intercellular space for which no geometry is apparently assigned at present.

For many years, there has been a fairly lively controversy concerning the state of water in biological systems, and whether or not water contributes to a gross crystal structure or has properties of a fluid structure one would find in ionic solutions. Jenin and Schwan [33] and Foster and Schwan [21] have examined the dielectric properties of red cells and barnacle muscle very carefully over a wide frequency and temperature range, and could find no properties not expected from those of an ionic solution of the same temperature. They were able to detect the water of hydration of major macromolecular components in red cells, estimating them to be about 5% of the total intracellular water. Their conclusions are reasonably consistent with the spin-lattice and spin-spin relaxation times measured by Fung and McGaughy [22], who also detected the hydration coats of muscle water, finding that the hydration water relaxation was fairly similar to the relaxation of some of the muscle protein groups.

Evidence that direct perturbation of the electronic structure of carbon chains of macromolecules occurred in RF fields was obtained by James *et al.* [31, 32], who used an off-resonant RF field to perturb the resonant structure of sickle cell hemoglobin. While done for analytical purposes with very careful NMR techniques, the direct demonstration of such a modulation of locally-generated magnetic fields by an external RF field is indeed interesting.

It is usually held that molecular disruptions not involving bond breakage or some high energy rearrangement are not capable of producing a lasting bioeffect. Recent studies of isolated liposomes, a dispersate of pure lecithin, show how fallacious such a conception can be. Kinskey and Nicolotti [36] have reviewed work in which glucose or other marker molecules are trapped, during the formation of liposomes, together with trapped purified specific antibody. Addition of the appropriate antigen and complement can cause complete lysis of the liposome and release of the trapped marker. Simply adding appropriate antigen is sufficient to cause release of the marker. In neither case is there a molecular degradation of any of the components present, but there is certainly a change in the state of the membrane lipid. Since the system is a model of immune responses at the cellular level, it can be seen that it is not necessary to have interactions producing molecular degradation in order to have far-reaching biological consequences.

With the above as background, it is profitable to examine the chief evidence for effects of fields which produce recognizable bioeffects under conditions where hyperthermia cannot be produced as a direct result of field interaction. Adey's group [1] has performed numerous experiments on calcium efflux from chick brain halves which show that fields of some

1 mW/cm^2 can produce alterations of calcium efflux when modulated at a particular frequency, in this case about 16 Hz. The group had previously shown that modulated VLF fields could produce errors in time estimation in primates. The effects of modulated fields on calcium efflux were subsequently confirmed by Blackman *et al.* [9] but questioned in a recent abstract by Albert [4]. EEG alterations by chronically applied RF fields were obtained by Takashima and Schwan [71] at levels comparable to those producing the behavioral effects noted by Adey's group. Taken together, Adey's group has found that the sign of the calcium runout effect reverses somewhere between 60 Hz and 147 mHz, but that higher frequencies don't seem remarkably sensitive to carrier frequency. Modulation frequencies are quite sharp. Albert *et al.* [3] found that there was a modulation dependent release of calcium from pancreatic slices, but that calcium dependent protein secretion wasn't affected. The discrepancy in brain slice findings hinges on two points: the small proportion of total calcium involved, and whether total loss or loss in the eluate is measured. Adey had already demonstrated that metabolically poisoned brains still showed modulation-dependent calcium release, and Grodsky's [25] hypothesis only required a cooperative binding in a surface sheet. Neither hypothesis requires interaction with secretory processes or predicts effects on them. This is one effect that certainly will be intensively studied in the future.

A number of systems have been reported which involve extremely sensitive methodologies for bacterial genetic exchange, the mating pairs of several microorganisms. Present genetic map calibrations for *E. coli* under standard mating conditions show that the entire bacterial genome is transferred in 109 minutes. This transfer occurs with the formation of a sex pylus and transfer of single-strand DNA. The genetic map is constructed with marker mutants and very careful genetic analysis of the crosses made. The conjugation pair is kept together by forces so tenuous that the process can be disarranged by gentle shaking of the incubation mixture, a regular procedure for genome dissection; Webb [76] reported disruption of such a transfer with reorientation of transferred characteristics by millimeter band fields.

This series of experiments excited considerable interest, as some of the resonances observed were less than 100 mHz wide at 60 gHz, counter to any known theory of interaction. Further, the type of analysis performed is not familiar to those usually adept at analysis of the circular map of *E. coli*. The experiments have not been repeated, and apparently no attempt has been made to replicate the precise experiments performed by Webb. Smolenskaya and Vilenskaya [67] studied colicin synthesis and sensitivity by apposing a synthetic strain and a sensitive strain of *E. coli*, irradiating the mixed culture with wavelengths from 6.5-6.58 mm. They detected sharp peaks in sensitivity at separations of about 0.02 gHz, with amplitudes of 0.01 mW/cm^2, but with a cutoff of effect at amplitudes above that. Similarly, Davyatkov [15] found effects on yeast cultures at 7.18 mm, but a depressed growth and clumping at 7.16 and 7.19 mm. Such effects are similar in direction and at frequencies similar to effects in tumor cells reported by Stamm *et al.* [69] and by Webb and Booth [78].

All of these experiments are marked by uncertainties in the exact nature of the exposure apparatus, and means have been taken to assure a phase-plane stability at the entrance to the sample when different sample loadings were added with phase-plane or mode stability in the sample compartment. Thus, the type of coupling into the sample is very much in doubt and the absolute SAR is very much open to question. At a much lower frequency (1.07 gHz), Moody *et al.* [52] showed that bacteriophage incorporation into resistant bacteria could be accomplished after pretreating the bacteria to microwave generated hyperthermia, but that a similar treatment delivered to the isolated bacteriophage decreased its infectivity at fairly low temperatures.

It is perhaps worthy of note that the experiments reporting low field effects, with caveat of the actual magnitude, involved transfer in a multicellular system or one with known phage activities. Replication of the analytical results reported by Stamm *et al.* has been unsuccessful, but very little work has been done on bacterial transfer systems as opposed to growth systems in this country. Since many common bacterial and mycotic species, as well as some protozoans, have viral infestations which confer several different genetic characteristics on the affected organism [40], including antibiotic resistance for instance, it would seem most sensible to study field effects on such relatively discrete systems. Blackman *et al.* [8] for instance, attempted to replicate some features of Webb's experiments, but were unsuccessful in spite of their great care in growth stage, sample exposure conditions, and determination of the thermal characteristics of their exposure system. Their power densities of 2.45 gHz spanned those reported by Smolanskaya and Vilenskaya [67] and those of Webb [76] at higher frequencies, but the sharp resonance peaks reported by these workers have not been replicated as yet.

Summary

Most of the reported effects of RF and microwave radiation in biological systems seem adequately explained as direct response to hyperthermia, or induction of synthetic systems to adapt to it. A series of experiments demonstrating calcium efflux and behavioral changes in fields modulated at specific frequencies determined by the radiated structure indicate there may be some nonlinear response with generation of bioeffect by modulating some internal system event. Work on multicellular systems where such a cooperative event could be analyzed is in its infancy.

J. W. Frazer, Ph.D.
Associate Editor

References and Bibliography

[1] W. R. Adey, "Evidence for cooperative mechanisms in the susceptibility of cerebral tissue to environmental and intrinsic electric fields," in *Functional Linkage in Biomolecular Systems*, F. O. Schmidt, D. M. Schneider, and D. M. Crothers, New York, NY: Rover Press, pp. 325-342, 1975.

[2] R. Albanese and E. Bell, *Radiofrequency Radiation and Living Tissue—Theoretical Studies*, *USAF SAM TR 79-41*, Dec. 1979.

[3] E. N. Albert, C. F. Blackman, and F. Slaby, "Calcium dependent secretory protein release and calcium efflux after UHF electro-

magnetic radiation of rat pancreatic slices," URSI Symposium, Paris, France, 1980.

[4] E. N. Albert, F. J. Slaby, K. Patumras, and Q. Balzane, "147 mHz RF irradiation does not increase calcium release from chick brains," Bioelectromag. Soc., San Antonio, TX, 1980.

[5] R. E. Amelunxen and A. L. Murdock, "Mechanism of thermophily," *CRC Critical Reviews in Microbiology*, vol. 6, no. 4, pp. 343-393, 1978.

[6] D. Auer, G. Brandner, and W. Bodemer, "Dielectric breakdown of the red blood cell membrane and uptake of SV40 DNA and mammallian cell RNA," *Naturwessenschaften*, vol. 63, p. 391, 1976.

[7] H. Biessman, S. Wadsworth, B. Levy, W. McCarthy, and B. J. McCarthy, "Correlation of structural changes in chromatin with transcription in the drosophila heat-shock response," *Cold Spr. Harbor Symp. Quant. Biol.*, vol. 42, no. 2, pp. 829-834, 1977.

[8] C. F. Blackman, S. G. Benane, C. M. Weil, and J. S. Ali, "Effects of non-ionizing electromagnetic radiation on single cell biologic systems," *Ann. N.Y. Acad. Sci.*, vol. 247, pp. 352-366, 1975.

[9] C. F. Blackman, J. A. Elder, C. M. Weil, S. G. Benane, D. C. Eilhinger, and D. E. House, "Induction of calcium ion efflux from brain tissue by radiofrequency radiation. Effects of modulation frequency and field strength," *Radio Sci.*, vol. 14, no. 65, pp. 93-98, 1979.

[10] A. W. Boddie, Jr., L. Booker, J. D. Mullins, C. J. Buckley, and C. M. McBride, "Hepatic hyperthermia by total isolation and regional perfusion *in vivo*," *J. Surg. Res.*, vol. 26, pp. 447-457, 1979.

[11] Dr. Gerald Braver, "Effect of 16.5 gHz microwaves on genetic characteristics of the fruit fly drosophila milanogoster," *4th OTP Report on Elect. Pollut. Envt.*, D-23, 1976.

[12] C. A. Cody, A. J. Modestino, P. J. Miller, and S. M. Klainer, Final report Contract F 41609-75-C-0043, Block Engr. Co., Cambridge, MA to USAFSAM RZ, Jan. 9, 1976.

[13] Robert H. Cole, "Dielectric theory and properties of DNA in solution," *Ann. N.Y. Acad. Sci.*, vol. 303, pp. 59-73, 1977.

[14] D. L. Conover, J. M. Lary, and E. D. Foley, "Induction of teratogenic effects in rats by 27.12 mHz RF radiation," *International Microwave Power Inst. Symposium Abstr.*, p. 31, Ottawa, 1978.

[15] M. D. Davyatkov, "Influence of millimeter-band electromagnetic radiation on biological objects," *Usp. Fiz Nauk*, vol. 110, pp. 452-469.

[16] J. W. Drake and R. H. Baltz, "The biochemistry of mutogenesis," *Annu. Rev. Biochem.*, vol. 45, pp. 11-37, 1976.

[17] S. J. Edelstein, "Cooperative interactions of hemoglobin," *Annu. Rev. Biochem.*, vol. 44, pp. 209-232, 1975.

[18] J. M. Eyster and E. W. Prohofsky, "Lattice vibrational modes of poly (rU) and poly (rA)," *Biopolymers*, vol. 13, p. 2505, 1974.

[19] J. M. Eyster and E. W. Prohofsky, *Biopolymers*, vol. 16, p. 965, 1977.

[20] J. M. Eyster and E. W. Prohofsky, *Phys. Rev. Lett.*, vol. 38, p. 371, 1977.

[21] K. R. Foster, J. L. Schepps, and H. P. Schwann, "Microwave dielectric relaxation in muscle—a second look," *Biophys. J.*, vol. 29, pp. 271-282, 1980.

[22] B. M. Fung and T. W. McGaughy, "Study of spin-lattice and spin-spin relaxation times of ^{1}H, ^{2}H, and ^{17}O in muscle water," *Biophys. J.*, vol. 28, pp. 293-304, Nov. 1979.

[23] B. R. Gelin and M. Karplus, "Lidechain tortional potentials and motion of amino acids in proteins: bovine pancreatic trypsin inhibitor," *Proc. Natn. Acad. Sci.*, USA, vol. 72, pp. 2002-2006, 1975.

[24] S. Glasstone, *Text Physical Chemistry*, 2nd ed., pp. 1220-1230, 1959. *Electrokinetic (Zeta) Potentials*, NY: D. Van Nostrand, 1959.

[25] I. T. Grodsky, "Possible physical substrates for the interaction of electromagnetic fields with biologic membranes," *Ann. N.Y. Acad. Sci.*, vol. 247, pp. 117-124, 1975.

[26] A. J. Grodzinsky and J. R. Melcher, "Electromechanical transduction with charged polyelectrolyte membranes," *IEEE Trans. Bio.-Med. Eng.*, vol. BME-23, no. 6, pp. 421-433, 1976.

[27] S. B. Hladky, "The carrier mechanism," *Current Topics Membranes and Transport*, vol. 12, pp. 53-164, 1979.

[28] R. Hober, *Physical Chemistry of Cells and Tissues*, New York, NY: McGraw-Hill (Blakiston), 1945.

[29] L. Hood, H. V. Herang, and W. J. Dreyer, "The area code hypothesis: the immune system provides clues to understanding the genetic and molecular basis of cell recognition during development," *J. Supramolecular Structure*, vol. 7, pp. 531-559, 1977.

[30] L. F. Jaffe, "Electrophoresis along cell membranes," *Nature*, vol. 265, pp. 600-602, 1977.

[31] T. L. James, G. B. Matson, and I. D. Kuntz, "Off-resonance RF field to resonant NMR expt," *J. Amer. Chem. Soc.*, vol. 100, pp. 3590-3594, 1978.

[32] T. L. James, R. Matthews, and G. B. Matson, "Hemoglobin aggregation in oxygenated sickle cells studied by carbon-13 rotating frame spin-lattice relaxation in the presence of an off-resonance radiofrequency field," *Biopolymers*, vol. 18, no. 7, pp. 1763-1768, 1979.

[33] P. C. Jenin and H. P. Schwan, "Some observations of the dielectric properties of hemoglobin's suspending medium inside human erythrocytes," *Biophys. J.*, vol. 30, pp. 285-294, May 1980.

[34] R.J.R. Johnson, J. R. Subjeck, D. Z. Moreau, H. Kowal, and D. Yakar, "Radiation and hyperthermia," *Ann. N.Y. Acad. Med.*, vol. 55, pp. 1193-1204, 1979.

[35] D. R. Justesen, "Method of equivalent temperatures using Bacellus sterothermophylus at 60°C," *Office of Telecommunications Policy 3rd Report, Appendix B*, p. 9.

[36] S. C. Kinsky and R. A. Nicolotti, "Immunological properties of model membranes," *Ann. Rev. Biochem.*, vol. 46, pp. 49-67, 1977.

[37] S. M. Klainer and J. W. Frazer, "Raman spectroscopy of molecular species during exposure to 100 mHz RF fields," *Ann. N.Y. Acad. Sci.*, vol. 247, pp. 323-326, 1975.

[38] H. L. Leffert, Ed., "Growth regulation by ion fluxes," *Ann. N.Y. Acad. Sci.*, vol. 339, 1980.

[39] D. L. Lelchior and J. M. Steim, "Thermotropic transitions in biomembranes," Mullins, *et al.*, Eds., *Ann. Rev. Biophys. Bioeng.*, vol. 5, pp. 205-238, 1976.

[40] P. A. Lemke, "Viruses of Eucaryotic microorganisms," *Ann. Rev. Microbiology*, vol. 30, pp. 105-145, 1976.

[41] R. M. Levy and M. Karplus, "Vibrational approach to the dynamics of an X helix," *Biopolymers*, vol. 18, pp. 2465-2495, 1979.

[42] R. P. Liburdy, "Radiofrequency radiation alters the immune system; modulation of T and B lymphocytes levels and cell mediated immunocompetence by hyperthermic radiation," *Radiation Research*, vol. 77, pp. 34-46, 1979.

[43] R. P. Liburdy, "Effects of radiofrequency radiation on inflammation," *Radio Sci.*, vol. 12, pp. 179-183, 1977.

[43(a)] G. H. Ling, *Physical Theory of the Living State*, New York, NY: Blaisdel, 1962.

[44] J. Lords, "Induced chronatropic effect on the heart," *Ann. Biomed. Engr. Proc.*, vol. 5, pp. 395-409, 1977.

[45] W. G. Lotz and S. M. Michaelson, "Temperature and corticosterone relationships in microwave-exposed rats," *J. Appl. Phys.: Respirat. Environ. Exercise Physiol.*, vol. 44, no. 3, pp. 438-445, 1978.

[46] K. C. Lu, L. L. Van Zandt, and E. W. Prohofsky, "Displacement of backbone vibrational modes of A-DNA and B-DNA," *Biophys. J.*, vol. 28, pp. 27-32, 1979.

[47] G. Maret, R. Oldenbourg, G. Wenterling, K. Dransfeld, and A. Rupprecht, "Velocity of high frequency sound waves in oriented DNA fibres and films determined by Brillouin scattering," presented at workshop on Microwaves Bioeffects, Maryland, May 17, 1979. To be published in *Biopolymers*.

[48] G. R. McArthur, J. L. Lords, and C. H. Durney, "Microwave radiation alters peristaltic activity of isolated segments of rat gut," *Radio Sci.*, vol. 12, no. 65, pp. 157-160, Dec. 1977.

[49] S. M. Michaelson, "The tri-service program, a tribute to George M. Knauff USAF(MC)," *IEEE Trans. Mic. Theory Tech.*, vol. MTT-19, no. 2, pp. 131-146, 1971.

[50] M. E. Mirault, M. L. Goldschmidt-Clermont, A. P. Arrigo, and A. Tissiere, "The effect of heat shock on gene expression in Drosophila Melanogoster," *Cold Spr. Harbor Symp. Quant. Biol.*, vol. 42, no. 2, pp. 819-827, 1977.

[51] A. Modak, W. Stavinoha, and A. Dean, "Effect of short electromagnetic pulses on brain acetylcholine content and spontaneous motor activity of mice," submitted to *Bioelectromagnetics*, June 1980.

[52] E. M. Moody, C. McLaren, J. W. Frazer, and V. A. Segreto, "Effects of 1.07 GHz RF fields on microbial systems," *Proc. Bioelectromagnetic Soc.*, p. 438, Seattle, WA, 1979.

[53] R. B. Olcerst, "A non-linear microwave radiation effect on the passive effects of Na^+ and Rb from rabbit erythrocytes," *URSI/ Bioelectromagnetic Soc. Meeting*, Abstr. NL2-6, Seattle, WA, p. 308, 1979.

[53(a)] F. J. Old and E. Stockert, "Immunogenetics of cell surface antigens of mouse leukemia," *Ann. Rev. Genetics*, vol. 11, pp. 127-160, 1977.

[54] J.A.F. Op Den Kamp, "Lipid asymmetry in membranes," *Ann. Rev. Biochem.*, vol. 48, pp. 47-71, 1979.

[55] J. E. Parrillo and A. S. Fauci, "Mechanisms of glucocorticoid action on immune processes," *Ann. Rev. Pharmacol. Toxicol.*, vol. 19, pp. 179-201, 1979.

[56] I. Prigogine and R. Le Fever, "Membranes, dissipative structures and evaluation," *Cado Chem. Phys.*, vol. 29, p. 1, 1975.

[57] W. L. Peticolas, "Mean-square amplitudes of the longitudinal vibrations of helical polymers," *Biopolymers*, vol. 18, pp. 747-755, 1979.

[58] M.-M. Poo and K. R. Robinson, "Electrophoresis of concanavallin A receptors along embryonic muscle cell membrane," *Nature*, vol. 265, pp. 602-605, 1977.

[59] S. Pyle, D. Nichols, F. Barnes, and A. Gamow, "Threshold effects of microwave radiation on embryo cell systems," *Ann. N.Y. Acad. Sci.*, vol. 247, pp. 401-407, 1975.

[60] J. R. Reed, III, J. L. Lords, and C. H. Durney, "Microwave irradiation of the isolated rat heart after treatment with ANS blocking agents," *Radio Sci.*, vol. 12, no. 65, pp. 161-166, 1977.

[61] F. Ritossa, "A new puffing pattern induced by heat shock and DNP in drosophila," *Experentia*, vol. 18, p. 571, 1962.

[62] M. M. Rubin, K. A. Puz, and A. Katchalsky, "Equilibrium mechanochemistry of collagen fibers," *Biochemistry*, vol. 8, pp. 3628-3637, 1969.

[63] T. Rupp, J. Montet, and J. Frazer, "A comparison of thermal and radiofrequency exposure effects on trace metal content of blood plasma and liver cell fractions of rodents," *Ann. N.Y. Acad. Sci.*, vol. 247, pp. 282-291, 1975.

[63(a)] F. O. Schmidt, D. M. Schneider, and D. M. Crothers, "Introduction: the role of functional linkage in molecular neurobiology," in *Functional Linkage in Biomolecular Systems*, NY: Raven Press, pp. 1-13, 1975.

[64] H. P. Schwan, "Field interactions with biological matter," *Ann. N.Y. Acad. Sci.*, vol. 303, pp. 198-213, 1977.

[65] S. J. Singer, "The molecular organization of membranes," *Ann. Rev. Biochem.*, vol. 43, pp. 805-833, 1974.

[66] S. J. Singer and G. L. Nicholson, "The fluid mosaic model of the structure of cell membranes," *Science*, vol. 175, pp. 720-731, 1972.

[67] A. Z. Smolenskaya and R. L. Vilenskaya, "Effects of millimeter band electromagnetic radiation on the functional activity of certain genetic elements of bacterial cells," *Usp. Fiz. Nauk*, vol. 110, pp. 452-469, 1973.

[68] S. Sokerov and G. Weil, "Polarized fluorescence in an electric field: comparison with other electrooptical effects for rod-like fragments of DNA and the problem of the saturation of the induced moment in polyelectrolytes," *Biophys. Chem.*, vol. 10, pp. 161-171, 1979.

[69] M. E. Stamm, W. D. Winters, D. L. Morton, and S. L. Warren, "Microwave characteristics of human tumor cells," *Oncology*, vol. 29, pp. 294-301, 1974.

[70] W. B. Stavinoha, J. Frazer, and A. Modak, "Microwave fixation for the study of acetylcholine metabolism in cholinergic mechanism and psychopharmacology," D. J. Jenden, Ed., *Adv. Behav. Biol.*, vol. 24, pp. 169-180, 1978.

[71] Takashima and Schwan, "Effect of modulated RF energy on the EEG of mammalian brain," *Radiat. Envt. Biophys.*, vol. 16, pp. 15-27, 1979.

[72] G. B. Thurston and D. I. Dowling, "The frequency dependence of the Kerr effect for suspensions of rigid particles," *J. Colloid. and Interface Sci.*, vol. 30, no. 1, pp. 34-45, 1969.

[73] V. V. Tyasolev, S. I. Alekseyev, L.K.H. Faijova, and V. V. Chertishchev, "Peculiarities of microwave effect on gramicidin-modified bilayers," *URSI/CMFRS Symposium, Electromagnetic Waves and Biology*, Paris, France, June 30-July 4, 1980.

[74] V. V. Tyasolev, S. I. Alekseyev, and P. A. Grigorev, "Change in the conductivity of phospholipid membranes modified by alamectricin on exposure to a high frequency electromagnetic field," *Biophysics*, vol. 23, pp. 750-751, 1979.

[75] J. J. Turner, "The effects of radar on the human body," Arlington, VA, Astia AD 273787, pp. 1-81, 1962.

[76] S. J. Webb, "Genetic continuity and metabolic regulation as seen by the effects of various microwave and black light frequencies on these phenomena," *Ann. N.Y. Acad. Sci.*, vol. 247, pp. 327-351, 1975.

[77] S. H. Wickner, "DNA replication proteins of *E. coli*," *Ann. Rev. Biochem.*, vol. 47, pp. 1163-1191, 1978.

[78] S. J. Webb and A. D. Booth, "Microwave absorption of normal and tumor cells," *Science*, vol. 174, p. 72, 1971.

[79] B. Zimm, "Fundamental principles of cooperative and transductive coupling," in *Functional Linkage in Biomolecular Systems*, D. M. Crothers, Ed., NY: Raven Press, pp. 17-41, 1975.

[80] C. Tanford, "Dielectric constants and dipole moments," in *Physical Chemistry of Macromolecules*, NY: John Wiley, pp. 105-116, 133-137, 1961.

[81] E. C. Pollard, "Thermal effects on protein, nucleic acid and viruses," in *Advances in Chemical Physics*, vol. 7, NY: John Wiley, pp. 201-237, 1964.

[82] Karl H. Illinger, Ed., *Biological Effects of Nonionizing Radiation*, ACS Symposium Series 157, Washington, DC: American Chemical Society, 1981.

Evidence for Cooperative Mechanisms in the Susceptibility of Cerebral Tissue to Environmental and Intrinsic Electric Fields

W. Ross Adey

The cerebral cortex of all vertebrates and the cerebral ganglia of invertebrates produce a rhythmic electrical activity that arises in their closely packed cellular elements. This activity has long been considered to be little more than a "noise" in cerebral tissue, having no direct physiological role in information processing, although the advent of sophisticated computer analyses of the EEG combined with pattern-recognition techniques has increasingly challenged this view. At the very least, it is now clear that fine correlates exist between the EEG and a wide range of behavioral states (Adey, 1974*b*). Nevertheless, adequate evidence has been lacking that would assign a causal role in information processing to the EEG.

How may we proceed further to answer the question of whether the EEG may have a physiological role in brain tissue? Do brain cells sense field potentials such as the EEG, and do such potentials modify their excitability in the genesis of propagated action potentials? What experimental paradigms and criteria should be used in testing for possible interactions between brain cells and extracellular fields? Very importantly, does the anatomical and physiological organization of brain tissue possess properties that might make possible interactions between weak intrinsic or environmental electric fields and the aggregate behavior of a domain of elements, thus precluding observation of these effects in so-called simpler systems?

Classical neuroanatomy and neurophysiology, particularly synaptic neurophysiology, have drawn heavily on simple systems, including spinal motoneurons, aplysian ganglion cells, and cerebellar Purkinje cells. It is doubtful that these are appropriate models for the intrinsic processes unique to cerebral tissue in transaction, storage, and recall of information. The disappointing but inescapable conclusion that we are rapidly approaching an impasse in following cerebral models that consider only synaptic mechanisms and axonal connectivity has invited investigation of all the possible ways in which interactions might occur in cerebral tissue. By experimental isolation of a tissue or its cellular elements, it was hoped to discern better certain properties which may be miniscule in individual elements but substantive in systems as a whole. Complexity may be an inherent and essential quality of cerebral tissue. There is a kinship with Heisenberg's uncertainty principle, not so much in the effects of measurement on the system being measured, but rather in the effects of its experimental isolation, or in the functional limitations inherent in most simple systems.

We have observed effects of weak electric and electromagnetic fields on the behavior of both humans and animals, and have correlated these effects with altered neurophysiological activity and modified brain chemistry. One striking conclusion of these studies is that mammalian central nervous functions can be modified by electrical gradients in cerebral tissue that are substantially smaller than those known to occur in postsynaptic excitation, as well as those presumed to occur with inward membrane currents at synaptic terminals during the release of transmitter substances.

Neither these observations nor models of cerebral organization which arise from them are nihilistic to the impressive body of synaptic physiology. Rather, they suggest the existence of hierarchies of excitatory organization, in which synaptic mechanisms represent but one level.

BRAIN AS A TISSUE IN THE PROCESSING OF INFORMATION

Information processing in brain tissue no longer is considered to involve only the nerve cells. There are strong physiological interactions between nervous and surrounding neuroglial cells, including concurrent electrical changes (Karahashi and Goldring, 1966) and simultaneous and sometimes reciprocal changes in chemical measures. The latter include enzyme activity and protein synthesis (Hydén, 1972), differential accumulation of the putative transmitter substances GABA (Henn and Hamberger, 1971), glycine (Ehinger and Falck, 1971), glutamic acid, aspartic acid, and taurine (Ehinger, 1973). By light microscopy, staining for calcium is low in neuronal cytoplasm, but appreciable in neuroglia (Tarby and Adey, 1967; Adey, 1970).

These interactions occur across the intervening intercellular space, which has been shown by rapid freezing techniques to be approximately 20% of the cerebral volume (van Harreveld, Crowell, and Malhotra, 1965), a figure confirmed by biochemical estimates (Reed, Woodbury, and Holtzer, 1964). The space is occupied by highly hydrated macromolecular material, forming a loose "fuzz." Although polyanionic, this material binds strongly to acidic solutions of phosphotungstic acid (Pease, 1966; Rambourg and Leblond, 1967). Despite its loose structure, it appears to blend with other glycoprotein material to form the outer coats or "glycocalyces" of cell membranes (Bennett, 1963).

Only a small portion of any extracellular current penetrates either neuronal or neuroglial membranes (Cole, 1940). Weak electric currents in cerebral tissue preferentially flow through the extracellular space; tissue

Reprinted with permission from *Functional Linkage in Biomolecular Systems*, edited by F. O. Schmitt, D. M. Schneider, and D. M. Crothers, pp. 325–342, 1975.

impedence measurements therefore appear to reflect primarily conductance in the extracellular space (Coombs, Curtis, and Eccles, 1959; Nicholls and Kuffler, 1964). Moreover, conductance changes in cortical and subcortical structures accompany a variety of learned responses, suggesting that the cell surface and intercellular macromolecular material may be one site of structural change during information storage and its retrieval (Adey, Kado, and Didio, 1962*b*; Adey, Kado, Didio, and Schindler, 1963; Adey, Kado, McIlwain, and Walter, 1966). As discussed below, these surface regions may have other quite special functions in the detection and transduction of weak chemical and electrical events at the membrane surface (Schmitt and Samson, 1969; Adey, 1972).

The hypothesis has been advanced elsewhere (Adey and Walter, 1963; Adey, 1974*b*) that the characteristic phenomenon of overlapping dendritic fields in the palisades of cells that characterize all vertebrate cerebral ganglia may be associated with the concurrent development of a rhythmic electric wave process. This in turn may be fundamentally related to the ability of these tissues to undergo permanent changes in excitability as a result of their prior participation in specified patterns of excitation. This dendritic organization appears to constitute a specific arrangement in cerebral tissue, and is discussed further below.

HOW MIGHT INFORMATION BE PROCESSED IN BRAIN TISSUE?

Although it is clear that fiber conduction and synaptic activation are essential elements in brain function, there are at least three other modes of information handling in cerebral neurons that merit equivalent attention. These include dendro-dendritic conduction, neuronal-neuroglial interactions across the intercellular space, and the detection of weak stimuli that modify the immediate environment of the neuron. The last would include sensitivity to weak electric (and perhaps magnetic) fields and to minute amounts of chemical substances, including drugs, hormones, and neurohumors. The susceptibility of brain tissue to drugs such as LSD in body fluid concentrations as low as 10^{-9} (Adey, Bell, and Dennis, 1962*a*) is well known and generally accepted, and hormone concentrations that predictably modify brain function are even lower.

It is therefore surprising that such scant consideration has been given to the possibility that brain tissue may be sensitive to field potentials in the environment of the neuron, including the intrinsic fields of the EEG. The EEG appears to be the summed contribution of waves from many neuronal generators, volume-conducted through the extracellular medium. This volume-conducted component will appear simultaneously at points quite distant from the original generators, including the brain surface and scalp. Unlike axonal spike conduction, it will not behave as a propagated disturbance with a measurable conduction velocity, but will exhibit a frequency-dependent attenuation of its high-frequency components. The extent of this attenuation will be determined by the impedance characteristics of the extracellular medium. Intracellular records from cortical neurons show waves with amplitudes up to 20 mV whose frequency spectrum resembles that of the EEG from the same region (Creutzfelt, Fuster, Lux, and Nacimiento, 1964; Elul, 1964, 1967*a*, 1972; Fujita and Sato, 1964). Evoked field potentials have also been correlated with membrane potential deflections in intracellular records (Purpura, Shofer, and Musgrave, 1964; Elul and Adey, 1965; Creutzfelt, Watanabe, and Lux, 1966). Approximately 100 μV of signal would be contributed to the volume-conducted signal by a neuronal wave 10 mV in amplitude in intracellular records.

How, then, could such a weak extracellular field influence neuronal excitability? With a membrane potential of 50 mV, a typical depolarization of at least several millivolts is necessary to initiate a propagated spike discharge. Nevertheless, field gradients comparable with the EEG alter firing thresholds in spinal motoneurons (Nelson, 1966).

We may consider a mechanism of "membrane amplification" to account for sensing these fields during the possible sequence of events activating a domain of cortical neurons, here considered to be a group of several hundred cells (Anninos, Beek, Csermely, Harth, and Pertile, 1970). Neuronal spikes would be generated in the axon hillock of the neuronal soma and then pass by short axons or axon collaterals to adjacent neurons in the same domain. In a similar fashion, interdendritic connections would provide paths for the spread of slower neuronal wave activity generated in the dendrites, which might sweep longitudinally toward the soma (Green, Maxwell, Schindler, and Stumpf, 1960). These processes would jointly generate an EEG in the extracellular space enclosing the cell. A system of macromolecular sensors on the membrane surface might transduce components of the extracellular fields by altered ion binding and an associated conformation change. This altered membrane surface would then trigger a transmembrane amplification of ionic movements, thus modifying both the initial events mediated by neuronal slow waves and concurrent membrane potential changes induced by synaptic activation.

We may summarize this working model:

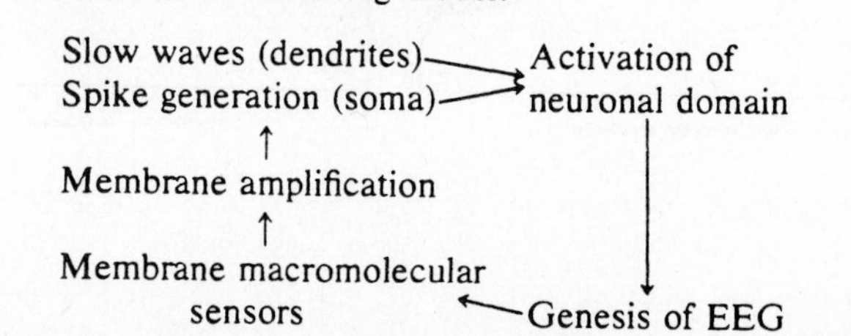

Such a model is hierarchically organized. Molecular events at the membrane surface would influence the excitability of a particular neuron, which in turn would influence others in its domain through conduction processes. The joint activity of other neurons would then produce a volume-conducted field through the domain, and this field would again modify the environment at each neuronal surface. This field is a remarkable "enchanted loom" when recorded at cellular dimensions, showing great differences in adjoining tripolar records along different electrical axes (Elul, 1962).

The following observations of central nervous interactions with weak electric and electromagnetic fields invite serious consideration of this type of model, since they occur at energy levels far below those required for classical synaptic activation, and will lead in turn to a more detailed consideration of membrane surface phenomena that might be involved.

EFFECTS OF LOW-LEVEL, LOW-FREQUENCY ELECTRIC FIELDS ON EEG AND BEHAVIOR

In 1960, Konig and Ankermüller reported altered human reaction times in response to low-level electric fields between 5 and 15 Hz; this was later confirmed by Hamer (1968) in our laboratory. Wever (1968) exposed human subjects to 10-Hz fields with an electric gradient of 2.5 V/m, and measured their patterns of circadian rhythms from sleep-wake cycles and from peaks and troughs in diurnal temperature cycles. After 7 to 10 days of control measurements, the fields were activated for an approximately equal period. Circadian rhythms were shortened by 1 to 2 hr in many subjects during field exposure.

After pilot studies in our laboratory suggested that subjective estimates of the passage of time were influenced by such fields, a detailed study was made of time estimation in the presence of weak electric fields in the pigtail macaque monkey (Gavalas, Walter, Hamer, and Adey, 1970).

Three monkeys were trained to estimate 5 sec between level presses. If the animals pressed with a 2.5-sec reward-enable interval, they were rewarded with apple juice. They were then exposed to the fields for 4 hr each day. In these initial experiments, the fields were generated between two large metal plates 50 cm square and 40 cm apart, and the field amplitude was 2.8 V peak-peak at 7 Hz. Each monkey was tested in two experiments of 20 exposures to 7-Hz fields and two comparable control experiments in the absence of any field. In summary, five of the six experiments showed a shift to significantly faster interresponse time in the 7-Hz fields compared with performance in the absence of fields; all mean differences were 0.4 sec or greater. Shifts in modal values also occurred in all five experiments, and were all 0.2 sec or greater. Although the total output of responses and the variability of those responses differed considerably from monkey to monkey (Fig. 1), the trend to shorter interresponse times in the presence of the fields was remarkably consistent, and the size of the shift was relatively large.

These initial findings have been confirmed and extended in five monkeys.

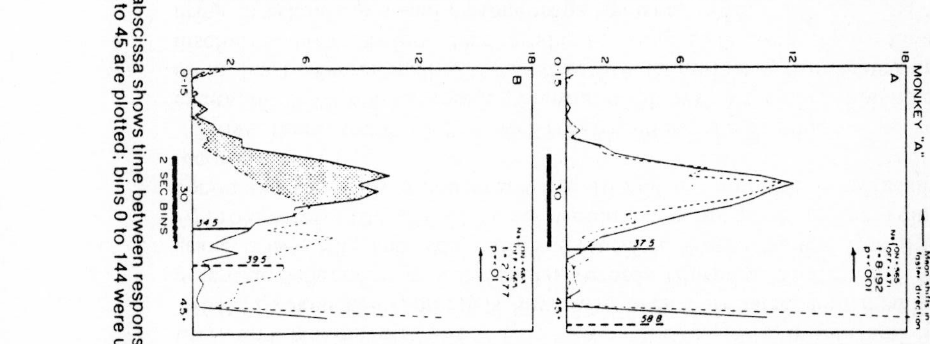

FIG. 1. Behavioral data showing shifts in interresponse time under 7-Hz fields. The abscissa shows time between responses in 0.2-sec bins; the ordinate shows percent of total responses at each interval. Note that only bins 15 to 45 are plotted; bins 0 to 144 were used in calculation of means and standard deviations. (From Gavalas et al., 1970.)

We have tested fields at 7, 10, 45, 60, and 75 Hz, at gradients ranging from 10 to 100 V/m. These later studies have indicated that higher voltage gradients decrease the variability in interresponse times.

Since questions of the possible production by implanted electrodes of antenna effects that might produce a focal field enhancement within the tissue have been widely raised, it is noteworthy that these thresholds and frequency differentials are similar for unimplanted and implanted subjects (Tables 1 and 2). Thus, it seems clear that the field is sensed directly by the organism without intercurrent effects attributable to an implanted electrode system.

TABLE 1. *Unimplanted monkey in a high-voltage field (56 V/m)**

Entire experiment			Hour 1			Hour 2			Hour 3		
Hz	V/m	$\bar{x}$IRT	Hz	V/m	$\bar{x}$IRT	Hz	V/m	$\bar{x}$IRT	Hz	V/m	$\bar{x}$IRT
7	56	5.347	7	56	5.440	7	56	5.697	75	56	5.659
75	56	5.576	75	56	5.498	75	56	5.922	60	56	5.784
60	56	5.756	60	56	5.698	0	0	5.976	7	56	5.819
0	0	5.960	0	0	5.883	60	56	5.987	0	0	6.121

* Rank-order weighted mean interresponse times (last bin excluded).

TABLE 2. *Unimplanted monkey in a high-voltage field (56 V/m)**

Entire experiment			Hour 1			Hour 2			Hour 3		
Hz	V/m	σ	Hz	V/m	σ	Hz	V/m	σ	Hz	V/m	σ
75	56	0.796	75	56	0.714	0	0	0.872	75	56	0.814
60	56	0.897	7.	56	0.821	7	56	0.929	60	56	0.858
7	56	0.913	60	56	0.826	75	56	1.050	0	0	0.943
0	0	1.016	0	0	1.027	60	56	1.081	7	56	1.039

* Rank-order weighted standard deviations (last bin excluded).

EEG records during field exposure were not markedly changed by the fields. All of the monkeys in the initial study showed an altered EEG power output at 6 to 8 Hz in the hippocampus, with less consistent alterations in the amygdala and nucleus centrum medianum with the 7-Hz fields.

EFFECTS ON BRAIN ELECTRICAL RHYTHMS AND BEHAVIOR OF LOW-LEVEL, VERY-HIGH-FREQUENCY (VHF) RADIO FIELDS AMPLITUDE MODULATED AT EEG FREQUENCIES

In a search for more definite evidence that the membrane surface might transduce weak extracellular fields as a step in the excitation process, we have examined the effects of VHF electromagnetic fields, amplitude modulated at EEG frequencies. This approach was based on the strong asymmetry of the fixed charge distribution on membrane surface glycoproteins with respect to extracellular fluid and to deeper layers of the membrane. Such a phase partition would be expected to demodulate the envelope of a high-frequency carrier wave, as in a semiconductor, although remaining unresponsive to the carrier frequency itself. If this hypothesis were correct, low-frequency modulation on the carrier would be a prerequisite for central nervous effects, and differential effects at specific brain sites would depend on particular modulating frequencies.

This hypothesis was strongly supported by our findings in cats exposed to a 147 MHz, 2.0 mW/cm^2 field, amplitude modulated at 0.5 to 30 Hz. Field modulation at frequencies identical with EEG "signatures" in single brain structures sharply reinforced the occurrence of these rhythms, in both conditional and unconditional behavioral situations (Bawin, Gavalas-Medici, and Adey, 1973). Animals trained to produce a specific brain rhythm in the presence of the field (or receive an aversive stimulus) continued to produce the rhythm without aversive stimulation (extinction trials) for 25 to 40 days, but ceased to perform within a few days in the absence of the fields. The fields also increased the rate of occurrence of spontaneous rhythms at specific locations, but only when modulated at frequencies close to the dominant biological frequency of the selected intrinsic EEG rhythm.

Spectral analysis of EEG records showed a concentration of energy around the imposed modulated frequency. However, daily controls with field amplitudes modulated at different frequencies for short periods failed to produce any changes or any artifactual patterns in these highly stable responses. If, despite their shielding up to the head, antenna effects on the electrode leads had induced even small potentials at the tip of the electrodes, spectral analysis would be expected to reveal voltages so induced and rectified at the electrode-tissue contact. The EEG changes were anatomically localized, highly specific in terms of frequency and associated

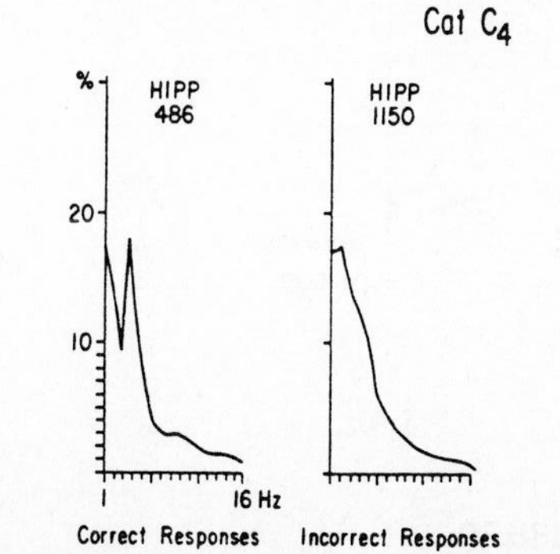

FIG. 2. Comparison of the EEG autospectra of hippocampal patterns in cases of correct and incorrect responses during exposures to VHF fields amplitude modulated at 4.5 Hz. Autospectra were calculated from averages of 20 epochs, each 2.5 sec in duration. (From Bawin et al., 1973.)

with transient patterns. Figure 2 compares the EEG autospectra in one animal accompanying correct and incorrect responses during irradiation with fields modulated at 4.5 Hz (the peak frequency of the response), and shows that the spectral peak shifted away from the imposed frequency when the animal was not performing. It therefore seems unlikely that the tissue effects can be attributed to direct injection of field voltages via the electrodes, a viewpoint strongly supported by the changes in chemistry induced in isolated brain tissue that are described below.

PARTICIPATION OF MEMBRANE SURFACE MACROMOLECULES IN EXCITATION; EVIDENCE FOR COOPERATIVE INTERACTIONS WITH CALCIUM IONS

Our studies strongly imply that the binding and release of calcium ions to membrane surface macromolecules is an important step in these field interactions. This surface glycocalyx greatly extends the effective membrane thickness, perhaps to as much as 2000 Å, and may play a role in the detection of hormones and neurohumoral substances that are effective in minute amounts. Initial conformational changes at the binding site may be followed by transmembrane effects, with molecular "switches" such as prostaglandins triggered in the presence of Ca^{2+} (Ramwell and Shaw, 1970), thus greatly "amplifying" the initial binding energies. We have followed three lines of related research.

The Role of Ca^{2+} in the Release of Ca^{2+} and GABA from Cat Cerebral Cortex

When cat cerebral cortex is equilibrated with $^{45}Ca^{2+}$ and ^{3}H-GABA in the absence of general anesthesia, a small increase in unlabeled Ca^{2+} in the solution bathing the cortex elicited a large release of labeled $^{45}Ca^{2+}$ and labeled GABA (Fig. 3). Moreover, the effect of a 1 mM increment in Ca^{2+} concentration was only slightly less than that of a 20 mM increment (Kaczmarek and Adey, 1973). Mg^{2+} did not trigger the release of either Mg^{2+} or Ca^{2+}. A possible mechanism for this highly nonlinear release of $^{45}Ca^{2+}$ may be the displacement of $^{45}Ca^{2+}$ bound to polyanionic sites on the membrane surface. To be consistent with the $^{45}Ca^{2+}$ efflux and net Ca^{2+} binding, this mechanism might take the form

$$Ca^{2+} + {}^{45}Ca\text{-}M^{n-} \rightarrow Ca - {}^{45}Ca\text{-}M^{(n-2)-}$$
$$Ca - {}^{45}Ca\text{-}M^{(n-2)-} \rightarrow {}^{45}Ca^{2+} + Ca\text{-}M^{n-}$$

where M represents a membrane anionic species. The efflux of Ca^{2+} ions from the membrane would then be proportional to a higher power of the bound Ca^{2+} ion concentration.

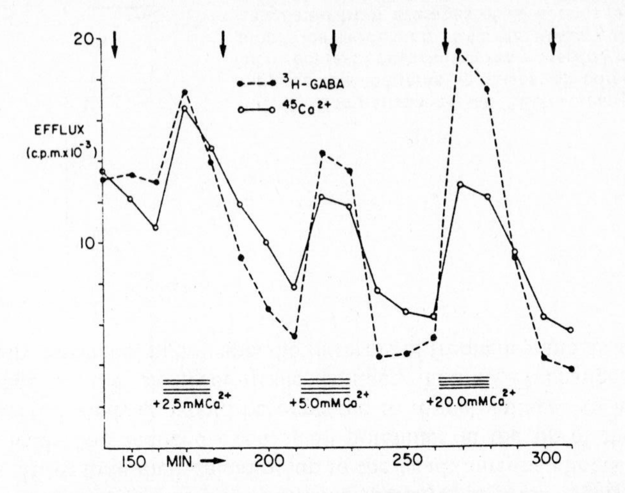

FIG. 3. The simultaneous efflux of $^{45}Ca^{2+}$ and ^{3}H-GABA from cat cortex in an experiment in which aminoxyacetic acid (AOAA), 5 mg/kg body weight, had been administered before incubation with ^{3}H-GABA. The superfusion medium contained 2.16 mM Ca^{2+} before increasing the Ca^{2+} concentration by the amount indicated. Time after the start of superfusion is shown on the abscissa, and the arrows indicate gallamine triethiodide administration. (From Kaczmarek and Adey, 1973.)

Effects of Weak Electrical Gradients on Ca ion and GABA Fluxes in Cortex

This sharp nonlinearity in the release of bound calcium by a small increase in extracellular Ca^{2+} suggested that Ca^{2+} release might be triggered by a weak electric gradient (Kaczmarek and Adey, 1974). Pulsed electrical stimulation of cat cortex with gradients in the range of 20 to 60 mV/cm increased the efflux of both $^{45}Ca^{2+}$ and ^{3}H-GABA (Fig. 4). The mean increase with a gradient of 50 mV/cm was a 1.29 ± 0.04 for ^{3}H-GABA, but these values may reflect larger changes in the rate of binding and release in the tissue.

Important questions may be raised by these findings in terms of classical processes of transmitter release. If a typical synaptic terminal is 0.5 μ in diameter, the extracellular gradient imposed by these fields is, at most, 2.5 μV across the terminal. It is unclear how such a weak stimulus would be able to modify a transmembrane potential of 50 mV sufficiently to influence transmitter release; the gradients in these experiments were more than four orders of magnitude less than the transmembrane gradient. Similar considerations apply to effects of the fields on postsynaptic excitability.

To what extent might the field generated by one cortical neuron influence a nearby cell? For the spinal motoneuron, extracellular gradients generated by the neurons exceed 50 mV/cm and do alter neuronal excitability (Nelson,

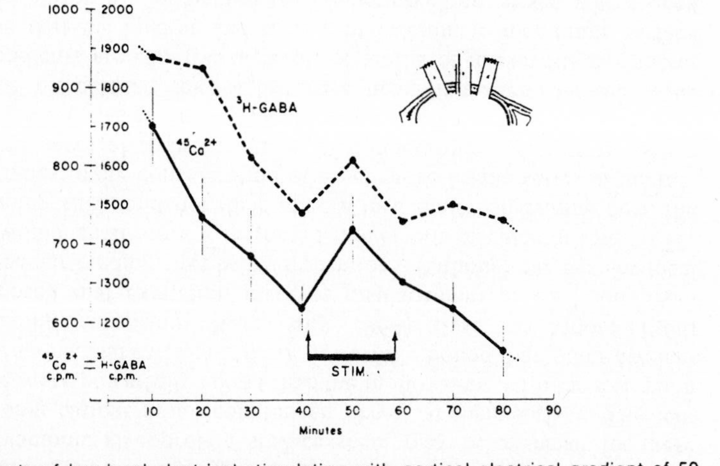

FIG. 4. Effects of low-level electrical stimulation with cortical electrical gradient of 50 mV/cm, 200 pulses/sec, 1.0 msec duration, on efflux of $^{45}Ca^{2+}$ and ^{3}H-GABA. (From Kaczmarek and Adey, 1974.)

1966). In our experiments, seizures induced by topical glutamate or intravenous thiosemicarbazide were as high as 37.5 mV/cm across a 1.0-mm dipole. Considerably higher gradients in the normal EEG can be recorded from microelectrode tips separated by cellular dimensions (Elul, 1962).

The applied fields in these experiments are therefore in the range of naturally occurring gradients, thus supporting the hypothesis that cortical neurons are sensitive to the natural electric field gradients which surround them.

Altered Ca Efflux in the Isolated Chick Brain with Modulated VHF Fields

The release of Ca^{2+} by weak electrical stimulation led us to test the effects of the modulated VHF fields described above on freshly isolated chicken brain (Bawin, Kaczmarek, and Adey, 1974). The cerebrum was placed in a $^{45}Ca^{2-}$ Ringer solution for 30 min and the efflux of Ca^{2+} subsequently observed for a 90-min period, with and without VHF field exposure. There was a remarkable "tuning curve" for different modulation frequencies, with increased Ca^{2+} efflux from the cortex at modulation frequencies between 9 and 20 Hz, but very little increase outside this frequency band (Fig. 5). Moreover, the results were identical in brains killed with 10^{-4} M potassium cyanide prior to equilibration with $^{45}Ca^{2+}$. Previous studies in our laboratory have shown the persistence of membrane fixed charges after cyanide poisoning of cultured neurons (Elul, 1967*b*), and it therefore seems reasonable to assume that the binding of Ca^{2+} and its subsequent efflux

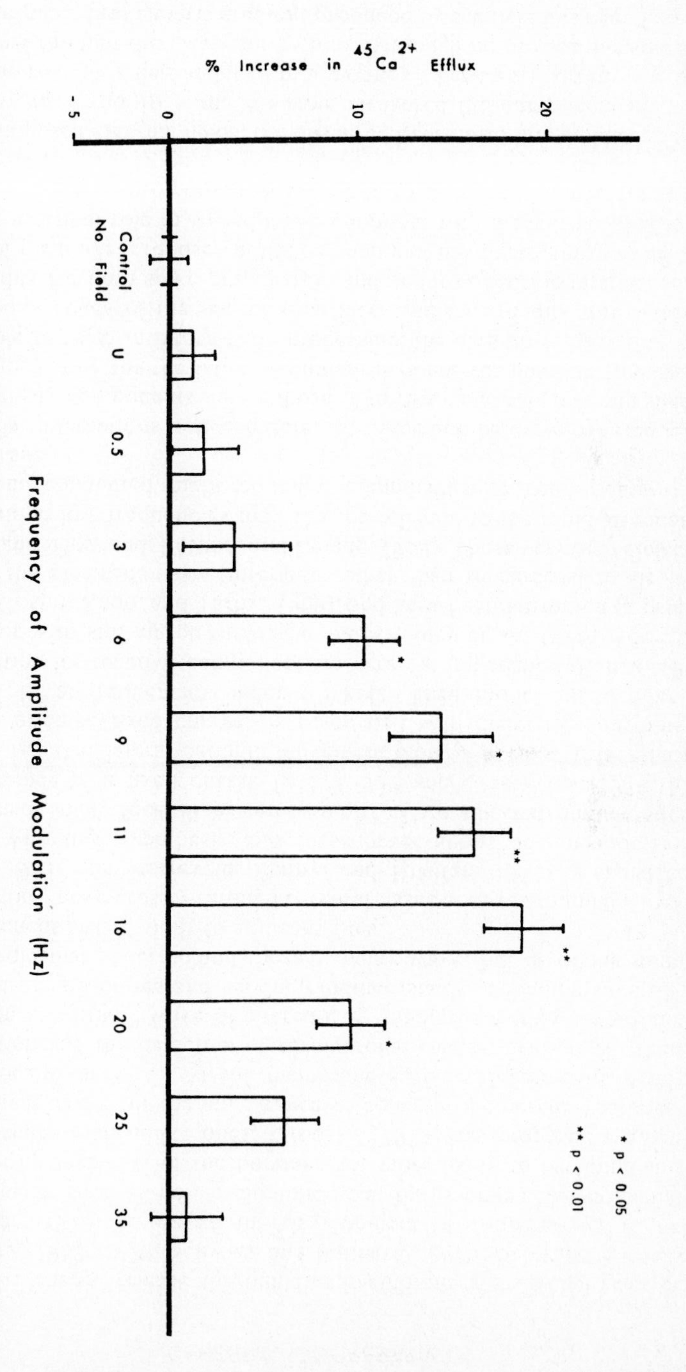

FIG. 5. Effects of amplitude-modulated VHF fields on the $^{45}Ca^{2+}$ efflux from the isolated brain of the neonate chick. The results, given ±S.E.M., are expressed as percent of increase of the calcium efflux, by comparison with control condition, in the absence of fields (From Bawin et al., 1974.)

relate to persisting properties of membrane surface polyanions. Identical exposures of isolated gastrocnemius muscle to these VHF fields have not elicited increased $^{45}Ca^{2+}$ efflux, suggesting that the phenomenon may be specific to brain tissue.

DISCUSSION

Three major points emerge from these studies. It seems clear that unequivocal interactions have been demonstrated between the brain tissue of mammals and birds and a variety of electric and electromagnetic fields, with tissue electric gradients and associated transmembrane ion fluxes far lower than those known to occur in the classical processes of excitation. Indeed, behavioral effects seen in our studies and those of others occur with such low tissue electric gradients that they may be best modeled from a consideration of cooperative processes at the membrane surface. A second basic consideration is the emergence of the calcium ion as a strong candidate for an essential role in these interactions. Third, the evidence suggests a functional role for intrinsic electric gradients, including the electroencephalogram, since behavioral electrophysiological and neurochemical changes have all followed the imposition of external electric fields with tissue components at or below levels of these intrinsic gradients.

We may model these membrane events in certain sequences and at differing levels of organization. The effectiveness of these low-level fields suggests an initial interaction in the long axis of the membrane, perhaps involving macromolecular conformation changes attributable to altered calcium binding, and acting as precursors to transmembrane responses. The broad polyanionic glycoprotein surface sheet may be a sensor for these fields, binding cations as a "counter-ion" layer at their surface. Divalent cations are more powerfully bound than the monovalent, with the exception of hydrogen ions, and calcium is more powerfully bound than other divalent ions, including magnesium (Katchalsky, 1964). Bass and Moore (1968) have proposed that excitation involves displacement of Ca^{2+} ions from macromolecular binding sites by hydrogen ions. Although the ensuing local alkalosis would entail only a restricted movement of calcium ions to adjacent binding sites, and not their release in a major diffusional flow, the present studies invite consideration of a yet more fragile series of interactions, at the level of transient states of cooperative organization among fixed charges on surface macromolecules.

Our data on calcium release by these fields indicates that the tissue electric gradients are effective at levels of tenths of microvolts per micrometer, and that the binding and release of calcium to membrane surface polyanions is probably in the class of "cooperative" processes, with a weak trigger at one point initiating macromolecular conformational changes over considerable distances along the membrane, and thereafter triggering

metabolic energy release through transmembrane signals. Schwarz et al. (Schwarz, 1967, 1970; Schwarz and Balthasar, 1970; Schwarz, Klose, and Balthasar, 1970) envisaged the development of cooperativity in linear biopolymers, such as poly-L-glutamic acid, by assuming that immediately neighboring segments of the polymer are more likely to be found in like charge states than unlike ones. Grodsky (1974) has proposed a quantum mechanical model for the appearance of coherent membrane "patches" in the glyocalyx at 37°C. At the membrane surface, decremental dendritic conduction and the detection of perineuronal electric fields might then be based on a "virtual" wave of altered Ca^{2+} binding, traveling longitudinally on dendritic structures and leaving modified states of binding sites on the macromolecular sheet behind the advancing wave, but involving minimal displacement of Ca^{2+} ions to adjacent sites.

Ca^{2+} ions move grossly through cerebral tissue at rates around 0.3 μM/sec (Adey, 1971; van Harreveld, Dafny, and Khattab, 1971), a speed compatible with this hypothesis, but these observations do not address the question of small, focal displacements of Ca^{2+} to adjacent binding sites in the presence of a fixed charge field. Einolf and Carstensen (1971) have pointed out that lateral cationic movement along a porous surface having radially oriented fixed charges is associated with dielectric constants as high as 10^6 at frequencies under 1.0 kHz. Their model would offer an explanation for reactive components in cerebral impedance (Ranck, 1963) that alter with shifting physiological states (Adey et al., 1966; Adey, Bystrom, Costin, Kado, and Tarby, 1969), and have been attributed to properties of the electrode-tissue interface rather than to cerebral tissue itself (Gesteland, Howland, Lettvin, and Levine, 1959). These very high dielectric constants at low frequencies may also be relevant to the rapid attenuation of volume-conducted high-frequency components of neural activity in brain tissue.

Finally, interactions between cerebral tissue and fields whose frequency and intensity components mimic those of intrinsic natural gradients suggest a functional role for the latter. Although it is not yet possible to assign a clear role to them in information processing, the data do suggest that they can modify states of the neuronal surface, and thus modify thresholds of excitability for both spike propagation and dendro-dendritic interaction. In terms of the model proposed at the beginning of this paper, they would thus play an essential role in transductive coupling, with these first weak interactions relating in a hierarchical sequence to later well-known transmembrane events.

For imposed external fields, the observed behavioral thresholds appear to lie between 1 and 10 V/m. We have measured the component of a 7 V peak-to-peak 7-Hz field induced in a monkey's head as a current to ground of 0.8 nA through the electrode system. Although no precise measurement of the intracerebral electric gradient produced by the field has been feasible, it would be about 0.02 μV/cm in a brain having a conducting cross section of approximately 10 cm^2 and a maximum linear dimension of 7 cm in the axis of the field, based on a specific brain impedance of 300 Ω-cm at these frequencies (Ranck, 1963). That such a weak field should be an effective stimulus is baffling in terms of classical synaptic excitation, in which an action potential transiently abolishes a gradient of about 1 kV/cm, some ten orders of magnitude larger than these weak fields. A significant release of transmitter substances from within the membranes of presynaptic terminals is equally difficult to explain. This disparity between observed field effects on the central nervous system as a whole and the improbability of a direct synaptic action surely invites fresh consideration of more subtle membrane sensitivities.

ACKNOWLEDGMENTS

These studies were supported by National Science Foundation grant GB-27740, U.S. Air Force contract F44620-70-C-0017, and ONR contract N00014-A-200-4037.

APPENDIX: A POSSIBLE MECHANISM FOR CALCIUM STORAGE AND RELEASE

David A. Rees

The polysaccharide-Ca^{2+} associations described in Chapter 1 suggested that similar types of interaction might be involved in Ca^{2+} sequestering and release within the intercellular space. Although little is yet known about the polysaccharide content of the intracellular space of brain tissue, carbohydrates known to be present in the intercellular space of other tissue include hyaluronic acid, keratan sulfate, and chondroitin sulfate, all of which contain repetitive regions that are capable of forming extended helices. For example, hyaluronic acid has been thought to form stable structures similar to those of the carrageenans (Dea, Moorhouse, Rees, Arnott, Guss, and Balazs, 1973).

The carbohydrates that are more intimately associated with cell membranes, particularly those that are covalently attached to polypeptide or lipid moieties anchored in the membrane bilayer, do not have regular sequences of sugar residues (Kiss, 1969; Hughes, 1973; Spiro, 1973), and are therefore unlikely to form regular structures such as the helix or ribbon. However, the conclusions that have emerged from studies with the long-chain repetitive molecules do suggest the possibility that certain kinds of interactions might have biological significance. For example, sialic acid is found in abundance as the terminal residue of the glycolipids and glycoproteins that are anchored in the bilayer. The disposition of carboxylate and other oxygen functions within this sugar carboxylic acid residue

suggests an analogy with the sugar carboxylate residues that form egg-box structures and cooperatively bind Ca^{2+} (see Chapter 1). In addition, sialic acid residues have been shown to be involved in important calcium binding sites on the external surfaces of some cells (Long and Mouat, 1971).

Gangliosides are major components of many cell membranes, and these glycolipids are particularly abundant in neuronal membranes. Since Ca^{2+} is intimately involved in nerve excitability, it is tempting to speculate that these gangliosides might interact specifically with Ca^{2+}. The covalent structures of many gangliosides are well known, and it was possible to derive tentative conformations for them using principles of conformational analysis that were recently developed for carbohydrate polymers (Rees, 1973; D. Thom, *personal communication*).

The conformations predicted for several of the gangliosides that contain consecutive sialic acid residues linked to position 3 of an interior galactose residue were of particular interest. The conformations predicted for these molecules contained an extensive system of long-range hydrogen bonding. This supported the plausibility of the structures, because the models were built only from a consideration of short-range interactions. In these models, a cavity is present that matches the ionic radius of Ca^{2+}, and that is lined with about nine oxygen atoms within a suitable distance for coordination, including one oxygen from each of the two carboxylate groups. The region opposite this site is a hydrophobic basin formed by the C—H bonds of the sugar rings and of the N-acetates; binding of an apolar species at this site might be coupled to Ca^{2+} binding and release.

REFERENCES

Adey, W. R. (1970): Cerebral structure and information storage. *Prog. Physiol. Psychol.*, pp. 181–200.

Adey, W. R. (1971): Evidence for cerebral membrane effects of calcium, derived from direct-current gradient, impedance, and intracellular records. *Exp. Neurol.*, 30:78–102.

Adey, W. R. (1972): Organization of brain tissue: Is the brain a noisy processor? *Int. J. Neurosci.*, 3:271–284.

Adey, W. R. (1974*a*): Extracellular microenvironment. In: *Dynamic Patterns of Brain Cell Assemblies. Neurosciences Res. Prog. Bull.*, 12:80–85.

Adey, W. R. (1974*b*): The influences of impressed electrical fields at EEG frequencies on brain and behavior. In: *Behavior and Brain Electrical Activity*, edited by H. Altshuler and N. Burch. New York: Plenum Press.

Adey, W. R., Bell, F. R., and Dennis, B. J. (1962*a*): Effects of LSD-25, psilocybin, and psilocin on temporal lobe EEG patterns and learned behavior in the cat. *Neurology*, 12:591–602.

Adey, W. R., Bystrom, B. G., Costin, A., Kado, R. T., and Tarby, T. J. (1969): Divalent cations in cerebral impedance and cell membrane morphology. *Exp. Neurol.*, 23:29–50.

Adey, W. R., Kado, R. T., and Didio, J. (1962*b*): Impedance measurements in brain tissue of animals using microvolt signals. *Exp. Neurol.*, 5:47–66.

Adey, W. R., Kado, R. T., Didio, J., and Schindler, W. J. (1963): Impedance changes in cerebral tissue accompanying a learned discriminative performance in the cat. *Exp. Neurol.*, 7:259–281.

Adey, W. R., Kado, R. T., McIlwain, J. T., and Walter, D. O. (1966): The role of neuronal elements in regional cerebral impedance changes in alerting, orienting and discriminative responses. *Exp. Neurol.*, 15:490–510.

Adey, W. R., and Walter, D. O. (1963): Application of phase detection and averaging techniques in computer analysis of EEG records in the cat. *Exp. Neurol.*, 7:186–209.

Anninos, P. A., Beek, B., Csermely, T. J., Harth, E. M., and Pertile, G. (1970): Dynamics of neural structures. *J. Theor. Biol.*, 26:121–148.

Bass, L., and Moore, W. J. (1968): A model of nervous excitation based on the Wien dissociation effect. In: *Structural Chemistry and Molecular Biology*, edited by A. Rich and C. M. Davidson, pp. 356–368. San Francisco: Freeman.

Bawin, S. M., Gavalas-Medici, R. J., and Adey, W. R. (1973): Effects of modulated very high frequency fields on specific brain rhythms in cats. *Brain Res.*, 58:365–384.

Bawin, S. M., Kaczmarek, L. K., and Adey, W. R. (1974): Effects of modulated VHF fields on the central nervous system. *Ann. N.Y. Acad. Sci. (in press).*

Bennett, H. S. (1963): Morphological aspects of extracellular polysaccharides. *J. Histochem. Cytochem.*, 11:14–23.

Cole, K. S. (1940): Permeability and impermeability of cell membranes for ions. *Cold Spring Harbor Symp. Quant. Biol.*, 8:110–122.

Coombs, J. S., Curtis, D. R., and Eccles, J. C. (1959): The electric constants of motoneurone membrane. *J. Physiol.*, 145:505–528.

Creutzfelt, O. D., Fuster, J. M., Lux, H. D., and Nacimiento, A. (1964): Experimenteller Nachweis von Beziehungen zwischen EEG-Wellen and Activtät corticaler Nervenzellen. *Naturwiss.*, 51:166–167.

Creutzfelt, O. D., Watanabe, S., and Lux, H. D. (1966): Relations between EEG phenomena and potentials of single cortical cells. II. Spontaneous and convulsoid activity. *Electroencephalogr. Clin. Neurophysiol.*, 20:19–37.

Dea, I. C. M., Moorhouse, R., Rees, D. A., Arnott, S., Guss, J. M., and Balazs, E. A. (1973): Hyaluronic acid: A novel, double helical molecule. *Science*, 179:560–562.

Ehinger, B. (1973): Glial uptake of taurine in the rabbit retina. *Brain Res.*, 60:512–516.

Ehinger, B., and Falck, B. (1971): Autoradiography of some suspected neurotransmitter substances: GABA, glycine, glutamic acid, histamine, dopamine, and L-DOPA. *Brain Res.*, 33:157–172.

Einolf, C. W., Jr., and Carstensen, E. L. (1971): Low-frequency dielectric dispersion in suspensions of ion-exchange resins. *J. Physical Chem.*, 75:1091–1099.

Elul, R. (1962): Dipoles of spontaneous activity in the cerebral cortex. *Exp. Neurol.*, 6:285–299.

Elul, R. (1964): Specific site of generation of brain waves. *Physiologist*, 7:125 (Abstr.).

Elul, R. (1967*a*): Statistical mechanisms in generation of the EEG. In: *Progress in Biomedical Engineering*, edited by L. J. Fogel and F. W. George, pp. 131–150. Washington, D.C.: Spartan Books.

Elul, R. (1967*b*): Fixed charge in the cell membrane. *J. Physiol.*, 189:351–365.

Elul, R. (1972): The genesis of the EEG. *Int. Rev. Neurobiol.*, 15:227–272.

Elul, R., and Adey, W. R. (1965): The intracellular correlates of gross evoked responses. *Proc. 23rd Int. Cong. Physiol. Sci.*, Tokyo, p. 434 (Abstr.).

Fujita, Y., and Sato, T. (1964): Intracellular records from hippocampal pyramidal cells in rabbit during theta rhythm activity. *J. Neurophysiol.*, 27:1011–1025.

Gavalas, R. J., Walter, D. O., Hamer, J., and Adey, W. R. (1970): Effect of low-level, low-frequency electric fields on EEG and behavior in *Macaca nemestrina*. *Brain Res.*, 18:491–501.

Gesteland, R. C., Howland, B., Lettvin, J. Y., and Levine, S. (1959): Comments on microelectrodes. *Proc. Inst. Radio Eng.*, 47:1856–1861.

Green, J. D., Maxwell, D. S., Schindler, W. J., and Stumpf, C. (1960): Rabbit EEG "theta" rhythm; its anatomical source and relation to activity in single neurons. *J. Neurophysiol.*, 23:403–420.

Grodsky, I. T. (1974): Possible physical substrates for the interaction of electromagnetic fields with biological membranes. *Ann. N.Y. Acad. Sci. (in press).*

Hamer, J. R. (1968): Effects of low level, low frequency electric fields on human reaction time. *Commun. Behav. Biol.*, 2(A):217–222.

Henn, F. A., and Hamberger, A. (1971): Glial cell function: Uptake of transmitter substances. *Proc. Nat. Acad. Sci.*, 68:2686–2690.

Hughes, R. C. (1973): Glycoproteins as components of cellular membranes. *Prog. Biophys. Mol. Biol.*, 26:191–268.

Hydén, H. (1972): Macromolecules and behavior. In: *The Arthur Thomson Lectures* (University of Birmingham, England), edited by G. B. Ansell and P. B. Bradley. London: Macmillan.

Kaczmarek, L. K., and Adey, W. R. (1973): The efflux of $^{45}Ca^{2+}$ and [^{3}H]-γ-aminobutyric acid from cat cerebral cortex. *Brain Res.*, 63:331–342.

Kaczmarek, L. K., and Adey, W. R. (1974): Weak electric gradients change ionic and transmitter fluxes in cortex. *Brain Res.*, 66:537–540.

Karahashi, Y., and Goldring, S. (1966): Intracellular potentials from "idle" cells in cerebral cortex of cat. *Electroencephalogr. Clin. Neurophysiol.*, 20:600–607.

Katchalsky, A. (1964): Polyelectrolytes and their biological interactions. In: *Connective Tissue: Intercellular Macromolecules. Proceedings of a Symposium sponsored by the New York Heart Association*, pp. 9–41. Boston: Little, Brown.

Kiss, J. (1969): Glycosphingolipids (sugar-sphingosine conjugates). *Adv. Carbohydrate Chem. Biochem.*, 24:381–433.

König, H., and Ankermüller, F. (1960): Über den Einfluss besonders niederfrequenter elektrischer Vorgänge in der Atmosphäre auf den Menschen. *Naturwiss.*, 47:486–490.

Long, C., and Mouat, B. (1971): The binding of calcium ions by erythrocytes and "ghost"-cell membranes. *Biochem. J.*, 123:829–836.

Nelson, P. G. (1966): Interaction between spinal motoneurons of the cat. *J. Neurophysiol.*, 29:275–287.

Nicholls, J. G., and Kuffler, S. W. (1964): Extracellular space as a pathway for exchange between blood and neurons in the central nervous system of the leech: Ionic composition of glial cells and neurons. *J. Neurophysiol.*, 27:645–671.

Pease, D. C. (1966): Polysaccharides associated with the exterior surface of epithelial cells: Kidney, intestine, brain. *J. Ultrastr. Res.*, 15:555–588.

Purpura, D. P., Shofer, R. J., and Musgrave, F. S. (1964): Cortical intracellular potentials during augmenting and recruiting responses. II. Patterns of synaptic activities in pyramidal and nonpyramidal tract neurons. *J. Neurophysiol.*, 27:133–151.

Rambourg, A., and Leblond, C. P. (1967): Electron microscope observations on the carbohydrate-rich cell coat present at the surface of cells in the rat. *J. Cell. Biol.*, 32:27–53.

Ramwell, P. W., and Shaw, J. E. (1970): Biological significance of the prostaglandins. *Rec. Prog. Horm. Res.*, 26:139–187.

Ranck, J. B. (1963): Specific impedance of rabbit cerebral cortex. *Exp. Neurol.*, 7:144–152.

Reed, D. J., Woodbury, D. M., and Holtzer, R. L. (1964): Brain edema, electrolytes, and extracellular space. Effect of triethyl tin on brain and skeletal muscle. *Arch. Neurol.*, 10:604–616.

Rees, D. A. (1973): Polysaccaride conformation. *M.T.P. Internat. Rev. Sci. Org. Chem.*, 7:251–283.

Schmitt, F. O., and Samson, F. E., Jr. (1969): Brain Cell Microenvironment. Neurosciences Res. Prog. Bull., 7:277–417.

Schwarz, G. (1967): A basic approach to a general theory for cooperative intramolecular conformation changes of linear biopolymers. *Biopolymers*, 5:321–324.

Schwarz, G. (1970): Cooperative binding to linear biopolymers. 1. Fundamental static and dynamic properties. *Eur. J. Biochem.*, 12:442–453.

Schwarz, G., and Balthasar, W. (1970): Cooperative binding to linear biopolymers. 3. Thermodynamic and kinetic analysis of the acridine orange-poly(L-glutamic acid) system. *Eur. J. Biochem.*, 12:461–467.

Schwarz, G., Klose, S., and Balthasar, W. (1970): Cooperative binding to linear biopolymers. 2. Thermodynamic analysis of the proflavine-poly(L-glutamic acid) system. *Eur. J. Biochem.*, 12:454–460.

Spiro, R. G. (1973): Glycoproteins. *Adv. Protein Chem.*, 27:349–467.

Tarby, T. J., and Adey, W. R. (1967): Cytological chemical identification of calcium in brain tissue. *Anat. Record*, 157:331–332 (Abstr.).

van Harreveld, A., Crowell, J., and Malhotra, S. K. (1965): A study of extracellular space in central nervous tissue by freeze-substition. *J. Cell. Biol.*, 25:117–137.

van Harreveld, A., Dafny, N., and Khattab, F. I. (1971): Effects of calcium on the electrical resistance and the extracellular space of cerebral cortex. *Exp. Neurol.*, 31:358–367.

Wever, R. (1968): Einfluss schwacher elektro-magnetischer Felder auf die circadiane Periodik des Menschen. *Naturwiss.*, 55:29–33.

COHERENT PHONONS AND EXCITONS IN BIOLOGICAL SYSTEMS

S.A. MOSKALENKO, M.F. MIGLEI, P.I. KHADSHI, E.P. POKATILOV and E.S. KISELYOVA
Institute of Applied Physics, Academy of Sciences of the Moldavian SSR, Kishinev, USSR

Received 28 May 1979

The possibility of Bose condensation of phonons in biological systems induced by millimeter electromagnetic waves is discussed. The existence of bistability and hysteresis is predicted and the possibility of Bose condensation of excitons, the creation of solitons and superfluid transfer of energy in such systems is also discussed.

1. In a series of papers [1–5] the possibility of Bose condensation of phonons in biological systems has been investigated. A biological system is considered as a macroscopic model consisting of separate units which oscillate with different frequencies due to the interaction among themselves and form a set of electric dipole oscillators. There is an individual finite group of modes among these units (they are called active phonons), the frequencies of which are split into a narrow band $\omega_0 \leqslant \omega_i \leqslant \omega_N$. The rest of the oscillators form a heat bath. If the energy is supplied to the active modes at a rate s beyond a critical rate s_0 the lowest mode is very strongly excited, i.e. the number of phonons with energy $\hbar\omega_0$ becomes macroscopically large. This phenomenon is equivalent to Bose condensation of phonons in a biological system. Excitation of such a group of oscillators has far-reaching biological consequences, for it would lead to a selective long-range interaction [6–8] and probably is responsible for growth control in normal tissue and the origin of enzyme activity [6,7,9]. The absence of active phonons might lead to cancer [8,9].

It should be noted that Bose condensation of excitons, photons and other elementary excitations in solids has been studied earlier [10,11]. From this point of view the phonon Bose condensation problem is not a new one, but it is unexpected that this phenomenon is actual in biological systems.

Conclusive experimental evidence of the existence of electric vibrations predicted by Fröhlich [12–14] in the region 10^{11}–10^{12} Hz has been obtained using the Raman effect [8,15–17]. Bacterial cells in a nutrient give a number of strong relatively sharp Raman lines in the region of 7×10^{10} and 5×10^{12} Hz [15]. In ref. [16] the ratio R of the intensities of the anti-Stokes and Stokes lines 124 cm^{-1} of *E. coli* B bacteria has been measured. The living cells of the green alga *Chlorella Pyrenoidosa* also give an enhancement of anti-Stokes Raman scattering [17]. The other conclusive experimental evidence for the existence of coherent electric vibrations has been obtained by the action of coherent millimeter waves [18,19]. The division rate of one type of cells and the steadiness against X-ray irradiation of another have been measured. Inspite of the complexities of the biological effects three common features emerge: (i) strong resonance dependence of the observed effects on frequency, (ii) existence of a sharp critical power supply threshold for the irradiation intensity, (iii) further independence of the effects with the growth of the power supply. These three effects stimulated Fröhlich to explain them by means of Bose condensation of phonons. This point of view is supported by other authors [4,5]. On the other hand, Livshits [3] has come to the conclusion that Bose condensation of phonons in biological systems, especially in Fröhlich's model, is impossible. Livshits has added the two-quantum terms in Fröhlich's rate equation. These terms describe the simultaneous creation and annihilation of two active phonons.

Utilizing microscopic techniques Wu and Austin [5] discussed Fröhlich's model of Bose condensation in biological systems and came to the conclusion that the criticism of Livshits is invalid.

Starting from the rate equation of Bogolyubov's

Reprinted with permission from *Phys. Lett.*, vol. 76A, no. 2, pp. 197–200, Mar. 17, 1980.

superfluidity theory [20] we have independently demonstrated [21] that in Fröhlich's model Bose condensation of phonons is possible. We have taken into account the terms describing: (a) the creation and annihilation of one biological active phonon due to interaction with two phonons of the heat bath, (b) the scattering of one active phonon due to the creation or annihilation of one phonon of the heat bath, (c) the simultaneous creation and annihilation of two active phonons due to the creation or annihilation of one heat bath phonon, (d) the processes with three and four active phonons without the participation of heat bath phonons. If we suppose that the constants of the triple anharmonicity for both processes (b) and (c) are of the same order it is easy to demonstrate that the average probability of a process of type (b) is some orders larger than that of type (c). In this case the Fröhlich approach is justified. If the mentioned constants differ essentially conditions may be obtained under which Livshits's suggestion is correct. Nevertheless it is necessary to take into account the processes of three and four active phonons.

2. In the present paper special attention is paid to the excitation level of a system with z H-bonds. The interaction between bonds produces a narrow band of frequencies $\Delta\omega$ ($\omega_0 \leqslant \omega_i \leqslant \omega_N$). The distances between the energy levels depend both on the dispersion law and on size quantization. Supposing that the levels in the band quadratically depend on the discrete value of the wave vector, and the frequency ω_{0+} next to the lowest level is given by $\omega_{0+} = \omega_0 + \pi^2\Delta\omega^2/z^2\omega_0$ when $\mu = \hbar\omega_0$ the authors of ref. [4] have calculated the occupation number of the ω_{0+} level:

$$N_1 = 2z^2\omega_0 kT/\hbar\pi^2\Delta\omega^2 .$$

It is clear that the total number N' on the levels above the ground level exceeds N_1. At $T = 300$ K, $z = 10^4$, $\omega_0 = 5 \times 10^{12}$ and $\Delta\omega = 2 \times 10^8$ Hz [18] we obtain $N_1 = 10^{17}$. If the modes in the band are spread at the same distances $\omega_n = \omega_0 + \Delta\omega\, n/z$ ($n = 0, 1, 2, ...$) then the number of excitations "out of the condensate" can be appreciated by the formula

$$N' = (kTz/\hbar\Delta\omega)\,\mathrm{Ei}\,(-\hbar\Delta\omega/kTz) ,$$

where $\mathrm{Ei}(-\hbar\Delta\omega/kTz)$ is the integral exponential function. For the above given values of the parameters $N' = 4.3 \times 10^{11}$.

We note that for the ideal bulk Bose gas the density of condensate particles is $N'/V = 2.612\,(mkT/2\pi\hbar^2)^{3/2}$. At $T = 300$ K and $m = m_p \approx 2000\, m_e$ (m_p and m_e are the proton and electron masses, respectively) $N'/V = 2.6 \times 10^{21}\ \mathrm{cm}^{-3}$, while the number of modes per unit crystal volume $z/V = 10^{23}\ \mathrm{cm}^{-3}$. The excitation order of one mode $N'/z = 10^{-2}$. From the above given estimations it follows that for observation of Bose condensation at room temperature the order of mode excitation of a linear chain of biological objects has to be 7–11 orders more than that of the bulk system. To reach such excitation levels the life times have to be 7–11 orders larger than the ones of phonons in crystals. However, at such a large density of elementary excitations they probably will destroy each·other and this may lead to the breaking of H-bonds. Therefore together with spontaneous Bose condensation it is worth discussing the induced Bose condensation of electric active vibrations under the action of millimeter electromagnetic waves. If the spectral width of the irradiation is less than the difference between the levels in the band then one level is strongly excited. We shall consider only this case. If the spectral width of the millimeter irradiation is of the same order as the width of the band $\Delta\omega$ then all the modes are macroscopically excited. This problem is analogous to that of exciton Bose condensation in the external field of laser irradiation [22] and belongs to the theory of nonlinear induced oscillations. A similar problem has been discussed in the theory of nutation in the excitonic region of the spectrum [23].

Assigning creation and destruction operators to the ground mode (a^+, a) and photons (c^+, c) it is possible to write the hamiltonian of biological systems as:

$$H = \hbar\omega_0 a^+a + \hbar c_0 k c^+c + \mathrm{i}\varphi(c^+a - a^+c) + \tfrac{1}{2}\nu a^+a^+aa ,$$

where ω_0 is the excited mode frequency, k is the photon wave vector, c_0 is the light velocity, φ and ν are the coupling constants for the photon–phonon and phonon–phonon interaction, respectively. The rate equations for the macroscopically large amplitudes a and c, taking into account the damping γ and the pump power p, are:

$$\mathrm{i}\hbar\dot{a} = \omega_0 a - \mathrm{i}\varphi c + \nu a^+aa - \mathrm{i}\gamma a ,$$

$$\mathrm{i}\hbar\dot{c} = c_0 kc - \mathrm{i}\varphi a + \mathrm{i}p\,\mathrm{e}^{-\mathrm{i}\omega t} .$$

Their solution leads to the following equation for the occupation number of the excited mode $N = a^+a \gg 1$:

$$\nu^2(\omega - c_0k)^2N^3 - 2\nu(\omega - c_0k)[(\omega - \omega_0)(\omega - c_0k) - \varphi^2]N^2 + \{[(\omega - \omega_0)(\omega - c_0k) - \varphi^2]^2 + \gamma^2(\omega - c_0k)^2\}N - \varphi^2p^2 = 0\,. \quad (1)$$

The equation $\partial p/\partial N = 0$ shows that if the pump power $p > p_c$ $[p_c^2 = 8\gamma^3(\omega - c_0k)^2/3\sqrt{3}\nu\varphi^2]$ then eq. (1) has two relative extrema N_1 and N_2 to which the power p_1 and p_2 correspond (fig. 1). When the pump power is increased above the critical value p_1 the number of excitations N of the coherent mode makes a rapid leap. This is equivalent to a first-order phase transition and to the origin of the condensate. When the pump power is decreased below p_2 the number of excitations sharply decreases. So there is a hysteresis in the dependence of the number of excitations N on the pump power p. Therefore it is of interest to repeat the experiments [18] and to measure the dependence both of the protective action of the millimeter electromagnetic field on the bone brain of mice and of the induction factor of colicin synthesis in *E.coli* bacteria not only in the direction of increasing power but also backwards (decreasing pump power). Thus the existence of the predicted hysteresis should be expected.

In all papers on phonon Bose condensation the repulsive phonon forces are supposed (not evidently) to prevail. This assumption has led to the conclusion of the possibility of phonon Bose condensation in biological objects. If the attractive phonon forces predominate the creation of biphonons [24] or phonon drops is possible. The problem of biphonon Bose condensation in biological objects is not solved yet.

Coherent phonons in biological systems can also appear in the regime of ultrashort pulses under the action of resonance maser irradiation or at Raman scattering of laser irradiation on phonons. Such experiments have been done for a number of crystals and liquids

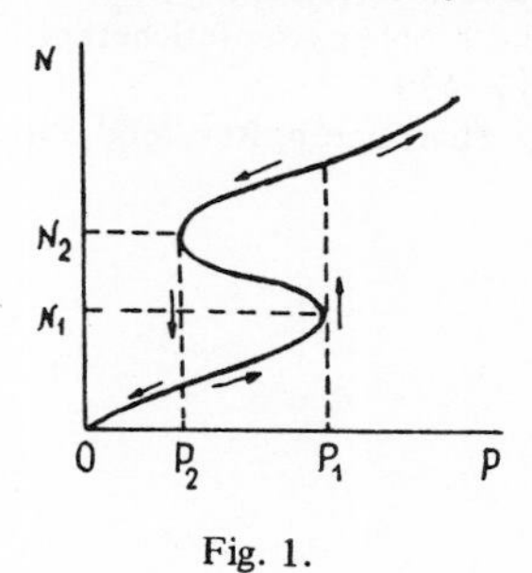

Fig. 1.

[26]. The coherent phonons and photons arising in the regime of ultrashort pulses or Bose condensation can propagate in the medium under some conditions in the form of steady state soliton wave packets [27]. If the pulse width is smaller than the relaxation time of the phonons, the propagation of solitons in biological systems without dissipation of energy will lead to the phenomenon of self-induced transparency discovered in gases by McCall and Hahn [28]. It is necessary to note that solitons in biological objects have already been discussed by Davydov and Kislucha [29]. However, in that paper the interaction of an electron excitation with the atom displacements in a linear molecular chain was considered and the propagation of autolocalized states of polaron type was investigated.

In our case the polariton consists of a phonon and a photon. It may be called a polariton soliton. Introducing the phonon and photon quasi wave number in a medium we can speak about the polariton dispersion law and its translational mass. The changes in the dispersion law are stronger in the crossing region of the dispersion branches of noninteracting phonons and photons. In that region the steepness of the polariton dispersion branches is maximal and the polariton mass can become equal to or less than the electromagnetic mass $\hbar\omega/c_0^2$. The parameters of polariton solitons in biological systems may be appreciated by the formulae of ref. [27].

3. Some remarks relative to the collective properties of excitons in biological objects should be made. Excitons are of great importance in the migration of energy in the process of photosynthesis. Their existence has been experimentally detected in investigations of energy migration in DNA [30] and other systems [31]. A high density of excitons can essentially change their own life times due to the interaction between themselves and lead to biexciton formation. Both these possibilities have been discussed in Borisov's report [32].

We would like to pay attention to the possibility of creation of high density coherent excitons in biological systems. They may be created in different ways. One of them is the thermalisation of excitons and establishment of quasi-equilibrium in the band, if the exciton life time is considerably larger than their relaxation time and the translational mass is not too large. Otherwise too large a concentration of excitons is needed. In this case the excitons destroy one another. The second way is Bose condensation of excitons in the field of an external electromagnetic wave [22]. And finally coherent excitons may be created in the regime of ultrashort pulses and may exist only during times less than the relaxation time. These problems are well known and have been discussed for a comparatively long time for semiconductors and other crystals [11]. It is possible that the idea of Bose condensation of excitons will be effective for the understanding of some consequences of laser action on biological structures and will open new perspective in this branch of science.

4. We want to pay attention to the possibility of propagation of the condensate wave in biological systems and the migration of energy over a large distance. This phenomenon has been discussed in the case of excitons and was called superfluidity [10,33]. We examine the superfluidity of coherent excitons, photons and phonons in biological systems under quasi-equilibrium weak nonstationary conditions. Superfluidity means nondissipative migration of energy. A more detailed analysis of the problem of exciton superfluidity in semiconductors was made by Keldysh [34]. He indicated that the excitons carry energy, momentum as well as electric and magnetic moments. The superfluidity of nonequilibrium excitons means the existence of energy and polarisation flows during the exciton life time. The damping time of superfluidity flows will be determined not by the exciton scattering time but by their life time which is some powers larger in semiconductors [34]. However, excitons do not carry mass as Kohn and Sherrington have concluded [35].

References

[1] H. Fröhlich, Phys. Lett. 26A (1968) 404.
[2] H. Fröhlich, Biofizika 22 (1977) 743.
[3] M.A. Livshits, Biofizika 17 (1972) 694; 22 (1977) 744.
[4] D. Bhaumik, K. Bhaumik and B. Dutta-Roy, Phys. Lett. 56A (1976) 145; 59A (1976) 77.
[5] T.M. Wu and S. Austin, Phys. Lett. 65A (1978) 74.
[6] H. Fröhlich, Phys. Lett. 39A (1972) 153.
[7] H. Fröhlich, Proc. Nat. Acad. Sci. US 72 (1975) 4211.
[8] H. Fröhlich, IEEE Trans. Microwave Theory Tech. MTT-26 (1978) 613.
[9] H. Fröhlich, Riv. Nuovo Cimento 7 (1977) 399.
[10] S.A. Moskalenko, Sov. Phys. Solid State 4 (1962) 276.
[11] S.A. Moskalenko, M.F. Miglei, M.I. Shmiglyk, P.I. Khadshi and A.V. Lelyakov, Zh. Eksp. Teor. Fiz. 64 (1973) 1786.
[12] H. Fröhlich, J. Quantum Chem. 2 (1968) 641.
[13] H. Fröhlich, Nature 228 (1970) 1093.
[14] H. Fröhlich, Phys. Lett. 51A (1975) 21.
[15] S.J. Webb and M.E. Stoneham, Phys. Lett. 60A (1977) 267.
[16] S.J. Webb, M.E. Stoneham and H. Fröhlich, Phys. Lett. 63A (1977) 407.
[17] F. Drissler and R.M. Macfarlane, Phys. Lett. 69A (1978) 65.
[18] N.D. Devyatkov, Usp. Fiz. Nauk 110 (1973) 453.
[19] W. Grundler, F. Keilman and H. Fröhlich, Phys. Lett. 62A (1977) 463.
[20] N.N. Bogolyubov, Zh. Eksp. Teor. Fiz. 18 (1948) 622.
[21] S.A. Moskalenko, E.P. Pokatilov, M.F. Miglei and E.S. Kiselyova, Int. J. Quantum Chem. 16 (1979), to be published.
[22] V.F. Elesin and Yu.B. Kopaev, Zh. Eksp. Teor. Fiz. 63 (1972) 1447.
[23] P.I. Khadshi, S.A. Moskalenko and S.N. Belkin, Pis'ma Zh. Eksp. Teor. Fiz. 29 (1979) 223.
[24] V.M. Agranovith, I.A. Efretov and I.K. Kobozev, Sov. Phys. Solid State 18 (1976) 3221.
[25] A.S. Kovalev and A.M. Kosevith, Sov. Phys. Low Temp. Phys. 2 (1976) 913.
[26] M.S. Pesin and I.L. Fabelinskii, Usp. Fiz. Nauk 120 (1976) 273.
[27] E.S. Moskalenko, Sov. Phys. Solid State 20 (1978) 2246.
[28] S.L. McCall and E.L. Hahn, Phys. Rev. 183 (1969) 457.
[29] A.S. Davydov and N.I. Kislukha, Phys. Stat. Sol. (b) 75 (1976) 735.
[30] S.L. Shapiro, A.J. Campillo, V.H. Koliman and W.B. Good, Opt. Commun. 15 (1975) 308.
[31] A.J. Campillo and S.L. Shapiro, Opt. Commun. 18 (1976) 142.
[32] A.Yu. Borisov and A.S. Piskarskas, Digest of reports of the IXth National Conf. on Coherent and nonlinear optics, dedicated to the memory of academician R.V. Khohlov (Leningrad, 1978) part II, p. 122.
[33] S.A. Moskalenko et al., Vsaimodeistvie eksitonov v poluprovodnikah (Shtiinza, Kishinev, 1974).
[34] L.V. Keldysh, in: Problemy teoreticheskoi fizika (Nauka, Moscow, 1972) p. 433.
[35] W. Kohn and D. Sherrington, Rev. Mod. Phys. 42 (1970) 1.

DIELECTRIC THEORY AND PROPERTIES OF DNA IN SOLUTION

Robert H. Cole

Department of Chemistry
Brown University
Providence, Rhode Island 02912

INTRODUCTION

In this paper, we discuss three problems in dielectrics that are relevant to the use of dielectric measurements as a tool for the better understanding of the behavior of biopolymers in solution. In each case, the work is unfinished and will be published in detail elsewhere, hence the paper is in the nature of a progress report presented in the hope that the incomplete results to date will be of interest.

The first problem concerns the effect of electrostatic dipole interaction forces on the static dielectric constant (permittivity) of polar liquids, especially in relation to predictions based on the Onsager and Kirkwood equations. The second concerns what conclusions can be drawn from the differences in Kerr effect response (time-dependent birefringence) of a polar liquid following application or removal of a strong electric field. The third part is a discussion of experimental problems in measuring low-frequency dielectric relaxation of charged biopolymers in solution, with a description of an improved four-terminal method and some results for DNA solutions obtained by M. S. Tung and R. J. Molinari.

DIPOLE INTERACTIONS

Onsager's equation[1] for the dielectric constant ϵ of a polar liquid can be written

$$(\epsilon - \epsilon_\infty)\left(\frac{2\epsilon + \epsilon_\infty}{3\epsilon}\right) = \left(\frac{\epsilon_\infty + 2}{3}\right)^2 \frac{4\pi n\mu^2}{3\text{kT}} \tag{1}$$

where μ is the permanent dipole moment of the polar molecule, ϵ_∞ is related to the polarizability α by $(\epsilon_\infty - 1)(\epsilon_\infty + 2) = 4\pi n\alpha/3$, n is the number density of molecules, and kT has the usual meaning. It has long been recognized that this result is, so to speak, the best that can be done with use of macroscopic continuum arguments and is a useful approximate formula that works surprisingly well. Kirkwood[2] obtained an equally celebrated, formally more general, expression, one form of which can be written

$$\frac{\epsilon - 1}{\epsilon + 2} - \frac{\epsilon_\infty - 1}{\epsilon_\infty + 2} = \langle\mu^2 + \boldsymbol{\mu}\cdot\Sigma_i(\boldsymbol{\mu}_i + e\mathbf{r}_i)\rangle\, \frac{4\pi n\mu^2}{9\text{kT}} \tag{2}$$

In this equation, the effect of dipole interactions appears in the average correlation of a permanent dipole $\boldsymbol{\mu}$ with the sum of permanent and induced moments $\boldsymbol{\mu}_i + e\mathbf{r}_i$ of other molecules in a finite spherical sample. If this correlation is approximated by a macroscopic continuum, Equation 2 reduces to Onsager's result.

It has often been tactily assumed that purely dipolar forces between polar molecules lead to no average correlation effects other than those imposed by macroscopic boundary conditions at the surface of the sample, and hence that Onsager's Equation takes account of the short-range effects of such forces. However, theoretical treatments of two different models by M. Lax[3] and by Wertheim,[4] taking explicit account of dipole forces, do not support this conclusion. These results seem to be unknown to most dielectricians, partly because the predicted dependences of dielectric constant on dipole moment, density, and temperature seem to have been given only in part or not at all. Our purpose here is to call attention to these theories and present numerical calculations of some of their predictions with an indication of their implications.

The Spherical Model

In the Toupin and Lax[5] rigid cubic lattice model of polarizable permanent dipoles with moments $\mathbf{m}_i = \mathbf{u}_i + e\mathbf{r}_i$, the potential energy U of the system is given by

$$U = -\sum_i \mathbf{m}_i\mathbf{E}_0 - \sum_i\sum_{j\neq 2} \mathbf{m}_i\mathbf{G}_{ij}\mathbf{m}_j + \frac{1}{2\alpha}\sum_i (e\mathbf{r}_i)^2 \tag{3}$$

where $\mathbf{E}_0$ is the uniform field of external charges. The basic statistical mechanical problem is then to evaluate the partition function Z and from it the polarization $P = k\text{T}\partial \ell n\, Z/\partial E_0$. The former is given by

$$Z = \prod_i \int_{-\mu}^{\mu} d\mu_i \prod_j \int_{-\infty}^{\infty} dr_i \exp(-U/\text{kT})$$

in which the phase-space integrations are over the restricted permanent dipole orientations, as symbolized by the limits $-\mu$ and μ, and induced dipole harmonic oscillator displacements r_i. Integration over the latter only presents no difficulty because, as Van Vleck[6] pointed out many years ago, one can reduce U to a sum of squares by changes of variable and then carry out the integrations over the new variables because their ranges are from $-\infty$ to $+\infty$. By this procedure, Van Vleck provided the first dynamic proof of the validity of the Clausius-Mossotti formula for this model of induced polarization.

Toupin and Lax accomplished the first part of a similar treatment with permanent dipoles included and approximated the 3n fold integrations over restricted orientations of n dipoles by the limits $-\infty$ to $+\infty$ and the "spherical" constraint $\sum_i \mu_i^2 = n\mu^2$, which, when represented by an integral delta function, permitted evaluation of Z and P. The normal coordinate transformation

$$\mu_i = \sum_k Y_k \exp(i\mathbf{k}\mathbf{R}_i)$$

Reprinted with permission from *Ann. N.Y. Acad. Sci.*, vol. 303, pp. 59-73, Dec. 30, 1977.

introduces "dipole wave sums" λ_k given by

$$\boldsymbol{\lambda}_k = \sum_i \sum_j \mathbf{G}_{ij} \exp\left[i\mathbf{k}(\mathbf{R}_i - \mathbf{R}_j)\right] \tag{4}$$

where $\mathbf{R}_i$ is the lattice vector position of dipole $\boldsymbol{\mu}_i$.

Evaluation of the polarization results in the dielectric constant formula

$$\sum_k \left[\frac{3}{\epsilon - 1} + 1 - \frac{3}{4\pi}\boldsymbol{\lambda}_k\right] = \frac{4\pi}{3}\frac{n\mu^2}{3\mathrm{kT}} \tag{5}$$

Here, for simplicity, only the permanent dipole terms have been included, but the complete formula for $\alpha \neq 0$ presents no difficulty.

A remarkable feature of this result is that it reduces to Onsager's formula if the $\boldsymbol{\lambda}_k$ are approximated by their values $-8\pi/3$ (longitudinal) and $+4\pi/3$ twice (transverse) in the limit $k \to 0$. (As expected, setting all $\boldsymbol{\lambda}_k = 0$ and thereby neglecting dipole coupling altogether gives the Clausius-Mossotti formula.) To determine the effect of dispersion of the $\boldsymbol{\lambda}_k$ for finite k, Toupin and Lax considered explicitly only a simple cubic lattice by complicated analytical evaluations of the $\boldsymbol{\lambda}_k$ and sums involving the $\boldsymbol{\lambda}_k$.

We have used Cohen and Keffer's[7] numerical calculations of $\boldsymbol{\lambda}_k$ to evaluate the k sums in Equation 4 for the more realistic body-centered lattice and for a dipole continuum outside an "Onsager cavity" of radius a satisfying $4\pi a^3 n/3 = 1$ with the correct domain in k space. These are given graphically in *Predictions of the Models* after a brief description of Wertheim's treatment and result for a liquid model.

Wertheim's Mean Spherical Model Theory

Wertheim[4] considered the problem of evaluating dipole correlation effects as expressed in Equation 2 for a model of a dipolar fluid. In terms of the pair distribution function $g(\boldsymbol{r}_{12})$, the contribution to be evaluated is proportional to the integral

$$\int d\Omega_1 \int dV_{12}\, \mu_1\, \mu_2\, e\, g(\boldsymbol{r}_{12})$$

where the integrations are over orientations and vector separation r_{12} of dipoles μ_1 and μ_2, induced dipoles and interactions being neglected. The problem solved exactly is a generalization of the "mean spherical model" for hard-sphere fluids, which takes account of the dipole interaction term in Equation 3 for U. In this model, the direct correlation function $C(\boldsymbol{r}_{12})$ is related to $g(\boldsymbol{r}_{12})$ by a generalized Ornstein-Zernike equation

$$g(\boldsymbol{r}_{12}) - 1 = C(\boldsymbol{r}_{12}) + n\int d\boldsymbol{r}_{13}\, C(\boldsymbol{r}_{13})[g(\boldsymbol{r}_{12}) - 1] \tag{6}$$

and the approximation of the model is taking $C(\boldsymbol{r}_{12}) = -U(\boldsymbol{r}_{12})/\mathrm{kT}$. Wertheim, by an elegant analysis, reduced the solution to that of the Percus-Yevick equation for hard spheres. The result for ϵ is given by the parametric equations

$$\frac{\epsilon - 1}{\epsilon + 2} = \frac{q(K\eta) - q(-K\eta)}{q(2K\eta) + 2q(-k\eta)} \tag{7}$$

$$\frac{4\pi}{3}\frac{n\mu^2}{3\mathrm{kT}} = \frac{1}{3}\left[q(2K\eta) - q(-K\eta)\right]$$

where the second equation determines K in terms of $q(\eta) = (1 + 2\eta)^2/1 - \eta)^4$, obtained from solution of the Percus-Yevick equation with $\eta = \pi R_0{}^3 n/6$, R_0 being the hard-sphere radius. The dielectric constant is then obtained as a function of the reduced dipole temperature $T/T_c = 9\mathrm{kT}/4\pi n\mu^2$ by assigning values of $K\eta$ in the range 0 to 1/2.

Predictions of the Models

In FIGURE 1, calculated values of ϵ from the results of the various theoretical models are plotted as a function of the reduced temperature T/T_c. It is seen that the dipole interaction models give values much closer to the Onsager result than the Lorentz field (Clansius-Mossotti) values. With the exception of the simple cubic lattice behavior at low temperatures (with a phase transition indicated by the arrow), the predicted values of ϵ are increasingly larger than from Onsager's Equation at lower temperatures. The similarity of the results for the Wertheim hard-sphere fluid and the "continuum" dipole lattice in the spherical model is also striking, and the body-center lattice values are also quite similar. (Incomplete calculations for the face-centered lattice indicate that the results will not be greatly different from those for the body-centered lattice).

The behavior of the various results at low temperatures is better seen from the plots of $1/(\epsilon - 1)$ against T/T_c in FIGURE 2. The continuum, body-, and face-

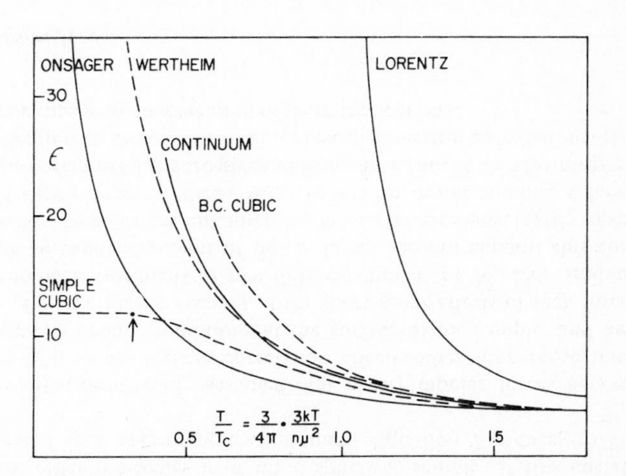

FIGURE 1. Dielectric constant ϵ versus reduced temperature T/T_c from theories of rigid dipole interaction effects discussed in text.

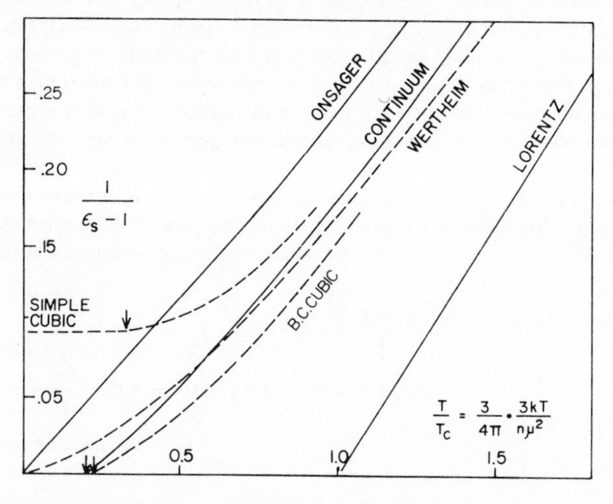

FIGURE 2. Reciprocal of $\epsilon - 1$ versus reduced temperature T/T_c from theories of rigid dipole interaction effects discussed in the text.

centered cubic lattice sums all indicate an infinite permittivity and phase transitions at $T/T_c \simeq 0.2$ to 0.25, i.e., at about one-fifth the Curie temperature for the Lorentz field "$4\pi/3$ catastrophe." These predictions and questions of the significance to be attached to them need further study. The hard-sphere fluid values show no transition; $\epsilon - 1$ approaches infinity only as $T \to 0$, and is more like the Onsager result in this respect. (But a little-known result of Pirenne[8] should be noted: below a critical temperature of order 0.1 T_c, the electrostatic equations of Onsager's model do admit a solution with finite spontaneous polarization in zero field). The simple cubic lattice behavior is quite different and associated with anomalous dispersion of the $\lambda(\mathbf{k})$ values and related lattice sums; this has been discussed by Fulton.[9] As the more closely packed structures are presumably more realistic for most systems, these peculiarities are unlikely to be of great interest.

For simplicity, we have given results only for rigid dipoles ($\alpha = 0$). Real molecules are, of course, polarizable with $(\epsilon_\infty - 1)/(\epsilon_\infty + 2)$ typically of order 0.25 to 0.5. The Toupin and Lax results for $\epsilon - 1$ with polarizability included is a bit tedious to work out but involves no real difficulty once the $\lambda(\mathbf{k})$ eigenvalues are known; calculations indicate that the principal effect is to increase the temperatures for a given ϵ by a factor $[(\epsilon_\infty + 2)/3]^2$ without major changes in shape of the curves in FIGURES 1 and 2. Necessary modifications to generalize Wertheim's analysis to include polarizability will require further study.

The principal conclusion from the approximate calculations of dipole interaction effects is seen to be that the interactions result in *larger* dielectric constants than predicted by Onsager's Equation or by Kirkwood's Equation if net correlations of a dipole with its neighbors are *assumed* to vanish. The Kirkwood g factor expressing the difference is correspondingly significantly greater than 1 for quite weak dipole forces, but detailed discussion of the implications should await more complete evaluations of the polarizability contributions. In any case, it should

be clear that Onsager's Equation, or modifications of it from similar electrostatic continuum models, should be used with caution in attempts at quantitative calculations of ϵ from dipole moments or vice versa.

TRANSIENT KERR EFFECT

Studies of anisotropy in electric polarization produced by a strong electric field can give useful information about molecular parameters and intermolecular forces, and if the field is time-dependent about the kinetics of molecular reorientation. One measure of the anisotropy is the difference in refractive indices parallel and at right angles to the orienting field $E(t)$. This can be determined by sending linearly polarized light through a sample in the x direction at right angles to $E(t)$ as shown in FIGURE 3, and evaluating the difference Δn of refractive in-

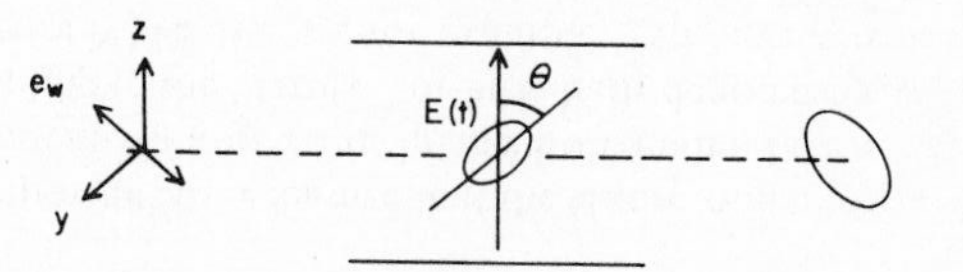

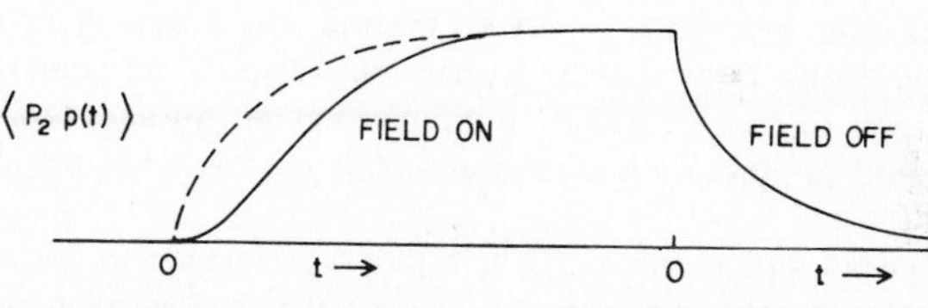

FIGURE 3. *Top:* Schematic arrangement of Kerr effect measurements. *Bottom:* Transient Kerr effect following application and removal of the polarizing field. The dashed curve is the mirror image of the decay function following removal of the field.

dices n_z and n_y from the birefringence (elliptical polarization) it produces. For axially symmetric molecules with a difference $\Delta\alpha = \alpha_p - \alpha_s$ of polarizabilities α_p parallel and α_s perpendicular to the molecular symmetry axis at an angle Θ with the orienting field $E(t)$ along z, the refractive index difference Δn is necessarily an even function of $E(t)$, i.e., independent of the sign of $E(t)$.

For not too intense fields, $\Delta n = KE^2$ and the Kerr "constant" K is related to $\Delta\alpha$ and degree of orientation induced by the field by an expression of the form

$$K = A\,\Delta\alpha \langle P_2(\cos\Theta)\rho(t)\rangle$$

In this expression, numerical factors, number density of molecules, and instrumental constants have been represented by the factor A. (See Beevers *et al.*,[10] for example, for the complete expression.) The quantity of interest here is the average

$<P_2(\cos\Theta)\rho(t)>$, where $P_2(\cos\Theta) = (3\cos^2\Theta - 1)/2$ is the second Legendre polynomial, resulting from the projection of the optical field components onto molecular axes and the projection of the moments they induce back onto laboratory axes, and $\rho(t)$, is the time-dependent distribution function.

In the equilibrium case for constant field E, the equilibrium distribution function ρ_E is given for polar molecules with dipole moment μ by

$$\rho_E = N \exp\left(-\beta\left[H_0 - \mu P_1(\cos\Theta)E - \frac{1}{3}\Delta\alpha P_2(\cos\Theta)E^2\right]\right) \quad \textbf{(8)}$$

where $\beta = 1/kT$, N is the normalization constant (partition function), H_0 is the Hamiltonian for $E = 0$, and for simplicity, intermolecular dipole forces have been neglected. The well-known solution[10] for the quadratic field dependence is

$$<P_2(\cos\Theta)\rho_E>$$

$$= \left(\frac{1}{3}\beta\Delta\alpha < P_2^2 \exp(-\beta H_0) > + \frac{1}{2}\beta^2 < P_2 P_1^2 \exp(-\beta H_0) >\right) E^2$$

$$= \frac{1}{15}\beta(\Delta\alpha + \beta\mu^2)E^2, \quad \textbf{(9)}$$

showing the dependence of the effect on both $\Delta\alpha$ and μ^2.

For time-dependent orienting fields $E(t)$, the appropriate time-dependent distribution function $\rho(t)$ must be determined if the relation of the time-dependent birefringence to molecular reorientations is to be used to learn about the latter. Benoit[11] solved this problem for the model of forced rotational diffusion when a constant field E is either applied or removed at time $t = 0$. His results are

(a) Field removed at $t = 0$,

$$<P_2\rho(t)> = \frac{1}{15}\beta(\Delta\alpha + \beta\mu^2)E^2 \exp(-6Dt) \quad \textbf{(10)}$$

(b) Field applied at $t = 0$

$$<P_2\rho(t)> = \frac{1}{15}\beta(\Delta\alpha + \beta\mu^2)E^2[1 - \exp(-6Dt)]$$

$$- \frac{1}{10}\beta^2\mu^2E^2[\exp(-2Dt) - \exp(-6Dt)]$$

where D is the rotational diffusion coefficient.

The remarkable feature of Benoit's results that has made the transient Kerr effect of such interest is that the response and decay curves lack the symmetry of linear polarization effects. As shown in FIGURE 3, the rising response is slower than the decay, and for the diffusion model the decay is associated entirely with the exponential loss of correlation of $P_2(\cos\Theta)$ with rate constant $6D$. The response, on the other hand, is not just the mirror image of the decay of P_2 correlation. The extra effect results from a difference of the P_2 and P_1 correlations, with the latter decaying more slowly (rate constant $2D$), and clearly arises from the second-order effect of dipole orientation energy $-\mu\cos\Theta E$ and corresponding torque, as it is proportional to μ^2.

The rotational diffusion model can thus be tested for a given system by determining whether the polarization does decay exponentially, and if it does, the relative importance of permanent and induced dipole polarizations can be determined if the difference between response and decay of polarization is consistent with the model. The questions we consider here concern what can be inferred more generally, without a specific model, from the observed transients. Thus if the decay function is not exponential, as it often is not, is it still true that this function is determined only by $P_2(\cos\Theta)$ time correlation, and is it possible to infer relative magnitudes of permanent and induced dipole contributions with useful information about $P_1(\cos\Theta)$ time correlation from the difference of response and decay functions? This sort of problem should be amenable to more general nonequilibrium theory, but presents difficulties not present in linear response theory because interactions of the molecules with the applied field and with the lattice or heat bath become coupled in second- and higher-order approximations that are necessary to take account of permanent dipole torques and energies. An analysis of this kind has not yet been fully carried out, but enough progress has been made to permit some useful conclusions, and a brief outline of the present state can be given.

Although either classical or quantum treatments should be possible, the latter has advantages, even for the classical limit, of greater ease of manipulation. The distribution function is then the density matrix $\rho(t)$ satisfying $d\rho(t)/dt = -(i/\hbar)[H(t), \rho(t)]$, where the bracket is the commutator and $H(t)$ the time-dependent Hamiltonian. This can be taken to be

$$H(t) = H_0 + H_c(t) - \mu P_1(\cos\Theta)E(t) - \frac{1}{3}\Delta\alpha P_2(\cos\Theta)E^2(t) \quad \textbf{(11)}$$

where it is important to recognize that the Hamiltonian for zero field is not just the time-independent H_0 appearing in the Boltzmann factor $\exp(-\beta H_0)$, but also includes coupling to the "lattice" or heat bath denoted by $H_c(t)$. This is assumed, as in spin-lattice NMR theory, for example,[12] to have zero-time average and only short-time correlation such that the time average $\overline{H_c(t)H_c(t+\tau)} \rightarrow$ constant for time $t \gg \tau$ of interest. General perturbation solutions for $\rho(t)$ through terms in $E_2(t)$ present considerable complication, but there is more hope for the simpler cases that a constant field E is applied or removed at $t = 0$.

If the field E is removed at $t = 0$, then $\rho(t)$ relaxes to the final equilibrium ρ_0 as

$$\frac{d}{dt}(\rho - \rho_0) = -(i/h)[H_0 + H_c(t), \rho - \rho_0] - (i/h)[H_c(t), \rho_0]$$

with the initial condition $\rho = \rho_E$ given above. Because the field-dependent terms in $H(t)$ are not involved in the rate equation, the solution is readily obtained and expressed in terms of the "natural motions" of $P_1(t)$ and $P_2(t)$ satisfying

$$\frac{dP}{dt} = +(i/\hbar)[H_0 + H_c(t), P].$$

The result for the relaxation function of interest is

$$<P_2\rho(t)> = \frac{1}{3}\beta\Delta\alpha <P_2P_2(t)\exp(-\beta H_0)> E^2$$

$$+ \frac{1}{2}\beta^2\mu^2 <P_2P_1^2(t)\exp(-\beta H_0)> E^2$$

$$= \frac{1}{15}\beta(\Delta\alpha + \beta\mu^2)\phi_2(t)E^2$$

where $\phi_2(t)$ is the normalized time-dependent correlation function for $P_2(\cos\Theta)$. The result is then just the generalization of the diffusion model solution (a) that one might expect from replacing $\exp(-6Dt)$ by $\phi_2(t)$.

If the field E is applied at $t = 0$, the relaxation of $\rho(t)$ to its final equilibrium value ρ_E is given by

$$\frac{d}{dt}(\rho - \rho_E) = -(i/\mathrm{h})[H_0 + H_c(t) - \mu P_1 E - \frac{1}{3}\Delta\alpha P_2 E^2, \rho - \rho_E]$$

$$-(i/h)[H_c(t), \rho_E]$$

with $\rho = \rho_0$ at $t = 0$. The motion is now perturbed by the field, complicating the second-order perturbation methods considerably. We have so far succeeded in obtaining a partial solution only of the form

$$<P_2\rho(t)> = \frac{1}{15}\beta(\Delta\alpha + \beta\mu^2)(1 - \phi_2(t)E^2$$

$$- \frac{1}{15}\beta\mu^2\left[\int_0^t dt'\phi_2(t')\phi_1(t - t') + F(t)\right]E^2$$

where the function $F(t)$ in the second term has so far resisted reduction to any simple form in terms of $\phi_2(t)$ and $\phi_1(t)$ for the time-dependent correlation of $P_1(\cos\Theta) = \cos\Theta$. Comparison with the diffusion solution (b) above shows that the first term is the generalized relaxation of $P_2(t)$, whereas the second involving both $P_2(t)$ and $P_1(t)$ arises from permanent dipole polarization only.

At this stage of development, the response-theory treatment thus shows that the decay function depends only on $P_2(t)$ relaxation, but the response function depends on both the mirror image of this relaxation plus terms depending on response of permanent dipole polarization. One can thus answer the question of whether permanent dipole polarization is present by comparing the response and decay curves, but the answer to the question of whether analysis of the difference between the two curves can give useful information about both kinds of relaxation must await further analysis of response-function integrals.

Low-Frequency Relaxation of Biopolymers in Solution

Although dielectric relaxation processes at audio- and subaudiofrequencies have been known or inferred for many biopolymers of high molecular weight in aqueous solution and may arise from the major contributions to orientational polarization, relatively little is known about these processes in many systems because of experimental difficulties.

There are four related kinds of difficulty. The first is that the dispersion and absorption currents for a given applied alternating field decrease with frequency of the field, requiring high-current sensitivity, low-frequency measurements. This problem is not too serious by itself, but becomes much more so if there is appreciable ionic conductance of the solution, since ion currents for a given field are independent of frequency and hence increasingly dominant dielectric currents by as much as a factor of a million at frequencies of a few hertz. The third complication is from space charge or electrode polarization if the metal electrodes are partly or wholly blocking for ion-electron charge transfer, as they usually are; this increasingly acts to reduce the polarizing field in the body of the solution at low frequencies. Finally, biopolymer solutions often fail to reach a completely steady state, with a consequent drift of real and apparent conductance with time.

Four-Terminal Null Methods

These various difficulties can be overcome, to a large extent, and the range of useful measurements extended to much lower frequencies by making use of AC four terminal methods. In these, currents are driven in the solution between two current electrodes across which an alternating field is applied, as shown in Figure 4, while the potentials at two points in the solution with negligible electrode space charge are picked up by probe electrodes, which ideally do not draw current

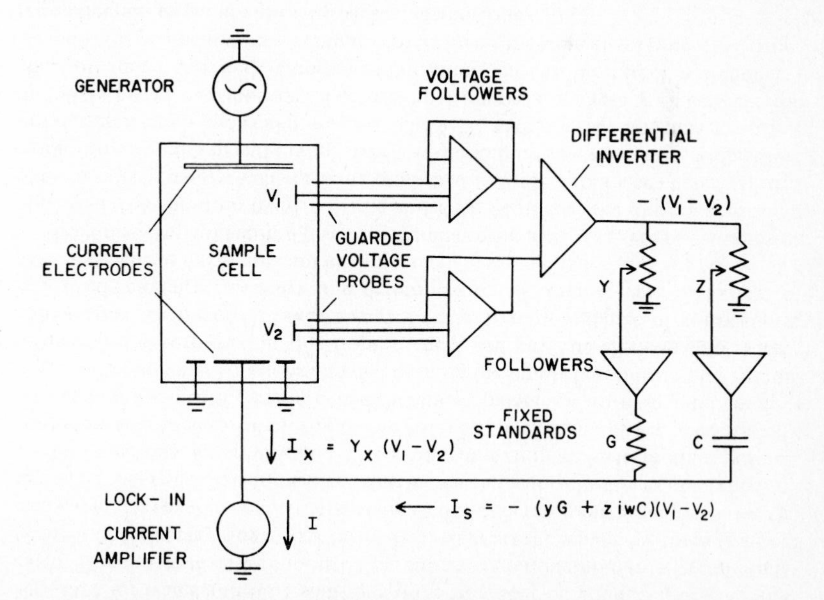

Figure 4. Simplified schematic diagram of cell and electronic instrumentation for four-terminal null measurements of dielectric properties of conducting solutions.

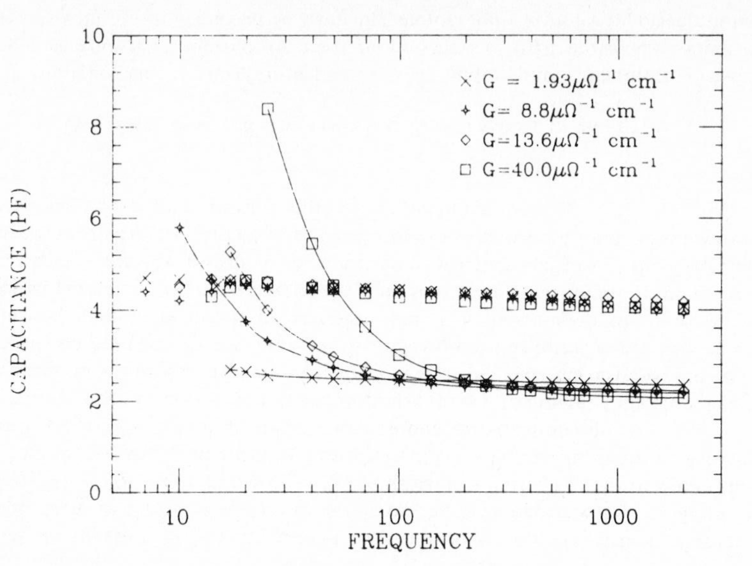

FIGURE 5. Uncorrected calibration data. Measured capacitances of acqueous KCl solutions of four specific conductances as a function of frequency. The frequency-independent values are from four-terminal null measurements; the curves from three-terminal bridge measurements show increases at low frequency as a result of electrode polarization.

or otherwise distort the current flow. For such conditions, the ratio of the current to the probe potential difference is directly proportional to the sample admittance $y = i\omega\epsilon_t^*$, where ϵ_t^* is the "total permittivity" usually and somewhat arbitrarily written as the sum of ohmic conduction and dielectric currents: $\epsilon_t^* = \epsilon^* + \sigma/i\omega E$, with E a conversion factor for consistent units.

The four-terminal method can reduce electrode polarization errors enormously, leaving the problem of measuring small currents of interest when much larger ionic currents are superposed. Various electronic null methods to accomplish this have been proposed, with sensitivity and resolution increasing as more sophisticated solid-state devices become available. A four-terminal "bridge" for the purpose, originally developed by Berberian and Cole[13] and since refined considerably by R. J. Molinari, is shown schematically in FIGURE 4. Ideal voltage probes are approximated by small platinum electrodes in the solution connected to voltage followers with very small FET input currents (10^{-13} amp) by coaxial lines with driven shields to virtually eliminate probe to ground stray capacitances. Variable fractions x and y of the inverted difference $V_2 - V_1$ of probe potentials are applied to a standard capacitance C and conductance G producing current I_s, which by adjustment of x and y balances the solution current I_x. For this condition, $I = I_x + I_s = 0$ as determined by the null detector and the sample admittance $Y = I/(V_1 - V_2)$ between the probes is given by $Y = x(i\omega C) + yG$.

For this scheme to work successfully down to a few Hertz or less, care must be taken to determine and compensate or correct for extremely small phase shifts. The degree to which this has been accomplished is shown by results plotted in FIGURE 5 for water and KCl solutions of several specific conductances, as indicated. The measured capacitances by the four-terminal method are virtually independent of frequency down to 5 Hz or less, whereas the values by three-terminal measurements (probe input circuits connected to generator and detector) show the familiar rapid rise at low frequencies and increasing with solution conductance.

The necessary sensitivity and phase discrimination in determining the null condition is realized satisfactorily by a PAR-124A lock-in amplifier, and the only serious limitation of the method is difficulty in balancing drifting ionic conductances of biopolymer solutions and determining the small components attributable to dielectric absorption. In this respect, the ingenious but electronically rather complicated scheme of Hayakawa *et al.*[14] for automatic balance of conductance at a fixed reference frequency and determination of conductance differences at other frequencies has significant advantages.

The four-terminal method just described has been used in a series of studies of aqueous DNA solutions by M. S. Tung and R. J. Molinari. We outline here some of the results to date, which are being published in more detail elsewhere.[15] In this work, three samples of calf thymus DNA as the sodium salt were used, with stated or inferred average molecular weights of 1.4, 2, and 9×10^6 and unknown but doubtless broad distributions. The principal interests in these studies were the effect of added salt and temperature variation on the limiting low-frequency, steady-state permittivity for comparison with the theoretical treatment of counterion fluctuation polarization by McTague and Gibbs.[16]

The observed dielectric behavior for a 3.8×10^{-4} solution of average molecular weight 2×10^6 at 25° is shown in FIGURE 6, together with the effect of added NaCl (4×10^{-5}N). It is seen that the relaxation is within experimental error described by the circular arc function $\epsilon^* = \epsilon_0 + (\epsilon_0 - \epsilon_\infty)/[1 + (i\omega\tau)^\beta]$[17] with β of order 0.75 and central relaxation frequency 10 Hz. These values are both lowered somewhat by added NaCl, but the principal effect is the decrease of extrapolated static permittivity ϵ_0 by about 40 percent. The effect of divalent positive ions is even more marked, a 1.5×10^{-4} normality of Mg^{++} ion reducing the increment in ϵ_0 over that of water virtually to zero.

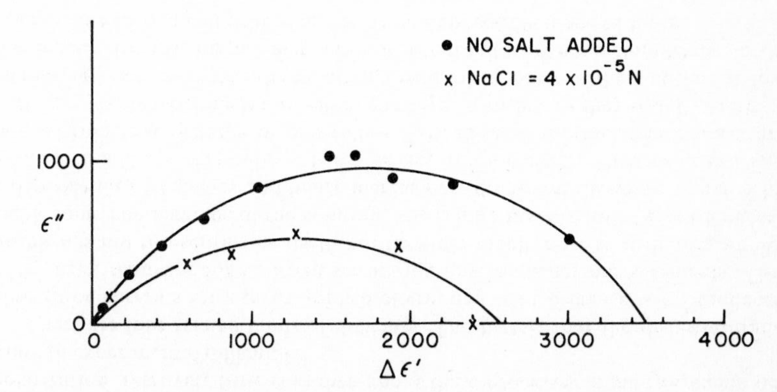

FIGURE 6. Complex plane loci of dielectric constants of salt-free DNA solution and DNA solution with added NaCl.

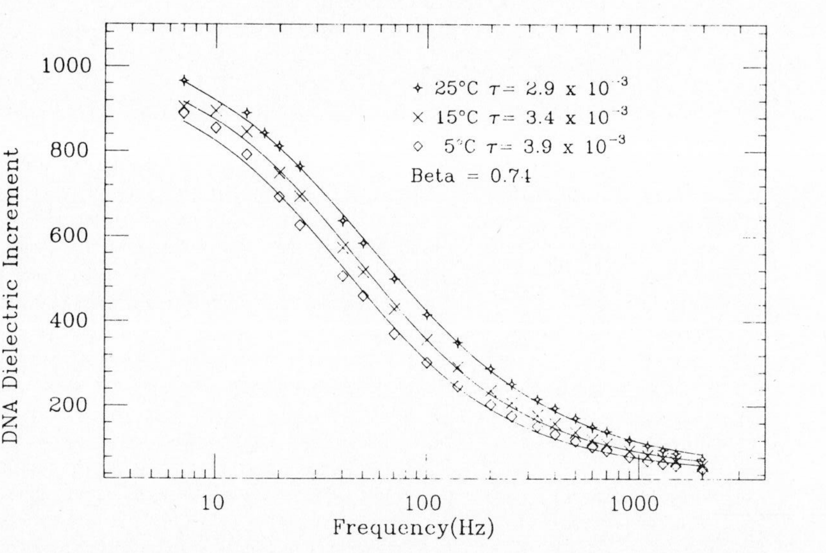

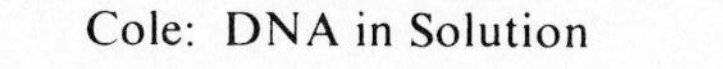

FIGURE 7. Dispersion curves of ϵ' versus frequency for DNA solutions (molecular weight 9×10^6, concentration 6.2×10^{-4} M) at 5°, 15°, and 25°. The solid curves are circular arc functions with $\beta = 0.74$.

The effect of temperature on the dispersion of ϵ' for solutions of concentration 6.2×10^{-4}M and average molecular weight 9×10^6 is shown in the plots of ϵ' versus frequency at 25°, 15°, and 5° in FIGURE 7. These have been computer least squares fitted by the solid curves for circular arc functions, with $\beta = 0.74$ having no significant temperature-dependence in this limited range. The static permittivities from the fitted curves are plotted against temperature in FIGURE 8 and show a significant *positive* temperature coefficient, in contrast to the negative coefficient ($1/T$ temperature-dependence) characteristic of permanent dipole polarization.

That these large low-frequency polarizations are characteristic of the helical form of DNA is shown by the fact, established by several other studies as well as these, that partial denaturation by heating to temperatures above 60° reduces their magnitude markedly, heating to 95° for 15 minutes sufficing to reduce the dielectric increments observed at room temperature virtually to zero.

The evidence just cited and other measurements are qualitatively consistent with explanations of the low-frequency polarization in terms of field-induced displacements of counterions attached to a fraction of available negative charge sites on the DNA helix. This fraction (one minus the charge fraction) is expected to be somewhat greater than one-half for the salt-free DNA samples at the concentrations used studied. The McTague-Gibbs Theory predicts, as one would expect intuitively, that the static polarization is a maximum for half the sites occupied, decreasing to zero for none or all filled. One then expects that added positive counterions will decrease the polarization as observed because more sites are occupied, whereas an increase in temperature will reduce the number and increase the polarization, again as observed.

The magnitude of polarization predicted by the McTague-Gibbs Theory is also in at least qualitative agreement with the observed values. For an estimated mean square end-to-end length of 3400 Å for DNA of molecular weight 2×10^6, assuming a contour length of 10,000 Å and persistence length of 600 Å, the theory predicts $\Delta\epsilon/c = 3.8 \times 10^6$, which is comparable with the observed value of 9.8×10^6 at a concentration of 0.93×10^{-3} molar.

The measurements of Tung and Molinari also indicate an increase in the specific dielectric increment $\Delta\epsilon/c$ as the concentration is reduced. This is consistent with a decrease in fraction of occupied sites on dilution and increase with polarization per molecule, but is somewhat ambiguous because the concentrations are rather high, with appreciable concentration-dependent solute interactions likely as a result and because slow denaturation after dilution was also observed.

The frequency dependence of the low-frequency relaxation process differs considerably from simple Debye behavior for all the DNA solutions so far studied by Tung, Molinari, and other workers. It does not seem possible to conclude with any assurance the extent to which the rather broad frequency range of relaxation is the result of heterogeneity in size, concentration-dependent solute interactions, cooperative counterion motions, or other effects, and more extensive studies will be required to throw light on such questions. The average or central relaxation frequencies in the subaudiorange are at least of the order predicted by kinetic theories of counterion fluctuations,[18-20,22] and one can secure agreement of their predictions with observed values by assigning plausible values to such parameters as counterion jump rates or mobilities appearing in the theories, but the question of whether real significance can be attached to such values or the models is unanswered as yet.

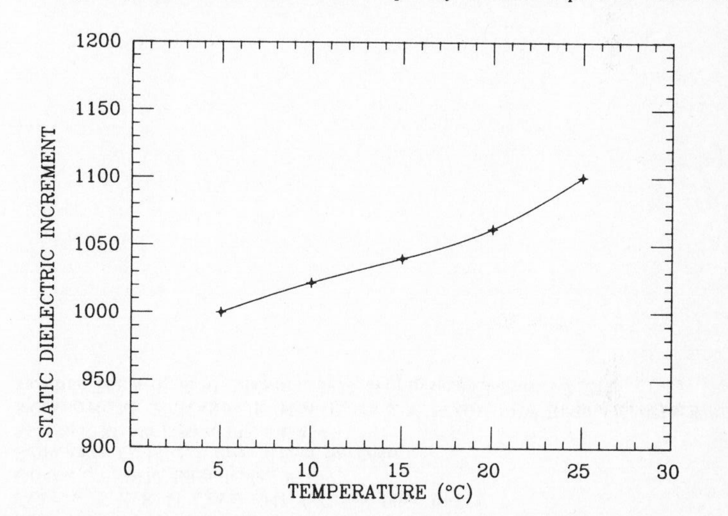

FIGURE 8. Static dielectric constants from the data plotted in FIGURE 7 as a function of temperature.

The results summarized above agree with measurements of other workers at higher frequencies and those of Sakamoto *et al.*[21] at room temperature to the extent comparison is possible for samples of different origins, molecular weights, and weight distributions. They are consistent with predictions from models of counterion fluctuation polarization, which appear to be more satisfactory from several points of view than alternative explanations that have been put forward. The principal difficulty in reaching definitive conclusions is that DNA samples of high molecular weights and broad distribution of molecular weights have been studied at rather high concentrations, with no clear indications of effects of polydispersity and solute interactions. Extensive measurements are needed on better and more sharply characterized samples over wide ranges of concentrations, added salt, temperature, and so on, if more quantitative conclusions are to be reached.

REFERENCES

1. ONSAGER, L. 1936. J. Am. Chem. Soc. **58:** 1486.
2. KIRKWOOD, J. G. 1939. J. Chem. Phys. **7:** 911.
3. LAX, M. 1952. J. Chem. Phys. **20:** 1351.
4. WERTHEIM, M. S. 1971. J. Chem. Phys. **55:** 4291.

5. TOUPIN, R. A. & M. LAX. 1957. J. Chem. Phys. **27:** 458.
6. VAN VLECK, J. H. 1937. J. Chem. Phys. **5:** 556.
7. COHEN, M. H. & F. KEFFER. 1954. Scientific Paper 60-94469-2P2, Westinghouse Research Laboratories.
8. PIERENNE, J. 1949. Helv. Phys. Acta **22:** 479.
9. FULTON, R. L. 1975. J. Chem. Phys. **62:** 3676.
10. BEEVERS, M., J. CROSSLEY, D. C. GARRINGTON, & G. WILLIAMS. 1976. J. Chem. Soc. Faraday II **72:** 1482.
11. BENOIT, H. 1951. Ann. Physique **6:** 561; 1952. J. Chim. Phys. **49:** 517.
12. SLICHTER, C. P. 1963. Principles of Magnetic Resonance, Chapter 5. Harper and Row. New York, N.Y.
13. BERBERIAN, J. G. & R. H. COLE. 1969. Rev. Sci. Inst. **40:** 811.
14. HAYAKAWA, R., H. KANDA, M. SAKAMOTO, & Y. WADA. 1975. Jap. J. App. Phys. **14:** 2039.
15. TANG, M. S., R. J. MOLINARI, R. H. COLE & J. H. GIBBS. Submitted to Biopolymers.
16. MCTAGUE, J. P. & H. H. GIBBS. 1966. J. Chem. Phys. **44:** 4295.
17. COLE, K. S. & R. H. COLE. 1941. J. Chem. Phys. **9:** 341.
18. OOSAWA, F. 1970. Biopolymers **9:** 677.
19. SCHWARTZ, G. 1962. J. Phys. Chem. **66:** 2646.
20. MANDEL, M. 1961. Mol. Phys. **4:** 489.
21. SAKAMOTO, M., H. KANDA, R. HAYAKAWA & Y. WADA. 1976. Biopolymers **15:** 879.
22. VAN DER TOUW, F. & M. MANDEL. 1974. Biophysical Chemistry **2:** 218.

FIELD INTERACTION WITH BIOLOGICAL MATTER

H. P. Schwan

Department of Bioengineering
University of Pennsylvania
Philadelphia, Pennsylvania 19174

INTRODUCTION

The interaction of electrical fields with macromolecules and biological membranes continues to be of interest to many biophysicists. It is the dominant topic of this meeting. Work on the giant squid axon as summarized by the Hodgkin Huxley Equations and their more recent modifications suggest that electrical currents will cause excitation if the induced membrane potential change is of the order of 10 mV. Orientation of polar macromolecules by electrical fields appears to require kilovolts/cm to become noticeable. Corresponding current densities in biological fluids are of the order of 10–100 amperes/cm^2 and cause rapid heating.

A third group of biophysically significant electrical phenomena is caused by electromechanical forces. These field-induced forces may cause orientation and "pearl chain" formation of cells, cellular deformation, or destruction or the movement of cells in inhomogeneous fields as they exist in tissues. These phenomena require a threshold field strength which is inversely related to particle size. This field strength is rather high for biopolymers and of the order of volts/cm for large cellular size.

These modes of field interactions do not provide for explanations of more subtle phenomena that have been reported in recent years. We shall first discuss the electrical properties of tissues and cells, since these properties contain relevant information.

TISSUE PROPERTIES

Suppose a step function potential is applied to a sample of matter. Then the time dependence of the accumulated charge can be frequently characterized by an exponential of the form a + b (1-exp($-t/T$). The corresponding behavior in the frequency domain is given by the Debye expression

$$\epsilon^* = \epsilon_\infty + \frac{\epsilon_0 - \epsilon_\infty}{1 + j\omega T} \qquad (1)$$

with $\epsilon^* = \epsilon' - j\epsilon''$ the complex dielectric constant; T the time constant that characterizes the exponential behavior in the time domain; $\omega = 2\pi f$ angular frequency; the subscripts 0 and ∞ indicate limit values of ϵ' for low and high frequencies. If the time dependence of the charge accumulation is not a simple exponential of time, it may be due to several processes characterized by several time constants. In all generality, any response may be written as a sum of exponentials and the corresponding frequency behavior in the frequency domain as a sum of "dispersions" of the type given in Equation 1.

The phenomena of dielectric dispersion is exhibited by practically all tissues following relaxation equations of the type:

$$\epsilon_0 - \epsilon_\infty + \frac{\epsilon_0 - \epsilon_\infty}{1 + (f/f_c)^2} \qquad (2)$$

$$\kappa = \kappa_0 + (\kappa_\infty - \kappa_0)\frac{(f/f_c)^2}{1 + (f/f_c)^2}. \qquad (3)$$

These equations are obtained from (1) by separation and introducing the conductivity $\kappa = \omega\epsilon''$ and allowing for a low-frequency limit conductivity κ_0

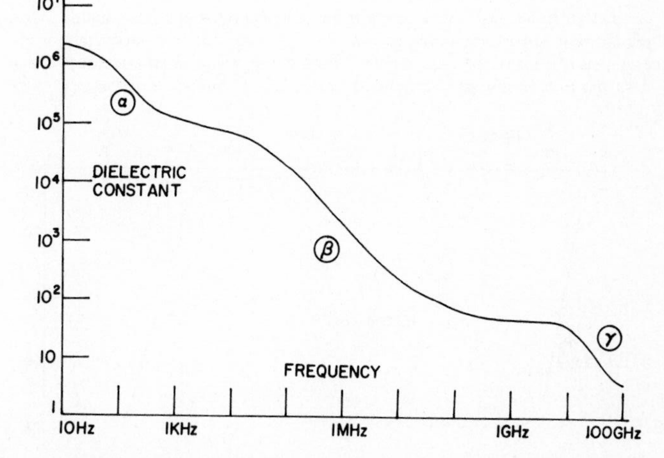

FIGURE 1. Dielectric constant of muscle tissue as a function of frequency. Three dispersion effects are centered near 0.1 kHz, 100 kHz, and 20 GHz. The data are characteristic for most tissues with high water content, even though detailed values for characteristic frequencies and dispersion magnitudes may vary. Room temperature. Data from Ref. 1.

different from zero. In (2) and (3) the "characteristic frequency" is $f_c = \frac{1}{2}\pi T$, ϵ stands for relative dielectric constant, and κ for conductivity. The dispersion magnitudes $\epsilon_0 - \epsilon_\infty$ and $\kappa_\infty - \kappa_0$ interrelated by

$$\kappa_\infty - \kappa_0 = (\epsilon - \epsilon_\infty)\epsilon_r \cdot 2\pi f_c \qquad (4)$$

which results from the Kronig-Kramer relationships for the case of one time constant and with ϵ_r dielectric constant of free space.

The electrical properties of tissues are characterized by three major dispersions labeled, α, β and γ in FIGURE 1. Each dispersion is fairly well defined by either a single relaxation time $T = \frac{1}{2}\pi f_c$ or a small spectrum of relaxation times. Major mechanisms responsible for these dispersions have been fairly well identified and include[1,2] the following:

Reprinted with permission from *Ann. N.Y. Acad. Sci.*, vol. 303, pp. 198–213, Dec. 30, 1977.

γ-dispersion: Dielectric relaxation of free water (f_c near 20 GHz);
β-dispersion: Maxwell-Wagner type relaxation resulting from the charging of cell membranes:
α-dispersion: Variability with frequency of the apparent outer-cell membrane capacitance. This membrane capacitance frequency dependence may result from

1) a frequency-dependent access to inner membrane systems that connect with the outer membrane;[3]
2) a frequency-dependent surface admittance tangential to the cell membrane caused by counter ion displacement about the charged cell and, possibly, fixed charge sites near the outer membrane;
3) a boundary potential-related capacitance element in series with the membrane capacitance;
4) a frequency-dependent membrane capacitance *per se* that results from ionic gating currents.[1,4,5]

Not entirely masked by these major mechanisms are secondary dispersions resulting from

1) Maxwell-Wagner relaxation effects resulting from the charging of membranes of subcellular organelles;
2) Debye type relaxations caused by the polar moments of biopolymers, primarily proteins;
3) relaxation of primarily protein-bound water occurring at some hundred MHz.[1,6-8]

These secondary effects may well be responsible for deviations from a single time-constant response for the β region, particularly at its high frequency tail. Other reasons relate to the complexity of cellular shape and density of packing. The relative contribution of all these secondary mechanisms is not yet entirely clarified.

A considerable amount of work has been done in order to identify the various mechanisms summarized above, particularly by Cole[9] and Schwan.[1] It is apparent that cellular and tissue properties are largely determined by the properties of biological membranes and tissue water. The contribution of biological macromolecules is probably largely masked by the strong Maxwell-Wagner polarization effects that result from the charging of membrane systems through the extracellular and intracellular fluids.

Water and Biopolymers

The interaction of alternating electrical fields with biological macromolecules has been extensively studied. (For summaries see Oncley,[10] and Takashima and Minakata.[11] Much of this work was done with macromolecules suspended in water or electrolyte. It has been suggested that tissue water may be differently structured than is free water.[12] It is, therefore, of interest to discuss the electrical properties of tissue water and compare them with normal water. Free-water estimates may be derived from microwave conductivity data obtained well above 1 GHz and the dielectric constant between β- and γ-dispersion. In addition, the mobility of ions may be compared with those in normal water from an analysis of tissue conductivity data obtained at frequencies between the β- and γ-dispersion ranges. We shall briefly discuss these three possibilities and conclude that in the few tissues studied so far water appears to be "free."

The dielectric properties of protein bound water have been investigated.[1,6-8] FIGURE 2 indicates the dispersions of the dielectric constant for bound water, ice and normal water. Ice relaxes at audio frequencies, bound water between 100 and 1000 MHz, and free water near 20 GHz. Normal or free water is for all practical purposes characterized by a single time constant.[13] Ice and particularly bound water relax over a broader frequency range, possibly integrating over various states.

The magnitudes of the dielectric constant dispersions $\epsilon_0 - \epsilon_\infty$ for ice, bound water, and free water are similar, as indicated in FIGURE 2. Hence the Debye contribution $\kappa - \kappa_0$ to the conductivity which may be calculated from Equations 2 and 3 is easily tenfold and if not hundredfold larger for free water than for protein-bound water. This strong contribution increases sharply with frequency for f-values below and approaching 20 GHz, as may be seen from the appropriate reduction of Equations 3 and 4.

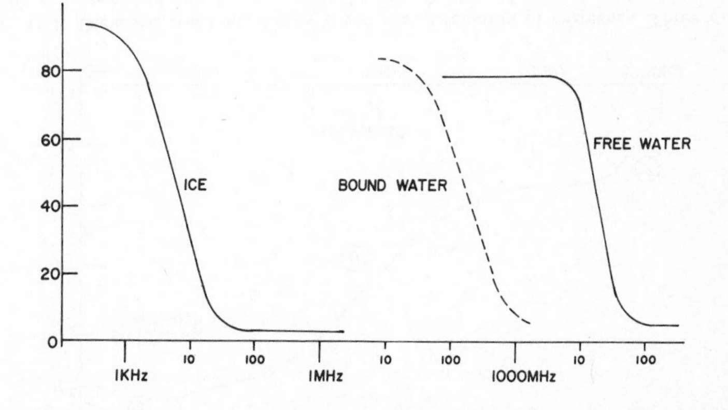

FIGURE 2. Dielectric constant of free water, protein-bound water, and ice as a function of frequency. Interpretation of available data suggests that all three forms of water have similar static dielectric constants near 80–100, but different relaxation frequencies. Protein-bound water appears to be located between ice and free water. Temperature near 23°C.

$$\kappa - \kappa_0 = (\epsilon_0 - \epsilon_\infty) 2\pi \epsilon_r f^2 / f_c. \tag{5}$$

This provides for a possibility to determine the amount of free water in tissue from microwave conductivity. FIGURE 3 shows the well-established conductivity of free water, that of bound water, and a few values obtained for liver tissue. The 10 GHz value for liver is about 30% lower than that of free water, indicating a free-water content in liver above 60%. A more detailed analysis of this sort has been carried out recently for muscle tissue.[14] The analysis yields a free water content equal to most muscle water.

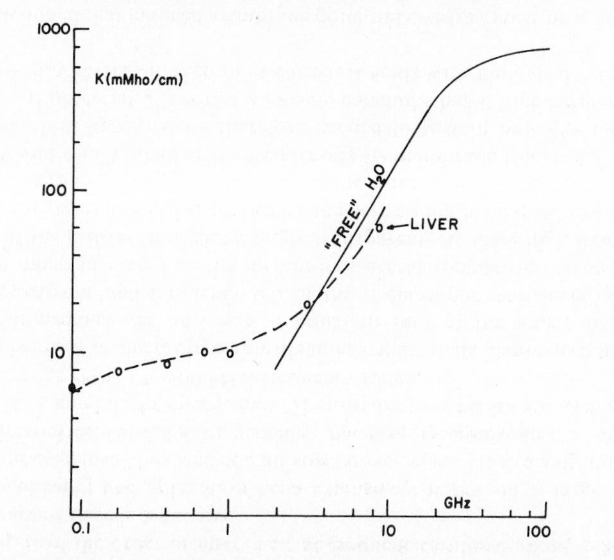

FIGURE 3. Conductivity of liver tissue is compared with that of water at microwave frequencies. The presence of substantial amounts of free water in liver tissue is indicated by the strong increase of κ (liver) above 2000 MHz. Liver data as quoted by Schwan.[1] The water conductivity is calculated by a relaxation frequency of 20 GHz and a static dielectric constant of water of 78 and the Debye Equation.[1] The very slight deviation of the behavior of water from that characterized by a single relaxation type[13] does not noticeably affect this presentation.

FIGURE 4 is chosen to demonstrate how the dielectric constant may be used to estimate the amount of free water from the relationship

$$\epsilon = \epsilon(H_2O) - \delta C \qquad (6)$$

where C is the volume fraction occupied by biopolymers, membranes, and bound water. This relationship applies between the end of the β-dispersion and beginning of the γ-dispersion. The dielectric decrement δ is primarily caused by the protein fraction and near 1, as indicated by theoretical estimates from mixture equations and experimental work with protein and larger particles.[1] The flat part of the muscle curve near 1000 MHz indicates $\epsilon = 50$, whereas free water has a value $\epsilon(H_2O) = 78$ at room temperature. A water content of 100–28—i.e., 72%—is indicated.

Ionic mobility in tissues and blood can be judged from microwave measurements near 100 MHz.[15] These mobilities compare with those in water fairly well if the inner frictional effect of macromolecular surfaces on the mobility is duly considered. Tissue water appears, therefore, largely of the normal type for most tissues investigated so far. However, an analysis of some initial results[16] indicates that the simple relationships that appear valid for most tissues with high water content may not apply to brain tissues, with a possibility for more structured water.

A major fraction of all biopolymers may therefore be anticipated in a normal

nonstructured water environment, and selective interactions at moderate or small field strength values are not indicated by these bulk dielectric data. But further refinements are recommended to elaborate on this approach.

MEMBRANE PROPERTIES AND EXCITATION

Membrane capacitance values are available for many cell types from bulk measurements. Here the applied current to a tissue or cell suspension is varied with frequency and analyzed with use of appropriate spherical, ellipsoidal, or cylindrical shape approximations for cells. Measurements are also available with electrode systems impaled into the cells and directly measuring across the membrane. Although this technique is restricted to larger cell size, it has provided membrane capacitance values in agreement with those obtained from the bulk measurements. Briefly, the capacitance is of the order of 1 $\mu F/cm^2$ with a range extending from 0.6 to 1.3 $\mu F/cm^2$ and an accuracy of about 20% or 30% in most cases. This capacitance appears frequency-independent for the frequency range from about 1 KHz to 100 MHz. It corresponds to a membrane dielectric constant of about 10 relative to free space, and this dielectric constant has been rationalized to reflect a mixture of values from the lipid and protein contributions to the membrane.[1]

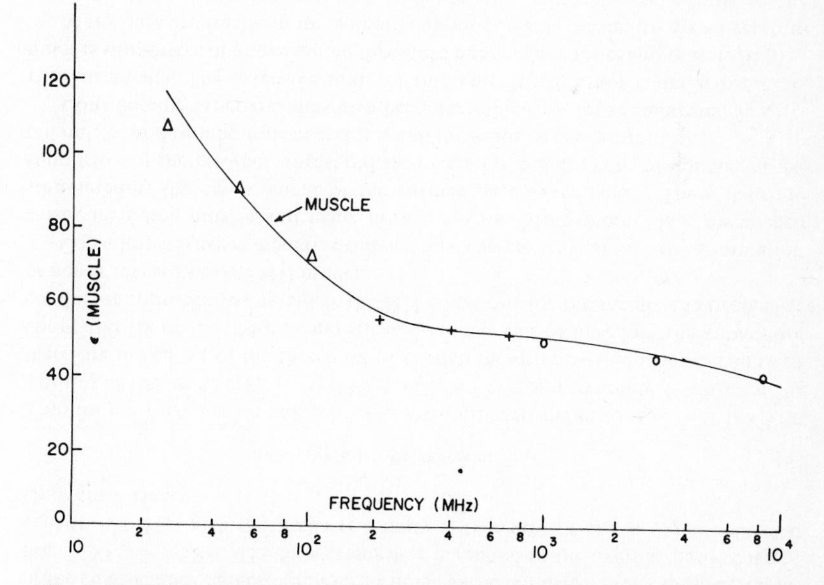

FIGURE 4. Dielectric constant of muscle at microwave frequencies at room temperature. The increase below 100 MHz is due to a Maxwell-Wagner type β-dispersion effect.[1] The decrease above 3000 MHz is due to water and the slow gradual change between 300 and 3000 MHz at least partially due to the δ-dispersion of protein-bound water.[1,6-8] The value between 1000 and 3000 MHz may be used to estimate the amount of free tissue water. The data are taken from Schwan.[1]

The membrane capacitance is polarized by an applied external field with cytoplasmic and extracellular tissue fluids serving as access impedance elements in series with the membrane capacitance. Thus the evoked membrane potential is given by the applied field E sampled over the cell dimension

$$\Delta V_m = 1.5RE \text{ (for spherical shape)} \quad (7)$$

provided that the applied frequency is low enough to make the membrane impedance large in comparison with that of the access impedance elements; i.e., f must be lower than the characteristic frequency $f_c = \frac{1}{2}\pi T$ where T is the time constant given by membrane capacitance, cellular size, and access impedance. For example, for cellular spherical shape[1]

$$T = RC_m(\rho_i + \tfrac{1}{2}\rho_a) \quad (8)$$

where R radius in cm, C membrane capacitance per cm^2, ρ_i internal and ρ_a external resistivity in Ohm-cm. It follows then that larger cells are more sensitive to an external field than smaller ones. This principle of sampling the field over a distance given by the cellular dimensions can explain the electrical sensitivities demonstrated by some large cells.[17,18] It appears also to be used effectively by the receptor organs (Ampullae Lorenzini) of electrosensitive fishes including sharks and rays.[19]

Membrane conductances cannot be extracted from bulk data of tissues or cell suspensions, since minor variations in the extracellular shunt path correspond to major membrane conductance values. Thus, precise knowledge of membrane conductances is limited to larger cells and presently excludes cellular organelles. The data available for large plant cells, nerve cells, and muscle indicate values of the order of 10 mMho/cm^2, but over a wide range. The membrane time constant $T = C_m/G_m$ is of the order of msec; i.e., at frequencies above about 1 KHz the membrane admittance is capacitive.

Above-indicated principles have been extensively tested on a large number of membrane systems. These include all sorts of sea eggs, muscle and most other tissues, erythrocytes, subcellular organelles, bacteria, synaptosomes, a variety of vesicles, PPLO, and bilayer membranes. They are responsible for the β-dispersion, which terminates at the 1 to 100 MHz frequency range.

The mechanism of the β-dispersion demands that at its center frequency the membrane impedance per unit area is equal to that of the access impedance $R(\rho_i + \frac{1}{2}\rho_a)$. At the high frequency tail of the β-dispersion the membrane impedance becomes smaller; i.e., the membranes are at frequencies corresponding to the end of the β-dispersion increasingly transparent. At these high frequencies, cell membranes therefore no longer receive the field strength sampled over cellular dimensions. Applied fields result in membrane potentials of the order of R/D times, 1000-fold smaller than at low frequencies (D membrane thickness). Rapidly increasing current densities are therefore needed to sustain any given imposed AC membrane potential. From this we would conclude that at high frequencies the likelihood of any direct interaction of electrical fields with biological membranes is small.

If it is assumed that induced membrane potential changes must be a noticeable fraction of the resting potential, 1–10 μV, bulk membrane current densities needed for excitation can be calculated. Such calculations have been carried out,[20] and indicate values ranging from 0.1 to 10 mA/cm^2, depending on cellular size, membrane conductance values, and extracellular space. Consider, for example, Equation 7, $\Delta V_m = 1.5RE$. The field strength E is related to the average tissue current density $j = E/\rho$ with the tissue resistivity ρ typically of the order of 600–700 Ohm-cm.[21] Hence

$$\Delta V_m = Rj\,(\text{mA/cm}^2). \quad (9)$$

Thus for $j = 1$mA/cm^2, a radius of 10 μm would correspond to $\Delta V_m = 1$ mV and a radius of 100 μm to $\Delta V_m = 10$ mV. A value of very approximately 1 mA/cm^2 is therefore indicative of the threshold of excitation with low-frequency currents of about 100 Hz or less. More sophisticated models can be based on the cable core conductor approach, using either the HH equations for the membrane properties or newer revisions of the HH model.

Considerable experimental evidence on current threshold for stimulation exists, including muscle and heart muscle tissues. More recent data have been stimulated by the development of the cardiac pacemaker field.[22] These data are supportive of the mA/cm^2 threshold for excitation but depend, of course, in detail on electrode configuration and size and the mode of excitation.

Considerable experience has also been assembled by those interested in electrical accidents. The extensive body of this knowledge is not summarized here, but it is supportive of above-stated threshold figures. (See for example Ref. 20.)

Many data are furthermore available about currents needed to evoke cerebral responses such as sleep, anesthesia, and narcosis. Currents are usually in the range between 10 and 100 mA, with current densities probably an order of magnitude lower. For details, see the report "An Evaluation of Electroanesthesia and Electrosleep," Div. Med. Sci. Nat. Res. Council.

All available knowledge from the more basic principles, literature on hazards, and more clinically oriented work appears internally consistent. But this internal consistency does not necessarily exclude the possibility of more subtle responses, particularly if they are masked by the more strong and readily apparent excitation phenomena discussed above.

Indeed, weaker interactions corresponding to membrane potentials of less than 1 mV have been reported. Terzuolo and Bullock[23] estimated the voltage gradient in the saline solution surrounding a neuron (stretch receptor of the crayfish) just sufficient to cause a noticeable change in the firing frequency to be of the order 0.01 V/cm. This corresponds to a current density in the external medium of about 0.02 mA/cm^2 and is almost one order of magnitude lower than the above postulated range of 0.1 to 10 mA/cm^2. It is not very different from the current densities which can significantly interact with cardiac tissues if applied with large electrodes.[22] Schmitt *et al.*[24] have recently pointed out that interactions in the central nervous system are mediated by graded electronic potential changes that are propagated through highly sensitive synapses. Corresponding membrane potentials are quoted at the submillivolt threshold level. There appears but little doubt that significant effects other than the propagation of spike potentials occur at current densities and corresponding potentials across membranes about tenfold lower than indicated by the above-quoted range extending down to 0.1 mA/cm^2

and 1–10 mV across the membrane. It must be pointed out, however, that the translation of membrane potentials into bulk current densities in the surrounding tissue or extracellular fluid is difficult and only approximate estimates can be made, as stated above. The real significance of the reported membrane sensitivities at the fractional mV level is that these values are at least tenfold lower than suggested by axonology so far.

Some Extraordinary Sensitivities to Electrical Fields

Probably the highest confirmed sensitivities to electrical fields have been reported for a variety of electrosensitive fish species. For sharks and rays, fields as low as 0.01 μV/cm can be perceived.[19,25] How can such an extraordinary sensitivity be achieved? Two principles come to mind. Sampling of the electrical field as suggested by cellular structures and expressed in Equation 7 and narrow bandwidth and, related, sampling over corresponding long periods of time. Equation 7 would suggest that large cellular structures or corresponding devices that sample the field strength over appreciable distances be utilized. Fish tissues have a higher resistivity than the surrounding sea water. Hence external field strength levels are of the same order of magnitude as *in situ* internal tissue values. One device used by electrosensing fish is long tubular structures, the ampullae of Lorenzini. These organs integrate the field strength over their total length at sufficiently low frequencies and apply this potential to their end receptor epithelium. These structures extend over more than 10 cm. Therefore, above-quoted field strength values result in receptor potentials of the order of 0.1 μV. Detection of these exceedingly small membrane potentials in the presence of a typical noise level of about 1 μV across the membrane is aided by the low band pass characteristics of the ampullae Lorenzini. The ability to detect the quoted low field values decreases rapidly and more than linearly with increasing frequencies above 8 or 10 Hz. The receptor potential of 0.1 μV may therefore correspond to sensitivities of the order of 0.01 mV on a broad band basis extending to 1000 Hz. This potential is about ten times less than potentials quoted above as involved in electrotonic conduction processes in the brain. The evolution of these highly sensitive field-integrating tubular structures, in combination with very sensitive receptor membrane structures, may well have been aided by the high conductivity of water in comparison to that of air, favoring adequate "coupling" of external fields. The receptor membrane structures operate more closely to noise level. Induced potentials are not buried in it and hence, in principle, detectable. By contrast, axons operate at a level about 1000-fold above noise.

A large number of reports have been published claiming that fairly low external fields of the order of a volt per cm in air can be perceived. Field-strength values in air "couple" fairly ineffectively into tissues. Consider, for example, a sphere of tissue surrounded by air. Then for low frequencies, the ratio of internal field E_i to external field in air E_a is

$$E_i/E_a = f_\rho/6 \cdot 10^{11} \quad \textbf{(10)}$$

(f (frequency in Herz, ρ tissue resistivity in Ohm-cm). For a tissue slab exposed to a vertical air field

$$E_i/E_a = f_\rho/18 \cdot 10^{11}. \quad \textbf{(11)}$$

Thus, for 60 Hz and a tissue resistivity of about 1000 Ohm-cm, E_i is about 10^7 times weaker than the air field strength E_a. External field levels of the order of 1 V/cm therefore result in internal field values about 10^{-7} V/cm, even though local fields may be higher because of tissue differences in electrical properties. Corresponding current densities are of the order of 10^{-7} mA/cm^2 and corresponding membrane potentials for 10-μm sized cell 10^{-7} mV or 10^{-4} μV. This is about 1000-fold less than the sensitivity of the end receptor membranes in the sharks, and 10,000 times below membrane noise. Clearly, exceedingly large distance sampling of the field strength such as achieved by the shark's tubular structures and very limited frequency response characteristics are needed to perceive signals of the order of 1 V/cm in air. Sharks perceive *in situ* tissue field-strength levels of 10^{-8} V/cm. But they sample the field over distances of 10 cm not demonstrated so far in other than fish species.

Most of the evidence for subtle effects of ELF fields on biological systems appears subject to inquiry. However, in recent years evidence in favor of such effects has been increasing and is well discussed in a text edited by Adey and Bawin[26] containing many pertinent references. A direct interaction with the central nervous system has been suggested by Adey. But the possibility of subtle peripheral transduction effects occurring at the surface of the body should also be considered.[27] The possibility is suggested by the fact that short-wave fields of a frequency near 150 MHz as well as ELF fields near 10 Hz have been noted as biologically effective by Bawin *et al.*[28,29] In both cases, similar external field-strength levels have been used, but internal field levels consequently differ by a factor of about 10^7. There appears to be agreement that if the reported sensitivities of nervous tissues can be confirmed and are not caused by surface transduction effects, exceedingly effective spatial and temporal integration mechanism must be involved. An example for spatial field integration was discussed above as it is utilized by the shark receptor organs. Another principle that concerns field-generated forces acting on particles is discussed below. It also demands larger size for smaller field threshold values to overcome thermal noise. In addition, several proposals have been made stating that highly cooperative molecular phenomena and metastable giant dipole effects may well provide a biophysical explanation of the claimed unusual sensitivities. These cooperative models either apply to exceedingly high frequencies above 100 GHz, where water loses its dampening properties,[30] or are so far rather qualitative and therefore not subject yet to verification.[31] Most of these models have one element in common. They demand interactions over extended regions or large volumes reacting to fields. There appears only little possibility that low-frequency fields of the mentioned low magnitudes can interact with individual macromolecules or membranes.

Some of the strongest evidence for subtle field effects, as reported by Adey and his associates and Wever's work on circadian rhythm as described in Reference 26, remains to be confirmed. If these findings can be confirmed and if it can be shown that they are caused by direct interaction with the central nervous system, entirely new modes of interaction of electrical fields with biological systems are indicated.

Field-Generated Force Effects and Membrane Destruction

Two classes of effects are less subtle than some of the previously discussed phenomena and occur at similar field levels. They will be discussed, since they probably account for a large number of experimental observations. The first deals with field-generated forces of sufficient magnitude to affect cellular characteristics, and the second deals with fields high enough to apply to membrane potentials that are sufficient to cause temporary or permanent damage.

Electrical fields can directly interact with matter and create forces that can act on molecules as well as on cellular and larger structures. Most of these interactions are reversible and do not necessarily yield demonstrable biological effects. An example is the movement of ions in an alternating field, which is inconsequential provided that the field is sufficiently weak to prevent undue heating from molecular collisons (i.e., below about 1 V/cm corresponding to 1 mA/cm^2 in a physiological medium). Another example is the orientation of polar macromolecules. For field-strength values of interest here, only a very minor preferential orientation with the field results. Complete orientation and consequent dielectric saturation require field strength values of the order of KV/cm. Changes of this magnitude occur in membranes upon depolarization. Hence field-induced orientation or changes in orientation of membrane molecules appear entirely possible. Corresponding current tissue current densities are of the order of mA/cm^2, as discussed above.

But electrical fields can just as well interact with nonpolar cells and organelles in the absence of any net charge. The application of the electrical field will create induced dipole moments in suspended cells or particles. These dipole moments can interact with each other or they may be responsible for orientation effects or cellular shape changes. Reported field interactions at the cellular level have been summarized by Schwan & Sher[32] and include a variety of effects such as:

Cellular and particle movement in inhomogeneous field;
orientation of particles and cells with or perpendicular to the field;
"pearl chain" formation;
cellular-shape changes;
AC-induced cytoplasmic streaming; and
erythrocyte and other cellular destruction in a nonthermal manner.

Common to all these effects appear to be field-induced mechanical forces which drive the system involved toward a state of lowered electrical energy. These forces are in competition with random thermal collisional forces. The response to the field emerges when the electrical potential energy or its change with the response (i.e., upon orientation, etc.) becomes larger than the thermal energy KT (K Boltzmann constant, T absolute temperature). Thus there exists a threshold field strength needed for the phenomena of interest to occur. Since the electrical potential energy is proportional to E^2 and the volume V of the responding particle,[33] the threshold field strength E_{th} must be given by the equation

$$E_{th}^2 V \sim aKT = A \rightarrow E_{th}^2 \sim 1/R^3 \qquad \textbf{(12)}$$

where $A = aKT$ is a constant for a given temperature. The constant a depends on

the particular phenomena of interest, electrical properties of suspended and suspending phase, shape of particles, and so on. Fairly abundant experimental evidence confirms that the principle and threshold field values for responses of 10-μm-sized cells are of the order of 10 V/cm. Figure 5 summarizes some of these results. But for 100-Å macromolecules they are about 10 KV/cm, and comparable to the fields needed for complete orientation due to the existence of a typical dipole moment of the order of 100 Debye units.

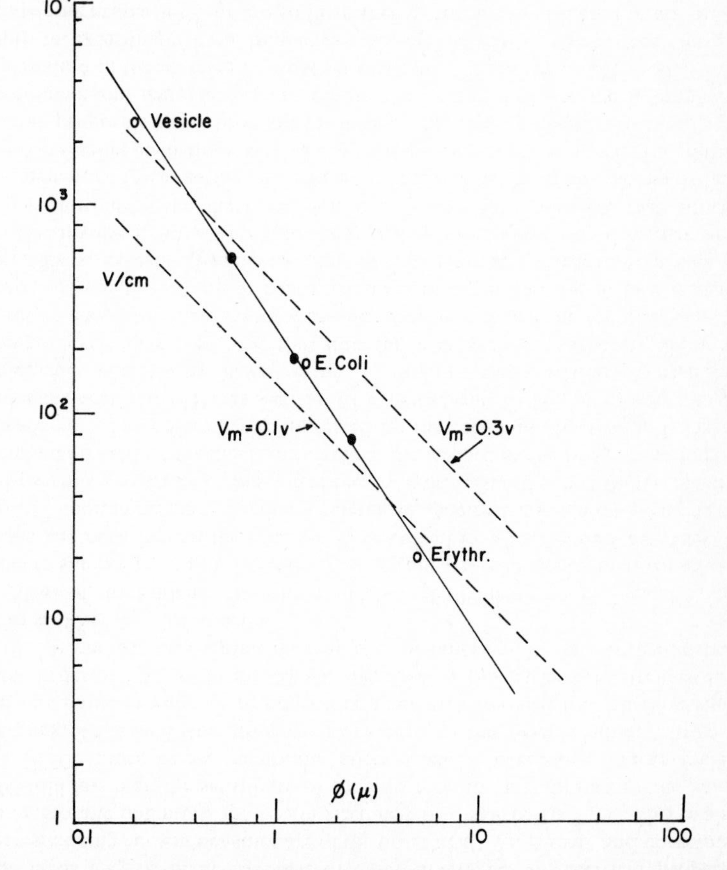

Figure 5. Threshhold of field effects on particles and cells. The solid curve gives the threshold for field-generated force effects. The dashed curves give the threshhold for dielectric membrane breakdown, assuming two different values for the membrane breakdown potential. Some typical membrane breakdown potentials have been listed by Schwan[27] and are of the order of 200 mV. For erythrocytes Zimmermann *et al.*[43] give a breakdown potential above 1000 mV. Data given are for permeability changes in chromaffin granule vesicles (Neumann & Rosenheck[42]), orientation and pearl chain formation effects in *E. coli* and erythrocytes (Sher & Schwan[32,38], and pearl-chain formation threshold data for silicone particles (solid points, Sher & Schwan[32,38]), axis: particle diameter.

The field effects may demonstrate themselves as movement or deformation of cells in an inhomogenous field, destructive effects on the cells, orientation, and alignment phenomena.[32] Particularly, the movement and deformation or destruction of cells in highly inhomogenous fields (for example, near the tip of a small electrode) can be very dramatic.[34] Cellular movement in inhomogeneous field (dielectrophoresis) has been used for purposes of cell sorting.[35] However, in physiological environments heat development is often pronounced, since high field-strength values are needed.

The usual random movement of a population of protozoa *Euglena* is replaced by movement in the direction of an applied alternating RF field or perpendicular to it, depending on frequency.[36] Orientation effects on *Fucus* eggs and *E. coli* have been observed by Novak and Bentrup[37] and by Sher,[38] respectively. The orientation of nonspherical particles has been treated theoretically by Schwarz *et al.*[39] and by Saito *et al.*[40] Further theoretical and experimental work is summarized by Schwan and Sher.[32] Experimental and theoretical evidence indicates that pulsed fields cannot be more effective than continuous fields of the same average power. Hence, modulation effects are not to be expected.[41] Field forces due to the induced dipole moment of the field have sometimes been listed as evidence of nonthermal action of electrical fields on biological systems. However, the effects require fairly large field-strength levels, frequently above those which give rise to undue heat development or stimulation of excitable tissues. The field forces are also strongly dependent on the electrical properties of the particle considered and its environment. Hence the threshold above noise is a strong function of frequency and has been proposed for purposes of cell classification.[35] In general, available evidence and present understanding indicate that significant effects with field-evoked forces above thermal noise require field-strength values above 1 V/cm in the medium unless cellular dimensions are well above 100 μm.

A second principle which relates to cellular responses to fields is also indicated in FIGURE 5. At sufficiently low frequencies the total potential applied across a cell by an external field is applied to the outer membranes, since the cytoplasmic phase is at an equipotential, as indicated by Equation 7. Let it be assumed that dielectric breakdown of membrane systems occur at a given field level V_m above that typical for the resting state. Then a threshold field strength E_{th} for cellular effects resulting from dielectric breakdown in membrane system is given by

$$E_{th} = V_m/1.5\mathrm{R} \qquad \mathbf{(13)}$$

where V_m is the induced membrane breakdown potential. The curves in FIGURE 5 demonstrate that both membrane breakdown as well as field-generated force cellular effects require external *in situ* field-strength levels of the order of 10–100 V/cm or more for cellular sizes and that the field strength for smaller dimensions is accordingly higher. On the other hand, large cells such as investigated by Friend *et al.*[17] can respond at field levels of some V/cm in the medium. FIGURE 5 also indicates that membrane-breakdown effects and effects caused by fields acting on the total cell can be expected at comparable field-strength levels in the medium. This makes it somewhat difficult to establish clearly what mechanism may be involved in any particular situation. But in any case, a perusal of the studies so far conducted on fields acting on cells and vesicles reveals experimental findings in accord with the classical principles stated.

SUMMARY

Electric field interactions with biological matter can be summarized as follows.

1. Rather strong field values of the order of KV/cm are probably needed to orient polar macromolecules significantly.

2. Various interactions at the cellular level depend strongly on size and may occur at field levels as low as 1 V/cm or less for unusually large unicellular organisms. They appear to be largely caused by field-generated forces and the dipole moments induced by the electrical field. These forces emerge above the thermal level at a threshold field-strength that is inversely related to size. For usual cellular size, required field values produce significant heating resulting from ionic movement.

3. Strong membrane interactions leading to membrane breakdown correspond to the order of some hundred mV across the membrane. Corresponding field-strength values in the medium are comparable to those which yield significant field-generated forces.

4. Classical membrane excitation phenomena occur at field levels of the order of 1 V/cm and current densities of very approximately 1 mA/cm^2. Corresponding membrane potentials are in the range of 1–10 mV.

5. Weak membrane interactions corresponding to a fraction of a mV have been reported for CNS tissues. Some fish species display still higher sensitivities corresponding to values below 1 μV/cm, with a highest sensitivity near 0.01 μV/cm in sea water. These highest sensitivities are achieved by a sampling of the field over large distances with special organs and use of narrow bandwidth characteristics. Corresponding membrane sensitivities are below 0.1 μV across the membrane and above noise level over a rather narrow frequency range of a few Hertz.

6. Extraordinary weak field effects have been reported corresponding to tissue field gradients of the order of 10^{-7} V/cm. These effects, if shown to be related to more complex interactions at the membrane level, would require the development of new concepts.

REFERENCES

1. SCHWAN, H. 1957. Electrical properties of tissue and cell suspensions. *In* Advances in Biological and Medical Physics. J. H. Lawrence and C. A. Tobias, Eds. Vol. **V:** 147–209. Academic Press. New York, N.Y.
2. SCHWAN, H. P. 1965. Biological impedance determinations. J. Cell. Comp. Physiol. **66:** 5–12.
3. FATT, P. 1964. An analysis of the transverse electrical impedance of striated muscle. Proc. Royal Soc. B. **159:** 606–651.
4. TAKASHIMA, S. & H. P. SCHWAN. 1974. Passive electrical properties of squid axon membrane. J. Membrane Biol. **17:** 51–68.
5. TAKASHIMA, S., K. S. COLE & H. P. SCHWAN. 1975. Membrane impedance of squid axon during hyper- and depolarization. Biophys. J. **15** (no. 2, part 2); 39a.
6. SCHWAN, H. P. 1965. Electrical properties of bound water. Ann. N. Y. Acad. Sci. **125:** 344–354.
7. PENNOCK, B. and H. P. SCHWAN. 1969. Further observations on the electrical properties of hemoglobin bound water. J. Phys. Chem. **73:** 2600–2610.
8. GRANT, E. H. 1965. The structure of water neighboring proteins, peptides and amino acids as deduced from dielectric measurements. Ann. N. Y. Acad. Sci. **125:** 418–427.

9. Cole, K. S. 1968. Membranes, Ions and Impulses, parts I and II. Univ. California Press. Berkeley, Calif.
10. Oncley, J. L. 1943. Chapter 22 *In* Proteins, Amino Acids and Peptides as Ions and Dipolar Ions. E. J. Cohn and J. T. Edsall, Eds. Reinhold. New York, N.Y.
11. Takashima, S. & A. Minakata. 1975. Dielectric properties of biological macromolecules. *In* Digest of Dielectric Literature **37:** 602.
12. Ling, G. N., C. Miller & M. M. Ochsenfeld. 1973. The physical state of solutes and water in living cells according to the association induction hypothesis. Ann. N. Y. Acad. Sci. **204.** C. F. Hazlewood, Ed.
13. Schwan, H. P., R. J. Sheppard & E. H. Grant. 1976. Complex permittivity of water. J. Chem. Phys. **64:** 2257–2258.
14. Schwan, H. P. & K. R. Foster. 1977. Microwave dielectric properties of tissues: Some comments on the rotational mobility of tissue water. Biophys. J. **17:** 193–197.
15. Pauly, H. & H. P. Schwan. Dielectric properties and ion mobility in erythrocytes. Biophys. J. **6:** 621–639. 1966.
16. Stoy, R. D. & H. P. Schwan. 1976. Dielectric properties of rabbit cerebral cortex in the radio frequency range. Physiologist **19**(3).
17. Friend, W. A., E. D. Finch & H. P. Schwan. 1974. Low frequency electric field induced changes in the shape and motility of amoebae. Science **187:** 357–359.
18. Goodman, E. M., B. Greenbaum & M. T. Marrow. 1976. Effects of ELF radiation on *Physarum polycephalum.* Technical Report N00014-76-C-0180. See also Nature. 1975. **254:** 66–67.
19. Kalmijn, A. J. 1966. Electroperception in sharks and rays. Nature **212:** 1232–1233.
20. Schwan, H. P. 1972. Biological hazards from exposure to ELF electrical fields and potentials. NWL Technical Report TR-2713. US Naval Weapons Lab. Dahlgren, Va.
21. Schwan, H. P. & C. F. Kay. 1957. Conductivity of living tissues. Ann. N. Y. Acad. Sci. **65:** 1007–1013.
22. Roy, O. Z., J. R. Scott & G. C. Park. 1976. 60 Hz ventricular fibrillation and pump failure thresholds versus electrode area. IEEE Trans. Biomed. Eng. **BME-23:** 45–48.
23. Terzuolo, C. A. & T. H. Bullock. 1956. Measurement of imposed voltage gradient adequate to modulate neuronal firing. Proc. Nat. Acad. Sci. USA **42:** 687–694.
24. Schmitt, F. O., P. Dev & B. H. Smith. 1976. Electronic processing of information by brain cells. Science **193:** 114–120.
25. Kalmijn, A. J. 1974. The detection of electrical fields from inanimate and animate sources other than electric organs. Chapter 5. *In* Handbook of Sensory Physiology. Vol. III/3. A. Fessard, Ed. Springer Verlag. New York, N.Y.
26. Adey, W. R. & S. M. Bawin. 1977. Brain interactions with weak electric and magnetic fields. Neurosci. Res. Prog. Bull. 15. MIT Press. Cambridge, Mass.
27. Schwan, H. P. 1977. Tissue determinants of interactions with electrical fields. *In* Brain Interactions with Weak Electric and Magnetic Fields. W. R. Adey and S. M. Bawin, Eds. Neurosci. Res. Prog. Bull. 15. MIT Press. Cambridge, Mass.
28. Bawin, S. M. & W. R. Adey. 1976. Sensitivity of calcium binding in cerebral tissue to weak environmental electric fields oscillating at low frequency. Proc. Nat. Acad. Sci. USA **73:** 1999–2003.
29. Bawin, S. M., L. K. Kaczmarek & W. R. Adey. 1975. Effects of modulated VMF fields on the central nervous system. Ann. N. Y. Acad. Sci. **247:** 74–81.
30. Fröhlich, H. 1977. Possibilities of long- and short-range electric interactions of biological systems. *In* Brain Interactions with Weak Electric and Magnetic Fields. W. R. Adey and S. M. Bawin, Eds. Neurosci. Res. Prog. Bull. 15. MIT Press. Cambridge, Mass.
31. Grodsky, I. T. 1977. Biophysical basis of tissue interactions. *In* Brain Interactions with Weak Electric and Magnetic Fields. W. R. Adey and S. M. Bawin, Eds. Neurosci. Res. Prog. Bull. 15. MIT Press. Cambridge, Mass.
32. Schwan, H. P. & L. D. Sher. 1969. Alternating current field induced forces and their biological implications. J. Electrochem. Soc. **116:** 170–174.
33. Schwarz, G. 1962. A theory of the low frequency dielectric dispersion of colloidal particles in electrolyte solution. J. Phys. Chem. **66:** 2636–2642.
34. Elul, R. 1967. Fixed charge in the cell membrane. J. Physiol. **189:** 351–365.
35. Pohl, H. A. 1973. Biophysical aspects of dielectrophoresis. J. Biolog. Phys. **1:** 1–16.
36. Teixeira-Pinto, A. A., L. L. Nejelski, J. L. Cutler & J. H. Heller. 1960. Exp. Cell Res. **20:** 548.
37. Novak, B. & F. W. Bentrup. 1973. Orientation of Fucus egg polarity by electric ac and dc fields. Biophysik **9:** 253–260.
38. Sher, L. D. 1963. Mechanical effects of AC fields on particles dispersed in a liquid. Biological implications. Ph.D. Thesis, Univ. Pennsylvania. Philadelphia, Pa.
39. Schwarz, G., M. Saito & H. P. Schwan. 1965. On the orientation of non-spherical particles in an alternating field. J. Chem. Phys. **43:** 3562–3569.
40. Saito, M., H. P. Schwan & G. Schwarz. 1966. Response of non-spherical biological particles to alternating electric fields. Biophys. J. **6:** 313–327.
41. Sher, L. D., E. Kresch & H. P. Schwan. 1970. On the possibility of nonthermal biological effects of pulsed electromagnetic radiation. Biophys. J. **10:** 970–979.
42. Neumann, E. & K. Rosenheck. 1972. Permeability changes induced by electric impulses in vesicular membranes. J. Membrane Biol. **10:** 279–290.
43. Zimmermann, U., G. Pilwat & R. Riemann. 1974. Dielectric breakdown of cell membranes. Biophys. J. **14:** 881–889.

Summaries of Selected Papers from USSR Academy of Sciences

Scientific Session of the Division of General Physics and Astronomy, USSR Academy of Sciences (17-18 January 1973)

Usp. Fiz. Nauk **110**, 452-469 (July 1973)

A scientific session of the Division of General Physics and Astronomy was held on January 17 and 18 at the conference hall of the P. N. Lebedev Physics Institute. The following papers were delivered:

1. K. I. Gringauz, The Formation of the Plasmapause and Its Influence on the Physics of the Earth's Ionosphere.

2. I. A. Zhulin, Dynamics and Physical Nature of Particles According to Coordinated Studies in the Magnetically Conjugate Regions.

3. A. D. Sytinskiĭ, The Relation of the Earth's Seismicity to Solar Activity.

4. N. D. Devyatkov, Influence of Millimeter-Band Electromagnetic Radiation on Biological Objects.

5. É. B. Bazanova, A. K. Bryukhova, R. L. Vilenskaya, E. A. Gel'vich, M. B. Golant, N. S. Landau, V. M. Mel'nikova, N. P. Mikaélyan, G. M. Okhokhonina, L. A. Sevast'yanova, A. Z. Smolyanskaya, and N. A. Sycheva (General Editor: N. D. Devyatkov), Certain Methodological Problems and Results of Experimental Investigation of the Effects of Microwaves on Microorganisms and Animals.

6. L. A. Sevast'yanova and R. L. Vilenskaya, A Study of the Effects of Millimeter-Band Microwaves on the Bone Marrow of Mice.

7. A. Z. Smolyanskaya and R. L. Vilenskaya, Effects of Millimeter-band Electromagnetic Radiation on the Functional Activity of Certain Genetic Elements of Bacterial Cells.

8. V. F. Konrat'eva, E. N. Chistyakova, I. R. Shmakova, N. B. Ivanova, and A. A. Treskunov, Effects of Millimeter-band Radio Waves on Certain Properties of Bacteria.

9. S. E. Manoĭlov, E. N. Chistyakova, V. F. Kondrat'eva, and M. A. Strelkova, Effects of Millimeter-band Electromagnetic Waves on Certain Aspects of Protein Metabolism in Bacteria.

10. N. P. Zalyubovskaya, Reactions of Living Organisms to Exposure to Millimeter-band Electromagnetic Waves.

11. R. I. Kiselev and N. P. Zalyubovskaya, Effects of Millimeter-band Electromagnetic Waves on the Cell and Certain Structural Elements of the Cell.

12. V. I. Gaĭduk, Yu. I. Khurgin, and V. A. Kudryashova, Outlook for Study of the Mechanisms of the Nonthermal Effects of Millimeter- and Submillimeter-band Electromagnetic Radiation on Biologically Active Compounds.

We publish below brief contents of some of the papers and an outline of remarks offered in discussion by D. S Chernavskiĭ.

N. D. Devyatkov. Influence of Millimeter-band Electromagnetic Radiation on Biological Objects

The development of microwave electronics is attended by steadily widening opportunities for use of its achievements not only in the areas that are traditional for it (transmission and reception of various types of information), but also in new areas, such as biology and medicine.

Study of the possible uses of coherent electromagnetic oscillations in recently opened bands has now acquired special importance. We refer here to the millimeter and submillimeter wavelength bands and shorter-wavelength regions of the electromagnetic spectrum—those in which lasers are now operating.

The present scientific session of the Division of General Physics and Astronomy, USSR Academy of Sciences, to which specialists working in the fields of biophysics, microbiology, biochemistry, and medicine have been invited, is devoted basically to experimental study of the effects of exposure of various biological objects to millimeter-band electromagnetic fields, on the molecular and cellular levels and at the level of more complex living organisms.

Several years ago, organizations of the USSR Ministry of the Electronics Industry and the Institute of Radiophysics and Electronics of the Ukrainian Academy of Sciences (Khar'kov) completed the development of millimeter- and submillimeter-band generators of the backward-wave-tube type, which make possible a broad range of continuous frequency variation of the waves generated. Special machines for biological studies were built on the basis of the newly developed millimeter-wave generators. Beginning around 1965, a number of organizations in the USSR began systematic research on the effects of millimeter waves on biological objects.

Experimental studies made in the millimeter band at very low microwave energy flux densities (no more than a few milliwatts per square centimeter) produced highly interesting specific effects of irradiation. It was found for almost all of the biological objects studied that:

a) the effect of irradiation depends strongly on the frequency of the microwaves;

b) in certain microwave-power ranges, the effect of exposure depends weakly on variation of the power through several orders of magnitude;

Reprinted with permission from *Sov. Phys. Usp.*, vol. 16, no. 4, pp. 568–579, Jan.–Feb. 1974.

c) the effects are observed to depend significantly on time of irradiation.

The results obtained are of great scientific and practical interest. For example, it was established that the vital activity of microorganisms is affected by millimeter-wave irradiation. The effect may be positive or negative, depending on the particular part of the band and the particular conditions of irradiation.

The effects obtained on irradiation of microorganisms may eventually form a basis for new methods of producing vaccines and increasing the productivity of antibiotic-production methods. Another possibility that had not been excluded is that of millimeter-wave irradiation to treat burns and other suppurating wounds in order to accelerate the healing process.

Using as an example one of the studies that we made jointly with K. S. Rozhnov and his staff at the Leningrad Electrical Engineering Institute, I should like to demonstrate graphically the strength of the effect of millimeter-wave electromagnetic irradiation on cell division. A machine for irradiation of microorganisms and direct observation of their behavior during and after irradiation was designed around an MIS-51 comparison microscope.

Various yeast cultures were irradiated. A resonant effect of millimeter-wave irradiation on the division rate of the cells being irradiated was observed. Thus, for example, irradiation of a culture of Rhodotorula rubra for 15 hours at wavelengths of 7.16, 7.17, 7.18, and 7.19 mm (10 experiments were performed for each frequency), showed a sharp frequency dependence (Fig. 1; the ordinate is the ratio of the number of cells in the experiment to the control): cell division is stimulated at 7.18 mm and slightly depressed at the other wavelengths.

Irradiation of a Candida culture caused a marked change in the nature of cell division as compared with the control. Figures 2—4 show clearly the difference between control and experiment at various stages during irradiation.

Figure 5 illustrates the behavior of irradiated and unirradiated (control) cultures after about 15 hours of irradiation (which was administered at 20—21°C. After the 15 hours of irradiation, the culture temperatures were about 16—17°C).

An explanation of the mechanism of the resonant effect of irradiation and some of its other properties would be of enormous interest from the scientific standpoint. As yet, we have no rigorous scientific explanations for the effects of millimeter-band electromagnetic waves. There have been only a few attempts to develop approximate hypotheses to account for the resonant effect, and they require further experimental and theoretical confirmation.

It would be desirable to have scientific manpower at the institutes of the Division of Biochemistry, Biophysics, and Chemistry of Physiologically Active Compounds of the USSR Academy of Sciences put to work on a scientific explanation of the observed phenomena.

In addition to the scientific groundwork, a more active search should be made for fields of practical application of the effects of millimeter-wave irradiation. In this matter again, we look for assistance to the competent specialists at the Institutes of the USSR Academy of Sciences.

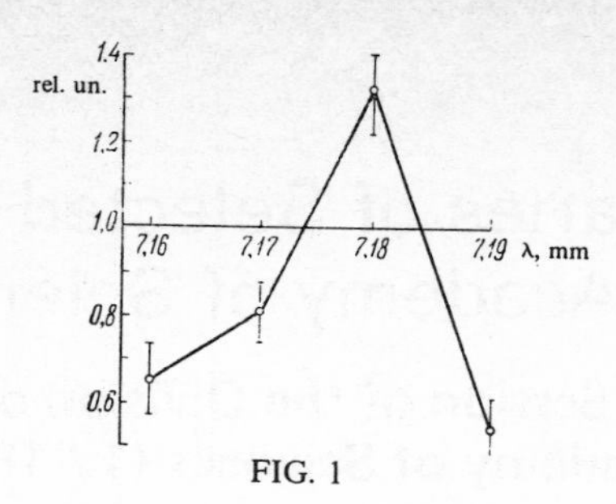

FIG. 1

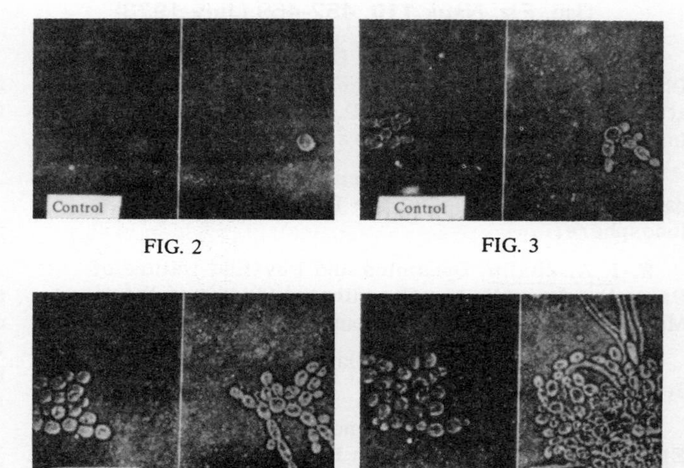

FIG. 2

FIG. 3

FIG. 4

FIG. 5

É. B. Bazanova, A. K. Bryukhova, R. L. Vilenskaya, É. A. Gel'vich, M. B. Golant, N. S. Landau, V. M. Mel'nikova, N. P. Mikaélyan, G. M. Okhokhonina, L. A. Sevast'yanova, A. Z. Smolyanskaya, and N. A. Sycheva (General Editor: N. D. Devyatkov), Certain Methodological Problems and Results of Experimental Investigation of the Effects of Microwaves on Microorganisms and Animals

1. Research methodology problems and general relationships

Since this was the first time that we had worked in the millimeter band, we began by giving serious attention to problems of the experimental method. A diagram of the experiment appears in the figure. The effects of frequency, power flux density, exposure time, ambient temperature, and the identity of biological conditions were investigated in detail and recorded. The polarization was held constant through all experiments. All were performed with continuous irradiation.

It was found in the course of the experiments that the observed effects are not very critical in regard to the incident power flux density. In particular, it was established for a wide variety of microorganisms and tests that, beginning at a certain threshold power flux density of about 0.01 mW/cm^2, the effects vary weakly over several (two and five) orders of magnitude when a marked thermal effect is already beginning to make its appearance. We note that heating by other sources did not produce the effects observed under exposure to microwaves at a low power level. Similar results are also observed in animals, except that the threshold power flux densities are much higher here.

The influence of irradiation time is of a different kind. The longer this time, the stronger is the observed effect, although saturation is noted after several hours. Thus, the total-dose concept that is often used in application to ionizing radiation (the total irradiation energy) does not give a definite answer in this case to the question as to the magnitude of the exposure: changes in power and time act differently.

The dependence of the effects of exposure on frequency is of acutely resonant nature: the width of the bands corresponding to a given exposure effect varies from a fraction of a percent to several percent, depending on the object and the test, and this dictated requirements as to the stability of the power sources and frequency meters used in the equipment.

Biological changes provided the indication of the effect. Investigation of the absorption or emission spectra by the methods conventional to radioelectronics and physics yielded nothing in this case. The losses in the biological systems that were studied were great; in addition, the bands of the individual resonances may overlap one another with the enormous number of degrees of freedom that are present. But when a particular biological test is analyzed, it is found to be influenced by a very limited number of the possible resonances in the object, and this offers an indicator that is highly sensitive and, at the same time, the only one necessary for investigation of the particular test.

The scatter of the results obtained with the method developed was much smaller than the quantity to be measured.

2. Results of exposure of certain biological objects to millimeter-band electromagnetic waves[2)]

a) Increase in production of proteases with fibrinolytic activity by the fungus Asp. aryzal (Moscow State University strain) under radiation exposure. The work was done in collaboration with Moscow State University. The original strain was irradiated at a wavelength of 6.6 mm. The power flux density was 0.1 mW/cm^2. Ten two-hour doses were administered. Irradiation caused an increase in the proteolytic activity of the aspergillis by a factor of 1.5—2. Six months of observations verified that the effect was inherited. The increase in fibrinolytic activity was not accompanied by an increase in biomass.

b) Effects of irradiation on staphylococci (strain 209). This study was made jointly with the Central Scientific Research Institute of Traumatology and Orthopedics in collaboration with the Leningrad Institute of Aviation Instruments and the Leningrad Chemical Physics Institute. The culture was irradiated repeatedly for one hour each day at 7.08 mm and a power flux density of 0.1 mW/cm^2. Over the course of the multiple exposures, hemolytic activity, the ability to coagulate plasma, lecithinase activity, and the gold pigment disappeared, in that order. In experiments on animals (rabbits) with 8 × 8-cm traumata on their backs that has been infected with staphylococci of the strain indicated, we established a decrease in the inoculability of the staphylococci after a series of daily radiation treatments of the wounds (for 20 minutes each day). The healing time of the wounds was reduced 20% as compared with the control. A normal regeneration process was observed on cytological examination of the secreta from the irradiated wounds. The peripheral blood showed a moderate leukocytosis of neutrophilic nature, monocytosis, and lymphopenia as compared with the control, indicating an augmented protective response of the organism.

This study was carried out under the auspices of the Ministry of the Electronics Industry jointly with the IÉKO, the Central Institute of Traumatology and Orthopedics, and the Moscow State University Biology Department.

L. A. Sevast'yanova and R. L. Vilenskaya. A Study of the Effects of Millimeter-Band Microwaves on the Bone Marrow of Mice

We have previously reported the results of studies of the bone marrow of animals that were irradiated with millimeter-band electromagnetic radiation (microwaves) and subsequently with x-rays[1,2]. Despite the fact that millimeter microwaves with $\lambda = 7.1$ mm are absorbed in the surface layer of the skin of the animals at a depth of approximately 3×10^{-2} cm[3], we observed a decrease in the number of bone-marrow cells that were damaged by the x-rays when the animals were first exposed to a microwave field. A similar effect was observed when animals were given microwave irradiation prior to administration of toxic antineoplastic substances used in chemotherapy that also destroy bone-marrow cells—such as chrysomallin and sarcolysin[4].

The present paper reports counts of mouse bone-marrow cells that remained undamaged by x-irradiation after prior irradiation with a microwave field in which the exposure time to the microwaves, the power density of the field, and wavelength were varied.

We felt that it would be interesting to investigate the protective effect of the microwave field at various power densities. Counts of the remaining undamaged bone-marrow cells, normalized to the control (N/N_0), were plotted against power flux density (P) in the range from 1 to 75 mW/cm^2. The exposure time was held constant at one hour. The x-ray dose was 700 rad. The results of these measurements and the increases in the skin temperature of the animals appear in Fig. 1. The figure shows that preliminary microwave irradiation of

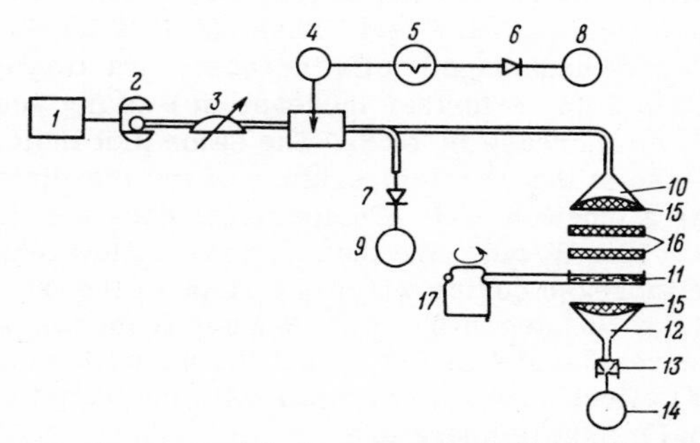

Diagram of experiment. 1—Power supply; 2—OV-612 backwardwave tube; 3—attenuator; 4—measuring line; 5—wavemeter; 6, 7—detector heads; 8—pointer-type indicator; 9—incident-power meter; 10, 12—horns; 11—object; 13—thermistor head; 14—transmitted-power meter; 15—correcting lens; 16—transformer; 17—electric motor to rotate and stir the medium.

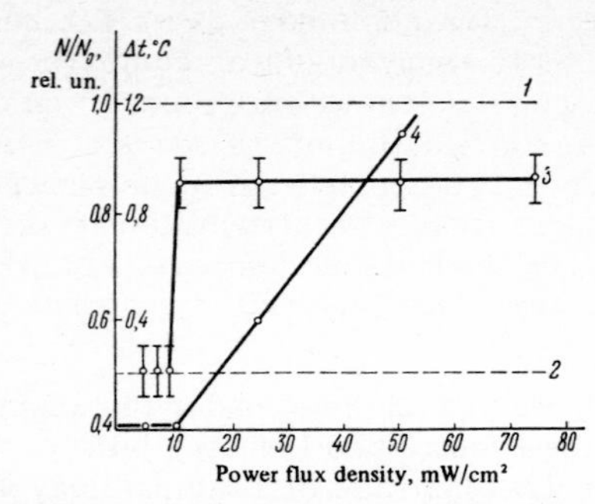

FIG. 1. Changes in number of bone-marrow cells (N/N_0) and skin temperature (Δt) of irradiated animal as functions of power flux density. 1—Number of bone-marrow cells (control); 2—exposure to x-radiation; 3—combined exposure to microwaves and x-rays; 4—change of skin surface temperature.

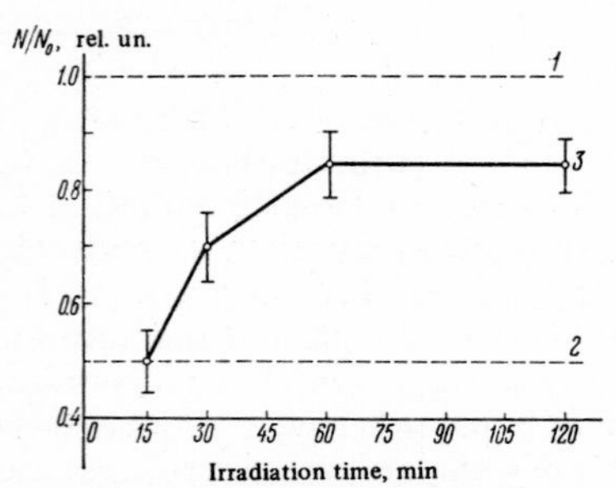

FIG. 2. Variation of number of bone-marrow cells with microwave irradiation time. 1—Control (unirradiated-animals); 2—x-irradiation; 3—microwave field and x-irradiation.

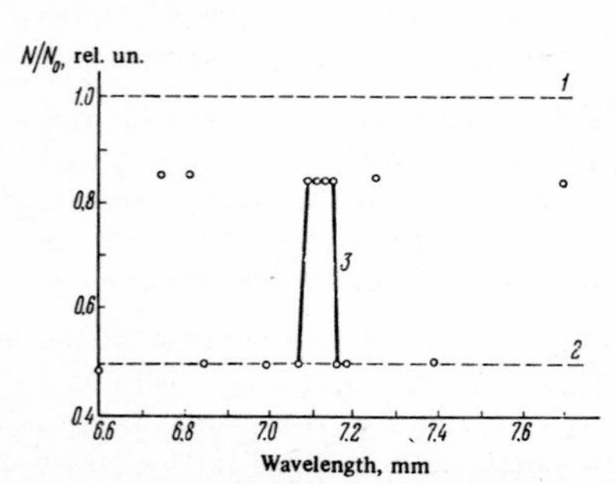

FIG. 3. Variation of number of bone-marrow cells with wavelength. 1—Control (unirradiated animals); 2—x-irradiation; 3—microwave field and x-irradiation.

the animals has no influence whatever on N/N_0 up to a power flux density $P = 9\ mW/cm^2$. Thus, there is a certain threshold power flux density below which the microwave field has no effect. Then, as P is increased, the number of undamaged cells increases practically jumpwise to 0.85. A further increase in P is not accompanied by an increase in N/N_0. The same plot indicates the increase in the skin temperature of the irradiated animal as a function of P. Temperature does not change below $P = 10\ mW/cm^2$. We then observe a slow temperature increase during which the slope of the Δt (°C) line is $2.5 \times 10^{-2}\ deg/mW \cdot cm^{-2}$. We see from comparison of curves 3 and 4 in Fig. 1 that the magnitude of the biological effect does not correlate with the variation of the animal's skin temperature.

Thus, the optimum power flux density, at which the microwaves can be observed to have a protective effect on bone marrow but do not cause heating of the skin, is around $10\ mW/cm^2$. It was this circumstance that dictated selection of a power density of $10\ mW/cm^2$.

When the microwave exposure times of the animals were varied, it was found that no microwave effect appears at all before $t = 30$ min (Fig. 2). As the irradiation time increases to 60 minutes, we observe an increase in the protective effect and N/N_0 reaches 0.8. Further increase of the exposure is not accompanied by any appreciable increase in the number of cells that remain undamaged by x-rays. Thus, the optimum irradiation time was found to be 60 minutes.

We were most interested in investigating N/N_0 as a function of microwave wavelength in experiments with combined exposure to microwaves and x-rays. The wavelength of the microwaves was varied from 6.6 to 7.7 mm. The results of these measurements appear in Fig. 3. It was found that the protective effect of preliminary microwave exposure of the animals is distinctly selective in nature. Thus, the undamaged-cell count rises from 0.5 to 0.85 at the wavelengths of 6.7 and 6.82 mm, in the range 7.09—7.16 mm, and at 7.26 and 7.7 mm, while there was no protective effect at all at the same microwave power density at the other wavelengths studied (6.6, 6.85, 7.0, 7.07, 7.17, 7.19, and 7.4 mm). This behavior of the $N/N_0(\lambda)$ relationship suggests a resonant mechanism for the action of the microwave field.

[1] L. A. Sevast'yanova, S. L. Potapov, V. G. Adamenko, and R. L. Vilenskaya, Biol. Nauki, No. 6, 46 (1969).

[2] L. A. Sevast'yanova, S. L. Potapov, V. G. Adamenko, and R. L. Vilenskaya, Changes in Hemopoiesis Under the Influence of Microwave and X-Radiation. Morphological and Hematological Aspects of the Biological Effects of Ionizing Radiation and Cytostatic Preparations, Fifth Conference of the Central Scientific Research Laboratory, Tomsk, 1970.

[3] R. L. Vilenskaya, L. A. Sevast'yanova, and A. S. Faleev, A Study of the Absorption of Millimeter Waves in the Skin of Experimental Animals. Elektronnaya Tekhnika, Ser. 1 (Elektronika SVCh), No. 7, 97 (1971).

[4] L. A Sevast'yanova, M. B. Golant, V. G. Adamenko, and R. L. Vilenskaya. Biol. Nauki, No. 6, 58 (1971).

[5] R. L. Vilenskaya, É. A. Gel'vich, M. B. Golant, and A. Z. Smolyanskaya, ibid., No. 7, 69 (1972).

A. Z. Smolyanskaya and R. L. Vilenskaya. Effects of Millimeter-band Electromagnetic Radiation on the Functional Activity of Certain Genetic Elements of Bacterial Cells

The effects of millimeter waves on intracellular systems responsible for lethal synthesis in bacteria, i.e., the synthesis of substances that result in the depth of the cell, were investigated. The colicinogenic factor of Bacillus coli was chosen as the test object. The col-factor is an extrachromosomic genetic element. The functional activity of this element is normally repressed. Suppression of the col-factor results in synthesis of a special proteic substance known as colicin; the cell then perishes. The colicin that it has produced has an antibacterial action with respect to other bacteria of the same or similar species.

We studied the influence of millimeter waves on colicin synthesis in the colicinogenic strain E. coli C600 (E_1) and in the strain E. coli K12S, which is sensitive to the colicin of the former. The activity of the colicin synthesis was determined by the method of lacunas[1], in which the numbers of individual colicin-synthesizing bacteria are counted. The effect was evaluated with the aid of the so-called induction coefficient, which is determined by the ratio of the lacuna-formation frequencies in the experiment and the control:

$$K_i = \frac{L_e K_c}{K_e L_c},$$

where L_e is the number of cells forming colicin in the experiment, K_e is the total number of colicinogenic cells in the experiment, L_c is the number of cells forming colicin in the control, and K_c is the total number of colicinogenic cells in the control.

It was found that the number of colicin-synthesizing cells increased sharply on irradiation of the colicinogenic strain with millimeter waves of certain wavelengths. Thus, the number of cells that synthesized colicin increased by an average of 300% on irradiation at wavelengths of 5.8, 6.5, and 7.1 mm. At the same time, neighboring wavelengths, 6.15 and 6.57 mm, showed no such effect. The results obtained were reproduced with high regularity.

Thus, it was observed that millimeter waves of certain wavelengths are capable of inducing the synthesis of colicin by colicinogenic bacteria. This indicates that these waves may be able to influence the regulation of functional activity in certain (in this case of extrachromosomic) genetic elements of bacterial cells.

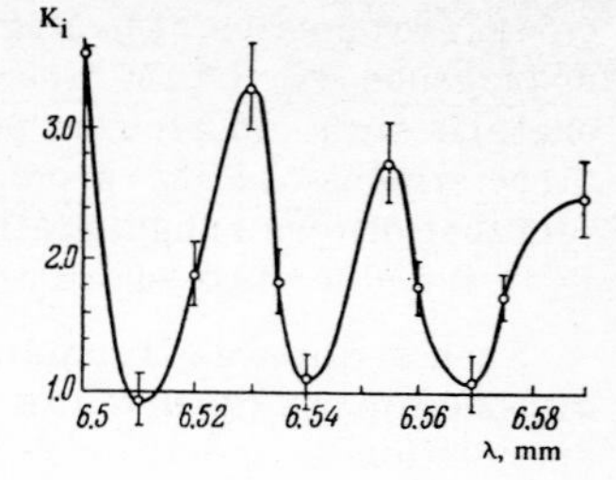

FIG. 1. Induction coefficient K_i of colicin synthesis as a function of wavelength.

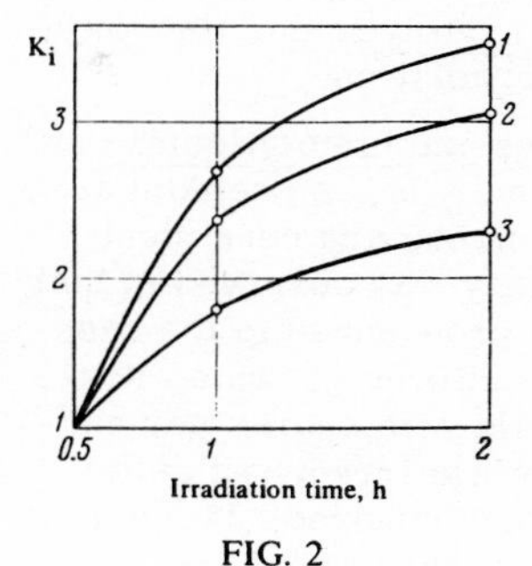

FIG. 2

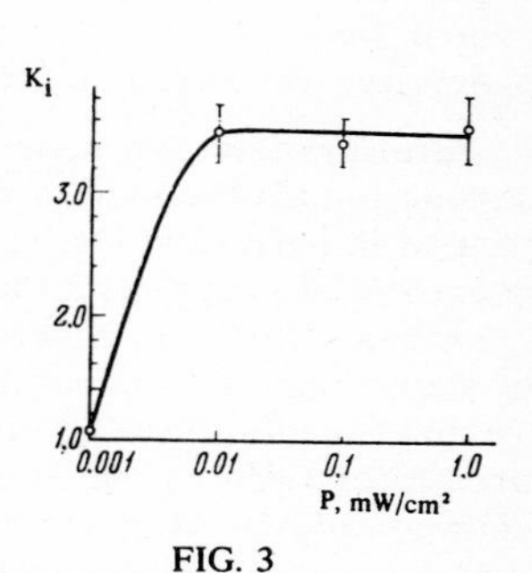

FIG. 3

FIG. 2. Induction coefficient of colicin synthesis as a function of irradiation time. λ(mm) = 6.5 (1), 5.8 (2), and 7.1 (3).

FIG. 3. Induction coefficient of colicin synthesis as a function of power flux density.

The behavior of the induction coefficient of the bacterial colicin synthesis as a function of wavelength was investigated in greater detail in the range 6.50—6.59 mm, because it was precisely in this range that an "active" wavelength (6.50 mm) and an "inactive" one (6.57 mm) had been detected.

Investigation of 11 points with the aid of a specially adapted wavemeter capable of measuring wavelengths with a resolution of 0.01% produced the curve in Fig. 1. It follows from the figure that colicin synthesis is a resonant function of wavelength. Note must also be taken of the high sensitivity of the biological system to variation of wavelength. The statistical significance ($P \leq 0.001$) of the differences between the comparison indicators in the control and experimental systems was demonstrated on statistical reduction of the results of repeated experiments (from 15 to 25 at each point). The effect was directly dependent on irradiation time. Irradiation for 30 minutes at $t = 20°C$ had no influence on colicin synthesis; the numbers of cells synthesizing colicin increased by a factor of 1.5—2 after irradiation for one hour and reached a maximum after 2 hours (Fig. 2). At 37°C, colicin synthesis was induced by as little as 30 minutes of irradiation. This would apparently be associated with the higher functional activity of all systems of the cell under these conditions.

We then studied the influence of the power flux density of the radiation on induction of the colicin synthesis. Variation of the power flux density through a factor of 100, from 0.01 to 1.00 mW/cm^2, had no influence on the induction coefficient, and only a further reduction of the power to 0.01 mW/cm^2 resulted in a sharp decrease in the biological effect (Fig. 3).

Thus, the magnitude of the biological effect is affected differently by variations of exposure time, power density, and temperature. While exposure time has a very strong influence, variation of the power of the radiation over a broad range leaves the magnitude of the effect practically unchanged. That the effect does not depend on power is another weighty argument in favor of the nonthermal effects of millimeter waves, since all thermal effects depend primarily on flux intensity. Direct temperature measurements with a thermocouple indicated that the bacterial suspensions in the experimental and control systems had practically the same temperature during irradiation.

Up to the present time, the ability of various agents (both physical and chemical) to induce the colicin synthesis, which is lethal to the bacterial cell, has been linked basically to the ability of these agents to disintegrate DNA or to block its synthesis. The classical inductors of the colicin syntheses of other similar genetic systems (for example, that of temperate phage)—UV irradiation or mitomycin C[2]—also exhibit these properties. As we know, both agents rupture chemical bonds in the DNA molecule, with formation of pyrimidine-base dimers. From this point of view, millimeter-band radiation can be regarded as a fundamentally new agent that disturbs the functional regulatory mechanism of genetic elements in the cell, and extrachromosomic elements in particular, without causing direct damage to the DNA molecule.

[1] H. Ozeki, B. A. D. Stocker, and H. Margeri, Nature **184**, 337 (1959).

[2] V. G. Likhoded, Mikrobiologiya, No. 7, 116 (1963); P. Amati, J. Mol. Biol. 8, 239 (1964); W. deWitt and D. Helinsky, ibid. 13, 692 (1965); P. Fredericq, J. Theor. Biol. 4, 159 (1963).

V. F. Kondrat'eva, E. N. Chistyakova, I. F. Shmakova, N. B. Ivanova, and A. A. Treskunov. Effects of Millimeter-band Radio Waves on Certain Properties of Bacteria

Over a number of years beginning in 1965, we investigated the influence of millimeter-band radio waves on certain properties of bacteria. Three strains of Cl. sporogenes, two of Cl. histolyticum—anaerobic spore bacteria with conspicuous proteolytic properties—and three strains of Bact. prodigiosum—an aerobic bacterium that differs substantially in a number of properties from the other two—were chosen as objects.

Each strain was irradiated 20 times at a wavelength of 7.20 mm for three hours at a time. Morphology, sporulation (in the anaerobes), the nature of growth on culture media, saccharolytic, proteolytic, and antigenic properties, and, in the case of hystolyticum, also pathogenicity, were studied after irradiation.

After irradiation, sporogenes and histolyticum shrank to half the size of the controls, and seldom appeared in pairs and chains. A strong and consistent decrease in spore-forming ability was observed in both anerobes. This was especially pronounced in the case of sporogenesis: unirradiated cultures grown on Kitt-Tarozzi medium had 50—54 cells with spores per hundred after twenty-four hours, while irradiated cells showed only 5—20 spore cells per hundred. Two cultures lost their ability to form spores altogether. It was not recovered after the cultures had been stored for a year and subcultured 20 times.

There were changes in growth on dense nutrient media. Rounder colonies with only slightly convoluted margins and smoother surfaces were encountered among the sporogenes colonies. Beginning at the 8th to 10th exposure, prodigiosum cultures began to grow in the form of pale pink colonies that did not redden in the light. Suspensions prepared from irradiated cultures that had grown for twenty-four hours were colorless or slightly pinkish, while their controls were deep pink or bright red.

The antigenic properties of the bacteria were affected. Irradiated sporogenes and histolyticum cultures began to agglutinate at titers $\frac{1}{2}$ to $\frac{1}{4}$ those of the controls. Antigens from the irradiated cultures had a positive precipitation reaction in agar gel in dilutions 1—3 smaller than the control dilutions. The intensity of the reaction was also weaker. The control cultures formed three lines of precipitation in the reaction between undiluted and 1 : 2 diluted sera and antigens, but there were seldom two lines, and usually only one, in irradiated cultures with the same antigen concentrations.

There was no change in the saccharolytic activity of the bacteria, but proteolytic activity declined. The irradiated bacteria began to peptonize milk 2—6 days later than the controls. Irradiated sporogenes cultures were slower to decompose the fragments of meat in the Kitt-Tarozzi medium. The pieces normally vanish completely within 20—30 days, but 35—52 days were required when irradiated bacteria were cultured.

The decrease in the ability to peptonize milk and decompose meat indicates a change in protein metabolism.

To investigate the virulence of histolyticum, rabbits and white mice were inoculated with cultures that had been growing for four days. The cultures were administered intramuscularly, undiluted and in successive twofold dilutions up to 1 :32, in 0.5-ml doses to the mice and in 1-ml doses to the rabbits. Eight irradiated and 3 unirradiated cultures were used in the experiment. The animals were observed for six days.

The animals inoculated with the unirradiated cultures perished during the first four days up to a maximum dilution of 1 :8.

Local effects were observed in the rabbits given the 1 :16 dilution, and also in one of the animals given the 1 :32 dilution.

Of the eight rabbits inoculated with the irradiated cultures, one perished at the 1 :8 dilution, and the rest after administration of undiluted or 1 :2 diluted cultures. Three rabbits showed local effects from the 1 :4 dilution, and one even from 1 :8.

We also studied the influence of microwaves on the survivability of the bacteria. It had been established originally that the 7.2-mm wavelength was most injurious. We investigated the effects of wavelengths λ = 7.1, 7.15, 7.16, 7.17, 7.18, 7.19, and 7.20 mm to define a narrower wavelength band between 7.1 and 7.2 mm.

In all experiments, the number of microorganisms was smaller after exposure than in the controls; the strongest effect was observed from the 7.15-mm wavelength.

Thus, millimeter waves have a substantial lethal effect on bacteria. Survival time was found to depend on wavelength. Sporogenesis, antigenic and proteolytic properties, and virulence are affected by microwave exposure.

S. E. Manoĭlov, E. N. Chistyakova, V. F. Kondrat'eva, and M. A. Strelkova. Effects of Millimeter-band Electromagnetic Waves on Certain Aspects of Protein Metabolism in Bacteria

It had been shown in earlier studies that this type of radiation has a very strong influence on the functions both of biologically active substances—hemoglobin, the cytochromes, etc.[1,2] and on microorganisms[3]. In the present study, we report material from an investigation of the effects of millimeter radio waves on certain aspects of the protein metabolism of anaerobic and aerobic bacteria and on fungi, whose protein metabolisms exhibit qualitative differences.

The following species were the objects of investigation: Cl. sporogenes, Cl. histolyticum (anaerobes), B. prodigiosum, Staphylococcus aureus (aerobes), Act. norsie, Pen. nigricans (fungi). The microbes were irradiated at wavelengths of 7.2 and 7.6 mm by a backward-wave-tube source at an average power flux density of 4—5 mW/cm^2 in the radiation incident on the object. Equal numbers of irradiated and unirradiated microorganisms were introduced into the nutrient medium, and the free amino acid contents were determined by Baudet's method in this medium after one day of growth.

All data were calculated as percentages of the amino-acid contents in the nutrient medium after unirradiated microorganisms had been cultivated on it. The amino-acid contents for all of the bacteria studied in the nutrient medium can be classified into three groups: in the first group, there is no difference between the quantities of amino acids in the nutrient medium between the irradiated and unirradiated bacteria; in the second group, the quantity of amino acids in the medium was larger when irradiated microbes were cultured than in the case of unirradiated microbes; in the third group, the amounts of amino acids in the medium were smaller for irradiated than unirradiated microorganisms.

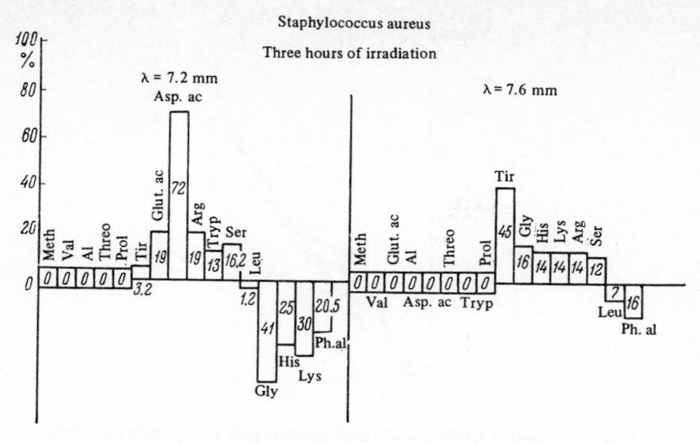

Several series of experiments were set up. In one series, we studied the effects of different wavelengths (7.2 and 7.6 mm) on the Staphylococcus, an aerobic microbe. In the second series, we investigated the protein metabolism of the individual amino acids in the various microbes after irradiation at the same wavelength (7.2 mm). The resulting data were processed statistically. As an example, we present data (see the figure) on the effects of various wavelengths on the Staphylococcus. As the figure shows, the following amino acids belong to the first group after irradiation at 7.2 mm: methionine, valine, alanine, threonine, and proline (five amino acids); on irradiation at 7.6 mm, methionine, valine, alanine, proline, tryptophan, glutamic acid, and ascorbic acid (seven amino acids). In the second group after irradiation at 7.2 mm: glutamic acid 19%, aspartic acid 72%, arginine 19%, tryptophan 13%, serine 16%, tirosine 3.2%. After irradiation at 7.6 mm: arginine 14%, serine 12%, tirosine 45%, glycine 16%, histidine 14%, lysine 14%, threonine 50%. In group 3 after 7.2-mm irradiation: leucine 1.2%, glycine 41%, lysine 30%, phenylalanine 20.5%, histidine 25%. After irradiation at 7.6 mm: leucine 7%, phenylalanine 15%. Consequently, there are qualitative and quantitative differences in the effects of the different wavelengths (7.2 and 7.6 mm).

In analysis of the factual material, our attention is drawn to two groups of amino acids: those with acidic properties (glutamic and aspartic acids) and those with alkaline properties (histidine, lysine, and arginine). While the number of "acidic" amino acids in the nutrient medium increases after growth of the microbes irradiated at 7.2 mm, no differences in their contents in the medium are detected after irradiation at 7.6 mm.

As for the "alkaline" amino acids, the amounts are smaller (group 3) during growth of bacteria irradiated at 7.2 mm and larger in the case of 7.6 mm. Changes in the contents of "acidic" and "alkaline" amino acids are also observed after irradiation of aerobes, anaerobes, and fungi at 7.2 mm. The metabolism of other amino acids is subject to substantial variations, both qualitative and quantitative. How can these facts be explained? We believe that electromagnetic radiation in the millimeter band has a definite influence on the protein metabolism of bacteria. This is manifested either in the form of activation or inactivation of proteolytic enzymes or in a change in the activity of enzymes participating in the metabolism of the individual amino acids.

[1] V. E. Manoĭlov, et al., Sb. Trudov LkhFI, No. 21, 1, 78 (1967).

[2] N. D. Devyatkov, Elektronnaya Tekhnika SVCh, Part 4, Sov. Radio, 1970, p. 190.

[3] V. F. Kondrat'eva and E. N. Chistyakova, in[1], pp. 1 and 83.

N. P. Zalyubovskaya. Reactions of Living Organisms to Exposure to Millimeter-band Electromagnetic Waves

We have been investigating the effects of millimeter-band electromagnetic waves on intact organisms, isolated cells, and cellular structures since 1966. To establish the biological effects of millimeter-band radiation, we studied the reactions of organisms in various stages of evolutionary development (viruses, microbes, insects, birds, and mammals).

Exposure of microorganisms (Staphylococcus, Streptococcus, B. coli, typhoid bacillus) to millimeter waves lowered their survival rates by 60% and more, affected the morphological, culturing, and biochemical properties, increased their sensitivity to antibiotics, and modified their antigenic properties. The infective activity of irradiated viruses was lowered.

The biological effects of the millimeter waves depended on wavelength and exposure time. The bactericidal action of millimeter waves was most pronounced at a wavelength of 6.5 mm. These studies permitted the conclusion that millimeter-band electromagnetic waves influence the viability of microorganisms.

In the experiments in which insects (Drosophila) were irradiated, we studied the influence of millimeter waves on the survival rates of the irradiated individuals, their ability to reproduce, and the influence of such irradiation on their offspring in the first and second generations.

Irradiation (for 15–60 min) of adult male and female Drosophila individuals was not lethal; they showed no externally evident changes, and breeding of such insects generally produced normal offspring. However, the offspring were fewer in number, and the fertility of the insects depended on the wavelength of the radiation to which they had been exposed (Fig. 1) and on the exposure time (Fig. 2).

Prolonged exposure to millimeter waves (for 3, 4, and 5 hours) resulted in significant changes in the first and second generations of Drosophila. Male individuals obtained from irradiated parents in the second generation were characterized by lower than normal viability; many perished 3–6 days after crossing. In most cases, female individuals laid no eggs.

Mutants seldom appeared in the first generation; most of them were observed in the second generation after prolonged exposure to radiation at 6.5 mm.

Thus, genetic changes were observed after exposure to millimeter waves in the insect experiments, and were manifested in lowered fertility and viability of the offspring. The observed changes apparently took place in reproductive cells, since the offspring inherited them. Individual genes obviously exhibited definite sensitivity to millimeter waves; this was indicated by multiple occurrences of the mutations in the offspring of the irradiated Drosophila.

With the object of studying the influence of millimeter-band microwaves on the formation, growth, and development of living organisms at a more advanced

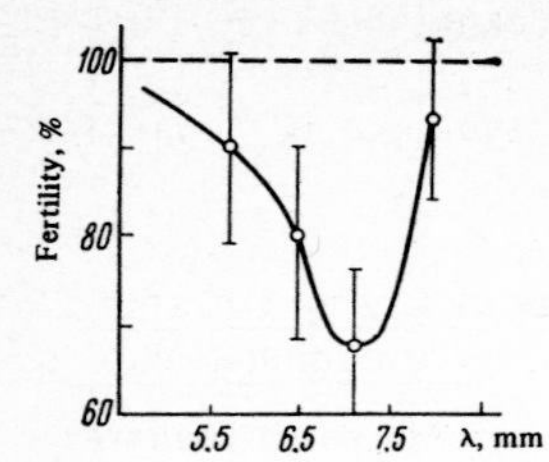

FIG. 1. Effects of various millimeter-band wavelengths on fertility of Drosophila in the second generation.

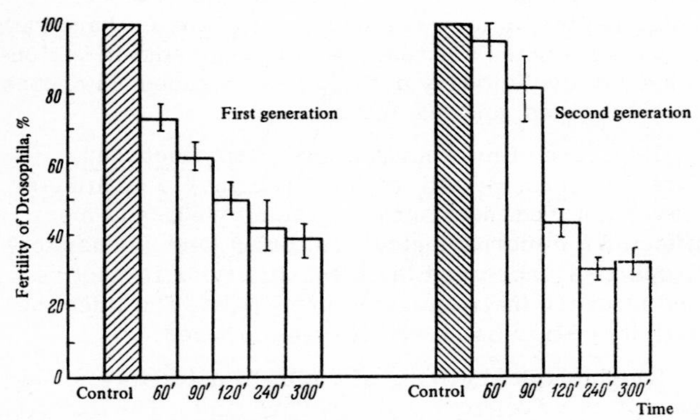

FIG. 2. Variation of fertility in first and second generations of Drosophila after irradiation for various exposure times at a wavelength of 6.5 mm.

stage in evolutionary development, we irradiated chick embryos and then followed them through their embryonal and post-embryonal states.

Beginning at the 7-th day, chick embryos were irradiated with millimeter waves five times for 30 minutes at a time. None of them perished after irradiation, and no loss of weight of the embryos as compared with the control was observed during the incubation period, but the incubation period was lengthened by 2—3 days.

The chicks that developed from the exposed embryos were found to be somewhat retarded in their development, especially when the irradiation had been at 6.5 mm. These chicks were unable to stand on their feet and began to peck at food later than the others; all of the irradiated chicks feathered out poorly.

The chicks that developed from the irradiated embryos were observed for 50 days. Up to the seventh day, their weights differed little from those of the controls, but they began to lag behind in this respect beginning at the 10th to 12th day. This decrease in weight continued to the last (50th) day (Fig. 3).

Thus, the results indicate a distinct effect of millimeter waves on the vital activity of chicks developed from irradiated embryos. The degree of the effect on post-embryonal development depended on wavelength.

Processes unfolding at all levels of organization of the particular organism take place in shaping the response reactions to millimeter-wave exposure in multicellular organisms.

In the experiments on mammals, we were interested in the response reactions associated with functional changes in the organism due to disturbances of the various complex functional systems. The tests showed that irradiation of the experimental animals (white rats and mice) over 40—50 days for 10—15 minutes per day was not lethal. However, these animals were sluggish and

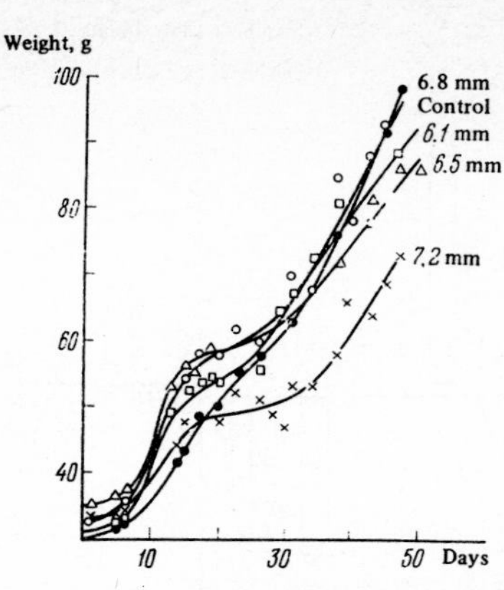

FIG. 3. Weight variations of chicks developed from embryos irradiated with millimeter waves.

their fur dishevelled; they refused food and drink for some time. The hair on shaved skin areas (under local irradiation) failed to grow back over the entire course of the exposures. Biopsies of irradiated skin areas showed atrophy of the Malpighian layer, sclerosis of the derma, and signs of accumulation of fat with penetration of fatty vacuoles into the derma, i.e., abnormalities in the outer layers of the skin and the underlying muscular layer were observed after irradiation.

The sensitivity of the organism's humoral system to millimeter irradiation was inferred from hematological indicators. The blood coagulation rate was higher after irradiation (68.2 ± 1.5 sec in the control, 35.0 ± 1.3 sec after irradiation at 6.5 mm; $p < 0.01$), and hemoglobin content decreased to 11.0 g-% as compared with 16.0 g-% in the unirradiated animals. Blood-serum albumin was lower by 30% and more in the irradiated animals as compared with the intact ones.

Decreases in total nucleic acids and albumins were observed in the blood-forming organs (liver and spleen) of the irradiated animals. Thus, while the total nucleic-acid content in the control was 312 ± 8.24 μg, with 94.8 ± 3.3 μg of RNA, 217.3 ± 7.2 μg of DNA, and 51.76 ± 1.1 mg of albumins, the respective concentrations decreased after irradiation at 6.5 mm: total nucleic acids to 250.0 ± 6.4 μg, RNA 109.0 ± 6.0 μg, DNA 140.0 ± 7.6 μg, and albumins 38.0 ± 2.2 mg; $p < 0.05$. The irradiated animals showed lower resistance to infection. Their antibody (agglutinin) and blood lysozyme levels were half those of the control. Irradiation of immunized animals had no effect on specific resistance; these animals remained stable against infection.

Thus, the studies showed that simple and highly organized animals are sensitive to electromagnetic waves in the millimeter band. This suggests that the action of millimeter waves is a general biological one and is not limited by phylogenetic differences between organisms. The effects of millimeter waves on the living organism were manifested in functional and systemic changes, although certain distinctive features of the reaction were noted in specific organisms. The biological effect of the millimeter waves depended on wavelength and exposure time. In the 5—8-mm band, 6.5-mm microwaves were characterized by higher biological effectiveness.

The biological effects of the millimeter waves were manifested in changes in many vital processes. The very diversity of these effects makes their investigation one of the problems whose solution may aid in understanding of other general biological processes associated with manifestation of the ultimate effects of irradiation.

R. I. Kiselev and N. P. Zalyubovskaya. Effects of Millimeter-band Electromagnetic Waves in the Cell and Certain Structural Elements of the Cell

Study of the mechanism by which electromagnetic waves in the millimeter band act on biological objects acquires substantial importance for the use of these waves in biology and medicine. During recent years, we have studied the influence of the millimeter band on isolated human and animal cells. Such cells offered a convenient model that enabled the experimentor to obtain individual cells in a monolayer form in which they were readily accessible to microwave exposure and subsequent study of its effects. In addition, structural elements of cells, viruses, and microorganisms were irradiated with microwaves. The basic criteria for evaluation of millimeter-wave effects were the morphological and biochemical indicators, survival rates, and changes in the antigenic, culturing, and virulence properties of the irradiated objects.

These studies indicated that millimeter-wave irradiation of isolated cells resulted in damage to the cell membrane, degeneration of protoplasm, and an increase in the sizes of the cells (control $5904 \pm 183\ \mu^3$, irradiated at 6.5 mm $6985 \pm 185\ \mu^3$; $p < 0.01$) and the nuclei (control $492 \pm 62\ \mu^3$, irradiated at 6.5 mm $590 \pm 43\ \mu^3$; $p < 0.01$).

The total nucleic acids and albumin contents of cells irradiated at 6.50 mm showed an increase. While the control had RNA $74.9 \pm 5.1\ \mu g$, DNA $96.8 \pm 9.4\ \mu g$, and albumins 109.8 ± 6.7 mg, the figures after irradiation were RNA $97.3 \pm 3.6\ \mu g$, DNA $137.7 \pm 6.2\ \mu g$, albumins 130 ± 8.6 mg; $p < 0.01$. It is possible that, by affecting cell metabolism, microwaves influence synthetic processes.

We noted a decrease in the number of viable cells after irradiation at the various wavelengths. In the range 5.90—7.50 mm, the 6.50-mm wavelength showed more conspicuous biological activity (Figs. 1 and 2).

After irradiation of red blood cells (erythrocytes) at 6.50 mm, we noted significant changes in hemolytic stability, an indication that these cells are sensitive to such radiation and that functional and structural changes occur under exposure to it.

Changes in the nucleic acid and albumin contents took place after irradiation in nuclei and mitochondria that were separated from liver cells. The extremely weak luminescence of the irradiated cell elements was down considerably in comparison with the control, and there was a sharp decrease in the rate of buildup of chemoluminescence intensity on heating (Fig. 3).

Millimeter-wave irradiation of various viruses (adenoviruses, measles virus, vesicular stomatitis virus, and others) resulted in a quantitative reduction of the virus particles (on irradiation of the whole virus) by a factor of 2—3. The lowered infectious activity of irradiated adenoviruses and measles virus was manifested in a delay of the cytopathogenic effect on a tissue culture.

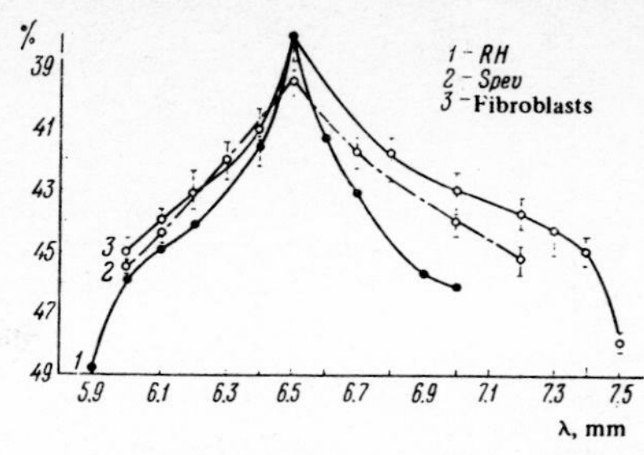

FIG. 1. Nature of millimeter-band microwave effect on various types of tissue cultures.

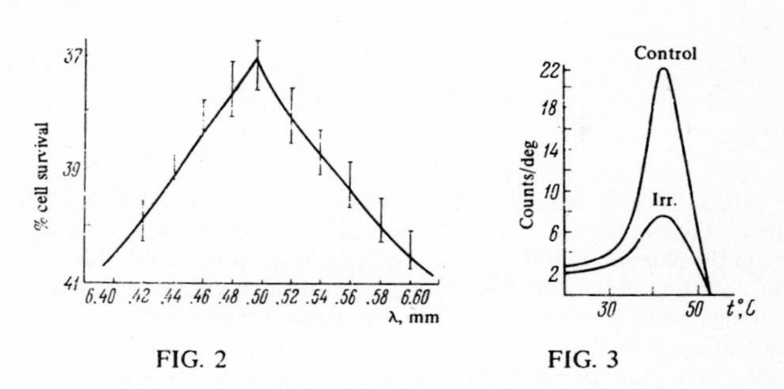

FIG. 2

FIG. 3

FIG. 2. Influence of millimeter-band microwave irradiation on survival rate of tissue culture.

FIG. 3. Growth rate of chemiluminescence intensity of cell nuclei after millimeter-wave irradiation.

A decrease in infectious activity was observed after irradiation of virus DNA preparations (isolated from adenoviruses) as compared to unirradiated specimens. While the cytopathogenic effect was observed on the 10th day in tissue cultures that had been treated with unirradiated DNA and was morphologically similar to the manifestations of adenovirus cell infection, the infectious activity in tissue cultures treated with DNA that had been irradiated at 6.50 mm appeared between days 15 and 16 and corresponded morphologically to a manifestation of the whole virus. It appears that millimeter-band microwave irradiation of the virus DNA resulted in this case in a partial loss of infectious activity, although transforming activity was not lost, as manifested in the later appearance of the cytopathogenic effect.

We judged the influence of microwave irradiation on the cellular genome from the increase in latent-phage and colicin productive activity after irradiation of lysogenic and colicinogenic microbe strains. After irradiation of the latter at 6.50 mm, the colicin titer increased to 320 conventional units. Increased production of phage particles as compared with the control was observed in the lysogenic microbe strains after millimeter-band microwave irradiation. Thus, while the number of phage particles was 1471 ± 152.0 in the control, it was 2934 ± 64.0 after irradiation at 5.9 mm, 4042 ± 152.0 after irradiation at 6.1 mm, 5725 ± 129.2 after irradiation at 6.50 mm, and 1296 ± 60.4 after irradiation at 7.5 mm; $p < 0.01$.

Thus, these studies indicated that millimeter-band electromagnetic waves affect both cells and cell structures.

The data obtained may serve as a basis for the use of millimeter-band electromagnetic waves in experiments toward controlled modification of viruses and microbes.

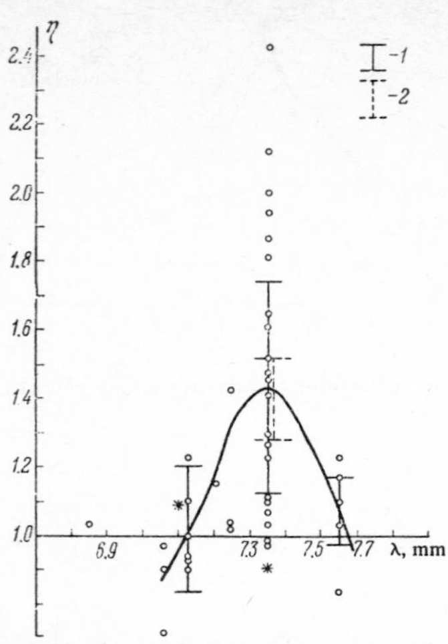

FIG. 1. Coefficient η of increase in Hb stability to acid splitting as a function of irradiation wavelength (according to [3]). $\eta = x_{contr}/x_{irrad}$, where the percentage splitting $X = 1 - C_{hydr}/C_{init}$ and C is the albumin concentration in the initial state (C_{init}) and the concentration of the remaining unsplit albumin after hydrolysis (C_{hydr}). The points indicate the scatter of the data from the various experiments; 1—rms scatter; 2—scatter after reduction of experimental data within the framework of a model taking account of the effect of microwaves on the kinetics of the oxy-met transition; asterisks indicate values of the coefficient η that we obtained for lyophylically dried Hb preparations.

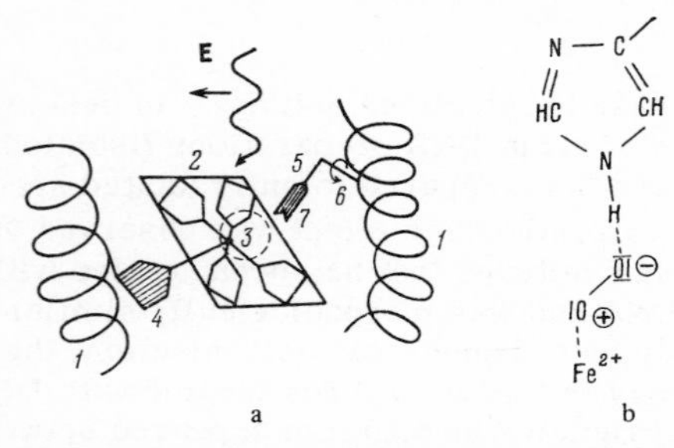

FIG. 2. Schematic representation of the structure of the active center of one of the four chains of the hemoglobin molecule (a) and the O-O bridge between the iron of the heme and the hydrogen of the distal histidine (b). 1—Spiral segments of the protein part of the micromolecule; 2—heme; 3—iron ion at center of heme; 4—proximal histidine F8 linking heme to protein part of molecule; 5—distal histidine E7, which is set in rotational rocking motion about the "axis" 6 by the microwaves; 7—conditional boundaries of region in which distal histidine interacts with iron ion.

V. I. Gaĭduk, Yu. I. Khurgin, and V. A. Kudryashova. Outlook for Study of the Mechanisms of the Nonthermal Effects of Millimeter- and Submillimeter-band Electromagnetic Radiation on Biologically Active Compounds

1. We irradiated aqueous solutions of hemoglobin (Hb) in the millimeter band. Not only was the complete molecular structure of this protein known[1], but Komov, Chistyakova, and Manĭolov had previously obtained an effect in which microwave exposure produced irreversible nonthermal changes in dried Hb preparations (see[2]). We had observed[3] a change in the chemical properties of 5% aqueous solutions of native Hb obtained from human erythrocytes. The material was irradiated at wavelengths around 7.35 mm for 5 hours at a radiation intensity on the order of 1 mW/cm^2 and temperatures of 37–40°C. On completion of the radiation exposures, the percentage acid hydrolysis of the Hb (splitting off of heme) was lower than that in the control (Fig. 1), i.e., exposure to microwaves resulted in increased strength of the bond between the heme and the protein. This can be interpreted as an increase in the stability of the Hb to the transition (during irradiation) from the active oxy form to the inactive met form, since the rate of acid hydrolysis of oxy-Hb is much lower than that for met-Hb[4]. Consideration[3] of the kinetics of this process makes it possible to improve the reliability of the experimental data.

2. According to the hypothesis of[5], a single molecular group in the active center can be regarded as responsible for the changes that take place; this is the distal histidine E7, which is essential to the Hb function[1] (Fig. 2). It can describe pivoting motions with frequencies on the order of a few cm^{-1}. The energy of the histidine oscillation and its average position with respect to the iron atom may be changed under the influence of the microwave field. Under certain conditions, the oxygen molecule in Hb forms[6] a bridge between the iron atom and the hydrogen of the E7 histidine, over which an electron is removed from the Fe atom and the Hb converted to the met form. Excitation of the histidine E7 oscillations can be inhibited by the formation of this bridge, thus preventing the inactivation of the Hb. However, the effect of the microwaves on the histidine E7 itself is not the only possibility; in principle, it can also be transferred to the histidine as a result of excitation of vibrations of the macromolecule as a single entity. This mechanism is based on the protein-machine hypothesis[7]. It is not now possible to choose between the two hypotheses or to synthesize them, and it has by no means been proven that changes in the disposition of the histidine E7 are responsible for the observed effects. Moreover, $kT/h\nu \approx 150$ under the conditions of our experiments, i.e., there are weak effects on the chemical behavior of the molecular systems (see, for example,[8]); it must be remembered that the walls of the "cavity" within which the E7 histidine is situated can fluctuate. The frequency and amplitude of these fluctuations (more precisely, their spectral distribution) are determined not only by temperature, but also by the physicomechanical properties of the macromolecule, which somehow transforms the energy of random collisions of molecules of the aqueous medium against the surface of the Hb. We may hope that microwave irradiation will assist us in finding the degrees of freedom through which most of the energy is "transferred," and that the importance of the nonlinearity[9,10] of the dipoles (which results from the energy dependence of their oscillation frequency) interacting with the microwave field will be established.

3. In principle, it would be interesting to use microwaves for resonance control of biological processes, e.g., of the addition of oxygen to Hb, in order to regulate enzyme reactions and to study the properties of the biopolymers themselves, e.g., their catalytic activity. The use of physicochemical methods to indicate microwave exposure is also promising. Selectivity of biopolymer absorption in aqueous media has not yet been observed in the microwave band. The presence of smooth dispersion and the influence of the type of solvent on absorption indicate that the interaction of the microwaves with matter is predominantly of relaxation nature as the wavelength is reduced all the way down to $\lambda \approx 0.1$ mm. In the submillimeter band, the inertia of the molecules in their oscillations about their temporary equilibrium positions begins to make itself felt[11]. Nevertheless, study of microwave absorption yields much information

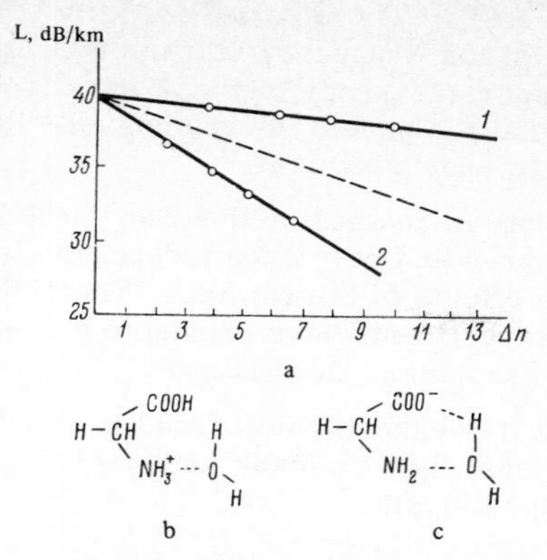

FIG. 3. Curves of absorption of aqueous glycine solution in acidic (1) and alkaline (2) media as a function of water content in the solution (a) and a schematic interpretation of the data shown (b, c). a) The dashed line indicates the decrease in absorption of pure water on a corresponding decrease in the amount of it ($\Delta n = 55.5 - n_1$, where n_1 is the water content (mole/liter) in the solution and Δn is the decrease of solution water content due to solution of the amino acid); b) cationic form (acidic solution of amino acid): the water molecule is not strongly bound and has freedom of displacement with respect to the hydrogen bond; c) anionic form (alkaline solution of amino acid); several water molecules are strongly bound and have lost freedom of displacement; for simplicity, the diagram shows only one strongly bound water molecule.

even now. For example, in the related compounds

$$CH_3-\underset{CH_3}{\overset{CH_3}{\overset{|}{\underset{|}{C}}}}-Cl$$

(tert-butyl chloride; $\mu \approx 2.14$ D) and $CH_3-CH_2-CH_2-CH_2-Cl$ (n-butyl chloride; $\mu \approx 2.08$ D), whose molecular dipole moments μ differ (in the gaseous phase) by only 3%, solutions of these compounds in nonpolar solvents give absorption values differing by a factor of about two, like the pure liquids—the explanation for which may be found in the difference between the potentials of the intermolecular interactions that results from the shape differences between the molecules and their polarization ellipsoids. Data on the influence of the interaction of amino acids with solvent (water) molecules as it affects absorption are highly interesting. Absorption decreases with decreasing amount of water in the solution, by approximately three times as much in an alkaline medium, in which the amino acids are in the cationic form, than in an acidic medium (Fig. 3). This is explained by strong bonding of about 2 water molecules by the amino acid in alkaline media, which is qualitatively consistent with the conclusions obtained in[12] by IR spectroscopy. Several strongly bound water molecules (and not only one, as in[12]) can be registered in the microwave band. These results indicate the possibility of developing a microwave method for measuring the degree of hydration of biopolymers in solution at different temperatures (NMR and microcalorimetric measurements yield the amount of free water when frozen solutions are melted). The proposed method would yield information on the interaction of amino acids and other biologically active compounds with water and could be used to construct a theory of resonant microwave effects on biopolymers. In addition, direct determination of the hydration number would be helpful to understanding of the nature of the reactivity of organic compounds in the liquid phase and certain specific properties of the macromolecules that are determined by their interaction with the aqueous medium.

[1] M. F. Perutz, et al., Nature 219 (5150), 131 (1968).
[2] N. D. Devyatkov, Elektronnaya Tekhnika, Ser. 1 (Elektronika SVCh), No. 4, 130 (1970).
[3] V. A. Kudryashova, S. A. Il'ina, A. S. Faleev, V. I. Gaĭduk, and V. V. Dementienko, in: Gigiena truda i biologicheskoe deystvie elektromagnitnykh voln radiochastot (Industrial Hygiene and the Biological Effects of Radio-Frequency Electromagnetic Waves), 1972; Preprint IRE Akad. Nauk SSSR No. 115, Moscow, 1972.
[4] V. P. Komov, Author's abstract of candidate's thesis (Leningrad, 1966).
[5] L. G. Koreneva and V. I. Gaĭduk, Dokl. Akad. Nauk SSSR 193, 465 (1970).
[6] Struktura i svyaz' (Structure and Bonds), (Russ. transl.), Mir, p. 318.
[7] D. S. Chernavskiĭ, Yu. I. Khurgin, and S. É. Shnol', Mol. Biol. 1, 419 (1967).
[8] L. A. Blyumenfel'd, V. I. Gol'danskiĭ, M. I. Podgoretskiĭ, and D. S. Chernavskiĭ, Zh. Strukt. Khim. 8, 854 (1967).
[9] V. I. Gaĭduk, R. F. Matveev, A. T. Fialkovskiĭ, and V. V. Dementienko, Radiotekh. i Elektron. 12, 749 (1973).
[10] V. I. Gaĭduk, Preprint IRE Akad. Nauk SSSR, No. 97, 1972.
[11] N. E. Hill, Proc. Phil. Soc. 5,(529), 723 (1963).
[12] N. G. Orlov, V. S. Markin, Yu. V. Moiseev, and Yu. I. Khurgin, Zh. Strukt. Khim. 8, 854 (1967).

D. S. Chernavskiĭ. I should like to make a few remarks on two possible mechanisms of the effect of microwaves on hemoglobin.

One of them relates to the direct action of the microwaves on the histidine, and the other to an effect on the molecule as a whole and the excitation of elastic oscillations of the entire structure of the protein. It seems to me that these mechanisms should be regarded not as alternatives, but instead as complementing one another, i.e., as two aspects of the same mechanism. To clarify, it might be appropriate to recall the hypothesis of the role of elastic deformations in enzymatic catalysis. According to this hypothesis, the energy needed to lower the activation barrier is stored in the polypeptide in the form of elastic deformations and released (or converted to another form) at the instant of the enzymatic event. It is important to stress that the deformations must be elastic, since otherwise the stored energy is dissipated. To resort to a simile: the enzyme-protein works like a machine: it contains a stressed region ("spring"), and at the proper moment the energy is transferred through a system of levers into the region of the active center, where the substrate molecule is located. As a result, a certain bond in the substrate molecule is ruptured and the desired reaction takes place.

Estimates have indicated[1] that the dimensions of the spring should be in the tens of angstrom units, i.e., of the same order of magnitude as the entire polypeptide globule. According to the estimates of[2], a system of these dimensions has a natural vibration frequency $\omega = 10^{11}$ Hz, i.e., of the same order of magnitude as the resonance frequency in the spectrum of the microwave effect. It appears that the elastic vibrations themselves do not play a substantial part in the enzymatic event. The conclusion that a preferred frequency exists follows from the hypothesis that the deformations are elastic in nature.

On the other hand, the very conception of the "protein machine" indicates that one generalized degree of freedom that encompasses the entire molecule and is characterized by a common frequency should be preferred in the macromolecule. It is therefore not surprising that the vibration frequencies of the several parts of the machine that participate in the enzymatic event (in particular, the histidine radical) are the same as the common frequency.

From this point of view, the direct action of microwaves in the histidine in hemoglobin should cause the entire "machine" to oscillate. Here the two mechanisms that were discussed represent different aspects of the same mechanism.

The special and highly important question as to the influence of temperature on the oscillation requires special discussion. However, it seems to me that this effect might not even be particularly strong. In classical physics, we have many examples of the excitation and survival of low-frequency vibrations against a background of higher-frequency thermal vibrations. We should note that the protein macromolecule is sufficiently massive and "classical."

I should note in conclusion that the "protein machine" concept appeared and developed independently of experiments on the effects of microwaves. The fact that the microwave experiments have proven to be related to it should be regarded as gift of nature.

I feel that this experimental trend is highly promising and may cast light on many fundamental problems of enzymatic catalysis.

---

[1] Yu. I. Khurgin, D. S. Chernavskiĭ, and S. É. Shnol', in: Kolebatel'nye protsessy v biologicheskikh i khimicheskikh sistemakh (Vibrational Processes in Biological and Chemical Systems), Nauka, 1967; Molek. Biol. 1, 419 (1967); D. S. Chernavskiĭ, Fizicheskiĭ modeli biologicheskogo kataliza (Physical Models of Biological Catalysis), Znanie, 1972.

[2] V. P. Komov, Yu. I. Shmelev, and D. S. Chernavskiĭ, Kr. soobshch. fiz. (FIAN Akad. Nauk SSSR), No. 9, 38 (1972).

Fundamental Principles of Cooperative and Transductive Coupling

FUNDAMENTAL PRINCIPLES OF COOPERATIVE PROCESSES:

Bruno Zimm

Many cooperative processes can be examined theoretically if one assumes that each object in an ensemble can exist in *only* two states. These processes include such basic biological phenomena as the alternation of enzyme subunits between two different conformations having different enzymic activities and the transformation of DNA base pairs from Watson-Crick helix to random coil.

The Four-Subunit Enzyme as an Illustrative Example

The differences between cooperative and noncooperative transitions can be illustrated with the simple example of a four-subunit enzyme in which each subunit may exist in either of two conformational states, referred to as (+) and (−), where the (+) form is enzymatically active. The mathematical analysis of this example is familiar in connection with the binding of small molecules to proteins, which has been widely discussed (see van Holde, 1971).

If there is no cooperativity among the four enzyme subunits, the probability of any subunit assuming the (+) or (−) state is independent of the conformation of the other subunits. It is therefore possible to define for each subunit a single equilibrium, or "stability," constant, s, for the ratio of subunits in the two states:

$$s = \frac{n_+}{n_-} \tag{1}$$

where n_+ and n_- are the average number of subunits in the (+) and (−) states, respectively, averaged over the ensemble of molecules. This stability constant can be related to the standard free-energy change for the reaction by the relationship

$$\Delta G^0 = G_+{}^0 - G_-{}^0 = -RT \ln s \tag{2}$$

In general, s will be a function of such parameters as temperature, pressure, pH, and allosteric effector concentration.

The quantity of interest is usually the fraction of subunits in the (+) state, θ:

$$\theta = \frac{n_+}{n_+ + n_-} \tag{3}$$

All possible states of the ensemble of four subunits must be considered in order to calculate θ, as shown in Table 1. The relative free energy of the (−)(−)(−)(−) state is arbitrarily taken as zero. There is one (−)(−)(−)(−) state, four states (of equal probability, since they have equal energy) having one subunit in the (+) state, and so on. θ can be calculated with this information and equation (1). A statistical calculation for all subunits in the (+) state shows that θ is given by (see van Holde, 1971)

$$\theta = \frac{0 + 4s + 12s^2 + 12s^3 + 4s^4}{Q} \tag{4}$$

and

$$\theta = \frac{1}{4}\frac{s}{Q}\frac{dQ}{ds} \tag{5}$$

where Q is the partition function and is given by

$$Q = 1 + 4s + 6s^2 + 4s^3 + s^4 \tag{6}$$

Q is a sum of the products of the number of ways a given energy state may be formed, multiplied by the stability constant raised to a power equal to

TABLE 1

		Free energy
(1)	(−)(−)(−)(−)	0
(2)	(+)(−)(−)(−) (−)(+)(−)(−) (−)(−)(+)(−) (−)(−)(−)(+)	ΔG^0
(3)	(+)(+)(−)(−) (+)(−)(+)(−) (+)(−)(−)(+) (−)(+)(+)(−) (−)(+)(−)(+) (−)(−)(+)(+)	$2\ \Delta G^0$
(4)	(−)(+)(+)(−) (−)(+)(−)(+) (−)(−)(+)(+) (−)(+)(+)(+)	$3\ \Delta G^0$
(5)	(−)(+)(+)(+)	$4\ \Delta G^0$

Reprinted with permission from *Functional Linkage in Biomolecular Systems*, edited by F. O. Schmitt, D. M. Schneider, and D. M. Crothers, pp. 17–41, 1975.

the number of (+) conformations for that state (see Table 1). Equation (6) for Q is equivalent to

$$Q = (1 + s)^4 \quad (7)$$

Q is thus the partition function of a system of four independent units, each having a (−) and a (+) state, with relative weights of 1 and s, respectively. Substitution of equation (7) into equation (5) yields

$$\theta = \frac{s}{1 + s} \quad (8)$$

as would be expected for independent subunits.

The transition between (−) and (+) states can be induced by varying the concentration of a ligand that binds only to the (+) state. The variation of the stability constant with the concentration of ligand, C_L, can be found by an elementary, though tedious, calculation, and is given by

$$s = s_0(1 + K_B C_L) \quad (9)$$

where K_B is the binding constant and s_0 is the value of s in the absence of ligand. s increases as C_L increases, encouraging the shift from (−) to (+) conformation. A calculation of θ versus C_L for a noncooperative transition will yield a hyperbolic curve of the form shown in Fig. 1.

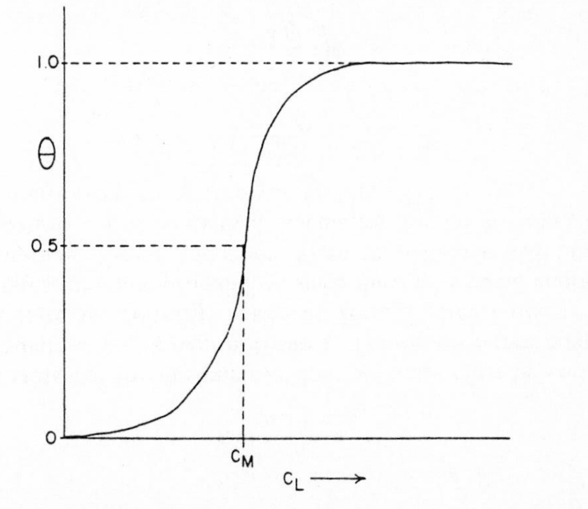

FIG. 1. The fraction of subunits in a four-subunit enzyme θ in the active (+) state as a function of ligand concentration (C_L) when no cooperativity exists between the subunits. C_M is the concentration at which $s = 1$ and θ is 0.5. When C is less than C_M, s is less than 1, and when C is greater than C_M, s is larger than 1 and θ is greater than 0.5.

A Cooperative Transition with No Intermediate States

In a transition with no intermediate states, in which only the (−)(−)(−)(−) and (+)(+)(+)(+) are possible, a shift in one subunit from the (−) to the (+) conformation forces the other three subunits to shift simultaneously to the (+) conformation. In this case

$$Q = 1 + s^4 \quad (10)$$

and

$$\theta = \frac{s^4}{1 + s^4} \quad (11)$$

and a plot of θ versus C_L is sigmoidal (Fig. 2).

Given the most general case of an enzyme having N subunits (Fig. 3) (van Holde, 1971)

$$\theta = \frac{s^N}{1 + s^N} \quad (12)$$

Cooperative Transition with Nucleation Parameters

Most real physical systems show cooperative behavior that is intermediate between the noncooperative case and a transition with no intermediate states. In terms of our simple four-subunit model, a conformational change in one subunit from (−) to (+) makes it easier (energetically more

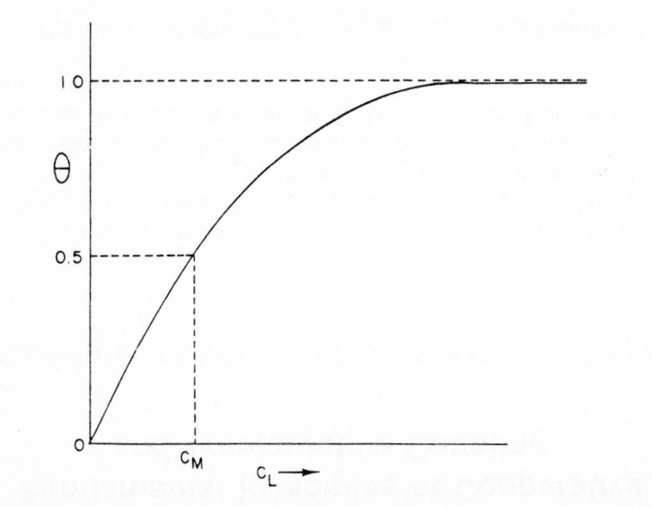

FIG. 2. The fraction of subunits in a four-subunit enzyme (θ) in (+) (+) (+) (+) state as a function of ligand concentration when no intermediate states are possible. As in the noncooperative case (Fig. 1), C_M is the value of C_L at which $\bar{s} = 1$; θ is still 0.5 at C_M.

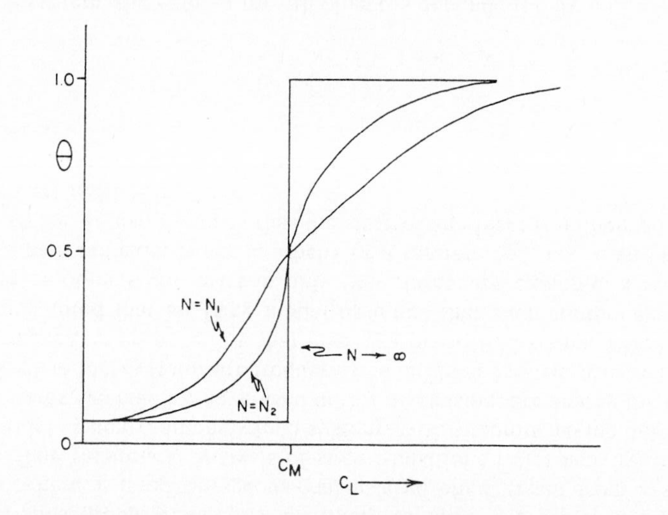

FIG. 3. The fraction of subunits in an N-subunit enzyme (θ) in the (+) (+) (+) (+) state as a function of C_L for a completely cooperative transition in which no intermediate states are possible. The steepness of the sigmoidal curve increases as N increases ($N_2 > N_1 > 1$); a true phase transition, indicated by a discontinuity in the θ versus C_L curve, is approached as $N \rightarrow \infty$.

favorable) for the other subunits to change conformation from (−) to (+). It is thus more difficult to shift the conformation of the first subunit than to shift the others. This can be expressed in mathematical terms in a manner analogous to the previous cases described, making the simplifying assumption that all other subunits are affected in the same way by the first conformational change, that is, that they have the same stability constant. The stability constant for a (−) to (+) conformational change for the first subunit is defined as σs, where s is the stability constant for the (−) to (+) conformational change for the succeeding subunits, and σ is less than unity, since the first conformational change is more difficult.

For this cooperative case, the partition function for the four-subunit example is

$$Q = 1 + 4\sigma s + 6\sigma s^2 + 4\sigma s^3 + \sigma s^4 \qquad (13)$$

Substitution into equation (4) will give an expression for θ in terms of σ and s. Again, plots of θ versus C_L will yield a sigmoidal curve.

A comparison of θ versus C_L for three values of σ are shown in Fig. 4; smaller values of σ yield more cooperative, sharper transitions. In physical terms, the transition becomes more cooperative as the difficulty of making the first conformational change, sometimes called the nucleation step, increases relative to subsequent steps. Once the difficult first conformational

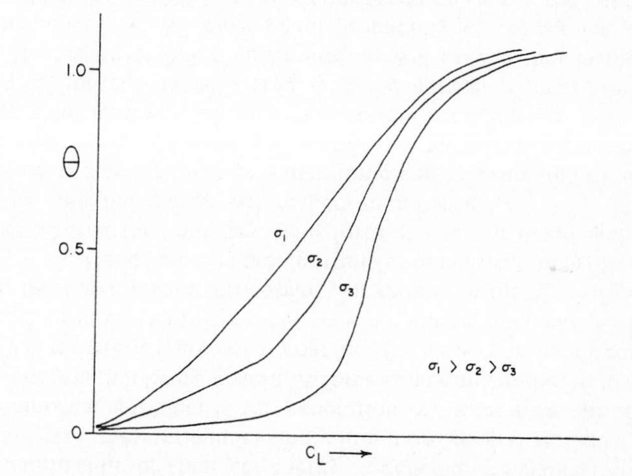

FIG. 4. The fraction of subunits in a four-subunit enzyme in the (+) state (θ) as a function of ligand concentration (C_L) for three values of σ, where σ is the relative degree of difficulty for the conformational change of the first subunit versus that of the other three. $\sigma_1 > \sigma_2 > \sigma_3$.

change is made, subsequent changes can occur with less difficulty given only a small change in experimental conditions, such as ligand concentration, thus giving rise to sharper, more sigmoidal curves.

Helix-coil transitions in nucleic acids and polypeptides can be examined theoretically in a similar manner. The simplest model for formation of a DNA helix by the association of two complementary random coils assumes only one type of base pair (Fig. 5). The most difficult step in forming a helix is "nucleation," through the formation of the first base pair. This step is difficult partly because of the translational entropy loss involved in bringing

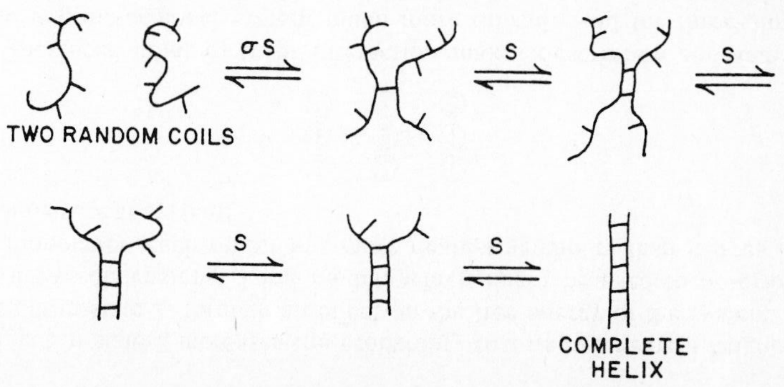

FIG. 5. The formation of a DNA helix having five base pairs.

the two strands together and partly because of the lack of vertical stacking energy with only one base pair. After the first base pair is formed, however, subsequent pairs have different stabilities, and a DNA molecule containing all four nucleotides has a broader, less sharp transition than one in which a single type of base pair predominates. Helix-coil transitions in nucleic acids are usually studied as a function of temperature, and can therefore be related to the stability constant, s, by the Van't Hoff relationship:

$$\frac{d \ln s}{dT} = \frac{\Delta H}{RT^2}$$

The relationship between θ and T is sigmoidal, and variations in the shape of the curves with variation in σ and helix length are similar to the simplified example of a helix with only one type of base pair.

It must be emphasized that sigmoidal curves in themselves do not *necessarily* imply cooperative transitions. The variation of θ with some thermodynamic variables to which s may be functionally related, especially when there is a logarithmic relationship, as with pH, may produce sigmoidal curves even if the transition is totally noncooperative. The sigmoidal shape is in this case an artifact of the method of plotting the data. This means that cooperativity must generally be defined with reference to some model noncooperative process, such as the titration of a monobasic acid in the case of a pH-driven transition.

Cooperation between subunits of macromolecular systems results in sharper transitions; in extreme cases the transitions approach all-or-none characteristics. Such transitions are likely to have evolved in living organisms, where it is advantageous that a large effect result from a small stimulus. The theoretical analysis of such transitions relies heavily on the use of model systems, among which systems with subunits having only two (+ or −) states, such as we have used in this discussion, are among the most useful. Such model systems are frequently called Ising models, after a model of ferromagnetism proposed by that investigator in 1925 (Ising, 1925).

L. Klotz noted that an Ising model assumes that each component of a collection of objects can exist in only two states; for example, a nucleic acid base pair can exist either in a helix or a random coil, not in any intermediate states. It also assumes that the state of an object is influenced only by its nearest neighbors.

Consider a one-dimensional Ising model in which each unit can exist in states 1 or 2:

$$\begin{array}{ccccc} ① & ① & ② & ② & ① \\ n-2 & n-1 & n & n+1 & n+2 \end{array}$$

In such a system, the state of the nth object is dependent only on the states of the $n-1$ and $n+1$ objects. Cooperativity means that the existence of an object in state 2 increases the probability that its two nearest neighbors will exist in state 2, through an effect on the free energy of the system.

In a two-dimensional Ising model, which might be used to describe a lipid monolayer, each object has more nearest neighbors than it does in a one-dimensional system:

$$\begin{array}{ccc} ② & ② & ① \\ ① & ② & ① \\ ① & ① & ② \end{array}$$

This increased complexity of interaction makes description and analysis of the two-dimensional system much more difficult, and the three-dimensional model is so complex that the equations have never been solved.

SOME BASIC PRINCIPLES OF TRANSDUCTIVE COUPLING:

Donald M. Crothers

In a transductive process, there is a transformation or conversion of energy between one mode and another, as in the transduction of a light impulse to an electrical current by a multiplier tube. In biological systems, the general modality of energy is chemical, and the problem is to understand how the energy of a chemical reaction is transduced to do various kinds of work. One must also deal with the reverse process in which other forms of energy are converted into chemical form, as in the transduction of light energy to a nerve impulse, possibly via intermediate processes that utilize chemical energy. For simplification, this section will deal primarily with the utilization of chemical energy to generate mechanical or osmotic work; however, these principles apply in a more general sense.

One important problem is the generation of forces from the molecular motions present in a living system. A force can be generated from molecular motion when there is a greater probability for motion in one direction than another; the simplest example is the pressure exerted by a gas. As shown in Fig. 6*a*, there is a greater probability for motion in the $P_1 \rightarrow P_2$ direction when $P_1 > P_2$. Hence there is greater transfer of momentum in this direction and a net force at the boundary. This force can be calculated as the derivative of the potential energy with respect to distance.

Diffusion is a second way in which force can be generated by molecular motion when there is a greater probability of molecular movement in one direction than another. Given a concentration gradient of a solute as a function of linear distance x (Fig. 6*b*), the number of molecules moving toward the region of lower concentration will exceed that in the reverse direction, since they all have equal probability of undergoing Brownian motion. As in the case of two gases at different pressures, the average force can be calculated from the gradient of the chemical potential with respect to

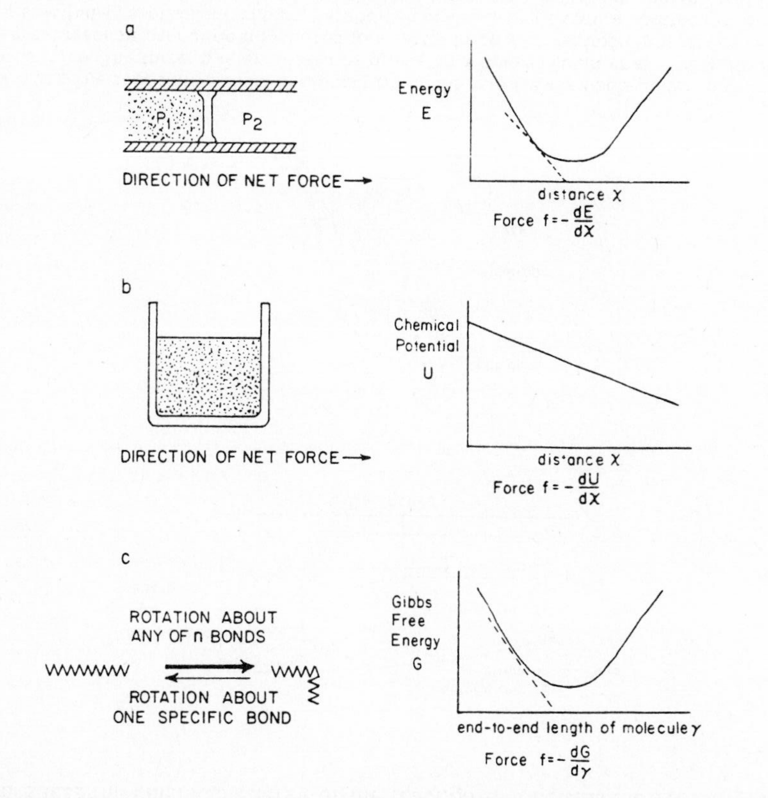

FIG. 6. Force generation as the result of molecular motion. A force will be generated: (*a*) when there is a greater probability for motion in one direction than another, as for two gases at different pressures, separated by an impermeable but moveable barrier; (*b*) as the result of diffusion; and (*c*) due to intramolecular elasticity (see text for discussion).

distance. Balancing this force against the viscous drag in the steady state yields the first law of diffusion. It is important to recognize the conceptual connection between molecular motion and the average force on the molecules, which arises from the greater probability of motion in one direction than another.

Similar forces can be produced by changing the internal structure of a macromolecule, as, for example, when rotation occurs about a carbon-carbon bond (Fig. 6*c*). Extension of the chain will be opposed by a restoring force against extension, since there is again a difference in probability between the two directions of movement. The fully extended chain can achieve a more folded state by rotation about any one of n bonds, while the production of a fully extended molecule requires rotation about a particular bond. There is therefore greater probability of motion to form the more

folded structure, and consequently a restoring force to assume this state results. As in the previous cases, the free energy can be plotted as a function of the end-to-end distance γ, and the force will be the gradient of free energy with respect to γ. The Gibbs free energy is enthalpy minus the temperature times entropy term ($G = H - TS$) and need not have any energy component: it can be comprised solely of entropy or probability. This means that a restoring force is present even if the two states of the molecular have identical energies, in which case the force derives solely from the greater probability of motion of one kind.

Molecular elasticity can also be coupled to a well-defined conformational change. This is illustrated in Fig. 7*a* by a helix-coil transformation. The free energy is a function of the end-to-end length of the molecule, and the force is therefore the gradient of free energy with respect to length. If the reaction conditions are adjusted so that the molecule is 50% helix in the absence of any applied force, the equilibrium can be displaced in either direction by applying a compressive or extensive force; the length of the macromolecule can thus be coupled to an applied force.

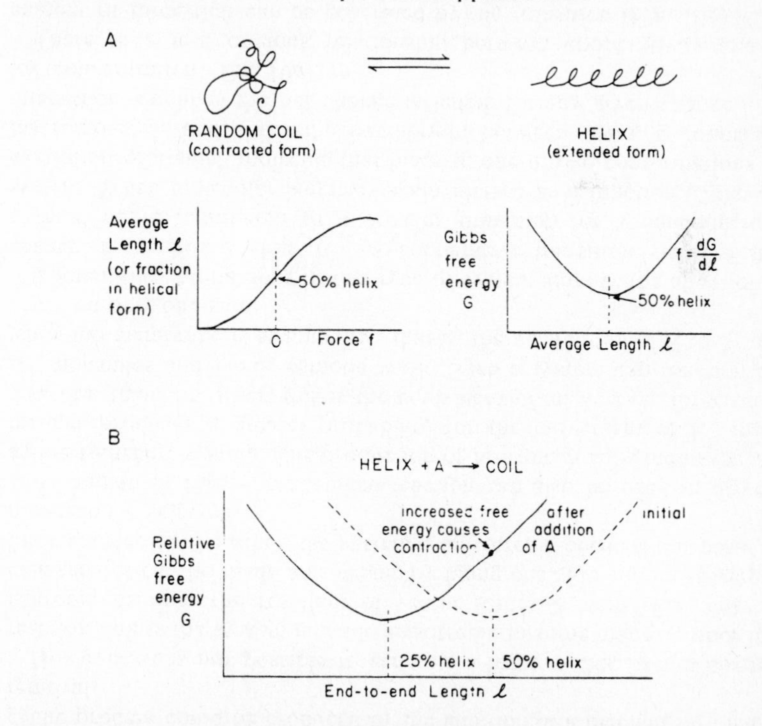

FIG. 7. (*a*) Elasticity coupled to a helix-coil transformation. (*b*) A chemical reaction can alter this helix-coil equilibrium. Addition of a small molecule generates a force (makes the probability of $h \rightarrow c$ greater than $c \rightarrow h$.)

The equilibrium between two forms can be altered by chemical reactions that produce a greater probability for movement in one direction. This is illustrated in Fig. 7*b* for a small molecule (A) that binds more readily to the coil than to the helix, and will therefore shift the equilibrium toward the coiled form under any specific reaction condition. The instantaneous effect of adding A is to increase the free energy relative to the value at equilibrium, thus producing a force (dG/dl) that will cause the molecule to contract. This force results from the greater probability for the helix to coil conversion than that for coil to helix.

An excellent illustrative example of force generation by a chemical reaction is found in the mechanochemistry of collagen fibers. The addition of LiBr to the fibers shifts the helix-coil equilibrium toward the coiled, contracted form, decreasing the fiber length and increasing tension; the binding of LiBr to the fiber thus does mechanical work (Rubin, Piez, and Katchalsky, 1969; Fig. 8*a*). In this example, the binding of LiBr to the collagen fibers transduces the chemical energy of the reaction to a contractile process. A

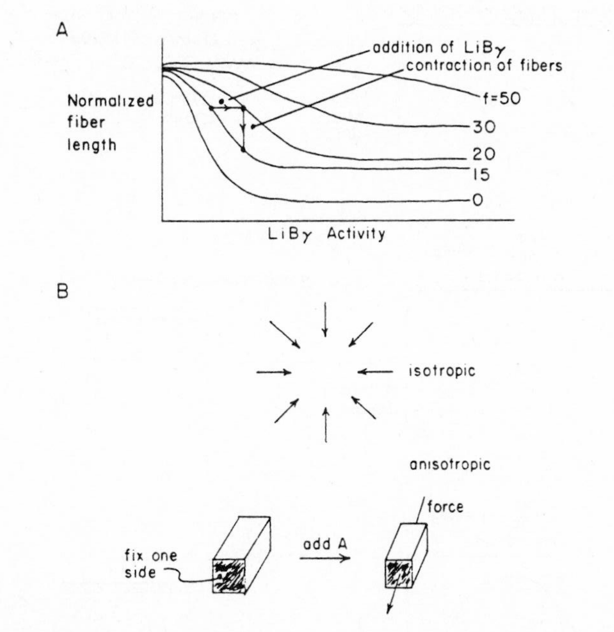

FIG. 8. (*a*). The cooperative effect of LiBr activity on the shrinkage of fibers under constant tensile force. The force, *f*, is expressed as grams on a pulley (Rubin et al., 1969). (*b*) A chemical reaction cannot be transduced to a vector force in an isotropic phase. For example, within a gel of collagen fibers, the addition of A (LiBr) will produce contraction. But the spherical symmetry within the isotropic gel will result in a net directional force of 0. Only when the medium is nonisotropic can a directional force be produced as in (*a*).

cyclic process could be produced by the addition of a mechanism for LiBr removal.

However, it is not possible to generate a force vector from a chemical reaction unless the system is made anisotropic in some manner, since in an isotropic medium the resultant of vector forces is zero (Fig. 8*b*). The collagen gel can be made anisotropic by fixing one side and measuring the force exerted on the other; the system is anisotropic because one particular dimension is selected.

As shown in Fig. 9, a chemical reaction can also be used to drive an active transport system. The production of X′ from X in some unspecified manner produces a greater probability for the movement of K^+ in one direction than the other; this is the equivalent of an average force on the K^+ molecules and yields osmotic work. Such a system can function only when the membrane is asymmetric; that is, the conversion of X to X′ must occur on only one side.

Cooperative interactions in this type of system amplify the effect of any change in conditions (Fig. 10). A cooperative transition yields a steep sigmoid curve, compared to a broader transition for a noncooperative system. When molecular length is again plotted as a function of the concentration of a small molecule that binds to one of two conformations of a macromolecule and shifts the equilibrium between them, a cooperative interaction amplifies the net change in length for any given change in the log (concentration) (Fig. 10*a*).

There is a less obvious relationship between cooperativity and the amount of force that can be generated by the transition (Fig. 10*b*). If the free energy is plotted as a function of the length of the molecule, which varies between the lengths of the coiled and helical forms, the free energy in a discontinuous phase transition will be the same for the two phases when they are at equilibrium. Similarly, at any given value of length displacement, the slope of free energy versus length will be greater for a noncooperative than a cooperative transition; that is, the cooperativity

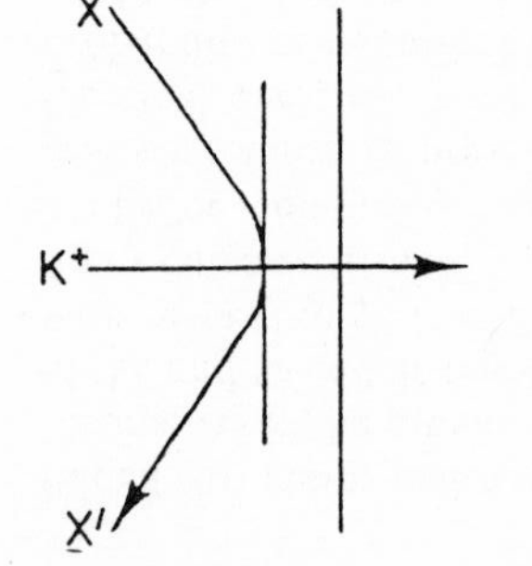

FIG. 9. The use of a chemical reaction, the conversion of X to X′, to drive the transport of K^+.

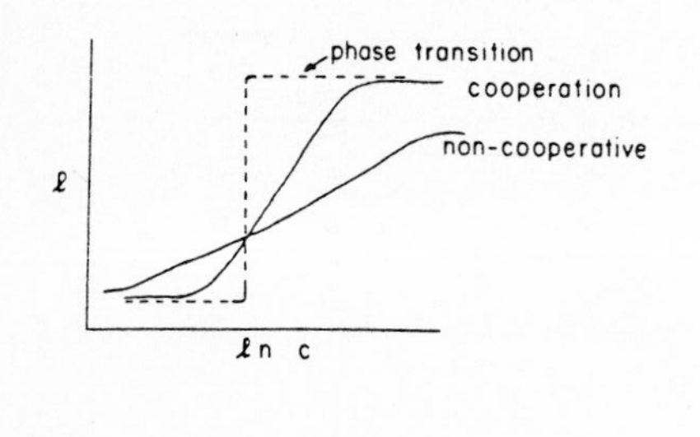

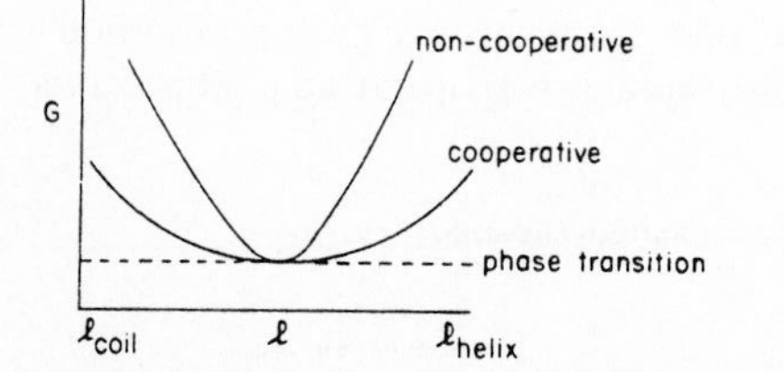

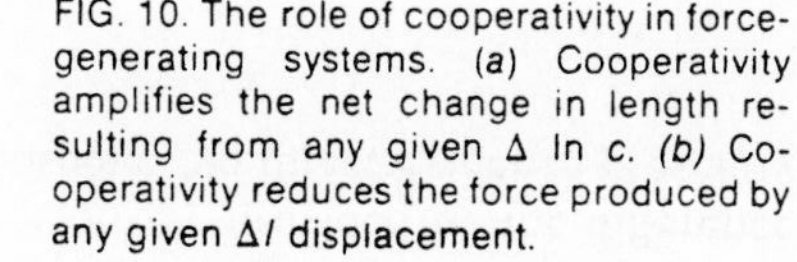

FIG. 10. The role of cooperativity in force-generating systems. (*a*) Cooperativity amplifies the net change in length resulting from any given Δ ln *c*. (*b*) Cooperativity reduces the force produced by any given Δ*l* displacement.

reduces the force generated for a given Δ*l* displacement. The noncooperative system is thus a superior force generator.

Another way of stating the same conclusion is that the elasticity of a cooperative system is greater than that of a noncooperative one. Elasticity arising from the order-disorder lipid phase transition could be important for the mechanical stability of cell membranes, since it would allow the membrane to respond to sudden osmotic changes by expanding or contracting.

A cooperative reaction also results in an increased available free energy for any given concentration perturbation Δ ln *c* (Fig. 11). When the fraction of molecules in one state (θ) is determined as a function of concentration, the available free energy produced by a change in concentration from C_1 to C_2 is calculated by

$$\Delta G = \frac{RT(\Delta \ln c)(\Delta \theta)}{2}$$

which is greater at a given Δ ln *c* for a cooperative than a noncooperative transition. Thus, the available free energy for any given concentration displacement is larger for a cooperative transition than for a noncooperative one.

Cooperativity also has a significant effect on the kinetics of a molecular transition, generally producing a net decrease in relaxation rates. In a noncooperative system, each unit is uncoupled from its neighbors and independently relaxes at its elementary rate. However, in a cooperative

FUNDAMENTAL PRINCIPLES OF COUPLING

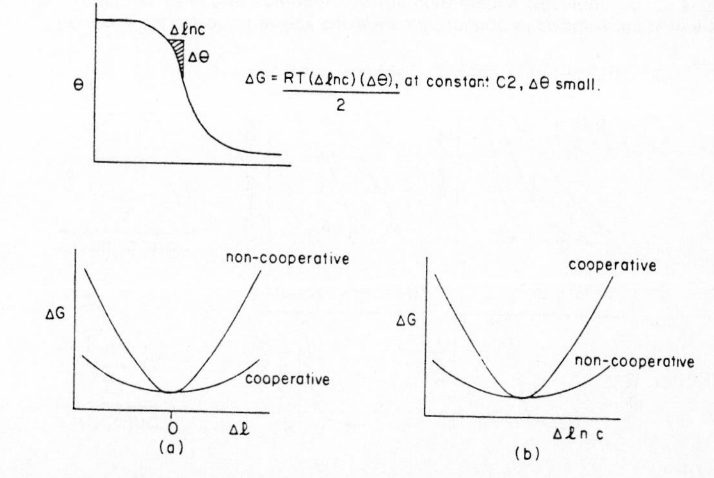

FIG. 11. A comparison of the amount of free energy available for a cooperative and noncooperative transition. Cooperativity reduces the force produced for a given Δ*l* (*a*) and increases the available free energy produced by any given Δ ln *c* (*b*).

system, reactions are more probable at boundary regions, and a sequence of steps must occur before a system has sampled all available configurations.

These principles may be illustrated by a simple model of messenger RNA (mRNA) translocation based on the mechanics of unfolding of transfer RNA (tRNA), whose structure has been determined crystallographically (Kim, Quigley, Suddath, McPherson, Sneden, Kim, Weinzierl, and Rich, 1973). A combination of relaxation kinetics and magnetic resonance methods has been used to document the sequence of steps involved in the unfolding of the macromolecule in response to increased temperature (Crothers, Cole, Hilbers, and Shulman, 1974). The initial step is transient unfolding of the dihydrouridine (DHU) helix at the junction between the two parts of the "L" structure (Fig. 12). This is followed by a full melting of the DHU helix and a tertiary interaction between the loop regions of the molecule, which are not involved in double-helix formation.

This highly specific series of molecular transformations may possibly be related to the generation of the force required to drive the movement of mRNA relative to the ribosome during the translation process, as illustrated in Fig. 13. One hypothesis is the following: First the peptide chain is transferred to the second (2) tRNA and the spent tRNA (1) is removed. The amino acid acceptor region of the L structure then moves simultaneously to the P site of the ribosome, with disruption of the DHU helix and tertiary structure. This could be thermodynamically coupled to removal of a bound Mg^{2+}, thus destablizing the helix. The ordered structure could then re-form and contract, producing movement of the mRNA and the

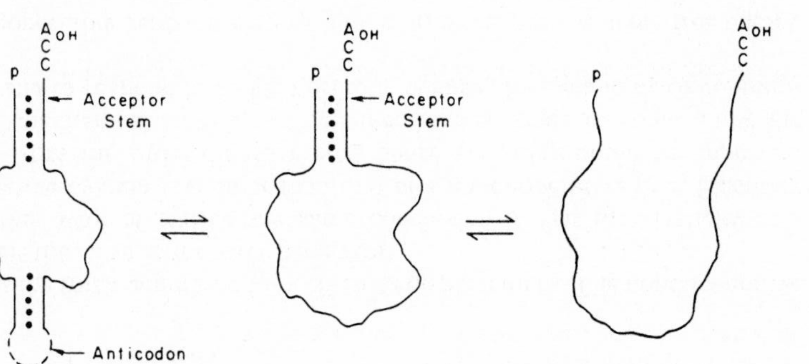

FIG. 12. Thermally induced unfolding of *E. coli* transfer RNA.

anticodon. The result is a simple transduction mechanism for driving the motion of an mRNA chain by the cyclic opening and closing of part of a macromolecule, with the necessary anisotropy provided by the difference between the A and P sites on the ribosome. No direct evidence presently

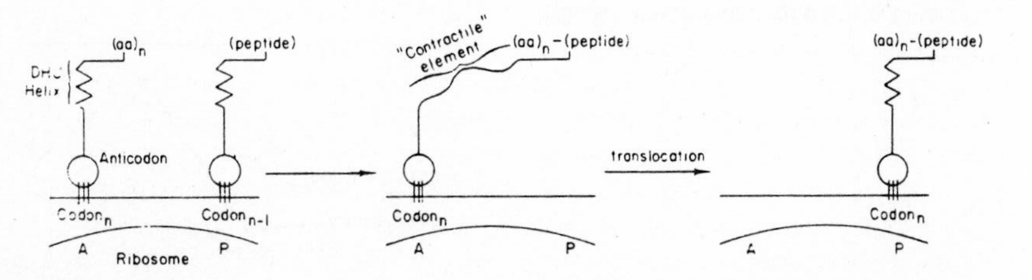

FIG. 13. A postulated mechanism for utilizing the opening and closing of the tRNA molecule in translation.

supports such an hypothesis; it is presented only to illustrate how cyclic conformational changes can be used to generate a force.

SHARPNESS AND KINETICS OF COOPERATIVE TRANSITIONS:

Gerhard Schwarz

When dealing with cooperativity in biological systems, one is generally most interested in the effect of a change in an external parameter such as ligand concentration on the equilibrium constant for a given reaction. Although the sharp transition from one state of high stability to another such state that is characteristic of a cooperative process can also be achieved by noncooperative means, much larger interaction energies would be required and the transition would therefore occur much more slowly. Thus, the sharp and fast transitions that are required in biological systems require the use of cooperative interactions, such as the individually weak forces involved in a series of hydrogen bonds or hydrophobic interactions. The essential features of cooperative transitions in biological systems are thus high stability below a certain threshold, but a sharp and rapid transition above it.

These cooperative transitions can be used for control and regulation in biological systems in the same way that electronic devices are used for these same purposes. The close similarity between the characteristics of electronic tubes and those of cooperative transitions is illustrated in Fig. 14.

Quantitative calculations involving cooperative transitions are usually

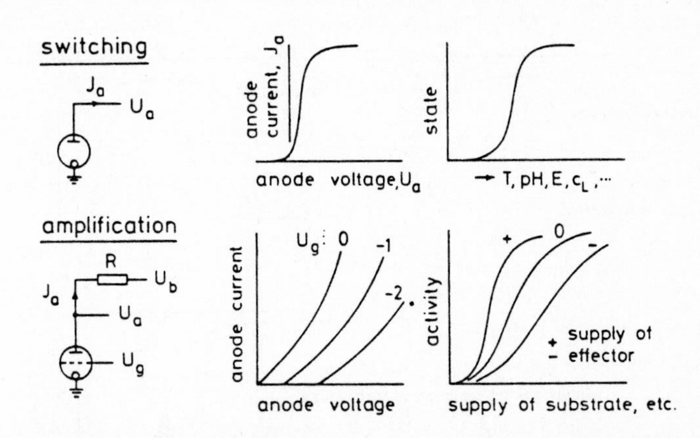

FIG. 14. The sharp transition of anode current as a function of anode voltage in a simple electronic switching device is analogous to the cooperative transition of the state of a biologically significant macromolecular structure as a function of external parameters. A close similarity to a simple electronic amplification device can also be demonstrated when grid voltage is compared to effector concentration (shift of threshold!).

based on the theoretical Ising model, in which cooperative interactions occur only between elementary reaction sites that are nearest neighbors, as discussed in the preceding section. Each of these sites is assumed to be able to exist in one of two states, for example, A or B. A plot of θ (the overall degree of transition) versus the stability constant s (the equilibrium constant for $A \rightleftharpoons B$ when half of the nearest neighbors are in state A, half in state B) always has the sigmoid shape characteristic of a cooperative reaction if the cooperative transition occurs in a system in which each site is associated with only two neighbors, as in a linear chain (Engel and Schwarz, 1970; Schwarz, 1970). The stimulus required to change this simplified system from one state to the other may be described by the necessary change of ln s. This critical stimulus is proportional to the square root of the general cooperativity parameter σ (see Fig. 18); the smaller σ, the smaller the change necessary to shift the state of the system (after the threshold has been reached). However, a true phase transition, having an infinite slope of θ versus s, can be achieved only when more than two neighbors are involved, as in the at-least-two-dimensional system indicated in Fig. 15, or when the interactions are of infinite magnitude or infinite range.

In view of the electrical excitability of nerve membranes, it is of particular interest that the application of an electric field strength E to a cooperative system alters the equilibrium constant according to the van't Hoff relationship:

$$\frac{\partial \ln s}{\partial E} = \frac{\Delta M}{RT}$$

where ΔM is the difference in the partial molar electrical moments (parallel to E) of reaction products and reactants (Bergmann, Eigen, and DeMaeyer, 1963; Schwarz, 1967). A change of ln s is therefore produced when an electric field is applied to a biological macromolecule. A much higher electric field than can be reached in biological systems would be required to induce a transition in a noncooperative system, whereas a much lower

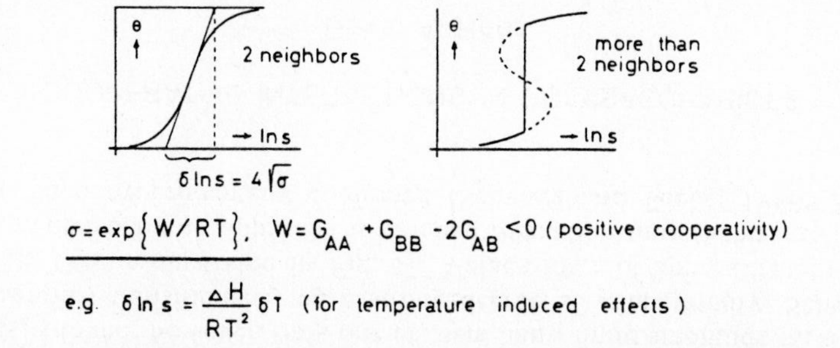

FIG. 15. Cooperative transitions based on the interaction of nearest neighbors (Ising model) with negligible end effects.

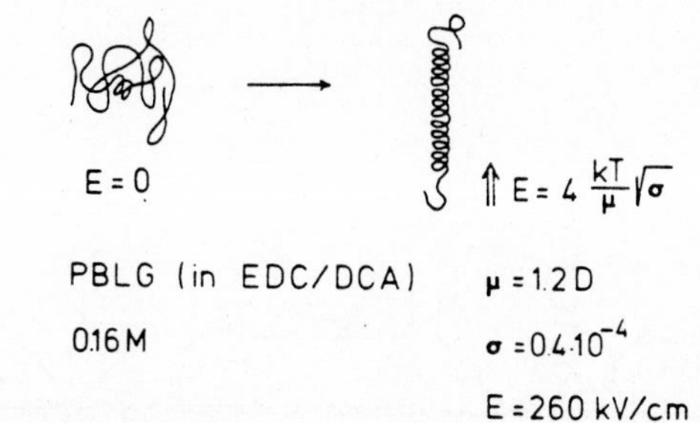

FIG. 16. A conformational helix-coil transition induced by an electric field in poly (γ-benzyl L-glutamate) (PBLG) in a mixture of ethylene dicholaride (EDC) (EDC)/dichloroacetic acid (DCA) (μ is the dipole moment per helical subunit, i.e., per hydrogen bond; it is parallel to the helix axis).

electric field density is sufficient to induce a transition in a system having a high degree of cooperativity. The well-known case of the helix-coil transition of polypeptides is shown in Fig. 16 (Schwarz and Seelig, 1968).

Although the existence of cooperativity permits rather fast transitions, they must nevertheless always be slower than the elementary processes for the individual sites. For example, the elementary process in an α-helix-coil transition is hydrogen bond formation, which has a reaction time of about 10^{-10} sec (Bergmann et al., 1963). The total reaction time is increased by the cooperative interactions, but to a much lower extent than if large interaction and activation energies were involved.

The kinetics of such conformational transitions of linear polymers can be examined in chemical relaxation experiments (Schwarz and Engel, 1972), which examine the system as it returns to a stable state following the application of a slight perturbation. The relaxation process is described by a number of relaxation times, τ, and respective amplitudes, the so-called relaxation spectrum (Eigen and DeMaeyer, 1963; Schwarz, 1968). Using the linear Ising model, without end effects, a spectrum of four relaxation

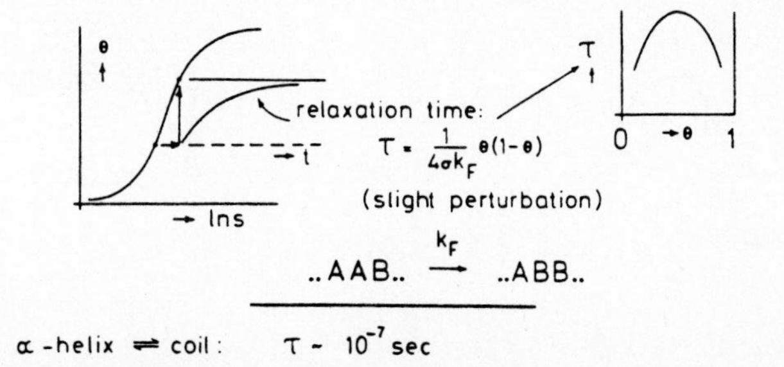

FIG. 17. Chemical relaxation of a conformational change in a one-dimensional system with strong cooperative interactions.

times is found; however, only one of them has a finite amplitude for a conformational transition having a strong degree of cooperativity (Schwarz, 1972). This is illustrated in Fig. 17. Values for τ of approximately 10^{-7} sec predicted by an approach to a mean relaxation time (Schwarz, 1965) have been experimentally confirmed (Schwarz and Seelig, 1968; Zana, 1972).

COOPERATIVE INTERACTIONS IN POLYSACCHARIDES:

David A. Rees

Long-chain polysaccharides of plant origin have been used as model systems to determine the types of interactions that might occur in the carbohydrate-containing molecules of vertebrates, such as those in the intercellular space. These plant polysaccharides form gels that contain as much as 99% water, whose structure results from the formation of noncovalent cross-links (Fig. 18*a;* Rees, 1969, 1972). The formation of these cross-links is a cooperative phenomenon similar to that shown by the nucleic acids, and a stable structural entity results from the accumulation of many hydrogen bonds and other weak forces (Fig. 18*b*). Sol-gel transitions show sigmoidal kinetics, and involve the same type of nucleation and propagation steps as those described by Zimm for nucleic acid transitions. Discontinuities in the repeating sequence of sugar moieties produce helix terminations, which make possible cross-linking between strands and the formation of a stable gel network (Rees, Steele, and Williamson, 1969; Fig. 18*a*).

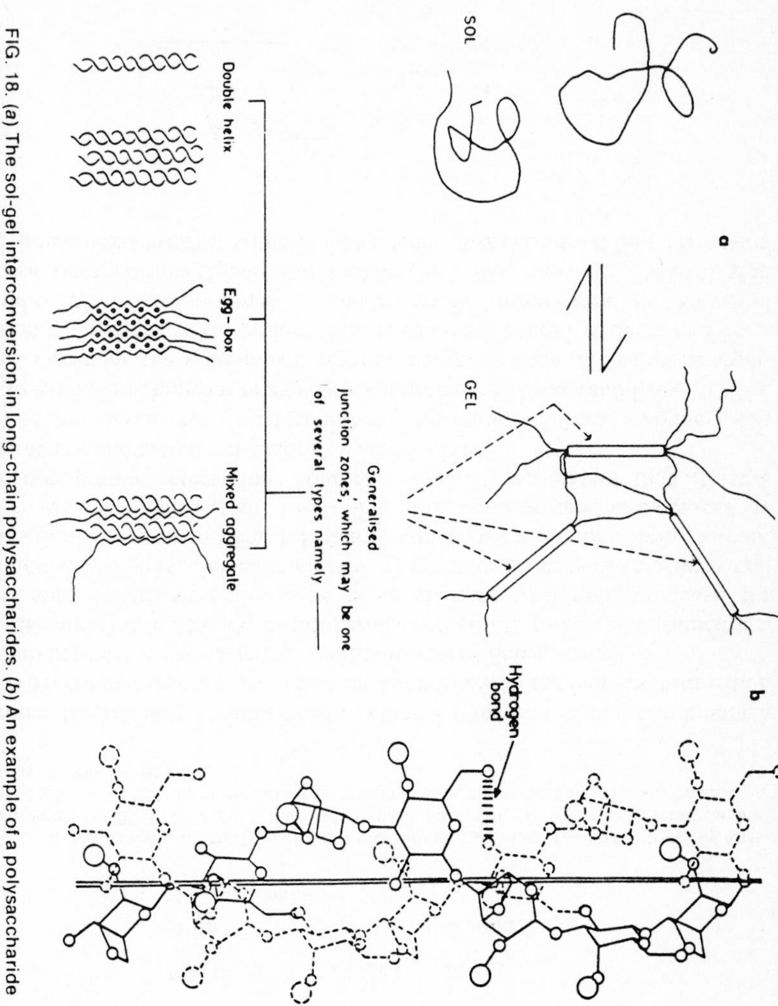

FIG. 18. (*a*) The sol-gel interconversion in long-chain polysaccharides. (*b*) An example of a polysaccharide double helix (for ι-carrageenan), stabilized by hydrogen bond formation. One strand is drawn in full lines, the other in broken lines. Large circles represent sulphate ester groups. The hydrogen bond indicated in the figure is repeated by helical symmetry, engaging all the sugar hydroxy groups to form a network of interstrand hydrogen bonding.

The formation of ordered polysaccharide structures is strongly dependent on the nature and sequence of the sugar residues, with the conformation determined by the bond torsion angles between the sugar rings. The two most important classes of polysaccharide conformation are the ribbon and helix (Rees and Scott, 1971; Rees, 1973), and these may interact to form ribbon sheets or multiple helices (Fig. 18*a*).

For example, the carrageenans, a class of polysaccharides that has a repetitive -A-B-A-B- structure, form stable double helices in solution (Rees et al., 1969; McKinnon, Rees, and Williamson, 1969; Bryce, McKinnon, Morris, Rees, and Thom, 1974). As in the nucleic acids, the helix-coil transition is a cooperative process. A sigmoidal relationship has been demonstrated for optical rotation as a function of temperature, indicating that melting is a cooperative process (Fig. 19; McKinnon et al., 1969), and the available evidence strongly suggests that the transition from coil to helix occurs by a switch mechanism, rather than involving a series of intermediate states (Bryce et al., 1974; Reid, Bryce, Clark, and Rees, 1974).

The alginates, whose sequences of acidic sugar residues are of the types

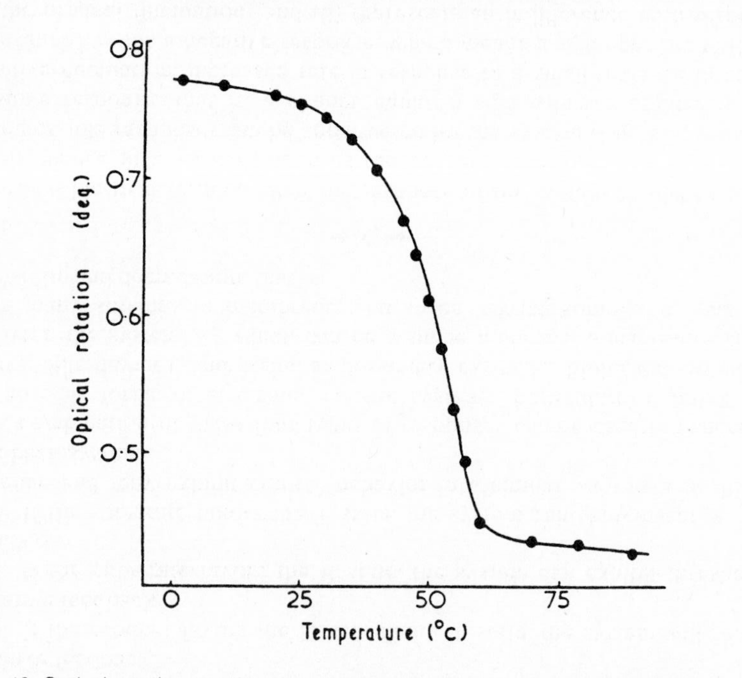

FIG. 19. Optical rotation as a function of temperature for ι-carrageenan segments.

-A-A-A, -B-B-B-, and -A-B-A-B-, undergo a Ca^{2+}-induced association into Ca^{2+}-containing "egg-box" structures (Kohn and Larsen, 1972; Figs. 18*a* and 20). Extensive evidence indicates that the formation of these structures is a cooperative process. A sharp transition to the egg-box structure is obtained only when the appropriate sequence within the polysaccharide chain is at least 16 to 20 units in length (J. Boyd, E. R. Morris, D. A. Rees, and D. Thom, *personal communication*). This minimal length requirement is comparable to that found for the nucleic acids, in which a minimum of 30 nucleotide units must be present for a cooperative transition to a stable

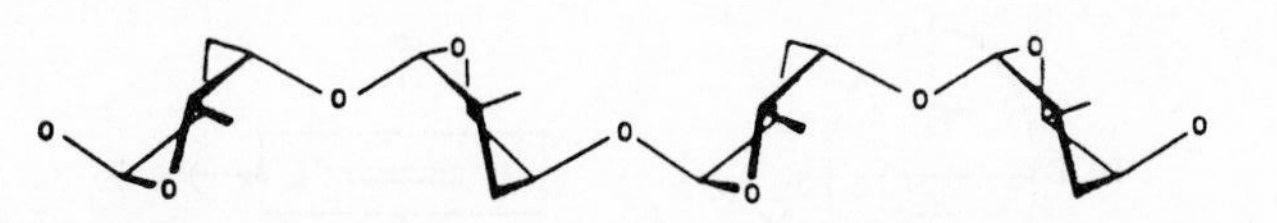

FIG. 20. Conformation of the poly (L-guluronate) sequences in alginates as they are presumed to exist in the proposed egg-box structure (Grant, Morris, Rees, Smith, and Thom, 1973). These buckled chains pack with Ca^{2+} ions in the manner shown schematically in Fig. 18*a*. Substituents on the sugar rings are not shown, except for the bond to the equatorial carboxylate (note that this projects toward the cavity that is occupied by Ca^{2+}). Ten sugar oxygen atoms are closely packed around Ca^{2+} to form a favorable coordination shell.

helix to occur. A sigmoid curve is observed for mixtures of Ca^{2+} and Na^+ when the binding of Ca^{2+} is determined as a function of the Ca^{2+}/Na^+ ratio (Kohn and Furda, 1968). A distinct transition point is also observed when bound Ca^{2+} is determined as a function of ester groups removed from pectin (Kohn and Furda, 1968).

Cooperative interactions are a general characteristic of any macromolecule that contains many similar repeating units. The principles of cooperativity apply to proteins, nucleic acids, and polysaccharides. This characteristic is thus a general biological principle.

CONSTRUCTION OF CONTROL AND SELF-ORGANIZING SYSTEMS:

Manfred Eigen

An extremely sensitive response system is produced when its components interact cooperatively and the system is maintained in the region of greatest sensitivity. An excellent example of such a biological system is the hemoglobin molecule.

Given that:

$$\frac{\partial \ln K}{\partial T} = \frac{\Delta H}{RT^2}$$

a steep response to a small change in temperature can be obtained if the enthalpy, ΔH, is large. Although this can occur in the absence of any cooperativity, as also noted by Schwarz, the appropriate entropy change, ΔS – defining a transition temperature T_t by $T_t \Delta S = \Delta H$ – cannot be achieved at room temperature unless the number of degrees of freedom in the system is severely restricted. This restriction can be achieved most readily in a cooperatively interacting system of repeating components, such as a nucleic acid. This type of repetitive system can undergo hysteresis, as can any structure containing large interacting domains. The advantages of hysteresis depend on evolutionary selection. For example, a nucleic acid chain 100 nucleotides in length may have a cooperativity factor σ of 10^4 and a reaction enthalpy per nucleotide of 10 kcal/mole. A helix-coil transition can be triggered in such a system by a $1\% \, \delta T$.

A conformational change in a protein subunit involves the cooperative interaction of many amino acids. A change in one subunit will force a change in others if there are many structural links between them. Such a system, involving steep transformations, can be used as control "switches" analogous to those in electronics.

The anode-cathode voltage across a vacuum tube (triode) or a corresponding transistor is analogous to a substrate-product reaction whose rate is controlled by the equilibrium between the inactive and active forms of an enzyme (Fig. 21). In the same way that a small change in grid current

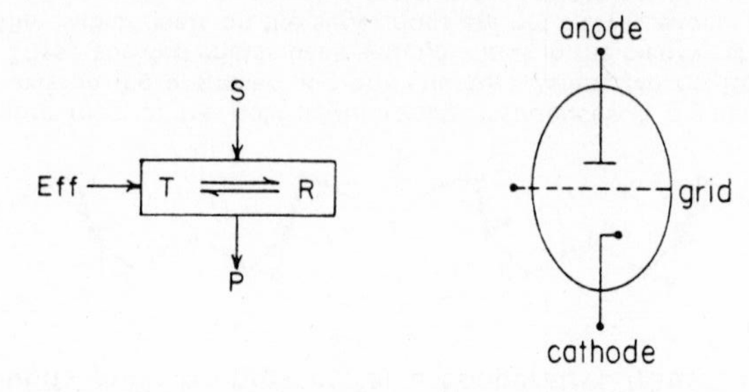

FIG. 21. Analogy between a vacuum tube and a substrate-product reaction whose rate is controlled by an effector (Eff) that alters the equilibrium between inactive (T) and active (R) forms of an enzyme.

produces a large change in the voltage across the tube, a small change in enzymic effector and the $T \rightleftharpoons R$ equilibrium produces a major alteration in the $S \rightarrow P$ conversion.

Such a system can be used either as an all-or-none switch or for amplification and control. This general design can be used as the basis for several types of systems:

1. If the product favors the higher-affinity R state, the system will show positive feedback.
2. If the product favors the lesser-affinity T state, the system will show negative feedback.
3. If the substrate favors the R state, the system can exhibit threshold behavior.
4. If the substrate favors the T state, the system can show negative resistance and may exhibit control behavior in a similar way as a positive feedback system.

A combination of these four types of responses can be used to generate all possible forms of electronic control systems, particularly if linked to irreversible devices, and a similar possibility exists for biological systems.

Given the system X_i, which can be a single molecule, a macromolecule with many subunits, a membrane, and so on, having some given rate of formation and degradation, that is:

$$\Rightarrow X_i \Rightarrow$$

it is possible to determine what the response of the system X_i will be to a small change in a variable such as the concentration of X_i. A matrix of nine possible responses can be constructed for the system (Fig. 22): (+) is a positive response, that is, a change equal in sign with the original (e.g., positive fluctuation, increased rate in response to a small increase in concentration); (−) is a negative response, which means a sign opposite to that of the original fluctuation; and (0) represents an indifference with respect to any change. Only two of these are potential control systems that could

FUNDAMENTAL PRINCIPLES OF COUPLING

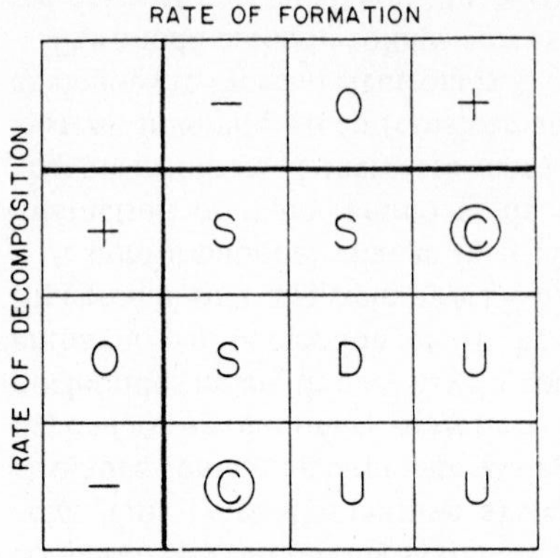

	−	0	+
+	S	S	©
0	S	D	U
−	©	U	U

FIG. 22. Possible response of a fluctuation of the concentration of X_i in the rate of formation or degradation. (+) is a response having the same sign as the fluctuation (e.g., increased rate in response to a small increase of concentration), (−) is a response with opposite sign (e.g., a decreased rate for an increase in concentration), and (0) represents a lack of change. S = stable, because after any change the system ultimately returns to the (reference) steady state; D = exhibits drift behavior if any noise is present; U = unstable; C = potential control system.

convert a cooperative structural behavior into a cooperative temporal one, as in an allosteric enzyme (Eigen, 1973). Either positive or negative feedback could be incorporated into such a system. Systems that utilize this type of control may have been selected for during the process of evolution. In biological systems, conversions of structural to temporal events are frequently mediated by conformational changes within allosteric proteins.

In many membrane-bound enzymes, conformational changes release Ca^{2+}, which in turn triggers or controls the subsequent reaction process. Every biological example of a functionally linked system discussed in subsequent chapters utilizes Ca^{2+} in some way. This almost universal use of Ca^{2+} leads to a consideration of what properties are responsible for its unique position.

The preference for Ca^{2+} over other divalent cations such as Mg^{2+} may be simply explained by its coordination chemistry. Although the stability constants for these ions do not differ significantly, their substitution rates are markedly different; the rates of Mg^{2-} and Ca^{2+} substitutions in water are 10^5/sec and 10^8/sec, respectively. A stability constant of 10^5 to 10^7 has been calculated as ideal for a trigger system operating within the physiological concentration range. If the "on" rate of Mg^{2+} is 10^5, its "off rate" is 1/sec, yielding a stability constant of 10^5 and a fast exchange. Mg^{2+} would be suitable only for a competitive system, where it is frequently found (Eigen and Hammes, 1963).

REFERENCES

Bergmann, K., Eigen, M., and DeMaeyer, L. (1963): Dielektrische Relaxation als Folge chemischer Relaxation. *Ber. Bunsenges. physik. Chem.*, 67:819–826.

Bryce, R. A., McKinnon, A. A., Morris, E. R., Rees, D. A., and Thom, D. (1974): Chain conformations in the sol-gel transitions for polysaccharide systems, and their characterization by spectroscopic methods. *Disc. Faraday Soc.* (*in press*).

Crothers, D. M., Cole, P. E., Hilbers, C. W., and Shulman, R. G. (1974): The molecular mechanism of thermal unfolding of *E. coli* $tRNA^{fMet}$. *J. Mol. Biol.*, 87:63–89.

Eigen, M. (1973): In: *The Physicist's Conception of Nature*, edited by J. Mehra, p. 594. Boston: Reidel Publishing Co.
Eigen, M., and DeMaeyer, L. (1963): Relaxation methods. In: *Technique of Organic Chemistry. Volume VIII. Investigation of Rates and Mechanisms of Reactions*, edited by S. L. Friess, E. S. Lewis, and A. Weissberger, pp. 895–1054. New York: Interscience Publishers.
Eigen, M., and Hammes, G. (1963): Elementary steps in enzyme reactions (as studied by relaxation spectrometry). *Adv. Enzymol.*, 25:1–38.
Engel, J., and Schwarz, G. (1970): Cooperative conformational transitions of linear biopolymers. *Angew. Chem. Int. Ed.*, 9:389–400.
Grant, G. T., Morris, E. R., Rees, D. A., Smith, P. J. C., and Thom, D. (1973): Biological interactions between polysaccharides and divalent cations: The egg-box model. *FEBS Letters*, 32:195–198.
Ising, E. (1925): Beitrag zur Theorie des Ferromagnetismus. *Z. Physik*, 31:253–258.
Kim, S. H., Quigley, G. J., Suddath, F. L., McPherson, A., Sneden, D., Kim, J. J., Weinzierl, J., and Rich, A. (1973): Three dimensional structure of yeast phenylalanine transfer RNA: Folding of the polynucleotide chain. *Science*, 179:285–288.
Kohn, R., and Furda, I. (1968): Binding of calcium ions to acetyl derivative of pectin. *Collect. Czech. Chem. Commun.*, 33:2217–2225.
Kohn, R., and Larsen, B. (1972): Preparation of water-soluble polyuronic acids and their calcium salts, and the determination of calcium ion activity in relation to the degree of polymerization. *Acta Chem. Scand.*, 26:2455–2468.
McKinnon, A. A., Rees, D. A., and Williamson, F. B. (1969): Coil to double helix transition for a polysaccharide. *Chem. Commun.*, 12:701–702.
Rees, D. A. (1969): Structure, conformation, and mechanism in the formation of polysaccharide gels and networks. *Adv. Carbohydrate Chem. Biochem.*, 24:267–332.
Rees, D. A. (1972): Polysaccharide gels. A molecular view. *Chem. and Ind.*, pp. 630–636.
Rees, D. A. (1973): *M. T. P. Internat. Rev. Sci., Org. Chem.*, 7:251.
Rees, D. A., and Scott, W. E. (1971): Polysaccharide conformation. Part VI. Computer model-building for linear and branched pyranoglycans. Correlations with biological function.

Preliminary assessment of inter-residue forces in aqueous solution. Further interpretation of optical rotation in terms of chain conformation. *J. Chem. Soc. B*, pp. 469–479.
Rees, D. A., Steele, I. W., and Williamson, F. B. (1969): Conformational analysis of polysaccharides. III. The relation between stereochemistry and properties of some natural polysaccharide sulfates. *J. Polymer Sci. C*, 28:261–276.
Reid, D. S., Bryce, T. A., Clark, A. H., and Rees, D. A. (1974): The helix-coil transition in gelling polysaccharides. *Disc. Faraday Soc. (in preparation)*.
Rubin, M. M., Piez, K. A., and Katchalsky, A. (1969): Equilibrium mechanochemistry of collagen fibers. *Biochemistry*, 8:3628–3637.
Schwarz, G. (1965): On the kinetics of the helix-coil transition of polypeptides in solution. *J. Mol. Biol.*, 11:64–77.
Schwarz, G. (1967): On dielectric relaxation due to chemical rate processes. *J. Phys. Chem.*, 71:4021–4030.
Schwarz, G. (1968): Kinetic analysis by chemical relaxation methods. *Rev. Mod. Phys.*, 40:206–218.
Schwarz, G. (1970): Cooperative binding to linear biopolymers. I. Fundamental static and dynamic Properties. *Eur. J. Biochem.*, 12:442–453.
Schwarz, G. (1972): Chemical relaxation of co-operative conformational transitions of linear biopolymers. *J. Theor. Biol.*, 36:569–580.
Schwarz, G., and Engel, J. (1972): Kinetics of cooperative conformational transitions of linear biopolymers. *Angew. Chem. Int. Ed.*, 11:568–575.
Schwarz, G., and Seelig, J. (1968): Kinetic properties and electric field effect of the helix-coil transition of poly (γ-benzyl L-glutamate) determined from dielectric relaxation measurements. *Biopolymers*, 6:1263–1277.
Van Holde, K. E. (1971): *Physical Biochemistry*, pp. 51–78. Englewood Cliffs, N.J.: Prentice-Hall.
Zana, R. (1972): On the detection of the helix-coil transition of polypeptides by ultrasonic absorption measurements in the megahertz range. Case of poly-L-glutamic acid. *J. Amer. Chem. Soc.*, 94:3646–3647.

DISPLACEMENTS OF BACKBONE VIBRATIONAL MODES OF A-DNA AND B-DNA

K.-C. LU, L. L. VAN ZANDT, AND E. W. PROHOFSKY, *Department of Physics, Purdue University, West Lafayette, Indiana 47907 U.S.A.*

ABSTRACT We display the displacement vectors or eigenvectors of calculations of the A- and B-DNA backbones. These calculations are based on a refinement scheme that simultaneously fit several backbone modes of A-DNA, B-DNA, and A-RNA. We discuss the role of symmetry operations in mode calculations and the relevance of these displacement vectors to the interpretation of linear dichroism measurements performed on the A- and B-DNA helix.

Linear dichroism data on nucleic acids have been interpreted in terms of structural details of the backbone phosphate oxygen displacements during normal mode vibrations (1, 2). The interpretation of the infrared dichroism has led to structural parameters disagreeing with x-ray scattering data (3).

The structural interpretation of the dichroism data has been criticized by Beetz et al (4). Their criticism notes that a PO_4 moiety, when incorporated into a nonsymmetric structure like a nucleic acid backbone, no longer vibrates in patterns displaying the high symmetry characteristic of the free ion or simple compounds. Furthermore, the mechanical coupling of the PO_4 to the rest of the nucleic acid molecule through the covalent bonds that bind the moiety into the backbone communicates vibrational energy to the rest of the polymer. Hence atoms other than those of the PO_4 group may participate in the vibrational motion excited by the IR photons, and the partial charges on these other atoms contribute to the observed dipole moment of the excited line. This additional contribution bears no simple relation to the PO_4 vibrations, either in magnitude or direction, and so serves to confuse the interpretation of the observed dichroism. We present the numerical results of a normal mode vibration calculation on the backbone of A-DNA and B-DNA. The eigenvectors so obtained show that both criticisms are well founded.

Double helical DNA composed of homopolymer chains—that is, all base pairs identical—and of infinite length possesses a screw-axis operator that is a true symmetry operator. The screw-axis operation rotates the helix and shifts the helix along its axis such that the nth unit cell of two bases and the associated backbone atoms assume the previous positions of the $n + 1$ unit. An additional c_2 operation exists that interchanges the backbones but also interchanges the complementary base pairs and is therefore not a true symmetry operation. The screw-axis operator and its higher powers form an Abelian group if periodic boundary conditions are assumed. No other true symmetry operators exist for the simple homopolymer double helix. Therefore, all the irreducible representations are one-dimensional and characterized by simple phase shifts of the displacement from one unit cell to the next. This is shown in detail by Higgs (5).

The absence of higher symmetry means that the actual displacements for each vibrational

Reprinted with permission from *Biophys. J.*, vol. 28, pp. 27–32, Oct. 1979.

mode must be determined by the diagonalization of the secular matrix. These displacements are found as eigenvectors of the diagonalization. No further symmetry-related block diagonalization is possible.

Polar moments are associated with these vibrational modes because the atoms have unbalanced charges; the polar moment interacts with electromagnetic fields giving rise to electric dipole IR absorption. This is usually the dominant term for IR absorption. These polar modes also can be analyzed with the help of group theory and must conform to irreducible basis functions. For these Abelian groups this means that both displacement eigenvectors and polar moment vector have the same phase shift, θ, from one unit cell to the other as this is the only symmetry-determined part of the problem. The vectors within a unit cell of the displacement and the vector within a unit cell of the polar moments need not be colinear and would only be colinear by some accident, as, for example, if all effective charges on the various atoms were equal.

The orientation of the polar moment within a unit cell of double helical DNA must be calculated rather than inferred from group theory. This is because no further symmetry information is available to reduce the matrix describing a single unit cell. The polar moments arise from the vibrational modes and have both the same frequency and θ dependence, but the polar moment is not necessarily a simple function of the vibrational displacements.

In Fig. 1 we show the DNA backbone-ribose model we adopted in this investigation. All the hydrogen atoms are assumed to be rigidly attached to their bonded atoms following common usage. The base atoms were neglected assuming relatively little interaction between the backbone and the bases for the relevant modes. More elaborate calculations (6) have confirmed the validity of this approximation.

We used the atomic coordinates as given by Arnott et al. (7) in evaluating the matrix, D, that transforms the mass-weighted cartesian coordinates into the internal coordinates. F is the force constant matrix in internal coordinates. The product matrix D^TFD was then diagonal-

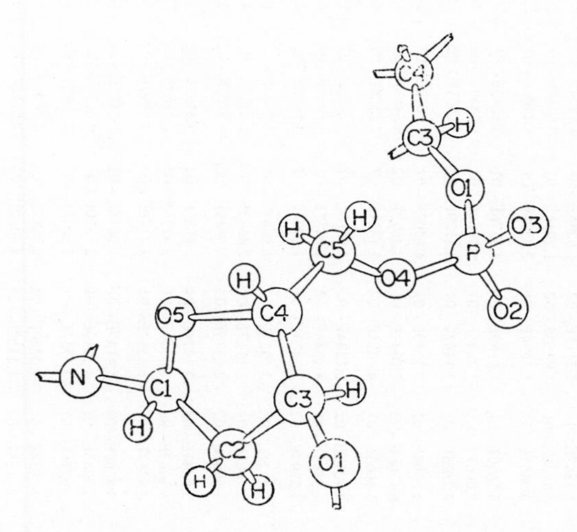

FIGURE 1 Atoms of the DNA backbone chain shown for one full unit and part of a second. Atom N is part of a base unit.

TABLE I

SOME EIGENVECTORS OF A-DNA BACKBONE

ν(Cm^{-1})....	8.0720E+02	8.3195E+02	8.8584E+02	9.7339E+02	9.8939E+02	1.0836E+03	1.1045E+03	1.1227E+03	1.1439E+03	1.1879E+03	1.2099E+03	1.2934E+03
						Eigenvectors						
C1* Δx	1.4450E-02	-1.3569E-02	-4.7680E-02	8.5710E-02	4.6536E-02	2.9552E-02	-2.6547E-02	1.7827E-01	1.0973E-02	1.9759E-02	-2.2492E-03	7.3358E-02
Δy	4.1311E-02	-1.2668E-01	-7.6843E-02	2.3099E-01	-2.9287E-01	-3.8620E-02	1.1512E-01	-5.3830E-01	1.6179E-01	-3.6042E-02	1.4477E-02	-2.1368E-01
Δx	4.7725E-02	2.4406E-01	1.9615E-01	-5.4874E-01	-2.7295E-02	-4.8437E-02	-2.3107E-03	-2.3194E-01	-1.6360E-01	-1.5175E-01	-2.5696E-02	-4.4160E-02
C2	-1.7642E-01	7.2111E-02	1.6288E-01	-4.5538E-02	2.1203E-01	-8.2059E-02	6.9384E-02	-1.9737E-01	5.2046E-02	8.9628E-02	2.4685E-02	-1.2303E-02
	2.8226E-01	1.9750E-02	-1.2214E-01	-7.6247E-02	-3.2003E-01	9.5751E-02	-9.9863E-02	2.2371E-01	-1.5184E-01	-1.1394E-01	-3.4108E-02	4.8621E-02
	2.0951E-02	-1.3494E-01	-2.0509E-01	5.2961E-01	5.9838E-02	4.0264E-02	2.2783E-02	1.4355E-01	1.5397E-01	2.2109E-02	-2.5434E-03	1.9671E-01
C3	2.5650E-01	-1.6160E-01	-3.2576E-01	-2.5245E-01	-2.7193E-01	6.7083E-03	-2.3181E-02	1.6936E-01	1.0816E-01	-1.3864E-01	-1.7903E-02	3.9657E-02
	1.0667E-01	-1.2555E-01	-9.8880E-02	-1.7469E-01	3.6457E-01	-2.3932E-01	1.7478E-01	-2.4614E-01	2.8173E-01	3.7654E-01	1.0659E-01	1.6787E-01
	-4.7128E-02	-3.1282E-01	-5.1229E-02	-3.6055E-02	-1.3467E-02	-6.0409E-02	-7.9411E-02	6.9416E-02	-3.4217E-01	3.6123E-01	7.9349E-02	-6.0039E-01
C4	4.5905E-02	2.7150E-01	4.6207E-01	1.5619E-01	-2.0103E-01	1.7509E-01	-3.4162E-02	-3.7226E-02	3.6409E-02	1.6163E-01	2.0623E-02	1.2694E-01
	-8.1706E-02	1.2041E-01	4.4670E-02	6.6406E-02	4.0217E-01	1.6084E-01	-6.1217E-02	2.6555E-02	4.1104E-02	-5.9149E-01	-1.1830E-01	-2.8743E-01
	-5.4663E-02	-4.4325E-02	-2.1312E-02	2.2528E-01	4.8913E-02	-8.0956E-02	2.1406E-03	-2.9375E-01	-5.0144E-01	-1.7285E-01	-5.6958E-02	4.3812E-01
O5	2.1015E-02	-5.1384E-02	-7.9687E-02	1.2020E-02	-5.1795E-02	-3.4547E-02	1.8566E-02	-1.3282E-01	-5.5635E-02	-6.4391E-02	-9.9555E-03	-3.7885E-04
	-1.2545E-01	-4.4046E-02	1.0451E-01	-1.5819E-01	2.3777E-01	-4.9594E-02	-6.9425E-02	4.0855E-01	-2.1261E-01	2.1730E-01	2.0111E-02	2.9228E-01
	-1.2793E-02	2.7863E-01	1.2794E-01	8.8552E-02	4.2638E-02	1.2553E-01	1.2762E-02	2.1625E-01	4.4686E-01	8.8742E-02	2.8641E-02	-1.7455E-01
C5	-9.2321E-02	-2.7727E-02	-1.3983E-01	3.9164E-02	1.2665E-01	-5.3684E-01	1.8811E-01	1.8131E-01	3.6230E-03	-2.6289E-01	-3.0069E-02	-1.2646E-01
	-2.5001E-01	3.1117E-01	-2.8622E-02	1.9267E-02	-4.2738E-01	-2.4548E-01	8.4741E-02	6.3979E-02	-9.4487E-04	1.2059E-01	5.5256E-02	4.0253E-02
	3.5664E-02	-2.2376E-02	-1.3719E-02	-5.5105E-02	-2.7268E-02	-3.2796E-02	6.7024E-03	9.4395E-02	4.7406E-02	7.2690E-03	-4.2060E-03	-4.9374E-02
P	5.9447E-02	-2.3924E-01	2.6083E-01	9.4132E-03	-5.5399E-02	1.3482E-02	1.4278E-01	5.3909E-02	-1.3137E-02	-1.2381E-01	4.1161E-01	1.3210E-02
	6.5087E-03	2.6939E-01	-2.4165E-01	-1.2582E-03	5.2995E-02	4.7976E-02	-6.8924E-03	-1.6739E-02	-4.4351E-02	-5.8354E-02	4.9455E-01	5.4285E-03
	2.6217E-01	1.3513E-01	-3.1176E-03	3.1392E-02	3.0172E-02	1.5367E-01	4.6894E-01	5.9223E-02	-1.1627E-01	4.0601E-02	-1.3321E-01	-3.1079E-02
O1	-9.8268E-02	5.4142E-02	1.0615E-01	1.1147E-01	5.5571E-02	1.0268E-02	-2.3578E-02	8.8747E-04	-1.5056E-01	8.2772E-02	-4.9386E-03	-1.3286E-01
	-4.3068E-01	-3.1510E-01	2.7270E-01	9.0134E-02	-1.6726E-01	7.2555E-02	-8.4111E-02	9.1140E-02	-3.1326E-02	-7.0302E-02	-4.1061E-02	-6.8375E-02
	-2.5248E-01	-3.6751E-01	1.1018E-01	-2.9041E-01	-1.5152E-01	-1.8082E-03	-2.1724E-03	-1.2795E-02	3.6392E-01	-1.9995E-01	-2.7069E-02	2.5124E-01
O2	-7.2974E-02	-2.3780E-02	-4.4887E-02	-4.3608E-02	-1.2532E-02	9.7546E-02	2.2072E-01	1.8604E-02	-1.5452E-02	4.2751E-02	-2.1579E-01	-1.5756E-03
	-1.4158E-01	-1.6350E-02	-4.4760E-02	-4.0544E-02	-2.4704E-02	1.2314E-01	3.3113E-01	3.4543E-02	-3.8317E-02	6.6498E-02	-3.1558E-01	-1.2436E-02
	1.1308E-01	7.0082E-03	6.3286E-02	6.3157E-02	3.0646E-02	-1.3477E-01	-3.7251E-01	-3.6315E-02	2.5741E-02	-6.6036E-02	3.4604E-01	2.3634E-03
O3	1.0282E-01	7.4694E-02	2.3507E-02	4.3140E-02	4.3346E-03	-2.1609E-01	-3.4848E-01	-4.4975E-02	6.8867E-02	7.3072E-02	-3.4532E-01	-4.9132E-03
	1.2758E-01	4.0209E-02	1.9699E-02	3.3911E-02	1.0355E-02	-1.7789E-01	-3.3506E-01	-4.8382E-02	6.5412E-02	7.2249E-02	-3.2543E-01	-2.3819E-03
	3.5224E-02	3.2352E-02	1.9280E-03	1.7545E-02	-4.0901E-04	-1.1266E-01	-1.7553E-01	-2.2849E-02	3.7337E-02	3.6155E-02	-1.7003E-01	-7.2049E-04
O4	-2.3711E-01	-8.4597E-03	-3.7140E-01	-1.3452E-01	9.7681E-02	5.2015E-01	-2.5471E-01	-1.6970E-01	3.8174E-03	1.3497E-01	-5.2098E-03	4.3663E-02
	4.3299E-01	-2.8871E-01	2.6025E-01	2.3723E-03	1.2694E-01	2.0832E-01	-3.1312E-02	-4.7355E-02	1.2974E-03	7.0858E-03	-4.3951E-02	7.6922E-03
	-2.5223E-01	1.1085E-01	-2.0572E-01	-4.2790E-02	-1.0158E-03	7.4025E-02	-7.0855E-02	-3.7862E-02	7.7723E-03	2.3991E-02	1.7043E-02	1.3157E-02

Theoretical eigenvectors (normalized) of backbone vibrations in A-DNA. Vibration amplitudes are expressed in cartesian coordinates as indicated for C1. Refer to Fig. 1 for atom labels and to the text for coordinate designations. The theoretical dioxy vibrations are at 1,210 cm^{-1} (antisymmetric) and 1,105 cm^{-1} (symmetric). A symmetric diester vibration falls at 807 cm^{-1}.

*Atom coordinate vibration amplitude (relative).

TABLE II
SOME EIGENVECTORS OF B-DNA BACKBONE

ν(Cm^{-1})....		7.8903E+02	8.3467E+02	8.8625E+02	9.5978E+02	9.9964E+02	1.0581E+03	1.0944E+03	1.1126E+03	1.1601E+03	1.2008E+03	1.2095E+03	1.2305E+03
							Eigenvectors						
C1*	Δx	2.8247E-02	-3.4235E-02	4.1185E-03	1.6064E-02	4.2118E-03	-1.0085E-01	-8.3462E-02	5.3188E-02	5.5689E-02	1.3131E-02	1.4573E-02	8.1501E-02
	Δy	-1.1451E-02	-3.4812E-01	1.5599E-01	5.7220E-01	-2.0644E-01	1.0607E-01	-1.5273E-02	3.3443E-02	-1.8789E-01	9.1982E-03	-2.6877E-03	1.1991E-01
	Δz	3.9054E-02	-4.0193E-02	6.6518E-02	1.5063E-01	2.0940E-02	-3.9657E-01	-3.3558E-01	2.1127E-01	3.0151E-01	6.9140E-02	6.0839E-02	2.9165E-01
C2		-1.6683E-01	-1.0096E-01	1.3959E-01	2.7866E-02	1.2729E-01	9.5586E-02	1.0280E-01	-6.0414E-02	7.8655E-02	-8.2599E-02	2.3721E-02	-4.6225E-02
		-1.3831E-02	2.9309E-02	1.1706E-02	-3.9285E-01	3.7158E-01	-8.4416E-02	1.1049E-02	-3.1122E-03	2.5883E-01	-6.6746E-02	5.3442E-02	3.7088E-02
		-2.4564E-01	-1.8845E-01	3.2410E-01	2.0542E-01	1.5896E-01	1.9683E-01	1.6266E-01	-9.8573E-02	4.4510E-02	6.7807E-04	-1.3568E-02	-1.8549E-01
C3		-1.7037E-01	9.0041E-02	4.9908E-02	-1.3820E-01	-1.4904E-02	-1.7182E-02	-1.9860E-01	1.2150E-01	-2.9517E-01	6.4220E-01	-1.9448E-01	-2.0811E-02
		1.3072E-01	4.9148E-01	-2.3020E-01	2.2086E-01	-3.1662E-01	-2.0873E-02	-6.5731E-02	2.1752E-02	-9.7200E-02	5.4244E-02	-5.9265E-02	-2.2239E-01
		6.2556E-02	-1.6096E-01	-1.5461E-01	-2.8004E-01	-3.0340E-01	-2.5895E-01	-1.0521E-01	4.6981E-02	-3.8126E-01	-2.7828E-01	1.8997E-02	2.9039E-01
C4		-1.8190E-01	2.5430E-01	-3.5187E-01	2.0794E-01	3.6653E-01	7.8536E-03	-1.2525E-01	1.4719E-01	-6.5673E-02	-2.7811E-01	4.4969E-02	-3.3370E-02
		-2.1856E-02	9.4973E-02	-4.4807E-02	-1.5863E-01	9.3926E-02	3.8778E-01	1.8872E-01	-4.6361E-02	-3.0057E-02	8.3819E-02	8.8325E-02	6.2706E-01
		-1.8874E-02	1.7978E-01	-1.9124E-01	-2.8137E-02	-2.9794E-01	1.8878E-01	7.0376E-02	-6.5200E-02	5.5239E-01	2.3493E-01	1.8375E-02	-9.8578E-02
05		3.7819E-02	-3.7907E-02	2.5305E-02	-5.0565E-02	-4.3325E-02	1.2325E-02	3.6112E-02	-3.4335E-02	-2.5829E-02	1.1836E-03	-1.5757E-02	-7.4551E-02
		6.1729E-03	-2.5673E-01	9.0214E-02	-2.5632E-01	-1.1156E-02	-3.1426E-01	-3.9564E-02	-5.1426E-02	3.8504E-02	-6.3309E-02	-6.2163E-02	-4.4319E-01
		1.1044E-01	5.7259E-02	-1.4681E-01	-1.9296E-01	2.5426E-02	2.3154E-01	2.2530E-01	-1.4122E-01	-3.7594E-01	-1.0987E-01	-5.6360E-02	-2.2753E-01
C5		2.2452E-01	-1.3806E-01	9.0269E-03	-6.2102E-02	-2.0740E-01	-1.9382E-01	3.0761E-01	-3.3889E-01	9.5611E-02	2.1328E-01	-1.0770E-02	1.1351E-01
		-4.5647E-02	7.9759E-02	2.8126E-02	6.2200E-02	3.0010E-02	2.1374E-02	-1.2864E-01	1.1601E-01	1.1608E-03	-8.9068E-02	-5.1968E-03	-8.9037E-02
		6.5190E-02	3.1018E-01	2.2690E-01	1.7636E-01	3.1470E-01	-3.1700E-01	1.4535E-01	-2.0458E-01	-1.8088E-01	5.7721E-02	-3.7465E-02	8.5453E-02
P		-1.1913E-01	-1.4067E-01	-2.1885E-01	4.3485E-02	5.7824E-02	1.3809E-02	-1.0380E-01	-1.7832E-01	6.6277E-02	-7.8656E-02	-4.7727E-01	8.0260E-02
		1.8909E-01	7.7604E-02	1.4937E-01	-2.6853E-02	-1.8290E-02	-2.1170E-02	2.2517E-01	3.0468E-01	5.4821E-02	-7.2288E-02	-4.0042E-01	4.8117E-02
		-2.2115E-01	2.4678E-01	2.7102E-01	-5.6616E-02	-9.6095E-02	4.7899E-02	-2.0311E-01	-2.2853E-01	1.9141E-02	-1.0492E-01	-1.5936E-01	4.5914E-02
01		1.6816E-01	1.0841E-01	5.6245E-02	1.1620E-01	-2.0626E-01	-3.1323E-02	1.3206E-01	-1.2757E-01	1.6908E-01	-4.1865E-01	1.4923E-01	2.4049E-02
		-7.4469E-02	-3.2091E-02	5.9835E-02	-3.3645E-02	1.3436E-02	-5.9134E-03	-5.2469E-03	-1.1994E-02	-6.9279E-03	4.5463E-02	-1.1105E-02	1.0118E-02
		5.9147E-01	-2.2854E-01	-2.7328E-01	1.6984E-01	3.5353E-01	8.3886E-02	1.9917E-02	1.0995E-01	2.4431E-02	1.8725E-01	-1.7067E-02	-8.1192E-02
02		8.8771E-02	-9.7541E-03	1.9331E-02	-1.0835E-02	2.6808E-02	5.3487E-02	-1.8012E-01	-1.6536E-01	-3.8604E-02	4.2419E-02	2.0658E-01	-2.7306E-02
		1.4506E-01	-1.1274E-02	3.4019E-02	-2.0313E-02	5.4009E-02	1.1269E-01	-3.9568E-01	-3.6764E-01	-8.4504E-02	8.8698E-02	4.4671E-01	-5.8898E-02
		2.2595E-02	-6.2724E-03	-6.1988E-03	7.0779E-03	-1.2315E-02	-1.1581E-02	4.5602E-02	3.5053E-02	1.4027E-02	-1.9868E-02	-3.9820E-02	6.3972E-03
03		-1.8070E-01	5.1328E-02	-7.2289E-03	-1.2915E-03	-5.0817E-02	-1.5770E-01	3.4109E-01	3.1977E-01	-7.1775E-02	1.0902E-01	4.1854E-01	-6.4661E-02
		-6.1112E-02	-4.5598E-03	-5.0892E-03	-5.7900E-03	-1.5228E-02	-1.8388E-02	5.8025E-02	7.0328E-02	-1.1840E-02	1.6748E-02	8.5682E-02	-1.4678E-02
		-1.3684E-01	8.9504E-03	-5.7553E-03	-1.6999E-02	-3.2165E-02	-8.9471E-02	2.0356E-01	2.1413E-01	-5.0652E-02	7.6161E-02	2.5986E-01	-4.1293E-02
04		2.5553E-01	4.9471E-02	3.6193E-01	-1.5923E-01	-6.5097E-02	2.8982E-01	-1.9920E-01	3.2536E-01	-9.3306E-03	-9.6173E-02	2.1348E-02	-5.5320E-02
		-2.9653E-01	-1.6073E-01	-3.4754E-01	7.9336E-02	1.7411E-02	-1.1743E-01	7.6600E-02	-1.6284E-01	1.3355E-02	4.6405E-02	9.6841E-03	2.0084E-02
		-1.8584E-01	-2.6949E-01	-2.0868E-01	-1.0327E-01	-1.2053E-01	2.5246E-01	-1.6630E-01	2.0996E-01	6.2863E-02	-6.5071E-02	3.4210E-02	-6.2670E-02

Theoretical eigenvectors (normalized) of backbone vibrations in B-DNA. Vibration amplitudes are expressed in cartesian coordinates as indicated for C1. Refer to Fig. 1 for atom labels and to the text for coordinate designations. The theoretical dioxy vibrations are at 1,210 cm^{-1} (antisymmetric) and 1,094 cm^{-1} (symmetric). A symmetric diester vibration falls at 789 cm^{-1}. The line at 835 cm^{-1} has a somewhat better defined antisymmetric diester character than any of the A-DNA lines.

*Atom coordinate vibration amplitude (relative).

ized numerically to obtain the vibrational modes of the backbone-ribose model. Comparisons of the calculated and observed frequencies were shown in an earlier paper (8).

The actual unbalanced charge on each atom will also affect the vibrational modes and this was estimated in previous calculations (8). This effect was found to be small for those modes in the frequency range considered in this paper.

Tables I and II give the maximum displacements of the backbone atoms in x, y, and z directions. The z axis is taken to be the helix axis, and the diad relating the strands of the double helix is along the x axis. Since in the long wavelength limit appropriate to IR absorption all displacements in different cells are in phase, we list in the table only 33 components that correspond to the displacements of the backbone and ribose atoms in one unit cell.

A number of interesting observations can be made about our eigenvalue solutions. The 1,209.9-cm^{-1} line for A-DNA has a large antisymmetric dioxy character, and has total displacements of the O_2---P---O_3 group, which are within 1° of the line connecting the O_2 and O_3 atoms in their unstretched positions. The same is true for the 1,209.5-cm^{-1} line calculated for B-DNA. The lines with symmetric dioxy character have displacements of the O_2PO_3 group lying within 3° of the O_2PO_3 bisector for A-DNA and within 4° of the bisector for B-DNA. The lines with symmetric diester character have displacements that are about 11 and 10° off from the O_1—P—O_4 bisector of A- and B-DNA, respectively. Although the displacements of the atoms in the PO_4 group are not far from those expected from symmetry arguments for an isolated PO_4, all the eigenvectors have considerable displacements over other atoms in the backbone.

The situation is especially unclear for the antisymmetric diester stretch modes. Referring to Table I, where we expect an antisymmetric stretch in the frequency range below 900 cm^{-1}, we see that there are two modes with substantial diester motion at 832 and 886 cm^{-1} but that neither is of markedly antisymmetric character. Furthermore, both modes show equally large or greater ion displacements in the ribose ring than in the backbone oxygens. Substantial displacement of C_1 in the 832 cm^{-1} line suggests that the base atoms may also share substantially in the motion.

Table II shows that this loss of antisymmetric character is not a fluke, but also appears in the B-conformation eigenvectors. The same spread of oscillation amplitude among the atoms of the ribose ring is also observed here, as well as the particular motion of the C_1 atom.

In even the most favorable case among the four lines studied here, that of the antisymmetric dioxy stretch, the characterization is no more than qualitative owing to the large amount of C_3 and C_4 motion in the eigenvector. In the worst case, the characterization is not even qualitative. An interpretation of optical absorption that made quantitative use of the "symmetric" or "antisymmetric" character of these resonances, but had no reference to quantitative experimental information about eigenvectors in these specific molecules, could only be correct by accident.

This work was supported by National Science Foundation grant DMR 74-14367 and National Institutes of Health grant GM 24443.

Received for publication 2 April 1979 and in revised form 15 May 1979.

REFERENCES

1. PILET, J., and J. BRAHMS. 1973. Investigation of DNA structural changes by infrared spectroscopy. *Biopolymers.* **12:**387.
2. KURSAR, T., and G. HOLZWARTH. 1976. Backbone conformational change in the A → B transition of deoxyribonucleic acid. *Biochemistry.* **15:**3352.
3. ARNOTT, S., and D. W. L. HUKINS. 1972. Optimized parameters for A-DNA and B-DNA. *Biochem. Biophys. Res. Commun.* **47:**1504. 1979.
4. BEETZ, C. P., G. ASCARELLI, and S. ARNOTT. 1979. A reinterpretation of the IR linear dichroism of oriented nucleic acid films and a calculation of some effective partial charges on the ribose phosphate backbone. *Biophys. J.* **28:**15.
5. HIGGS, P. W. 1953. The vibration spectra of helical molecules: infra-red and Raman selection rules, intensities and approximate frequencies. *Proc. Roy. Soc. Lond. A. Math. Phys. Sci.* **220:**472.
6. EYSTER, J. M., and E. W. PROHOFSKY. 1974. Lattice vibrational modes of poly(rU) and poly(rA). *Biopolymers.* **13:**2505.
7. ARNOTT, S., P. J. CAMPBELL SMITH, and R. CHANDRASEKARAN. 1976. Atomic coordinates and molecular conformation for DNA-DNA, RNA-RNA and DNA-RNA helices. *In* Handbook of Biochemistry, Molecular Biology. 3rd Ed. Nucleic Acids. G. Fasman, editor. Vol. 2. 411.
8. LU, K-C., E. W. PROHOFSKY, and L. L. VAN ZANDT. 1977. Vibrational modes of A-DNA, B-DNA, and A-RNA backbones: an application of a green-function refinement procedure *Biopolymers.* **16:**2491.

Introduction: The Role of Functional Linkage in Molecular Neurobiology

HISTORICAL PERSPECTIVES

The concept of functional linkage in biomolecular systems resulted from the fusion of several streams of scientific thought. In biology, epoch-making advances culminated in the great revolution of molecular biology; in this volume we are particularly concerned with concepts involving the highly dynamic, "cooperative" interactions within and between certain macromolecules and with their neurobiological consequences. In physics and chemistry, developments during the same period in the areas of thermodynamics, kinetics, statistical mechanics, and instrumentation provided theoretical and technical bases for understanding these dynamic intramolecular and intermolecular interactions or linkage properties that molecular biologists have demonstrated to be fundamental to life processes.

The nervous system is highly dependent on fast biophysical and biochemical processes that are electrochemical and chemo-electrical in nature. Neuroscience is therefore the branch of life science in which the concepts of functional linkage are of most direct and immediate application.

The concepts and processes dealt with in this book developed gradually over a period of a half-century, beginning in the mid-1920s, by which time the microscopic anatomy of brain tracts had been extensively characterized. The neuron doctrine, formulated earlier in the last century, had been developed to a point where the function of the nervous system was visualized as resulting from series and parallel interactions of nets of neurons in which an impulse arriving at a dendritic terminal was transmitted down the axon to synaptic terminals; neurons were thus viewed as contiguous, but not continuous as reticularist theory had claimed. In the late nineteenth century and early years of the twentieth, neurophysiology was dominated by Sherrington (see Swazey, 1969). His investigations and publications on the integrative action of the nervous system led to a concept of the interaction, via nerve impulses, of numerous neuronal centers in various parts of the central nervous system. Since the work of DuBois-Reymond (1848), these impulses were known to be based on bioelectrical properties; the propagating impulse was successfully modeled with an iron-wire model by Lillie (1923). Three decades later the basis of the nerve impulse was explained in the ionic diffusion theory of Hodgkin and Huxley (1952).

The application of the techniques of electronic amplification and oscilloscopic recording, initiated by Erlanger and Gasser (1937), introduced a new era, since it was now possible to record with temporal fidelity even the fastest (ca. 10^{-4} sec) bioelectric events in the nervous system. This permitted the mapping of action waves (fast spike potentials and slower afterpotentials) over axons, synapses, and extensive neuronal nets. Slower electrical phenomena of the whole brain (EEG) were also discovered during this period. The ability to record from microelectrodes placed inside neuronal cell bodies (unit recording) introduced a still more dynamic phase of neurophysiology, in which function could be referred to individual neurons. The time was now ripe for the development of molecular neurobiology and the application of the concepts of functional linkage that had been developing simultaneously in physics and chemistry.

However, let us first trace the origins of dynamic neurochemical concepts. For convenience these may also be traced to the starting point of our narrative, half a century ago, in the work of Loewi and Dale, which led to the acceptance of the concept that the nerve impulse is conducted across the synaptic gap between axon terminals and the postsynaptic membrane by chemical mediators or *transmitters*, not by electrical means, as had until then been supposed.

The search was now on for substances that could survive the rigid requirements for qualification as "Grade-A-Certified" excitatory or inhibitory neurotransmitters. Seven such substances have so far been identified, and others wait to be fully tested and accredited. The role of transmitters such as norepinephrine in complex behavioral patterns, such as affective behavior, soon became evident, and the now thriving fields of molecular pharmacology and psychopharmacology were established. Neurotransmitters can produce fast and biologically significant changes when present in extremely low concentrations, in the range of 10^{-8} to 10^{-12} M. How is such action accomplished?

Also beginning roughly in the mid-1920s were investigations with X-ray diffraction and polarization optics that, together with the development of high-resolution electron microscopy in the 1940s, induced revolutionary advances in our knowledge of tissue and molecular ultrastructure. Technical developments led to the widespread use of apparatus such as the ultracentrifuge and other equipment for characterizing macromolecules and ushered in the period of molecular biology, whose most spectacularly successful achievement was molecular genetics. Molecular neurobiology, which had developed substantial impetus during the rise of molecular biology, profited by the great increase in the popularity of neuroscience and mounted to a sustained crescendo in the late 1960s.

From the biological side, the stage was now set for the profitable investigation of the role of functional linkage in biomolecular systems.

THE CONCEPT OF FUNCTIONAL LINKAGE

In parallel with the revolution in molecular biology, biophysical chemists developed the concepts and techniques necessary for studying the func-

Reprinted with permission from *Functional Linkage in Biomolecular Systems*, edited by F. O. Schmitt, D. M. Schneider, and D. M. Crothers, pp. 1–13, 1975.

tional linkage between intra- and intermolecular parameters and applied these principles and techniques to cell physiology. Included in this category, for convenience if not rigorous conceptual usage, are the concepts of cooperativity, allosteric interaction, transductive coupling, and amplification. Many vital processes of the cell depend on such functional linkages: the action of many enzymes and enzyme systems, hormonal effects, metabolic and homeostatic control mechanisms, gene expression, biosynthetic mechanisms, the conversion of metabolic chemical energy into mechanical, electrical, osmotic, and other forms of work, and the transduction from one energy modality to another, as in sensory transduction.

The application of these concepts converted the highly descriptive, relatively static field of neurochemistry, founded by Thudichum (see Drabkin, 1958), into a new science whose *leitmotiv* is molecular dynamism. Like the revolution in neurophysiology that was initiated by the introduction of electronic amplifiers and oscilloscopes, the consideration of fast reactions (10^{-7} to 10^{-9} sec) within and between macromolecules opened up the possibility of new insights into the mechanisms of brain and mind.

The theme that runs through many of the phenomena of functional linkage is that of *cooperativity*, a term that refers to the ways in which the components of a macromolecule, a system of macromolecules, or, in the most general case, of any system of components, act together so as to switch from one stable state of a molecule to another. Frequently involved are phase transitions (as in helix-coil transitions), hysteresis, and a one-to-many input-output relationship.

A sigmoidal relationship is frequently shown between the input parameter and the output response of a cooperative system. Sigmoidicity alone is neither necessary nor sufficient evidence that a process is cooperative; it is important to determine whether the parameters can vary independently or whether they tend to change in a concerted manner. Suppose that a protein contains several subunits, and binding to one subunit has no effect on the affinity of other subunits for the ligand. It is then proper to say that binding is noncooperative with respect to interactions between the subunits, or between the ligands, since they act independently. However, if this noncooperative ligand binding converts a subunit from one conformational structure to another, then the reaction can be said to be cooperative *with respect to the interactions between the protein dihedral angles,* since these change in a concerted way. In summary, one must always specify the parts considered to be the elemental units of the system in connection with any statement about the cooperativity of a reaction.

Allosteric interactions in proteins and their role in the control of cellular metabolism by conformational changes represent one type of functional linkage. Enzymes frequently contain two kinds of sites, those that activate or inactivate the enzyme and those that exert the specific catalytic action. The influence of binding at the control site on the activity of the catalytic site is said to be an allosteric interaction (Greek, *allo* = other, *stereo* = space). Various combinations of positive and negative feedback by products of the enzyme action and other metabolites can exert control to produce a homeostatic kinetic process. Allosteric interactions may also occur between subunits of structures such as membrane receptors, adjacent enzymes, and macromolecular assemblies. The first enzyme in a self-regulating enzyme system is frequently inhibited by products of the action of the last enzyme in the pathway. The first enzyme in the pathway is called a regulatory or *allosteric* enzyme. The known allosteric enzymes have more than one, sometimes many, polypeptide chains. Such systems have an atypical dependence of reaction velocity on substrate concentration, and hyperbolic Michaelis-Menten kinetic relationships do not apply.

Another type of functional linkage is *transmodal transduction*. In the most general case, that in which the chemical free energy available in molecules such as ATP or cyclic AMP is coupled to the macromolecular machinery of tissues to do various kinds of work, the "high-energy" molecule must be covalently linked with the effector lattice of the cell. The best-known example of such transmodal energy conversion is muscle contractility. In this process, cooperative interactions of Ca^{2+} with troponin and of one tropomyosin molecule with seven actin subunits has been demonstrated (see Chapter 14), as has the transductive process proper, in which heavy meromyosin interacts via nucleotide triphosphate with actin and its complex with tropomyosin and troponin. As our knowledge has grown concerning receptors and their coupling with the systems that link the molecular complex of the receptor (for vision, audition, smell, taste, etc.) to the cellular systems that transduce sensory inputs to action potentials in sensory nerves, an appreciation has developed concerning the importance of functional linkage in the process. These matters are discussed in Chapters 1, 10, 11, 12, and 13.

Another form of functional linkage is that in which a signal gives rise to an amplified response, frequently larger than the input signal by orders of magnitude. One example of such *amplification* is the relatively large change in cyclic AMP concentration induced by small amounts of hormones such as glucagon (Chapter 9). More complex phenomena that include amplification are the immune response (Chapter 8) and the dramatic alteration of swimming behavior in chemotactic bacteria produced by a small concentration gradient of an attractant (Chapter 13). Amplification in biochemical systems can be achieved by controlling the activity of enzymatic catalysts. There are a number of well-documented cases of enzymes whose activity is controlled by ligands; this topic is considered at length in Chapter 2. Cooperativity plays an important role in amplification. An enzyme whose subunits interact cooperatively can be converted from an inactive to an active form over a narrow range of ligand concentration, thus increasing the amplifier gain. In such cases, the "amplified" response may not be chemically identical to the stimulus, or even in the same modality, but the output is much larger quantitatively than the input, in terms of energy or the potential to do work. This concept is of practical utility, if not theoretical rectitude. One case of amplification in the strictest sense is electrosensing in certain fish, in which extremely weak electrical signals are converted into action potentials in sensory nerves.

Cooperativity plays an important role in the transductive coupling of different energy modes. An example considered in Chapter 1 is the transduction of chemical to mechanical energy through a cooperative helix-coil transition in which cooperativity influences the rate of the process. The most spectacular example of transductive coupling in neurobiology is in sensory transduction, in which the activity of the receptor cell can be altered by the absorption of one photon (in vision) or of one molecule of pheromone attractant (in olfaction).

THE DEVELOPMENT OF CONCEPTS OF FUNCTIONAL LINKAGE

The earliest biochemical observations of phenomena much later to be termed cooperative were those of Christian Bohr (Bohr, 1903*a,b;* Bohr, Hasselbalch, and Krogh, 1904), the father of Niels Bohr, who discovered that O_2 is dissociated from hemoglobin as the CO_2 tension is increased. Another way of defining the "Bohr effect," formulated much later, is that the binding of oxygen by hemoglobin results in the discharge of protons. Ten years later, Haldane and his collaborators (Christiansen, Douglas and Haldane, 1914) demonstrated the converse effect, that oxygenation of hemoglobin drives off CO_2. Bohr had sought for the reciprocal effect, but failed to find it, although it could have been predicted mathematically.

The discovery of the adaptation of the chemical structure of hemoglobin to its function is delightfully and authoritatively told by Edsall (1972).[1] He relates how Bohr, in collaboration with Hasselbalch and Krogh (1904), critically analyzed the curves of O_2 binding by hemoglobin and concluded that their sigmoid shape (as compared with the hyperbolic shape that was expected from the mass action law for a single binding site) was not due to error. The controversy provoked much work to explain why the curve was not hyperbolic. Bohr's data showed that, after the first oxygen molecule was bound, succeeding ones bound to hemoglobin more easily. Unfortunately, he held to the view that hemoglobin solutions contain a heterogeneity of hemoglobin molecules and that this was responsible for the sigmoid curves; thus he missed intramolecular cooperativity.

During this period of uncertainty about the molecular status of hemoglobin, A. V. Hill (1910) derived a mathematical approximation which, a half-century later, is proving valuable as a measure of cooperativity between

[1] The editors thank Prof. J. T. Edsall for his helpful criticism in the preparation of this chapter.

ligand-binding sites. Hill proposed that the hemoglobin monomers, each with a single heme group, could aggregate reversibly to varying extents, and that the aggregates would bind oxygen in a fashion that we now term "cooperative." He expressed the percent saturation function y of hemoglobin as

$$y = \frac{100Kx^n}{1 + Kx^n}$$

where K is a constant and x represents the activity (concentration or partial pressure of the ligand); n, the Hill coefficient as it has come to be known, was originally conceived by Hill to be a measure of the degree of aggregation of hemoglobin. Subsequently, Wyman (1948, 1963) showed that the Hill coefficient, under certain conditions, measures the free energy of interaction between sites.

Adair, who, according to Edsall, was unique among the biochemists of his time in that he studied Gibbsian thermodynamics, used improved osmotic measurements to establish that hemoglobin is a tetramer weighing 67,000 daltons (Adair, 1925). He formulated a general equation for ligand binding of hemoglobin in terms of four successive binding constants. If these constants progressively increase as more oxygen is bound, the process will be cooperative. Adair's work had far-reaching significance not only for the study of oxygen carrying by hemoglobin, but also for the study of many enzymes.

Cooperativity theory really became established as a powerful concept from the landmark papers of Wyman (1948) and Wyman and Allen (1951), who proposed that cooperative ligand interaction has its origin in conformational changes, that is, in fast alterations in the three-dimensional tertiary structure of the protein molecule. The section of Wyman's paper on "linked functions" was remarkably perceptive. The thesis of cooperativity through conformational change in hemoglobin was confirmed by the crystallographic work of Perutz, Rossmann, Cullis, Muirhead, Will, and North (1960), who showed that the distance between the four heme groups is too great for direct interaction; the interaction must therefore be indirect, presumably through conformational changes. This was confirmed by the finding that the distances between the heme groups in hemoglobin are shifted substantially by oxygen binding.

During the next decade, the concept of cooperativity and the broader one of functional linkage received important and insightful contributions from the work of Novick and Szilard (1954), Umbarger (1956), Yates and Pardee (1956), and others concerning feedback (end-product) inhibition of bacterial biosynthetic pathways. Koshland (1958) suggested that binding to substrate involves an induced fit, produced by alteration of the conformation of the catalytic site.

However, it was not until 1963 that the word "allosteric" was used, in a paper by Monod, Changeux, and Jacob that further developed the concept of the control of enzymic—hence of cellular—activity through protein conformational changes. In their model, proteins were assumed to have two nonoverlapping receptor sites: a catalytically active site that binds and reacts with substrate, and an *allosteric* site that specifically and reversibly binds an allosteric effector. Binding of the latter produces a reversible alteration in the conformation of the protein—an allosteric transition—that changes the active site and certain kinetic parameters characteristic of the enzyme. These investigators concluded that there need be no direct interaction between the substrate and the metabolic ligand that activates the allosteric protein; the effect is due entirely to reversible conformational changes induced when the effector ligand is bound by the allosteric protein.

This conclusion had been foreshadowed by the theoretical work of Wyman 15 years earlier and was further enlarged by his later work in collaboration with Monod and Changeux (Monod, Wyman, and Changeux, 1965). Their main purpose was to develop and justify the idea that a general and simple relationship between symmetry and function may explain the emergence, evolution, and properties of oligomeric proteins as molecular "amplifiers" of random and structural accidents or of highly specific organized, metabolic interactions. In a systematic and remarkably lucid survey, they listed the general properties of allosteric proteins: They are oligomers with several identical structures; allosteric interactions involve quaternary conformational changes; and homotropic interactions, that is, between identical ligands, according to their model, are always cooperative, while heterotropic interactions, that is, between different ligands, may be either cooperative or antagonistic. This ability to mediate cooperative interactions between stereospecific ligands is the physiologically most important property of allosteric proteins. Protein monomers ("protomers") recognize and associate with like monomers with extreme specificity; specific association occurs even at high dilution and in the presence of other proteins, that is, in crude extracts. The authors concluded that cooperativity is a decisive factor in the emergence and selective maintenance of symmetrical oligomeric proteins.

The model of Monod et al. (1965) assumed that allosteric proteins exist in only two states, T (tensed) and R (relaxed), the latter having a much higher affinity for ligand than the former. Thus, all the subunits of the oligomeric protein had to change their conformation simultaneously in the $T \rightleftharpoons R$ transition. An alternative theory was developed by Koshland, Nemethy, and Filmer (1966), who proposed that each subunit of the oligomer could undergo a conformational transition, induced by the binding of ligand. This could then induce conformational changes in neighboring subunits. However, the transition was not all or none, as in the theory of Monod, Wyman, and Changeux. Either of the two theories was capable of explaining many of the experimental data on cooperativity. Tests to distinguish the two theories from each other, or indeed from other possible interpretations,

required new approaches, involving rapid kinetic methods and other techniques.

In addition to thermodynamics, the approach of statistical mechanics was of great importance for the theoretical development of the functional linkage concept. This approach had its basis in physics rather than in chemistry. Of particular historical interest in this field is the theoretical work of Ising (1925), who developed a nearest-neighbor interaction theory for transformations in one-dimensional ferromagnets. The Ising model, which exemplifies the simplest system of interacting particles that retains features of physical reality, raises the fundamental question of whether the formulations of statistical mechanics can predict phase transitions and, if so, how. For almost half a century, various forms of Ising models have formed the basis for theoretical constructs concerning interaction coupling, the aim being to express the partition function in a tractable form, thus facilitating the derivation of exact analytical expressions for thermodynamic quantities. A good review of the theory of the Ising model was provided by Newell and Montroll (1953).

The work of Schellman (1955) foreshadowed the application of statistical mechanical methods to helix-coil transitions. Using the heat and the entropy of adding a segment to the helix as basic parameters, he considered special nucleating end effects. In his simple theory, helix-coil transitions are sharp and dependent on chain length. A major advance in the statistical mechanical development of linkage theory was made by Zimm and Bragg (1959), who showed that helix-coil phase transitions in polypeptide chains are sharp, and are produced by changes of only a few degrees in temperature or a few percent of solvent concentration. The formation of the first turn of a helix is difficult and involves a large decrease of entropy. But once formed, this turn serves as a template for others; such nucleation is typical of sharp transitions. The finite chain problem was attacked with the matrix method, developed by Kramers and Wannier (1941*a,b*), which considers interactions between distant segments as well as nearest neighbors. In the 15 years since the work of Zimm and Bragg, there have been many applications of theories of cooperativity and other linkage phenomena to systems of biological interest, including polypeptides, nucleic acids, polysaccharides, and lipids.

These developments in concepts of functional linkage were aided by great advances in determining the kinetics of fast reactions. These were discussed at length at an International Colloquium for Fast Reactions in Solutions (see Eigen, 1960), with contributions representing every type of related technique then available. Developments proceeded along two lines, from biochemistry and from physics.

From the biochemical approach, the work of Hartridge and Roughton (1923; see also Roughton, 1960) pioneered the measurement of reaction velocities in the millisecond-to-second range, in contrast to previous methods. which had been limited to processes taking minutes or hours. Their apparatus permitted the rapid mixing of reactants in a restricted space, the composition of the streaming fluid being determined by fast—for example, spectroscopic—methods at varying distances from the mixing chamber or, in work done many years later, after sudden stoppage of the fluid after mixing (hence the appellation "stopflow" method). Types of chemical reactions amenable to study by rapid flow technique were reviewed by Chance (1960).

On the physical side, early formulations concerned the relaxation kinetics of reactions in gases (Einstein, 1920; Meixner, 1943*a,b*). The development of various relaxation techniques, especially those of Eigen and DeMaeyer (1963), permitted measuring the kinetics of fast (to 10^{-7} sec) and very fast (to 10^{-9} sec) reactions, whose rates are close to those for diffusion-controlled kinetics (10^{-10} sec). Additional methods for measuring fast reactions include those of flash photolysis (see particularly Porter, 1963), photochemistry, fluorescence, and electrochemistry, as well as the related methods of electron spin resonance and nuclear magnetic resonance. For a survey of these methods see Caldin (1964).

The availability of methods for measuring fast state transitions made it possible to investigate cooperative, allosteric, transductive, and other types of functional linkage. It became obvious that the attack on major problems of molecular neurobiology, involving bioelectric events in the range of 10^{-3} to 10^{-6} sec, would receive powerful support from biophysical chemists, particularly those concerned with fast reactions; conversely, problems of neuroscience, looming ever larger in importance among the problems of life science, indeed of all science, might greatly interest these same biophysical chemists. These factors were crucial in the decision in 1962 to organize a multidisciplinary, multiuniversity attack on basic problems of neuroscience. The Neurosciences Research Program (NRP), sponsored by the Massachusetts Institute of Technology, was founded from the impetus of the developments recounted above. Manfred Eigen and Leo DeMaeyer, leading investigators in the field of fast reactions, helped to organize the NRP and were among its charter members.

THE APPLICATION OF FUNCTIONAL LINKAGE CONCEPTS TO MOLECULAR NEUROBIOLOGY

It may be useful at this point to list some of the major problems of molecular neurobiology in which functional linkage is a key factor. Their treatment in the three NRP Intensive Study Programs (Quarton, Melnechuk, and Schmitt, 1967; Schmitt, 1970; Schmitt and Worden, 1974), and in the Work Sessions reported in the *NRP Bulletin*, as well as in the rapidly growing literature of neuroscience, made it reasonable to organize a conference on the subject of functional linkage.

Membrane processes are of primary interest and are currently being investigated from the viewpoint of molecular organization and composition, including the chemical characterization of membrane-borne molecular machines (channels, ionophores, and other mediators of excitability, pumps, permeases, receptors, adenylate cyclase, and other enzymes), and with regard to the dynamic interactions and coupling that subserve brain function. Bilayer systems and similar models have proved extremely valuable in analyzing functional linkage.

The synapse presents important opportunities and challenges, particularly with regard to its ultrastructure and the dynamics of transmitter action: Functional linkage is involved in the release and reabsorption of transmitter, receptor binding of the transmitter ligand, and the resulting excitation or inhibition of postsynaptic neurons. The properties of dendrodendritic synapses between "local circuit" interneurons, characteristic of the retina and the olfactory bulb, suggest fascinating new concepts of reciprocal linkage and information processing; the possible existence and function of such connections in other brain regions, such as the cerebral cortex, is now being investigated in various laboratories.

Recent studies have begun to elucidate the mechanisms that couple metabolic processes with the synthesis of transmitters and neurohormones and their regulation according to functional requirements. Coupling has been demonstrated between the postsynaptic membrane and the genetic expression of the messenger RNAs that code for the enzymes needed for transmitter synthesis. Linkage also exists between the cell soma and the cellular membrane to regulate the synthesis of membrane material and its incorporation via the Golgi apparatus into the cell membrane.

Neuroplasm is constantly synthesized and transferred down the axon. Its synthesis is regulated by linkage with the cell membrane, in which are located receptors and other transductive mechanisms. The role of microtubules in the fast transport of materials from the cell soma through its axonal and dendritic extensions is thought by many to be a cooperative process involving the microtubular subunits and calcium ions.

Mechanisms by which organisms sense visual, olfactory, gustatory, hormonal, neurotransmitter, mechanical, and electrical stimuli operate at or near the limits of sensitivity. The typical system utilized for transmodal transduction includes an input-specific receptor and a molecular amplifier system that alters the polarization of the postsynaptic membrane (for stimulation or inhibition) or gives rise to action potential waves in sensory axons.

Many processes that occur at the cellular, tissue, and whole-brain level may, in a formal sense, be considered cooperative in nature: a small input may generate a large output. All-or-nothing, phase-like transitions that are sensitive to environmental parameters are characteristic of brain function, and changes at one locus may be rapidly transmitted to a distant locus or to an entire surround (volume or holistic effect). Dynamic patterns may arise when chemical reactions in a distributed system are coupled by diffusion; for example, under nonequilibrium thermodynamic conditions, functional linkage processes are involved in cooperative pacemaker assemblies, oscillating fields, pulse distribution, and other coupled phenomena (see Katchalsky, Rowland, and Blumenthal, 1974). The dense interactions of neurons in some brain centers has led to the consideration of domains of cooperative neural activity (Freeman, 1975).

The detection of weak electrical fields by the brain is another field or volume effect that displays highly cooperative characteristics, according to Adey (Chapter 15).

At the perceptual level, the phenomenon of visual depth perception (stereopsis) illustrates cooperativity in an experiential mode that is grasped more vividly than can be conveyed to most by mathematical analysis (Julesz, 1974; see also Szentágothai and Arbib, 1974).

Reserved for a possible future conference is a consideration of other aspects of neuroscience, particularly at the level of brain regions or of the brain as a whole. It is hoped that this book will stimulate interest in functional linkage, especially among neuroscientists who may see in the phenomenology and the physicochemical principles new and productive ways of approaching neuroscience problems, including those of higher brain functions.

REFERENCES

Adair, G. S. (1925): The hemoglobin system. *J. Biol. Chem.*, 63:493–545.

Bohr, C. (1903*a*): Die Sauerstoffaufnahme des genuinen Blutfarbstoffes und des aus dem Blute dargestellten Hamoglobins. *Zentr. Physiol.*, 17:688–711.

Bohr, C. (1903*b*): Theoretische Behandlung der quantitativen Verhaltnisse bei der Sauerstoffaufnahme des Hamoglobins. *Zentr. Physiol.*, 17:682–688.

Bohr, C., Hasselbalch, K., and Krogh, A. (1904): Ueber einen in biologischer Beziehung wichtigen Einfluss, den die Kohlensäurespannung des Blutes auf dessen Sauerstoffbindung übt. *Skand. Arch. Physiol.*, 16:402–412.

Caldin, E. F. (1964): *Fast Reactions in Solution.* New York: John Wiley & Sons.

Chance, B. (1960): Catalysis in biochemical reactions. *Z. Elektrochem.*, 64:7–13.

Christiansen, J., Douglas, C. G., and Haldane, J. S. (1914): The absorption and dissociation of carbon dioxide by human blood. *J. Physiol.*, 48:245–271.

Drabkin, D. L. (1958): *Thudichum, Chemist of the Brain.* Philadelphia: University of Pennsylvania Press.

DuBois-Reymond, E. (1848): *Untersuchungen über thierische Elektrizität.* Berlin: Reimer.

Edsall, J. T. (1972): Blood and hemoglobin. *J. Hist. Biol.*, 5:205–257.

Eigen, M., ed. (1960): Bericht über das Internationale Kolloquium über schnelle Reaktionen in Lösungen. Hahnenklee/Harz. 14–17 September 1959. *Z. Elektrochem.*, 64:1–204.

Eigen, M., and De Maeyer, L. (1963): Relaxation methods. In: *Investigations of Rates and Mechanisms of Reaction*, edited by S. L. Friess, E. S. Lewis, and A. Weissberger, pp. 895–1054. *Technique of Organic Chemistry, Vol. 8, Part II.*

Einstein, A. (1920): Schallausbreitung in teilweise dissoziierten Gasen. *Sitz. ber. Preus. Akad. Wiss., Physik. Math. Kl.*, Berlin, pp. 380–385.

Erlanger, J., and Gasser, H. S. (1937): *Electrical Signs of Nervous Activity.* Philadelphia: University of Pennsylvania Press.

Freeman, W. J. (1975): *Mass Action in the Nervous System.* New York: Academic Press (*in press*).

Hartridge, H., and Roughton, F. J. W. (1923): A method of measuring the velocity of very rapid chemical reactions. *Proc. Roy. Soc. A*, 104:376–394.

Hill, A. V. (1910): The possible effects of the aggregation of the molecules of haemoglobin on its dissociation curves. *J. Physiol.*, 40:iv–vii.

Hodgkin, A. L., and Huxley, A. F. (1952): A quantitative description of membrane current and its application to conduction and excitation in nerve. *J. Physiol.*, 117:500–544.

Ising, E. (1925): Beitrag zur Theorie des Ferromagnetismus. *Z. Physik.*, 31:253–258.

Julesz, B. (1974): Cooperative phenomena in binocular depth perception. *Amer. Sci.*, 62:32–53.

Katchalsky, A. K., Rowland, V., and Blumenthal, R. (1974): *Dynamic Patterns of Brain Cell Assemblies.* MIT Press, Cambridge.

Koshland, D. E., Jr. (1958): Application of a theory of enzyme specificity to protein synthesis. *Proc. Nat. Acad. Sci.*, 48:98–104.

Koshland, D. E., Jr., Nemethy, A., and Filmer, D. (1966): Comparison of experimental binding data and theoretical models in proteins containing subunits. *Biochemistry*, 5:365–385.

Kramers, H. A., and Wannier, G. H. (1941*a*): Statistics of the two-dimensional ferromagnet. Part I. *Phys. Rev.*, 60:252–262.

Kramers, H. A., and Wannier, G. H. (1941*b*): Statistics of the two-dimensional ferromagnet. Part II. *Phys. Rev.*, 60:263–276.

Lillie, R. S. (1923): *Protoplasmic Action and Nervous Action.* Chicago: University of Chicago Press.

Meixner, J. (1943*a*): Zur Thermodynamik der irreversible Prozesse in Gase mit reagierenden dissoziierenden und anregbaren Komponenten. *Ann. Physik.*, 43:244–270.

Meixner, J. (1943*b*): Absorption und Dispersion des Schalles in Gasen mit chemisch reargierenden un anregbaren Komponenten. *Ann. Phys.*, 43:470–487.

Monod, J., Changeux, J.-P., and Jacob, F. (1963): Allosteric proteins and cellular control systems. *J. Mol. Biol.*, 6:306–329.

Monod, J., Wyman, J., and Changeux, J.-P. (1965): On the nature of allosteric transitions: A plausible model. *J. Mol. Biol.*, 12:88–118.

Newell, G. F., and Montroll, E. W. (1953): On the theory of the Ising model of ferromagnetism. *Rev. Mod. Phys.*, 25:353–389.

Novick, A., and Szilard, L., (1954): Experiments with the chemostat on the rates of amino acid synthesis in bacteria. In: *Dynamics of Growth Processes*, edited by E. J. Boell, pp. 21–32. Princeton, N.J.: Princeton University Press.

Perutz, M. F., Rossmann, M. G., Cullis, A. F., Muirhead, H., Will, G., and North, A. C. T. (1960): Structure of haemoglobin. A three-dimensional Fourier synthesis at 5.5 Å resolution, obtained by x-ray analysis. *Nature*, 185:416–422.

Porter, G. (1963): Flash photolysis. In: *Investigation of Rates and Mechanisms of Reactions*, edited by S. L. Friess, E. S. Lewis, and A. Weissberger, pp. 1055–1106. *Technique of Organic Chemistry, Vol. 8, Part II.*

Quarton, G. C., Melnechuk, T., and Schmitt, F. O., eds. (1967): *The Neurosciences—A Study Program.* New York: Rockefeller University Press.

Roughton, F. J. W. (1960): The origin of the Hartridge-Roughton rapid reaction method and its applications to the reactions of haemoglobin in intact red blood corpuscle. *Z. Elektrochem.*, 64:3–4.

Schellman, J. A. (1955): The stability of the hydrogen-bonded peptide structures in aqueous solution. *Compt. Rend. Trav. Lab Carlsberg, Ser. Chim.*, 29:230–259.

Schmitt, F. O., editor-in-chief (1970): *The Neurosciences—Second Study Program.* New York: Rockefeller University Press.

Schmitt, F. O., and Worden, F. G., eds. (1974): *The Neurosciences—Third Study Program.* Cambridge, Mass.: M.I.T. Press.

Swazey, J. P. (1969): *Reflexes and Motor Integration: Sherrington's Concept of Integrative Action.* Cambridge, Mass.: Harvard University Press.

Szentágothai, J., and Arbib, M. A. (1974): *Neurosciences Res. Prog. Bull.* (*in press*).

Umbarger, H. E. (1956): Evidence for a negative-feedback mechanism in the biosynthesis of isoleucine. *Science*, 123:848.

Wyman, J. (1948): Heme proteins. *Adv. Prot. Chem.*, 4:407–531.

Wyman, J. (1963): Allosteric effects in hemoglobin. *Cold Spring Harbor Symp. Quant. Biol.*, 28:483–489.

Wyman, J., and Allen, D. W. (1951): The problem of the heme interactions in hemoglobin and the basis of the Bohr effect. *J. Polymer Sci.*, 7:499–518.

INTRODUCTION

Yates, R. A., and Pardee, A. B. (1956): Control of pyrimidine biosynthesis in *Escherichia coli* by a feedback mechanism. *J. Biol. Chem.*, 221:757–770.

Zimm, B. H., and Bragg, J. K. (1959): Theory of the phase transition between helix and random coil in polypeptide chains. *J. Chem. Phys.*, 31:526–535.

Part III
Effects of Radio Fields on the Central Nervous System and Behavior

The central nervous system (CNS) is the object of intensive laboratory investigations at many levels of analysis, from the microscopic realm of ions and organelles to the organismic vista of behavior. Certainly the least understood of the body's organ systems, the CNS is associated as much with controversy as with enlightenment from the vantage of nonionizing radiobiology. Much of the controversy attaches to clinically or industrially based reports from Eastern Europe of reversible neurasthenic disorders in individuals with a history of prolonged occupational exposure to microwaves and other radio fields. The inherent looseness of scientific control in the epidemiological or clinical-case study has placed a burden of verification of claims of subtle neuropathies on the laboratory scientist. But how is this scientist objectively to discern, in his animal models, the human subjective complaints of headache, irritability, insomnia, and lethargy sometimes interpreted [1] as sequelae of shortwave or microwave irradiation? Since the CNS embraces subcellular, cellular, multicellular, subsystem, and organismic vistas, at which level of analysis can the scientist most fruitfully mount an attack? These questions, at the time of this writing, are unfortunately largely rhetorical. The contents of this chapter are offered, in part, as a stimulus to impel members of the engineering and scientific communities to participate in enlarging the store of pertinent data and theory. The primary aim, of course, is to celebrate the achievements of investigators whose works have already contributed to that store.

In this chapter, I shall avoid my holistic bias toward the study of the intact, freely-behaving animal [2], but not because of a change of scientific perspective. I remain a fervent believer in the Doctrine of Emergence—that phenomena emerge from a complex system that are not evidenced by its isolated parts—but I am just as committed to the view that integrative experimentation at all levels of organization is critical to the ultimate business of science: comprehensive explanation. Accordingly, I have selected a bibliography of key works which bear on any level of analysis at which an important finding on, or technique for studying, the CNS has been reported and confirmed in independent study. I offer, too, a caveat. My selection of reprints* and my drafting of this chapter took place in early 1979. Doubtless, some of the voids in research and confirmatory reports noted in the following paragraphs will have been filled by the time this volume reaches the reader's hands.

No works have been selected from studies at the subcellular, *in vitro* level of analysis; none of moment have been reported, (so, too, for cell-culture studies of neurons). The report by Lebovitz and Seaman [3], which dealt with single-cell responses of neurons to pulsed microwaves in the brain of the intact feline is not included, but would have been if their work were not so recent that independent verification has yet to be reported.

The first key paper recounts work performed in the laboratories of William Ross Adey (see *Bawin, Kaczmarek, and Adey*, 1975) [4]. Susan Bawin and colleagues observed a neurochemical reaction in populations of neurons in isolated hemispheres of the avian (Leghorn-chick) brain. The finding of an exodus of radiolabeled calcium ions from brain tissue, which exodus is dependent on frequency of *sinusoidal* modulation of an applied VHF (147 MHz) carrier, was recently confirmed by Blackman and colleagues [5]. One import of this finding is that relatively weak fields, that is, fields at strengths well below those that are capable of direct stimulation of spike activity through depolarization of neurons, can apparently initiate activity along the surface of the greater membrane of the neuron. By analogy, much more force is required to lift and hurl a domino than to tip it onto a row of dominos and thereby initiate a cascade of significant activity. In a manner of speaking, Bawin and her colleagues have uncovered evidence of a *biological amplifier* sensitive to shortwave fields that wax and wane at certain sub-ELF frequencies, within which band (0–30 Hz) are identified prominent components of the electroencephalogram (EEG).

The EEG, in its crude form, provides a measure of whole- or part-brain neuroelectric activity, and, in the more sophisticated form of computer-summed sensorily-evoked potentials, provides fine-grained, time-locked data on multi-neuronal activity. Unfortunately, nearly all reports based on the EEG contain data that might be confounded because recording electrodes and conductive leads were in electrical contact with the scalp or brain during shortwave or microwave irradiation. Their presence in the field raises the twin problems of passive and active artifact (recording of spurious signals, stimulation of the preparation by demodulated currents). Several intriguing reports on the EEG response of small animals to continuous and pulsed microwaves have appeared in the Eastern European literature and would be worthy of inclusion in this volume were it not for the question of artifact.

Work on visually evoked, computer-summed electrocortical responses of guinea pigs *after* bouts of irradiation has been performed [6, 7]; this work confirmed the increase of neuronal conduction velocity that occurs in single-nerve preparations under conventionally induced elevations of brain temperature

---

Revised manuscript received July 15, 1981.

*Papers included in this volume are referenced by the authors' names (in italics) and year of publication; other papers are referenced by numerals only.

(see the review of classical single-nerve studies by Bruce-Wolfe and Justesen [7]). This work also indicates that both basal and experimentally elevated temperatures of the infraprimate neocortex are consistently 1-2 °C below those of the body's core. Both findings await independent confirmation. Microwave-pulse-evoked electrical responses have been recorded from the brain of the intact feline by *Frey, Fraser, Siefert, and Brish* (1968) [8], who used a coaxial electrode to minimize interactions between it and the field; the requirement for independent, if somewhat indirect, confirmation was recently met in reports by American [3] and Soviet [9] investigators.

I searched in vain for a reprint to showcase the important area of neurohistology. Unfortunately, the barrier of language and the absence of confirmatory reports preëmpted consideration of several papers by foreign authors. Elegant neurohistological work has been performed in the United States in the laboratory of Ernest Albert [10], but he is singular in bringing the electronmicroscope to bear on nervous materials of the microwave-irradiated animal. The absence of a confirmatory report also preëmpted selection of a work performed in my laboratories in collaboration with Kenneth Brizzee and Nancy King. Reference is to the finding [11] of increased packing density of neurons and, especially, of ectodermally derived glial cells in sensorimotor cortex of adult rats after repeated exposure to 2450-MHz radiations.

Moving on to behavioral studies, I have assorted these into three categories: 1) strong disturbances (work stoppage and convulsive activity), 2) sensory detection (perception of pulsed and nonpulsed radiations), and 3) weak disturbances (affective changes and alterations of learning and discrimination during exposure to relatively weak fields).

A paper that Nancy King and I presented at the so-called Richmond Symposium [2] in 1969 addressed work stoppage and other behavioral end points in the stead of several conceptual and technical innovations: dose-determinate exposures of small animals freely behaving *in* a microwave field, free-operant and Pavlovian conditioning in the multi-mode cavity, use of fiber optics to record and program reinforcement of operant responses, development of a means of providing appetitive reinforcement that does not perturb the field, a unit-mass dosimetry that is steadily gaining the acceptance of North American scientists, and an *a priori* attempt (in retrospect, a bit naive) to define quantitatively thresholds of behavioral change in terms of the important but imperfect correlation between dosimetric and densimetric measures, that is, between the W/kg of absorbed energy and the W/m^2 of incident energy. Given these innovations, most of which have now been confirmed, adopted, or extended by other investigators, why not include the paper in this volume? The reason is sheer length; the reader's interests will be better served by reprints of shorter papers whose authors have improved upon the earlier work.

King and I found that 2450-MHz irradiation at a whole-body-averaged dose rate approximating 9 mW/g (during 5-min periods of radiation alternating with 5-min intervals of no radiation) typically resulted, within 20 minutes, in cessation of work for food by hungry, freely-moving rats. More recently, *Lin, Guy, and Caldwell* (1977) [12] reported confirmatory data; specifically, they found for comparable durations of exposure that the dose-rate threshold of work stoppage occurs near 8 mW/g in the hungry, restrained rat when irradiated in the near zone of a 918 MHz CW field. The comparability of threshold dose-rates, given the differences in carrier frequency, in mode of irradiation (unidirectional near-field vs. multipath), and in behavioral management (corporal restraint vs. freedom to locomote) argues strongly for a thermal basis of work stoppage. Supportive of this interpretation are the sizable elevations of core temperature (>1 °C) observed by investigators of both groups. Neither we nor Lin and colleagues employed pulsed radiations, having respectively used sinusoidally modulated and CW fields; unknown but of considerable interest are the averaged power-density and dose-rate thresholds of work stoppage to pulsed radiations of high peak power.

The strongest biological disturbances are obviously associated with lethal or near-lethal intensities of radiation. The induction of epileptiform activity, a sometime precursor of lethality by a host of chemical, electrical, and thermogenic agents, is a neuroelectric storm readily evidenced by convulsive behavior. A study of microwave-induced convulsions in rats by *Phillips, Hunt, and King* (1975) [13] revealed that 2450-MHz irradiation at a dose rate of 100 mW/g resulted in convulsive activity at an averaged latency near four minutes. The total spread of latencies was small (±15%), in spite of body masses that ranged from 190 to 560 grams. This work was recently confirmed in my laboratories [14].

Under the rubric of sensory detection is the RF-hearing phenomenon, perception of clicking or popping sounds to pulsed microwaves, which has been an object of widespread interest for two decades [15, 16]. I have recounted elsewhere [15] the skepticism of many microwave investigators regarding an early report of the phenomenon by Allan Frey [14], and include among the reprinted papers a report from my laboratories (see *King, Justesen, and Clark*, 1971) [17] that was pivotal in changing many minds. This changing of minds, in retrospect, was well justified in the light of the human and infrahuman data—pulsed microwaves *can* be heard—but the change occurred for the wrong reason. Our finding that the rat's averaged dose-rate threshold of detection of 2450-MHz energy is near 600 μW/g (for 60-s exposures) is probably based on perception of warming, not on the thermoelastically induced waves of pressure that are now believed responsible for the auditory response. Our field was modulated as a 60-Hz half-wave sinusoid; the rise time and peak density of such an envelope of radiation fall short of those required to launch a perceptible wave of pressure. Viewed from hindsight, the merit of our paper lies in the implication that microwave irradiation at a dose rate of 600 μW/g is near the rat's absolute threshold of sensory detection of warming by nonpulsatile, 12-cm microwaves. Subsequently, an important study by *Frey and Messenger* (1973) [18] revealed that the energy-dose threshold for human detection of a pulse of microwave radiation is on the order of 20 μJ/g, which is three orders of magnitude below the threshold ostensibly associated with perception of warmth, about 36 mJ/g.

A subsequent paper by Foster and Finch [19] was pivotal,

and deservedly so, in directing the attention of investigators to the thermoelastically induced waves of pressure that attend the capture of a pulsed radiation. The first report of transient thermodynamic effects (thermoelastic expansion) of pulsed radiation was authored by White [20] who, like Foster and Finch, performed studies on nonbiological materials. Thermoelastic expansion occurs when any object, animate or inanimate, undergoes a rapid elevation of temperature—if its density is thermally dependent. The resulting waves of pressure are detectable by the irradiated observer and require little absorbed energy. It is estimated that human beings can hear an absorbed pulse that elevates the average temperature of the brain by as little as five one-millionths of a Celsius degree [21]. An elegant empirical study of guinea pigs (*Chou, Guy, and Galambos*, 1977) [22] indicates that a pulse of microwave radiation is converted to waves of pressure that arise in the skull or brain and flow to and excite the cochlea. One implication of downstream cochlear stimulation by transduced pressure waves (as opposed, say, to a mechanism based on direct activation of neurons of auditory cortex) is that cochlear microphonics should be generated, at least for pulses of relatively high energy content. The work by Chou and colleagues confirmed the existence of the cochlear microphonic.

The last category of behavior to be addressed, weak disturbances, includes measures of aversion—in effect, tests of the motivational sign of irradiation by radio fields at low-to-moderate power densities. In 1975, *Frey and Feld* [23] reported studies of rats in a shuttlebox, one side of which was shielded from, and the other irradiated by, CW or pulsed microwaves. The rats were free to choose between sides, their momentary preference being recorded continuously during experimental and control sessions. A reliable *tendency* to remain on the shielded side was demonstrated by the rats under pulsed but not under CW radiation. The pulse-irradiated rats spent approximately 70 percent of their time on the shielded side; controls and CW-irradiated rats, about 50 percent. One suspects that unpleasant auditory affect, such as that which accompanies unremitting tinnitus in human beings, was responsible for the shift of preference. Subsequently, Hjeresen and colleagues confirmed [24] the rat's tendency to escape from pulsed microwave fields, then found the same tendency when rats were subjected to pulsed sound waves.

There are other reports of aversive behavior [25, 26], for example, when each of a succession of mice was irradiated in a waveguide by 2450-MHz CW microwaves at a fixed level of forward power, the animal was observed to reduce its rate of energy absorption about 30 percent over 5–10 minute periods [25]. Reliable reductions of energy absorption occurred at initial dose rates as low as 60 μW/g when the temperature in the waveguide was very high (35 $^{\circ}$C). At normal environmental temperatures (20–25 $^{\circ}$C), reduction of dose rate occurred only at high initial rates near 40 mW/g. These findings were based solely on power-meter readings—a mouse was hidden from view in a waveguide—and may have reflected attempts by mice to escape from the field, but it is my suspicion that hyperpyrexia from a high ambient temperature or from excessive irradiation occurred and resulted in prostration. Thermally stressed rats assume a posture characteristic of heat prostration [2, 27]; perhaps the mice in the waveguide were also rendered prostrate, although the authors interpreted their power-meter datum as evidence of attempted (intentional) avoidance of the field. My interpretation hinges on the assumption that thermal collapse on the base of a waveguide under TE_{10} excitation reduces the animal's absorptive cross section and thus the rate of energy absorption, a testable but as yet dosimetrically untested assumption.

The final exemplar of weak disturbances lies in discrimination learning; specifically, the case in which microwave irradiation is presented continuously during measures of learning as a concurrent event, not as a cue to be discriminated or a stimulus to be avoided. King and I found evidence [2] that the efficiency of rats' discrimination of an acoustic signal that signalled the availability of a food reward was considerably augmented by concurrent microwave radiation at thermally significant (but not over-burdening) levels. The finding of enhanced discriminative efficiency was later confirmed and extended by *Nealeigh, Garner, Morgan, Cross, and Lambert* (1971) [28], who found that performance by rats in a Y-maze was considerably facilitated by moderately intense microwave irradiation. I selected the paper by Nealeigh and colleagues for inclusion here not only because they confirmed an interesting datum, but also because their discussion exemplifies an interpretative gap that currently divides behavioral investigators.

Nealeigh and colleagues, no doubt influenced by the reports of Eastern European investigators, suggested that the enhanced performance of their rats could be explained by microwave-induced *headache* (and, presumably, other accompaniments of the neurasthenic syndrome). Upon first examination, one's subjective reaction to a better-performance-from-pain interpretation may be one of skepticism, but the literature of experimental psychology contains many reports of studies in which presumably annoying stimulation of animals has indeed motivated behavior, not only as a negative but as a positive reinforcer. I reject the interpretation of Nealeigh and colleagues, preferring the more parsimonious explanation of simple acceleration of biological rate processes by increased temperature, but both interpretations (and several others as well) are candidates and will remain so until the closely knit integrative studies are performed that can link behavioral and neurological events into an understandable relation.

I shall conclude the chapter on the CNS and behavioral effects by giving numbers, where possible, which reflect threshold doses of shortwave or microwave energy. Power densities and dose rates in the absence of specified exposure durations are without substantive meaning; moreover, since transient thermodynamic effects (for example, hearing of pulsed microwaves) are not contained by averaged rates of energy absorption, there is definite utility in reckoning thresholds in terms of the energy dose [29]. In addition, because standards to prevent harmful exposures to plane-wave radio fields have been and probably will be couched in terms of averaged power densities of incident radiation, I shall accommodate the need for this datum by reporting it parenthetically, along with the minimal duration of exposure associated with a threshold response.

Highly elusive numbers are those associated with the VHF

field-induced exodus of calcium ions from the *in vitro* avian brain [4], because of the high degree of shortwave energy scatter by a relatively small target [30]. The power-density threshold of incident radiation was low, and the duration of treatment, although short, may have actually been longer than necessary (respectively, 1 to 2 mW/cm^2 and 20 minutes); the estimated total dose is ~9 mJ/g. A 918-MHz, microwave-pulse-evoked response from a single neuron of the intact cat [3] occurred at 4 μJ/g (~13 mW/cm^2 avg./s, 25-μs duration). The threshold of a multi-neuronal, pulse-evoked response from the intact cat [22] was somewhat higher, ~6 μJ/g (~20 mW/cm^2 given a 1-μs pulse and time averaging over 1 s).

Threshold doses for the shuttle-box performances of the rats that preferred the shielded to the pulse-irradiated side [23] cannot be estimated accurately from extant data because of the highly variable periods of time the rats spent on either side. (The minimal averaged power density associated with a reliable preference for the shielded side was 800 μW/cm^2.)

Detection by rats of sinusoidally modulated microwaves [17] (the thresholds of which, presumably, would hold for CW fields) occurred to a dose somewhat less than 35 mJ/g; for example, the threshold dose-rate is near 600 μW/g for exposure durations of 60 s, but the animals reacted to irradiation before termination of a 60-s exposure; the estimated equivalent plane-wave power density of incident radiation is near 3.5 mW/cm^2 at the threshold of detection. This power-density number is of interest because it is not too far from the number reflecting black-body emittance of radiant energy by the adult rat.

Microwave-induced work stoppage by hungry rats [2] under normal environmental conditions occurred to a microwave dose between 8 and 10 J/g, for example, after irradiation for 20 minutes at an averaged dose rate of 8 mW/g, which is close to the specific metabolic rate of the highly active rat. Irradiation of a rat by CW microwaves near its resonant frequency (600 MHz) and at a power density of 20 mW/cm^2 produced work stoppage after 23 minutes; at an off-resonant frequency of 400 MHz, time to work stoppage doubled at the same power density [27]. At both frequencies, the threshold energy doses were approximately the same (~8 J/g), which reflects the increase of energy absorption that defines the frequency-dependent phenomenon of resonance.

The acutely-delivered microwave dose resulting in a convulsively-indexed epileptic seizure [13] is close to 25 J/g. For example, at an averaged dose rate of 100 mW/g, rats convulsed after an average of 225 seconds of 2450-MHz irradiation (the extrapolated average power density at this frequency, which is off resonance for the rat, is about 500 mW/cm^2).

By way of final comment, I direct the reader's attention to several notable hiatuses in the literature on behavioral effects of microwaves, which have a bearing on exposure standards. The first is the virtual void of information on thresholds of pulsed radiation for work stoppage and epileptiform activity. Second is the paucity of information on endurance. King and colleagues developed [31] and validated [32] a technique—an automated swimming alley—for determining the ability of rats to perform sustained work after acute bouts of irradiation, and data based on the technique were subsequently reported by Hunt and colleagues [33], (theirs is the only study in which this important dimension of behavior is addressed). Third is the absence of data on mating behaviors, a critical end point that might hold the key to evaluation of the neurasthenic syndrome, primary manifestations of which are impotence and loss of libido. Finally, there is a notable absence of dosimetrically anchored experimental data on thresholds of discomfort or pain as wrought either by CW or by pulsed radiation in part- or whole-body exposures. These voids need to be filled, since they limit rational use in the clinic of shortwave and microwave energy.

* * * * * *

Addendum

The published reports by Allan Frey that were originally selected by the editor for inclusion in this volume are not included. Although Mr. Frey gave written permission in March of 1979 to republish his papers, he later requested that his papers be withheld. The reason for this reversal is unknown, but the omitted works obviously possess both scientific merit and historical value, hence their attempted inclusion. References to these works have been retained so that the interested reader may pursue other avenues of accession.

Don R. Justesen
Associate Editor

References

[1] M. S. Tolgskaya and Z. V. Gordon, *Pathological Effects of Microwaves*, New York: Consultants Bureau, 1973.

[2] D. R. Justesen and N. W. King, "Behavioral effects of low-level microwave irradiation in the closed space situation," in *Biological Effects and Health Implications of Microwave Radiation—Symp. Proc.*, S. F. Cleary, Ed., United States Public Health Service Publication no. BRH/DBE 70-2, Washington, DC: United States Government Printing Office, pp. 154–179, 1970.

[3] R. M. Lebovitz and R. L. Seaman, "Microwave hearing: the response of single auditory neurons in the cat to pulsed microwave radiation," *Radio Sci.*, vol. 12, no. 6S, pp. 229–236, 1977.

[4]* S. M. Bawin, L. K. Kaczmarek, and W. R. Adey, "Effects of modulated VHF fields on the central nervous system," *Ann. N.Y. Acad. Sci.*, vol. 247, pp. 74–80, 1975.

[5] C. F. Blackman, J. A. Elder, C. M. Weil, G. Benane, and D. C. Eichinger, "Modulation-frequency and field-strength dependent induction of calcium-ion efflux from brain tissue by radio-frequency radiation," *Radio Sci.*, vol 14, no. 6S, pp. 93–98, 1979.

[6] D. R. Justesen and V. Bruce-Wolfe, "Microwave hyperthermia and the visually evoked potential: preliminary observations of the guinea pig," *Proc. Symp. on Biological Effects and Measurement of RF/Microwaves*, HEW Publication (FDA) 77-8026, pp. 54–61, 1977.

[7] V. Bruce-Wolfe and D. R. Justesen, "Microwave-induced hyperthermia and the visually evoked electrocortical response of the guinea pig," *Radio Sci.*, vol. 14, no. 6S, pp. 187–191, 1979.

[8]* A. H. Frey, A. Fraser, Siefert, and T. Brish, "A coaxial pathway for recording from the cat brain stem during illumination with UHF energy," *Physiology and Behavior*, vol. 3, pp. 363-364, 1968.

[9] V. V. Tyazhelov, R. E. Tigraniam, and E. P. Khizhniak, "New artifact-free electrodes for recording of biological potentials in

*Asterisk indicates papers that were originally selected for reprinting in this volume. All papers so identified, save those withdrawn by Allan Frey, are reprinted in the following pages.

strong electromagnetic fields," *Radio Sci.*, vol. 12, no. 6S, pp. 121-123, 1977.

[10] E. N. Albert and M. DeSantis, "Do microwaves alter nervous system structure?" *Ann. N.Y. Acad. Sci.*, vol. 247, pp. 87-106, 1975.

[11] K. R. Brizzee, D. R. Justesen, and N. W. King, "Increased cell-packing density in neocortex of rats following irradiation by 12.25 cm microwaves," *Proc. Dept. of Defense Electromag. Workshop*, P. E. Tyler, Ed., United States Navy Bureau of Medicine and Surgery, Washington, DC, pp. 353-360, 1971.

[12]* J. C. Lin, A. W. Guy, and L. R. Caldwell, "Thermographic and behavioral studies of rats in the near field of 918-MHz radiations," *IEEE Trans. Microwave Theory Tech.*, vol. MTT-25, pp. 833-836, 1977.

[13]* R. D. Phillips, E. L. Hunt, and N. W. King, "Field measurements, absorbed dose, and biological dosimetry of microwaves," *Ann. N.Y. Acad. Sci.*, vol. 247, pp. 499-509, 1975.

[14] D. R. Justesen, R. A. Morantz, M. Clark, D. L. Reeves, and M. A. Mathews, "Effects of handling and surgical treatment on convulsive latencies and mortality of tumor-bearing rats to 2450-MHz microwave radiation," *Abstracts*, 1977 USNC/URSI Symposium on Biological Effects of Electromagnetic Waves, Washington, DC: National Academy of Sciences, p. 40, 1977.

[15] A. H. Frey, "Auditory system response to radio frequency energy," *Aerosp. Med.*, vol. 32, pp. 1140-1142, 1961.

[16] D. R. Justesen, "Microwaves and behavior," *Amer. Psychologist*, vol. 30, pp. 391-401, 1975.

[17]* N. W. King, D. R. Justesen, and R. L. Clarke, "Behavioral sensitivity to microwave irradiation," *Science*, vol. 172, pp. 398-401, 1971.

[18]* A. H. Frey and R. Messenger, "Human perception of illumination with pulsed ultra-high frequency electromagnetic energy," *Science*, vol. 181, pp. 356-358, 1973.

[19] K. R. Foster and E. D. Finch, "Microwave hearing: evidence for thermoacoustic auditory stimulation by pulsed microwaves," *Science*, vol. 185, pp. 256-258, 1974.

[20] R. M. White, "Generation of elastic waves by transient surface heating," *J. Appl. Phys.*, vol. 34, pp. 3559-3567, 1963.

[21] A. W. Guy, C.-K. Chou, J. C. Lin, and D. Christensen, "Microwave-induced acoustic effects in mammalian auditory systems and physical materials," in *Biologic Effects of Nonionizing Radiation*, P. E. Tyler, Ed., *Ann. N.Y. Acad. Sci.*, vol. 247, pp. 194-218, 1975.

[22]* C.-K. Chou, A. W. Guy, and R. Galambos, "Characteristics of microwave-induced cochlear microphonics," *Radio Sci.*, vol. 12, no. 6S, pp. 221-227, 1977.

[23]* A. H. Frey and S. R. Feld, "Avoidance by rats of illumination with low power non-ionizing electromagnetic energy," *J. Comparative and Physiological Psychology*, vol. 89, pp. 183-188, 1975.

[24] D. Hjeresen, S. R. Doctor, and R. L. Sheldon, "Shuttlebox side preference during pulsed microwave and conventional auditory cueing," in *Electromagnetic Fields in Biological Systems*, S. S. Stuchly, Ed., Edmonton, Alberta: International Microwave Power Institute, pp. 194-214, 1979.

[25] J. C. Monahan and H. S. Ho, "Microwave-induced avoidance behavior in the mouse," in *Biological Effects of Electromagnetic Waves, Selected Papers of the USNC/URSI Annual Meeting*, 1975, vol. I., C. C. Johnson and M. L. Shore, Eds., HEW Publication (FDA) 77-8010, Washington, DC: United States Government Printing Office, pp. 274-283, 1977.

[26] J. C. Monahan and H. W. Ho., "The effect of ambient temperature on the reduction of microwave energy absorption by mice," *Radio Sci.*, vol. 12, no. 6S, pp. 257-262, 1977.

[27] J. A. D'Andrea, O. P. Gandhi, and J. L. Lords, "Behavioral and thermal effects of microwave radiation at resonant and nonresonant wavelengths," *Radio Sci.*, vol. 12, no. 6S, pp. 251-256, 1977.

[28]* R. C. Nealeigh, R. J. Garner, R. J. Morgan, H. A. Cross, and P. D. Lambert, "The effect of microwaves on Y-maze learning in the white rat," *J. Microwave Power*, vol. 6, pp. 49-54, 1971.

[29] D. R. Justesen, "Toward a prescriptive grammar for the radiobiology of nonionizing radiations: quantities, definitions and units of absorbed electromagnetic energy," *J. Microwave Power*, vol. 100, pp. 343-356, 1975.

[30] C. H. Durney, C. C. Johnson, P. W. Barber, H. Massoudi, M. F. Iskander, J. L. Lords, D. K. Ryser, S. J. Allen, and J. C. Mitchell, Eds., *Radiofrequency Radiation Dosimetry Handbook*, second ed., Prepared for the USAF School of Aerospace Medicine by the Department of Bioengineering, University of Utah, Salt Lake City, UT, 1976.

[31] N. W. King, E. L. Hunt, R. D. Castro, and R. D. Phillips, "An automated swim-alley for small animals: I—instrumentation," *Behavior Research Methods and Instrumentation*, vol. 6, no. 6, pp. 531-534, 1974.

[32] N. W. King, E. L. Hunt, R. D. Castro, and R. D. Phillips, "An automated swim-alley for small animals: II—training and procedure," *Behavior Research Methods and Instrumentation*, vol. 6, no. 6, pp. 535-540, 1974.

[33] E. L. Hunt, N. W. King, and R. D. Phillips, "Behavioral effects of pulsed microwave radiation," in *Biologic Effects of Nonionizing Radiation*, P. E. Tyler, Ed., *Ann. N.Y. Acad. Sci.*, vol. 247, pp. 440-453, 1975.

*Asterisk indicates papers that were originally selected for reprinting in this volume. All papers so identified, save those withdrawn by Allan Frey, are reprinted in the following pages.

EFFECTS OF MODULATED VHF FIELDS ON THE CENTRAL NERVOUS SYSTEM*

S. M. Bawin, L. K. Kaczmarek, and W. R. Adey

Space Biology Laboratory
Brain Research Institute and Department of Anatomy
University of California
Los Angeles, California 90024

Introduction

The existence of brief epochs in which electroencephalographic and neuronal activities are strongly correlated has been repeatedly established in different areas of the brain. For example, periodic slow oscillations of neuronal membrane potentials, frequency related to the concomitant local electroencephalogram (EEG), have been recorded in various cortical areas[1–4] and in hippocampal and thalamic sites.[5,6]

Therefore, the EEG appears to reflect the attenuated undulations of the membrane potential of a surrounding population of neurons, and rhythmic electroencephalographic patterns could be generated by extracellular summation of simultaneous transient slow electrical events in a population of cells.

On the other hand, weak extracellular voltage gradients (1–5 mV/mm) have been shown to significantly affect the excitability, or firing thresholds, of stretch receptor neurons in the crayfish and spinal motor neurons in the cat.[7,8] As pointed out earlier by Nelson,[8] complex structural organization of brain tissues, as seen in the cerebrum, should be highly favorable for multiple electric field interactions, both in the intricate rate of overlapping dendritic trees and between the macromolecules of the extracellular space and the glycoproteins of the outer surface of the cell membrane.[9–14]

Indeed, extremely weak vhf fields [147 MHz, 1 mW/cm^2], amplitude modulated at brain wave frequencies, have been shown to strongly influence spontaneous and conditioned EEG patterns in the cat.[14,15] The hypothesis was offered that the weak electrical forces induced in the brain were modifying the excitability of the central neurons and that these changes were reflected in the recorded transient EEG episodes.

The extracellular electrical gradients could exert their forces on the multiequilibrium system that exists in the outer zone of the neuronal membrane, where mono- and divalent cations compete for binding sites on polyanionic macromolecules and polar ends of intrinsic membrane proteins.[16–20]

Local variations in the surrounding electric field could result in slight modifications of the negative binding sites, either by triggering configurational changes of the surface macromolecules or by inducing small displacements of the surface-bound cations. These alterations would, in turn, influence the activity of adjacent ions, which would disturb further the molecular arrangement of the surface macromolecules, thus propagating and multiplying the initial electrical disturbance.

Concurrent work in this laboratory[21] indicated that weak pulsed electric currents (2–5 mV/mm, 200 pulses/sec) applied across the cat cortex were able to trigger the release of previously bound radioactive calcium ($^{45}Ca^{2+}$). This, with earlier findings[22] that a local increment in extracellular calcium induces a local increase in the efflux of membrane-bound $^{45}Ca^{2+}$, appears to support the hypothesis of an intrinsic membrane amplification mechanism.

Intracranial injection of Ca^{2+} or Mg^{2+} [20 μl, 40 mM] in chronically implanted neonatal chicks resulted in an almost immediate synchronization of the hyperstriatal EEG, accompanied by behavioral depression.[23] During successive testing days, the animals appeared to recover behaviorally but never showed any sustained EEG arousal. By contrast, animals treated with sodium chloride recovered completely within the first hour after the injection.

The chick forebrain, being so highly sensitive to small perturbations of the extracellular concentrations of either divalent cations, was therefore chosen for investigating, *in vitro*, the possible interactions between extracellular weak voltage gradients, induced by vhf radiations, and ionic movements in cerebral tissue. In the present experiment, $^{45}Ca^{2+}$ fluxes from irradiated brains are compared at various frequencies of amplitude modulation of the carrier wave.

Materials and Methods

The experiments were conducted in an environmental chamber (ambient temperature, 37°C; relative humidity, 35%) specially adapted for the use of vhf fields. The screening procedures and implementation of the 147-MHz fields have been described elsewhere.[9,15] Briefly, the feedline from the transmitter was applied, via an antenna coupler, to the narrow apices of two large triangular aluminum plates (4100 cm^2). The applied power, monitored by two in-line wattmeters and a radio frequency microampmeter, was adjusted to provide field intensities of 1 to 2 mW/cm^2. Sinusoidal modulating frequencies at 0.5–35 Hz were fed into a power supply specifically designed to provide modulated high voltages to the transmitter. Modulation depths were kept between 80 and 90%.

Five hundred neonatal chicks, which ranged in age from 2 to 7 days, were used in these experiments. The animals were sacrificed by decapitation. The forebrains were rapidly dissected from the cranial cavities, and each cerebral hemisphere was incubated for 30 min in a polyallomer test tube that contained 1 ml of physiologic medium [37°C, 155 mM NaCl, 5.6 mM KCl, 2.16 mM $CaCl_2$, 24 mM $NaHCO_3$, and D-glucose (2 g/liter)] together with 0.2 ml of saline that contained 0.2 μCi of $^{45}Ca^{2+}$ (sp act, 1.39 Ci/mmole).

After incubation, the samples were washed three times with nonradioactive solution. The brains were than bathed in 1 ml of physiologic medium for 20-min epochs each, during which time they were exposed to one of the experimental conditions. Assay of radioactivity was by liquid scintillation counting with 0.2 ml of the bathing solution diluted in 11 ml of Packard "Instagel."

The experiment included irradiation with sinusoidal modulation of the vhf field at 0.5, 3, 6, 9, 11, 16, 20, 25, and 35 Hz, also with an unmodulated carrier wave. Controls were run in the absence of fields. Ten half brains were tested simultaneously for each field condition and control. Every field condition was tested at least three times, to provide sufficiently large populations for further statistical analyses. In each experiment, three series of 10 samples were irradiated with the vhf fields modulated at three different frequencies, and one series of 10 half brains served as the control (no field condition). The order of presentation of the four conditions was randomized for each daily experiment. The radioactivities (cpm) of all samples were related to the mean value of the counts obtained with the 10 control samples, taken

*Supported in part by National Science Foundation Grant GB-27740 and by Office of Aerospace Research Contract F44620-70-C-0017.

Reprinted with permission from *Ann. N.Y. Acad. Sci.*, vol. 247, pp. 74–81, Feb. 28, 1975.

as 100. All normalized data within each condition were then statistically compared with matched samples of control values.

An additional series of 40 chicks was used to compare field effects on the $^{45}Ca^{2+}$ efflux from brains bathed in physiologic medium and brains poisoned, after incubation, with 10^{-4} M NaCN. Five poisoned samples were tested simultaneously with five normal brains for three field exposures (unmodulated radiation and fields modulated at 0.5 and 16 Hz) and control condition. Each condition was tested twice, and the data were normalized to the control mean values.

Possible field effects on the $^{45}Ca^{2+}$ efflux from a variety of biologic tissues are now being investigated. Preliminary experiments have been conducted with skeletal muscles (lateral head of the gastrocnemius, 25 animals). Muscle and cerebrum dissected from the same animal have been tested simultaneously for two field exposures (6-Hz modulation, 10 samples; 16-Hz modulation, 15 samples) and no field, 20-min epochs, for control.

Results

$^{45}Ca^{2+}$ *Efflux from Brain Tissues Bathed in Physiologic Medium*

Unmodulated radiations and fields modulated at 0.5 and 3 Hz failed to induce any significant changes in the $^{45}Ca^{2+}$ efflux, by comparison with unirradiated control brains (three repetitions, 30 samples for each condition). By contrast, there was a progressive increase in the $^{45}Ca^{2+}$ efflux from the brains exposed to the fields modulated at 6 Hz (40 samples, 10.1%, $p < 0.05$), 9 Hz (30 samples, 14.3%, $p < 0.05$), 11 Hz (50 samples, 16.0%, $p < 0.01$), and 16 Hz (80 samples, 18.5%, $p < 0.01$). These effects gradually decline at higher frequencies. Exposures to 20-Hz sinusoidal modulation lead to a small increase of the $^{45}Ca^{2+}$ efflux (30 samples, 9.5%, $p < 0.05$). The results obtained with 25-Hz modulation (30 samples, 6%) were not statistically significant, and the fluxes observed with 35-Hz modulation did not differ from the controls.

These findings are illustrated in Figure 1. The data are expressed as percentage of increase of the $^{45}Ca^{2+}$ efflux for the various experimental conditions, by comparison with the no field condition.

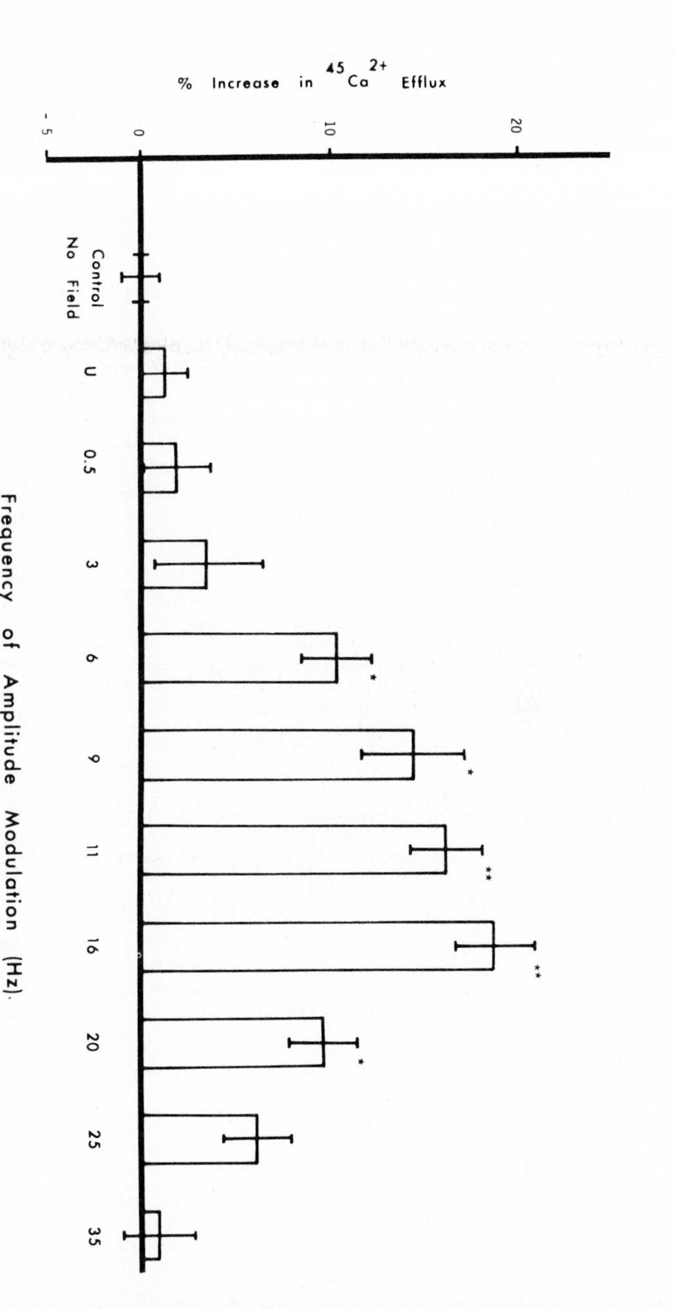

Figure 1. Effects of amplitude-modulated 147 MHz vhf fields on the $^{45}Ca^{2+}$ efflux from the isolated forebrain of the neonatal chick. The results, given ± SEM, are expressed as percentage of increase of the calcium efflux, by comparison with control condition, in the absence of fields. *, $p < 0.05$; **, $p < 0.01$.

$^{45}Ca^{2+}$ *Efflux from Brain Tissues Poisoned with 10^{-4} M NaCN*

The results with poisoned brains were identical to those observed simultaneously for the samples bathed in physiologic solution. The field effects observed previously were not altered by the cyanide treatment, which strongly suggests that the $^{45}Ca^{2+}$ effluxes from the cerebral tissues are independent of any ongoing metabolism. The $^{45}Ca^{2+}$ fluxes from poisoned brains and normal samples are compared in Table 1. The data have been normalized to the mean value of the radioactivity counts obtained in the absence of the vhf field with unpoisoned brains.

$^{45}Ca^{2+}$ *Efflux from Skeletal Muscle*

There exists a much greater variability in the radioactivity counts obtained with muscular tissue than with cerebral tissue, within each experimental condition. This

TABLE 1

EFFECTS OF AMPLITUDE-MODULATED VHF FIELDS ON $^{45}Ca^{2+}$ EFFLUX FROM THE ISOLATED CHICK BRAIN*

	Relative $^{45}Ca^{2+}$ Efflux†	
	Normal Brain	Poisoned Brain
Control, no field	100.0 ± 4.0 (15)	96.7 ± 3.6 (15)
Unmodulated field	103.7 ± 6.0 (10)	102.3 ± 5.3 (10)
0.5-Hz modulation	100.5 ± 4.6 (10)	98.7 ± 5.0 (10)
16-Hz modulation	114.2 ± 6.4 (10)‡	118.9 ± 7.7 (10)‡

*Effects of 10^{-4} M NaCN at spot frequencies of the modulation.

†The number of determinations is given in parentheses. The data referred to the mean control value of 100 are given ± SEM.

‡$p < 0.05$.

TABLE 2

EFFECTS OF AMPLITUDE-MODULATED VHF FIELDS ON $^{45}Ca^{2+}$ EFFLUX FROM BIOLOGIC TISSUES*

	Relative $^{45}Ca^{2+}$ Efflux†	
	Brain	Muscle
Control, no field	100.0 ± 5.4 (10)	100.0 ± 5.8 (10)
6-Hz modulation	115.0 ± 4.2 (10)‡	105.1 ± 9.8 (10)
16-Hz modulation	119.1 ± 6.2 (10)‡	100.1 ± 3.9 (15)

*Comparison of brain and skeletal muscle at two spot frequencies of the modulation.

†The number of determinations is given in parentheses. The results referred to the mean control value of 100 are given ± SEM.

‡$p < 0.05$.

variability may be due to small differences in the weight of the samples, and the results would probably be more homogeneous if expressed in radioactivity counts per minute per gram of tissue. However, the preliminary results obtained after exposures to the vhf radiations modulated at 6 and 16 Hz do not indicate any changes in the $^{45}Ca^{2+}$ efflux, by comparison with unirradiated muscles. The brain samples tested concurrently with the muscular tissues gave results identical to these reported above.

Comparisons between the $^{45}Ca^{2+}$ effluxes from both tissues are given in TABLE 2, for the two field exposures tested so far. The experimental data have been related to the mean value of the radioactivity counts obtained with unirradiated muscles and brains.

DISCUSSION

The experimental data indicate that weak vhf fields, amplitude modulated at brain wave frequencies, are able to increase the calcium efflux from the isolated brain of the neonatal chick.

The need for additional investigation of calcium fluxes in a variety of tissues is obvious. Nevertheless, it is worth noting that the two frequencies of modulation tested in the preliminary experiment reported here did not trigger significant changes in calcium efflux from striated muscle, even though both frequencies were effective for the brain samples tested at the same time.

The $^{45}Ca^{2+}$ fluxes from the brain were not influenced by cyanide treatment. The results obtained with the poisoned control samples, in the absence of field, as well as with the poisoned irradiated samples, were not different from those obtained with brains bathed in physiologic medium. Therefore, it may be assumed that the ionic exchanges observed in this experiment were independent of ongoing metabolic processes. This finding seems to favor the divalent cation, present in the highly complex border zone where the cell membrane is in contact with extracellular macromolecular material, as a possible mediator of the observed phenomena.

As mentioned in the introduction, the electrochemical equilibrium that exists in cerebral tissues between ions, polyanionic macromolecules, and glycoproteins of the cell surface can be disrupted by small variations of either surrounding ionic concentrations or local electrical gradients. Thus, rhythmic modulations of the radio frequency energy, reflected as slow undulations of the extracellular electric field, could easily affect the binding of calcium to the neuronal membrane.

A small displacement of these calcium ions, which would result in cooperative interaction between modified adjacent binding sites, could play an important role in the propagation and amplification of local electrical events.

The fact that unmodulated radio frequency energy failed to induce any change in the fluxes, in addition to the findings that the changes observed at 11 and 16 Hz progressively decreased in magnitude for ascending and descending frequencies of modulation, seems to indicate that the calcium ion movements under observation were critically related to very specific slow components of the irradiation.

This finding is in accordance with many previous studies,[24–28] in which the pulse repetition rate of radio frequency energies has been shown to be of critical importance in eliciting specific biologic field effects, and it supports further the hypothesis that the specific frequencies of modulation were responsible for the specific changes seen in the cat EEG. In this context, it might be of significance that the frequencies that lead to the largest calcium efflux (11 and 16 Hz) are constituents of the EEG of the aroused neonatal chick.[29,30]

Finally, because the binding and release of calcium has been linked to inhibition and excitation in the cerebral cortex of the cat,[22] the mode of interaction between external electric fields and central neurons proposed here could explain, at least partially, our previous electroencephalographic results.

ACKNOWLEDGMENTS

We wish to thank Mrs. I. Sabbot for her advice and help during this investigation, Ms. S. Skiles for preparation of the illustrations and editing of the manuscript, and Mr. D. Olsen and Mr. S. Huston for their engineering support.

REFERENCES

1. ELUL, R. 1962. Dipoles of spontaneous activity in the cerebral cortex. Exp. Neurol. **6:** 285–299.
2. ELUL, R. 1964. Specific site of generation of brain waves. Physiologist **7:** 125.
3. CREUTZFELDT, O. D., S. WATANABE & H. D. LUX. 1966. Relations between EEG phenomena and potentials of single cortical cells. II. Spontaneous and convulsoid activity. Electroencephalogr. Clin. Neurophysiol. **20:** 19–37.
4. JASPER, H. & C. STEFANIS. 1965. Intracellular and oscillatory rhythms in pyramidal tract neurons in the cat. Electroencephalogr. Clin. Neurophysiol. **18:** 541–553.

5. FUJITA, Y. & T. SATO. 1964. Intracellular records from hippocampal pyramidal cells in rabbit during theta rhythm activity. J. Neurophysiol. **27:** 1011–1025.
6. PURPURA, D. P. & B. COHEN. 1962. Intracellular recording from thalamic neurons during recruiting responses. J. Neurophysiol. **25:** 621.
7. TERZUOLO, C. A. & T. H. BULLOCK. 1956. Measurement of imposed voltage gradient adequate to modulate neuronal firing. Proc. Nat. Acad. Sci. USA **42:** 687–694.
8. NELSON, P. G. 1966. Interaction between spinal motoneurons of the cat. J. Neurophysiol. **29:** 275–287.
9. ADEY, W. R. 1974. The influences of impressed electrical fields at EEG frequencies on brain and behavior. *In* Behavior and Brain Electrical Activity. H. Eltshuler & N. Burch, Eds. Plenum Publishing Corporation. New York, N.Y. In press.
10. LEHNINGER, A. L. 1968. The neuronal membrane. Proc. Nat. Acad. Sci. USA **60:** 1055–1101.
11. LENARD, J. & S. J. SINGER. 1966. Protein conformation in cell membrane preparations as studied by optical rotatory dispersion and circular dichroism. Proc. Nat. Acad. Sci. USA **56:** 1828–1835.
12. WALLACH, D. F. H. & P. H. ZAHLER. 1966. Protein conformation in cellular membranes. Proc. Nat. Acad. Sci. USA **56:** 1552–1559.
13. SCHMITT, F. O. 1969. Brain cell membranes and their microscopic environment. Neurosci. Res. Progr. Bull. **7:** 281–300.
14. BAWIN, S. M. 1972. Cat EEG and behavior in very high frequency electric fields amplitude-modulated at brain wave frequencies. Ph.D. Dissertation. University of California, Los Angeles, Calif.
15. BAWIN, S. M., R. J. GAVALAS-MEDICI & W. R. ADEY. 1973. Effects of modulated VHF fields on specific brain rhythms in cats. Brain Res. **58:** 365–384.
16. CARVALHO, A. P., H. SANUI & N. PACE. 1963. Calcium and magnesium binding properties of cell membrane. J. Cell. Comp. Physiol. **62:** 311–317.
17. SIMPSON, L. A. 1968. Review Article: The role of calcium in neurohumoral and neurohormonal extrusion processes. J. Pharm. Pharmacol. **20:** 889–910.
18. TASAKI, I., T. TAKEWA & S. YAMAGISHI. 1968. Abrupt depolarization and bi-ionic action potentials in internally perfused squid giant axons. Amer. J. Physiol. **215:** 152–159.
19. RAMWELL, P. W. & J. E. SHAW. 1970. Bioloical significance of the prostaglandins. Recent Prog. Hormone Res. **26:** 133–187.
20. ADEY, W. R. 1971. Evidence for cerebral membrane effects of calcium, derived from direct-current gradient, impedance, and intracellular records. Exp. Neurol. **30:** 78–102.
21. KACZMAREK, L. K. & W. R. ADEY. 1974. Weak electric gradients change ionic and transmitter fluxes in cortex. Brain Res. **66:** 537–540.
22. KACZMAREK, L. K. & W. R. ADEY. 1973. The efflux of $^{45}Ca^{2+}$ and $[^{3}H]$ γ aminobutyric acid from cat cerebral cortex. Brain Res. **63:** 331–342.
23. BAWIN, S. M., L. K. KACZMAREK, S. R. MAGDALENO & W. R. ADEY. 1974. Effects of intracranial injection of ions on the electrocortical activity of the chick. In preparation.
24. GAVALAS, R. J., D. O. WALTER, J. HAMER & W. R. ADEY. 1970. Effect of low-level, low-frequency electric fields on EEG and behavior in *Macaca nemestrina*. Brain Res. **18:** 491–501.
25. HAMER, J. 1968. Effects of low level, low frequency electric fields on human reaction time. Commun. Behav. Biol. **2**(A): 217–222.
26. KORBEL, S. F. & H. L. FINE. 1967. Effects of low intensity UHF radio fields as a function of frequency. Psychonom. Sci. **9:** 527, 528.
27. SUBBOTA, A. G. 1958. The effect of a pulsed super-high frequency SHF electromagnetic field on the higher nervous activity of dogs. Bull. Exp. Med. **46:** 1206–1211.
28. TANNER, J. A., C. ROMERO-SIERRA & S. J. DAVIE. 1967. Nonthermal effects of microwave radiation on birds. Nature (London) **216:** 1139.
29. CORNER, M. A., J. J. PETERS & P. RUTGERS VAN DER LOEFF. 1966. Electrical activity patterns in the cerebral hemisphere of the chick during maturation, correlated with behavior in a test situation. Brain Res. **2:** 274–292.
30. SCHULMAN, A. H. 1971. The effects of imprinting training on hyperstriatal EEG in young domestic chicks [*Gallus domesticus*]. Develop. Psychobiol. **4**(4): 353–362.

DISCUSSION

DR. J. C. SHARP (*Walter Reed Army Institute of Research, Washington, D.C.*): Can an interaction from the radio-frequency field be picked up by the electrode on the brain tissue?

DR. BAWIN: We thought that we might be able to induce some current in the brain through the electrode. Nevertheless, as the cable is shielded up to the head, most of the radio-frequency energy picked up on the cable shielding is returned to ground. Only a short length (about 3 cm) of unshielded electrode was present beyond the connection to the shielded cable. This was only a small fraction of a wavelength at 147 MHz. Our best argument exists in the EEG records. We found no propagation phenomena between electrodes; that is, the field effect that we describe as "sharpening" of a response toward certain frequencies was only associated with a very short transient rhythm similar to that in the on-going EEG. The evoked rhythm in the conditioned cat subsided after a long time in extinction trials. It vanished completely from the EEG while the animal was in extinction trials in the presence of radio-frequency fields. It would not be reasonable to assume that we are inducing a current in the brain that will only materialize when the animal is responding to the conditioning stimulus. There was no cross talk between one electrode and another. I generally monitored four or five electrode sites simultaneously, and I conditioned rhythms in only one or two specific locations. I detected no rhythm propagation into any other leads.

DR. J. A. ELDER: Upon turning the modulated vhf fields off, does the calcium again become bound to the brain tissue? Is the effect reversible or irreversible?

DR. BAWIN: I don't know. The calcium is constantly exchanged back and forth, and I only measured at a particular point in time how much calcium had been exchanged from the brain to the saline milieu.

Characteristics of microwave-induced cochlear microphonics

Chung-Kwang Chou, Arthur William Guy

Bioelectromagnetics Research Laboratory, Department of Rehabilitation Medicine RJ-30, University of Washington School of Medicine, Seattle, Washington 98195

Robert Galambos

Department of Neurosciences A-012, University of California, San Diego, School of Medicine, La Jolla, California 92093

Cochlear microphonics (CM) have been recorded from guinea pigs and from cats of differing body mass during irradiation by 918- and by 2450-MHz pulsed microwaves. Both horn applicators and a cylindrical waveguide exposure system were used to radiate the animals. The CM frequency and duration were similar irrespective of carrier frequency and mode of application. Parameters of the CM (except amplitude) were not influenced by orientation of the body axis to the electrical field or by pulse width of microwaves for pulses less than 30 μsec wide. The frequency of the CM correlated well with the longest dimension of the brain cavity and poorly with several dimensions of the head. Extrapolations of these animal findings to the human being indicates that the frequency of the microwave-induced CM in man should be between 7 and 10 kHz.

1. INTRODUCTION

In our previous investigations [*Chou et al.*, 1975, 1976, 1977], we have shown that with proper recording techniques, a cochlear microphonic (CM) can be recorded from guinea pigs and from cats during pulsed irradiation by 918-MHz microwaves. Recordings of the pulse-evoked CM indicate that the microwaves initiate mechanical events that activate the auditory system at the cochlear level. Other studies [*Foster and Finch*, 1974; *Lebovitz and Seaman*, 1977; *Lin*, 1977] provide additional evidence in support of such an electromechanical interaction, one that presumably involves thermal expansion, which would contradict an earlier hypothesis of direct neural stimulation [*Frey*, 1971].

In this paper, we present more data on the microwave-induced CM in an attempt to answer the following questions: What dimensional parameters of the skull determine the frequency of CM? Do 918- and 2450-MHz microwave pulses produce differing CMs? Is the CM waveform dependent upon the polarization of the electrical field? Does the CM change as a function of source of radiation? What is the threshold of the microwave-induced auditory response? What is the expected frequency of CM in the human being?

2. METHODS

Guinea pigs and cats of different body masses were anesthetized with pentobarbital sodium (35-40 mg/kg IP). Either the right or left bulla was opened and a fine (microwave transparent) carbon lead was placed against the round window and cemented to nearby bone. An indifferent electrode was connected to proximal tissue. The heads of the guinea pigs and small cats were then placed inside and radiated within a cylindrical waveguide [*Chou et al.*, 1975]. The inserted head was parallel to the electrical field of the TE_{11} mode waves in the waveguide. For the large adult cats, a 13 × 13 cm aperture-source was loaded with dielectric material and was used for the 918-MHz exposure; a standard horn applicator was used to apply 2450-MHz microwaves. All animals, their exposure systems, the microwave generator (Applied Microwave Laboratory PH 40K), and two power meters were enclosed in an anechoic chamber, an isolation procedure that minimized microwave artifact in physiological recordings. Conventional physiological amplifiers (gain 2×10^4, bandpass 1 Hz to 100 kHz) processed the CM signal. The animals were radiated intermittently for 1.5 minutes at a time by 918- or by 2450-MHz microwave pulses of 10-μsec duration (rise time less than 1 μsec) and 100-pps repetition rate at various peak-power levels below 10 kW. The responses were recorded on a magnetic tape system (Honeywell 7600) with a frequency response of 80 kHz. For recordings made using a 100-μsec time base, a Honeywell SAI-43A correlation and probability analyzer was used to enhance and visualize the induced responses. The averaged responses were then plotted by an *x-y* recorder on graphic paper.

Two types of dosimetric measurements were made to calculate the quantity of absorbed energy per pulse. The parameters are stated in terms of energy instead of power, since the CM response is produced by microwave pulses of the same energy for all pulses of durations less than 30 μsec [*Guy et al.*, 1975]. For the guinea pigs and small cats that were radiated in the cylindrical waveguide, the averaged absorbed energy per pulse was calculated by dividing the total energy of the microwave pulse by

Reprinted with permission from *Radio Sci.*, vol. 12, pp. 221–227, Nov.–Dec. 1977.

the mass of the head, since more than 99% of the microwave energy was absorbed by the head [*Chou et al.*, 1977]. In order to use existing thermographic data [*Johnson and Guy*, 1972], a 3.1-kg cat was held in a stereotaxic instrument and was radiated by microwaves from a square-aperture source that was located 8 cm from the occipital pole of the cat. For cats radiated with the head in close proximity to the applicators, dosimetric measures were not made since the major interest was to elicit a CM response of reasonably high amplitude.

In order to obtain dimensions of the skull, animals were decapitated after the radiation experiments; the mass of the heads was determined and soft tissues were then removed from the skull. The following measurements were then made -- for terminology, refer to *Crouch* [1969] and *Cooper and Schiller* [1975]:

Skull mass (without lower mandible)
Skull size: L = between incisors and lambdoidal ridge
W = between right and left outside squamous portions
H = between outside vertex and basisphenoid in cats and spenoid in guinea pigs
Skull thickness (at vertex)
Brain-cavity dimensions

Cat: L = between upper transverse ridge and base of *tentorium cerebelli*
W = between right and left interior of squamous portions
H = between interior of vertex and occipital bone

Guinea Pig: L = between cribriform plate and occipital bone
W = same as in cat
H = between inside vertex and sphenoid

Brain-cavity volume
Bulla size L × W × H
Cerebellar-cavity dimensions

Cat: L = between intersection of sagittal and lambdoidal suture and basisphenoid
W = between widest points of cerebellum
H = between *tentorium cerebelli* and margin of skull

Cerebellar-cavity volume

3. RESULTS

Typical microwave-induced CM responses are shown in Figure 1. These responses occurred in every animal that exhibited an acoustically induced CM that was greater than 0.5 mV in amplitude. The amplitude of the CM varied with the placement of the electrode but the time course of the response did not. Therefore, only the time course of the responses will be mentioned in the following discussion. We have shown previously that microwave-induced CMs and auditory neural responses are independent of the pulse width for pulses of duration less than 30 μsec [*Chou et al.*, 1975; *Guy et al.*, 1975]. After considering the energy per pulse and the lifetime of the microwave generator tubes, we chose 10 μsec for all the experiments.

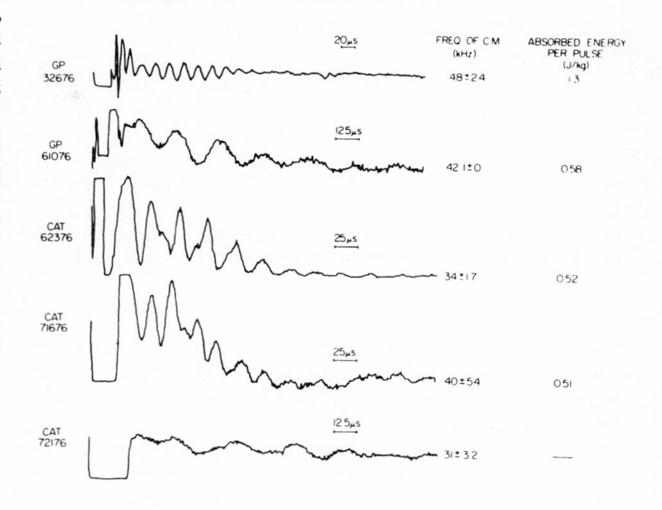

Fig. 1. Microwave-induced cochlear microphonics in guinea pigs and cats. Pulsed microwave: 918 MHz, 10 μs width.

There are two major differences in Figure 1. The frequencies of the CM are different and the waveforms of the CM of cats are more complicated than those of guinea pigs. The following experimental parameters were changed in an attempt to explain the above results.

Pulse width. Same frequency and duration of microwave-induced CM were produced by pulsed microwaves of 10, 5 and 1 μsec pulse width [*Chou et al.*, 1975]. The CM response is time-locked to the onset of the microwave pulses.

Polarization of electrical field. Figure 2 illustrates that there was no difference in the CM when the head was irradiated with the electrical field of TE_{10} mode waves either parallel or perpendicular to the anterior-posterial body axis.

Carrier frequency. Figure 3 shows that the same frequency and duration of CM were induced by 918- and 2450-MHz microwaves as delivered by horn applicators. The amplitude of the CM response was different because of the differing coupling efficiencies of the two horn applicators. When these horn applicators were used, it was necessary to place a cat's head in close proximity in order to record a detectable CM.

Horn applicator versus waveguide exposure system. For small cats exposed to the horn applicator and then to the cylindrical waveguide, the frequency and duration of CM were similar. The amplitude of the CM was much greater, however, when exposures were made in the cylindrical waveguide due to more efficient coupling of energy.

Body mass. Guinea pigs and cats of differing body

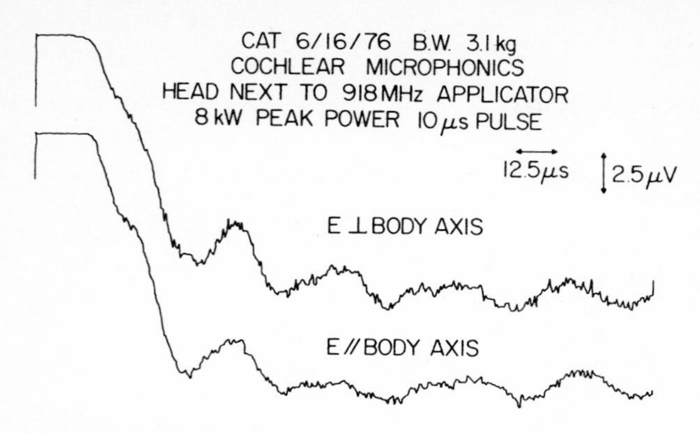

Fig. 2. Comparison of cochlear microphonics induced by pulsed microwaves at different polarizations of electrical fields.

mass were used. Tables 1 and 2 show the frequency of CM, the number of oscillations of CM, and $t_{1/e}$ (time required for the amplitude of the CM to drop to $1/e$). Pulsed microwaves of 10 μsec duration at 918 MHz were delivered at an averaged absorbed energy per pulse as indicated. Due to the large quantity of reflected power, which would limit the life of the microwave generator, exposure in the cylindrical cavity was made only once (GP 30675).

For guinea pigs, Table 1 shows that the larger the mass of the animal, the lower the frequency of the CM. Except for GP 32875, the number of oscillations decreased as body mass increased. The parameter $t_{1/e}$ varied between 31.6 and 55 μsec. Table 2 shows that the frequency of the CM varied between 29.6 and 32.5 kHz, with less than five oscillations, for adult cats exposed to the applicator. For kittens, (body mass < 2 kg) that were irradiated either in the cylindrical waveguide or by the horn applicator, the frequency varied between 31 to 40 kHz and the number of oscillations were between 5 and 10. There was no consistent relationship between the frequency of the CM and the body mass of the cats and kittens. It is also noted that the $t_{1/e}$ for cats is about two to three times longer than for guinea pigs.

Physical dimensions of the head. In an effort to isolate the variables that determine the CM, numerous measurements of dimensions of the animals' heads were made and are listed in Table 3. Note that as the body mass increases, so too does the head mass, skull mass, skull dimension, skull thickness and dimension of the cerebellar cavity; the brain cavity and bulla dimensions, however, increase only slightly. Plots of CM frequency versus each of the measured parameters as shown in Table 3 did not provide a consistent explanation for the variation in the frequency of CM in cats. However, the longest dimension in the brain cavity (for cats, that distance between the anterior orbital gyrus and the posterior tip of the posterior composite gyrus), shows a consistent relationship to frequency of the CM both for cats and for guinea pigs. As shown in Figure 4, it is clear that the greater the length of the brain cavity, the lower the frequency of the microwave-induced CM.

Concerning threshold of energy, Figure 5 illustrates the intensity-function for a small cat that was exposed in the cylindrical waveguide. This function is similar to that of a guinea pig studied by *Chou et al.* [1975]. For a larger, 3.1-kg adult cat, Figure 6 shows the relationship between N_1 (nerve response) amplitude and latency versus the peak value of absorbed energy per pulse. Since the amplitude of the CM was smaller than 5 μV under this exposure condition, the amplitude is not shown in the figure. For a 918-MHz plane wave that is incident on a spherical model of the brain (3 cm radius), the averaged specific absorption rate (SAR) in W/kg is about 3/8 of its peak SAR in the brain [*Johnson and Guy*, 1972]. As is seen in Figures 5 and 6, the efficiency of energy coupling in the cylindrical waveguide is about ten times higher than that of the aperture-source of radiation. The lowest energy required to induce a response (CM or N_1) was 10 mJ/kg peak in adult cats, 2.5 mJ/kg average in kittens, and 7.5 mJ/kg average in the adult guinea pigs. The peak absorbed energy density per pulse of guinea pigs as exposed in a cylindrical waveguide is about ten times higher than the averaged absorbed energy per pulse [*Chou et al.*, 1977]. The above results are consistent with our previous data for an estimated peak threshold of energy of 16 mJ/kg for hearing in human beings and an evoked response in the medial geniculate in cats of 10 mJ/kg. Comparable results have been reported by *Lebovitz and Seaman* [1977].

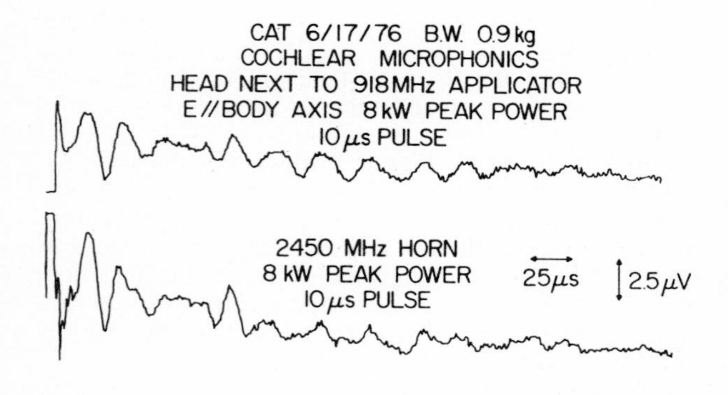

Fig. 3. Comparison of cochlear microphonics induced by pulsed microwaves at 918 and 2450 MHz.

4. DISCUSSION

Cochlear microphonics are electrical potentials that mimic the sonic waveforms of acoustic stimuli. An understanding of the microwave-induced CM should shed more light on the mechanisms of microwave-induced hearing. It was shown that the characteristics of CM (except amplitude) do not depend on carrier frequency, mode of application, field polarization and pulse width of the applied microwave pulses. Instead, the frequency of CM correlates well with the length of the brain cavity and poorly with other measurements made upon the head and the skull. These results provide more evidence that the microwave auditory effect is mechanical in nature.

TABLE 1. Characteristics of microwave-induced cochlear microphonics in guinea pigs.

Animal Number	Body Mass (kg)	Frequency ± SD (kHz)	No. of Oscillations	$t_{1/e}$ (μs)	Exposure Apparatus	Averaged Absorbed Energy per Pulse (J/kg)
GP 30675	∿ 0.4	∿ 50	8	--	Cylindrical Cavity	--
GP 32575	∿ 0.4	50 ± 0	11	--	Cylindrical Waveguide	1.4
GP 32675	∿ 0.45	48 ± 2.4	12	31.6	Cylindrical Waveguide	1.3
GP 32875	∿ 0.45	50 ± 0	5	--	Cylindrical Waveguide	1.3
GP 61076	1.10	42.1 ± 0	8	36.25	Cylindrical Waveguide	0.73
GP 61476	1.07	46.1 ± 2.5	10	55	Cylindrical Waveguide	0.8
GP 61576	1.13	39.2 ± 4	6	49	Cylindrical Waveguide	0.6

TABLE 2. Characteristics of microwave-induced cochlear microphonics in cats.

Animal Number	Body Mass (kg)	Frequency ± SD (kHz)	No. of Oscillations	$t_{1/e}$ (μs)	Exposure Apparatus	Averaged Absorbed Energy per Pulse (J/kg)
Cat 12076	2.9	∿ 33	3	--	Applicator	--
Cat 52176	3.2	31.7 ± 7.9	3	91	Applicator	--
Cat 52576	2.7	31.6 ± 4.3	5	--	Applicator	--
Cat 60376	2.6	32.5 ± 2.8	5	--	Applicator	--
Cat 60476	2.9	29.6 ± 1.3	4	140	Applicator	--
Cat 61676	3.1	30.5 ± 5.0	4	96.25	Applicator	--
Cat 61776	0.9	36 ± 3.3 35.5 ± 4.8	9 10	117 120	Applicator Cylindrical Waveguide	-- 0.8
Cat 62376	1.0	35 ± 2.7 34 ± 1.7	6 7	97.5 107.5	Applicator Cylindrical Waveguide	-- 0.52
Cat 71676	1.45	40 ± 5.4	8	100	Cylindrical Waveguide	0.51
Cat 71976	1.78	32.2 ± 1.7	5	110	Cylindrical Waveguide	0.43
Cat 72076	1.92	31 ± 3.0	8	---	Cylindrical Waveguide	0.51
Cat 72176	4.3	31 ± 3.2	4	137	Applicator	---

TABLE 3. Physical dimensions of animal heads.

ANIMAL Number	Body Mass (kg)	Head Mass (g)	Skull Mass (g)	Skull Dimensions L x W x H (cm)	Skull Thickness (cm)	Brain Cavity Dimensions L x W x H (cm)	Brain Volume (cc)	Bulla Dimensions L x W x H (cm)	Cerebellar Cavity Dimensions L x W x H (cm)	Cerebellar Volume (cc)
Guinea pig 61076	1.09	133	9.49	6.77 x 2.65 x 1.65	0.135	3.70 x 2.57 x 1.29	4.86	1.29 x 0.95 x 0.3		
Guinea pig 82476	0.57	70	6.39	6.18 x 2.48 x 1.63	0.08	3.46 x 2.26 x 1.27	4.6	1.17 x 0.88 x 0.26		
Cat 61676	3.13	304	28.2	9.6 x 4.28 x 3.5	0.195	3.55 x 3.96 x 2.85	22.4	1.89 x 1.24 x 0.68	3.01 x 2.39 x 2.84	6.8
Cat 62376	1.01	124	9.9	6.75 x 4.05 x 3.05	0.105	3.57 x 3.79 x 2.70	18.4	1.72 x 1.16 x 0.72	2.44 x 1.65 x 2.44	3.6
Cat 71676	1.45	152	13.5	7.24 x 4.13 x 3.08	0.1	3.4 x 3.86 x 2.70	18.4	1.66 x 1.14 x 0.73	2.62 x 1.94 x 2.80	5.1
Cat 71976	1.78	182	15.7	7.54 x 4.20 x 3.20	0.11	3.6 x 4.0 x 2.88	21	1.72 x 1.18 x 0.675	2.65 x 1.96 x 2.82	5.6
Cat 72076	1.29	154	13.8	7.31 x 4.11 x 3.22	0.1	3.6 x 3.87 x 2.84	20	1.69 x 1.16 x 0.65	2.59 x 1.92 x 2.69	5.0
Cat 72176	4.3	494	39.4	9.65 x 4.30 x 3.62	0.275	3.58 x 4.19 x 3.04	23.6	1.98 x 1.39 x 0.55	2.90 x 2.24 x 3.10	7.6
Human	---	---	847	18.4 x 14 x 12.4	0.8	16.8 x 11.4 x 13.0	1410			

Foster and Finch [1974] showed that a reflected pressure wave was measured in a lucite tank filled with KCL solution when exposed to pulsed microwaves. The time between pressure waves corresponded to the propagation time of acoustic waves in the solution. It is postulated that the induced pressure on the skull can be transmitted to the cochlea through bone conduction. Based on this hypothesis and the data from guinea pigs and cats, the extrapolated frequency for human beings lies somewhere between 7 and 10 kHz. Dimensions of the human skull are listed at the bottom of Table 3. Our results are consistent with theoretical predictions of *Lin* [1977].

In Figure 1, the shape of the CM induced in cats is more complicated than that of guinea pigs. The variation of the periods between CM peaks is also larger. These differences are probably due to differing brain-cavity structures in the two animals. In the cat, the cerebral cavity is separated from the cerebellum by a bony partition, the *tentorium cerebelli* (see Figure 7). The shape of the cerebral cavity is close to a quarter portion of a sphere. In the guinea pig (Figure 8) the brain cavity resembles an elongated ellipsoid and there

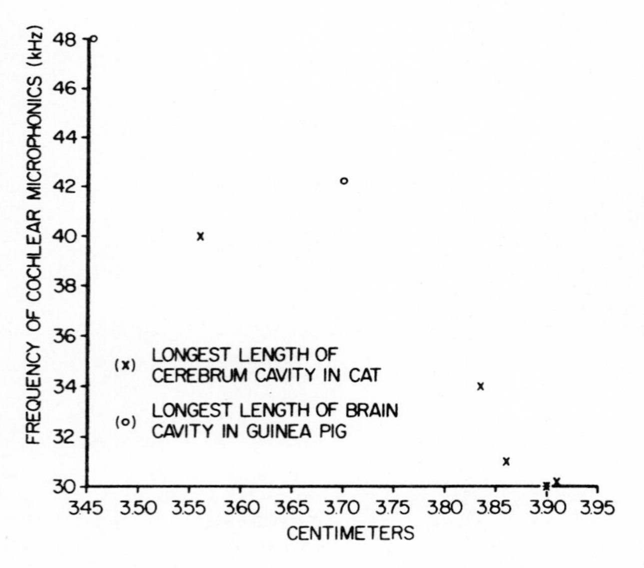

Fig. 4. Frequency of microwave-induced cochlear microphonics versus longest length in brain cavity.

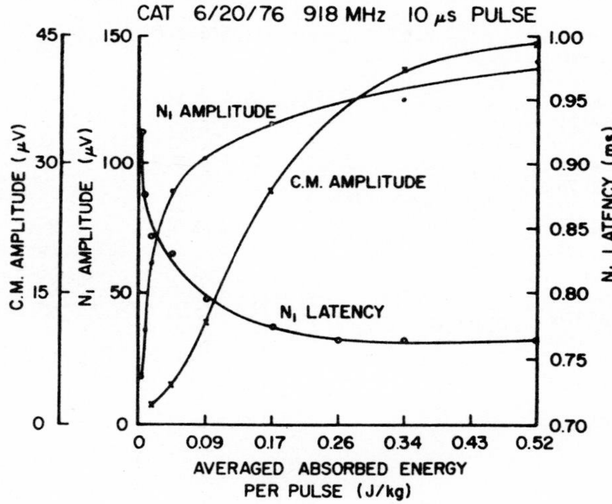

Fig. 5. Amplitude of microwave-induced CM and N_1, and latency of N_1, as a function of the averaged absorbed energy per pulse.

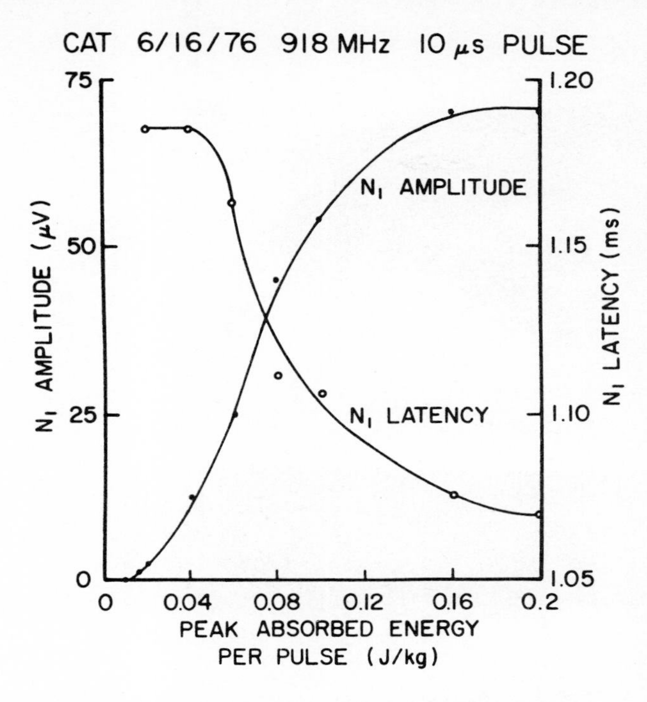

Fig. 6. Amplitude of microwave-induced N_1 and latency of N_1, as a function of the peak absorbed energy per pulse.

is no bony partition separating cerebrum from cerebellum. Thus, the temporal bone, where the cochlea is embedded, forms the wall of the cerebellar cavity in cats and differs from the whole-brain cavity of guinea pigs. The two separate compartments in cats may have caused the observed difference in waveforms.

The number of oscillations, as shown in Tables 1 and 2, seems to relate to the size of the head, presumably to the thickness of the skull. Although the number of oscillations depends on the intensity of microwave radiation, it appears to be similar for the two different exposure apparatus (cat 61776 and 62376). Based on the thickness of the human skull (0.8 cm at vertex), the number of oscillations may be below three for an absorbed energy per pulse that is less than 0.5 J/kg. The $t_{1/e}$ of guinea pigs is about half of that in cats. The implication of this difference is not clear.

Acknowledgments. This work was supported in part by the Bureau of Radiological Health Grant No. FD 00646-06; the Office of Naval Research Contract No. N00014-75-C-0464; General Medical Sciences Fellowship Award No. 1 F32 GM05681-01; and the Rehabilitation Services Administration Grant No. 16-P-56818. We thank Ms. Carrol Harris and Mr. Stephen Barnes for assistance.

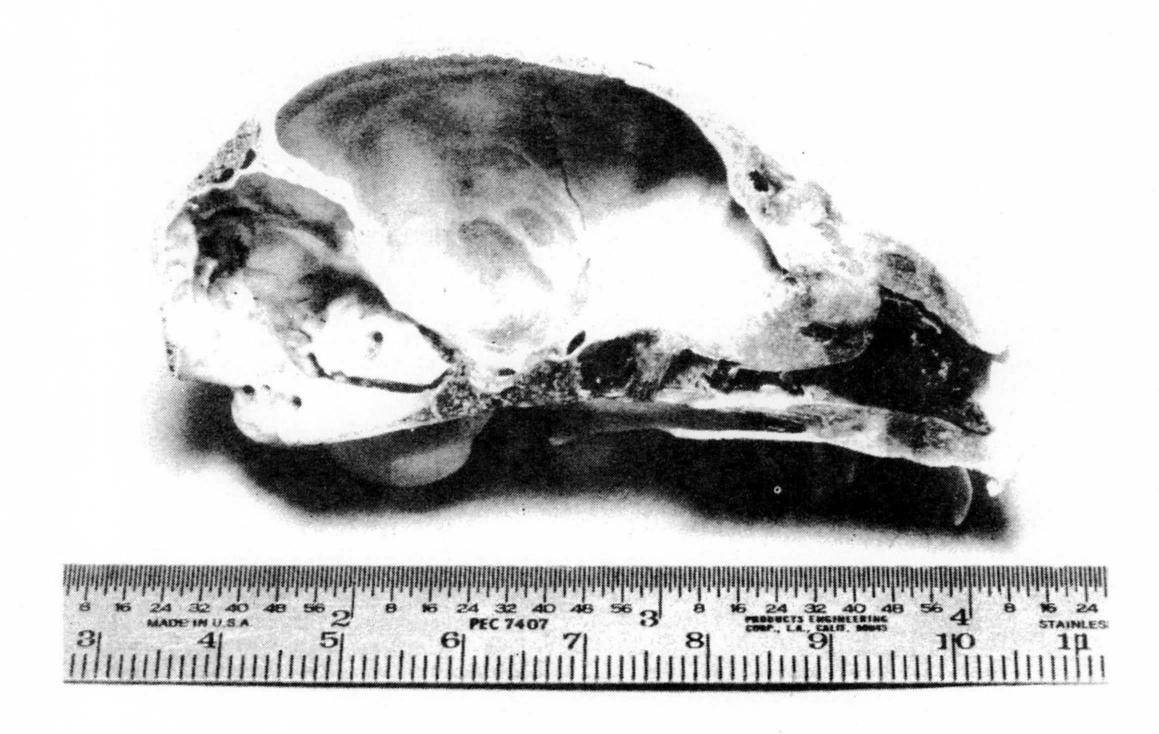

Fig. 7. Photograph of cat skull in sagittal plane.

Fig. 8. Photograph of guinea pig skull in sagittal plane.

REFERENCES

Chou, C. K., R. Galambos, A. W. Guy, and R. H. Lovely (1975), Cochlear microphonics generated by microwave pulses, *J. Microwave Power, 10*(4), 361-367.

Chou, C. K., A. W. Guy, and R. Galambos (1976), Microwave-induced cochlear microphonics in cats, *J. Microwave Power, 11*(2), 171-173.

Chou, C. K., A. W. Guy, and R. Galambos (1977), Microwave-induced auditory response - Cochlear microphonics, in *Biological Effects of Electromagnetic Waves,* edited by C. C. Johnson and M. L. Shore, pp 89-103, Bureau of Radiological Health, FDA (77-8010), Rockville, Maryland.

Cooper, G., and A. L. Schiller (1975), *Anatomy of the Guinea Pig*, Harvard University Press, Cambridge, Massachusetts.

Crouch, J. E. (1969), *Atlas of Cat Anatomy*, pp. 10-32, Lea & Febiger, Philadelphia, Pennsylvania.

Foster, K. R., and E. D. Finch (1974), Microwave hearing: evidence for thermoacoustic auditory stimulation by pulsed microwaves, *Science, 185*, 256-258.

Frey, A. H. (1971), Biological function as influenced by low-power modulated RF energy, *IEEE Trans. Microwave Theory Tech., 19*(2), 153-164.

Guy, A. W., C. K. Chou, J. C. Lin, and D. Christensen (1975), Microwave-induced acoustic effects in mammalian auditory systems and physical materials, in *Biologic Effects of Non-ionizing Radiation,* edited by Paul E. Tyler, 194-218, *Ann. N. Y. Acad. Sci.*, New York.

Johnson, C. C., and A. W. Guy (1972), Nonionizing electromagnetic wave effects in biological materials and systems. *Proc. IEEE, 60*(6), 692-718.

Lebovitz, R., and R. Seaman (1977), Microwave hearing: the response of single auditory neurons in the cat to pulsed microwave radiation. *Radio Sci.*, this issue.

Lin, J. C. (1977), Theoretical calculation of frequencies and thresholds of microwave induced auditory signals, *Radio Sci.*, this issue.

Behavioral Sensitivity to Microwave Irradiation

Abstract. Rats assayed by the technique of conditional suppression were able to detect the presence of 12.25-centimeter microwaves at doses of power approximating 0.5 to 6.4 milliwatts per gram. The assay, which controlled for sensitization, for pseudo and temporal conditioning, and for several possible sources of artifactual cueing, revealed that irradiation by microwaves, although lacking the saliency of an auditory stimulus, can function as a highly reliable cue. Efficiency of detection was strongly and positively related to the amount of microwave energy to which the rats were exposed.

Nearly a decade ago, Frey (*1*) reported that human beings can detect pulse-modulated electromagnetic energy at wavelengths of 10 to 70 cm and at average power densities of 0.4 to 2.1 mw/cm^2. The sensations reported were usually auditory in character and were often described as "hissing, buzzing, and clicking sounds." Although no confirmatory studies have been published since Frey's reports, three instances of verification have been communicated to him (*2*), and he has referred to successful use of microwave energy as a signaling stimulus in cats (*3*). Two directly related studies yielded negative results. Jones (*4*) reported that none of 20 college students could discriminate presence from absence of 30- or 60-cm microwaves. Justesen and King (*5*) intermittently presented 12.25-cm microwave energy to each of six rats as a cue for obtaining sugar water, but none of the rats discriminated the cue. Since unmodulated energy was used in Jones's study, its negative findings comport with Frey's belief (*2*) that modulation is necessary for perception of microwaves. Modulated energy was used by Justesen and King, but the assay for perception was based upon appetitive rather than aversive motivation and may have lacked sensitivity. Much indirect evidence of relevance to detection of microwaves has been published, particularly in the Soviet literature (*6*); altered thresholds to physiologically adequate stimulation have been reported as sequelae of microwave irradiation in olfactory (*7*), auditory (*8*), visual (*9*), and cutaneous (*10*) modalities. Other aftereffects include cardiovascular changes (*11–13*), irritability and irascibility (*11*), neurasthenia (*13*), and headache and disturbance of sleep (*11*). Acute responses observed during irradiation include changes of blood pressure (*14*), heart rate (*15*), and cortical and subcortical electrophysiological activity (*3*, *16*).

Although the mechanisms responsible for chronic and acute changes are unresolved and much debated (*3*, *17*), the evidence suggests that microwaves at densities below the safety limit of 10 mw/cm^2 observed in the United States (*18*) can affect nervous activity and could, therefore, possess stimulus properties. We report here attempts to assess in rats the efficacy and the reliability of modulated 12.25-cm microwaves as a warning stimulus for impending electrical shock.

Six male albino rats of common age were obtained from the Simonsen Company of Minnesota. Three randomly selected rats (R-1, R-2, and R-5) served as subjects for irradiation; control rats (R-3, R-4, and R-6) were never irradiated, but were maintained in their home cages with unrestricted access to Purina Lab Chow and water. Although irradiated rats had the same access to water, they were partially deprived of food until their individual weights fell to 75 percent of that before experimentation; a diet that led to 75 percent of normal gains in weight was thereafter instituted on the basis of data on weight gained by the control animals. Weights of R-1, R-2, and R-5 at the commencement of experimentation were, respectively, 409, 455, and 427 g.

A highly sensitive measure of conventional sensory stimulation, the technique of conditional suppression (*19*), was used. With this technique, a subject is reinforced after making an operant response; then reinforcement is scheduled intermittently until the response occurs frequently and consistently. Finally, a Pavlovian conditioning regime is superimposed in which a warning signal is presented from time to time, always terminating in a brief, but aversive, unconditional stimulus (US). After repeated presentations of the warning signal and the US, a subject will respond stably except when the warning signal is being presented; that is, operant behavior is conditionally suppressed. The operant response required of our rats was the tongue lick, which was detected photoelectrically and reinforced by discrete volumes (30 μl) of sugar water (dextrose, 16 g/100 ml). A radiolucent ensemble by which licks were detected and reinforced is described elsewhere (*20*). An aperiodic schedule of reinforcement was used during all experiments; the passage of each 2-second interval after reinforcement led to availability of another reinforcer with a probability (*P*) of .25, .125, or .0625. The *P* value was not changed during a given experiment, but was varied across experiments to maintain stable responding. The US was unavoidable electrical shock to the feet presented by a radiopaque floor-grid of aluminum rods (*21*). A conventional warning stimulus, with which microwave irradiation was compared for cueing efficacy, was a 525-hz tone, and was produced by a 3-inch, 4-ohm loudspeaker (Jensen VK-300) driven by a sinusoidal current at 800 μw of continuous power.

Because conventional (open space) methods of exposure to microwaves require immobilization of a subject in order to preserve constancy of incident energy (*5, 22, 23*), we irradiated our animals in the closed space of a multimodal exposure cavity, a modified Tappan R3L microwave oven (*5, 22*). Microwaves (at 2450 ± 50 Mhz) were generated by a QK707-A magnetron and were doubly modulated at 60 and 12 hz. The exposure cavity was fitted with a Plexiglas conditioning chamber (internal dimensions, 26 by 37 by 24 cm) and with the radiolucent ensemble by which licks were detected and reinforced (*20*). Dosimetry was accomplished by measuring available microwave power (*22*) within the chamber by water calorimetry (*24*), then dividing obtained wattage values by mean weight of the three rats to yield average doses within ± 15 percent of 6.4, 4.8, 2.4, and 1.2 mw/g. The smallest dose was not based upon calorimetry but was estimated to be 500 ± 90 μw/g on the basis of the level of focusing current used to control and monitor the output power of the magnetron (*25*). A shift from zero to a preset level of available power in the exposure cavity was accomplished by applying 5 kv of 60-hz a-c voltage to the anode of the magnetron.

The exposure cavity was cooled and sounds transmitted to the conditioning chamber were masked by fans that provided a continuous flow of air from an external, thermostatically controlled source. Temperatures within the operant chamber were maintained at $24° \pm 2°$C; relative humidity was between 20 and 40 percent. One-minute periods of tonal stimulation or of microwave irradiation and 0.5-second periods of electrical shock (averaging 790 μa root-mean-square) were programmed for automatic presentation by a punched-tape control system. The number of licks that occurred during 60-second control ("safe") periods, and during ensuing 60-second periods of warning stimulation (which usually terminated in shock), were tallied by digital counters and cumulative recorders. A rat's discriminative efficiency (that is, the degree to which an animal's operant responding was suppressed during warning stimulation) was quantified by the formula $[(S - W)/S] \times 100$ (where S is the number of responses made during safe periods and W is the number made during periods of warning stimulation). Each of the three rats was tested in a total of 14 experiments based upon 62 2-hour sessions. The 62 sessions were interspersed with another 87 sessions (without irradiation) in which baselines of operant responding were established or reestablished, all 149 sessions being conducted within 6 months. Each session was 120 minutes long, and the minimum interval between sessions was 22 hours. During each session conditioning stimuli were presented eight times, and the length of intervals between presentations was made random to control for temporal conditioning. Each of the 14 experiments comprised a set of two or more contiguous sessions (Fig. 1).

None of the rats exhibited signs of spurious (unconditional) suppression when the tonal stimulus was presented without shock to the feet (session set 1); but during sessions of the second set,

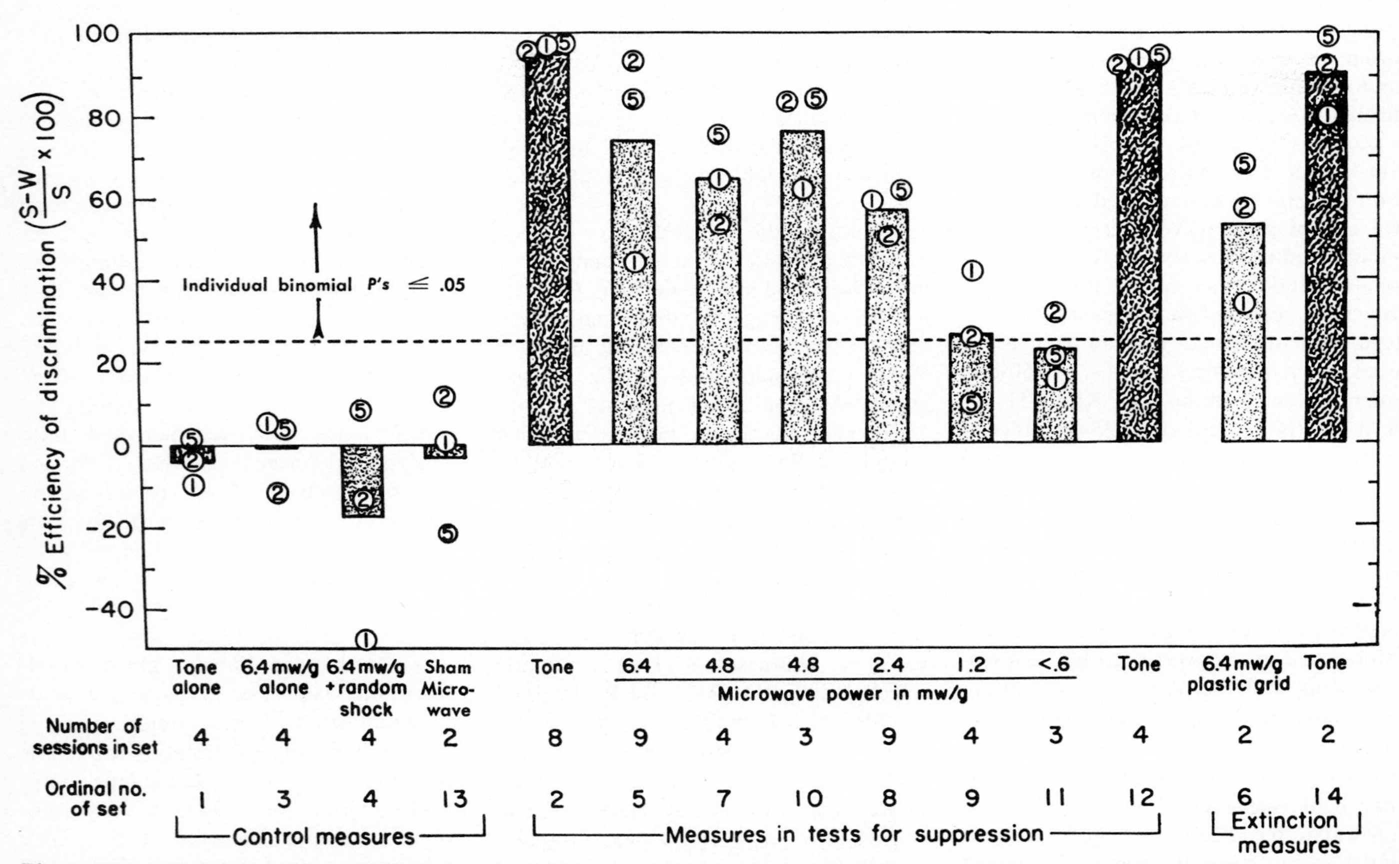

Fig. 1. Mean efficiencies of rats discriminating a 525-hz tone or microwave irradiation presented as a warning stimulus that usually preceded unavoidable footshock. Means of individual subjects (rats 1, 2, and 5) are shown by encircled numbers; overall means are illustrated by verticle bars. A total of 14 experiments was ordered across time in sets of two or more contiguous sessions as indicated in the two rows of entries at the bottom of the figure. Conditions are explained in text.

when shock did follow the tone, suppression of responding was quickly conditioned. When periods of microwave irradiation were presented alone (session set 3) or made random in time with respect to an equivalent number of shocks (session set 4), neither spurious suppression nor aversive sensitization, respectively, was observed. During the fourth set, one rat, R-1, responded reliably more often during periods of irradiation than it did during control periods ($P < .01$); since shocks never coincided with irradiation R-1 apparently learned that it was "safe" to respond when irradiated. During sessions of the fifth set, shock was presented at the termination of each period of irradiation, and the rats were again irradiated at 6.4 mw/g. All rats suppressed reliably, R-1 less efficiently than the others. During sessions of the sixth set, a conditioning chamber formed entirely of Plexiglas was inserted into the exposure cavity. Only microwaves were presented. This arrangement permitted assessment of resistance to extinction (which was relatively high in all animals), but was primarily used to control for (and was found to eliminate) the possibility that the suppression observed during the prior set of sessions was produced by demodulated microwave energy that could have been wave-trapped by the aluminum shock-grid and could have led to artifactual electrical stimulation of the footpad. Periods of microwave irradiation terminating in shock were programmed during several subsequent sessions (sets 7 through 11), but at lower doses. During the 12th set of sessions the 525-hz tone again preceded shock; efficiency of discrimination was almost as high for all three rats as during the second set. In order to test for spurious auditory or vibratory cueing by relays and switches controlling presentations of microwave energy, neither irradiation nor shock was presented during the 13th set of sessions. The source of current to the anode of the magnetron was interrupted, preventing generation of microwave energy, but all control relays and switches were operated as in the fifth set. Spurious suppression was not observed. During the 14th and final set of sessions, resistance to extinction to the tonal stimulus was measured. All three rats continued to suppress during tonal stimulation in spite of the absence of shock.

Detection of microwaves was generally less efficient than detection of the tonal cue, but was highly reliable at all but the two lowest doses: binomial P values ranged between 10^{-7} at 6.4 mw/g for the best performing rat to $\sim .10$ at 1.2 mw/g for the worst performing (*26*). Even at the lowest dose, one of the rats discriminated reliably (R-2, $P < .001$). The appearance of a strong relation between dose and response was tested by plotting mean efficiency of discrimination against average dose as presented during session sets 5, 7, 8, 9, 10, and 11. The product-moment r is .95 (at 4 d.f., $P < .01$).

Under the conditions described, the albino rat can either detect microwave energy or is sensitive to some concomitant of irradiation. Because of the strong dose-response relation, an artifactual cue would have varied in intensity with the amount of available power. X-irradiation is a possibility and is not only generated by high-voltage, thermionic devices such as the magnetron, it has also been demonstrated to function as a signaling stimulus (*27*). However, a metallic wall of the exposure cavity and the 1-cm-thick Plexiglas sheet from which the conditioning chamber was fabricated separated the rats from the magnetron; x-rays at the low photon energies developed in a magnetron would have little probability of penetrating the metallic wall and the Plexiglas sheet. Another possible cue derives from the small portion of microwave energy that is absorbed by the metallic walls of a multimodal cavity. The loudspeaker used to present the tonal cue was located on the external surface of the cavity and could have trapped demodulated microwave energy and translated it into an audible vibration of the speaker-cone. Even though our experiments controlled for this possibility (compare session sets 2 and 3), we took the additional precaution of removing the loudspeaker from the apparatus during most of the tests for microwave cueing.

The only unargued fate of absorbed microwave energy is thermalization. However, rats exposed to 6.4 mw/g and lesser doses of irradiation for 60-second periods have never exhibited reliable elevations of whole-body temperature in our laboratories, even when measured by expanded-scale electronic thermometers with a resolution of 0.05°C. Some investigators (*28*) would conclude that nonthermal effects are implicated; others (*29*) would argue with some cogency that the rats were simply more sensitive to a weak thermal stimulus than were our thermometers.

Whatever the mechanism of detection (*30*) we offer our data as evidence that confirms and extends the generality of Frey's findings: mammals are sensitive to something that inheres in or accompanies illumination by microwaves at low levels of available power.

NANCY WILLIAMS KING
Neuropsychology Laboratories, Veterans Administration Hospital, 4801 Linwood Boulevard, Kansas City, Missouri 64128

DON R. JUSTESEN
Department of Psychiatry, Kansas University Medical Center, Kansas City

REX L. CLARKE
Department of Psychology, University of Kansas, Lawrence

References and Notes

1. A. H. Frey, *Aerosp. Med.* **32**, 1140 (1961); *J. Appl. Physiol.* **17**, 689 (1962).
2. ———, *Psychol. Bull.* **63**, 322 (1965).
3. ———, *J. Appl. Physiol.* **23**, 984 (1967).
4. I. A. Jones, thesis, Baylor University, Waco, Texas (1966).
5. D. R. Justesen and N. W. King, in *Biological Effects and Health Implications of Microwave Radiation Symposium Proceedings*, No. BRH/DBE 70-2, S. F. Cleary, Ed. (U.S. Public Health Service, Rockville, Md., 1970), p. 154.
6. Several English translations of reports by Soviet investigators are available in the United States; for references see A. H. Frey (*2*) and C. Dodge and S. Kassel [*Soviet Research on the Neural Effects of Microwaves*, Air Technology Division (A.T.D.) Report 66-133 (1966)]. See also a review of the Soviet literature by W. D. Thompson and A. E. Bourgeois [*Effects of Microwave Exposure on Behavioral and Related Phenomena: an Annotated Bibliography* (Aeromedical Research Laboratory, Holloman Air Force Base, New Mexico, 1965)].
7. Ye. A. Lobanova and Z. V. Gordon, in *Biological Action of UHF* [translated from Russian, Joint Publication Research Study (JPRS) 12471], A. A. Letavet and Z. V. Gordon, Eds. (Academy of Medical Sciences USSR, Moscow, 1960), p. 50.
8. I. A. Kitsovsakaya, *ibid.*, p. 75; N. N. Livshits, *Biofizika* **2**, 198 (1957).
9. N. I. Matzuov, *Byull. Eksp. Biol. Med.* **48**, 816 (1959).
10. A. G. Grinbarg, *Kazan Med. Zh.* **40**, 63 (1959).
11. A. A. Kevork'ian, *Institute of Work Hygiene of Professional Diseases* [translated from Russian, Office of Technical Services (OTS) 59-21098] (Academy of Medical Sciences USSR, Moscow, 1948).
12. M. N. Sadchikova, in *Biological Action of UHF* (translated from Russian, JPRS 12471), A. A. Letavet and Z. V. Gordon, Eds. (Academy of Medical Sciences USSR, Moscow, 1960), p. 25.
13. M. N. Sadchikova and A. A. Orlova, *Industrial Hygiene and Occupational Diseases* (translated from Russian, OTS 59-11437) **2**, 18 (1958).
14. E. Pflomm, *Arch. Klin. Chir.* **166**, 1 (1931); Z. V. Gordon, in *Biological Action of UHF* (translated from Russian, JPRS 12471), A. A. Letavet and Z. V. Gordon, Eds. (Academy of Medical Sciences USSR, Moscow, 1960), p. 18; V. A. Baronenko and K. F. Timofeeva, *Fiziol. Zh. SSSR im. I. M. Sechenova* **45**, 184 (1959).
15. A. S. Presman and N. A. Levitina, *Byull. Eksp. Biol. Med.* **53**, 39 (1962); *ibid.* **53**, 41 (1962); N. N. Livshits, *Biofizika* **2**, 378 (1957).
16. Z. M. Gvozdikova, V. M. Ananieb, I. N. Zenina, V. I. Zak, *Byull. Eksp. Biol. Med.* **57**, 63 (1964); Y. Kholodov, *ibid.* **56**, 42 (1963); *ibid.* **57**, 98 (1964).
17. H. C. Sommer and H. C. Von Gierke, *Aerosp. Med.* **35**, 834 (1964); A. D. McAfee, *Amer. J. Physiol.* **203**, 374 (1962); A. S. Presman, *Usp.*

Sovrem. Biol. (translated from Russian, OTS 61-31472) **56**, 161 (1963); N. N. Livshits, *Biofizika* **2**, 378 (1957).
18. Currently instituted by the United States of America Standards Institute; see W. M. Mumford, in *Biological Effects and Health Implications of Microwave Radiation Symposium Proceedings*, No. BRH/DBE 70-2, S. F. Cleary, Ed. (U. S. Public Health Service, Rockville, Md., 1970), p. 21.
19. W. K. Estes and B. F. Skinner, *J. Exp. Psychol.* **29**, 390 (1941).
20. N. W. King, D. R. Justesen, A. D. Simpson, *Behav. Res. Method Instrum.* **2**, 125 (1970).
21. D. R. Justesen, N. W. King, R. L. Clarke, *ibid.*, in press.
22. J. H. Vogelman, in *Biological Effects of Microwave Radiation*, M. F. Peyton, Ed. (Plenum, New York, 1961), p. 23.
23. W. Moore, *Biological Aspects of Microwave Radiation* (U.S. Public Health Service, Rockville, Md., 1968).
24. A volume of distilled water approximating the gram weight of a mature rat is placed in a thick-walled vessel of foamed polystyrene at the ambient temperature of the exposure cavity. The water (which is continuously agitated or stirred) is then irradiated and the average available power is obtained by the formula $w = T_{\Delta}V/kt$, where w is the average available (thermalized) power in watts, T_{Δ} is the temperature increment in degrees Celsius, V is the volume of water in milliliters (weight in grams), k is Joule's conversion factor (0.239), and t is the duration of irradiation in seconds. Temperatures are measured either by expanded-scale spirit thermometers or by electronic thermometers with thermistor sensors.
25. Since the unit-mass (watts per gram) dose is an estimate of absorbed energy, and the conventional unit-surface (watts per square centimeter) dose is a planar index of field density, the two doses cannot be precisely equated. Maximum limiting values of the unit-surface dose can be approximated (*5*) and for the unit-mass doses as given are approximately 20, 15, 7.5, 3.75, and <2 mw/cm^2.
26. Individual binomial probabilities were derived as follows: (i) The number of responses generated by an animal during an *S* interval was compared to the number generated during the succeeding *W* interval; (ii) if the former number was higher, an instance of cueing was noted; if equal or lower, an instance of no cueing was noted; and (iii) frequencies of positive and negative instances were cumulated across a total set of sessions and evaluated for reliability by use of the binomial theorem.
27. D. D. Morris, *J. Exp. Anal. Behav.* **9**, 29 (1966).
28. Frey (*2*) has discussed the practice by some investigators of assuming that irradiation which does not lead to measurable thermalization is ipso facto "nonthermalizing." Such an operational fiat may lead one to overlook the possibility of nonlinear heating of tissues of a biological preparation or of relative insensitivity of thermal measurements.
29. See the discussion by L. D. Sher (*5*, p. 192).
30. The only creature suspected of possessing a specialized sensory-motor apparatus capable of receiving and transmitting microwaves is the corn earworm; see the *J. Microwave Power* **5**, 149 (1970).
31. Supported by 8200 Research Funds from the U.S. Veterans Administration and by contract funds (DADA17-68-C-8021) from the Surgeon General, U.S. Army, to D.R.J. We thank E. L. Wike, C. L. Sheridan, H. F. Fisher, and D. G. Cross for technical advice and criticism.

4 March 1971

Thermographic and Behavioral Studies of Rats in the Near Field of 918-MHz Radiations

JAMES C. LIN, SENIOR MEMBER, IEEE, ARTHUR W. GUY, FELLOW, IEEE, AND LYNN R. CALDWELL

***Abstract*—Patterns of thermalized energy of rat carcasses exposed to 918-MHz CW radiation in the near zone have been determined using a computerized thermograph. Peak absorption of energy in the body was estimated to be 0.9 W/kg per mW/cm² of incident energy. Operant responses of irradiated rats to schedules of fixed-ratio (food) reinforcement under the same conditions as the dosimetric test were observed to occur at averaged power densities of 30–40 mW/cm². This range of densities corresponds to absorbed peaks of energy of 27–36 W/kg. No change in behavior was observed for incident power densities and peaks of absorbed energy to 20–30 mW/cm² and to 18–27 W/kg, respectively, and all changes at higher values were reversible.**

Introduction

THE ABUNDANT literature on biological effects of microwave radiation divides into two categories, one pertaining to well-defined thermal effects on living tissues at averaged power densities of radiation above 100 mW/cm², the other pertaining to "nonthermal" or "low-level" effects of CW or pulsed energy, usually at densities below 20 mW/cm². The effects at high densities are probably related to thermalization of tissues by absorbed microwave energy but concurrent nonthermal effects that might be masked or complexed by heating should not be ruled out. The mechanisms underlying low-level effects such as the RF hearing phenomenon [1] are unknown but probably are due to thermoelastic transduction. There are several reports [2]–[6], that confirm alteration of behaviors of small animals during or after exposure to low to moderate densities of microwave energy. Specifically, radiation has been found to facilitate the performance of rats in a Y maze [2], and to influence the spontaneous activity and emotionality of animals as well as their sensitivity to electroconvulsive shock [4]. A decrease in the frequency of bar pressing for food and lessened endurance as indicated by swimming tasks have also been found in irradiated animals [5]. In addition, birds have exhibited escape reactions when exposed to microwave fields [6]. Although these subtle behavioral changes may be produced by seemingly thermally insignificant fields, the lack of information on the spatial distribution of absorbed energy in the target animal makes a nonthermal interpretation somewhat suspect. Given fields of very low density (i.e., $< 500\ \mu W/cm^2$), there is still the possibility that concentrations of thermalized energy—"hot spots"—may conspire to produce thermal stimulation. For instance, it was demonstrated [7] that by granting the animals freedom of movement in a large, electrically shielded enclosure the behavioral changes observed at 0.1 mW/cm² averaged density of microwave energy were associated with peak absorptions of 140 W/kg in tissues of radiated animals. Thus free movement can seriously compromise control of exposure in a closed space or in the free field.

A specially designed apparatus that restricted the body movement of rats but permitted simultaneous exposure and monitoring of performances was assembled. Measures of incident and absorbed energy were made in an effort to determine levels that disrupt the operant behavior.

Experiment

The subjects were 200 (± 25) gram female white rats (Sprague–Dawley). The animals were partially deprived of food until their body masses fell to 80 percent of those before deprivation. They were placed in a plastic body-movement restrainer and were then trained to perform a head-raising response for a food pellet. The criterion was 30 rapidly and regularly executed head movements for each food pellet (fixed ratio, or FR-30) during 30-min sessions. After stabilizing, the rats typically responded 2000 times during each session. Following measures of base-line performance, the animals were exposed to increasing levels of radiation while performance was monitored.

Behavior Task

The general arrangements for the behavioral test are shown in Fig. 1. The rat holder shown in Fig. 2 was designed to provide necessary restriction of body movement to facilitate determination of absorbed energy during experimentation while permitting sufficient movement of the animal's head and neck for collection of behavioral data. The holder is constructed of acrylic to reduce the amount of distortion of the incident microwave field. The spaced-bar construction provided adequate ventilation for control of the animal's surface temperature and permitted easy placement of the animal. After the first few sessions, the rats learned to position themselves in the holder by running into it and extending their heads through its opening. The holder

Manuscript received October 10, 1974; revised March 17, 1977. This work was supported in part by the Office of Naval Research under Contract N00014-67-A-0103, Task Order 0026, in part by the Social and Rehabilitation Service under Grant 16-P-56818/0-12, and in part by the National Science Foundation under Grant Eng 75-15227.

J. C. Lin is with the Department of Electrical Engineering and Center for Bioengineering, Wayne State University, Detroit, MI 48202.

A. W. Guy and L. R. Caldwell are with the Department of Rehabilitation Medicine, Bioelectromagnetics Research Laboratory, University of Washington School of Medicine, Seattle, WA 98195.

Reprinted from *IEEE Trans. Microwave Theory Tech.*, vol. MTT-25, pp. 833–836, Oct. 1977.

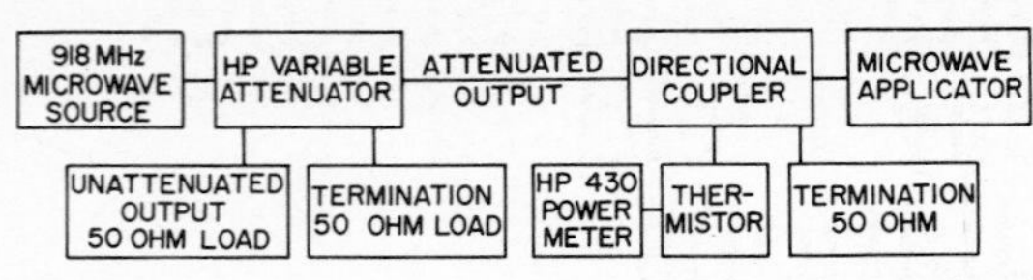

Fig. 1. Schematic of microwave exposure facilities for small animal behavioral investigation.

Fig. 2. Rat holder.

Fig. 3. Rat holder and receiver.

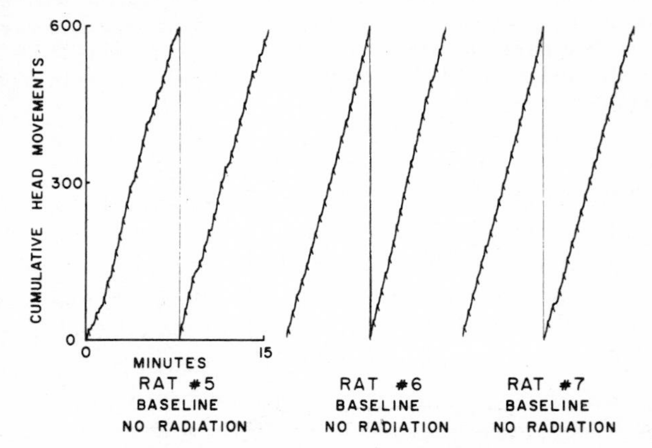

Fig. 4. Cumulative records of fixed-ratio performance of three animals during a pre-irradiation session.

with the rat enclosed was then placed in a receiver as shown in Fig. 3. The holder and receiver were placed in a modified cooler chest that was lined with microwave absorbing material. In addition, a small fan attached to a wall at one end of the chest provided a forced flow of air to keep the environmental conditions within reasonable limits. The receiver positioned the rat in such a way that it allowed the rat to interrupt a light beam by moving its head in a short vertical arc. The small head movement, which interrupted the light beam, constituted the operant behavior. After 30 interruptions of the beam of light a switch closed leading to the delivery of food. An external feeder caused a small 45-mg food pellet to be delivered via a polyethylene tube to a receptacle, which was constructed of the same material as the holder, directly below the rat's head. The rat was able to eat the food pellet with only a slight downward movement of its head. This system provided a consistent means of investigating behavior that is adaptable to the special requirements of microwave radiation in the free field. Examples of baseline performances of animals in the apparatus are shown in Fig. 4. A slash on the cumulative records indicates that 30 head raises had been performed and that a food pellet had been delivered.

Exposure and Dosimetry

After base-line performances were established, the rats were exposed to four levels of 918-MHz radiation (0, 10, 20, and 40 mW/cm^2). The technique for exposing the animals is shown in Fig. 5. Microwave energy was emitted by a cavity-backed applicator with a square aperture. The maximum density of the radiated energy was directed toward the longitudinal midpoint of the rat's body. Immobilization of the tail was necessary to minimize a hot spot that developed near its point of attachment to the body and was observed during preliminary studies. The distance between the proximal surface of the animal's body and the radiator was 8 cm.

Two principal methods were used for estimating absorbed energy: 1) the energy density at the same position as the axis of the rat's body inside the holder in the empty exposure chamber and directly under the center of the radiator was calibrated by a National Bureau of Standards electric energy density meter; 2) thermographic measurements were made of the patterns of thermalized energy in exposed rat carcasses as described previously [8]. The animal was killed by an overdose of Nembutal, was subsequently frozen in the holder, was cast in a polyfoam block, and was then dissected along with its holder. After return to room temperature, the reassembled rat carcass and the holder were then taken to the exposure chamber and were radiated for 30 s at high power (1500 mW/cm^2) under the same geometrical arrangement as during the session of behavioral testing. The thermographically recorded temperature patterns over the midsagital plane of the rat were processed digitally using a Honeywell DDP 516 computer to obtain isothermal contours. The tissue properties were assumed to be homogeneous and equal to that of muscle material.

The isothermal contours inside the rat along the midsagital plane are shown in Fig. 6. Observable peaks of absorbed energy occurred both in the body and in the tail. The peak absorption at the tail corresponds to 0.9 W/kg per 1 mW/cm^2 of incident energy while the value for the body is 0.8 W/kg. The position of the tail had considerable influence

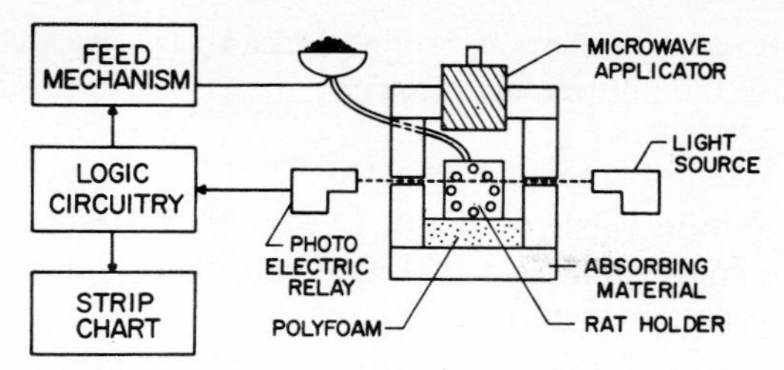

Fig. 5. Block diagram for irradiation of rats undergoing behavioral tests.

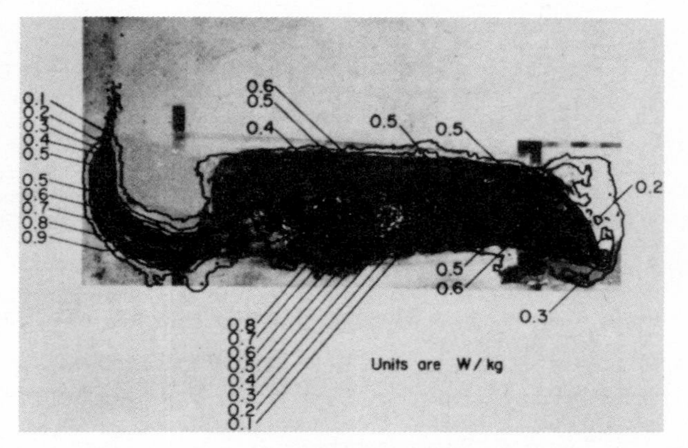

Fig. 6. Absorbed energy patterns (midsagittal plane) in a rat exposed to 918-MHz radiation at 8 cm; values are for 1 mW/cm² applied power.

on the absorbed energy. Prior dosimetric studies of phantom models of the rat indicated that absorbed energy at the base of the tail could increase by an order of magnitude when it was extended straight back [9]. Enhanced absorption was presumably due to an increase of current density that resulted from sharp changes in the cross-sectional area of the tail and a standing-wave maximum that resulted from resonance, since the rat model used in the previous study [9] was approximately one wavelength long.

Both visual inspection of cumulative records and measures of total responses across sessions revealed no noticeable effects of exposure until the incident energy power at densities of 40 mW/cm² was applied (Fig. 7). At this level the animal showed physiological changes, i.e., panting, fatigue, and foaming of the mouth, which are indicative of general hyperthermia (heat stress). Rat no. 5 was exposed during each of five consecutive days in order to observe gross cumulative effects of irradiation and to provide some control for unobserved sources of variability. Rat no. 6 was exposed on alternate days, Monday, Wednesday, and Friday. Rat no. 7 was sham-exposed. Recovery of the base-line performance for both exposed animals on the first day following exposure at 40 mW/cm² is readily observed in Fig. 7.

Another experiment in which the incident power densities were raised in 3-mW/cm² increments indicated a slight decrease in responding during exposures to densities of 32 mW/cm² (Fig. 8). The corresponding peak in the rate of absorbed energy was 29 W/kg. Immediately after irradiation, the rat exhibited symptoms similar to the rat exposed to densities of 40 mW/cm². As shown in Fig. 8, the deterioration of responding did not occur in this animal until 20 min after initiation of exposure. In contrast, the response of

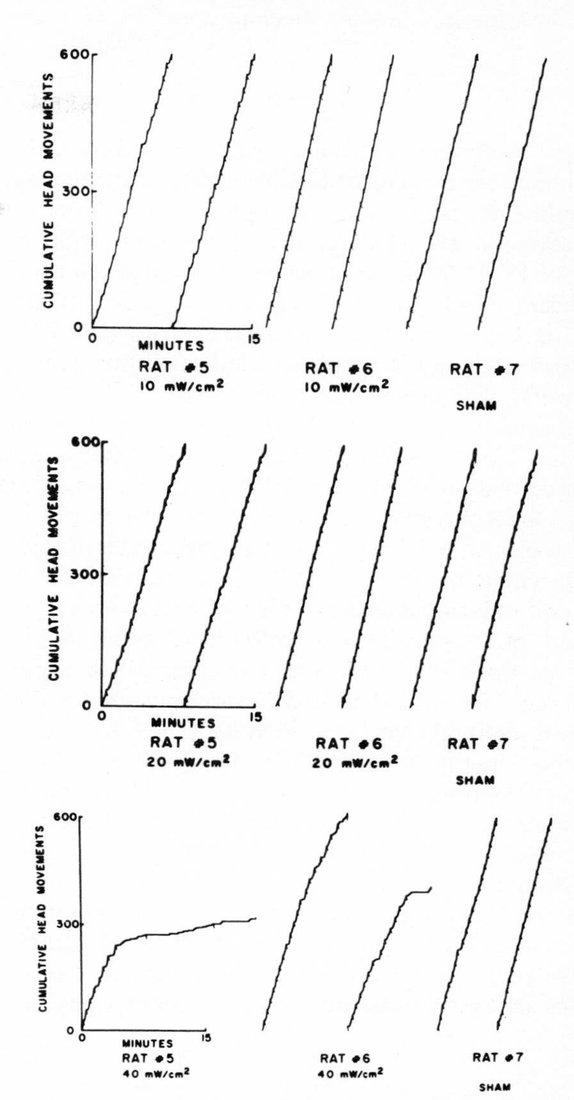

Fig. 7. Cumulative records of FR-30 responding during a microwave irradiation session.

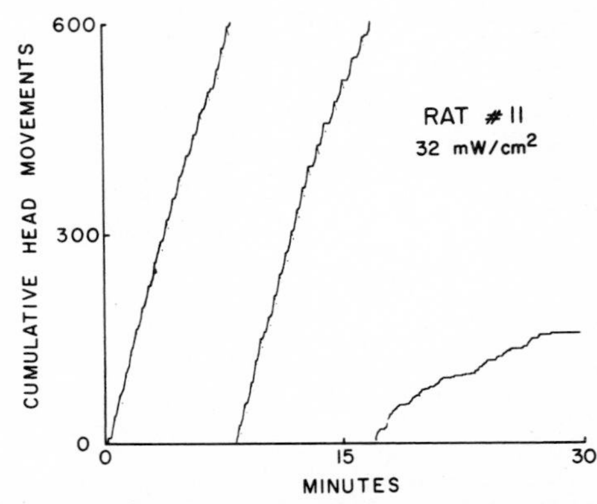

Fig. 8. Response record of a rat exposed to 32-mW/cm² indicent power density with a peak rate of absorbed energy of 28.8 W/kg.

animal no. 5 exposed to 40 mW/cm^2 decreased dramatically only 5 min into the period of irradiation. These data clearly point to a thermal loading mechanism of interaction.

Summary

These preliminary studies with reasonably accurate determinations of incident and absorbed energy suggest that incident CW microwaves at power densities of 20–30 mW/cm^2 and averaged rates of energy absorption in the body of 18–27 W/kg, after 30-min exposures, do not have significant effects on the fixed ratio behavior of the rat. Control performances one day after exposure to fairly high densities of energy (40 mW/cm^2) indicated that behavioral disruption, when occurring, is reversible.

It should be noted that the bodily dimensions of the rat should render it a resonant structure to 918 MHz of energy. Based on mass-equivalent muscle spheres, theoretical calculations show maximum peak absorption occurring in a body the size of a rat to have peak and averaged rates of absorbed energy equal to 0.85 and 0.21 W/kg, respectively, per 1 mW/cm^2 of incident energy. This indicates that energy at an incident power density of 40 mW/cm^2 is associated with a peak absorption of 34 W/kg and an averaged absorption of 8.4 W/kg. The measured peak-absorption rate for the rats exposed to 40 mW/cm^2 was 36 W/kg. The 8.4-W/kg rate of absorbed energy is particularly significant in light of the reported depressions in general activity of rats immediately following exposure to far-zone 2450-MHz microwaves that resulted in an average absorption rate of 8.0 W/kg [10].

It should also be mentioned that this study indicates that animals (at least the rat) can be trained to accept immobilization and nonetheless to perform well on an appetitively motivated operant task. Thus dose-determinate behavioral studies of the rat can take place in almost any kind of exposure environment if proper precautions are taken to habituate the animals to restraint.

Acknowledgment

The authors wish to thank Dr. D. R. Justesen for his valuable suggestions and J. McDougall, D. Maloney, and T. Sparks for their assistance.

References

[1] A. H. Frey and R. Messenger, Jr., "Human perception of illumination with pulsed ultra-high frequency electromagnetic energy," *Science*, vol. 181, pp. 356–358, 1973.

[2] R. C. Nealeigh, R. J. Garner, R. J. Morgan, H. A. Cross, and P. D. Lambert, "The effect of microwave in Y-maze learning in the white rat," *J. Microwave Power*, vol. 6, pp. 49–54, 1971.

[3] D. R. Justesen and N. W. King, "Behavioral effects of low-level microwave irradiation in the closed space situation," in *Biological Effects and Health Implications of Microwave Radiation* (Symp. Proc. Bureau of Radiological Health), Tech. Rep. BRH/DBE 70-2, 1970, pp. 154–179.

[4] S. F. Korbel, "Behavioral effects of low intensity UHF radiation," in *Biological Effects and Health Implications of Microwave Radiation* (Symp. Proc. Bureau of Radiological Health), Tech. Rep. BRH/DBE 70-2, 1970, pp. 180–184.

[5] E. L. Hunt, N. W. King, and R. D. Phillips, "Behavioral effects of pulsed microwave radiation," *Annals of N.Y. Acad. Science*, vol. 247, pp. 440–452, 1975.

[6] J. A. Tanner, C. Romero-Sierra, and S. J. Davie, "The effects of microwaves on birds: Preliminary experiments," *J. Microwave Power*, vol. 4, pp. 122–128, 1969.

[7] A. W. Guy and S. F. Korbel, "Dosimetry studies on a UHF cavity exposure chamber for rodents," in *1972 Microwave Power Symp., Summaries of Papers* (Ottawa, Canada), pp. 180–193.

[8] C. C. Johnson and A. W. Guy, "Nonionizing electromagnetic wave effects in biological materials and systems," *Proc. IEEE*, vol. 60, pp. 692–718, 1972.

[9] A. W. Guy, "Quantitation of induced electromagnetic field patterns in tissue and associated biological effects," in *Biologic Effects and Health Hazards of Microwave Radiation*, P. Czerski *et al.*, Eds. Warsaw, Poland, 1974.

[10] E. L. Hunt, R. D. Phillips, R. D. Castro, and N. W. King, "General activity of rats immediately following exposure to 2450 MHz microwaves," in *1972 Microwave Power Symp., Summaries of Papers* (Ottawa, Canada), p. 119.

The Effect of Microwave on Y-Maze Learning in the White Rat

Roger C. Nealeigh†,
R. John Garner†,
R. John Morgan‡,
Henry A. Cross††
and
Paul D. Lambert†

ABSTRACT

The performance of white rats in a Y-maze learning task was altered by exposures of 2.45 GHz microwaves at a measured maximum level of 50 mw/cm².

Introduction

The expanding use in recent years of microwave heating devices in industry and microwave ovens in the home has increased the possibility of exposure of workers and of the general population to microwave radiation. It has been suggested that microwaves directly affect neural processes thus inducing transient changes in behavior.[1,2,3] Retrograde amnesia and depressed learning have been described in rats exposed to microwaves[4,5] but the field intensity in these studies was evidently quite high since tissue temperatures were elevated within several seconds. However, behavioral effects have also been demonstrated with low intensity fields.[6,7,8]

This experiment was designed to study the effects of 2.45GHz microwaves on Y-maze learning in white rats. The microwave exposure was given just prior to the test situation in an attempt to produce retrograde amnesia. Behavioral changes under these conditions would be helpful in designing neurophysiological experiments to study microwave and neural interactions.

Materials and Methods

Twenty female Sprague-Dawley rats, about 90 days old, were randomly assigned to two equal groups, a control group and a microwave-irradiated group. The microwave exposures consisted of 2.45 GHz continuous wave irradiation for 20 minutes in a field having a maximum power density of 50 mw/cm² (Fig. 1). The field in Fig 1 was produced by a 2.45 GHz Scintillonics, Inc. microwave generator, Model HV15A 100-watt CW variable output oscillator used in conjunction with a Scientific Atlanta standard gain horn Model SGH 2.60 (Frequency range: 2.60 - 3.95 GHz) mounted one meter above the center of the body of a standing animal in the start box. The arrangement of horn, start box and Eccosorb blocks can be seen in Fig. 2. A forward power setting of 100 watts was used to feed the horn. The power densities were measured with a Narda Model 8100 Electromagnetic Radiation Survey Meter (Narda Microwave Corp.). The Narda probe was positioned above the Eccosorb CV-B (−40 db) microwave absorbing blocks which were used beneath the start box. A simulated start box without a floor did not affect the readings appreciably.

* This investigation was supported by U.S. Public Health Service Contract No. CPE-R-70-0001. Original manuscript received October 27, 1970; in final form May 19, 1971.

† Collaborative Radiological Health Laboratory, Colorado State University, Fort Collins, Colorado. 80521.

†† Department of Psychology, Colorado State University, Fort Collins, Colorado 80521.

‡ Department of Electrical Engineering, Colorado State University, Fort Collins, Colorado 80521.

Reprinted with permission from *J. Microwave Power*, vol. 6, no. 1, pp. 49–54, Mar. 1971.

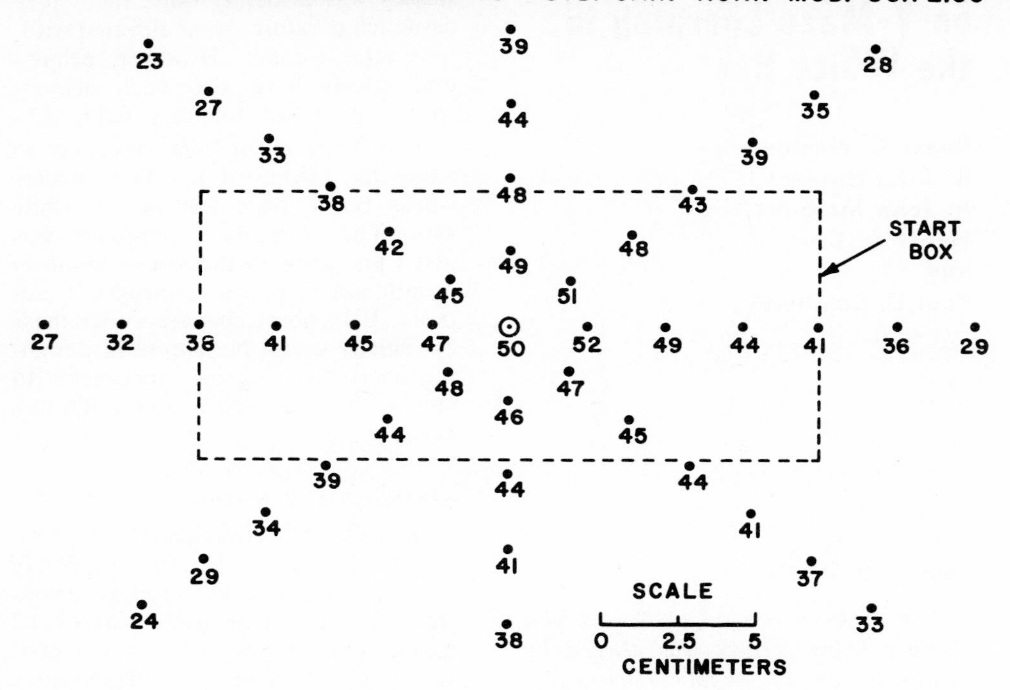

Fig. 1 The power density distribution measured in air in a plane one meter from the neck of the standard gain horn. The dashed lines indicate the walls of the start box.

Each irradiated rat received three exposures, one each on the first, second and third day just prior to running the maze (Fig. 2). The control group was sham irradiated using the same procedure and apparatus. The Y-maze arms were 24 inches long with a 24 inch runway to the choice point. All of the alleys were 4 inches wide by 5 inches high and the complete maze was painted black with a non-metallic carbon black pigment paint. There were restraining doors on the runway and each alley. The sections of the maze were covered with plexiglass to contain the animal, and yet allow it to be clearly seem from above. The maze was situated in a hooded ventilator with the lights and fan on to mask extraneous noise and movements of the experimenter.

The animals were placed on a 23½ hour deprivation schedule for 13 days prior to the experiment. During this time they were handled by the experimenter and placed in the maze to become familiar with it. This regimen has been shown to produce a high degree of motivation in the subjects for this type of testing.[9]

The testing paradigm consisted of placing the animals in the start box of the maze and holding them there for 20 minutes of irradiation or sham irradiation. They were then released to run the maze selecting either the right or left arm. As soon as the animal had entered one arm to the base of its tail the restraining door was closed. The animal was kept in the arm for about 20 seconds to allow time for eating the reward when it was

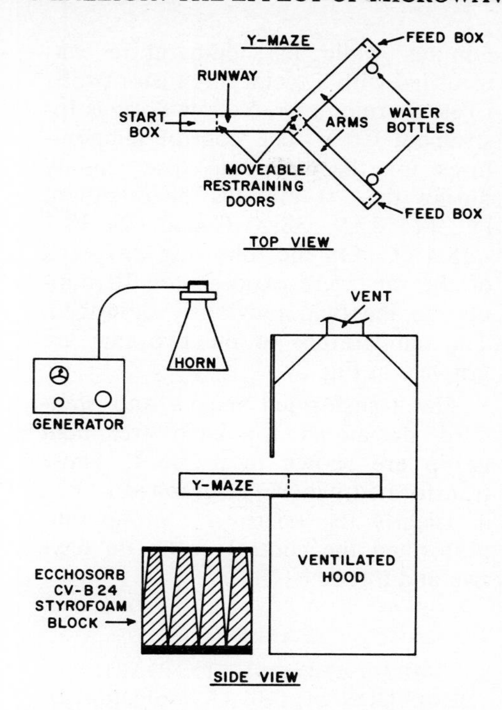

Fig. 2 Schematic diagram of Y-maze in position under the microwave horn.

present. Trial one was completed when the rat found the reward for the first time. This allowed the animal to become oriented with the reward. After the correct arm was chosen the next correct response was the alternate arm. Reinforcement was a 20-50 mg piece of Sugar Rice Krinkles (General Foods). Two control and two irradiated rats were tested each day and fifty trials per rat were run with alternate right and left responses rewarded. This procedure was followed for three consecutive days. The percentage of correct responses were recorded for each rat's 50 trial session. These values were transformed using arcsin $\sqrt{\text{percentage}}$ before testing for significance by means of analysis of variance and trend analysis.[10]

Experimental Findings

The control and irradiated animals showed no difference in behavior during the microwave exposure. The irra-

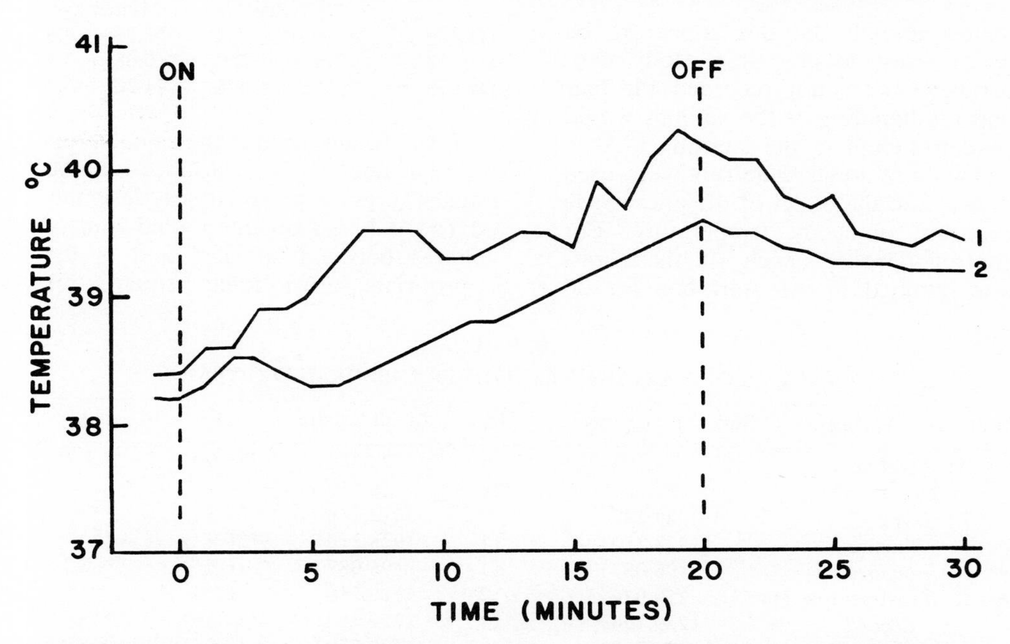

Fig. 3 The increase in body temperature of two male white rats exposed to 2450 MHz microwaves. The body temperature was allowed to stabilize for two minutes before the generator was turned on.

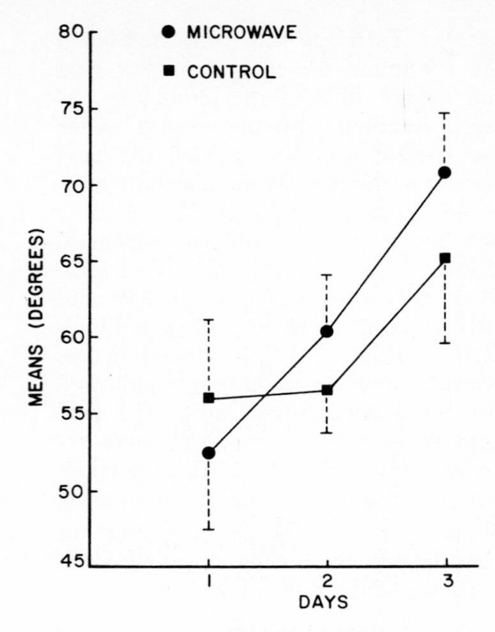

Fig. 4 The average transformed percentage of correct responses plotted against trial days. The transformation of the response scores was arcsin $\sqrt{\text{percentage}}$.

diated animals did not appear to be under stress at any time. Body temperatures were not recorded for fear that the handling of the animals would be detrimental to the learning task.

Two additional male rats were used to evaluate the effect of the microwave exposure on body temperature. For control purposes each of these rats was retained in the start box for 20 minutes while his temperature was recorded with a rectal thermistor probe (Tele-thermometer, Yellow Springs Instrument Co.). The control temperatures for the two rats rose slightly during the 20 minutes of retention, i.e., (1) 37.9 - 38.2° C and (2) 38.1 - 38.4° C. On the following day each of the rats were exposed for 20 minutes to the field previously described. The temperature of the two rats are graphed in Fig. 3.

The transformed means and standard deviations for each treatment group are shown in Table I. These transformed means are graphed in Fig. 4. Clearly the microwave group outperformed the control group on days two and three.

TABLE I

MEAN AND ONE STANDARD DEVIATION OF THE TRANSFORMED DATA (ARCSIN $\sqrt{\text{PERCENTAGE}}$) FOR THE TWO GROUPS OF RATS

	Control	Microwave
Day 1	56.10 ± 5.2	52.53 ± 4.8
Day 2	56.47 ± 2.9	60.68 ± 3.9
Day 3	65.20 ± 6.8	71.20 ± 3.2

Table II summarizes the calculations for the analysis of variance of the data. The error mean square designated (a) is based on the pooled sum of squares between subjects and is the appropriate error mean square for

TABLE II

ANALYSIS OF VARIANCE OF THE TRANSFORMED DATA

Source of Variation	Sum of Squares	d.f.	Mean Square	F	Sig.
A: Treatment	73.394	1	73.394	2.01	N.S.
Error (a)	657.350	18	36.519		
B: Days	2024.297	2	1012.148	67.83	**
linear	(1928.210)	(1)	(1928.210)	(130.99)	(**)
quadratic	(96.087)	(1)	(96.087)	(6.53)	(N.S.)
A x B: Treatment x Days	259.034	2	129.516	8.68	**
linear	(229.154)	(1)	(229.154)	(15.56)	(**)
quadratic	(29.880)	(1)	(29.880)	(2.03)	(N.S.)
Error (b)	529.935	36	14.720		
Total	3544.011				

testing the significance of the A or irradiation treatment effect. The error mean square designated (b) is based upon the pooled subjects x days interactions and is the appropriate mean square for testing the significance of the B or day effect and the A x B or treatment x days interaction.

The nonsignificant F for the A effect indicates that the average performance over 3 days is not significantly different for the two treatments, control and microwave. The significant value for the B effect shows that the average value for each of the three days differs significantly.

Most important of all, from the experimental point of view, is the significant F for the A x B interaction which means that the two learning and/or performance curves were not of the same form. Our primary interest was, as is usually the case in this type of experimental design, with the B (days) effect and the A x B interaction, both of which were highly significant. Trend analysis was indicated in order to best account for the difference between the two learning curves.

Two orthogonal comparisons, linear and quadratic, were made of the over-all day means and the treatment x days interaction. The linear comparison regarding the over-all trend of the day means was significant $P < 0.01$, from which one must conclude that the levels of B were affected differently by the two treatments, i.e., the microwave irradiation enhanced performance on the 2nd and 3rd days. The quadratic comparison of the A x B interaction was not significant. This would indicate that the quadratic components of the learning curves were not significantly different.

Discussion

The results of our experiment suggest that microwave irradiation may produce effects similar to a variety of CNS stimulants which act on the CNS by facilitating consolidation of the memory trace. Drugs such as caffeine, physostigmine, amphetamine, nicotine, and the convulsants, picrotoxins, strychnine and pentylenetetrazol have been shown to improve learning in animals when administered immediately before or after the learning sessions.[11] In our experiment the irradiated rats had a learning curve significantly different from that of the controls. It would appear then that 2.45 GHz microwaves affected performance at the power density level used.

Although there was an attempt to produce anterograde amnesia with microwaves, just the opposite effect was encountered. This was not entirely unexpected since Bryan[4] in his rat experiment, using 12-cm microwaves fifteen seconds after the training trial, found initially that the treated animals retained the conditioned avoidance response better than the controls.

From this experiment one could conclude that the effect observed would be detrimental in persons exposed to extended periods of microwave irradiation. It is understandable that a continuing stimulation of the CNS could result in headaches and other ill-defined effects that supposedly have been produced in workers exposed to microwaves. Other experiments are obviously indicated, e.g., studies involving lower power density levels and longer exposures to microwaves.

References

1 Frey, A. H., "Behavioral Biophysics," Psychol. Bull. 5:322-337, 1965.

2 Kholodov, Y., "The Influence of Electromagnetic and Magnetic Fields on the Central Nervous System," Moscow: Soviet Acad. Sci., 1966.

3 Presman, A. S., "The Role of Electromagnetic Fields in Life Activity Processes," Biofizica 9:131-134, 1964.

4 Bryan, R. N., "Retrograde Amnesia: Effects of Handling and Microwave Radiation," Sci. 153:897-899, 1966.

5 Justensen, D. R., Pendleton, R. B. and Porter, P. B., "Effects of Hyperthermia on Activity and Learning," Psychol. Repts. 9:99, 1961.

6 Korbel, S. F. and Fine, H. L., "Effects of Low Intensity UHF Radio Fields as a Function of Frequency," Psychonomic Science 9: 527-528, 1967.

7 Korbel, S. F. (Eakin) and Thompson, W. D., "Behavioral Effects of Stimulation by UHF Radio Fields," Psychol. Repts. 17:595-602, 1965.

8 Tanner, J. A., Romero-Sierra, C. and Davie, S. J., "The Effects of Microwaves on Birds: Preliminary Experiments," J. Microwave Power 4:122-128, 1969.

9 Gay, W. I., "Methods of Animal Experimentation," Vol. III:1-25, 1968.

10 Edwards, A. L., "Experimental Design in Psychological Research," Revised Edition, Holt, Rinchart and Winston, New York, pp. 130, 131 and pp. 224-253, 1960.

11 Ganong, W. F., "Review of Medical Physiology," 4th Edition, Large Medical Publication, Los Altos, California, p. 212, 1969.

FIELD MEASUREMENTS, ABSORBED DOSE, AND BIOLOGIC DOSIMETRY OF MICROWAVES*

Richard D. Phillips, Edward L. Hunt,† and Nancy W. King

Biology Department
Battelle, Pacific Northwest Laboratories
Richland, Washington 99352

INTRODUCTION

A variety of exposure arrangements have been used for irradiating experimental animals with microwaves to study the biologic effects. Methods of dosimetry and ways of expressing radiation dose have been equally diverse. Without a common dose unit, it has been very difficult, if not impossible in many cases, to compare the biologic effects produced with different exposure arrangements. There is an obvious need to establish a uniform method of expressing radiation dose if we are to be able to compare results between studies.

Various solutions to the dosimetry problem have been proposed and investigated,[1-6] but the problem of measuring the amount of energy absorbed by biologic materials continues to plague investigators. Several years ago, we developed a technique for measuring the amount of energy an animal absorbs when exposed to microwaves in a cavity arrangement.[7,11] Recently, we have applied this technique to determine absorbed doses with far-field exposures of rats in an anechoic chamber and with exposures in the multimodel resonating cavity. We compared the biologic effects produced by these two treatment arrangements at equivalent radiation doses. In addition, we determined the accuracy with which physical models of rats could be used for estimating energy absorption by the whole animal in both irradiation arrangements.

MICROWAVE EXPOSURE SYSTEMS

Two microwave exposure arrangements were used: a multimodal resonating cavity system, in which the animal serves as the load and is exposed multilaterally; and a far-field exposure system in an anechoic chamber, in which the animal is exposed unilaterally to a well-defined incident field.

Multimodal Resonating Cavity Exposure System

A block diagram of the system used for exposing animals to microwaves in the cavity is shown in FIGURE 1. The cavity is powered by a magnetron source operated at 2450 MHz with a pulse repetition rate of 120 Hz. In this study, the output power of the source was adjusted to 175 W, which produced pulses of 2.5 msec duration. A three-port circulator is located in the wave guide to protect the magnetron from

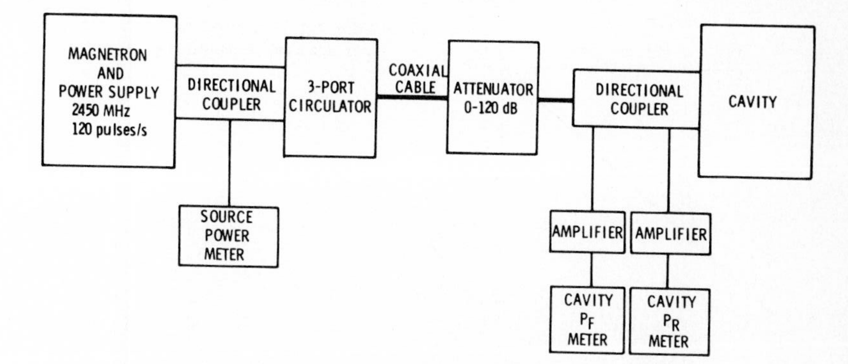

FIGURE 1. Block diagram of multimodal resonating cavity exposure system.

reflected power and reduce moding of the magatron. Power delivered to the cavity is adjusted with the variable attenuator. Crystal detectors in the bidirectional coupler and calibrated power meters were used to measure forward and reflected powers at the cavity input. Temperature and relative humidity within the cavity were maintained at approximately 24° C and 20–40%, respectively, by a forced-air ventilation system. In this exposure system, the animal serves as the working load in the cavity, and when the animal is positioned in a standard location, it absorbs energy in proportion to net power delivered to the cavity.[7,11]

Far-Field Exposure System

The arrangement employed for exposing animals in the far field of a transmitting antenna in an anechoic chamber is shown in FIGURE 2. The source is an APS-20E radar transmitter with a coaxial magnetron. It generates 2880 MHz microwaves at 925 or 308 pulses/sec, with pulse widths of approximately 0.8 and 2.4 μsec, respectively. A circulator has been added to the system to protect the magnetron from reflected power; a variable attenuator is used for adjusting the power level to the transmitting antenna; and calibrated power meters connected to directional couplers are utilized to measure forward and reflected powers in the wave guide. The transmitting antenna, located in the anechoic chamber, is a horn with a 16.1-dB gain. The transmission system is pressurized with nitrogen gas at 10 psig to reduce the possibility of arcing within the system.

The anechoic chamber consists of a radio-frequency-shielded room lined with microwave-absorbent material. The anechoic, or specimen exposure, space is located in the far field of the antenna and extends from 4 to 10 ft from the aperture of the horn antenna. The anechoic space is 2 ft wide by 1 ft high, with its geometric center located on the beam axis. Reflectivity measurements were achieved with the free-space voltage standing wave ratio technique. It was found that the reflected power into this space from the ceiling, floor, side walls, and back wall is decreased at least 45 dB from the main beam power. The cross polarization of the chamber is −38.5 dB, which shows that the room does not appreciably alter the plane of polarization of the beam from the transmitter antenna. Temperature within the chamber was maintained at about 24° C by a ventilation system that exchanges the room air approximately 60 times per minute. The relative humidity during these experiments ranged from 20 to 40%.

*Supported by Office of Naval Research Contracts N00014-70-C-0197 and N00014-70-C-0332 with funds provided by the Bureau of Medicine and Surgery, Department of the Navy.

†Present address: Department of Microwave Research, Walter Reed Army Institute of Research, Washington, D.C. 20012.

Reprinted with permission from *Ann. N.Y. Acad. Sci.*, vol. 247, pp. 499–509, Feb. 28, 1975.

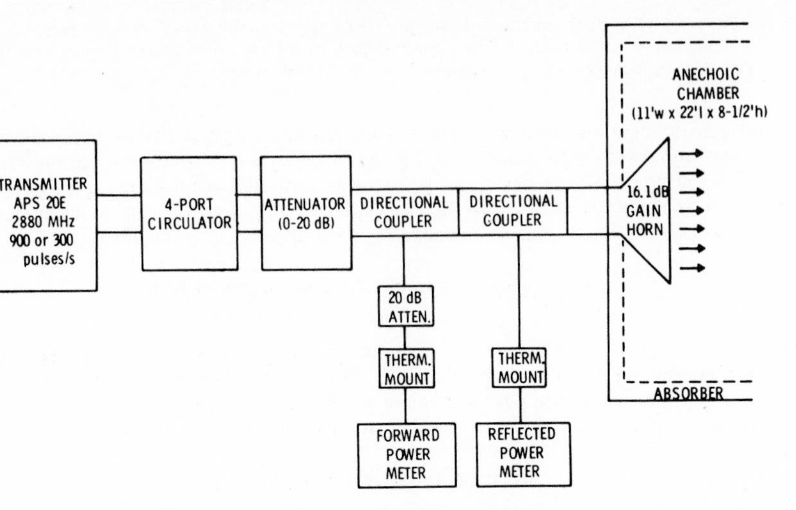

FIGURE 2. Block diagram of far-field exposure system.

Power densities were measured in the far field by a standard procedure.[8] A 16.1-dB receiving antenna, identical to the transmitting antenna, was placed in the far field at the location of interest. Both antennas were oriented with their electric fields horizontal. A given power level, measured in the wave guide at the directional coupler, was delivered to the transmitting antenna. The power absorbed by the receiving antenna was conducted through wave guide and coaxial cable of known characteristics to a thermistor power bridge. The power density was calculated from the following equation:[8]

$$P_d = \left(\frac{4\pi}{9 \times 10^8}\right) f^2 \left(\frac{A}{G}\right) P_t,$$

where P_d represents the power density (mW/cm²), f equals the frequency of radiation (MHz), A is the attenuation of the coupling circuit, G denotes gain of receiver antenna, and P_t represents the power registered on the thermistor power bridge (mW). For the arrangement employed, the attenuation of the coupling circuit was 24.85 dB, the gain of the receiver antenna was 16.1 dB, and the frequency of the source was 2880 MHz. Insertion of these values in the equation gives the result, $P_d = 0.88\ P_t$.

We made a mathematic analysis of the conjugate mismatch uncertainties of the coupling circuit used for the power density measurements.[9,10] The lower and upper conjugate mismatch loss limits were −0.2 and 0.8 dB, respectively, which yields an uncertainty range of 1 dB. With this potential error source in the power density determinations, and all other potential instrument errors included, the total uncertainty of the absolute accuracy of the power density measurement was −11 to 32% of the stated values.

Power density measurements were performed 7 ft from the transmitting antenna with different forward powers in the wave guide (FIGURE 3). A linear relationship exists between the power density measurements and the forward power measured in

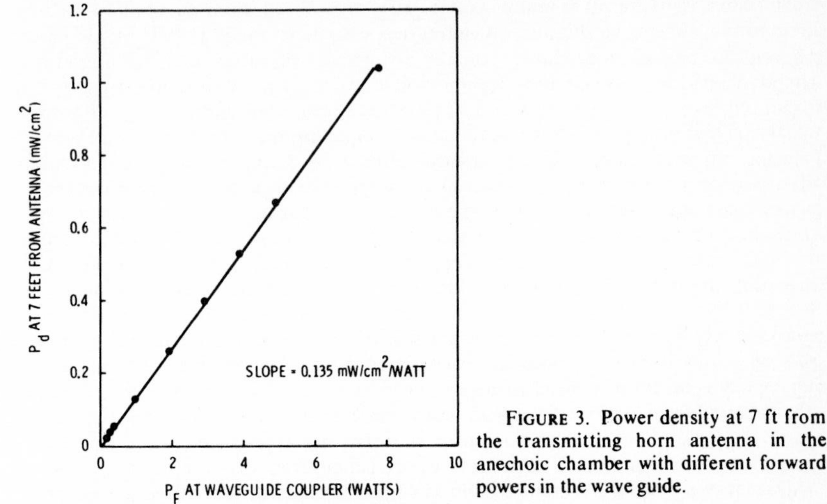

FIGURE 3. Power density at 7 ft from the transmitting horn antenna in the anechoic chamber with different forward powers in the wave guide.

the wave guide at the directional coupler. Reflected power is minimal in this system, less than 0.2% of the forward power, and can be ignored (FIGURE 4).

Power densities were also determined in the far field at different distances from the transmitting antenna. FIGURE 5 depicts the power densities in the far field as a function of distance from the receiving antenna, with the transmitting antenna operated at a forward power of 352 W. The data fit the inverse square law, with the isotropic reference point located 18 in. behind the horn antenna aperture.

ABSORBED DOSE MEASUREMENTS

We have previously described the technique utilized for determining the absolute amount of energy absorbed by animals exposed to microwaves in a cavity arrangement.[7,11] The same technique is used for determining the amount of energy absorbed

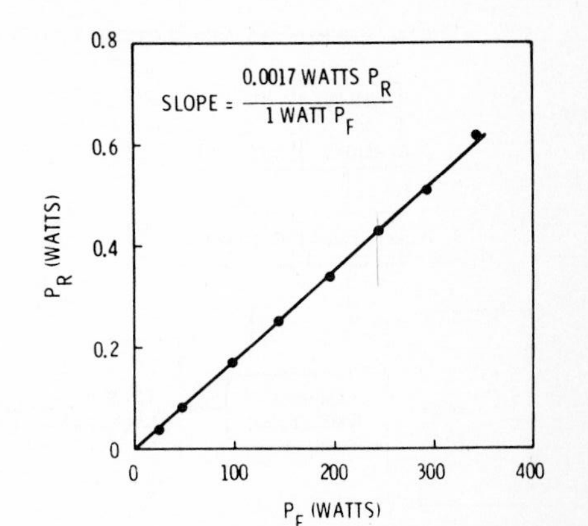

FIGURE 4. Relationship between forward and reflected powers in the wave guide.

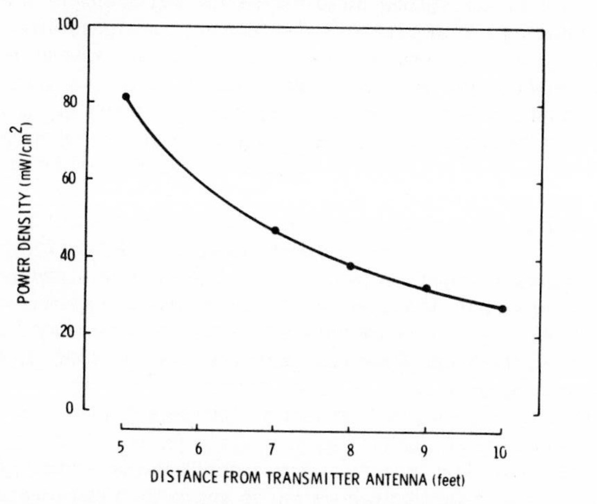

FIGURE 5. Power densities in the far field at different distances from the transmitting antenna operated at a forward power of 352 W.

with far-field exposures. Briefly, this value is determined by measuring the amount of heat generated in a fresh rat carcass irradiated with microwaves by use of a differential, or twin-well, calorimeter. The general procedure consists of placing a pair of freshly killed rats of equivalent body weight into a twin-well calorimeter and calculating the differential body heat content of the pair. Both carcasses are then removed from the calorimeter and placed in insulated containers constructed of expanded bead polystyrene. One animal is then exposed briefly to microwaves in either the cavity or in the far field in the anechoic chamber. The other animal serves as a sham-exposed control and reference heat source. Immediately after treatment, both animals are put back into the calorimeter, and the heat added to one animal's body heat by irradiation is measured. Use of dead animals eliminates the problems of physiologic heat production and loss and requires only the assumption that death does not alter the lossiness of animal tissues for absorption of microwave energy. The accuracy of this technique for determining the absolute amount of energy absorbed by rats has been calculated to be 0.8%.[12]

Absorbed Doses with Cavity and Far-Field Exposures

By means of the calorimetric technique, we measured the absorbed dose in rats exposed to microwaves in the cavity and in the far field of a transmitting antenna in the anechoic chamber. Determinations were made for three different sizes of animals in each treatment arrangement.

The amounts of energy absorbed by rats irradiated in the cavity are listed in TABLE 1. The mean and standard error values given in the Table are based on determinations from five animals in each weight group. The total amount of energy absorbed in a 1-min exposure for each watt of net cavity power increased with body mass. The absorption efficiencies can be calculated by comparing the total amount of energy absorbed to that which is available. The absorption efficiencies of 190-, 289-, and 375-g rats were 43.3, 48.2, and 51.6%, respectively. When the absorbed

TABLE 1

ENERGY ABSORBED (MEAN ± SE) BY RATS EXPOSED IN A CAVITY TO 2450 MHz MICROWAVES FOR 1 MIN FOR EACH WATT OF NET POWER AT CAVITY

Body Mass (g)	*N*	Total Surface Area* (cm^2)	Total Energy Absorbed (J)	Mass Absorption Density (mJ/g)	Total Area Absorption Density (mJ/cm^2)
189.6 ± 1.1	5	357	26.0 ± 0.5	137.2 ± 2.8	78.2 ± 1.5
288.6 ± 3.0	5	437	28.9 ± 0.5	100.1 ± 1.7	66.1 ± 1.1
374.5 ± 2.8	5	565	31.1 ± 0.5	82.9 ± 1.3	55.0 ± 0.8

*Altman and Dittmer.[13]

energies are expressed per unit of body mass, or as mass absorption densities, the average energy absorbed per gram decreased with an increase in body mass. The area absorption densities, which reflect average energy absorption per unit total surface area of the animals,[13] also declined with an increase in body mass.

TABLE 2 gives the absorbed dose values for rats exposed in the far field of the transmitting antenna in the anechoic chamber. The values shown are means and standard errors and are based on measurements from six animals in each weight group. The absorbed energies are normalized for a 1-min exposure for each mW/cm^2 of field power density. As with cavity irradiation, the total energy absorbed increased with body mass. The mass absorption densities were the same for the two heaviest weight groups and slightly higher for the lightest weight group. The silhouette area absorption densities, which reflect the average absorption per unit of shadow cross-sectional area of the animals,[3] did not change with animal size. If the total energy absorbed is compared to the amount of energy available for absorption on the basis of the silhouette surface area, it is found that the animals absorb about 67% of the available incident energy.

Comparison of Energy Absorbed by Models and Rats

Many investigators use biophysical models for estimating absorbed doses in cavity irradiations and far-field exposures.[3–6,14–18] Such models usually simulate the weight and volume of the animal and are shaped like spheres or cylinders. They are composed of water or a homogeneous mixture of materials that simulate the average dielectric properties of the whole animal or particular tissues. Because ani-

TABLE 2

ENERGY ABSORBED (MEAN ± SE) BY RATS EXPOSED UNILATERALLY IN THE FAR FIELD TO 2880 MHz MICROWAVES PER MINUTE PER mW/cm^2 FIELD POWER DENSITY

Body Mass (g)	*N*	Silhouette Area* (cm^2)	Total Energy Absorbed (J)	Mass Absorption Density (mJ/g)	Silhouette Area Absorption Density (mJ/cm^2)
196.6 ± 3.6	6	85	3.43 ± 0.15	17.5 ± 0.8	40.4 ± 1.8
296.7 ± 2.8	6	105	4.13 ± 0.08	13.9 ± 0.3	39.5 ± 0.9
378.0 ± 4.4	6	125	5.17 ± 0.08	13.7 ± 0.2	41.5 ± 0.6

*Justesen and King.[3]

TABLE 3

COMPARISON OF ENERGY ABSORBED (MEAN ± SE) BY RATS AND WATER MODELS EXPOSED IN A CAVITY TO 2450 MHz MICROWAVES FOR 1 MIN PER WATT NET POWER AT CAVITY

Mass (g)	Rats	Models	Difference (%) (Models − Rats)
	Mass Absorption Density (mJ/g)		
190	137.2 ± 2.8	145.8 ± 1.8	+6.2†
290	100.1 ± 1.7	104.4 ± 1.2	+4.4†
	Area Absorption Density (mJ/cm²)		
190	78.2 ± 1.1*	212.1 ± 2.6‡	+170†
290	66.1 ± 1.1*	178.9 ± 2.1‡	+170†

*Total surface areas of 190- and 290-g rats were 357 and 437 cm², respectively.[13]
†$p < 0.05$.
‡Total surface areas of 190- and 290-g models were 129 and 166 cm², respectively.

mals are heterogeneously structured of differently shaped tissues that vary in their composition and dielectric properties, it is not known how accurately simplified models estimate energy absorption by the whole animal.

We tested the accuracy of one type of model for estimating absorbed energy by the whole animal. The model is a cylindrical vessel composed of expanded bead polystyrene lined with a thin layer of Silastic® and filled with water.[7] The amounts of energy absorbed by two sizes of models, 190 g (17 cm long by 3.8 cm diameter) and 290 g (19.6 cm long by 4.5 cm diameter), were compared to those of rats of the same weights. The procedures used for irradiation and calorimetry were the same for the rats and models. Absorbed dose measurements were conducted on five water models and five rats, for each of two sizes, in each treatment arrangement.

TABLE 3 summarizes the results for rats and models exposed to microwaves in the cavity. The models only slightly overestimated the mass absorption densities of rats of comparable weight. In terms of the total area absorption densities, there were marked differences between models and rats. The values obtained with water models were 170% higher than those obtained with rats of the same weight.

A comparison of energy absorbed by rats and models exposed unilaterally to microwaves in the far field is illustrated in TABLE 4. In terms of the mass absorption densities, the water models underestimated the energy absorbed by rats by about 13%. This finding contrasts with the results obtained with cavity irradiations, where values based on models overestimated energy absorption of rats by about 5%. In terms of the silhouette area absorption densities, the values obtained with water models did not differ reliably from those of rats.

It is evident that extreme care must be used when energy absorption for the whole animal is estimated from models. With cavity irradiations, the mass absorption density appears to be the appropriate unit for estimating absorbed doses with models. With unilateral far-field exposures, the silhouette area absorption density appears to be useful for estimating absorbed doses with models. It is important to note that the results of these comparisons between models and rats only provide information as to the relative importance of mass and area for determining absorbed doses with cavity and far-field exposures and indicate nothing about the biologic relevance of either the mass absorption density or the area absorption density.

TABLE 4

COMPARISON OF ENERGY ABSORBED (MEAN ± SE) BY RATS AND WATER MODELS EXPOSED UNILATERALLY IN THE FAR FIELD TO 2880 MHz MICROWAVES PER MINUTE PER mW/cm² FIELD POWER DENSITY

Mass (g)	Rats	Models	Difference (%) (Models − Rats)
	Mass Absorption Density (mJ/g)		
190	17.5 ± 0.8	15.2 ± 0.7	−13.1*
290	13.9 ± 0.3	12.0 ± 0.3	−13.7*
	Silhouette Area Absorption Density (mJ/cm²)		
190	40.4 ± 1.8†	44.3 ± 3.0‡	ns§
290	39.5 ± 0.9†	40.5 ± 0.3‡	ns

*$p < 0.05$.
†The silhouette areas of 190- and 290-g rats were 85 and 105 cm², respectively.[3]
‡The silhouette areas of 190- and 290-g models were 66 and 85 cm², respectively.
§ns, not significant.

BIOLOGIC EFFECTS OF CAVITY AND FAR-FIELD EXPOSURES

We have obtained some biologic data that indicate that the average absorption of energy per unit mass is a useful parameter for expressing dose with cavity irradiations. In one experiment, we measured the latency for inducing convulsive seizures in rats during cavity exposures. With four different sizes of male rats of the Wistar strain, which ranged in weight from 190 to 562 g, a family of dose-response curves was obtained when latency to convulsions was plotted in terms of net power delivered to the cavity (FIGURE 6). As the weight of the animal increased, so did the latency at a given power level, even though the heavier animal absorbed more total energy. When the same data are replotted according to the rate at which energy is

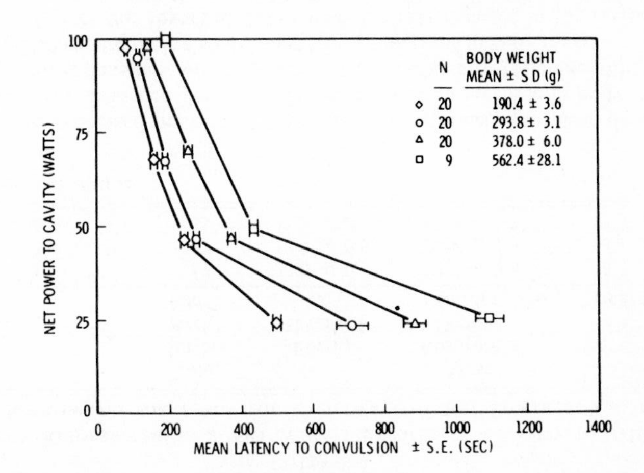

FIGURE 6. Latency to convulsions in rats during microwave irradiation with different net powers delivered to the cavity.

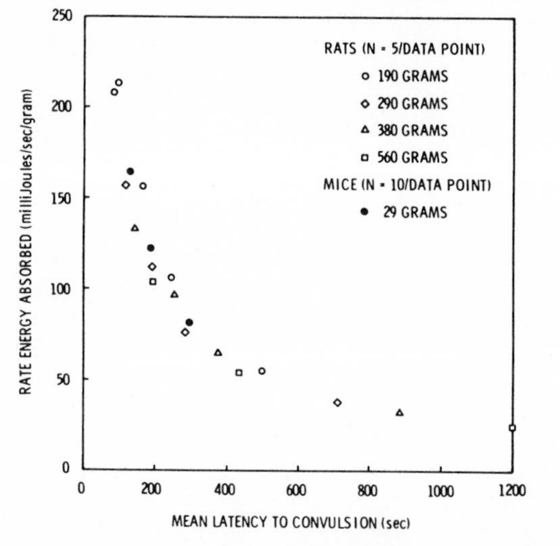

FIGURE 7. Latency to convulsions in rats and mice during microwave irradiation as a function of the rate at which energy is absorbed per gram of tissue.

absorbed per gram of tissue, based on the calorimetric measurements, the curves are no longer separated (FIGURE 7), and the data will fit a single curve. A similar study has been conducted in C57BL/6J mice (cf. FIGURE 7), and their responses fit the dose-response curve of rats. For this one biologic endpoint, latency to thermally induced convulsions, the average energy absorbed per unit mass appears to be a meaningful and useful parameter.

We do not know if the same relationship between absorbed dose rate and latency to convulsions would be obtained in rats and mice exposed unilaterally in the far field in the anechoic chamber. Power of sufficient magnitude to induce convulsions in rats cannot be generated in most far-field exposure systems, including the one used in this study. It may be possible to investigate this problem with mice as the test subjects. Because mice have a mass absorption density approximately 10 times that of rats, it should be possible to produce convulsions at much lower power densities in the far field.

We have initiated a series of experiments to compare the biologic responses of rats exposed in the cavity to those of rats irradiated in the far field in the anechoic chamber. Although we have obtained only preliminary data at this time, it is apparent that the biologic responses are different under these two treatment arrangements. Both qaauntitative and qualitative differences in responses have been observed over an equivalent range of absorbed doses.

Rats exposed to microwaves in the cavity for 30 min at an absorbed dose rate of 11.5 mW/g die within 24 hr after irradiation. A similar response is observed in rats irradiated unilaterally in the far field at an absorbed dose rate of only 8.3 mW/g for 30 min. Marked differences also occur in physiologic responses between animals irradiated in both exposure arrangements. With cavity irradiation, an array of functional changes occurs within 1–3 hr after a 30-min exposure at a dose rate of 6.5 to 11.0 mW/g.[19] Typically, the irradiated animal will exhibit thermoregulatory overcompensation, a lowered metabolic rate, bradycardia, irregular heart rate, and

incomplete heart blockage. None of these functional alterations have been observed in rats exposed to microwaves in the far field for 30 min at comparable dose rates.

There are several possible explanations for these differences in biologic responses to irradiation under these two exposure arrangements. With a cavity exposure the animal is irradiated multilaterally, while with far field exposures the animal is irradiated unilaterally. This would be expected to result in marked differences in distribution of absorbed energy that could lead to different biologic responses. Also, the frequency and pulsing characteristics of the microwave sources for the cavity and far-field exposure systems were different. The appropriate experiments need to be made to eliminate these differences. We have recently modified our exposure systems, so that the same source can be used for either the cavity or the anechoic chamber.

CONCLUSIONS

Four major implications of this study are:

The mass absorption density appears to be a useful dose unit for comparing biologic effects produced with different exposure arrangements. Other measures, such as silhouette area absorption density, may be useful for a particular geometry but not for others.

The use of models for estimating energy absorption by animals must take into consideration the geometry of the exposure arrangement. With cavity irradiation, the mass of the load appears to be the important factor. In unilateral far-field irradiations, the silhouette surface area of the object is the important factor.

It may not be possible to generalize information derived from cavity irradiations to far-field exposure conditions.

A safety standard based on incident field densities and based on data from experiments with far-field exposures may not be applicable under conditions where multilateral irradiations occur.

REFERENCES

1. ANNE, A., M. SAITO, O. M. SALATI & H. P. SCHWAN. 1961. Relative microwave absorption cross sections of biological significance. *In* Biological Effects of Microwave Radiation. M. F. Peyton, Ed. Vol. 1: 153–176. Plenum Publishing Corporation. New York, N.Y.
2. VOGELMAN, J. H. 1961. Microwave instrumentation for the measurement of biological effects. *In* Biological Effects of Microwave Radiation. M. F. Peyton, Ed. Vol. 1: 23–31. Plenum Publishing Corporation. New York, N.Y.
3. JUSTESEN, D. R. & N. W. KING. 1969. Behavioral effects of low level microwave irradiation in the closed space situation. *In* Biological Effects and Health Implications of Microwaves. S. F. Cleary, Ed. Report BRH/DBE 70-2: 154–184. Department of Health, Education and Welfare, Bureau of Radiological Health. Washington, D.C.
4. JUSTESEN, D. R., D. M. LEVINSON, R. L. CLARKE & N. W. KING. 1971. A microwave oven for behavioral and biological research: electrical and structural modifications, calorimetry, dosimetry, and functional evaluation. J. Microwave Power 6: 237–258.
5. GUY, A. W. & S. F. KORBEL. 1972. Dosimetry studies of a uhf cavity exposure chamber for rodents. *In* Microwave Power Symposium, Summaries of Presented Papers.: 180–193. International Microwave Power Institute. Edmonton, Alberta, Canada.
6. LIN, J. C., A. W. GUY & G. H. KRAFT. 1974. Microwave selective brain heating. J. Microwave Power 8: 275–286.

7. HUNT, E. L. & R. D. PHILLIPS. 1972. Absolute dosimetry for whole animal experiments. *In* Joint Army/Georgia Institute of Technology Microwave Dosimetry Workshop.: 74–77. Walter Reed Army Institute of Research. Washington, D.C.
8. GRISSOM, J. L. 1970. Determination of radiated rf power density. Electr. Instr. Dig. **6**: 61–63.
9. BEATTY, R. W. 1963. Intrinsic attenuation. IEEE Trans. Microwave Theory Tech. MTT-**11**(3): 179–182.
10. BEATTY, R. W. 1964. Insertion loss concepts. Proc. IEEE **52**(6): 663–671.
11. PHILLIPS, R. D. & E. L. HUNT. 1971. Problems of physical and biological dosimetry of microwave irradiation. J. Microwave Power **6**:90 (abs.).
12. HUNT, E. L., R. D. PHILLIPS, D. M. FLEMING & R. D. CASTRO. Dosimetry for whole-animal microwave irradiation. Health Phys. Submitted for publication.
13. ALTMAN, P. L. & D. S. DITTMER. (Eds.) 1962. *In* Growth.: 538. Federation of American Societies for Experimental Biology and Medicine. Washington, D.C.
14. MERMAGEN, H. 1961. Phantom experiments with microwaves at the University of Rochester. *In* Biological Effects of Microwave Radiation. M. F. Peyton, Ed. Vol. 1: 143–152. Plenum Publishing Corporation. New York, N.Y.
15. JOHNSON, C. C. & A. W. GUY. 1972. Nonionizing electromagnetic wave effects in biological materials and systems. Proc. IEEE **60**: 692–718.
16. HO, H. S., A. W. GUY, R. S. SIGELMANN & J. F. LEHMANN. 1971. Microwave heating of simulated human limbs by aperture sources. IEEE Trans. Microwave Theory Tech. MTT-**19**(2): 224–231.
17. GUY, A. W. 1971. Analyses of electromagnetic fields induced in biological tissues by thermographic studies on equivalent phantom models. IEEE Trans. Microwave Theory Tech. MTT-**19**(2): 205–214.
18. SCHWAN, H. P. 1971. Interaction of microwave and radiofrequency radiation with biological systems. IEEE Trans. Microwave Theory Tech. MTT-**19**(2): 146–152.

19. PHILLIPS, R. D., E. L. HUNT & N. W. KING. 1973. Physiologic response of rats to hyperthermia induced by exposure to 2450 MHz microwave radiation. Physiologist **16**(3): 423 (abs.).

DISCUSSION

DR. C. WILEY (*Environmental Protection Agency, Research Triangle Park, N.C.*): What was the shape of the model that you used?

DR. PHILLIPS: The model was a cylinder-shaped vessel.

DR. WILEY: Did you use any form of cooling in the cavity?

DR. PHILLIPS: The cavity has a forced-air ventilation system. We monitored but did not control the temperature. Typically, the temperature after the equipment had been running for awhile would be about 73° F. The temperature would increase by about 0.5° F during a 30-min exposure because of the heat generated by the equipment.

DR. A. W. GUY: We have verified that you do obtain hot spots in the rat with average absorbed power densities as you are measuring. The spherical model theory seems to also confirm this finding. You can obtain peak absorbed power densities up to approximately five times the average. The value is in the range of typical localized diathermy treatments, between 50 and 100 mW/g.

BEHAVIORAL EFFECTS OF PULSED MICROWAVE RADIATION*

Edward L. Hunt,† Nancy W. King, and Richard D. Phillips

Battelle, Pacific Northwest Laboratories
Richland, Washington 99352

INTRODUCTION

Research on microwave bioeffects at our Institution has included investigations of prompt behavioral effects of exposure to pulsed microwaves in the 10-cm region of the spectrum. One series of these experiments was designed to determine the dosage levels that, in association with mild to severe microwave heating of the experimental animal, produce prompt degradations in its performance. This paper summarizes the results obtained with three widely divergent forms of behavior: exploratory activity, swimming, and discrimination performance on a vigilance task.

Young adult male rats of the Wistar strain (Hilltop Laboratories) were used as subjects in all of these experiments. After suitable preliminary handling and training, each animal was irradiated or sham irradiated with pulsed 2.45-GHz microwaves for 30 min in a multimodal resonating cavity. The performance tests were initiated within 2 min postirradiation and 1 or 24 hr postirradiation.

MICROWAVE IRRADIATION PROCEDURES

The animals were exposed to microwaves in a multimodal resonating cavity system similar to that developed by Justesen and associates.[1] A block diagram of the cavity exposure system is shown elsewhere[2] in this monograph (see FIGURE 2, p. 501). The 2.45-GHz source (Varian, PPS-2.5A) produced pulses of quasi-sinusoidal shape with a half-amplitude duration of approximately 2.5 msec and a repetition rate of 120/sec. This energy was delivered to the cavity (Varian, TCS-2.5A) through a circulator (Varian, EW3-CL3) to protect operation of the magnetron from reflected power, then through a variable attenuator (Narda, W22912) to provide reductions in pulse amplitude and of power delivered to the cavity without altering pulse modulation, and finally through a bidirectional coupler located at the cavity input to sample and measure the levels of forward power (P_F) directed toward the cavity and reflected power (P_R). The net power delivered to the cavity was calculated as the difference value, $P_F - P_R$. The cavity was equipped with a mode stirrer, mounted on the back wall, that increased the number of resonance modes and enhanced field homogeneity.

The experimental animal was located in a standard position within the cavity and was exposed multilaterally to a complex, rapidly shifting incident energy field. The efficiency with which energy available in the cavity field is absorbed by the animal in this arrangement is determined by the electrical characteristics, size and shape of the animal, and by its location. The system was calibrated to provide a known rate of energy absorption for rats of various sizes by means of absolute dosimetry measurements based on the whole-animal calorimetry method developed previously.[3–5] Characteristics of absorption by rats are provided elsewhere in this monograph (see TABLE 2, p. 504).

A modified form of the Bollman holder,[6] made of expanded bead polystyrene (Polyfoam®) and Lucite® rods (0.6 cm diam), was used for confining the animal during exposure. Two Polyfoam end blocks were joined by the rods pressed through the blocks to form a cylinder-shaped space for the rat. A Polyfoam base plate served as a spacer between the end blocks and provided a floor support surface for the rat. This tended to reduce struggling and rotational movements during confinement and irradiation. The rats readily habituated to this confinement.[7] This holder would be expected to produce minimal distortions in the incident field, except for some reflections from the Lucite rods. Because it was of open construction, it would interfere little with normal heat dissipated by the animal. The temperature and relative humidity within the cavity were maintained at 24°C and at 20–40% by a ventilation system that exchanged air with the surrounding laboratory at a high rate (7.4 m³/min).

All animals were repeatedly confined to the holder, and sham irradiated when appropriate, prior to their experimental treatment. All exposures and sham exposures were 30 min in duration. For irradiation, power to the cavity was adjusted according to each animal's weight to produce an absorbed dose rate just above 6 mW/g and also, in the swimming and discrimination investigations, 11 mW/g. No power was delivered to the cavity during sham irradiations.

EXPLORATORY ACTIVITY INVESTIGATION

Experimental Design and Methods

To test for effects of microwave radiation on exploratory activity, rats were exposed and then placed in the activity apparatus for the first time in their lives and allowed to explore freely for 1 or 2 hr. In three replications, the activity test was initiated immediately after irradiation. In parallel replications, the animal was held undisturbed in a metal cage similar to its home cage for 1 hr postirradiation before the activity test was started. This procedure was employed to determine whether the animals' exploratory activity was related to the transient elevation in body temperature produced by the radiation. In the third replication of the delay series, measurements of deep colonic temperature were made with a thermistor probe (Yellow Springs) immediately after exposure and again just before the activity test was initiated.

A commercial activity meter (Columbus Instruments, model O) was used to index exploratory activity. The animal was confined on the movement-sensitive top surface of this unit in a Lucite chamber with base dimensions of 25 × 37.5 cm and 15-cm high walls. To promote exploratory activity, several head-sized holes were made in the outside walls, and two equal interior areas were formed by a partial wall divider. The unit was located inside a sound-deadened room and illuminated from overhead by a 15-W lamp. Electromechanical counters, located outside the room, accumulated counts for each 30-min period in the test. With spare rats of the size used in the experiments, the sensitivity of the meter was adjusted to extract counts primarily from whole-animal movements and relatively fewer from minor move-

*Supported by Office of Naval Research Contracts N00014-70-C-0197 and N00014-70-C-0332 with funds provided by the Naval Bureau of Medicine and Surgery.

†Present address: Department of Microwave Research, Walter Reed Army Institute of Research, Washington, D.C. 20012.

Reprinted with permission from *Ann. N.Y. Acad. Sci.*, vol. 247, pp. 440–453, Feb. 28, 1975.

ments, such as scratching, licking, and grooming. Naive rats placed in this apparatus typically spent much of their time perambulating between the interior sections and sticking their heads out through the holes in the outside walls. These activities diminished with time, and by the end the test, the animals were relatively inactive.

Results

FIGURE 1 shows the mean activity counts obtained from each 30 min of testing in the three replications in which the exploratory activity test was begun immediately after treatment. Because of the variation among experiments in the general level of activity counts, a comparison among experiments was made from the calculation of departures of irradiation group means from control group means within each experiment. Control group variances were pooled to obtain a standard error value for the control mean (SEM). The results of this analysis are provided in FIGURE 2; the shaded area represents the ±1 SEM limits.

FIGURES 3 and 4 similarly illustrate the experimental results and graphic analysis, respectively, for the three replications in which the exploratory activity test was initiated after a postirradiation delay of 1 hr. Immediately after treatment in the last replication, the exposed group had a mean temperature of 40.3 ±0.1° (SEM),

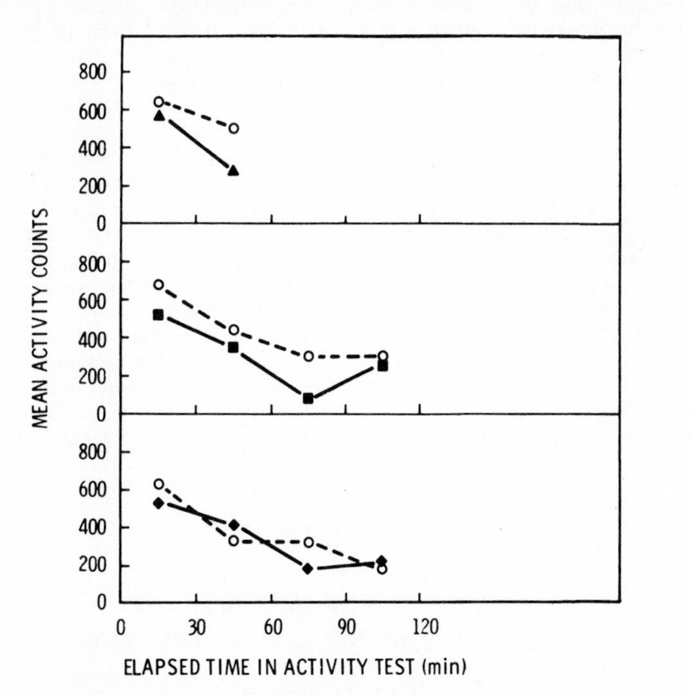

FIGURE 1. Activity of rats tested immediately after a 30-min irradiation at a dose rate of 6.3 mW/g. Three replications. Control (---): 10 (*top*), 5 (*middle* and *bottom*); irradiated (—): 10 (*top*), 5 (*middle* and *bottom*).

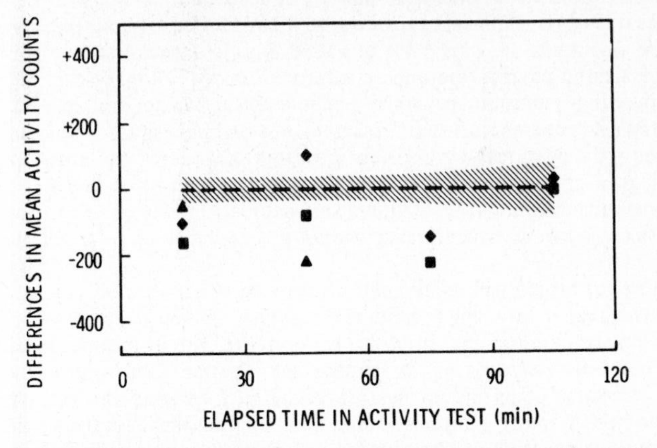

FIGURE 2. Departures of activity level of irradiated rats from controls (20 or 10), combined from three replications shown in FIGURE 1 (symbols are same as in previous Figure). Shaded area shows ±1 SEM (standard error of mean) limits for pooled controls.

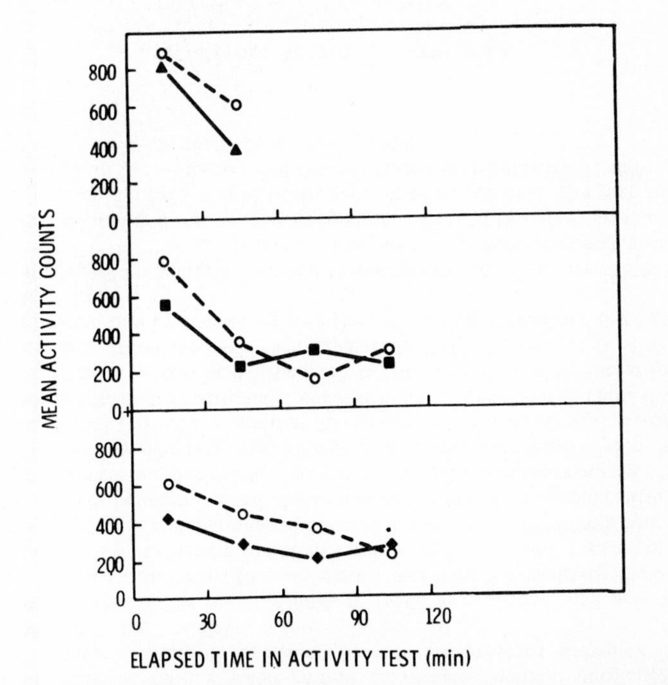

FIGURE 3. Activity of rats tested after delay of 1 hr postirradiation following a 30-min exposure at a dose rate of 6.3 mW/g. Three replications. Control (---): 5 rats each; irradiated (—): 5 rats each.

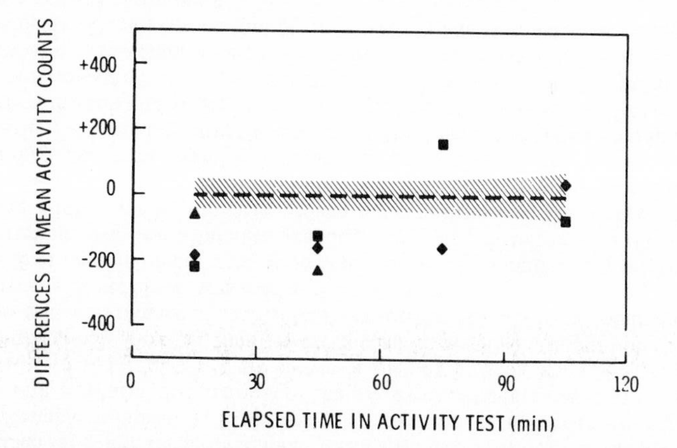

FIGURE 4. Departures of activity level of irradiated rats from controls (15 or 10), combined from three replications shown in FIGURE 3 (symbols are same as in previous Figure). Shaded area shows ±1 SEM limits for pooled controls.

and the controls a temperature of 38.6 ± 0.2° C. One hour later, their respective values were 37.8 ± 0.2 and 38.0 ± 0.2° C, essentially identical and within normal range.

The general tendency in both series of experiments was for irradiated animals to exhibit less activity than controls during most of the test period. The activity level of controls decreased finally to about the same level as that of exposed animals. These results were consistent with observations made through the window of the room. Irradiated animals during the middle portion of the test period were more likely to be seen in a sleeping position. There was no evidence that a postirradiation delay of 1 hr to the start of the activity test attenuated the radiation effect.

Discussion

The decreased level of exploratory activity by irradiated animals in these experiments evidently was not directly related to the transient hyperthermia produced by microwave heating. The exposed animals failed to exhibit a greater reduction in activity during the first 30 min postirradiation, when the hyperthermia was greatest. Also, the effect on activity was as evident after imposition of the delay, when recovery from the hyperthermia was complete, as it was when tested promptly.

The microwave effect found in this investigation does not appear to be comparable to effects on activity previously reported. Korbel[8] noted a reduction in rodent activity produced by exposure at a field power density estimated to be less than 1 mW/cm^2, which would not heat animals appreciably. This decline in activity first appeared only after several weeks of continuous exposure and was not a prompt effect of an acute irradiation. Furthermore, it was produced with microwaves in a continuously swept range of frequencies from 0.3 to 0.9 GHz, with wavelengths 3–10 times longer than that used in our investigation.

Justesen and King,[9] investigating effects of microwave radiation on a conditioned operant behavior, utilized 2.45-GHz microwaves in a multimodal resonating cavity

system. They used a recurrent cycle of exposure, 5 min on and 5 min off, and found that a stoppage of performance frequently occurred before the end of a 1-hr test session with a dose rate of 6.2 mW/g (3.1 mW/g average over each cycle.) At the highest dose rate used, 9.2 mW/g, this effect occurred much earlier and very abruptly. The inactivity induced in their animals was characterized by a flaccidity from which the animal could be aroused by handling, and it appeared to be directly related to the body heat load produced by irradiation. In contrast with their findings, the relative inactivity displayed postexposure in our investigation was not distinctively different from that displayed by the controls later in the test. Also, it was not directly related to the microwave-induced hyperthermia.

In some other experiments, Phillips and associates used irradiation methods and levels identical to those employed with this study to investigate physiologic effects in rats.[7] It was found that a 30-min treatment at a dose rate of 6.5 mW/g, but not 4.3 mW/g, provoked compensatory reactions to the initial heating, which included a relative hypothermia and bradycardia that persisted for 3–4 hr postirradiation. The loss in vigor and interest by the exposed animals in continuing exploratory activity during this period might represent one of the costs to the animal of making active metabolic adjustments to the thermal loading from microwave irradiation.

SWIMMING PERFORMANCE INVESTIGATION

Experimental Design and Methods

To test for effects of microwave radiation on the highly-motivated animal's work performance on a physically demanding task, trained rats were tested on a repetitive swim task immediately after a 30-min irradiation at dose rates of 6.3 or 11 mW/g and one day after treatment at the higher rate.

A straight alley swim apparatus and a training and testing procedure were developed for use in this investigation.[10,11] FIGURE 5 schematically depicts a side view of the swim alley. Water at 24° C, regulated within ±0.5° C, was continuously supplied through a manifold inlet in the center bottom and drained from both ends through the stand pipes. The swimming course between the end platforms was 6 m long and 0.3 m wide. After swimming the length of the alley, the animal was allowed to rest on the raised platform for a fixed time interval, 20 or 30 sec, and then the next traverse was started by the platform being lowered from under the animal. Simultaneously, the platform at the other end was raised and its position was illuminated to provide the next resting place. The platform movements were driven

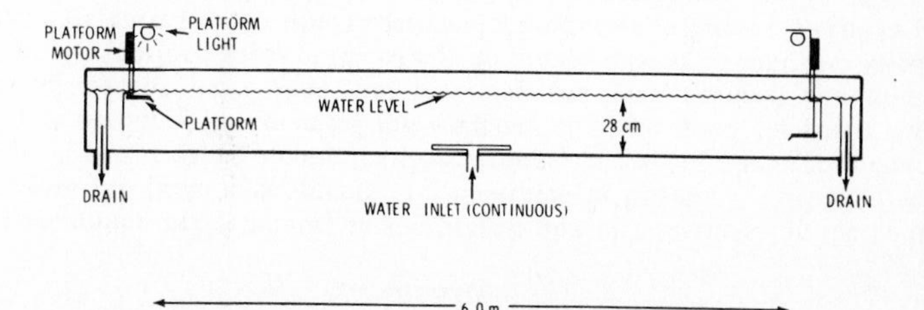

FIGURE 5. Schematic diagram of swim alley apparatus, side view section.

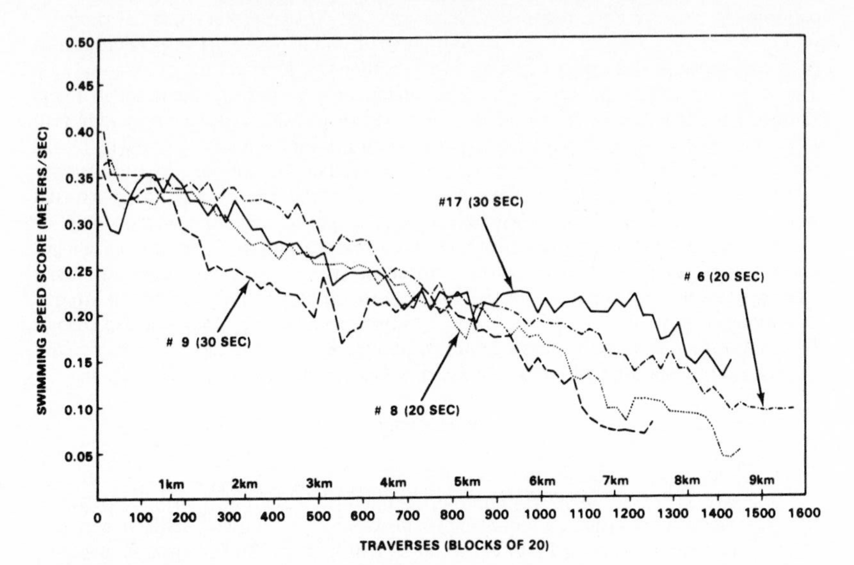

FIGURE 6. Performances of individually trained rats in 24-hr swim test. Platform rest times of 20 and 30 secs.

pneumatically. Automatic control equipment was used to switch and time apparatus operations and to record traverse times.

Although many animals performed reliably without specific training, others were erratic and would sometimes circle or even just float. Continuous directional swimming was readily established in all animals early in their training by punishing specifically all other behavior. This was accomplished by submerging the animal by hand momentarily and releasing it under water with the rat oriented in the proper direction. Reliable proficient performance on this task was established in four training sessions with a total of 200 traverses. The performance characteristics of trained animals in a long-duration test are shown in FIGURE 6. Even after 24 hr of swimming, these animals were paddling toward the raised platform, though very weakly. To follow the performance of each animal, its median velocity in each block of traverses, usually 20 per block, was used as its speed score.

In each of the microwave irradiation experiments, the animals were subjected to a pretreatment swim test after their training. They were distributed on the basis of performance into equally proficient groups for exposures and sham exposures. In the experiment that tested for prompt effects of a 30-min irradiation at a dose rate of 6.3 mW/g, the pretest was 3 hr long and the postirradiation test was made 1 day later, starting immediately after the irradiation treatment, and lasted 5 hr. In the experiments with radiation at the 11-mW/g rate, the pretest was set at 200 traverses, and the postexposure test was set at 400 traverses and was made 2 days later, immediately after irradiation in one experiment and after a 1-day delay in the other. Measurements of deep colonic temperature were made routinely after irradiation and before and after swimming.

Results

The swimming performances in the pretest and test sessions for the control groups from the three experiments are illustrated in FIGURE 7. Although some variations in performance among the three groups are evident, the overall characteristics are similar. After an initial few traverses at a high speed, the performance level plateaued for more than 100 traverses and then declined continuously thereafter. The performances of the irradiated groups relative to controls, based on the mean difference scores within experiments, are shown in FIGURE 8. The largest SEM value among control groups in each block of traverses was used to depict the ± 1 SEM limits (shaded area) for control group performance.

The performance of animals tested immediately after irradiation at the dose rate of 6.3 mW/g was essentially similar to that of their sham-exposed controls for approximately 200 traverses (1.2 km integral distance). They then slowed to a velocity that their controls approached only later. A similar performance was exhibited by animals irradiated at 11 mW/g that were tested not immediately but after a 24-hr delay. The animals tested immediately after exposure at the higher rate, however, displayed an observably gross impairment in performance for a few initial traverses, which was followed by a period of apparent recovery to the controls' level of proficiency. After about 100 traverses, their performance was again degraded to a level that only much later was approached by the controls.

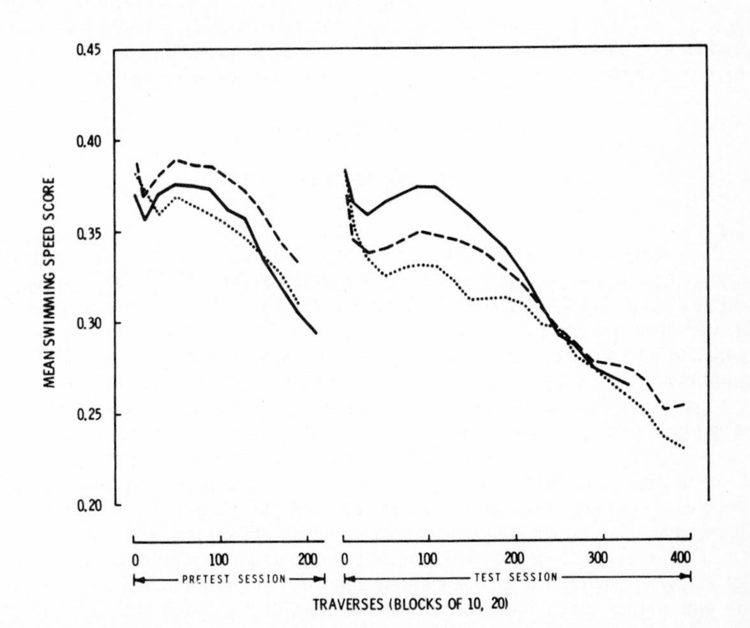

FIGURE 7. Performances on swim alley in preirradiation and postirradiation tests of control groups in three microwave studies. Speed score in meters/sec. (—) 6.3 mW/g, no delay, test-rest interval 30 sec, $N = 7$; (---) 11 mW/g, no delay, test-rest interval 20 sec, $N = 10$; (······) 11 mW/g, 24-hr delay, test-rest interval 20 sec, $N = 9$.

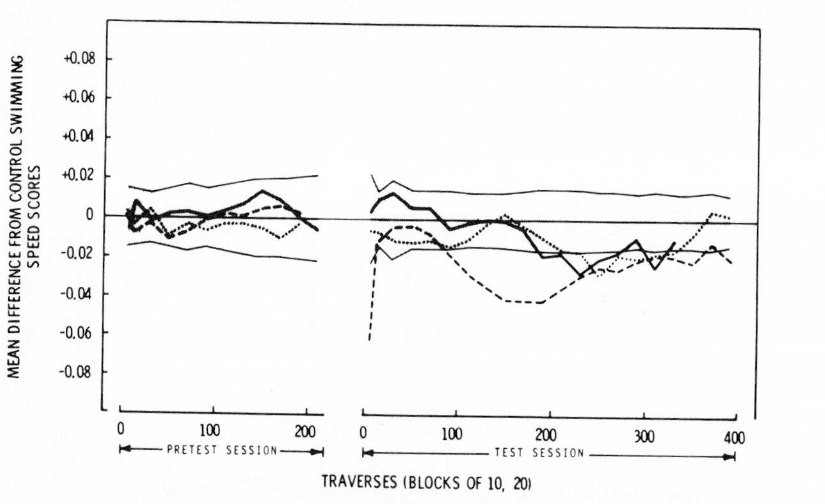

FIGURE 8. Departures in performance of irradiated groups from respective control groups shown in FIGURE 7. Shaded area shows ±1 SEM limits for controls, based on largest value among control groups in each block of traverses. Symbols and conditions are the same as in previous Figure.

Microwave irradiation for 30 min at the rate of 11 mW/g produced a severe hyperthermia, characterized by deep colonic temperatures of 41°C and higher. In other investigations, it was found that such animals subsequently exhibit thermoregulatory overcompensation, a lowered metabolic rate, bradycardia, irregular heart rates, and incomplete heart blocks.[7] Prompt immersion in 24°C water in the swim test might not protect against all of these sequellae but would promptly reduce the hyperthermia. In another experiment in which animals were irradiated at the 11-mW/g rate, or sham irradiated, and then immediately swum for 30 min, the mean temperatures of rats in both groups were comparable, 31.1 ±0.7 and 32.0 ±0.7°C, respectively, when they were removed from the water.

Although repetitive swimming in 24°C water resulted in hypothermia, with body temperatures as low as 26°C after 24 hr of swimming,[11] no systematic differences in body temperatures were found in this investigation between irradiated and control animals after they had swum for comparable periods of time.

Discussion

All of the microwave-exposed groups in this investigation exhibited a moderate reduction in swimming speed late in the test, after they had been swimming for a considerable distance with apparently normal proficiency. Reductions in speed late in the test session, after a steady level (plateau) of performance had been exhibited, very probably resulted from a loss of capacity due to fatigue.[11] The hypothermia produced by swimming in 24°C water was established early in the test, during the period when all animals were performing proficiently, and since it did not increase differentially for the irradiated animals, the hypothermia probably contributed little, if at all, to their subsequent losses in performance.

Prompt immersion in 24°C water was obviously therapeutic for the animals made severely hyperthermic by the 11-mW/g rate of irradiation; they quickly recovered and performed at their trained level of proficiency. Their subsequent abrupt loss in speed, though perhaps due to early fatigue effects, might also have reflected metabolic adjustments and cardiovascular complications that would have occurred without imposition of the swim task.[7] In similarly exposed animals that were given a 1-day delay in which to recover before the test was imposed, the only apparent residual effect of the irradiation was the relative slowing exhibited later in the test session.

An adverse effect on swimming endurance has been reported to occur in rats after a single exposure to 10-cm wavelength microwaves at field power densities as low as 10 mW/cm^2.[12] The tank-swimming method was used for measuring how long the animal, with an additional 10-g weight attached, could safely be left in the water. In the absence of information on some critical test factors, such as the criterion for removal from the water, the animal's prior experience with the test, and the water temperature,[10,11] a comparison of our results with those reported cannot be made. Our investigation, furthermore, made use of a specific form of swimming for which the proficiency measure of velocity was available, and this was used to assess changes in performance during the test and not simply the animal's vitality under stress.

VIGILANCE DISCRIMINATION INVESTIGATION

Experimental Design and Methods

To test for prompt effects of microwave radiation on the animal's performance on a complex discrimination task, rats trained to perform accurately on a vigilance task were tested for a 30-min period immediately after each treatment at dose rates of 0, 6.5, and 11 mW/g. The order in which these tests were imposed was varied among animals.

The vigilance discrimination task was based on a paradigm developed by Kornetsky and Bain[13] for testing effects of various drugs on the capacity of an animal to attend to a brief stimulus presented randomly and infrequently.[13–15] Standard operant conditioning equipment (Lehigh Valley) and procedures were used. For ease of training, a cross-modal discrimination was employed. A light flash signaled availability of a single reinforcement, and a brief burst of sound indicated that a "time-out" punishment would follow a lever response. One or the other cue was presented at the start of each interval of 5 sec. This signal repetition interval was externally clocked with the positive light (S+) presented randomly at a relative frequency of 12.5%, or on the average of 45 times in the 30-min test. The rats were maintained on a 23-hr water-deprivation cycle and were reinforced for correct responding by delivery of 0.08 ml of saccharin-flavored water. Failures to respond in time to the S+ signal constituted errors of omission. The negative sound (S−) was presented at all other intervals, and responses in these intervals resulted in a 15-sec time-out period in which the house light was turned off and all signal presentations ceased. Responses in the S− intervals constituted errors of commission. Multiple responses in the S+ intervals also earned time-out punishment, but with extensive training and practice these errors of commission virtually disappeared.

In trained animals, the omission error rate in a 30-min test, based on the relative frequency of available reinforcements that were missed, seldom exceeded 15%. The commission error rate, based on the relative frequency of earned time-out punishments to S− presentations, typically was less than 2%. But, this rate usually was exceeded during the first few minutes of each test session, when, in starting to work, the animals seemed particularly eager.

In each irradiation and sham-irradiation test session, the animals first were given a 10-min practice on the discrimination task, then were confined to the exposure holder and placed in the cavity for 30 min, and finally were subjected to the 30-min discrimination test. After they were sufficiently practiced on this testing schedule to exhibit stabilized performances, the animals were tested over a 5-day period in which each was subjected once to each level of microwave radiation. All rats were given a sham-irradiation test on the first day and the series of irradiation tests thereafter, with two or three rats assigned to each of four test sequences. These sequences, identified by mW/g dose rate, were: 0.0, 6.5, and 11.0; 11.0, 0.0, and 6.5; 0.0, 11.0, and 6.5; and 6.5, 0.0, and 11.0. A sham-irradiation test session was interposed on the day after each actual exposure to verify recovery. One of the animals failed to recover adequately from the 11 mW/g treatment and consequently was deleted from the 6.5-mW/g test condition. All sham-irradiation tests, with the cavity unpowered, were identical to the 0.0-mW/g tests, except that the performance scores from the former tests were not included in analysis.

Measurement was made routinely of deep colonic temperature in each rat in the short interval between each irradiation treatment and the performance test.

Results

The omission and commission error rates in the discrimination test after each level of microwave radiation treatment are shown in FIGURE 9. It is evident that the rats were missing or ignoring presentations of the S+ signal at the outset of testing

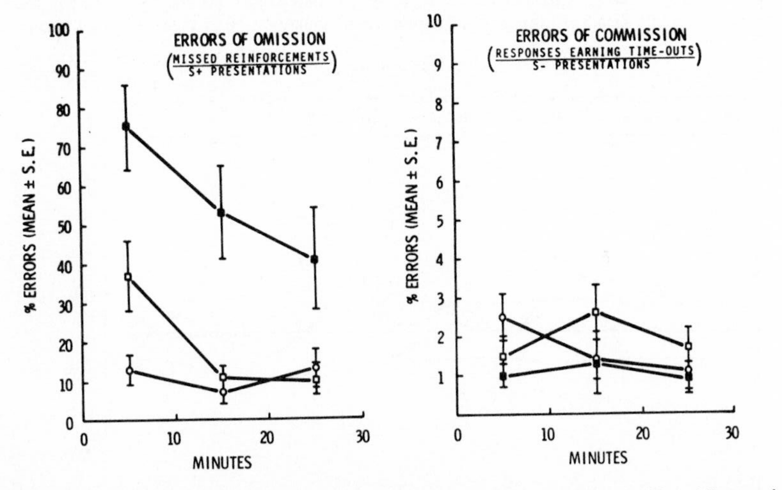

FIGURE 9. Performances on vigilance discrimination task for trained rats (N = 10) tested after each of three irradiation treatment conditions. Colonic temperatures were measured in the interval (1–2 min) between exposure and onset of discrimination test. (○) 0 mW/g, 38.89±0.10° C; (□) 6.5 mW/g, 40.47±0.11° C; (■) 11.0 mW/g, 41.60±0.19° C.

after an irradiation at either dose rate. The effect was greater after exposure at the higher rate and was related to the body temperatures measured at the beginning of the test (FIGURE 9 legend). The omission error rate during the first 10 min of the test was sharply increased, when the animal's temperature was greater than 40° C, although a few animals maintained an omission error rate below 20% with a body temperature as high as 41° C.

A rapid rate of recovery of discrimination responding was evident also from the start of and continued throughout the test. Recovery was apparently complete by the middle of the test period in the 6.5-mW/g condition.

There was no evidence of a change in commission error rate after irradiation. The slight increase exhibited by the animals in the middle of the test under the 6.5-mW/g condition corresponded in time to their resumption of discrimination responding and in rate to their normal level at the start of testing.

Discussion

The performance losses produced by microwave exposures in this investigation are probably closely related to the inactivity and work stoppages that have been reported to occur during irradiation.[9] The degradation in both situations appeared to be directly related to the microwave-induced hyperthermia.

The rapid recovery in discrimination performance occurred with no evident loss in accuracy and at rates roughly similar for both radiation levels. In other experiments,[7] rats similarly irradiated and then confined to holders for physiologic testing displayed similarly rapid rates of recovery from their hyperthermia. The animals' rapid behavioral recovery might directly reflect this adjustment to thermal loading.

CONCLUSIONS

The use of a common form and schedule of microwave irradiation in these investigations provided the means for comparing effects on a wide variety of behaviors and for relating these effects to the level of heat loading produced by the radiation. It is noteworthy that some type of prompt effect was produced on each of the behaviors tested after a 30-min exposure at an absorbed dose rate of about 6 mW/g, which produced a moderate hyperthermia.

The characteristics of the microwave effects produced on the three behaviors were largely dissimilar. Only on the psychologically most complex task, vigilance discrimination, was the effect apparently related in a direct manner to the induction of and recovery from hyperthermia. During the same postirradiation period in which the vigilance discrimination performance was most severely affected, the least motivated, most random behavior tested, exploratory activity, was only moderately degraded. This effect on activity persisted for 2 hr or more. It did not appear to be related to the hyperthermia directly, but perhaps it was related to a metabolic overcompensation to the heat load that in other, related investigations had been found during this postirradiation period.[7] Immersion in water for the arduous swim test provided prompt relief from the hyperthermia, but a delayed effect indicative of early fatigue occurred after more than 2 hr of proficient swimming was displayed. A nearly lethal irradiation at the rate of 11 mW/g produced a marked performance degradation that occurred after these animals first had recovered from an initial incapacitation and then had displayed their trained level of proficiency for more than 1 hr.

The microwave effects on the behavior of trained animals evidently reflected alterations in performance factors rather than interference with trained skills. Both proficiency in swimming and accuracy in discrimination responding were retained without retraining after a nearly lethal irradiation.

References

1. Justesen, D. R., D. M. Levinson, R. L. Clarke & N. W. King. 1971. A microwave oven for behavioural and biological research: electrical and structural modifications, calorimetry, dosimetry, and functional evaluation. J. Microwave Power **6:** 237–258.
2. Phillips, R. D., E. L. Hunt & N. W. King. This monograph.
3. Phillips, R. D. & E. L. Hunt. 1971. Problems of physical and biological dosimetry of microwave irradiation. J. Microwave Power **6:** 90 (Abs.).
4. Hunt, E. L. & R. D. Phillips. 1972. Absolute dosimetry for whole animal experiments. *In* Joint Army/Georgia Institute of Technology Microwave Dosimetry Workshop, Digest of Papers.: 74–77. Walter Reed Army Institute of Research. Washington, D.C.
5. Hunt, E. L., R. D. Phillips, D. M. Fleming & R. D. Castro. 1975. Dosimetry for whole-animal microwave irradiation. In preparation.
6. Bollman, J. L. 1948. A cage which limits the activity of rats. J. Lab. Clin. Med. **33:** 1348.
7. Phillips, R. D., E. L. Hunt & N. W. King. 1975. Thermoregulatory, metabolic and cardiovascular response of rats to microwaves. J. Appl. Physiol. In press.
8. Korbel, S. F. 1969. Behavioral effects of low intensity uhf radiation. *In* Biological Effects and Health Implications of Microwaves. S. F. Cleary, Ed. Report BRH/DBE 70–2: 180–184. Bureau of Radiological Health. Washington, D. C.
9. Justesen, D. R. & N. W. King. 1969. Behavioral effects of low level microwave irradiation in the closed space situation. *In* Biological Effects and Health Implications of Microwaves. S. F. Cleary, Ed. Report BRH/DBE 70–2: 154–179. Bureau of Radiological Health. Washington, D. C.
10. King, N. W., E. L. Hunt, R. D. Castro & R. D. Phillips. 1975. An automated swim alley for small animals: I. Instrumentation. Behav. Res. Methods Instr. In press.
11. King, N. W., E. L. Hunt, R. D. Castro & R. D. Phillips. 1975. An automated swim alley for small animals: II. Training and procedures. Behav. Res. Methods Instr. In press.
12. Lobanova, Y. A. 1960. Survival and development of animals with various intensities and durations of the influence of uhf. *In* The Biological Action of Ultrahigh Frequencies. A. A. Letavet & Z. V. Gordon, Eds.: 60–63. Academy of Medical Sciences USSR. Moscow, USSR. (Engl. transl. JPRS: 12471; OTS: 62-19175. U. S. Joint Publications Research Service, Washington, D. C.)
13. Kornetsky, C. & G. Bain. 1965. The effects of chlorpromazine and pentobarbital on sustained attention in the rat. Psychopharmacology (Berlin) **8:** 277–284.
14. Kornetsky, C. & M. Eliasson. 1969. Reticular stimulation and chlorpromazine: an animal model for schizophrenic overarousal. Science **165:** 1273, 1274.
15. Eliasson, M. & C. Kornetsky. 1972. Interaction effects of chlorpromazine and reticular stimulation on visual attention in rats. Psychon. Sci. **26:** 261, 262.

Discussion

Dr. H. S. Ho: In exposing animals in the multimode cavity, have you investigated whether changes occur in rate of energy absorption as a function of animal movement? In exposing animals in a wave guide, we have found that even though the animal's range of movement is sharply confined, limited movement significantly perturbs the field.

Mr. Hunt: Usually, our animals are held in a fixed position within the cavity, but we do have indirect evidence that bears on your question. We have forward and reflected power meters by which we index energy absorption by an animal. Variations of reflected power, which, of course, indicate alterations in coupling between the wave guide and the cavity, do occur and they correlate with postural changes made by an animal, such as tail flipping or head movements. We have not conducted specific studies to quantify energy absorption changes as a function of animal orientation.

Dr. Ho: Since temperature, especially colonic, is apparently increased in animals under restraint and will also be influenced by ambient conditions, did you measure this parameter? Also, did you record humidity, which may also be a factor in affecting the thermal response of the animal?

Mr. Hunt: We are careful to habituate animals to restraint before irradiating them. We monitor ambient temperature and relative humidity, which usually remains near 30–35%.

Dr. D. R. Justesen: A noble aspect of Mr. Hunt's presentation lies in the precision with which he *doses* his animals with microwave energy. Mr. Hunt and his colleagues are among the few investigators who distinguish between measurements of energy dose and energy density. Contrary to widespread (mis-)use, neither Poynting's vector nor its integral are doses or dose rates. They are measures of the amount of energy that flows in time through, respectively, a two- or three-dimensional space. Both are conceptually independent of any target or object that may inhabit the space transited by the energy. In fact, it has been demonstrated that a salient target will likely severely distort the field.[1] It would be folly, of course, to argue that there is no utility in measurements of energy flux density. For a precise scientific reckoning, we have as much need to know how much energy is *incident upon* a target as to know how much energy is *absorbed by* it. However, to confuse the operations and meanings of densitometry with those of energy dose is to court egregious errors that may extend to orders of magnitude, as was so elegantly demonstrated in earlier work by Guy and Korbel[2] and, more recently, by Gandhi.[3] Whenever possible, measurements or careful estimates of density and of dose should be performed in studies of the biologic response to microwaves and other nonionizing radiations.

References

1. Reno, V. R. 1974. Microwave reflection, diffraction and transmission studies of man. Naval Aerospace Med. Res. Lab. Tech. Rep. 1199. Pensacola, Fla.
2. Guy, A. W. & S. F. Korbel. 1972. Dosimetry studies on a uhf cavity exposure chamber for rodents. Proc. Microwave Power Symp., Ottawa, Ontario, Canada. Microwave Power Institute. Edmonton, Alberta, Canada.
3. Gandhi, O. P. 1974. Polarization and frequency effects on whole animal absorption of rf energy. Proc. IEEE **62:** 1166–1168.

On Microwave-Induced Hearing Sensation

JAMES C. LIN, MEMBER, IEEE

***Abstract*—When a human subject is exposed to pulsed microwave radiation, an audible sound occurs which appears to originate from within or immediately behind the head. Laboratory studies have also indicated that evoked auditory activities may be recorded from cats, chinchillas, and guinea pigs. Using a spherical model of the head, this paper analyzes a process by which microwave energy may cause the observed effect. The problem is formulated in terms of thermoelasticity theory in which the absorbed microwave energy represents the volume heat source which depends on both space and time. The inhomogeneous thermoelastic motion equation is solved for the acoustic wave parameters under stress-free surface conditions using boundary value technique and Duhamel's theorem. Numerical results show that the predicted frequencies of vibration and threshold pressure amplitude agree reasonably well with experimental findings.**

Manuscript received July 12, 1976; revised October 5, 1976. This work was supported in part by the National Science Foundation under Grant ENG 75-15227.

The author is with the Department of Electrical and Computer Engineering, Wayne State University, Detroit, MI 48202.

I. Introduction

IT HAS BEEN demonstrated that sound can be generated in laboratory animals by the absorption of microwave energy in the head [1]–[3]. These reports indicate that auditory activities may be evoked by irradiating the heads of cats, chinchillas, and guinea pigs with pulsed microwave energy [1], [4]–[6]. Responses elicited in cats by both conventional acoustic stimuli and by pulsed microwaves disappear following destruction of the round window of the cochlea [4], and following death [3]. This suggests that microwave-induced audition is transduced by a mechanism similar to that responsible for conventional acoustic reception, and that the primary site of interaction resides peripherally with respect to the cochlea. More recently [6], sonic oscillations at 50 kHz have been recorded from the round window of guinea pigs during irradiation by pulsed

Reprinted from *IEEE Trans. Microwave Theory Tech.*, vol. MTT-25, pp. 605–613, July 1977.

TABLE I
ELASTIC AND THERMAL PROPERTIES OF BRAIN MATTER

Specific heat, c_h	0.88 cal/gm-°C
density, ρ	1.05 gm/cm^3
coefficient of thermal expansion, α	4.1 x 10^{-5}/°C
Lame's constant, λ	2.24 x 10^{10}dyn/cm^2
Lame's constant, μ	10.52 x 10^{3}dyn/cm^2
Bulk velocity of propagation, c_1	1.460 x 10^{5}cm/sec

microwaves at 918 MHz. The oscillations promptly followed onset of radiation, preceded the nerve responses, and disappeared after death. It is therefore reasonable to conclude that the microwave-induced auditory effect is a cochlear response to acoustic signals that are generated, presumably in the head, by pulsed microwaves.

When human subjects are exposed to pulsed microwave radiation, an audible sound occurs which appears to originate from within or immediately behind the head. The microwave-generated sound has been described as clicking, buzzing, or chirping depending on such factors as pulsewidth and repetition rate [2], [4], [5], [7], [8]. The effect is of great significance since the average incident power densities required to elicit the response are considerably lower than those found for other microwave biological effects and the threshold average power densities are many orders of magnitude smaller than the current safety standard of 10 mW/cm² [9].

Although the effect is widely accepted as a genuine biologic effect occurring at low average power densities, there exists some controversy regarding the mechanism by which pulsed microwave energy is converted to sound [1], [4], [7], [10]–[13]. This paper analyzes the acoustic wave generated in the heads of animals and man exposed to pulsed microwave radiation as a result of rapid thermal expansion.

We assume that the auditory effect arises from the minuscule but rapid rise of temperature in the brain as a result of absorption of microwave energy. The rise of temperature occurring in a very short time is believed to create thermal expansion of the brain matter which then launches the acoustic wave of pressure that is detected by the cochlea [13].

We consider the head to be perfectly spherical and consisting only of brain matter. The impinging radiation is assumed to be a plane wave of pulsed microwave energy. Our approach is first to obtain the absorbed microwave energy inside the head. The accompanying temperature rise is then derived, and finally the inhomogeneous thermoelastic motion equation is solved for the acoustic wave generated in the head.

The relevant physical parameters of brain matter are listed in Table I. All except one are typical values obtained from the literature [14]–[16]. For the coefficient of thermal expansion, which does not seem to have been measured in the past, we assume a value equal to 60 percent of the corresponding value for water. These values will be useful for quantitative estimations of the frequency and threshold of pulsed microwave-induced hearing.

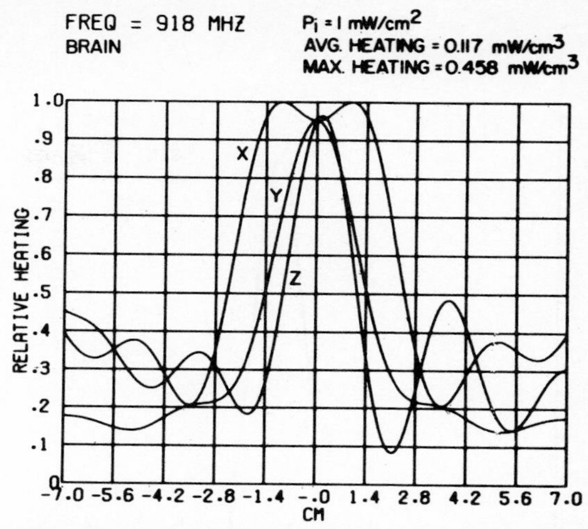

Fig. 1. Absorbed energy distribution in a 7-cm-radius spherical model of the head exposed to 918-MHz plane wave. The incident power density is 1 mW/cm² [20].

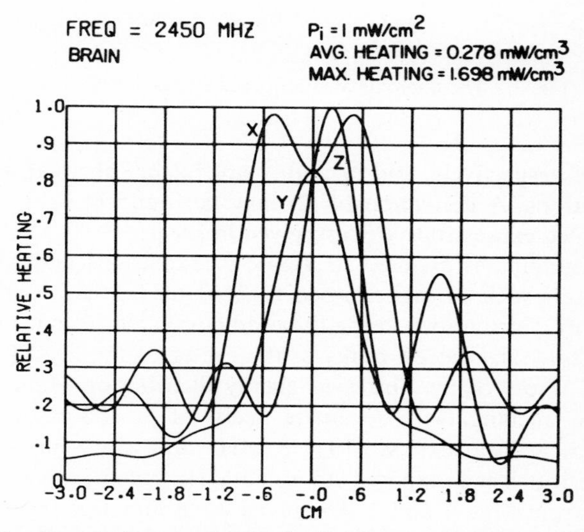

Fig. 2. Absorbed energy distribution in a 3-cm-radius spherical model of the head exposed to 2450-MHz plane wave. The incident power density is 1 mW/cm² [20].

II. THEORETICAL FORMULATION

A. Microwave Absorption

Let us consider a homogeneous spherical model of the head exposed to a plane wave of pulsed microwave energy. The absorbed microwave energy $I(r,t)$ at any point inside the head is given by

$$I(r,t) = \tfrac{1}{2}\sigma|\bar{E}|^2 \quad (1)$$

where σ is the electrical conductivity of brain matter. The induced electric field $\bar{E}$ is given by

$$\bar{E} = E_0 e^{-i\omega t} \sum_{j=1}^{\infty} i^j \frac{2j+1}{j(j+1)} [a_j \bar{M}_{o1j} - i b_j \bar{N}_{e1j}] \quad (2)$$

where E_0 is the incident electric field strength, $\omega = 2\pi f$, f is frequency, a_j and b_j are magnetic and electric oscilla-

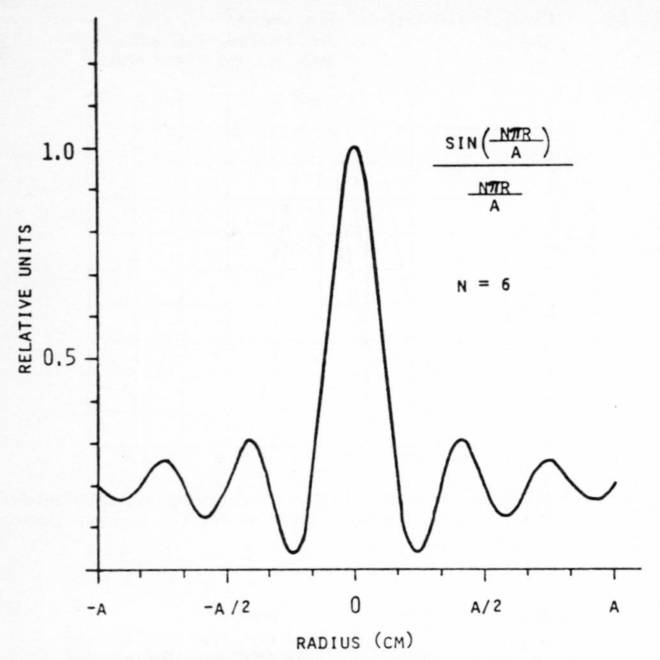

Fig. 3. The approximated absorbed energy distribution.

tions, respectively, and $\overline{M}$ and $\overline{N}$ are vector spherical wave functions. A derivation of (2) may be found in [17]. The detailed expressions are also given in [18].

For humans exposed to 918-MHz radiations and small animals such as cats exposed to 2450 MHz, the absorbed energy distributions inside the head computed from (1) and (2) show absorption peaks in the center of the head [19], [20]. Plots of the absorbed energy distribution along the three rectangular coordinate axes of a 7.0-cm-radius spherical head exposed to 918 MHz and a 3.0-cm-radius spherical head exposed to 2450-MHz plane waves are shown in Figs. 1 and 2. The plane wave impinges from the negative z direction and is polarized in the x direction. Note that in both cases the absorbed energy along the three coordinate axes exhibits characteristic oscillations along the outer portion of the spherical head and reaches a maximum near the center.

Although the detailed absorption along the three axes is not the same, we will assume a spherically symmetric absorption pattern and approximate the absorbed energy distribution inside the head by the spherically symmetric function

$$W(r,t) = I_0 \sin\left(\frac{N\pi r}{a}\right) \Big/ \left(\frac{N\pi r}{a}\right) \tag{3}$$

where I_0 is the peak absorbed energy per unit volume, r is the radial variable, and a is the radius of the spherical head. The parameter N specifies the number of oscillations in the approximated spatial dependence of the absorbed energy. Fig. 3 shows the approximated energy absorption pattern for $N = 6$ and is particularly suited for the cases shown in Figs. 1 and 2. For some frequencies and sphere sizes, the integer N may be changed to account for the difference in absorption patterns. For instance, $N = 3$ may be chosen to approximate the absorption pattern inside a 5-cm-radius spherical head exposed to 918-MHz radiation [20]. For other frequencies and sphere sizes, a different function will be required to describe the absorbed energy distribution.

B. Temperature Rise

We take advantage of the symmetry of the absorbed energy pattern by expressing the heat conduction equation as a function of r alone [21]. That is,

$$\frac{1}{r^2}\frac{\partial}{\partial r} r^2 \frac{\partial v}{\partial r} - \frac{1}{\kappa}\frac{\partial v}{\partial t} = \frac{-W(r,t)}{K} \tag{4}$$

where v is temperature, κ and K are, respectively, the thermal diffusivity and conductivity of brain matter, and W is the heat production rate, which is the same as the absorbed microwave energy pattern and is assumed for the moment to be constant over time.

Because microwave absorption occurs in a very short time interval, there will be little chance for heat conduction to take place. We may therefore neglect the spatial derivatives in (4) such that

$$\frac{1}{\kappa}\frac{dv}{dt} = \frac{W}{K}. \tag{5}$$

Equation (5) may be integrated, directly, to give the change in temperature by setting the initial temperatures equal to zero. Thus

$$v(r,t) = \frac{I_0}{\rho c_h} \frac{\sin(N\pi r/a)}{N\pi r/a} t \tag{6}$$

where ρ and c_h are the density and specific heat of brain matter, respectively, and $\rho c_h = K/\kappa$.

In biological materials, the stress-wave development times are short compared with temperature equilibrium times. The temperature decay is therefore a slowly varying function of time and becomes significant only for times greater than milliseconds. We may thus assume for a square pulse of microwave energy, immediately after termination of radiation, that

$$v(r,t) = \frac{I_0}{\rho c_h} \frac{\sin(N\pi r/a)}{N\pi r/a} t_0 \tag{7}$$

where t_0 is the pulsewidth.

C. Sound Generation

We now consider the spherical head with homogeneous brain matter as a linear, elastic medium without viscous damping. The thermoelastic equation of motion in spherical coordinates [22] is then given by

$$\frac{\partial^2 u}{\partial r^2} + \frac{2}{r}\frac{\partial u}{\partial r} - \frac{2}{r^2} u - \frac{1}{c_1^2}\frac{\partial^2 u}{\partial t^2} = \frac{\beta}{\lambda + 2\mu}\frac{\partial v}{\partial r} \tag{8}$$

where u is the displacement of brain matter, $c_1 = [(\lambda + 2\mu)/\rho]^{1/2}$ is the velocity of bulk acoustic wave propagation, $\beta = \alpha(3\lambda + 2\mu)$, α is the coefficient of linear thermal expansion, and λ and μ are Lame's constants. It should be noted that the curl of u equals zero since u is in the radial direction only. The right-hand side of (8) is the change in temperature which gives rise to the displacement. We first write

$$\frac{\beta}{\lambda + 2\mu}\frac{\partial v}{\partial r} = u_0 F_r(r) F_t(t). \tag{9}$$

Hence

$$u_0 = \frac{I_0}{\rho c_h}\frac{\beta}{\lambda + 2\mu} \tag{10}$$

and

$$F_r(r) = \frac{d}{dr}\left[\sin\left(\frac{N\pi r}{a}\right)\Big/\left(\frac{N\pi r}{a}\right)\right]. \tag{11}$$

From (6) and (7), we have

$$F_t(t) = \begin{cases} t, & 0 \le t \le t_0 \\ t_0, & t \ge t_0. \end{cases} \tag{12}$$

If the surface of the sphere is stress free, then the boundary condition at $r = a$ is

$$(\lambda + 2\mu)\frac{\partial u}{\partial r} + 2\lambda\frac{u}{r} = \beta v = 0. \tag{13}$$

The initial conditions are

$$u(r,0) = \frac{\partial u(r,0)}{\partial t} = 0. \tag{14}$$

Our approach in the following derivations is first to obtain a solution for the case of step of microwave energy, $F_t(t) = 1$, at some instant $t = 0$ and then to extend the solution to a rectangular pulse using Duhamel's theorem [23].

1) Unit Step: If we write the displacement $u(r,t)$ as

$$u(r,t) = u_s(r) + u_t(r,t) \tag{15}$$

and substitute (15) into (8), the equation of motion becomes two differential equations: a stationary one and a time-varying one. Thus

$$\frac{d^2u_s(r)}{dr^2} + \frac{2}{r}\frac{du_s(r)}{dr} - \frac{2}{r^2}u_s(r) = u_0 F_r(r) \tag{16}$$

and

$$\frac{\partial^2 u_t(r,t)}{\partial r^2} + \frac{2}{r}\frac{\partial u_t(r,t)}{\partial r} - \frac{2}{r^2}u_t(r,t) = \frac{1}{c_1^2}\frac{\partial^2 u_t(r,t)}{\partial t^2}. \tag{17}$$

The corresponding boundary conditions at $r = a$ are

$$(\lambda + 2\mu)\, du_s/dr + 2\lambda u_s/r = 0 \tag{18}$$

and

$$(\lambda + 2\mu)\, \partial u_t/\partial r + 2\lambda u_t/r = 0. \tag{19}$$

To obtain $u_s(r)$, we assume a solution of the form

$$u_s(r) = u_p(r) + D_1/r^2 + D_2 r \tag{20}$$

where $u_p(r)$ is a particular solution of (16). We now rewrite the left-hand side of (16) as follows:

$$\frac{d}{dr}\left[\frac{1}{r^2}\frac{d(r^2u_p)}{dr}\right] = u_0 F_r(r). \tag{21}$$

We then integrate (21) from 0 to r to get the expression

$$u_p(r) = u_0\left(\frac{a}{N\pi}\right)j_1\left(\frac{N\pi r}{a}\right). \tag{22}$$

Since $u_s(r)$ must remain finite as $r \to 0$, D_1 reduces immediately to zero. The coefficient D_2 is obtained by applying the boundary condition of (18), and it is

$$D_2 = \pm u_0\left(\frac{1}{N^2\pi^2}\right)\frac{4\mu}{3\lambda + 2\mu}, \qquad N = \begin{cases}1,3,5\cdots \\ 2,4,6\cdots\end{cases}. \tag{23}$$

The solution of (16) is therefore given by

$$u_s(r) = u_0\left[\frac{a}{N\pi}j_1\left(\frac{N\pi r}{a}\right) \pm \frac{4\mu}{3\lambda + 2\mu}\frac{r}{N^2\pi^2}\right], \qquad N = \begin{cases}1,3,5\cdots \\ 2,4,6\cdots\end{cases} \tag{24}$$

where $j_1(N\pi r/a)$ is the spherical Bessel function of the first kind and first order.

Now we let

$$u_t(r,t) = R(r)T(t) \tag{25}$$

and use the method of separation of variables to solve (17) for the time-varying component. Inserting (25) into (17) yields the two ordinary differential equations

$$\frac{d^2R}{dr^2} + \frac{2}{r}\frac{dR}{dr} + \left(k^2 - \frac{2}{r^2}\right)R = 0 \tag{26}$$

$$\frac{d^2T}{dt^2} + k^2c_1^2T = 0 \tag{27}$$

where k is the constant of separation to be determined. Equation (26) is Bessel's equation and its solution is [17]

$$R(r) = B_1 j_1(kr) + B_2 y_1(kr) \tag{28}$$

where $j_1(kr)$ and $y_1(kr)$ are the spherical Bessel functions of the first and second kind of the first order. Since $R(r)$ is finite at $r = 0$, B_2 must be zero. Combining (28) and the boundary condition of (19), we obtain a transcendental equation for k, the constant of separation,

$$\tan(ka) = (ka)/[1 - (\lambda + 2\mu)(ka)^2/(4\mu)]. \tag{29}$$

The solution of (29) is an infinite sequence of eigenvalues k_m; each corresponds to a characteristic mode of vibration of the spherical head. It can be shown that, using the values for brain matter given in Table I, $k_m a = m\pi$, $m = 1,2,3\cdots$ to within an accuracy of 10^{-7}. Moreover, since (27) is harmonic in time, a general solution for $u_t(r,t)$ may be written as

$$u_t(r,t) = \sum_{m=1}^{\infty} A_m j_1(k_m r)\cos\omega_m t \tag{30}$$

where

$$\omega_m = k_m c_1 = m\pi c_1/a \tag{31}$$

and ω_m is the angular frequency of vibration of the sphere. Note that the frequency of vibration is independent of the absorbed energy pattern. It is only a function of the spherical head size and the elastic properties of the medium.

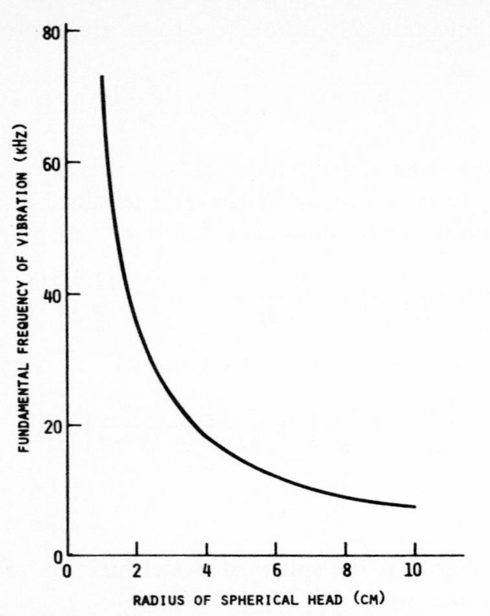

Fig. 4. The fundamental frequencies of sound generated inside the head as a function of spherical head radii.

The fundamental frequency of sound generated inside the spherical head is therefore given by

$$f_1 = c_1/2a. \tag{32}$$

Fig. 4 is a plot of the fundamental frequency of sound generated in the head as a function of head radii. The frequency varies from above 80 kHz for mice ($a \cong 1$ cm) to about 8 kHz for humans ($a = 7$–10 cm).

We evaluate the constants A_m by using the initial conditions in (14) to obtain

$$A_m = -u_0 \left\{ \frac{a}{N\pi} \int_0^a r^2 j_1(k_m r) j_1 \left(\frac{N\pi r}{a} \right) dr \pm \frac{4\mu}{3\lambda + 2\mu} \left(\frac{1}{N\pi} \right)^2 \int_0^a r^3 j_1(k_m r)\, dr \right\} \Big/ \left\{ \int_0^a r^2 [j_1(k_m r)]^2\, dr \right\}, \qquad N = \begin{cases} 1,3,5,\cdots \\ 2,4,6,\cdots \end{cases} \tag{33}$$

The integrals in (33) may be evaluated [24] to give

$$\int_0^a r^2 j_1(k_m r) j_1 \left(\frac{N\pi r}{a} \right) dr = \left[\frac{-a^3}{(k_m a)^2 - (N\pi)^2} \right] \left[\pm \frac{ka}{N\pi} \right] j_0(k_m a), \qquad N = \begin{cases} 1,3,5,\cdots \\ 2,4,6,\cdots \end{cases} \tag{34}$$

$$\int_0^a r^3 j_1(k_m r)\, dr = \left(\frac{a^2}{k_m a} \right)^2 [3 j_1(k_m a) - k_m a j_0(k_m a)] = \frac{a^3}{k_m} j_2(k_m a) \tag{35}$$

$$\int_0^a r^2 [j_1(k_m r)]^2\, dr = \frac{a^3}{2} \{ [j_1(k_m a)]^2 - j_0(k_m a) j_2(k_m a) \} \tag{36}$$

where $j_2(k_m a)$ is the spherical Bessel function of the first kind and second order.

Using these values (33) becomes

$$A_m = \mp u_0 a \left(\frac{1}{N\pi} \right)^2 \left[\frac{2}{[j_1(k_m a)]^2 - j_0(k_m a) j_2(k_m a)} \right] \cdot \left\{ \frac{4\mu}{3\lambda + 2\mu} \left(\frac{1}{k_m a} \right) j_2(k_m a) - k_m a j_0(k_m a) \frac{1}{(k_m a)^2 - (N\pi)^2} \right\}, \qquad N = \begin{cases} 1,3,5,\cdots \\ 2,4,6,\cdots \end{cases} \tag{37}$$

For $k_m a = m\pi = N\pi$, (37) simplifies to

$$A_m = -u_0 a \left(\frac{1}{N\pi} \right) \left[1 + \frac{24\mu}{3\lambda + 2\mu} \left(\frac{1}{N\pi} \right)^2 \right]. \tag{38}$$

The displacement response of the sphere to a step input of microwave energy is now given by introducing (37) in (30) and then combining (24) and (30) in (15). We have

$$u(r,t) = u_0 Q + \sum_{m=1}^{\infty} A_m j_1(k_m r) \cos \omega_m t \tag{39}$$

$$Q = \frac{a}{N\pi} j_1 \left(\frac{N\pi r}{a} \right) \pm \frac{4\mu}{3\lambda + 2\mu} \frac{r}{N^2 \pi^2}, \qquad N = \begin{cases} 1,3,5,\cdots \\ 2,4,6,\cdots \end{cases} \tag{40}$$

The radial stress can be deduced from the displacement solution using [22] and (13):

$$\sigma_r(r,t) = (\lambda + 2\mu) \frac{\partial u}{\partial r} + 2\lambda \frac{u}{r} - \beta v. \tag{41}$$

We have, therefore, by substituting (6), (10), and (39) into (41),

$$\sigma_r(r,t) = 4\mu u_0 S + \sum_{m=1}^{\infty} A_m k_m M_m \cos \omega_m t \tag{42}$$

$$S = \pm \left(\frac{1}{N\pi} \right)^2 - j_1 \left(\frac{N\pi r}{a} \right) \Big/ \left(\frac{N\pi r}{a} \right), \qquad N = \begin{cases} 1,3,5,\cdots \\ 2,4,6,\cdots \end{cases} \tag{43}$$

$$M_m = [(\lambda + 2\mu) j_0(k_m r) - 4\mu j_1(k_m r)/(k_m r)]. \tag{44}$$

2) *Rectangular Pulse:* We now can obtain the displacement and radial stress for a rectangular pulse of microwave energy by applying Duhamel's theorem [23] to the solutions expressed by (39) and (42). That is,

$$u(r,t) = \frac{\partial}{\partial t} \int_0^t F_t(t - t') u'(r,t')\, dt' \tag{45}$$

where $u'(r,t)$ is the solution given by (39) for the case of a sudden application of microwave radiation. An equivalent expression can, of course, be written for the radial stress. Therefore, by substituting (12) and (39) into (45), we have for the displacement

$$u(r,t) = u_0 Q t + \sum_{m=1}^{\infty} A_m j_1(k_m r) \frac{\sin \omega_m t}{\omega_m}, \qquad 0 \le t \le t_0 \tag{46}$$

$$u(r,t) = u_0 Q t_0 + \sum_{m=1}^{\infty} A_m j_1(k_m r) \left[\frac{\sin \omega_m t}{\omega_m} - \frac{\sin \omega_m (t - t_0)}{\omega_m} \right], \qquad t \ge t_0. \tag{47}$$

Similarly, we have for the radial stress

$$\sigma_r(r,t) = 4\mu u_0 St + \sum_{m=1}^{\infty} A_m k_m M_m \frac{\sin \omega_m t}{\omega_m}, \qquad 0 \leq t \leq t_0 \tag{48}$$

$$\sigma_r(r,t) = 4\mu u_0 St_0 + \sum_{m=1}^{\infty} A_m k_m M_m \cdot \left[\frac{\sin \omega_m t}{\omega_m} - \frac{\sin \omega_m (t - t_0)}{\omega_m}\right], \qquad t \geq t_0. \tag{49}$$

Equations (46)–(49) represent the general solution for the displacement and radial stress in a spherical head exposed to pulsed microwave radiation as a function of the microwave, thermal, elastic, and geometric parameters of the model.

Since u_0 and A_m are directly proportional to I_0, both the displacement and the radial stress are proportional to the peak absorbed power density. It is easy to see that the displacement and radial stress also depend linearly on the peak incident power density.

At the center of the sphere, $r = 0$, both (46) and (47) reduce to zero, and (48) and (49) become

$$\sigma_r = 4\mu u_0 \left[\pm \left(\frac{1}{N\pi}\right)^2 - \frac{1}{3}\right] t + \sum_{m=1}^{\infty} A_m k_m \left(\lambda + \frac{2}{3}\mu\right) \frac{\sin \omega_m t}{\omega_m}, \qquad 0 \leq t \leq t_0 \tag{50}$$

and

$$\sigma_r = 4\mu u_0 \left[\pm \left(\frac{1}{N\pi}\right)^2 - \frac{1}{3}\right] t_0 + \sum_{m=1}^{\infty} A_m k_m \left(\lambda + \frac{2}{3}\mu\right)\left[\frac{\sin \omega_m t}{\omega_m} - \frac{\sin \omega_m (t - t_0)}{\omega_m}\right],$$

$$t \geq t_0, \; N = \begin{cases} 1,3,5, \cdots \\ 2,4,6, \cdots \end{cases} \tag{51}$$

The radial stress is therefore given by (50) and (51), and there is no displacement at the center of the model. On the other hand, at the surface ($r = a$), (43) becomes naught. The radial stress is given by the summation of the harmonic time functions alone.

III. Displacement and Sound Pressure

Using the parameters for brain matter given in Table I, we can compute the effect of microwave pulses on spherical models of the head from the solutions derived above. Fig. 5 shows the results of pressure computations in a 7-cm spherical head exposed to 918-MHz radiation with pulsewidth ranging from 0.1 to 100 μs while keeping the peak incident (or absorbed) power density constant. The relations between peak incident and absorbed power density are obtained from Figs. 2 and 3. The sound pressure amplitudes clearly depend on the pulsewidth of the impinging radiation. Moreover, there seems to be a minimum pulsewidth around 2 μs. The sound pressure amplitude rises rapidly first to a maximum and then alternates around

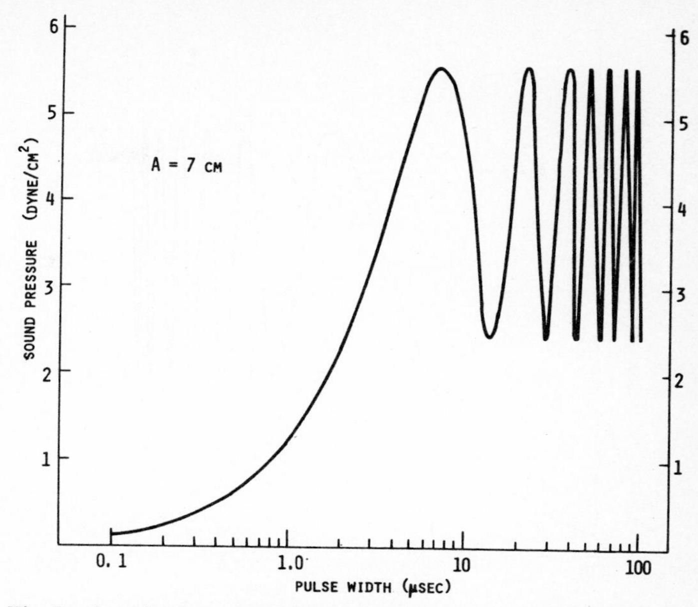

Fig. 5. Sound pressure amplitude generated in a 7-cm-radius spherical head exposed to 918-MHz plane wave as a function of pulsewidth. The peak absorbed energy is 1000 mW/cm³.

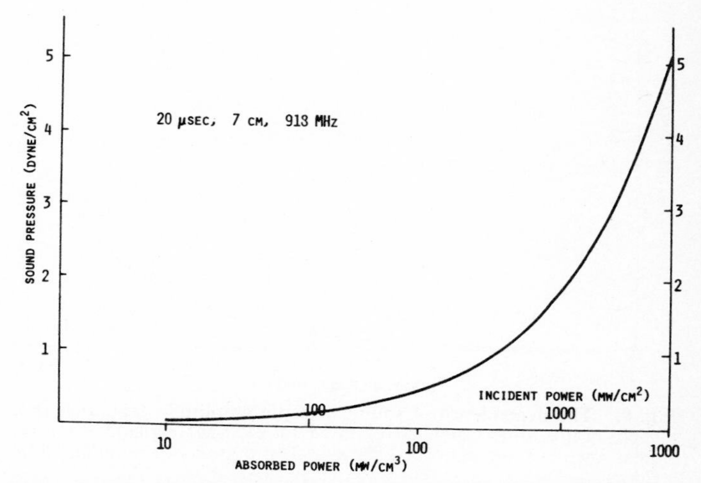

Fig. 6. The dependence sound pressure amplitude generated in a 7-cm-radius spherical head exposed to 918-MHz plane wave on peak incident and absorbed powers. The pulsewidth is taken to be 20 μs.

a constant average amplitude. The dependence of sound pressure amplitude on peak powers is illustrated in Fig. 6. The pulsewidth is taken to be 20 μs. It is therefore apparent that the sound pressure amplitude depends upon peak power as well.

Fig. 7 gives the computed pressures in a 3-cm-radius sphere exposed to 2450-MHz radiation. It is readily seen that microwave-induced sound is a function of both pulsewidth and peak powers (Fig. 8). The minimum pulsewidth for efficient sound generation by 2450-MHz microwaves impinging on a 3-cm-radius spherical head is around 1 μs.

Figs. 9 and 10 depict typical displacements of brain matter in spherical models of the head exposed to pulsed microwaves. The pulsewidth used for computing Figs. 9 and 10 is 20 μs. These are representative graphs and are shown for $r = 0$, $a/2$, and a, where a is the radius of the sphere. As

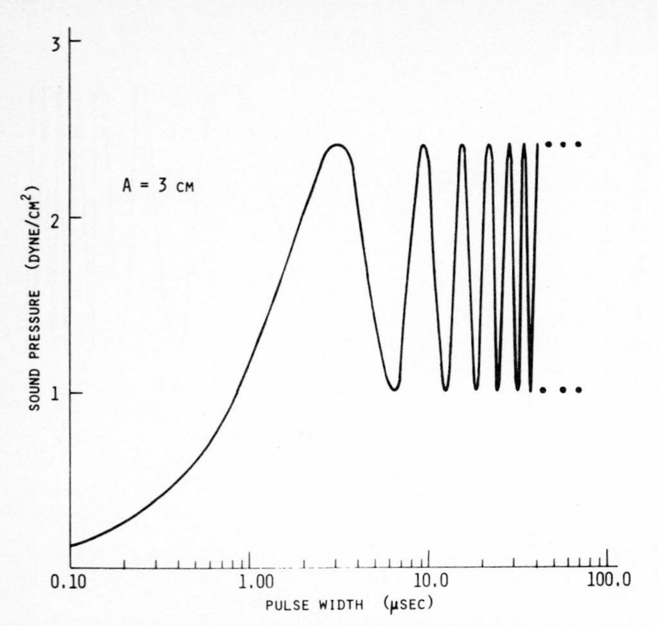

Fig. 7. Sound pressure amplitude generated in a 3-cm-radius spherical head exposed to 2450-MHz plane wave as a function of pulsewidth. The peak absorbed energy is 1000 mW/cm^3.

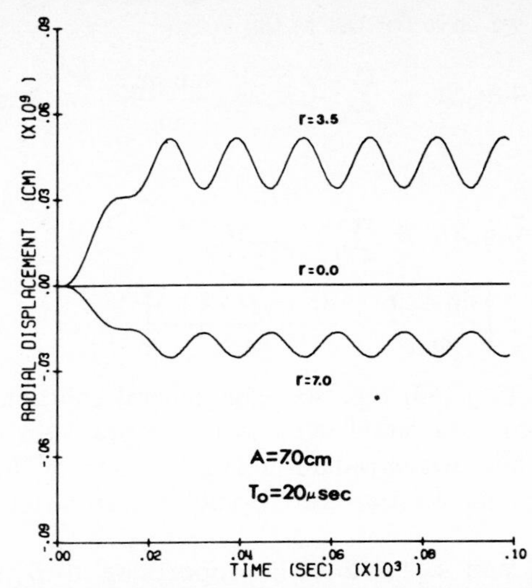

Fig. 9. Radial displacement as a function of time a 7-cm-radius spherical head exposed to 918-MHz plane wave. The peak absorption is 1000 mW/cm^3.

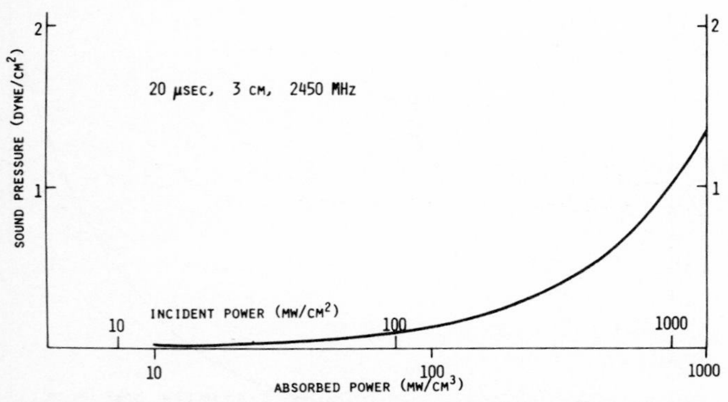

Fig. 8. The dependence of sound pressure amplitude generated in a 3-cm-radius spherical head exposed to 2450-MHz plane wave on peak incident and absorbed powers. The pulsewidth is taken to be 20 μs.

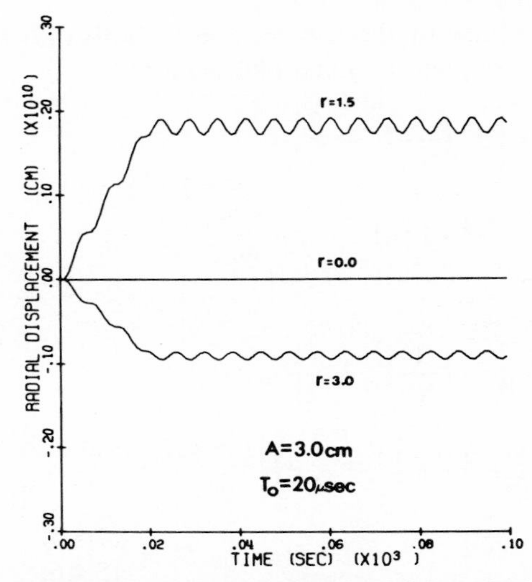

Fig. 10. Radical displacement as a function of time of a 3-cm-radius spherical head exposed to 2450-MHz plane wave. The peak absorption is 1000 mW/cm^3.

expected, the displacement at the center of the sphere is zero. At other locations the displacement increases almost linearly as a function of time until $t = t_0$, the pulsewidth, and then starts to oscillate around the value attained at $t = t_0$. In both cases, the maximum displacements are on the order of 10^{-11} cm. The displacements stay constant after a transient buildup because of the lossless assumption for the elastic media. The apparent higher frequency of oscillation seems to stem from the contribution of higher order modes. But how these frequencies are chosen over all others is not clear. Further investigations are currently in progress.

The sound pressures (radial stresses) in the spherical head models are shown in Figs. 11 and 12 for the corresponding cases shown in Figs. 9 and 10. It is interesting to note that the sound pressure begins with zero amplitude and then grows to an intermediate value. With a sudden rise of amplitude the main body of the pressure wave arrives, oscillating at a constant pressure level in the absence of elastic loss. The final jump in amplitude is marked by $t = t_0$.

IV. Conclusions

We have presented a model for sound wave generation in spheres simulating heads of laboratory animals and human beings by assuming a spherically symmetric microwave absorption pattern. The impinging microwaves are taken to be plane wave rectangular pulses. The problem has been formulated in terms of thermoelastic theory in which the absorbed microwave energy represents the volume heat source. The thermoelastic equation of motion is solved for the sound wave under stress-free boundary conditions using boundary value technique and Duhamel's theorem. The extension to constrained surface is currently under investigation. It may be noted that the related case of micro-

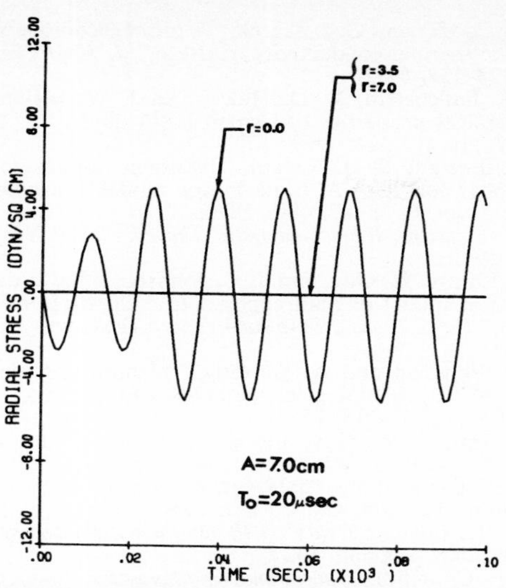

Fig. 11. Radial stress (sound pressure) generated in a 7-cm-radius spherical head exposed to 918-MHz plane wave. The peak absorption is 1000 mW/cm³.

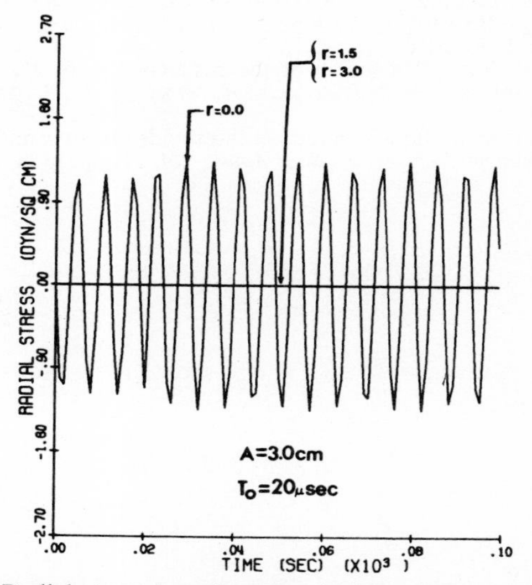

Fig. 12. Radial stress (sound pressure) generated in a 3-cm-radius spherical head exposed to 2450-MHz plane wave. The peak absorption is 1000 mW/cm³.

wave pulses impinging on a semi-infinite medium of absorbing material has been given previously [25], [26].

Examination of the numerical results given in the last section indicates that pulsed microwave-induced sound pressure amplitude depends upon both pulsewidth and peak power density. In addition, there is apparently an optimal pulsewidth for maximum sound pressure generation which varies according to the sphere size and the frequency of the impinging radiation. As shown in Tables II and III, for a peak absorbed power density of 1000 mW/cm³ (which corresponds to 600 mW/cm² incident power at 2450 MHz impinging on a 3-cm spherical head, and to 2200 mW/cm² incident power at 918 MHz impinging on a 7-cm spherical head), the pressure amplitudes generated at the center of the sphere are 15–30 dB above the reported threshold of hearing by bone conduction (60 dB, Re 0.0002 dyne/cm², 5–10 kHz) [27], [28] for pulses between 1 and 50 μs wide. The incident power required compares favorably with that reported previously [2], [4], [5]. At an absorbed power density of 1000 mW/cm³, the corresponding rate of temperature rise at the center of both spheres, $r = 0$, is 0.258°C/s in the absence of heat conduction. The temperature rise in 20 μs is 5.2×10^{-6}°C.

TABLE II
SOUND PRESSURE IN A MAN-SIZED (a = 7 cm) SPHERICAL HEAD EXPOSED TO 918-MHz RADIATION

Pulse width (μs)	Incident power mW/cm²	Absorbed power mW/cm³	Pressure (dyne/cm²)	db re 0.0002 dyne/cm²
0.1	2200	1000	0.12	55.5
0.5	2200	1000	0.60	69.5
1.0	2200	1000	1.19	75.5
5.0	2200	1000	4.90	87.8
10.0	2200	1000	4.70	87.4
20.0	2200	1000	5.10	88.1
30.0	2200	1000	2.80	82.9
40.0	2200	1000	4.10	86.2
50.0	2200	1000	5.40	88.6

TABLE III
SOUND PRESSURE IN A CAT-SIZED (a = 3 cm) SPHERICAL HEAD EXPOSED TO 2450-MHz RADIATION

Pulse width (μs)	Incident power (mW/cm²)	Absorbed power (mW/cm³)	Pressure (dyne/cm²)	db re 0.0002 dyne/cm²
0.1	600	1000	0.12	55.6
0.5	600	1000	0.59	69.4
1.0	600	1000	1.15	75.2
5.0	600	1000	1.40	76.9
10.0	600	1000	2.30	81.2
20.0	600	1000	1.35	76.6
30.0	600	1000	1.50	77.5
40.0	600	1000	2.2	80.8
50.0	600	1000	1.2	75.6

Estimations of the fundamental sound frequency generated inside the head show that the frequency varies from about 8 kHz for a man-sized sphere to approximately 80 kHz for a small animal's, such as a mouse's, head. Assuming an equivalent radius of 1.5 cm for the brain of a guinea pig, Fig. 4 indicates a fundamental sound frequency of 48 kHz, which is in reasonably good agreement with the 50-kHz cochlea microphonic oscillations recorded from the round window of guinea pigs [6], which also happens to be the only available data in the literature,

Finally, it should be mentioned that the numerical results presented in this paper should be interpreted as giving estimates of the sound waves expected to be produced in mammalian heads by microwave pulses, subject to our ability to describe microwave, thermal, elastic, and geometric properties of mammalian cranial structures. In general, the results of this analysis indicate that thermoelastically generated stresses, resulting from microwave absorptive heating inside the head, represent a highly possible mechanism for sound generation.

ACKNOWLEDGMENT

The author wishes to thank P. M. Nefcy and C. K. Lam for their assistance.

References

[1] A. H. Frey, "Auditory system response to radio frequency energy," *Aerospace Medicine*, vol. 32, pp. 1140–1142, 1961.

[2] A. W. Guy, E. M. Taylor, B. Ashleman, and J. C. Lin, "Microwave interaction with the auditory systems of humans and cats," *IEEE/MTT Symposium Digest*, pp. 321–323, 1973.

[3] E. M. Taylor and B. T. Ashleman, "Analysis of the central nervous involvement in the microwave auditory effect," *Brain Research*, vol. 74, pp. 201–208, 1974.

[4] A. W. Guy, C. K. Chou, J. C. Lin, and D. Christensen, "Microwave induced acoustic effects in mammalian auditory systems and physical materials," *Annals N.Y. Acad. Sciences*, vol. 247, pp. 194–215, 1975.

[5] W. J. Rissman and C. A. Cain, "Microwave hearing in mammals," *Proc. Nat. Elect. Conf.*, vol. 30, pp. 239–244, 1975.

[6] C. K. Chou, R. Galambos, A. W. Guy, and R. H. Lovely, "Cochlea microphonics generated by microwave pulses," *J. Microwave Power*, vol. 10, pp. 361–367, 1975.

[7] A. H. Frey, "Human auditory system response to modulated electromagnetic energy," *J. Appl. Physiol.*, vol. 17, pp. 689–692, 1962.

[8] A. H. Frey and R. Messenger, Jr., "Human perception of illumination with pulsed ultra-high frequency electromagnetic energy," *Science*, vol. 181, pp. 356–358, 1973.

[9] ANSI Standard, "Safety level of electromagnetic radiation with respect to personnel," *ANSI C95.1*, 1974.

[10] H. C. Sommer and H. E. von Gierke, "Hearing sensations in electric field," *Aerospace Med.*, vol. 35, pp. 834–839, 1964.

[11] J. C. Sharp, H. M. Grove, and O. P. Gandhi, "Generation of acoustic signals by pulsed microwave energy," *IEEE Trans. Microwave Theory Tech.*, vol. 22, pp. 583–584, 1974.

[12] K. R. Foster and E. D. Finch, "Microwave hearing: evidence for thermoacoustical auditory stimulation by pulsed microwaves," *Science*, vol. 185, pp. 256–258, 1974.

[13] J. C. Lin, "Microwave auditory effect—A comparison of some possible transduction mechanisms," *J. Microwave Power*, vol. 11, pp. 77–81, 1976.

[14] T. E. Cooper and G. J. Trezek, "A probe technique for determining the thermal conductivity of tissue," *J. Heat Transfer* vol. 94, pp. 133–138, 1972.

[15] G. T. Fallenstein, V. D. Hulce, and J. W. Melvin, "Dynamic mechanical properties of human brain tissue," *J. Biomechanics*, vol. 2, pp. 217–226, 1969.

[16] Y. C. Lee and S. H. Advani, "Transient response of a sphere to torsional loading—A head injury model," *Mathematical Bioscience*, vol. 6, pp. 473–486, 1970.

[17] J. A. Stratton, *Electromagnetic Theory*. New York: McGraw-Hill, 1941.

[18] J. C. Lin, A. W. Guy, and C. C. Johnson, "Power deposition in a spherical model of man exposed to 1–20 MHz electromagnetic fields," *IEEE Trans. Microwave Theory Tech.*, vol. 21, pp. 791–797, 1973.

[19] C. C. Johnson and A. W. Guy, "Nonionizing electromagnetic wave effects in biological materials and systems," *Proc. IEEE*, vol. 60, pp. 692–718, 1972.

[20] J. C. Lin, A. W. Guy, and G. H. Kraft, "Microwave selective brain heating," *J. Microwave Power*, vol. 8, pp. 275–286, 1973.

[21] H. S. Carslaw and J. C. Jaeger, *Conduction of heat in solids*, 2nd Edition. London: Oxford Univ. Press, 1959.

[22] A. E. H. Love, *A treaty on the mathematical theory of elasticity*. Cambridge, England, 1927.

[23] R. V. Churchill, *Operational mathematics*, 2nd edition. New York: McGraw-Hill, 1958.

[24] E. Jahnke and F. Emde, *Tables of functions*, 4th edition. New York: Dover, 1945.

[25] R. M. White, "Generation of elastic waves by transient surface heating," *J. Appl. Phys.*, vol. 34, pp. 3559–3569, 1963.

[26] L. S. Gournay, "Conversion of electromagnetic to acoustic energy by surface heating," *J. Acoust. Soc. Am.*, vol. 40, pp. 1322–1330, 1966.

[27] J. Zwislocki, "In search of the bone-conduction threshold in a free sound field," *J. Acous. Soc. Amer.*, vol. 29, pp. 795–804, 1957.

[28] J. F. Corso, "Bone-conduction thresholds for sonic and ultrasonic frequencies," *J. Acous. Soc. Amer.*, vol. 35, pp. 1738–1743, 1963.

Part IV
Pathophysiologic Aspects of Microwave/Radiofrequency Energy Exposure

Introduction and Bibliography

Extensive investigations into radiofrequency/microwave bioeffects during the last quarter century indicate that exposure to power density of 100 mW/cm^2 for several minutes or hours can result in pathophysiologic manifestations of a thermal nature in laboratory animals. Such effects may or may not be characterized by a measurable temperature rise, which is a function of the thermal regulatory processes and active adaptation of the animal. The end result is reversible or irreversible change, depending on the conditions of irradiation and the physiologic state of the animal. At lower power densities, evidence of pathologic changes or physiologic alteration is nonexistent or equivocal. A great deal of discussion has nevertheless been engendered concerning the relative importance of thermal or nonthermal, as well as low-level, field effects of radiofrequency (RF) and microwave energy absorption.

The results of some *in vitro* studies have been considered as evidence of nonthermal effects of RF radiation. Although some investigators and reviewers still question the interpretation of these so-called nonthermal effects, several support nonthermal interactions between tissues and electric and magnetic fields.

Temperature increase during exposure to microwaves depends on: (a) the specific area of the body exposed and the efficiency of heat elimination, (b) intensity or field strength, (c) duration of exposure, (d) specific frequency or wavelength, and (e) thickness of skin and subcutaneous tissue. These variables determine the percentage of radiant energy absorbed by various tissues of the body.

In partial body exposure under normal conditions, the body acts as a cooling reservoir, which stabilizes the temperature of the exposed part. The stabilization is due to an equilibrium established between the energy absorbed by the exposed part of the body and the amount of heat carried away from it. This heat transport is due to increased blood flow to cooler parts of the body, maintained at normal temperature by heat-regulating mechanisms such as heat loss due to evaporation, radiation, and convection. If the amount of absorbed energy exceeds the optimal amount of heat energy that can be handled by the mechanisms of temperature regulation, the excess energy will cause continuous temperature rise with time. Hyperthermia and, under some circumstances, local tissue destruction can result.

Manuscript received July 15, 1981.
This paper is based on work performed under Contract No. DE-AC02-76EV03490 with the U.S. Department of Energy at the University of Rochester, Department of Radiation Biology and Biophysics, and has been assigned Report. No. UR-3490-2266.

Elucidation of the biologic effects of microwave exposure requires a careful review and critical analysis of the available literature. Such review requires an appreciation as well as differentiation of the established effects and mechanisms from speculative and unsubstantiated reports. Although most of the experimental data support the concept that the effects of microwave exposure are primarily, if not only, a response to hyperthermia or altered thermal gradients in the body, there are large areas of confusion, uncertainty, and actual misinformation.

Basic considerations—Illumination of biological systems with microwave/radiofrequency (MW/RF) energy leads to temperature elevation when the rate of energy absorption exceeds the rate of energy dissipation. Whether the resultant temperature elevation is diffuse or confined to specific anatomical sites depends on the electromagnetic field characteristics and distributions within the body, as well as to the passive and active thermoregulatory mechanisms available to the particular biological entity.

Experiments with small animals, such as mice and rats, to evaluate the potential effects of MW/RF energy must be carefully designed and performed. The responses may be the result of another unrelated agent, inadvertently introduced into the experimental design, rather than the factor intended to be studied. The fact that a living organism responds to many stimuli is a part of the process of living; such responses are examples of biological "effects." Since animal organisms have considerable tolerance to change, these "effects" may be well within the capability of the organism to maintain a normal equilbrium or condition of homeostasis. If, on the other hand, an effect is of such an intense nature that it compromises the individual's ability to function properly or overcomes the recovery capability of the individual, then the "effect" should be considered a "hazard." In any discussion of the potential for biological "effects" from exposure to electromagnetic energies, we must first determine whether any "effect" can be demonstrated, and then determine whether such an observed "effect" is "hazardous."

In an analysis of scientific literature to determine the probability of a biological response from exposure to a noxious agent, we must consider the consistency of experimental results claimed, both the nature of the response and the biological system involved, the ability to replicate the results of studies with consistency, and whether the results claimed and observations reported can be explained by accepted biological principles.

When assessing the results of research on biological effects of MW/RF exposure, it is important to note whether the techniques used are such that possible effects of intervening factors, for example, noise, vibration, chemicals, variation in temperature, humidity, and air flow are avoided and care is taken to avoid population densities that perturb the field to the extent that measurements become meaningless. The sensitivity of the experiment should be adequate to ensure a reasonable probability that an effect would be detected if indeed any exists. The experiment and observational techniques should be objective. Data should be subjected to acceptable analytical methods with no relevant data deleted from consideration. If an effect is claimed, it should be demonstrated at an acceptable level of statistical significance by application of appropriate tests. A given experiment should be internally consistent with respect to the effect of interest. Finally, the results should be quantifiable and susceptible to confirmation by other investigators.

Proper investigation of the biologic effects of electromagnetic fields requires an understanding and appreciation of biophysical principles and "comparative biomedicine." Such studies require interspecies "scaling," the selection of biomedical parameters which consider basic physiological functions, identification of specific and nonspecific reactions, and differentiation of adaptational or compensatory changes from pathological manifestations.

Much of the research on biological effects of MW/RF has been done with small rodents that have coefficients of heat absorption, field concentration effects, body surface areas, and thermal regulatory mechanisms significantly different from man. Adverse reaction in animals does not prove adverse effect in man, and lack of reaction in animals does not prove that man will not be affected. Even closely related species can differ widely in their response. The literature is replete with "anomalous" reactions. Thus, results of exposure of common laboratory animals cannot be readily extrapolated to man unless some form of "scaling" among different animal species, and from animal to man, can be invoked in an accurate way to obtain a quantitatively valid extrapolation from the actual data observed.

The physical factors that must be considered include: frequency of radiation, intensity, animal orientation with respect to the source, size of animal with respect to the wavelength, portion of the body irradiated, exposure time-intensity factors, environmental conditions (temperature, humidity, air flow), and absorbed energy distribution in the body. In addition, variables such as restraint, metabolic rate, ratio of body volume/surface area, and thermoregulatory mechanisms will affect the biological response to microwave/radiofrequency energies.

The need for proper dosimetry in experimental procedures and the importance of realistic scaling factors required for extrapolation of data obtained with small laboratory animals to man are clearly required. Maximum absorption in man occurs at 80 MHz and falls off at higher frequencies. Formulas for scaling factors among species are available. Five milliwatts per centimeter square, 2450 MHz, exposure of a small animal, such as a mouse or rat, can result in a thermal effect that could influence the central nervous system or elicit behavioral and other physiologic responses in that animal, but not necessarily in a larger animal or man. One, therefore, has to be very circumspect in assessing the results of animal experiments for predictive value insofar as man is concerned.

A number of retrospective studies have been done on human populations exposed or believed to have been exposed to MW/RF energies. Those performed in the United States and Poland have not revealed any relationship of altered morbidity or mortality to MW/RF exposure. Nervous system and cardiovascular alterations in humans exposed to microwaves have been reported in Soviet and other Eastern European literature. Most of the reported effects are subjective and reversible, and pathological damage is insignificant. There is considerable difficulty in establishing the presence of, and quantifying the frequency and severity of, "subjective" complaints. Individuals suffering from a variety of chronic diseases may exhibit the same dysfunctions of the central nervous and cardiovascular systems as those reported to be a result of exposure to microwaves; thus, it is extremely difficult, if not impossible, to rule out other factors in attempting to relate microwave exposure to clinical conditions especially in the environment or industrial setting.

The publications included in this bibliography were selected on the basis of their impact on the assessment of the pathophysiologic aspects of microwave/radiofrequency energy exposure. Although some of the papers do not meet the criteria of sound scientific publications, they were included because they have been utilized by various individuals or groups in their assessment of biologic effects and health implications of radiofrequency/microwave energy exposure.

Sol M. Michaelson
Associate Editor

Reviews

[1] S. Baranski and P. Czerski, *Biological Effects of Microwaves*, Stroudsburg, PA: Dowden, Hutchinson & Ross, 234 pp., 1976.

[2] R. L. Carpenter, "Microwave radiation," in *Handbook of Physiology*, no. 9—*Reactions to Environmental Agents*, D.H.K. Lee, H. L. Falk, and S. D. Murthy, Eds., pp. 111–125, Bethesda, MD: American Physiological Society, 1977.

[3] S. F. Cleary, "Biological effects of microwaves and radiofrequency radiation," in *CRC Critical Reviews in Environmental Health*, C. Staub, Ed., vol. 7, p. 121–165, 1977.

[4] Z. V. Gordon, A. V. Roscin, and M. S. Byckov, "Main directions and results of research in the USSR on the biologic effects of microwaves," in *Biologic Effects and Health Hazards of Microwave Radiation, Proc. International Symposium*, Warsaw, Oct. 15–18, 1973, Warsaw: Polish Medical Publishers, pp. 22–35, 1974.

[5] C. C. Johnson and A. W. Guy, "Non-ionizing electromagnetic wave effects in biological materials and systems," *Proc. IEEE*, vol. 60, pp. 692–718, 1972.

[6] H. Kalant, "Physiologic hazards of microwave radiation, survey of published literature," *Canada Med. Ass. J.*, vol. 81, pp. 575, 1959.

[7] K. Marha, J. Musil, and H. Tuha, *Electromagnetic Fields and the Living Environment*, Prague: State Health Publishing House, 1968. Translation—SBN 911302-13-7, San Francisco Press, 1971.

[8] D. I. McRee, "Environmental aspects of microwave radiation," *Environ. Health Perspect.*, vol. 5, p. 41, 1972.

[9] S. M. Michaelson, "Biological effects of microwave exposure," in *Biological Effects and Health Implications of Microwave Radiation, Symp. Proc.*, S. F. Cleary, Ed., United States Dept. Health,

Education, and Welfare, Public Health Service BRH/DBE 70-2, p. 35, 1970.

[10] S. M. Michaelson, "The tri-service program—a tribute to George M. Knauf, USAF (MC)," *IEEE Trans. Microwave Theory Tech.*, vol. MTT-19, pp. 131-146, 1971.

[11] S. M. Michaelson, "Human exposure to non-ionizing radiant energy—potential hazards and safety standards," *Proc. IEEE*, vol. 60, pp. 389-421, 1972.

[12] S. M. Michaelson, "Effects of exposure to microwaves: problems and perspectives," *Environmental Health Perspectives*, vol. 8, pp. 133-156, 1974.

[13] S. M. Michaelson, "Radiofrequency and microwave energies, magnetic and electric fields," in *The Foundations of Space Biology and Medicine*, M. Calvin and O. G. Gazenko, Eds., Washington, DC: NASA, vol. II, Book 2, chapt. 1, pp. 409-452, 1975.

[14] S. M. Michaelson, "Biologic and pathophysiologic effects of exposure to microwaves," in *Microwave Bioeffects and Radiation Safety Trans.*, M. A. Stuchly, Ed., Edmonton, Alberta, Canada: The International Microwave Power Institute (IMPI), vol. 8, pp. 55-94, 1978.

[15] S. M. Michaelson, "Microwave biological effects: an overview," *Proc. IEEE*, vol. 68, pp. 40-49, 1980.

[16] S. M. Michaelson and C. H. Dodge, "Soviet views on the biologic effects of microwaves: an analysis," *Health Phys.*, vol. 21, pp. 108-111, 1971.

[17] S. M. Michaelson, R. A. E. Thomson, and J. W. Howland, "Physiologic aspects of microwave irradiation of mammals," *Amer. J. Physiol.*, vol. 201, p. 351, 1961.

[18] S. M. Michaelson, R. A. E. Thomson, and J. W. Howland, "Comparative studies on 1285 and 2800 mc/sec pulsed microwaves," *Aerosp. Med.*, vol. 36, p. 1059, 1965.

[19] S. M. Michaelson, R. A. E. Thomson, and J. W. Howland, "Biologic Effects of Microwave Exposure," Griffiss Air Force Base, Rome Air Development Ctr., Rome, NY, ASTIA Doc. No. AD 824-242, 138 pp., 1967.

[20] W. C. Milroy and S. M. Michaelson, "Biological effects of microwave radiation," *Health Phys.*, vol. 20, p. 567, 1971.

[21] B. A. Minin, "Microwaves and Human Safety—Part I and II," *JPRS* 65506-1 and 2, 1975. Transl. from Russian "Svch i Bezopasnost Cheloveka," *Ivol*, 342 pp., Moscow: Izdatel'stvo Sovetskoye Radio, 1974.

[22] Y. I. Novitskiy, Z. V. Gordon, A. S. Presman, and Yu. A. Kholodov, "Radio Frequencies and Microwaves, Magnetic and Electrical Fields," Washington, DC:, NASA, NASA TT F-14.021, 1971.

[23] *Influence of Microwave Radiation on the Organism of Man and Animals*, I. R. Petrov, Ed., Leningrad: Meditsina Press, NASA TT F-708, 1970.

[24] A. S. Presman, *Electromagnetic Fields and Life*, Moscow: Izd-vo Nauka, 1968. Translation, New York: Plenum Press, 1970.

[25] H. P. Schwan, "Radiation biology, medical applications and radiation hazards," in *Microwave Power Engineering*, E. C. Okress, Ed., vol. 2, New York: Academic Press, pp. 213-243, 1968.

[26] H. P. Schwan, "Interaction of microwave and radio frequency radiation with biological systems," *IEEE Trans. Microwave Theory Tech.*, vol. MTT-19, p. 146, 1971.

[27] H. P. Schwan, and G. M. Piersol, "The absorption of electromagnetic energy in body tissues, a review and critical analysis, Part I, Biophysical aspects," *Amer. J. Phys. Med.*, vol. 33, pp. 371-404, 1954.

[28] H. P. Schwan and G. M. Piersol, "The absorption of electromagnetic energy in body tissues, a review and critical analysis, Part II, Physiological and clinical aspects," *Amer. J. Phys. Med.*, vol. 34, pp. 425-448, 1955.

[29] C. Silverman, "Nervous and behavioral effects of microwave radiation in humans," *Amer. J. Epidemiol.*, vol. 97, p. 219, 1973.

[30] *Biological Effects of Nonionzing Radiation*, P. E. Tyler, Ed., *Ann. N.Y. Acad. Sci.*, vol. 247, 1975.

[31] Special issues on nonionizing radiation:
(a) *Bull. NY Acad. Med.*, vol. 55, no. 11, Dec. 1979.
(b) *Proc. IEEE*, vol. 68, no. 1, Jan. 1980.

Biochemistry and Metabolism

[1] E. N. Albert, G. McCullars, and M. Short, "The effect of 2450 MHz microwave radiation on liver adenosine triphosphate (ATP)," *J. Microwave Power*, vol. 9, pp. 205-211, 1974.

[2] J. W. Allis, "Irradiation of bovine serum albumin with a crossed-beam exposure-detection system" in *Biologic Effects of Nonionizing Radiation*, P. E. Tyler, Ed., *Ann. N.Y. Acad. Sci.*, vol. 247, pp. 312-322, 1975.

[3] M. L. Belkhode, D. L. Johnson, and A. M. Muc, "Thermal and athermal effects of microwave radiation on the activity of glucose-6-phosphate dehydrogenese in human blood," *Health Phys.*, vol. 26, p. 45, 1974.

[4] M. Bini, A. Checcucci, A. Ignesti, L. Millanta, N. Rubino, S. Camici, G. Manao, and G. Ramponi, "Analysis of the effects of microwave energy on enzymatic activity of lactate dehydrogenase (LDH)," *J. Microwave Power*, vol. 13, pp. 96-99, 1978.

[5] P. E. Hamrick, "Thermal denaturation of DNA exposed to 2450 MHz CW microwave radiation," *Radiat. Res.*, vol. 56, p. 400, 1973.

[6] H. S. Ho and W. P. Edwards, "Oxygen-consumption rate of mice under differing dose rates of microwave radiation," *Radio Sci.*, vol. 126(S), p. 131, 1977.

[7] D. E. Janes, W. M. Leach, W. A. Mills, R. T. Moore, and M. L. Shore, "Effects of 2450 MHz microwaves on protein synthesis and on chromosomes in Chinese hamsters," *Non. Ioniz. Radiat.*, vol. 1, pp. 125-130, 1969.

[8] J. H. Kinoshita, L. D. Merola, E. D. Dikmak, and R. L. Carpenter, "Biochemical changes in microwave cataracts," *Doc. Ophthalmol.*, vol. 20, p. 91, 1966.

[9] L. O. Merola and J. H. Kinoshita, "Changes in the ascorbic acid content in lenses of rabbit eyes exposed to microwave radiation," in *Biological Effects of Microwave Radiation*, M. F. Peyton, Ed., vol. 1, New York: Plenum Press, p. 285, 1961.

[10] L. Miro, R. Loubiere, and A. Pfister, "Effects of microwaves on the cell metabolism of the reticulo-endothelial system," in *Biologic Effects and Health Hazards of Microwave Radiation*, P. Czerski et al., Eds., Warsaw: Polish Medical Publishers, p. 89, 1974.

[11] S. V. Nikogosyan, "Influence of UHF on the cholinesterase activity in the blood serum and organs in animals," in *The Biological Action of Ultrahigh Frequencies*, A. A. Letavet and Z. V. Gordon, Eds., JPRS-12471, p. 83, 1962.

[12] R. B. Olcerst and J. R. Rabinowitz, "Studies on the interaction of microwave radiation with cholinesterase," *Rad. and Environm. Biophys.* vol. 15, p. 289, 1978.

[13] T. Rupp, J. Montet, and J. W. Frazer, "A comparison of thermal and RF exposure effects on trace metal content of blood plasma and liver cell fractions of rodents," *Ann. N.Y. Acad. Sci.*, vol. 247, pp. 282-290, 1975.

[14] K. D. Straub and P. Carver, "Effects of electromagnetic fields on microsomal ATPase and mitochondrial oxidative phosphorylation," in *Biological Effects of Nonionizing Radiation*, P. E. Tyler, Ed., *Ann. N.Y. Acad. Sci.*, vol. 247, pp. 292-300, 1975.

[15] R. T. Wangemann and S. F. Cleary, "The *in vivo* effects of 2.45 GHz microwave radiation on rabbit serum components," *Radiat. Environ. Biophys.*, vol. 13, pp. 89-103, 1976.

Cellular/Molecular Biology

[1] S. Baranski, H. Debiec, K. Kwarecki, and T. Mezykowski, "Influence of microwaves on genetical processes of Aspergillus nidulans," *J. Microwave Power*, vol. 11, pp. 146-147, 1976.

[2] S. Baranski, S. Szmigielski, and J. Moneta, "Effects of microwave irradiation *in vitro* on cell membrane permeability," in *Biological Effects and Health Hazards of Microwave Radiation*, P. Czerski et al., Eds., Warsaw: Polish Medical Publishers, pp. 173-177, 1974.

[3] C. F. Blackman, S. G. Benane, C. M. Weil, and J. S. Ali, "Effects of nonionizing electromagnetic radiation on single-cell biologic systems," in *Biological Effects of Nonionizing Radiation*, P. E. Tyler, Ed., *Ann. N.Y. Acad. Sci.*, vol. 247, pp. 352-365, 1975.

[4] C. F. Blackman, M. C. Surles, and S. G. Benane, "The effects of microwave exposure on bacteria mutation reduction," *Symp. Biol. Eff. of E.M. Waves*, United States Dept. Health, Education, and Welfare publication, (FDA) 77-8010, vol. 1, pp. 406-413, 1976.

[5] J. C. Correlli, R. J. Gutmann, S. Kohazi, and J. Levy, "Effects of 2.6-4.0 GHz microwave radiation on E.-coli B," *J. Microwave Power*, vol. 12, p. 141, 1977.

[6] J. E. Prince, L. H. Mori, J. W. Frazer, and J. C. Mitchell, "Cytologic aspect of RF radiation in the monkey," *Aerosp Med.*, vol. 43, no. 7, pp. 759-761, 1972.

[7] S. J. Webb, "Genetic continuity and metabolic regulation as seen by the effects of various microwave and black light frequencies on these phenomena," in *Biologic Effects of Nonionizing Radiation*, P. E. Tyler, Ed., *Ann. N.Y. Acad. Sci.*, vol. 247, pp. 327–351, 1975.
[8] K. T. S. Yao, "Microwave radiation-induced chromosomal abberrations in corneal epithelium of Chinese hamsters," *J. Heredit.*, vol. 69, pp. 409–412, 1978.
[9] H. Fröhlich, "The biological effects of microwaves and related questions," in *Advances in Electronics and Electron Physics*, vol. 53, pp. 85–152, NY: Academic Press, 1980.

Genetics

[1] S. Mittler, "Failure of 2 and 10 meter radio waves to induce genetic damage in Drosophila melanogaster," *Environmental Res.*, vol. 11, pp. 326–330, 1976.
[2] T. L. Pay, E. C. Beyer, and C. F. Reichelderfer, "Microwave effects on reproductive capacity and genetic transmission in Drosophila melanogaster," *J. Microwave Power*, vol. 7, p. 75, 1972.

Gonads and Reproduction

[1] A. N. Bereznitskaya, "The effect of 10-centimeter and ultrashort waves on the reproductive function of female mice," *Gig. Tr. Prof. Zabol.*, no. 9, p. 33, 1968.
[2] A. N. Bereznitskaya, "Research on the reproductive function in female mice under the impact of low-intensity radio waves of different ranges," in *Industrial Health and Biological Effects of Radio Frequency Electromagnetic Waves*, material of the Fourth All-Union Symposium, October 17–19, 1972, Moscow, 51 pp., 1972.
[3] A. N. Bereznitskaya and I. M. Kazbekov, "Studies on the reproduction and testicular microstructure of mice exposed to microwaves," in *Biological Effects of Radiofrequency Electromagnetic Fields*, Z. V. Gordon, Ed., no. 4, Moscow, pp. 221–229, 1973. JPRS-63321, 1974.
[4] T. S. Ely, D. Goldman, J. Z. Hearon, R. B. Williams, and H. M. Carpenter, *Heating Characteristics of Laboratory Animals Exposed to Ten Centimeter Microwaves*, Bethesda, Md: US Nav. Med. Res. Inst., Res. Rep. Proj. NM 001-056.13.02, 1957. *IEEE Trans. Bio-Med. Eng.*, vol. 11, pp. 123–137, 1964.
[5] S. F. Gorodetskaya, "The effect of centimeter radio waves on mouse fertility," *Fiziol. Zh.*, vol. 9, p. 394, 1963.
[6] S. A. Gunn, T. C. Gould, and W. A. D. Anderson, "The effect of microwave radiation (24000 mc) on the male endocrine system of the rat," in *Biological Effects of Microwave Radiation*, M. F. Dayton, Ed., New York: Plenum Press, vol. I, pp. 99–115, 1961.
[7] C. J. Imig, J. D. Thomson, and H. M. Hines, "Testicular degeneration as a result of microwave irradiation," *Proc. Soc. Exp. Biol. Med.*, vol. 69, pp. 382–386, 1948.
[8] I. Lancranjan, M. Maicanescu, E. Rafaila, I. Klepsch, and H. I. Popescu, "Gonadic function in workmen with long-term exposure to microwaves," *Health Phys.*, vol. 29, p. 381, 1975.
[9] G. J. Muraca, Jr., E. S. Ferri, and F. L. Buchta, "A study of the effects of microwave irradiation of the rat testes," in *Biological Effects of Electromagnetic Waves*, C. C. Johnson and M. Shore, Eds., United States Dept. Health, Education, and Welfare, (FDA) 77-8010, Rockville, MD, vol. I, pp. 484–494, 1977.
[10] D. S. Rosenthal and S. C. Beering, "Hypogonadism after microwave radiation," *J. Amer. Med. Ass.*, vol. 205, p. 245, 1968.
[11] M. M. Varma, E. L. Dage, and S. R. Joshi, "Mutagenicity induced by nonionizing radiation in Swiss male mice," in *Biological Effects of Electromagnetic Waves*, C. C. Johnson and M. L. Shore, Eds., United States Dept. Health, Education, and Welfare, Food and Drug Administration, Rockville, MD, (FDA) 77-8010, vol. 1, pp. 397–405, 1976.
[12] M. M. Varma and E. A. Traboulay, Jr., "Evaluation of dominant lethal test and DNA studies in measuring mutagenicity caused by non-ionizing radiation," in *Biological Effects of Electromagnetic Waves*, C. C. Johnson and M. L. Shore, Eds., United States Dept. Health, Education, and Welfare, Food and Drug Administration, publication (FDA) 77-8010, Rockville, MD, vol. 1, pp. 386–396, 1976.

Development

[1] E. Berman, J. B. Kinn, and H. B. Carter, "Observations of mouse fetuses after irradiation with 2.45 GHz microwaves," *Health Phys.*, vol. 35, pp. 791–801, 1978.
[2] R. L. Carpenter and E. M. Livstone, "Evidence for nonthermal effects of microwave radiation: abnormal development of irradiated insect pupae," *IEEE Trans. Microwave Theory Tech.*, vol. MTT-19, p. 173, 1971.
[3] M. E. Chernovetz, D. R. Justesen, N. W. King, and J. E. Wagner, "Teratology, survival, and reversal learning after fetal irradiation of mice by 2450 MHz microwave energy," *J. Microwave Power*, vol. 10, p. 391, 1975.
[4] M. E. Chernovetz, D. R. Justesen, and A. F. Oke, "A teratologic study of the rat: microwave and infrared radiations compared," *Radio Sci.*, vol. 12, no. 6S, p. 191, 1977.
[5] J. Daels, "Microwave heating of the uterine wall during parturition," *Obstet. Gynecol.*, vol. 42, pp. 76–79, 1973.
[6] J. Daels, "Microwave heating of the uterine wall during parturition," *J. Microwave Power*, vol. 11, pp. 166–168, 1976.
[7] F. Dietzel, "Effects of non-ionizing electro-magnetic radiation on the development and intrauterine implantation of the rat," in *Biologic Effects of Nonionizing Radiation*, P. E. Tyler, Ed., *Ann. N.Y. Acad. Sci.*, vol. 247, p. 367, 1975.
[8] F. Dietzel and W. Kern, "Abortion following ultra-shortwave hyperthermia animal experiments," *Arch. Gynakol.*, vol. 209, p. 445, 1970.
[9] F. Dietzel, W. Kern, and R. Steckenmesser, "Deformity and intrauterine death after short-wave therapy in early pregnancy in experimental animals," *Munch. Med. Wschr.*, vol. 114, p. 228, 1972.
[10] R. Guillet and S. M. Michaelson, "The effect of repeated microwave exposure on neonatal rats," *Radio Sci.*, vol. 12, no. 6(S), pp. 125–130, 1977.
[11] R. P. Jensh and J. Ludlow, "Behavioral teratology: application in low dose chronic microwave irradiation studies," chapter 8 in *Advances in the Study of Birth Defects*, T.V.N. Persand, Ed., vol. 4, Neural and Behavioral Teratology, Lancaster, England: MTP Press Ltd., pp. 135–162, 1980.
[12] G. A. Lindauer, L. M. Liu, G. W. Skewes, and F. J. Rosenbaum, "Further experiments seeking evidence of nonthermal biological effects of microwave radiation," *IEEE Trans. Microwave Theory Tech.*, vol. 22, pp. 790–793, 1974.
[13] L. M. Liu, F. J. Rosenbaum, and W. F. Pickard, "The relation of teratogenesis in *Tenebrio molitor* to the incidence of low level microwaves," *IEEE Trans. Microwave Theory Tech.*, vol. 23, pp. 929–931, 1975.
[14] D. I. McRee and P. E. Hamrick, "Exposure of Japanese quail embryos to 2.45 GHz microwave radiation during development," *Radiat. Res.*, vol. 71, pp. 355–366, 1977.
[15] D. I. McRee, P. E. Hamrick, J. E. Zinkl, P. Thaxton, and C. R. Parkhurst, "Some effects of exposure of the Japanese quail embryo to 2.45 GHz microwave radiation," in *Biologic Effects of Nonionizing Radiation*, P. E. Tyler, Ed., *Ann. N.Y. Acad. Sci.*, vol. 247, pp. 377–390, 1975.
[16] S. M. Michaelson, R. Guillet, and F. W. Heggeness, "The influence of microwave exposure on functional maturation of the rat," *Proc. 17th Hanford Biology Symposium—Developmental Toxicology of Energy-Related Pollutants*, Department of Energy Symposium Series 47, D. Mahlum et al., Eds., pp. 300–316, CONF 771017 TIC, USDOE NTIS, Springfield, VA, 1978.
[17] R. G. Olsen, "Insect teratogenesis in a standing-wave irradiation system," *Radio Sci.*, vol. 12, no. 6S, pp. 199–208, 1977.
[18] R. Rugh, E. I. Ginns, H. S. Ho, and W. M. Leach, "Are microwaves teratogenic?" in *Biological Effects and Health Hazards of Microwave Radiation*, P. Czerski, et al., Eds., Warsaw: Polish Medical Publishers, p. 98, 1974.
[19] R. Rugh, E. I. Ginns, H. S. Ho, and W. M. Leach, "Responses of the mouse to microwave radiation during estrous cycle and pregnancy," *Radiat. Res.*, vol. 62, p. 225, 1975.
[20] R. Rugh and M. McManaway, "Anesthesia as an effective agent against the production of congenital anomalies in mouse fetuses exposed to electromagnetic radiation," *J. Exp. Zool.*, vol. 197, p. 363, 1976.
[21] R. Rugh and M. McManaway, "Mouse fetal sensitivity to microwave radiation," *Congenital Anomalies*, vol. 17, pp. 39–45, 1977.

[22] W. B. Stavinoha, M. A. Medina, J. Frazer, S. T. Weintraub, D. H. Ross, A. T. Modak, and D. J. Jones, "The effects of 19 megacycle irradiation on mice and rats," in *Biolgical Effects of Electromagnetic Waves*, C. C. Johnson and M. L. Shore, Eds., United States Dept. Health, Education, and Welfare, publication (FDA) 77-8010, Rockville, Md., vol. I, pp. 431–448, 1976.

[23] C. A. Van Ummersen, "The effect of 2450 mc radiation on the development of the chick embryo," in *Biological Effects of Microwave Radiation*, M. F. Peyton, Ed., New York: Plenum Press, vol. 1, p. 201, 1961.

Central Nervous System

[1] *Brain Interactions with Weak Electric and Magnetic Fields*, W. R. Adey and S. M. Bawin, Eds., Neurosciences Res. Program Bull. 15, MIT Press, 129 pp., 1977.

[2] E. N. Albert, "Light and electron microscopic observations on the blood brain barrier after microwave irradiation," *Symp. Biol. Eff. and Measurements of Radiofrequency/Microwaves*, D. S. Hazzard, Ed., United States Dept. Health, Education, and Welfare, publication (FDA) 77-8026, Rockville, MD, pp. 294–304, 1977.

[3] E. N. Albert, L. Grau, and J. Kerns, "Morphologic alterations in hamster blood-brain barrier after microwave irradiation," *J. Microwave Power*, vol. 12, pp. 43–44, 1977.

[4] E. N. Albert and M. DeSantis, "Do microwaves alter nervous system-structure?" in *Biologic Effects of Nonionizing Radiation*, P. E. Tyler, Ed., *Ann. N.Y. Acad. Sci.*, vol. 247, pp. 87–108, 1975.

[5] M. S. Baldwin, S. A. Bach, and S. A. Lewis, "Effects of radio frequency energy on primate cerebral activity," *Neurol.*, vol. 10, pp. 178–187, 1960.

[6] S. Baranski, "Histological and histochemical effects of microwave irradiation on the central nervous system of rabbits and guinea pigs," *Amer. J. Phys. Med.*, vol. 51, pp. 182–191, 1972.

[7] S. Baranski and Z. Edelwejn, "Electroencephalographic and morphological investigations on the influence of microwaves on the central nervous system," *Acta Physiol. Pol.*, vol. 18, p. 423, 1967.

[8] S. Baranski and Z. Edelwejn, "Studies on the combined effect of microwaves and some drugs on bioelectric activity of the rabbit CNS," *Acta Physiol. Pol.*, vol. 19, pp. 37–50, 1968.

[9] S. Baranski and Z. Edelwejn, "Experimental morphologic and electroencephalographic studies of microwave effects on the nervous system," in *Biologic Effects of Nonionizing Radiation*, P. E. Tyler, Ed., *Ann. N.Y. Acad. Sci.*, vol. 247, pp. 109–116, 1975.

[10] S. M. Bawin and W. R. Adey, "Sensitivity of calcium binding in cerebral tissue to weak environmental electric fields oscillating at low frequency," *Proc. Natl. Acad. Sci.*, vol. 73, pp. 1999–2003, 1976.

[11] S. M. Bawin and W. R. Adey, "Calcium binding in cerebral tissue," in *Symposium on Biological Effects and Measurement of Radio Frequency/Microwaves*, D. G. Hazzard, Ed., United States Dept. Health, Education, and Welfare, publication (FDA) 77-8026, Washington, DC, p. 305, 1977.

[12] C. F. Blackman, J. A. Elder, C. M. Weil, S. G. Benane, D. C. Eichinger, and D. E. House, "Induction of calcium ion efflux from brain tissue by radio-frequency radiation: effects of modulation frequency and field strength," *Radio Sci.*, vol. 14, pp. 93–98, 1979.

[13] Z. M. Gvozdikova, V. M. Anan'yev, I. N. Zenina, and V. I. Zak, "Sensitivity of the rabbit central nervous system to a continuous (non-pulsed) ultrahigh frequency electromagnetic field," *Biul. Eksp. Biol. Med.*, Moscow, vol. 58, p. 63, 1964.

[14] Yu. A. Kholodov, "Changes in the electrical activity of the rabbit cerebral cortex during exposure to a UHF-HF electromagnetic field—Part 2—the direct action of the UHF-HF field on the central nervous system," *Biul. Eksp. Biol. Med.*, Moscow, vol. 56, p. 42, 1963.

[15] Yu. A. Kholodov "The influence of a VHF-HF electromagnetic field on the electrical activity of an isolated strip of cerebral cortex, *Biul. Eksp. Biol. Med.*, Moscow, vol. 57, p. 98, 1964.

[16] Yu. A. Kholodov, *The Effect of Electromagnetic and Magnetic Fields on the Central Nervous System*, Moscow: Nauka Press, 283 pp., NASA TT-F-465, 1966.

[17] I. A. Kitsovskaya, "An investigation of the interrelationships between the main nervous processes in rats on exposure to SHF fields of various intensities," *Tr. Gig. Tr. Prof. AMN SSSR*, no. 1, p. 75, 1960.

[18] M. N. Livanov, A. B. Tsypin, Yu. G. Grigoriev, U. G. Kruschev, S. M. Stepanov, and A. M. Anen'yev, "The effect of electromagnetic fields on the bioelectric activity of cerebral cortex in rabbits," *Byull. Eksp. Biol. i Med.*, vol. 49, p. 63, 1960.

[19] E. A. Lobanova and A. V. Goncharova, "Investigation of conditioned-reflex activity in animals (albino rats) subjected to the effect of ultrashort and short radio-waves," *Gig. Tr. Prof. Zabol.*, vol. 15, no. 1, pp. 29–33, 1971.

[20] Y. Maruyama, R. Nakamura, and K. Kobayashi, "Effect of microwave irradiation on brain structure and catecholamine distribution," *Psychopharmacol.*, vol. 67, pp. 119–123, 1980.

[21] J. H. Merritt, A. F. Chamness, and S. J. Allen, "Studies on blood-brain barrier permeability after microwave-radiation," *Rad. and Environm. Biopys.*, vol. 15, pp. 367–377, 1978.

[22] W. H. Oldendorf, "Focal neurological lesions produced by microwave irradiation," *Proc. Soc. Ex. Biol. Med.*, vol. 72, p. 432, 1949.

[23] K. J. Oscar and T. D. Hawkins, "Microwave alteration of the blood-brain barrier system of rats," *Brain Res.*, vol. 126, pp. 281–293, 1977.

[24] B. Roberti, G. H. Heebels, J. C. M. Hendricx, A. H. A. M. de Greef, and O. L. Wolthuis, "Preliminary investigations of the effects of low-level microwave radiation on spontaneous motor activity in rats," in *Biological Effects of Nonionizing Radiation*, P. E. Tyler, Ed., *Ann. N.Y. Acad. Sci.*, vol. 247, pp. 417–423, 1975.

[25] B. Servantie, G. Bertharion, R. July, A. M. Servantie, J. Etienne, P. Dreyfus, and P. Escoubet, "Pharmacologic effects of a pulsed microwave field," in *Biologic Effects and Health Hazards of Microwave Radiation.*, P. Czerski et al., Eds., Warsaw: Polish Med. Publ., pp. 36–45, 1974.

[26] B. Servantie, A. M. Servantie, and J. Etienne, "Synchronization of cortical neurons by a pulsed microwave field as evidenced by spectral analysis of EEG from the white rat," in *Biological Effects of Nonionizing Radiation*, P. E. Tyler, Ed., *Ann. N.Y. Acad. Sci.*, vol. 247, p. 82, 1975.

[27] W. D. Thompson and A. E. Bourgeois, "Effects of Microwave Exposure on Behavior and Related Phenomena," Primate Behavior Lab., Aeromedical Research Lab. Rep., (ARL-TR-65-20; AD 489245) Wright-Patterson AFB, OH, 1965.

[28] M. S. Tolgskaya and Z. V. Gordon, "Changes in the receptor and interoreceptor apparatuses under the influence of UHF," in *The Biological Action of Ultrahigh Frequencies*, A. A. Letavet and Z. V. Gordon, Eds., Moscow: Academy of Medical Science, p. 104, 1960.

[29] M. S. Tolgskaya and Z. V. Gordon, "Comparative morphological characterization of action of microwaves of various ranges," *Tr. Gig. Tr. i Prof., AMN SSSR*, no. 2, p. 80, 1964.

[30] M. S. Tolgskaya and Z. V. Gordon, *Pathological Effects of Radio Waves*, Moscow: Meditsina Press, 1973. Translation, New York: Consultants Bureau, 1973.

[31] M. S. Tolgskaya, Z. V. Gordon, and Ye. A. Lobanova, "Morphological changes in experimental animals under the influence of pulsed and continuous wave SHF-UHF radiation," *Tr. Gig. Tr. i Prof., AMN SSSR*, no. 1, p. 90, 1960.

Perception—Thermal/Auditory

[1] C. K. Chou, R. Galambos, A. W. Guy, and R. H. Lovely, "Cochlear microphonics generated by microwave pulses," *J. Microwave Power*, vol. 10, p. 361, 1975.

[2] H. F. Cook, "The pain threshold for microwave and infra-red radiations," *J. Physiol.*, vol. 118, p. 1, 1952.

[3] K. R. Foster and E. E. Finch, "Microwave hearing; evidence for thermoacoustical auditory stimulation by pulsed microwaves," *Science*, vol. 185, p. 256, 1974.

[4] A. H. Frey, "Auditory system response to rf energy," *Aerosp. Med.*, vol. 32, p. 1140, 1961.

[5] A. H. Frey, "Human auditory system response to modulated electromagnetic energy," *J. Appl. Physiol.*, vol. 17, p. 689, 1962.

[6] A. W. Guy, C. K. Chou, J. C. Lin, and D. Christensen, "Microwave induced acoustic effects in mammalian auditory systems and physical materials," *Ann. N.Y. Acad. Sci.*, vol. 247, pp. 194–215, 1975.

[7] E. Hendler, "Cutaneous receptor response to microwave irra-

diation," in *Thermal Problems in Aerospace Medicine*, J. D. Hardy, Ed., Surrey: Unwin, Ltd., pp. 149–161, 1968.

[8] E. Hendler, J. D. Hardy, and D. Murgatroyd, "Skin heating and temperature sensation produced by infra-red and microwave irradiation," in *Temperature Measurement and Control in Science and Industry*, Part 3, Biology and Medicine, J. D. Hardy, Ed., New York: Reinhold, pp. 221–230, 1963.

[9] J. C. Lin, "On microwave-induced hearing sensation," *IEEE Trans. Microwave Theory Tech.*, vol. MTT-25, p. 605, 1977.

Neuroendocrine

[1] S. Baranski, K. Ostrowski, and W. Stodolnik-Baranska, "Functional and morphological studies of the thyroid gland in animals exposed to microwave irradiation," *Acta Physiol. Pol.*, vol. 23, p. 1029, 1972.

[2] R. Guillet, W. G. Lotz, and S. M. Michaelson, "Time-course of adrenal response in microwave-exposed rats," in *Proc. 1975 Annual Meeting of USNC/URSI*, University of Colorado, Boulder, CO, Washington, DC: National Academy of Sciences, p. 316, 1975.

[3] J. Lenko, A. Dolatowski, L. Gruszecki, S. Klajman, and L. Januszkiewicz, "Effects of 10-cm radar waves on the level of 17-ketosteroids and 17-hydoxycorticosteroids in the urine of rabbits," *Przegl. Lek.*, vol. 22, p. 296, 1966.

[4] W. G. Lotz and S. M. Michaelson, "Temperature and corticosterone relationship in microwave exposed rats," *J. Appl. Physiol: Respirat. Environ. Exercise Physiol.*, vol. 44, pp. 438–445, 1978.

[5] W. S. Lotz and S. M. Michaelson, "Effects of hypophysectomy and dexamethasone on rat adrenal response to microwaves," *J. Appl. Physiol: Respirat. Environ. Exercise Physiol.*, vol. 47, pp. 1284–1288, 1979.

[6] S. T. Lu, W. G. Lotz, and S. M. Michaelson, "Advances in microwave-induced neuroendocrine effects: The concept of stress," *Proc. IEEE*, vol. 68, pp. 73–77, 1980.

[7] R. L. Magin, S. T. Lu, and S. M. Michaelson, "Stimulation of dog thyroid by local application of high intensity microwaves," *Amer. J. Physiol.*, vol. 233, pp. E363–368, 1977.

[8] R. L. Magin, S. T. Lu, and S. M. Michaelson, "Microwave heating effect on the dog thyroid," *IEEE Trans. Bio-Med. Eng.*, vol. 24, pp. 522–529, 1977.

[9] H. Mikolajczyk, "Microwave irradiation and endocrine functions," in *Biologic Effects and Health Hazards of Microwave Radiation*, P. Czerski et al., Eds., Warsaw: Polish Medical Publishers, p. 46, 1974.

[10] H. Mikolajczyk, "Microwave-induced shifts of gonadotrophic activity in anterior pituitary gland of rats," in *Biological Effects of Electromagnetic Waves*, vol. I, C. C. Johnson and M. L. Shore, Eds., United States Dept. Health, Education, and Welfare, publication (FDA) 77-8010, Rockville, MD, pp. 377–383, 1977.

[11] W. C. Milroy and S. M. Michaelson, "Thyroid pathophysiology of microwave radiation," *Aerosp. Med.*, vol. 43, p. 1126, 1972.

[12] L. N. Parker, "Thyroid suppression and adrenomedullary activation by low-intensity microwave radiation," *Amer. J. Physiol.*, vol. 224, p. 1388, 1973.

[13] M. I. Smirnova and M. N. Sadchikova, "Determination of the functional activity of the thyroid gland by means of radioactive iodine in workers with UHF generators," in *The Biological Action of Ultrahigh Frequencies*, A. A. Letavet and Z. V. Gordon, Eds., Moscow, p. 50, 1960.

Behavior

[1] E. R. Adair and B. W. Adams, "Microwaves induce peripheral vasodilation in squirrel monkey," *Science*, vol. 207, pp. 1381–1383, 1980.

[2] J. A. D'Andrea, O. P. Gandhi, and J. L. Lords, "Behavioral and thermal effects of microwave radiation at resonant and nonresonant wave lengths," *Radio Sci.*, vol. 12, no. 6S, pp. 251–256, 1977.

[3] J. de Lorge, *Operant Behavior and Colonic Temperature of Squirrel Monkeys (Saimiri sciureus) During Microwave Irradiation*, NAMRL-1236, Naval Aerospace Medical Res. Lab., Pensacola, FL 32508, 1977.

[4] J. de Lorge, "Disruption of behavior in mammals of three different sizes exposed to microwaves: Extrapolation to large mammals," in *Proc. 1978 Symposium on Electromagnetic Fields in Biological Systems*, S. S. Stuchly, Ed., Ottawa, Canada, p. 215–228, 1978.

[5] J. de Lorge, "Operant behavior and rectal temperature of squirrel monkey during 2.45 GHz microwave irradiation," *Radio Sci.*, vol. 14, pp. 217–225, 1979.

[6] M. Gage, "Behavior in rats after exposure to various power densities of 2450 MHz microwaves," *Neurobehavioral Toxicology*, vol. 1, 1979.

[7] W. D. Galloway, "Microwave dose-response relationships on two behavioral tasks," in *Biological Effects of Non-Ionizing Radiation*, P. E. Tyler, Ed., *Ann. N.Y. Acad. Sci.*, vol. 247, p. 410, 1975.

[8] W. D. Galloway and M. Waxler, "Interaction between microwave and neuroactive compounds," in *Symp. on Biol. Eff. and Measurement of Radio Frequency/Microwaves*, D. Hazzard, Ed., United States Dept. Health, Education, and Welfare publication (FDA) 77-8026, Rockville, MD, p. 66, 1977.

[9] E. L. Hunt, N. W. King, and R. D. Phillips, "Behavioral effects of pulsed microwave radiation," in *Biological Effects of Nonionizing Radiation*, P. E. Tyler, Ed., *Ann. N.Y. Acad. Sci.*, vol. 247, pp. 440–453, 1975.

[10] D. R. Justesen and N. W. King, "Behavioral effects of low level microwave irradiation in the closed space situation," in *Biological Effects and Health Implications of Microwave Radiation*, S. F. Cleary, Ed., Symposium Proceedings, United States Dept. Health, Education, and Welfare, PHS, BRH/DBE 70-2, p. 154, 1970.

[11] J. C. Lin, A. W. Guy, and L. T. Caldwell, "Thermographic and behavioral studies of rats in the near field of 918-MHz radiations," *IEEE Trans. Microwave Theory Tech.*, vol. MTT-25, pp. 833–836, 1977.

[12] J. N. Sanza and J. de Lorge, "Fixed interval behavior of rats exposed to microwaves at low power densities," *Radio Sci.*, vol. 12, no. 6S, pp. 273–277, 1977.

[13] S. Stern, "Behavioral effects of microwaves," *Neurobehavioral Toxicology*, vol. 2, pp. 49–58, 1980.

[14] S. Stern, L. Margolin, B. Weiss, S. T. Lu, and S. M. Michaelson, "Microwaves: effect on thermoregulatory behavior in rats," *Science*, vol. 206, pp. 1198–1201, 1979.

[15] J. R. Thomas, E. D. Finch, D. W. Fulk, and L. S. Burch, "Effects of low level microwave radiation on behavioral baselines," in *Biologic Effects of Nonionizing Radiation*, P. E. Tyler, Ed., *Ann. N.Y. Acad. Sci.*, vol. 247, p. 425, 1975.

[16] J. R. Thomas, L. S. Burch, and S. S. Yeandle, "Microwave radiation and chlordiazepoxide. Syncryistic effects on fixed-internal behavior," *Science*, vol. 203, pp. 1357–1358, March 30, 1979.

[17] J. R. Thomas and G. Maitland, "Combined effects on behavior of low-level microwave radiation and dextroamphetamine," in *Abstracts of Scientific Papers, URSI 1977 Internat. Symp. Biol. Eff. Electromagnetic Waves*, Airlie, p. 121, 1977.

[18] W. D. Thompson and A. E. Bourgeois, *Effects of Microwave Exposure on Behavior and Related Phenomena*, Primate Behavior Lab., Aeromedical Res. Lab. Rep., (ARL-TR-65-20; AD 489245) Wright-Patterson AFB, OH, 1965.

Cardiovascular

[1] T. Cooper, M. Jellinek, T. Pinakatt, and A. W. Richardson, "The effects of pyridoxine and pyridoxal on the circulatory responses of rats to microwave radiation," *Experientia*, vol. 21, p. 28, 1965.

[2] T. Cooper, T. Pinakatt, M. Jellinek, and A. W. Richardson, "Effects of adrenalectomy, vagotomy and ganglionic blockade on the circulatory response to microwave hyperthermia," *Aerosp. Med.*, vol. 33, p. 794, 1962.

[3] T. Cooper, T. Pinakatt, and A. W. Richardson, "Effect of microwave induced hyperthermia on the cardiac output of the rat," *The Physiologist*, vol. 4, p. 21, 1961.

[4] R. D. Phillips, E. L. Hunt, R. D. Castro, and N. W. King, "Thermoregulatory, metabolic and cardiovascular response of rats to microwaves," *J. Appl. Physiol.*, vol. 38, pp. 630–635, 1975.

[5] T. Pinakatt, T. Cooper, and A. W. Richardson, "Effect of ouabain on the circulatory response to microwave hyperthermia in the rat," *Aerosp. Med.*, vol. 34, p. 497, 1963.

[6] T. Pinakatt, A. W. Richardson, and T. Cooper, "The effect of digitoxin on the circulatory response of rats to microwave irradiation," *Arch. Int. Pharmacodyn. Ther.*, vol. 156, p. 151, 1965.

Hematopoiesis and Hematology

[1] S. Baranski, "Effect of chronic microwave irradiation on the blood forming system of guinea pigs and rabbits," *Aerosp. Med.*, vol. 42, pp. 1196–1199, 1971.

[2] S. Baranski, "Effect of microwaves on the reactions of the white blood cell system," *Acta Physiol. Pol.*, vol. 23, p. 685, 1972.

[3] P. Czerski, "Microwave effects on the blood-forming system with particular reference to the lymphocyte," in *Biologic Effects of Nonionizing Radiation*, P. E. Tyler, Ed., *Ann. N.Y. Acad. Sci.*, vol. 247, pp. 232–242, 1975.

[4] P. Czerski, E. Paprocka-Slonka, M. Siekierzynski, and A. Stolarska, "Influence of microwave radiation on the hematopoietic system," in *Biologic Effects and Health Hazards of Microwave Radiation*, P. Czerski et al., Eds., Warsaw: Polish Medical Publishers, pp. 67–74, 1974.

[5] P. E. Czerski, E. Paprocka-Slonka, and A. Stolarska, "Microwave irradiation and the circadian rhythm of bone marrow cell mitosis," *J. Microwave Power*, vol. 9, pp. 31–37, 1974.

[6] A. S. Hyde and J. J. Friedman, "Some effects of acute and chronic microwave irradiation of mice," in *Thermal Problems in Aerospace Medicine*, J. D. Hardy, Ed., Surrey: Unwin, Ltd., pp. 163–175, 1968.

[7] A. I. Ivanov, "Changes of phagocytic activity and mobility of neutrophils under the influence of microwave fields," in *Summaries of Reports, Questions of the Biological Effect of a SHF-UHF Electromagnetic Field*, Leningrad: Kirov Order of Lenin Military Medical Academy, p. 24, 1962.

[8] I. A. Kitsovskaya, "The effect of centimeter waves of different intensities on the blood and hemopoietic organs of white rats," *Gig. Tr. Prof. Zabol.*, vol. 8, p. 14, 1964.

[9] J. C. Lin, J. C. Nelson, and M. E. Ekstrom, "Effects of repeated exposure to 148 MHz radiowaves on growth and hematology of mice," *Radio Sci.*, vol. 14, pp. 173–179, 1979.

[10] S. M. Michaelson, R. A. E. Thomson, M. Y. E. Tamami, H. S. Seth, and J. W. Howland, "Hematologic effects of microwave exposure," *Aerosp. Med.*, vol. 35, pp. 824–828, 1964.

[11] J. M. P. Rotkovska and D. R. A. Vacek, "Effect of high-frequency electromagnetic fields upon haemopoietic stem cells in mice," *Folia Biol. (Praha)*, vol. 18, p. 292, 1972.

Immunology

[1] R. P. Liburdy, "Radiofrequency radiation alters the immune system: modulation of T- and B-lymphocyte levels and cell-mediated immunocompetence by hyperthermic radiation," *Radiat. Res.*, vol. 77, pp. 34–36, 1979.

[2] R. J. Smialowicz, "The effect of microwaves (2450 MHz) on lymphocyte blast transformation *in vitro*," in *Biological Effects of Electromagnetic Waves*, C. C. Johnson and M. L. Shore, Eds., United States Dept. Health, Education, and Welfare, publication (FDA) 77-8010, Rockville, MD, pp. 472–483, 1976.

[3] R. W. Smialowicz, J. B. Kinn, and J. A. Elder, "Prenatal exposure of rats to 2450 MHz (CW) microwave radiation: Effects on lymphocytes," *Radio Sci.*, vol. 14, no. 6S, p. 147, 1979.

[4] R. J. Smialowicz, M. M. Riddle, P. L. Brugnolotti, J. M. Sperazza, and J. B. Kinn, "Evaluation of lymphocyte function in mice exposed to 2450 MHz (CW) microwaves," *Proc. 1978 Symposium on Electromagnetic Fields in Biological Systems*, S. S. Stuchly, Ed., Ottawa, Canada, p. 122–152, 1979.

[5] R. J. Smialowicz, C. M. Weil, J. B. Kinn, and J. A. Elder, "Exposure of rats to 425 MHz (CW) microwave radiation: effects on lymphocytes," *J. Microwave Power*, 1980.

[6] W. Stodolnik-Baranska, "The effects of microwaves on human lymphocyte cultures," in *Biologic Effects and Health Hazards of Microwave Radiation*, P. Czerski et al., Eds., Warsaw: Polish Medical Publishers, pp. 189–195, 1974.

[7] W. Wiktor-Jedrzejczakz, A. Ahmed, P. Czerski, W. M. Leach, and K. W. Sell, "Immunologic response of mice to 2450 MHz microwave radiation: overview of immunology and empirical studies of lymphoid spleen cells," *Radio Sci.*, vol. 12, no. 6S, pp. 209–219, 1977.

[8] W. Wiktor-Jedrzejcazk, A. Ahmed, K. W. Sell, P. Czerski, and W. M. Leach, "Microwaves induce an increase in the frequency of complement receptor-bearing lymphoid spleen cells in mice," *J. Immunol.*, vol. 118, pp. 1499–1502, 1977.

Pathology

[1] L. Minecki and R. Bilski, "Histopathological changes in internal organs of mice exposed to the action of microwaves," *Medycyna Pracy*, Poland, vol. 12, p. 337, 1961.

[2] L. Miro, R. Loubiere, and A. Pfister, "Studies of visceral lesions observed in mice and rats exposed to UHF waves. A particular study of the effects of these waves on the reproduction of these animals," *Rev. Med. Aeronaut.*, Paris, vol. 4, p. 37, 1965.

Cataract and Other Ocular Effects

[1] B. Appleton, *Results of Clinical Surveys for Microwave Ocular Effects*, United States Dept. Health, Education, and Welfare publication (FDA) 73-8031, BRH/DBE 73-3, 1972.

[2] B. Appleton, "Experimental microwave ocular effects," in *Biologic Effects and Health Hazards of Microwave Radiation*, P. Czerski et al., Eds., Warsaw: Polish Medical Publishers, pp. 186–188, 1974.

[3] B. Appleton, "Microwave cataracts," *J. Amer. Med. Ass.*, vol. 229, p. 407, 1974.

[4] B. Appleton, S. Hirsch, and P. V. K. Brown, "Investigation of single-exposure microwave ocular effects at 3000 MHz," in *Biologic Effects of Nonionizing Radiation*, P. E. Tyler, Ed., *Ann. N.Y. Acad. Sci.*, vol. 247, pp. 125–134, 1975.

[5] B. Appleton, S. Hirsch, R. O. Kinion, M. Soles, G. C. McCrossen, and R. M. Neidlinger, "Microwave lens effects in humans," *Arch. Ophthalmol.*, vol. 93, p. 257, 1975.

[6] B. Appleton, and G. C. McCrossen, "Microwave lens effects in humans," *Arch. Ophthal.*, vol. 88, p. 259, 1972.

[7] E. Aurell and B. Tengroth, "Lenticular and retinal changes secondary to microwave exposure," *Acta Opthal.*, vol. 51, p. 764, 1973.

[8] R. L. Carpenter, D. K. Biddle, and C. A. van Ummersen, "Biological effects of microwave radiation with particular reference to the eye," *Proc. Third Int. Conf. Med. Electronics*, London, England, vol. 3, p. 401, 1960.

[9] R. L. Carpenter and C. A. van Ummersen, "The action of microwave radiation on the eye," *J. Microwave Power*, vol. 3, p. 3, 1968.

[10] A. W. Guy, J. C. Lin, P. O. Kramar, and A. F. Emery, "Measurement of absorbed power patterns in the head and eyes of rabbits exposed to typical microwave sources," in *Proc. 1974 Conf. on Precision Electromagnetic Measurements*, London, England, p. 255, 1974.

[11] A. W. Guy, J. C. Lin, P. O. Kramar, and A. F. Emery, "Quantitation of microwave radiation effects on the eyes of rabbits at 2450 MHz and 918 MHz," Scientific Rep. No. 2, Univ. Washington, Seattle, WA, Jan. 1974.

[12] A. W. Guy, J. C. Lin, P. O. Kramar, and A. F. Emery, "Effect of 2450 MHz radiation on the rabbit eye," *IEEE Trans. Microwave Theory Tech.*, vol. MTT-23, p. 492, 1975.

[13] S. E. Hirsch, B. Appleton, B. S. Fine, and P. V. K. Brown, "Effects of repeated microwave irradiations to the albino rabbit eye," *Invest. Ophthalmol. Vis. Sci.*, vol. 16, p. 315, 1977.

[14] P. O. Kramar, A. F. Emery, A. W. Guy, and J. C. Lin, "The ocular effects of microwaves on hypothermic rabbits: a study of microwave cataractogenic mechanism," *Ann. N.Y. Acad. Sci.*, vol. 247, pp. 155–165, 1975.

[15] P. O. Kramar, A. W. Guy, A. F. Emery, J. C. Lin, and C. A. Harris, "Quantitation of microwave radiation effects on the eyes of rabbits and primates at 2450 MHz and 918 MHz," Univ. Washington, Bioelectromagnetics Res. Lab. Scientific Rep. No. 6, Seattle, WA, 1976.

[16] B. Tengroth and E. Aurell, "Retinal changes in microwave workers," in *Biological Effects and Health Hazards of Microwave Radiation*, P. Czerski, et al., Eds., Warsaw: Polish Medical Publishers, p. 302, 1974.

[17] J. J. Weiter, E. D. Finch, W. Schultz, and V. Frattali, "Ascorbic acid changes in cultured rabbit lenses after microwave irradiation," in *Biologic Effects of Nonionizing Radiation*, P. E. Tyler, Ed., *Ann. N.Y. Acad. Sci.*, vol. 247, pp. 175–181, 1975.

[18] M. M. Zaret, S. Cleary, B. Pasternack, M. Eisenbud, and H. Schmidt, "A Study of Lenticular Imperfections in the Eyes of a Sample of Microwave Workers and a Control Population," New York Univ., New York, NY, Fina. Rep. RADC-TDR-63-125, ASTIA Doc. AD 413-294, 1963.

[19] M. M. Zaret, I. T. Kaplan, and A. M. Kay, "Clinical microwave cataracts," in *Biological Effects and Health Implications of Microwave Radiation*, S. Cleary, Ed., United States Dept. Health, Education, and Welfare, PHS, BRH/DBE 70-2, p. 82, 1970.

[20] S. Zydecki, "Assessment of lens translucency in juveniles, microwave workers and age-matched groups," in *Biologic Effects and Health Hazards of Microwave Radiation*, P. Czerski et al., Eds., Warsaw: Polish Medical Publishers, p. 306, 1974.

[21] E. S. Ferri and G. J. Hagan, "Chronic low-level exposure of rabbits to microwaves," *Biological Effects of Electromagnetic Waves*, vol. 1, United States Dept. of Health, Education, and Welfare, publication FDA 77-8010, pp. 129–142, Dec. 1976.

[22] S. Cleary, "Microwave cataractogenesis," *Proc. IEEE*, vol. 68, pp. 49–55, Jan. 1980.

Epidemiology

[1] R. M. Albrecht and E. Landau, "Microwave radiation: an epidemiologic assessment," *Rev. Environ. Health*, vol. 3, pp. 44–58, 1978.

[2] C. I. Barron and A. A. Baraff, "Medical considerations of exposure to microwaves (radar)," *J. Amer. Med. Ass.*, vol. 168, p. 1194, 1958.

[3] C. I. Barron, A. A. Love, and A. A. Baraff, "Physical evaluation of personnel exposed to microwave emanations," *J. Aviat. Med.*, vol. 26, p. 442, 1955.

[4] B. Cohen, A. M. Lilienfeld, S. Kramer, and L. C. Hyman, "Parental factors in Down's syndrome. Results of the second Baltimore case-control study," in *Population Cytogenetics: Studies in Humans*, E. B. Hook and I. H. Porter, Eds., New York: Academic Press, p. 301, 1977.

[5] P. Czerski, M. Siekierzynski, and A. Gidynski, "Health surveillance of personnel occupationally exposed to microwaves. I. Theoretical considerations and practical aspects," *Aerosp. Med.*, vol. 45, p. 1137, 1974.

[6] P. Czerski and M. Piotrowski, "Proposals for specification of allowable levels of microwave radiation," *Medcyna Lotnicza*, Polish, no. 39, pp. 127–139, 1972.

[7] P. Czerski and M. Siekierzynski, "Analysis of occupational exposure to microwave radiation," in *Fundamental and Applied Aspects of Non-Ionizing Radiations*, S. M. Michaelson, M. W. Miller, R. Magin, and E. L. Carstensen, Eds., New York: Plenum Press, p. 367, 1975.

[8] L. Daily, "A clinical study of the results of exposure of laboratory personnel to radar and high frequency radio," *U.S. Nav. Med. Bull.*, vol. 41, p. 1052, 1943.

[9] C. H. Dodge and Z. R. Glaser, "Trends in nonionizing radiation bioeffects research and related occupational health aspects," *J. Microwave Power*, vol. 12, p. 319, 1977.

[10] E. A. Drogichina, M. N. Sadchikova, M. N. Snegova, G. V. Konchalovskaya, and K. T. Glotova, "Autonomic and cardiovascular disorders during chronic exposure to super-high frequency electromagnetic fields," *Gig. Tr. Prof. Zabol.*, USSR, vol. 10, p. 13, 1966.

[11] Z. V. Gordon, *Biologic Effect of Microwaves in Occupational Hygiene*, Leningrad: Izd. Med., (TT 70-50087, NASA TT F-633, 1970) 165 pp., 1966.

[12] Z. V. Gordon, "Occupational health aspects of radio-frequency electromagnetic radiation," in *Ergonomics and Physical Environmental Factors*, Geneva: International Labour Office, p. 159, 1970.

[13] A. K. Gus'kova and Ye. H. Kochanova, "Some aspects of etiological diagnostics of occupational diseases as related to the effects of microwave radiation," *Gig. Truda I. Prof. Zabol.*, Moscow, vol. 3:14, 1975; JPRS L/6135, June 1976.

[14] A. M. Lilienfeld, J. Tonascia, S. Tonascia, C. H. Libauer, G. M. Canthen, J. A. Markowitz, and S. Weida, "Foreign Service and Other Employees from Selected Eastern European Posts Final Report, July 31, 1978, contract no. 6025-619073, Dept. of Epidemiol., Johns Hopkins Univ., Baltimore, MD, 1978. *NTIS PB-288*, 1963.

[15] Yu. A. Osipov, *Occupational Hygiene and the Effect of Radiofrequency Electromagnetic Fields on Workers*, Leningrad: Izd. Meditsina Press, pp. 78–103, 1965.

[16] C. D. Robinette and C. Silverman, "Cause of death following occupational exposure to microwave radiation (Radar) 1950–1974," in *Symposium on Biological Effects and Measurement of Radio Frequency/Microwaves*, United States Dept. Health, Education, and Welfare, publication (FDA) 77-8026, pp. 337–344, 1977.

[17] M. N. Sadchikova, "Clinical manifestations of reactions to microwave irradiation in various occupational groups," in *Biologic Effects and Health Hazards of Microwave Radiation*, P. Czerski et al., Eds., Warsaw: Polish Medical Publishers, p. 261, 1974.

[18] M. N. Sadchikova and A. A. Orlova, "Clinical picture of the chronic effects of electromagnetic microwaves," *Ind. Hyg. Occupat. Dis.*, USSR, vol. 2, pp. 16–22, 1958.

[19] M. Siekierzynski, "A study of the health status of microwave workers," in *Biologic Effects and Health Hazards of Microwave Radiation*, P. Czerski et al., Eds., Warsaw: Polish Medical Publishers, p. 273, 1974.

[20] M. Siekierzynski, P. Czerski, H. Milczarek, A. Gidynski, C. Czarnecki, D. Dziuk, and W. Jedrzescak, "Health surveillance of personnel occupationally exposed to microwaves. II. Functional disturbances," *Aerosp. Med.*, vol. 45, p. 1143, 1974.

[21] A. T. Sigler, A. M. Lilienfeld, B. H. Cohen, and J. E. Westlake, "Radiation exposure in parents of children with mongolism," *Bull. Johns Hopkins Hosp.*, vol. 117, p. 374, 1965.

[22] C. Silverman, "Epidemologic studies of microwave effects," *Proc. IEEE*, vol. 68, pp. 78–84, 1980.

[23] C. Silverman, "Epidemiologic approach to the study of microwave effects," *Bull. N.Y. Acad. Med.*, vol. 55, pp. 1166–1181, Dec. 1979.

Heating Characteristics of Laboratory Animals Exposed to Ten-Centimeter Microwaves

T. S. ELY, D. E. GOLDMAN, AND J. Z. HEARON

Summary—**Experimental animals were exposed to a 10-cm microwave field in order to study the heating and cooling characteristics of the entire animal and localized sensitive structures. The flanks of rats, rabbits and dogs were exposed and the whole body heating was observed. After heating, the cooling curve was determined. Similarly, restricted area fields were used to study heating and cooling of eyes and testes. Data on the heating and cooling rates were used to determine the most sensitive structures. The experimental findings, together with the values for some pertinent related factors from the literature, provide the basis for an estimation of the possible risks to man from exposure to microwaves.**

This paper originally appeared as a Naval Medical Research Institute Report (NM 001 056.13.02, dated March 21, 1957). It was prepared in response to a practical need for estimating hazards to personnel from exposure to microwave fields. Since that time the use of high powered microwave devices has continued to increase and, although there have been a number of studies directed to special aspects of the biological effects of microwaves, there has been little work published which deals directly with the over-all hazard problem. The authors feel that the report is somewhat "dated," but the editors hope that reprinting it will make some useful material available to a wider circle of readers than could have been reached with the limited Navy Report distribution.

Introduction

THE INCREASED use of microwave diathermy and of high field strength radar installations make it desirable to investigate some of the physiological effects of this form of energy. Although considerable work has been done on this subject, much of it does not lend itself to quantitative evaluation. The primary effort of this study has been directed towards the delineation of a hazard, and therefore certain facets of the problem have been given emphasis, while other areas, however interesting, have been bypassed.

The structures most likely to be damaged appear to be the body as a whole, the lens of the eye, and the testis. They have already been subjected to some study, and have been mentioned by Clark [1], Schwan and Piersol [4] and others. The lens and the testis owe their special sensitivity to their physical location relative to the body surface, their poor ability to dissipate heat, and in the case of the testis, high sensitivity to temperature increase. The body as a whole can tolerate only a moderate temperature increase, and has a limited ability to lose heat. It is less able to lose heat by convection, radiation, and evaporation to the surroundings, for instance, than a small, well vascularized area of that body would by conduction and bloodstream convection to the remainder of the same body. Although the lens tolerates relatively high temperatures, the eye is completely avascular in the lens and vitreous, therefore leaving conduction through the humors essentially the only route of heat loss. The testis is extremely sensitive to elevated temperature. Other localized areas of the body are relatively less sensitive to 10-cm microwaves by reason of being less temperature sensitive, being deep enough to avoid most of the energy, having a better vascular or conductive pathway for the loss of heat, or a combination of these factors. The undesirable effect of excess temperature in each of these three sensitive structures, the whole body, the eye and the testis, is, respectively, heat disablement, lenticular opacity and tubular injury.

As the investigation progressed, it became apparent that in all but rather restricted circumstances, the testis was the limiting factor with regard to hazard. Therefore the greatest effort was made on the study of this organ.

The 10-cm wavelength is important because of its biologically "in-between" position in the electromagnetic spectrum of radiant energy. Substantially longer wavelengths penetrate deeper, and are absorbed more diffusely by animal tissues, and are therefore less likely to produce small regions of differential heating. Energy of significantly shorter wavelength has very little penetration, and approaches the infrared in behavior. Radiant energy in the 10-cm region of the spectrum can penetrate tissues sufficiently to avoid concentrating heat production in the region of skin receptors, and may produce undesirable temperature elevations below the surface.

For the purposes of this study, we have made the assumption that the entire effect of the microwave radiation on biological material is the production of heat. While not excluding the possibility that some "athermal" process may exist, we have had no reason, in any of the work, to believe that anything but a thermal effect is involved. This view is supported by other workers in the field, *e.g.*, Clark [1], and Schwan and

Manuscript received June 9, 1964. The opinions or assertions contained herein are the private ones of the authors and are not to be construed as official, or reflecting the views of the Navy Department or the Naval service at large.

T. S. Ely is with the University of Rochester, Rochester, N. Y. He was formerly with the Naval Medical Research Institute, Bethesda, Md.

D. E. Goldman is with the Naval Medical Research Institute, Bethesda, Md.

J. Z. Hearon is with the National Institutes of Health, Bethesda, Md.

Reprinted from *IEEE Trans. Bio-Med. Eng.*, vol. BME-11, pp. 123–137, Oct. 1964.

Piersol [2]. Williams, *et al.*, point out that the photon energy in this band of microwaves is only about 1/3,000,000 that required to produce ionization in tissue [3].

It has been assumed that for a given experimental setup, the heating rate is proportional to the field intensity, and independent of time or temperature. Heating rate change would depend on a change in dielectric characteristics which would be minor. In the case of the eye and the testis, the further assumption is made that cooling rate is proportional to the temperature rise of the part, as opposed to other limited areas of the body, (*e.g.*, the thigh), where Herrick and Krusen have found that cooling is markedly influenced by a thermally induced increase in blood flow [5]. Since the eye has no vascularity in the region of the greatest heating, alteration of cooling rate by blood flow change is remote and secondary. The testis is limited in blood cooling capability. In addition, for testicular temperatures below body core temperature, an increase in blood flow would heat rather than cool the organ. Experimental support for these two assumptions is illustrated by Figs. 9 through 13, and is explained in the Appendix.

The whole body, which undergoes a considerably more complex physiological adjustment with increased temperature, follows a much more involved cooling pattern, and in addition experiences a change in metabolic heat contribution.

The cooling time constant (τ) was estimated usually by replotting the cooling curve on log-linear paper. It was checked in some cases by plotting the heating curve in a similar fashion, and in some cases by timing the heating interval required to produce a given temperature rise. The Appendix develops the relationship

$$\ln\left(1 - \frac{I'}{I}\right) = \frac{-t}{\tau}$$

which is used in estimating τ by this last method. Since this more closely represents the practical situation, it was used as a check.

Much of the quantitative data of the study involved the determination of the field required to maintain a prescribed temperature elevation of the structure under consideration under steady-state conditions (I'). Field intensity could have been controlled in several ways. A method which proved very convenient, mainly because it lent itself readily to automatic control and accurate measurement, was that of "on and off" cycling in response to the temperature of the subject. Since the temperature variation was very small with respect to total temperature rise (ΔT), the cycle time was much shorter than the time constant of cooling and thus the time average field could be used as the field intensity.

A portion of this study concerns itself with the determination of the microwave power absorbed by the animal to a profile exposure. For the purpose here, it is represented by the percentage of the power in the area of the animal profile determined under free field conditions which is absorbed and appears as heat in the animal. This absorption is determined from the initial slope of the heating curve. The values obtained are not directly comparable to true absorption efficiencies, such as those of Schwan and Li [6] principally because less than the entire animal surface presented to the microwave beam was perpendicular to it, and because of the appreciable effect of beam diffraction where the subject's dimensions are not very much greater than the wavelength used. Hence, the term *absorption "efficiency,"* as used here, is a qualified figure.

Because one of the principal factors in this study was the rate of heat dissipation, the ambient conditions were of direct importance. During the experimentation, dry bulb temperatures were approximately 24°C, radiant environment that of the dry bulb, relative humidity about 50 per cent and air velocity essentially that due to convection only. In the application of the data to practical situations, ambient conditions form a direct and important factor.

In the evaluation of the critical temperature and critical field intensities, *any demonstrable damage* was used as the criterion. It was felt that if this resulted in an error in the estimation of threshold values, the error would be in the safe direction. In the case of the eye, opacity formed the basis of calculating a critical temperature, even though a small opacity may be asymptomatic. For the testis, mild tubular epithelial injury is certainly a reversible process. However, the study of damage rate, recovery rate and their interrelationships is another large field, and was not included in the scope of the study.

Most of the results have been presented in the form of a relationship between effect and field intensity, *i.e.*, power per unit area or milliwatts per square centimeter. This seemed to offer the most direct relationship of cause and effect. Heat gain and loss are both closely related to surface area, consequently this area factor is largely cancelled out. In a practical situation also, field intensity is the quantity which is measured, and biological result the quantity of interest.

Materials and Methods

Fig. 1 is a block diagram of the major equipment used. An electronically regulated power supply provided a transmitter input sufficiently stable to achieve an average microwave output which was constant to within one per cent with time. The microwave generator was a high power pulsed military radar transmitter, having a frequency of 2880 Mc/sec, or 10.4-cm wavelength, which is in the "S" band. The microwave power output was passed through a waveguide into a directional coupler which had a coaxial output attenuated 40 db. This greatly reduced power was then further attenuated by a coaxial attenuator pad, and measured with a Hewlett-

Packard 430-C microwave wattmeter. Total power output of the transmitter could therefore be determined from the wattmeter reading. The directional coupler led into a power divider, which permitted continuously variable division of the power between the transmitting horn and a water cooled load. Field strength could be reduced to any desired degree, or the entire power could be shifted into the water load for calibration purposes. The horn radiated the microwave power into space providing a divergent field with vertical polarization. In the illustration, the subject is shown much closer to the horn than was usually the case.

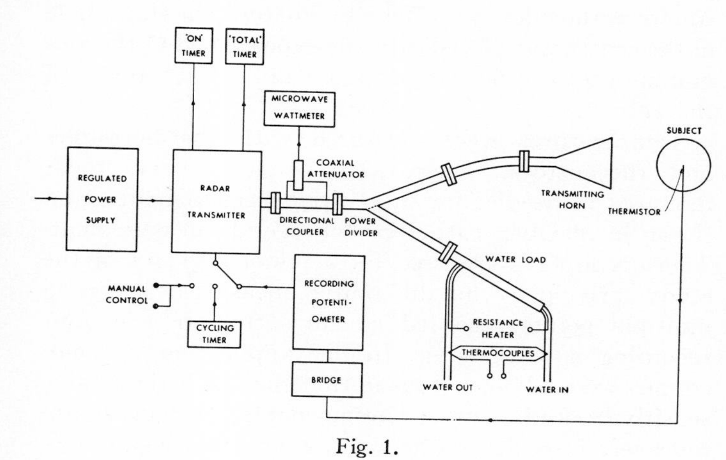

Fig. 1.

Three methods were used for on and off control of the transmitter, manual, time cycling and temperature cycling. The great majority of the work was done with the last named method of control. For some work where it was desired to reduce the field rapidly by an accurately known amount, time cycling by a synchronous variable sector cam timer with a speed of three rpm was used.

Two synchronous timers were used to determine the time average of transmitter "on" time. One determined the total time during a specific exposure interval, and the other, which operated only with the microwave power, indicated the accumulated "on" time. Temperature controlled cycling was relatively slow, so that small starting and stopping errors of the timer were negligible. However, for the time controlled cycling, where very numerous and frequent starts and stops occurred, in order to prevent a cumulative starting and/or stopping error, the individual microwave pulses were counted electronically, and the time average power computed from the total pulses counted.

All of the microwave equipment and the subject were contained in a commercial prefabricated double copper screen shielded room. A microwave absorber lining the room provided a relatively anechoic chamber in order that the field pattern be undistorted by reflections. Measurements of standing wave ratios in the room demonstrated a satisfactorily low reflection of about ± 0.1-db power standing wave ratio.

Field Calibration

The water load was provided with a resistance heater for use on low-frequency power, and thermocouples in the water inlet and outlet in order that the temperature rise of the water passing through could be measured. A determination of the standing wave ratio of the complete water load indicated that it was an efficient absorber of microwave power. The first phase of field calibration was the determination of microwave power output of the transmitter. For this purpose, a constant flow of water was passed through the water load. The water was heated by one of two sources. First, heating was accomplished by diverting all of the microwave power into the water load. Temperature rise was recorded on a recording potentiometer. Microwave power was then turned off, and simultaneously house current supplied to the resistance heater. This was varied by an adjustable transformer until the temperature rise matched that produced by the microwaves, and the corresponding power level was read on a wattmeter in the line. The actual temperature rise of the water and its rate of flow provided a third determination of power. Measurements made by the three methods were in good agreement.

In order to find field strengths, it was necessary either to measure the field directly, or to determine the field distribution for a known total radiated power. The former method is beset with many difficulties, and it proved more practical to use the latter. This was done using a small receiving horn, attenuators, the microwave wattmeter and a recorder. The system, while it did not yield absolute values, did behave as a linear system, and was entirely satisfactory for relative measurements. The receiver, automatically scanning the beam in a plane perpendicular to its axis, recorded families of curves which were then transposed to the form of contour plots of field intensity in planes at several ranges. A sample distribution plot is shown in Fig. 2. The field distributions so obtained were in excellent agreement with those calculated from horn geometry. Absolute field strengths were then calculated from the total microwave power, field distribution, and of course, average "on" time when cycling was used.

The animal exposures can be divided into two groups, those in which a large area of the animal was exposed to the beam, and those in which a relatively restricted area was exposed. The "whole body" exposures constitute the first type, and the eye and testicular exposures the second. *Profile area*, that is, the lateral area of most experimental animals, and the frontal area of a human, probably represents the largest area likely to be exposed in the practical case. Greater area is possible, but not without multiple sources, or unusual reflection. Five rats having a weight in the range of 200–300 grams, 10 rabbits in the range of 3–4 kilograms, and 9 dogs in the range of 8–20 kilograms were used in these large area exposures.

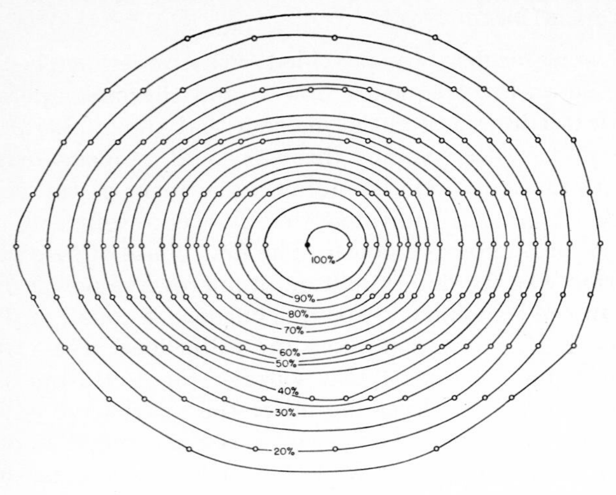

Fig. 2.

For the restricted area, or localized exposures, radiation was limited as far as possible to the area of direct interest, in order not to confuse the determination with any more general body heating than possible. During these limited area exposures, the axis of the animal was oriented parallel to that of the beam, thus reducing the intercepting area considerably, and the appropriate end of the animal passed through a hole in a batt of microwave absorber. This limited diffraction of the beam, and absorption of any irregular field by the remainder of the animal. "On" field strengths were limited to about 100/mw/cm² in most cases, in order to minimize differential heating. Eight rabbits and one dog were used in eye heating experiments. Fifty dogs and three rabbits were used in the testis experiments.

Animal support structures were constructed of low-loss materials of thin dimensions. Plastics, glass, cotton and nylon were the principal materials used. Wood, although having greater loss, was used sparingly in regions of lower field intensity. For the whole body exposures, the rats and rabbits were supported in a tube of plastic insect screen suspended from a polystyrene rod, which was supported on glass tubes. The dogs required a somewhat stouter support consisting of a canvas sling, a polystyrene rod and wood uprights. These uprights were in a region of reduced field intensity, and were oriented to present only the edge to the beam.

General anesthesia would have been useful in the whole body exposures in order to provide immobility. However, it was found that any of several general anesthetic agents tried caused a profound fall in body temperature which badly confused the experiment. The rats and rabbits proved relatively unaggressive during exposure, and were exposed without medication. Dogs, however, if exposed without anesthesia, often became quite active, with a resulting additional metabolic heating which confused or completely obliterated data from microwave heating. It was then found that premedication with 6 mg/kilogram of chlorpromazine 1 hour before exposure reduced the activity of the dogs and permitted the collection of useful data. This drug did not result in temperature depression sufficient to influence the data significantly. Results with the chlorpromazine compared favorably with what useful data were obtained from the unmedicated animals.

For the localized exposures, general anesthesia could be used, since exposure times were, in general, shorter, and body temperature was not directly the quantity of interest. The limited area of exposure and the use of general anesthesia simplified the support problem. After the animal was anesthetized, he was placed parallel to the beam axis on a table of appropriate height, with the proper end directed towards the antenna. A 2-foot square of the microwave absorber material having a hole in the center was placed over the exposed end of the animal in order to shield the support and the remainder of the animal.

Temperatures were measured with thermocouples and thermistors. These were of two types, those mounted in needles for interstitial measurements, and those in flexible catheters for rectal measurement. Thermocouples were used for a minor segment of the study, principally for differential measurements using multiple points recorded concurrently, on a 6-point recording potentiometer. In this application, thermocouples were advantageous in that they all had the same sensitivity and range. Thermocouples, however, are relatively insensitive, and require a reference junction. Also, it was inconvenient to change the range and sensitivity of the associated recording equipment.

Thermistors were used for the bulk of the work, since most of it consisted of single point recording on a continuous line recorder. With thermistors, range and sensitivity could be changed easily and rapidly, and the high output permitted much higher sensitivities and accuracies. A small thermistor bead sealed in a short 2-mm diameter glass rod was used in a catheter for rectal temperature measurement, and a very small thermistor mounted inside the tip of a 24-gauge needle for the eye and testicular determinations.

The thermistor was connected as one arm of a direct current unbalanced bridge, the output of which drove a 3-mv recording potentiometer. Precision variable resistors in the bridge permitted the rapid changing of range and sensitivity. Bridge balance was set at the hot end of the scale, and maximum unbalance at the cold. By this mechanism, the nonlinearity of the thermistor was almost completely compensated by the nonlinearity of the bridge, resulting in a recorder scale which was linear with temperature within very small error.

Thermistors were calibrated against a National Bureau of Standards certified mercury in glass thermometer. The thermistor recording system had a sensitivity of about 0.01°C and an accuracy of about 0.05°C.

Whole body temperature was measured rectally. This is the best single index of the core temperature according to Hardy [7]. Although the average body temperature (which would have been the most desirable quantity to determine) is lower than the rectal temperature by a significant amount over the relatively large temperature increases used, and since it was an increase rather than a decrease where greater body differentials would be expected, it was assumed that a change in rectal temperature reflected a change in average body temperature sufficiently closely.

During the localized eye exposures, temperature was measured with the 24-gauge needle thermistor at the posterior pole of the lens. For this wavelength band, this location has been shown to be that of theoretical and actual maximum temperature development in the eye and the site of opacity production by Salisbury, *et al.* [8], Richardson *et al.* [9], and others. Rabbits were used for all but one of the eye experiments. This species was used for the same reasons of others in the field, mainly, the similarity of size and shape between the rabbit eye and the human eye.

Testicular work was carried out mainly on the dog, since this species was available, and had testes comparable in size and shape to the human organ. Testicular temperature was measured about 1 cm from that surface nearest the microwave source. By experimentation with high field intensities it was found that this was approximately the location of greatest temperature development. Testicular temperatures were found to equalize throughout the organ with relative rapidity, and with the low field intensities used it was found, by multiple point recording, that differential temperature development within one testis was small. It was also found, that with symmetrical orientation of the testes, temperature rise in each was essentially the same. This fortunate fact allowed the simultaneous heating of both testes, temperature measurement in one, and biopsy of the other which was untraumatized by the thermistor needle.

Experimental Procedure

For the bulk of the experimental runs, the following general sequence was used. The animal was weighed and premedicated (in the whole body exposures), and placed in the appropriate support. The thermistor or thermocouples were inserted and, in the case of the whole body exposures, the profile determined. Temperature recording was started. Fig. 3 illustrates the series of events. After a period sufficient to give a good initial temperature value, the microwave power was turned on which caused heating of the particular structure under study.

When the temperature reached a predetermined level, automatic cycling occurred in response to temperture. This was allowed to continue until many cycles had been produced, and a stable steady-state field intensity could be calculated. Usually, then, the control point was raised by an integral number of degrees, causing further rise in temperature to the new control level. Following the determination of steady-state field intensities at one or more temperatures, the microwave power was shut off completely, and cooling was allowed to take place.

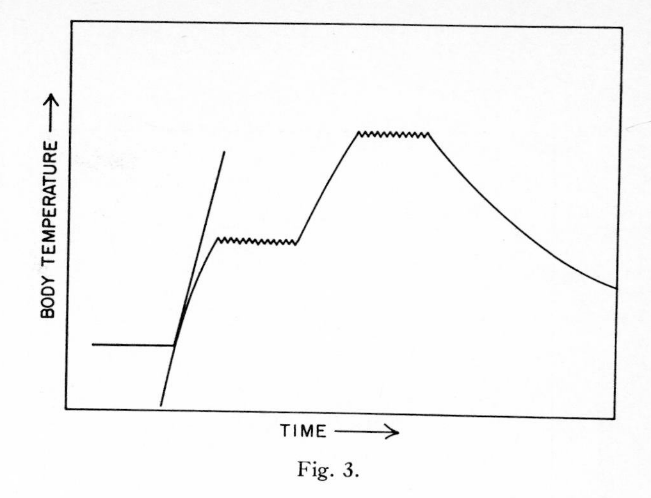

Fig. 3.

Results

Fig. 4 shows the absorption "efficiencies" calculated for several animals of three species, and the mean of the values for each species. In order to obtain a rough evaluation of the effect of fur on the absorption of microwave energy, a small series of dogs was exposed before and after clipping. Average "efficiency" before clipping was 33 per cent and after clipping was 35 per cent, this difference not being significant.

In Figs. 5 through 7 are plotted the steady-state values of field intensity required to maintain a given body temperature for whole body exposures of rats, rabbits and dogs.

Fig. 8 shows approximate composite minimum field intensities for each of the three species.

Most animals of all three species survived a maximum temperature of 42°C, for this relatively short exposure, which was generally less than an hour at maximum temperature. Roughly half survived 43°C maximum, while 44°C proved uniformly fatal. It was our experience that if the animal did not die during or within a very few minutes of the exposure, survival was the rule.

Fig. 9 shows the steady-state field intensities of the eye exposures. One dog eye is represented, the reminder being rabbit eyes.

Cooling time constants (τ) for the eyes were all within the range of 100 to 180 sec. In several cases, τ was also computed from the heating curve, and agreement was good in most cases. Two eye cooling curves are shown in Fig. 10.

Fig. 11 shows steady-state field intensities obtained on a very limited number of rabbit testes. Although the results are entirely consistent with those of the dogs, rabbit work was not continued since the size of the rabbit testis is much smaller than the human. Fig. 12 shows the dog testis steady-state field intensity results. Since 35 dogs were used in this particular testicular series, the graphs are plotted with mean, standard deviation and minimum field for each temperature rather than individual lines. Of course, in the actual temperature plot, the first set of values represents the mean, standard deviation and maximum initial temperature. Cooling

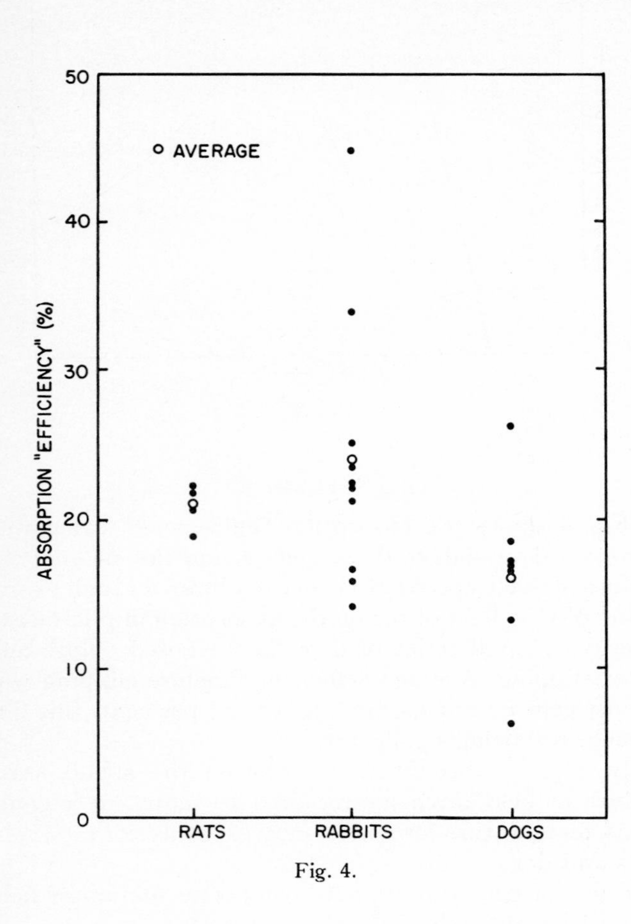

Fig. 4.

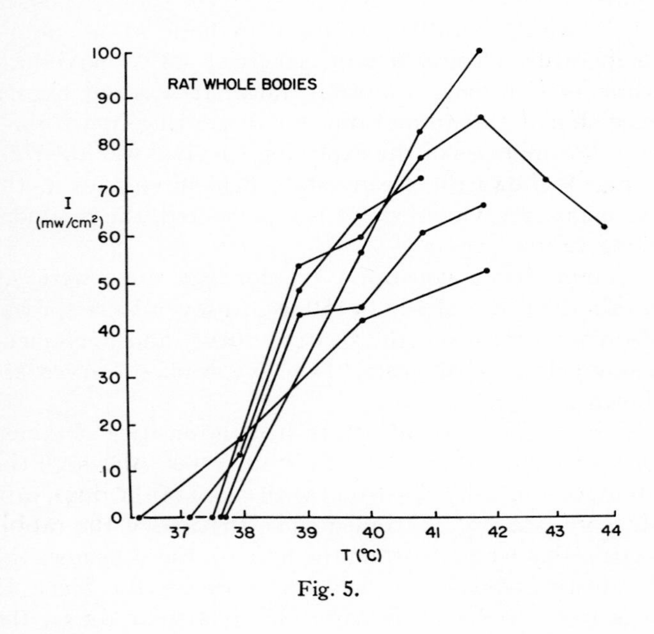

Fig. 5.

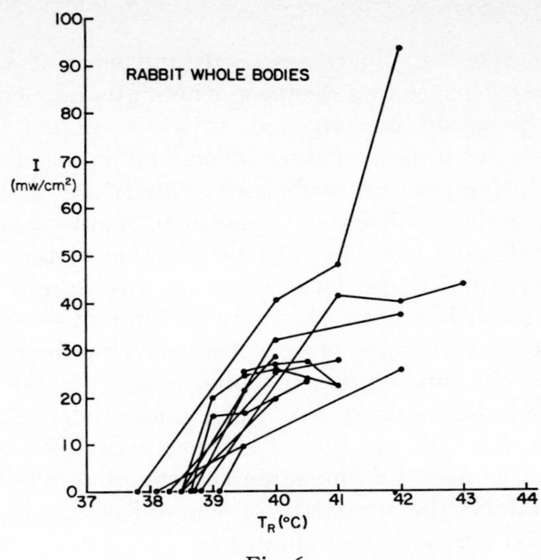

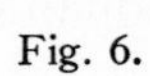

Fig. 6.

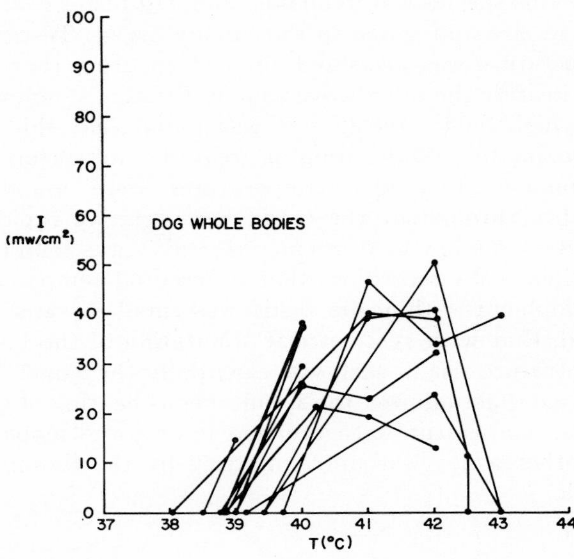

Fig. 7.

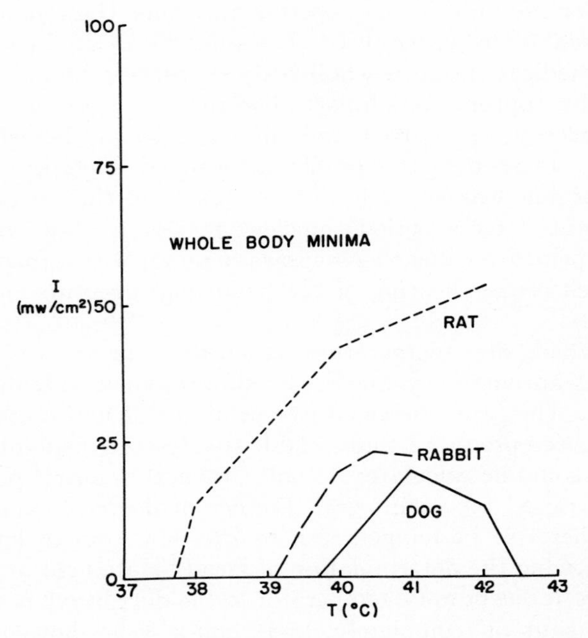

Fig. 8.

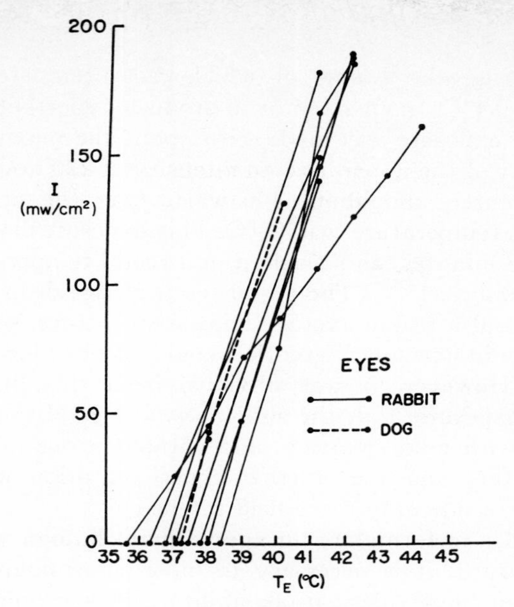

Fig. 9.

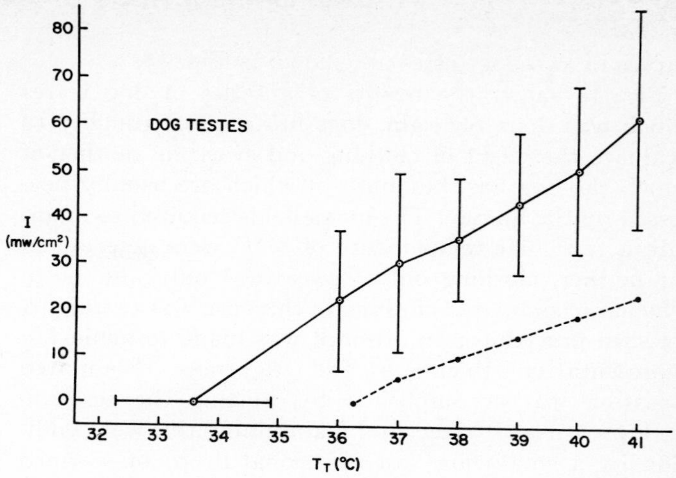

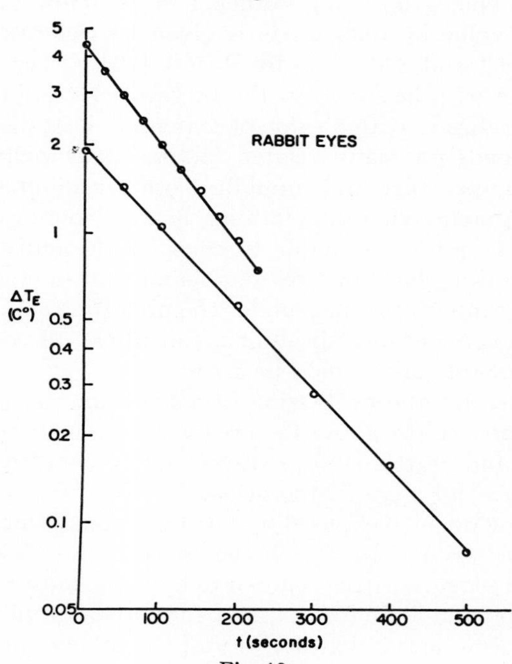

Fig. 10.

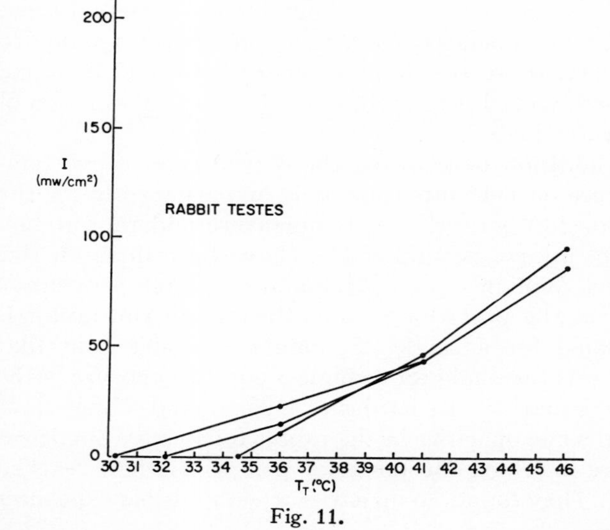

Fig. 11.

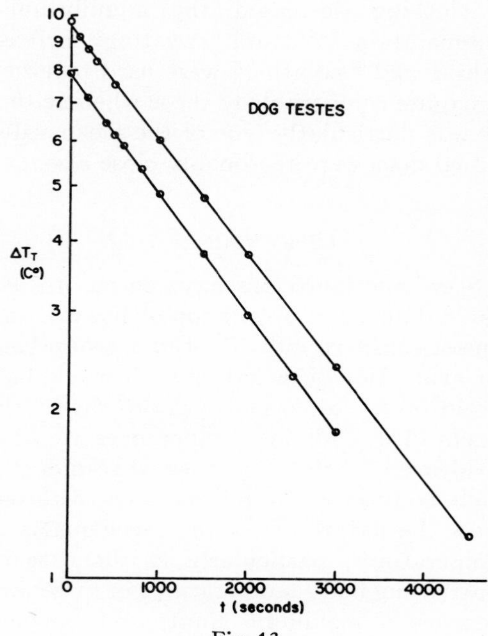

Fig. 13.

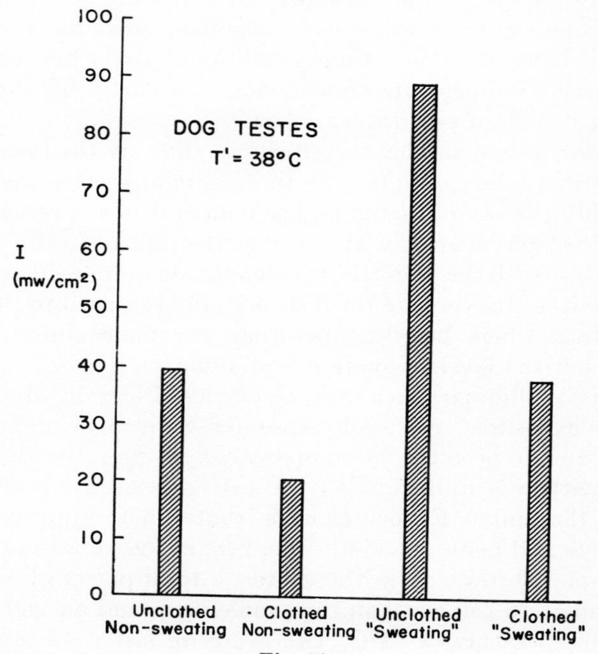

Fig. 14.

Fig. 12.

curves of two dog testes are shown in Fig. 13.

Fig. 14 shows the results of a series of dog testes which was done on eight dogs in an effort roughly to evaluate the effect of clothing and sweating neither of which the dog has, but both of which are usually possessed by the human. The mean fields required to maintain a testicular temperature of 38°C were determined for neither, clothing only, "sweating" only and both. The use of simulated clothing in this case was a straightforward procedure; an attempt was made to achieve a representative thickness and tightness. Simulated sweating was accomplished by wetting the scrotum with prewarmed water, and maintaining this wet condition by a small flow (an occasional drop) of warmed water through a small polyethylene tube. As would be expected, clothing decreased the equilibrium field required to maintain 38°C and "sweating" increased it. When clothing and "sweating" were used together, the results were quite comparable to those when neither was used. This was particularly true of the mean value but the individual dogs were reasonably close also.

Discussion

As has been mentioned, we have chosen to use any demonstrable damage as a criterion of hazard. In addition, the most sensitive individual in a group has been used in the evaluation of results, since it would be desirable to avoid injury to en entire group rather than to some segment of it. Individual differences are considerable, as evidenced by the experimental results.

The whole body exposure results (Figs. 5 through 7) clearly show the failure of cooling mechanisms at the higher temperatures, particularly in the case of the dogs. At increasing body temperatures in these animals, cooling reaches a maximum limit, and beyond this undergoes a failure presumably due primarily to a central failure of the respiratory mechanism. Also, metabolism is increased under these conditions, partly because of specific temperature dependence, and partly because of an increased respiratory effort.

Examination of the curves shows that, at the lower elevations, the slopes for the three animal species were roughly the same. At the higher temperatures, a reduction in slope occurred with a few of the rats and rabbits whereas, with the dogs, this tendency was quite evident. The curves for three of the dogs actually returned to the abscissa, where body temperature was maintained at the elevated level by metabolism alone.

As a preliminary to a proposed series of human whole body exposures, two such exposures were made. Since the human profile was comparatively large, the field intensity was quite low at the extremes as compared with the center. Exposures were frontal, with appropriate eye and testis shielding. The first exposure was to a field of 100 mw/cm^2 in the center, a total power of approximately 250 watts in the profile area, and an ambient temperature of 24°C. The exposure lasted 48 minutes, during the course of which rectal temperature dropped 0.4°C. In an attempt to produce hyperthermia, a second exposure was made with about the maximum capability of the apparatus, an intensity of 220 mw/cm^2 in the center, and about 400 watts over the profile. Ambient temperature was 25°C. This exposure likewise lasted 48 minutes, and resulted in a rectal temperature drop of only 0.15°C. The rectal temperature drop does not rule out a rise in average body temperature, since a reduction in normal differentials could account for both events. However, it was obvious, from this limited human exposure, that the subject was in good thermal control with mild sweating, and without a rise in core temperature and that further experimentation would require considerably more field intensity.

In order to formulate an approximate human whole body hazard, it is necessary to have some figure for maximum heat dissipation ability. This would, of course, depend on many other factors. However, a nominal value of 1000 watts is given by Schwan and Piersol [4], and 800 watts by Bazett [10]. Other similar values may be found in the literature. It is important to realize that this value of maximum heat dissipation depends on many related factors, principally air speed, temperature and humidity, other radiant environment, metabolic rate, clothing, beam, geometry and time. It is quite reasonable to consider the individual who is working hard in a reasonable amount of clothing on a hot, humid, still day under the noontime sun, who is on the verge of heat disability, and whose microwave heat input tolerance would be zero.

Schemes for the evaluation of a particular situation with respect to the above factors have been formulated. Belding and Hatch [11] provide a set of charts which can be used for a graphic solution.

The one dog eye exposed was at least consistent with the rabbit results. In Fig. 9, the rabbit eye which appears the most sensitive is noted to have the most experimental points. The entire experiment for this rabbit took considerably longer than for any of the others, and the apparent sensitivity is probably due to a concurrent heating of the head of the animal. In this animal then, sensitivity appears to be a function of heating time to some extent. In the human, this factor would be much less pronounced because of the better heat dissipation of the whole body.

In addition to knowing the dependence of eye temperature on field intensity, it is necessary to know the relationship between eye temperature and the production of injury. Several studies have been done on this subject. Williams, *et al.* [3], found that a temperature of 49°C at the posterior pole of the lens in the rabbit is threshold for a single 25 minute exposure, and that 53°C was threshold for a single 5 minute exposure in the same situation. Richardson, Duane, and Hines [12] found some opacities in the rabbit following a single exposure in which the posterior pole of the lens reached 52°C. They found no opacities when multiple exposures were given under conditions which produced a temperature of 47.1°C. After a review of these and other investigators, Schwan and Piersol [4] have suggested a human threshold eye temperature of 45°C.

A preliminary evaluation of the relative sensitivites of the three structures in the human will be made now. Experimental animal data are used, and the pitfalls of this maneuver are acknowledged. Estimates of the initial temperature, the maximum permissible temperature, the corresponding temperature rise, the field strength necessary to maintain that temperature rise and the time constant of each of the structures in the human are listed in Table I. Initial rectal temperature is considered 37.0°C. Maximum permissible rectal temperature is listed at 39.0°C (102.2°F), although this is somewhat arbitrary. This represents a temperature rise of 2.0°C. From the whole body experimental animal data this amount of temperature rise can be maintained by a field intensity of about 20 mw/cm² in the most sensitive individuals. However, under average conditions the human is able to dissipate heat at a considerably greater rate than these laboratory animals, principally because of the sweating mechanism. Field strength required to maintain a 2.0°C temperature rise in the human has been estimated at 100 mw/cm². This is based on a somewhat standardized man of 70-kilogram weight, 2.0 square meter total body area, 1.0 square meter profile area, 100 per cent absorption efficiency, 0.83 specific heat, 1000 watts maximum heat dissipation ability above metabolic rate, and essentially even distribution of temperature increase, at rest in average ambient conditions.

TABLE I

Structure	Initial Temp. (°C)	Maximum Temp. (°C)	Temp. Rise (°C)	Steady-State Field (mw/cm²)	Cooling Time Constant (sec)
Whole body	37.0	39.0	2.0	100	(50 joules/cm²)
Eye	37.0	45.0	8.0	155	100
Testis	35.6	37.0	1.4	5	250

The maximum profile area estimate is the limiting case, and is used only for simplicity. A somewhat closer estimate might be $1/\pi$ of the total area. The 100 per cent absorption efficiency is also a limiting case, but Schwan and Li [6] have shown that absorptions of close to this are possible theoretically. The maximum heat dissipation ability is a simplified round number mentioned previously. Although direct use of the fact is not made in the curve to follow, it is significant that this level of heat dissipation is associated with a rectal temperature of about 39°C in the steady state under average circumstances [10]. A time constant could not be used for the human whole body, since cooling rate is a complex function, only partially dependent on rectal temperature. This is true of the human, with his very effective sweating mechanism even more than of the non-sweating animals used in the study. The thermal mass of the body in terms of profile intake intensity has therefore been given rather than τ. This also is based on the assumptions above.

Human eye temperature at the posterior pole of the lens has been estimated as equal to the rectal temperature. This certainly is a maximum, and is probably somewhat lower in actuality, as evidenced by the rabbit data. The maximum permissible eye temperature given is the 45°C value suggested by Schwan. For the resulting temperature rise of 8.0°C, a field intensity of 155 mw/cm² has been chosen, based on the most sensitive rabbit eye in the present study. The time constant τ was also chosen from the present study, and represents the most sensitive condition (smallest value).

For the testis, an initial temperature of 35.6°C was used, as this is the maximum human normal found by both Badenoch [13], and Newman and Wilhelm [14], of which more will appear later. The threshold temperature of 37°C was the lowest damaging temperature found in the present study. The field intensity required to maintain a threshold temperature was chosen from the most sensitive of the dog series at 5 mw/cm². It is of considerable significance that this is about the same figure that would result if computation were based on the 1.4°C temperature rise rather than the actual temperature of 37°C. Although time constants varied widely, and bore some relationship to I', a time constant of 250 sec was used as representative for this preliminary evaluation. Further evaluation of the testicular sensitivity will be made later in this report.

Fig. 15 is a plot of whole body, eye and testis thresholds from the values in Table I. Any point on the line for each structure represents the time necessary to produce the specified temperature rise at a particular field intensity under the conditions already stated, or conversely, the intensity necessary to produce the temperature rise in a given time interval. The early, short time, portion of each curve approaches the conditions of no cooling, approaches the hyperbola which is proportional to the heat capacity. The long time condition represents the steady state where heat input is balanced by heat output, and becomes a constant value of field intensity. The central portion of the curves represents the transition between these two conditions, where significant cooling occurs, but before this cooling reaches the rate of heat input.

Graphic presentation for the eye and testis is directly from formula (13) of the Appendix. Since heat loss from the whole human body is not a simple function of its temperature rise, a complete smooth curve was not drawn. The hyperbolic portion was taken from the 50 joules/cm² figure of thermal mass, and the late portion of constant field intensity from the heat dissipation ability. These curves were extended with dashed lines to intersection although, in fact, there would probably be a smooth transition as in the case of the eye and the testis, somewhat above the dashed portion. This dashed line, then, is in error in the safe direction.

A factor of significance at the shorter times, and affecting mainly the consideration of whole body heating, is that of the time of heat distribution. Since absorption is essentially all in the first few centimeters of tissue, and the above calculations are on the basis of even distribution of heat throughout the body, the whole body curve at short times would undoubtedly shift to another hyperbolic segment with smaller constant representing the heat capacity of the volume which is heated directly, and probably somewhat larger permissible temperature rise. This consideration applies to a much less extent to the localized exposures, where biological effect is essentially at or near the site of heat generation, and heat distribution is consequently much less a factor.

These curves are simplified, depend on many assumptions and are only approximate. They are, however, bases for some general observations. It appears that for a short exposure the eye is more sensitive than the whole body because of its smaller thermal mass. At longer times, the whole body appears to be more sensitive than the eye principally because of its lower critical temperature. It also seems justified to assume that the testis is considerably more sensitive than either of the other structures.

To develop information of significance to the human species, a further evaluation of the testicular situation was indicated. Two unknowns immediately present themselves; what is "normal" human testicular temperature, and what is a threshold hazardous human testicular temperature? The literature on this subject was reviewed in an effort to supply an answer for each of these questions. An experimental approach to either was outside the scope of the study.

The most useful studies of "normal" temperature were those of Badenoch [13], and Newman and Wilhelm [14]. In both studies general, spinal and local anesthesia were used alone or in combination. Only those cases under local anesthesia were used for the present evaluation in order to avoid the likely effects of spinal or particularly general anesthesia on the measured temperatures. For the former study this involved 16 young men with a mean age of about 27 years and, for the latter study, 11 older men with a mean age of 61 years.

In Badenoch's series, peritoneal and scrotal temperatures were measured, rather than rectal and testicular. It is felt, however, that these should be roughly equivalent. Data from these two series, plus that of the present study, are plotted in Fig. 16. For each item, the range from minimum to maximum, the standard deviation and the mean are indicated.

It appears that the ambient temperature of the present study was between that of the other two series. The deep temperatures measured in the two human series are essentially the same, while that of the dogs naturally is higher. The mean testicular temperature of the dogs,

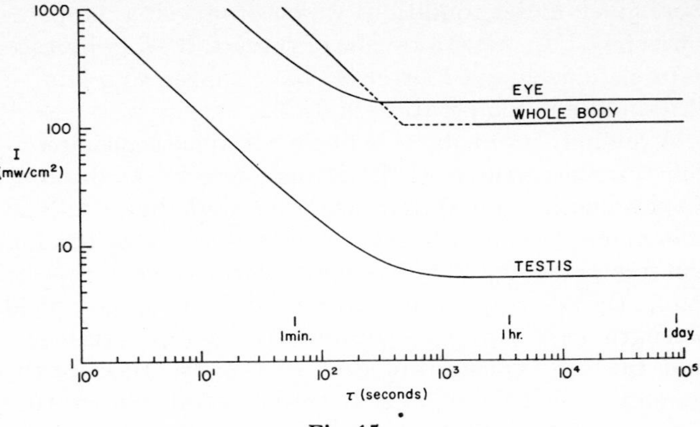

Fig. 15.

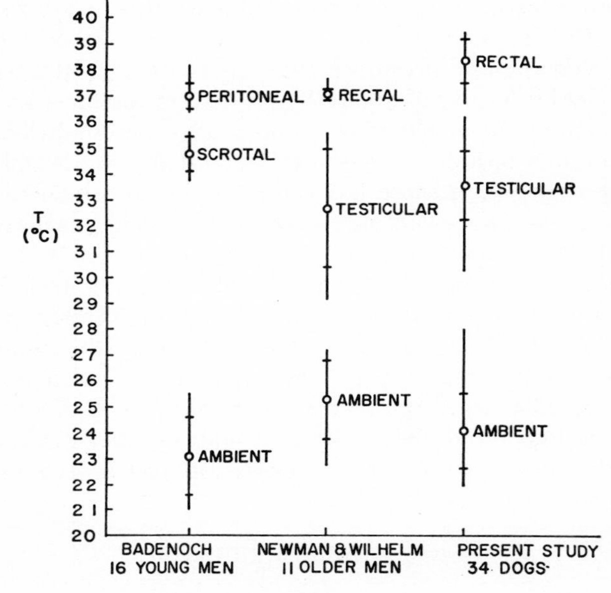

Fig. 16.

however, falls between the testicular temperature of the Newman and Wilhelm series, and the scrotal temperature of the Badenoch series. It is interesting to note that the maximum temperature of both of these two items in the human series was 35.6°C. It is also interesting that in the series of Badenoch, the individual having the highest scrotal temperature had the lowest peritoneal temperature, with a difference of only 0.8°C. The highest initial dog testis temperature is higher than that of either human series.

"Normal" testicular temperature would appear to vary considerably from individual to individual and, despite the thermoregulatory ability of the scrotum, even in one individual from time to time. Such factors of environment as confinement by adjacent areas of the body or clothing, hot and humid air, and hot water could prevent the testes from being cooler than the body, and even raise their temperature above that of the body. Fukui [15] discusses the effects of clothing and hot water, and points out that the propensity of the Japanese for taking hot baths involves an exposure of about 30 minutes to water of from 40° to 46°C. Testicular temperature surely approaches water temperature in this case. The only significant cooling is by the meager circulation to the rest of the body, which itself undergoes considerable hyperthermia under these conditions. It may be that such factors cause human testes to sustain a certain amount of thermal damage and repair repeatedly throughout life, as a normal course.

The evaluation of a temperature threshold for testicular damage is a rather difficult problem. It is undoubtedly a function of exposure time, and probably also age. Several authors have demonstrated testicular damage following experimental cryptorchidism which resulted in testicular temperature equal to core temperature. Moore [16], [17] found the tubular epithelium of the guinea pig highly disorganized after retention in the abdominal cavity for a period of 5 days, and completely disorganized in 7 days. He also [18] found complete degeneration of the ram's germinal epithelium 80 days after testicular cooling was prevented by enclosure of the scrotum in a loose fitting thermal insulating wool bag. Wangensteen [19] found considerable destruction of the dog's tubular epithelium after 5 days of cryptorchidism. Knaus [20] produced cryptorchidism in rabbits and found 100 per cent fertility 2 days later, 60 per cent at 3 days, 33 per cent at 4 days and 0 per cent at 5 days.

It appears that normal body core temperature is damaging to the testes of many species. Although species normal core temperatures differ, that of the rat is close to the human. It is also known that human cryptorchidism is incompatible with normal testicular function, even though a well defined time factor is not known for the species.

In the present study, tubular degeneration in the dog testis was seen as low as 37°C for a 1-hour exposure. MacLeod and Hotchkiss [21] found a marked decrease in sperm counts in six normal humans 40 to 50 days after a 30-minute period of testicular temperature of about 40.5°C, caused by diathermy and warm air. Bauer and Gutman [22] produced necrospermia in five human subjects by two to six 20-minute diathermy treatments to the testes. Temperature in this series was not reported. Williams [23] reports degeneration in a rat testis which had been exposed to S-band microwaves for a total of 5 minutes, reaching a maximum temperature of 38.2°C at the end of the exposure. Imig, Thompson and Hines [24] found damage in the rat testis following a 5-minute exposure to infrared radiation which produced a testicular temperature of 40°C. The same authors found damage to rat testes caused by a 5-minute exposure to S-band microwaves at as low a testicular temperature as 31°C. This is difficult to reconcile with the infrared results and the fact that 31°C is within or below the range of normal rat testicular temperatures.

It is apparent that data on threshold temperature sensitivity with time is meager at present. The minimum temperatures which are suggested from the above review are roughly 37°C for 5 days, 37°C for 1 hour and 38.2°C for 1 minute. Although the last, from Williams, was a 5-minute exposure, the temperature was increasing throughout, reaching the maximum figure at the end of exposure. This should be equivalent to a somewhat shorter exposure at that temperature, say, 1 minute. We have then, a suggestion of threshold temperature according to time of exposure. At exposures of less than 1 minute, the threshold would probably be higher. For exposures of more than 5 days, it would undoubtedly be reduced.

The approximate curve of microwave field intensity and time of exposure for the testis may be refined somewhat. Time constant data must be brought into the calculations. In the series of dog testes, time constants were found in the rather large range of from 110 to over 2000 sec. This represents principally the wide individual difference in scrotal structure and, to a lesser extent, possibly circulation. The time constants of the dog testes were plotted with respect to their maintenance field intensities at each of several integral temperatures. Figs. 17 and 18 show these plots for 37°C and 38°C. Since the short time constant and the low equilibrium field represent the more sensitive condition, the suggested line was drawn on that side of the field of points. Several points along these lines were substituted in formula (13) of the Appendix, and the resulting curves plotted as in Fig. 19, which is the 38°C condition. In Fig. 20, the curves for 37°, 38°, 39° and 40°C are presented with respect to time and field intensity combination which would produce the indicated temperatures as a maximum for any testis having constants falling within the field of experimental points. Each line represents a line through the minima of the solid curve families as in Fig. 19. The dashed line in this figure is a hypothetical curve using the minimum I' and the minimum $I'\tau$. It could be used in the interest of some simplification but is somewhat in error on the safe side in the central portion. Although tempting, a hypothetical combination of minimum I' and minimum τ would be greatly misleading at the shorter times. A set of curves, such as those in Fig. 20, together with information on threshold temperatures according to time of exposure, could be used in the estimation of testicular hazard.

Since heat development from microwave exposure in the testis is nonuniform with respect to volume, differential temperatures are produced. These differentials are small with respect to the over-all temperature rise for intensities of 100 mw/cm^2 and below, as used in this study. At much greater intensities, localized heating results in a zone having a very short time constant by reason of cooling to the surrounding testicular tissue. The same mechanism, however, is responsible for a high I' for this zone, and a sensitivity which is not much greater than that which had been based on the low intensity field data.

This point was checked experimentally by exposing

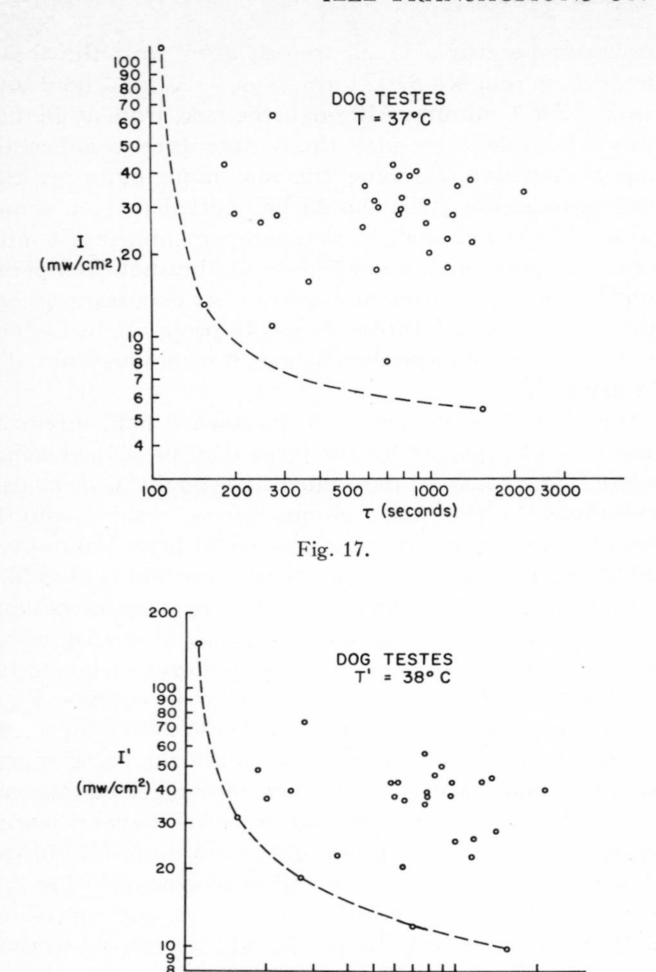

Fig. 17.

Fig. 18

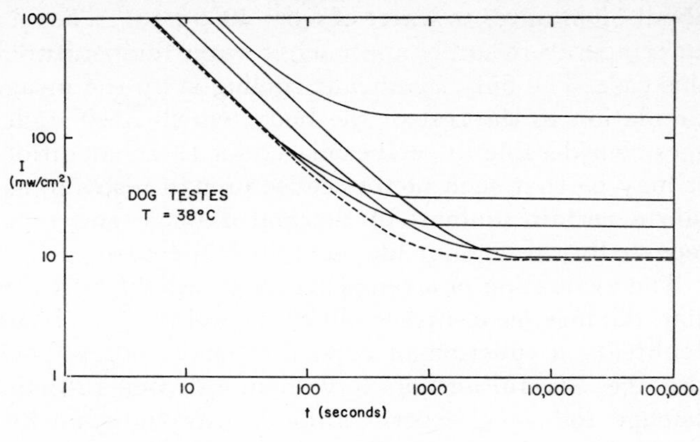

Fig. 19.

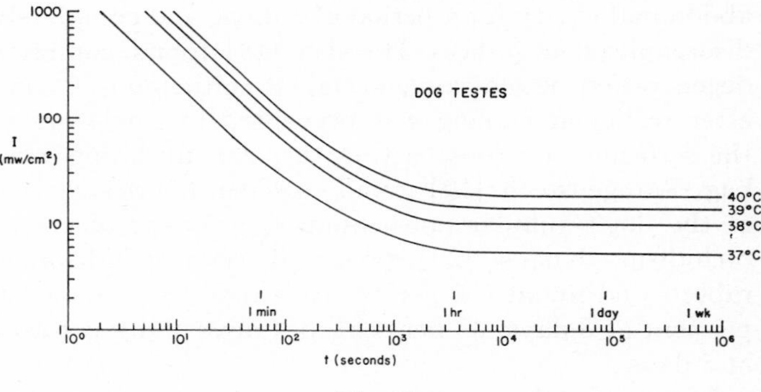

Fig. 20.

dog testes with the thermistor in the location of greatest heat production and time cycling the transmitter on a 3-sec cycle with a variable segment. This mechanism allowed reducing a very intense field of several watts per square centimeter, which produced significant differentials, to a much lower value in the order of less than 100 mw/cm², which produced negligible differentials. In this way the ratio between the field strengths could be determined accurately by pulse counting. Change of field strength by other means would have required measurement of actual field intensity, which is a procedure permitting of less accurate determination.

Essentially, one may conclude that as field strengths increase, testicular heating changes from a pattern of uniform temperature rise to that of temperature rise according to region of heat production. Since this latter does not vary with heating rate, it is the limiting factor in the production of differential temperatures. It would be expected then, that at the higher field intensities, corresponding to the shorter times, the curve would shift to a somewhat lower value, also having hyperbolic form.

An additional factor important in the evaluation of hazard is repetitive exposure. The irregular exposures cannot be treated in a general way, and would have to be evaluated on an individual basis. Regular exposures are to be seen in three main practical situations; those resulting from individual pulses of pulsed radar, which are repeated many times per second; those resulting from a scanning antenna, where repetition rate may be between 360 and $\frac{1}{2}$ per minute and those resulting from work schedules such as a 4-hour watch or an 8-hour shift. The principal reference in evaluating these repeated exposures is the time constant of the structure exposed. Pulsed radar pulse repetition rates are all much faster than any of the time constants demonstrated, as are the faster scanning rates, and time average field intensity is the proper value to be used in evaluation. This opinion is supported by Salisbury, Clark and Hines [8], and others. Repeated exposures occurring as a result of a daily shift or similar circumstance in which the repetition rate is several hours at least and is long with respect to tissue thermal time constants, should be treated as individual exposure incidents. The complex case occurs when the exposure repetition rate is of the same order of magnitude as the time constant of the structure in question. It would appear that the main example of this condition is with the slowly scanning radar antennas.

Evaluation of the hazard under conditions of a regularly repetitive exposure can be done by a reasonably straightforward mathematical analysis. Equations for the general case can also be developed.

Although the criterion of hazard used in this study has been the least demonstrable damage, other factors should be considered in the over-all viewpoint. The minimal testicular damage is almost certainly completely reversible. Even considerably more severe testicular insult will probably be reversible, with the only finding being a temporary sterility. An even greater injury can result in permanent sterility, which result would evoke varying reaction.

In the case of the eye, a small, asymptomatic lens opacity would be of more concern than transient testicular damage. A disabling cataract would require only a moderately greater exposure, and would be considered a serious event.

From the whole body standpoint, minimum damage from generalized hyperthermia is difficult to evaluate. However, the large exposure resulting in death represents the ultimate effect.

Summary

Experimental animals were exposed to 10-cm microwave fields. From simultaneous temperature recordings of the animal, an evaluation was made of the heating effects. Cooling data enabled the formulation of curves of the time and field intensity required to achieve a given temperature. Of the three main sensitive structures, the whole body, the eye and the testis, the last was found to be an order of magnitude more sensitive than the other two. It was found that in the three structures, heating rate is essentially proportional to field intensity, and in the case of the eye and the testis, cooling rate was essentially proportional to the temperature elevation of the part. The whole body cooling rate is a complex function only partly dependent on the temperature elevation. Threshold damage temperatures were taken from the literature and, in the case of the testis, also from a histological evaluation which was part of this study. The heating, cooling and threshold temperature data were put into a somewhat general form, which may be useful in the estimation of hazards to humans.

The limitations of this study, such as those imposed by numbers of animals, restricted parameters and extrapolation from lower animals to humans, should be emphasized. Also, it should be stressed that this study attempts to provide but one link in the body of information which would be necessary to evaluate a hazard, and, therefore, of itself does not implicate or absolve any particular situation.

Acknowledgment

The authors express gratitude to Dr. R. G. Fellers and J. A. Kaiser of the Naval Research Laboratory for their frequent technical assistance on matters concerning the radar transmitter, and to L. T. Letchworth, Aviation Electronics Technician, First Class USN, for his capable operation of the transmitter and over-all assistance with the project.

APPENDIX

Some Mathematical Considerations

J. Z. HEARON

Consider a body of heat capacity C (cal/degree) which is being heated by an internal source S (cal/sec) and is cooling according to the law

$$\text{cal lost/sec} = \lambda a(T - T_0)$$

where λ (cal/sec cm^2 degree) is the loss per unit time per unit area for a unit temperature differential and a is the surface area of the body. Then

$$(\text{cal gained in } dt) = CdT = Sdt - \lambda a(T - T_0)dt$$

where dT is the temperature change in dt. Or, $T(t)$ is governed by

$$\frac{dT}{dt} = \frac{S}{C} - k(T - T_0) \qquad (1)$$

where $k(\text{sec}^{-1}) = \lambda a/C$. In the above it is assumed that S and T are *spatially uniform* and that T_0 is the external or environmental temperature. If spatial uniformity does not obtain, it can be shown that (1) still applies if I and S are replaced by their spatial or volume means and k has a different physical connotation (*i.e.*, $k \neq \lambda a/C$) but still represents a formal "cooling constant" in terms of the (spatial) average or mean temperature.[1,2]

In the situation at hand, T and S are not uniform but we wish to show that $T(t)$, measured at a fixed point in the body (*e.g.*, point of maximum temperature) does in

[1] N. Rashevsky, "Mathematical Biophysics," University of Chicago Press, Ill.; 1948.

[2] J. Z. Hearon, *Bull. Math. Biophys.*, vol. 15, p. 23–31; March, 1953.

fact obey an equation of type (1) wherein a specific assumption is made relating S/C to a known external energy field, and T_0 is taken to be the "normal initial temperature" (*i.e.*, the temperature in the absence of the external field).

In particular, assume

$$S/C = K'I \tag{2}$$

where I is the field intensity and K' is of the nature of a "pick-up" or "dissipation" coefficient, and independent of T and of the explicit time t. Then (1) is

$$\frac{dT}{dt} = K'I - k(T - T_0). \tag{3}$$

Denote the steady-state temperature by T_s and define the temperature rise at any t by

$$\Delta T(t) = T(t) - T_0 \tag{4}$$

and in the steady state

$$\Delta T_s = T_s - T_0. \tag{5}$$

Then (3) can be written

$$\frac{d\Delta T}{dt} = k(KI - \Delta T) \tag{6}$$

where $K = K'/k$.

The general integral of (6) is

$$\Delta T(t) = KI(1 - e^{-kt}) + \Delta T(0)e^{-kt}. \tag{7}$$

It is clear from (7), and otherwise obvious from (6) with $d\Delta T/dt = 0$, that

$$\Delta T_s = KI. \tag{8}$$

Thus (7) may be written as

$$\Delta T(t) = \Delta T_s(1 - e^{-kt}) + \Delta T(0)e^{-kt}. \tag{9}$$

The assumptions embodied in (2) can be checked through (8) which predicts the steady-state rise in temperature to be proportional to I. In particular the slope of the line (8) estimates K. The result (8) depends, of course, upon the assumed form $k(T - T_0)$ for the cooling law, but this can be validated separately and independently of assumption (2), as follows. If the body is heated, heating is stopped (*i.e.*, I made zero), and time is reckoned from the instant of cessation of heating, then from that instant on, $\Delta T(t)$ obeys [from (7) with $I = 0$]

$$\Delta T(t) = \Delta T(0)e^{-kt} \tag{10}$$

and under these conditions a semi-log plot of the temperature rise should be linear. The slope of this line estimates[3] $-k$. It is seen from (10) that beginning with any arbitrary initial temperature rise $\Delta T(0)$, the time required for the temperature rise to fall to $1/e$ of this initial value is $\tau = 1/k$. Or, beginning with an initial rise $\Delta T(0)$ of zero [*i.e.*, $T(0) = T_0$, see (4)] it follows from (9) that at time $\tau = 1/k$ the temperature rise achieves the fraction $1 - 1/e \cong 0.67$ of its final value.[4] It is convenient to deal with this characteristic time τ. We consider now *only* situations in which $\Delta T(0) = 0$. Then (9) is

$$\frac{\Delta T(t)}{\Delta T_s} = 1 - e^{-t/\tau} = \frac{\Delta T(t)}{KI}\cdot \tag{11}$$

Consider a system being heated under field intensity I, its temperature rise being given by (11). Then for any t, and corresponding $\Delta T(t)$, there exists an intensity $I' \leq I$ which will produce a steady-state temperature rise equal to the observed temperature rise $\Delta T(t)$ at that time. It is clear from (7) that the inequality $I' \leq I$ is not valid unless $\Delta T(0) \leq KI$. But the agreement to consider $\Delta T(0) = 0$ entitles us to speak of the *reduced intensity* I' such that if the system is heated till time t under intensity I and at that time the intensity is instantaneously reduced to I', the temperature rise $\Delta T(t)$ obtained at that time will be maintained constantly and indefinitely. Plainly I' is a function of t and I but we do not display that fact in the notation. According to the argument leading to (8), I' is given by

$$KI' = \Delta T(t) \tag{12}$$

and (12) into (11) yields

$$\frac{I'}{I} = 1 - e^{-t/\tau} \tag{13}$$

or

$$\ln\left(1 - \frac{I'}{I}\right) = -t/\tau. \tag{14}$$

According to (11), if we prescribe a temperature rise, say $\Delta T(t) = A \leq KI$, then given A, there is, for each time t, an intensity I such that the temperature rise is precisely A at time t. These times and intensities are related by

$$A = KI(1 - e^{-t/\tau}) \tag{15}$$

for small t, *i.e.*, $t \ll \tau$,

$$A = \frac{K}{\tau}It = K'I\cdot t. \tag{16}$$

Thus over the range of sufficiently small t (and large I) the hyperbolic relation 16 holds. The slope of the initial linear portion of the log I vs log t plot is -1, the ordinate intercepts log (A/K').

[3] Also, k can be estimated in a slightly different fashion, see (14) and compare text.

[4] Stated otherwise, at time $\tau = 1/k$ the fractional displacement of the temperature rise from its steady-state value is $1/e$, *i.e.*,

$$\frac{\Delta T_s - \Delta T(t)}{\Delta T_s} = 1/e.$$

References

[1] J. W. Clark, "Effects of intense microwave radiation on living organisms," Proc. IRE, vol. 38, pp. 1028–1032; September, 1950.

[2] H. P. Schwan and G. M. Piersol, "The absorption of electromagnetic energy in body tissues." *Am. J. Phys. Med.*, vol. 33, pp. 371–404; 1954.

[3] D. B. Williams, J. P. Monahan, W. J. Nicholson, and J. J. Aldrich, "Biologic Effects Studies on Microwave Radiation," U. S. Air Force School of Aviation Medicine, Rept. No. 55-94.

[4] H. P. Schwan and G. M. Piersol, "The absorption of electromagnetic energy in body tissues. II. Physiological and clinical aspects," *Am. J. Phys. Med.*, vol. 34, pp. 424–448; 1955.

[5] J. F. Herrick and F. H. Krusen, "Certain physiologic effects of microwaves," *Elec. Engrg.*, vol. 72, pp. 239–244; 1953.

[6] H. P. Schwan and K. Li, "The mechanism of absorption of ultrahigh frequency electromagnetic energy in tissues, as related to the problem of tolerance dosage," IRE Trans. on Medical Electronics, vol. PGME-4, pp. 45–49; February, 1956.

[7] J. D. Hardy, "Summary review of heat loss and heat production in physiologic temperature regulation," U. S. Naval Air Dev. Ctr., Johnsville, Pa., Rept. No. NADC-MA-5431; October 14, 1954.

[8] W. W. Salisbury, J. W. Clark, and H. M. Hines, "Exposure to microwaves," *Electronics*, pp. 66–67; May, 1949.

[9] A. W. Richardson, T. D. Duane, and H. M. Hines, "Experimental cataract produced by three cm. pulsed microwave irradiations," *AMA Arch. of Ophth.*, vol. 45, pp. 352–356; 1951.

[10] H. C. Bazett, "The regulation of body temperatures," in "Physiology of Heat Regulation and the Science of Clothing," L. H. Newburgh, Ed., W. B. Saunders, Philadelphia, Pa., pp. 109–192; 1949.

[11] H. S. Belding and T. F. Hatch, "Index for evaluating heat stress in terms of resulting physiological strains," *Heating, Piping Air Conditioning*, pp. 129–136; August, 1955.

[12] A. W. Richardson, T. D. Duane, and H. M. Hines, "Experimental lenticular opacitites produced by microwave irradiations," *Arch. Phys. Med.*, vol. 29, pp. 765–759; 1948.

[13] A. W. Badenoch, "Descent of the testis in relation to temperature," *Brit. Med. J.*, vol. 2, pp. 601–603; 1945.

[14] H. F. Newman and S. F. Wilhelm, "Testicluar temperature in man," *J. Urol.*, vol. 63, pp. 349–352; 1950.

[15] N. Fukui, "On the action of heat rays upon the testicle: an histological, hygienic and endocrinological study," *Acta Scholae Med. Univ. Kioto*, vol. 6, pp. 225–260; 1923.

[16] C. R. Moore and W. J. Quick, "The scrotum as a temperature regulator for the testes," *Am. J. Phys.*, vol. 68, pp. 70–79; 1924.

[17] C. R. Moore, "Biology of the testes," in "Sex and Internal Secretions," Allen, Danfurth and Doisy, Eds., Williams and Wilkins Co., Baltimore, Md., ch. 7; 1739.

[18] C. R. Moore and R. Oslund, "Experiments of the sheep testis—cryptorchidism, vasectomy and scrotal insulation," *J. Physiol.*, vol. 67, pp. 595–607; 1924.

[19] W. H. Wangensteen, "The undescended testis," *Arch. Surg.*, vol. 14, pp. 663–731; 1927.

[20] H. Knaus, "Thermosensibilita die testicoli e degli spermatozio," *Minerva Med.*, vol. 1, pp. 322–323; 1940.

[21] J. MacLeod and R. S. Hotchkiss, "The effects of hyperpyrexia upon spermatozoa counts in man," *Endocrinology*, vol. 28, pp. 780–784; 1941.

[22] J. Bauer and G. Gutman, "The effect of diathermy on testicular function," *Urol. and Cutan. Rev.*, vol. 44, pp. 64–66; 1940.

[23] D. B. Williams, personal communication.

[24] C. J. Imig, J. D. Thomson, and H. M. Mines, "Testicular degeneration as a result of microwave irradiation," *Proc. Soc. Exper. Biol. and Med.*, vol. 69, pp. 382–386; 1948.

16729

Testicular Degeneration as a Result of Microwave Irradiation.

C. J. Imig, J. D. Thomson, and H. M. Hines.

From the Department of Physiology, State University of Iowa.

The mammalian scrotum has been established as being a local thermoregulator for the testes. Moore and Quick[1] found the scrotal temperature of white rats to be from 2° to 8°C lower than the temperature of the abdominal cavity. A sub-abdominal temperature in the scrotum has been shown to be necessary for the continuance of spermatogenesis. Moore[2] confined the testes of guinea pigs in the abdominal cavity for varying periods of time and found that an abdominal retention of seven days resulted in a complete disorganization of the germinal epithelium of the seminiferous tubules. He considered the cause of this degeneration to be due to the higher temperature of the abdominal cavity.

Investigations of Moore[3] showed that testicular degeneration resulted from a single application of heat at approximately 7°C above body temperature for a 15-minute period. The heating devices which were used consisted of hot water baths, electric stoves, electric light bulbs, and hot water pads. Testicular degeneration was visible histologically within four to six days following the heat application, and was entirely similar in type to that resulting from early experimental cryptorchidism.

Fukui[4] exposed the scrotum of rabbits to sunlight and to warm air, and found a definite relationship between the temperature and the time required to cause regressive changes in the germinal cells. The minimum scrotal temperature at which he was able to produce testicular damage was 40°C. The time of required exposure at this temperature was more than one hundred hours.

It was the purpose of this study to deter-

1 Moore, C. R., and Quick, W. J., *Am. J. Physiol.*, 1924, **68**, 70.

2 Moore, C. R., *Am. J. Anat.*, 1924, **34**, 269.

3 Moore, C. R., *Am. J. Anat.*, 1924, **34**, 337.

4 Fukui, N., *Japan M. World*, 1923, **3**, 27.

mine the effect of 12 cm electromagnetic radiations of testicular tissue. In addition, attempts were made to confirm the results of previous investigators relative to the effects of infra red irradiations on testes.

Procedure. Male albino rats of the Sprague-Dawley strain, ranging in age from 120 to 200 days were employed in this study. In preparation for irradiation the animals were anesthetized with ether and the scrotum swabbed with 95% alcohol. Each animal was then arranged on a platform behind a copper shield, and the scrotum was inserted through an opening provided in the shield. Thus, the remainder of the animal was protected from the radiations. An iron-constantan thermocouple needle, of the type described by Tuttle and Janney,[5] was then inserted into the center of one of the testes. This served to register the degree of temperature produced. The thermocouple had been calibrated previously with a Bureau of Standards thermometer. Thermocouple potentials were measured by a Leeds and Northrup potentiometer. Temperatures were read to the nearest tenth of a degree centigrade. The temperature of the testes was allowed to fall to approximately 29°C before the irradiations were started. Because of possible damage to the testes resulting from thermocouple needle puncture, one testis was employed for temperature measurements and the contralateral testis was used to study the histologic effects of the radiation. Care was taken to align both testes at an equal distance from the source of radiation. In preliminary experiments it was found that if these precautions were taken the temperature rise due to the radiations was the same in both testes.

A Raytheon Microtherm generator (model CMD4) which produced a wave length of approximately 12 cm was used to apply the high frequency radiations. A variac was provided by means of which the power output was regulated. The corner type reflector was used. The infra-red source rated at 600 watts was of the non-luminous type, and a 9-inch hemispherical reflector was employed. In all cases irradiation was applied to the testis through the scrotum.

Two series of experiments were performed with 12 cm electromagnetic waves. In the first of these series the procedure consisted of elevating the testicular temperature to levels of 47, 46, 45, 44, 43, 40, 37, 35, 34, 33, 32, 31 and 30°C. For each temperature level 4 animals were used, and the testicular temperature was maintained at the selected temperature for a single period of 5 minutes. The temperature was maintained by varying the power output of the microwave machine. From each group of 4, an animal was sacrificed at 4, 8, 12 and 16 days following the exposure and the testes were prepared for histologic examination.

In the second series of irradiations 2 groups of 10 animals were given single 15-minute exposures at temperature levels of 33° and 34°C. A third group of 10 animals was exposed to 35° C for a period of 10 minutes. All animals were sacrificed for histologic studies 4 days following the exposures. The testes were fixed in Bouin's solution and the sections stained in hematoxylin-eosin.

Infra-red irradiation was used in two series of experiments. In the first series testicular temperatures were raised by irradiation to levels of 33, 37, 40, and 43°C for periods of 5 minutes. Four animals were exposed at each temperature level and a histologic examination of the testes was made at 4, 8, 12, and 16 days after the exposures. The procedure in the second series of infra-red irradiations was to expose two groups of 10 animals to temperature levels of 38° and 40°C for 10 minutes. Four days after exposure the animals were sacrificed and the testes were removed for histologic study.

The testes of 12 normal non-irradiated males were employed as histologic controls.

Results. The results of experiments with 12 cm electromagnetic irradiations are listed in Table I. These results show that in all cases when the testicular temperature was raised to 35°C or higher there was evidence of testis tissue damage. At temperatures from 31° to 35°C approximately 50% of the testes showed signs of degenerative changes. The

[5] Tuttle, W. W., and Janney, C. D., *Arch. Phys. Med.*, 1948, **29**, 416.

TABLE I.
Effect of 12 Cm Irradiation on Male Gonads.

No. animals exposed	Exposure temp. °C	Time of exposure min.	Animals with testicular damage	% with testicular damage
4	47	5	4	100
4	46	5	4	100
4	45	5	4	100
4	44	5	4	100
4	43	5	4	100
4	40	5	4	100
4	37	5	4	100
4	35	5	4	100
4	33	5	2	50
4	32	5	1	25
4	31	5	1	25
4	30	5	0	0
13	35	10	13	100
4	33	10	1	25
4	31	10	1	25
3	35	15	3	100
7	34	15	3	43
16	33	15	9	56
4	31	15	0	0

TABLE II.
Effect of Infra Red Irradiation on Male Gonads.

No. animals exposed	Exposure temp. °C	Time of exposure min.	Animals having testicular damage	% with testicular damage
3	43	5	2	67
4	40	5	2	50
4	37	5	0	0
4	33	5	0	0
8	40	10	3	38
10	38	10	0	0

testes of animals exposed to 30°C were not affected.

The results of experiments with infra red irradiation (Table II) show that testes from 67% of the animals exposed to 43°C were damaged. Testicular degeneration was not found in the experiments where the temperature of the testes was maintained at 38°C for 10 minutes by infra red. No damage was found in the testes of control animals.

A typical histological picture of testicular degeneration following exposure to electromagnetic waves shows an area of degeneration along the side of the testis nearest to the source of radiation (area A, Fig.1). Continuing from this area of degeneration to the opposite side of the testis, all gradations from completely degenerated to normal tubules can be found (areas B and C, Fig. 1). Fig. 2 clearly shows the transition of degeneration from area A to area B of Fig. 1. These areas are well defined. A detailed picture of a degenerating tubule in area B is shown in Fig. 3. In area A, which was nearest to the source of radiation, complete coagulation of the tubules was found (area A of Fig. 2). This area showed an absence of germ cell nuclear material, the cytoplasm appearing much more refractile than normal cytoplasm. The tissue resembled that seen in burn necrosis. In general the tubules in area B show sloughing of degenerating germinal elements into the lumen, multinuclear masses termed "giant cells" which apparently consist of fused spermatid nuclei, and usually absence of spermatozoa (area B of Fig. 2 and 3). Sertoli and interstitial cells apparently remain intact. Leucocytic infiltration of the

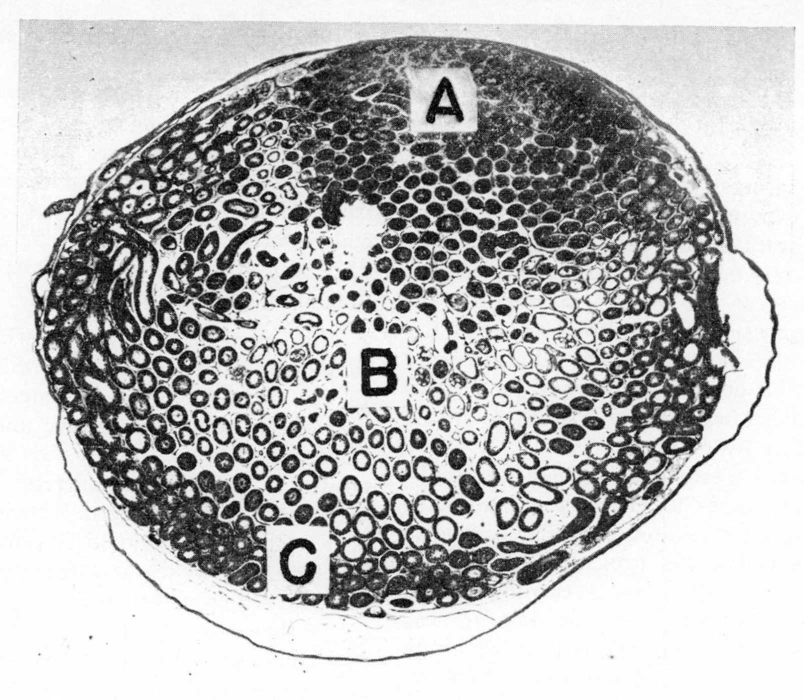

FIG. 1.
Cross-section of entire testis removed 4 days after a single irradiation with 12 cm microwaves at 35°C for 10 minutes (× 10). See text for explanation of the areas indicated.

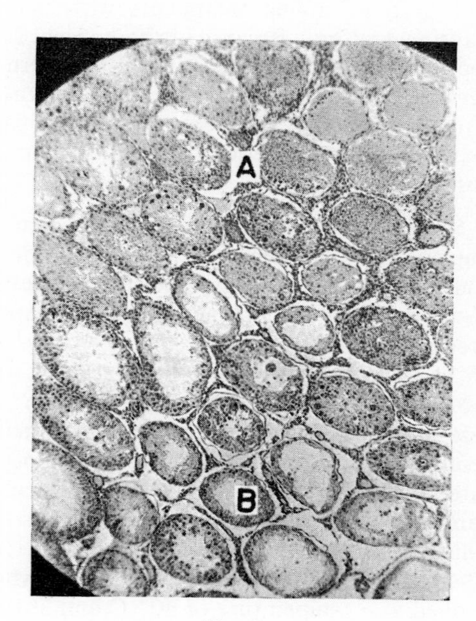

FIG. 2.
Seminiferous tubules from areas A and B of Fig. 1. (× 40).

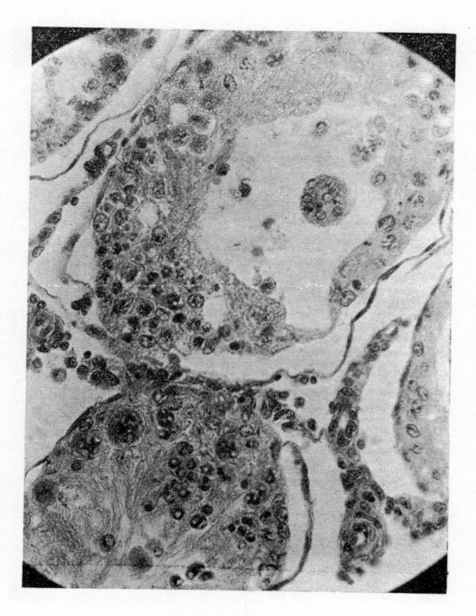

FIG. 3.
High magnification of a degenerating seminiferous tubule from area B of Fig. 2. (× 200).

tissue was evidenced by the presence in some cases of intertubular polymorphonuclear leucocytes, in the areas of degeneration. In area C of Fig. 1, the tubules were essentially normal.

Fig. 4 is a cross section through the center of a testis exposed for 10 minutes at 40° C to infra red irradiations. Area A, the side most directly exposed, shows coagulation of the tubules, whereas the tubules in area B show varying degrees of degeneration. The tubules farthest from the source of radiation were normal. The general picture of the degenerative changes was similar to that produced by microwave irradiation.

Discussion. Testicular degeneration resulting from exposures to microwaves and infra red irradiations presented a similar histologic appearance which was typical of the degeneration seen in experimental cryptorchidism.

The temperatures at which damage was noted from infra red irradiations were approximately 3 to 5°C lower than those reported by Moore and Chase. These investigators placed the bulb of a thermometer close to the scrotum to register the degree of heat applied. With the needle thermocouple method of temperature measurement used in the experiments reported here it was possible to register the temperature within the testes.

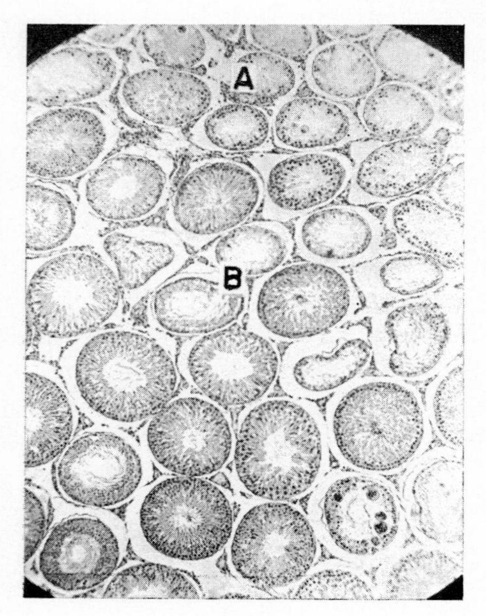

FIG. 4.
Degenerating tubules of a testis 4 days after a single irradiation with infra red at 40°C for 10 minutes. (× 40).

Following electromagnetic irradiation testicular degeneration was found at temperatures below those at which damage occurred from infra red irradiations. All of the testes which were elevated to a temperature level of 35°C and above with microwaves were found to contain degenerated tubules.

The outcome of this experiment clearly shows that testicular damage will result from 12 cm irradiations at a temperature below that of the abdominal cavity and below that necessary to cause injury by infra red exposures. This finding suggests that damage may result in part from factors other than heat. However, it should be pointed out that measurements of temperature were made only near the center of the testes and the possibility exists that areas adjacent to the field of irradiation may have been subjected to temperatures somewhat in excess of those recorded.

These findings suggest that precautions should be taken by those working in the field of high frequency electromagnetic generators and to those giving treatments with microwave generators. Because of the unusual susceptibility of testicular tissue to thermal agents, it seems desirable to shield these structures from high frequency electromagnetic waves during periods of treatment or exposure.

Summary. A study was made concerning the effects of 12 cm electromagnetic waves and of infra red irradiations upon the testes of adult albino rats. A single ten-minute exposure to microwaves at a temperature of 35°C as measured in the central areas of the testes caused testicular degeneration in all cases. In some experiments testicular damage resulted from a single exposure at temperatures between 30° and 35°C. Testicular damage was not found in experiments in which infra red irradiation was applied for 10 minutes at 38°C but was observed when applied at a temperature of 40° C and above. The type of degeneration resulting from microwave exposure could not be distinguished from that produced by infra red.

Studies on Blood-Brain Barrier Permeability After Microwave-Radiation

J. H. Merritt, A. F. Chamness, and S. J. Allen*

Radiation Sciences Division, USAF School of Aerospace Medicine,
Brooks AFB, Tex. 78235, USA

Summary. Since the reported alterations of permeability of the blood-brain barrier by microwave radiation have implications for safety considerations in man, studies were conducted to replicate some of the initial investigations. No transfer of parenterally-administered fluorescein across the blood-brain barrier of rats after 30 min of 1.2-GHz radiation at power densities from 2–75 mW/cm^2 was noted. Increased fluorescein uptake was seen only when the rats were made hyperthermic in a warm-air environment. Similarly, no increase of brain uptake of ^{14}C-mannitol using the Oldendorf dual isotope technique was seen as a result of exposure to pulsed 1.3-GHz radiation at peak power densities up to 20 mW/cm^2, or in the continuous wave mode from 0.1–50 mW/cm^2. An attempt to alter the permeability of the blood-brain barrier for serotonin with microwave radiation was unsuccessful. From these studies it would appear that the brain must be made hyperthermic for changes in permeability of the barrier induced by microwave radiation to occur.

Introduction

The mammalian blood-brain barrier is a selectively permeable system which exerts extraordinary fine control over the penetration of plasma-borne solutes into the brain extracellular fluid. As Oldendorf (1975) has stated, the function of the blood-brain barrier is to facilitate "optimization of the fluid environment of the brain cells" by passing some molecules but excluding others. In this manner, precise control is exerted over the brain chemical milieu. It is the loss of this precise control, through alterations of the blood-brain barrier, that has led to the concern over reports of microwave-induced changes in the barrier system.

The first study which purported to show alterations of the blood-brain barrier was reported in 1971 by Polyashchuck (1971). Frey et al. (1975) then reported penetration of fluorescein-labeled albumin across the rat blood-brain barrier as a result of pulsed 1.2-GHz radiation at power densities down to 0.2 mW/cm^2. Oscar and Hawkins (1977) have shown that single 20-min exposures of rates to pulsed or continuous wave 1.3-GHz radiation increased the permeability of the blood brain for the saccharides mannitol and inulin.

It is the concern for changes in the barrier system by microwaves that prompted this work to replicate some of the reported studies in an attempt to independently verify the results. This is in consonance with the generally held view in the nonionizing radiation bioeffects community that studies showing significant effects should be replicated. The studies of Frey et al. (1975) and Oscar and Hawkins (1977) were selected for replication.

Materials and Methods

Male Sprague Dawley rats, 150–225 g, were used throughout in these studies. They were maintained in a vivarium on water and commercial chow ad libitum.

Radiation and Dosimetry

All exposures were made in an anechoic chamber. The rats were anesthetized with sodium pentobarbital (50 mg/kg intraperitoneally) and positioned on a styrofoam block in front of the horn as seen in Figure 1. In the terminology of Johnson et al. (1976), this orientation is KEH. Radiation was produced by a Cober Model 1831 microwave generator and transmitted to an American Electronics Laboratory Model H5001 horn by a flexible cable. Power density measurements were made with a National Bureau of Standards EDM-1B probe. Specific absorption rates (SAR) were obtained calorimetrically by the method of Allen and Hurt (1978).

At a frequency of 1200 MHz, the radiation was delivered at a pulse repetition rate of 1000 pps, 0.5 ms pulse width, and 2 mW/cm^2 peak power to reproduce the parameters cited by Frey et al. (1975). For radiation at 1300 MHz, the radiation was pulsed at 1000 pps, 10 μs pulse width, with various peak power densities (Oscar and Hawkins, 1977). In addition, some animals were radiated in the continuous wave mode.

Some sham-irradiated animals were heated in a laboratory drying oven to achieve approximately equivalent brain and rectal temperature for the highest power used (75 mW/cm^2).

Brain Slice Studies

After 30 min of radiation, the femoral vein was exposed, and 0.2 ml of a 4-% sodium fluorescein solution were injected. 5 min later the brains were perfused with heparinized saline through the left side of the heart, removed, and embedded in gelatin. The

* The research reported in this paper was conducted by personnel of the Radiation Sciences Division, USAF School of Aerospace Medicine, Brooks AFB, Tex. 78235. The animals involved in the study were procured, maintained, and used in accordance with the Animal Welfare Act of 1970 and the "Guide for the Care und Use of Laboratory Animals" prepared by the Institute of Laboratory Animal Resources, National Research Council

Fig. 1. The anesthetized rat was placed on a styrofoam block in the anechoic chamber. The orientation is KEH (Johnson et al., 1976)

brains were then frozen on the stage of a tissue chopper, cut into approximately 600-μm coronal sections, and placed on glass slides. The slices were then viewed under ultraviolet light in a darkened room. For photography, the slides were illuminated with ultraviolet light from a Leitz xenon arc light source and photographed on high-speed daylight Ectachrome film using an ultraviolet filter. In addition, some animals were injected with 1 ml of 2-% Evan's blue and treated similarly, with viewing under white light.

Brain Uptake Studies

The method of Oldendorf (1970) was used to assess brain uptake of ^{14}C-labeled mannitol relative to ^{3}H-water. A mixture of ^{14}C-labeled mannitol and ^{3}H-water in Ringer's solution (pH 7.5) was injected as a bolus into the common carotid artery of the anesthetized rat. 15 s later the animal was decapitated and the brain removed. After dissection into regional areas following the scheme of Glowinski and Iversen (1966), the tissue was dissolved in a quaternary ammonium hydroxide solubilizer (Soluene 350, Packard). 1-ml aliquots of the solubilized tissue were placed in vials and 10 ml of a scintillation cocktail (Dimilume-30, Packard) added. The samples were then counted in a Beckman LS-230 liquid scintillation instrument equipped for dual isotope counting.

The brain uptake index (BUI) is derived from the counts of ^{14}C and ^{3}H in brain after 15 s relative to the counts of these isotopes in the original injectate, and is calculated as follows:

$$BUI = \frac{^{14}C/^{3}H\ \text{(brain tissue)}}{^{14}C/^{3}H\ \text{(injectate)}} \times 100\,.$$

Immediately after 20 min of irradiation, heat, or sham-irradiation, the common carotid artery was exposed and the mixture injected. In experiments in which the barrier was opened osmotically, 0.2 ml of 10 M urea was injected. After about 1 min, the ^{14}C-^{3}H mixture was injected and the procedure above followed.

In some instances, the brain tissue used was as described in the original Oldendorf method (1970). The half of the brain ipsilateral to the injection side and rostral to the midbrain was dissected out. Portions of tissue (approximately 250 mg) were extruded from a syringe through a 20-gauge needle and solubilized and counted as already described.

Serotonin Uptake Studies

After radiation or sham-radiation, anesthetized rats were given 50 mg serotonin/kg body weight intraperitoneally. 30 min after administration of serotonin, the animals were euthanized and the brains removed. Brain serotonin was assayed by the method of Snyder et al. (1965).

Rectal and Brain Temperature Measurements

Temperature measurements were made immediately after exposure to the radiation conditions. A Yellow Springs Model 73 Telethermometer was used to make the measurements. The probes and instrument were calibrated against a National Bureau of Standards mercury thermometer. Rectal temperatures were made with the probe inserted to a depth of 4 cm. Brain temperature measurements were made after euthanization of the rats (by cervical dislocation) by inserting the probe through a puncture made in the skull. Rectal and brain temperature measurements were made on separate animals. This was necessary since the individual measurements had to be made immediately upon removal from the chamber, and could not be made simultaneously.

Results

Brain Slice Studies

No fluorescence could be seen in any of the slides that could be ascribed to penetration into the extracellular fluid as a result of exposure to any of the radiation conditions (Figs. 2 and 3). Occasionally the pineal gland was included when the brain was

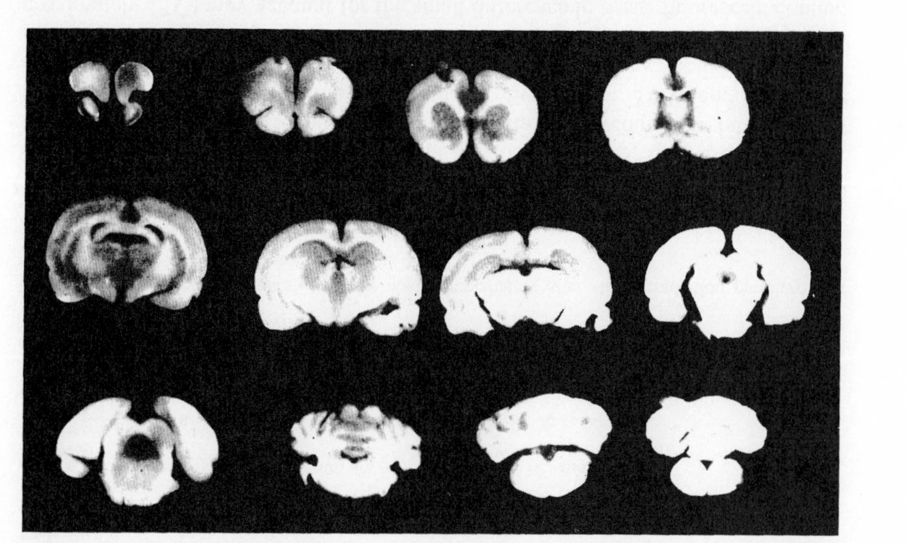

Fig. 2. Brain slices from sham-irradiated rat. Fluorescein was infused into the femoral vein. 5 min later the brain was perfused and frozen and then 600-μm coronal slices were cut on a tissue chopper. Slices were photographed under ultraviolet light

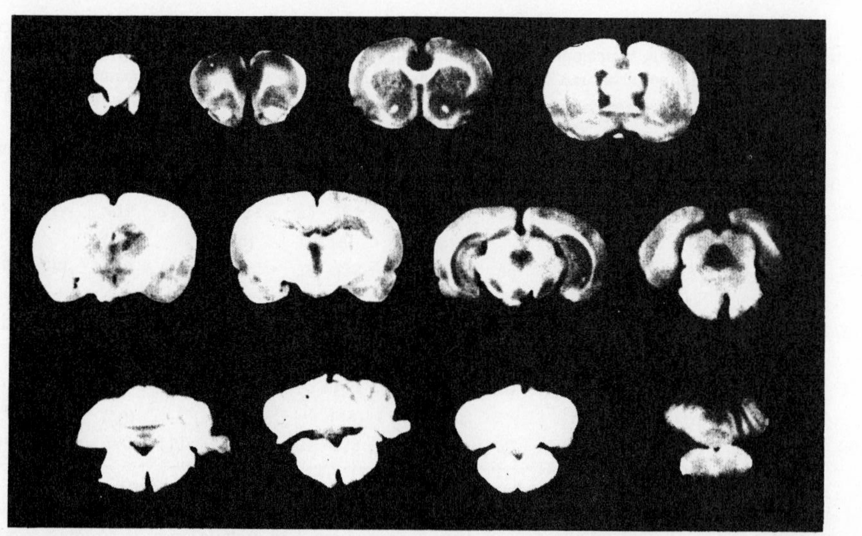

Fig. 3. Brain slices from irradiated rat. Fluorescein was infused into the femoral vein. 5 min later the brain was perfused and frozen and then 600-μm coronal slices were cut on a tissue chopper. Slices were photographed under ultraviolet light

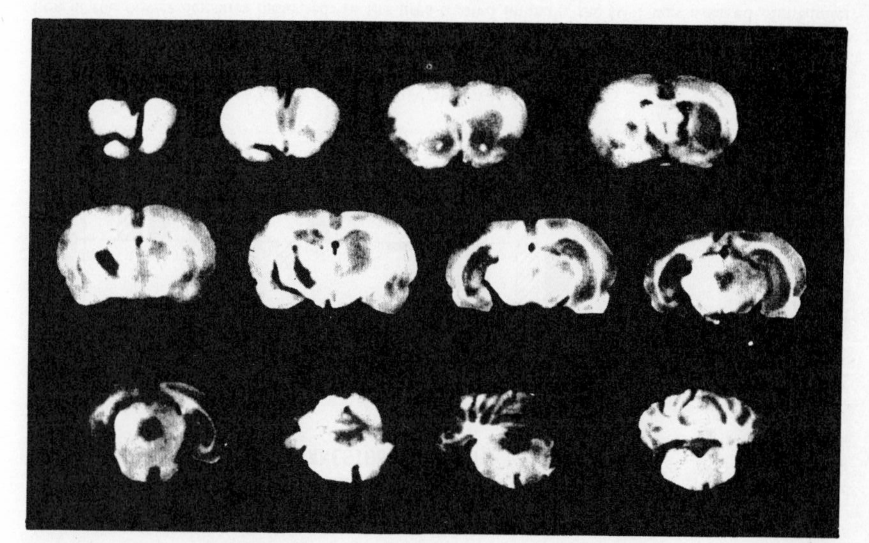

Fig. 4. Brain slices from a urea-treated rat. Fluorescein was infused into the femoral vein. 5 min later 10 M urea was injected into the right common carotid artery. 1 min later the brain was perfused and treated as in Figures 2 and 3. Large amounts of fluorescein-labeled albumin have penetrated across the opened blood-brain barrier into the brain parenchyma

removed, and it fluoresced in both the control and irradiated animals. Because Frey et al. (1975) considered their field measurements order of magnitude, the actual exposure power density could have been higher. For this reason, some rats were exposed to a power density of 75 mW/cm^2 peak (38 mW/cm^2 average), and the brain sections were examined for fluorescence. Again, fluorescein in the brain parenchyma could not be visually detected.

In contrast, penetration of fluorescein into the brain as a result of opening the blood-brain barrier with hypertonic urea (10 M) was plainly visible (Fig. 4). The studies using Evan's blue instead of fluorescein yielded identical results.

In order to confirm these observations, rat brain areas were chemically analyzed for fluorescein. Instead of injecting a 4-% solution of sodium fluorescein, 0.1 ml of 0.5-% fluorescein/kg of body weight was injected into the exposed femoral vein. The rat brain was then perfused as indicated in the brain slice study, and the brain areas were homogenized in butanol. After centrifugation, an aliquot of the clear organic phase was shaken with borate buffer (pH 10) and the buffer phase read in a spectrophotofluorometer at an activation wavelength of 482 nm and a fluorescence wavelength of 518 nm.

The results of this study are seen in Table 1. Though there tended to be an increase in fluorescein as power increased above 15 mW/cm^2, the only statistically significant values were those in the animals made hyperthermic in a warm-air environment. An interesting observation is that, in every case, the tissue fluorescein concentration in the 2 mW/cm^2 was insignificantly smaller than in the sham group.

Table 1. Fluorescein in rat brain in ng/g brain tissue ± S.D.

	Hypothalamus	Striatum	Midbrain	Hippocampus	Cerebellum	Medulla	Cortex
Sham	260 ± 111 (20)	194 ± 103 (20)	239 ± 82 (20)	198 ± 72 (20)	299 ± 100 (20)	301 ± 82 (20)	221 ± 86 (20)
2 mW/cm²	255 ± 86 (11)	142 ± 62 (12)	197 ± 37 (11)	167 ± 50 (12)	261 ± 88 (12)	257 ± 61 (12)	181 ± 35 (12)
15 mW/cm²	298 ± 66 (6)	166 ± 46 (6)	251 ± 68 (6)	157 ± 39 (6)	307 ± 36 (6)	315 ± 67 (6)	227 ± 30 (6)
25 mW/cm²	406 ± 162 (6)	185 ± 73 (6)	238 ± 86 (6)	202 ± 99 (6)	346 ± 142 (6)	294 ± 115 (6)	228 ± 88 (6)
75 mW/cm²	421 ± 233 (9)	235 ± 72 (9)	264 ± 102 (9)	246 ± 78 (9)	379 ± 115 (9)	375 ± 136 (9)	294 ± 102 (9)
Heated	490 ± 164[a] (6)	252 ± 93 (6)	310 ± 75 (6)	291 ± 118 (6)	441 ± 94[a] (6)	457 ± 99[a] (6)	321 ± 113 (6)

[a] $P < 0.01$ Numbers in parentheses are numbers of animals. Animals were exposed for 30 min to pulsed 1.2 GHz radiation (1000 pps, 0.5 ms pulse width) at the power densities indicated. Student's t-test was used for calculations of p values

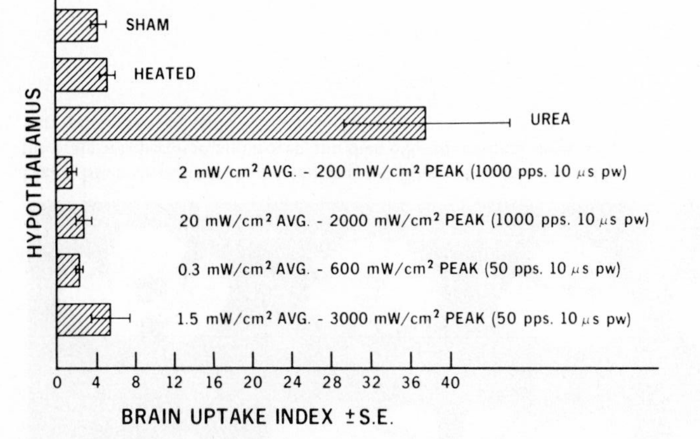

Fig. 5. Brain and rectal temperatures were made immediately after sham-irradiation or microwave irradiation at the power densities indicated

An appreciable hyperthermia was seen only in the animals exposed at a power density of 75 mW/cm² and in animals kept for 30 min in a 43° C environment (Fig. 5). Note that the sham-irradiated animals have markedly reduced rectal and brain temperatures due to the 30 min of anesthesia. Even when 2 mW/cm² of radiant energy is imposed, the barbiturate-induced hypothermia persists, though not as profound as in the sham group. This difference in degree of brain hypothermia (approximately 1° C) may account for the small difference in tissue fluorescein content between the sham and 2 mW/cm² noted above.

Brain Uptake of Mannitol

Figure 6 shows the results of 20 min of 1.2-GHz radiation on uptake of ^{14}C-mannitol in the hypothalamus. None of the radiation regimens increased the uptake of ^{14}C-mannitol into the brain parenchyma. In fact, there were decreases, if anything. For instance, the BUI at 2 mW/cm² is smaller than the sham-irradiated control, though these decreases are small and of doubtful significance. Oldendorf and Braun (1976) have pointed out the large variances obtained using an almost completely extracted reference (^3H_2O) when studying a minimally extracted test substance.

On the other hand, opening of the endothelial tight junctions with hypertonic urea greatly increases extraction of ^{14}C-mannitol. These results were true of the four other brain areas studied, namely hippocampus, cerebellum, medulla, and cortex (Fig. 7).

Oscar and Hawkins (1977) showed a "power window" for alteration of uptake of ^{14}C-mannitol in the brain; i.e., uptake increased up to approximately 1 mW/cm², leveled off, and declined when power density reached 2–3 mW/cm². Table 2 shows the results of our attempt to replicate these findings. Rats were radiated at 1.3 GHz for 35 min at 0.1, 1, 10, and 50 mW/cm² in the CW mode, and brain uptake of ^{14}C-mannitol was assessed. There was no difference in uptake at any of the power densities used.

Serotonin Uptake Study

Serotonin does not cross the intact blood-brain barrier. The results of an attempt to alter uptake of serotonin by microwave radiation are shown in Table 3. There was no uptake of the neurotransmitter except in the animals made hyperthermic in a warm-air environment.

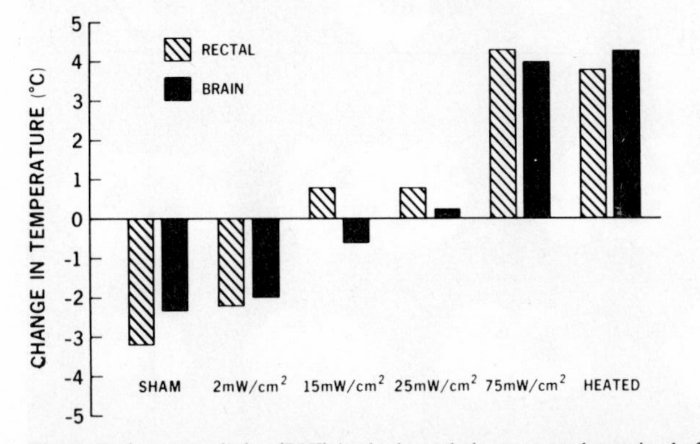

Fig. 6. Brain uptake index (BUI) in the hypothalamus was determined after 20 min of 1.3 GHz radiation at the power densities indicated. In the urea-treated animals, the BUI was assayed immediately after opening the blood-brain barrier with 10 M urea. The heated animals were placed in a warm-air environment for 20 min

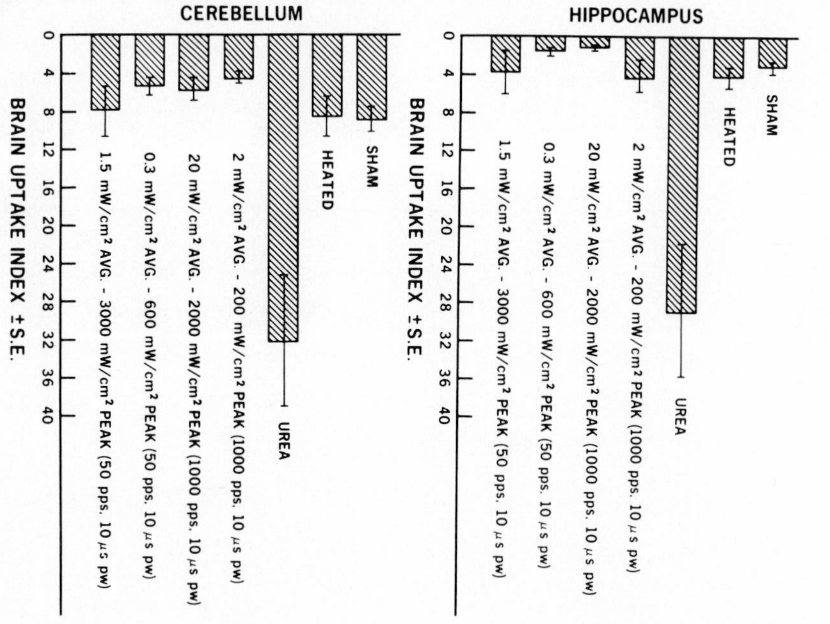

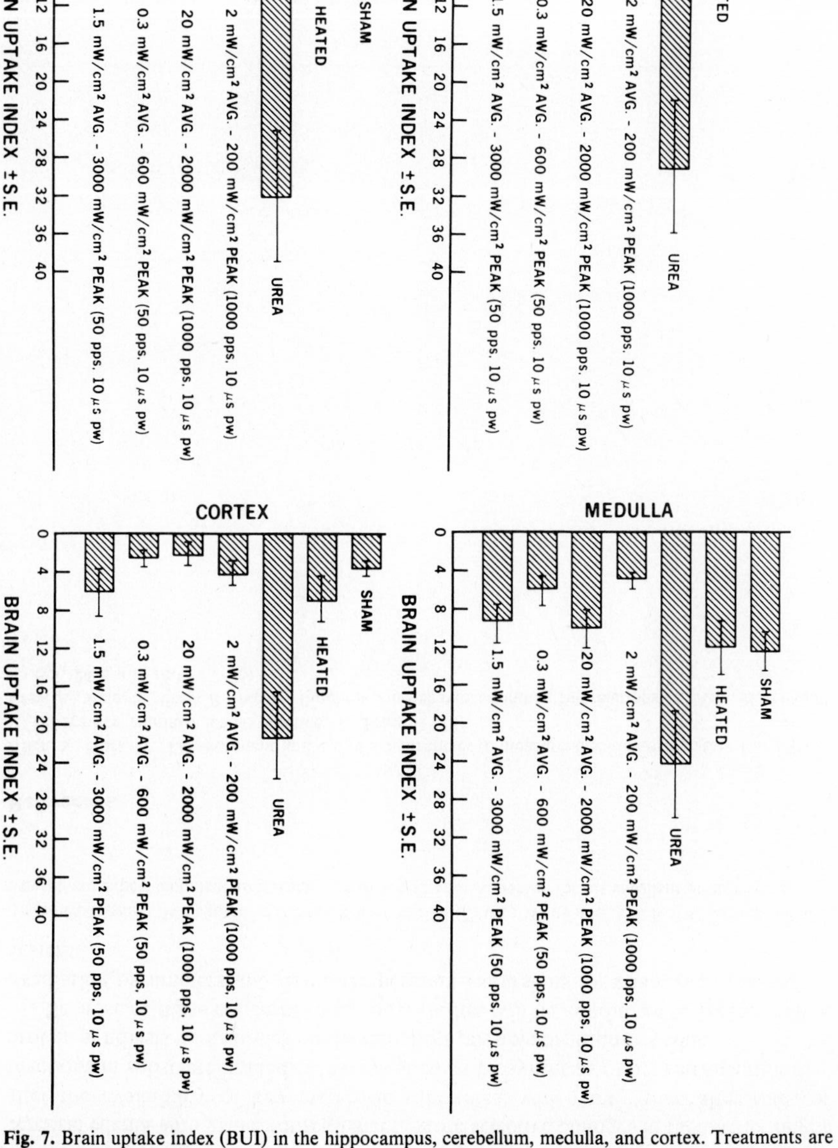

Fig. 7. Brain uptake index (BUI) in the hippocampus, cerebellum, medulla, and cortex. Treatments are as described in Figure 6

Table 2. Effect of 1.3 GHz radiation on permeability of blood-brain barrier for ^{14}C-mannitol

	Brain uptake index (BUI) ± S.E.
Sham	1.2 ± 0.1
0.1 mW/cm²	2.3 ± 0.6
1.0 mW/cm²	1.5 ± 0.2
10 mW/cm²	1.3 ± 0.1
50 mW/cm²	1.3 ± 0.1

Table 3. Effect of 1.3 GHz radiation on permeability of blood-brain barrier for serotonin

	ng serotonin/g brain ± S.D.
Sham	484 ± 118
Sham + serotonin	432 ± 84
Irradiated – 75 mW/cm²	430 ± 122
Irradiated – 75 mW/cm² + serotonin	507 ± 82
Heated	533 ± 75
Heated + serotonin	708 ± 56[a]

[a] $P < 0.01$ Student's *t*-test was used for calculation of *p* values

Discussion

The reasons for lack of agreement of these results and previous reports on microwave-induced alterations of the blood-brain barrier (Frey et al., 1975; Oscar and Hawkins, 1977) are not apparent. Interlaboratory replications of microwave exposure parameters are fraught with difficulties and uncertainties. Differences in measurement techniques, distance from emitters, emitter characteristics, and the like can all additively confound the replication attempt.

These data indicate that brain tissue must be made hyperthermic by microwave irradiation for changes in the blood-brain barrier permeability for either small or large molecules to occur. From these data, it would appear that alterations in the human blood-brain barrier are unlikely for exposures at 10 mW/cm². "Hot spots" in the brain may occur under certain conditions of frequency and power density in which local temperature increases may be several times the mean body temperature increase. Presumably, man can thermoregulate in 10-mW/cm² fields so that such "hot spots" would be trivial, and the temperature elevations incurred probably would not produce alteration of the blood-brain barrier.

Since the blood-brain barrier operates to maintain an optimal chemical environment for neurons, alterations of this environment should affect neuronal function. The significance of such alterations is not completely characterized. Of course, generalized breakdown of the barrier, with the possibility of free diffusion, could cause cerebral edema with grave consequences. More selective changes in the barrier might alter the availability of key metabolic substrates, which may then alter rates of reactions in substrate-limited pathways such as biosynthesis of neurotransmitters or protein synthesis with subtler but nevertheless possible profound results.

Because of these concerns, additional studies will be conducted to further define irradiation parameters and tissue temperature excursions for changes in the barrier system.

Acknowledgement. We express our appreciation to Sgt. Charles White and AlC Roy Gresham for their careful and expert technical assistance, and to Mr. Orville Anderson for his excellent photographs.

References

Allen, S., Hurt, W.: Development and use of a calorimeter to measure specific absorption rates in small laboratory animals. Radio Science (In Press, 1978)

Frey, A., Feld, S., Frey, B.: Neural function and behavior: defining the relationship. Ann. N.Y. Acad. Sci. **247,** 433–439 (1975)

Glowinski, J., Iversen, L.: Regional studies of catecholamines in the rat brain. J. Neurochem. **13,** 655–669 (1966)

Johnson, C., Durney, C., Barber, P., Massoudi, H., Allen, S., Mitchell, J.: Radiofrequency radiation dosimetry handbook. USAF School of Aerospace Medicine Report SAM-TR-76-35, Brooks AFB, Texas, 1976

Oldendorf, W.: Measurement of brain uptake of radiolabeled substances using a tritiated water internal standard. Brain Res. **24,** 372–376 (1970)

Oldendorf, W.: Permeability of the blood-brain barrier. In: The Nervous System, Vol. 1, The Basic Neurosciences, p. 279 (Brady, R., Ed.). New York: Raven Press 1975

Oldendorf, W., Braun, L.: (^{3}H) tryptamine and ^{3}H-water as diffusible internal standards for measuring brain extraction of radiolabeled substances following carotid injection. Brain Res. **113,** 219–224 (1976)

Oscar, K., Hawkins, T.: Microwave alterations of the blood-brain barrier system of rats. Brain Res. **126,** 281–293 (1977)

Polyashchuck, L.: Changes in permeability of histo-hematic barriers under the effect of microwaves. Dokl. Akad. Nauk Ukrain. **8,** 754–758 (1971): Transl. in JPRS 58203 (1973)

Snyder, S., Axelrod, J., Zweig, M.: A sensitive and specific assay for tissue serotonin. Biochem. Pharmacol. **14,** 831–835 (1965)

Received March 2, 1978

Advances in Microwave-Induced Neuroendocrine Effects: The Concept of Stress

SHIN-TSU LU, W. GREGORY LOTZ, AND SOL M. MICHAELSON

Abstract—Recent evidence indicates that neuroendocrine effects are induced by microwave exposure with a threshold intensity required for the onset of the response. The level of that threshold is dependent upon intensity and duration of exposure. The threshold can vary with the given endocrine parameter studied. The level of that threshold is yet unclear due to conflicting reports of effect in chronic or repeatedly exposed populations of man or experimental animals. The response of the endocrine systems appears to be a nonspecific stress reaction in the case of adrenocortical and growth hormone changes, but it is apparently a metabolically specific response to increased energy input in the case of pituitary-thyroid changes.

Introduction

NUMEROUS biological effects of microwaves have been reported in the literature and have been the subject of several reviews. The validity as well as the actual significance of many of these reported microwave bioeffects have elicited controversies regarding the setting of exposure standards. The suggestion of direct action by low-level microwave exposure on the central nervous and the endocrine systems apart from the well-established heating effect of microwaves has raised uncertainties in the characterization of the general effects of exposure to microwave energy.

At all levels of mammalian biological organization adverse environment elicits a complex array of nervous, endocrine, neurohumoral, and motor reactions to adjust body fluid balance, energy metabolism, and behavior to the needs concomitant with survival in a changed environment. The neuroendocrine system, a complex of hormone secreting glands, and the central nervous system function as a chemical regulatory system in mammals to control and regulate metabolism and growth and to protect the body from endogenous and exogenous alterations in homeostasis.

Neuroendocrine function is of considerable importance in the response of an organism to microwave exposure. However, the information available at the present time is not sufficient to clarify microwave-induced neuroendocrine effects, due in part to insufficient documentation of some of the available data. Since other reviews [1]–[4] have provided a general survey of this subject area, we will not attempt to reiterate what has been covered in those treatises. The objective of this presentation is to evaluate recent progress in the area of

Manuscript received May 8, 1979; revised August 7, 1979.

This paper is based on work performed with the U.S. Department of Energy at the Department of Radiation Biology and Biophysics University of Rochester under contract UR-3490-1694, and the U.S. Navy, Naval Aerospace Medical Research Laboratory.

S. T. Lu and S. M. Michaelson are with the Department of Radiation Biology and Biophysics, School of Medicine and Dentistry, University of Rochester, Rochester, NY 14642.

W. G. Lotz is with the Naval Aerospace Medical Research Laboratory, Naval Air Station, Pensacola, FL 32508.

Reprinted from *Proc. IEEE*, vol. 68, pp. 73–77, Jan. 1980.

microwave-induced neuroendocrine effects with respect to the current concepts of neuroendocrine control mechanisms.

STRESS

In its medical sense, *stress* is essentially the rate of wear and tear on the body and can be defined as a nonspecific response of the body to any demand. In any event, wear and tear is only the end result. The pathophysiologic picture of stress has been characterized as the General Adaptation Syndrome, which develops in three stages, the alarm reaction, the stage of resistance, and the stage of exhaustion [5]. The classic triad of alarm reaction (adrenocortical stimulation, thymicolymphatic hypotrophy, and gastrointestinal ulcer) denotes the stereotyped response of the body to any demand [5] that severely taxes the regulatory processes. The traid of the alarm reaction also points out the involvement of the hypothalamo-hypophyseal-adrenocortical (HHA) system and autonomic control. Only recently has the secretory pattern of adenohypophyseal hormones (other than adrenocorticotropin) been found to be involved in the nonspecific stress responses. It is now well established that in both female and male rats acute stress inhibits growth hormone (GH) secretion [6]-[11] and stimulates adrenocorticotropic hormone (ACTH) [12], [13] and prolactin release [11], [14]-[17]. Some discrepancies still exist, regarding the magnitude and pattern of the stress response of luteinizing hormone, follicle-stimulating hormone, and thyroid-stimulating hormone (TSH) response [18].

The most important aspect of the stress-induced secretory pattern of these adenohypophyseal hormones is the nonspecific character of the response. In general, these stress-induced hormonal changes are not related to the nature, but rather to the intensity and duration of the stressing agent. Standard laboratory stressors can easily be found in the experimental procedures that are used to investigate the biological effects of microwaves just as in other biological studies. Such commonplace procedures as handling, novelty of experimental environment and procedures, extreme environmental temperatures, forced muscular exercise, immobilization, transportation, noise, electrical shock, ether anesthesia etc. can act as stressors under certain conditions. Great care should thus be exercised to ensure that the changes in hormone levels are the response to a specific stressor in question, (i.e., microwaves in this case) rather than to some extraneous factors.

HYPOTHALAMO-HYPOPHYSEAL-ADRENOCORTICAL AXIS

Based on physiologic responses indicative of adrenocortical stimulation in dogs, Michaelson *et al.* [19] suggested that microwaves of high power density can act as a *stressor* affecting regulatory and integrative homeokinetic activity resulting in an alteration in homeostasis. In recent studies plasma corticosterone (CS) levels in rats exhibited a variable power density/threshold pattern of response, with a different threshold for 120-min exposure (20 mW/cm^2) than for 30- or 60-min exposure (50 mW/cm^2) to 2450 MHz CW [20]. For all these durations of exposure, a strong correlation was evident between mean colonic temperature and mean plasma corticosterone levels. These thresholds occurred with whole-body specific absorption rates (SAR) of 3.2 and 8.0 W/kg, respectively. An independent study confirmed the 50 mW/cm^2 as a threshold power density for 60-min exposure of 300-g rats (Lu *et al.*, unpublished observation). In other experiments that used sequential sampling techniques it was shown that plasma CS increases within 15-30 min of the start of exposure and falls sharply within 15-30 min after termination of exposure [21]. The response of the adrenal cortex is transient in all cases.

In contrast to the pronounced adrenocortical response observed in intact rats, plasma CS levels in acutely hypophysectomized rats exposed to 60 mW/cm^2 for 60 min were below control levels. The CS response to microwaves at 50 mW/cm^2 for 60 min was completely suppressed by pretreatment of normal rats with 3.2 μg dexamethasone/100 g body weight [22]. These results indicate that the microwave-induced CS response observed in rats is dependent upon adrenocorticotropin secretion by the pituitary, i.e., the adrenal gland is not directly stimulated by microwave exposure.

The involvement of higher organizational levels of control in the HHA axis response to microwaves was also indicated by Novitskii *et al.*, [23] who studied corticotropin releasing factor (CRF) levels in the median eminence, pituitary ACTH content and plasma 11-oxycorticosteroid (11-OCS) levels in 180-230 g Wistar rats whole-body exposed for 30 min to 0, 0.01, 0.1, 10, and 75 mW/cm^2 of 12.6 cm (2.6 GHz) microwaves. In these experiments the threshold intensity was 0.1 mW/cm^2 for increases in CRF, ACTH, and 11-OCS; the largest increases were seen at 1 mW/cm^2. The findings confirmed that the adrenocortical stimulation was a process mediated by the central nervous system. However, the dose rate-response relation was suggestive of a bell-shaped curve and a much lower threshold than the reports cited previously. The reasons for these discrepancies are uncertain at present.

When adult rats were exposed to 2450 MHz CW, 20 mW/cm^2 for 8 h, serum CS increase was significantly inhibited from the expected circadian elevation [24]. This inhibition of CS circadian elevation was also noted in rats exposed to 0.1 and 1 mW/cm^2 for 4 h [25]. Thus a dual action of microwaves on the HHA axis was demonstrated, in which low-intensity exposure (< 10 mW/cm^2) inhibited CS levels during the peak period of the CS circadian rhythm, while higher intensity exposures (> 25 mW/cm^2) stimulated CS secretion during any interval of the circadian periodicity. These divergent responses may indicate that microwave exposure actually represents two different types of action, depending upon intensity, that elicit different responses relative to the timing of the exposure with respect to circadian rhythms. Such periodic sensitivities have been noted with other stimuli used in the study of the pituitary-adrenal response [26]. Some reports indicate that there is a dissociation of the modes of pituitary-adrenal activation for the stress response and the rhymicity at the level of the hypothalamic CRF neurons. The divergent nature of the HHA axis responses to low- and high-intensity microwaves may be somewhat related to the dissociation of these two types of neural control over pituitary-adrenal function.

Adrenocortical stimulation is significant because it points to an influence of unfavorable conditions, so called *stress stimuli*, acting upon the whole organism [27]. Enhancement of corticosteroid activity during and after microwave irradiation could be an adaptive reaction [23], [28]. On the other hand, the less pronounced activation of the HHA system when exposed to microwaves (2.6 GHz) of 10 to 75 mW/cm^2 may indicate a pathological reaction to the effect of the microwaves [23]. The findings in repeated 30-min exposures to 2.6 GHz at 1 mW/cm^2 was considered to be evidence of *cumulative effect* and *exhaustion* by microwave exposure [23]. However, the pattern of this HHA reaction could very well fit into the stage of resistance of the general adaptation syndrome [5].

The few reports on the HHA axis response to chronic or repeated exposures are not all in agreement. Dumanskij and Sandala [27] reported adrenocortical stimulation in animals repeatedly exposed (8 h/day, 120 days) to power density well below 1 mW/cm^2 at 0.5 and 2.5 GHz. It is interesting to note that a threshold of exposure intensity was observed for the response, which is consistent with the majority of reports of an acute stimulation of pituitary-adrenal secretion by microwave exposure, but disagrees sharply in the intensity of the exposure at which such stimulation occurs. In contrast, Lenko *et al.* [29] found that rabbits exposed to 3 GHz (50 to 60 mW/cm^2) 4 h daily for 20 days tended to show an initial decline in urinary 17-hydroxy-corticosteroids, followed by a general return to normal. In between the above reports are the observations of chronic exposure studies done by Moe *et al.* [30] and Lovely *et al.* [31], in which no differences were noted in basal or *ether-stressed* levels of plasma corticosterone between control rats and rats exposed to 918 MHz for 3 weeks at 10 mW/cm^2 (3.6 W/kg), 10 h/day or for 13 weeks at 2.5 mW/cm^2 (0.9 W/kg), 10 h/day.

Adrenocortical stimulation has been generally accepted to be a result of a stressor stimulus, i.e. a level of stimulation that requires bodily adjustment to counteract the insult. There is consistent evidence that microwave exposure above 25 mW/cm^2 (~4 W/kg) stimulates the HHA axis in the rat, and that this stimulation is modulated by the central nervous system. The existing evidence is contradictory for exposures below 25 mW/cm^2, suggesting stimulation in some cases, inhibition in others, and no change in others. It may be that alterations in adrenocortical function at low-exposure intensity are smaller than the magnitude of the daily oscillation of this system and are modified by their timing with respect to the normal biological periodicities.

Hypothalamo-Hypophyseal-Thyroid (HHT) Axis

Among the functions of thyroid hormones are their effect on metabolic rate. Thyrotropin secretion is under the control of the central nervous system, as are the secretions of the adenohypophyseal hormones. Pituitary secretion of thyrotropin (TSH) has been shown to respond in a specific metabolic pattern to extreme environmental temperature [18], [32] and appears to respond in a nonspecific manner to other stressful stimuli [18]. Enlargement of the thryoid gland and increased radioactive iodine uptake, in some cases without clinical symptoms of hyperfunction has been reported among microwave workers [33]. Baranski *et al.* [34] showed a stimulatory influence of repeated microwave exposures (3 GHz, 5 mW/cm^2, 3 h daily for 4 months) on trapping and secretory activity of the rabbit thyroid gland.

Lu *et al.* [24] reported that serum thyroxine levels are transiently elevated in rats after exposure to 2.45 GHz at 1 mW/cm^2 for 4 h. This transient increase was not accompanied by changes in serum TSH [35]. Magin *et al.* [36] demonstrated that localized thyroid exposure (2.45 GHz) which resulted in thyroid temperature elevation can stimulate thyroid secretion in the absence of pituitary influence. Stimulation of thyroid function was also noted in rats exposed to a high power density (2.45 GHz, 70 mW/cm^2) for 1 h (Lu *et al.*, unpublished observations) or in rats exposed to 2.45 GHz at 40 mW/cm^2 for 2 h in which TSH had been suppressed at the time of exposure by pretreatment with triiodothyronine [38].

In contrast to the above described reports of thyroid stimulation, rats exposed to 15 mW/cm^2, 2.45 GHz for 60 h had a depression of serum thyroxine and protein-bound iodide levels and thyroidal radioiodine-concentrating ability [37]. Similarly, levels of thyroid hormone were found by Vetter [39] to decrease as power density of 2.45-GHz microwaves increased from 5 to 25 mW/cm^2. Lu *et al.* [24] also noted decreased thyroxine levels in rats exposed to 2.45 GHz at 20 mW/cm^2 for 4 to 8 h. The thyroid depression apparently reflects the inhibition of hypophyseal TSH secretion as evidenced by decreased circulating TSH prior to and accompanied by the decreases in serum thyroxine in rats exposed to 2.45-GHz microwaves at 10 mW/cm^2 for 1 and 2 h and 20 mW/cm^2 for 2 and 8 h [35].

Lotz *et al.* (unpublished observations) investigated the TSH levels in rats exposed to 2.45 GHz at 13 to 60 mW/cm^2 for 30, 60, and 120 min. A 30-min exposure did not affect the TSH levels; depressed TSH levels were noted in rats exposed at 30 mW/cm^2 (SAR = 4.8 W/kg) or higher for 60 min and 13 mW/cm^2 (SAR = 2.1 W/kg) or higher for 120 min. A high correlation between decreases in serum thyroxine and TSH was also reported in rats exposed to 2.45 GHz, 8 mW/cm^2, 8 h daily up to 21 days [40].

Thus microwaves may act on the HHT axis, by local thyroid stimulation and/or axial inhibition. The local thyroid stimulation is in contrast to the ACTH dependent adrenocortical stimulation caused by high intensity microwave exposure. The inhibition of the HHT axis by thermogenic microwave exposure is homeostatically appropriate to the increased heat load, requiring a lower level of metabolism. The response can be viewed as a specific stress response.

Growth Hormone (GH)

In addition to its known, but unclear role in mediating somatic growth of the organism, GH is an important component of the endocrine control of circulating metabolities. It is, like the other adenohypophyseal hormones, under the control of the central nervous system mediated through the hypothalamic inhibiting hormone, somatostatin, and a hypothalamic releasing factor. As noted earlier, changes in growth hormone secretion during stress is considered to be a nonspecific response to a stimulus. Unlike those for TSH or ACTH, the GH stress response is somewhat species dependent, with a decrease in rodents, and an increase noted in dogs and primates including man.

In the young rat (8 to 12 weeks) exposed to 10 mW/cm^2 (2.45 GHz) there is an increase in GH, whereas at 36 mW/cm^2 rat serum GH drops after 60 min of exposure, to significantly low levels. Lotz *et al.* [41] noted the threshold intensity for GH inhibition is 50 mW/cm^2 (SAR = 8.0 W/kg) for rats exposed to 2.45 GHz, for 30- and 60-min. For 2-h exposure, GH levels were lower than among sham-exposed controls after exposure at 13 mW/cm^2 with progressively lower GH at each successively higher power density. Part of this sensitivity may be attributed to the higher GH levels in the 2-h sham-exposed rats than 30- or 60-min shams. The significance of this report [41] is that the GH levels were determined in the same rats as those reported for the CS study [20]. While GH and adrenocortical responses to stress often occur together, the dissociation of these two responses in primates has been observed for certain stimuli [42].

Neuroendocrine/Metabolic Correlation

The acute reaction of the HHT to thermogenic levels of microwaves is to lower the hormone affecting resting meta-

TABLE I

r	Probability (P)	Functions
0.89	< 0.001	CT versus PD
0.85	< 0.001	CS versus PD
0.84	< 0.001	CS versus CT
−0.70	< 0.001	TSH versus PD
−0.74	< 0.001	TSH versus CT
−0.43	< 0.02	GH versus PD
−0.46	< 0.02	GH versus CT

CT = Colonic temperature, CS = Corticosterone, TSH = Thyrotropin, GH = Growth hormone, PD = Power density.

bolic rate. Phillips *et al.* [43] have shown that the decreases in the resting metabolic rate of male rats exposed to microwaves was dependent upon the quantity of absorbed energy. The threshold at 6.5 W/kg was approximately double the resting metabolic rate. The decreased resting metabolic rate is considered to reflect physiologic adjustment in contrast to the van't Hoff relationship describing the speed of chemical reactions, including those of metabolism, affected by temperature in such a way that the rate of reaction doubles with each 10°C rise in temperature.

Using the Pearson's moment–product coefficient r, certain correlations among hormone levels, colonic temperatures (CT), and power denisty (PD) have been noted (Lu *et al.*, unpublished observation). A significant correlation between colonic temperature and CS level was found in sham-exposed rats sacrificed between 1230 and 1930 h ($r = 0.36$; $p < 0.05$) or rats exposed to 1 to 70 mW/cm^2 for 1 h and sacrificed at 1230 h ($r = 0.84$; $p < 0.001$); 0.1 to 40 mW/cm^2 for 4 h and sacrificed at 1530 h ($r = 0.52$; $p < 0.001$). The temperature coefficient decreased in the following sequence: 20.2 µg/dl/°C sham; 15.3 µg/dl/°C 1 h exposed and 8.8 µg/dl/°C 4-h exposed rats. The temperature coefficient of CS was significantly lower in the 4-h exposed rats than the sham-exposed rats ($t = 1.70$; $df = 110$; $p < 0.05$; 1 tail) or 1-h exposed rats ($t = 3.11$, $df = 86$; $p < 0.005$ 1 tail). Other correlations in the rat exposed to 2.45 GHz (CW), 1–70 mW/cm^2 for 1 h are listed in Table I.

In sham exposed rats, the correlation was significant between CS and colonic temperature ($r = 0.36$; $p < 0.05$) and between TSH and colonic temperature ($r = -0.41$; $p < 0.02$); the correlation between GH and colonic temperature was not significant. The implications of these findings are several fold.

1) Dose-response curves of measured parameters can be mutually correlated in either sham or exposed animals.

2) Investigators have to be aware of the effects of body temperature on pituitary hormones in sham or exposed animals, so that precise biological effects can be allocated specifically to microwave exposure. Physical activity and environmental factors should be controlled to avoid variations affecting body temperature or thermoregulation between sham and exposed animals.

3) Threshold intensities can vary with how one accepts the normal range in sham-exposed or unexposed populations of animals.

deLorge [44] plotted a relationship between power densities at which disruption of ongoing operant behavior occurred with rats, squirrel monkeys, and rhesus monkeys to use as a basis for interspecies comparison. Although comparable comparisons for neuroendocrine parameters have not been made, Lotz [45] found that rats responded to 1.29-GHz microwaves (2-µs pulse 0.001-duty cycle) by increased corticosterone levels at 15 mW/cm^2 for 30 min or longer. The same pulsed-microwave exposure (1.29 GHz) was without effect on adrenocortical and GH secretion in the male adult rhesus monkey [45], [46] exposed at 20, 28, and 38 mW/cm^2 for 8 h. The increases in rectal temperature were 0.6, 0.6, and 1.5°C, respectively. These differences may suggest that neuroendocrine function in the rhesus monkey is more stable than in the rat during thermogenic microwave exposure. This illustrates the difference in the response of a given species of animal to a different frequency and the resulting different energy deposition and distribution.

Conclusion

The acute effects of microwaves on hypothalamo-hypophyseal function are increased adrenocorticotropic secretion, decreased thyrotropic secretion, and decreased GH secretion. These stereotyped changes can be observed simultaneously in rats acutely exposed to 2.45-GHz microwaves at 50, 60, or 70 mW/cm^2 for 1 h (Lu *et al.*, unpublished observation). The characteristics of these changes of hypo-physeal hormones constitute the pattern of stress reactions of animals. Because of their physiologic significance, these biological endpoints of microwave effects can serve as meaningful criteria for hazard evaluation if sufficient care has been incorporated into the design of chronic or repeated exposure experiments.

In the present state of the art, endocrine activity cannot be separated from the functional state of the neural network. The influence of endocrine function on body metabolism is also longer lasting than that due to neural disturbances. Nonspecific stress reactions to microwave exposure have to be isolated from extraneous factors that are usually associated with experimental procedures. Furthermore, evaluation of given endocrine parameters involves not only its perturbation, but also its recovery or manifestation of delayed response if such should occur.

For the most part, research on the biological effects of microwaves relates to the problem of hazard assessment for man. This requires determination of the absorbed energy required to cause deleterious changes in the body functions of experimental animals and extrapolation of the results of animal experiments to the exposure conditions of man—a process referred to as *scaling*.

Different body functions may be affected at different levels of microwave exposure. The most sensitive function could be taken as the determinant of a hazard level, but the possible differentiation between an effect and a deleterious change in function is an important question in hazard assessment. The sensitivity of neuroendocrine reactions to microwave/RF exposure can be indicative of unfavorable conditions. Neuroendocrine function can thus fulfill the requirements for a meaningful biological endpoint in hazard assessment.

The neuroendocrine data are consistent with the hypothesis that the adenohypophyseal responses are the integral result of CNS processing of multiple signals from many body locations, such that no single localization of absorbed energy is pivotal to the onset of a response. Factors, such as circadian rhythmicity, stimulus intensity, and perhaps interspecies differences are important in determining the pattern of these responses. Thus in addition to further studies to characterize the basic neuroendocrine response to microwave exposure, studies are needed to determine the physiological mechanism or mechanisms by which this regulatory system is affected during microwave exposure.

References

[1] S. M. Michaelson, W. M. Houk, N. J. A. Lebda, S.-T. Lu, and R. Magin, "Biochemical and neuroendocrine aspects of exposure to microwaves," *Ann. NY Acad. Sci.*, vol. 247, pp. 21–45, Feb. 1975.

[2] S. M. Michaelson, "Endocrine and biochemical effects," in *Microwave and Radiofrequency Radiation.* Copenhagen, Denmark; World Health Organization, Regional Office for Europe, 1977, sect. 7, pp. 18–23.

[3] S. Baranski and P. Czerski, "Endocrine and metabolic effects of microwave exposure," in *Biological Effects of Microwaves.* Stroudsburg, PA: Dowden, Hutchinson and Ross, Inc., 1977, ch. 4, pp. 122–126.

[4] S. F. Cleary, "Biological effects of microwave and radiofrequency radiation," *CRC Crit. Rev. Environ. Contr.*, vol. 7, pp. 121–166, June 1977.

[5] H. Selye, "The general adaptation syndrome and the diseases of adaptation," *J. Clin. Endocrinol.*, vol. 6, pp. 117–230, Feb. 1946.

[6] N. Kokka, J. F. Garcia, R. George, and H. W. Elliott, "Growth hormone and ACTH secretion: Evidence for an inverse relationship in rats," *Endocrinology*, vol. 30, pp. 735–743, Mar. 1972.

[7] R. Collu, J. C. Jequier, J. Letarte, G. Leboeuf, and J. R. Ducharme, "Effect of stress and hypothalamic deafferentiation on the secretion of growth hormone in the rat," *Neuroendocrinology*, vol. 11, pp. 183–190, 1973.

[8] G. M. Brown and J. B. Martin, "Corticosterone, prolactin and growth hormone responses to handling and new environment in the rat," *Psychom. Med.*, vol. 36, pp. 241–247, May-June 1974.

[9] J. D. Dunn, W. J. Schindler, M. D. Hutchins, L. E. Scheving, and C. Turpen, "Daily variation in rat growth hormone concentration and the effect of stress on periodicity," *Neuroendocrinology*, vol. 13, pp. 69–78, 1974.

[10] L. Krulich, E. Hefco, P. Illner, and C. B. Read, "The effects of acute stress on the secretion of LH, FSH, prolactin and GH in the normal male rat, with comments on their statistical evaluation," *Neuroendocrinology*, vol. 16, pp. 293–311, 1974.

[11] R. Collu and J. C. Jequier, "Pituitary response to auditory stress: Effect of treatment with α-methyl-p-tyrosine. Usefulness of a factorial mixed design for statistical analysis," *Can. J. Physiol. Pharmacol.*, vol. 54, pp. 596–602, Aug. 1976.

[12] H. Matsuyama, A. Ruhmann-Wemhold, and D. H. Nelson, "Radioimmunoassay of plasma ACTH in intact rats," *Endocrinology*, vol. 88, pp. 692–695, Mar. 1971.

[13] J. C. Buckingham and J. R. Hodges, "Hypothalamo-pituitary adrenocortical function in the rat after treatment with betamethasone," *Brit. J. Pharmacol.*, vol. 56, pp. 235–239, Feb. 1976.

[14] J. D. Neill, "Effect of stress on serum prolactin and luteinizing hormone levels during the estrus cycle of the rat," *Endocrinology*, vol. 87, pp. 1192–1197, Dec. 1970.

[15] I. Wakabayshi, A. Arimura, and A. V. Schally, "Effect of pentobarbital and ether stress on serum prolactin levels in rats," *Proc. Soc. Exp. Biol. Med.*, vol. 137, pp. 1189–1193, Sept. 1971.

[16] K. Ajika, S. P. Kalra, C. P. Fawcett, L. Krulich, and S. M. McCann, "The effect of stress and nembutal on plasma levels of gonadotropins and prolactin in ovariectomized rats," *Endocrinology*, vol. 90, pp. 707–715, Mar. 1972.

[17] J. D. Dunn, A. Arimura, and L. E. Scheving, "Effect of stress on circadian periodicity in serum LH and prolactin concentration," *Endocrinology*, vol. 90, pp. 29–33, Jan. 1972.

[18] P. Du Ruisseau, Y. Taché, P. Breau, and R. Collu, "Pattern of adenohypophyseal hormone changes induced by various stressors in female and male rats," *Neuroendocrinology*, vol. 27, pp. 257–271, 1978.

[19] S. M. Michaelson, R. A. E. Thomson, M. Y. El Tamami, H. S. Seth, and J. W. Howland, "The hematologic effects of microwave exposure," *Aerosp. Med.*, vol. 35, pp. 824–829, Sept. 1964.

[20] W. G. Lotz and S. M. Michaelson, "Temperature and corticosterone relationships in microwave-exposed rats," *J. Appl. Physiol.*, vol. 44, pp. 438–445, Mar. 1978.

[21] R. Guillet, W. G. Lotz, and S. M. Michaelson, "Time course of adrenal response in microwave-exposed rats," in *Proc. Annu. Meeting USRI* (Boulder, CO, Oct. 20–23, 1975), p. 316.

[22] W. G. Lotz and S. M. Michaelson, "Effects of hypophysectomy and dexamethasone on the rat's adrenal response to microwave irradiation," in *Abstracts of Scientific Papers, 1977 Int. Symp. Biol. Effects of Electromagnetic Waves* (Airlie, VA, Oct. 30–Nov. 4, 1977), p. 38.

[23] A. A. Novitskii, B. F. Murashov, P. E. Krasnobaev, and N. F. Markozova, "The functional condition of the system hypothalamus-hypophysis-adrenal cortex as a criterium in establishing the permissible levels of superhigh frequency electromagnetic emissions," *Voenn. Med. Zh.*, vol. 8, pp. 53–56, Aug. 1977.

[24] S.-T. Lu, N. Lebda, S. M. Michaelson, S. Pettit, and D. Rivera, "Thermal and endocrinological effects of protracted irradiation of rats by 2450 MHz microwaves," *Rad. Sci.*, vol. 12, supp., pp. 147–156, Nov.-Dec. 1977.

[25] S.-T. Lu, S. Pettit, and S. M. Michaelson, "Dual actions of microwaves on serum corticosterone in rats," *presented at Bioelectromagnetics Symp.*, (Seattle, WA, June 18–22, 1979).

[26] E. S. Redgate, "Central nervous system mediation of pituitary adrenal rhythmicity," *Life Sci.*, vol. 19, pp. 137–146, July 1976.

[27] J. D. Dumanskij and M. G. Sandala, "The biological action and hygienic significance of electromagnetic fields of superhigh and ultrahigh frequencies in densely populated areas," in *Biological Effects and Health Hazards of Microwave Radiation.* Warsaw, Poland: Polish Medical Publishers, 1974, pp. 289–293.

[28] I. R. Petrov and V. A. Syngayevskaya, "Endocrine glands," in *Influence of Microwave Radiation on the Organism of Man and Animals*, I. R. Petrov, Ed. Lenningrad, U.S.S.R.: Meditsina Press, 1970, ch. 2, pp. 31–41. (NASA TT F-708.)

[29] J. Lenko, A. Dolatowski, L. Gruszeki, S. Klajman, and L. Januszkiewicz, "Effect of 10-cm radarwaves on the level of 17-ketosteroids and 17-hydroxy-corticosteroids in the urine of rabbits," *Przeglad Lekarski*, vol. 22, pp. 296–299, 1966.

[30] K. E. Moe, R. H. Lovely, D. E. Meyers, and A. W. Guys, "Physiological and behavioral effects of chronic low level microwave radiation," in *Biological Effects of Electromagnetic Waves.* Rockville, MD: Bureau Radiol. Health, 1977, pp. 248–256 (HEW Pub. (FDA) 77-8010.)

[31] R. H. Lovely, D. E. Meyers, and A. W. Guy, "Irradiation of rats by 918 MHz microwaves at 2.5 mW/cm^2: Delineating the dose-response relationship," *Rad. Sci.*, vol. 12, supp., pp. 139–146, Nov.-Dec. 1977.

[32] S. Reichlin, J. B. Martin, M. Mitnick, R. L. Boshans, Y. Grimm, J. Bollinger, J. Gordon, and J. Malacara, "The hypothalamus in pituitary-thyroid regulation," *Rec. Prog. Horm. Res.*, vol. 28, pp. 229–286, 1972.

[33] M. I. Smirnova and M. N. Sadchikova, "Determination of the functional activity of the thyroid by means of radioactive iodine in workers with UHF generators," in *The Biological Action of Ultrahigh Frequencies*, A. A. Letavet and Z. V. Gordon, Eds. Moscow, U.S.S.R.: Acad. Medical Science, 1960, pp. 47–49. (JPRS 12471.)

[34] S. Baranski, K. Ostrowski, and W. Stodolnik-Baranska, "Functional and morphological studies of the thyroid gland in animals exposed to microwave irradiation," *Acta Physiol. Polonica*, vol. 23, pp. 1029–1039, 1972.

[35] S.-T. Lu, N. Lebda, and S. M. Michaelson, "Effects of microwave radiation on the rat's pituitary-thyroid axis," in *Abstracts of Scientific Papers, 1977 Int. Symp. Biol. Effects of Electromagnetic Waves* (Airlie, VA, Oct. 30–Nov. 4, 1977), p. 37.

[36] R. L. Magin, S.-T. Lu and S. M. Michaelson, "Stimulation of the dog thyroid by the local application of high intensity microwave," *Amer. J. Physiol.*, vol. 233, pp. E363–E368, Nov. 1977.

[37] L. N. Parker, "Thyroid suppression and adreno-medullary activation by low-intensity microwave radiation," *Amer. J. Physiol.*, vol. 224, pp. 1388–1390, June 1973.

[38] S.-T. Lu, N. Lebda, S. Pettit, and S. M. Michaelson, "Modification of microwave biological end-points by increased resting metabolic heat load in rats," *presented at Bioelectromagnetics Symp.* (Seattle, WA, June 18–22, 1979).

[39] R. J. Vetter, "Neuroendocrine response to microwave irradiation," *Proc. Nat. Electron. Conf.*, vol. 30, pp. 237–238, 1975.

[40] W. D. Travers and R. J. Vetter, "Low intensity microwave effects on the synthesis of thyroid hormones and serum proteins," *presented at 22nd Annu. Meeting, Health Physics Soc.* (Atlanta, GA, July 3–8, 1977).

[41] W. G. Lotz, S. M. Michaelson, and N. J. Lebda, "Growth hormone levels of rats exposed to 2450-MHz (CW) microwaves," in *Abstracts of Scientific Papers, 1977 Int. Symp. Biol. Effects of Electromagnetic Waves* (Airlie, VA, Oct. 30–Nov. 4, 1977), p. 39.

[42] G. M. Brown, J. A. Seggie, J. W. Chambers, and P. G. Ettigi, "Psychoendocrinology and growth hormone: A review," *Psychoneuroendocrinology*, vol. 3, pp. 131–153, 1978.

[43] R. D. Phillips, E. L. Hunt, R. D. Castro, and N. W. King, "Thermoregulatory, metabolic, and cardiovascular response of rats to microwaves," *J. Appl. Physiol.*, vol. 38, pp. 630–635, Apr. 1975.

[44] J. deLorge, "Disruption of behavior in mammals of three different sizes exposed to microwave: Extrapolation to larger mammals," in *Abstracts of Scientific Papers, 1978 Symp. Electromagnetic Fields Biol. Syst.* (Ottawa, Canada, June 27–30, 1978), p. 37.

[45] W. G. Lotz, "Endocrine function in rhesus monkeys and rats exposed to 1.29 GHz microwave radiation," *presented at Bioelectromagnetics Symp.*, (Seattle, WA, June 18–22, 1979).

[46] W. G. Lotz, "Neuroendocrine function in rhesus monkeys exposed to pulsed microwave radiation," in *Abstracts of Scientific Papers, 1978 Symp. on Electromagnetic Fields in Biol. Syst.* (Ottawa, Canada, June 27–30, 1978).

Microwaves Induce Peripheral Vasodilation in Squirrel Monkey

Abstract. *Vasomotor activity in cutaneous tail veins was indexed by changes in local skin temperature during exposure of the whole body to 12.3-centimeter continuous microwaves. At an ambient temperature (26°C) just below that at which tail vessels normally vasodilate, criterion dilation was initiated by 5-minute exposures to a microwave power density of 8 milliwatts per square centimeter. This intensity deposits energy equivalent to approximately 20 percent of the monkey's resting metabolic rate but produces no observable change in deep body temperature. Intensity increments of 3 to 4 milliwatts per square centimeter for 1°C reductions in ambient temperature below 26°C produced identical responses. That no vasodilation occurred during infrared exposures of equivalent power density suggests that noncutaneous thermosensitive structures may mediate microwave activation of thermoregulatory responses in the peripheral vasomotor system.*

In thermally neutral environments, the peripheral vasomotor response of warm-blooded (endothermic) species continuously provides fine control of body temperature. Physiological regulation of heat flow into or out of the body depends largely on autonomically controlled changes in the volume, rate, and distribution of blood supplied to the skin. The stimulus to constriction or dilation of cutaneous vessels is often peripheral as, for example, when localized or whole-body changes occur in the temperature of the skin (*1*). The stimulus can also originate centrally, in the absence of peripheral thermal events, when the deep body temperature rises (*2*). Rapid changes in peripheral vasomotor state have been produced in a variety of experimental animals by altering the temperature of the anterior hypothalamus with stereotaxically implanted thermode devices (*3*). In such experiments, dilation or constriction in highly vasoactive skin areas such as ears, tail, or extremities is often indexed by abrupt increases or decreases in local skin temperature.

Under specific exposure conditions and at relatively low intensities, electromagnetic energy of the microwave frequency range can produce body heating, often favoring deep tissues over the skin (*4, 5*). Exposure to intense microwaves raises the body temperature of animal subjects and interferes significantly with ongoing behavioral and physiological processes, thereby upsetting thermal homeostasis (*6*). On the other hand, investigations into the biological effects of low-intensity microwaves (often dubbed "nonthermal") have largely ignored the thermoregulatory consequences of such exposure, although subtle thermal effects caused by exposure to power densities of 1 to 10 mW/cm^2 have been suspected but not demonstrated (*7*). We now report that monkeys in a cool environment can be induced to vasodilate by brief whole-body exposures to microwaves at intensities that produce no ob-

Reprinted with permission from *Science*, vol. 207, no. 4437, pp. 1381–1383, Mar. 21, 1980.

servable change in deep body temperature. The characteristics of the evoked response resemble those produced by direct experimental heating of thermosensitive tissue in the hypothalamus.

Three adult male squirrel monkeys (*Saimiri sciureus*) served as subjects. They were restrained in a chair, one at a time, in the far field of a horn antenna inside an air-conditioned electromagnetically anechoic chamber (*8*). Rectal and four representative skin temperatures (abdomen, tail, leg, and foot) were monitored continuously with small copper-constantan thermocouples (*9*, *10*). After minimum 2-hour equilibration to a cool environment of constant temperature (range, 22° to 26.5°C), at which tail and extremities were fully vasoconstricted (*11*, *12*), the monkey underwent 5-minute exposures to 2450 ± 25 MHz continuous microwaves. Microwave power density (*10*), initially at a low level of 2.5 to 4 mW/cm^2, was increased at each successive exposure until a criterion tail vasodilation occurred. This criterion was defined by an abrupt and rapid rise in temperature of the tail skin that exceeded any increase in air temperature and that persisted after the end of the microwave exposure. Figure 1A shows an example of criterion tail vasodilation in one monkey equilibrated to an ambient temperature of 25°C. The microwave power density producing the response in this case was 10 mW/cm^2, which represents a whole-body energy absorption rate of 1.5 W/kg (*13*) or roughly 25 percent of the resting metabolic heat production of the squirrel monkey (*11*, *14*).

Microwave power densities below that which initiated tail vasodilation often increased the temperature of the air or skin areas other than the tail. Control experiments (Fig. 1B) demonstrated that tail vasodilation was not initiated passively as a result of elevated air temperature. This monkey, also equilibrated to a 25°C environment, failed to exhibit any change in peripheral vasomotor state when exposed to infrared radiation of equivalent power density to the microwaves (*15*). Taken together, these results suggest that thermosensitive structures other than those in the skin may be responsible for altered thermoregulatory responses during microwave exposure.

The microwave power density required to stimulate criterion tail vasodilation was directly related to the environmental temperature in which the monkey was restrained. Figure 2A summarizes the data from 16 experiments on three animals at discrete ambient temperatures that range downward from 26.5°C, the temperature at which the tail vessels of a sedentary monkey may dilate spontaneously (*11*, *12*). The plotted points describe a linear relationship which reveals that, to initiate a criterion tail vasodilation response above threshold, an increase of 3 to 4 mW/cm^2 in microwave power density is required for every 1°C reduction of ambient temperature. A second abscissa relates the data to absorbed microwave energy based on our dosimetry (*13*). Thus, in a 23°C environment, when the animal's metabolic heat production is elevated 2.5 to 3 W/kg above the resting level (*11*, *14*), microwave energy deposited at a rate approximating this metabolic elevation will

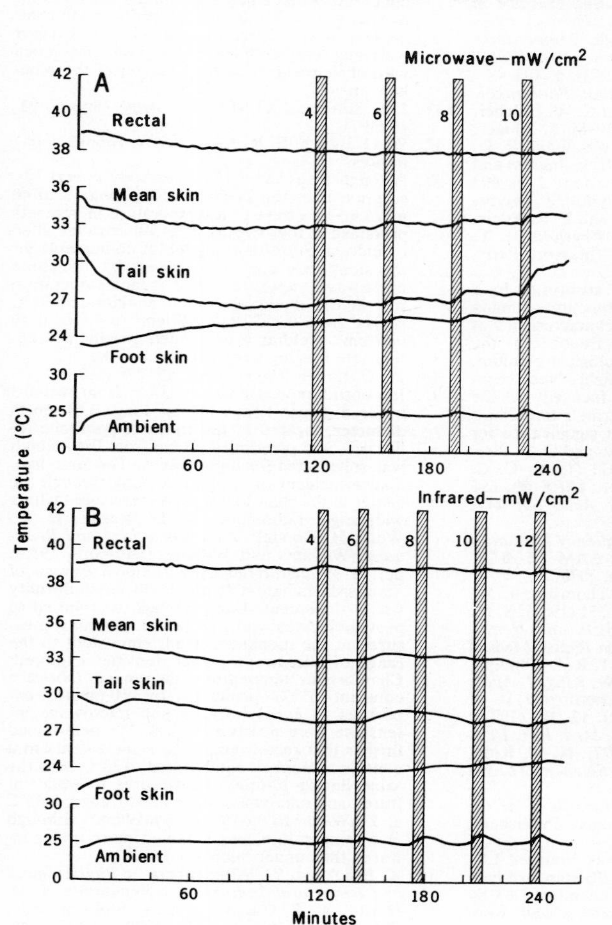

Fig. 1. Representative experiments showing effects of 5-minute exposures (hatched bars) to 2450-MHz continuous microwaves (A) or infrared radiation (B) on skin and deep body temperatures of squirrel monkeys equilibrated to a 25°C environment. (A) Exposure to microwaves of increasing power density (4, 6, 8, and 10 mW/cm^2) were separated by 15-minute recovery periods. Criterion vasodilation of tail skin, evidenced by rapid and persistent rise in tail temperature, occurred at 10 mW/cm^2 and prevented a rise in deep body temperature. (B) Control exposures to infrared radiation of increasing incident power density failed to induce tail vasodilation despite significant elevations in ambient temperature.

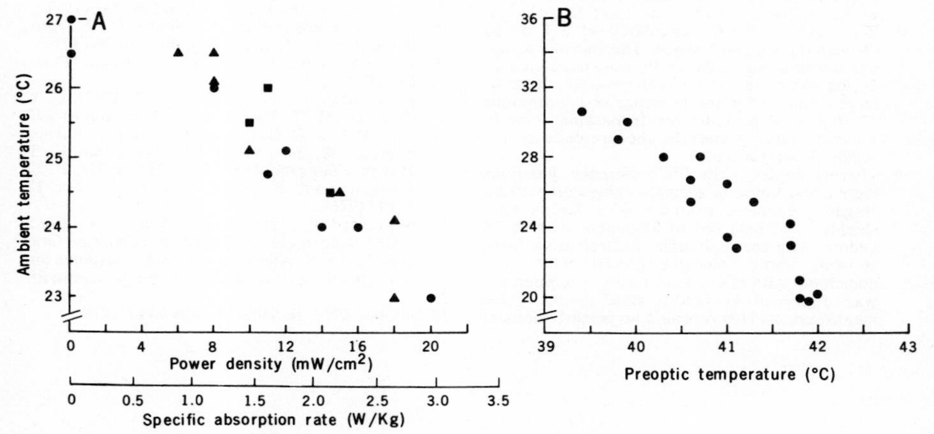

Fig. 2. Threshold functions for vasodilation of the squirrel monkey tail (A) produced by 5-minute whole-body exposures to 2450-MHz continuous microwaves and foot (B) produced by 5-minute periods of localized heating of the anterior hypothalamic-preoptic area through four implanted water-perfused thermodes. (A) Each data point represents the least microwave power density or absorbed microwave energy required to induce criterion tail skin warming in three monkeys at the ambient temperature indicated. (B) Each data point represents the least increase in preoptic temperature, measured bilaterally 2 mm from the thermodes, required to induce criterion foot skin warming in two monkeys at the ambient temperature indicated.

vasodilate the tail. Stability of the internal body temperature is thereby assured within the limits possible through changes in vasomotor state (uppermost tracing in Fig. 1A).

Thresholds for initiation of other thermoregulatory effector processes, such as shivering and panting, have been demonstrated to vary with both the ambient (skin) temperature and the local temperature either of the preoptic hypothalamus (*16*) or of other thermosensitive sites such as the spinal cord (*17*) as controlled by implanted thermodes. The form of such functions often resembles the relation presented in Fig. 2A. Recent research in our laboratory has determined how tail and foot vasodilation can be triggered by heating thermodes implanted in the hypothalamus of squirrel monkeys restrained in cool environments (*12*). Some of these results appear in Fig. 2B in a form that facilitates direct comparison with the adjacent microwave data. The striking resemblance lends credence to the hypothesis that low-intensity microwaves, absorbed in the vicinity of thermosensitive neural tissue in the hypothalamus and elsewhere (for example, posterior hypothalamus, midbrain, spinal cord, or deep viscera), can provoke immediate and dramatic changes in thermoregulatory effector response systems. Theoretical analyses (*18*) suggest that internal hot spots could occur under our experimental conditions (10 mW/cm², 2450-MHz microwaves) that would locally elevate temperature as much as 0.5°C. A possible neural mechanism would integrate many small afferent signals from diverse structures throughout the body into a strong effector command. The thermoregulatory neural substrate exhibits the diversity and integrative function appropriate to such a mechanism (*19*). Further, researches into the consequences of multiple thermal inputs confirm that the magnitude of the thermoregulatory effector response can be directly related to the number and sign of localized temperature changes occurring at discrete thermosensitive sites within the body (*20*).

ELEANOR R. ADAIR
BARBARA W. ADAMS
John B. Pierce Foundation Laboratory
and *Yale University,*
New Haven, Connecticut 06519

References and Notes

1. A. B. Hertzman and J. B. Dillon, *Am. J. Physiol.* **127**, 671 (1939); A. Hemingway and L. A. French, *ibid.* **174**, 264 (1953); A. Tholozan and E. Brown-Séquard, *J. Physiol. (Paris)* **1**, 497 (1858); G. W. Pickering, *Heart* **16**, 115 (1932); S. Robinson, in *Physiology of Heat Regulation and the Science of Clothing*, L. H. Newburgh, Ed. (Saunders, Philadelphia, 1949), p. 203; J. A. J. Stolwijk and J. D. Hardy, *J. Appl. Physiol.* **21**, 967 (1966).
2. This situation often occurs during exercise in cool environments. For a discussion of other relevant variables, see J. Bligh, *Temperature Regulation in Mammals and Other Vertebrates* (North-Holland, Amsterdam, 1973), p. 103.
3. H. G. Barbour, *Arch. Exp. Pathol. Pharmakol.* **70**, 1 (1912); A. Hemingway and C. W. Lillehei, *Am. J. Physiol.* **162**, 301 (1950); M. J. Kluger, *ibid.* **226**, 817 (1974); F. H. Jacobson and R. D. Squires, *ibid.* **218**, 1575 (1970); D. L. Ingram and K. F. Legge, *J. Physiol. (London)* **215**, 693 (1971); B. Kruk and A. F. Davydov, *J. Therm. Biol.* **2**, 75 (1977); W. C. Lynch and E. R. Adair, in *New Trends in Thermal Physiology*, Y. Houdas and J. D. Guieu, Eds. (Masson, Paris, 1978), p. 130.
4. The rate of microwave energy absorption by a biological target is a complex function of many factors including the physical characteristics of the radiation (particularly its frequency), the size and complexity of the biological medium, and body orientation in the field. Near resonance, microwaves may be focused by the body's curved surfaces to generate internal hot-spots that may have profound significance for thermoregulation [H. P. Schwan and G. P. Piersol, *Am. J. Phys. Med.* **33**, 371 (1954); C. C. Johnson and A. W. Guy, *Proc. IEEE* **60**, 692 (1972); O. P. Gandhi, *Ann. N.Y. Acad. Sci.* **247**, 532 (1975)].
5. C. H. Durney *et al.*, *Radiofrequency Radiation Dosimetry Handbook* (Report SAM-TR-78-22, Brooks Air Force Base, Texas, 1978).
6. S. M. Michaelson, R. A. E. Thomson, J. W. Howland, *Am. J. Physiol.* **201**, 351 (1961); S. M. Michaelson, in *Biological Effects and Health Hazards of Microwave Radiation* (Polish Medical Publishers, Warsaw, 1974), p. 1; R. D. Phillips, E. L. Hunt, R. D. Castro, N. W. King, *J. Appl. Physiol.* **38**, 630 (1975); M. E. Chernovetz, D. R. Justesen, A. F. Oke, *Radio Sci.* **12**, 191 (1977); J. deLorge, *U.S. Nav. Aerosp. Med. Res. Lab. (Pensacola) NAMRL 1236* (1977); N. W. King, D. R. Justesen, R. L. Clarke, *Science* **172**, 398 (1971).
7. *Biological Effects and Health Hazards of Microwave Radiation* (Polish Medical Publishers, Warsaw, 1974).

The Lucite restraining chair was mounted 1.85 m from the front edge of a 15-dB standard-gain horn antenna inside a lighted chamber 1.83 by 1.83 by 2.45 m. The interior chamber walls were covered with 20-cm pyramidal microwave absorber (Advanced Absorber Products type AAP-8) to minimize reflections. The long axis of the monkey's body was aligned with the electric vector of the incident plane wave (*E* polarization). Air (± 0.5°C) circulated at 1.1 m/sec through the anechoic space. The animal was under constant surveillance by television camera during the 4- to 5-hour test sessions; sessions were conducted in the presence of a continuous 73-dB (sound pressure level) masking noise to prevent auditory cues to the presence or absence of microwaves.

9. Thermocouples with 0°C reference junctions were constructed in special configurations from 36-gauge copper-constantan wire. Leads were shielded and held out of alignment with the *E* vector. Any thermocouple electromotive force showing abrupt changes greater than 4-μV coincident with microwave onset or termination was discarded as inadmissible datum. Field measurements (*10*) revealed no perturbations of the microwave field by the fine wires at the monkey's location.
10. Microwaves generated by a Cober (model S2.5W) source were fed to the antenna through standard waveguide components. Calibrations to determine far-field uniformity were made with a broadband isotropic radiation detector (Narda model 8306 B). Field intensity was mapped at 12-cm intervals across a 1 by 1.5 m plane passing through the center of the restraining chair location orthogonal to the incident microwave. The maximum nonuniformity of the central 50 by 50 cm of this plane, 8 percent with the chair absent, increased an additional 5 percent with the chair present. Power densities specified in this report were measured with the Narda probe positioned with chair present, at the location of the monkey's head.
11. J. T. Stitt and J. D. Hardy, *J. Appl. Physiol.* **31**, 48 (1971).
12. W. C. Lynch, E. R. Adair, B. W. Adams, *ibid.*, in press.
13. A rough assessment of whole-body energy absorption over the power density range 5 to 40 mW/cm² was based on temperature increments produced at four depths in a 1.1-liter saline-filled cylindrical Styrofoam model (of comparable dimensions to a squirrel monkey) by 10-minute microwave exposures. The mean temperature rise in the liquid above an equilibrated 35°C ranged from 0.1°C at 5 mW/cm² to 0.6°C at 40 mW/cm², yielding a calculated specific absorption rate ranging from 0.5 to 5.8 W/kg.
14. W. C. Lynch, *Physiologist* **19**, 279 (1976).
15. In control experiments, radiation from two T-3 infrared quartz lamps (41 cm long, 0.64 cm in diameter, located at the focus of parabolic reflectors, and positioned 60 cm from the animal) was substituted for microwaves. The lamp irradiance incident on a plane passing through the center of the chair location was measured with a wide-angle radiometer [J. D. Hardy, H. C. Wolff, H. Goodell, *Pain Sensations and Reactions* (Williams and Wilkins, Baltimore, 1952), pp. 73–79] calibrated by a National Bureau of Standards radiation lamp. Field nonuniformity was < 1 percent. Lamp voltage was varied to provide incident infrared power densities, measured at the monkey's head, equivalent to the range of microwave power densities explored. Chamber air temperature increments (above a constant 35°C) produced by 10-minute exposures to equal infrared and microwave intensities were nearly identical. We determined further that rectal temperature increments in a conscious monkey equilibrated to 33°C were the same during 10-minute exposures to equal infrared and microwave intensities (range, 0.05°C at 5 mW/cm² to 0.65°C at 20 mW/cm²) although skin temperature was elevated more under infrared than under microwaves.
16. K. Brück and W. Wünnenberg, in *Physiological and Behavioral Temperature Regulation*, J. D. Hardy, A. P. Gagge, J. A. J. Stolwijk, Eds. (Thomas, Springfield, Ill., 1970), p. 777; M. Cabanac, J. Chatonnet, R. Philipot, *C. R. Acad. Sci.* **260**, 680 (1965); J. Chatonnet, M. Cabanac, M. Mottaz, *C. R. Soc. Biol.* **158**, 1354 (1964).
17. C. Jessen, *J. Physiol. (London)* **264**, 585 (1977).
18. H. N. Kritikos and H. P. Schwan, *IEEE Trans. Biomed. Eng.* **23**, 168 (1976); *ibid.* **26**, 29 (1979).
19. J. D. Guieu and J. D. Hardy, *J. Physiol (Paris)* **63**, 253 (1971).
20. E. R. Adair, *Physiol. Behav.* **7**, 21 (1971); C. Y. Chai and M. T. Lin, *J. Physiol. (London)* **225**, 297 (1972); J. D. Guieu and J. D. Hardy, *J. Appl. Physiol.* **28**, 540 (1970); C. Jessen and E. T. Mayer, *Pfluegers Arch.* **324**, 189 (1971); R. O. Rawson and K. P. Quick, *Israel J. Med. Sci.* **12**, 1040 (1976).
21. Supported by grant 77-3420 from the Air Force Office of Scientific Research. We thank H. Graichen and S. J. Allen for valuable assistance and J. D. Hardy for his continuing encouragement.

15 October 1979; revised 10 December 1979

DISRUPTION OF BEHAVIOR IN MAMMALS OF THREE DIFFERENT SIZES EXPOSED TO MICROWAVES: EXTRAPOLATION TO LARGER MAMMALS*

J. de Lorge
Naval Aerospace Medical Research Laboratory
Pensacola, FL 32508, USA

ABSTRACT

In three separate studies albino rats, squirrel monkeys, and rhesus monkeys were exposed to 2.45-GHz irradiation (100% amplitude modulated at 120 Hz) under far field conditions in anechoic chambers. Incident radiation at power densities from 0 to 75 mW/cm^2 was measured in the absence of the animal. All animals were performing on operant schedules for food reinforcement during the microwave exposures. Rats worked unrestrained in a response chamber of Styrofoam while monkeys worked restrained in Styrofoam chairs. Rectal temperature was measured continuously during exposures of the monkeys and measured immediately after exposures in rats. Exposure sessions lasted 60 min and were repeated on

*Opinions or conclusions contained in this paper are those of the author and do not necessarily reflect the views of the endorsement of the Navy Department. This series of studies was conducted under work unit number 62758N, ZF51.524.0150037 sponsored by the Naval Medical Research and Development Command. Mr. C. S. Ezell was responsible for the running of the monkey studies and construction of associated equipment and much appreciation goes to him for this assistance. Thanks are also extended to Mrs. K. Venner for her typing assistance. The animals used in this study were handled in accordance with the Principles of Laboratory Animal Care established by the Committee on the Guide for Laboratory Animal Resources, National Academy of Science - National Research Council.

a daily basis. Stable performance on the operant schedules was achieved but was disrupted by irradiation in all three species: behavior of rats was perturbed at lower power densities; of squirrel monkeys at intermediate levels; and of the rhesus at higher levels. When the averages of the thresholds of disruption (28, 45, and 67 mW/cm^2) are plotted as a function of body mass (0.3, 0.7, and 5 kg) a semilog relationship becomes evident. Extrapolation along the resulting curve allows one to predict the power densities needed to disrupt ongoing operant behavior in larger animals such as man. In addition, the power densities associated with behavioral disruption approximate those power densities that produce an increase in rectal temperatures of at least 1°C (above control levels) in the corresponding animals. The conclusion is that under the environmental conditions of these studies a well controlled behavior is not disrupted in most mammals by microwaves until the energy deposited in the subject produces a measurable and sustained increase in core temperature.

INTRODUCTION

Although much research has been conducted on the biological effects of microwave irradiation little effort has been directed at predicting effects on man. The work that has concerned man traditionally dealt with temperature elevations or areas suggesting hazards to health. Exceptions to this general trend are found in the Russian literature (see for example Czerski and Baranski, 1976). If effects other than those caused by heat generation are elicited by microwaves, models predicting something other than thermal deposition or health hazards must be developed.

Our laboratory has taken a many-faceted approach to the problem of predicting microwave biological effects. Part of the approach has been to establish, in mammals of three distinctly different body masses, a well-controlled operant behavior requiring similar performances. The assumption is made that the chosen behavioral tasks should make demands on comparable elements of the central nervous system of each species and also from larger animals such as man. When animals of differing mass are engaged in such behavior, one should then be able to determine the minimum power density at a given set of microwave parameters that is needed to disrupt or otherwise influence performance. Having identified such thresholds one should be able to construct a scale and then extrapolate to animals of larger body mass. The purpose of this paper is to describe the generation of such a scale and to present a formula for predicting the power density necessary to disrupt behavior when animals, including man, are exposed to 2450-MHz microwave irradiation. Each of the three contributing experiments has been published previously (Sanza and de Lorge, 1977; de Lorge, 1975, 1977).

METHOD

SUBJECTS

Four male rats of the Sprague-Dawley strain with an average body mass of 315 g were used in one experiment (Sanza and de lorge, 1977). Three male squirrel monkeys (*Saimiri sciureus*) with an average body mass of 697 g (de Lorge, 1977) and three male rhesus monkeys (*Macaca mulatta*) with an average body mass of 4.8 kg (de Lorge, 1975) were subjects in the other experiments.

APPARATUS

A Holaday magnetron, model HI-1200, was the source of the 2450-MHz microwaves which were amplitude modulated at 120 Hz. The microwaves were transmitted into anechoic exposure chambers via waveguide and a 15-dB standard gain horn, Narda model 644. The polarization of the électric field was parallel to the long axis of the rat operant conditioning chamber, parallel to the front-back axis of the squirrel monkey chair and parallel to the long axis of the sitting rhesus monkey. The rat and squirrel monkey studies occurred in the identical anechoic chamber with the horn above the animal whereas the rhesus monkey was exposed from the front in a larger anechoic chamber. Field measurements were obtained with a National Bureau of Standards probe, EDM 1C, at various points within the space to be occupied by the animal. The areas corresponding to the center of each animal's head were used as reference points. All exposures and measurements were in the far field.

The rats worked unrestrained in a Styrofoam operant conditioning box and the monkeys worked while seated in restraint chairs of Styrofoam.

The anechoic chambers were ventilated and equipped with masking noise. Room temperature and humidity were continuously recorded during the experiments. Room temperature averaged 24 °C for the rat study and 23 °C for the squirrel monkey study while the averaged relative humidity for all three studies was 70 percent. The room temperature for the rhesus monkey study averaged 22.5 °C.

Rectal temperatures were obtained during experiment sessions from the monkeys by inserting a Yellow Springs Instruments' rectal probe into the anus of the sitting animal. Reference probes were located below the chairs. Similar temperatures were obtained from similar-sized rats in the same exposure chamber immediately after exposure.

Irradiation at power densities of 8.8, 18.4 and 37.5 mW/cm^2 was used in the rat experiment. In the squirrel monkey study, power densities of 10, 20, 30, 40, 50, 60, 65, 70, and 75 mW/cm^2 were used. In the rhesus monkey study power densities of 4, 16, 32, 42, 52, 62, and 72 mW/cm^2 were used. All data in this report were obtained during exposure sessions of 60-min.

PROCEDURE

All animals were trained to respond by pressing or pulling levers for food pellets on operant schedules of reinforcement. All animals were deprived of food prior to training and generally worked at weights lower than their normal ad lib feeding body weight. The rats worked at 80 percent of their free-feeding body mass. The squirrel monkeys worked at 76 percent of their free-feeding body mass and the rhesus monkeys worked at 95 percent of their free-feeding body mass.

The rats performed on a fixed-interval 50-sec (FI 50-sec) schedule of food reinforcement. On this schedule a rat is required to press the lever at least once after 50 sec have transpired. When the lever press occurs at the end of 50 sec a 45-mg food pellet (Noyes) is dispensed into a food aperture next to the response lever.

The squirrel monkeys performed on a variable-interval 1-min (VI 1-min) schedule of food reinforcement. This schedule required the monkey to respond on one lever on its right which produced either a 0.5-sec illumination of a red light, most of the time, or a 10-sec illumination of a blue light on the average of once a minute. When the blue light appeared the monkey had to press a lever on its left to obtain a 190-mg (Noyes) monkey pellet that was delivered into a recess on the chair top. Responses on the left lever in the absence of the blue light were not reinforced.

The rhesus monkeys performed on a VI 1-min schedule of food reinforcement (a 0.86-g Purina Monkey Chow pellet). This schedule was the same as that utilized with the squirrel monkey except that the stimuli were two different tones instead of lignts. A response by a rhesus on a lever to its right produced either a 0.5-sec tone of 1070 Hz, most of the time, or a 2740 Hz tone on the average of once a minute, which remained on until a lever at the left was pressed. A response on the left lever delivered a food pellet into a recess on top of the restraint chair. A response on the left lever in the absence of the higher tone was not reinforced.

The animals were all trained to respond stably prior to exposure to microwaves. Sham exposures were given between exposures and occurred aperiodically in all studies. Sessions were usually scheduled during the 5-day work week. Water was not available during experimental sessions.

RESULTS AND DISCUSSION

The typical result of microwave exposure was a decrement in response rate and an increment in the pause time following a reinforcement. These effects were clearly seen in the rat and rhesus during exposure and in the squirrel monkey following exposure. One of the squirrel monkeys actually increased his response rate at the highest power densities. Other effects were transient and occurred at either the beginning or the end of irradiation. On these occasions the animals would generally pause in the midst of responding. Details of these results may be obtained from the original articles (Sanza and de Lorge, 1977; de Lorge, 1975, 1977).

The effect of microwaves on the operant tasks used in this series of studies tended to be rather discrete in most individual animals. That is, with the exception of the squirrel monkeys, the behavior of an individual animal tended to be either unperturbed or greatly disturbed when the microwaves reached a threshold of perturbation. The same was not true of body temperature. With animals of all three species there were gradual increases of rectal temperature as the power density increased, as shown in Figure 1. The rat data are symbolized by filled circles in Figure 1 and are from similar-sized rats irradiated in a similar chamber and with a source similar to that used by Lotz and Michaelson (1977). The other rat data were obtained by William Houk in an unpublished study

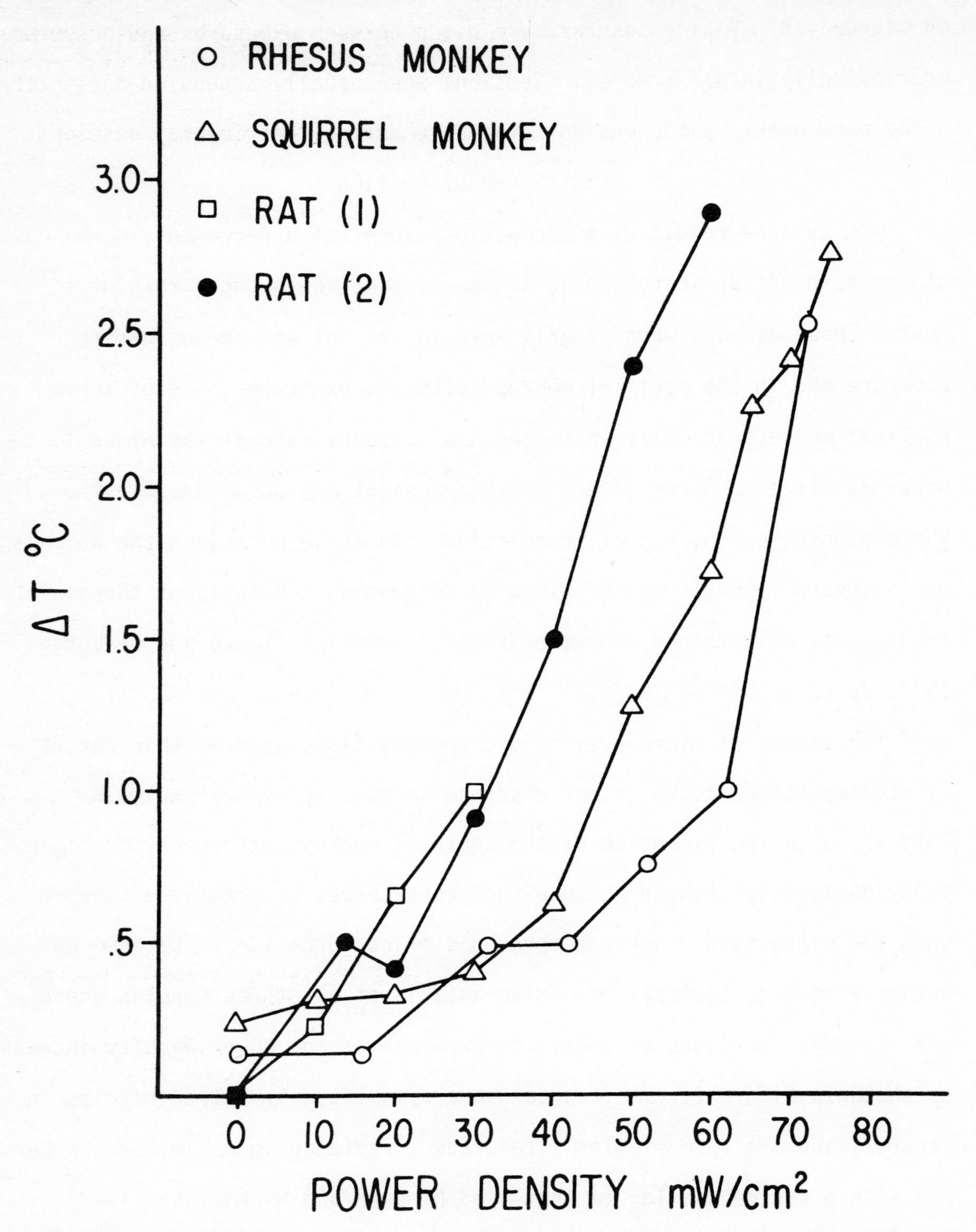

Fig. 1 The change in rectal temperature of various groups of animals as a function of 60-min exposures to 2450 MHz microwaves.
Rat (1) from our laboratory. Rat (2) from a study by Lotz and Michaelson (1978).

that was conducted in our laboratory in the same experimental arrangement as the rat behavioral study.

By repeatedly exposing the animals to radiation at various power densities, it was possible to calculate an average power density where disruption of responding occurred in the case of each separate animal. The medians of these power densities were then used to calculate disruption thresholds in each experiment. The threshold for disruption of lever responding during 60-min exposures to 2450 MHz was at 28 mW/cm^2 in rats, 45 mW/cm^2 in squirrel monkeys and 67 mW/cm^2 in rhesus monkeys. These thresholds are the median power densities between those levels where no effects on responding were seen and those levels where effects were first seen, hence the numbers are interpolations. An approximation of the average specific absorption rate (SAR) in W/kg for these thresholds is obtained from Durney et al (1978). The average SAR corresponding to the threshold of disruption for the rat is 5.8 W/kg; that for the squirrel monkey ranges from 2.5 to 4.5 W/kg depending upon whether or not the monkey is sitting with its back straight (2.5 W/kg) or curled over in its typical sitting position (4.5 W/kg); that for the rhesus monkey is 4.7 W/kg. Although more accurate approximations of SAR's for the monkeys are not available Lotz and Michaelson (1978) determined calorimetrically an SAR of 4.8 W/kg in similarly sized rats exposed to almost identical conditions (30 mW/cm^2). Hence, the present thresholds of disruption in all three species probably occur at an average SAR between 4 and 5 W/kg. These approximations are somewhat lower than the values obtained by Lin, Guy and Caldwell (1977) using rats exposed to near-field radiation at 918-MHz. They observed no disruption of operant responding until a power density of 40 mW/cm^2 was applied. This level corresponded to an average SAR of 8.4-W/kg.

When the threshold values in mW/cm^2 are plotted on semilog paper with power density on the linear x axis and body mass on the logarithmic y axis a linear function results. The equation for this curve is P (power density) = 14.14 $\log_e$ W(body mass) + 46.42. Such a plot is shown by the circles in Figure 2. With this equation one would predict that a 70-kg man would exhibit behavioral disruption of a simple operant task at a power density of 106 mW/cm^2 when exposed to 2450-MHz microwaves for a period of one hour. The effect should occur before the 60 minutes has elapsed.

Extrapolation to man under these circumstances is questionable but no more so than that from the heat-deposition models. In the present studies a comparative approach, as recommended by Michaelson (1974), has been used. Animals that are of the same order as man have been used. And an animal, the rhesus monkey--one that is considered to be an excellent analogue for investigating thermoregulation in man (Smiles, Elizondo, and Barney, 1976)--has been used to form one point on the scale. Finally, disruption of a well-controlled behavior, similar to behavior a human could be engaged in, was used as the dependent variable on which to base the scale.

As seen in Figure 2 the points also correspond well with the points generated by plotting the power density needed to raise the rectal temperature of these same animals at least 1 °C above baseline (sham-exposure) levels. These temperatures are indicated by the triangles in

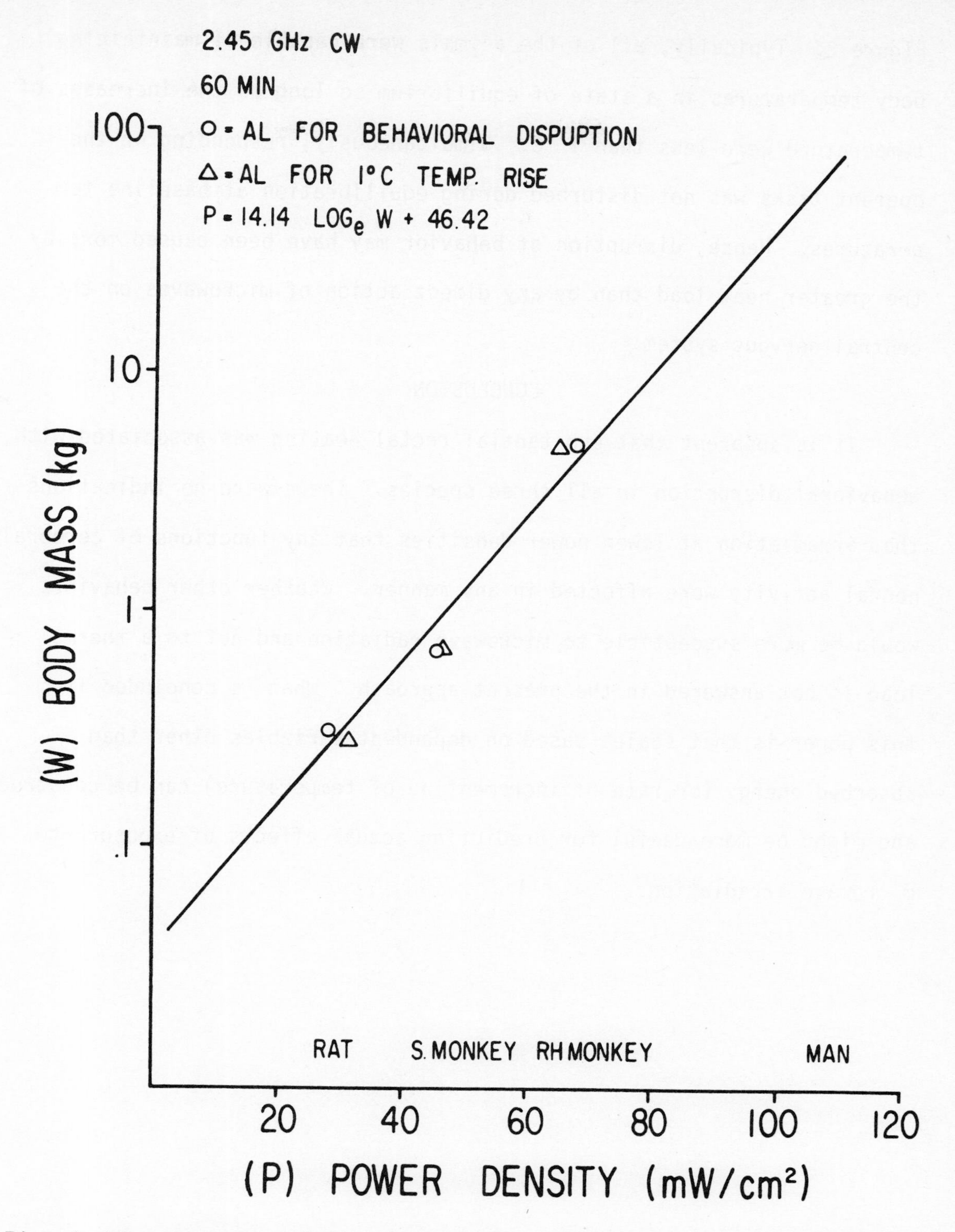

Fig. 2 The thresholds (AL) of behavioral disruption of an operant task for three different sized animals as a function of power density and body mass. The body mass is plotted on a logarithmic scale.

Figure 2. Typically, all of the animals were capable of maintaining body temperatures in a state of equilibrium so long as the increases of temperature were less than 1 °C. Simultaneously, responding on the operant tasks was not disturbed during equilibration at baseline temperatures. Hence, disruption of behavior may have been caused more by the greater heat load than by any direct action of microwaves on the central nervous system.

CONCLUSION

It is apparent that substantial rectal heating was associated with behavioral disruption in all three species. There were no indications that irradiation at lower power densities that any functions of cerebral neural activity were affected in any manner. Whether other behaviors would be more susceptible to microwave radiation and not to a thermal load is not answered in the present approach. What is concluded from this paper is that scales based on dependent variables other than absorbed energy (or rate of incrementing of temperature) can be constructed and might be more useful for predicting actual effects of exposure to microwave irradiation.

REFERENCES

Czerski, P. and Baranski, S. _Biological Effects of Microwaves_, Strouds burg: Dowden, Hutchinson and Ross, 1976.

de Lorge, J. The effects of microwave radiation on behavior and temperature in rhesus monkeys, In: _Biological Effects of Electromagnetic Waves-Selected Papers of the USNC/URSI Annual Meeting_, edited by C.C. Johnson and M.L. Shore. Washington, D.C.: HEW, 1975, vol. 2, p. 158-174.

de Lorge, J. Operant behavior and colonic temperature of squirrel monkeys _(Saimiri sciureus)_ during microwave irradiation. NAMRL-1236. Pensacola, Florida: Naval Aerospace Medical Research Laboratory, June 1977. (AD A043706)

Durney, C.H., Johnson, C.C., Barber, P.W., Massoudi, H., Iskander, M.F., Lords, J.L., Ryser, D.K., Allen, S.J., and Mitchell, J.C., (1978), Radiofrequency radiation dosimetry handbook, 2nd edition, SAM-TR-78-22. Brooks Air Force Base, Texas 78235: USAF School of Aerospace Medicine.

Lin, J.L., Guy, A.W., and Caldwell, L.R. Thermographic and behavioral studies of rats in the near field of 918-MHz radiation. _IEEE Transactions on Microwave Theory and Techniques_, _MTT-25_, 1977.

Lotz, W.G., and Michaelson, S.M. Temperature and corticosterone relationships in microwave-exposed rats. _Journal of Applied Physiology_, _44_: 438-445, 1978.

Michaelson, S.M. Thermal effects of single and repeated exposures to microwaves - a review. In: _Biologic Effects and Health Hazards of_

Microwave Radiation, edited by P. Czerski, K. Ostrowski, M.L. Shore, C. Silverman, M.J. Suess, and B. Waldeskog. Warsaw: Polish Med. Publishers, 1974, p. 1-14.

Sanza, J.N. and de Lorge, J. Fixed interval behavior of rats exposed to microwaves at low power densities. *Radio Science*, *12* (6S), 273-277, 1977.

Smiles, K.A., Elizondo, R.S., and Barney, C.C. Sweating responses during changes of hypothalamic temperature in the rhesus monkeys. *Journal of Applied Physiology*. *40*: 653-657, 1976.

Microwaves: Effect on Thermoregulatory Behavior in Rats

Abstract. *Rats, with their fur clipped, pressed a lever to turn on an infrared lamp while in a cold chamber. When they were exposed to continuous-wave microwaves at 2450 megahertz for 15-minute periods, the rate at which they turned on the infrared lamp decreased as a function of the microwave power density, which ranged between 5 and 20 milliwatts per square centimeter. This result indicates that behaviorally significant levels of heating may occur at an exposure duration and intensities that do not produce measurable changes in many other behavioral measures or in colonic temperature. Further study of how microwaves affect thermoregulatory behavior may help us understand such phenomena as the reported "nonthermal" behavioral effects of microwaves.*

Reports of "nonthermal" behavioral effects of microwaves have contributed to debates about the safety of microwaves. There is no question that microwaves can affect behavior (*1*). Microwaves heat tissue, and heat itself can affect behavior (*2*). When microwaves increase colonic temperature, concomitant behavioral changes are often attributed to the thermal burden (*3*). When microwaves produce no observable changes in colonic temperature, concomitant behavioral changes are sometimes attributed to "nonthermal" actions of microwaves, especially by Soviet and Eastern European investigators (*4*). Numerous biological processes, however, are affected by local temperatures that are not highly correlated with the core temperature (*5*); indeed, many help ensure its constancy. Thermoregulatory behaviors, those behaviors that directly affect, and are often controlled by, the thermal environment of the subject, generally respond to skin and hypothalamic, rather than colonic, temperatures (*6–8*). It seems plausible, then, that microwaves might alter behavior as a consequence of thermal stimulation in the absence of measurable core temperature changes.

Current techniques for measuring temperature changes in animals exposed to microwaves are inadequate for several reasons: (i) sensitivity is limited to about 0.1°C; (ii) ongoing thermoregulation serves to dissipate heat; (iii) "hotspots," that is, localized increases in temperature, can occur at locations not being monitored; (iv) most sensors distort the microwave field. Finally, since any absorption produces some temperature increase, it is still necessary to determine its functional significance. One way to avoid these difficulties is to use the organism's behavior as the thermometer, so to speak.

Six male Long-Evans hooded rats (325 to 450 g) were individually trained to press a small lever in order to turn on an infrared lamp for 2 seconds. Responses made during the 2-second period produced no programmed consequences. Test sessions, each of approximately 24 hours, started late in the afternoon. At this time, the fur of the rat was clipped, and the rat was placed in a chamber (*9*) located in a dark, refrigerated room (see Fig. 1). After a few such sessions, the rat generally pressed the lever at a nearly constant rate for several hours. This performance provided a baseline for study-

Reprinted with permission from *Science*, vol. 206, no. 4423, pp. 1198–1201, Dec. 7, 1979.

ing the effect of 2450 MHz continuous-wave (CW) microwaves on thermoregulatory behavior. For sessions in which microwave exposure (*10, 11*) was scheduled, the first control period began in the morning after the rat had been in the chamber for several hours. Alternating control and exposure periods lasted 15 minutes each. Within a single session, across different exposure periods, the microwaves illuminated the chamber either in an ascending followed by a descending series or in a descending followed by an ascending series of power densities.

Figures 2 and 3 show the results of exposing rats to 2450-MHz CW microwaves (*12*). Figure 2a shows that the proportion of time during which the heat lamp was kept on decreased as the microwave power density increased. Figure 2b confirms this relationship by showing a decrease in the ratio of the lamp-on time during an exposure period to the lamp-on time during the preceding control period. Figure 3 shows cumulative records of lever presses that turned on the heat lamp. The slopes of the records remained fairly constant within individual exposure and control periods, demonstrating that the consequences of presentation or removal of microwaves were immediate and constant throughout the respective condition. These data indicate that rats turn on the heat lamp less frequently during a 15-minute period of exposure to microwaves than during control periods; that the decrease is a direct function of power density; that it appears almost immediately; that it occurs at 5 mW/cm^2; and that recovery follows immediately after each exposure.

Statistical analyses support these conclusions. A linear regression model provided a close fit to the data (*13*). Individual regression lines for the five rats with complete data had nonzero slopes ($P < .01$, t-test). By calculating the coefficient of determination (r^2; the percentage variation determined by power density) we obtained the following values for each rat: MW14, 63; MW17, 81; MW21, 70; MW22, 76; and MW24, 83 percent. The usual assumptions of homogeneity of variance and normal distribution of errors were checked. An analysis of covariance demonstrated that the individual regression lines of heat lamp on time to power density had different intercepts, $F(4,50) = 11.54$, $P < .01$, but not different slopes, $F(4,54) = 0.79$. By analysis of r^2 we found that this model accounted for 79 percent of the variance. The common slope (− 0.0062) indicates that if one views microwaves as replacing heat from the lamp, then the amount of heat replaced by a given power density is the same for each rat (*14*).

The most reasonable interpretation of these data is that the rat responds to maintain a nearly constant thermal state (*15*). When heat from one source, microwaves, is introduced, the rat compensates by reducing the heat contributed by another source, the infrared heat lamp. The sensitivity of the rat to relatively small changes in power densities under this procedure, compared to most others used to study behavioral effects of microwaves (*16*), as well as the immediate recovery during the control periods, support this view. Thermoregulatory behavior, therefore, may provide an index of the thermal burden contributed by microwaves. Even if measurable colonic temperature change may sometimes occur at 5 mW/cm^2, the observed time course of the behavior change does not parallel it (*11, 17*). Thermoregulatory behavior responded immediately to thermal change (*18*).

In this experiment we measured one thermoregulatory behavior directly, but obviously such behaviors occur whether or not one measures them. Furthermore, they occur concurrently with other behaviors (*19*). Changes in the frequency of thermoregulatory behaviors, such as reduced activity, sprawling, and saliva spreading, could provoke changes in the frequencies of these other behaviors. Interpretations of reported "nonthermal"

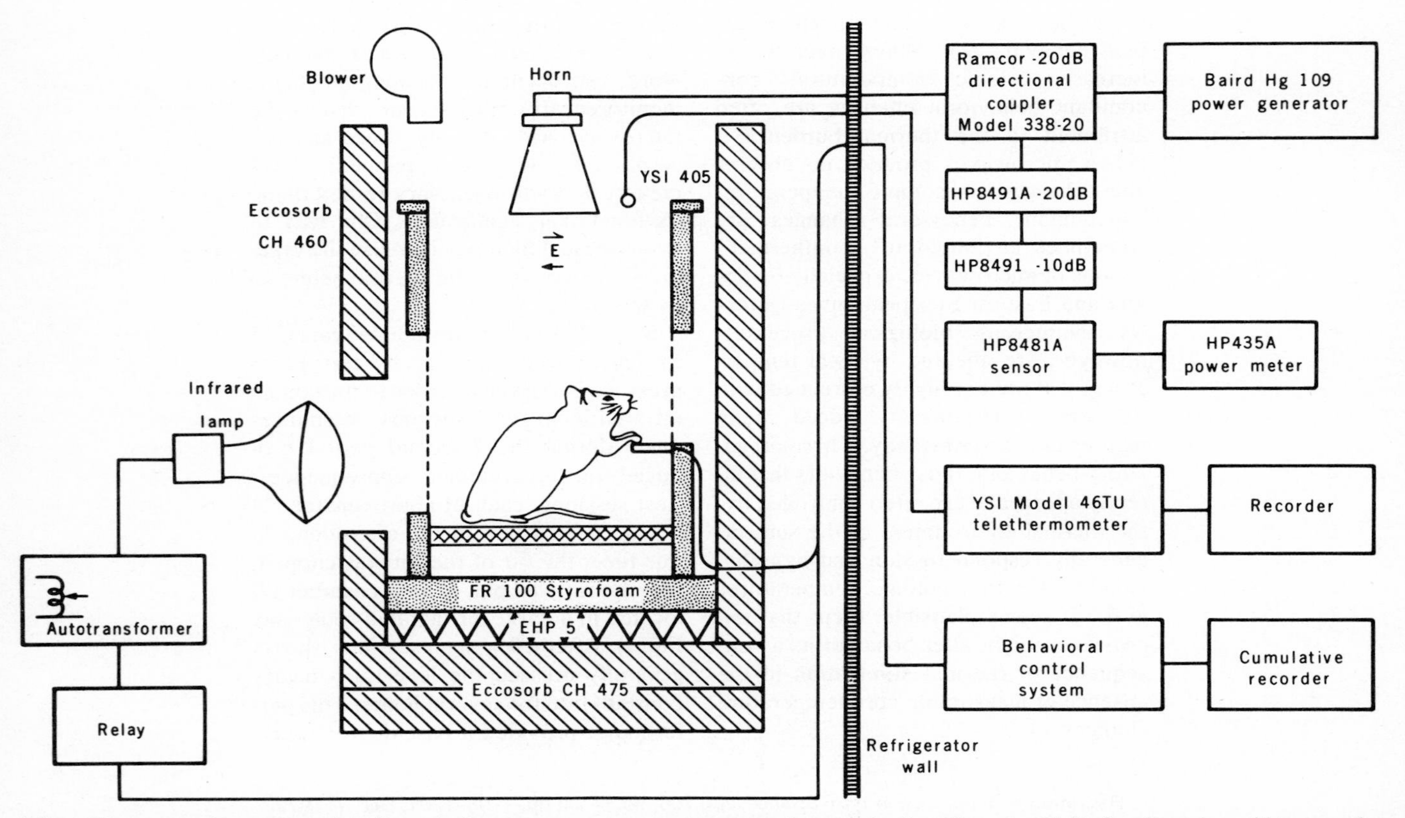

Fig. 1. The system for studying the effect of microwave exposure on thermoregulatory behavior. The rat, with its fur clipped, could turn on the infrared lamp for 2 seconds by pressing the lever. Microwaves were presented on a schedule independent of the rat's behavior.

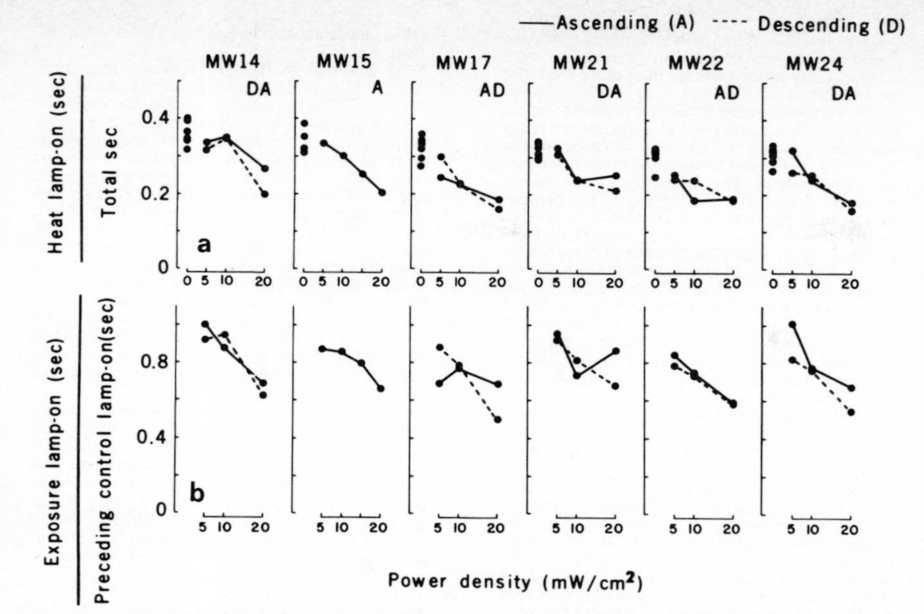

Fig. 2. (a) The proportion of each 15 minutes that the infrared lamp was on during each control period and during each exposure to 2450 MHz CW microwaves. The order of the letters *A* and *D* designates the sequence of the ascending (*A*) and descending (*D*) series of power densities. (b) The ratio of the proportion of each 15-minute microwave-exposure period that the infrared lamp was on to the proportion of the immediately preceding 15-minute control period that the infrared lamp was on. A ratio of less than 1.0 indicates that the rat turned the infrared lamp on less frequently during exposure to microwaves than during the preceding control period (MW14, MW15, and so on, refer to individual rats).

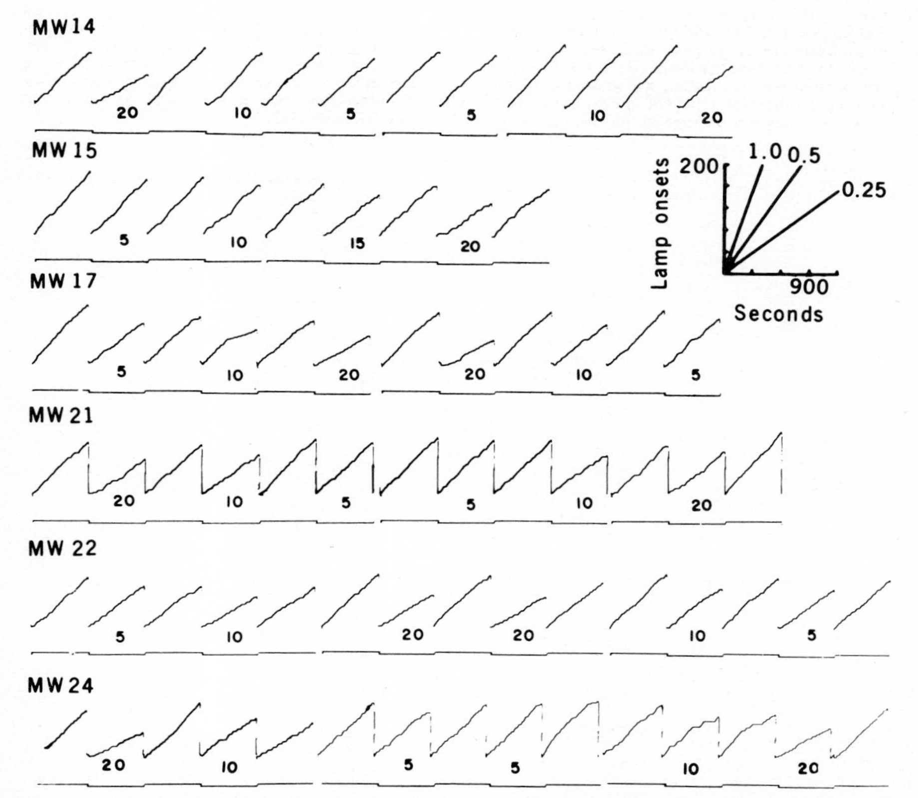

Fig. 3. Cumulative records showing the rate at which rats turned on the infrared lamp during the same exposure periods and in the immediately preceding control periods, from which the data of Fig. 2, a and b, were obtained. Records from additional periods after exposure also show the recovery of responding. The chart speed was constant. Each lever press that turned on the infrared lamp moved the upper pen vertically. Since the lamp could be turned on at a maximum rate of once every 2 seconds, the slope of the record shows the proportion of time the lamp was on. The inset shows the proportion of selected slopes. The upper pen was reset at the end of each 15-minute period. The lower pen was deflected downward during the exposure period. The power density of each exposure is indicated above the lower channel. Note that rat MW15 was exposed only to an ascending series of power densities.

behavioral effects of microwaves might be clarified if concurrent thermoregulatory behaviors were recorded.

SANDER STERN
LEONID MARGOLIN
BERNARD WEISS
SHIN-TSU LU
SOL M. MICHAELSON

Department of Radiation Biology and Biophysics and *Environmental Health Sciences Center, School of Medicine and Dentistry, University of Rochester, Rochester, New York 14642*

References and Notes

1. For a recent review, see S. F. Cleary, *CRC Crit. Rev. Environ. Control* **7**, 121 (1977). See also D. R. Justesen and A. W. Guy, Eds., *Radio Sci.* **12**, No. 6(S), Suppl. (1977); C. C. Johnson and M. L. Shore, Eds., *Biological Effects of Electromagnetic Waves, Selected Papers of the USNC/URSI Annual Meeting, Boulder, Colorado, October 20–23, 1975* [Government Printing Office, HEW Publ. (FDA) 77-8010, Washington, D.C., 1977], vol. 1.
2. E. Satinoff and R. Hendersen, in *Handbook of Operant Behavior*, W. K. Honig and J. E. R. Staddon, Eds. (Prentice-Hall, Englewood Cliffs, N.J., 1977), p. 153.
3. D. Justesen and N. W. King, in *Biological Effects and Health Implications of Microwaves*, S. F. Cleary, Ed. (Government Printing Office, U.S. Public Health Service Publ. No. PB193-898, Washington, D.C., 1970), p. 154; J. de Lorge, in *Biological Effects of Electromagnetic Waves, Selected Papers of the USNC/URSI Annual Meeting, Boulder, Colorado, October 20–23, 1975*, C. C. Johnson and M. L. Shore, Eds. [Government Printing Office, HEW Publ. (FDA) 77-8010, Washington, D.C., 1977], vol. 1, p. 158.
4. For discussions see A. S. Presman, *Electromagnetic Fields and Life*, F. L. Sinclair, Transl. (Izd-vo Nauka, Moscow, 1968; Plenum, New York, 1970); S. Baranski and P. Czerski, *Biological Effects of Microwaves* (Dowden, Hutchinson, Ross, Stroudsburg, Pa., 1976); D. R. Justesen, *Radio Sci.* **12**, 355 (1977); S. M. Michaelson, *Microwave and Radiofrequency Radiation* (World Health Organization, Rep. ICP/CEP 803, Copenhagen, 1977).
5. T. H. Benzinger and G. W. Taylor, in *Temperature: Its Measurement and Control in Science and Industry*, J. D. Hardy, Ed. (Reinhold, New York, 1963), vol. 3, p. 111.
6. Thermoregulatory behavior is sensitive to numerous variables including the following: ambient temperature; internal temperature; magnitude or duration of reinforcement, or both; species; age; metabolic state; drugs; chemical or electrolytic brain lesions; and behavioral requirement for temperature change. For reviews, see Weiss and Laties (*7*); Carlisle (*8*); J. D. Corbit, in *Physiological and Behavioral Temperature Regulation*, J. D. Hardy, A. P. Gagge, J. A. J. Stolwijk, Eds. (Thomas, Springfield, Ill., 1970), p. 777; D. Murgatroyd and J. D. Hardy, in *ibid*. p. 874; Satinoff and Henderson (*2*).
7. B. Weiss and V. G. Laties, *Science* **133**, 1338 (1961).
8. H. J. Carlisle, in *Animal Psychophysics: The Design and Conduct of Sensory Experiments*, W. C. Stebbins, Ed. (Prentice-Hall, Englewood Cliffs, N.J., 1970), p. 211.
9. The chamber (20.4 by 20.8 by 40.5 cm) was constructed out of FR 100 Styrofoam foamed polystyrene that is transparent to 2450 MHz microwaves [V. R. Reno and J. O. de Lorge, *IEEE Trans. Biomed. Eng.* **24**, 201 (March 1977)]. Windows (15 by 25 cm) on three sides, each lined with 0.0127-cm cellulose acetate, allowed transmission of radiant energy. The fourth side, which contained the response lever, was lined with Mylar to reduce damage by gnawing and scratching. Nylon mesh in a Styrofoam frame served as the cover. An acrylic lever (2.6 cm wide and 0.32 cm thick) protuded 5 cm into the chamber 4 cm above the polystyrene grid floor. The lever, mounted on a metal base attached to the outside wall, contained a microswitch for detecting lever presses. The inside of that wall was located 44 cm from the front of the 250 W infrared lamp that was operated at 130 V a-c. The temperature decreased intermittently to approximately 1.1°C. It recovered quickly to approximately 3.6°C and over the next 2 to 3 hours in-

creased at an approximately linear rate to 5.5°C. Control and exposure data were obtained between 3.9° and 5.3°C. A blower (100 ft³/min) provided airflow from above. The airflow, measured below the empty chamber with a hot-wire anemometer (Datametrics Airflow Multimeter Model 800 VTP), was approximately 6 m per minute in the region of the lever and increased to approximately 18 m per minute at the opposite side of the chamber.

10. Microwaves were transmitted to the horn via a coaxial cable to a coaxial-waveguide adapter. The feeder horn with a 6.5 by 7.5 cm rectangular aperture directed the microwaves toward the chamber floor 44.5 cm below with the E field parallel to the axis established by the response lever and infrared lamp. The designated power density specifies that value at the lever. A Narda Model 8315 probe calibrated against an NBS XD-1 probe was used to map the field. With the back wall removed, the field was measured at the level of the lever at nine locations in a 10 by 10 cm grid with loci 5 cm apart covering the central area of the chamber. The distribution of power densities varied within 11 percent of the mean value. The field was also mapped with a smaller probe fabricated by J. Ali (U.S. Environmental Protection Agency) which enabled measurements to be made at loci close to the chamber walls. There was close correspondence between the two sets of measurements. The energy absorption rate per unit mass was estimated according to the relationship $P = 4.186\ C\Delta T/t$, where C is the tissue specific heat in calories per gram per degree Celsius (in this analysis, $C = 0.83$), ΔT is the temperature increase in degrees Celsius, t is the duration of exposure in seconds, and P is watts per kilogram [C. C. Johnson, *J. Microwave Power* **10**, 249 (1975)]. A YSI 423 probe inserted 6 cm into the colon measured the temperature of a pentobarbital-anesthetized rat encased in a Styrofoam block during a brief exposure to microwaves [Lu *et al.* (*11*)]. Delta-*t*, the rate of temperature change, included a correction for the rate of temperature change immediately preceding the exposure. The absorption rate was approximately 8.4 W/kg at a power density of 41 mW/cm², resulting in a specific absorption rate of 0.20 W/kg per milliwatt per square centimeter.
11. S.-T. Lu, N. Lebda, S. Michaelson, S. Pettit, D. Rivera, *Radio Sci.* **12(S)**, 147 (1977).
12. These data are the results of the second exposure session. After studying three rats, we found that the results from the first session were similar to but more variable than those from the second. Since such an outcome might be attributable to the novelty of the microwaves, an effect often seen during initial exposure to drugs and other stimuli, we decided to focus on the data from the second sessions of those three rats plus three others. The following are the means (standard errors in parentheses) of the proportion of time the heat lamp remained on at each power density for all six rats. For the first exposure session: 0 mW/cm², 0.364 (0.017); 5 mW/cm², 0.333 (0.040); 10 mW/cm², 0.324 (0.032); and 20 mW/cm², 0.273 (0.029). For the second exposure session: 0 mW/cm², 0.326 (0.009); 5 mW/cm², 0.298 (0.015); 10 mW/cm², 0.264 (0.020); and 20 mW/cm², 0.199 (0.011). Thus the two functions are similar, but the variability from the second session is less, as would be expected, after previous experience.
13. Serial correlation within individual rats was examined by the Durbin-Watson test [J. Neter and M. Wasserman, *Applied Linear Statistical Models* (Irwin, Homewood, Ill., 1974)], which confirmed the absence of an effect from the previous 15-minute exposure and the suitability of the linear regression model.
14. Using randomization tests [A. R. Feinstein, *Clinical Biostatistics* (Mosby, St. Louis, 1977)] we compared results for the six rats at each power density with all control periods, with only the preceding control period, and with the adjacent exposure to a different and nonzero power density. Two-sided P values were less than $P = .003$ except for 5 mW/cm² against all zero exposures and 5 mW/cm² against adjacent 10 mW/cm² ($P = .02$). These results provide additional confirmation of the sensitivity of the procedure as well as an independent check of the regression analysis.
15. That interpretation is consistent with the conclusion reached in numerous studies of behavioral thermoregulation, including those of Weiss and Laties (*7*) and Carlisle (*8*).
16. See N. W. King, D. R. Justesen, and R. L. Clarke [*Science* **172**, 398 (1971)] for an example of a different sensitive procedure. We did not attempt to determine the limits of the sensitivity in the present study; instead, we wanted to determine if the thermal action of microwaves could produce behavioral change in the absence of reliable, measurable changes in colonic temperature.
17. S. M. Michaelson, W. M. Houk, N. J. A. Lebda, S.-T. Lu, R. L. Magin, *Ann. N.Y. Acad. Sci.* **247**, 21 (1975). In addition, using the exposure system and behavioral contingencies of the present experiment, we measured the rectal temperature of two rats at 0900, 1200, and 1500 hours immediately after they were exposed to 0, 5, 10, and 20 mW/cm² for 15 minutes each. The respective mean temperatures were 37.4°, 37.7°, 37.7°, and 37.4°C for one rat and 37.7°, 37.7°, 38.0°, and 37.5°C for the other, indicating that the microwaves did not produce ordered changes in rectal temperature but did produce such changes in thermoregulatory behavior.
18. Although a compensatory change in behavior occurs, we do not know if the overall absorbed heating is constant across conditions. Certainly, the spatial distributions of absorbed energies differ between microwaves and infrared heat. The relationship between electromagnetic energy and transduced thermal energy is complex. For example, see R. A. Tell, *An Analysis of Radiofrequency and Microwave Absorption Data with Consideration of Thermal Safety Standards* (U.S. Environmental Protection Agency Tech. Note No. ORP/EAD 78-2, Washington, D.C., April 1978). For present purposes it is sufficient to assume at least an ordinal relationship between the two.
19. V. G. Laties, *J. Physiol. (Paris)* **63**, 315 (1971). See also A. C. Catania [in *Operant Behavior: Areas of Research and Application*, W. K. Honig, Ed. (Prentice-Hall, Engelwood Cliffs, N.J., 1966), p. 213] and P. de Villiers [in *Handbook of Operant Behavior*, W. K. Honig and J. E. R. Staddon, Eds. (Prentice-Hall, Engelwood Cliffs, N.J., 1977), p. 233] for a comprehensive introduction to the study of concurrent schedules.
20. We thank C. Cox and H. T. Davis for the statistical analyses, and C. A. Wallen for criticism of the manuscript. We thank J. Ali and EPA for assistance in mapping the microwave field. This paper is based on work supported in part by postdoctoral fellowship F22 ES01804 (awarded to S.S.) and grant ES-01247, both from the National Institute of Environmental Health Sciences, and in part by a contract with DOE at the Department of Radiation Biology and Biophysics, University of Rochester, and is Report No. UR-3490-1537.

22 January 1979; revised 13 August 1979

Immune response of mice to 2450-MHz microwave radiation: Overview of immunology and empirical studies of lymphoid splenic cells

Wieslaw Wiktor-Jedrzejczak, Aftab Ahmed, Przemyslaw Czerski, William M. Leach, and Kenneth W. Sell

Cellular Immunology Division, Clinical and Experimental Immunology Department, Naval Medical Research Institute, Bethesda, Maryland 20014 and Division of Biological Effects, Bureau of Radiological Health, Food and Drug Administration, Rockville, Maryland 20857

Male CBA/J mice were exposed to 2450-MHz microwave radiation (each exposure: 30 minutes at an averaged dose rate near 14 mW/g). The mice were tested later to determine effects of the radiation on (1) the relative frequency of T and B cells; (2) the functional capacity of spleen cells from irradiated mice to respond to T- and B-cell-specific membrane stimuli; and (3) the ability to respond to sheep red blood cells and dinitrophenyl-lyslyl-Ficoll. Results demonstrated that microwave radiation has weak stimulatory effects on B- but not T-lymphoid cells in the spleen. Single exposures to radiation produced an increase in the incidence of cells with complement receptor on the cell's surface. Three exposures to radiation induced increases in the total number of splenic cells, in the incidence of immunoglobin-positive cells, and in the incidence of complement receptor-positive cells. The total number of T cells was unaffected by single or triple exposures of mice to the radiation.

1. INTRODUCTION

Rapid increases in the number of devices that emit potentially hazardous radio-frequency radiations prompted us to carry out investigations on effects (harmful or beneficial) that exposure to microwaves may have on the immune system. The immune system produces specific cellular and humoral responses to microörganisms and to other intrinsic and extrinsic agents. Among the major elements of the immune system are the lymphoid cells. These cells are highly similar morphologically (relatively small, spherical, with a relatively large nucleus in relation to the cytoplasm) and are present in the bone marrow, thymus, spleen, peripheral blood, lymph nodes, and Peyer's patches. Despite their morphological similarity, as determined by classical hematological and histological criteria, the lymphoid cells are composed of a number of functionally distinct subpopulations. All of the subpopulations originate from a common lymphoid stem cell (Figure 1). One of the progeny of the stem cells, which infiltrates, differentiates, and matures in or under the influence of the thymus, thereafter recirculating to the peripheral lymphoid tissues, is called the thymus-derived lymphoid (T) cell. This cell interacts with a foreign antigen by direct cellular contact and the immune response that follows is termed "the cellular immune response." Other descendants of the lymphoid stem cell are present in the bone marrow -- and from the marrow, directly infiltrate peripheral organs such as the spleen, lymph nodes, and Peyer's patches -- and have been termed the bone-marrow-derived lymphoid (B) cells. These cells interact with foreign antigens by secreting specific proteins called immunoglobulins (or antibodies). Thus, the two major subclasses of lymphoid cells that are part of the immune system are the T cells and the B cells.

The T cells are identified by the use of a cytotoxic antisera against a unique marker on their surface termed the theta isoantigen [*Reif and Allen,* 1966], whereas B cells possess surface immunoglobulin (Ig), which is easily detected by membrane immunofluorescence through the use of a fluorescent-conjugated, anti-immunoglobulin antisera. Further, there is considerable heterogeneity among T cells and B cells. This heterogeneity is based on the observation that there are other markers that are unique to, but are not present on all, T and B cells [*Raff et al.,* 1970]. Thus, a fraction of adult B cells, besides expressing surface Ig, has been shown to express a receptor for activated complement (C'3),which may be detected by its ability to form rosettes with a reagent consisting of antibody and C'3-coated sheep erythrocytes (EAC reagent) as described by *Bianco et al.* [1970]. Cells of this fraction are not present in neonatal mice. Thus, it can be theorized that stem cells give rise to precursor B cells that, when mature, manifest surface Ig, and then further mature into B cells that express the complement receptor (CR) (Figure 2). After interaction with an antigen, e.g., sheep erythrocytes, adult B cells develop into plasma cells and produce antibody, which may be quantitated by a local, hemolysis-in-gel (plaque-formation) technique of *Jerne and Nordin* [1963]. Functionally, T and B cells can be assessed by their ability to respond to T-cell- and B-cell-specific mitogenic agents. Thus, T cells preferentially undergo DNA synthesis in response to the mitogens phytohemagglutinin and concanavalin A, and the B cells preferentially undergo DNA synthesis in response to the mitogenic action of bacterial lipopolysaccharide and polyinosinic-polycytidylic acid. The magnitude of DNA synthesis can be assessed and used as an indicator both of the presence and of the frequency of functional T or B cells. Therefore, at present, several techniques are available that not only permit the enumeration of the total number of T and B cells but also the frequency and functional capacity of each subpopulation in a given mixture of cells.

The purpose of our investigation was to learn whether

Reprinted with permission from *Radio. Sci.*, vol. 12, no. 6(S), pp. 209-219, Nov.-Dec. 1977.

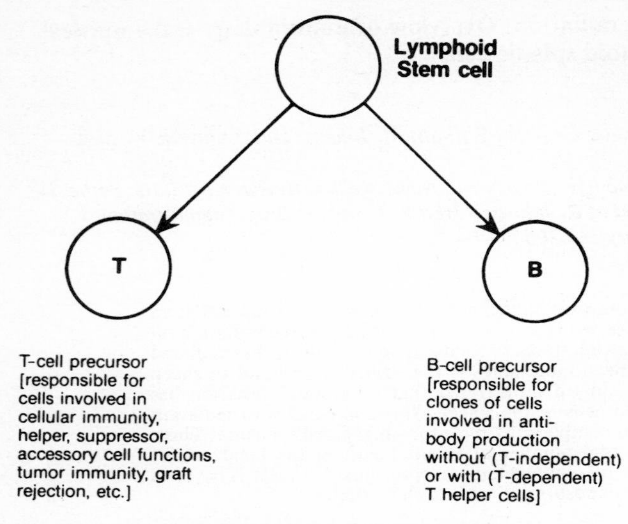

Fig. 1. Lymphoid stem cells are primitive constituents of the immune system that undergo a series of transformations during maturation. The initial transformation is that to precursor T cell (for thymus) or precursor B cell (historically identified with the bursa of an avian species). During subsequent maturation, T and B cells perform a diversity of specialized functions.

exposure of mice to microwaves has an effect on their immune function by determining, not only the relative frequency of each subpopulation, but also, through further study, to determine the functional capacity of the T and B cells to respond to T-cell-specific and B-cell-specific membrane stimuli after immunization of microwave- and sham-exposed mice against T-cell-dependent and T-cell-independent antigens.

According to contemporary opinion, the immune system is essentially the same among differing mammalian species and, therefore, the choice of an experimental animal model in studies involving microwave radiation is dependent mostly on the amenities of the radiation facility. An advanced apparatus, an environmentally controlled waveguide system for exposure of mice, is available at the Bureau of Radiological Health, Food and Drug Administration, Rockville, Maryland [*Ho et al.*, 1973; *Christman et al.*, 1974]. In the mouse, a lymphoid organ that is widely used for experimental purposes is the spleen. It is an organ that contains both categories of differentiated T and B cells in sufficient quantities to allow simultaneous performance of different assays with the same suspension of cells.

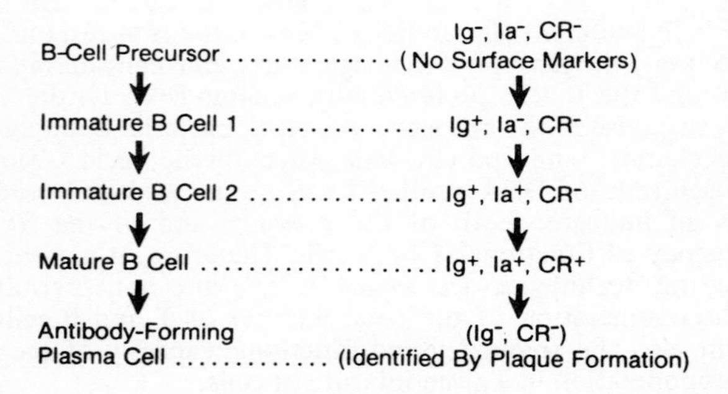

Fig. 2. Sequential development of surface markers on B-lymphoid cells of the mouse.

In this paper, we shall describe the effect of a single microwave exposure and of three such exposures on (a) the frequency of various subpopulations of T and B cells; (b) the functional capacity of splenic cells from exposed mice to respond to T- and B-cell-specific membrane stimuli; and (c) the ability of microwave- and sham-exposed mice to mount an antibody response to SRBC (which requires the presence of T cells) and to DNP-lys-Ficoll (which does not require the presence of T cells and is therefore T-cell independent).

2. MATERIALS AND METHODS

2.1. *Animals.*

Highly inbred CBA/J male mice (8-12 weeks of age) were purchased from the Jackson Laboratory, Bar Harbor, ME, and were used in all experiments.

2.2. *Exposure conditions and protocol.*

Each mouse was individually positioned in the waveguide with its head towards the source of radiation while restrained in a plastic polystyrene holder. A mouse was exposed to 2450-MHz microwaves at a forward power of about 0.6 watts. The microwave source was a Hewlett-Packard 8616A signal generator that was coupled to a 491C amplifier; the amplifier was tuned to 2450 MHz ± 500 Hz by a coherent synchronizer (Sage Laboratories Model 251). The body mass of all animals was measured before exposure and ranged from 15 to 20 grams. Rectal temperatures were determined with a Yellow Springs Model 46 TUC telethermometer before and after exposures. During exposures, an ambient temperature of 25 ± 0.5 °C and a relative humidity of 50 ± 5% were maintained as was a constant flow of air in the waveguide (38 liters/min). The duration of exposure was always 30 minutes and, during exposures, forward, reflected, and transmitted powers were measured with power meters. Animals were exposed either once or three times. The averaged dose rate for the single exposures was 13 ± 2.5 (SD) mW/g and was associated with a mean ΔT of rectal temperature of –0.05 K. The averaged dose rates for the animals that were exposed three times were, respectively, for the first, second, and third exposure, 15.6 ± 1.2, 15.5 ± 0.9, and 13.3 ± 0.6 mW/g, which were associated with mean ΔTs of –0.10, –0.20, and +0.05 K. The dose rates were based on the body mass of individual animals (in grams), which was divided by the computer-extrapolated, averaged rate of energy absorption (in watts), as previously described [*Youmans and Ho*, 1975; *Justesen*, 1975; *Johnson*, 1975]. Sham-exposed animals served as controls. They were handled in the same manner as microwave-exposed animals and were kept in the same polystyrene holder in the waveguide for 30-minute periods, but were not exposed to radiation.

2.3. *Statistics.*

Only groups of six MW-exposed and six sham-exposed mice could be treated during any given experimental run. Data from each mouse were recorded individually and the arithmetic mean (± SE) of each determination was calculated with the help of a Wang 700 C calculator. Differences between means were analyzed by Student's *t* test.

2.4. *Suspensions of Cells.*

During specified days, as indicated in Section 3, mice were anesthetized with methoxyflurane (Penthrane®, Abbott Laboratories) and were euthanized by cervical dislocation; their spleens were removed under aseptic conditions. Cells were obtained by gentle teasing of the splenic capsule with the aid of a rubber spatula in Hanks Balanced Salt Solution (Microbiological Associates), as described in detail elsewhere [*Strong et al.*, 1975]. Cells in suspension were washed in media and were resuspended in RPMI 1640 supplemented with penicillin (100 units/ml), streptomycin (100 μg/ml), L-glutamine (2mM), 25mm Hepes (N-2-hydroxyethylpiperazine N'-2-ethane sulfonic acid) and 10% heat-inactivated (56 °C, 30 min) fetal bovine serum (FBS) (Microbiological Associates). Cell counts were determined with a Fisher Scientific Co. autocytometer.

2.5. *Antisera.*

Anti-T-cell-specific sera (anti-theta or anti-Thy 1.2 sera) was prepared according to the method of *Reif and Allen* [1966]. Fluorescent-conjugated goat antimouse-polyvalent IgG was obtained from Meloy Laboratories.

2.6. *Mitogen Assay.*

An assay of the ability of the splenic cells to respond *in vitro* to T- and B-cell-specific mitogenic agents was performed with a microculture system described previously [*Strong et al.*, 1973]. Essentially, cells in 100-μl suspensions (8×10^6 cells/ml) were cultured with 100 μl of media (control) or 100 μl of media containing an (optimal) concentration of each mitogenic agent. The T-cell mitogens were phytohemagglutinin-P (PHA-P) obtained from Difco Laboratories, and concanavalin A (con A) obtained from Calbiochem. Pokeweed mitogen (PWM), which induces proliferation of both T and B cells, was purchased from Grand Island Biological Co.

Various mitogens that specifically induce B-cell activation were used in these studies. These consisted of dextran sulphate (DxS), which has been shown previously to stimulate immature B cells [*Gronowicz et al.*, 1974; *Gronowicz and Coutinho*, 1975; *Ahmed et al.*, 1976], bacterial lipopolysaccharide of *E. coli* (LPS), and the homoribopolynucleotide, polyinosinic polycytidylic acid (Poly I•C), and purified protein derivative (PPD). The DxS was purchased from Pharmacia, (Uppsala, Sweden); LPS from Difco Labs; PPD from Parke Davis; and Poly I•C from P. L. Biochemicals.

Cultures were performed in triplicate and were incubated at 37 °C in a humidified atmosphere of 5% CO_2 for a total of 72 hours. Eighteen hours prior to harvest each culture received 1 μCi of methyl-^{3}H-thymidine (Schwarz-Mann) in 20 μl of media. Cultures were harvested using the Multiple Automated Sample Harvester as previously described [*Thurman et al.*, 1973].

2.7. *Enumeration of complement-receptor-bearing lymphoid cells.*

The number of CR+ cells was determined by the technique of *Strong et al.* [1975]. Briefly described, splenic cells were treated with tris-buffered NH_4Cl for red-cell lysis, were washed, and then were resuspended to 5×10^6 cells/ml. One ml of the suspension was incubated with 1 ml of EAC reagent at 37 °C for 45 minutes. After incubation, a drop of the mixture was observed under a microscope and the number of lymphoid cells that formed rosettes (at least five erythrocytes adhering to one lymphocyte was enumerated. A total of 300 or more lymphoid cells was counted and the percentage of CR+ cells was determined. The EAC reagent was made of sheep red blood cells (E) and a 19S fraction of human antisheep red-blood-cell antibody (A); fresh mouse serum was used as a source of complement (C). The EAC reagent also contained 0.01 M EDTA to prevent binding of EAC complexes to granulocytes or macrophages.

2.8. *Immunization of mice.*

Groups of mice were given an immunization treatment of saline (control), of sheep red blood cells, or of DNP-lys-Ficoll, all intraperitoneally. The SRBC was used as an antigen that requires the presence of functional T cells for the B cells to make antibody. The DNP-lys-Ficoll was used as an antigen that has been characterized previously as not requiring T cells for the B cells to make antibody. Thus, SRBC was used as a T-dependent antigen and DNP-lys-Ficoll was used as a T-independent antigen [*Sharon et al.*, 1975]. After immunization, the mice were divided into two groups; one half was exposed to microwaves and the other half was sham exposed.

2.9. *Antigens*

Sheep red blood cells (SRBC) were collected in Alsever's solution, were washed several times in isotonic saline and, finally, were resuspended to give a 10% suspension. The washed SRBC were kept at 4 °C and were never more than four weeks old prior to use. Each mouse received 0.2 ml of the washed SRBC suspension. For the other antigenic treatment, dinitrophenol (Eastman Kodak, Rochester, NY) was conjugated to L-lysine and Ficoll (both from Sigma Chemicals). Each mouse was injected with 100 μg of DNP-lys-Ficoll in a volume of 0.5 ml (Figure 3).

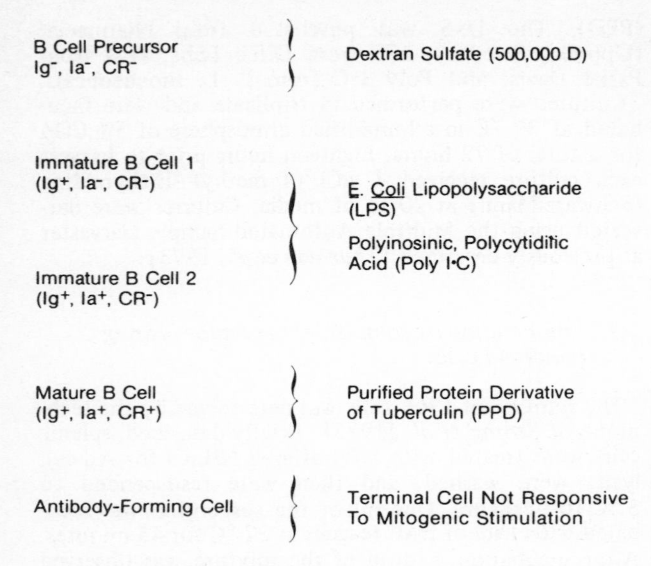

Fig. 3. Correlation of maturational stage with specificity of response of B-lymphoid cells to mitogenic stimulation. A mitogen not only induces proliferation of cells but drives the cell to the indicated stage of maturation.

2.10. *Direct plaque-forming-cell (PFC) assay.*

The splenic cells from microwave- and sham-exposed mice were assayed for presence of immunoglobulin-M (IgM) antibody-forming cells by a modification [*Mosier*, 1969] of Jerne's direct PFC assay [*Jerne and Nordin*, 1963]. The choice of the indicator cells depended on the type of antigen that was used for immunization, as is summarized in Table 1. Cells producing IgM anti-SRBC antibody in response to SRBC immunization were assayed using SRBC as indicator cells. Cells producing IgM anti-DNP antibody in response to DNP-lys-Ficoll were assayed using SRBC conjugated with trinitrophenyl-haptenated SRBC as indicator cells. This light haptenation of SRBC was accomplished by the method of *Rittenberg and Pratt* [1969]. Splenic cells from saline-treated controls, and from microwave-exposed and sham-exposed animals were assayed against both kinds of indicator cells. The data are expressed as the PFC per 10^6 spleen cells and represent the mean and the standard error of triplicate samples.

TABLE 1. Antigens and appropriate indicator cells that were used to evaluate effects of 2450-MHz microwave radation on IgM response of mice to T-dependent and T-independent antigens.

Antigenic Treatment	Indicator Cells for PFC-Assay
0.5-ml saline i.p. (control)	SRBC, TNP-SRBC
0.2-ml 10% SRBC in saline i.p.	SRBC
100-mg DNP-lys-Ficoll in 0.5-ml saline i.p.	TNP-SRBC

Abbreviations: SRBC, sheep red blood cells; PFC, plaque-forming cells; TNP-SRBC, SRBC coated with trinitrophenyl sulfonic acid; DNP-lys-Ficoll, conjugation of dinitrophenyl with lysine and Ficoll, known to be a T-independent antigen.

3. RESULTS

3.1. *Effect of a single microwave exposure of mice on frequency of T- and B-lymphoid-cell subpopulations in the spleen.*

Mice in groups of six were individually exposed to a single dose of 2450-MHz radiation for 30 minutes in the waveguide. Spleens were removed six days after the exposure. Spleen-cell suspensions were then prepared and assayed for the total number of T cells (by the use of cytotoxic anti-Thy-1.2 sera and C'), and the total number of B cells (by the use of fluorescent antimouse Ig sera). Sham-exposed mice were used as controls. As is seen in Table 2, a single exposure to 2450-MHz microwaves failed to produce any significant increase or decrease in the total number of relative frequency of thymus $(1.2\text{-T})^+$ cells or IG+ B cells. Interestingly enough, when the frequency of CR+ cells was determined, there were marked net increases of 24 to 29%.

3.2. *Effect of three exposures of mice to microwaves on the frequency of T- and B-lymphoid-cell subpopulations in the spleen.*

Groups of six mice were given three, 30-minute doses of 2450-MHz radiation, one dose each on days 0, 3, and 6; six days after the last exposure their spleens were removed and assayed for the total number of T cells, B cells, and CR+ cells. Sham-exposed mice served as controls. As is seen in Table 3, there was no significant change in the total number of $(\text{thy-}1.2)^+$-bearing T cells, but there was a significant increase in the total number of Ig-bearing cells and further net increases (52-64%) in the frequency of CR+ cells as compared with values obtained for spleen cells from sham-exposed mice.

TABLE 2. Effect of a single, 30-minute exposure of CBA/J mice to microwaves (2450-MHz; ~ 13 mW/g) on the frequency of subpopulations of T- and B-lymphoid cells (Mean ± SE).

Surface Marker	Type of Cell Detected	Frequency of Positive Cells[1] Sham-Exposed	Microwave-Exposed
Theta	T	37.8 ± 1.7	34.7 ± 2.3
Ig	B	51.6 ± 2.9	53.9 ± 2.7
CR	B	22.5 ± 2.0	30.6 ± 1.5*

[1]Data were determined on spleen-cell suspensions from six sham-exposed and six microwave-exposed mice.
*P < .01

TABLE 3. Effect of three, 30-minute exposure of CBA/J mice to 2450-MHz microwaves (~ 14 mW/g) on the frequency of subpopulations of T- and B-lymphoid cells (Mean ± SE).

Surface Marker	Type of Cell Detected	Frequency of Positive Cells[1]	
		Sham-Exposed	Microwave-Exposed
Theta	T	35.5 ± 1.4	30.9 ± 1.8
Ig	B	48.1 ± 1.4	59.4 ± 2.4*
CR	B	23.4 ± 2.1	36.7 ± 3.8*

[1]Data were determined on spleen-cell suspensions from six sham-exposed and six microwave-exposed mice.
*$P < .05$.

3.3. *Effect of a single exposure of mice to 2450-MHz microwaves on the response of splenic cells in vitro to T-cell mitogens.*

Splenic cells from microwave and sham-exposed mice were cultured *in vitro* in microculture plates in the presence of media (control) or of various dilutions of either PHA-P or con-A, two plant lectins that induce blastic transformation of T cells but not B cells. As is seen in Table 4, no significant difference ($P > 0.05$) occurred in the uptake of ^{3}H-TdR in splenic cells of microwave- and sham-exposed mice. The six spleen-cell suspensions from microwave-exposed mice and the six spleen-cell suspensions from sham-exposed mice were assayed individually.

3.4. *Effect of a single exposure of mice to 2450-MHz microwaves on the response of splenic cells in vitro to various B-cell mitogens.*

Groups of six mice were exposed once to microwaves and 3, 6, or 9 days thereafter their spleens were removed. Splenic cells were assayed for ability to undergo DNA synthesis in the presence of a variety of B-cell-specific, mitogenic agents. Splenic cells from sham-exposed mice were used as controls. Various concentrations of each mitogen were used in this assay in efforts to determine if exposure to microwaves caused a change in the optimal mitogenic dose. For reasons of brevity, data from cells that exhibited incorporation of the largest quantities of ^{3}H-TdR are reported here. As is seen in Table 5, splenic cells from both microwave-exposed and sham-exposed mice responded well to all four mitogens. There was no major difference in the response to the various B-cell mitogens. However, there was a systematic trend toward an increasing incorporation of ^{3}H-TdR by splenic cells from microwave-exposed mice in response to LPS, Poly I•C, and PPD. Although most of the differences are statistically significant, as evaluated by *t* test, further experiments are necessary to confirm the validity of the data.

3.5. *Effect of three exposures of mice to 2450-MHz radiation on the response to T-cell-specific mitogens.*

A group of 18 CBA/J mice were exposed for 30 minutes to 2450-MHz microwave radiation on days 0, 3, and 6. Sham-exposed mice served as controls. Three, 6, and 9 days after the last exposure, splenic cells from six microwave-exposed and six sham-exposed mice were cultured *in vitro* with varying concentrations of phytohemagglutinin-P and con A, two plant lectins that specifically induce the activation of T cells but not B cells. As seen in Table 6, no significant differences were noted in the ability of splenic cells from either microwave- or sham-exposed mice to respond to PHA-P or to con A.

3.6. *Effect of three exposures of mice to 2450-MHz microwaves on the response of splenic cells to a variety of B-cell mitogens.*

Aliquots of the same splenic cells that had been tested for ability to respond to the T-cell-specific mitogens were also tested for their ability to respond to a variety of B-cell mitogens. Again, splenic cells from these mice responded quite well to all of the B-cell-specific mitogens (Table 7) and, similar to the results obtained with cells from mice exposed to a single dose of microwave radiation, the cells showed a slight augmentation in their response to LPS and to Poly I•C. The augmentation to LPS is statistically significant, but the biological significance is unknown.

3.7. *Effect of three exposures of mice to 2450-MHz microwaves on the response to thymus-dependent (SRBC) and thymus-independent (DNP-lys-Ficoll) antigens.*

Mice in groups of four were given intraperitoneal immunizing treatments either with 0.2 ml of a 10% suspension of SRBC, with 50 μg of DNP-lys-Ficoll, or with 0.2 ml of saline (control). After the treatment, they were all exposed individually for 30 minutes to 2450-MHz radiation on days 1, 2, and 3. On the 4th day, their spleens were removed and cells were assayed for ability to form antibody to SRBC or TNP-SRBC by the plaque-formation assay. As is seen in Table 8, while there was a modest but insignificant increase in the response of saline controls (from 713 ± 308 to 1219 ± 274 PFC per spleen) there was a significant decrease in the response of mice immunized with SRBC (from 24,625 ± 4288 to 11,000 ± 2121 PFC per spleen). Similarly, there was a considerable, if statistically insignificant, decrease in the response of mice immunized with DNP-lys-Ficoll (Table 9) from 24,375 ± 4399 PFC per spleen to 17,000 ± 5700 PFC per spleen. The decrease was not secondary to an increase in numbers of splenic cells since the decrease was also of the same magnitude when data were calculated on the basis of plaque-forming cells per 10^6 splenic cells.

4. DISCUSSION

These studies demonstrate that exposure of mice to 2450-MHz microwaves in an environmentally controlled waveguide has weak stimulatory effects on B- but not

TABLE 4. The effect of a single exposure of mice[1] to 2450-MHz microwaves on response of splenic cells *in vitro* to T-cell mitogens, PHA and con A.

		Mean Uptake[2] of ^{3}H-Thymidine cpm ±SE					
		Time Following Exposure					
		Three Days[2]		Six Days[2]		Nine Days[2]	
Mitogen	Concentration per Culture[3]	Sham	Exposed	Sham	Exposed	Sham	Exposed
None	Media control	2415 ± 165	2423 ± 198	5077 ± 325	4563 ± 434	2118 ± 555	2081 ± 462
PHA-P	0.1%	100158 ± 5671	108051 ± 3149	99386 ± 6116	108421 ± 1312	54385 ± 7773	46551 ± 482
Con A	0.25 μg	208616 ± 7695	193889 ± 9302	218837 ± 6503	215914 ± 4255	192687 ± 25302	185333 ± 15564

[1]Six animals per group.
[2]72-hr. culture evaluated by the pulse with 1 μCi of ^{3}H-TdR 18 hr prior to harvest.
[3]Optimal concentration.

on T-lymphoid cells of the spleen. A single exposure of mice neither increased the total number of splenic cells nor the frequency of either theta-bearing (T) cells or Ig+ (B) lymphoid cells. However, the single exposure caused a highly significant increase in the frequency of cells possessing complement receptors (CR) on their surface, which is a characteristic of mature B cells. In contrast, three 30-minute exposures of mice to 2450-MHz radiation induced an increase in the total number of splenic cells, in the total frequency of Ig+ (B cells), and further increased the frequency of CR+ B cells, but had no significant effect on the total number of T cells. The results indicate, at least at the dosages used, that microwaves act primarily on B-lymphoid cells and produce quantitative changes in the absolute frequency of B-cell subpopulations.

TABLE 5. The effect of a single exposure of mice[1] to 2450-MHz microwaves on the response of splenic cells *in vitro* to B-cell mitogens.

		Mean Uptake[2] of ^{3}H-Thymidine cpm ±SE					
		Time following exposure					
		Three Days[2]		Six Days[2]		Nine Days[2]	
Mitogen	Concentration per Culture[3]	Sham	Exposed	Sham	Exposed	Sham	Exposed
None	Media control	2415 ± 165	2423 ± 198	5077 ± 325	4563 ± 434	2118 ± 555	2081 ± 262
DxS	50 μg	Not performed		12307 ± 1074	9515 ± 1784	Not performed	
LPS	25 μg	47496 ± 3313	49823 ± 1882	72385 ± 3299	80151 ± 4297	60121 ± 2252	75119 ± 1844*
Poly I-C	50 μg	50523 ± 4950	63577 ± 3432*	70851 ± 5413	93465 ± 3032*	48358 ± 8629	58873 ± 3667
PPD	100 μg	36898 ± 4408	46763 ± 1275*	34254 ± 1156	43406 ± 1735*	53436 ± 2972	60002 ± 1279*

[1]Six animals per group.
[2]72-hr culture evaluated by a pulse with 1 μCi of ^{3}H-TdR 18 hr prior to harvest.
[3]Optimal concentration
*$P < .05$

TABLE 6. The effect of three exposures of mice[1] to 2450-MHz microwaves on the response of splenic cells *in vitro* to T-cell mitogens, PHA-P and con A.

Mitogen	Concentration per Culture[3]	Mean Uptake[2] of ^{3}H-Thymidine cpm ±SE — Time following first exposure — Nine Days[2]: Sham	Nine Days[2]: Exposed	Twelve Days[2]: Sham	Twelve Days[2]: Exposed
None	Media control	6539 ±1109	5051 ±574	4409 ± 795	3830 ± 796
PHA-P	0.1%	98949 ± 7980	108337 ± 5930	66848 ± 10724	79035 ± 22031
Con A	0.25 µg	225033 ± 38054	237792 ± 38645	276672 ± 34783	298164 ± 32991

[1]Six animals per group.
[2]72-hr. culture evaluated by a pulse with 1 µCi of ^{3}H-TdR 18 hr prior to harvest.
[3]Optimal concentration.

TABLE 7. The effect of three exposures of mice[1] to 2450-MHz microwaves on response of splenic cells *in vitro* to B-cell mitogens, LPS, POLY I•C, PPD.

Mitogen	Concentration per Culture[3]	Mean Uptake[2] of ^{3}H-Thymidine cpm ±SE — Time following first exposure — Nine Days[2]: Sham	Nine Days[2]: Exposed	Twelve Days[2]: Sham	Twelve Days[2]: Exposed
None	Media control	6539 ± 1109	5051 ± 574	4409 ± 795	3830 ± 796
LPS	25 µg	60613 ± 4683	61720 ± 3266	58641 ± 5020	76727 ± 5693*
Poly I-C	50 µg	73336 ± 6521	71356 ± 8561	57960 ± 7158	70789 ± 11314
PPD	100 µg	43160 ± 3283	46459 ± 4271	41761 ± 6692	46128 ± 5695

[1]Six animals per group.
[2]72-hr. culture evaluated by a pulse with 1 µCi of ^{3}H-TdR 18 hr prior to harvest.
[3]Optimal concentration only.
*$P < .05$.

TABLE 8. Effect of three exposures to 2450-MHz microwaves on response of mice to a thymus-dependent antigen (SRBC).

Immunizing Treatment	Mean D-PFC/spleen[1]		Mean D-PFC/10^6 splenic cells[1]	
	Sham-Exposed	Microwave-Exposed	Sham-Exposed	Microwave-Exposed
Saline	713 ± 308	1219 ± 274	5.3 ± 2.2	9.7 ± 3.0
SRBC	24625 ± 4258	11000 ± 2121*	204.4 ± 22.2	112 ± 16.8*

[1]Splenic cells from six mice were assayed in duplicate for ability to form plaque-forming cells against sheep erythrocytes. Arithmetic means were calculated and from the data obtained the mean of means of direct IgM plaque-forming cells (PFC) per 10^6 splenic cells (±S.E.) was calculated per group of six mice.
*$P<.05$.

In general, two questions regarding the mechanisms of these effects should be considered. First, does exposure to microwave radiation result in proliferation of certain subpopulations of B cells? Such an occurrence would be in agreement with the data of *Stodolnik-Baranska* [1967] and of *Czerski* [1975]. Second, the same effects might be obtained by stimulation of immature B cells that are already present in the spleen and that subsequently express properties of more highly differentiated cells. A lack of change in the total number of splenic cells after exposure of mice to microwaves, which was found in an initial pilot study, would indicate that the latter possibility was the case. Further experiments were carried out by which to evaluate the rate of DNA synthesis in splenic cells of mice at various intervals following exposure to microwaves. Cells were cultured *in vitro* for various periods of time (4 hours to 7 days) and incorporation of a specific DNA-precursor (tritiated thymidine) was measured. No differences in uptake of thymidine were observed between cells from microwave-irradiated and sham-exposed mice. It was concluded that neither a single nor a triple dose of microwave radiation produces detectable proliferation of cells [*Wiktor-Jedrzejczak,* unpublished observations]. The possibility that B cells may proliferate elsewhere (e.g., in the bone marrow) from which they may migrate to the spleen, also seems to be unlikely since no increase was found in the rate of DNA synthesis after short-term (4-hour incubation of bone marrow cells or peripheral blood lymphocytes in culture [*Wiktor-Jedrzejczak,* unpublished observations].

There are two major ways of studying the effects of microwave radiation on the functional capacity of lymphoid cells: (1) evaluate the ability of the cells to undergo specific activation by antigens, and (2) evaluate the ability of the cells to be stimulated by nonspecific activators -- by mitogens. In any mammal, there are groups of cells, T and B, that are genetically preprogrammed to respond to antigens that possess unique structural properties. Such groups of cells are derived originally from single precursors and, therefore, are called "clones." Mitogens can stimulate many clones of lymphoid cells that have similar general characteristics apart from their specific antigenic reactivities

TABLE 9. Effect of three exposures to 2450-MHz microwaves on response of mice to a thymus-independent antigen (DNP-lys-Ficoll).

Immunizing Treatment	Mean D-PFC/spleen[1]		Mean D-PFC/10^6 splenic cells[1]	
	Sham-Exposed	Microwave-Exposed	Sham-Exposed	Microwave-Exposed
Saline	713 ± 308	1219 ± 274	5.3 ± 2.2	9.7 ± 3.0
DNP-lys-Ficoll	24375 ± 4399	17000 ± 5700	163 ± 23	152 ± 42

[1]Splenic cells from six mice were assayed in duplicate for ability to form PFC against sheep erythrocytes. Arithmetic means were calculated and from the data obtained the mean of means of direct IgM plaque-forming cells (PFC) per 10^6 splenic cells (±S.E.) was calculated per group of six mice.

[*Nowell,* 1976]. Moreover, some routinely used mitogens stimulate clones of T-cells only, while others only stimulate clones of B cells. For example, phytohemagglutinin (PHA) and concanavalin A (con A) only activate clones of T-lymphoid cells [*Greaves and Bauminger,* 1972; *Andersson et al.,* 1972]. On the other hand, dextran sulfate, the lipopolysaccharide of *E. coli* (LPS), polyinosinic polycytidylic acid (poly I•C), and purified protein derivative of tuberculin (PPD) only stimulate B cells [*Gronowicz and Coutinho,* 1975; *Scher et al.,* 1973]. There are also mitogens such as pokeweed (PWM) that stimulate both T- and B-lymphoid cells [*Janossy and Greaves,* 1971]. Further, there are indications that difference subpopulations of T or B cells are stimulated by difference T- and B-cell-specific mitogens. These differences correlate with the functional (maturational) level of the cells. For example, con A was found primarily to stimulate T-helper cells [*Hirst et al.,* 1975]; that is, cells that help B cells "recognize" certain antigens. In the case of B-cell-specific mitogens, a good correlation exists between stage of maturation of the B cells and their response to a given mitogen (Figure 3). Dextran sulfate stimulates precursors of B cells [*Gronowicz and Coutinho,* 1975]; LPS and Poly I•C stimulate immature B cells [*Gronowicz and Coutinho,* 1975; *Ahmed et al.,* 1976]; and PPD stimulates mature B cells [*Gronowicz and Coutinho,* 1975]. These mitogens stimulate cells at a given stage not only to proliferate but also to make the transition to the next maturational stage. The process whereby B cells from many different clones are stimulated nonspecifically is called "B-cell polyclonal activation" [*Gronowicz and Coutinho,* 1975]. The ability of lymphoid cells to undergo mitogenic stimulation is, therefore, a measure of the functional capacity of subpopulations of lymphoid cells. The assays for mitogenic activity are partly automated [*Strong et al.,* 1973], which allows simultaneous performance of several assays at various concentrations of cells and mitogens. The "mitogen assays" are useful tools, therefore, in the screening of functional changes in lymphoid cells that are produced by various factors; the assays were used by us to characterize the effects of microwave radiation on the function of lymphoid cells.

Neither a single exposure nor a triple exposure of mice to microwaves affected the response of their splenic cells to the T-cell-specific mitogens. Also, the response to PWM, which stimulates both T and B cells, was unchanged. The response to the B-precursor-cell-specific mitogen, dextran sulfate, was unaltered, while the response to the B-cell-specific mitogens LPS and Poly I•C was significantly increased, both after single and triple exposures. The response to the mature B-cell mitogen PPD was also significantly increased after a single exposure. The results were clearly correlated with previously discussed changes in the proportion of cells that bear appropriate surface markers. Microwave irradiation did not stimulate lymphoid cells to proliferate; the irradiation probably increased the number of cells that are responsive to stimulation by the various B-cell mitogens. The radiation apparently acted as a polyclonal B-cell activator without mitogenic properties. The data obtained on mitogens also provided further evidence that the microwaves did not affect the T-lymphoid cells.

Another important approach to the study of the effects of microwaves on the function of the immune system is the evaluation of T- and B-cell responses to specific antigens. Studies of the specific response require the choice of an antigen, the response to which can serve as an example of the immune response to a given category of antigens. For this purpose, one must distinguish between antigens that activate appropriate clones of T cells and antigens that stimulate antibody-producing B cells. The structural properties of an antigen determine the clone of T- and B-lymphoid cells with which it will interact [*Feldmann et al.,* 1975; *Coutinho,* 1975]. The main structural criterion for the selection of the type of clone of responsive cells is the difference between the structure of the antigens that are recognized by lymphoid cells as foreign and the structure of autologous antigens of the organism. In essence, B cells by themselves respond to antigens that are clearly different structurally from autologous antigens, e.g., bacterial antigens. The T cells, on the other hand, are able to recognize antigens that are highly similar to, yet different from, those normally occurring in the body, e.g., antigens of another animal of the same species or antigens of autologous, malignant tumors. There are also antigens that B and T cells respond to cooperatively. As already mentioned, an antigen must be highly different from an autologous antigen in order to be recognized as foreign for B cells. In the case of antigens where structural differences are of the intermediate type, B cells recognize a part of them as foreign. The other part is recognized as foreign by a special subpopulation of T cells, called "T-helper cells," and these T-helper cells, in turn, specifically activate maturation of B cells and lead to the production of antibody that is specific to the antigen. The type of antigens that require help of T cells to engender a B-cell response are called "T-dependent," while other antigens, for which B cells do not require T-helper cells, are called "T-independent."

In order to determine if exposure of mice to microwaves induces a specific effect on their ability to respond to thymus-dependent and thymus-independent antigens, groups of mice were immunized with a well-known and established thymus-dependent antigen, sheep red blood cells (SRBC) and a well-known and documented thymus-independent antigen (DNP-lys-Ficoll). The studies indicate that while three small doses of microwave radiation produce polyclonal stimulation of B-lymphoid cells in the spleen of exposed mice, they may simultaneously decrease the number of antibody-forming cells that are induced specifically by injection of thymus-dependent or thymus-independent antigens.

The apparent discrepancy between the two effects may be explained as follows: injection of antigen produces a stimulus for activation of a specific clone of B cells which, in the course of the reaction to antigen, generates antibody-producing plasma cells. The magnitude of the response depends to some extent on the number of available antigen-reactive B cells. Microwaves may nonspecifically stimulate some of the cells to mature before they are activated by antigen. When signals produced by injection of an antigen reach them, at least some of them will be already in a late, unresponsive stage. The discrepancy between our data and observations previously reported by *Czerski* [1975], who found an increase in the IgM response to SRBC immunization after long-term exposure to microwaves, may be explained by differences both in the exposure-immunization schedule and in the exposure conditions of the two studies. Czerski first exposed animals, then immunized them and assayed for response, while in our study the animals were first immunized and then exposed before the response was evaluated. Nevertheless, the discrepancy between Czerski's results and ours suggests that the decrease in the number of antibody-forming cells that was produced in response to antigens in our microwave-exposed mice may be a temporary phenomenon. We cannot stress too strongly the importance of the immunization-exposure schedule for interpreting the interaction between microwave radiation and cells of the immune system.

The CR+ lymphocytes represent in mice a maturational stage of B-lymphoid cells that precede antibody-forming cells [*Hämmerling et al.*, 1976]. Therefore, studies of the changes in the subpopulations of CR+ B cells should explain some of the observed changes in the number of direct PFC and should also relate the effects observed in our study to previously reported changes in the maturation of B cells in microwave-exposed mice [*Wiktor-Jedrzejczak et al.*, 1976, 1977]. Contrary to our expectations, no simple conclusions can be drawn. The immunization alone, microwaves alone, and both in combination were found to alter the frequency of CR+ cells in the murine spleen. Moreover, the differences were observed between T-dependent and T-independent antigens, both in sham- and microwave-exposed animals, suggesting that the relation between CR+ B cells and different clones of antibody-producing cells is more complex than previously suspected.

Our studies do indicate that exposure to microwave radiation can produce a number of effects on the immune system of mice and that some immunization-exposure relationshps implicate an inhibition of the immune response both to T-dependent and to T-independent antigens.

5. FINAL REMARKS

We have not discussed the important role played in the immune system by macrophages and granulocytes and, since some important effects of microwave radiation on these cells have been reported, we refer readers to the previously published papers by *Mayers and Habeshaw* [1973] and by *Szmigielski et al.* [1975]. Beyond the scope of these papers are the effects of microwaves on T-killer cells and other cytotoxic mechanisms by which lymphoid cells specifically eliminate antigens. Whether the studies of mice reported here can be extrapolated directly to man is questionable, since the frequency of radiation we used (2450-MHz) allows for almost complete penetration of microwaves through the body of the mouse but would allow for only superficial penetration of the tissues in man [*Baranski and Czerski*, 1976]. Studies on the effects of microwaves on the immune system are in their infancy; until experiments are performed in different laboratories, under different exposure conditions, and with different species of animals, firm conclusions cannot be drawn.

REFERENCES

Ahmed, A., I. Scher, and K. W. Sell (1976), Functional studies of the ontogeny of the M-locus product: A surface antigen of murine B lymphocytes, in *Leukocyte Membrane Determinants Regulating Immune Reactivity*, edited by V. P. Eijsvogel, D. Roos, and W. P. Zeijlemaker, pp. 703-709, Academic, New York.

Anderson, J., G. M. Edelman, G. Moller, and O. Sjoberg (1972), Activation of B lymphocytes by locally concentrated concanavalin A, *Eur. J. Immunol.*, *2*, 233-235.

Baranski, S., and P. Czerski (1976), *Biological Effects of Microwaves*, 234 pp., Dowden, Hutchinson & Ross, Stroudsburg.

Bianco, C., R. Patrick, and V. Nussenzweig (1970), A population of lymphocytes bearing a membrane receptor for antigen-antibody-complement complexes. I. Separation and characterization, *J. Exp. Med.*, *132*, 702-720.

Christman, C. L., H. S. Ho, and S. Yarrow (1974), A microwave dosimetry system for measured sampled integral-dose rate, *IEEE Trans. Microwave Theory Tech.*, *MTT-22*, Pt. II, 1267-1272.

Coutinho, A. (1975), The theory of the "one nonspecific signal" model for 3 cell activation, *Transplant. Rev.*, *23*, 49-65.

Czerski, P. (1975), Microwave effects on the blood-forming system with particular reference to the lymphocyte, *Ann. N. Y. Acad. Sci.*, *247*, 232-242.

Feldmann, M., M. G. Howard, and C. Desaymard (1975), Role of antigen structure in the discrimination between tolerance and immunity by B cells, *Transplant. Rev.*, *23*, 78-97.

Greaves, M. F., and S. Bauminger (1972), Activation of T and B lymphocytes by insoluble phytomitogens, *Nat. New Biol.*, *235*, 67-70.

Gronowicz, E., and A. Coutinho (1975), Functional analysis of B cell heterogeneity, *Transplant. Rev.*, *24*, 3-40.

Gronowicz, E., A. Coutinho, and G. Moeller (1974), Differentiation of B cells: Sequential appearance of responsiveness to polyclonal activators. *Scand. J. Immunol.*, *3*, 413-421.

Hämmerling, U., A. F. Chin, and J. Abbott (1976), Ontogeny of murine B lymphocytes: Sequence of B-cell differentiation from surface-immunoglobin-negative precursors to plasma cells, *Proc. Natl. Acad. Sci. U.S.A.*, *73*, 2008-2012.

Hirst, J. A., P. C. L. Beverley, P. Kisielow, M. K. Hoffman, and H. F. Oettgen (1975), Ly antigens: Markers of T cell function on mouse spleen cells, *J. Immunol.*, *115*, 1555-1557.

Ho, H. S., E. I. Ginns, and C. L. Christman (1973), Environmentally controlled waveguide irradiation facility, *IEEE Trans. Microwave Theory Tech.*, *MTT-21*, 837-840.

Janossy, G., and M. F. Greaves (1971), Lymphocyte activation. I. Response of T and B lymphocytes to phytomitogens, *Clin. Exp. Immunol.*, *9*, 483-498.

Jerne, N. E., and A. A. Nordin (1963), Plaque formation in agar by single antibody-producing cells., *Science*, *140*, 405.

Johnson, C. C. (1975), Recommendations for specifying EM wave irradiation conditions in bioeffects research, *J. Microwave Power*, *10*, 249-250.

Justesen, D. R. (1975), Toward a prescriptive grammar for the radiobiology of non-ionising radiation: Quantities, definitions, and units of absorbed electromagnetic energy–an essay, *J. Microwave Power*, *10*, 343-356.

Mayers, C. P., and J. A. Habeshaw (1973), Depression of phagocytosis: a nonthermal effect of microwave radiation as a potential hazard to health, *Int. J. Radiat. Biol.*, *24*, 449-459.

Mosier, D. E. (1969), Cell interactions in the primary immune response *in vitro:* A requirement for specific cell clusters. *J. Exp. Med.*, *129*, 351-361.

Nowell, P. C. (1976), Mitogens in immunobiology: Introduction, in *Mitogens in Immunobiology*, edited by J. J. Oppenheim and D. L. Rosenstreich, pp. 3-9. Academic, New York.

Raff, M. C., M. Sternberg, and R. B. Taylor (1970), Immune immunoglobulin determinants on the surface of mouse lymphoid cells, *Nature (London)*, *225*, 553-554.

Reif, A. E., and J. M. Allen (1966), Mouse thymic iso-antigens. *Nature (London)*, *209*, 521-523.

Rittenberg, M. B., and K. L. Pratt (1969), Antitrinitrophenyl (TNP) plaque assay. Primary response of Balb/c mice to soluble and particulate immunogen, *Proc. Soc. Exp. Biol. Med.*, *132*, 575-581.

Scher, I., D. M. Strong, A. Ahmed, R. C. Knudsen, and K. W. Sell (1973), Specific murine B-cell activation by synthetic single- and double-stranded polynucleotides, *J. Exp. Med.*, *138*, 1545-1563.

Sharon, R., P. R. B. McMaster, A. M. Kask, J. D. Owens, and W. E. Paul (1975), DNP-Lys-Ficoll: A T-independent antigen which elicits both IgM and IgG anti-DNP antibody secreting cells, *J. Immunol.*, *114*, 1585-1589.

Stodolnik-Baranska, W. (1967), Lymphoblastoid transformation of lymphocytes *in vitro* after microwave irradiation, *Nature (London)*, *214*, 102-103.

Strong, D. M., A. Ahmed, G. B. Thurman, and K. W. Sell (1973), *In vitro* stimulation of murine spleen cells using a microculture system and a multiple automated sample harvester, *J. Immunol. Methods*, *2*, 279-291.

Strong, D. M., J. N. Woody, M. A. Facktor, A. Ahmed, and K. W. Sell (1975), Immunological responsiveness of frozen-thawed human lymphocytes, *Clin. Exp. Immunol.*, *21*, 442-455.

Szmigielski, S., J. Jeljaszewicz, and M. Wiranowska (1975), Acute straphylococcal infections in rabbits irradiated with 3 GHz microwaves, *Ann. N. Y. Acad. Sci.*, *247*, 305-311.

Thurman, G. B., D. M. Strong, A. Ahmed, and K. W. Sell (1973), Human mixed lymphocyte cultures. Evaluation of a microculture technique utilizing the Multiple Automatic Sample Harvester (MASH), *Clin. Exp. Immunol.*, *5*, 289-302.

Wiktor-Jedrzejczak, W., A. Ahmed, P. Czerski, W. M. Leach, and K. W. Sell (1976), The effects of microwaves (2450 MHz) on the immune system in mice: Increase in complement receptor bearing lymphocytes (CRL). *Exp. Hematol. 4* (Suppl.), 73.

Wiktor-Jedrzejczak, W., A. Ahmed, K. W. Sell, P. Czerski, and W. M. Leach (1977), Microwaves induced an increase in the frequency of complement receptor-bearing lymphoid spleen cells in mice, *J. Immunol.*, *118*, 1499-1502.

Youmans, H. D., and H. S. Ho (1975), Development of dosimetry for RF and microwave radiation. I. Dosimetric quantities for RF and electromagnetic fields, *Health Phys.*, *29*, 313-316.

THE ACTION OF MICROWAVE RADIATION ON THE EYE†

Russell L. Carpenter* and Clair A. Van Ummersen**

ABSTRACT

Microwave power can cause formation of opacities in the lens of the rabbit eye exposed to continuous wave or pulsed wave radiation at frequencies from 2.45 GHz to 10 GHz. When the eye is irradiated in a free field, the opacity (cataract) develops in the posterior part of the lens; in location, form and growth, it resembles cataracts caused by ionizing radiation. When the eye is irradiated at the same frequencies as part of a 'closed' waveguide system, the cataract develops in the anterior part of the lens, like those caused by infrared radiation. Although for every power level there is a minimal exposure period which will cause an opacity, repeated shorter exposures can have a cumulative effect, the main determining factor being the time interval between successive exposures. Experimental evidence suggests that microwave cataracts are not simply a result of microwave heating, but are caused by some other property of the radiation.

INTRODUCTION

Reports presented before the International Microwave Power Institute are mainly concerned with what microwave power can do *for* man, - how it can be employed to serve his purposes. In our laboratory, we have been looking at the other side of the coin to consider the question of what microwave power can do *to* man. That certain regions of the electromagnetic spectrum directly and profoundly affect man and other living organisms is common knowledge. Not only the visible spectrum but also the invisible ultra-violet, infrared, radio-frequency and ionizing radiations exert an influence upon living tissue. Sometimes it is a beneficial influence; sometimes it is harmful. Indeed, what may at one time be beneficial may, at another time, through higher intensity or longer period of action, become harmful radiation.

Microwave radiation, properly applied in diathermy to living muscles and joints, induces beneficial deep heat. It can also, in a microwave oven, thoroughly cook the muscles and joints of broiler chickens. It is quite true that 'microwave power is not a new form of energy, but it is being used in inventive ways to make man's tasks easier' Jolly (1967). As the useful applications increase, it becomes increasingly important to discover how it may affect man's body and to learn, so far as possible, how much is too much where living tissue is concerned.

Over the past two decades, many investigators have studied the effect of radio-frequency energy upon living things. Attempts have been made to assess the relative roles of frequency, single or repeated exposures to radiation, continuous or pulsed wave, whole body or localized irradiation, and differences related to animal species and age. These various studies have recently been summarized in tabular form by Carpenter and Clark, (1966).

The possibility of hazard to personnel through exposure to radio-frequency radiation has led to experimental investigation of the effect of irradiating the whole body. In general, whole body irradiation induces an elevation of body temperature (hyperthermia) and there occur changes in blood pressure, rate of heart beat and rate of respiration, as well as changes in blood constituents. When the power level is sufficiently high or the duration of exposure to radiation is sufficiently long, the induced hyperthermia causes death of the experimental animal and consequent

† Paper received December 28, 1967. Based on the original paper presented by one of the authors (RLC) at the 1967 Symposium on Microwave Power, Stanford University, March 30, 1967, entitled 'The Action of Microwave Radiation on the Eye'.

*Professor of Zoology; **Lecturer in Biology and Research Associate; Department of Biology, Tufts University, Medford, Massachusetts.

Reprinted with permission from *J. of Microwave Power*, vol. 3, no. 1, pp. 3–19, Mar. 1968.

termination of the experiment. As a result, it becomes impossible to identify any long range effects or effects which might be of a non-thermal nature. On the other hand, such effects may not be evoked in the case of whole body irradiation if either the power or the duration are reduced to non-lethal levels.

It seemed to us that a fruitful approach to the problem would be to apply the power to a limited region of the body and then thoroughly study the effect upon a single irradiated organ. This is the procedure we have followed in ascertaining how microwave radiation can affect the mammalian eye. In presenting here some of the evidence not previously reported as to how microwave power can affect the eye, it seems advisable also to summarize some work already reported. This communication will therefore be partly in the nature of a review.

The first evidence that radio-frequency power could harmfully affect the eye came from experiments on the dog by Daily et al (1948) and on the rabbit by Richardson et al (1948). In both instances, microwave radiation caused the development of small opaque areas, or cataracts, in the crystalline lens of the eye. Similar results were subsequently obtained by these and other investigators (Richardson et al, 1951; Daily et al, 1952; Williams et al, 1955; Carpenter, 1958; 1959; Carpenter et al, 1959, 1960.) Still others reported no effect upon the eye (Howland and Michaelson, 1959; Cogan et al, 1958; and Osborne and Addington, 1959). From these various and varied experiments, however, certain general conclusions were warranted:

1. Microwave radiation at frequencies of 2.45 GHz to 10 GHz, either continuous wave or pulsed, caused development of lens opacities in the eyes of dogs and rabbits.

2. At low frequencies, such as 200 or 400 MHz, there was no lens damage.

3. In those experiments in which measurements of intraocular temperatures were made, exposure of the eye to microwave radiation caused an increase in intraocular temperature.

4. Because there existed no common yardstick for measuring power absorbed by the eye, it was not possible to correlate lens damage with power density and duration of exposure. Neither could results obtained with different frequencies be compared, since data on power levels were not translatable into some common unit of measurement.

5. Development of cataracts could be provoked not only by exposure to a single dose of radiation but also by multiple exposures at intervals of a day (Daily et al, 1948, 1959, 1952; Richardson et al, 1948; Belova and Gordon, 1956). Although these experiments raised the possibility of a cumulative effect of microwave power, they provided no firm evidence, for in no instance had it been established that a single one of the multiple exposures was of itself incapable of causing opacity formation.

When we undertook for the U.S. Air Force a study of how much hazard to the eye microwave power might be, we were concerned with resolving several fundamental questions:

1. For each level of microwave power incident on the eye, is there a minimal single period of exposure (a threshold dose) which will cause an opacity to develop in the lens?

2. Is there a cumulative effect if the eye is exposed repeatedly to less than threshold doses? *This is the condition more likely to be encountered with industrial, commercial and military applications of microwave power and cases of microwave cataract incurred by personnel have been reported (Hirsch et al, 1952).*

3. Inasmuch as microwave power causes both a rise in intraocular temperature and the formation of lens opacities, are these two effects causally related or are they independent and merely coincident?

4. Radiation dose is determined by the intensity of the power absorbed by the eye and by the duration of the period during which it is absorbed. The latter is easily measured. Is it possible to devise a method by which the amount of power absorbed can be measured with a fair degree of accuracy?

As our experimental subject, we used the New Zealand White rabbit, a strain whose eye has been used often in experiments on the effect of radiated energy. The rabbit eye is three-fourths the diameter of the human eye and is structurally similar. Any changes in the transparency of the lens can easily be identified by examination in the living animal. Rabbit body temperature (38.7° C) is only 1.2 degrees higher than human body temperature (37.5° C).

The general structure of the mammalian eye is shown in Figure 1. It functions as the organ of sight only when light rays form an image upon and stimulate certain photosensitive cellular elements present in its innermost layer, the retina. The formation of a clear image on the retina depends to a considerable degree upon maintenance of transparency in the several media through which light rays must pass to reach the retina. In their consecutive anatomical order, these are the cornea, the fluid aqueous humour in the anterior chamber, the crystalline lens, and the vitreous body. The chief light refracting surface is the cornea; the lens gives sharp focus to the image. Because blood

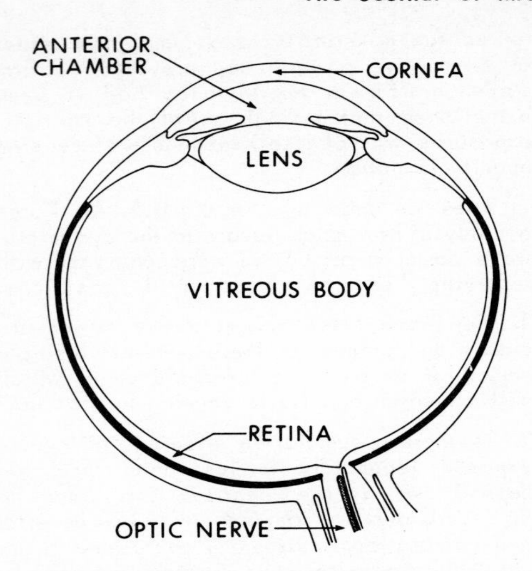

Fig. 1. Diagrammatic horizontal section of a mammalian eye.

flowing through vessels obstructs light rays, it is advantageous to have the light transmitting media unencumbered by blood vessels. Thus we find that the cornea, the lens and the gel-like vitreous body are without blood supply. That the retina itself has blood vessels is an optical disadvantage which we must of necessity tolerate.

The lens is biconvex in form (Fig. 2). The central point of its front (anterior) surface is its anterior pole; the center of its back (posterior) surface is the posterior pole. A line joining both poles is its axis. Its rounded edge, where anterior and posterior surfaces join, is its equator. In its composition, the lens has an outer thin capsule which is elastic. This encloses the lens substance, which consists of many lens fibers, arranged in concentric layers as the outer lens cortex and more firmly packed centrally to form the lens nucleus. The fibers are long prismatic cells which extend from the equator forward and backward toward the poles, their tapered ends meeting the ends of similar fibers reaching from the opposite side of the lens. Where the many fiber ends meet, their junctions form a linear suture. In the rabbit eye, the suture line is horizontal on the posterior side of the lens and vertical on the anterior side.

Directly under the anterior lens capsule, a single layer of cube-shaped cells constitutes the lens epithelium. These epithelial cells, in a zone toward the equator, undergo cell division to produce new cells which move to the equator and there grow in length to become the new lens fibers. Thus, throughout life, new layers of young fibers continually are being differentiated, each additional layer going to form new lens cortex directly beneath the epithelium in front and under the lens capsule behind. The lens is therefore always growing, most actively before birth and at a diminishing rate thereafter. So long as this growth proceeds normally, the lens maintains its transparency (Plate 1a) but if some agency, either external or internal, interferes even temporarily with the normal pattern of growth, there may result areas of lens fibers which are misformed and lack transparency. These opacities in the lens substance are termed cataracts. Depending upon their size and position, they may seriously interfere with sight.

Because dividing and differentiating cells are known to be susceptible to harm by radiated energy (e.g., infrared, ultraviolet and ionizing radiations), the lens epithelium must be suspect as the possible site of action of microwave radiation.

DETERMINATION OF THRESHOLD DOSES

Although Williams et al (1955) reported time and power thresholds for the production of lens opacities by 2.45 GHz radiation, it was necessary for us similarly to establish thresholds for our equipment operating under the conditions obtaining in our laboratory. Our power source initially was a modified microwave diathermy instrument, the Raytheon CMD4 Microtherm. Operating at 2.45 GHz, its maximal CW output was approximately 85 watts. Power was monitored through a direc-

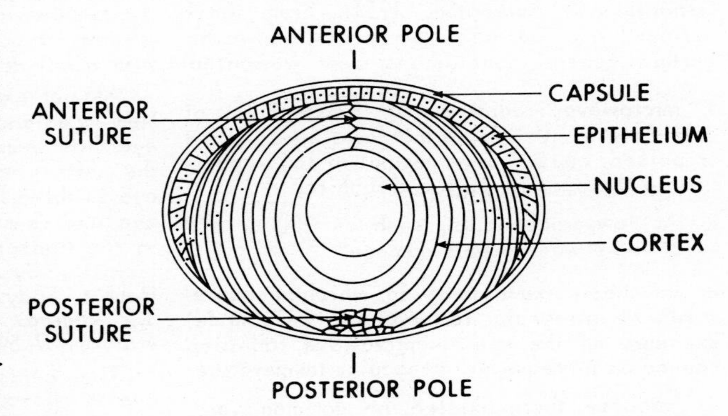

Fig. 2. Diagrammatic horizontal section of the rabbit lens.

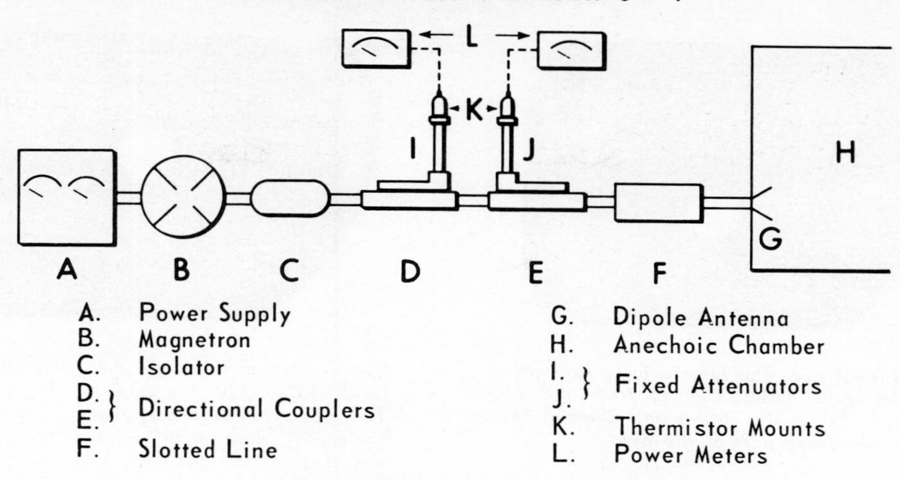

Fig. 3. System for irradiating eyes at 2.45 GHz in an anechoic chamber.

tional coupler inserted in the coaxial cable to the dipole antenna and leading to a thermistor and power meter (Fig. 3). The eye of the anesthetized rabbit was positioned exactly opposite the dipole crossover and two inches from the plastic housing which covered it. Irradiating was done in an anechoic chamber lined by microwave absorbent.

If we were to establish for several power levels threshold doses for induction of cataracts,a uniform method for measuring power density became essential. A method to be preferred would be one measuring the power absorbed by the eye. That not being feasible, we has to settle for calorimetric measurements made at the position of the eye. The calorimeter, a thin-walled sphere made of plastic of low dielectric constant and having a known volume and profile area, was the size of the rabbit eye. It was filled with a saline solution having a dielectric constant similar to that of the eye. Temperature changes reflecting energy absorbed or lost were measured by a thermistor-bridge circuit, with the thermistor enclosed in the tip of a 24 gauge hypodermic needle inserted to the center of the sphere. They were recorded on a strip-chart recorder. Because the measurements thus made were for any power level consistent and reproducible, they permitted us to establish power levels at the position of the rabbit eye which bore a constant relation to each other, assuming consistent perturbation of the microwave field due to the presence of the calorimeter and assuming also that perturbation of the field by the head of one rabbit would be the same as that caused by the head of another rabbit, so long as their heads were similar in size and shape. That the calorimeter and the rabbit head perturbed the field differently was immaterial.

By this calorimetry, the maximal continuous wave output of our generator was determined to be 400 mw/cm^2 at the two inch position of the eye. In all experiments, the body of the animal was behind a barrier of microwave absorbent material, so that only the head was exposed. The non-irradiated left eye served as a control. Following irradiation, both eyes were examined regularly by ophthalmoscope and slit-lamp biomicroscope.

Figure 4 summarizes the results of 141 experiments in which right eyes were subjected to single exposures at power levels varying from 400 mw/cm^2 down to 80 mw/cm^2 The threshold curve defines the minimal exposure period at each power level which caused development of an opacity in the lens. Shorter subthreshold exposure periods had no apparent effect. In general, a threshold exposure evoked a minimal response, such as small granules or vesicles on or near the the posterior suture. Always they were located just beneath the capsule in the posterior cortex and probably were too slight to interfere with the animal's vision.

With longer exposures, minimal changes of this type were usually the first to appear but from day to day they increased progressively to become obvious opacities (Plate 1; (b), (c), and (d)). Some were of circumscribed character and were sharply delineated in the posterior lens cortex, on or near the lens axis or above and below the suture. Other cataracts were of a diffuse nature, involved more extensive areas and penetrated more deeply into the posterior cortex (Plate 1; (i)).

These microwave cataracts were in many respects similar to those caused by ionizing radiation. In both types, the opacities develop in the cortex beneath the posterior capsule and the first changes occur in the region of

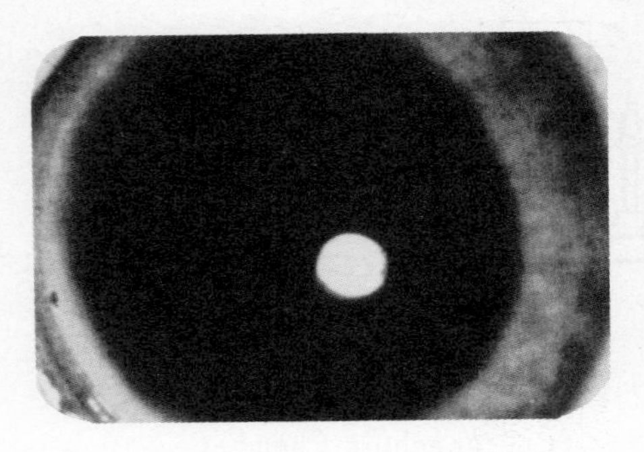

(a) Normal eye. The lens is fully transparent, the pink color coming from the blood vessels of the retina and choroid layers in the back part of the eye.

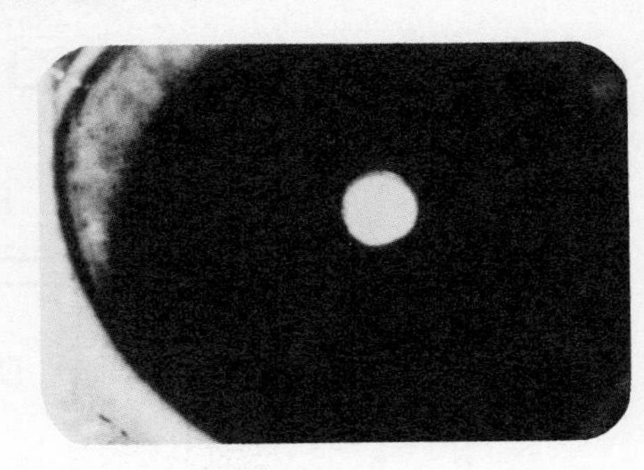

(b) Eye 5 days after single dose irradiation (2.45 GHz, 280 mw/cm2 for 12 min.). The posterior suture appears as a hazy horizontal line with an arc of small vesicles at its right end.

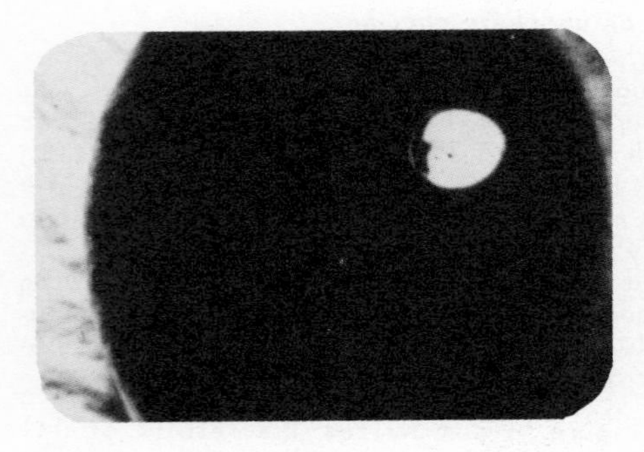

(c) The same eye a day later. A string of large vesicles has formed near the equator on the left side. An opacity is developing in the area above and below the posterior suture.

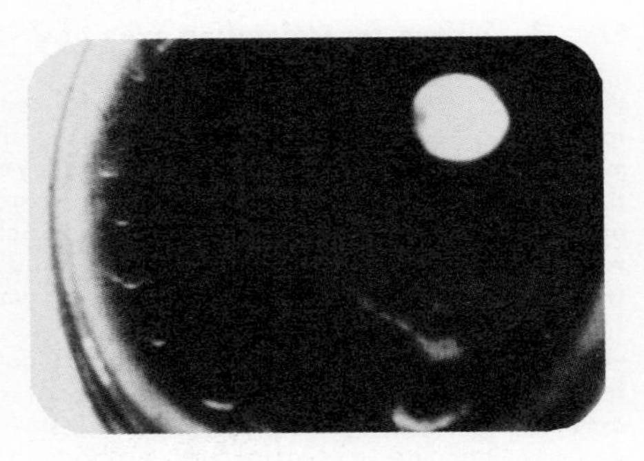

(d) The same eye one week after Photograph (c). The opacity around the suture is now well developed.

(e) Eye 15 days after last of 5 daily sub-threshold exposures (2.45 GHz, 280 mw/cm2 for 4 min.). The well defined opacity includes an area of the posterior sub-capsular cortex above and below the suture.

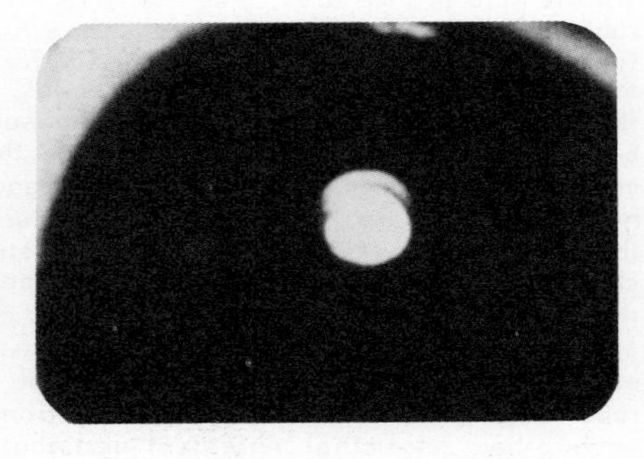

(f) The same eye as shown in Photograph (e) but 14 months later. The cataract has not changed but transparent new lens fibers have been added behind it.

***PLATE 1.** Kodachrome photographs of microwave irradiated right eyes of New Zealand White rabbits. The pupils have been dilated to permit maximal viewing of the lens. The white disc is a reflection of the electronic flash by which the photograph was taken in the non-anesthetized animal. Although in black and white, these photographs are included for the information of readers of this volume.*

(g) Eye 6 days after last of 15 daily subthreshold exposures (2.45 GHz, 80 mw/cm^2 for 1 hour). Clumps of granules and vesicles form an interrupted string along the upper part of the lens in the posterior cortex. An opacity is forming around the suture and is especially dense above and below its left end.

(h) The same eye as shown in Photograph (g) but 3 weeks later. The opacity has become large and fully formed in the posterior subcapsular cortex.

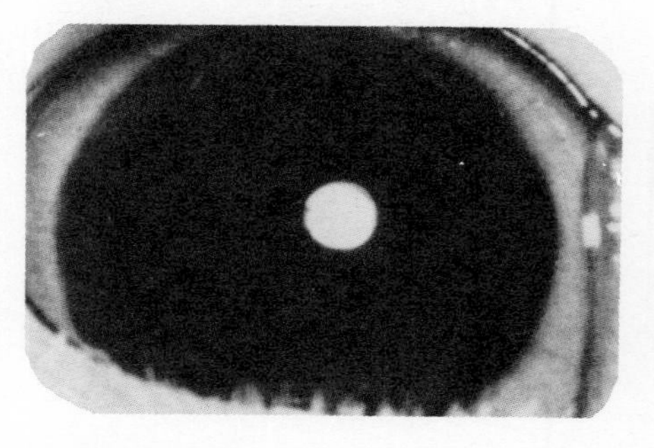

(i) Eye 50 days after single dose irradiation (2.45 GHz, 240 mw/cm^2 for 25 min.). The cataract is diffuse, covering most of the posterior subcapsular cortex.

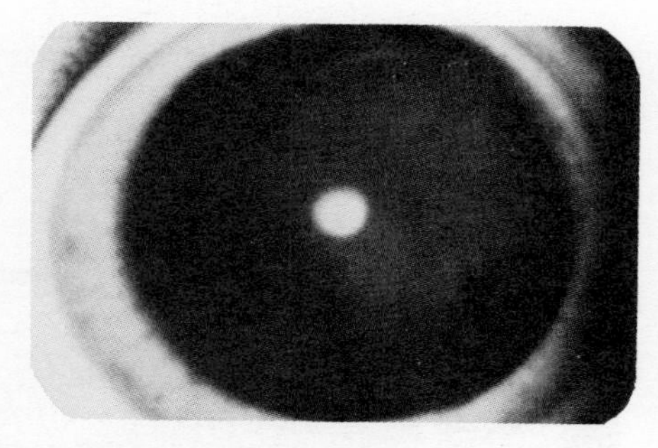

(j) Eye 52 days after single dose irradiation employing 'closed' waveguide system (2.45 GHz, 2.67 watts for 5 min.). The cataract is in the anterior cortex of the lens.

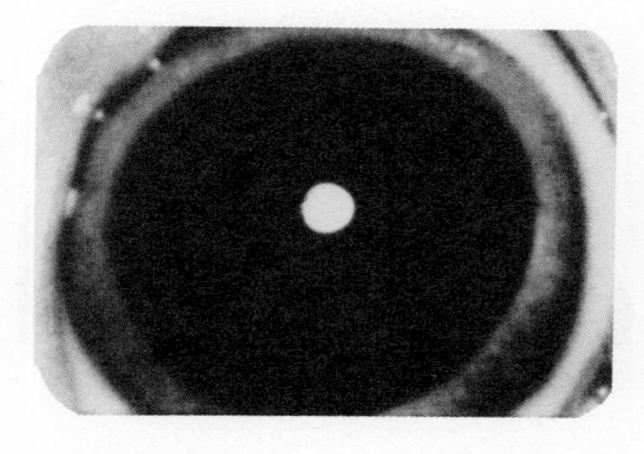

(k) Eye 58 days after single dose irradiation in anechoic chamber by pulsed wave (2.45 GHz, 0.5 duty cycle, 1,000 pps. pulse repetition rate, average power 140 mw/cm^2, exposure period of 20 min.). The cataract is in the posterior subcapsular cortex.

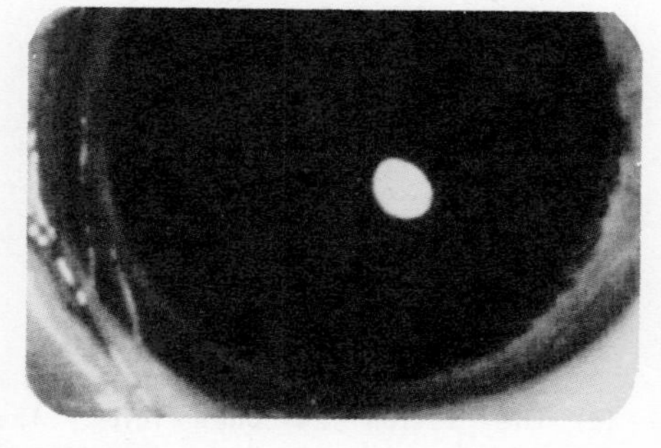

(l) Eye 10 days after single dose irradiation by pulsed wave and 'closed' waveguide system (9.375 GHz, average power 900 mw, exposure period of 4 min.). The cataract is in the anterior cortex along the vertical suture.

PLATE 1. (Continued) Kodachrome photographs of microwave irradiated right eyes of New Zealand White rabbits. The pupils have been dilated to permit maximal viewing of the lens. The white disc is a reflection of the electronic flash by which the photograph was taken in the non-anesthetized animal. Although in black and white, these photographs are included for the information of readers of this volume.

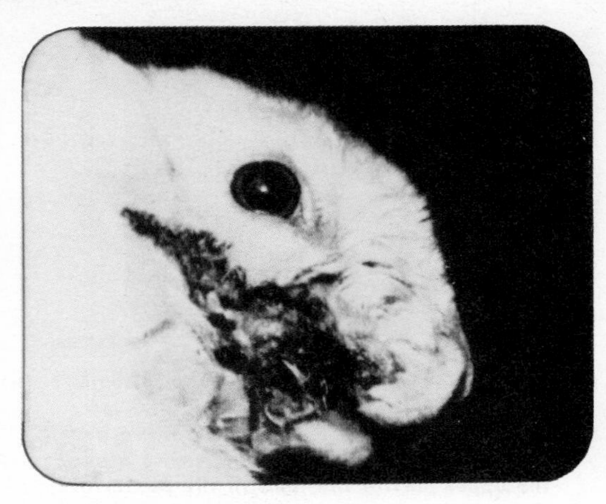

(a) Rabbit from which an area of facial skin has just sloughed off, 23 days after incurring subcutaneous microwave burn.

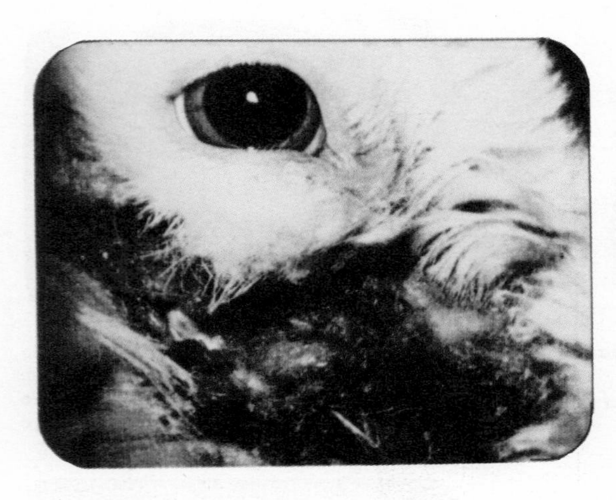

(b) Magnified view of the lesion. There is no infection and healing is under way. Note the triangle-shaped cataract in the eye.

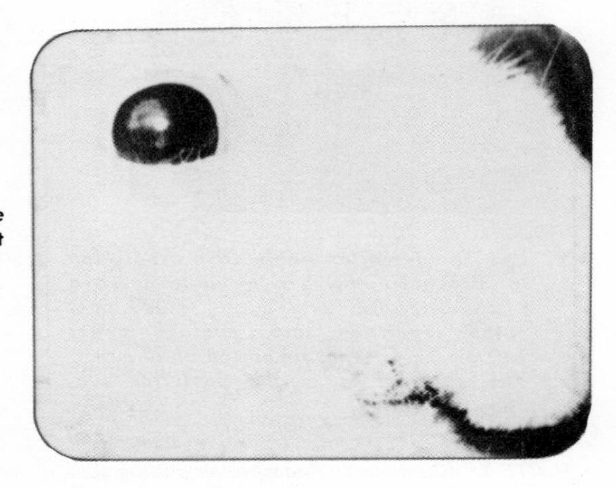

(c) The lesion 24 days later. The burned area has healed and is almost completely covered by new skin.

PLATE 2

Although in black and white, these photographs are included for the information of readers of this volume.

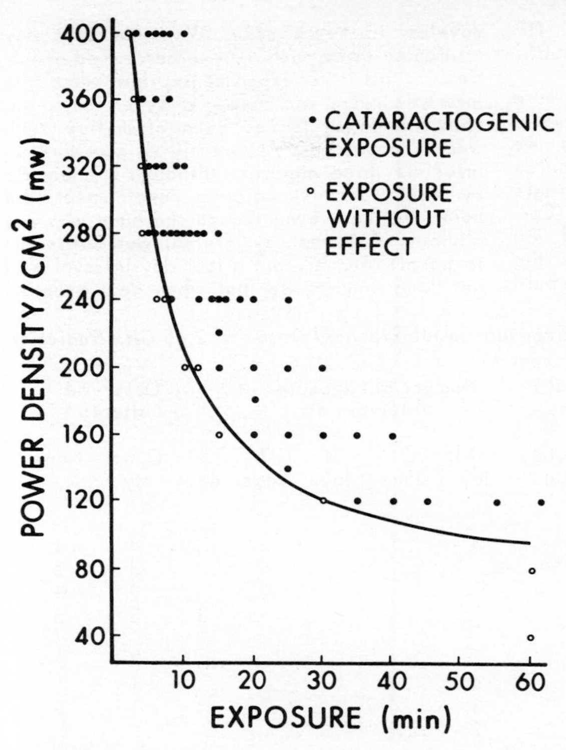

Fig. 4. Time and power thresholds for induction of lens opacities in the rabbit eye by single dose irradiation at 2.45 GHz. Solid circles represent irradiations which induced opacities; open circles represent irradiations without effect. Each circle may represent a number of experiments; altogether, 140 experiments are represented. Power densities do not denote power absorbed by the eye but are values obtained by calorimetric measurements at the position of the irradiated eye.

the posterior suture. A distinct difference of the microwave cataract, however, is that it usually develops within one to eight days after irradiation, the average latent period being 3 1/2 days. Cataracts caused by single doses of X-ray develop only after latent periods of 25 to 30 days, according to Cogan and Donaldson (1951).

INTRAOCULAR TEMPERATURE DURING IRRADIATION

In microwave diathery and microwave cooking, biological tissues - whether living or dead - become heated as they absorb the radiated energy. The amount of heating depends partly on how much radiation the tissue absorbs and partly on its ability to dissipate heat, a process more effectively carried out by living tissues through which circulating blood forms a cooling system. Measurements of temperature rise in the eye during irradiation have been made by several investigators (Richardson et al, 1948; Osborne and Frederick, 1948; Daily et al, 1950; Williams et al, 1955; Ely and Goldman, 1957; and Carpenter, 1958), with reported increases ranging from 4°C to 16°C.

Our measurements, shown in Figure 5, were made by means of a shielded 22 gauge hypodermic needle thermistor probe inserted into the eye so that the thermistor bead was positioned in the vitreous body directly adjacent to the posterior pole of the lens. Covering a range of eight power levels, the temperature graphs illustrate that the higher the power, the more rapidly does the temperature rise and the higher the level it reaches before tending to stabilize.

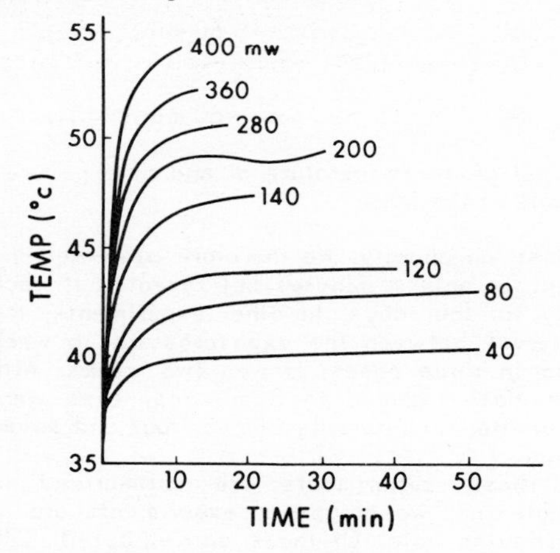

Fig. 5. Intraocular temperatures during irradiation of the eye at power levels from 40 mw/cm² to 400 mw/cm². Each curve is based on recordings from three or more eyes.

IS THERE A CUMULATIVE EFFECT OF REPEATED SUBTHRESHOLD DOSES?

Having established threshold exposure times for the induction of cataracts at several reproducible microwave power levels, we then tested the effect of repeated exposures of shorter duration than the minimal single exposure required to cause a cataract (Carpenter et al, 1960a). At our designated 280 mw/cm^2 power density, 5 minutes was the minimal single exposure period which would develop in every case. Given at four day intervals, three such exposures caused opacities in all five experiments, but when the intervals were increased to seven days, no opacities were formed even after five such weekly exposures. Apparently an exposure as brief as three minutes, although inadequate by itself to cause cataract development, did harm the lens, even though the harm was not obvious. A seven day interval was sufficient to permit recovery but a four day interval was not long enough, so that when next exposed

Table I Cumulative Effect of Repeated Exposure to Subthreshold Doses of 2.45 GHz Radiation

Power density in mw/cm^2	Single dose threshold	Subthreshold exposures		Number of exposures at intervals of:					Observed effects	
		Duration	Ocular temp.*	1 day	2 days	4 days	7 days	14 days	Opacity	None
				4	....	....	....	....	4	4
280	5 min.	4 min.	48.5°				4	....	3	2
							2	....	2	0
								3	3	5
				5	....	....	....	....	7	0
				4	....	....	....	....	1	1
280	5 min.	3 min.	47.2°	3	....	....	....	....	1	3
				2	....	....	....	....	2	0
						3	....	....	5	0
							5	....	0	5
120	35 min.	25 min.	44.0°	5	....	....	....	....	5	0
120	35 min.	30 min.	44.0°		3	....	....	....	4	0
				15	....	....	....	....	3	0
80	....	60 min.	43.3°	10	....	....	....	....	2	0
				1	....	....	....	....	0	5
40	....	60 min.	....	15	....	....	....	....	0	2

*Centigrade temperature at end of exposure period, recorded from vitreous body at posterior pole of the lens.

cause an opacity. We therefore irradiated the eye for only 4 minutes but repeated it each day for four days. In other experiments, the interval between the exposures was a week and in three cases, it was two weeks. With the dose reduced to 3 minutes, eyes were irradiated at intervals of one, four and seven days.

These experiments are summarized in Table 1. Two groups of experiments are of particular note. Of those carried out at 280 mw/cm^2, with a subthreshold exposure of 4 minutes, opacities resulted when a day (Plate 1, (e) and (f))or even a week or two weeks intervened between successive exposures. Daily three minute exposures for five consecutive days caused opacities to to radiation, the lens had not recovered from the damage previously inflicted and multiple exposures therefore had a cumulative effect. With the power reduced to 80 mw/cm^2, a single hour long exposure had no observable effect in five cases but when the same dose was given daily for fifteen or even ten consecutive days, a cataract developed in every instance. Plate 1, (g) and (h)).

It was obvious that the effect of microwave power on the eye can be cumulative, so that single episodes of exposure to radiation which are not of themselves harmful may become hazardous if they are repeated sufficiently often. Our experiments suggested that the interval between subthreshold doses may be an important factor.

In view of the experimental results, we suggested (Carpenter et al, 1960a) that 'the cataractogenic effect of microwave radiation involves initiation of a chain of events in the lens, the visible and end result of which is an opacity, and that this chain of events must be initiated by an adequate power density acting for a sufficient duration of time if it is to progress to the development of an opacity. If either the power density or the duration of the irradiation are below a certain threshold value, then the damage done to the lens is not irreparable and recovery can occur, provided sufficient time elapses before a subsequent similar episode'.

The microwave cataract represents a limited reaction of the lens to the action of an external agency. This response occurs in a relatively short time and the affected portion of the lens remains permanently opaque. However, unless further exposure to radiation occurs, the incident is a closed one and new lens fibers which form later on are not affected. This is well illustrated by Plate 1 (e) with Plate 1 (f) showing the same lens 13 months later. It is apparent that the cataract has not changed and that the newer lens cortex formed later is fully transparent.

ARE MICROWAVE CATARACTS PRODUCED AS A THERMAL EFFECT?

Microwave radiation causes a rise in intraocular temperature and, after a time interval, a lens opacity. It is a simple exercise to conclude that the former is the cause of the latter and this assumption has, indeed, been commonly made. It has been reinforced by the argument that the lens should be expected to be particularly susceptible to microwave heating because, as a structure lacking a blood supply, it is less able to dissipate heat. This view was presented by Ely and Goldman (1957); and Schwan and Piersol (1955) suggested that 45°C may be the intraocular temperature above which damage to the human lens can result.

On the other hand, critical examination of our data led us to suggest (Carpenter, 1958, Carpenter et al, 1960a) the possibility that some factor other than a thermal one might be operative in the production of opacities. If the higher intraocular temperature is the determinant, one should expect evidence of a critical temperature above which opacities would be induced and below which none would occur.

Our experiments yielded no such evidence. On the contrary, they demonstrated that if induction of lens opacities is indeed a thermal effect, it is one which is in no way dependent upon critical temperature, as Figures 4 and 5 together show. At the 280 mw/cm^2 power, the minimal opacity-producing exposure period was five minutes, during which the intraocular temperature was raised to 49.3° C. At the same power level, a three minute exposure raised eye temperature to 47.2° C but caused no opacity. However, if the three minute exposure was repeated at four day intervals, three such exposures caused opacity formation, although the temperature within the eye did not exceed 47.2°. Five such exposures, spaced a week apart, did not affect the lens. If the power was reduced by more than a half, to 120 mw/cm^2, a 35 minute exposure period raised intraocular temperature to 44.2° and invariably caused formation of a lens opacity, but if the exposure period was shortened to 25 or 30 minutes, the intraocular temperature reached 44° but no opacity resulted.

If lens opacities are in fact caused by microwave heating, then the experiments just cited raised two provocative questions: (1) Is lens transparency so dependent upon a fixed temperature limit that an increase in intraocular temperature from 44° to 44.2° is enough to alter it? (2) If a rise in intraocular temperature to 44.2° is sufficient to cause an opacity to develop in the lens, why is it that a rise 5 degrees greater - to 47.2°, as is the case during a single three minute exposure at 280 mw/cm^2 - leaves the transparency of the lens unimpaired?

At the lower power level of 80 mw/cm^2, irradiation for one hour brought a rise in intraocular temperature to only 43.3° C and no apparent harm to the lens. Yet daily exposures of the eye to the same dose for 10 consecutive days caused lens opacities to form. It is of interest to note that it was not necessary to employ anesthesia at this low power, for the animals remained positioned in front of the antenna without moving during the hour. They gave no sign of discomfort or annoyance. In fact, we had the distinct impression that they were particularly comfortable. Inasmuch as the temperature increase in the eye was the same for each exposure, it became difficult to explain formation of a cataract as a thermal effect. The response of the lens was independent of: (a) any fixed temperature limit; (b) the amount of temperature rise in the tissue; and (c) the total duration of the periods during which the elevated temperature was maintained.

We therefore concluded that microwave cataracts are not merely the result of microwave heating but are caused by some other property of this radiation. Further evidence of a non-thermal biological effect of microwave power has been provided by the experiments of Gunn et al (1960) on the rat tests, Van Ummersen (1963) on developing chick embryo, and Livstone (1965) on insect pupae.

LENS OPACITIES INDUCED BY X-BAND RADIATION IN A 'CLOSED' WAVEGUIDE SYSTEM

We were interested in comparing the cataractogenic effect of a frequency higher than 2.45 GHz. Richardson et al (1951) had reported occurrence of opacities in the rabbit lens following a five minute exposure to pulsed wave radiation at 10 GHz, but the power level was not specified. With no other information and with the probability that frequencies higher than this, because of limited penetration, might not reach the lens, we decided to investigate the effect on the eye of 10 GHz radiation.

For experiments on laboratory animals, the investigator is usually restricted by circumstances to employing microwave generating equipment not powerful enough to provide adequate radiated power except in the near field, where slight differences in placement of animals in the field can provide the basis for pronounced differences in results. In the experiments already described, we routinely took particular care in positioning the eye with respect to the dipole antenna and in checking this position during irradiation. It was our experience that any movement of the head during this time could affect the field to such an extent that the power meter reading changed sufficiently to render that particular experiment invalid.

It would be distincly advantageous if we could deliver microwave power to the eye in a manner permitting us to measure the amount of power delivered and the amount reflected, the difference comprising the power absorbed by the eye. Equally effective would be a method which permitted tuning out all reflections, leaving only the delivered power to be measured. The method finally chosen was one which allowed us to transmit microwave power directly into an eye which, together with a thin brass iris, formed a wave-guide termination. Variability introduced by transmission through air was thereby eliminated. The system, which was capable of transmitting frequencies in the range of 8.2 to 12 GHz, is shown in Fig. 6. It was designed and constructed by Professor George Hammond of the Tufts Department of Electrical Engineering. Its chief innovation lay in making the eye itself part of a continuous closed system beginning with the power tube and terminating in the eye. The animal's head was placed so that the right eye was firmly applied to and closed a 1 cm. circular aperture in a thin brass iris stretched across the terminal section of wave-guide. The eye and the iris together comprised the final unit of the system. The impedance mismatch introduced by the iris-eye unit was assessed by means of a directional coupler and crystal detector sampling the reflected power and reading out on a microammeter. With suitable tuning provided by an E-H tuner, reflection could be reduced routinely to 0.1% of forward-going power, equivalent to a residual VSWR of 1.01. It could therefore be assumed for all practical purposes that the power directed eyeward in the system was the same as the power ab-

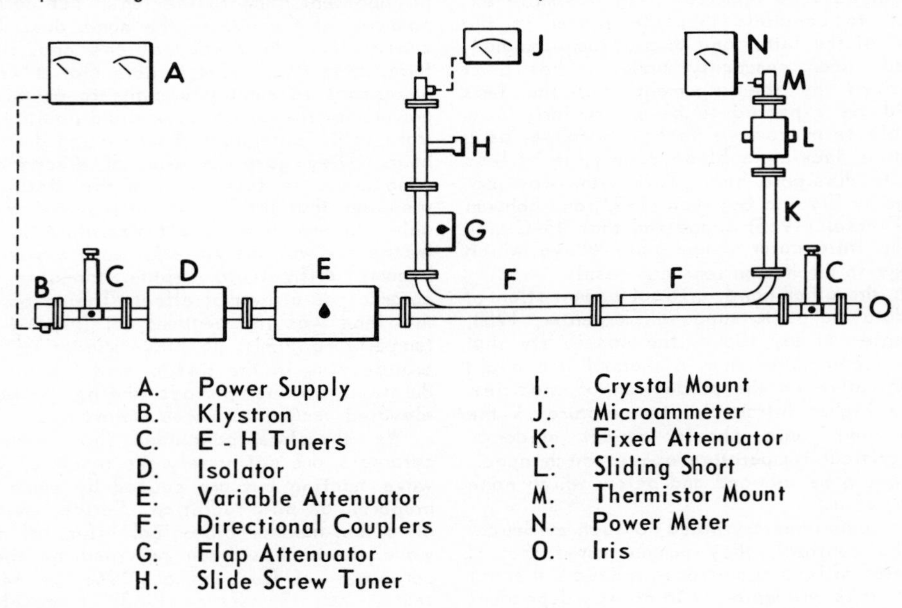

Fig. 6. 'Closed' waveguide system for irradiating eyes at X-band frequencies.

sorbed by the iris and eye together. Since the amount of power absorbed by the iris was probably negligible, the forward-going power could therefore be measured as a close approximation of the power absorbed by the eye. This was done by a second directional coupler, thermistor and power meter.

It should be noted that this was a measure to the total microwave power delivered to the eye through the iris aperture. It was not a measure of power density nor could it be converted into one.

Employing a frequency of 10.05 GHz, this 'closed' system was used in irradiation of eyes at five levels of delivered power, from 150 mw. to 1090 mw. At each power, exposure times were varied in order to determine thresholds for the induction of lens opacities. The results are summarized in Figure 7. Although this threshold curve is similar in shape to that previously shown for 2.45 GHz radiation, the power values cannot be compared, because of their entirely different derivations.

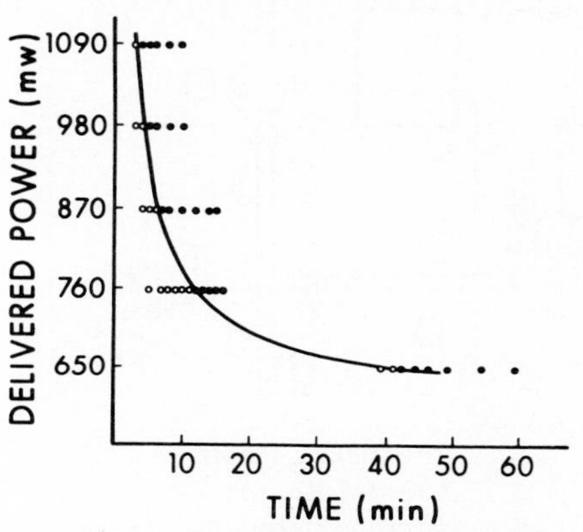

Fig. 7. Time and power thresholds for induction of opacities at 10.05 GHz. Solid circles represent irradiations which induced opacities; open circles represent irradiation without effect.

The cataracts which developed were in one respect strikingly different from those caused by the lower frequency (2.45 GHz). It will be recalled that in the latter case, the opacities were located typically in the posterior cortex of the lens. At the higher frequency of 10 GHz, they developed always in the anterior cortex region as localized granular opacities just under the epithelium. This was the case in 107 experiments in which cataracts were formed. In this respect, they were similar to cataracts caused by infrared radiation, which likewise appear typically in the anterior cortex.

Although damage to the lens was not apparent until the second or third day following irradiation, a granular opacity was commonly seen in the cornea within 24 hours but it usually disappeared within four days.

IRRADIATION AT FREQUENCY OF 8.236 GHz

To compare the effect of another frequency, a second group of animals was exposed to 8.236 GHz radiation. The same 'closed' system was used but with appropriate corrections for the altered frequency-dependent response of the various components. Eyes were irradiated at delivered power levels of 650 mw, 760 mw and 870 mw. Exposure periods corresponded to those employed at the higher frequency. As Figure 8 shows, a curve describing the time-power threshold for induction of opacities at 8.236 GHz does not differ from that for 10.05 GHz. The response of the rabbit lens to each of the two frequencies was also identical with respect to latent period, degree of reaction, form of opacity, and its location in the anterior subcapsular cortex.

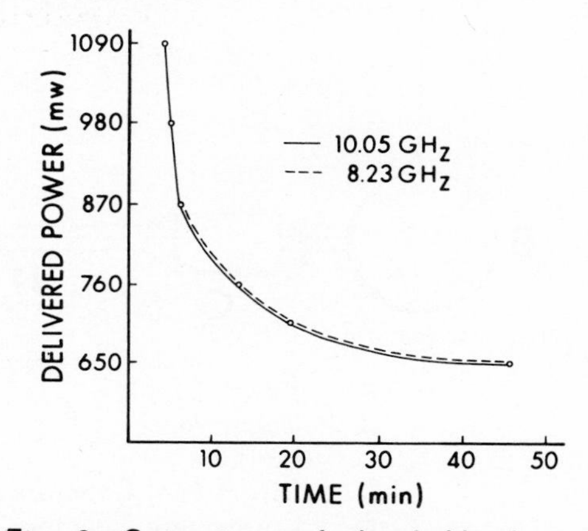

Fig. 8. Comparison of thresholds for induction of opacities by single exposures at 8.236 GHz and 10.050 GHz.

These experiments, employing high frequencies and the 'closed' system of delivering power to the eye, yielded distinctive and consistent results. The change of location of the cataract, as compared to those caused by lower frequencies, could be attributed either to the frequency difference or to the different system of transmitting power to the eye. It was possible that the anterior site of the cataract might be related to the limited tissue penetration which characterizes the higher microwave frequencies. On the other hand, there was also the possibility that with the eye pressed against the aperture in the terminal iris of the wave-guide, the interface created a condition whereby the radiation was concentrated on the anterior part of the lens.

The question could not be answered until we learned what would be the results of irradiating the eye: (1) in a free field at 10 GHz and (2) as part of a 'closed' wave-guide system transmitting 2.45 GHz radiation. We therefore undertook to perform both experiments.

X-BAND IRRADIATION IN A FREE FIELD

To irradiate the eye in a free field, experiments were carried out in an anechoic chamber 14 feet long and 6 feet in width and height. It was completely lined with microwave absorbent material*. Our power source was a magnetron tube generating up to 200 watts at 10.165 GHz. The system is shown schematically in Figure 9. The terminal wave-guide section passed through an opening in the front wall of the anechoic room to a transmitting horn having a gain of 20 db.

The rabbit, supported on a small styrofoam couch, was placed on a styrofoam stand in the room and oriented so that its right eye was centered in the microwave beam. This was accomplished by lining up the eye with the image of cross hairs projected by an optical pointer centered within the transmitting horn so that its optical axis coincided with the axis of the horn. When the eye had been correctly positioned, the pointer was removed from the horn and the eye was then irradiated. The amount of power supplied to the horn in all experiments was 120 watts. With the eye placed 50 cm. away from the horn, lens opacities developed from exposure periods of 10 to 20 minutes. At a distance of 62.5 cm., exposures of 18 to 40 minutes caused opacities. In all cases, they developed in the posterior subcapsular cortex of the lens and were in every way similar to those which formed when the eye was subjected to 2.45 GHz radiation from an antenna in an anechoic chamber. It seemed quite clear, therefore, that in the experiments done at X-band frequencies with the 'closed' wave-guide system, the anterior location of the cataracts was due not to the particular frequency but to the method by which the power was transmitted to the eye.

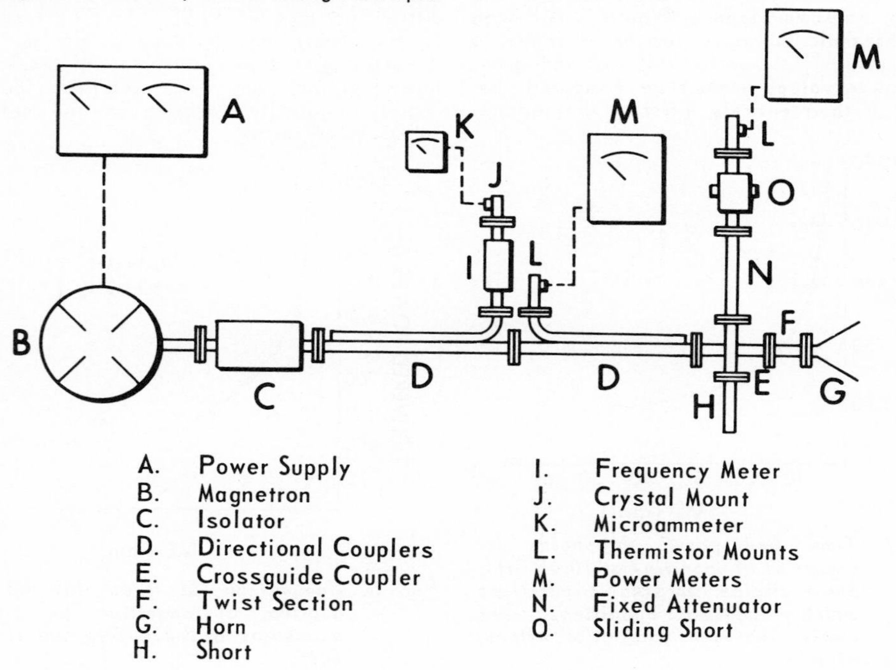

A. Power Supply
B. Magnetron
C. Isolator
D. Directional Couplers
E. Crossguide Coupler
F. Twist Section
G. Horn
H. Short
I. Frequency Meter
J. Crystal Mount
K. Microammeter
L. Thermistor Mounts
M. Power Meters
N. Fixed Attenuator
O. Sliding Short

Fig. 9. System for irradiating eyes at 10.165 GHz in an anechoic room.

It should be mentioned that although the results were consistent with regard to the posterior location of the opacities, there was considerable variability in the response of the eye and of the animal itself to the radiation. Unlike irradiation at 2.45 GHz in the near field, those at 10 GHz in the far field were unpredictable as to result. At cataractogenic field intensities, the animal's temperature increased enough to be lethal, until we provided protection for its body. This was done by shielding all but the head behind a barrier of 2-inch thick microwave absorbent through which a strong current of air was continually

*Eccosorb FR330, FR340 and CV6. From Emerson & Cumming, Inc., Canton, Massachusetts.

passed to carry away the heat. A stream of air was also directed on the animal's head to facilitate heat loss from the ears, the rabbit's principal heat dissipators.

Under these conditions, irradiations were successfully carried out but the results varied, even under conditions which seemed to be identical. A given exposure would cause an opacity in one case and not in another. Some animals experienced burns of the cornea or eyelids; others had hemorrhage into the anterior chamber or detachments of the retina. In many cases, the radiation did not harm the eye but did cause burns on the face. Sometimes the burns were evident as swollen areas the following day but quite often they gave no sign until several days later, when an area of dead skin was sloughed from the face, revealing a sterile subcutaneous burn which had already begun to heal. This occurred most frequently in the region of the face just beneath and behind the eye (Plate 2). The burned area seldom became infected and eventually healed and was covered by new skin. This phenomenon had been reported by Howland and Michaelson (1959b), who described and pictured microwave burns on the chest and flank regions of dogs subjected to 2.88 GHz radiation from a military radar. They termed them 'reverse third degree burns' (occurring from the inside out).

The variability of these reactions could not be due to age differences, for Van Ummersen and Cogan (1965) have shown that age of the animal is not a factor in susceptibility of the lens to damage by microwave power. We were led to speculate either that rabbits varied widely in their individual responses to radiation - a supposition unsupported by our previous experiments - or that power distribution in the microwave field varied from one experiment to another. The latter explanation proved to be correct. Using a scatter technique, Fisher (1966) plotted equal power contours in the free field of our anechoic chamber. The highest intensity was in the center of the field and it diminished uniformly as the scattering probe was moved in any radius away from the center. He then plotted the same field with a rabbit placed immediately behind the plane of the measured field and with its eye centered on the beam axis. Under these circumstances, equal power contours were extremely distorted and the region of highest intensity was displaced from the center of the field to a position on the rabbit's face just beneath and behind the eye.

It became obvious that the presence of an animal in the microwave field perturbed the field to such an extent that conditions could not have been uniform in the 102 experiments we performed. This helped to explain the facial burns and the inconsistency of lens responses, although it had no bearing on the particular question being studied; namely, the location of the cataract in the lens under the conditions of irradiation.

'CLOSED' WAVEGUIDE IRRADIATION AT 2.45 GHz.

Although the question had been answered clearly, we nevertheless thought it worthwhile to approach the problem from the other direction by employing the closed' waveguide system to irradiate eyes at 2.45 GHz (Fig. 10). We encountered several technical problems related chiefly to the larger size of the waveguide. Although the ridged terminal section of the waveguide was tapered as much as was feasible, the brass iris closing it was still slightly over two inches square. The large

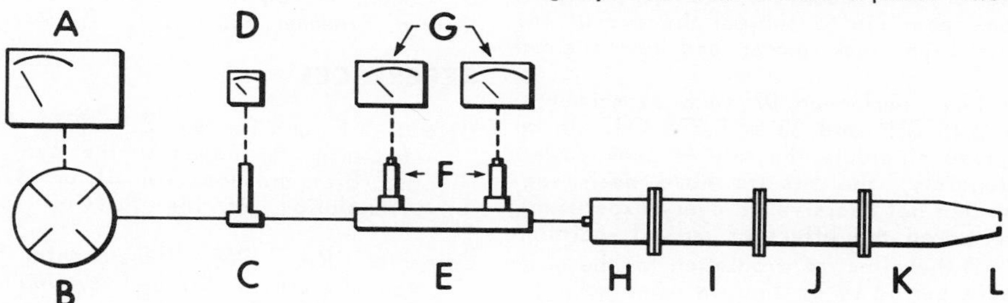

A. Power Supply
B. Magnetron
C. Adjustable Attenuator
D. Wave Meter
E. Dual Directional Coupler
F. Thermistor Mounts
G. Power Meters
H. Coax-Waveguide Transition
I. Slotted Line
J. Slug Tuner
K. Ridged & Tapered Termination
L. Iris

Fig. 10. 'Closed' waveguide system for irradiating eyes at 2.45 GHz.

iris and the shape of the rabbit skull made it difficult to place the head close enough to the flat brass surface so that the eye could be applied snugly to the 1 cm. central aperture. If fitted loosely, the cornea became burned and the experiment had to be discarded. Fortunately, the eye of the anesthetized rabbit can be made to protrude from its bony recess and in this proptosed position, we found that we were able to fit it firmly against the aperture while still leaving clearance between the cheek and the iris.

No attempt was made to establish time and power thresholds. Lens opacities developed following five minute irradiation at 2.67 watts of power delivered to the eye and in all cases the opacities were situated in the anterior cortex of the lens, exactly as they were after X-band irradiation of the eye in the same manner. We could therefore conclude that, in the range of frequencies studied, microwave power transmitted directly to the eye by waveguide causes cataracts in the anterior cortex of the lens which are similar to those caused by infrared radiation. When microwave power is transmitted to the eye through the air, the cataracts which develop are situated in the posterior subcapsular cortex and are similar to those caused by ionizing radiations.

THE QUESTION OF PULSED WAVE AND CONTINUOUS WAVE RADIATION

Microwave power, whether pulsed or continuous, can cause cataracts in the eye. We cannot say, in the case of pulsed waves, whether the induction of cataracts is related to average power or to peak power. With pulsed radiation, the eye can be subjected to rapidly repeated peaks of high energy while the average power during the exposure remains low. Inasmuch as the thermal response of tissues depends upon the average power, it becomes possible to subject the eye at one time to high peak power and low thermal flux.

We have performed 97 such experiments, 48 at 2.45 GHz and 33 at 9.375 GHz, in an attempt to elucidate the role of peak power. Unfortunately, the results have been suggestive but not decisive. In every experiment, we compared the effect of pulsed radiation with: (1) the effect of irradiation for the same exposure period by continuous wave having a power equal to the average power of the pulsed wave; and (2) the effect of continuous wave irradiation equal in intensity to the peak power and for an exposure period equalling the total 'on' time of the pulsed wave radiation.

The results were indecisive. At 2.45 GHz, the pulsed waves caused a cataract (Plate 1, (k)) in almost half of the experiments, with exposure periods significantly shorter than the minimal period required to induce cataracts by continuous wave radiation equal to the average power of the pulsed waves. At 9.375 GHz, we employed the 'closed' waveguide system, with 0.05 duty cycle, pulsed repetition rate of 1280 pps, pulse width of 0.4 microseconds, and peak power of 2600 mw. Average power was 760 mw. The comparable continuous wave radiation was likewise 760 mw. and the frequency was 10.05 GHz. Lens opacities developed in half of the animals exposed to continuous waves; pulsed radiation of equal average power caused opacities in two-thirds of the irradiated eyes. (Plate 1 (l)).

Our experiments did not permit definite conclusions to be drawn as to the relative potencies of continuous wave and pulsed wave radiation in the induction of cataracts. Our present view is that it would be unwise to ignore the factor of peak power in assessing pulsed radiation as an ocular hazard.

ACKNOWLEDGEMENTS

The researches reported and reviewed in this paper were supported from 1956 through 1961 by the U.S. Air Force under Contract AF41(657) - 86, administered by the Rome Air Development Center, Griffiss Air Force Base, N.Y., and since May 1, 1962, by U.S. Public Health Service Grant No. GM 09495 from the National Institute of General Medical Sciences. We gratefully acknowledge this support.

During the period covered by this work, valuable contributions to it were made by several investigators who were for a time our colleagues in the Microwave Radiobiology Laboratory at Tufts University. We wish to acknowledge in particular the contributions of Professor George H. Hammond, Dr. Frances C. Cogan, Mr. David K. Biddle, Dr. H. MacKenzie Freeman and Dr. C. P. Mangahas.

REFERENCES

Belova, S.F. and Gordon, Z.V. 1956; 'Action of Centimeter Waves on the Eye', Bull. Exp. Biol. and Med., Vol. 41. pp. 327-330. (Translation of the Russian journal)

Carpenter, R.L. 1958; 'Experimental Radiation Cataracts Induced by Microwave Radiation', Proc. Second Annual Tri-Service Conf. on Biol. Effects of Microwave Energy. Rome Air Dev. Ctr., Air Res. and Dev. Command, Rome, N.Y. ARDC-TR-58-54, ASTIA Doc. No. AD-131-477, pp. 146-166.

Carpenter, R.L. 1959; 'Studies on the Effects of 2450 Mc. Radiation on the Eye of the Rabbit', Proc. Third Annual Tri-Service

Conf. on Biol. Effects of Microwave Radiating Equipments, Rome Air Dev. Ctr., Air Res. and Dev. Command. at Berkeley, Calif. RADC-TR-59-140, pp. 279-290.

Carpenter, R.L. 1965; 'Suppression of Differentiation in Living Tissues Exposed to Microwave Radiation', Dig. of Sixth. Int. Conf. on Med. Elect. and Bio. Eng., Tokyo, pp. 573-574.

Carpenter, R.L., Biddle, D.K., Van Ummersen, C.A., Mangahas, C.P. and Freeman, H.M. 1959; 'Experimental Radiation Cataracts Induced by Microwave Radiation', Amer. J. Ophth. Vol. 47, p. 94.

Carpenter, R.L., Biddle, D.K. and Van Ummersen, C.A. 1960a; 'Opacities in the Lens of the Eye Experimentally Induced by Exposure to Microwave Radiation', IRE Trans. on Medical Electronics, Vol. ME7, pp. 152-157.

Carpenter, R.L., Biddle, D.K. and Van Ummersen, C.A. 1960b; 'Biological Effects of Microwave Radiation with Particular Reference to the Eye', Proc. Third Int. Conf. on Med. Elect., London, pp. 401-408.

Carpenter, R.L. and Clark, V.A. 1966; 'Responses to Radio-frequency Radiation', Table 31 in 'Environmental Biology' edited by Altman, P.L. and Dittmer, D.S., Bethesda, Md., pp. 131-138.

Cogan, D.G. and Donaldson, D.D. 1951; 'Experimental Radiation Cataracts: Cataracts in the Rabbit Following Single X-ray Exposure', A.M.A. Arch. Ophth., Vol. 45, pp. 1-15.

Cogan, D.G., Fricker, S.J., Lubin, M., Donaldson, D.D. and Hardy, H. 1958; 'Cataracts and Ultra-high Frequency Radiation', A.M.A. Arch. Indust. Health, Vol. 18, pp. 299-302.

Daily, L., Wakim, K.G., Herrick, J.F. and Parkhill, E.M. 1948; 'The Effect of Microwave Diathermy on the Eye', Amer. J. Physiol., Vol. 155, p. 432.

Daily, L. Wakim, K.G., Herrick, J.F., Parkhill, E.M. and Benedict, W.L. 1950; 'The Effects of Microwave Diathermy on the Eye', Amer. J. Ophth., Vol. 33, pp. 1241-1245.

Daily, L., Wakim, K.G., Herrick, J.F., Parkhill, E.M. and Benedict, W.L. 1952; 'The Effects of Microwave Diathermy on the Eye of the Rabbit', Amer. J. Ophth., Vol. 35, pp. 1001-1017.

Ely, T.S. and Goldman, D.E. 1957; 'Heating Characteristics of Laboratory Animals exposed to 10 cm. Microwaves', Research Report, Project NM 001 056.13.02, Naval Medical Research Institute.

Fisher, L.J. 1966; 'Microwave Field Measurements in the Vicinity of a Biological Specimen', M.S. Thesis, Tufts University.

Gunn, S.A., Gould, T.C. and Anderson, W.A.D. 1960; 'The Effect of Microwave Radiation (24,000 Mc) on the Male Endocrine System of the Rat', Proc. Fourth Annual Tri-Service Conference on Biol. Effects of Microwave Radiations, Plenum Press, N.Y., Vol. 1, pp. 99 - 115.

Hirsch, F.G. and Parker, J.T. 1952; 'Bilateral Lenticular Opacities Occurring in a Technician Operating a Microwave Generator', A.M.M. Arch. Indust. Health, Vol. 18, p. 299.

Howland J.W. and Michaelson, S. 1959a; 'Studies on the Biological Effects of Microwave Radiation on the Dog and Rabbit', Rome Air Dev. Ctr., Air Res. and Dev. Command, Rome, N.Y. ASTIA Doc. No. AD-212110.

Howland, J.W. and Michaelson, S. 1959b; 'Studies on the Biological Effects of Microwave Irradiation of the Dog and Rabbit', Proc. Third. Annual Tri-Service Conf. on Biol. Effects of Microwave Radiating Equipments, Rome Air Dev. Ctr., Air Res. and Dev. Command, at Berkeley, Calif. RADC-TR-59-140, pp. 191-238.

Jolly, J.A. 1967; Guest Editorial, Jour. Mic. Power, Vol. 2, p. 65.

Livstone, E.M. 1965; Unpublished undergraduate research report, Dept. of Biology, Tufts University.

Osborne, C.M. and Addington, C.H. 1959; 'Studies on the Biological Effects of 200 Megacycles: Part III - Ophthalmological Studies', Investigators' Conf. on Biol. Effects of Electronic Radiating Equipments, Rome Air Dev. Ctr., Air Res. and Dev. Command, Rome, N.Y. ASTIA Doc. No. AD-214693, pp. 24-25.

Osborne, S.L. and Frederick, J.N. 1948; 'Microwave Radiations; Heating of Human and Animal Tissues by Means of High Frequency Current with Wave Length of

12 cm. (microtherm),' J. Amer. Med. Assn., Vol. 137, p. 1036.

Richardson, A.W., Duane, T.D. and Hines, H.M. 1948; 'Experimental Lenticular Opacities Produced by Microwave Irradiations', Arch. Phys. Med., Vol. 29, pp. 765-769.

Richardson, A.W., Duane, T.D. and Hines, H.M. 1951; 'Experimental Cataract Produced by Three Centimeter Pulsed Microwave Irradiation', A.M.A. Arch. Ophth. Vol. 45, pp. 382-386.

Schwan, H.P. and Piersol, G.M. 1955; 'The Absorption of Electromagnetic Energy in Body Tissues. II. Physiological and Clinical Aspects', Amer. J. Phys. Med., Vol. 34, pp. 424-448.

Van Ummersen, C.A. 1963; 'An Experimental Study of Developmental Abnormalities Induced in the Chick Embryo by Exposure to Radio-frequency Waves', Ph.D. Thesis, Dept. of Biology, Tufts University.

Van Ummersen, C.A. and Cogan, F.C. 1965; 'Experimental Microwave Cataracts. Age as Factor in Induction of Cataracts in the Rabbit', A.M.A. Arch. Ind. Health, Vol. 11, pp. 177-178.

Williams, D.B., Monahan, J.P., Nicholson, W.J. and Aldrich, J.J. 1955; 'Biologic Effects of Microwave Radiation: Time and Power Thresholds for the Production of Lens Opacities by 12.3 cm. Microwaves', U.S.A.F. School of Aviation Medicine Report No. 55-94

Long-Term 2450-MHz CW Microwave Irradiation of Rabbits: Methodology and Evaluation of Ocular and Physiologic Effects*

A.W. Guy, P.O. Kramar, C.A. Harris, and C.K. Chou[1]

ABSTRACT

In order to assess the biological effects of long-term microwave radiation, special exposure systems were developed and used to expose four rabbits to 10-mW/cm² microwave radiation (maximum 17 W/kg SAR) for 23 h per day for 180 days. Comparisons with four sham-exposed rabbits revealed no significant effects in terms of eyes, body mass, urinary output, rectal temperature, hematocrit, hemoglobin, white cell count, and basic blood-coagulation studies. After the experiment, the animals were sent to the National Institute of Environmental Health Sciences for additional analyses which revealed biochemical effects as reported in a companion paper in this issue.

INTRODUCTION

One of the more striking differences in methodologies of assessing the biological response to electromagnetic fields between Western and Eastern bloc countries lies in the relative preponderance of work in the Eastern bloc devoted to long-term exposure of animals at relatively low intensities of radiation; in contrast, most study in the Western bloc has been devoted to acute, short-term exposures of animals at much higher intensities [1]. A few Western authors, however, have reported that cataractogenic or other changes indicative of ocular pathology occur in eyes of workers exposed chronically to microwave radiation at low intensities [2,3].

In order to answer some of the questions raised by the differences in methodology between the various countries and to determine whether low-level microwave radiation can produce cataractogenic effects, a special exposure system was developed and used chronically to irradiate a small sample of rabbits at a power density of 10 mW/cm² for a period of six months.

New Zealand white rabbits were chosen as subjects for the chronic exposures based on several considerations. A large amount of data exists for 2450 MHz acute exposures of this species. Most of the data relate to the effect of the radiation on the eyes of the animal. The first work by Richardson [4] and Daily [5,6] demonstrated that cataracts would result from acute exposures localized to the region of the eye by the application of the energy from a standard clinical diathermy applicator. Since the diathermy devices were operated at or near full power with the applicator within a few inches of the eye, it is estimated that the power density for the exposure fields was on the order of several hundred mW/cm².

Williams [7] was the first to establish a power density vs. time dependency for the threshold of cataractogenesis in the eyes of rabbits under such exposure conditions. He found that cataracts could result from exposures as short as 5 minutes at exposure power densities of 590 mW/cm², while, on the other hand, power densities of 220 mW/cm² required 4.5 h of exposure to produce the cataracts. No cataracts could be produced at exposure level of 120 mW/cm².

*Manuscript received September 20, 1979; in revised form January 24, 1980.

[1]Bioelectromagnetics Research Laboratory, Department of Rehabilitation Medicine, University of Washington School of Medicine, Seattle, Washington 98195, USA.

Reprinted with permission from *J. of Microwave Power,* vol. 15, no. 1, pp. 37-44, Mar. 1980

Carpenter* (8-10) established a very precise time-power density dependency for the same exposure conditions based on a large number of acute experiments showing cataract formation for exposures varying from 600 mW/cm² for 3 min to 180 mW/cm² for 60 min. One of the important problems that needed to be considered was the question of cumulative effects from longer term or repeated exposures at lower levels. Carpenter* [8] found that exposures repeated from 4-15 times over intervals from 1-14 days resulted in a lower cataractogenetic threshold of only 120 mW/cm².

Kramar et al. [12] and Guy et al. [13] replicated Carpenter's work for single acute exposures with essentially the same results. They were also able to quantify the threshold of cataractogenesis in terms of the specific absorption rate (SAR) and temperature distribution in the eye [14]. Guy et al. [14], however, were not able to demonstrate the formation of cataracts with repeated subthreshold power density exposures but were able to produce vacuoles in the lenses of the rabbits with repeated exposures at threshold power density levels applied over subthreshold time periods.

Another important problem requiring attention was the question of possible cataractogenesis resulting from whole body exposures. Appleton [15,16] demonstrated that acute whole body exposures at 3000 MHz would be lethal to the rabbit at exposure levels insufficient to produce cataracts or other ocular effects. Though considerable data now exist on cataractogenesis due to short-term exposures of the rabbits, the possibility of cumulative effects for long-term subthreshold power densities for cataractogenesis also remained an important question. Therefore, within the constraints of our research resources, we chose to direct attention to this problem, with the rabbit as a subject. These resources allowed a maximum exposure duration of 6 months. Thus, to take full advantage of the time, we chose a maximum daily exposure period of 23 h (a maximum of 1 h allowed for cleaning cages and attending animals) applied 7 days a week for the full 6-month period, with food and water available ad lib. Whole-body exposure conditions were selected to allow maximum freedom of movement of the animal. Since Appleton (15) had demonstrated that the maximum lethal whole-body exposure level for a rabbit was between 25-50 mW/cm², we chose an exposure level of 10 mW/cm² to reduce the possibility of any type of acute effects in the animal.

Most chronic studies performed in Eastern Europe have involved exposure of animals in a compact group to a plane-wave field in an anechoic chamber or to standing-wave fields in a metallic cavity. Both of these methods of exposure present serious dosimetric problems. Coupling of fields to each animal is a function of the number of animals in the group, the spacing between animals, the orientation of each animal, and the presence, composition, and orientation of water and food dispensers. It has been shown under a fixed set of exposure conditions that the peak densities of absorbed energy in a particular animal of a group exposed in a metal cavity can vary over a range of three orders of magnitude with normal changes of posture and position [17]. The dosimetric problem could be eliminated in part by restricting exposures to a single restrained animal, but the cost of resources and the time required to perform even simple experiments involving chronic exposures with conventional systems now in use would be prohibitive. An economical exposure system has been designed that eliminates some of these problems.

We developed an inexpensive set of exposure systems by which to irradiate individually a number of animals. The system was developed to irradiate four experimental animals under simulated free-field conditions; four control animals can be treated under near identical conditions, but without irradiation. The entire assembly of systems was designed to fit within a small vivarium room (248 x 419 cm²).

In order to eliminate effects of mutual coupling, it was necessary to design a radiation chamber for each animal. Experimental animals are typically exposed with the k vector directed laterally, anteriorly, or posteriorly to the animal. A more compact arrangement of multiple exposure chambers can be implemented, however, if the animal is irradiated from above — with

* Cited power density values have been corrected based on more recent data reported by Carpenter [11].

the k vector directed toward the animal's dorsum. In order to test the equivalence of dorsal and lateral irradiation in transmitting microwave energy to the head and eyes of the animal, the SAR was measured in the eye under both conditions of exposure, through deployment of thermocouple methodology [12,13,18]. The exposure set-up is shown in Fig. 1. Standard gain horns, in which the aperture was placed one meter from the primary axis of an animal's body, were used both for horizontal and for vertical irradiation. The polarization of the electric field was parallel to the long axis of the body to facilitate thermographic measurements of SARs for vertical irradiation and perpendicular to the axis for horizontal irradiation. Incident radiation at a power density of 525 mW/cm^2 was applied for 15 seconds at the position of the eye. Figure 2 reveals that higher peaks of absorption were found with the vertically oriented horn, but the averaged SARs along the axis in the eyes under the two conditions were similar (0.679 W/kg per mW/cm^2 for the vertical and 0.481 W/kg per mW/cm^2 for the horizontal exposure condition).

Figure 1 A rabbit is shown in relation to 2450-MHz horn applicators located both horizontally and vertically. Simultaneous irradiation did not take place; different groups of rabbits were assessed dosimetrically for the difference in modes of exposure.

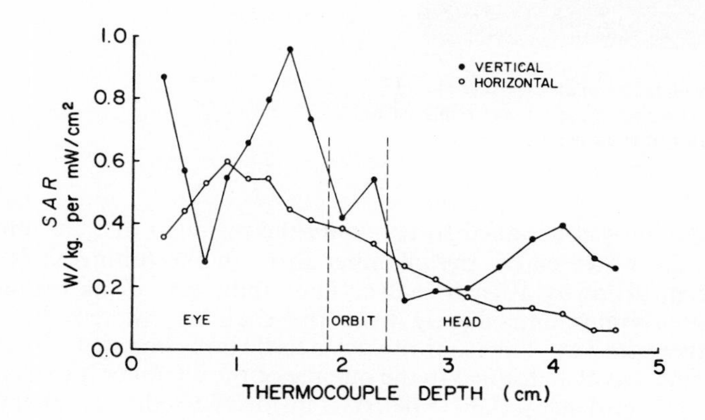

Figure 2 Averaged SAR patterns based on two rabbits exposed to 2450-MHz microwaves with respect to horizontal vs. vertical conditions in which the aperture of a horn was one meter from the mid line of the rabbit.

Exposure Conditions

A total of eight rabbits was studied simultaneously; four received 2450-MHz CW radiation, and four served as controls. Each rabbit was kept in an acrylic cage (Fig. 3) surrounded by microwave-absorber walls and a floor that formed a chamber with a 61 x 61 cm cross section and 122 cm in height. A standard microwave horn was placed one meter from the center line of the body of the rabbit, which is 16 cm from the floor of the chamber. The arrangement of the eight chambers is shown in Fig. 4. Square panels (61 x 61 x 20.3 cm) of Emerson Cuming Ch 490 microwave absorbing hair were used to fabricate the chambers of the four experimental rabbits. The chambers for control rabbits were constructed with thinner panels of packing material of the same size and similar composition as the microwave absorber. In all cases, the absorber and packing material panels were covered with a thin plastic film to block outgassing that might arise from the material. With the exception of the absence of the microwave horns, the topography of the control chambers was similar to that of exposure chambers. The ambient temperature within each chamber was maintained at 24 ± 2°C by a thermostatically-controlled air conditioner. The air velocity and humidity were not measured during the course of the experiment.

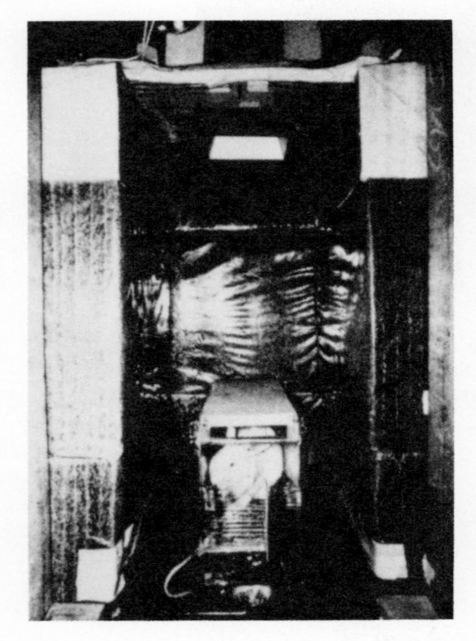

Figure 3 Photograph of rabbit in an exposure chamber. (The front section of absorber material was removed to permit viewing of interior of chamber.)

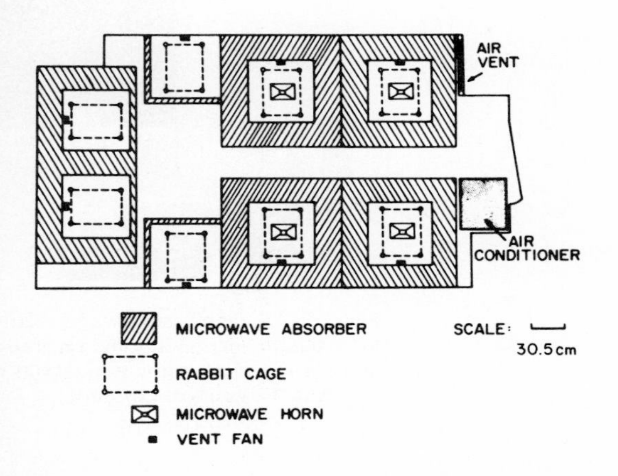

Figure 4 View from the overhead of chronic-exposure laboratory showing four experimental chambers and four control chambers.

A watering system was designed to eliminate the problem of field intensification during drinking (Fig. 5). As water enters the chamber through the tubing, it is separated into individual, radiolucent drops by a needle valve. The rabbit may drink by catching drops in its mouth; excess water drains immediately from the chamber, thereby breaking up any continuously conductive pathway. The floor of each acrylic box consisted of 1 cm diam. glass rods spaced at 2 cm. Urine was also drained to the outside of the chamber into a collection beaker.

The effective-power-density (EPD) patterns inside the radiation chambers were mapped with an NBS EDM-IC probe. The electric-field vector of the radiation field was polarized parallel to the long axis of the cage in each chamber. The animal remained in a position with his body extended along the long dimension of the cage during most of the radiation. The fields along the axis of the horn in both the E-plane and H-plane fields inside the cages were measured

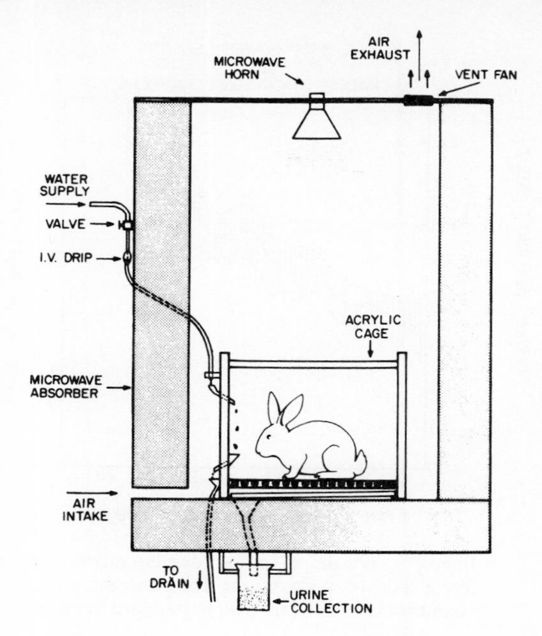

Figure 5 Schematic of chronic-exposure chamber showing water-supply system.

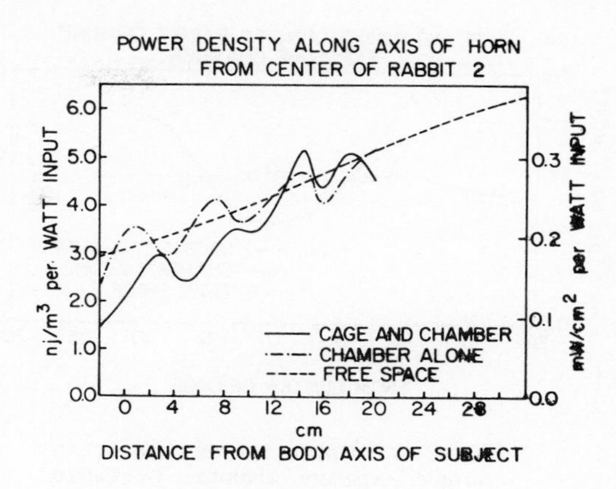

Figure 6 Power-density values along axis of horn in exposure chamber (subject not present).

as shown in Figs. 6-8. Means and standard deviations of power density along the axis of the horn are shown in Fig. 9. The field, expressed in terms of both nanojoules/m³ as measured by the NBS probe and the equivalent power density in mW/cm² per watt input to the horn, are plotted as a function of distance from the body axis of the subject in a resting position.

Fields inside the radiation chambers were mapped with the NBS EDM-IC probe. Due to standing waves and divergence of the fields from the feed horn, the maximal power densities of incident radiation were found to be 29.1, 23.4, 30.4, and 25.3 mW/cm² for the four exposure chambers (Fig. 4) at the dorsal aspect of the rabbit's head in contrast to the value of 10 mW/cm² predicted at the body axis of the animal from free-space standard-gain-horn characteristics. This prompted a series of field measurements to be made in all of the chambers; the results are given in Fig. 9.

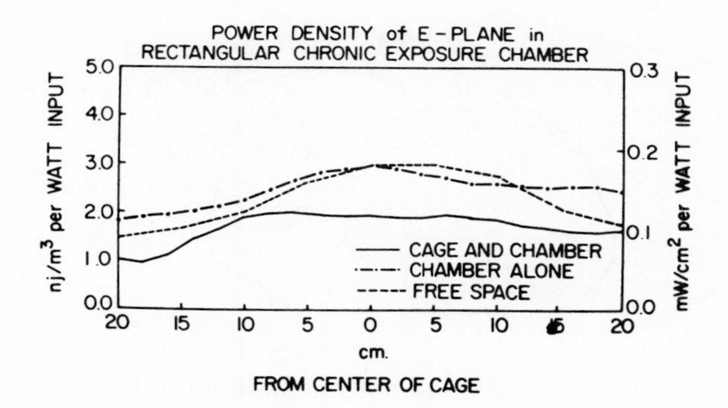

Figure 7 E-plane power density in chronic exposure chamber (measured along position of body axis of animal with animal not present).

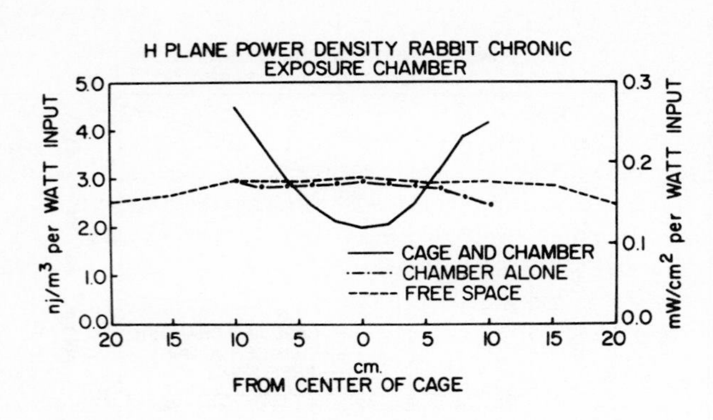

Figure 8 H-plane power density in chronic exposure chamber (measured along position of body axis of animal with animal not present).

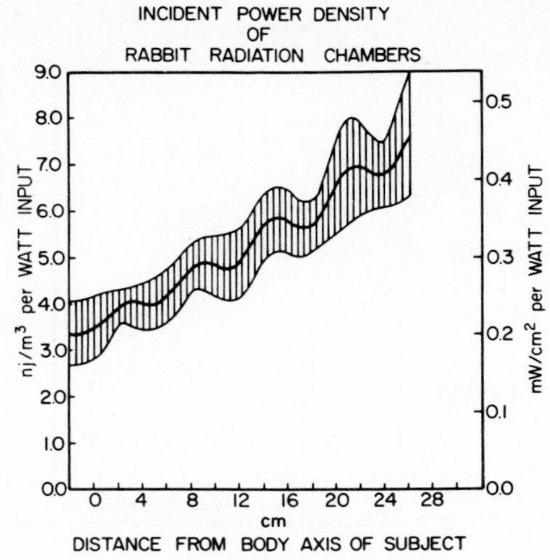

Figure 9 Means of power density measurements along axis of horn in the four chronic exposure chambers (vertical lines denote standard deviation).

During the long-term exposure of the final group of test animals, the input power for each standard-gain horn was set at 32 W, thereby providing a mean value of EPD of 7 mW/cm² at the position of the body axis of the animal (measured with the animal absent). According to Fig. 9, this corresponds to a mean EPD of 10 mW/cm² at the usual position of the animal's head, i.e., 16 cm above the body's long axis, and 14.5 mW/cm² at the top of the cage. Lines of constant specific absorption rates (SARs) were measured and plotted for rabbits exposed in the chamber. The results, shown in Fig. 10 and normalized to 1 mW/cm² of EPD at the body axis of the animal, were obtained by observation of temperature increases in the exposed rabbit carcasses by thermography (19). The images recorded by the thermograph were transmitted through an A-D converter to a digital computer for subsequent computation and plotting of the constant SAR lines. The results indicate that one could expect a peak SAR of 17 W/kg in the head of the rabbit when an animal was in the normal resting position in the cage with the head under the 10 mW/cm² exposure, in contrast to an estimated maximum average SAR of 1.5 W/kg based on theoretical data for a prolate spheroid model of the rabbit exposed to a plane wave [20].

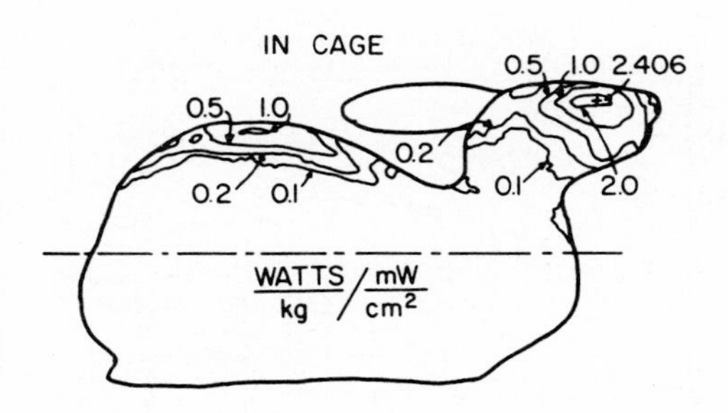

Figure 10 Pattern of SARs in rabbit exposed to 2450-MHz fields as normalized to measured power density at the body axis of the rabbit.

RESULTS

Four male New Zealand albino rabbits weighing approximately four kilograms were exposed in the system seven days a week, 23 h/day for a period of six months. A separate group of four rabbits was maintained under similar conditions except that radiation was not applied. Periodic examinations of the eyes with a Nikon slip lamp microscope were made and the following parameters were monitored throughout the exposure period: body mass, urinary output, rectal temperature, hematocrit, hemoglobin, white cell count with differential, platelet count, and basic blood-coagulation studies. No significant differences other than a decrease in percentage of eosinophils were found between experimental and control animals, as shown in the tabulation of results in Table 1. It should be noted, however, that eosinophil percentage varies widely among animals.

Table 1 Summary of Hematological Evaluations

Test		Pre-Irradiation	1 1/2 Months	3 Months	4 1/2 Months	6 Months
Lymphocyte (%)	R*	45.5± 15.4	42.8± 14.4	38.3± 17.2	41.0± 12.6	35.8± 22.2
	C*	60.3± 1.5	57.0± 11.4	48.8± 22.7	44.7± 15.9	50.0± 14.4
Neutrophil (%)	R	45.5± 12.3	53.0± 12.3	55.0± 13.1	51.8± 10.6	58.8± 22.5
	C	31.3± 6.4	36.3± 8.6	45.3± 21.1	47.0± 13.1	42.0± 13.6
Monocyte (%)	R	2.8± 1.7	2.0± 1.4	3.5± 3.1	2.8± 1.3	3.0± 1.8
	C	3.8± 2.1	2.5± 1.7	4.3± 1.7	5.0± 1.0	3.8± 3.0
Eosinophil (%)	R	2.5± 2.4	0.8± 1.0	1.0± 0.8	2.0± 0.8	0.3± 0.5
	C	1.0± 1.4	0.8± 1.0	1.0± 0.8	1.3± 1.5	1.8± 1.0
Basophil (%)	R	3.8± 1.3	1.5± 0.6	2.5± 1.3	2.8± 1.7	2.5± 1.9
	C	4.0± 2.6	3.5± 2.4	1.3± 0.5	2.0± 1.7	1.3± 1.0
White blood count (thousand/cumm)	R	7.6± 2.7	8.1± 0.8	8.2± 2.4	8.5± 0.9	7.7± 0.8
	C	7.7± 0.9	10.0± 2.3	8.4± 0.8	8.9± 1.7	9.1± 1.7
Hemoglobin (g/dl)	R	13.0± 1.4	13.2± 0.5	13.7± 1.2	13.0± 0.6	13.2± 0.1
	C	14.0± 0.7	14.0± 0.7	14.5± 0.9	13.9± 0.4	13.6± 0.7
Hematocrit (%)	R	37.3± 2.9	37.8± 1.5	39.3± 3.0	36.8± 1.5	37.8± 0.5
	C	39.0± 1.2	40.3± 1.5	41.8± 2.2	40.0± 0.8	38.5± 1.3
Platelet count (x 1000/cumm)	R	582.5± 280.1	295.0± 109.6	432.0± 116.9	387.0± 105.9	369.0± 74.5
	C	384.5± 117.9	266.0± 53.0	354.0± 139.2	387.0± 97.9	366.8± 104.5
Fibrinogen clauss (mg/dl)	R	316.8± 32.1	210.5± 61.4	295.3± 85.2	291.8± 63.2	224.3± 54.3
	C	300.0± 41.3	256.5± 60.0	314.3± 122.1	313.0± 99.6	281.3± 78.3
Prothrombin time (sec)	R	12.8± 1.3	12.1± 1.6	11.6± 1.9	11.5± 0.7	11.3± 1.3
	C	11.4± 0.8	11.8± 1.0	11.8± 1.0	10.9± 0.4	11.2± 0.6
Partial Thrombo-plastin (sec)	R	24.3± 2.6	25.5± 3.1	23.3± 1.3	23.3± 1.3	21.0± 2.2
	C	33.3± 2.5	20.8± 2.8	22.5± 3.7	22.8± 2.6	20.8± 2.6

*NOTE: R = the average of 4 radiated rabbits
C = the average of 4 control rabbits
All values are mean ± SD

Although inspection of the sixth line in Table 1 (WBC) suggests that there was a slight but consistently lower value of white cell production in the exposed group, an analysis of variance for trend revealed that the two groups were not significantly different ($F = 1.59$, $df = 1/6$). Aside from normal aging changes found in the lenses of the 8 animals, no differences were noted in the eyes of the two groups. Additional tests, however, in the National Institute of Environmental Health (NIEHS) laboratory indicated a significant decrease in albumin and calcium levels and albumin/total globulin ratio in exposed animals as compared to control animals. Analysis in the NIEHS laboratory also showed that there was a significant increase in the myeloid/erythroid rate in the bone marrow from the sternum of the exposed animals. These results of these NIEHS tests are discussed in the companion paper by McRee, et al. [21].

ACKNOWLEDGEMENT

This work was supported in part by the Association of Home Appliances Manufacturers and in part by the Rehabilitation Services Administration Grant No. 16-P-56818/0.

REFERENCES

1 Biologic Effects and Health Hazards of Microwave Radiation, Proceedings of an International Symposium, Warsaw, Poland, October 15-18, 1973, Polish Medical Publishers, Warsaw, 1974.

2 Zaret, M.M.: Selected cases of microwave cataract in man associated with concomitant annotated pathologies. In Biologic Effects and Health Hazards of Microwave Radiation, Proc. of an Internat'l Symp., Warsaw, Poland, October 15-18, 1973, Polish Medical Publishers, Warsaw, 1974, p. 294.

3 Tengroth, B., and E. Aurell: Retinal changes in microwave workers. In Biologic Effects and Health Hazards of Microwave Radiation, Proc. of an Internat'l Symp., Warsaw, Poland, October 15-18, 1973, Polish Medical Publishers, Warsaw, 1974, p. 302.

4 Richardson, A.W., T.D. Duane, and H.M. Hines: Experimental lenticular opacities produced by microwave irradiations. Arch. Phys. Med. 765-769, Dec., 1948.

5 Daily, L., K.G. Wakim, J.F. Herrick, E.M. Parkhill, and W.L. Benedict: The effects of microwave diathermy on the eye. Amer. J. Ophthal. 33:1241-1254, 1950.

6 Daily, L., K.G. Wakim, J.F. Herrick, E.M. Parkhill, and W.L. Benedict: The effects of microwave diathermy on the eye of the rabbit. Amer. J. Ophthal. 35:1001-1017, 1952.

7 Williams, D.B., J.P. Monahan, W.J. Nicholson, and J.J. Aldrich: Time and power thresholds for the production of lens opacities by 12.3 cm microwaves. IRE Trans. Bio-Med. Electron. 4:17-22, 1956.

8 Carpenter, R.L., and C.A. Van Ummersen: The action of microwave power on the eye. J. Microwave Power 3:3-19, 1968.

9 Carpenter, R.L., E.S. Ferri, and G.J. Hagan: Lens opacities in eyes of rabbits following repeated daily irradiation at 2.45 GHz. In: Summaries of Presented Papers, Microwave Power Symposium, Ottawa, Canada, May 24-26, 1972 (IMPI), p. 143.

10 Carpenter, R.L., E.S. Ferri, and G.J. Hagan: Assessing microwaves as a hazard to the eye — progress and problems. In Biologic Effects and Health Hazards of Microwave Radiation, Proc. International Symposium, Warsaw, October 15-18, 1973, Polish Medical Publishers, Warsaw, 1974, pp. 178-185.

11 Carpenter, Russel L.: Ocular effects of microwave radiation. Bull. N.Y. Acad. Med. 55:1048-1057, 1979.

12 Kramar, P.O., A.F. Emery, A.W. Guy, and J.C. Lin: The ocular effects of microwaves on hypothermic rabbits: a study of microwave cataractogenic mechanisms. Ann. N.Y. Acad. Sci. 247:155-165, 1975.

13 Guy, A.W., J.C. Lin, P.O. Kramar, and A.F. Emery: Effect of 2450-MHz radiation on the rabbit eye. IEEE Trans. MTT-23:492-498, 1975.

14 Guy, A.W., P.O. Kramar, A.F. Emery, J.C. Lin, and C.A. Harris: Quantitation of microwave radiation effects on the eyes of rabbits and primates at 2450 MHz and 918 MHz. Scientific Report No. 8, Office of Naval Research, Contract N00014-75-C-0414, March, 1976.

15 Appleton, B.: Experimental microwave ocular effects. In Biologic Effects and Health Hazards of Microwave Radiation, Proc. of an International Symp., Warsaw, October 15-18, 1973, Polish Medical Publishers, Warsaw, 1974, pp. 186-188.

16 Appleton, B., S.E. Hirsch, and P.V.K. Brown: Investigation of single-exposure microwave ocular effects at 3000 MHz. Ann. N.Y. Acad. Sci. 247:125-132, 1975.

17 Guy, A.W., and S.F. Korbel: Dosimetry studies on UHF cavity exposure chamber for rodents," 1972 Microwave Power Symposium, Ottawa, Canada, May 24-26, 1972.

18 Emery, A.F., P. Kramar, A.W. Guy, and J.C. Lin: Microwave induced temperature rises in rabbit eyes in cataract research. J. Heat Trans. 2/75:123-128.

19 Guy, A.W.: Quantitation of induced electromagnetic field patterns in tissue and associated biologic effects. In Biological Effects and Health Hazards of Microwave Radiation, Proc. of an International Symp., Warsaw, Poland, October 15-18, 1973, Polish Medical Publishers, Warsaw, 1974.

20 Durney, C.H., C.C. Johnson, P.W. Barber, H. Massoudi, M.F. Iskander, J.L. Lords, D.K. Ryser, S.F. Allen, and J.C. Mitchell: Radiofrequency Radiation Dosimetry Handbook, Second Ed., Report No. SAM-TR-78-22, prepared by the Univ. of Utah for USAF School of Aerospace Med., Brooks, AFB, TX, 1978.

21 McRee, D.I., R. Faith, E.E. McConnell, and A.W. Guy: Long-term 2450 MHz CW microwave irradiation of rabbits: evaluation of hematological and immunological effects. J. Microwave Power, this issue.

MEDICAL CONSIDERATIONS OF EXPOSURE TO MICROWAVES (RADAR)

Charles I. Barron, M.D.
and
Albert A. Baraff, M.D., Burbank, Calif.

Apprehension over the possibility of injury to man by microwaves is based largely on the fact that sufficiently intense and prolonged radiation of this frequency has caused severe injuries to experimental animals. Several hundred workers who have been occupied about radar installations or have been exposed to radar beams have therefore been observed in a comprehensive medical surveillance program. This has been in progress for four years. The initial study was a comparison of 88 nonexposed persons with 226 radar-exposed employees, some of whom had worked with radar as long as 13 years. No acute, transient, or cumulative physiological or pathological changes attributable to microwaves have been revealed by this study in people working with high-power radar transmitters and frequently exposed to their output. These subjects are free to heed the warning sensation of heat and are advised to avoid exposure to any firing beam when in a zone defined by a minimum power density of 0.0131 watts per square centimeter. Since the development of increasingly high-powered transmitters is to be anticipated, the need for more precise and refined statements of human tolerance is evident.

Considerable interest in the biological aspects of exposure to radar beams has been generated during the past year by widespread publicity of an alleged case of human death occurring after a brief exposure to an unknown quantum of microwaves.[1] The incident served to direct attention to this relatively new agent, and questions naturally arose concerning the extent of the hazard, if any, to persons working with radar transmitters and to those who might be exposed in some manner to the energized beam. Many of these questions have subsequently been answered, and much needed research is being conducted by military and civilian agencies to more closely explore the multifaceted disciplines associated with this complex problem. Unfortunately, these less sensational but more scientifically conducted efforts have received little recognition or dissemination beyond the scientific world.

It is not generally known that apprehension over the biological potentials of microwaves dates back to the early days of World War II, when Daily[2] performed his original studies on U. S. Navy personnel engaged in the operation and testing of relatively low-powered radars. Although this study revealed no evidence of radar-induced pathology in human beings, numerous reports have since appeared indicating that tissue injury and animal death can occur under certain experimental conditions. These studies indicate that cataracts, corneal opacities, testicular degeneration, and hemorrhagic phenomena have been induced in anesthetized, small, furry, test animals by exposure to microwaves in the frequency range of 2,800 to 9,000 megacycles for various time exposures. Boysen,[3] using a transmitter with a frequency of 300 megacycles, exposed rabbits in a wave guide and produced damage to the central nervous system, degenerative changes in the kidneys, heart, liver, and gastrointestinal tract, and hemorrhagic changes in the respiratory tree. The power density measured in the wave guide was in excess of 0.1 watts per square centimeter. The animals were exposed for periods of 7 to 10 minutes, and all whose rectal temperatures exceeded 44.5 C (112.1 F) died. Boysen was of the opinion that the pathology and death were causally related to the hyperthermia.

Because of these rather startling findings and the apprehension engendered by their publication in scientific journals, the medical department of the airframe manufacturer, Lockheed Aircraft Corporation, coincidentally installing, testing, and servicing the most powerful airborne transmitters, early in 1954 instituted a comprehensive medical surveillance program for its several hundred employees working with radar or those who might be exposed to the energized beam. This program has now been in progress for four years and constitutes one of the longest continuous medical surveys of radar-exposed personnel in the United States. It is because of this that we believe the results of our efforts and observations should be presented.

It is not our intent to discuss either the physics of microwave propagation or its ignition hazards. This presentation will be limited to a brief discussion of the objectives, methodology, findings, and interpretation of our program.

From the Medical Department, Lockheed Aircraft Corporation.

Read before the Joint Meeting of the Section on Preventive Medicine and the Aero Medical Association at the 107th Annual Meeting of the American Medical Association, San Francisco, June 25, 1958.

Our objectives were threefold: (1) to detect any cumulative biological effects of long-time exposure to microwaves of varying frequency and power output in persons who had taken minimal precautions; (2) to observe possible effects on persons working for short periods of time with or near extremely high-powered airborne radar with pulsed wave emissions; and (3) to establish correlation between objective findings and units of exposure expressed in time-power density factors with the highly idealized objective of establishing safe maximum exposure standards.

Effects of Long Periods of Exposure

Our initial study included 226 radar-exposed employees and 88 nonexposed control subjects. The examination program was designed to detect in our subjects pathology similar to that observed in exposed animals and included procedures and laboratory studies previously performed by other investigators. Several additional procedures were used in an attempt to duplicate findings alleged to have been observed in a study of human subjects.

Examination in every case included an extensive system and organ inventory, with emphasis on the ocular structures, central nervous system, gastrointestinal and urinary tracts, hematopoietic system, and skin. Imbedded metallic foreign bodies were identified; a careful marital and fertility history was elicited; and duration and manner of exposure to radar was identified.

Physical examination was extensive with respect to the body systems as outlined above. In addition, each subject was inspected for manifest hemorrhagic phenomena. A modified test for Rumpel-Leede phenomenon was then performed by means of placing the blood pressure cuff on the arm and maintaining pressure midway between systolic and diastolic pressure for three minutes. The appearance of more than 10 fresh petechiae in a circle 4 cm. in diameter below the cuff was considered a positive result.

The second phase consisted of an ocular examination, including a slit-lamp study performed with the subject subjected to cycloplegia by a competent ophthalmologist; complete blood cell and platelet counts; chest x-rays; and urinalyses.

Among the radar groups were 83 with 2 to 5 years of exposure and 37 with 5 to 13 years. Many of them were exposed while in military service or with other companies prior to their employment at Lockheed. Few had observed any precautions whatsoever prior to 1953. Despite this, no pathology or adverse physiological effects unequivocally attributable to microwave exposure could be demonstrated. Minor variations in the red and white blood cell counts were comparable with those of the controls. An apparent decrease in polymorphonuclear cells and increase in eosinophils and monocytes was later found to be due to a variation of interpretation by a laboratory technician. There were no significant variations in blood platelet counts. Abnormal urinalyses were found to be proportionate in both groups.

Physical abnormalities revealed a higher percentage of circulatory and gastrointestinal diseases in the controls, with a higher incidence of jaundice, severe headaches, and bleeding phenomena. Ocular pathology was considerably greater in the radar group. However, with the single exception of a small, solitary retinal hemorrhage, which absorbed completely within three months, all abnormalities were causally related to diseases or conditions not generally associated with microwave exposure. Chest x-rays were noncontributory. Modified tests for Rumpel-Leede phenomenon were interesting in that 8% of the control group showed positive results, compared to 2% of the radar group. There was no apparent correlation between positive findings and reduced platelet counts. Findings were negative in all but one subject with platelet counts below 200,000 per cubic millimeter. The positive results were, with this single exception, in persons with normal platelet counts.

Fertility studies revealed essentially the same percentage of offspring in both groups, when correcting for the larger number of unmarried men in the radar group. The percentage of childless marriages attributable to unknown causes was comparable, and in not a single instance in the radar group could an admission of male sterility be elicited.

On the basis of these findings we concluded that no person in this study had sustained any acute or chronic injury secondary to radar exposure. Reference is made to an earlier report describing our results in greater detail.[4]

Effects of Short Periods of Exposure

The second objective of our program is currently being accomplished. Having established base-line or reference criteria, we proceeded to reexamine our personnel, first at 6-month, then at 12-month, and finally at 24-month intervals, approximately four years after the original study. This latter program is now in progress. As a result of our original study and findings it was decided to modify our procedures and eliminate several of the more costly, time-consuming, and noncontributory tests. An extensive medical questionnaire was prepared, and each subject was interviewed by a physician. Physical examinations were performed only when indicated on the basis of the medical history or laboratory studies. Ocular and slit-lamp studies were repeated, and complete blood cell counts and urinalyses were performed. Blood platelet studies were repeated on alternate years. A limited number of electrophoretic serum protein patterns were made in an attempt to validate significant changes claimed to have been observed by another investigator.

Routine control studies were discontinued during the second and third examinations but were resumed in a limited degree during the present program. Despite a small sampling of radar-exposed personnel to date (49) it is believed that this group, under surveillance for four years and with exposure to identifiable transmitters, will provide valuable information for comparative purposes. Additional information obtained includes number of days of sick leave, leaves of absence, and other health statistics for the year 1957. Also, a large number of tests for Rumpel-Leede phenomenon were performed on applicants for employment and employees seeking treatment for routine ailments. None of these subjects had had any known exposure to radar emanations.

TABLE 1.—*Subjects in Medical Surveillance Program*

Age, Yr.	Exposed Personnel by Study Group, No.					Controls, No.
	Single Test	1-Yr.	2-Yr.	4-Yr.	Total	
<20	1	0	0	0	1	0
20-29	62	15	30	5	112	12
30-39	46	24	55	24	149	52
40-49	14	8	22	16	60	30
50-58	5	1	3	4	13	6
Total	128	48	110	49	335	100

The results of our studies are graphically presented in the accompanying tables. Table 1 identifies radar-exposed personnel as to numbers, age group, and years under medical surveillance and compares them to the control group. The total exposure group has increased to 335 by the addition of newly hired or reclassified employees. Persons in the one-year study generally have had two examinations, in the two-year study two or three examinations, and in the four-year study three or four examinations. Only the results of the last examinations are included in subsequent tables.

TABLE 2.—*Fertility Data for Subjects in Medical Surveillance Program*

	Exposed Personnel by Study Group							Controls		
	Single Test	1-Yr.	2-Yr.	4-Yr.	Total	Av. No. of Children	% Without Children	Total	Av. No. of Children	% Without Children
Children, no.	112	36	158	76	382	1.94	...	152	2.11	...
Fathers, no.	72	18	77	30	197	...	...	72	...	...
No children										
Cause known,* no.	40	14	25	17	96	...	27	...	...	10
Cause unknown, no.	16	16	8	2	42	...	13	...	...	18
Total	128	48	110	49	335					

* Major known causes: unmarried, wife sterile, birth control.

Table 2 reveals the fertility history of the radar-exposed and control groups. The slightly lower percentage in the first group who have fathered children is reflected in the larger number of unmarried men in this younger age group. Of those who are married and for unknown reasons have no offspring, the control group shows a somewhat higher percentage. The average number of children per family for each group is approximately two.

Table 3 lists comparative pathology or major subjective complaints for the various groups. These conditions were present at the time of the last examination or had been present the previous year. Among the radar-exposed group, sinus, gastrointestinal, genitourinary, and dermatological complaints were most prevalent. Headaches and nervousness were the commonest subjective complaints. The control group exhibited sinus, allergic, gastrointestinal, joint, and genitourinary disease prevalence, with fewer headaches and skin and respiratory complaints. There were no marked deviations or trends from the common disorders and no unusual or unexplained hemorrhagic phenomena. Bleeding was primarily nasal, rectal, or urinary in origin, and generally of known causation.

TABLE 3.—*Comparative Pathology and Number of Subjects with Major Complaints in Medical Surveillance Program*

Pathology	Exposed Personnel by Study Group (335)					Controls (100)
	Single Test	1-Yr.	2-Yr.	4-Yr.	Total	
Nervousness	4	1	2	3	10	4
Headache	5	...	5	3	13	6
Sinusitis	6	...	5	2	13	14
Hay fever	2	...	...	2	4	10
Respiratory disease or asthma	1	...	1	...	2	8
Circulatory and coronary disease	3	...	1	1	5	4
Peptic ulcer	...	1	4	2	7	...
Other gastrointestinal disorder	2	2	7	3	14	8
Urinary tract infection	2	...	5	1	8	8
Skin disorder	4	...	1	2	7	6
Arthritis, bursitis	3	...	2	2	7	10
Bleeding (nose, rectal, urinary)	3	3	2	...	8	4
Other	2*	1†	...	1‡	4	3§

* Mumps, deafness.
† Diabetes.
‡ Ruptured appendix with peritonitis (subject died).
§ Diabetes, hypothyroidism, hernia.

Table 4 reveals ocular findings for the radar-exposed groups only. In our opinion not a single finding can be attributable to radar exposure. There were no cataracts characteristic of those experimentally induced in animals by hyperthermia, and the corneal scars were, in the main, associated with other known causative agents. There were no tendencies toward progressive ocular diseases, and the four-year group revealed no pathology significantly different from that of the other groups. Congenital sutural cataracts are commonly seen in the general populace and are of no significance.

Sick leave for the 49 subjects who were in the four-year group averaged 3.0 days for the year 1957, compared to 3.1 days for the entire factory. In addition, there were five absences of more than 30 days in the radar-exposed group for the following reasons: hemorrhoidectomy, herniorrhaphy, appenectomy, nervous breakdown, and skull fracture. These longer absences represent a rate of 10% for this small group, compared to a plant-wide rate of 8%. One death occurred among the radar-exposed group; it was attributable to complications after a ruptured appendix. The diagnosis was confirmed on autopsy, and the pathological findings were typical of the disease process. Coincidentally, there were over 200 appendectomies, for acute appendicitis, performed on company personnel during the past year. In 1957 there were 113 deaths of company employees, including several cases of leukemia and plasma cell (multiple) myeloma. There was no known exposure to microwaves in any of these cases.

Table 4.—*Number of Ocular Findings in Radar-Exposed Groups*

Pathology	Single Test (128)	1-Yr. (48)	2-Yr. (110)	4-Yr. (49)	Total (335)
Pterygium	1	1	2	1	5
Esotropia or exotropia	5	2	3	3	13
Cornea					
Old scars	4	1	4	1	10
Keratitis bullosa	1	...	...	...	1
Lens					
Cataract, sutural (congenital)	6	2	9	2	19
Cataract, traumatic	1	1	1	...	3
Vitreous "floaters"	1	1	1	1	4
Drüsen	4	...	3	...	7
Other*	6	4	3	4	17

* Papilloma, acute trauma, telangiectasis, blepharitis; leukoma (2), episcleritis, pigment change in iris, allergic conjunctivitis; eclipse burn of macula, small retinal hemorrhage, medullated nerve fibers (3), macular scar (cause unknown); diabetic retinopathy, healed chorioretinitis, surgical ophthalmostcresis.

Table 5 reveals blood findings in the radar-exposed groups and in the 100 control subjects. Only the last blood cell counts for the radar group are shown. It is apparent that the blood picture of the radar-exposed and control groups is comparable in most respects. An unusually high incidence of increased monocytes and eosinophils is noted. However, it is somewhat higher in the control group, and the relative increase in these cells can be traced to the interpretation of one laboratory technician.

Blood platelet determinations revealed counts of less than 200,000 per cubic millimeter in only 2 of 243 subjects tested, despite a higher incidence of reduced counts in our original study. Of the 49 subjects studied over four years, only one had a reduced count. This was unusually revealing, in view of the attempts of at least one investigator to use this as an index of exposure. Also, since the same investigator associated exposure with positive findings on tests for Rumpel-Leede phenomenon, we decided to perform these tests on persons who routinely came to the attention of our medical department. Accordingly, 145 tests were performed on employees seeking minor first-aid care, management personnel undergoing annual health examinations, and a small number of applicants for employment. No subject was included in this study if he had knowingly been exposed to microwaves during military service or employment. Of the 20 positive findings obtained, 14 were in men and 6 in women; these represented 14% of the group. Of these subjects, two presented histories of hemorrhagic tendencies, two had been exposed to considerable ionizing radiation (x-ray), three had undergone recent major abdominal or pelvic surgery, and one was taking sedative medication. Of the 88 subjects used in our original control group (in 1954), positive results were noted in 8%.

In 26 cases selected at random, electrophoresis of serum proteins was performed at a hospital laboratory. Many of these specimens were from subjects in the two-year and four-year groups. Results in 16 tests were reported within normal range for all component proteins; 10 revealed deviations as follows: elevation of gamma globulin level, 4; depression of beta globulin level, 3; elevation of beta globulin level, 1; depression of gamma globulin level, 1; and depression of alpha globulin level, 1. In only one subject was the deviation more than slight or considered significant, and this was partially reversed within two months after elimination of an active known infection. Deviations described are generally associated with dietary deficiencies, infections, or obesity. One subject had been exposed to ionizing radiation, two had known active infections, and two had undergone recent surgery.

In no case was there any significant decrease in serum albumin or total protein levels. The albumin-globulin ratio was within normal limits in all persons.

Maximum Exposure Standards

Our third objective was to delineate safe maximum exposure standards. Obviously, this was contingent on the detection of pathological changes in our subjects and determination of the exposure parameters with respect to frequency or wave length, field power density, exposure time, and total test environment.

Table 5.—*Comparative Blood Cellular Findings in Subjects in Medical Surveillance Program*

	Exposed Personnel by Study Group, No.						Controls	
	Single Test	1-Yr.	2-Yr.	4-Yr.	Total	% of 335	Total	% of 100
Red blood cell count, per cu. mm.								
<4,500,000	...	9	1	1	11	3.3	0	0
4,500,000-5,000,000	20	9	11	2	42	12.5	10	10
White blood cell count, per cu. mm.								
Total								
< 5,000	...	...	...	...	0	0	0	0
>10,000	18	7	21	11	57	17.0	16	16
Differential count								
Polymorphonuclear cells								
<55%	24	4	15	16	59	17.0	19	19
>70%	18	3	14	3	38	11.3	12	12
Monocytes (>6%)	13	10	10	18	51	15.4	21	21
Eosinophils (>4%)	16	4	12	16	48	14.3	16	16

It soon became apparent that this objective could not be achieved in our study. We uncovered no pathology caused by either single or repeated exposure, and consequently we cannot speak authoritatively of so-called hazardous exposure conditions. The majority of our personnel had been exposed to radars of the following types: AN/APS-20A, B, C, and E; AN/APS-28, 30, 31, 33, and 45; AN/APG-40, SG, SX, and SR; AN/APS-6 and 7; and AN/APS 70. These transmitters operate in a frequency range of 400 to 9,000 mc and include powerful "S" band components. It was impossible to obtain precise data covering exposure time and average field power density, since these often were unknown. Exposure, however, varied from an occasional incidental contact with the beam to as much as four hours daily close exposure for periods up to four years. Exposures of several minutes a day at distances of less than 10 ft. from the radars were not uncommon.

Protective clothing was not worn by any of our subjects while in the radar beam. For some time precautions have been exercised in testing of our equipment and in exposure of personnel. In general, ground testing of high-powered, aircraft-mounted transmitters involves scanning of a very limited sector, with the antenna pointed toward an open, uninhabited field and rotating at 2 or 6 rpm, usually firing at reduced power.

Personnel were advised to avoid exposure to any firing beam when in a zone defined by a minimum power density of 0.0131 watts per square centimeter. A second zone, extending from the area previously defined to that with a minimum power density of 0.0039 watts per square centimeter, was deemed acceptable for occasional pass-through but no constant exposure, and, finally, there was a third limitless zone in which exposure was not deemed biologically significant.

Unfortunately, because most persons are exposed to radar emanations while on the ground and frequently within the so-called near radar field, it is extremely difficult to evaluate biological effects and hazards in relation to absolute power levels without accurate measurements. The need for such accuracy in quantitative determinations of exposure is obvious and can be achieved by the development of exposure meters reflecting absorption in quantum units of radar energy.

It has been suggested that the sensation of heat is almost universal on exposure to radar and that this in itself is indicative of an overexposure. In our study, only 17% of the 335 subjects experienced heat sensation and frequently only when in close proximity to "X" band radars. Almost 6% were aware of a buzzing or pulsating sensation when in an "S" band field. Less than 1% experienced other sensations or warning phenomena, such as sparking between dental fillings or a peculiar metallic taste. Eight subjects gave a history of metallic implants, such as bullets, buckshot, steel pins, and plates. None experienced any unusual reaction attributable to the metal. There were no complaints of heat directed to rings, wrist watches, or bracelets.

Comment

During the past 18 years thousands of persons, in the course of their employment or while in military service, have been exposed to microwaves, many without protection. Concern over the effects of such exposure is natural and to be expected. The majority of radars in common use today are relatively low powered, with the exception of some military transmitters which exceed one megawatt in peak power output. Radars with many times this power will be operational in due course and may radically change our entire concept of the biological potentials of this form of energy.

Since microwaves of varying frequency and power output are also being used to provide television display, for diathermy, and in electronic ovens, the personal safety problem is one of general public interest.

Experiments to date have been conducted primarily on small fur-bearing animals and under unusual test conditions. It is generally accepted that the modus of injury by microwaves is a hyperthermia produced by absorption of this form of energy by the body. Extreme caution must be exercised in attempting to extrapolate the results of small animal responses to heat to those of the human body. Small fur-bearing animals have a high coefficient of heat absorption, a small body surface, and a relatively poor heat regulating system. The human body, by comparison, has one of the best and can readily adjust and maintain thermal homeostasis under severe stress conditions. Adequate physiological function can be maintained in environments of 240 F for 23 minutes if the humidity is low, and at least one subject has been exposed to a temperature of 400 F, for a period of approximately one minute, without tissue injury.

Conditions of radar operation and testing vary from experimental conditions. Humans are generally exposed while in free air and rarely to a stationary energized beam. Some radar beams are extremely narrow, and only a small portion of the human body is instantaneously exposed. The body can dissipate heat readily to the environment between such exposures. One is reminded of a similar problem associated with exposure of personnel to the thermal effects of ultrasonic energy. In an analogous situation, small fur-bearing animals were destroyed by hyperthermia when placed in a jet engine noise field, yet there is no evidence of any adverse heating effects on man when exposed to the same environment. It has been estimated that it would require many million times the ultrasonic energy of that generated by any current jet engine to produce these effects in human beings.

There is reason to believe that the dramatic effects observed in small animals exposed to whole-body radiation will not be reproduced in larger, live animals under identical test conditions and that the human body will be the most resistant of all. This is not to imply that localized application of heat cannot injure human tissue. We have witnessed one case of accidental 15-second exposure, at a 6-to-10-in. distance, to an "X" band radar of over 100,000 watts in peak power output, with resultant erythema and a sensation of warmth for an hour but with full and uneventful recovery. Unless carefully controlled and operated, microwave diathermy with use of "S" band frequencies can cause local tissue damage.

In our study we have failed to detect any acute, transient, or cumulative physiological or pathological changes in subjects working with and frequently exposed to high-power radar transmitters. It would therefore appear extremely unlikely that there exists a biological hazard to the radar technician observing reasonable precautions or that the general public, exposed to greatly attenuated and intermittent doses of microwaves in the environment, is in any danger of body injury. We can see no causal relationship betwen microwave exposure and any increase in such conditions as coronary heart disease, leukemia, bone and lung cancer, and degenerative diseases of the nervous system.

There is need for additional research to explore the effects on living tissue of extended wave lengths and frequencies of microwaves and transmitters of higher energy, and military research is being directed in this area.[5] Col. G. M. Knauf has reported on the progress of this research at the American Medical Association's Annual Meeting in San Francisco. It is hoped this study will provide the basis for establishment of a realistic safety program acceptable to all scientists.

Finally, a plea is made for deductive rather than inductive research in this difficult field. With the increasing exposure to microwaves in and around the home, as well as in industry, careless and scientifically uncorroborated reports of human injury and death cannot avoid receiving dramatic and widespread dissemination. Such reports should not appear unless sufficient scientific data are included to support the conclusions and unequivocally establish the modus of injury. If radar is incriminated, the report must contain a definite history of exposure, including proper identification of the transmitter, wave length, power density, exposure time, symptomatology, laboratory data, pathological findings, and other factors.

Summary

In 1954, a medical surveillance program was instituted, covering 335 employees working with or exposed to microwaves in an airframe manufacturing company. Examinations have been performed at intervals of 6, 12, and 24 months in an effort to detect acute or cumulative biological effects of exposure at various intervals to energized radar beams in the 400-to-9,000-mc range and with peak power output exceeding one megawatt. Whenever possible, identical examinations were also accomplished on a nonexposed control group.

The examinations have failed to detect any significant changes in the physical inventories of the subjects. The incidence of death and chronic disease, sick leave, and subjective complaints was comparable in both groups. A high percentage of eye pathology was identified, but none with causal relation to the hyperthermia produced by microwave absorption. Fertility studies revealed essentially the same findings for both groups.

Laboratory studies for total red and white blood cell counts and differential counts revealed no significant changes above those noted in the control group. Urinalyses and chest x-rays were noncontributory with respect to radar exposures. Electrophoretic serum protein level determinations were performed on 26 subjects, with insignificant or accountable deviations in 10. Platelet counts and controlled capillary fragility studies for Rumpel-Leede phenomenon revealed the fallacy of using either to identify radar exposure. In addition, only a small percentage of the exposed subjects had been aware of heat or other subjective warning phenomena. Neither these tests nor subjective complaints were considered reliable indexes of exposure.

Absolute or safe maximum exposure standards were impossible to define, inasmuch as no radar-induced pathology could be identified. Subjects had been exposed for various periods, at indefinite distances, to a multitude of radars under flexible test conditions. The need for more precise and refined exposure data is indicated.

On the basis of these studies there appears to be no justification for public concern about the effects of greatly attenuated microwave energy in the environment. It would seem, therefore, that one may continue to enjoy his television without undue apprehension.

References

1. McLaughlin, J. T.: Tissue Destruction and Death from Microwave Radiation (Radar), California Med. **86:** 336-339 (May) 1957.

2. Daily, L. E.: Clinical Study of Results of Exposure of Laboratory Personnel to Radar and High Frequency Radio, U. S. Nav. M. Bull. **41:**1052-1056 (July) 1943.

3. Boysen, J. E.: Hyperthermic and Pathologic Effects of Electromagnetic Radiation (350 Mc), A.M.A. Arch. Indust. Hyg. **7:**516-525 (June) 1953.

4. Barron, C. I.; Love, A. A.; and Baraff, A. A.: Physical Evaluation of Personnel Exposed to Microwave Emanations, J. Aviation Med. **26:**442-452 (Dec.) 1955.

5. Knauf, G. M.: Biological Effects of Microwave Radiation on Air Force Personnel, A.M.A. Arch. Indust. Health **17:**48-52 (Jan.) 1958.

ANALYSIS OF OCCUPATIONAL EXPOSURE TO MICROWAVE RADIATION

P. Czerski and M. Siekierzyński

Natl. Res. Inst. of Mother and Child, Warsaw
and Inst. for Postgraduate Study, Military Medical
Academy, Warsaw, Poland

ABSTRACT

Principles of analysis of environmental conditions and of comparison of various microwave worker groups are discussed. Early and actual findings concerning the health status of microwave workers, published by various Polish authors, are compared and discussed in the light of personal experiences. Advantageous effects from enforcement of safety rules may be documented by the comparison of present results with those published ten years ago.

Introduction

The present paper is an attempt to summarize briefly personal experience and that of our colleagues, who worked in close contact with us during the last sixteen years. All the comments, however, express personal opinions of the present authors only. The list of references was restricted to the essential minimum, as an almost complete list of pertinent papers can easily be had by referring to the bibliographies in references 4, 13, 14, 24, and 29.

Analysis of occupational exposure to microwave radiation is fraught with many difficulties, the main being the assessment of the relationship between the microwave exposure levels and the health status of the examined groups of workers. The possible role of other environmental factors and of socio-economic conditions must be taken into account. As it often happens in clinical work, it is difficult to demonstrate a causal relationship between a disease and the influence of environmental factors, at least in individual cases. Large groups must be observed to obtain statistically significant epidemiological data. The problem of adequate control groups is controversial and hinges mostly on what one considers "adequate."

Analysis of Environmental Conditions

Precise quantitation of human exposure is possible only in the case of therapeutic or diagnostic applications of microwaves. In view of the lack of adequate instrumentation, especially of individual dosimeters, the quantitation of occupational exposure is extremely doubtful. This is particularly true where personnel move around in the course of their duties and are exposed to non-stationary feilds (i.e., moving beam or antenna), as well as to near- and far-fields alternatively. It is impossible to quantitate the exposure over a period of several years within reasonable limits. Attempts to present detailed data as to the source of microwave radiation, effective area of irradiation, position of the body in respect to the field, etc. for an individual worker for a period of several years would be misleading to an extreme degree. In the present authors' opinion, it is far better to present approximate evaluations, than to create an impression of accuracy, where none can be had.

Gordon (14) divided the microwave exposed workers examined by her into 3 groups, according to exposure levels:

1. periodic exposure to "high energy density" levels, i.e. 0.1-10 mW/cm^2,
2. periodic exposure to "low energy density" levels, i.e. 0.01-0.1 mW/cm^2,
3. systematic exposure to low energy density levels.

The first group consisted of technical maintenance personnel and workers of repair shops and certain factories (montage). This group consisted of production, montage, technical maintenance and repair of microwave equipment personnel. It should be mentioned that a large part of this personnel was periodically exposed to near-zone fields. The second group consisted of technical maintenance personnel as well as certain categories of personnel engaged in the use of microwave apparatus, research workers and others. The third group consisted of personnel engaged in the use of various microwave equipment, mainly radar.

In our investigations we adopted a similar rule of division of the personnel examined into high, mean or low exposure groups, or later, into two groups - high and low exposure. In regard to environmental conditions, analysis of such factors as air temperature, movement and humidity, noise, lighting and exposure to ionizing radiation generated incidentally by electronic equipment were taken into account. It should be stressed that exposure to these factors is carefully controlled, according to Polish laws.

Safe microwave exposure limits and regulations enforcing safety measures and precautions were introduced in Poland in 1961.

The exposure limits adopted at this time were identical to those introduced two years earlier in the USSR (0.01 mW/cm^2 unlimited exposure, 0.1 mW/cm^2 permissible for 2 hr per day and 1 mW/cm^2 for 20 min per day). It was possible to enforce this law only gradually. In view of this, all examinations of personnel carried out before 1962 concerned persons subjected to uncontrolled exposure. The probable exposure levels could be evaluated only ex post, and only approximately. In the period 1962-1968 data on exposure levels based on power density measurements and analysis of working conditions became available; most examined individuals had, however, a shorter or longer uncontrolled exposure history.

In view of this, the Polish publications concerning the health status of personnel professionally exposed to microwaves may be divided into papers concerning:

1. persons having a history of longer or shorter periods of work under uncontrolled conditions, exposure levels undetermined or calculated retrospectively;
2. persons with a history as above and a period of work in controlled environment;
3. persons who were examined medically before work was undertaken, found fit for work under microwave exposure conditions, and working under controlled conditions.

Numerous Polish papers concern the first and second groups (1, 6, 9, 11, 15, 17, 19, 23) or the second group and third groups jointly (3, 12, 16, 18, 20, 30) and only very few exclusively the third group (7, 8, 26, 27).

Early Investigations

Selected groups of microwave workers were first examined at intervals of 3 months, later 6 months, and, finally, annually. All the persons examined were usually divided into 3 groups:

1. low-level exposure of the order of tens of μW/cm^2 usually in far-field conditions in open space or in complex fields in closed rooms where only very low power equipment was installed;
2. "mean" exposure levels in far- and near-field zones; the measured levels being of the order of hundreds of μW/cm^2 up to about 1 mW/cm^2;
3. high level exposures of the order of 1 mW/cm^2 up to 10 mW/cm^2; in certain instances even more.

No attempt was made to differentiate between exposures at various microwave frequencies, as most individuals were exposed at one time or another over the entire microwave range. No adequate control group could be found because of difficulties in finding individuals working under sufficiently similar conditions (temperature, noise, humidity, time of day, etc.) of work and of sufficiently similar economic and social position, having similar everyday living habits, as well as belonging to the same age group. It was decided, therefore, to analyse the material within the group according to the total period of work and according to the exposure level by making comparisons between groups. It was possible to collect such groups which were reasonably similar in all respects (socio-economical and psychological) except the level of exposure.

All cases, where pathology related to any known etiology could be found, were excluded from the analysed material. Medical documentation on earlier examinations and physical check-ups was obtained in all cases and compared with actual findings. This documentation was nevertheless in most instances unsatisfactory in respect to eye examination and laboratory data.

The results obtained confirmed the findings of USSR authors (14, 24, 29). Complaints were analogous and demonstrated a periodicity of occurrence in relation to the duration of occupational exposure, stressed in the Soviet literature (24).

The presence of complaints was characteristic persons subjected to periods of uncontrolled exposure, before safety rules were introduced. Headaches and fatigue disproportionate to effort occurred respectively in 47% and 45% in group 3, 30% and 34% in group 2, and 30% and 30% in group 1 during the first year of work. These complaints disappear for two years, recur during the period 3-5 years of work and may reappear in certain individuals after 5 or 10 years. Abnormally excessive sweating (during the night) had a similar time dependence and was found in 68.8% in group 3, in 33.4% in group 2 and in 22.5% in group 1 during the first year of work; in the period of 3-5 years of work, the respective values being 14.5%, 15%, and 7%. Later on this symptom was not observed. Changes in blood pressure occurred only in group 3; the percent of hypotonia being 18 during the first year, 14 in the period 1-3 years, 6 in the period 3-5 years, 8 in the period 5-10 years and 11 after 10 years of work. In the remaining groups this percentage was less than 1. It should be added that no correlation with changes in heart rate could be demonstrated.

The peripheral blood picture did not demonstrate any abnormalities. In group 1 diverse WBC responses were seen during the first year of work. After ten years of work in the same group 10.5% of workers show absolute lymphocytosis usually accompanied by monocytosis; the total WBC being over 10,000 per mm^3.

Neurological examinations are difficult to evaluate. Many of the physicians, who carried out these examinations, differed in the evaluation of reflexes, dermographism, signs of irritability, etc. In view of this, the only means of obtaining objective results are electroencephalographic studies. These do not demonstrate any abnormalities in group 1. In groups 2 and 3 depending on duration of work and degree of exposure, a definite decrease in the number and amplitude of alpha waves occurs. Theta and delta waves and spike discharges may occur. The response to photostimulation is decreased. The most impressive finding is the poorly expressed bioelectric activity after more than 5 and, especially after 10 years of work.

It should be stressed that a rather specific phenomenon occurs in microwave workers (3, 12). Intravenous administration of cardiazole (metrazole) may be used for provocation of discharges (preconvulsive discharges) in the EEG, convulsions or shock. According to the literature this phenomenon is dose dependent and a cardiazole threshold exists; the dose of 7 mg/kg body weight being without any effect. Intravenous administration of 500 mg cardiazole in 10 ml (1 ml/30 s at 30 s intervals) does not provoke any effects in the normal adult male. In microwave workers with greater than 3 years of exposure, theta waves, theta discharges, spike discharges and even convulsions occurred. Twelve persons were examined, in 8 the test could not be completed. The study was discontinued, as it was considered dangerous for the patient. It should be pointed out that this phenomenon was studied extensively in rabbits (2, 3), and a decrease of cardiazole tolerance in irradiated animals may be considered as established.

Cases of what was considered a "microwave sickness" were reported in the Polish literature (5, 6). It may be doubted if a specific nosologic entity may convincingly be demonstrated in the present state of understanding of microwave bioeffects. The abnormal findings consisted mainly of a syndrome of autonomic nervous system disturbances with bradycardia, hypotonia and a disabling neurotic syndrome with typical EEG changes as described above.

In the course of work connected with health surveillance and risk analysis, the present authors encountered "clusters" of certain deviations from normal in factories or other working places, where exposure levels were exceptionally high, i.e. about 10 mW/cm^2 or more during about 1 hr/day. In such places also uncontrolled exposure by leakage could be expected. The abnormalities were present in 0.5-2% of otherwise healthy persons with deep bradycardia (less than 50/min) and signs of impairment of myocardial conductivity as represented by ECG, variable percentage of workers with gastric ulcers or peripheral blood picture changes (slight anaemia, lymphocytosis, granulocytopenia, or persisting unexplained granulocytosis). Such groups were usually too small to draw any valid conclusions, so only an impression remains that working conditions had "something to do" with these phenomena. This impression is strengthened by the fact that after introduction of rigorous health and microwave exposure surveillance, as well as partial exchange of personnel, no further cases were noted.

It should be added that no cases of "microwave cataracts" were described in the Polish literature or found by us. A higher incidence of lenticular opacities was reported in groups with histories of uncontrolled exposure periods and may possibly be related to poorly controlled exposure conditions (15, 17, 30).

These findings were used to develop the principles of medical examinations for selection of candidates for work under microwave exposure (8), and as an empirical basis for setting up the new Polish microwave safe exposure limits (7), which were introduced in 1972 (25).

Actual Findings

A special study was undertaken to determine if work under conditions conforming to actual Polish safe exposure limits (25) may be considered as truly safe. Detailed results are to be published shortly in English (8, 26, 27); therefore, the results will be presented very briefly.

An analysis of the incidence of disorders considered as contraindications for occupational microwave exposure among 841 males aged 20 to 45 years and exposed occupationally to microwaves for various periods was made. The analysed population was subdivided into two groups differing only in respect to microwave exposure - low i.e. below 0.2 mW/cm^2 and high i.e. between 0.2 mW/cm^2 and 6 mW/cm^2. No dependence of the incidence of disorders, considered contraindications for occupational microwave exposure, on the exposure level or duration of occupational exposure could be demonstrated (8). The incidence of lenticular opacities was compared between both these groups, as well as analysed within each group, subdivided according to age or duration of occupational exposure. No dependence of the incidence of lenticular opacities on the exposure level, nor on duration of occupational exposure was found. Significant correlation with age was demonstrated (27).

The incidence of functional disturbances (neurotic syndrome, gastro-intestinal tract disturbances, cardiovascular disturbances with abnormal ECG) was also analysed and no dependence on the exposure level or duration of occupational exposure (years) could be demonstrated (26).

Conclusions

Uncontrolled occupational exposure leads to the appearance of autonomic and central nervous system disturbances, asthenic syndromes and other chronic (prolonged exposure) effects, which are well documented by early Soviet, Polish and Czech reports. Similar observations were made by Miro (22) and Deroche (10) in France and in the United Kingdom and USA, according to a personal communication made by Mumford to Seth and Michaelson (28).

Controlled occupational exposure of healthy adults seems to have no untoward effects if the (new) Polish safe exposure limits (25) or, even more so, the conservative Soviet standard (24) are observed.

The last point, which is most important and cannot be sufficiently emphasized, is that all available data concern healthy human adult exposure, mostly men. The effects of intermittent or continuous exposure of children living near radar installations or TV transmitters is completely unexplored. Children may be expected, because of body size and geometry, to absorb microwave energy differently from adults. Exposures to 4-8 min/day microwave irradiation at low mean (tens or hundreds microwatts per cm^2), and very high peak power densities are sufficiently real for children, as to cause concern. This is one of the many reasons to adopt lower safe exposure limits for the general public. It seems reasonable to differentiate between occupational exposure (adult, healthy men under medical surveillance) and general public exposure (the aged, the sick, pregnant women and children). It seems evident that in the former case higher values are acceptable, in the latter - a safety margin must be introduced.

REFERENCES

(Russian names, book-titles and journals were transliterated according to International Organization for Standardization recommendation IO/RG-1954 E)

1. BARANSKI, S.: Badania biologicznych efektów swoistego oddzialywania mikrofal, Inspektorat Lotnictwa. Warsaw (1967).

2. BARANSKI, S. and EDELWEJN, Z.: Acta Physiol. Polonica 19 (1968) 37.

3. BARANSKI, S. and EDELWEJN, Z.: Pharmacologic analysis of microwave effects on the central nervous system in experimental animals, In "Biologic Effects and Health Hazards of Microwave Radiation" (P. Czerski **et al**. eds). Polish Medical Publishers. Warsaw (1974) in press.

4. BARANSKI, S., and CZERSKI, P.: Biological Effects of Microwaves. Polish Medical Publishers. Warsaw (1975) in press.

5. CIESLIK, Z., KORZENIOWSKI, K. and KAFLIK, I.: Problemy Lekarskie 12 (1973) 483.

6. CZERSKI, P., HORNOWSKI, J. and SZEWCZYKOWSKI, J.: Medycyna Pracy 15 (1964) 251.

7. CZERSKI, P. and PIOTROWSKI, M.: Medycyna Lotnicza 39 (1972) 104.

8. CZERSKI, P., SIEKIERZYNSKI, M. and GIDYNSKI, A.: Aerospace Med. (in press).

9. DENISIEWICZ, R., DZIUK, E. and SIEKIERZYNSKI, M.: Polski Archiwum Med. Wewn. 15 (1970) 19.

10. DEROCHE, M.: Arch. des Malashes Professionnelles 62 (1971) 679.

11. DZIUK, E., DENISIEWICZ, R., SIEKIERZYNSKI, M. and SYMONOWICZ, N.: Lekarz Wojskowy 42 (1970) 20.

12. EDELWEJN, Z. and BARANSKI, S.: Lekarz Wojskowy 46 (1970) 781.

13. GLASER, Z.R.: Bibliography of Reported Phenomena (Effects) and Clinical Manifestations Attributed to Microwave and Radio-frequency Radiation. BUMED Report, National Technical Information Service, Springfield Va., A.D. 750271, 1972 and Bureau of Medicine and Surgery, Dept. of the Navy, Washington (1973).

14. GORDON, Z.V.: Voprosy Gigieny Truda i Biologiceskogo deistvija elektromagnitnyh polei sverhvysokih castot. Medicina, Moskva (1966).

15. HORNOWSKI, J., MARKS, E., CHMURKO, E. and PANNERT, L.: Medycyna Pracy, 17 (1966) 213.

16. KOLAKOWSKI, Z.: Lekarz Wojskowy 47 (1971) 309.

17. MAJEWSKA, K.: Klinika Oczna 38 (1968) 323.

18. MAZURKIEWICZ, J., MILCZAREK, H., ZALEJSKI, S. and SIEKIERZYNSKI, M.: Lekarz Wojskowy 42 (1966) 9.

19. MILCZAREK, H.: Lekarz Wojskowy 47 (1971) 442.

20. MILCZAREK, H., ZALEJSKI, S., MAZURKIEWICZ, J. and SIEKIERZYNSKI, M.: Polski Tygodnik Lekarski 50 (1967) 1924.

21. MINECKI, L.: Medycyna Pracy 12 (1961) 329.

22. MIRO, L.: Rev. Med. Aeronaut 1 (1962) 16.

23. ORNOWSKI, M.: Medycyna Lotnicza 20 (1967) 47.

24. PETROV, I.R. (Ed.): Influence of Microwave Radiation on the Organism of Man and Animals. NASA TT-F-708. Natl. Technical Information Service. Springfield (1972).

25. Rozporzadzenie Rady Ministrów z 25.05, 1972. Dziennik Ustaw PRL 21, poz. 153.(Order on Polish safe exposure limits in Polish statute journal).

26. SIEKIERZYNSKI, M., CZERSKI, P., MILCZAREK, H., GIDYNSKI, A., CZARNECKI, C., DZIUK, E. and JEDRZEJCZAK, W.: Aerospace Med. (in press).

27. SIEKIERZYNSKI, M., CZERSKI, P. ZYDECKI, S., CZARNECKI, C., DZIUK, E. and JEDRZEJCZAK, W.: Aerospace Med. (in press).

28. SETH, H.S. and MICHAELSON, S.M.: Aerospace Med. 35 (1966) 734.

29. TJAGIN, N.V.: Kliniceskie Aspekty Oblucenija SVC-diapazona. Medicina. Leningrad (1971).

30. ZYDECKI, S.: Assessment of Lens Translucency in Juveniles, Microwave Workers and Age - Matched Groups. In "Biological Effects and Health Hazards of Microwave Radiation" (P. Czerski et al. Eds). Polish Medical Publishers. Warsaw (1974) in press.

-DISCUSSION-

McAFEE - Whenever electronic devices operate at high voltages, there is the possibility of significant deleterious by-products, emanating in the vicinity of the tubes. These by-products include X-radiation, ultraviolet radiation, ozone and oxides of nitrogen. Is it possible that chronic exposure to these toxic substances may explain the results of the provocative cardiazol tests as well as the altered blood and neurologic picture reported among microwave workers?

CZERSKI - Certainly not. Why? Because as I tried to say in the introduction, these things were controlled in human exposure. As concerns Cardiazol tests, the reaction of this drug was explored under experimental conditions. Rabbits were exposed in an anechoic chamber in far field conditions of course in the absence of X-rays. These animals demonstrated the reaction to Cardiazol. So I feel that I can easily say that none of the factors mentioned by Dr. McAfee could influence the response.

KAUFMAN - In talking about subjective complaints, as you are aware, it is often very hard to determine the cause of such complaints. I wonder whether complaints that occur in the first year and then go away might be due to the psychological stress associated with starting a new job, particularly if these people were told that their work might be hazardous.

CZERSKI - Such a possibility certainly exists, but we try to explain that really there is no work which is not hazardous,

CZERSKI - because as it is said "whatever you do - work is dangerous to your health." Seriously speaking, I think one question has to be considered. We had some effect at very low levels as I tried to show in the example of the influence of microwaves on the circadian rhythm of adrenalin-noradrenalin levels. Then you have an appearance and disappearance of complaints in the same individual. They complained the first year and then in the second and third year they had no complaints, and they were perfectly happy. Afterward, they started the complaining once more about the fifth year.

I am just wondering if we shouldn't try to evaluate the whole thing according to the classical concept of stress, adaptation, and fatigue, and not look for any more complicated explanations when we may find an answer in well-known physiologic mechanisms.

POSTOW - Is there a specific reason why tuberculosis is considered a contraindication for occupational exposure to microwaves?

CZERSKI - There is really a non-specific reason on general grounds. This is a socially dangerous disease from a transmission point-of-view. This is a precaution. Especially, perhaps under Polish conditions, where just after the war for many years tuberculosis was a serious problem. I am happy to say it is no longer; we are apt, however, still to remain oversensitive just in this respect.

TOMPKINS - In your chamber exposure situation at the end, would you repeat the exposure conditions? I am sorry; I missed them.

CZERSKI - You mean those I showed on the slides. This is really no anechoic chamber exposure, but simply a real open-space exposure. The situation was such that there were some people and installations at a certain distance in the open space.

Measurements were made at these points where the examined individuals worked and the exposure level was 0.2 mW/cm^2 and at certain times up to 6 mW/cm^2 during two minutes per 4 hours. What I said is based on actual timing. The men worked four hours and rested four hours, respectively during 24 hours.

We found another situation where another group of men had done exactly the same work, but there were no microwaves detectable by measuring equipment. We looked especially for any microwave irradiation at the second location, and none could be detected. And, as I said, the men performed exactly the same work in the same cycle, and the adrenalin-noradrenalin

CZERSKI - secretion was compared.

THOMSON - Were there any differences in the incidence of absence from work in the people exposed occupationally to microwaves?

CZERSKI - No, but this is another question because when we introduced this very strict medical surveillance, in certain instances we had such a situation, where we had no complaints, no absence from work, and so on. After medical surveillance and strictly controlled exposure conditions were introduced, the people were, I would say, more healthy, in comparison to matched groups. Of course, this was a result of selection and good medical care.

SOME ASPECTS OF ETIOLOGICAL DIAGNOSTICS OF OCCUPATIONAL DISEASES AS RELATED TO THE EFFECTS OF MICROWAVE RADIATION

Moscow GIGIYENA TRUDA I PROFESSIONAL'NYYE ZABOLEVANIYA in Russian No.3, 1976, pp. 14-17.

[Article by A. K. Gus'kova and Ye. M. Kochanova (Moscow), Institute of Industrial Hygiene and Occupational Pathology, USSR Academy of Medical Sciences, submitted 13 Sept. 1975]

[Text] One of the industrial factors, the effects of which on the cardiovascular system are reflected, according to the results of some studies of recent years, in the incidence of cardiovascular disease, is microwave radiation (B. V. Il'inskiy, et al.; V. P. Medvedev). It is very difficult to make etiological diagnoses of pathology of the circulatory system in the group of workers dealing with sources of superhigh frequency radiation, whether it involves tuning radio equipment or work as operators of radar stations, since such work requires a high degree of nervous and emotional tension and involves other deleterious factors.

According to the data of different authors, the changes in activity of the cardiovascular system occuring in individuals working with sources of microwave radiation form the syndrome of neurocirculatory dystonia of the hypotensive or hypertensive type (Ye. V. Gembitskiy, N. V. Tyagin, K. V. Glotova and M. N. Sadchikova), and they could lead to development of essential hypertension and cardiac ischemia (B. V. Il'inskiy, et al.; V. P. Medvedev).

If we compare the data over a period of several years pertaining to the incidence of arterial hypertension and hypotension among those working with SHF [superhigh frequency] radiation, we can see that there has been a perceptible decline in incidence of arterial hypotension in the last few years, from 38% according to 1948 data (A. A. Kevorkyan) to 7%, according to 1971 data (M. N. Sadchikova and K. V. Nikonova). The higher incidence of arterial hypotension, according to the results of examining workers exposed to the most deleterious conditions in 1948-1964 (22% according to Yu. A. Osipov, 26% according to N. M. Konchalovskaya, et al.), when safety devices were not adequately used, is indicative of the significant probability of a link between neurocirculatory hypotension and microwave radiation. As for arterial hypertension, the incidence thereof has been increasing in the last few years among those exposed to UHF radiation: 5.8-7.6% according to 1963-1964 publications, and 28% according to those of 1971-1972. In spite of the fact that many investigators do not indicate the criteria they used to assess high arterial pressure, on the whole the incidence of hypertension with chronic exposure to SHF fields coincides, according to the results of studies pursued in 1970-1972, with the findings of a screening of the nonorganized population of Moscow and use of WHO criteria (V. I. Metelitsa): 28% in males 35-64 years of age.

The wide distribution of functional cardiopathies is indicative of a need for comprehensive analysis of each specific case of determining the occupational etiology of a cardiovascular disease.

At the present time, there has been comprehensive investigation of so-called risk factors in development of essential hypertension and cardiac ischemia.

The occurrence of the five main risk factors and their involvement in development of cardiac ischemia are shown in Table 1, which we took from the works of V. I. Metelitsa.

Table 1. Occurrence of the five main risk factors and their influence on incidence of cardiac ischemia.

Risk factor	Occurrence %	Occurrence mean age (years)	Occurrence Criteria	Increased risk of cardiac ischemia
Hypercholesterolemia	23 or more	35-64	Over 259 mg%	2.2-5.5 times greater, firmly established
Arterial hypertension	28	35-64	Arterial pressure, 140 mm Hg or higher systolic, 90 mm Hg or higher diastolic	1.5-6 times greater, firmly established
Smoking	50	30-64	Regular smoking	1.5-6.5 times greater, firmly established
Inactive life style	--	---	---	1.4-4.4 times greater, inadequately investigated
Obesity	22	35-64	9 kg above average weight	1.3-3.4 times greater, poorly documented

In addition to those listed in Table 1, the following are proven risk factors in development of cardiac ischemia: genetic, emotional stress, psychological personality factors that determine an individual's reactivity to environmental conditions, etc.

The objective of our work was to make a comprehensive analysis of the possible risk factors in development of cardiovascular pathology in 21 patients with the diagnosis of radiowave disease.

The patients were hospitalized a second time in 1973-1974. All of the subjects had previously worked under adverse conditions, with exposure to radiation in excess of the permissible level. By the time they were rehospitalized, 2-3 years had elapsed after discontinuing work involving radiation. All of the subjects were men ranging in age from 35 to 54 years; most (19) were 40 to 54 years old and had been in contact with radiation for over 10 years.

Table 2. Quantitative distribution of some deleterious factors, with reference to development of cardiovascular disease, in 21 patients with the diagnosis of microwave sickness.

Factors	Number of patients related to the factor
Nervous and emotional stress	14
Combining work with studies	4
Night shifts	17
Regular smoking	7
Moderate drinking	7
Hereditary burden	5
History of skull trauma	3
Obesity	5
Hypercholesterolemia	12
Arterial hypertension	10

The following cardiovascular diseases were demonstrated on the basis of general clinical diagnostic criteria: neurocirculatory dystonia of the hypertensive type in 7 cases; essential hypertension in 10; cardiac ischemia in 2; hypothalamic deficiency in 2. According to our findings individuals 40-49 years of age are the most susceptible to these diseases.

An interval of 2-3 years after discontinuing contact with radiation did not lead to a perceptible change in course of cardiovascular disease in any of the subjects. It was possible to normalize high arterial pressure (over 160/100 mm Hg) in the clinic only by means of complex hypotensive therapy in the cases of essential hypertension. In three cases, in spite of therapy, hypertensive crises developed in the clinic, with transient signs of impaired cerebral circulation.

During the examination, two patients with cardiac ischemia reported pain in the region of the heart and retrosternally, with irradiation under the left scapula and left arm, which was curbed by validol or nitroglycerin. Electrocardiographic examination of these patients revealed changes of the 4-1, 4-2, 5-1 and 5-2 type, according to the Minnesota code, which are typical for cardiac ischemia.

Table 2 shows that nervous and emotional stress, and bad habits were the main exogenous risk factors in the 21 patients with the diagnosis of radiowave sickness. Hypercholesterolemia, arterial hypertension, obesity and hereditary burden were the most common endogenous factors.

Table 3 illustrates combinations of different risk factors in 10 patients with arterial hypertension, with the diagnosis of radiowave sickness. In 5 out of the 10 cases, we observed a combination of at least 3 deleterious factors, and in the other 5 there were 4 to 6 risk factors. Thus, all of the usual risk factors retain their significance in onset of cardiovascular pathology in these 10 cases.

For this reason, when diagnosing cardiovascular pathology in individuals exposed to UHF, in addition to assessing working conditions, the general rules should be followed in analyzing etiological and pathogenetic causes. In each case, it is also imperative to take into consideration such very frequently encountered causes of cardiopathy as changes in the cervicothoracic spine, intercostal neuralgia, cardiospasm, spastic or atonic colonopathy. When investigating working conditions, this should not be limited to demonstration of a high level of UHF fields; one must also pay attention to the

Table 3. Occurrence of deleterious factors in 10 patients with arterial hypertension and the diagnosis of radiowave sickness.

Name of patient	Age group (years)	Nervous & emotional stress	Work & studies	Night shifts	Heredit. burden	Skull trauma	Obesity	Hypercholesterolemia
Zhad-v	35-39 (37)	+	+		+			+
R-k		+	+	+			+	+
A-n		+		+				+
Zh-v	40-44	+		+				+
Chev-ov		+	+	+	+	+	+	
Chi-ov		+		+		+		+
T-n	45-49	+		+	+			
Vl-v		+		+				+
T-ko	50-54	+		+				+
Sh-v		+		+		+	+	+

extent of nervous and emotional stress, combination of work and studies, working in night shifts.

As for the effects of microwave radiation on onset of cardiovascular pathology, in our opinion it may take a certain place among other deleterious factors. However, for the time being there is no reason to relate development of disease to it alone, This question can be definitively answered only on the basis of the result of well-organized epidemiological screening of the relevant groups, with investigation of the incidence among them of proven risk factors.

The results of such an investigation, in conjunction with information concerning the incidence of such disease in the nonorganized population, will help single out the role of microwave radiation as a more or less significant etiological factor in development of cardiovascular pathology.

BIBLIOGRAPHY

1. Glotova, K. V., and Sadchikova, M. N., GIG. TRUDA [Industrial Hygiene] No.7, 1970, pp. 24-27.

2. Gembitskiy, Ye. V., TRUDY VOYEN.-MED. AKAD. IM. S. M. KIROVA [Works of the Military Medical Academy imeni S. M. Kirov], Vol. 166, 1966, pp. 121-129.

3. D'yachenko, N. A., VOYEN.-MED. ZH. [Military Medical Journal], No.2, 1970, pp. 35-37.

4. Istamanova, T. S. "Functional Internal Organ Disorders in Neurasthenia," Moscow, 1958.

5. Ionescu, V., "Cardiovascular Disorders on the Borderline Between Normal and Pathology," Bucharest, 1973.

6. Il'inskiy B. V., et al., in "Tezisy dokladov 2-go Vsesoyuzn. s"yezda kardiologov" [Summaries of Papers Delivered at the 2nd All-Union Congress of Cardiologists], Moscow, Vol. 1, 1973, pp. 34-36.

7. Kevork'yan, A. A., GIG. TRUDA, No.3, 1948, pp. 26-30.

8. Medvedev, V. P., Ibid, No.3, 1973, p.6.

9. Metelitsa, V. I., in "Epidemiologiya ishemicheskoy bolezni serdtsa i arterial'noy gipertonii" [Epidemiology of Cardiac Ischemia and Arterial Hypertension], Moscow, 1971, pp. 11-35.

10. Osipov, Yu. A., " Industrial Hygiene and Effects on Workers of Radio-Frequency Electromagnetic Fields, " Leningrad, 1965.

11. Ryvkin, I. A., et al., in " Epidemiologiya arterial'noy gipertonii i koronarnogo ateroskleroza " [Epidemiology of Arterial Hypertension and Coronary Atherosclerosis], Moscow, 1969, pp. 123-162.

12. Stamler, J., et al., KARDIOLOGIYA [Cardiology], No. 5, 1971, pp. 26-39.

13. Sadchikova, M. N., and Nikonova, K. V., GIG. TRUDA, No. 9, 1971, pp. 10-13.

14. Tyagin, N. V. "Klinicheskiye aspekty oblucheniy SVCh-diapazona" [Clinical Aspects of Irradiation in the Ultrahigh Frequency Range], Moscow--Leningrad, 1971.

15. Khanina, S.B., and Shirinskaya, G. I. "Funktsional'nyye kardiopatii" [Functional Cardiopathies], Moscow, 1971.

16. Shkhvatsabaya, I. K., KARDIOLOGIYA, No. 8, 1972, pp. 5-13.

Epidemiologic Studies of Microwave Effects

CHARLOTTE SILVERMAN

Abstract—**This is a selective review of human epidemiologic studies and related information concerning biologic and health effects of microwave radiation. Following a description of the objectives and methods of epidemiology, the approach to microwave effects is considered and two recent but not yet published studies are described, namely, a study of U.S. naval personnel occupationally exposed to radar, and a study of American Embassy personnel in Moscow. Investigations of several reported or suspected adverse effects are assessed: ocular effects, nervous and behavioral effects, congenital anomalies, and cancer. Suggestions are offered for further epidemiologic research.**

INTRODUCTION

EPIDEMIOLOGY is the study of both the distribution and the determinants of disease (or any health-related condition) in defined groups of individuals or the population [1]. The study of the distribution of disease, in terms of age, sex, race, geography, environment, occupation, socioeconomic status, and other characteristics is essentially descriptive (descriptive epidemiology); investigation of the influence of these and other factors on disease patterns is analytic, a search for causes (analytic epidemiology).

Epidemiology is one of many disciplines that search, each in its own fashion, for causal factors in health and disease. Since there are many levels of influences which lead to disease, and disorders rarely are due to one cause alone, many types of knowledge are needed. In epidemiology, a science basic to public health and important to medicine, the search is for determinants that can be modified so as to prevent disease or ill health. Interest covers the whole spectrum of disease from the inapparent or subclinical state to frank illness to death. Epidemiologic evidence about etiology may be incomplete or inconclusive but still provide a reasonable basis for preventive action as well as for further study aimed at improvement of our understanding of causation.

In contrast to clinical medicine, which deals with individual cases or a series of cases, epidemiology is concerned with all cases, or a representative sample of them, in defined groups of individuals. Cases are related to the population group and the time period in which they occur; the reference population is as necessary as the cases of interest. Several basic rates measure the frequency of disease and deaths. For example, morbidity rates, in terms of occurrence of cases (incidence) or prevailing frequency of cases at a particular time (prevalence), and mortality rates, make it possible to compare the health experience of different groups of individuals. And, of great importance in epidemiologic studies, the same quality of information is needed for cases and controls.

In the main, methods used in search of etiologic factors are those for observational rather than experimental studies. A study may be suggested by the findings of descriptive epidemiology, clinical investigation, experimental work or other sources. A tentative hypothesis is followed by demonstrating a statistical relationship between the condition being studied and certain individual or group characteristics. Then, the meaning of the relationship must be established. If causal inferences are derived, they may be tested in studies of individuals with the condition or characteristic compared to those without it.

Two principal approaches are used: studies can start either with a given population in which cases are sought over time (cohort, prospective), or with cases which are referred back to their population groups (case-control, retrospective). In epidemiologic studies, *prospective* and *retrospective* do not have the usual connotation of future and past time. They refer to two different basic methods of study and each can be concerned with past or current events. The essential feature of prospective studies is the classification and forward (longitudinal) followup of groups of individuals with and without the characteristic of interest (exposed and not exposed, for example) to determine if adverse effects (related to exposure) develop over a period of time. In retrospective studies, one starts with cases of a suspected adverse effect (e.g., cataract) and with controls without the adverse effect, and follows back in their histories for evidence of exposure in the cases and not in the controls. There are advantages and disadvantages to both types of studies, as well as different indications for their use and differences in the type and quality of information produced.

The strength of epidemiology is its ability to provide a direct measure of risk in humans. Risk estimates from nonhuman experimental studies may require epidemiologic confirmation and for some diseases experimental studies cannot be done at all because there are no suitable animal models. Adequate risk information cannot be furnished by clinical medicine alone.

The limitations of epidemiology are those of a population-based life science. Information must come from many other disciplines such as clinical medicine, pathology, environmental sciences, and statistical and social sciences. More often than not, data sources have been developed for other purposes. There can be considerable variation not only in the quality and quantity of data but also in the availability and completeness of records. And the epidemiologic method may be impractical for detecting some low-level risks because of large sample size requirements and problems of identifying and controlling confounding variables.

II. EPIDEMIOLOGIC APPROACH TO MICROWAVE EFFECTS

Epidemiologic studies of microwave effects have been few in number and generally limited in scope. Persons occupationally exposed in the military services or in industrial settings have been the principal groups studied. A few other populations living or working near generating sources or exposed to medical diathermy have been or are being investigated.

Information about health status has come from medical records, questionnaries, physical and laboratory examinations,

Manuscript received June 6, 1979; revised July 30, 1979.

The author is with the U.S. Department of Health, Education, and Welfare, Public Health Services, and with the Division of Biological Effects, Bureau of Radiological Health, Food and Drug Administration, 5600 Fishers Lane, Rockville, MD 20857.

Reprinted from *Proc. IEEE*, vol. 68, pp. 78–84, Jan. 1980.

and vital statistics. Sources of exposure data include personnel records, questionnaires, environmental measurements, equipment emission measurements, and (assumed adherence to) established exposure limits. Although there have been recent advances in dosimetry, microwave dosimetry presents formidable problems for meaningful assessment in most epidemiologic studies. There is at present no practical way of determining exposure to large numbers of individuals.

Two cohort studies completed during the past year but not yet available in published form are described below.

U.S. Naval Personnel Occupationally Exposed to Radar [2]

A pilot study by the Bureau of Radiological Health under the sponsorship of the Office of Naval Research determined that it was feasible to study a variety of long-term health effects among enlisted men occupationally exposed to different levels of microwave radiation during military service. Due to fiscal restraints, the endpoints of the study had to be reduced to those which could be investigated through the use of largely automated military and veterans' records, namely, mortality, hospitalized illness in naval and veterans' hospitals, and disability. The study was conducted by the National Academy of Sciences.

The study population was drawn from the thousands of enlisted men trained in the use and maintenance of radar equipment for navigation and gunfire control in technical schools maintained by the Navy since World War II. Other technical schools also graduated large numbers of technicians similarly selected for educational achievement, general intelligence and aptitude, thus offering the possibility of selecting valid comparison groups. The men selected for this study graduated during the period 1950 through 1954. The Korean War period was chosen for two reasons: wartime service ensured virtually complete ascertainment of deaths, and exposure in the 1950's provided a sufficient period of time for long-term effects to develop.

Measurements made by the Navy offered a guide for selecting the most highly exposed occupations. On the basis of a consensus decision by Navy personnel involved in training and operations, occupational groups were classified as probably maximally exposed (those repairing radar equipment) and probably minimally exposed (those operating radar equipment). Technical occupational groups probably nonexposed (such as engine room personnel) were considered unsuitable because of their exposure to heat and humidity stress. Men selected for the study were drawn from six Naval Enlisted Classifications of occupations.

The high-exposure cohort is made up of Electronics Technicians (13 078), Fire Control Technicians (3298), and Aircraft Electronics Technicians (3733). The groups in the low-exposure cohort, consisting of men trained in equipment operations, are classified as Radiomen (9253), Radarmen (10 116), and Aircraft Electrician's Mates (1412). The study population of approximately 40 000 consists of about 20 000 in each of the two cohorts. The groups were composed predominantly but not exclusively of young men who entered service shortly after graduating from high school; the high exposure groups had more older men who were veterans of World War II and had re-enlisted. The mean age in 1952 of the total low-exposure men was 20.7, whereas the average age of the high-exposure group was 22.1 years. The airmen in both the high (mean age 23.4) and low (mean age 24.6) exposure groups were older than the men who served on ships.

Follow-up medical information was derived from search and linkage of Navy and Veterans Administration (VA) records. The death of almost every war veteran is a matter of record in VA files, since applications for burial benefits are made for about 98 percent of war veterans. The application usually includes a copy of the death certificate, from which the certified cause of death may be obtained. The Navy's records of hospital admissions were searched, and records of admissions to VA hospitals were available for computer search, as were current awards for disability compensation. The cohorts of over 40 000 men were followed through extant records for the following endpoints and time periods of ascertainment:

Mortality	1955–1974
Morbidity (in-service hospitalization)	1950–1959 (excluding 1955)
Morbidity (VA hospitalization)	1963–1976
Disability compensation	1976.

The lack of dosimetry of occupational exposure did not permit the assignment of exposure doses to any individuals in this study. The only measurements possible were environmental or arose out of efforts to reconstruct the circumstances of an accidental overexposure. There have been enough accidental exposures at estimated levels exceeding 100 mW/cm^2 to indicate that there are occupations in which some men at some times on certain classes of ships have been exposed well in excess of the 10 mW/cm^2 limit [3]. Shipboard monitoring programs in the Navy since 1957 show that men in other occupations rarely, if ever, were exposed to doses in excess of this limit [4]. Radiomen and radar operators whose duties keep them far from radar pulse generators and antennae were generally exposed to levels well below 1 mW/cm^2, whereas gunfire control technicians and electronics technicians were exposed to higher levels in the course of their duties.

In addition to occupation *per se* other relevant elements of exposure were included in the analysis, namely, length of time in the occupation, class of ship, and power of equipment on the ship at the time of exposure. An index of *potential* microwave exposure to individuals called the *hazard number* was constructed for a sample of men in the high exposure group. This consisted of the sum of the power ratings of all gunfire control radars aboard the ship or search radars aboard the aircraft to which the technician was assigned, multiplied by the number of months of assignment. Data on navigational radars, which had very-low-power output by comparison, were not available. A technician with a low hazard number had little opportunity for substantial exposure to microwaves, while men with large hazard numbers may have had substantial opportunity for such exposure. The distribution of hazard numbers by specific occupational rating showed that within the high exposure group the Fire Control Technicians and the Aircraft Electronics Technicians had much larger proportions of men with large hazard numbers than did the Electronics Technician group. Study constraints prevented the determination of hazard numbers for the low exposure groups.

Because no measures of *actual* as opposed to *potential* exposure were available, the so-called *high exposure* rosters were made up of a mixture, in unknown proportions, of men whose actual exposures varied from high to negligible. If a large proportion of the men had, in fact, very small exposures, the consequence would have been to obscure by dilution any differences which might have been found had it been possible to study a large group of men who actually received high exposures. Further, it is possible that effects involving the cardiovascular, endocrine, and central nervous systems are transient, disappearing with the termination of exposure or soon thereafter, or are not perceived to be of sufficient consequence to result in admission to hospital.

It was not possible in this study to determine hospitalization outside the Navy and VA systems, nonhospitalized medical conditions during and after service, reproductive performance and health of offspring, or employment history after discharge from service. A subsample of living men with presumed high and low exposure patterns during service, however, can be identified for intensive individual followup. This would make it possible to obtain additional information about occupational exposure, by reviewing individual service files and by making direct inquiries of the men.

B. American Embassy Personnel in Moscow [5]

The long-term microwave irradiation by the Russians of the American Embassy Buildings in Moscow was highly publicized in 1976 and led to a two-year study of possible adverse health effects. The purpose of the study, sponsored by the U.S. Department of State and conducted by the Johns Hopkins University, School of Hygiene and Public Health, was to compare the morbidity and mortality experience of U.S. Government employees at the Moscow Embassy during the period 1953 to 1976 with the experience of employees who had served in other selected, nonirradiated Eastern European embassies or consulates during the same time period. Eight comparison posts were selected for their similarity to Moscow in climate, diet, geographic location, disease problems and general social milieu: Budapest, Leningrad, Prague, Warsaw, Belgrade, Bucharest, Sofia, and Zagreb.

Primary attention was given to the employees but spouses and children (whether or not at the embassies) and other dependents who had resided in the embassies during the study period were included in the investigation. A major effort was required to construct a basic list of all personnel who had served in any of the selected posts at any time during the 23-yr study period and to identify their dependents who might have been with them during their tours of duty at any study post. The final identified study population consisted of 1827 employees at the Moscow Embassy and over 3000 of their dependents, and 2561 employees at the eight comparison posts and 5000 of their dependents. The total study group was comprised of 4388 employees and 8283 dependents.

Microwave exposure at the Moscow Embassy varied during the period of the study. The direction and intensity of the microwave signal changed in 1975 but it was always directed toward the upper floors of the Chancery. Maximum exposure and exposed areas by time period were estimated by the State Department [5] as follows:

Time Period	Exposed Area of Chancery	Maximum Exposure
1953 to May 3, 1975	West Facade	Maximum of 5 μm/cm^2, 9h/day
June 1975 to February 7, 1976	South and East Facade	15 μm/cm^2, 18h/day
Since February 7, 1976	South and East Facade	Fractions of 1 μm/cm^2, 18h/day

Measurements of maximum exposure were made at or near the windows of the upper central building. Exposures according to time period were determined for individual floors in the living and working areas. Apartment complexes in Moscow distant from the Chancery were monitored every few months and only background levels were found. Tests for microwave radiation (between frequencies of 0.5 GHz and 10 GHz) at all Eastern European comparison posts were made periodically. Only background levels were detected at these Eastern European embassies.

Once a study member was identified, tracing was done by means of questionnaire and telephone interviews, with nearly complete success despite the extraordinary mobility of the study population. Medical information was abstracted from the records of the numerous physical examinations of the foreign service employees and their dependents. Additional health data, as well as family and detailed occupational information, were sought by means of an extensive Health History Questionnaire mailed to all employees and certain dependents, followed by telephone interviews. Information on mortality experience (living or dead) was reasonably complete because of the high tracing success; specific causes of death were obtained from death certificates and from other sources.

For purposes of the study, persons in the Moscow population were divided into three exposure subgroups: the exposed (to other than background levels of 1 μm/cm^2), the unexposed, and those with questionable exposure. Information about individual working and living locations was not available in personnel records and had to be obtained from the participants by the mailed Health History Questionnaire and personal telephone interviews. Exposure data, by location during assignments in the Moscow Embassy, were provided by the State Department for only two time periods: before and after May 1975. Many individuals remained in the questionable category due to the nature of their employment at the embassy or because they could not remember exactly where they worked or lived in the embassy compound.

About one-third of the employees were followed by means of records and inquiries for 15-20 years and over half for longer than 10 years. There were approximately 50 000 person-years of observation of the total employee group and 18 000 person-years of observation of employees who served in Moscow at any time.

Limitations of the study have been noted by the investigators: identification of the entire study population was thought to be very nearly complete but confirmation of completeness was not possible; compilation of dependents was incomplete, to an unknown degree; death certificates could not be obtained during the limited time period of the study for approximately one-third of the employees known to have died; the response rate to the Health History Questionnaire was low, no higher than 59 percent for the most responsive subgroup; classification of exposure status was inadequate; the highest exposure levels were the most recent; and the size of the study population was not sufficient to detect excess risks less than twofold for many of the medical conditions studied, while for some diseases only a larger excess could be detected.

Hundreds of factors were examined in terms of two basic comparisons: Moscow post versus comparison post individuals, and the Moscow population divided into subgroups by various measures of exposure to microwave radiation. No differences between the Moscow and comparison populations were found with respect to various components of the study such as success of tracing the ascertained study population, abstracting the medical records, response to the Health History Questionnaire, validation of the conditions and diseases reported on the Questionnaire, and ascertainment of deaths and acquisition of death certificates. Exhaustive comparative analyses were made of all symptoms, conditions, diseases, and causes of death.

Neither of the two studies demonstrated differential health risks associated with presumed exposure to microwave radiation but the limitations of the studies preclude conclusions at this time about possible microwave-related health effects.

III. Some Specific Health Effects

A. Ocular Effects

In some laboratory animals there is clear evidence that exposure of the head to microwaves under controlled environmental conditions will cause not only minor lens changes but also cataracts. Cataracts develop following a single high threshold dose or repeated subthreshold doses to the eye [6]. In man, over 50 cases of alleged microwave-induced cataract have been accumulated, mainly by a single investigator [7].

Numerous surveys of ocular effects in man have been made, especially in the U.S. Most investigations have involved service personnel and civilian workers at military bases and in industrial settings. The principal subject of interest has been the significance of minor lens changes in the cataractogenic process; cataracts (opacities impairing vision) have been infrequently investigated; and only recently have retinal changes been sought.

1) Minor Lens Changes: Lenticular defects too minor to affect visual acuity have been studied as possible early markers of microwave exposure or precursors of cataracts. The studies have been mainly prevalence surveys although the time periods are often variable or not specified; re-examination data rarely permit estimates of incidence. These occupational studies have generally emphasized careful clinical eye examinations including the use of slit-lamp biomicroscopy and photographs, without comparable attention to study design and follow-up plans for exposed and comparison groups.

The following generalizations can be made about observations of lens changes in microwave workers and comparison groups:

a) Lens imperfections occur normally and increase markedly with age among employed males studied. There is evidence that lens changes increase with age even during the childhood years [8]. By about age 50, lens defects have been reported in most comparison subjects. This is illustrated in Fig. 1 based on data from various studies.

b) Although a few suggestive differences have been reported [8]-[10], there is no clear indication that minor lens defects are a marker for microwave exposure in terms of type or frequency of changes, exposure factors or occupation. Inspection of Fig. 1 suggests possible earlier appearance of lens defects in microwave workers than in comparison groups, but there is considerable variation in the type, number and size of defects recorded, in the scoring methods used by different observers, and in the numbers examined.

c) Clinically significant lens changes, which would permit selection of individuals to be followed, have not been identified [11].

d) There is no evidence from ophthalmic surveys to date that minor lens opacities are precursors of clinical cataracts.

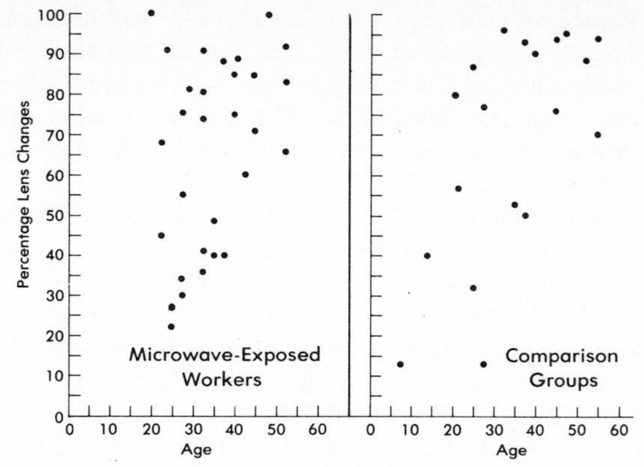

Fig. 1. Percent lens changes (all types) by age: Various studies.

2) Cataracts: Although there has been much interest in the cataractogenic effect of microwave radiation, only a minimal effort has been made to investigate cataracts as such, as distinct from their precursors. The only epidemiologic study of cataracts in microwave workers, a case-control study of World War II and Korean War veterans with negative findings, was reported in 1965 [12]. A recent statement that "not one epidemiologic study ... suggests even a slight excess of cataracts in microwave workers" [13] is certainly true—since there has been only one study.

Neither definitions nor methods of detection of cataract are standardized. The common meaning of cataract, a lens opacity that interferes with visual acuity, is open to many interpretations as to degree and nature of the opacity and loss of visual acuity. Specific disorders, physical agents and injuries are known to cause cataracts but many cataracts are loosely called *senile* when they occur after middle age, implying they result solely from aging of the lens. Microwave cataracts are not distinguishable from other cataracts in the opinion of most observers.

The most prominent characteristic of cataracts is their age distribution. Although estimates of frequency are not comparable because of differences in population groups surveyed, as well as nonuniform methods of detection and definition, all point to low frequencies until about the fifth decade of life when sharp increases occur.

Preliminary national estimates by age of the total prevalence of cataracts in the civilian noninstitutionalized population 1-74 years of age in the U.S. have just been made available by the National Center for Health Statistics [14]. They are based on diagnoses by ophthalmologist examiners in the Health and Nutrition Examination Survey of 1971-1972 and are plotted in Fig. 2.

One or more cataract conditions were found in 9 percent of the population. For the various age groups under 45, frequency of the condition increases gradually from 0.4 percent in those 1-5 years of age to 4 percent in the 35-44 year age group. The marked increase that occurs after age 45 reaches a maximum in the oldest group examined: of those 65-74 years of age, more than half had cataracts.

Cataract data for personnel on active duty in the armed services (who are mainly healthy, relatively young males) are available as incidence rates which show similar age dependence up to about age 55. Although not comparable with general population figures, recorded mean annual incidence rates are extremely low, of the order of 2/100 000 [15].

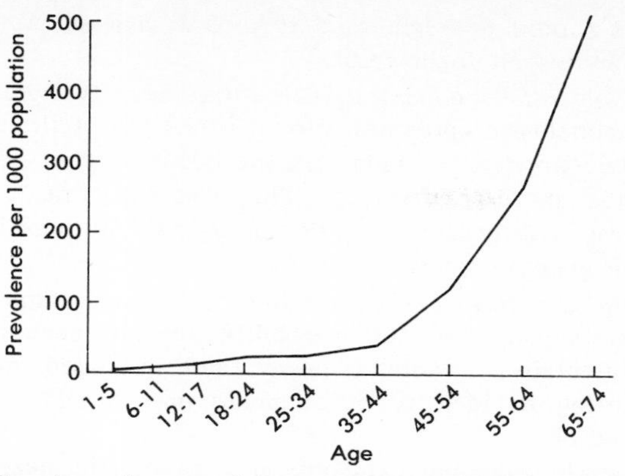

Fig. 2. Prevalence by age of one or more types of cataracts among persons 1-74 yrs of age, U.S., 1971-72. Preliminary national estimates. (Source: Division of Health Examination Statistics, National Center for Health Statistics, Medical Statistics Branch, Mar. 2, 1979.)

In the one case-control study of cataracts mentioned earlier [12], armed services personnel with cataracts which were ascribable to a nonmicrowave factor were eliminated from the sampling plan—that is, all congenital, traumatic, diabetic, and other specified types. The sampling plan also eliminated veterans 55 years of age and over to minimize dilution of the study group with senile cataracts.

Studies of microwave workers have been designed to find out whether cataract formation is accelerated in younger persons. It is necessary to look for possible increases in cataracts by age, as well as in high risk groups excluded from the study described above, to detect possible heightened microwave-induced susceptibility. The determinants of the microwave cataractogenic effect are not fully understood [6] and the epidemiology of cataracts has been inadequately studied [16].

3) Retinal Lesions: Until recently retinal lesions have not been considered a possible microwave effect. There is some reason to think that oculists examining microwave workers have observed but not reported retinal changes, because of no known relation to microwave exposure [17]. A small Swedish study reported in 1973 [18], which included examinations for retinal as well as lens changes, was prompted by preliminary findings of paramacular and macular pathology in industrial radar workers. A significantly higher proportion of retinal lesions in the central part of the fundus was found in microwave workers aged 26–40 years than in controls. The retinal lesions had resulted in decreased vision in two cases. No further reports are available.

B. Nervous and Behavioral Effects

Nervous system and behavioral changes in experimental animals following low-level exposures have been reported for many years from Eastern Europe [19]. Clinical laboratory studies of groups employed in the operation, testing, maintenance and manufacture of microwave-generating equipment have also been reported in large number from the USSR and other Eastern European countries [17], [20], [21]. Exposures have been mainly low-level (microwatts up to a few mW/cm^2) and long term.

With few exceptions [22], functional disturbances of the central nervous system have been described as a typical kind of radiowave sickness, the neurasthenic or asthenic syndrome. The symptoms and signs include headache, fatigability, irritability, loss of appetite, sleepiness, sweating, thyroid gland enlargement, difficulties in concentration or memory, depression and emotional instability. This clinical syndrome is generally reversible if exposure is discontinued.

Another frequently described manifestation is a set of labile functional cardiovascular changes including bradycardia (or occasional tachycardia), arterial hypertension (or hypotension) and changes in cardiac conduction. This form of neurocirculatory asthenia is also attributed to nervous system influence. More serious but less frequent neurologic or neuropsychiatric disturbances have occasionally been described as a diencephalic syndrome.

The only American epidemiologic study to date of some of these effects is the cohort study of American Embassy employees in Moscow and other Eastern European capitals [5]. Although much symptomatology was found, no differences were attributable to microwave exposure at the intensities measured outside the Moscow embassy. These levels, however, were even lower than exposures reported in the Russian occupational studies.

The identification and assessment of poorly defined nonspecific complaints, symptom-complexes and illnesses is extremely difficult [23], [24]. For the future, in addition to medical examinations, consultation with specialists in the behavioral sciences is needed. The use of health questionnaires designed to detect emotional ill health, and objective psychologic tests for specific types of symptomatology, can provide relevant information. Useful data may also come from attendance rates at clinics or physicians' offices, absentee rates due to illness, accident liability and job performance.

C. Congenital Anomalies

Microwave radiation can be teratogenic in experimental animals exposed at specific times during gestation to relatively low average doses [25]. In humans there is no evidence of anomalies following exposure *in utero*.

A case-control study of *Down's syndrome* in Baltimore [26] yielded an unexpected finding regarding paternal exposure to microwave radiation. Fathers of children with mongolism gave more frequent unsolicited histories of occupational exposure to radar during military service than did fathers of unaffected children, a difference that was of borderline statistical significance. Exposure during military service occurred prior to the birth of the affected child. After publication of the first report in 1965, expansion of the study group, followup of all fathers to obtain more detailed information about radar exposure, and search of available armed forces records were undertaken. The suggestive excess of radar exposure of fathers of Down's syndrome cases was not confirmed on further study but occupational exposures were difficult to document [27]. A chromosome study of peripheral blood of exposed and unexposed fathers showed some suggestive but inconclusive changes; the findings are to be reported.

A study of congenital anomalies in Alabama [28] showed that during the 3-year period 1968–1971 the adjoining counties of Dale and Coffee, in which Fort Rucker was located, had a reported number of *clubfoot* cases among white babies that greatly exceeded the expected number (based on birth certificate notifications for the State). A more detailed investigation revealed that in the 6-county area surrounding Fort Rucker, there was, during the same time period, a considerably higher rate of anomalies (diagnosed within 24 h after birth) among births to military personnel than in the State as a whole. Fort Rucker was a training base for fixed-wing and helicopter aircraft, situated within 35 mi of dozens of radar stations. Errors in malformation data on birth certificates and probable overreporting from Fort Rucker led to the conclusion that convincing evidence was lacking that radar exposure was related to congenital malformations [29]. The high malformation rate across a group of counties of the State was presumably environmentally induced but no specific agent was suggested. It was not possible to do a more detailed study at this or at another military base.

The use of microwave heating as diathermy to relieve the pain of uterine contractions during labor was reported from Belgium in 1973 [30]. The analgesic effect was found helpful in 1000 selected patients without obstetric pathology, and the babies were born healthy with good circulation. By 1976 [31], 2000 microwave-exposed patients and 2000 controls had been observed. No untoward immediate or long-term effects on the fetus are known from such exposure shortly before delivery. The only possible congenital defect so late in gestation would involve the *central nervous system.* Systematic follow-up examinations have not yet been made.

D. Cancer

Microwave-induced cancer has not been reported experimentally or suspected in medical surveillance examinations of microwave workers or service personnel. Only the two recently completed prospective epidemiologic studies [1], [3] have looked into the question systematically. Neither has revealed an excess of any form of cancer to date that can be interpreted as microwave-related.

IV. Suggestions for Further Research

In view of the exceptional difficulty in extrapolating microwave effects from experimental animals to man, epidemiologic studies, including appropriate clinical and laboratory examinations, are essential to improve our understanding of possible hazards to health.

1) Studies of identified populations should be improved, continued or expanded wherever possible and useful. It is difficult to identify exposed populations, select suitable controls and obtain exposure data. Some study groups already characterized can be improved by the acquisition of additional exposure data, some groups should be followed for longer periods of time, and some should be investigated for additional endpoints.

2) Additional study populations should be sought. Exposure to microwave radiation has been experienced increasingly for more than three decades. A careful search should be made for exposed groups not yet investigated or considered for study. In epidemiologic studies, as in experimental or clinical work, there is rarely a single study, positive or negative, that can be accepted as definitive. Replication and validation are needed at all exposure levels.

3) Some specific endpoints should be studied further.

a) Cataracts: It is reasonable to hypothesize and feasible to investigate a synergistic or additive action of microwaves in cataract formation. Comparative frequencies of cataracts from all causes in exposed and nonexposed occupational groups can be investigated by following study subjects into the older age periods when cataracts increase in frequency.

b) Mental and Behavioral Changes: The numerous reports from Eastern Europe of a wide variety of functional changes and possible nervous system effects have yet to be confirmed or rejected. In appropriate epidemiologic studies, medical reports should be augmented to include assessment of emotional and psychologic status.

c) Congenital Anomalies: There have been two preliminary and inconclusive epidemiologic investigations of the effect of *paternal* exposure to radar on the occurrence of congenital anomalies in their offspring. The subject is in need of more intensive investigation in both fathers and mothers to assess possible genetic effects.

d) Malignancies: There is no direct evidence that microwaves are carcinogenic but the possibility has not been excluded experimentally. More intensive and extended morbidity monitoring to identify malignancies may provide pertinent information.

4) Newly available exposure and dosimetric measurement prediction techniques should be used wherever applicable.

To clarify the complexities of the biologic and health effects of microwave radiation, all possible study approaches are needed. Epidemiologic research, which is in an early state of development, should be broadened.

References

[1] A. M. Lilienfeld, *Foundations of Epidemiology.* New York: Oxford Univ. Press, 1976.

[2] "Occupational exposure to microwave radiation (radar)," Final rep. (Contract No. FDA 223-76-6003) to Food and Drug Administration, National Academy of Sciences-National Research Council, Dec. 27, 1978.

[3] Z. R. Glaser and G. M. Heimer, "Determination and elimination of hazardous microwave fields aboard naval ships," *IEEE Trans. Microwave Theory Tech.*, vol. MIT-19, p. 232, Feb. 1971.

[4] G. M. Heimer, personal communication.

[5] A. M. Lilienfeld, J. Tonascia, S. Tonascia *et al.*, "Foreign service health status study-evaluation of health status of foreign service and other employees from selected Eastern European posts," Final rep. (Contract No. 6025-619073) to U.S. Dep. of State, July 31, 1978.

[6] R. L. Carpenter, "Microwave radiation," in: *Handbook of Physiology #9—Reactions to Environmental Agents*, D. H. K. Lee, H. L. Falk, and S. D. Murthy, Eds. Bethesda, MD: Amer. Physiol. Soc., pp. 111-125, 1977.

[7] M. M. Zaret, "Selected cases of microwave cataract in man associated with concomitant annotated pathologies," in: *Biologic Effects and Health Hazards of Microwave Radiation*, P. Czerski, K. Ostrowski, C. Silverman, *et al.*, Eds. Warsaw, Poland: Polish Medical Publishers, pp. 294-301, 1974.

[8] S. Zydecki, Assessment of lens translucency in juveniles, microwave workers and age-matched groups, in: *Biologic Effects and Health Hazards of Microwave Radiation*, P. Czerski, K. Ostrowski, C. Silverman, *et al.*, Eds. Warsaw, Poland: Polish Medical Publishers, pp. 306-308, 1974.

[9] S. F. Cleary and B. S. Pasternack, "Lenticular changes in microwave workers, a statistical study," *Arch. Environ. Health*, vol. 12, pp. 23-29, 1966.

[10] K. Majewska, "Study of effects of microwaves on visual organs," *Klin. Oczna* (Polish) vol. 38, pp. 323-328, 1968.

[11] M. M. Zaret, S. F. Cleary, B. Pasternack, *et al.*, A study of Lenticular imperfections in the eyes of a sample of microwave workers and a control population," Final Contract Rep. for Rome Air Development Center, RADC-TDR-6310125, Mar. 15, 1963.

[12] S. F. Cleary, B. S. Pasternack, and G. W. Beebe, "Cataract incidence in radar workers," *Arch. Environ. Health*, vol. 11, pp. 179-182, 1965.

[13] J. A. Hathaway, reply to Dr. Zaret (letter to Ed.), *J. Occup. Med.*, vol. 20, pp. 316-317, 1978.

[14] U.S. Dep. Health, Education, and Welfare, National Center for Health Statistics, Div. Health Examination Statistics, Medical Statistics Branch, Mar. 2, 1979.

[15] L. T. Odland, "Observations on microwave hazards to USAF personnel," *J. Occup. Med.*, vol. 14, pp. 544-547, 1972.

[16] A. Sommer, "Cataracts as an epidemiologic problem," *Amer. J. Ophthalmol.*, vol. 83, pp. 334-339, 1977.

[17] S. Baranski and P. Czerski, *Biological Effects of Microwaves.* Stroudsburg, PA: Dowden, Hutchinson, and Ross, Inc. 1976.

[18] E. Aurell and B. Tengroth, "Lenticular and retinal changes secondary to microwave exposure," *Acta Ophthal.*, vol. 51, pp. 764-771, 1973.

[19] I. R. Petrov, Ed., *Influence of Microwave Radiation on the*

Organism of Men and Animals. Leningrad, USSR: Meditsina Press, 1970, NASA TTF-708, 1972.

[20] Z. V. Gordon, *Biological Effect of Microwaves in Occupational Hygiene*, 1966, translated from Russian, NASA TTF-633, 1970.

[21] M. N. Sadchikova, "Clinical manifestations of reactions to microwave irradiation in various occupational groups," in *Biologic Effects and Health Hazards of Microwave Radiation*, P. Czerski, K. Ostrowski, C. Silverman, *et al.*, Eds. Warsaw, Poland: Polish Medical Publishers, 1974, pp. 261-267.

[22] M. Siekierzynski, C. Czarnecki, E. Dziuk, *et al.*, "Microwave radiation and other harmful factors of working environment in radiolocation: Method of determination of microwave effects," *J. Microwave Power*, vol. 11, pp. 144-145, 1976.

[23] C. Silverman, "Nervous and behavioral effects of microwave radiation in humans," *Amer. J. Epidemiol.*, vol. 97, pp. 219-224, 1973.

[24] —, *The Epidemiology of Depression.* Baltimore, MD: Johns Hopkins Press, 1968.

[25] R. Rugh, E. I. Ginns, H. S. Ho, *et al.*, Responses of the mouse to microwave radiation during estrous cycle and pregnancy," *Rad. Res.*, vol. 62, pp. 225-241, 1975.

[26] A. T. Sigler, A. M. Lilienfeld, B. H. Cohen, *et al.*, Radiation exposure in parents of children with mongolism (Down's Syndrome), *Bull. Johns Hopkins Hosp.*, vol. 117, pp. 374-399, 1965.

[27] B. H. Cohen, A. M. Lilienfeld, S. Kramer *et al.*, "Parental factors in Down's Syndrome-results of the Second Baltimore Case-control Study," in *Population Cytogenetics, Studies in Humans*, E. B. Hook, and I. H. Porter, Eds. New York: Academic, 1977, pp. 301-352.

[28] P. B. Peacock, J. W. Simpson, C. A. Alford *et al.*, "Congenital anomalies in Alabama," *J. Med. Ass. State AL.*, July 1971, pp. 42-50.

[29] J. A. Burdeshaw and S. Schaffer, "Factors associated with the incidence of congenital anomalies: A localized investigation. Environmental Protection Agency, Final Rep. Contract No. 68-02-0791, Mar. 31, 1976.

[30] J. Daels, "Microwave heating of the uterine wall during parturition," *Obstet-Gyn.*, vol. 42, pp. 76-79, 1973.

[31] —, "Microwave heating of the uterine wall during parturition," *J. Microwave Power*, vol. 11, pp. 166-168, 1976.

The Tri-Service Program—A Tribute to George M. Knauf, USAF (MC)

SOL M. MICHAELSON

Abstract—During World War II, the Department of Defense medical services became interested and concerned about possible hazards associated with the development, operation, and maintenance of the increasing numbers of radars and other radio-frequency emitting electronic equipment. After some investigations by the U. S. Navy and the U. S. Air Force, responsibility for research on the biomedical aspects of microwave radiation was delegated in July, 1957, as a tri-service arrangement to Rome Air Development Center, Griffiss Air Force Base, N. Y. Primary responsibility for coordination of the program rested with Dr. George M. Knauf, USAF (MC). The Tri-Service Program included investigation of effects of exposure of the whole-body, selected organs and tissues, single cells, and enzyme systems using various power levels, pulsed and continuous wave, in the frequency spectrum from 200 through 24 500 MHz under acute, subacute, and chronic conditions. The most important contribution of the Tri-Service Program was the validation of the 10-mW/cm² safety standard. The Tri-Service Program is to date the only large-scale coordinated effort in the Western world to elucidate and understand some of the basic mechanisms of microwave bioeffects and to assess possible health implications of this form of energy. This paper constitutes an overview of the Tri-Service Program to provide some historical insight into the significance of the program and its contributions to our understanding of the biologic effects of microwaves. The initiative, foresight, and drive of Colonel Knauf was an immeasurable contribution to the success of the program.

Historical Perspective

BEFORE 1956, research on the biological effects of microwave energy was delegated the role of a "stepchild" among other research problems competing for support within the Armed Services. In 1956, however, the Department of Defense assigned to the Air Force the responsibility of tri-service coordination of this research to ensure that all three departments, Army, Navy, and Air Force, would be well informed. To implement this responsibility, the Rome Air Development Center and the Air Research and Development Command Headquarters held a Tri-Service Conference on July 15–16, 1957, to bring together key investigators in the area of microwave bioeffects.

At the time the Tri-Service Program was initiated, a review of existing information on microwave bioeffects indicated the following.

1) Microwave radiation injury had been qualitatively demonstrated in animals but had not been observed clinically in radar personnel.

2) In experimental animals the testes were found to be vulnerable to the shorter wavelengths.

3) Experimentally induced injury appeared to be thermal in nature; i.e., the temperatures induced in the affected regions were sufficiently high to account for injury on a thermal basis.

4) Reliable information on power densities for radar beams in use was not readily available, and the parameters of injurious exposure were unknown [82].

A Tri-Service Ad Hoc Committee was formed in 1958 consisting of military representatives and a number of civilian advisors in the fields of physics, physiology, ophthalmology, microwave engineering, biophysics, and pathology. Representatives from the participating groups met regularly together with the investigators from the various universities, critically analyzed experimental results, made recommendations, and suggested sound approaches to meeting the requirements of the military services. The research program was coordinated by the Rome Air Development Center (RADC); primary responsibility for coordination of the program rested with Colonel Knauf.

Annual Tri-Service Conferences were initiated in 1956 by RADC as the means for reporting to the military services. These annual conferences were attended by key investigators in the field of biological effects and related scientific fields, and by representatives of other interested Government agencies, universities, and industry, to discuss and contribute knowledge towards a better understanding of the biological effects of microwave energy. The findings were reported through the Tri-Service Publications, which represent a valuable data source for investigation of problems related to the field of microwave bioeffects.

In an effort to establish a safe exposure level to microwaves, many variables were considered, such as the frequency of the energy to which an individual might be exposed and the nature of the exposure including time of exposure, field strength, and other aspects. A survey made at the Rome Air Development Center of earlier investigations concluded that the minimum power density[1] for injurious microwave effects was about 0.2 W/cm² [65]. Knauf [27] pointed out that the reports which were available concerning tolerance levels for microwave exposure were widely divergent.

Manuscript received June 1, 1970; revised September 8, 1970. This work was supported by the U. S. Atomic Energy Commission under the University of Rochester Atomic Energy Project (Rep. UR-49-1237).

The author is with the Department of Radiation Biology and Biophysics, University of Rochester, Rochester, N. Y. 14620.

[1] In general the power density referred to in the discussion of hazards and safety standards is the power density present in the absence of an animal or subject.

Reprinted from *IEEE Trans. Microwave Theory Tech.*, vol. MTT-19, pp. 131–146, Feb. 1971.

Sufficient factual data were not available to determine the "safe" exposure level for each frequency throughout the spectrum; therefore, it was decided to select one level satisfactory for all radio frequencies. The accuracy of the methods and instrumentation used was somewhat questionable, and possibly some cases of reported damage were no doubt caused by power densities of approximately 0.1 W/cm². Also, the expanded use of electronics equipment had resulted in the addition of microwave energy from incidental sources at many different frequencies. Since it is impractical to measure the power density at each of these frequencies separately, and since the sum of all these assorted RF sources would normally be extremely small, a safety factor of 10 was decided upon, and a safe exposure level of 0.01 W/cm² was established [78].

As Colonel Knauf first envisioned it in July 1957, the Tri-Service Program was to include a study of the frequency spectrum from 200 through 24 500 MHz, investigating effects of whole-body exposure on selected organs and tissues, single cells, and enzyme systems using various power levels, pulsed and continuous wave, under conditions of acute, subacute, and chronic exposure.

In summary, the program aimed at attaining the following objectives.

1) Determination of adequate mechanisms, techniques, instrumentation, and information for the protection of personnel from propagating energies when engaged in the development, test, and operational use of military electromagnetic radiation devices.

2) Establishment of valid safety criteria and instrumentation techniques for identifying electromagnetic hazard areas.

3) Formulation of recommendations concerning the means for recognizing the biological effects by medical surveillance.

4) Evaluation of means by which microwave energy may be put to use to serve other military needs.

Research proposals were developed to investigate whether there are effects other than thermal, whether the biological effects are dependent on the frequency of the microwave radiation, and whether there are regions in the frequency spectrum which can be expected to cause specific molecular resonance phenomena in structural elements of the body or possibly cause specific enzyme changes. There was also interest in whether there are "biological windows" in the frequency spectrum or regions of the spectrum which give a more pronounced effect than others or regions which give no effect at all. There was also the desire to discover whether the repetition frequency in pulsed radiation was of any importance from the biological point of view.

The frequencies chosen were 200, 1280, 2450, 2800, 3000, and 10 000 MHz to cover the equipment in use at the time. Investigation at 24 500 MHz was incorporated into the program because of its biological interest in relation to peak absorption.

Work was supported at the following institutions:

1) University of Pennsylvania: to investigate the penetration and thermal dissipation characteristics of microwave energy of 100 to 10 000 MHz in tissues by using phantom models of the human torso;

2) University of Miami: to obtain reliable and detailed information on the effects of 24 500 MHz on whole animals, cells, and enzymes using both CW and pulsed emissions;

3) University of California: to study effects at *X*-band frequencies (pulsed) on small laboratory animals and on cellular organisms as well as related microwave engineering aspects;

4) State University of Iowa: to investigate the effects of microwave radiation on the brain and hollow viscera of dogs [frequency 2450 MHz (CW)];

5) St. Louis University: to study the physiologic effects of 2450 MHz (CW);

6) University of Rochester: to a) detect and characterize damaging effects of acute and chronic nature in animals exposed to microwaves and b) determine the effect of combined radiation from X-ray similar to that produced by transmitter tubes and microwave sources [frequency 1280, 2800 MHz (pulsed)];

7) University of Buffalo: to study the effects of exposure to to 200 MHz (CW);

8) Tulane University: for the investigation of microwave irradiation on peripheral nerve systems to determine if any neural response can be isolated [frequency 3000 and 10 000 MHz (pulsed)];

9) Tufts University: to determine the effect of microwave radiation on the eyes of rabbits [frequency 2450 and 9100 MHz (CW and pulsed)];

10) New York University: to assess the incidence of lenticular imperfections in radar exposed personnel.

The First Annual Tri-Service Conference on Biological Hazards of Microwave Radiation[2] was held at Griffiss Air Force Base, Rome, N. Y., on July 15–16, 1957. "The very capable and energetic host was Colonel G. M. Knauf, who presided at the sessions and contributed to the success of the meeting by delivering two papers in addition to his appropriate commentaries and introductions [42]." The Second Annual Tri-Service Conference on Biologic Effects of Microwave Energy[3] was held at Griffiss Air Force Base, Rome, N. Y., on July 8–10, 1958. "Colonel Knauf was again the genial host and chairman in addition to presenting three significant papers [42]." The Third Annual Tri-Service

[2] E. G. Pattishall, Ed., *Proc. Tri-Service Conf. Biol. Hazards of Microwave Radiation* (George Washington Univ., Washington, D. C., July 15–16, 1957), ASTIA Doc. AD 11 5603.

[3] E. G. Pattishall and F. W. Banghart, Eds., *Proc. 2nd Annu. Tri-Service Conf. Biol. Effects of Microwave Energy* (University of Virginia, Charlottesville, July 8–10, 1958), ASTIA Doc. AD 131 477.

Conference[4] was held at the University of California, Berkeley, August 25–27, 1959. It should be noted that the data presented at this meeting substantiated the previous choice of 10 mW/cm² for the upper limit of microwave exposure. The Fourth (and last) Tri-Service Conference was held at the New York University Medical Center, August 16–18, 1960. The proceedings of this meeting have been published in book form.[5] Perhaps the singular most important contribution of the Tri-Service Program was the validation of the 10 mW/cm² safety standard.

The following excerpts from the opening remarks by Colonel Knauf at the Second Tri-Service Conference in July 1958 provide the historical perspective of the Tri-Service Program and give eloquent testimony to the foresight of this man and his associates.

> When this program was first conceived at Rome, we limited our thinking to the problem of hazards that might result from exposure to this energy. . . . When we looked at the whole research effort in its new setting it became obvious that it could not any longer be properly called a hazards investigation. We were truly looking at the broad spectrum of possible microwave effects with a view toward being ready should our development people be successful in coming up with the higher power radar they were dreaming about. We dropped the word "hazards" from the title of our effort and replaced it with the word "effects". . . . We first are interested in any possible effect on man which might be produced by the power produced by radiating equipment in use today. Then next we want to get some idea what lies ahead biologically if we continue to increase the power output of our equipment. . . . I think this might be a good time to say that up to date there has not been any effect produced or even hinted at at power levels which remotely approach our established maximum safe exposure level. An effect does not become a hazard until man is exposed to such an effect. . . . It is sufficient to say that the power of this proposed equipment is much greater than anything we have dealt with before. The problem of accidental exposure to this higher power becomes a possibility. The need for protective clothing to cope with certain operating and maintenance problems appears inevitable as does the need for more attention to shielding for buildings and passageways in the operational areas. . . . It is with the thought that we must be ready with answers for the questions that will certainly be asked, that we are going ahead with this forward looking research effort. . . .

A decade has now elapsed since the last Tri-Service Conference was held. The Tri-Service Program and the Tri-Service Conferences served as a landmark in investigations and dissemination of data concerning the biologic effects of microwaves and contributed immensely to our understanding of these effects. After the demise of the Tri-Service Program, coordinated research into the biologic effects of microwaves came to a virtual standstill with very little active research going on in this area in the United States.

The Tri-Service Program

A. Electrical Properties of Living Tissues and Absorption Characteristics of Microwaves

Schwan and his associates at the University of Pennsylvania continued and expanded on their earlier studies of the electrical properties of living tissue and absorption characteristics of microwaves in the body.

All these studies were carried out over a broad frequency spectrum, in some instances from 100 to 10 000 MHz; in others over the total range of radio and radar frequencies from 1 to 10 000 MHz (Table I).

These studies were essential 1) to determine what percentage of airborne radiation is absorbed by the human head, eye, and total body; 2) to judge how deep the radiation penetrates; and 3) to determine how and where the radiant energy is transformed into heat and establishes a major thermal load. Possession of such knowledge makes it possible to state whether head exposure establishes a more serious hazard than the one predicted for the case of "total body" irradiation. It was, furthermore, anticipated that it would be possible to state whether and what frequencies are particularly dangerous for the human eye [61].

This work has been the subject of numerous papers including [4], [5], [56]–[58], [60], [61], [63], [64], and [70].

As a result of these studies, it was determined that on the basis of electrical properties, the tissues of living organisms can be divided into three groups according to their water content: suspensions of cells and protein molecules of liquid consistency (blood, lymph), similar suspensions in a condensed state (muscle, skin, liver, etc.), and tissues with a low water content (fat, bone).

At frequencies of the order of several megahertz, the mechanism of dispersion, due to cell-wall relaxation, is supplemented by the comparatively weaker effects of the relaxation of dipole protein molecules (characteristic frequencies close to 10 MHz) and the structural relaxation of subcellular components [62].

It has been suggested [59], [62] that in the frequency range of 100 MHz to 1 GHz, the electric properties of tissues with a high water content are due to an electrolytic medium containing suspended protein molecules with a lower dielectric constant.

Theoretical investigations by Saito and Schwan [56] indicated that pearl-chain formation is due to the attraction between particles in which dipole changes are induced by the EMF.

[4] C. Susskind, Ed., *Proc. 3rd Annu. Tri-Service Conf. Biol. Hazards of Microwave Radiating Equipments* (University of California, Berkeley, Aug. 25–27, 1959).

[5] M. F. Peyton, Ed., *Proc. 4th Annu. Tri-Service Conf. Biol. Effects of Microwave Radiating Equipments.* New York: Plenum Press, 1961; RADC-TR-60-180.

TABLE I

BIOLOGICAL CHARACTERISTICS OF MICROWAVES

<table>
<tr><th rowspan="2">Frequency (MHz)</th><th rowspan="2">Wavelength (cm)</th><th rowspan="2">Energy (eV—approximate)</th><th colspan="3">Relative Absorption Cross Section Percent[a]</th><th rowspan="2">Tolerance Levels (incident power density)</th></tr>
<tr><th>Fat Thickness (1–3 cm)</th><th>Skin Thickness (0.2–0.4 cm)</th><th>Body</th></tr>
<tr><td rowspan="2">400</td><td rowspan="2">75</td><td rowspan="2">10^{-7}</td><td colspan="2" rowspan="2">43–73</td><td rowspan="3">50–100</td><td rowspan="4">0.01 W/cm²</td></tr>
<tr></tr>
<tr><td rowspan="2">900</td><td rowspan="2">33</td><td rowspan="2">10^{-6}</td><td colspan="2" rowspan="2">51–89</td></tr>
<tr><td rowspan="2">20–100</td></tr>
<tr><td rowspan="2">3000</td><td rowspan="2">10</td><td rowspan="2">10^{-5}</td><td colspan="2" rowspan="2">18–50</td><td rowspan="4">0.02 W/cm²</td></tr>
<tr><td rowspan="3">50</td></tr>
<tr><td rowspan="2">10 000</td><td rowspan="2">3</td><td rowspan="2">10^{-4}</td><td colspan="2" rowspan="2">41–50</td></tr>
<tr></tr>
</table>

[a] Adapted from [4].

"Dielectric saturation" in solutions of proteins and other biological macromolecules due to intense EMFs of superhigh frequencies was considered by Schwan [61]. He suggests that such fields cause all the polarized side chains of the macromolecules to become oriented in the direction of the electric lines of force and that this can lead to breakage of hydrogen bonds and other secondary intra- and intermolecular bonds and to alteration of the hydration zone (on which the solubility of molecules depends).

B. 24 000 MHz

The investigations into the effects of 24 000 MHz by Deichmann and his associates at the Uuiversity of Miami School of Medicine are noteworthy; they constitute the only multispecies broad-scale studies available at this frequency. These studies have been reviewed in several publications [11]–[18], [20], [21].

These investigations were performed with a pulsed-modulated magnetron operated at 24 000 MHz (1.25-cm wavelength) with an average power output of 20 W. The studies were designed to determine the biological effects of this type of radiation, utilizing various experimental animals.

Exposure of rats, mice, and chicks indicated that there is an inverse relationship between power density and lethal exposure time (rat and mouse, prone position, total body exposure). For the following power densities (W/cm²), the minimum lethal exposure periods were as follows: (0.15) rats 35 min, mice 5 min; (0.08) rats 56 min, mice 13 min; (0.05) rats 80 min, mice 35 min; (0.03) rats 135 min, mice 140 min.

The local effects produced in the rat were in relation to the severity of exposure and ranged from hyperemia to first, second, or third degree burns.

When compared with continuous exposure (0.30 W/cm²), which killed a rat in 15 min, intermittent exposure to 50 percent of this energy per unit of time over a period of 31 min (generator 1 min on, 1 min off) killed in 16 min of actual exposure time, while 17 percent of the above microwave energy (1 min on, 5 min off) killed in 34 min of actual exposure. These studies suggest that microwave energy might induce subacute or chronic effects possibly for no other reason than prolonged mild hyperpyrexia [13].

The environmental temperature was found to influence the systemic effects of microwave radiation. The period of survival of a rat was more than doubled (from 17.4 to 37.0 min) by a drop in environmental temperature from 35°C (95°F) to 15°C (59°F). A most remarkable prolongation of life was brought about by an effective air volume exchange. Rats exposed continuously to 250 mW/cm² at 15°C lived for 47 min (40 to 63.5 min). Rats exposed similarly but aided in losing microwave-induced heat by air (15°C) from a blower survived for eight to 24 h.

Osborne–Mendel, CFN, and Fischer strain rats showed evidence of leukopenia, lymphopenia, and neutrophilia after 7 h of continuous exposure to 20 mW/cm² with recovery in 1 week. Hematologic recovery occurred in 2 days after 10 min of continuous exposure to 20 mW/cm², or 3 h of continuous exposure to 10 mW/cm². Effects on erythrocytes, hemoglobin, and hematocrit differed in the three strains used. In Osborne–Mendel and CFN rats all values increased; in Fischer rats they decreased [16].

In an ancillary study an attempt was made to determine whether exposure of the testes of rats to microwaves of 1.25-cm wavelength would produce a functional disturbance in the male endocrine system, even though there might be no morphological evidence of damage to the testes [20], [21]. Rat testes were exposed to 250 mW/cm² for 5, 10, and 15 min. This investigation indicated that output of androgen may have been affected.

Deichmann [12] suggests that microwave radiation effects are not directly related only to the level of power density. There are a number of factors involved in determining radiation effects of a certain dose measured in mW/cm². These include the following.

1) Wave length of frequency: absorption increases with reduction in wave length.

2) Time: absorption increases as the duration of exposure is increased.

3) Radiation cycle rate: per unit of time (1 min), the periods of exposure and nonexposure are important, even when the total exposure for this unit of time (1 min) is kept constant.

4) Air motion: with an increase in the velocity of air movement the effects of radiation decrease.

5) Humidity: radiation effects increase with an increase in the humidity.

6) Clothing: in general, radiation effects are more marked in those parts of the body covered by clothing. Certain types of clothing offer protection, acting as a shield or reflector.

7) Environmental temperature: Radiation effects increase with an increase in environmental temperature.

8) Body tissues: injury from a certain dose of radiation can be expected, particularly in organs or tissues with poor blood supply (eyes).

9) Body weight, type, or mass.

10) Position in the field: variables exist because of an exposed subject's position or orientation in the radar beam.

C. 10 000 MHz

At the University of California, Berkeley, investigations were related to the effects of 10 000-MHz microwave radiation on laboratory animals and cellular organisms, the principles and instrumentation problems relating to dosimetry, the hazards to personnel in the vicinity of radiators, the employment of electromagnetic radiation as a tool for biophysical research, and the measurement of the dielectric properties of biological solutions by cavity-perturbation and other methods. This group also studied the effect of intermittent chronic irradiation on the pathophysiology and longevity of the mouse.

Temperature regulation was studied in mice and rats. The trials were carried out partly as continuous long-term irradiation with a low intensity, and partly in the form of intermittent irradiation with intensities varying between 0.009 and 0.438 W/cm^2. The rise in body temperature reached a constant value below the temperature at which death took place when mice were irradiated with an intensity of less than 0.060 W/cm^2. The constant temperature level is a function of the intensity. A constant raised temperature level did not usually occur during exposure of the animals to low intensities, but the body temperature presented a fluctuating curve with rises and falls of short duration. The same phenomenon appeared in normal rats but not in hypophysectomized rats.

The factors which determine the rate of cooling following heating of an experimental animal with microwaves were also studied. The intensity of the radiation in these trials with mice varied between 0.156 and 0.438 W/cm^2. The average rise in temperature was 5°C above the normal body temperature. The rate of cooling was the same in all cases. Irradiation with an intensity of 0.04 W/cm^2 gave a "steady-state" rise in temperature of 2.3°C. A study was also made of the dependence of the effect of previous irradiation on subsequent exposure to microwaves. The initial exposure seems to stimulate the cooling mechanism of the body so that it functions more effectively during the subsequent exposure [51], [71], [72].

In the longevity study, 200 male Swiss Albino mice were exposed to 0.109 W/cm^2 for 4.5 min (LD_{50}=11.5 min), producing an average body temperature rise of 3.5°C. The animals were irradiated, 10 at a time, 5 times a week for 59 weeks. Longevity did not appear to be affected by irradiation at this level [52], [73].

Of still more long-term significance were the investigations made at Berkeley on cellular and molecular samples. The first doctoral dissertation to result from the program [81] was concerned with the dielectric properties of water and their role in enzyme–substrate interactions, and was to point the way to extensive bioengineering studies. From the instrumentation viewpoint, perhaps the most important contribution was the development of a cavity-perturbation method of making quick determinations of both real and imaginary components of the dielectric constants of proteins and other lossy materials containing water [74].

D. 3000 MHz

Investigators at Tulane University studied the effect of microwave exposure on growth rate of mice. Preliminary experiments [44] had suggested early decrease followed by increase in growth rate. In later studies, however, in which larger numbers of mice and improved statistical analyses were utilized, it was shown that the growth pattern in mice exposed to 10-cm microwaves, 10 mW/cm^2, 2 min/h each day for 36 days, was no different from that of the controls [6].

E. 2800 MHz

At the University of Rochester, Howland and his associates performed studies to recognize the general biologic responses caused by irradiation from a 2800-MHz pulsed radar unit. This provided baseline data for comparison with studies at other frequencies. Animal response following 1240-MHz pulsed and 200-MHz (CW) exposure was later investigated. The scope of the investigation was extended to include the interaction of microwave and ionizing radiation in mammals.

The results of these studies have been described in several reports and publications: [22], [23], [33], [35]–[41], [68], [69], [75]–[77].

These investigations revealed that animals exposed to microwaves of high power density experience thermal stress. As long as the thermal burden is not excessive, the animal can cope with it. Once it exceeds this level, homeokinetic capability is impaired, and difficulty in

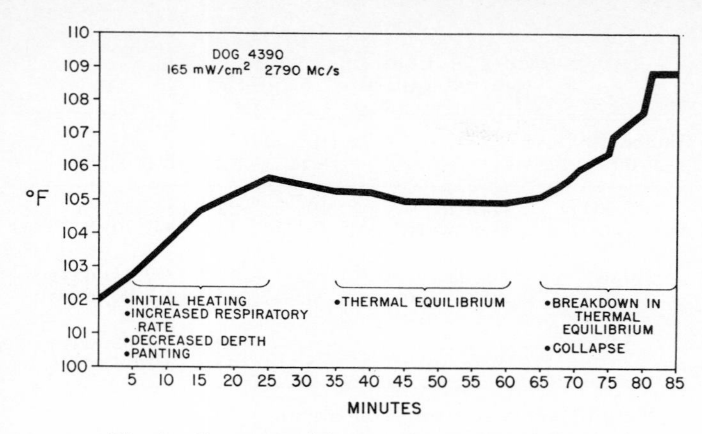

Fig. 1. Response of dog to microwave exposure.

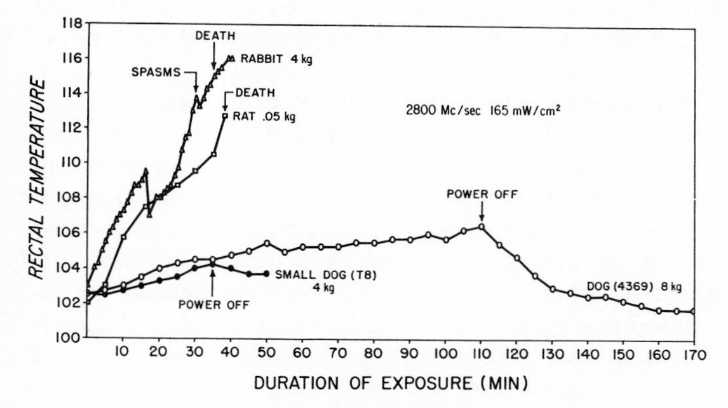

Fig. 2. Thermal response of various species of animals exposed to microwaves.

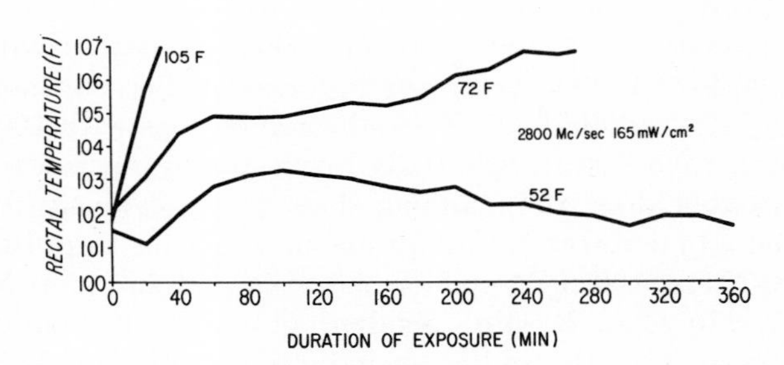

Fig. 3. Response of dogs exposed to microwaves at various environmental temperatures.

maintaining normal function occurs. Superficial or deep burns in various portions of the body, particularly over the thoracic cage, may develop in some dogs.

The thermal response in the dog exposed to 165 mW/cm², 2800-MHz pulsed microwaves at 30-percent humidity consists of three distinct phases. In phase 1, the initial thermal response, body temperature increases by 2–3°F 1/2 h after onset of exposure. In phase 2, the period of thermal equilibrium, rectal temperature stabilizes. This may last 1 h, during which the temperature will cycle between 105° and 106°F. In phase 3, the period of thermal breakdown, the temperature rises above 106°F, and continues increasing rapidly until a critical

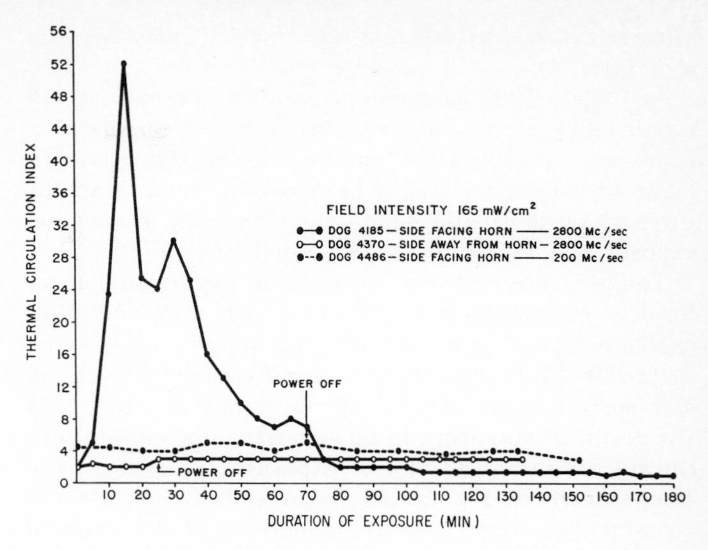

Fig. 4. Thermal circulation index for dogs exposed to microwaves.

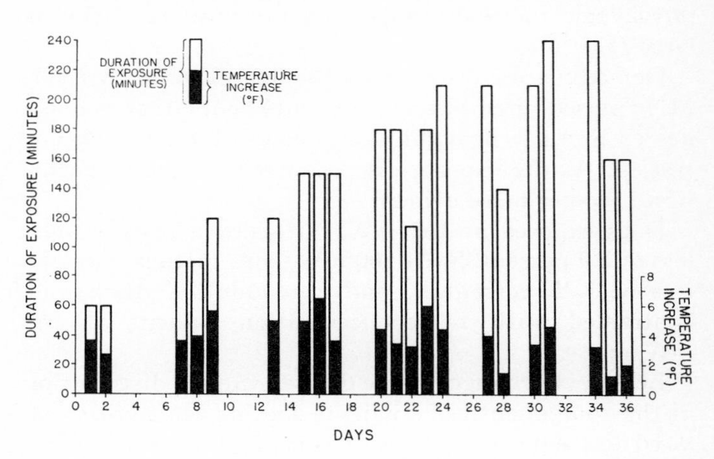

Fig. 5. Development of microwave tolerance at 2800 MHz, 165 mW/cm².

temperature of 107°F, or greater, is reached. If exposure is not stopped, death will occur (Fig. 1).

A critical rectal temperature with no equilibration is reached in 10 min in the rabbit and 20 min in the rat when exposed at 165 mW/cm². The influence of body size on the thermal response at this frequency is negligible. Fox terriers weighing 4 kg, which is equivalent to the weight of the rabbit, respond the same as medium-sized dogs ranging in weight from 8 to 20 kg (Fig. 2).

Exposure of dogs at 100 mW/cm² for periods up to 6 h does not cause a critical rectal temperature. Initial heating is slight, the animal remaining in thermal equilibrium during the remainder of exposure.

Exposure of rabbits at 165 mW/cm² produces an extremely violent reaction. Within 5 min, desperate attempts are made to escape from the cage. Peripheral engorgement of all vessels yields an acrocyanotic picture. Forty min of exposure results in death. When rabbits are exposed at 100 mW/cm² for 1 h, they become prostrate.

At 103.5–106°F, 20-percent humidity, thermal regulation is adequate to maintain normal body temperature. Exposure to 2800-MHz microwaves at this environmental temperature results in increased rectal temperature, which is greater than that seen with microwave exposures at 72°F, 30-percent humidity (Fig. 3).

Temperature recovery assumes an exponential form for dogs exposed to 165 mW/cm² or 100 mW/cm². Pre-exposure level is reached within an hour.

At 2800 MHz the temperature difference between the skin surface of the side facing the radiating source and the rectal temperature in anesthetized dogs diminishes during the initial 30 min of exposure. Thereafter, the temperature of the exposed skin stabilizes while the rectal temperature continues to increase. On the exposed side, the physiologic thermal gradient (difference between skin and rectal temperature) is small while the physiologic thermal gradient on the unexposed side is large (Fig. 4).

Hemodynamic response of the dog exposed to 2800-MHz pulsed resembles that of acute heat stress as manifested by early hemodilution followed by hemoconcentration. As the exposure is prolonged, hemoconcentration becomes more evident.

Dogs exposed at 165 mW/cm² show a body weight loss of 2.0 percent/h. At 100 mW/cm², there is a weight loss of 1.25 percent/h, and hemodilution occurs, as contrasted with hemoconcentration evident at 165 mW/cm².

Leukocyte changes in the dog reflected in distribution of the component cells indicate specific sensitivities related to frequency, power density, and duration of exposure. Lymphocytes and eosinophiles are decreased after 6 h of exposure at 100 mW/cm², 2800 MHz. Total leukocytes and neutrophils are slightly increased at 24 h.

After 2 h of exposure to 165 mW/cm², the total leukocyte count is slightly decreased. When the exposure is prolonged to 3 h, leukocyte increase is observed, which is more pronounced at 24 h. Lymphocyte and eosinophil changes are variable but are decreased immediately or 24 h after exposure.

Dogs anesthetized with sodium pentobarbital or premedicated with chlorpromazine or morphine are more sensitive to the thermal effects of 2800-MHz microwaves than untreated animals.

Repeated exposure to 165 mW/cm², 2800 MHz results in physiologic adaptation as evidenced by the ability to prolong exposure time successively with minimal increase in temperature and progressive depression of "basal" temperature. No overt cumulative effect of microwaves is demonstrated if maintenance of adequate thermal regulation is used as the criterion. Intermittent exposure of the normal animal can be tolerated for extended periods of time and is related to the interval between exposures, which permits the animal time to recover (Fig. 5).

TABLE II

SURVIVAL TIME FOR DOGS THAT RECEIVED IONIZING RADIATION TO THE HEAD

Midline Air Dose (r)		Number of Dogs	Survival Time
25 000	normal	10	22.0 (±1.37)[a] h
	microwave	10	43.1 (±6.70) h
10 000	normal	5	13.8 (±0.73) days
	microwave	5	15.4 (±1.03) days
5 000	normal	5	40.6 (±4.68) days
	microwave	5	43.6 (±15.4) days

[a] Mean (± standard error of the mean).

All animals returned to normal after exposure. Periodic physical and eye examination by slit lamp and ophthalmoscope up to 1 year after exposure failed to reveal any physiologic decrements or lenticular changes.

Latent residual effects of ionizing radiation were revealed by microwave exposure of dogs that had survived ionizing irradiation. This was evidenced by increased susceptibility to the microwave induced hyperthermia. This response indicates a lowered threshold to the stress of microwaves among ionizing radiation survivors and/or inefficient physiologic function in the X-irradiated animal. There was a significant prolongation in survival time among dogs that previously received whole-body microwave exposure when irradiated with 25 000 R of 1000 kVp X-rays to the head. Percent mortality after whole-body X-irradiation of dogs with 1000 kVp X-rays was substantially lower among microwave pretreated dogs than among dogs not previously exposed to microwaves. Simultaneous exposure to pulsed 2800-MHz, 100 mW/cm² microwaves and 250 kVp X-rays, 720 R (2 R/min) resulted in earlier leukocytic recovery from the radiation induced leukopenia. Such enhanced recovery was not observed in comparable exposures at the higher rate of 4.6 R/min (Table II, Fig. 6).

F. 2450 MHz

Investigators at Iowa State University studied the effects of 2450 MHz (CW) microwaves on various biologic systems [24], [67]. One study was to determine whether any functional changes occur in dogs after chronic exposure of the head to 2450 MHz (CW). While the primary purpose was to observe the animals for functional disturbances associated with the exposure to microwaves, an index of temperatures reached at various points in the head and body was also determined.

These studies revealed that temperatures in the cisterna magna, midbrain, frontal lobe, fourth ventricle,

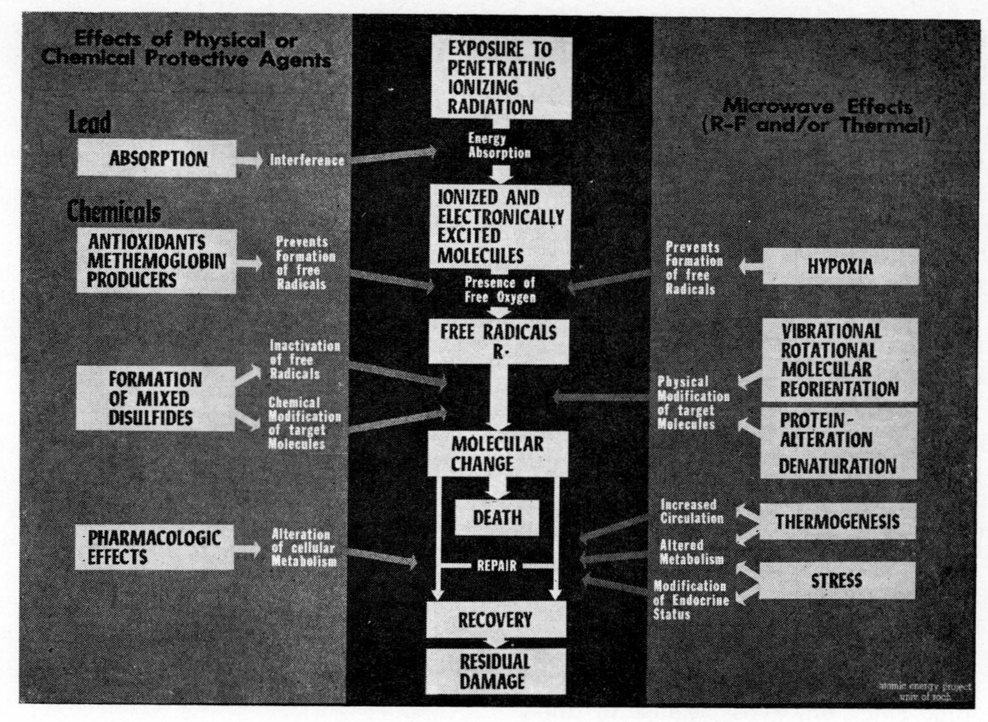

Fig. 6. Modification of sensitivity to ionizing radiation.

and rectum were significantly raised by microwave irradiation of the top of the head in anesthetized dogs. The generator (125-W output) was varied so that the skin damage threshold (42°C) was approached but not exceeded. At no time did the calculated power density exceed 0.8 W/cm² in the center of the field. No gross damage to the brain or its membranes and no clinical evidence of neurologic involvement resulted from either single or repeated 22 periods of exposure. In unanesthetized dogs subjected to a similar pattern of irradiation, the rectal temperature did not rise significantly throughout the exposure, but a panting response was exhibited.

Temperatures were measured in the skin and certain visceral tissues of larger dogs, smaller dogs, and rabbits during two different patterns of microwave exposure. With both of these procedures the amount of temperature developed in the visceral tissues during the exposure was inversely related to animal size, being greatest in the rabbits, next highest in the smaller dogs, and lowest in the larger dogs. It appeared, therefore, that the amount of temperature developed in the visceral organs during microwave irradiation might be a function of animal size as well as duration and intensity of radiation.

Larvae of *Drosophila melanogaster* were exposed to a power density as high as approximately 1.0 W/cm² and induced-heat dissipated. Growth rates during irradiation were not significantly different from control values. When heat was not dissipated, a marked mortality resulted. The results seemed to indicate the absence of nonthermal effects upon the organism.

Richardson [53][6] studied the effect of exposure to 2450-MHz (CW) microwave radiation on body weight in rats. With repeated exposures the animals lost weight immediately after irradiation for a short time, but during periods between exposures they gained weight more rapidly reaching or even surpassing the control level. Lethality occurred in rats exposed to 70 mW/cm² for 15 min.

Richardson [53], [54], was also interested in developing a suitable dosimeter for microwave exposure.

It is of interest to note that Richardson and his associates [55] performed some of the earliest studies on experimental microwave cataractogenesis. Since these studies were not part of the Tri-Service Program, they will not be reviewed in this paper.

The effect of microwave exposure on another biological system characterized by cell differentiation and proliferation was investigated by Van Ummersen [79].[7] All experiments were carried out on chick embryos which had been incubated at 39°C to approximately the 48-h stage of development and exposed to 200–280 mW/cm².

The radiation caused a significant temperature increase in the egg. Abnormalities were produced which included inhibition or retardation of differentiation in the eye, brain, and heart, as well as suppression of development in the posterior regions of the embryos. The allantois failed to develop.

Carpenter [8] suggests two possibilities for microwave

[6] St. Louis University.
[7] Tufts University.

action which should be considered in this investigation: "(1) it may act directly on the differentiating cells, which because of special conditions of their metabolism may be peculiarly susceptible to the radiation; (2) it may affect the environment of the cells and thus alter the organic milieu on which their growth and development depend. In the growing chick embryo, for example, it might alter or perhaps inactivate one or more of the enzymes responsible for metabolism of yolk, the only available source of the energy required for the processes of growth and differentiation."

Since protein denaturation takes place in the neighborhood of 40°C, anything which elevates the temperature to this region could start degeneration of cell structures. The high power densities that were used in this study might be sufficient to cause protein denaturation on a thermal basis. Osborne [47], [48] found that at no stage of development was the chick embryo affected by long exposure to 200-MHz microwaves. In these experiments the temperature rise within the egg was never more than 1°C. One cannot, however, compare 200-MHz with 2450-MHz microwaves in their effects on developing chick embryos because of differences in wavelength and resultant energy absorption by an object the size of a hen's egg.

G. 1240 MHz[8]

Rectal temperature response in dogs exposed to 1240 MHz was comparable to those exposed to 2800 MHz, at similar microwave power densities. The reactions of the animal varied little from the preexposure state until critical rectal temperature level was approached, at which time a moderate increase in panting was noted. The animal appeared comfortable, or slightly agitated. Salivation was normal or slightly increased [40].

While exposed to 1240 MHz, 100 mW/cm², dogs were more relaxed than when exposed to 2800 MHz. Manifestations of a critical rectal temperature (106°F or greater), which were readily apparent at 2800 MHz, were negligible at 1240 MHz. Agitation of the animal was rarely seen. Weakness of the hind quarter was not prominent during exposure but was readily apparent when the animal was removed from the exposure compartment. Panting during exposure at 1240 MHz was less marked than at 2800 MHz.

Daily exposures at 1240 MHz for 6 h/day, 5 days/week, for 4 weeks revealed that at 100 mW/cm² there was an increase in rectal temperature during each exposure for the first week. During the subsequent three weeks, temperature increases were moderate. A progressive lowering of preexposure temperature was evident as the number of exposures was increased. Exposure to 50 mW/cm² initially resulted in slight temperature increase. Progressive lowering of the preexposure temperatures was noted as the exposures were repeated. Exposure to 20 mW/cm² resulted in slight temperature decrease. When compared with sham-exposed dogs, evidence of a thermal regulatory response to the procedure was apparent as the nonirradiated dogs showed a greater decrease in body temperature.

[8] Studies performed by the University of Rochester group.

H. 200 MHz

The uniqueness of these studies performed by Osborne and associates at the University of Buffalo is that they are the only ones of a broad scope performed at this frequency [1]–[3], [47], [48].

Addington *et al.* [2] studied the importance of the position of the body in the microwave field for the occurrence of harmful effects from primary heating. It was found in trials with dogs that the rate and degree of temperature increase was greater when the body was positioned so that its longitudinal axis was parallel to the polarization plane of the microwave field. It was also found that qualititave differences as regards to the form of the head, length of tail, etc., in the animal seemed to be factors in the thermal response.

An approximate median lethal dose for dogs was obtained with 15-min exposure at 330 mW/cm² at perpendicular orientation [3]. Markedly lower lethal rates were produced with slightly longer duration irradiation at 220 and 194 mW/cm², and no deaths resulted from exposures of $\frac{1}{2}$- to 1-h duration at power densities of 165 mW/cm² and less (all at perpendicular polarization). All of the dogs that died had rectal temperature rises of 7.5°F or more. Because of the small number of animals used (37) the data have little statistical significance; however, they give some idea of the range of lethality for 200 MHz.

Studies were also carried out with various lower animals and plants such as protozoa (amoeba, paramecium) worms, flies, bacteria, fungi, etc., with power densities of under 0.04–0.05 W/cm² to find an organism to serve as a standard in microwave irradiation. It was not possible, however, to find an organism which would give a response so unequivocal and reproducible that it could be used for this purpose.

Using the same equipment, the University of Rochester investigators found that at 200 MHz, 165 mW/cm², the dog equilibrates later than at 2800 MHz and remains in thermal equilibrium for a longer period before breakdown. Body size of the animal does not influence response. Equilibration is minimal in the rabbit, and critical rectal temperature occurs within 30 min. Rats exposed for 1 h do not show evidence of thermal breakdown. A dog exposed at 200 MHz, 165 mW/cm² showed little change from its initial thermal circulation index. This difference in thermal circulation index from that which was found with 2800-MHz exposure reflects the difference in depth of microwave energy absorption between 200 MHz and 2800 MHz [40] (Fig. 4).

I. Neural Effects

The group at Tulane University studied the effects of 1.25-, 3.0-, and 10-cm microwave irradiation on the nervous system of both warm- and cold-blooded animals.[9] Experiments were designed to separate thermal from nonthermal neural effects. Refrigerated cold-blooded animals were irradiated with 3-cm microwaves at a power level of 45 mW/cm^2 for various lengths of time. In addition, experiments at the same wavelength and power density were conducted on isolated nerve preparations.

Results from the peripheral nerve experiments indicate that previous reports of neural effects may be explainable as effects due to local heating of peripheral nerves rather than to excitation of the central nervous system [28]. The principal effect observed (viz., increased activity of the experimental animal) was found to be temperature dependent. The most important result of these experiments was that every observed effect of microwaves on isolated nerves can be duplicated by use of convective heat or infrared radiation.

In experiments on steady potentials of the central nervous system (CNS), concurrent rectal temperature measurements indicated a relationship between temperature and the steady potentials. In these experiments measurements were included to develop information on the relation between power and steady-potential measurements in the CNS. Network theory applied to the analysis of measured data provided results reaffirming the reported nonlinear characteristics of neural tissue [19], [43], [45].

The propagation properties of electromagnetic waves in brain material and surrounding tissues described by Nieset *et al.* [46] and Fleming *et al.* [19] showed large values of attenuation for the electric field in the brain and its surrounding tissues of skin, fat, muscle, and bone. Their results indicate a maximum penetration into the brain at about 1000 MHz, with a frequency band of greatest penetration from 300 MHz to 1500 MHz. Much of the incident power is reflected at the lower frequencies while the propagating power is more readily absorbed by the various tissues at higher frequencies.

Nieset *et al.* [46] showed by means of irradiating isolated frog's skin that exposure for 10 to 20 min to 3-cm microwave radiation (0.042 W/cm^2) usually greatly reduced ion transport. It was later found, however, that raising the skin temperature to 38°C usually blocked Na^+ transport.

Pinneo [49] studied direct current potentials of the CNS. The normal brain of the mammal is electrically polarized. Two forms of this steady potential are evident: a "resting" potential present in the absence of stimuli and a second potential that is apparently due to incoming sensory volleys. The importance of these potentials is emphasized by the dc shifts associated with external and locally applied stimulation; these shifts are also intrinsically linked with such phenomena as paroxysmal seizures, spreading depression, and cortical–reticular formation "arousal."

[9] Portions of these studies were also supported by the Veterans Administration Hospital, Radioisotopes Research, New Orleans, La.

Pinneo *et al.* [50] point out that while peripheral nerves may be thermally activated by absorption of electromagnetic radiation in the 10- to 25-GHz frequency band, CNS structures and systems are effectively shielded at these frequencies by skin, muscle tissue, and skull. Thus, they suggest that frequencies in the range of 400 MHz to 2 GHz be used in the investigation of the effects of microwave energy on neural function.

McAfee [28], [31], using pulsed 3-cm microwave radiation with 1000 pulses per s and an intensity of about 200 mW/cm^2, demonstrated the occurrence of a nociceptive reflex (i.e., a reflex which occasions a movement in the animal to shield or defend itself against harmful influence) in decerebrate cats during microwave radiation. This reflex comprises a sharp rise in blood pressure and pulse frequency, a change in the respiration rate and depth, and movements of the extremities and of the body in decerebrate cats. These symptoms occur when an exposed sciatic, radial, facial, or trigeminus nerve reaches a temperature between 42° and 47°C.

The studies by McAfee [28]–[32] provide convincing evidence that the presumed nonthermal effect of microwaves on the CNS is a result of thermal stimulation of peripheral nervous structures. Studies claiming CNS effects of microwaves should at least include controls for the possible peripheral nervous system effects described by McAfee.

J. Cataracts

Carpenter and his associates at Tufts University carried out some of the most extensive and detailed studies made on the effects of microwaves on the rabbit lens. The radiation source operated at 2450 MHz, CW or pulsed. Except at certain low powers, exposure of the animals was carried out under sodium pentobarbital anesthesia. Measurements were made of the intraocular temperature during exposure. Following irradiation, the eyes were examined regularly by ophthalmoscope and by slit-lamp microscope. In many cases, changes in the appearance of a lens opacity were recorded by stereo-color photography [7], [9].

Carpenter was also interested in comparing the cataractogenic effect of 10 000 MHz [10]. The method of exposure permitted transmission of microwave power directly into an eye which, together with a thin brass iris, formed a waveguide termination.

Lens changes which were observed by the slit-lamp microscope in 136 cases form the basis for Carpenter's threshold curve (Fig. 7). The latency for opacity development was 1 to 6 days after the exposure; the average latent period was 3.5 days [9], [10].

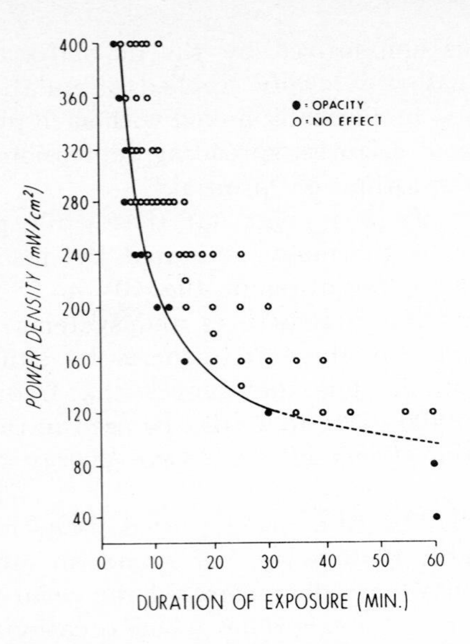

Fig. 7. Time and power thresholds for induction of lens opacities in the rabbit eye from single exposure at 2450 MHz [7], [10].

In further investigations, Carpenter obtained data which he considers as evidence of nonthermal and a cumulative effect of microwaves [7], [9], [10].

At 280 mW/cm² power density the minimal exposure period that caused an opacity was 5 min, by which time the temperature of the vitreous body had risen to 49.3° C. At this power density there was no observable effect on the lens from a 3-min exposure, which raised the vitreous temperature to 47.2°C. However, when the power density was reduced to 120 mW/cm², a 35-min exposure period induced lens opacities while the vitreous temperature was increased to 44°C.

It was also found that rabbits' eyes exposed to 140 mW/cm² with 2450 MHz (CW) for 20 min did not develop opacities, but that rabbits' eyes exposed to 140 mW/cm² pulsed power for 20 min did develop opacities.

Having established threshold cataractogenic exposure periods for the induction of opacities, Carpenter tested the effect upon the lens of repeated exposures of briefer duration than the minimal exposure required for causing an opacity to form. It was reasoned that this procedure should cause less temperature change in the eye than is associated with an exposure at the cataractogenic threshold, and that if any cumulative effect were to occur, it would suggest the operation of some influence other than a thermal one.

It was found that a 4-day interval was not long enough to permit the lens to recover from the damage inflicted upon it by radiation, but that a 7-day interval was sufficiently long for such recovery to take place.

Carpenter has suggested that the cataractogenic effect of radiation may involve "initiation of a chain of events in the lens, the visible and end results of which is an opacity, and that this chain of events must be initiated by an adequate power density acting for a sufficient duration of time if it is to progress to the development of an opacity. If either the power density or the duration of the irradiation are below a certain threshold value, the damage done to the lens is not irreparable and recovery can occur, provided that sufficient time elapses before a second similar episode [9]."

These results do not necessarily indicate a nonthermal cumulative effect. Acute injury of the lens leads first to hydration, and this is reversible providing no lens protein denaturation has taken place despite the fact that banding, striations, and opacification are evident. Hydration of lens fibers may last for many days. If the excess water leaves the lens before denaturation has occurred, no permanent residua results. If another thermal injury intervenes, however, at a time when the lens is partially damaged, there may be a summation of effect [83].

Most investigators agree that there is a critical intraocular temperature which must be reached before opacities develop. This temperature, as reported by various authors, ranges from 45° to 55°C. Obviously, no cumulative rise in temperature can occur if the intervals between exposures exceed the time required for the tissue to return to normal temperature. The cumulative effect to be anticipated, therefore, is the accumulation of damage resulting from repeated exposures, each of which is individually capable of producing some degree of damage [25].

Using 10 000 MHz, eyes were exposed in a "closed system" at five levels of delivered power, from 150 mW/cm² to 1090 mW/cm². At each power exposure times were varied in order to determine thresholds for the induction of lens opacities. Although this threshold curve was similar in shape to that previously shown for 2450 MHz, the power values could not be compared because of their entirely different derivations [10].

The opacities which developed were always in the anterior cortex region. In this respect they were similar to cataracts caused by infrared radiation, which likewise appear typically in the anterior cortex.

Although damage to the lens was not apparent until the second or third day following irradiation, a granular opacity was commonly seen in the cornea in 24 h, but it usually disappeared within four days. Similar results were obtained with 8000-MHz exposure. Exposure at 10 000 MHz in the far field resulted in opacity in the posterior subcapsular cortex of the lens similar to cataracts which resulted when the eye was subjected to 2450-MHz radiation from an antenna in an anechoic chamber. It seemed quite clear, therefore, that in the experiments done at 10 000 MHz, with the "closed" waveguide system, the anterior location of the cataracts was due not to the particular frequency but to the method by which the power was transmitted to the eye [10].

Attempts at a comparison of the cataractogenic relationship of pulsed versus CW radiation were inconclusive. Carpenter and Van Ummersen [10] suggested that it would be unwise to ignore the factor pf peak power in assessing pulsed radiation as an ocular hazard.

Van Ummersen and Cogan [80] showed that the age of the animal was not a factor in susceptibility of the lens to damage by microwave power.

In other studies, attempts were made to determine the biochemical changes in lenses of rabbits exposed to 2450 MHz, 280 mW/cm². The duration of exposure was varied depending on the severity of opacification desired [34], [26].

The results of these experiments indicated that the factors which influence the cation distribution of the lens were not particularly sensitive to microwave irradiation. A number of other constituents of the lens were also equally unaffected. The levels of protein, thiol groups, ammonia, glucose, and lactate fell only after an opacity had developed. These results suggested that the first sign of damage of the lens by microwave radiation was the formation of opacities rather than any change in the chemical constituents.

Ascorbic acid and glutathione in the lens appeared to be more sensitive to microwave radiation than any of the other chemical constituents studied. The investigators suggest that ascorbic acid is the most sensitive chemical constituent to be affected by microwave irradiation of the lens.

The University of Rochester group also studied microwave-induced cataractogenesis [40]. In unanesthetized rabbits exposed in the far field to 2800 MHz (CW), lenticular changes in the form of rapid and complete opacification developed after exposure to 220–240 mW/cm² for 1 h. Transient tumescence of the lens fibers was apparent at the lower exposure levels. Daily exposures for short periods did not produce any cumulative effect [69].

Studies of protein metabolism in the lens of rabbits' eyes exposed to 160–170 mW/cm² (2800 MHz, CW) for 30–40 min indicated that the only discernible alteration occurred in the posterior lens capsule. A significant increase in the permeability of the capsule was shown, as well as an impairment in the turnover of labeled sulfate. Studies on protein and nucleic acid metabolism failed to reveal any significant differences between the experimental and control lenses.

Whole-body exposure of dogs to a single treatment of 3-h duration at 2800 MHz (pulsed) microwaves, 165 mW/cm² or for as long as 6 h daily for a 3-week period (2800 MHz, 165 mW/cm²) did not produce any lenticular changes. These animals were examined with a slit-lamp periodically for one year after exposure. It should be pointed out that in these exposures, the dogs could move around freely in their cages and their eyes were never exposed directly to the microwave energy for long periods of time.

In dogs, single or fractionated exposure of the eyes to 2800-MHz (pulsed) microwaves, 350 mW/cm² for 20 min did not cause permanent lenticular alteration. However, single or fractionated exposure of the eyes to 700 mW/cm² at 2800-MHz (pulsed) microwaves for 20 min caused lens opacification which involved the posterior lens capsule and the posterior subcapsular cortex (Fig. 8) [40].

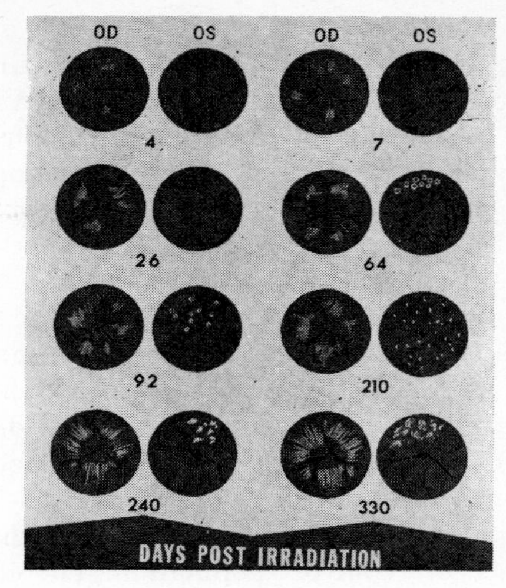

Fig. 8. Lens changes in dogs as a function of post irradiation time; 2800 MHz, 700 mW/cm².

K. Lenticular Effects in Man

These studies by investigators of New York University are no doubt the most extensive performed on individuals exposed to microwaves up to and including the period of the Tri-Service Program.

Zaret *et al.* [86] assessed the frequency of occurrence of lenticular imperfections in the eyes of a sample of 736 microwave workers engaged in installation, operation, and development of microwave equipment and a control sample of 559 individuals of similar age grouping. This study was the first intensive effort undertaken to study the effects of microwave energy on the human eye and to relate ophthalmological findings to actual conditions of exposure.

Analysis of the data revealed that the lenses of both microwave workers and controls had defects and that the number of defects increase linearly with age. Although there was an apparent statistical difference in eye score between the exposed and control groups, the difference was demonstrated to be clinically insignificant. The extent of minor lenticular imperfection does not serve as a clinically useful indicator of cumulative exposure to microwave radiation. No relationship between lens imperfection and microwave cataract was found. A peculiar finding was that the difference in eye scores appeared to be constant for all age groups. The clinical impression gained from the detailed microscopic lenticular examinations was that the defects present in the lens are not generally an indicator of cumulative chronic exposure to RF radiation. However, the effect of microwave exposure was demonstrable by standard statistical treatment of the data.

It was determined that irreversible injury to the human lens can occur when the magnitude of exposure reaches levels of several hundred milliwatts per square centimeter and there are multiple exposures at least of several minutes duration. Continuous exposure to permissible levels of microwave radiation (10 mW/cm²) did not appear to have any significant clinical effect on the human lens insofar as the sample studies appeared to indicate.

It was also noted that exposures to levels of microwave radiation in excess of 10 mW/cm² but below 300 mW/cm² may produce reversible injury of the lens. This condition is termed intumescence and is recognized as a swelling of the lens and it is probably due to hydration.

An important aspect of this study was the determination that, in areas where accepted safety practices were observed, microwave workers did not reveal lenticular changes significantly different from those exhibited by an unexposed or "control" population. The etiology of microwave cataract appears to be of a thermal nature [86].

This study had additional implications for the conduct of follow-on animal research. For example, it was shown that small lenticular imperfections such as granules, vacuoles, and minute opacities are not necessarily precursors of damaging microwave effects. The observation that marked species differences exist with respect to the type location and induced lenticular injuries imposed the additional requirement that both criteria and examination techniques demonstrate the feasibility of extrapolating animal data to man [84]–[86].

Epilog

For those of us who participated in the exciting days of the Tri-Service Program, we can only look back with a feeling of nostalgia and respect for the stimulus which Colonel Knauf provided in his sometimes authoritative, but always congenial manner. Although the Tri-Service Program has been criticized for the lack of quantitative data produced, one must not lose sight of the fact that this program was the only large-scale coordinated effort in the Western world to elucidate and understand some of the basic mechanisms of microwave bioeffects and to assess possible health implications of this form of energy. It must be realized that, at the time the studies were initiated, the art of microwave dosimetry was just emerging. Also of great importance is the fact that there were very few groups available at the time that were oriented toward the type of study required for elucidation of microwave bioeffects. Any thorough and objective review of the proceedings of the Tri-Service Conferences reveals the wealth of information that became available in the short period during which this program was in effect.

We have had a chance to look back and reflect on the results of the Tri-Service Program, the period of apathy since the Tri-Service Program ceased, and the recent resurgence of interest in the question of microwave hazards since the enactment of Public Law 90-602 in October 1968. We should now be in a better position than ever to engage in productive investigations into the biologic effects of microwaves based on well-designed, adequately controlled, conceptually valid investigations in this obviously growing and very important field of microwave bioeffects investigation. We can appreciate the remarks made by Colonel Knauf at the Second Tri-Service Conference in July 1958, which are as pertinent today as they were twelve years ago:

> You must remember that this subject has an enormous public appeal and lends itself readily to sensational headlines in the press. I know. We are presenting the whole story honestly and factually in keeping with our promise of last year. This is done because we feel that a technical group such as this will be able to accurately interpret the results of our work in terms of hazards. We feel this group represents the folks who have a need to know what we are learning. This might be looked upon as a report to the stockholders.

Appendix

George M. Knauf, M.D.

Dr. George M. Knauf was born in Camden, N. J., October 17, 1906. He received the B.S. degree from the Hahnemann College of Science in 1930 and the degree of Doctor of Medicine from the Hahnemann Medical College, Philadelphia, Pa., in 1934. After a rotating internship, Dr. Knauf entered the private practice of medicine in 1935 at Rising Sun, Md. He continued in private practice until he entered the military service in 1940.

In the military service Dr. Knauf was assigned a wide variety of military medical duties during the period from 1940 until 1952. These duties included such assignments as Wing Surgeon, Hospital Commander, and Center Surgeon both in this country and overseas. Following a tour in the Far East during the Korean hostilities, Dr. Knauf was assigned to duty at the Rome Air Development Center. Here he occupied two positions, one as Base Surgeon and the other as Deputy Director of Technical Services in the Electronics Research Center. In this latter capacity he was responsible for the Tri-Service Program.

George Knauf in his enthusiastic, convincing, tireless, and friendly personal way, brought together and stimulated a group of investigators which included engineers, physicists, physicians, physiologists, and pharmacologists. He helped each investigator to organize his own group of technical experts, paved the way, when paving was needed, to expedite the project, and never was too busy to make the rounds at the institutions involved for the purpose of correlating data or answering questions. All of us who got to know George, and who had the coffee pot always ready whenever he appeared on the scene, know the role that he played in putting microwave radiation research on a sound basis and in separating microwave fiction from facts.

It was he who was largely responsible for the organization of the Tri-Service Conferences, which brought the research teams into intimate contact with the military—conferences which met at Rome, N. Y., in 1957 and 1958, in Berkeley, Calif., in 1959, and in New York, N. Y., in 1960. The research carried out during approximately ten years resulted in the establishment of the current standard accepted in the United States and abroad.

In September 1958, Colonel George Knauf was reassigned to the Air Force Missile Test Center at Patrick Air Force Base, Fla. for duty as Center Surgeon. Here, his responsibilities included the supervision of a broad industrial medical program associated with problems resulting from missile and booster launch operations. He organized and conducted courses of instruction in the "Medical Aspects of Missile Operations," for medical service officers of all three medical departments.

Shortly after the activation of Project Mercury by the newly formed NASA, he was appointed Assistant for Bioastronautics to the Department of Defense Representative for Project Mercury Support and charged with responsibility for all DOD medical support of Project Mercury Space Flight Operations. In January 1962, Colonel Knauf was detailed to NASA to serve as the Deputy Director, Space Medicine, Manned Space Flight.

On October 31, 1962, Colonel Knauf retired from the Air Force, and immediately accepted an appointment to the same NASA position as a civilian. In May 1963, Dr. Knauf was appointed Acting Director, Space Medicine. In this capacity his major contributions were coordination of the NASA Space Medicine Program with Air Force and other DOD elements, provision of medical administrative guidance and council to the Directors of Manned Space Flight Centers, recognition of NASA's overall medical needs, and the establishment of finite program definition between the life science elements of manned space flight and advanced research and technology.

On December 31, 1964, Colonel G. Knauf resigned his position with NASA and accepted a position with the Boeing Company, Seattle, Wash., as Chief, Aeromedical Support for the Boeing Supersonic Transport Branch. In this assignment he reviewed the SST program at Boeing to identify aeromedical considerations which could influence the design of the aircraft. He also worked closely with special airline and Federal Aviation Administration committees assigned to investigate the aeromedical aspects of the U.S. SST design.

George Knauf passed away on June 12, 1968, after suffering an acute illness.

George was a proud man; he was sincere and honest; he was dynamic and a tireless individual whose driving force and enthusiasm for assigned duties and tasks flowed over and influenced and stimulated those around him. The world has become a better one because of his contributions to mankind. Those of us who had the privilege of working with him shall never forget him.

WILLIAM B. DEICHMANN[10]

[10] *Ind. Med. Surg.*, vol. 37, Sept. 1968, pp. 645, 650.

References

[1] C. H. Addington, F. Fischer, R. Neubauer, C. Osborn, Y. T. Sarkees, and G. Swartz, "Studies on the biological effects of 200 megacycles," in *Proc. 2nd Annu. Tri-Service Conf. Biol. Effects of Microwave Energy*, E. G. Pattishall and F. W. Banghart, Eds. (Univ. Virginia, Charlottesville, July 8–10, 1958); ASTIA Doc. AD 131 477.

[2] C. H. Addington, C. Osborn, G. Swartz, F. Fischer, and Y. T. Sarkees, "Thermal effects of 200 megacycles (CW) irradiation as related to shape, location and orientation in the field," in *Proc. 3rd Annu. Tri-Service Conf. Biol. Hazards of Microwave Radiating Equipments*, C. Susskind, Ed. (Univ. California, Berkeley, Aug. 25–27, 1959), p. 10.

[3] C. H. Addington, C. Osborn, G. Swartz, F. Fischer, R. A. Neubauer, and Y. T. Sarkees, "Biological effects of microwave energy at 200 megacycles," in *Proc. 4th Annu. Tri-Service Conf. Biol. Effects of Microwave Radiating Equipments: Biological Effects of Microwave Radiations*, M. F. Peyton, Ed. New York: Plenum Press, 1961; RADC-TR-60-180, p. 177.

[4] A. Anne, "Scattering and absorption of microwaves by dissipative dielectric objects: The biological significance and hazard to mankind," Ph.D. dissertation, Univ. Pennsylvania, Philadelphia, Cont. Nonr 55105, 1963; ASTIA 408 997 ONRTR 36.

[5] A. Anne, M. Saito, O. M. Salati, and H. P. Schwan, "Relative microwave absorption cross sections of biological significance," in *Proc. 4th Annu. Tri-Service Conf. Biol. Effects of Microwave Radiating Equipments; Biological Effects of Microwave Radiations*, M. F. Peyton, Ed. New York: Plenum Press, 1961; RADC-TR-60-180, p. 153.

[6] R. Baus, and J. D. Fleming, "Biological effects of microwave radiation with limited body heating," in *Proc. 3rd Annu. Tri-Service Conf. Biol. Hazards of Microwave Radiating Equipments*, C. Susskind, Ed. (Univ. California, Berkeley, Aug. 25–27, 1959), p. 292.

[7] R. L. Carpenter, "Studies on the effects of 2450 megacycle radiation on the eye of the rabbit," in *Proc. 3rd Annu. Tri-Service Conf. Biol. Hazards of Microwave Radiating Equipments*, C. Susskind, Ed. (Univ. California, Berkeley, Aug. 25–27, 1959), p. 279.

[8] ——, "Suppression of differentiation in living tissues exposed to microwave radiation," *Dig. 6th Int. Conf. Med. Elect. Biol. Eng.* (Tokyo), 1965, p. 573.

[9] R. L. Carpenter, D. K. Biddle, and C. A. Van Ummersen, "Biological effects of microwave radiation with particular reference to the eye," *Proc. 3rd Int. Conf. Med. Electron.* (London), vol. 3, 1960, p. 401.

[10] R. L. Carpenter and C. A. Van Ummersen, "The action of microwave radiation on the eye," *J. Microwave Power*, vol. 3, 1968, p. 1.

[11] W. B. Deichmann, "Annual report of microwave radiation research," Griffiss AFB, N. Y., Contract AF 30 (602)-1753, RADC Tech. Rep., 1959, vol. 59, p. 228.

[12] ——, "Biological effects of microwave radiation of 24 000 megacycles," *Arch. Toxika*, vol. 22, 1966, p. 24.

[13] W. B. Deichmann, E. Bernal, and M. Keplinger, "Effects of environmental temperature and air volume exchange on survival of rats exposed to microwave radiation of 24 000 megacycles," in *Proc. 3rd Annu. Tri-Service Conf. Biol. Hazards of Microwave Radiating Equipments*, C. Susskind, Ed. (Univ. California, Berkeley, Aug. 25–27, 1959), p. 62; *Ind. Med. Surg.*, vol. 28, p. 535.

[14] W. B. Deichmann, M. Keplinger, and E. Bernal, "Relation of interrupted pulsed microwaves to biological hazards," in *Proc. 3rd Annu. Tri-Service Conf. Biol. Hazards of Microwave Radiating Equipments*, C. Susskind, Ed. (Univ. California, Berkeley, Aug. 25–27, 1959); *Ind. Med. Surg.*, vol. 28, p. 212.

[15] W. B. Deichmann, E. Bernal, F. Stephens, and K. Landeen, "Effects on dogs of chronic exposure to microwave radiation," *J. Occup. Med.*, vol. 5, 1963, p. 418.

[16] W. B. Deichmann, J. Miale, and K. Landeen, "Effect of microwave radiation on the hemopoietic system of the rat," *Toxicol. Appl. Pharmacol.*, vol. 6, 1964, p. 71.

[17] W. B. Deichmann and F. H. Stephens, "Microwave radiation of 10 mW/cm^2 and factors that influence biological effects at various power densities," *Ind. Med. Surg.*, vol. 30, 1961, p. 221.

[18] W. B. Deichmann, F. H. Stephens, M. Keplinger, and K. F. Lampe, "Acute effects of microwave radiation on experimental animals (24,000 Mc)," *J. Occup. Med.*, vol. 1, 1959, p. 369.

[19] J. Fleming, L. Pinneo, R. Baus, and R. McAfee, "Microwave radiation in relation to biological systems and neural activity," in *Proc. 4th Annu. Tri-Service Conf. Biol. Effects of Microwave Radiating Equipments: Biological Effects of Microwave Radiations*, M. F. Peyton, Ed. New York: Plenum Press, 1961; RADC-TR-60-180, p. 229.

[20] S. A. Gunn, T. C. Gould, and W. A. D. Anderson, "The effect of microwave radiation (24,000 Mc) on the male endocrine system of the rat," in *Proc. 4th Annu. Tri-Service Conf. Biol. Effects*

of Microwave Radiating Equipments: Biological Effects of Microwave Radiations, M. F. Peyton, Ed. New York: Plenum Press, 1961; RADC-TR-60-180, p. 99.

[21] ——, "The effect of microwave radiation on morphology and function of the rat testis," *Lab. Invest.*, vol. 10, 1961, p. 301.

[22] J. W. Howland and S. Michaelson, "Studies on the biological effects of microwave irradiation of the dog and rabbit," in *Proc. 3rd Annu. Tri-Service Conf. Biol. Hazards of Microwave Radiating Equipments*, C. Susskind, Ed. (Univ. California, Berkeley, Aug. 25–27, 1959), p. 191.

[23] J. W. Howland, R. A. E. Thomson, and S. M. Michaelson, "Biomedical aspects of microwave irradiation of mammals," in *Proc. 4th Annu. Tri-Service Conf. Biol. Effects of Microwave Radiating Equipments: Biological Effects of Microwave Radiations*, M. F. Peyton, Ed. New York: Plenum Press, 1961; RADC-TR-60-180, p. 261.

[24] C. J. Imig and C. W. Searle, "Studies on organisms exposed to 2450 Mc-CW microwave irradiation," Griffiss AFB, N. Y., Contract AF 41 (657) 113, RADC-TDR-62-358, 1962.

[25] H. Kalant, "Physiological hazards of microwave radiation: A survey of published literature," *Can. Med. Ass. J.*, vol. 81, 1959, p. 575.

[26] J. H. Kinoshita, L. D. Merola, E. D. Kammak, and R. L. Carpenter, "Biochemical changes in microwave cataracts," *Doc. Ophthalmol.*, vol. 20, 1966, p. 91.

[27] G. Knauf, "Review of the biological effects program," in *Proc. 2nd Annu. Tri-Service Conf. Biol. Effects Microwave Energy*, E. G. Pattishall and F. W. Banghart, Eds. (Univ. Virginia, Charlottesville, July 8–10, 1958); ASTIA Doc. AD 131 477.

[28] R. D. McAfee, "Neurophysiological effects of microwave irradiation," in *Proc. 3rd Annu. Tri-Service Conf. Biol. Hazards of Microwave Radiating Equipments*, C. Susskind, Ed. (Univ. California, Berkeley, Aug. 25–27, 1959), p. 314.

[29] ——, "Neurophysiological effect of 3-cm microwave radiation," *Amer. J. Physiol.*, vol. 200, 1961, p. 192.

[30] ——, "Physiological effects of thermode and microwave stimulation of peripheral nerves," *Amer. J. Physiol.*, vol. 203, 1962, p. 374.

[31] ——, "Microwave stimulation of the sympathetic nervous system," *Biomed. Sci. Instrum.* vol. 1. New York: Plenum Press, 1963, p. 167.

[32] ——, "The neural and hormonal response to microwave stimulation of peripheral nerves," presented at the Symp. Biol. Eff. Health Implic. Microwave Radiat., Richmond, Va., Sept. 1969.

[33] H. Mermagen, "Phantom experiments with microwaves," in *Proc. 4th Annu. Tri-Service Conf. Biol. Effects of Microwave Radiating Equipments: Biological Effects of Microwave Radiations*, M. F. Peyton, Ed. New York: Plenum Press, 1961; RADC-TR-60-180, p. 143.

[34] L. O. Merola and J. H. Kinoshita, "Changes in the ascorbic acid content in lenses of rabbit eyes exposed to microwave radiation," in *Proc. 4th Annu. Tri-Service Conf. Biol. Effects of Microwave Radiating Equipments: Biological Effects of Microwave Radiations*, M. F. Peyton, Ed. New York: Plenum Press, 1961, RADC-TR-60-180, p. 285.

[35] S. M. Michaelson, J. W. Howland, R. A. E. Thomson, and H. Mermagen, "Comparison of responses to 2800 Mc and 200 Mc microwaves or increased environmental temperature," in *Proc. 3rd Annu. Tri-Service Conf. Biol. Hazards of Microwave Radiating Equipments*, C. Susskind, Ed. (Univ. California, Berkeley, Aug. 25–27, 1959), p. 161.

[36] S. M. Michaelson, R. A. E. Thomson, and J. W. Howland, "Physiologic aspects of microwave irradiation of mammals," *Amer. J. Physiol.*, vol. 201, 1961, p. 351.

[37] S. M. Michaelson, R. A. E. Thomson, L. T. Odland, and J. W. Howland, "The influence of microwaves on ionizing radiation exposure," *Aerosp. Med.*, vol. 34, 1963, p. 111.

[38] S. M. Michaelson, R. A. E. Thomson, M. Y. El Tamami, H. S. Seth, and J. W. Howland, "Hematologic effects of microwave exposure," *Aerosp. Med.*, vol. 35, 1964, p. 824.

[39] S. M. Michaelson, R. A. E. Thomson, and J. W. Howland, "Comparative studies on 1185 and 2800 Mc/sec pulsed microwave," *Aerosp. Med.*, vol. 36, 1965, p. 1059.

[40] ——, "Biologic effects of microwave exposure," 138 p. Griffiss AFB, N. Y., Rome Air Development Center, UR-49-976, Unclas. H69-25367; ASTIA Doc. No. AD 824-242.

[41] S. M. Michaelson, R. A. E. Thomson, and W. J. Quinlan, "Effects of electromagnetic radiations on physiologic responses," *Aerosp. Med.* vol. 38, 1967, p. 293.

[42] W. W. Mumford, "Some technical aspects of microwave radiation hazards," Bell Telephone Syst., Tech. Publ. 3865; *Proc. IRE*, vol. 49, Feb. 1961, pp. 427–447.

[43] R. T. Nieset and R. Baus, Jr., "Investigation of the biological effects of microwave irradiation," Biophysical Laboratory, Tulane Univ., New Orleans, Contract NONR 47503, Ann. Prog. Rep., June 11–Nov. 15, 1957.

[44] R. T. Nieset, R. Baus, Jr., R. D. McAfee, J. J. Friedman, A. S. Hyde, and J. D. Fleming, Jr., "Review of the work conducted at Tulane University. Investigations of the biological effects of microwave irradiation," in *Proc. 2nd Annu. Tri-Service Conf. Biol. Effects of Microwave Energy*, E. G. Pattishall and F. W. Banghart, Eds. (Univ. Virginia, Charlottesville, July 8–10, 1958); ASTIA Doc. AD 131 477.

[45] R. T. Nieset, R. Baus, Jr., R. D. McAfee, J. D. Fleming, Jr., and L. R. Pinneo, "The Neural effects of microwave irradiation," Quart. Rep. RADC-TN-59-311, June, 1959.

[46] R. T. Nieset, L. R. Pinneo, R. Baus, Jr., and J. D. Fleming, Jr., "The neural effects of microwave irradiation," Biophysics Laboratory, Tulane University, New Orleans, La., Contract AF 30(602) 965, Ann. Rep., 1960.

[47] C. Osborne, "Studies on the biological effects of 200 mc," in *Proc. 2nd Annu. Tri-Service Conf. Biol. Effects of Microwave Energy*, E. G. Pattishall and F. W. Banghart, Eds. (Univ. Virginia, Charlottesville, July 8–10, 1958); ASTIA Doc. AD 131 477.

[48] ——, "Studies on the biological effects of 200 mc," *Invest. Conf. Biol. Effects of Electronic Radiating Equipments*, Rome Air Development Center, Air Research and Development Command, Rome, N. Y., ASTIA Doc. AD 214 693, 1959, p. 20.

[49] L. R. Pinneo, "Direct current potentials of the central nervous system," RADC-TN-59-137, 1959.

[50] L. R. Pinneo, R. Baus, Jr., R. D. McAfee, and J. D. Fleming, Jr., "The neural effects of microwave radiation," ASTIA Doc. AD 277 684, 1962.

[51] S. Prausnitz and C. Susskind, "Temperature regulation in laboratory animals irradiated with 3-cm microwaves," in *Proc. 3rd Annu. Tri-Service Conf. Biol. Hazards of Microwave Radiating Equipments*, C. Susskind, Ed. (Univ. California, Berkeley, Aug. 25–27, 1959), p. 33.

[52] ——, "Effects of chronic microwave irradiation on mice," *IRE Trans. Bio-Med. Electron.*, vol. BME-9, Apr. 1962, pp. 104–107.

[53] A. W. Richardson, "Review of the work conducted at the St. Louis University School of Medicine," in *Proc. 2nd Annu. Tri-Service Biol. Effects of Microwave Energy*, E. G. Pattishall and F. W. Banghart, Eds. (Univ. Virginia, Charlottesville, July 8–10, 1958); ASTIA Doc. AD 131 477, p. 169.

[54] ——, "New microwave dosimetry and the physiological need," in *Proc. 3rd Annu. Tri-Service Conf. Biol. Hazards of Microwave Radiating Equipments*, C. Susskind, Ed. (Univ. California, Berkeley, Aug. 25–27, 1959), p. 244.

[55] A. W. Richardson, T. D. Duane, and H. M. Hines, "Experimental lenticular opacities produced by microwave irradiations," *A. M. A. Arch. Phys. Med.*, vol. 29, 1948, p. 765.

[56] M. Saito and H. P. Schwan, "Time constants of pear 1-chain formation," in *Proc. 4th Annu. Tri-Service Conf. Biol. Effects of Microwave Radiating Equipments: Biological Effects of Microwave Radiations*, M. F. Peyton, Ed. New York: Plenum Press, 1961; RADC-TR-60-180, p. 85.

[57] M. Saito, H. P. Schwan, and G. Schwarz, "Response of nonspherical biological particles to alternating electric fields," *Biophys. J.*, vol. 6, 1966, p. 313.

[58] O. M. Salati, and H. P. Schwan, "A technique for relative absorption cross-section determination," in *Proc. 3rd Annu. Tri-Service Conf. Biol. Hazards of Microwave Radiating Equipments*, C. Susslomd. Ed. (Univ. California, Berkeley, Aug. 25–27, 1959), p. 107.

[59] H. P. Schwan, "Electrical properties of tissues and cell suspensions," in *Advances in Biological and Medical Physics*, vol. 5, J. H. Lawrence and C. Tobias (Eds.) New York: Academic Press, 1957.

[60] ——, "Biophysics of diathermy," in *Therapeutic Heat*, S. H. Licht (Ed.) New Haven: E. Licht, 1958, p. 55.

[61] ——, "Survey of microwave absorption characteristics of body tissues," in *Proc. 2nd Annu. Tri-Service Conf. Biol. Effects Microwave Energy*, E. G. Pattishall and F. W. Banghart, Eds. (Univ. Virginia, Charlottesville, July 8–10, 1958); ASTIA Doc. AD 131 477, p. 126.

[62] ——, "Alternating current spectroscopy of biological substances," *Proc. IRE*, vol. 47, Nov. 1959, pp. 1841–1855.

[63] ——, "Absorption and energy transfer of microwaves and ultrasound in tissues; Characteristics," *Medical Physics*, vol. 3, Glasser, Ed. Chicago: Year Book Publishers, 1960, p. 1.

[64] ——, "Radiation biology, medical applications and radiation hazards," in *Microwave Power Engineering*, vol. 2, E. C. Okress, Ed. New York: Academic Press, 1968, p. 215.

[65] H. P. Schwan and K. Li, "Hazards due to total body irradiation by radar," *Proc. IRE*, vol. 44, Nov. 1956, pp. 1572–1581.

[66] H. P. Schwan, H. Pauly, J. Twisdom, and E. Frazer. "Effects of microwaves on mankind," Electromedical Laboratory, Moore School Elec. Eng. Dep. Physical Medicine, School of Medicine, Univ. Pennsylvania, Pa., 1st Ann. Prog. Rep., Mar. 1958.

[67] G. W. Searle, R. W. Dahlen, C. J. Imig, C. C. Wunder, J. D. Thomson, J. A. Thomas, and W. J. Moressi, "Effects of 2450 Mc microwaves in dogs, rats and larvae of the common fruit fly," in *Proc. 4th Annu. Tri-Service Conf. Biol. Effects of Microwave Radiating Equipments: Biological Effects of Microwave Radia-*

tions, M. F. Peyton, Ed. New York: Plenum Press, 1961, RADC-TR-60-180, p. 187.
[68] H. S. Seth and S. M. Michaelson, "Microwave hazards evaluation," *Aerosp. Med.*, vol. 35, 1964, p. 734.
[69] ——, "Microwave cataractogenesis," *J. Occup. Med.*, vol. 7, 1965, p. 439.
[70] L. D. Sher and H. P. Schwan, "Mechanical effects of AC fields on particles dispersed in a liquid: Biological implications," Ph.D. dissertation, Univ. Penn., Philadelphia, Contract AF 30(602), ONR Tech. Rep. 37, 1963.
[71] C. Susskind, "Biological effects of microwave radiations," Univ. California, Inst. Eng. Res., Ann. Sci. Rep., Ser. 60(205), RADC-TR-298, June 30, 1958.
[72] ——, "Microwave radiation as a biological hazard and tool," Univ. California, Berkeley, Ann. Sci. Rep. RADC-TR-60-122, 1960.
[73] ——, "Longevity study of the effects of 3-cm microwave radiation on mice," Univ. California, Berkeley, Ann. Sci. Rep. RADC-TR-61-205, 1961; ASTIA 269 385.
[74] C. Susskind and P. O. Vogelhut, "Cavity-perturbation measurement of the effects of microwave radiation on proteins," *Proc. Inst. Elec. Eng.*, vol. 109(B), Suppl. 23, 1961, p. 668 and p. 682.
[75] R. A. E. Thomson, S. M. Michaelson, and J. W. Howland, "Modification of X-irradiation lethality in mice by microwaves (radar)," *Radiat. Res.*, vol. 24, 1965, p. 631.
[76] ——, "Leukocyte response following simultaneous ionizing and microwave (radar) irradiation" *Blood*, vol. 28, 1966, p. 157.
[77] ——, "Microwave radiation and its effect on response to X-irradiation," *Aerosp. Med.*, vol. 38, 1967, p. 252.
[78] U. S. Air Force, "Electromagnetic radiation hazards," T. O.31Z-10-4, 1966; rev. 1967.
[79] C. A. Van Ummersen, "The effect of 2450 Mc radiation on the development of the chick embryo," in *Proc. 4th Annu. Tri-Service Conf. Biol. Effects of Microwave Radiating Equipments: Biological Effects of Microwave Radiations*, M. F. Peyton, Ed. New York: Plenum Press, 1961; RADC-TR-60-180, p. 201.
[80] C. A. Van Ummersen and F. C. Cogan, "Experimental microwave cataracts. Age as factor in induction of cataracts in the rabbit," *Arch. Environ. Health*, vol. 11, 1965, p. 177.
[81] P. O. Vogelhut, "The dielectric properties of water and their role in enzyme-substrate interactions," Univ. California Electron. Res. Lab., Berkeley, Calif., Rep. 60–476, Aug. 24, 1962.
[82] D. B. Williams and R. S. Fixott, "A summary of the School of Aviation Medicine, U. S. Air Force program for research on the biomedical aspects of microwave radiation," in *Proc. Tri-Service Conf. Biol. Hazards of Microwave Radiation*, G. Pattishall, Ed. (George Washington Univ., Washington, D. C., July 15–16, 1957); ASTIA Doc. AD 11 5603.
[83] M. M. Zaret, "Comments on papers delivered at Third Tri-Service Conference on Biological Effects of Microwave Radiation," in *Proc. 3rd Annu. Tri-Service Conf. Biol. Hazards of Microwave Radiating Equipments*, C. Susskind, Ed. (Univ. California, Berkeley, Aug. 25–27, 1959), p. 334.
[84] ——, "An experimental study of the cataractogenic effects of microwave radiation," Griffiss AFB, N. Y., Contract AF30 (602) 3087, RADC-TDR-64-273, Final Rep., 1964; ASTIA 608 746.
[85] M. M. Zaret, S. Cleary, B. Pasternack, and M. Eisenbud, "Occurrence of lenticular imperfections in the eyes of microwave workers and their association with environmental factors," New York University, New York, N. Y., Progr. Rep. Contract AF 30(602)2215, RADC-TN-61-226, 1961; ASTIA Doc. AD 266 831.
[86] M. M. Zaret, S. Cleary, B. Pasternack, M. Eisenbud, and H. Schmidt, "A study of lenticular imperfections in the eyes of a sample of microwave workers and a control population," New York University, New York, N. Y., AD 413 294, Final Rep. Contract AF 30(602) 2215, RADC-TDR-63-125, 1963; ASTIA Doc. AD 413 294.
[87] M. M. Zaret and M. Eisenbud, "Preliminary results of studies of the lenticular effects of microwaves among exposed personnel," in *Proc. 4th Annu. Tri-Service Conf. Biol. Effects of Microwave Radiating Equipments: Biological Effects of Microwave Radiations*, M. F. Peyton, Ed. New York: Plenum Press, 1961; RADC-TR-60-180, p. 293.

Part V
Medical Applications of Electromagnetic Fields

There is a great deal of anticipation about the beneficial applications of electromagnetic fields in medicine. The use of electromagnetic heating in diathermy is well established [1-6]; a wide variety of applications, particularly in rehabilitative medicine, have been discussed [1-3].

The most interesting and advanced potential application area is electromagnetic hyperthermia (local, regional, or whole-body) as an adjutant to cancer therapy [7-20]. Other promising areas for electromagnetic technology applications are: noninvasive biomedical imaging [21-26], electromagnetic heating for revival of hypothermic subjects and for organ and blood thawing [26, 27], microwave radiometry [28-31], and several miscellaneous medical applications [32, 33].

Electromagnetic energy offers several distinct and unique advantages of in-depth heating and rapid control. The engineering design, however, is a challenging electromagnetics problem, in that inhomogeneous lossy dielectric properties of the tissues coupled to complex exposure profiles must be solved in order to obtain accurate rates of heating for the various regions of the exposed body—a problem that is just beginning to be met because of significant recent advances in near-field inhomogeneous modeling techniques [34-36]. Because of these developments, there are expectations for even bigger strides in the future [36].

By far the most exciting new medical application is the use of electromagnetic hyperthermia as a supplement in cancer therapy, since malignant cells are generally more sensitive to thermal damage at temperatures in the range of 41-45°C than are normal cells [9]. Advances in RF, microwave, and electronic control technologies have led to studies at a number of laboratories in which electromagnetic hyperthermia is being used as an adjunct to ionizing radiation or chemotherapy for treatment of localized tumors [10-13, 15, 17, 18]. A number of clinicians have also begun to investigate the use of electromagnetic energy for regional hyperthermia [11, 13, 16] and there is a rapidly increasing interest in whole-body hyperthermia [14, 19, 20] with and without regional boosting, particularly for patients with advanced cancers involving widely spread metastases. Clinical studies of whole-body hyperthermia have used only conventional heating techniques to date (melted paraffin immersion with or without inserted hot water sacks for regional boosting [14], hot water "space suits" [19, 20], and extracorporeal heat exchange by perfusion). Even though electromagnetic heating has not yet been used for whole-body hyperthermia, it may be that frequencies on the order of 50-150 MHz offer the advantages of deep heating, precise control, and a time reduction to about one-half hour to reach the final hyperthermic temperature of 41.8°C, as compared to one and a half to three hours by current techniques [20] which rely upon surface heating and blood mediation to elevate body temperature. Work is currently in progress [35] to account for the inhomogeneous dielectric properties of the body in properly designed multielement electromagnetic applicators in order to obtain physician-prescribed subregional rates of heating over the various parts of the body, which may be relatively uniform (2 : 1) or highly focalized, if needed, over certain regions. Because of the *in situ* heating of the tissues and lesser reliance on blood conduction, whole-body electromagnetic hyperthermia may be less stressful to the patient with the possibility of getting by with analgesia alone. In present methods, whole-body anesthesia is administered for up to the six to nine hours needed for the desired four to six hours of chemotherapy sessions at elevated temperatures [19-20].

An exciting development in the field of biomedical imaging is the possibility of using nuclear magnetic resonance (NMR) to discriminate between various tissues [37-45]. The method is based on the different relaxation times, T_r, of the proton spins of water in normal and malignant tissues. T_r ranges from 0.3 to 0.6 s for water in normal tissues, whereas for tumors the corresponding range is from 0.5 to over 0.8 s. From early pioneering research by Damadian [37] and Lauterbur [38], the method has developed to a stage where it has been applied for biomedical imaging [39-45]. Although the quality of NMR images is no match to that of X-ray images, many investigators feel the method would be useful medically because of its correlation with the status of tissue water.

Pilot studies have recently been started in several laboratories [21-26] to evaluate the use of electromagnetic scattering from the human body or parts thereof for biomedical imaging. Even though serious questions have been raised about the spatial resolution capabilities of electromagnetic methods because of the larger wavelength of incident radiation (as compared to, say, ultrasonic waves), Larson and Jacobi [21, 22] have conducted experiments to try to obtain 4 GHz microwave images of an isolated canine kidney. Using moment methods, exploratory work is also being done at the University of Utah [24] to obtain complex dielectric properties of the tissues at 300-1000 locations from the RF E-field measurements at several locations surrounding the body. With this approach, it is envisaged that some of the receiving antennas could also be interchangeably used as transmitting antennas.

Whereas in X-ray imaging techniques the propagating beam is well collimated and therefore easy to manipulate, the electromagnetic propagation and scattering from inhomogeneous dielectric bodies is quite complicated. The vector fields are subject to diffraction and scattering. Complete Maxwell's equations therefore need to be solved to obtain the lossy dielectric properties from the measurements. In spite of the

Manuscript received July 15, 1981.

many difficulties, the promise of electromagnetic biomedical imaging remains high because of the highly pertinent nature of the information one would get—dielectric properties which are very sensitive to the water content of the tissues. Another use of the information so obtained would be its pertinence and therefore direct applicability to individualizing the electromagnetic hyperthermia regimens.

Microwave radiometry [28–31] relies upon detection of the intensity of black-body radiation from various parts of the body in the frequency band on the order of hundreds of MHz to several tens of GHz. The premise here is that since this radiation is not as highly absorbed by the skin as is the infrared (thermal emission), one should be able to detect somewhat deeper (on the order of centimeters) thermal anomalies that may therefore help in many diagnostic applications similar to those for conventional thermography. Examples here are detection of the location and extent of breast cancer and tumors of the thyroid, determination of the extent of burns and frostbite, and detection of rheumatoid arthritis and placental location. A great deal of work has been done by Barrett and Myers at MIT (frequencies 1–4 GHz) and by Edrich and colleagues at Denver Research Institute using higher frequencies in the millimeter-wave band. Even though substantial progress has been made in obtaining true positive detections, the image quality leaves much to be desired. Considerably more data processing work and proper electromagnetic modeling of the tissues and pick-up antenna(s) is needed before the perceived advantages of microwave techniques over infrared thermography can be realized.

In the field of neurobiology, rapid microwave heating [46–51] of the animal head (to temperatures in the range of 55°–90°C) is being used increasingly as a technique for inactivating brain enzymes in animals, thereby eliminating postmortem changes in many heat-stable neurochemicals such as acetylcholine, L-DOPA, gamma aminobutyric acid, cyclic AMP, and cyclic GMP. The determination of the distribution of these neurochemicals is an indicator of the central nervous system (CNS) and is consequently an important tool for neurobiological research (for example, in the evaluation of the effects of hormones or drugs on the CNS). The microwave technique offers the advantage of inactivating brain enzymes in milliseconds as compared to much longer durations for the competing freezing methods. Rapid freezing methods such as the freeze-blowing technique [52], though offering relatively fast inactivation times, suffer from the loss of anatomical features. This does not permit regional determination. There are also concerns that enzyme systems may recover during thawing, leading to additional artifactual changes.

Electromagnetic energy is also being evaluated for use in several miscellaneous biomedical applications and these are discussed at length in a review article by Iskander and Durney [32]. Some other highly promising, but not yet completely evaluated, applications of electromagnetic energy are, for example, in blood and organ thawing [26, 27], warming of neonatal infants post-open-heart surgery, wound healing [33], and for enhanced rates of post-surgical healing.

From the foregoing, it is obvious that there are several potential and highly useful biomedical applications of electromagnetic energy. It is necessary, however, that competent interdisciplinary teams consisting of physicians, biologists, and engineers work in these areas to convert the hopes into realities. It is unfortunate that the level of funding is totally inadequate and most of these highly likely applications are therefore not receiving a proper and thorough evaluation. Recent advances in electromagnetic bioengineering, including the proper modeling of the inhomogeneous tissue properties and the realistic near-field exposure conditions, have poised the field for major advances in the foreseeable future.

Om P. Gandhi
Associate Editor

References

[1] S. Licht, Ed., *Therapeutic Heat and Cold*, New Haven, CT: Licht, 1965.

[2] J. F. Lehmann, *Diathermy Handbook of Physical Medicine and Rehabilitation*, Krusen, Kottke, and Elwood, Eds., Philadelphia, PA: Saunders, 1971.

[3] A. W. Guy, J. F. Lehmann, and J. B. Stonebridge, "Therapeutic application of electromagnetic power," *Proc. IEEE*, vol. 62, pp. 55–75, 1974.

[4] A. W. Guy, J. F. Lehmann, J. B. Stonebridge, and C. C. Sorensen, "Development of a 915 MHz direct-contact applicator for therapeutic heating of tissues," *IEEE Trans. Microwave Theory Tech.*, vol. MTT-26, pp. 550–556, 1978.

[5] A. W. Guy, "Analyses of electromagnetic fields induced in biological tissues by thermographic studies of equivalent phantom models," *IEEE Trans. Microwave Theory Tech.*, vol. MTT-19, pp. 205–214, 1971. (See pp. 23–32, Part I, this volume.)

[6] J. F. Lehmann, A. W. Guy, J. B. Stonebridge, and B. J. DeLateur, "Evaluation of a therapeutic direct-contact applicator for therapeutic heating of tissues," *IEEE Trans. Microwave Theory Tech.*, vol. MTT-26, pp. 556–563, 1978.

[7] Two recently held conferences: Third International Symposium on Cancer Therapy by Hyperthermia, Drugs, and Radiation, Colorado State University, Fort Collins, CO, June 22–26, 1980; Conference on Thermal Characteristics of Tumors: Applications in Detection and Treatment, NY Acad. Sci., New York, NY, March 14–16, 1979.

[8] A. Y. Cheung and G. M. Samaras (Guest Eds.), *J. Microwave Power*, special issue on "Microwave and Radiofrequency Hyperthermia for Cancer Therapy," vol. 16, 1981.

[9] J. G. Short and P. F. Turner, "Physical hyperthermia and cancer therapy," *Proc. IEEE*, vol. 68, pp. 133–142, 1980.

[10] R. J. R. Johnson, J. R. Subjeck, D. Z. Moreau, H. Kowal, and D. Yakar, "Radiation and hyperthermia," *Bull. N.Y. Acad. Med.*, vol. 55, pp. 1193–1204, 1979.

[11] H. H. Leveen, S. Wapnick, V. Piccone, G. Falk, and N. Ahmed, "Tumor eradication by radio frequency therapy," *J. Amer. Med. Ass.*, vol. 235, pp. 2198–2200, 1976.

[12] J. H. Kim and E. W. Hahn, "Clinical and biological studies of localized hyperthermia," *Cancer Res.*, vol. 39, pp. 2258–2262, 1979.

[13] F. K. Storm, W. H. Harrison, R. S. Elliott, and D. L. Morton, "Normal tissue and solid tumor effects of hyperthermia in animal models and clinical trials," *Cancer Res.*, vol. 39, pp. 2245–2251, 1979.

[14] H. T. Law and R. T. Pettigrew, "New apparatus for the induction of localized hyperthermia for treatment of neoplasia," *IEEE Trans. Bio-Med. Eng.*, vol. BME-26, pp. 175–177, 1979.

[15] J. Mendecki, E. Friedenthal, C. Botstein, F. Sterzer, R. Paglione, M. Nowogrodzki, and E. Beck, "Microwave induced hyperthermia in cancer treatment: apparatus and preliminary results," *Int. J. Radiat. Oncol., Biol., and Phys.*, vol. 4, pp. 1095–1103, 1978.

[16] R. Paglione, F. Sterzer, J. Mendecki, E. Friedenthal, and C. Botstein, "27 MHz ridged waveguide applicators for localized hyperthermia treatment of deep-seated malignant tumors," *Microwave J.*, vol. 24, pp. 71–80, 1981.

[17] L. S. Taylor, "Implantable radiators for cancer therapy by microwave hyperthermia," *Proc. IEEE*, vol. 68, pp. 142–149, 1980.

[18] L. S. Taylor, "Brain cancer therapy using an implanted microwave radiator," *Microwave J.*, vol. 24, pp. 66-71, 1981.
[19] J. M. Larkin, "A clinical investigation of total-body hyperthermia as cancer therapy," *Cancer Res.*, vol. 39, pp. 2252-2254, 1979.
[20] J. Bull, "Combined modalities," *Cancer Res.*, vol. 39, p. 2337, 1979.
[21] L. E. Larson and J. H. Jacobi, "Microwave scattering parameter imagery of an isolated canine kidney," *Med. Phys.*, vol. 6, pp. 394-403, 1979.
[22] L. E. Larson and J. H. Jacobi, "The use of orthogonal polarizations in microwave imagery of isolated canine kidney," *IEEE Trans. on Nucl. Sci.*, vol. NS-27, pp. 1184-1191, 1980.
[23] R. Maini, M. F. Iskander, and C. H. Durney, "On the electromagnetic imaging using linear reconstruction techniques," *Proc. IEEE*, vol. 68, pp. 1550-1552, 1980.
[24] M. J. Hagmann, O. P. Gandhi, and D. K. Ghodgaonkar, "Application of moment-methods to electromagnetic biomedical imaging," *Digest of Papers*, 1981 Int. Microwave Symp., Los Angeles, CA, p. 482, June 15-19, 1981.
[25] M. F. Iskander, R. Maini, and C. H. Durney, "Microwave imaging: numerical simulation and results," *Digest of Papers*, 1981 Int. Microwave Symp., Los Angeles, CA, pp. 483-485, June 15-19, 1981.
[26] R. V. Rajotte, J. B. Dossetor, W. A. G. Voss, and C. R. Stiller, "Preservation studies on canine kidneys recovered from the deep frozen state by microwave thawing," *Proc. IEEE*, vol. 62, pp. 76-85, 1974.
[27] W. A. G. Voss, R. V. Rajotte, and J. B. Dossetor, "Applications of microwave thawing to the recovery of deep frozen cells and organs: a review," *J. Microwave Power*, vol. 9, pp. 181-194, 1974.
[28] P. C. Myers, N. L. Sandowsky, and A. H. Barrett, "Microwave thermography: principles, methods, and clinical applications," *J. Microwave Power*, vol. 14, pp. 105-115, 1979.
[29] J. Edrich, "Centimeter- and millimeter-wave thermography—a survey on tumor detection," *J. Microwave Power*, vol. 14, pp. 95-104, 1979.
[30] A. H. Barrett, P. C. Myers, and N. L. Sandowsky, "Detection of breast cancer by microwave radiometry," *Radio Sci.*, vol. 12(S), pp. 167-171, 1977.
[31] M. Gautherie, J. Edrich, R. Zimmer, J. L. Guerguin-Kern, and J. Robert, "Millimeter-wave thermography—application to breast cancer," *J. Microwave Power*, vol. 14, pp. 123-129, 1979.
[32] M. F. Iskander and C. H. Durney, "Electromagnetic techniques for medical diagnosis: a review," *Proc. IEEE*, vol. 68, pp. 126-132, 1980.
[33] C. Romero-Sierra, S. Halter, J. A. Tanner, M. W. Roomi, and D. Crabtree, "Electromagnetic fields and skin wound repair," *J. Microwave Power*, vol. 10, pp. 59-70, 1975.
[34] I. Chatterjee, M. J. Hagmann, and O. P. Gandhi, "Electromagnetic energy deposition in an inhomogeneous block model for near-field irradiation conditions," *IEEE Trans. Microwave Theory Tech.*, vol. MTT-28, pp. 1452-1459, 1980.
[35] O. P. Gandhi, M. J. Hagmann, and A. Riazi, "Electromagnetic applicators for whole-body and regional hyperthermia," Bioelectromagnetics Society Meeting, San Antonio, TX, September 14-18, 1980.
[36] O. P. Gandhi, "Electromagnetic absorption in an inhomogeneous model of man for realistic exposure conditions," presented at H. P. Schwan's 65th Birthday Symp., University of Pennsylvania, Philadelphia, PA, November 24, 1980; *Bioelectromagnetics*, vol. 3, pp. 81-90, 1982.
[37] R. Damadian "Tumor detection by nuclear magnetic resonance," *Science*, vol. 171, p. 1151, 1971.
[38] P. C. Lauterbur, "Image formation by induced local interactions: examples employing nuclear magnetic resonance," *Nature*, vol. 242, p. 190, 1973.
[39] P. C. Lauterbur, "Medical imaging by nuclear magnetic resonance zeugmatography," *IEEE Trans. Nucl. Sci.*, vol. NS-26, pp. 2808-2811, 1979.
[40] P. G. Morris, P. Mansfield, I. L. Pykett, R. J. Ordidge, and R. E. Coupland, "Human whole-body line scan imaging by nuclear magnetic resonance," *IEEE Trans. Nucl. Sci.*, vol. NS-26, pp. 2817-2820, 1979.
[41] J. L. Marx, "NMR opens a new window into the body," *Science*, vol. 210, pp. 302-305, October 17, 1980.
[42] W. S. Hinshaw, E. R. Andrew, P. A. Bottomley, G. N. Holland, and W. S. Moore, "Display of cross sectional anatomy by nuclear magnetic resonance imaging," *Brit. J. Radiol.*, vol. 51, pp. 273-280, 1978.
[43] R. Damadian, M. Goldsmith, and L. Minkoff, "NMR in cancer: XVI. FONAR image of the live human body," *Physiol. Chem. and Phys.*, vol. 9, pp. 97-100, 1977.
[44] P. Mansfield and I. L. Pykett, "Biological and medical imaging by NMR," *J. Magn. Resonance*, vol. 29, pp. 355-373, 1978.
[45] O. Nalcioglu (Guest Ed.), "Papers presented at the IEEE short course on nuclear magnetic resonance imaging for physicians and engineers," *IEEE Trans. Nucl. Sci.*, Special Issue on Nuclear Medicine, vol. NS-27, pp. 1220-1254, 1980.
[46] W. B. Stavinoha, S. T. Weintraub, and A. T. Modak, "The use of microwave heating to inactivate cholinesterase in the rat brain prior to analysis for acetylcholine," *J. Neurochem.*, vol. 20, pp. 361-371, 1973.
[47] R. H. Lenox, O. P. Gandhi, J. L. Meyerhoff, and H. M. Grove, "A microwave applicator for *in vivo* rapid inactivation of enzymes in the central nervous system," *IEEE Trans. Microwave Theory Tech.*, vol. 24, pp. 58-61, 1976.
[48] W. B. Stavinoha, "Microwave fixation for the study of acetylcholine metabolism," in *Cholinergic Mechanisms and Psychopharmacology*, D. Jenden, Ed., New York, NY: Plenum Press, pp. 169-179, 1978.
[49] R. H. Lenox, J. L. Meyerhoff, O. P. Gandhi, and H. L. Wray, "Microwave inactivation: pitfalls in determination of regional levels of cyclic AMP in rat brain," *J. Cyclic Nucleotide Res.*, vol. 3, pp. 367-369, 1977.
[50] J. H. Merritt and J. W. Frazer, "Microwave fixation of brain tissue as a neurochemical technique—a review," *J. Microwave Power*, vol. 12, pp. 133-139, 1977.
[51] J. L. Meyerhoff, R. H. Lenox, P. V. Brown, and O. P. Gandhi, "The inactivation of rodent brain enzymes *in vivo* using high-intensity microwave irradiation," *Proc. IEEE*, vol. 68, pp. 155-159, 1980.
[52] R. L. Veech, R. L. Harris, D. Veloso, and E. H. Veech, "Freeze-blowing: a new technique of the study of brain *in vivo*," *J. Neurochem.*, vol. 20, pp. 183-188, 1973.

Evaluation of a Therapeutic Direct-Contact 915-MHz Microwave Applicator for Effective Deep-Tissue Heating in Humans

JUSTUS F. LEHMANN, ARTHUR W. GUY, FELLOW, IEEE, JERRY B. STONEBRIDGE, AND BARBARA J. DELATEUR

***Abstract*—A 13-cm square direct-contact microwave applicator which operates at 915 MHz was evaluated in tissue models and human volunteers to determine its therapeutic effectiveness. It was found that the applicator with radome- and forced-air cooling selectively elevates temperatures in muscles (1–2 cm) to 43–45°C. At this higher range of temperature, certain physiologic responses such as an increase in blood flow are produced. The applicator may also be used to heat malignant tumors of muscle.**

I. Introduction

IT HAS BEEN demonstrated that microwave radiation at 915 MHz can produce local vigorous therapeutic responses when used clinically to heat deep tissue [1]. In the clinical use of diathermy, it is essential that the highest temperatures in the distribution occur at the anatomical site to be treated. Thus the heating modality to be used—which may be microwave, shortwave, or ultrasound energy—and the technique of application must be selected according to the specific site of pathology to be treated and heating pattern desired [1]. These criteria are well documented for therapeutic application in the realm of rehabilitative medicine [2]. It is also likely that these heating modalities will be useful in treating cancer.

To ensure that vigorous physiological responses are elicited, temperatures on the order of 43–45°C must be produced in the tissue. When these temperatures are pro-

Manuscript received July 15, 1977; revised September 22, 1977. This work was supported in part by Rehabilitation Services Administration under Grant 16-P56818.

The authors are with the Department of Rehabilitation Medicine, University of Washington School of Medicine, Seattle, WA 98195.

Reprinted from *IEEE Trans. Microwave Theory Tech.*, vol. MTT-26, pp. 556–563, Aug. 1978.

duced, increased blood flow is usually triggered, causing a decline in tissue temperature despite an unchanged rate of energy absorption in the tissues. Thus the occurrence of blood-flow changes demonstrates that therapeutic temperatures have been achieved [3], [4].

It has also been demonstrated that these temperatures should be achieved when heat and stretch are used in combination to produce an increase in the extensibility of collagenous tissue in the treatment of contractures [5], [6]. Hyperthermia with temperatures of 43–45°C has also been used in conjunction with radiation and chemotherapy for treatment of cancerous tumors [7]–[9].

Microwave energy is selected as the modality of choice when its heating pattern in tissue causes the highest temperatures to be produced at the site of pathology. From assessment of the electrical properties of tissues and the selective absorption and reflection of microwave energy in tissue and at tissue interfaces, it was found that microwaves are most useful for selectively heating muscle [10]–[14]. That such selective muscle heating with microwave energy in human tissue is effective has been demonstrated by deLateur [15]. The optimal frequency range was also determined [16]; however, within this range, only one frequency (915 MHz) is presently available in this country for medical and scientific use because of FCC regulations.

The distribution of temperatures produced by microwave radiation is also modified by the preexisting distribution of temperature in human tissue, i.e., warmer at the core and cooler on the outside. This distribution is further modified by changes in blood flow produced by the therapeutic agent.

II. Purpose

The purpose of this study was to evaluate a new direct-contact microwave applicator to determine the temperature distributions produced in tissue models and in the anterior thigh of human volunteers. The distributions were determined from the skin surface through subcutaneous layers of fat and muscle to the bone. The applicator was tested to determine whether it could readily raise and maintain the tissue temperature in the therapeutic range (40–45°C). A determination was also made of the rate of energy absorption required to achieve these desired temperatures. The effects of physiologic factors such as the initial distribution of temperature in tissues and changes in blood flow triggered by microwave radiation were also studied.

III. Methodology

The applicator tested was a 13-cm square direct-contact 915-MHz microwave applicator with stripline feed [17]. It was designed by A. W. Guy to produce an even temperature distribution with the highest temperatures in the muscle. A lightweight dielectric matching material was used which was porous and allowed for air flow (2.5-cm thick Echo foam, HiK Flex Dielectric, 4.0). The applicator housing was equipped with an air inlet so that cooled air could be passed through the applicator onto the skin surface to cool the skin and immediately underlying subcutaneous tissues.

The applicator was first tested on a plane parallel-layered model which had 2 cm of fat equivalent over 10 cm of muscle equivalent. The model, measuring $30 \times 30 \times 12$ cm, was irradiated with the E field parallel to the plane of separation. Thermographic scanning of the split surface was performed 5–10 s after irradiation.

The applicator was then tested on a thigh model of subcutaneous fat, muscle, and bone, which was previously described [18], [19]. Three models were used, with fat layers of 0.5, 1.0, and 2.0 cm, respectively. The models were irradiated with microwaves for 15 s at an average power input to the applicator of 520 W. Based on results obtained from the experiments on models, design modifications were made on the applicator before human subjects were used.

In the human experiments, seven volunteers who had subcutaneous fat layers $\leqslant 1$ cm participated along with six volunteers whose subcutaneous fat layers were $\geqslant 2$ cm. Before each experiment was conducted, the location of the fat–muscle interface and the thickness of the subcutaneous fat layer of the anterior thigh were determined by roentgenography. The volunteers, who were given neither local nor general anesthetics, were surgically scrubbed over the area of needle insertion and were placed in a prone position on a special table constructed without metal parts. A sterilized needle guide fabricated from phenolic and Rexolite plastics was placed against the lateral aspect of the middle third of the thigh. The configuration of this needle guide is similar to that previously described [20]. Sterilized thermistor probes which had previously been tested and calibrated were used to measure tissue temperatures. To facilitate smooth insertion of the thermistor probes, a stab incision was first made with a 15-gauge cutting needle through a hole in the needle guide. The needle was withdrawn, and a sleeve was placed in the needle guide hole to reduce the gauge to 20. Then, a 20-gauge Teflon catheter, stiffened by a trochar, was inserted into the tissue. X-rays were taken to verify the positions of the probes in the tissues. The trochars were then withdrawn and the thermistor probes were inserted into the catheters. The thermistor probes were aligned vertically in the area of peak intensity of the applicator from the superficial tissues to the periosteum of the middle third of the thigh. Temperatures were recorded continuously on an oscillograph recorder before, during, and after exposure to microwave radiation. Immediately after each exposure, the thermistor probes were calibrated.

The probes, which were designed to minimize disturbances in the field, were carefully oriented parallel to the magnetic field. This was done to minimize induction of high-frequency currents. When testing the probes at 915 MHz under the conditions of the experiment [21], they were found to be accurate within ± 0.2°C.

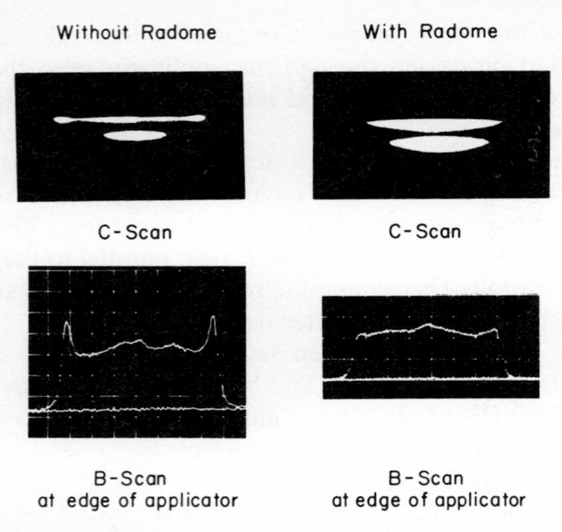

Fig. 1. Thermographs produced with a 13-cm square direct-contact microwave applicator operating at 915 MHz with and without radome on plane layered model.

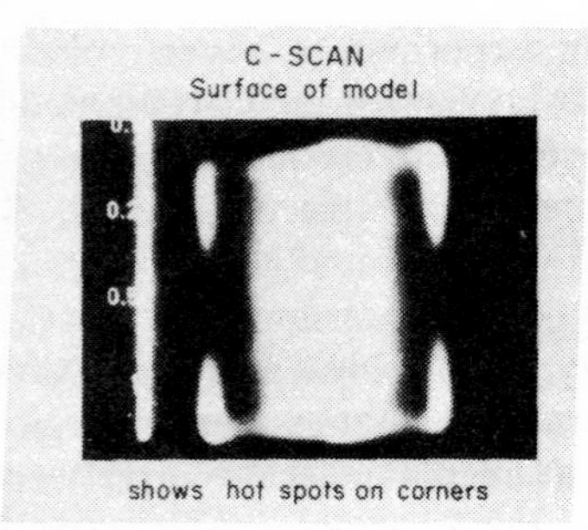

Fig. 2. *C* scan of the surface of a plane parallel-layered model after exposure with a 13-cm square direct-contact microwave applicator operating at 915 MHz without radome.

IV. Experimental Results

The applicator was first tested with the plain parallel-layered tissue model. The results of the test are shown in Figs. 1, 2, and 3. The applicator was placed in direct contact with the tissue model, first with and then without the radome. The models were then placed in front of the thermographic camera for display and recording of the distribution of heat produced in the models after microwave radiation. The *C* scan shows the total heating pattern within the model, while the *B* scan shows surface heating at the edge of the applicator. (See Figs. 2 and 3.) It can be seen that the use of the radome, without forced air cooling, significantly reduced the hot spots at the corners of the applicator.

The radome consisted of 4-mm Rexolite drilled with a matrix of holes terminating in grooves that allowed cooled air to flow along the surface of the skin. To equalize surface temperatures across the skin, hole spacing was closer at the center of the radome and at the corners [17].

Fig. 4 shows thermograms of the plain parallel-layered tissue model at various depths after a 15-s exposure. The applicator was used with radome and cooling as previously described. It can be seen that the highest temperatures in the distribution occurred along the centerline of the applicator just beneath the skin and at the muscle–fascia interface.

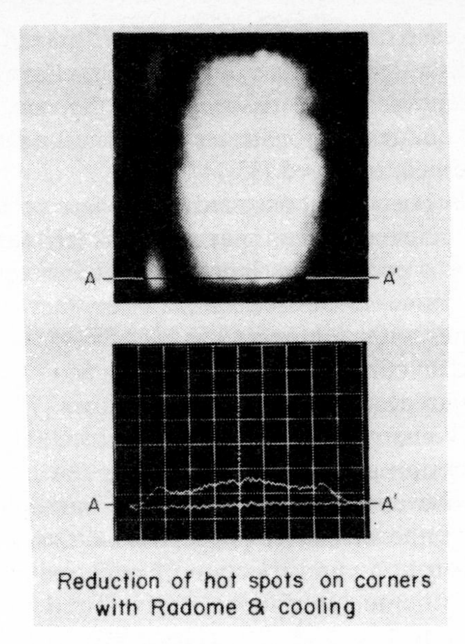

Fig. 3. *C* and *B* scans of the surface of a plane parallel-layered model after exposure with a 13-cm square direct-contact microwave applicator operating at 915 MHz with radome and cooling.

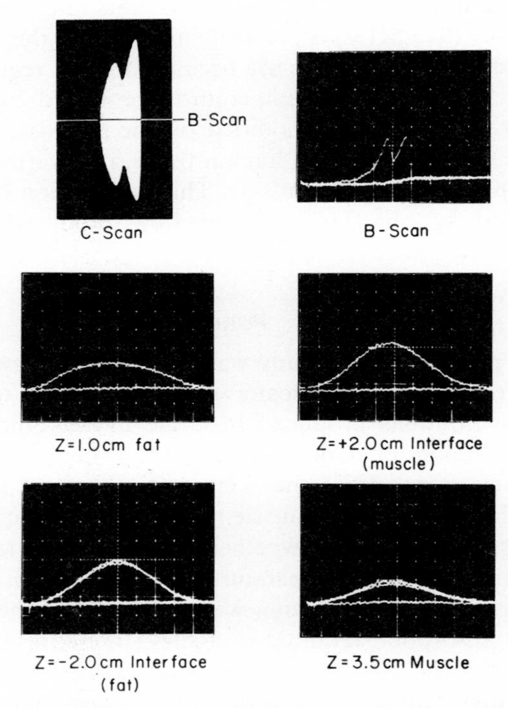

Fig. 4. *X–Z* plane thermograms taken of a plane parallel-layered model with a fat thickness of 2.0 cm after exposure to a 13-cm square direct-contact microwave applicator operating at 915 MHz with radome and cooling.

The applicator was next tested with the three models of the human anterior thigh as previously described. The model shown in Fig. 5 had a 2-cm fat-layer thickness, a 3-cm muscle layer, and a 3-cm bone thickness. The isotherms and *B* scans were calculated and plotted by an on-line computer connected to the thermograph. Fig. 6 shows the isotherms when the applicator was used with the radome without cooling. The temperature distribution shows a marked peak in the subcutaneous fat.

Fig. 5. Phantom thigh model with 13-cm square direct-contact microwave applicator with radome.

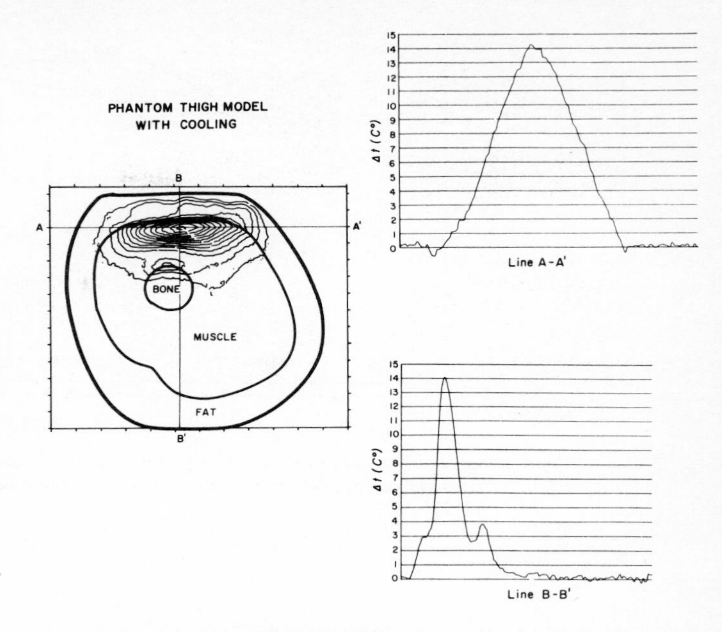

Fig. 7. Isotherms produced in phantom thigh model after exposure to a 13-cm square direct-contact microwave applicator operating at 915 MHz with radome and cooling.

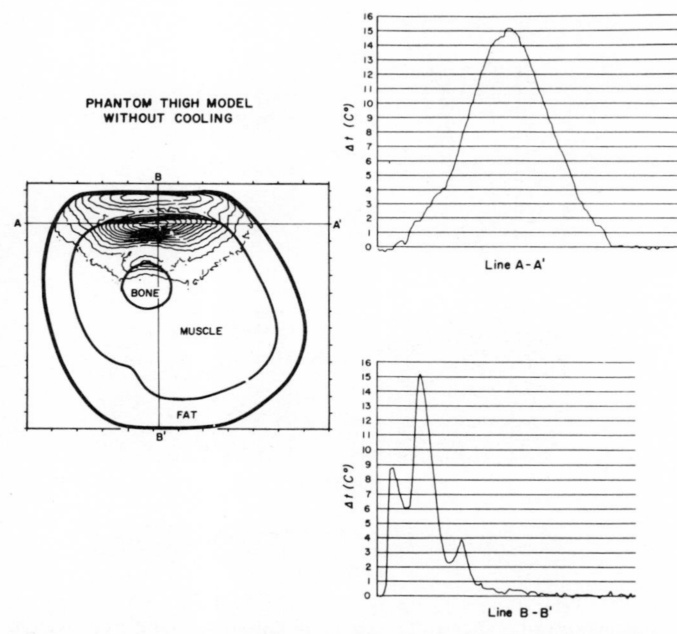

Fig. 6. Isotherms produced in the phantom thigh model after exposure to a 13-cm square direct-contact microwave applicator operating at 915 MHz with radome and without cooling.

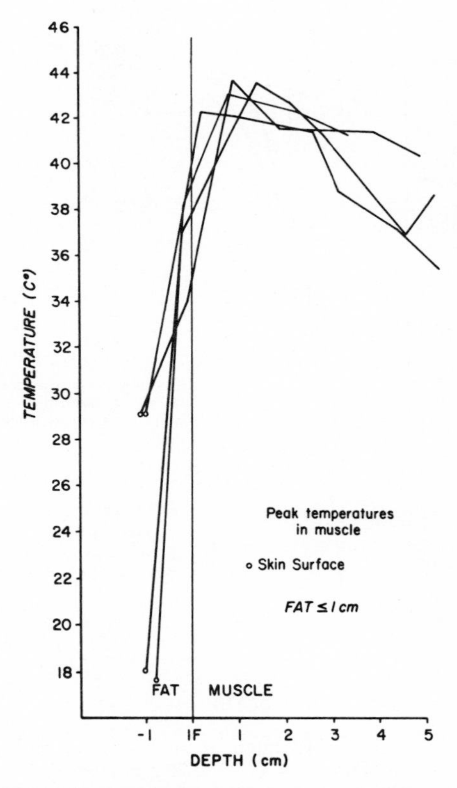

Fig. 8. Temperature distribution in all volunteers with ≤1 cm of subcutaneous fat after an average of 6.5 min of microwave application with the 13-cm square direct-contact microwave air-cooled applicator operating at 915 MHz.

When the run was repeated with cooling, as shown in Fig. 7, the temperatures produced in the subcutaneous fat were significantly lower. In both of these tests, a secondary peak of temperature occurred at the muscle–bone interface. This peak, which resulted from energy reflected at the interface, was not noticeable in thigh models which had more than 6 cm of simulated muscle.

From these experiments with tissue models, it was concluded that the applicator produces significant heating within the musculature, that a radome is necessary to prevent hot spots in the superficial tissues, and that air cooling is necessary to prevent excessive heating of superficial tissues.

The applicator was then tested on human volunteers who were divided into two groups, the first having an anterior thigh fat layer ≤1 cm, and the second having a fat layer ≥2 cm. The thermistor probes were inserted as previously described. The subjects were then irradiated

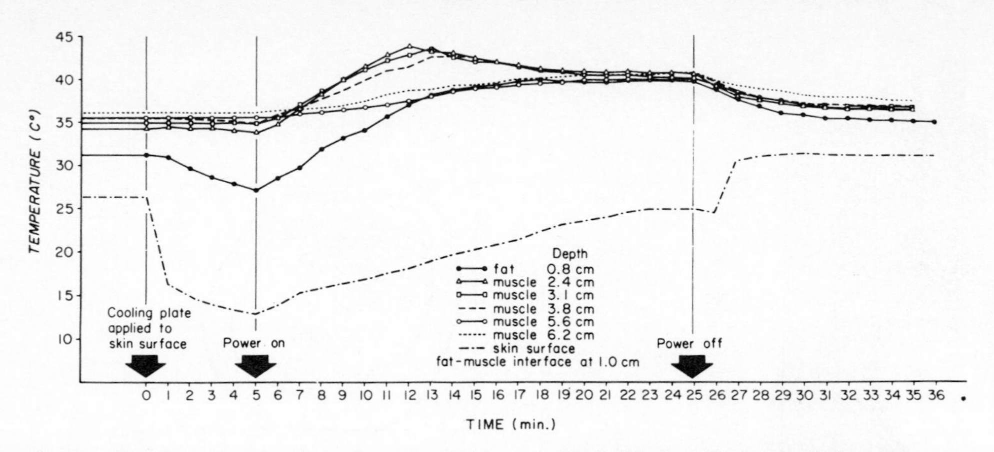

Fig. 9. Temperatures at various tissue depths in one individual before, during, and after exposure to microwave application with a 13-cm square direct-contact air-cooled microwave applicator operating at 915 MHz.

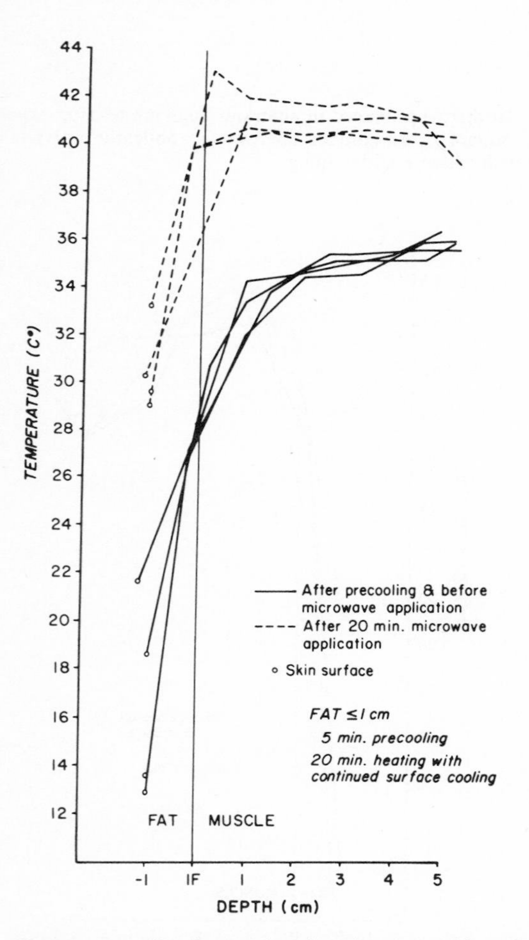

Fig. 10. Temperature distribution in all volunteers with ⩽ 1 cm of subcutaneous fat before (———) and 20 min after (------) microwave application with 13-cm square direct-contact air-cooled microwave applicator operating at 915 MHz.

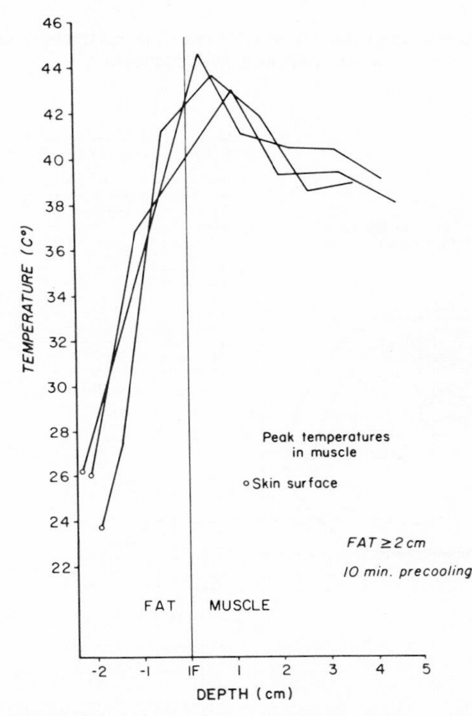

Fig. 11. Temperature distribution in all volunteers with ⩾ 2 cm of subcutaneous fat after an average of 10 min of microwave application with the 13-cm square direct-contact air-cooled microwave applicator operating at 915 MHz.

with the radome in place, with and without cooling. It was found that when cooling was not used, none of the three volunteers in the first group could tolerate more than 5 min of the planned 20-min exposure; concurrently, temperatures reached or exceeded the upper limit of the therapeutic range (45°C). The highest temperatures occurred in the muscle, 0.5–0.6 cm from the fat–muscle interface, and a sharp temperature drop was observed in deeper muscle. All of these subjects complained of surface and deep-muscle pain which was referred to the region of the knee. When cooling was used, with identical net power levels, a 20-min exposure was tolerated by all four volunteers. No pain necessitating discontinuation of treatment was reported. Peak temperatures produced in the muscle are shown in Fig. 8. These temperatures occurred at an average time of 6.5 min after onset of radiation. Fig. 9 gives a complete thermal distribution for one subject, showing all temperatures recorded and the timing of the experimental procedure. In this case, the peak temperatures occurred 7 min after onset of radiation. After this time, the increased flow of blood caused temperatures in

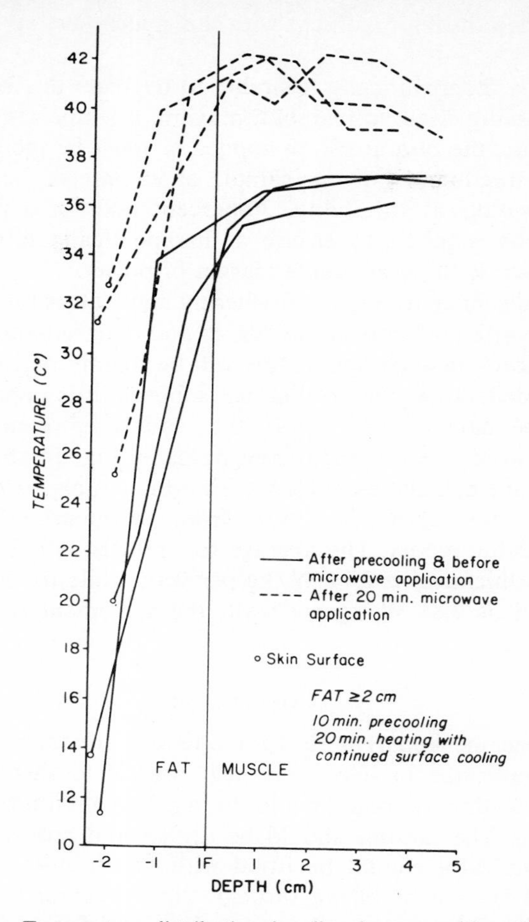

Fig. 12. Temperature distribution in all volunteers with ⩾2 cm of subcutaneous fat before (———) and 20 min after (------) microwave application with the 13-cm square direct-contact air-cooled microwave applicator operating at 915 MHz.

the tissue to decline despite unchanged power input. At the end of the 20-min treatment, the temperatures throughout the muscle tissue had stabilized at a therapeutic level. The final temperatures after 20 min of treatment for the ⩽1-cm fat subjects are shown in Fig. 10. The temperature curves in muscle are more uniform when cooling is used because of the cooling produced by blood flow.

The three volunteers with a thicker layer of fatty tissue were then tested in the same manner. Again, when cooling was not used, the subjects were unable to tolerate a 20-min treatment because of the higher temperatures produced at the skin surface. Again, the upper limit of therapeutic temperatures was reached or exceeded (45°C). Peak temperatures occurred on the skin surface, causing pain at this site. When cooling was used, the peak temperatures occurred at an average time of 10 min from onset of radiation (Fig. 11). The final distributions of temperature in the three volunteers that were cooled are shown in Fig. 12. Thus with cooling, excessive heating of the skin was prevented and effective heating of muscle was achieved. Once again the distribution of temperature was more uniform.

A schematic heating curve for human tissue is shown in Fig. 13. The rate of energy absorption can be calculated

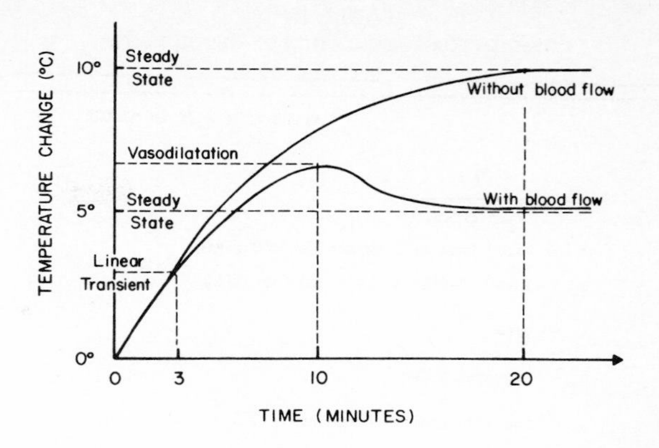

Fig. 13. Schematic representation of transient and steady-state temperature for a typical tissue under diathermy exposure.

TABLE I
FORMULA FOR CALCULATION OF ABSORBED POWER

MASS-NORMALIZED RATE OF ENERGY ABSORPTION = SPECIFIC ABSORPTION RATE (SAR)

$W_a = \frac{k.c.\Delta T}{t}$

W_a = Watts/kg

k = 4.186×10^3 (Joules/Calorie)

c = specific heat - kcal/kg °C

ΔT = °C

t = time (sec.)

TABLE II
CALCULATED VALUES OF ABSORBED POWER IN HUMAN MUSCLE

SPECIFIC ABSORPTION RATE (SAR) in W/kg[1] in MUSCULATURE (1-2cm)

RUN		SAR
1	-	121.60
2	-	78.17
3	-	118.70
4	-	75.27
5	-	167.93

1 at 555.55 mW/cm^2 maximum power density of incident radiation.

from the slope of the initial transient according to the formula taken from Table I ($W_a = k \cdot c \cdot \Delta T/t$) [22]. Typical specific heats for human muscle and fat tissue are 0.86 and 0.30 kcal/kg°C, respectively [23]. By using this formula, the rate of absorption was calculated for five subjects who participated in the experiment. These data are shown in Table II. The power levels were calculated at one of the thermistor locations 1–2 cm from the fat–muscle interface in the musculature. The total power input to the applicator was 40 W, thus the averaged specific absorption rate (SAR) was 2.81 W/kg per watt.

In a similar manner, it is possible to estimate the flow rate of blood in the musculature from the maximum slope

TABLE III
FORMULA FOR CALCULATION OF BLOOD FLOW

BLOOD-FLOW RATE IN MUSCLE

$$m = \frac{W_b}{k_2 \cdot c \cdot \Delta T' \cdot \rho_b}$$

m = blood flow millimeters per 100 g/min

W_b = power dissipated by blood flow W/kg

ρ_b = g/cm^3

k_2 = constant 0.698

c = specific heat of blood kcal/kg°C

ΔT = T - T_a

T_a = arterial temperature

T - tissue temperature

TABLE IV
CALCULATED VALUES FOR BLOOD FLOW IN HUMAN MUSCLE

CALCULATED BLOOD-FLOW RATE IN MUSCLE

RUN		ml/100 g/min
1	-	28.90
2	-	28.91
3	-	25.00
4	-	23.64
5	-	29.69

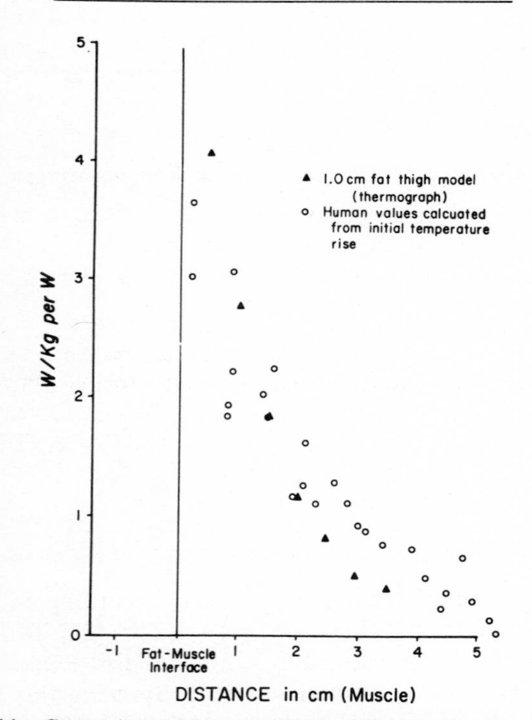

Fig. 14. Comparison of the calculations of the specific absorption rate (SAR) in thighs of human beings and models.

of temperature decline as produced by blood-flow cooling. This formula $m = W_b / k_2 \cdot c \cdot \Delta T' \rho_b$ is given in Table III [22]. The blood flow was calculated from the tissue-heating curves of the same five subjects and is given in Table IV. Since the maximum blood-flow rate in musculature during exercise is 30–33 ml per 100 g of tissue per minute [24], these values show that maximal blood-flow levels can be reached during treatment with this microwave applicator.

If it is therapeutically desirable to override this blood-flow cooling effect to maintain maximum temperatures in the tissue, the output of the applicator must be increased during treatment. For our sample of volunteers, because of physiological variability, 30-percent additional power would be required to ensure a therapeutically effective applicator with an adequate margin of power.

Finally, since testing the diathermy applicators for compliance with performance standards and determinations of the efficacy of such applicators will be mainly performed on models rather than on human subjects, it is important that the data obtained from the models represent that which would be obtained from humans. Fig. 14 shows a plot of the calculations of SAR from human experimentation as compared with those from thermograms of a comparable model. The average for models with 1–2-cm muscle thickness is 2.74 W/kg per watt while for human beings it is 2.81 W/kg per watt; the agreement is excellent.

V. CONCLUSIONS

It became apparent that for effective treatment using this applicator to selectively heat muscle to therapeutic levels, a thin radome should be used to minimize edge heating. The radome should be drilled and grooved and the applicator should be fitted with an air inlet so that cooled air can be blown through the applicator and dispersed by the radome for precooling and for continued cooling of the skin surface and subcutaneous tissues. This arrangement ensures that the highest temperatures in the distribution will be reached in muscle tissue.

It is evident from the experiments carried out on human volunteers that the 13-cm square direct-contact microwave applicator with radome and cooling selectively heats musculature, that temperature can be brought to a level between 43 and 45°C at 1–2 cm in muscle, and that by increasing power levels after blood flow has cooled this tissue, it may be feasible to push these temperatures back to 43–45°C and hold them at these levels for longer periods of treatment. Thus the possibility exists for using this applicator to heat cancerous tumors in muscle. Indeed, some physicians are presently using microwave applicators for this purpose.

From comparisons of calculations of SAR in human thighs with those in models it was found that the models are highly reliable predictors of temperature distributions produced in human tissues.

REFERENCES

[1] J. F. Lehmann and C. G. Warren, "Therapeutic heat and cold," *Clinical Orthopaedics*, vol. 99, no. 3, pp. 207–245, Mar./Apr. 1974.
[2] J. F. Lehmann, "Diathermy," in *Handbook of Physical Medicine and Rehabilitation*, Krusen, Kottke, and Elwood, Eds. Philadelphia, PA: Saunders, 1971, pp. 273–345.
[3] J. F. Lehmann, G. D. Brunner, D. R. Silverman, and V. C. Johnston, "Modification of heating patterns produced by microwaves at the frequency of 2456 and 900 MC by physiologic factors in the human," *Arch. Phys. Med. Rehab.*, vol. 45, no. 11, pp. 555–563, Nov. 1964.
[4] B. J. deLateur, J. F. Lehmann, J. B. Stonebridge, and C. G.

Warren, "Muscle heating in human subject with 915 MHz microwave contact applicator," *Arch. Phys. Med. Rehab.*, vol. 51, no. 3, pp. 147–151, Mar. 1970.

[5] J. F. Lehmann, A. Masock, C. G. Warren, and J. Koblanski, "Effect of therapeutic temperatures on tendon extensibility," *Arch. Phys. Med. Rehab.*, vol. 51, no. 8, pp. 482–487, Aug. 1970.

[6] C. G. Warren, J. F. Lehmann, and J. Koblanski, "Elongation of rat tail tendon effect of load and temperature," *Arch. Phys. Med. Rehab.*, vol. 51, no. 10, pp. 466–475, Oct. 1971.

[7] J. F. Lehmann and F. Krusen, "Biophysical effects of ultrasonic energy on carcinomas and their possible significance," *Arch. Phys. Med. Rehab.*, vol. 36, no. 7, pp. 452–459, 1955.

[8] J. E. Robinson, M. J. Weizenberg, and W. A. McCready, "Radiation and hyperthermal-response of normal tissue *in situ*," *Radiology*, vol. 113, no. 10, pp. 165–198, Oct. 1974.

[9] W. G. Connor, E. W. Serner, R. C. Miller, and M. L. M. Boone, "Implications of biological and physical data for applications of hyperthermia to man," *Radiology*, vol. 123, no. 2, pp. 497–503, May 1977.

[10] H. P. Schwan and G. M. Piersol, "The absorption of electromagnetic energy in body tissues," *Amer. J. Phys. Med.*, vol. 33, no. 12, pp. 370–404, Dec. 1954.

[11] H. P. Schwan, E. Carstensen, and K. Li, "The biophysical basis of physical medicine," *J. Amer. Med. Ass.*, Jan. 21, 1956.

[12] H. P. Schwan, "Electrical properties of tissues and cells," *Advan. Bio. Med. Phys.*, vol. 5, pp. 147–209, 1957.

[13] ——, "Survey of microwave absorption characteristics of body tissues," in *Proc. 2nd Tri-Serv. Conf. on Biological Effects of Microwave Energy*, pp. 126–145, 1958.

[14] C. C. Johnson and A. W. Guy, "Non-ionizing electromagnetic wave effects in biological materials and systems," *Proc. IEEE*, vol. 60, pp. 692–718, June 1972.

[15] B. J. deLateur, J. B. Stonebridge, and J. F. Lehmann, "Treatment of fibrous muscular contractures with a new direct contact applicator operating at 915 MHz," *Arch. Phys. Med. Rehab.*, unpublished.

[16] A. W. Guy and J. F. Lehmann, "On the determination of an optimum microwave diathermy frequency for a direct contact applicator," *IEEE Trans. Bio-Med. Eng.*, vol. BME-13, pp. 76–87, Apr. 1966.

[17] A. W. Guy, J. F. Lehmann, J. B. Stonebridge, and C. C. Sorensen, "Development of a 915-MHz direct-contact applicator for therapeutic heating of tissues," this issue, pp. 550–556.

[18] A. W. Guy, "Electromagnetic fields and relative heating patterns due to a rectangular aperature source in direct contact with biological tissue," *IEEE Trans. Microwave Theory Tech.*, vol. MTT-90, pp. 214–223, Feb. 1971.

[19] A. W. Guy, J. F. Lehmann, J. A. McDougall, and C. C. Sorenson, "Studies on therapeutic heating by electromagnetic energy," in *Thermo Problems in Biotechnology*. New York: ASME, 1968, pp. 26–45.

[20] J. F. Lehmann, B. J. deLateur, and J. B. Stonebridge, "Selective muscle heating by shortwave diathermy with a helical coil," *Arch. Phys. Med. Rehab.*, vol. 50, no. 3, pp. 118–123, Mar. 1969.

[21] J. F. Lehmann, A. W. Guy, B. J. deLateur, J. B. Stonebridge, and C. G. Warren, "Heating patterns produced by shortwave diathermy using helical induction coil applicator," *Arch. Phys. Med. Rehab.*, vol. 49, no. 4, pp. 193–198, Apr. 1968.

[22] A. W. Guy, J. F. Lehmann, and J. B. Stonebridge, "Therapeutic application of electromagnetic power," *Proc. IEEE*, vol. 62, pp. 55–75, Jan. 1974.

[23] D. Minard, *Body Heat Content in Physiological and Behavioral Temperature Regulation*. J. Hardy, A. Sagge, and J. Stoluijh, Eds. Springfield, IL: Thomas, 1978, pp. 345–357.

[24] C. A. Keele and E. Neil, Eds., *Samson Wrights Applied Physiology*, 10th ed. London, England: Oxford, 1961.

Physical Hyperthermia and Cancer Therapy

J. GORDON SHORT AND PAUL F. TURNER

Abstract—In general, malignant cells are more sensitive to heat than are normal cells in the range of 41–45°C. In addition, most clinically apparent tumors (above 1-cm diameter) have blood perfusion rates less than $\frac{1}{5}$ that of surrounding normal tissue, meaning that they may be preferentially heated. Hyperthermic treatment may be local (tumor only), regional (e.g., a limb), or whole body. Physical techniques for hyperthermia include metabolic heat containment, conduction through the skin (e.g., hot water bath), perfusion of externally heated blood, heated intravenous fluids and anesthetic gases, ultrasound, and electromagnetic EM coupling modalities. EM modalities include capacitive, inductive, and UHF-microwave radiative techniques, and may be invasive or noninvasive. Hyperthermia is effective against malignant cells not successfully attacked by ionizing radiation and also shows synergism with X-irradiation and some forms of chemotherapy making combination therapy attractive. Thermometric requirements vary with the treatment modality and clinical situation. Characteristics of tumors which may influence the choice of treatment are discussed. Local heating of deep-seated tumors with appropriate thermometry remains a technical challenge. Thermal dose requirements for various tumors and optimal protocols for adjuvant therapy are biological challenges.

Introduction

The history of heat therapy is both ancient and extensive. In areas of many thermal springs, such as the Mediterranean countries, balneotherapy was—and is—popular and widespread. Three millenia ago hot water and steam baths were praised by both Homer and Herodotus. The historical record attests the beneficence of the hot bath in ancient Israel, Egypt, North and tropical Africa, Java, and in America (among the Indians) as well as in the Roman Empire [46]. It remained to the Romans to organize the program into three operations, each with its special chamber: 1) sweating in hot air or steam, *sudatorium;* 2) hot bath *caldarium;* and 3) cold bath, *frigidarium.* In recent years it has been noted that populations using hot baths or saunas (Japanese and Finns) have a lower incidence of cancer of the skin, breast, testis and penis then their *not-so-hot* neighbors [45].

The modern era of hyperthermia in cancer therapy is considered to have begun in 1866 when W. Busch in Germany reported a spontaneous regression of multiple sarcomas of the face in a 43-year-old woman following a streptococcal skin infection (erysipelas) with fever. In this country the individual who made the most impact on fever therapy for cancer was W. B. Coley, whose professional life spanned the years from 1891–1936. Coley, picking up on the work of Busch and others, treated cancers by injecting virulent streptococci [9], [10]. Later, to avoid a treatment which, before antibiotics, was potentially as hazardous as the disease, he switched to toxins derived from killed organisms [11]. Although Coley had many notable successes with this therapy, he was unsure whether the fever (which was highly variable) or the stimulation of the immune system (which was unmeasurable) was more important. The question remains unanswered—especially since fevers produced rarely exceeded 104°F (40°C).

Meanwhile, a number of investigators in the first quarter of this century who were studying cancer in the laboratory found that tumors were killed by heat more rapidly than were normal benign tissues and that heat acted synergistically with X-irradiation to destroy cancer [8], [34], [43], [44]. Based on this knowledge, Warren introduced the era of physical hyperthermia for cancer therapy in 1935. He treated 32 hopeless tumor cases with diathermy or radiant energy to a rectal temperature of 41.5°C. There were no cures, but in all except 3 cases, there was immediate improvement lasting from 1 to 6 months [65].

At this point, hyperthemia should be defined. While any temperature above normal—37°C for humans—could be considered *excessive heat*, hyperthermia has come to mean temperature in humans above 41°C (106°F) induced with therapeutic intent. This distinguishes it from *normal* fever (pyrexia) of up to 41° and from hyperpyrexia, which is high fever such as might occur in heatstroke. Fever is internally induced temperature elevation resulting from an increase in the thermoregulatory setpoint of the body, while hyperthermia is externally induced temperature elevation in spite of a normal thermoregulatory setpoint [Kluger].

In the early part of this century the major interest in hyperthermia was for the treatment of infectious diseases, particularly syphilis and gonorrhea, although other diseases such as arthritis, asthma, and multiple sclerosis were also treated.

Neymann, in his monograph *Artificial Fever* (1938), reviews the techniques of *electopyrexia* [46]. Diathermy utilized contact applicators at frequencies around 1 MHz and currents from 3 to 6.5 A. Radiothermy utilized noncontact plates at frequencies between 10 and 100 MHz. Electric blankets were tried but Neymann's preference was for EM induction (at 12 MHz). Each method had its drawbacks, as a few quotations will show.

> Long experience with the use of diathermy has shown that if burns result due to the faulty concentration of the current in a circumscribed area, the subcutaneous adipose tissue is liquefied at a time when the skin is barely reddened. This melted fat later forms hard indurated tumors [masses] which in time may become infected (p. 94).

Neymann then used a radiotherm for 18 months:

> However, small burns under the armpits and wherever large amounts of perspiration gather are so frequent that they are a constant annoyance to the patient and operator. There are many objections to the practical use of the apparatus. First, it is difficult to raise the temperature quickly and with precision when excitable patients are being treated. Second, much heat is lost while lifting the patient out of the radiotherm to his bed. Third, the machine frequently catches on fire (p. 90).

Neymann's preference was for EM induction because of its

Manuscript received August 1, 1979; revised September 5, 1979.
The authors are with the BSD Corporation, University of Utah Research Park, Salt Lake City, UT 84108.

Reprinted from *Proc. IEEE*, vol. 68, pp. 133–142, Jan. 1980.

minor disadvantages:

> ... One must remove all metal springs from the bed since they will be heated by electromagnetic induction. The bed can actually be set on fire by the metal springs which may become red hot (p. 106).

Electric blanket technology was found wanting also:

> During the treatment, free perspiration, a good conductor, seeps into the blanket which is wrapped around the patient. Short circuits result. The blanket catches fire and the patient is stewed in his own juice. We are glad to report that we did not electrocute anyone with this appliance, though we did produce burns two inches deep and a foot in diameter (p. 98).
>
> Hyperthermia indeed!

Since physical hyperthermia was viewed primarily as a treatment for certain infectious and some degenerative diseases and had barely been tried clinically for human cancer, it seems in retrospect not unusual that hyperthermic research should have gone into eclipse in the forties with the advent of antibiotics as the *magic bullets* to fight infections and the advent of chemotherapy–hopefully *magic bullets* for cancer.

Two factors have stimulated the recent resurgence of interest in thermotherapy–fallout from recent technological advances in controlling *nonmagic bullets* and the relative failure of surgery, X-ray, and chemotherapy to control cancer adequately.

The clinical objective of the hyperthermic treatment of cancer is to produce sufficient heat to kill cancer cells without significantly damaging normal benign cells, and to treat all malignant cells existing within the host. For a variety of reasons this objective is often not attainable; therefore, hyperthermia is likely to be combined with (partial) surgical extirpation of the tumor(s), radiation therapy, chemotherapy, and immunotherapy. Most, if not all, malignant cells are more sensitive to heat than are benign cells in the range of about 41–45°C.

Thermal dose has been defined as the number of degrees above 37 or 40°C times the time in minutes, but Atkinson has proposed a dose based on biological effect and using a nonlinear weighting factor [2]. It is significant that as a rule of thumb the time necessary to produce a bioeffect (e.g., cell killing) is halved for each degree rise in temperature above 42.5°C [52], [53]. Above 46°C the times for cell killing become quite short (minutes) and the differential sensitivity of malignant and benign cells is lost or at least obscured. Above 50°C all cells are killed very rapidly.

Regardless of adjuvant therapies there are two approaches to choosing an adequate thermal dose. In some cases it may be possible to heat a tumor to 50°C or higher without significantly damaging surrounding normal tissue. This may be done by accurate and adequate heat injection into the tumor or by depending on impaired egress of heat from the tumor or, more commonly, by both.

The other alternative is to heat the tumor in the range of 41–45°C (for from 4h+ at 41°C to about $\frac{1}{4}$ h at 45°C) and depend on differential heat sensitivity rather than on preferential tumor heating. Clinically, the picture is seldom this clearcut and one would prefer to keep normal tissues at a temperature lower than the tumor. Furthermore, if the whole body is being heated (in order to treat known or suspected multiple tumors) a temperature above 42°C is hazardous, partly because it is difficult to control quickly and precisely and partly because of physiological stress.

While physical hyperthermia can be accomplished without electronics, recent advances in electronic technology have encouraged a new look at EM coupling modalities, which may offer the advantage of direct rather than blood-mediated heating of tissue. At frequencies below 50–100 MHz, capacitive and inductive techniques are utilized, while above 100-MHz, radiating applicators are used. Invasive as well as noninvasive applicators have been designed, and nonperturbing but invasive temperature probes are available. Noninvasive thermometry appears possible but a practical system has yet to be developed. Finally, microcomputer technology is being used to attain fully automatic temperature control of the subject tissue.

Anatomic Extent of Treatment

In principle one would prefer to heat all of the malignant cells just enough to kill them and spare normal tissues. But most malignancies have the capacity to spread to distant sites–metastasize–as well as to expand locally, and to do both in often unpredictable fashion. The rate of growth of tumor nodules may be highly variable depending on location, as well as on histological type and host defenses, and subclinical latent periods of many years are not uncommon. For this reason, as well as the fact that many patients have too many known tumors to be treated one-by-one, there has been a great deal of interest in whole-body hyperthermia [28], [35], [36], [48]–[50], [63], [64]. The target temperature most frequently mentioned is 41.8°C. This temperature is well tolerated by sensitive benign tissue, such as liver, but requires long treatment times to do significant damage to malignant cells–commonly several weekly sessions of 4–6 h each. The induction period is long–$1\frac{1}{2}$ to 3 h–and the patient requires close monitoring for the entire duration of both the treatment and the recovery. A major limitation is the cardiovascular status of the patient. The heart rate increases around 10 beats/degree Celcius rise in temperature. A pulse rate of 140 beats/min for 4–6 h is not tolerated well by some elderly and debilitated patients and fatalities have occurred. Whole-body hyperthermia by EM coupling has the potential advantage of reducing the warm-up period to about 30 min, since the energy penetrates into the body directly rather than requiring convection by blood perfusion to the interior from the skin as in most present methods [20].

Factors Influencing Choice of Treatment

In order to understand the technical challenge of thermotherapy, it is necessary to have some appreciation of the anatomic, physiologic, and biologic characteristics of the group of diseases we refer to collectively as *cancer* or *malignant neoplasms*.

It is difficult to define *tumor* in an unexceptionable way but the definition of Willis is acceptable:

> A tumour is an abnormal mass of tissue, the growth of which exceeds and is uncoordinated with that of the normal tissues, and persists in the same excessive manner after cessation of the stimuli which evoked the change [66].

Neoplasms are divided into 2 categories–benign and malignant–which in reality tend to form a continuum with no sharp dividing line. In general, however, benign tumors tend to be more or less spheroidal, encapsulated, and slow-growing. By contrast, malignant tumors, although they may seem rounded, have irregular and invasive advancing margins, are faster growing, and have the ability to form daughter colonies in distant locations (metastasis).

The vast majority of cancers are derived from the epithelial cells of surfaces and glands and are called carcinomas. They may spread along the surfaces where they originate, surrounding hollow organs as they also invade outward (e.g., esophagus, stomach, rectum, bronchus). They may also spread along serous surfaces such as the pleura or peritoneum. Metastasis is usually to local lymph nodes before distant spread.

Sarcomas derive from nonepithelial tissues such as muscle, fat, and connective tissue. They are more likely to start as spheroidal tumors. They have more of a tendency to spread early to distant locations (especially lung) via the blood stream. Leukemias are diffuse neoplasms in which white blood cells become malignant. Little attention has so far been directed to human leukemia in hyperthermia clinical research.

Cancer may occur anywhere in the body. Centrally located tumors, as in the lung or pancreas, may be 15 cm from the nearest skin surface, making noninvasive energy application difficult. However, none of the pancreas is more than 4 or 5 cm away from the stomach or duodenum, at least suggesting the possibility of access through the upper gastrointestinal tract. Tumors in the lung present special problems because of the high air content of surrounding tissue. Intracranial neoplasms present unique challenges because of potential reflections and resonances in the skull and because of integral fluid reservoirs in the ventricular system which might be subject to over-heating. Breast tumors may be surrounded by adipose tissue with its lower heat loss characteristics, which may be advantageous in microwave heating. Abdominal heating may be complicated by differential heating of intestinal contents.

One of the most important considerations in the hyperthermic treatment of cancer is the vascularity of tumors and its relation to the rate of blood perfusion and heat transfer [16]-[18], [25], [26], [41], [61]. It has been known for over half a century that tumors have an impaired blood circulation and, therefore, tend to concentrate applied heat because of reduced heat transfer capabilities [21]. In addition to the vital functions of supplying oxygen and nutrients to the tissues and removing carbon dioxide and metabolic wastes, the blood circulatory system has the equally vital function of transferring metabolic heat from the interior of the body to the skin and lungs, whence it may be dissipated to the environment. As a tumor grows, it must form new blood vessels. In a series of elegant experiments, Folkman has shown that cancers secrete a *tumor angiogenesis factor*, which stimulates the growth of new capillaries [16]. Although tumors may grow to encompass arteries and veins, it is felt that these larger vessels are not formed by the tumor. Thus a diffuse plexus of capillaries forms, some of which may dilate to form larger blood sinuses.

A tumor expands predominantly by the growth of cells at the advancing margins. Here the new capillaries are formed and here they are closely related to their conjunctive arteries and veins. Capillaries in the center of the tumor are connected only to other capillaries; thus blood flow becomes quite sluggish. At the same time, the poorly controlled growth of new capillaries at the margins of the tumor may produce what are functionally arteriovenous anastomoses—the equivalent of an electrical short circuit.

Thus the center of a tumor characteristically has poor oxygenation and nutrition with decreased metabolism and intrinsic heat production. Central necrosis (cell death) is common. Externally injected energy is poorly dissipated and the temperature increases above that of surrounding normally perfused tissue, typically from 1-5°C [32], [33], [39], [58]. By contrast, the advancing margins of the tumor have good perfusion, perhaps even better than normal, with high oxygenation, good nutrition, and rapid heat transfer. Since these marginal tumor cells are the most rapidly growing part of the cancer and have the best cooling, it would seem, in principle, best to target them for whatever thermal dose is considered necessary, with proper regard for surrounding benign tissue.

At least in the idealized case, the height of the increased core temperature is probably of little concern. Every malignant tumor has a benign cellular component—capillaries, connective tissue, inflammatory cells—and the temperature in the core of a tumor whose margins are heated to 43-44°C may rise to 48-50°C or even higher. This will likely kill all the benign cells with resultant thrombosis of all the vascular channels. This might be advantageous in delaying absorption of tumor breakdown products with a decrease in the possibility of producing disseminated intravascular coagulation (an uncommon but feared complication), and in increasing the immune stimulation of the host. Parenthetically, it should be mentioned that the well-oxygenated rapidly growing marginal cells of a tumor are more X-ray sensitive than are the central hypoxic cells, providing one rationale for combining hyperthermia with X-ray therapy [12], [13].

Although infiltration is the hallmark of cancer, it tends to predominate along paths of least resistance, which may modify the tumor's shape. The typical cancer is not a smooth-surfaced spheroid. Instead, fingers of malignant cells reach out into the adjacent tissue, markedly increasing the surface area of the tumor and allowing for improved heat transfer both by conduction and by perfusion. This complicates the job of the thermotherapist.

Metastases provide another complication. Their locations may be predicted statistically, but in individual cases they have the annoying propensity to pop up capriciously in unexpected locations. Futhermore, small metastatic deposits of cancer—e.g., less then 5-mm diameter—have a high surface-to-mass ratio with all that implies for rapid heat egress, both by conduction and convection. Preferential heating of these small tumors, therefore, is unlikely and one may have to rely only on the increased heat sensitivity of malignant cells as compared with benign tissue.

Cancer cells are abnormal not only in their escape from normal growth controls, but also in being more easily damaged by adverse environmental influences than are normal cells. These adverse conditions include ionizing radiation, certain chemicals, and hyperthermia. Radiosensitivity is associated in general with rapidly growing, primitive (dedifferentiated), well-oxygenated cancer cells. There is considerable variability among cancers of different histologic origin. Sensitivity to various types of metabolic inhibitors (chemotherapy) is also highly variable among tumors of different origins. Observations over the last century indicate that thermosensitivity is also variable. However, at the present time there is no great body of knowledge on thermal doses required for lethality in human cancers of various histologic types, nor on the synergism with radiotherapy and chemotherapy. This is a major research challenge for the immediate future.

It has been shown that cells, both normal and malignant, develop thermotolerance [29]. How important this is from a practical standpoint remains to be determined. The degree of thermotolerance appears to decay logarithmically with time, most being gone in 48 h and almost all having disappeared in 72 h. For this reason most hyperthermia treatments have been given at not less than 48-h intervals.

Almost every host develops resistance to the growth of its tumor. Specific immune resistance is a function of both the tumor type and the host's ability to respond. Immunotherapy is attracting considerable attention as an adjuvant treatment modality of some promise, but as of now one of less potency [27]. Nevertheless, there is evidence that immunogenicity may be stimulated if a tumor grows in a host, dies in the host, and is allowed to be absorbed slowly [57]. Although this precept has obvious theoretical implications for surgical excision vis-à-vis treatment *in situ*, the practicalities have yet to be worked out. At least in some cases and under the right conditions, small secondary tumors can be seen to regress after successful treatment of the primary cancer.

In addition to the foregoing general characteristics of tumors that may have a bearing on the choice of treatment, there may be other considerations for each individual case. The size of the tumor, its depth and location (especially in relation to delicate, vital structures) are important. Large tumors may need to be treated fractionally to prevent massive necrosis with rapid resorption of toxic breakdown products, expecially in a debilitated patient. A deep-seated tumor—10-15 cm—is a poor candidate for noninvasive local EM radiative heating because of attenuation, even at low frequencies. A lung tumor surrounded by aerated tissue is not a candidate for ultrasound. Heavy fat layers make Radio-frequency (RF) capacitive coupling unattractive because of the high resistance heating of fat with possible burns and necrosis. A limb tumor may be a suitable subject for regional hyperthermic perfusion.

The known or probable extent of the tumor is important. Local lymph node metastases may be included in the field of local or regional treatment, but distant metastases suggest the desirability of whole body hyperthermia unless the metastases are considered accessible for treatment on a one-by-one basis. However, the presence of known multiple metastases is almost sure to mean multiple small, subclinical metastases and one-by-one treatment is bound to be frustrating unless the patient is one of those fortunate people who possess a good immune response.

Another factor of importance is the vascularity and perfusion of the tumor. Vascularity is a static factor and can be roughly assessed by the histologic appearance of a biopsy. Perfusion is dynamic but could probably be evaluated, at least roughly, by studying development and disappearance of tumor stains in X-ray and/or isotope studies. Well-perfused tumors may not be differentially heated, but these appear to be a small minority.

Whole-body hyperthermia requires a patient whose physiological condition, particularly in relation to the cardiovascular system, will take the stress that may produce a heart rate of 140 beats/min for 6 h. Adequately trained personnel must monitor therapy continually. EM heating has the potential to make possible the treatment of several patients in a working day by a single treatment team because of a shorter induction period and more rapid control of energy input [20].

It would be helpful to know in advance the thermosensitivity of the tumor in order to know whether the thermal dose required to be effective can indeed be administered under the prevailing circumstances. There is some hope that cell cultures *in vitro* of the patient's tumor can help answer this question. Transplantation of human tumors into congenitally athymic nude mice would probably be even better for testing, since there is a more realistic simulation of the tumor in the original host. (Nude mice are a special strain that produce essentially no antibodies and thus will not reject tumors from a different species [23], [24].)

The compatibility of the hyperthermic treatment modality with other adjuvant therapies also needs to be considered. A combination with X-ray therapy is most likely to be successful if hyperthermia and X-irradiation are temporally as close as possible—hopefully within a $\frac{1}{2}$ h. Some even advocate giving the X-ray treatment during the hyperthermia [62]. Furthermore, radiography may be useful in the positioning of invasive or orificial applicators and of temperature probes. Intervals to allow for the disappearance of thermotolerance must be coordinated with radiotherapy protocols. Some forms of chemotherapy may be useful adjuncts including hyperthermic perfusion of cytocidal drugs [55], [56].

Although many lines of evidence point to the importance of tumor immunity, it has been difficult to manipulate immunity reliably for the benefit of the patient. There is a little suggestive evidence that whole-body hyperthermia may be immunosuppressive, as are ionizing radiation and chemotherapy [12], [13]. Local and regional hyperthermia, leaving a necrotic tumor to be slowly resorbed, should be stimulatory. For human tumors, the evidence on this subject is clearly very incomplete.

Finally, a most important consideration in treatment is the technical ability of the staff. What equipment is available and how well qualified is the treatment team to use it? Because hyperthermic techniques are not trivial, at the present time few people are working with more than one or two modalities, as indicated in Table I.

Energy Coupling Modalities

Only non-EM modalities have been used extensively for whole-body hyperthermia (with the partial exception of the Siemens heat cabinet) [42]. Conduction heating is combined with metabolic heat confinement. Heat production by the body amounts to about 75-100 kcal/h, equivalent to about 100 W. If this heat could be entirely confined, it would by itself raise the temperature of a 70 kg man 5°C in 4 h [50]. This would be a long induction period, but it at least indicates the importance of metabolic heat. To speed up the process and to allow for temperature control in the patient, two approaches have been used—immersion and temperature-controlled *space suits.*

Several media have been used for immersion heating. Hot air cabinets may utilize infrared radiation from light bulbs [42], [65] but humidity must be controlled to prevent undue evaporative cooling. Quick temperature control is difficult, access to the patient is limited, and the method has not been popular. Hot water baths eliminate the evaporative cooling from perspiration, but otherwise the same objections hold. Melted paraffin (MP 43-46°C) has been used in a technique refined over a number of years by Pettigrew and co-workers in Edinburgh [50]. Briefly their system works as follows: The patient is anesthetized and a urinary catheter and nasogastric tube are inserted. Temperature is monitored continuously from rectal, esophageal, and tracheal thermometers. Also monitored are electrocardiogram, pulse, blood pressure, and central venous pressure. An insulated endotracheal tube delivers oxygen-enriched air at 80°C. Intravenous infusions

TABLE I
REPORTED HYPERTHERMIA MODALITIES

Anatomic Extent	Method	Temperature	Time	Adjuvant Therapy	Comments and Considerations	References
Whole Body	Hot air and/or radiant heat	41–41.5	5–21 h	X-ray	Cumbersome–limited access to patient	Warren
	Hot water by immersion	40	2½ h	*	Slow temperature control	Von Ardenne
	Hot water by *space suit*	39.5–42	90 min-5 h	Systemic chemotherapy	*	Larkin, Bull
	Melted paraffin by immersion	41–41.8	90 min-4 h	Cytotoxic drugs	Limited access to patient	Pettigrew
	Perfusion–extracorporeal heat exchange	41.5–42	5 h	Cytotoxic drugs	Surgical procedure	Parks
	Microwave and hot air cabinet (Siemens)	43–43.6	40 min-1 h	*	*	Moricca
Regional	Hot water by immersion	46–47	67 min	X-ray, cytotoxic drugs	Limited to extremities	Crile
	Radiant heat–visible and infrared	43–45	20 min	*	*	Lehmann
	Perfusion	38–45	30 min-8 h	X-ray, chemotherapy, cytotoxic drugs	Surgical procedure	Cavaliere, Stehlin, Shingleton, Woodhall
	Capacitive RF	42–50	30 min-3 h	*	Skin and fat burns have been a problem.	LeVeen, Storm
	Inductive RF	41–63	35 min	*	*	Storm
Localized	Irrigation	41–45	1–3 h	*	Bladder tumors, peritoneal cavity	Hall *et al.* Ludgate
	Capacitive RF	48.4	30 min-3 h	*	*	LeVeen, G. Hahn
	Inductive RF	42.8–45	30 min-1 h	X-ray	*	Kim and Hahn, Von Ardenne
	Ultrasound	43.5	30 min	*	Best focusing. Reflections, caviation, and bone heating are problems.	G. Hahn
	Microwave	42–44	12–20 min	*	Deep heating (10 cm) difficult with noninvasive (NI) applicators.	Sandhu, Dickson, Mendecki, Samaris, *et al.*

are preheated to 45°C. The patient is wrapped in a double envelope of polyethylene, and paraffin at 50°C is pumped into the bath surrounding the enclosed patient. Normal heat loss is reversed, and energy is pumped into the patient at the rate of about 3250 cal/min (195 kcal/h), sufficient to raise the body temperature approximately 5°C in 1 h to the targeted treatment value of 41.8°C, which is maintained for 4 h. When the temperature reaches 41.8°C, all but a thin layer of solidified wax is removed and body temperature coasts up to 41°C, where it is maintained by peeling the wax layer back as necessary to allow evaporative cooling of perspiration. Energy sources are identified as follows: 1) from body metabolism–1250 cal/min; 2) from wax (including latent heat of solidification)–1500 cal/min; 3) from respiration of heated gases–500 cal/min [50].

In this country, the use of hot suits has been favored by a number of investigators as being less cumbersome and allowing access for X-rays [4], [35], [36]. Two commercially available circulating hot water suits are available under the trade names Blanketrol and K-Thermia.

Hyperthermic Perfusion

Stehlin has used hyperthermic perfusion for the treatment of melanomas and sarcomas of the limbs. The circulation to the limb is isolated and blood is perfused at a temperature of 43.3°C. The chemotherapeutic drug is added to the blood when the temperature of the limb is from 38–40°C. The blood is heated in a coiled tube passing through a carefully controlled water bath [55], [56].

Taking this method one step farther, Parks has developed a device for producing whole-body hyperthermia by perfusion [47], [48]. Blood circulates through an extracorporeal heat exchanger and mixer apparatus. There are two reservoirs, one for heated and one for chilled blood. The hot reservoir is kept at a temperature of 48°C, and the cool, at 30°C. With appropriate valves and pumps, the temperature and flow rate are closely controlled. The main advantage of the reservoirs is that they allow for quick temperature control.

Ultrasound

Ultrasound by definition is above 20 kHz, but as used for diathermy is usually around 1 MHz. At this frequency the wavelength is about 1.5 mm in tissue. Since many anatomic structures are large as compared with this wavelength, reflection of ultrasound from tissue interfaces is a problem. Up to 30 percent of energy is reflected from a muscle/bone interface, while bone absorbs ultrasonic energy 10 times more readily than muscle [37]. Even less energy penetrates the lung. An advantage of ultrasound, however, is that since absorption is proportional to the protein content of tissue, relatively little ultrasonic energy is absorbed in fat. Another advantage of ultrasound is its small angle of divergence–6.5° for a 2-cm diameter transducer, less for larger transducers. This makes it possible to focus energy into small tissue volumes, especially with multiport applicators. Multiple applicators also decrease the hazard of gaseous cavitation in the intervening normal tissues.

Since the velocity of sound is a function of the temperature of the medium, this makes possible the use of ultrasound for noninvasive thermometry according to S. Johnson. A diagnostic ultrasonic transducer has been fabricated by Filly to fit around a biopsy needle using imaging techniques for directing the biopsy needle to the tissue of interest. Presumably a similar system could be utilized to improve the accuracy of placement of invasive temperature probes and microwave or other applicators.

EM Coupling Modalities

EM coupling modalities have perhaps attracted the most interest in recent years. There have been significant refinements with the advent of solid-state electronics and with computerized control technology, as well as advances in thermometry. These advances, coupled with those in therapeutic radiology and chemotherapy, have evoked a profound new interest in the possibilities of thermotherapy.

Coupling of EM energy is through one of three modalities: capacitive, inductive, or radiative (microwave). Ultra-shortwave diathermy at the ISM (industrial, scientific, medical) frequencies of 13.56 and 27.12 MHz has been used with contact electrodes or with capacitor plates separated from the patient in order to produce more parallel field patterns and eliminate edge-heating effects [46]. The alternating current produced by this form of heating may produce heating patterns that are difficult to predict because the current will follow paths of least resistance, perhaps bypassing intended target tissues if these have low conductivity when parallel to adjacent high conductivity tissue. Also, the subcutaneous fat layer is a high resistance pathway in series which produces more heat—and dissipates it poorly because of low thermal conductivity and low vascular perfusion. Fat heating may be minimized either by choosing a pathway through the body where the subcutaneous fat layer is thin, or else by using invasive or orificial electrodes in or near the target tissue and large external electrodes. Specially designed electrodes for intravaginal or rectal insertion were used before the antibiotic era to treat pelvic inflammatory disease and prostatitis. Heat may also be conducted out of the fat by surface cooling techniques.

Inductive heating was demonstrated in 1893 by d'Arsonval, who elevated the temperature of an animal by placing it within a solenoid through which a high-frequency current was passing [40]. Flat applicators containing a pancake coil of several turns use conventional ISM frequencies, as in capacitive coupling (13.56 and 27.12 MHz). Although the problem of overheating subcutaneous fat is eliminated, a pancake coil has little ability to penetrate more than 1 to 2 cm of muscle beneath the fat.

Storm places the patient within a coil or coils [58]. In this situation, deeper heating can be produced in a segment of the thorax and/or abdomen, the length depending on the number of coils used. Surface cooling is used on the underside of the patient. Any preferential heating of a tumor depends on a decreased blood perfusion of the tumor. According to Storm, about 80 percent of tumors can be heated significantly above the level of the surrounding tissue with this method or regional hyperthermia. Induction has poor coupling efficiency, but the treatment appears to be well tolerated by patients and presumably should be able to produce heating of deep tumors, such as in lung and pancreas, with less trauma and stress to the patient than whole-body hyperthermia. Note, however, that inductive heating of the body tends to favor superficial tissues with attenuation at increasing depths, as Cetas has shown and we have confirmed.

Noninvasive applicators at UHF-microwave frequencies are not required to be in contact with the body. Small animal experimentation is usually performed with noncontact horn antennas. Contact applicators are usually preferred for human use, and coupling efficiencies of approximately 95 percent are expected. Low water content tissues such as fat and bone absorb relatively little microwave energy as compared with skin, muscle, and other wet tissues. As frequency increases, the depth of penetration decreases until at 2450 MHz, for an assumed 1-cm fat thickness, the energy is reduced to 40 percent of the incident level in 2 to 3 mm of the skin thickness [19]. Although the subcutaneous fat is relatively transparent to plane waves, it is clear that little energy is available for heating of deep tissues at this frequency.

The situation improves considerably at the next ISM frequency of 915 MHz. In fact, the frequency range of greatest clinical interest may be from about 100 to 1000 MHz. Unfortunately, in this country the next available band is at 40.68 MHz. (Outside of the U.S. an ISM band at 434 MHz is allowed.) At 40.68 MHz, the penetration depth in infinitely thick wet tissue (for energy reduction to 13.5 percent of the incident value) is 11.2 cm for plane waves, the body becoming increasingly penetrable [30]. Even though for a single applicator the pentration depth is likely to be somewhat lower on account of the multilayered and curved-shape nature of the tissues, the body is nevertheless more penetrable than at higher frequencies. Regardless of the frequency chosen, the energy density in the locally applied EM beam decays rapidly because of absorption, divergence, and fringe effects of the fields.

Several stratagems are used to counteract the unavoidable surface heating, particularly at UHF-microwave frequencies, when it is desired to drive heat patterns deeper into the body. Chilled distilled water is circulated through a chamber on the front of contact applicators to draw off excess heat from the skin. If monitoring of skin surface temperature is desired, it is important to insulate the temperature probe from the applicator (e.g., by covering it with a small piece of foam tape). In analogy with X-ray therapy, multiple entry ports have also been used [39], [59]. An array of applicators encircling a body part can be activated sequentially by switching mechanisms. Although this clearly spreads out the skin heating over a wider area for better heat dissipation, it also spreads out the applicators; and for deep tumors in the trunk of a large person, this may make the method inapplicable.

An alternative approach is to use invasive applicators. These are discussed in the paper in this volume by Taylor. Since attenuation occurs rapidly both from divergence and absorption, asynchronous time sharing has also been used here [60]. In addition to these interstitial antennas, applicators have been designed for insertion into natural body orifices such as esophagus, rectum, vagina, etc. These may be made to radiate from a considerable length of the applicator (up to 10 cm), and, with the increased diameter available, may incorporate surface cooling or temperature probes.

Surgical exposure of tumors presents another level of accessibility. More or less conventional applicators operating at

915 or even 2450 MHz could probably be designed to meet the requirements of the operating room with regard to size, sterility, coupling efficiency, electrical noninterference with monitoring equipment, anesthetic safety, etc. There is also the possibility of implanting active, passive, or active/passive radiators.

There are practical frequency considerations in addition to those mentioned regarding depth of penetration. For contact applicators in general, the minimum dimensions of the radiating face are inversely proportional to the frequency. Tricks of geometry and of dielectric loading can make low frequencies radiate from small applicators and broad-band applicators may have a range of up to 3 octaves, but there is always a trade-off penalty—loss of efficiency, variable and perhaps undesirable radiative patterns, special coupling requirements, etc. Large applicators—for deep penetration at low frequencies—may not necessarily be disadvantageous.

Recently one of us (P.F.T.) has done some work on synchronous phased arrays. If 2 or more applicators—either invasive or noninvasive—radiate EM energy toward a central volume of tissue in phase and with identical polarity, the E vectors from each applicator will add. Since power is proportional to the square of the voltage, the center will experience a power deposition which is proportional to the square of the number of applicators. In an asynchronous phased array with sequential firing of the applicators, the power deposition is only equal to the sum of the power from each applicator. In a synchronous array of 6 invasive monopole antenna applicators in a circle of 3-cm diameter in skeletal muscle of a living anesthetized pig, Turner was able to demonstrate a temperature in the center of the array that was 2°C higher than directly between 2 adjacent applicators using a frequency of 915 MHz.

Finally, EM coupling modalities must be used with extreme caution, if at all, in patients with cardiac pacemakers.

Metrology

There is probably a considerable difference between optimal and clinically necessary temperature measurement. While sensitivity of 0.01°C with accuracy and precision of 0.05° are desirable and perhaps even necessary in the laboratory, the clinical requirements are much less stringent for several reasons.

1) For any particular human cancer, no one yet knows the optimal thermal dose (i.e., the minimal dose necessary to guarantee tumor sterility), or if this is even achievable with acceptable levels of damage to normal tissue.

2) The temperature in the absence of therapy is not uniform throughout a typical tumor and becomes highly variable with local forms of hyperthermia for circulatory reasons previously discussed. Marked inhomogeneities of temperature may occur.

3) It should be remembered that many malignant tumors are far from spheroidal, implying an increased surface area for conductive heat transfer.

4) It is impractical to place more than a few temperature probes in a typical patient for reasons varying from the physical well-being of the patient to the psychological well-being of the physician. Since the domain of existing invasive temperature probes is only of the order of a couple of millimeters radius, sampling errors can be expected. Furthermore, placement of a probe in any but superficial tumors is likely to be inexact (with reference to whether the probe is in the tumor core or near a margin).

5) If the treatment is fractionated and at least partially successful, the thermal dynamics of the tumor will be changing as portions necrose and are resorbed, thus adding another source of inhomogeneity.

Although it has been shown that the time required to produce a given bioeffect—such as cell death—is halved for each degree Celsius rise above 42.5°C, for local hyperthermia, a typical thermotherapist would be satisfied if he could be assured that the temperature throughout the tumor(s) was within ±1.0°C of the target temperature. Thermometric systems operating within a tolerance of ±0.2°C will obviously be satisfactory and the onus will be on the therapist to place the probes in the appropriate locations.

Whole-body hyperthermia requires a higher degree of accuracy for the safety of the patient where the target temperature is usually 41.8° ± 0.2°C and where temperatures above 42°C may be life-threatening (e.g., from liver damage, or from the cardiovascular stress of a high cardiac output for 3-4 h in an older patient with coronary atherosclerosis, or from tachyarrhythmias). Clearly, whole-body hyperthermia has the greatest likelihood of producing homogeneous tumor temperatures. With local hyperthermia it would usually be most desirable if the margins of the tumor could be raised to a minimum target temperature—e.g., 43°C for 1 h—while allowing the core to reach whatever higher temperature it achieves under the circumstances.

Temporal resolution of several seconds is adequate for continuous reading nonperturbing probes. With thermocouple or thermistor probes having metallic leads, temperature readings can only be relied on when the RF power is off. Here a much faster response time is required in order to see when the temperature reading has stabilized—often several seconds. Frequency of temperature reading can only be determined by experience with due regard for patient safety, especially during induction of hyperthermia. Temperature probes must be isolated so as not to present a shock hazard to the patient and to avoid interference with any monitoring equipment being used on the patient, such as an electrocardiogram.

Calibration of invasive temperature probes involves considerations of stability and sterility. For many situations treatment times may extend to 4 h. If an hour is added for calibration and other set-up time, it is clear that drift in 5 h should be negligible (±0.1°C). For human use, it must be possible to sterilize the probes by a method known to kill hepatitis viruses and bacterial endospores (e.g., autoclave, ethylene oxide, formaldehyde); it must then be possible to maintain sterility during calibration. The protective sheath must have good heat transfer characteristics and must be impervious to water if a water bath is used for calibration. Since errors may be additive, it is important to depend ultimately on an NBS traceable thermometer. Thermometric fixed points, such as the ice point or the melting point of gallium (29.78°C), are also valid calibration checks.

Thermometry is hardly a new science. Nevertheless, the special demands of hyperthermia by EM modalities have produced many novel concepts for temperature measurement. The art of thermometry in EM fields is therefore in a fluid state with many promising methods in an investigational or prototype stage. The subject is well reviewed by Cetas and Connor and by Christensen [6], [7].

Currently available temperature probes are invasive and may be either perturbing or nonperturbing. The inexpensive and

TABLE II
(ACS 1979)
SELECTED CANCER STATISTICS AND HYPERTHERMIA CONSIDERATIONS

Site	Estimated New Cases	Estimated Deaths	Possible Regional or Local Hyperthermic Approaches ULT	CAP	IND	MIC	Problems and Considerations
Buccal cavity and pharynx	24 400	8650	NI	NI	NI	NI	Proximity to eyes, complex geometry–reflections, uneven heat deposition. Specialized applicators. High vascular perfusion.
Esophagus	8400	7500	–	OR	NI	OR	Local metastases at first diagnosis. Large blood vessels adjacent. Difficult thermometry (danger of mediastinitis).
Stomach	23 000	14 100	–	NI OR	NI	OR	Local metastases at first diagnosis. Difficult geometry and placement of applicators. Some success with WBH and hyperthermic irrigation.
Colon	77 000	42 800	–	NI	NI	NI OR OP	Right colon not accessible by orificial applicator (too deep). Adjacent loops of bowel containing liquids and gases. Thermometry difficult.
Rectum	35 000	9100	–	OR	NI	OR IN	Metastases in regional lymph nodes may be difficult to heat.
Liver and biliary	11 600	9200	NI	NI	NI	NI IN	Most sensitive tissue to heat. Large organ. Usually good differential heating of tumors. Gall bladder contents may overheat.
Pancreas	23 000	20 200	NI	NI	NI	NI OR OP	Deep, inaccessible. Intestinal mucosa are heat sensitive. Thermometry difficult. May require surgical implantation–active and passive radiators and temperature probes.
Larynx	10 400	3500	NI	–	–	NI OR	Good candidate for hyperthermia–accessible, early symptoms, second chance laryngectomy if hyperthermia fails.
Lung	112 000	97 500	–	NI	NI	NI OP	Often far advanced at first diagnosis, metastasis common. Geometry difficult–air, bone, major vessels. Thermometry tricky.
Bone and connective tissue	6400	3350	NI	–	NI	NI	Commonly metastasize to lung. Problems relate to size and location, etc.
Breast	106 900	34 500	NI	NI	NI	NI IN	Deep surface against muscle may be harder to heat. Metastases to axillary lymph nodes often small (1 mm). Synchronous phased array?
Melanoma	13 600	4300	NI	–	–	NI	Metastasize widely.
Uterus	53 000	10 700		OR IN	NI	OR IN OP	Nonsurgical candidates have local extension which is deep for external applicators, distant from intrauterine cavity.
Ovary	17 000	11 100	–	–	NI	NI OP	Many tumors cystic, large, easily ruptured–thermometry difficult. Tumors deep in pelvis. Disseminate through peritoneal cavity.
Prostate	64 000	21 000	NI	OR	–	NI OR	Metastasize to lower spine.
Brain	11 600	9500	IN		NI	IN OP	Reflections, resonance, fluid-filled ventricles, proximity to eyes. Surgical procedure for invasive applicators and temperature probes. Hyperthermic chemotherapeutic perfusion?
Thyroid	9000	1000	NI	–	NI	NI	Adjacent tissue has many large blood vessels and important nerves.
Leukemia	21 500	15 400	–	–	–	–	Diffuse disease. Whole body hyperthermia with chemotherapy and/or X-irradiation? Extracorporeal hyperthermia of blood?
Lymphoma	38 500	20 300	NI	NI	NI	NI	Often diffuse. May be difficult to heat preferentially. Dosimetry may be difficult.
Bladder	35 000	10 000			NI	OR	Temperature probe placement tricky in invasive tumors. Some success with hyperthermic irrigation.
Kidney	16 200	7500	NI	NI	NI	NI IN	May require X-ray placement of temperature probes and IN applicators.
All sites	765 000	395 000					

Abbreviations: ULT–Ultrasound, CAP–Capacitive RF, IND–Inductive RF, MIC–Radiative UHF, microwave, NI–Noninvasive, IN–Invasive, OR–Natural orifice (orificial), OP–Operative (surgical exposure), WBH–Whole body hyperthermia.

commonly available thermocouples and thermistors have metallic leads which act as antennas in an RF field. EM energy is both absorbed and reradiated. This effect can be minimized, in theory, by orienting the leads perpendicular to the electric field vector, but this is not always possible. Such probes would consequently give unreliable readings until the RF power is turned off.

In order to eliminate this interaction between the probe and the RF field, there has been a great deal of effort put into developing nonperturbing probes. One developed by Bowman utilize a thermistor and nonmetallic high resistance leads [3]. Since at about 5 MΩ the lead resistance is of the same order as that of the thermistor, 2 leads are required to bias the thermistor and a second pair to read the voltage drop across the thermistor. The diameter of the probe is currently about 1 mm, with a possible minimum said to be about 0.6 mm.

Recently developed equipment utilizes these nonperturbing probes to produce a feedback loop to a microprocessor based control and data acquisition system. The system is versatile, being adaptable to all forms of EM coupling over a frequency range of 10–2450 MHz. Both graphic and digital temperature versus time readouts appear on CRT monitors, and these and other experimental parameters are also recorded on magnetic discs and on hardcopy printouts [5].

CONCLUSIONS

With the emergence from the laboratory into the commercial arena of advanced equipment for physical hyperthermia, it can be expected that hyperthermic cancer research will proceed with great rapidity. The current state of the art will allow studies on large animals and humans of thermal doses for various cancers, thermotolerance, protocols for combination with radiotherapy, chemotherapy, and immumotherapy, etc. Considerable work remains to be done to develop applicators that will reach tumors in all their multifarious sizes, shapes and locations, as indicated in Table II.

Old techniques of cancer therapy which were inadequate by themselves will need to be re-evaluated. F. M. Allen suggested temporary strangulation of a tumor-containing area or injection of ergot into a tumor percutaneously or via an intra-arterial catheter. Hypotensive drugs and IV glucose (to serum levels of 400 mg/dl) have also been suggested to decrease tumor blood flow [13], [38]. Tumor hyperthermia superimposed on whole-body hypothermia may offer possibilities, especially combined with chemotherapy as per Shingleton.

The art of microencapsulation offers possibilities for adjuvant therapy. Microspheres of Gelfoam or silicone latex can produce artificial tumor embolization [R. D. Turner]. Liposomes can deliver drugs that will be released in a heated tumor [67]. Microspheres may be directed to tumors with tumor specific antibodies or with included magnetic particles (e.g., Fe_3O_4) which may also preferentially heat by induction according to Gilchrist. Radioactive microspheres are another possibility.

Although cancer prevention is our ultimate goal, hyperthermia has the potential to be a large step forward in cancer therapy.

Give me power to produce fever and I will cure all disease.
Hippocrates [22].

References

[1] J. H. Anderson, S. Wallace, C. Gianturco, and L. P. Gerson, "Mini gianturco stainless steel coils for transcatheter vascular occlusion," *Diag. Radiol.*, pp. 301-303, Aug. 1979.

[2] E. R. Atkinson, "Hyperthermia dose definition," pp. 251-253.

[3] R. Bowman, personal communication.

[4] J. M. Bull, D. Lees, W. Schuette, J. Whang-Peng, R. Smith, G. Bynum, E. R. Atkinson, J. S. Gottdiener, H. R. Gralnick, T. H. Shawker, and V. T. DeVita, Jr., "Whole body hyperthermia: A phase-1 trial of a potential adjuvant to chemotherapy," *Ann. Int. Med.*, vol. 90, pp. 317-323, Mar. 1979.

[5] BSD Corporation. An instrument specifically designed for conducting accurate hyperthermic research, 420 Chipeta Way, Salt Lake City, UT, 1979.

[6] T. C. Cetas and W. G. Connor, "Thermometry considerations in localized hyperthermia," *Med. Phys.*, vol. 5, no. 2, pp. 79-91, Mar./Apr. 1978.

[7] D. A. Christensen, "Thermal dosimetry and temperature measurements," *Cancer Res.*, vol. 39, pp. 2325-2327, June 1979.

[8] G. H. A. Clowes, "A study of the influence exerted by a variety of physical and chemical forces on the virulence of carcinoma in mice," *Brit. Med. J.*, pp. 1548-1554, Dec. 1906.

[9] W. B. Coley, "Contribution to the knowledge of sarcoma," *Ann. Surg.*, vol. 14, pp. 119-220, July-Dec. 1891.

[10] —, "The treatment of malignant tumors by repeated inoculations of erysipelas: with a report of ten original cases," *Amer. J. Med. Sci.*, vol. 105, no. 5, pp. 487-511, May 1893.

[11] —, "The therapeutic value of the mixed toxins of the streptococcus of erysipelas and bacillus prodigious in the treatment of inoperable malignant tumors," *Amer. J. Med. Sci.*, vol. 112, no. 3, pp. 251-281, Sept. 1896.

[12] J. A. Dickson, "The effects of hyperthermia in animal tumour systems," *Recent Results Cancer Res.*, vol. 59, pp. 43-108, 1977.

[13] J. A. Dickson and S. K. Calderwood, "Temperature range and selective heat sensitivity of tumors," *NYAC, Conf. Thermal Characteristics of Tumors: Applications in Detection and Treatment*, p. 13, Mar. 1979.

[14] J. A. Dickson and D. S. Muckle, "Total-body hyperthermia versus primary tumor hyperthermia in the treatment of the rabbit VX-2 carcinoma," *Cancer Res.*, vol. 32, pp. 1916-1923, Sept. 1972.

[15] F. Dietzel, "A summary and overview of cancer therapy with hyperthermia in Germany," *Proc. Int. Symp. Cancer Therapy by Hyperthermia and Radiation*, Apr. 1975.

[16] J. Folkman, "Tumor angiogenesis," *Harvard Med. School*, pp. 331-356, 1974.

[17] —, "The vascularization of tumors," *Sci. Amer.*, pp. 59-73, May 1976.

[18] J. Folkman and R. Cotran, "Relation of vascular proliferation to tumor growth," *Int. Rev. Exp. Pathol.*, pp. 207-248, 1976.

[19] O. P. Gandhi, *Bulletin N.Y.A.M.*, Special Issue *1979 meeting on Health Aspects of Nonionizing Radiation.*

[20] —, personal communication.

[21] A. C. Geyser, "Diathermia and the physiological treatment of cancer," *Fischer's Mag.*, pp. 6-9, Dec. 1921.

[22] U. Giles, "The historic development and modern application of artificial fever," *New Orleans Med. Soc. J.*, vol. 91, pp. 655-670, Nov. 1938/39.

[23] B. C. Giovanella, L. J. Williams, J. S. Stehlin, Jr., and A. C. Morgan, "Selective lethal effect of supranormal temperatures on human neoplastic cells," *Cancer Res.*, vol. 36, pp. 3944-3950, Nov. 1976.

[24] B. C. Giovanella, J. S. Stehlin, and L. J. Williams, Jr., "Heterotransplantation of human malignant tumors in nude thymusless mice. II. Malignant tumors induced by injection of cell cultures derived from human solid tumors," *J. Nat. Can. Inst.*, vol. 52, no. 3, pp. 921-927, Mar. 1974.

[25] P. M. Gullino, "Angiogenesis and oncogenesis," *J. Nat. Can. Inst.*, vol. 61, no. 3, pp. 639-643, Sept. 1978.

[26] P. M. Gullino and F. H. Grantham, "Studies on the exchange of fluids between host & tumor. II. The blood flow of hepatomas & other tumors in rats & mice," *J. Nat. Can. Inst.*, vol. 27, no. 6, pp. 1465-1484, Dec. 1961.

[27] J. Harris, "Tumor immunology," *Clin. Oncol.*, pp. 193-221, 1977.

[28] M. A. Henderson and R. T. Pettigrew, "Induction of controlled hyperthermia in treatment of cancer," *Lancet*, pp. 1275-1277, June 1971.

[29] K. J. Henle and L. A. Dethlefsen, "Heat fractionation and thermotolerance: A review," *Cancer Res.*, vol. 38, pp. 1843-1851, July 1978.

[30] C. C. Johnson and A. W. Guy, "Nonionizing electromagnetic wave effects in biological materials and systems," *Proc. IEEE*, vol. 60, pp. 692-718, June 1972.

[31] R. J. R. Johnson, "Radiation and hyperthermia," *Cancer Therapy by Hyperthermia and Radiation, Proc. 2nd Int. Symp.*, pp. 89-95, June 1977.

[32] J. H. Kim and E. W. Hahn, "Clinical and biological studies of localized hyperthermia," *Cancer Res.*, vol. 39, pp. 2258-2262, June 1979.

[33] J. H. Kim, E. W. Hahn, and N. Tokita, "Combination hyperthermia and radiation therapy for cutaneous malignant melanoma," *Cancer*, vol. 41, pp. 2143-2148, 1978.

[34] R. A. Lambert, "Demonstration of the greater susceptibility to heat of sarcoma cells as compared with actively proliferating connective-tissue cells," *J. Amer. Med. Ass.*, vol. 59, no. 24, pp. 2147-2148, Dec. 1912.

[35] J. M. Larkin, "A clinical investigation of total-body hyperthermia as cancer therapy," *Cancer Res.*, vol. 39, pp. 2252-2254, June 1979.

[36] J. M. Larkin, W. S. Edwards, D. E. Smith and P. J. Clark, "Systemic thermotherapy: Description of a method and physiologic tolerance in clinical subjects," *Cancer*, vol. 40, pp. 3155-3159, May 1977.

[37] J. F. Lehmann, "Diathermy," in *Handbook of Physical Medicine and Rehabilitation*, pp. 273-345, 1971.

[38] LeVeen, U.S. Patent 3,991,770, Nov. 16, 1976.

[39] H. H. LeVeen, S. Wapnick, V. Piccone, G. Falk, and N. Ahmed, "Tumor eradication by radiofrequency therapy," *J. Amer. Med. Ass.*, vol. 235, pp. 2198-2200, May 1976.

[40] S. Licht, "History of therapeutic heat," *Therapeutic Heat Cold*, vol. 2, no. 6, pp. 196-229, 1972.

[41] A. G. H. Lindgren, "The vascular supply of tumors with special reference to the capillary angioarchitekture," *Acta Pathal. Microwbiol. Scand.*, vol. 22, pp. 493-523, 1945.

[42] G. Moricca, R. Cavaliere, A. Caputo, A. Bigotti, and F. Colistro, "Hyperthermic treatment of tumours: Experimental and clinical applications," in *Selective Heat Sensitivity of Cancer Cells*, A. Rossi-Fanelli, R. Cavaliere, B. Mondovi, and G. Moricca, Eds. Berlin, Germany: Springer-Verlag, 1977, ch. 6, pp. 153-170.

[43] C. Muller, "Therapeutische erfahrungen an 100 mit kombination von rontgenstrahlen und hochfrequenz, resp. diathermie behandelten bosartigen newbildungen," *Muenchen. Med. Wochensch.*, vol. 28, pp. 1546-1549, July 1912.

[44] —, "Die Kresbskrankheit und ihre Behandlung mit Rontgenstrahlen und hochfrequenter Elektrizitat resp. Diathermie," *Strahlentherapie*, vol. 2, pp. 170-191, 1913.

[45] H. C. Nauts, "Pryogen therapy of cancer: a historical overview and current activities," *Int. Symp. Cancer Therapy by Hyperthermia and Radiation*, pp. 239-248, Apr. 1975.

[46] C. A. Neymann, "Historical development of artificial fever in the treatment of disease."

[47] L. C. Parks, personal communication and REAC, Inc., product information.

[48] L. C. Parks, D. Minaberry, D. P. Smith and W. A. Neely, "Treat-

ment of far advanced bronchogenic carcinoma by extracorporeal induced systemic hyperthermia," University of Mississippi.

[49] R. T. Pettigrew, J. M. Galt, C. M. Ludgate, and A. N. Smith, "Clinical effects of whole-body hyperthermia in advanced malignancy," *Brit. Med. J.*, vol. 4, pp. 679-682, 1974.

[50] R. T. Pettigrew and C. M. Ludgate, "Whole-body hyperthermia. A systemic treatment for disseminated cancer," *Recent Results Cancer Res.*, vol. 59, pp. 153-170, 1977.

[51] G. M. Samaras, J. E. Robinson, A. Y. Cheung, T. Prempree and R. G. Slawson, "Production of controlled hyperthermal fields for cancer therapy," *Cancer Therapy by Hyperthermia and Radiation, Proc. 2nd Int. Symp.*, June 1977.

[52] S. A. Sapareto, L. E. Hopwood, and W. C. Dewey, "Combined effects of X-irradiation and hyperthermia on CHO cells for various temperatures and orders of application," *Rad. Res.*, vol. 73, pp. 221-233, 1978.

[53] S. A. Sapareto, L. E. Hopwood, W. C. Dewey, M. R. Raju, and J. W. Gray, "Effects of hyperthermia on survival and progression of Chinese hamster ovary cells," *Cancer Res.*, vol. 38, pp. 393-400, Feb. 1978.

[54] P. R. Stauffer, T. C. Cetas, R. C. Jones and M. R. Manning, "A system for producing localized hyperthermia in brain tumors through magnetic induction heating of ferromagnetic implants," *Nat. Rad. Sci. Meeting Bioelectromagnet. Symp.*, pp. 420, June 1979.

[55] J. S. Stehlin, Jr., B. C. Giovanella, P. D. de Ipolyi, L. R. Meunz, and R. F. Anderson, "Results of hyperthermic perfusion for melanoma of the extremities," *Surgery, Gyn. and Obstetrics*, vol. 140, no. 3, pp. 339-348, Mar. 1975.

[56] J. S. Stehlin, Jr., P. D. de Ipolyi, B. C. Giovanella, A. E. Gutierrez, and R. F. Anderson, "Soft tissue sarcomas of the extremity," *Amer. J. Surg.*, vol. 130, pp. 643-646, Dec. 1975.

[57] H. B. Stone, R. M. Curtis, and J. H. Brewer, "Can resistance to cancer be induced?" *Ann. Surg.*, vol. 134, pp. 519-527, 1951.

[58] F. K. Storm, W. H. Harrison, R. S. Elliott, and D. L. Morton, "Normal tissue and solid tumor effects of hyperthermia in animal models and clinical trials," *Cancer Res.*, vol. 39, pp. 2245-2251, June, 1979.

[59] S. Sugaar and H. H. LeVeen, "A histopathologic study on the effects of radiofrequency thermotherapy on malignant tumors of the lung," *Cancer*, vol. 43, pp. 767-783, 1979.

[60] L. S. Taylor, "Implantable radiators for cancer therapy by microwave hyperthermia," *Proc. IEEE*, this issue.

[61] F. Urbach, "The blood supply of tumors," *Adv. Biol. Skin*, no. 9, pp. 123-149, 1961.

[62] D. Van Echo, personal communication.

[63] M. von Ardenne, "On a new physical principle for selective local hyperthermia of tumor tissues," *Cancer Therapy by Hyperthermia and Radiation, Proc. 2nd Int. Symp.*, pp. 96-103, June 1977.

[64] M. von Ardenne, "Selective multiphase cancer therapy: Conceptual aspects and experimental basis," *Advan. Pharmacol. Chemother.*, vol. 10, pp. 339-380, 1972.

[65] S. L. Warren, "Preliminary study of the effect of artificial fever upon hopeless tumor cases," *Amer. J. Roentg. Radium Ther.*, vol. 33, pp. 75-87, Jan. 1935.

[66] S. Warren and W. A. Meissner, "Neoplasms," *Pathology, Anderson*, vol. 1, pp. 440-429, 1966.

[67] J. N. Weinstein, R. L. Magin, M. B. Yatvin, and D. S. Zaharko, "Liposomes and local hyperthermia: Selective delivery of methotrexate to heated tumors," *Science*, vol. 204, pp. 188-191, Apr. 1979.

27 MHz Ridged Waveguide Applicators for Localized Hyperthermia Treatment of Deep-Seated Malignant Tumors

R. PAGLIONE and F. STERZER
Microwave Technology Center
RCA Laboratories, Princeton, NJ

J. MENDECKI, E. FRIEDENTHAL and C. BOTSTEIN
Department of Radiotherapy
Montefiore Hospital and Medical Center, Bronx, NY

Ridged waveguide applicators suitable for localized hyperthermia treatment of deep-seated malignant tumors are described. When driven with several hundred watts of 27 MHz RF power, these applicators can raise the temperature of deep-seated tumors to the hyperthermic range (42.5-43.5°C), i.e., the temperature range which appears to be optimum for the treatment of cancer. In initial clinical trials encouraging results were obtained with no discernible side effects.

INTRODUCTION

Localized hyperthermia has been shown to be effective in the treatment of a variety of malignant tumors, either as a stand-alone therapy, or more often in conjunction with radiation therapy.[1,2] A typical treatment with localized hyperthermia consists of raising the temperature of the tumor mass to about 42.5-43.5°C, taking care to minimize the heating of the surrounding healthy tissues. The tumor is maintained at the high temperature for approximately one-half to one hour at a time. Multiple treatments are usually given. Ionizing radiation when added is usually administered in reduced dosages either during the hyperthermia treatment, or immediately before or after the treatment.

One of the most useful methods of producing localized hyperthermia in tumors is dielectric heating with radio-frequency (RF) radiation.[1] Here, power from an RF generator is transmitted into the tissue volume to be heated by an antenna or applicator. The RF travels through the tissues of the body in the form of an exponentially decaying wave, giving up energy to the tissues (dielectric heating) as it traverses them.

The depth to which RF waves can penetrate into tissues and produce heating is primarily a function of the dielectric properties of the tissues and of the RF frequency.[3,4] In general, the lower the water content of the tissue, the deeper a wave at a given frequency can penetrate into it. Thus, for example, RF waves can penetrate much deeper into fat (low water content) than into muscle (high water content). Also, at the RF frequencies of interest, the lower the RF frequency the deeper the depth of penetration into a tissue with a given water content. This is illustrated in **Figure 1**, which shows the results of calculating the heat produced by plane waves at the five lowest ISM frequencies* in a simple tissue model (infinite layer of fat 2 cm thick followed by an infinite layer of muscle infinitely thick).[3] The values of the complex dielectric constant used

* ISM frequencies are frequencies set aside by the Federal Communications Commission for *I*ndustrial, *S*cientific and *M*edical applications.

Reprinted with permission from *Microwave J.*, vol. 24, no. 2, pp. 71, 72, 74, 75, 76, 79, 80, Feb. 1981.

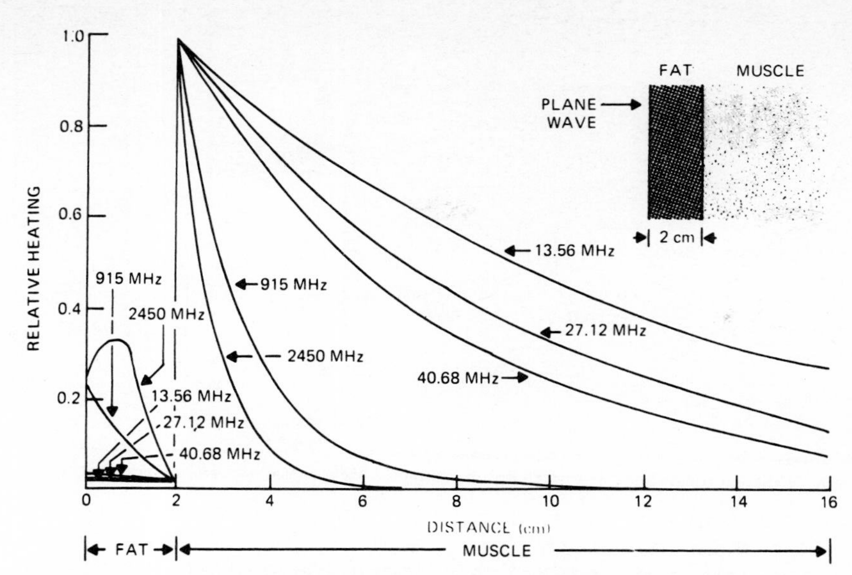

Fig. 1 Calculated relative heating in fat and muscle as a function of distance for five different ISM frequencies.

in the calculations were taken from Reference 9. Note the relatively small amount of RF power that is dissipated in the fat layer. In particular, note that virtually all of the energy at 13.56 MHz, 27.12 MHz and 40.68 MHz is transmitted through the fat into the muscle where it is dissipated.

In the January, 1980 issue of the *Microwave Journal* we described the theory and construction of hyperthermia applicators designed for operation at 915 and 2450 MHz.[5] Applicators at these frequencies are useful for noninvasive hyperthermia treatments of cutaneous and subcutaneous tumors, tumors located within or in the vicinity of natural body cavities, and tumors located in the breasts. On the other hand, tumors that are shielded from an accessible body surface by more than about 2 cm of tissue with high water content are usually difficult to treat with 915 or 2450 MHz radiation because of the high absorption of 915 and 2450 MHz radiation in tissues of this type (see **Figure 1**).

In the present paper, we describe applicators designed for operation with 27 MHz RF radiation. Unlike 915 or 2450 MHz radiation, 27 MHz radiation can penetrate deeply into tissues with high water content (see **Figure 1**), and can therefore be used to noninvasively heat deep-seated tumors that are not accessible with either 915 or 2450 MHz radiation. The major limitations of heating with 27 MHz are comparatively poor focusing (wavelengths in tissues are greater than 1 meter), and the large size of the present applicator designs. Also, accurate temperature measurements with thermocouples on thermistors are difficult in the presence of large 27 MHz fields.

The paper is divided into three parts: The first part describes the construction and principles of operation of the 27 MHz applicators. Despite their low frequency of operation, the design of these applicators is based on the microwave concept of dielectrically-loaded waveguide radiators. In the second part of the paper, the results of calculations of temperature profiles generated with plane waves at 27 MHz in simple living tissue models are presented. The third part covers the use of the applicators in animal experiments and as therapeutic tools in clinical trials involving various types of malignant tumors.

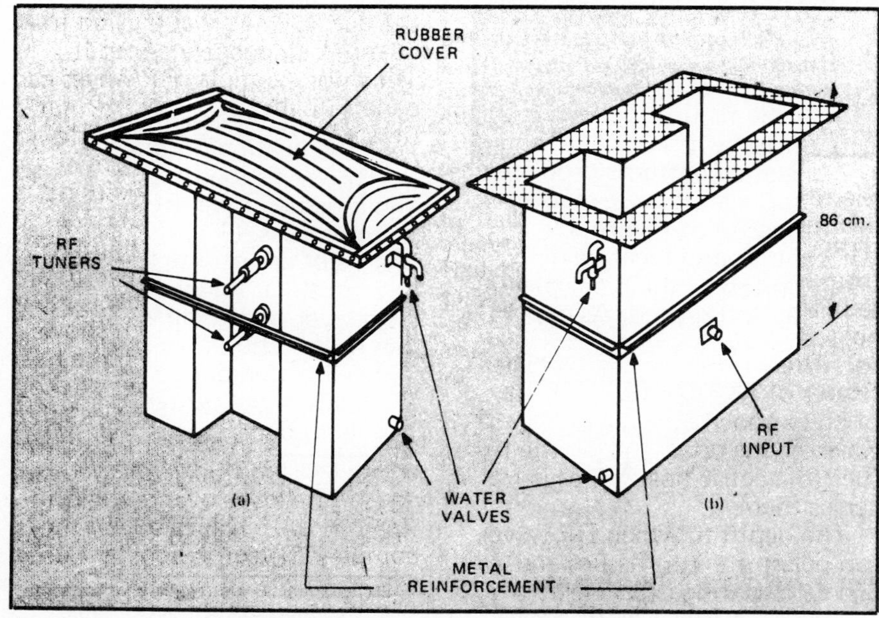

Fig. 2 Sketch of a 27 MHz ridged-waveguide applicator with (a) rubber cover in place and (b) with rubber cover removed.

DESCRIPTION OF 27 MHz APPLICATORS

Several considerations must be kept in mind when designing applicators for producing localized hyperthermia with RF radiation.

- The applicator must be able to handle the RF power required to raise the temperature of the tumor or tumors to be treated to the hyperthermic range. Power levels as high as several hundred watts CW are often required.
- The design of the applicator must minimize the amount of RF power being delivered to healthy tissues.
- The applicator design must be consistent with the physical comfort of a patient who has to undergo one or more treat-

ments at a time, each treatment lasting up to one hour, or in some instances even longer.

- Radiation into free space from the applicator must be kept to a minimum to protect the patient and the technicians administering the treatment from unnecessary exposure to RF.
- The applicator must be rugged, its cost must be moderate, and its size must be consistent with the space available in a typical hospital treatment room.

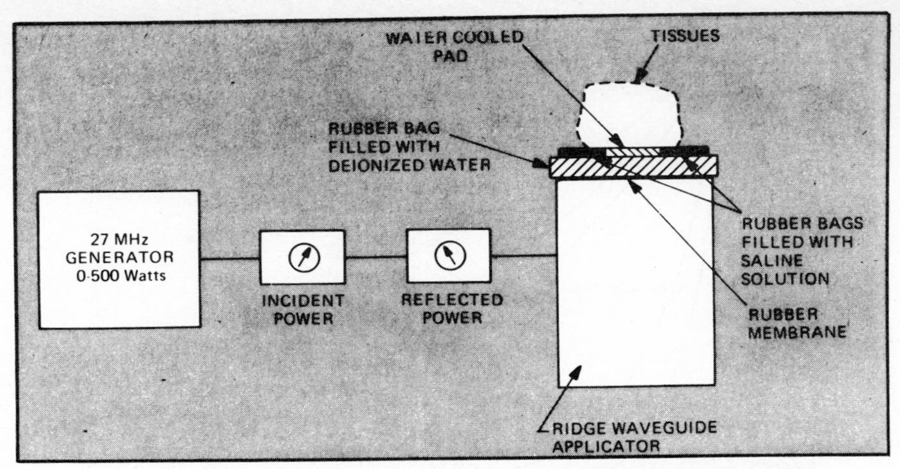

Fig. 3 Setup for hyperthermia treatment using 27 MHz ridged-waveguide applicator.

Based on the above design considerations, we have developed 27 MHz ridged waveguide applicators that have proven to be satisfactory for both animal studies and clinical work. **Figure 2** is a sketch of a typical applicator. The applicator is built from sheet metal in the shape of a shorted section of ridged waveguide. The open end of the guide is covered with a rubber membrane and the guide is filled with deionized water (ordinary water is too lossy). The 27 MHz power is introduced into the applicator via a coax-to-waveguide transition and is radiated from the applicator through the rubber sheet. Reflections of 27 MHz power back into the coaxial RF input port are minimized by means of three capacitive tuners.

The dimensions of the ridged waveguide are chosen to allow propagation of the TE_{10} mode at 27 MHz, typical cutoff frequencies for the TE_{10} mode being on the order of 20 MHz. Such low cutoff frequencies can be achieved with reasonable guide dimensions because: (1) the high dielectric constant of deionized water ($\epsilon \sim 81$) reduces the linear dimensions of the guide for a given cutoff frequency by a factor of approximately 9 over an air-filled guide and (2) ridged waveguides have lower cutoff frequencies than rectangular guides of the same outer dimensions. Suitable design equations for calculating the cutoff frequencies of ridged waveguides are given in References 6 and 7.

Figure 3 is a diagram showing a complete setup for inducing hyperthermia with a 27 MHz ridged-waveguide applicator. The applicator is driven by an RF generator whose power output can be varied between 0 and 500 watts. Incident and reflected powers are measured with conventional RF power meters. A rubber bag large enough to cover the entire rubber membrane is filled to a thickness of about 5 cm with deionized water and is placed on top of the applicator. The purpose of this bag is to spread the high electric fields present at the edge of the applicator; this prevents any excessive local heating of the patient. On top of the 5 cm thick water bag is a thin water-cooled pad for cooling the skin of the patient, and one or more small bags filled with saline solution—the saline solution is a good absorber of 27 MHz radiation and, therefore, it protects the tissues that one does not want to heat.

The heating patterns produced by the applicators were studied with the setup of **Figure 3**. Blocks of ground meat were placed on top of the 5 cm thick water bag and heated. As expected from standard ridged-waveguide theory, most of the heating took place in the volume of meat placed above and between the ridge and the wall opposite the ridges, i.e., above the shaded area of **Figure 4**. The hottest point was usually in the center of the shaded area (point A of **Figure 4**). This fall off in heating with distance from the applicator along a line perpendicular to the applicator and going through point A followed approximately the calculated curve for 27 MHz radiation shown in **Figure 1.**†

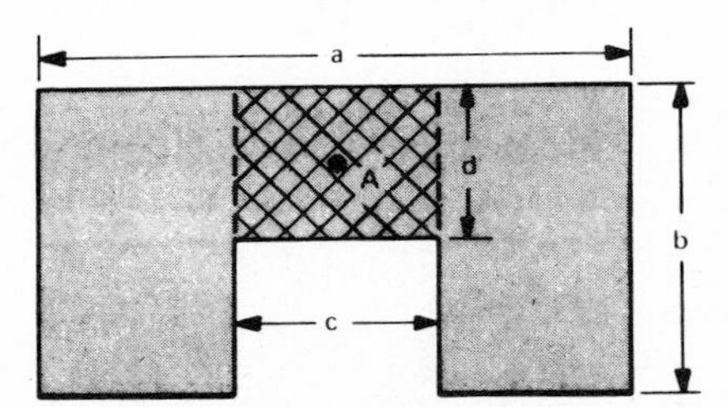

Fig. 4 Cross-section of ridged-waveguide applicator. Most of the RF energy in the guide is concentrated in the cross-hatched area.

While heating with one applicator is adequate in a number of clinical situations, it is often desirable to use two applicators in a cross-fire arrangement, since two applicators in such an arrangement can often produce deeper and more uniform noninvasive heating than a single applicator. This is illustrated in **Figure 5** which shows the results of calculating the heating produced by two RF waves that impinge on a simple tissue model from opposite directions. (The tissue model consists of an infinite layer of fat 2 cm thick followed by an infinite layer of muscle whose thickness is chosen so that the relative heating due to the two RF waves in the center of the muscle is approximately 80% of

† Figure 1 can be used to determine the maximum depth of penetration of waves emitted from symmetrical apertures (round, square, rectangular, etc.) into the tissue model, since the center ray emitted from any symmetrical aperture parallel to the surface of the model must, from symmetry considerations, always propagate in the same direction as the plane wave of Figure 1.

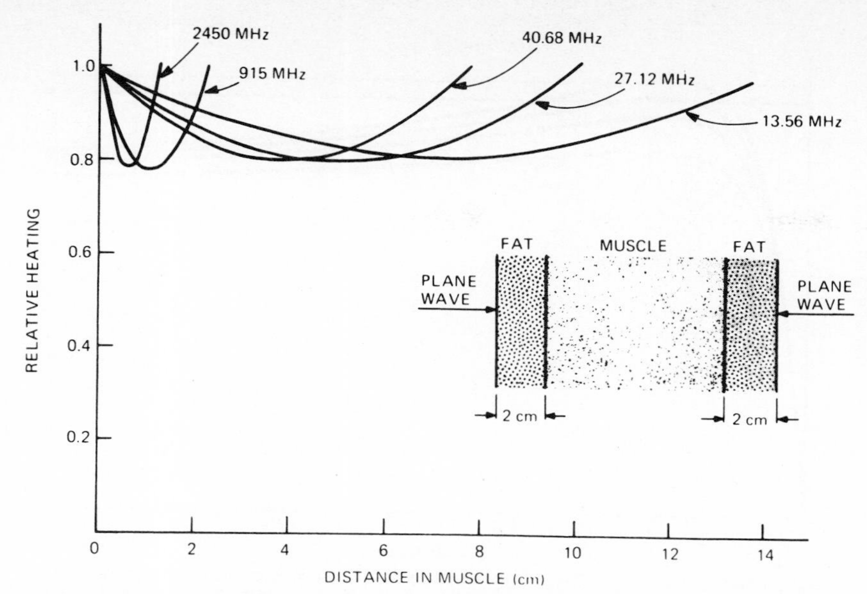

Fig. 5 Calculated relative heating in muscle as a function of distance for five different ISM frequencies due to two noncoherent plane waves of approximately the same frequency traveling in opposite directions.

the heating at the fat-muscle interface, followed again by an infinite layer of fat 2 cm thick.) The two plane waves incident on the tissue layers are assumed to be of approximately the same frequency but uncorrelated, and any reflections from the second tissue interface (muscle-fat) are neglected. **Figure 5** indicates that 10 cm of muscle can be heated nearly uniformly with two 27 MHz waves, a significant improvement over heating with a single 27 MHz wave (**Figure 1**).

We have built several ridged-waveguide applicators that can be used in "cross-fire" arrangements. The inside of these applicators is compartmentalized with solid plastic sheets, and each compartment is individually filled with deionized water. Compartmentalized applicators can be placed in a horizontal position without danger of rupturing the rubber membrane, since the plastic sheets protect the membrane from most of the water pressure. Two cross-fire arrangements are possible with these applicators: two applicators in horizontal positions facing each other, or one applicator in a horizontal position and the other in a vertical position.

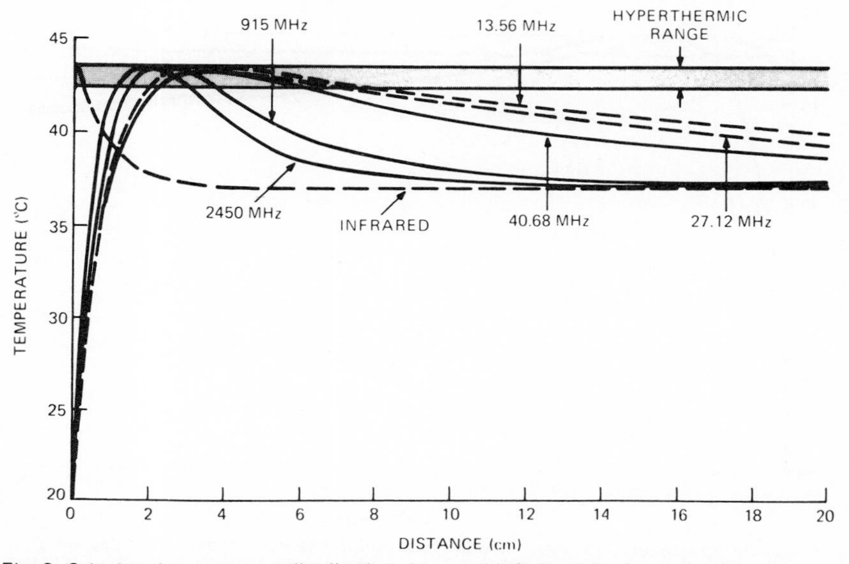

Fig. 6 Calculated temperature distributions in a semi-infinite slab of muscle when the muscle is heated with infrared radiation or with radiation at any one of five ISM frequencies.

CALCULATIONS OF TEMPERATURE DISTRIBUTIONS PRODUCED IN LIVING TISSUES

The temperature distributions produced by RF waves propagating through living tissues can be calculated by determining the complex propagation constants of the waves and then solving the heat transport equation. Such calculations are too difficult for the actual tissue geometries encountered when treating malignancies with RF hyperthermia; however, much information of practical value can be obtained by calculating temperature distributions in simplified tissue models.

Figures 6-10 are plots of calculated temperature distributions in semi-infinite slabs of muscle based on steady-state solutions to the one-dimensional heat transport equation in living tissues that are heated by RF waves. These solutions, which were given by Foster, Kritikos and Schwan,[8] assume a uniform plane wave incident on a homogeneous semi-infinite layer of tissue, and take into account blood flow (assumed to be temperature independent), tissue heat conductivity, and temperature difference between the surface of the tissue and the environment. In calculating these graphs, it was assumed that the arterial blood entering the tissues is at a temperature of 37°C. The values of the complex dielectric constants needed in the calculations were taken from Reference 9.

Figure 6 is a plot of the calculated temperature distributions in a semi-infinite slab of muscle when the muscle is heated with infrared radiation (assumed depth of penetration $\sim$0.001 cm), or with waves at one of five ISM frequencies. The following values of power densities transmitted into the muscle were used in the calculations: 0.05 W/cm^2 (infrared), 0.27 W/cm^2 (2450 MHz), 0.3 W/cm^2 (915 MHz), 0.56 W/cm^2 (40.68 MHz), 9.7 W/cm^2 (27.12 MHz), and 0.87 W/cm^2 (13.56 MHz). These values just raise the temperature in part of the muscle to the hyperthermic range (42.5-43.5°C). A value of $\lambda = 1.3\ cm^{-2}$ was assumed in the calculations for this figure and for **Figures 7** and **8**. (λ is the product of flow and heat capacity of blood divided by the coefficient of tissue heat conductance.)

Figure 6 shows that waves at the five ISM frequencies can produce temperatures in the hyperthermic range at distances of several centimeters into muscle without causing excessive temperature increases near the surface. On the other hand, most of the temperature increase produced by infrared radiation is confined to a few millimeters near the surface. As expected from the curves of **Figure 1**, the three lower ISM frequencies can produce hyperthermic temperatures at a significantly greater depth than 915 or 2450 MHz.

Figure 7 illustrates the dependence of temperature rise on the power density of 27 MHz radiation transmitted into muscle. Note that with an RF power density of 0.7 W/cm², temperature rises to the upper limit of the hyperthermic range are obtained; while with RF power densities about one-third smaller (0.5 W/cm²) the maximum temperature is well below the hyperthermic range.

Figure 8 compares active cooling of the surface to 20°C with passive cooling in room temperature surroundings (22°C). Significantly deeper heating is obtained with active cooling, since more power per unit area can be transmitted into the muscle without causing excessive heating near the surface.

Figures 9 and **10** illustrate the dependence of temperature profiles on blood flow. **Figure 9** shows heating to the hyperthermic temperature range with 27 MHz radiation for various values of blood flow. It can be seen from the figure that the lower the blood flow, the smaller the power density required to raise the temperature of the muscle to the hyperthermic range, and the greater the depth at which one obtains hyperthermic temperatures. **Figure 10** illustrates the temperature distribution due to heating with 27 MHz radiation in a model consisting of a 4 cm thick tumor buried at a depth of 6 cm in muscle. The tumor is assumed to have properties similar to normal muscle except that the blood flow in the tumor is assumed to be 1/4 that of the flow in the muscle. (Blood flow in malignant tumors is usually much lower than the blood flow in healthy tissues.)[10] Note that part of the tumor can be raised to the hyperthermic range without significantly increasing the temperature of the normal muscle. This phenomenon is of the greatest importance in hyperthermia treatment with RF because, generally, solid malignant tumors have poor blood circulation and therefore it is possible to selectively raise the temperature of such tumors with RF heating.

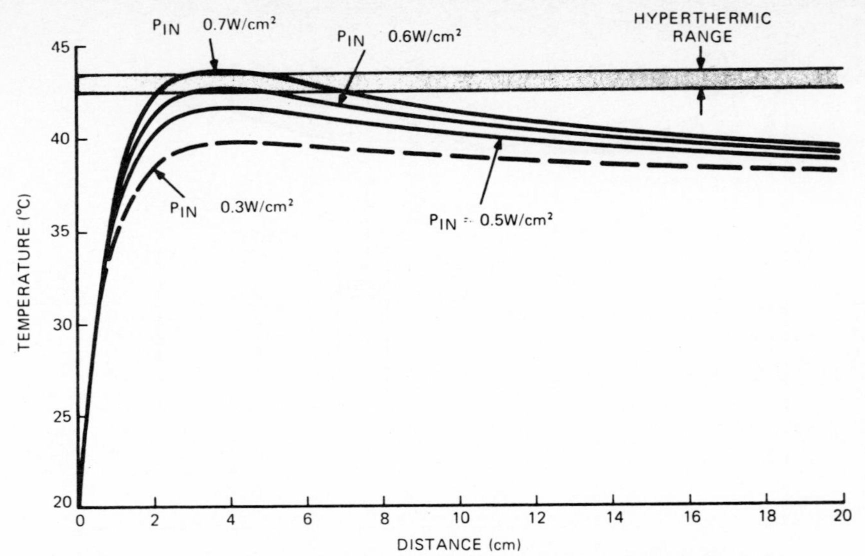

Fig. 7 Calculated temperature distributions in a semi-infinite slab of muscle for four different power densities when the muscle is heated with RF radiation at a frequency of 27 MHz and the surface of the muscle is maintained at 20°C by active cooling. (P_{in} = RF power density transmitted into muscle.)

ANIMAL EXPERIMENTS AND CLINICAL TRIALS

We have used the experimental arrangement shown in **Figure 3** in both animal experiments

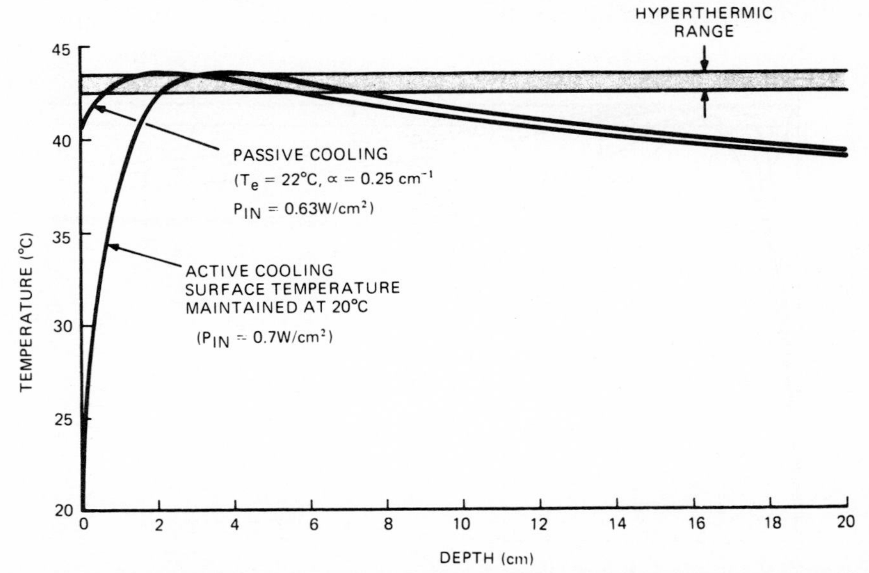

Fig. 8 Calculated temperature distributions in a semi-infinite slab of muscle for passive and active cooling when the muscle is heated with RF radiation at a frequency of 27 MHz. (T_e = temperature outside of tissue plane, α = heat loss coefficient from surface of muscle, P_{in} = RF power density transmitted into the muscle.)

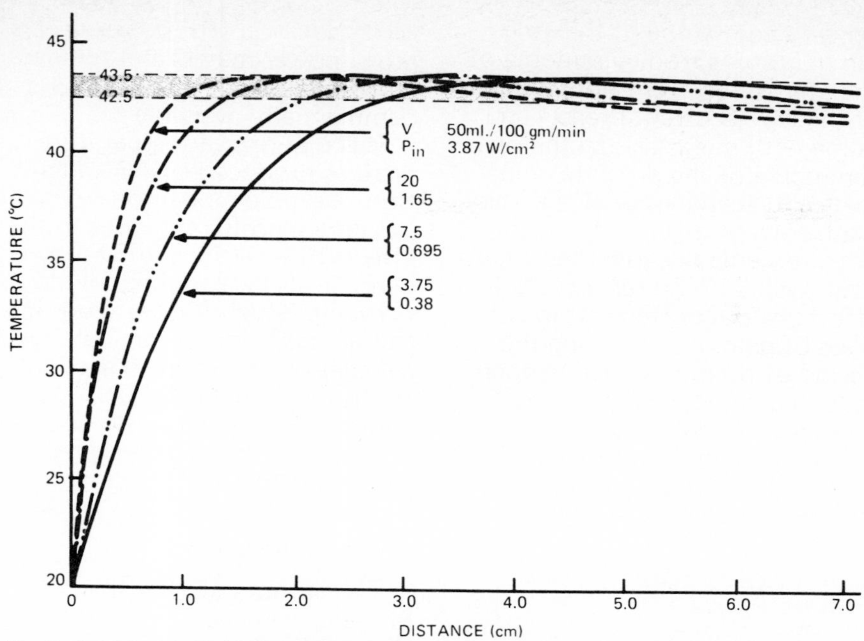

Fig. 9 Calculated temperature distributions in a semi-infinite slab of muscle for four different values of blood flow when the muscle is heated to the hyperthermic temperature range with RF radiation at a frequency of 27 MHz, and the surface of the muscle is maintained at 20°C by active cooling. P_{in} = RF power density transmitted into muscle.

and clinical trials. Tissue temperatures in animals or patients were measured with 3-mil diameter thermocouples immediately before and after heating. The thermocouples were removed during RF heating because 27 MHz RF fields induce currents in the thermocouple wires. These currents cause heating of the thermocouple junction and, therefore, produce misleading temperature readings. Various arrangements were used to support the animals or patients during RF heating. For example, in many of our animal experiments the animals were supported on a wooden table with a cutout for the ridged-waveguide applicator. Patients with tumors in the rectal area were treated while sitting on a specially constructed chair with a cutout for the applicator.

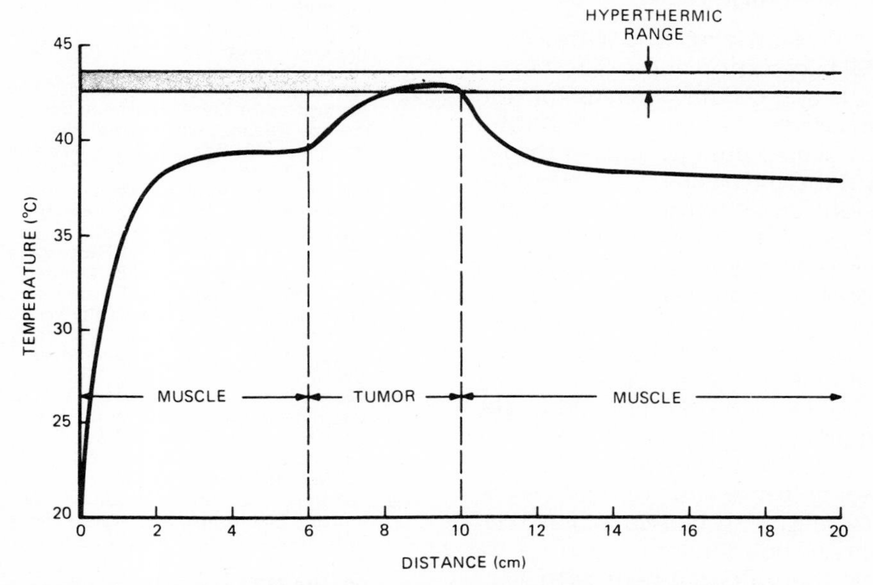

Fig. 10 Calculated temperature distribution in a layered one-dimensional structure of muscle-tumor-muscle when heated with RF power at a power density of 0.25 W/cm² at 17 MHz and the surface of the muscle is maintained at 20°C by active cooling. [λ (muscle) = 1. λ (muscle) = 1. λ (tumor) = 0.325 cm^{-2}.]

Patients who needed treatments in chest areas or in the extremities, etc., were treated lying down on stretchers built around applicators. Applicators of various dimensions were used. The dimensions of the smallest applicator were (see **Figure 4**): a = 50.8 cm, b = 22.9 cm, c = 25.4 cm, d = 7.6 cm, height = 38.1 cm; the dimensions of the largest applicator were a = 58.4 cm, b = 26.3 cm, c = 29.2 cm, d = 13.7 cm, height = 86.4 cm. When properly matched, all applicators could raise well-vascularized muscle to the hyperthermic range within several minutes. The required RF input power was typically 300-400 watts. Assuming that about 80% of the RF input power was concentrated across the ridge of the guide (i.e., the shaded area of **Figure 4**), this corresponds to power densities of roughly 0.8 W/cm², which is in good agreement with the calculated values of the previous section.

The ability of the waveguide applicators to produce uniform temperature distributions in healthy tissues was checked by heating the gluteal region of a 35 kg pig. The pig was anesthetized with IV nembutol and positioned on one of its sides with the gluteal region resting on a cooling pad placed on top of an applicator. Four plastic angiocaths were introduced into the muscle mass at distances of 2, 5, 7 and 9 cm from the skin in contact with the cooling pad. The pig was then heated with 250 watts of 27 MHz power. Muscle temperature was measured by inserting thermocouples at five-minute intervals into the angiocaths. During these temperature measurements, the RF power was interrupted for 20-30 seconds. **Figure 11** is a plot of the measured temperatures as a function of time. Note that after 20 minutes of heating the temperature distribution in the muscle is reasonably uniform, the peak temperature occurring 5 cm from the skin.

In order to test the concept of differential heating of poorly vascularized deep-seated tumors with 27 MHz radiation (see **Figure 10**), the following experiment was performed: A 60 g piece of meat was formed into an egg-shaped mass and placed in a rubber balloon. This simulated tumor was then inserted into the rectum of a 40 kg anesthetized male dog. The dog was positioned on its side with the ***gluteus maximus*** resting on a cooling pad that is placed on top of the rubber membrane of an applicator. Plastic angiocaths were placed in (1) the center of the "tumor," (2) 3 cm above the "tumor," and (3) 3 cm below the "tumor." The "tumor" itself was approximately 6 cm above the applicator.

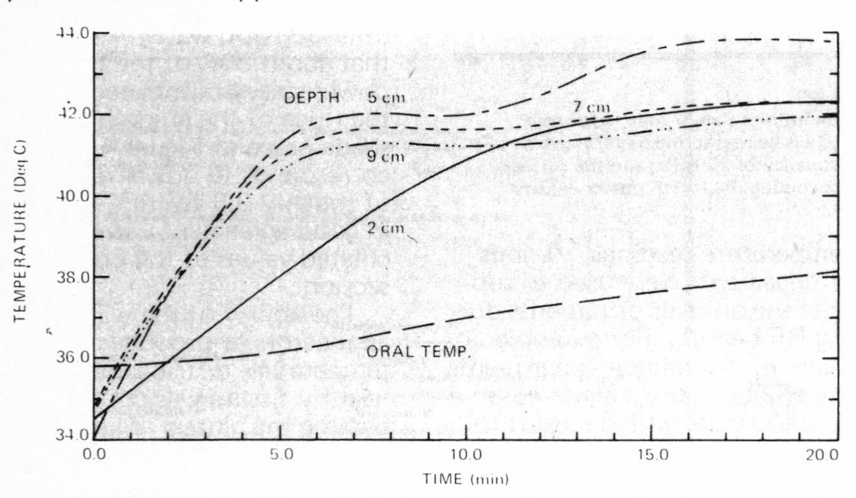

Fig. 11 Measured temperatures as a function of time in a gluteal muscle of a pig. Temperatures at 2, 4, 5, 7 and 9 cm from the skin are plotted.

Temperatures were measured at four-minute intervals by inserting thermocouples into the embedded angiocaths.

With 300 watts of 27 MHz power the non-vascularized "tumor" could be heated to 45-48°C within eight minutes. At the same time the normal tissues above and below the "tumor" remained below the hyperthermic range. The systemic temperature of the dog as measured by a fourth thermocouple inserted into a remote muscle rose to 39°C, indicating active dissipation of heat from the directly heated area. Also, the respiratory and cardiac rates of the dog increased markedly during the localized hyperthermia.

A number of patients with advanced malignancies were offered treatments of 27 MHz hyperthermia. Tumors treated included primary carcinoma of the lung, hip metastasis from carcinoma of the uterus, recurrent liposarcoma of the leg, recurrent breast carcinoma with metastasis to the ribs, carcinoma of the prostate, and metastatic carcinoma of the scapula from a primary lung cancer. Patients generally tolerated treatment well with no serious side effects, and reported varying degrees of pain relief. During the period of treatment oral temperature increased, usually to 39°C, accompanied by moderate sweating and acceleration of heart rate. The medical aspects of these studies will be published elsewhere.

CONCLUSIONS

Dielectric heating with 27 MHz radiation appears to be a safe and efficient means for noninvasively raising the temperature of deep-seated tumors to the hyperthermic range. 27 MHz radiation can penetrate through fat layers with little loss, can traverse anatomical features such as bones‡ or air spaces, and can heat tumors that are buried several centimeters deep inside muscle.

The water-filled ridged-waveguide applicators that are described in this paper are of reasonable size and are relatively inexpensive to fabricate. The rest of the equipment needed for hyperthermia treatment with 27 MHz radiation (RF power generators, power meters and temperature-measuring instruments) is commercially available.

Calculations of temperature profiles produced by RF radiation in simple models of living tissues are useful as guides to the effects of variations in RF power density, surface cooling, blood flow, etc. Much further work in this area is needed, particularly calculations taking into account finite aperture and tissue sizes, and the nonlinearity of blood flow as a function of temperature in normal and malignant tissues.

‡ Heating of bones can be minimized by orienting the electric fields in the RF wave perpendicular to the bones in the tissue being heated.

ACKNOWLEDGMENT

The authors wish to thank Elvira E. Petersack, Markus Nowogrodzki, Steven Weber and Francis J. Wozniak for their invaluable help during the course of this work.

REFERENCES

1. Mendeci, J., E. Friedenthal, C. Botstein, F. Sterzer, R. Paglione, M. Nowogrodzki and E. Beck, "Microwave-Induced Hyperthermia in Cancer Treatment: Apparatus and Preliminary Results," *International Journal of Radiation Oncology,* Biology and Physics, Vol. 4, Nov./Dec. 1978, pp. 1095-1103.
2. Short, J. G., and P. F. Turner, "Physical Hyperthermia and Cancer Therapy," *Proc. IEEE*, Vol. 68, January 1980, pp. 133-142.
3. Schwan, P., E. L. Curstensen and K. Li, "Heating of Fat-Muscle Layers by Electromagnetic and Ultrasonic Diathermy," American Institute of Electrical Engineers, Vol. 72, Pt. 1 *Communications and Electronics*, Sept. 1953, pp. 483-488.
4. Schwan, H. P., and G. M. Piersol, "The Absorption of Electromagnetic Energy in Body Tissue," *Am. Journal Phys. Med.*, Vol. 33, Dec. 1954, pp. 370-404.
5. Sterzer, F., et al., "Microwave Apparatus for the Treatment of Cancer by Hyperthermia," *Microwave Journal*, Vol. 23, Jan. 1980, pp. 39-44.
6. Cohn, S. B., "Properties of Ridge Wave Guide," *Proc. IRE*, Vol. 35, August 1947, pp. 783-788.
7. Chen, T. S., "Calculation of the Parameters of Rdige Waveguides," *IRE Trans. on Microwave Theory & Techniques*, Vol. MTT-5, Jan. 1957, pp. 12-17.
8. Foster, K. R., et al., "Effect of Surface Cooling and Blood Flow on the Microwave Heating of Tissue," *IEEE Trans. on Biomedical Engineering*, Vol. BME-25, May 1978, pp. 313-316.
9. Johnson, C. C., and A. W. Guy, "Nonionizing Electromagnetic Wave Effects in Biological Materials and Systems," *Proc. IEEE*, Vol. 60, June 1972, pp. 692-718.
10. Peterson, Hans-Inge, *Tumor Blood Circulation: Angiogenesis, Vascular Morphology and Blood Flow of Experimental and Human Tumors*, CRC Press, Inc., 1979.

New Apparatus for the Induction of Localized Hyperthermia for Treatment of Neoplasia

H. T. LAW AND R. T. PETTIGREW

***Abstract*—A method of providing heat to body cavities is described. Although developed in conjunction with a program investigating the effects of hyperthermia in the treatment of malignant disease, it may have other applications, for example resuscitation of hypothermic patients.**

INTRODUCTION

The effect of hyperthermia in the treatment of malignancy has been studied by several workers (1)–(3). Experiments *in vivo* and *in vitro* have shown that heat, applied alone or in conjunction with cytotoxic drugs, is an effective agent in bringing about selective destruction of malignant cells with relatively little effect on normal cells or tissue. Hyperthermia thus seems today to be a promising therapy for the treatment of malignant disease.

Henderson and Pettigrew (4) have described a method for the induction of whole-body hyperthermia, a later paper by Pettigrew *et al.* (5) summarizes the results obtained with this treatment in fifty-one patients, almost all of whom were in an advanced stage of their disease when treatment by hyperthermia was given.

The effect of the elevated temperature on central organs, notably the liver, is commented upon. Because of the probability of liver damage the core temperature in whole-body hyperthermia has been limited to 41.8°C.

The observations in these cases suggest strongly that a high tumor temperature, or extended duration of treatment, would produce a better result. Treatment time already extends over some 4 to 5 hours and it is difficult to foresee extension of this by any substantial factor. The desirability of localized hyperthermia, in which the temperature of the tumor is increased significantly above that of the remainder of the body, is thus highlighted.

Techniques which have already been applied to achieve this include perfusion [Cavaliere (1) Stehlin (6)], and irrigation of body cavities, e.g., bladder, stomach, by hot water [Hall (7)]. Consideration has also been given to local heating of the tumor by diathermy or microwave irradiation.

DESCRIPTION OF APPARATUS

The present method is an extension of the irrigation technique offering several important advantages. These are best demonstrated by considering some of the disadvantages of the irrigation method as hitherto applied, viz.:,

(1) The location of the inlet with respect to the outlet tube is known only approximately, so that the circulation of fluid within the cavity is very poorly defined. Remoter parts of the cavity volume, in which the local velocity is low, may have a much lower temperature than that of the fluid leaving the outlet tube. The temperature of the fluid in contact with the tumor is thus very uncertain.

(2) The power required to maintain any significant temperature difference between the irrigating fluid and the body (core) temperature is substantial and it is difficult to provide this amount of power by transfer of heated fluid from outside without causing scalding of the entrance passage (urethra, oesophagus etc.).

In this present apparatus:

(1) The required power is generated (electrically) within the cavity, and it is easy to provide a very high power by this means.

Manuscript received March 7, 1978; revised October 25, 1978.
H. T. Law is with Ferranti, Ltd., Edinburgh, Scotland.
R. T. Pettigrew is with Western General Hospital, Edinburgh, Scotland.

(2) There is no appreciable power transfer to the walls of the entrance passage, the temperature of the tube in contact with the walls is insignificantly above the whole-body temperature.

(3) Very vigorous mixing of the fluid within the cavity is provided ensuring that the temperature in all parts of the fluid volume filling the cavity is very uniform and the temperature of the fluid in contact with the tumor is accurately measured and controlled.

Figure 1 shows the essentials of the apparatus. For the present, the apparatus and procedure appropriate to use within the stomach is described. A stainless steel nozzle A, 14 mm in diameter, containing the heater, temperature control, and temperature measuring devices, is connected by a flexible P.V.C. tube (12 mm outer diameter, 8 mm inner diameter) to the pump B. Wires supplying the heater and connecting to the temperature-sensing elements are contained entirely within

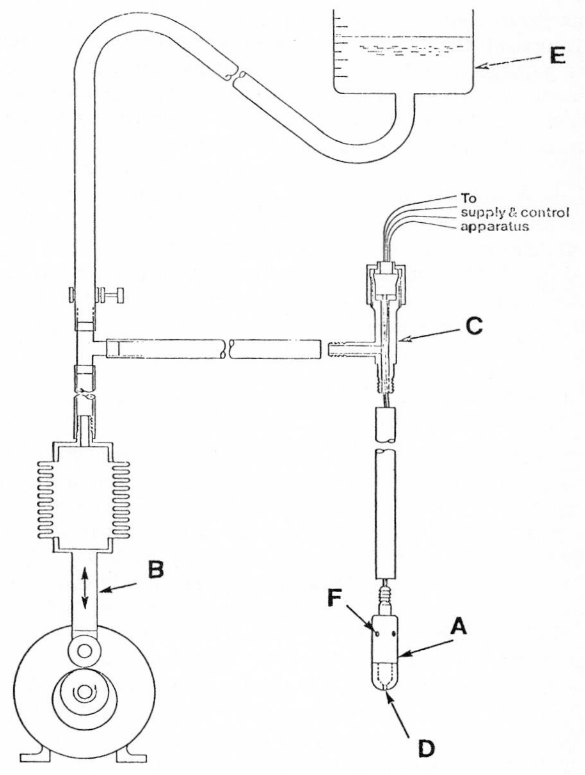

Fig. 1.

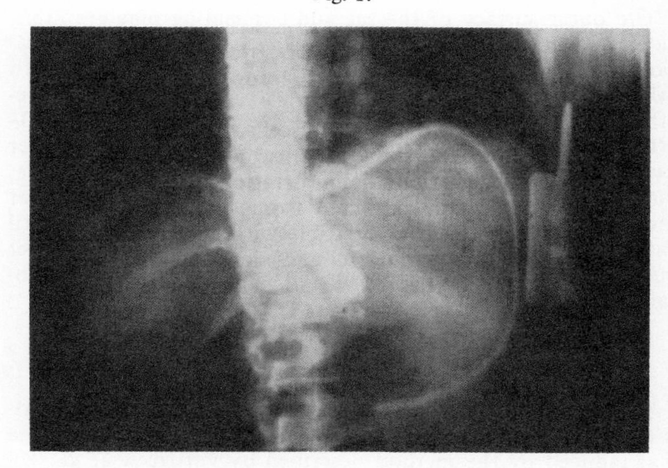

Fig. 2. Radiograph of the apparatus *in situ* in the stomach. (The smaller diameter tube is an independent thermometer.)

Reprinted from *IEEE Trans. Bio-Med. Eng.*, vol. BME-26, pp. 175–177, Mar. 1979.

the oesophageal part of the tube and leave the water system (external to the patient) via the seal assembly C. The pump has a swept volume of 10 ml, which generates a to-and-fro, tidal oscillation of the water in the connecting tube by about 10 cm.

The nozzle is valved so that water exits (on the delivery stroke of the pump) via the axial hole D, and enters at lower velocity (on the intake stroke) mainly via the several radially disposed holes (F). In this manner the pump maintains a vigorous agitation and mixing of the fluid, while the total volume of fluid within the cavity oscillates by ±5 ml or so about its central, constant value.

For use in the stomach, absorption or onward passage of the water is prevented by enclosing the nozzle in a thin-walled rubber sac (not shown in Fig. 1) sealed around the neck of the nozzle. The volume of water in the system is altered by opening the pinch clamp to the reservoir E.

Throughout the tidal pumping process the water is heated by its passage over the immersed heating element, the power to which is automatically regulated in response to measurements made on the temperature sensor (thermistor) contained in the nozzle near the exit hole. A second thermistor is fastened to the heater as a safety measure; the control equipment contains an audible alarm which operates if the heater temperature, as measured by this thermistor, exceeds 70°C. A chromel-alumel thermocouple is also included in the nozzle assembly. This is not interconnected electrically with the control equipment but is used purely as an independent monitor of the fluid temperature.

Brief Details of Procedure (In Treatment of Gastric Carcinoma)

After the rubber sac is securely fastened round the neck of the nozzle it is dilated by admitting 200–300 ml of water from the reservoir. The reservoir is then pinched off and the operation of the pump, heater, sensors, and control circuits is checked. In this condition it is relatively easy to "capture" any bubbles remaining within the sac (which might present an interface of poor thermal conductivity between the heated water and the tumor surface). The bubbles may be passed up the tube and out of the system via the reservoir line.

The sac is then collapsed by lowering the reservoir and opening the pinch clamp. Radio opaque dye may be added to the reservoir at this stage if desired, due allowance being made for the volume of the connecting tubes relative to the intended *in situ* volume of the sac.

The outer surface of the sac and connecting pipe are lubricated and passed via the oesophagus into the stomach. Location in the stomach may be verified by dilating the sac with 50–100 ml of fluid and gently attempting to withdraw. Thereafter the sac is dilated by admitting the desired volume (500–600 ml), the reservoir pinched off and the pump and heater switched on. The water warms rapidly to body temperature, mainly by abstraction of heat from the patient, thereafter more slowly under the action of the immersed heater, until the desired, pre-set temperature is achieved. The controller automatically adjusts the input power to the heater to maintain this temperature for the duration of treatment.

Results and Discussion

The apparatus has been used on two occasions in the treatment of carcinoma of the stomach. Whole-body hyperthermia was induced by the method described by Pettigrew *et al.* (8). When a temperature of 41.5°C was attained the patient was removed from the wax bath and placed in a warm-water-heated blanket. The nozzle and sac were passed, 600 ml of water introduced, and a temperature of 43.5°C attained within 15 minutes. This temperature was maintained for a further 80 minutes, after which the bag was drained and the nozzle withdrawn.

The patient recovered rapidly and was fully ambulant four days after treatment. Gastroscopic examination on the fourth day showed a sub-acute inflammatory reaction in the neoplastic mass. The appearance of the antrum and pylorus was normal. There was no evidence of increase of plasma bilirubin above normal levels, or other symptoms of liver damage.

On the second occasion the procedure was similar, except that the temperature of the water within the sac was raised slightly to 43.8°C (it is estimated that a temperature difference of 0.3°C is set up across the rubber of the bag at the power density prevailing) and the duration of treatment extended to 3 hours.

The power requirements were surprisingly high. At one stage of the treatment over 30 W was required to maintain the temperature of the water filling the sac at 43.8°C, when the rectal temperature of the patient had fallen to 40.0°C. The required power was variable and some time later had fallen to 22 W, for the same temperatures. This decrease was observed to correlate with a fall in the systolic blood pressure and pulse rate in response to renewed administration of epidural block. In view of the magnitude of this power requirement it is difficult to avoid the conclusion that previous methods of irrigation of the stomach have been ineffective in creating a significant temperature rise. Consider the required flow rate to attain 43.5°C at the stomach wall with a water inlet temperature of, say, 48°C. With perfect mixing, the water being removed at a temperature no higher than 43.5°C, each milliliter of water provides 4.5 cal. The power requirement is 30 W, i.e., 7.2 cal/s, giving the required flow rate of about 100 ml/minute. This is for a whole-body temperature of 40°C, at normal body temperature the required flow rate is roughly doubled.

Although we have referred throughout this paper to apparatus, procedures and requirements appropriate to heating within the stomach we consider that the technique is applicable to treatment of other body cavities, rectum, bladder, vagina, oesophagus and, with modification, peritoneal and pleural cavities.

Acknowledgment

We gratefully acknowledge the stimulus towards the evolution of this technique provided by discussions with Mr. M. A. Henderson, Consultant Surgeon, Royal Infirmary, Dumfries and Mr. A. N. Smith, Reader in Clinical Surgery, University of Edinburgh, and the enthusiastic and prompt participation of personnel of Ferranti Ltd., Edinburgh, notably Messrs. W. R. McGill and J. M. Morrison in the construction of the prototype apparatus.

References

1. R. Cavalier, et al. Selective heat sensitivity of cancer cells. Biochemical and clinical studies. *Cancer* (Philadelphia), 20, 1351. 1967.
2. F. M. Vermel and L. B. Kuznetsova. Klinicheskoe izuchenie N-nitrozome-tilmocheviny. *Voprovy Onkologii*, 16, 96. 1970.
3. K. Overgaard and J. Overgaard. Investigations of the possibility of a thermic tumour therapy. 1. Short-wave treatment of a transplanted isologous mouse mammary carcinoma. *European Journal of Cancer*, 8, 65. 1972.
4. M. A. Henderson and R. T. Pettigrew. Induction of controlled hyperthermia in treatment of cancer. *Lancet*, 1, 1275. 1971.
5. R. T. Pettigrew, et al. Clinical effects of whole-body hyperthermia in advanced malignancy. *British Medical Journal*, 4, 679. 1974.
6. J. S. Stehlin. Hyperthermic perfusion with chemotherapy for cancers of the extremities. *Surgery, Gynaecology and Obstetrics*, 129, 305. 1969.
7. R. R. Hall, R. O. K. Schade and J. Swinney. Effects of hyperthermia on bladder cancer. *British Medical Journal*, 2, 593. 1974.
8. R. T. Pettigrew, et al. Circulatory and biochemical effects of whole body hyperthermia. *British Journal of Surgery*, 61, 727. 1974.

Combined Modalities[1]

Joan Bull

Clinical Oncology, National Cancer Institute, NIH, Bethesda, Maryland 20205

Rather than discuss further current investigations in systemic hyperthermia, it may be more useful to project needs for future developments. This projection assumes that whole-body hyperthermia will prove to be a clinically useful tool and an effective therapy in some types of metastatic cancer. It assumes that the optimal use of heat as adjuvant to standard therapy may increase survival in solid tumors such as pancreas, colorectal, gastric, and metastatic breast cancer.

There are major questions that must be addressed for which there is little or no information at present. Regarding systemic hyperthermia as a single modality: what are the duration and the frequency of heating that will optimize tumor kill while sparing normal tissue? The frequency of repeated treatments and for which tumor type (if tumor type is a critical factor), have not been examined. Are these parameters different for different cancers (for example, gastric cancer *versus* melanoma) or for different types of a cancer (*e.g.*, different melanomas)? While 41.8°C may be the maximal tolerance of the whole organism for 1 to 4 hr, possibly brief temperature bursts to 43°C combined with 1 to 2 hr at 41.8°C would be tolerable and would accomplish more of a therapeutic advantage. A full Phase I exploration of heat alone has not yet been done.

When one considers the addition of chemotherapeutic agents to systemic hyperpyrexia, the question of temperature-chemical sequencing is of critical importance. It is likely that each class of agents may have different optimal sequencing times with heat that will increase tumor cell kill without lethal injury to normal cells.

One of the major problems in examining drug-heat interactions, as well as heat-time interactions, is the lack of suitable small animal models. The murine systems, useful in other analyses, do not tolerate systemic temperatures as high as 41.8°C. If an appropriate animal model could be developed, it is obvious that many basic questions of timing and sequencing could be worked out more rapidly than by relying totally on the clinical model.

While *in vitro* systems are useful in examining these questions of heat-drug interactions, the complex metabolism of pharmacological agents is made even more complex at higher temperatures by changes in hepatic and renal blood flow, as well as by metabolic changes.

An *in vivo* model that answers questions of drug-heat interactions is the murine model for local hyperthermia plus systemic chemotherapy. This model, described by Dr. Marmor, may be useful in developing heat-drug sequencing, although changes in drug pharmacokinetics and metabolism in the hyperthermia state are not examined.

Thus, major biological issues to examine are questions of timing, duration, extent of heat, and heat-drug sequencing, as well as variable tumor susceptibility to heat and heat plus drug.

Practical problems of application of systemic heat are also difficult if one considers using heat for large patient populations. Major problems are the time consumed and the manpower used to apply all techniques of whole-body hyperthermia. These cumbersome aspects apply to warm-water suit, hot paraffin wax, and external blood shunt methodologies. Each procedure occupies the time and manpower efforts comparable to renal hemodialysis.

A major factor is the long induction time required for existing techniques; 1.5 to 3 hr is not only wasted time, but it is dangerous cardiovascularly for the patients. Acute arrhythmia or cardiac failure is as likely or more likely during induction temperature than during plateau temperature. During plateau, the skin is cooled to maintain a constant core temperature at 41.8°C. An improvement of the external shunt method or use of microwaves or ultrasound could conceivably decrease this problem of slow core heat induction.

By whatever method of heat induction and maintenance, the need to monitor body temperature accurately is essential. There is a need to develop noninvasive temperature monitors to assess hepatic, brain, interabdominal, and intertumor temperatures.

Thus, improvement of systemic hyperthermia will require both hardware and technique development and improvements, as well as the development of biological insights to guide its optimal use.

[1] Presented at the Conference on Hyperthermia in Cancer Treatment, September 15 and 16, 1978, San Diego, Calif.

MEDICAL IMAGING BY NUCLEAR MAGNETIC RESONANCE ZEUGMATOGRAPHY

Paul C. Lauterbur*

Abstract

Both tomographic sections and complete three-dimensional images may be generated from nuclear magnetic resonance signals by a wide variety of techniques. In favorable situations, contrast and resolution may be comparable to those obtained by other imaging techniques, although scanning times will in general be longer. The practical problems encountered in the construction and use of head, limb, and whole-body systems are reviewed, progress toward useful whole-body imaging evaluated, and the effects of flow and motion, the sources and control of contrast, and the prospects for diagnostic tissue characterization discussed.

Introduction

Nuclear magnetic resonance zeugmatographic imaging has some of the characteristics of transmission computed tomography, ultrasonic imaging, and radioisotope emission imaging, as well as a number of unique features. Although it is too soon to be certain of the relative importance of NMR imaging techniques in medical research and diagnosis, it seems likely that they will eventually become of comparable importance to the earlier methods. Research in this field has been concentrated on the development of schemes for transforming NMR signals into images, upon the design and construction of large magnets and radiofrequency transmitter-receiver coils, and upon the relationships between NMR properties of tissues characteristic of disease and the distinctive contrast obtainable in the images. Secondary considerations, which may become of critical importance as practical solutions to the basic problems are devised and clinical studies begin, include the effects of flow and motion on the images, and the effects of the static and modulated magnetic fields and the continuous and pulsed radiofrequency electromagnetic fields on human beings. The effects of blood flow and of organ motion on the images can also be used, however, to obtain useful information, and it seems likely at this time that the hazards associated with NMR imaging can be made negligible if some extreme values of the field parameters are avoided.

Instrumentation

Two large NMR zeugmatographic imaging systems are in use or under construction in this laboratory. One, with a 42 cm bore, 1000 gauss (0.1 T) magnet, has been used for objects up to 15 cm in diameter, and will soon be used for 20 cm objects. A magnetic field uniformity of approximately 1 part in 10^5 has been achieved over a region about 18 cm in diameter and 15 cm long by careful positioning of the four independent coils of the sixth order electromagnet, empirical adjustment of the currents in the individual coils, and shimming of the field with a ferromagnetic plate and current-carrying coils to compensate for the field distortion produced by ferromagnetic structures in the building. Proton NMR zeugmatograms are generated by reconstruction from projections obtained by the Fourier transformation of free induction decays following 4 MHz rf pulses. The differently-oriented projections, in any directions in space, are made possible by the electrical reorientation of a linear magnetic field gradient of about 5×10^{-6} T cm^{-1} produced by three sets of gradient coils energized by a microprocessor-controlled current source.[1] Either 2D or 3D images may therefore be generated without any mechanical motion of the apparatus. A number of studies with phantoms and animals are under way in this system. The second system, with a 62 cm bore and the same magnetic field, will be used for objects up to about 40 cm in diameter, including the whole human body. A field uniformity similar to that in the 42 cm magnet has been achieved over a diameter of more than 35 cm, and imaging experiments will be carried out in the near future.

*Department of Chemistry
State University of New York at Stony Brook
Stony Brook, New York 11794

Techniques

We have emphasized the development of those NMR zeugmatographic imaging techniques that are based upon reconstruction from projections for several reasons. Full three-dimensional images may be obtained efficiently and directly, instead of being assembled slice by slice in a series of experiments, and the resolution in such 3D images may be the same in all directions. Rapidly switched or modulated magnetic field gradients are not required, which simplifies the construction of the apparatus and eliminates the possibility of biological effects from induced currents. Furthermore, relatively accurate measurements of standard NMR parameters, such as relaxation times, that are useful in tissue characterization may be made, and the sensitivity may be readily optimized. It is probable that those techniques that make use of the phase as well as amplitude information in the NMR signals that is present in rapidly changing gradients will ultimately prove to be more rapid and sensitive, however, especially if they can be used practically for direct 3D imaging.[2]

One area of application that may be more effectively served by those techniques that monitor the behavior of an NMR signal at a single location[3] is the study of time-dependent phenomena, such as pulsatile flow, heart motion, the movement of injected substances through the vascular system, and the rates of deposition and elimination of exogenous materials in organs. Preliminary studies of arterial phantoms by the sensitive point method have been successful,[4] and measurements of the effects of injected paramagnetic ions on relaxation times, as a function of time after injection, should be possible. The latter application should be of interest in studies of liver and kidney function, for example.

Diagnostic Applications

In addition to the imaging of anatomical structure, which depends for contrast upon the concentrations of water in various tissues, and upon the NMR relaxation times, which can change the image intensities under suitable conditions, it is possible to image the distribution of tissue abnormalities. For example, edema changes signal intensities directly by changing water concentrations, and also changes the tissue water NMR relaxation times by dilution effects and by producing other more subtle alterations. Detailed studies of these phenomena have been made, for example, on tissue samples in connection with experimentally produced edema in dog lungs[5,6] and of myocardial infarction in dog and pig hearts, especially with the help of paramagnetic contrast agents. Measurements on malignant tumors have also been made, and suggest that NMR

Reprinted from *IEEE Trans. Nucl. Sci.*, vol. NS-26, pp. 2808–2811, Apr. 1979.

imaging techniques may be quite useful in tumor detection.[7,8] The possibility of providing quantitative information on flow and motion has already been mentioned above.

Research Applications

Studies requiring the use of the much weaker NMR signals from substances other than water are unlikely to be competitive in straightforward diagnostic applications in the foreseeable future because of the unfavorable combination of imaging time, sensitivity, and resolution.[9] In more specialized studies, however, there may be some very interesting possibilities. Changes in phosphorus metabolites in intact cells, tissues and organs have been studied extensively during the past few years in several laboratories without imaging.[10] Recently, we have succeeded in generating separate images of the different spatial distributions of major phosphorus metabolites by ^{31}P NMR zeugmatography at a very high magnetic field. Only tissue-equivalent phantoms have been used so far, and it is not yet known how practical such measurements will be with the same substances in perfused organs or entire animals. Extensions of ^{31}P NMR imaging experiments to human beings will require the use of high-field whole-body superconducting magnets, and a number of technical questions must be answered before the eventual practicality of such measurements can be determined.

Acknowledgements

This investigation was supported by Grant No. CA-153000, awarded by the National Cancer Institute, DHEW, by Contract NO1-HV-5-2970, awarded by the National Heart, Lung and Blood Institute, DHEW, and by a grant from the Northport Veterans Administration Hospital.

References

1. C.-M. Lai, J.W. Shook and P.C. Lauterbur, "Microprocessor-Controlled Reorientation of Magnetic Field Gradients for NMR Zeugmatographic Imaging", Chem. Biomed. Environ. Instr., (in press).

2. A.Kumar, D. Welti, and R.R. Ernst, "NMR - Fourier - Zeugmatography", J. Mag. Res. 18, 69-83 (1975).

3. W.S. Hinshaw, "Image Formation by Nuclear Magnetic Resonance: The Sensitive Point Method", J. Appl. Phys. 47, 3709-3721 (1976).

4. P.C. Lauterbur and C.-M. Lai, "Feasibility Study of Nuclear Magnetic Resonance Zeugmatography for Use in Detecting Atherosclerosis", in Proc. of the NHLBI Division of Heart and Vascular Diseases, Devices and Technology Branch Annual Contractors Meeting 1977, pp. 158-159.

5. P.C. Lauterbur, J.A. Frank, and M.J. Jacobson, "Water Proton Spin-Lattice Relaxation Times in Normal and Edematous Dog Lungs", Physics in Canada 32, Special July Issue: Digest of the Fourth International Conference on Medical Physics, Abstract 33.9 (1976).

6. J.A. Frank, M.A. Feiler, W.V. House, P.C.Lauterbur, and M.J. Jacobson, "Measurement of Proton Nuclear Magnetic Longitudinal Relaxation Times and Water Content in Infarcted Canine Myocardium and Induced Pulmonary Injury", Clinical Research 24, 217A (1976).

7. P.C. Lauterbur, C.-M. Lai, J.A. Frank, and C.S. Dulcey, Jr., "In Vivo Zeugmatographic Imaging of Tumors", Physics in Canada 32, Special July Issue: Digest of the Fourth International Conference on Medical Physics, Abstract 33.11 (1976).

8. I.L. Pykett and P. Mansfield, "A Line Scan Image Study of a Tumorous Rat Leg by NMR", Phys. Med. Biol. 23 (5), 961-967 (1978).

9. P.C. Lauterbur, "Spatially-Resolved Studies of Whole Tissues, Organs and Organisms by NMR Zeugmatography", in NMR in Biology, R.A. Dwek, I.D. Campbell, R.E. Richards and R.J.P. Williams, Eds., Academic Press, London, 1977, pp. 323-335.

10. D.P. Hollis, R.L. Nunnally, G.J. Taylor IV, M.L. Weisfeldt, and W.E. Jacobus, "Phosphorus Nuclear Magnetic Resonance Studies of Heart Physiology", J. Mag. Res. 29, 319-330 (1978).

Appendix

A Bibliography on NMR Zeugmatographic Imaging and Related Techniques.

1. P.C. Lauterbur, "Image Formation by Induced Local Interactions: Examples Employing Nuclear Magnetic Resonance", Nature 242, 190 (1973).

2. P.C. Lauterbur, "Stable Isotope Distributions by NMR Zeugmatography", Proc. First International Conference on Stable Isotopes in Chemistry, Biology and Medicine, May 9-11, 1973, AEC, CONF-730525.

3. P. Mansfield, P.K. Grannell, A.N. Garroway, and D.C. Stalker, "Multi-pulse Line Narrowing Experiments: NMR 'Diffraction' in Solids?", Proc. First Specialized "Colloque Ampere", Institute of Nuclear Physics, Krakow, Poland, p. 16, 1973.

4. P. Mansfield and P.K. Grannell, "NMR 'Diffraction' in Solids?", J. Phys. C: Solid State Phys. 6, L422 (1973).

5. P.C. Lauterbur, "Magnetic Resonance Zeugmatography", Pure Appl. Chem. 40, 149 (1974).

6. P.C. Lauterbur, "Reconstruction in Zeugmatography-The Spatial Resolution of Magnetic Resonance Signals", in Techniques of Three-Dimensional Reconstruction: Proceedings of an International Workshop, July 16-19, 1974, Brookhaven National Laboratory, R.B. Marr, Ed., Publicationa BNL 20425, pp. 20-22.

7. W.S. Hinshaw, "Spin Mapping: The Application of Moving Gradients to NMR", Phys. Letters 48A, 87 (1974).

8. A.N. Garroway, "Velocity Measurements in Flowing Fluids by NMR", J. Phys. D: Appl. Phys. 7, L159-163 (1974).

9. K. Tanaka, Y. Yamada, T. Shimizu, F. Sano and Z. Abe, "Fundamental Investigations (in vitro) for a Non-Invasive Method of Tumor Detection by Nuclear Magnetic Resonance", Biotelemetry 1, 337-350 (1974).

10. A.N. Garroway, P.K. Grannell, and P. Mansfield, "Image Formation in NMR by a Selective Irradiation Process", J. Phys. C: Solid State Phys. 7, L457-L462 (1974).

11. P.C. Lauterbur, C.S. Dulcey, Jr., C.-M. Lai, M.A. Feiler, W.V. House, Jr., D.M. Kramer, C.-N. Chen, and R. Dias, "Magnetic Resonance Zeugmatography", in Magnetic Resonance and Related Phenomena, Proceedings of the 18th Ampere Congress, Nottingham Sept. 9-14, 1974, P.S. Allen, E.R. Andrew, and C.A. Bates, Eds., North-Holland, Amsterdam, 1975, Vol. 1, pp. 27-29.

12. P. Mansfield, P.K. Grannel, and A.A. Maudsley, "Diffraction and Microscopy in Solids and Liquids by NMR", in Magnetic Resonance and Related Phenomena, Proceedings of the 18th Ampere Congress, Nottingham, Sept. 9-14, 1974, P.S. Allen, E.R. Andrew, and C.A. Bates, Eds., North-Holland, Amsterdam, 1975, Vol. 2, pp. 431-432.

13. W.S. Hinshaw, "The Application of Time Dependent Field Gradients to NMR Spin Mapping", in Magnetic Resonance and Related Phenomena, Proceedings of the 18th Ampere Congress, Nottingham, Sept. 9-14, 1974, P.S. Allen, E.R. Andrew, and C.A. Bates, Eds., North-Holland, Amsterdam, 1975, Vol. 2, pp. 433-434.

14. A.N. Garroway, "Velocity Profile Measurements by NMR", in Magnetic Resonance and Related Phenomena, Proceedings of the 18th Ampere Congress, Nottingham Sept. 9-14, 1974, P.S. Allen, E.R. Andrew, and C.A. Bates, Eds., North-Holland, Amsterdam, 1975, Vol. 2, pp. 435-436.

15. J.M.S. Hutchison, J.R. Mallard, and C.C. Goll, "In-Vivo Imaging by Body Structures Using Proton Resonance", in Magnetic Resonance and Related Phenomena, Proceedings of the 18th Ampere Congress, Nottingham, Sept. 9-14, 1974, P.S. Allen, E.R. Andrew, and C.A. Bates, Eds., North-Holland, Amsterdam, 1975, Vol. 1, pp. 283-284.

16. P. Mansfield and P.K. Grannell, "'Diffraction' and Microscopy in Solids and Liquids by NMR", Phys. Rev. B 12, 3618-3634 (1975).

17. A. Kumar, D. Welti, and R.R. Ernst, "Imaging of Macroscopic Objects by NMR Fourier Zeugmatography", Naturwiss. 62, 34 (1975).

18. A. Kumar, D. Welti, and R.R. Ernst, "NMR - Fourier-Zeugmatography", J. Mag. Res. 18, 69-83 (1975).

19. P.C. Lauterbur, W.V. House, Jr., D.M. Kramer, C.-N. Chen, F.W. Porretto, and C.S. Dulcey, Jr., "Reconstruction from Selectively-Excited Signals in Nuclear Magnetic Resonance Zeugmatography", in Image Processing for 2-D and 3-D Reconstruction from Projections: Theory and Practice in Medicine and the Physical Sciences, Opt. Soc. Am., pp. MA10-1-MA10-3 (1975).

20. P.C. Lauterbur, D.M. Kramer, W.V. House, Jr., and C.-N. Chen, "Zeugmatographic High Resolution Nuclear Magnetic Resonance Spectroscopy. Images of Chemical Inhomogeneity within Macroscopic Objects", J. Am. Chem. Soc. 97, 6866-6868 (1975).

21. P.K. Grannell and P. Mansfield, "Microscopy in vivo by Nuclear Magnetic Resonance", Phys. Med. Biol. 20, 477-482 (1975).

22. P. Mansfield, A.A. Maudsley, and T. Baines, "Fast Scan Proton Density Imaging by NMR", J. Phys. E: Sci. Instrum. 9, 271-278 (1976).

23. P.C. Lauterbur, C.-M. Lai, J.A. Frank, C.S. Dulcey, Jr., "In Vivo Zeugmatographic Imaging of Tumors", Physics in Canada 32, Special July Issue: Digest of the Fourth International Conference on Medical Physics, Abstract 33.11 (1976).

24. R. Damadian, L. Minkoff, M. Goldsmith, M. Stanford, and J. Koutcher, "Tumor Imaging in a Live Animal by Field Focusing NMR (FONAR)", Physiol. Chem. & Phys. 8, 61-65 (1976).

25. W.S. Hinshaw, "Image Formation by Nuclear Magnetic Resonance: The Sensitive Point Method", J. Appl. Phys. 47, 3709-3721 (1976).

26. P. Mansfield and A.A. Maudsley, "Line Scan Proton Spin Imaging in Biological Structures by NMR", Phys. Med. Biol. 21, 847-852 (1976).

27. E.R. Andrew, W.S. Hinshaw, and W.S. Moore, "Spin Mapping", Brit. J. Rad. 49, 1052-1061 (1976).

28. T. Baines and P. Mansfield, "An Improved Picture Display for NMR Imaging", J. Phys. E: Sci. Instrum. 9, 809-811 (1976).

29. P. Mansfield and A.A. Maudsley, "Planar Spin Imaging by NMR", J. Phys. C: Solid State Phys. 9, L409-411 (1976).

30. P. Mansfield, "Proton Spin Imaging by NMR", Contemp. Phys. 17, 553-576 (1976).

31. P. Mansfield and A.A. Maudsley, "Planar and Line-Scan Spin Imaging by NMR", Proc. XIXth Congress Ampere, Heidelberg, 1976, pp. 247-252.

32. R. Damadian, L. Minkoff, M. Goldsmith, M. Stanford and J. Koutcher, "Field Focusing Nuclear Magnetic Resonance (FONAR): Visualization of a Tumor in a Live Animal", Science 194, 1430-1432 (1976).

33. P. Mansfield, "Multi-planar Image Formation Using NMR Spin Echoes", J. Phys. C: Solid State Phys. 10, L55-58 (1977).

34. J.M.S. Hutchison, "Imaging by Nuclear Magnetic Resonance" in Medical Images: Formation, Perception and Measurement, G.A. Hay, Ed., Wiley, 1977, pp. 135-141.

35. D.I. Hoult, "Zeugmatography: A Criticism of the Concept of a Selective Pulse in the Presence of a Field Gradient", J. Mag. Res. 26, 165-167 (1977).

36. P. Mansfield and A.A. Maudsley, "Medical Imaging by NMR", Brit. J. Rad. 50, 188-194 (1977).

37. G.J. Bene, B. Borcard, E. Hiltbrand, P. Magnin, and R. Sechehaye, "Magnetographie nucleaire en champ faible. Resultats preliminaires", C.R. Acad. Sc. Paris B, 284, 141-143 (1977).

38. P. Mansfield and A.A. Maudsley, "Planar Spin Imaging by NMR", J. Mag. Res. 27, 101-119 (1977).

39. E.R. Andrew, "Body-Scanning by Nuclear Spin", Spectrum 150, 2-6 (1977).

40. E.R. Andrew, "Zeugmatography", Proceedings of the IV Ampere International Summer School, Pula, Yugoslavia, Sept. 13-23, 1976, R. Blinc and G. Lahajuar, Eds., J. Stefan Institute, Ljubljana, Yugoslavia, 1977.

41. G.N. Holland and P.A. Bottomley, "A Colour Display Technique for NMR Imaging", J. Phys. E: Sci. Instrum. 10, 714-716 (1977).

42. E.R. Andrew, "Imaging by Nuclear Magnetic Resonance", Phys. Bull. 323 (1977).

43. G.N. Holland, P.A. Bottomley, and W.S. Hinshaw, "^{19}F Magnetic Resonance Imaging", J. Mag. Res. 28, 133-136 (1977).

44. P.C. Lauterbur, "Spatially-Resolved Studies of Whole Tissues, Organs and Organisms by NMR Zeugmatography", in NMR in Biology, R.A. Dwek, I.D. Campbell, R.E. Richards, and R.J.P. Williams, Eds., Academic Press, London, 1977, pp. 323-335.

45. E.R. Andrew, P.A. Bottomley, W.S. Hinshaw, G.N. Holland, W.S. Moore, and C. Simaroj, "NMR Images by the Multiple Sensitive Point Method: Application to Larger Biological Systems", Phys. Med. Biol. 22, 971-974 (1977).

46. R. Damadian, M. Goldsmith, and L. Minkoff, "NMR in Cancer: XVI. FONAR Image of the Live Human Body", Physiol. Chem. and Phys. 9, 97-100 (1977).

47. L. Minkoff, R. Damadian, T.E. Thomas, N. Hu, M. Goldsmith, J. Koutcher, and M. Stanford, "NMR in Cancer: XVII. Dewar for a 53-Inch Superconducting NMR Magnet", Physiol. Chem. and Phys. 9, 101-104 (1977).

48. M. Goldsmith, R. Damadian, M. Stanford, and M. Lipkowitz, "NMR in Cancer: XVIII. A Superconductive NMR Magnet for a Human Sample", Physiol. Chem. and Phys. 9, 105-107 (1977).

49. P.C. Lauterbur and C.-M. Lai, "Feasibility Study of Nuclear Magnetic Resonance Zeugmatography for Use in Detecting Atherosclerosis", Proc. of the NHLBI Division of Heart and Vascular Diseases, Devices and Technology Branch Annual Contractors Meeting 1977, pp. 158-159.

50. C.-M. Lai, W.V. House, Jr., and P.C. Lauterbur, "Nuclear Magnetic Resonance Zeugmatography for Medical Imaging", Proc. of the IEEE Electro/78 Conference, Session 30, "Technology for Non-Invasive Monitoring of Physiological Phenomena", May 25, 1978, paper 2.

51. W.S. Hinshaw, P.A. Bottomley, and G.N. Holland, "Radiographic Thin-Section Image of the Human Wrist by Nuclear Magnetic Resonance", Nature 270, 722-723 (1977).

52. L.E. Crooks, T.P. Grover, L. Kaufman, and J.R. Singer, "Tomographic Imaging with Nuclear Magnetic Resonance", Invest. Radiol. 13, 63-66 (1978).

53. P.A. Bottomley, W.S. Hinshaw, and G.N. Holland, "A Computer Driven Photoscanner for Medical Imaging", Phys. Med. Biol. 23, 309-317 (1978).

54. W.S. Hinshaw, E.R. Andrew, P.A. Bottomley, G.N. Holland, W.S. Moore, and B.S. Worthington, "Display of Cross Sectional Anatomy by Nuclear Magnetic Resonance Imaging", Brit. J. Radiol. 51, 273-280 (1978).

55. P.A. Bottomley and E.R. Andrew, "RF Magnetic Field Penetration, Phase Shift and Power Dissipation in Biological Tissue: Implications for NMR Imaging", Phys. Med. Biol. 23, 630-643 (1978).

56. P. Mansfield,and I.L. Pykett, "Biological and Medical Imaging by NMR", J. Mag. Res. 29, 355-373 (1978).

57. R.J. Sutherland and J.M.S. Hutchison, "Three-dimensional NMR Imaging Using Selective Excitation", J. Phys. E.: Sci. Instr. 11, 79-83 (1978).

58. J.M.S. Hutchison, R.J. Sutherland, and J.R. Mallard, "NMR Imaging: Image Recovery under Magnetic Fields with Large Non-uniformities", J. Phys. E.: Sci. Instrum. 11, 217-221 (1978).

59. R. Damadian, L. Minkoff, M. Goldsmith, and J.A. Koutcher, "Field-Focusing Nuclear Magnetic Resonance (FONAR)", Naturwis. 65, 250-252 (1978).

60. H.R. Brooker and W.S. Hinshaw, "Thin-Section NMR Imaging", J. Mag. Res. 30, 129-131 (1978).

61. E.R. Andrew, "L'Exploration du Corps par le Spin Nucléaire", Médecine et Hygiène 36, 1862-1869(1978).

62. P. Mansfield, I.L. Pykett, and P.G. Morris, "Human Whole Body Line-Scan Imaging by NMR", Brit. J. Radiol. 51, 921-922 (1978).

63. I.L. Pykett and P. Mansfield, "A Line Scan Image of a Tumorous Rat Leg by NMR", Phys. Med. Biol. 23, 961-967 (1978).

HUMAN WHOLE BODY LINE SCAN IMAGING BY NUCLEAR MAGNETIC RESONANCE

Peter G. Morris*, Peter Mansfield*, Ian L. Pykett*, Roger J. Ordidge* and Rex E. Coupland**

Abstract

An NMR imaging apparatus, capable of whole human body studies, is described. The resolution and spatial distortions of the system are discussed and illustrated. Representative images of the human anatomy are presented.

Introduction

Conventional nuclear magnetic resonance (NMR) techniques utilize a large static magnetic field of high homogeneity. The superposition of magnetic field gradients[1,2], however, allows the NMR parameters to be spatially localized and therefore imaged. A number of small-scale (< 10 cm) images have been reported, including a live human finger[3,4] and wrist[5]. However, the extension to whole human body dimensions has only recently been accomplished[6-8] and it is the purpose of this paper to describe our apparatus and to illustrate its performance. A number of other approaches to NMR imaging are being developed and are discussed and compared elsewhere[9,10].

Human Tissue Characteristics

NMR is applicable to any nucleus which possesses a magnetic moment. However, since human tissue contains an average of 75% water, proton imaging is an obvious first candidate for an imaging experiment. The variation in water content between the various healthy tissues is about 15% and therefore allows imaging of morphology on this basis alone. In addition, NMR is sensitive to the nature of the water: the manner in which it is bound to the tissue and the presence or otherwise of dissolved paramagnetic impurities. These properties manifest themselves through their effects on the relaxation times T_1 and T_2 of the protons. The spin-lattice relaxation time T_1 is also strongly temperature and frequency dependent, and, although at present it has not been well characterized for human tissue, the variation for the different human organs and that between normal and diseased states is likely to greatly exceed the variation attributable to water content alone. These differences may be as much as a factor of two or three. NMR experiments all depend in some way on T_1 and it may be desirable, particularly for tumour discrimination[11], to accentuate the effect of this parameter.

In addition to spin density and relaxation times, other NMR parameters which one could envisage measuring in an imaging experiment include: diffusion coefficients, macroscopic flow, and chemical shift.

Line Scanning

The technique of line scanning used in these experiments derives from the early work of Garroway et al.[12] and has been largely described elsewhere[13]. Only a brief account is therefore given here.

Plane selection is achieved with a linear magnetic field gradient. An orthogonal gradient is applied to the sample which is then irradiated with a radiofrequency (RF) pulse tailored in such a manner that only the spins lying along one particular line receive a 90° nutation. Immediately following the pulse, the field gradient is switched through 90° and the Fourier transform of the ensuing free induction decay yields the spin density profile along a line. This process is repeated to achieve an acceptable signal-to-noise ratio, the averaged data are stored and the line off-set is incremented. A complete image is thus built up line by line.

Apparatus

A block diagram of the apparatus is given in Figure 1.

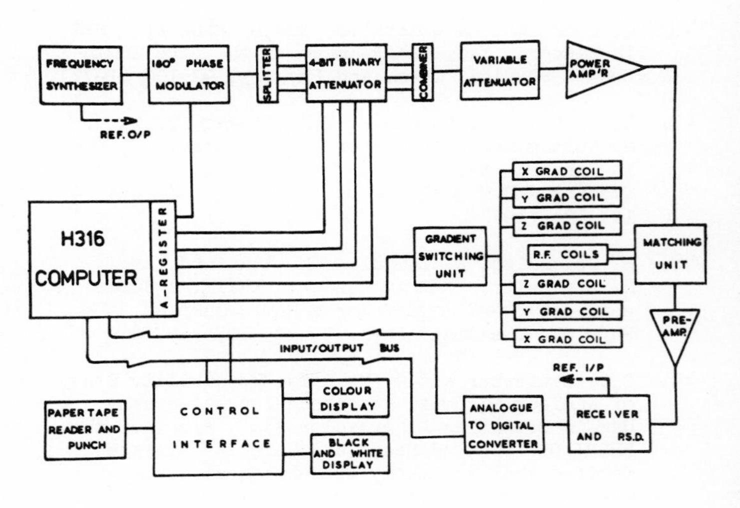

Figure 1: *Schematic diagram of the NMR imaging apparatus.*

The RF and gradient coils lie in a static field of 1 kG which is produced by an Oxford Instruments 4-coil 25½ inch bore electromagnet. The homogeneity is about 1 part in 10^4 over typical whole body dimensions though this figure should improve to 1 part in 10^5 with the removal of the structural ironwork in our present laboratory.

The gradient coils are of two types: one is a reversed Helmholtz pair, and the remaining two utilize line currents in a novel manner to be described in a later publication[14]. They all produce field gradients of approximately 0.04 G cm^{-1}, linear to within about 3%. Switching takes about 50 μs and is triggered from the computer (Honeywell 316).

The RF probe is a crossed-coil system with a 4-turn split-pair rectangular configuration (Q = 9, L = 18 μH) for the transmitter and a 5-turn elliptic coil (Q = 34, L = 9.6 μH) for the receiver. The magnet and coil systems are surrounded by an RF shielding cage constructed from aluminium sheet.

The low level RF is routed through a 180° phase modulator and binary attenuation gate which are triggered

*Department of Physics, University of Nottingham, University Park, Nottingham NG7 2RD, England.

**Department of Human Morphology, School of Medicine, University of Nottingham, University Park, Nottingham NG7 2RD, England.

Reprinted from *IEEE Trans. Nucl. Sci.*, vol. NS-26, pp. 2817-2820, Apr. 1979.

by the computer according to the calculated irradiation profile. A linear amplifier increases the peak power to about 10 W.

The nuclear signal is amplified, phase sensitively detected, and then sampled by a 10-bit analogue-to-digital converter (Data Laboratories ADC 710), again triggered from the computer. Signal averaging, Fourier transformation and storage of 8-bit data are all performed by the computer. During a scan each profile is normally monitored on a storage oscilloscope. Black and white display of the developing image is also available.

Final image data are generally scaled to 4 bits (with shift and off-set controls) and then displayed on either the black and white monitor or in colour on a domestic television set. Interpolation and data smoothing routines are also available.

Picture Quality

Figure 2 illustrates a phantom consisting of a supporting matrix containing $\frac{3}{4}$ inch water-filled tubes placed at 1 inch intervals to form the letters IP. The NMR image obtained from this phantom is shown in Figure 3 and illustrates the degree of resolution and some of the spatial distortions obtained with our present imaging system. The bright areas correspond in general to

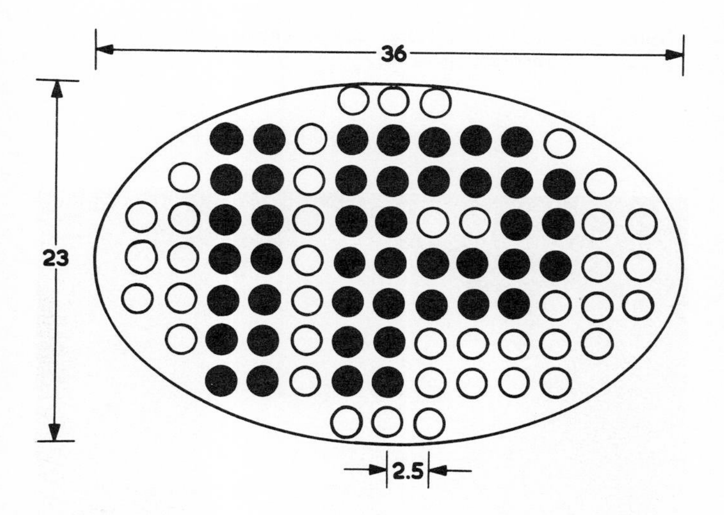

Figure 2: *Test-tube matrix (dimensions in cm). Shaded holes contained water-filled test-tubes*

areas with high mobile proton density. Some distortion is attributable to the gradient coils. However, the major distortions arise from static field inhomogeneity. The curvature in the plane is particularly apparent near the top of the 'I'. Also, curvature of the imaging plane itself can be inferred from the reduced intensities in some regions of the image. The slice thickness was defined to be about 3 to 4 cm. The test tubes in the phantom are 4.5 cm long. Curvature of the imaging plane causes resonance shifts of the water in the tubes. If sufficiently large, these shifts can lead to diminished resonance signal intensity from particular regions. This effect is in evidence in Figure 3.

In an NMR imaging experiment the resolution does not depend directly on the wavelength of the RF irradiation (for a frequency of 4 MHz this is about 75 m in free

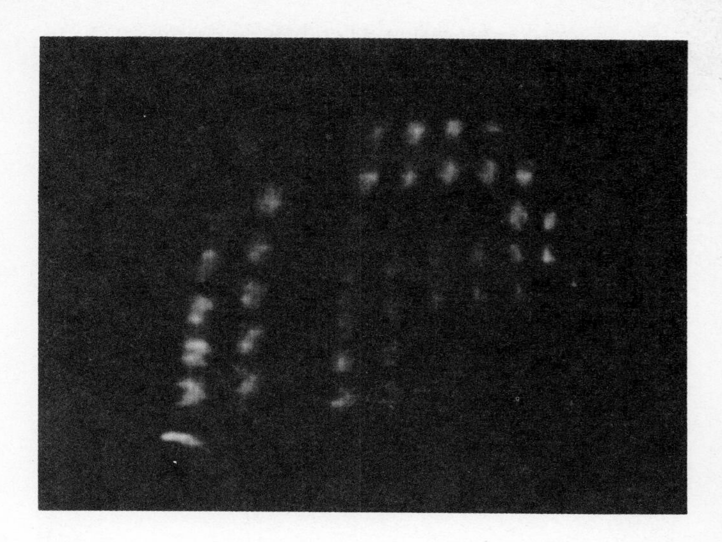

Figure 3: *NMR image of the 'IP' phantom shown in Figure 2.*

space!). Rather, it depends on the strength of the field gradients used and on the amounts of computer memory available. For the present system we have arranged for each line of 90 points to span the imaging field. The full image array is 90^2 points. This figure was dictated by memory limitations (8K presently) and was envisaged to be independent of image size whether it be whole human body, leg, finger, or even cell, the increased resolution necessary for scanning smaller objects being obtained by a proportional increase in the field gradient magnitude. However, in the present system no such facility has been incorporated and the resolution cannot be improved beyond that displayed by the 'IP' phantom. The test tubes are separated by $\frac{1}{4}$ inch so by inspection the resolution along a line would appear to be about 1/8 inch. Between lines (a function of the efficiency of the selective irradiation technique) it is somewhat worse, probably about $\frac{1}{4}$ inch. It should be stressed that this resolution is maintained over an object comparable in dimensions to a human body cross-section.

It has been possible to confirm the efficiency of the selection to a plane by the use of phantoms consisting of perspex rods of lengths in the range 1 to 5 cm.

Skin Depth Effects

It is possible to improve the signal-to-noise ratio of an NMR experiment and therefore to reduce the imaging time by working at higher frequencies. This is clearly the way to proceed if small-scale imaging, of cells say, is envisaged. However, for medical applications high fields have a number of disadvantages.

1. They require the use of expensive high field magnets.
2. It becomes increasingly difficult to tune RF coils of whole body dimensions.
3. Problems arise from lack of complete penetration of the human body by the RF field.

The skin depth at 10 MHz for instance is about 10 cm. Higher frequencies may be useful for scans of heads or limbs, but for whole body work the optimal frequency seems to be around 4 MHz ($\sim$ 1 kG for protons).

Representative Images

Figure 4 shows a cross-section through a cadaver thigh. The data were averaged 128 times and took 26 minutes to acquire. We emphasize that the apparatus

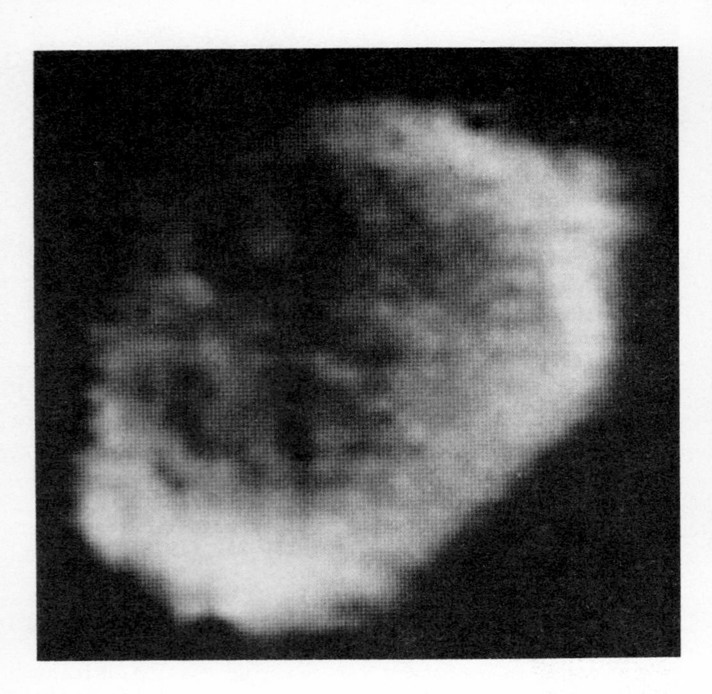

Figure 4: *Cross-sectional image through a human cadaver thigh.*

was set up for whole body work, that no compensatory increase in gradient strength was applied and therefore the resolution is correspondingly low - only 35 real points across the sample. Nevertheless clearly resolved anatomical detail is visible, for example, the femur, femoral vessels and profunda femoris vessels. In contrast with X-ray techniques the low and high signals come from bone and soft tissue respectively. The image shown is unsmoothed, but has been linearly interpolated three times.

Figure 5 shows a whole body cross-section through the abdomen of PM at the level of the third and lower part of the second lumbar vertebrae. This 75 x 90 point image was averaged 96 times and took 40 minutes to produce. The posterior is to the bottom of the picture and the liver is the large bright mass to the right. Other details visible include: abdominal muscle and retroperitoneal fat, the vertebra, pancreas, kidneys, stomach, and intestines, etc. For a full anatomical assignation we refer the reader to Reference 8. The missing anterior section is partly due to a progressively increasing phase error, but is largely ascribed to motional artefacts caused by the subject's breathing. The original data showed diminished signal from the interior caused by RF penetration and/or non-uniformity of receiver coil response. This image was corrected for these effects by first smoothing and then applying an elliptically varying enhancement function to be described elsewhere. It should be added that no sensations were experienced during the scan and no ill effects have since appeared.

The possible use of T_1 discrimination in tumour detection was mentioned above. Such discrimination is implemented in the line scanning experiment by simply decreasing the delay between scans during signal averaging. Figures 3 to 5 were obtained with a delay of approximately 0.3 s, whereas a value of 0.15 s was used for Figure 6. This shows a coronal projection of a

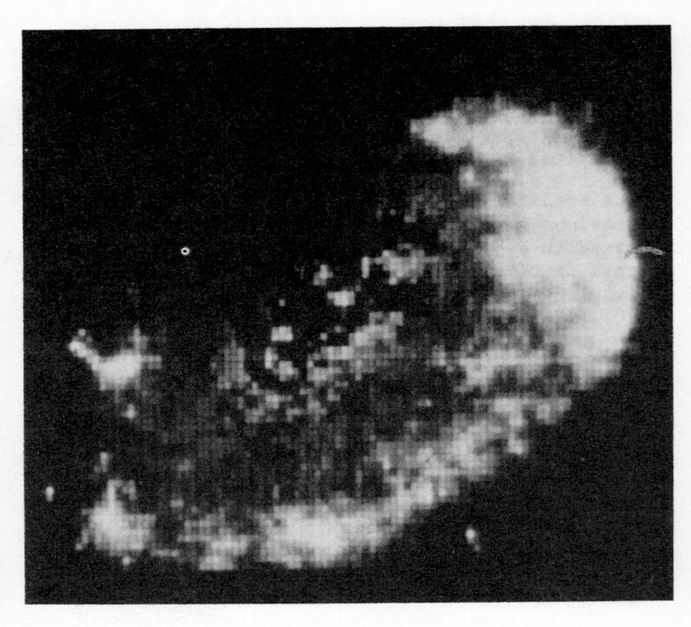

Figure 5: *Cross-sectional image through the human abdomen at L2-3.*

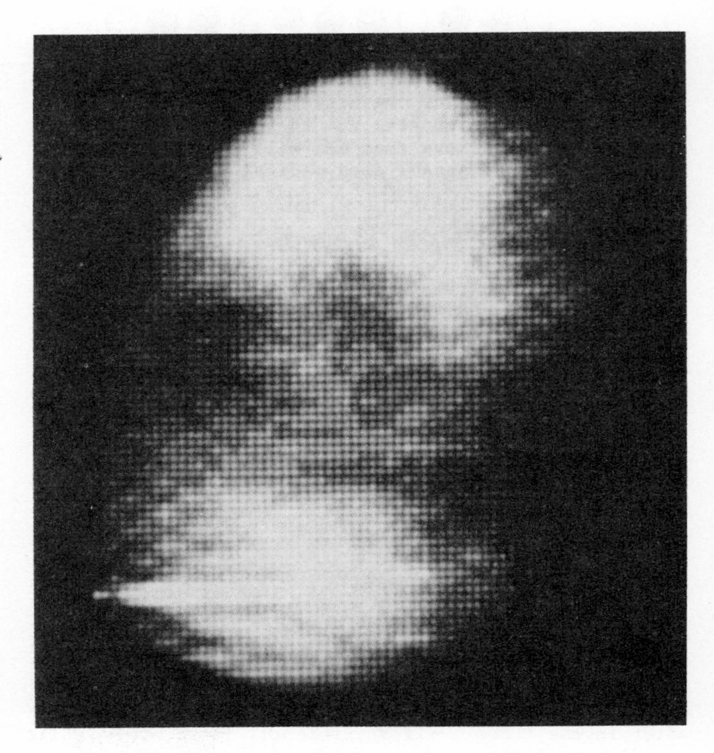

Figure 6: *Coronal projection of an excised human breast containing a scirrhous carcinoma.*

whole tumourous breast 1½ hours following simple mastectomy. The nipple is visible to the left and a scirrhous carcinoma to its right. In a normal breast the largest signal would be expected from the central region (corresponding to the greatest thickness). However, in this case, the long T_1's of the tumourous regions have caused the signal to become reduced in intensity thereby enhancing the image contrast. The tumour size and position correspond well with the paraffin sections and histological examination. The reader is referred to Reference 15 for further details.

Using the same technique of two-dimensional imaging it has been possible to locate a fibroadenoma in an excised breast specimen. Cerebral tumours have also been observed in post-mortem specimens.

The breast image has the same spatial resolution as all previous images. This resolution could be greatly increased in a purpose-built instrument. We anticipate that *in vivo* imaging of breasts is one area in which NMR imaging might find clinical application.

Timing

The advantages of the non-invasive and, so far as is known, hazardless nature of NMR imaging are somewhat offset by the inherent insensitivity of the NMR experiment. Scan times will thus be of major concern. Present computer restrictions have limited the imaging time for a 100 averaged 90 x 90 matrix to about 40 minutes. Simple modifications should reduce this to a few minutes. However, for high quality images such a technique would be suited only to regions of the body where motion is relatively unimportant. Real time images of the heart, for example, require imaging times much less than one second. A scheme capable of such speed has recently been proposed by Mansfield and Pykett[16].

Conclusion

In conclusion, NMR imaging should be seen in context with other medical imaging schemes where recent developments have set very high standards. NMR is unlikely to compete in terms of pure morphology, rather it holds the promise that physiological information may be obtained leading to the characterization of diseased states. Such studies have yet to receive clinical evaluation but we believe that NMR will shortly join the expanding range of established medical imaging techniques.

Acknowledgements

We are grateful to the MRC for support of the whole body NMR imaging project, the SRC for staff support and the use of the Appleton Laboratory's photowriter, and to D. Kerr for assistance in the construction of some of the apparatus. We are especially indebted to T. Baines for the design and construction of the electronic apparatus and V. Bangert for his assistance in the design of the gradient coils.

***

References

1. P. C. Lauterbur, 'Image formation by induced local interactions: Examples employing nuclear magnetic resonance', Nature, 242, 190 (1973).

2. P. Mansfield, P. K. Grannell, A. N. Garroway and D. C. Stalker, 'Multi-pulse line narrowing experiments: NMR diffraction in solids?', Proc. First Specialized 'Colloque Ampere', Institute of Nuclear Physics, Krakow, Poland, p. 16 (1973).

3. P. Mansfield and A. A. Maudsley, 'Planar and line-scan spin imaging by NMR', Proc. XIX Congress Ampere, Heidelberg, pp. 247-252 (1976).

4. P. Mansfield and A. A. Maudsley, 'Medical imaging by NMR', Brit. J. Radiol., 50, 188-194 (1977).

5. W. S. Hinshaw, P. A. Bottomley and G. N. Holland, 'Radiographic thin-section image of the human wrist by nuclear magnetic resonance', Nature, 270, 722-723 (1977).

6. R. Damadian, M. Goldsmith and L. Minkoff, 'NMR in cancer: XVI FONAR image of the live human body', Physiol. Chem. and Phys., 9, 97-100 (1977).

7. R. Damadian, L. Minkoff, M. Goldsmith and J. A. Koutcher, 'Field-focusing nuclear magnetic resonance (FONAR)', Naturwiss., 65, 250-252 (1978).

8. P. Mansfield, I. L. Pykett, P. G. Morris and R. E. Coupland, 'Human whole body line-scan imaging by NMR', Brit. J. Radiol., 51, 921-922 (1978).

9. P. Mansfield, 'Proton spin imaging by NMR', Contemp. Phys., 17, 553-576 (1976).

10. P. Brunner and R. R. Ernst, 'Sensitivity and performance time in NMR imaging', J. Mag. Res. (in press).

11. R. Damadian, 'Tumour detection by nuclear magnetic resonance', Science, 171, 1151 (1971).

12. A. N. Garroway, P. K. Grannell and P. Mansfield, 'Image formation in NMR by a selective irradiation process', J. Phys. C: Solid State Phys., 7, L457 (1974).

13. P. Mansfield, A. A. Maudsley and T. Baines, 'Fast scan proton density imaging by NMR', J. Phys. E: Scientific Instruments, 9, 271-278 (1976).

14. V. Bangert and P. Mansfield (in preparation).

15. P. Mansfield, P. G. Morris, R. J. Ordidge, R. E. Coupland, H. Bishop and R. Blamey, 'Carcinoma of the breast imaged by nuclear magnetic resonance (NMR)', Brit. J. Radiol. (in press).

16. P. Mansfield and I. L. Pykett, 'Biological and medical imaging by NMR', J. Mag. Res., 29, 355-373 (1978).

THE USE OF ORTHOGONAL POLARIZATIONS IN MICROWAVE IMAGERY OF ISOLATED CANINE KIDNEY

L. E. Larsen and J. H. Jacobi*

Abstract

A method of imaging biological targets using microwave radiation at a frequency of 4 GHz is presented. Linearly polarized radiation is transmitted through an isolated canine kidney and received with co-polarized and cross-polarized antennas. Images are displayed as the spatial variation of the magnitude of the transmission scattering parameter S_{21} for each mode of polarization. The relationship between the spatial variation of the magnitude of S_{21} and canine renal anatomy is discussed. It is shown that within the kidney the cross-polarized image tends to emphasize linear or piecewise linear structures, whereas the co-polarized image balances renal cortical lobulations.

Introduction

The microwave region of the electromagnetic spectrum has many advantages as a choice of radiant energy for the interrogation of biological systems. Primary among these is the fact that photon energy is low enough to interact prominently at the level of intra- and intermolecular forces rather than at the nuclear, atomic, and electronic levels as is the case with more energetic gamma, x-ray and optical photons, respectively. The measurement of bulk dielectric properties at microwave and radio frequencies allows inferences to be made concerning the tertiary structure of biomolecules, at least with respect to hydration, and their associated function (1, 2). At microwave frequencies, the dominant molecular species contributing to the bulk dielectric constant of biosystems is water. Likewise, water and electrolytes dominate the bulk dielectric losses in biosystems.

The spatial distribution of water and electrolytes may reasonably be predicted to provide a means for functional assessment in most biosystems since these are often attributes of the physiological activity and pathophysiological status of major organ systems. This is most certainly true in the case of the kidney where water and electrolyte processing (filtration, secretion and reabsorption) comprise the primary function of the organ. The kidney offers further advantages as a target organ to demonstrate the advantages of microwave imagery in that its microscopic unit of function, the nephron, is generally arrayed in a radial fashion such that macroscopic (anatomical) regions correspond to microscopic functonal specialization. Lastly, the organ displays a rough bilateral symmetry about the plane of the hilus. This is an advantage since none of the images thus far produced with transmitted microwave energy have the advantage of tomographic reconstruction in the plane of propagation.

Prior microwave imagery may be classified into two technologies based on the distinction of passive as contrasted with active sources of microwave energy. The former method is radiometric and makes no explicit use of applied microwave energy (3, 4). It is basically observational in character. On the other hand, active microwave imagery is more amenable to experimental manipulation. Among the many possibilities that become attainable with active radiation are precise control of frequency, phase and amplitude. These key words suggest a network analysis approach to microwave imagery. Prior publications have demonstrated images formed on the basis of the magnitude and phase of the scattering parameter S_{21} (5, 6, 7). The scattering parameter S_{21} is a complex number that represents the magnitude and phase of the insertion loss of a two port microwave network (8). In this case, that is the insertion loss between the transmitting and receiving antennas between which the kidney is interposed. Another attribute that was explored in earlier work was the effect of polarization on the magnitude of the scattering parameter S_{21} (5). In general, insertion losses in heterogeneous dielectrics depend upon frequency, coupling geometry and spatial distribution of complex permittivity within the object. In the prior work, circularly polarized radiation was launched from an Archimedean spiral antenna and received by a dielectrically loaded circular waveguide antenna with two orthogonal probes (5). The two probes were located in an air-filled section following a transition from the dielectrically loaded aperture of the receiving antenna. The deviation of the outputs of these two probes from mutual equality as a target was scanned past the apertures served to indicate a marked dependence of insertion loss upon polarization, geometry, and dielectric properties. In the case of polarization dependence, the magnitude of S_{21} was more sensitive than the phase of S_{21} at orthogonal positions within the circular waveguide when linear targets with various dielectric compositions were introduced.

In general, the expectation is that linear structures in the order of one-half the wavelength of the incident energy in the dielectric, $\lambda\kappa'/2$, would most effectively illustrate polarization dependent insertion loss (9). Thus, linear or quasilinear structures in the order of 4 mm would be most effective given the conditions of f_o=4 GHz, and a relative dielectric constant, κ', of 81. Raleigh scattering is known to take place when target dimensions diminish from resonant, $\ell \simeq \lambda\kappa'/2$, to ca. one-tenth of that condition, but aperture limitations prevent that argument from reaching its full realization. So called optical scattering takes place when the target dimension exceeds $10\lambda\kappa'/2$. This would be predicted to present a relatively constant scattering cross-section as frequencies increased and/or larger targets are involved.

The present report explores the use of polarization rotation as a means for microwave imagery to enhance linear or quasilinear discontinuities in heterogeneous biological dielectrics. More specifically, microwave

---

This material has been reviewed by the Walter Reed Army Institute of Research, and there is no objection to its presentation and/or publication. The opinions or assertions contained herein are the private views of the authors and are not to be construed as official or as reflecting the views of the Department of the Army or the Department of Defense.

The procedures used within this paper conform to the standards for the use of laboratory animals established by the Institute of Laboratory Animal Resources, U. S. National Academy of Sciences "Guide for the Care and Use of Laboratory Animals," DHEW Publication No. 74-23, U.S. Government Printing Office, Washington DC 20402.

*Department of Microwave Research, Walter Reed Army Institute of Research, Walter Reed Army Medical Center, Washington DC 20012.

Reprinted from *IEEE Trans. Nucl. Sci.*, vol. NS-27, pp. 1184-1191, June 1980.

images of isolated canine kidneys are collected under two conditions of polarization of the double ridged, water coupled transmitting and receiving antennas. In one case, known as co-polarization, both antennas are linearly polarized with vertical polarization. In the other case, known as cross-polarized, the receiving antenna is rotated 90° with respect to the transmitting antenna.

Methods

Specimen Preparation

A test target composed of a dielectric material simulating brain was prepared in a 3.6x10.6x10.6cm lucite box. In the center of the 3.6cm dimension, 3 polystyrene rods, 6.4mm in diameter, were placed at 15mm & 30mm separations, center-to-center. The polystyrene and brain phantom were isodense to within ca. 1%.

The specimens for the co-polarized and cross-polarized microwave scans were obtained from mongrel dogs of 10-15kg weight. The renal organs were removed under pentobarbital anethesia of surgical level prior to sacrificing the animals. The kidney was removed with care to retain an intact capsule and to retain blood volume that existed in the in situ condition. The hilar vessels and ureter were sectioned between ligatures. The specimen was quickly transported to the scanner at which time the ureter and pelvis were flushed with saline under a water seal to prevent introduction of air into the ureter or pelvis. The hilar structures were then attached to a ceramic rod by suture material. The organs were approximately 8cm in length, 4.5cm in height and 4cm in the transverse direction. Approximately 7 grams of weight were suspended from the now dependent portion of the organ by a small purse-string suture of 4-0 nylon in the capsule at a distance of ca. 10cm from the nearest margin of the capsule. The weight is needed to counter the bouyancy of the kidney when it is placed in the water bath. The water bath comprises a cubic volume of about 1 meter on a side. The water was continuously exchanged at a rate of 6-8 liters per hour, externally filtered and treated by iodination as a bacteriocidal measure prior to processing by ion exchange columns. The water was returned to a tank in the corner to minimize vibration and bubbling near the two antennas.

Antennas

The transmitting and receiving antennas are constructed of double-ridged waveguide (WRD180C24) with a co-axial feed. The air filled frequency range for dominant mode propagation is 18-36 GHz. The antenna is operated with water immersion to reduce its frequency of operation by the square root of the relative dielectric constant of water, i.e. a factor of $\sqrt{77} \approx 9$. In this way a small aperture permits high spatial resolution, comparable to the dimensions of the aperture in the case of near field imagery while the transmission losses remain manageable at those of 4 GHz. Water coupling has many advantages which are presented in greater detail in reference 10.

Data Acquisition

The kidney specimen was immersed in the cubical water tank approximately 25cm below the surface of the water. The transmitting and receiving antennas were located in the water on opposite sides of the specimen as close as possible to the specimen surface without actual contact occurring. The distance between the two antennas was fixed throughout a given scan at a value of ca. 5cm. The antennas were moved together in spatial synchronism in a raster scan pattern containing 64 lines and 64 points per line. The sampling increment in both the horizontal and vertical direction was 1.4mm. The position of the antennas was measured by external optical transducers to an accuracy of ±0.01mm. At each point in the scan, the magnitude and phase of S_{21} was measured at a frequency of 3.9 GHz using a Hewlett Packard 8542A Automatic Network Analyzer, then corrected for mismatch, frequency tracking and directivity, then recorded on a disc file. The disc file was later transferred to magnetic tape for offline image processing. A more detailed description of the data acquisition system is given in reference number 6.

Image Processing

The data provided by the scanner for both the copolarized and cross-polarized cases comprised two real valued arrays for each which were the magnitude and phase of the scattering parameter S_{21}. Further processing was limited to the magnitude of S_{21} hereafter denoted by $|S_{21}|$. The data was 15 bits plus sign for each of 4096 discrete locations on the plane of the 64 by 64 sampling raster. The data was truncated to 64 by 64 by 7 bits deep prior to image display on Comptal terminals of the Digital Image Analysis Laboratory (DIAL) at the U.S. Army Engineer Topographic Laboratory. The 64 by 64 by 7 bit image was then interpolated by a cubic spline to produce a 256 by 256 by 7 bits deep image (11). Subsequently, the interpolated image was processed in the amplitude domain to more effectively map that portion of the pixel distribution due to the organ onto the dynamic range of the video display. The mapping function consisted of two segments. The lower valued segment was fixed at the maximum brightness and the intersection point was interactively altered to make maximum use of the display range. A typical mapping function is shown in Fig. 1. Gradients were enhanced with a 3 by 3 Laplacian operator (12).

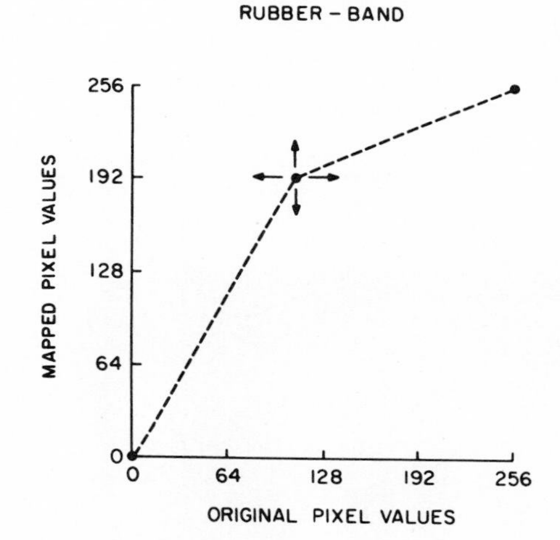

Figure 1 The grey scale mapping function with an interactively alterable hinge point.

Subsequently, frequency domain processing took place by means of a two dimensional, circularly symmetric band pass filter (12). The low frequency cutoff was set to reject low spatial frequency variations due to the thickness of the organ, and the high spatial frequency cutoff was set at $\pm f_{max}$ in azimuth and elevation to reject high spatial frequency noise introduced by the cubic spline interpolation. An example of the two dimensional band pass digital filter is shown in Fig. 2 overlaid on a typical power spectral density.

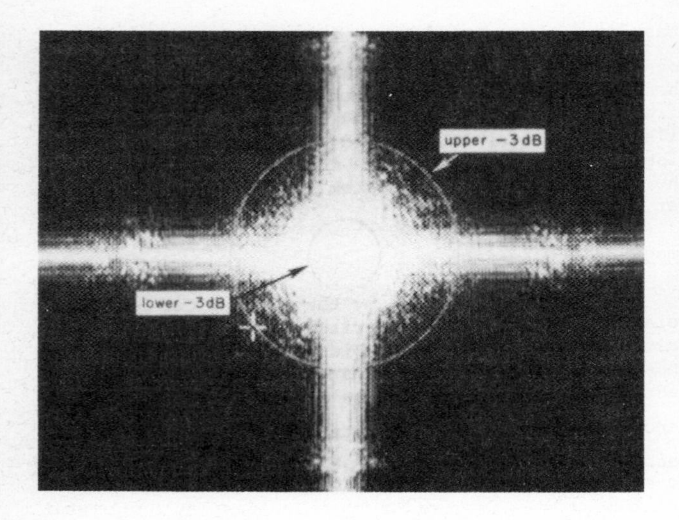

Figure 2 The half power points of the magnitude of the transfer function of the two dimensional band pass filter are shown as a graphic overlay on a typical psd function.

A cube root compression step preceded the power density calculation. The dampening factors of the digital filters were those of the second order Butterworth case. The digital filters were constrained to contain only cosine terms. The two dimensional Fourier transform of the image was multiplied by the specified transfer function and the product was then inverse transformed to the amplitude domain. The amplitude domain data was rescaled to ± 3σ range prior to display.

Results

The co-polarized image of the brain phantom is shown in Fig. 3a. The processing steps were limited to interpolation. The corresponding power spectrum is shown in Fig. 3b.

Figure 3a The co-polarized $|S_{21}|$ image of the brain phantom.

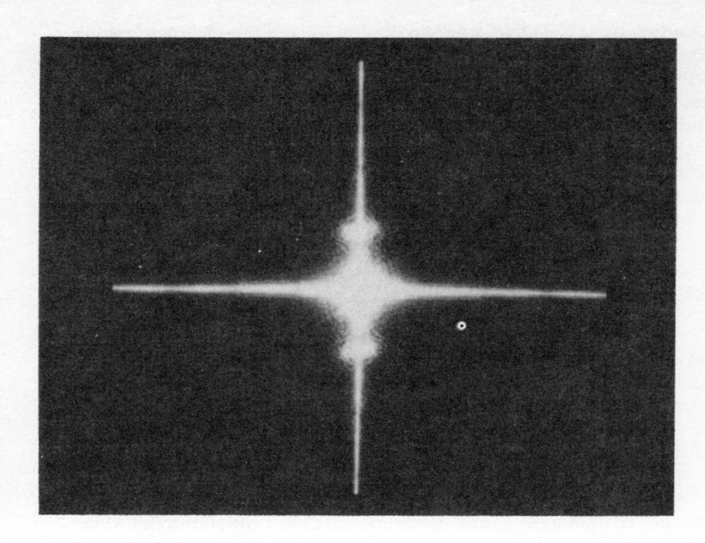

Figure 3b The psd function corresponding to the image of figure 3a. Note the asymmetry between the spatial frequencies in elevation and azimuth.

The co-polarized $|S_{21}|$ image of the first kidney is shown in Fig. 4a with processing steps limited to interpolation, grey scale mapping and contour enhancement by use of the Laplacian operator.

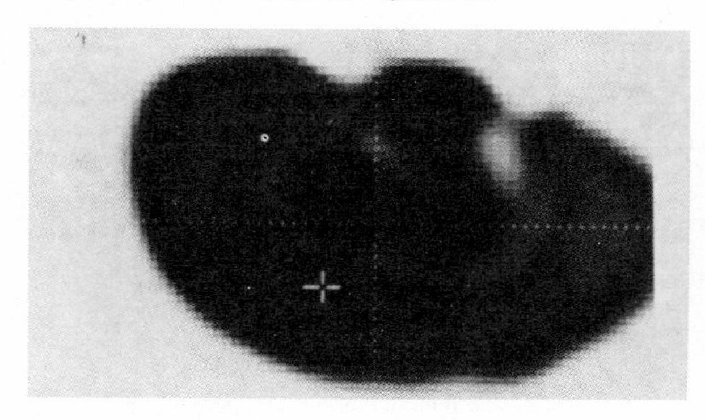

Figure 4a The co-polarized $|S_{21}|$ image of an isolated canine kidney. Processing has been limited to the amplitude domain.

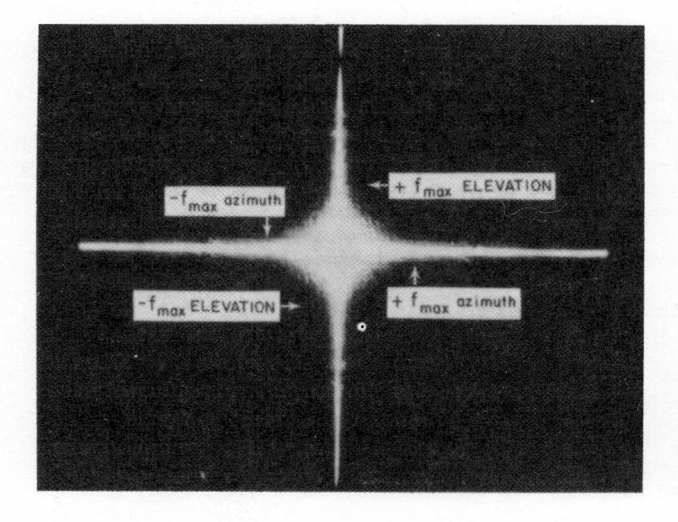

Figure 4b The psd function corresponding to the co-polarized image of Fig. 4a.

The corresponding two dimensional power spectral density (psd) function is shown in Fig. 4b. Note the smooth envelope of the psd at $\pm f_{max}$ in elevation.

The cross-polarized $|S_{21}|$ image of the second kidney is shown in Fig. 5a with processing steps limited to interpolation and amplitude domain mapping. The corresponding two dimensional power spectral function is shown in Fig. 5b. Note that the cross-polarized $|S_{21}|$ image and its psd contain broadened energy at $\pm f_{max}$ in elevation on comparison with the co-polarized case. Note also the general enhancement of borders in the image and the marked increase in artifact due to the support rod.

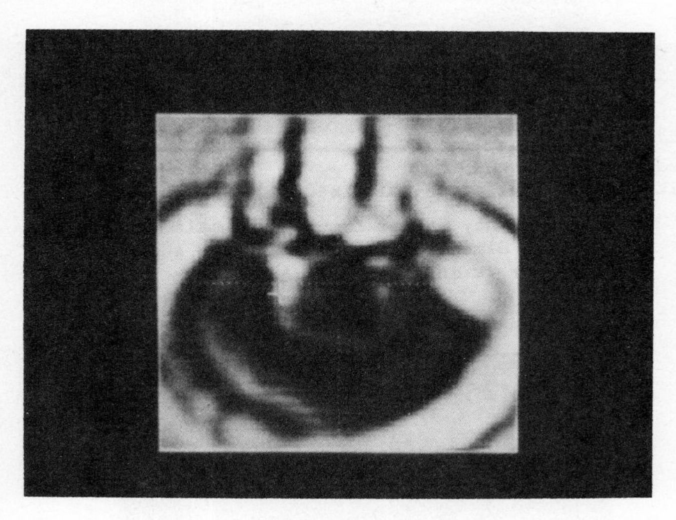

Figure 5a The cross-polarized $|S_{21}|$ image of an isolated canine kidney. Processing has been limited to the amplitude domain.

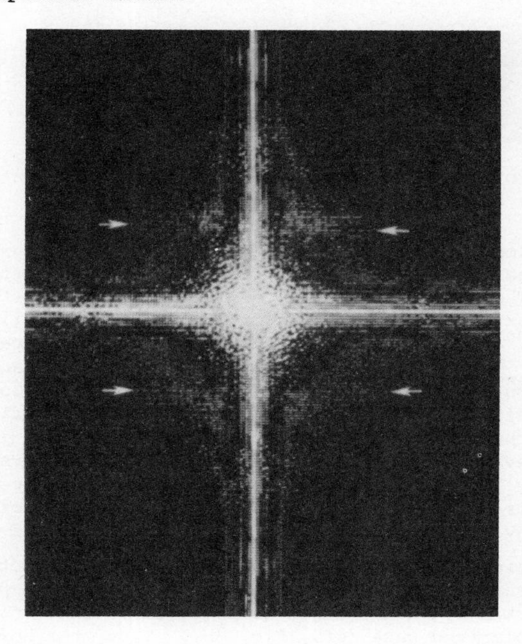

Figure 5b The psd function corresponding to the cross-polarized image of Fig. 5a. The arrows indicate asymmetry at $\pm f_{max}$ between spatial frequencies in elevation (shown) and azimuth.

The cross-polarized $|S_{21}|$ data when presented as an image derives a perceptual enhancement when the grey scale map is replaced by its complement to allow bright rather than dark contrast between the interior of the organ and the surrounding coupling medium. The grey scale complement of the cross-polarized $|S_{21}|$ kidney image is shown in Fig. 6. Note that this display technique alters the perceptual effect of the horizontal rastering and enhances the border areas of the organ.

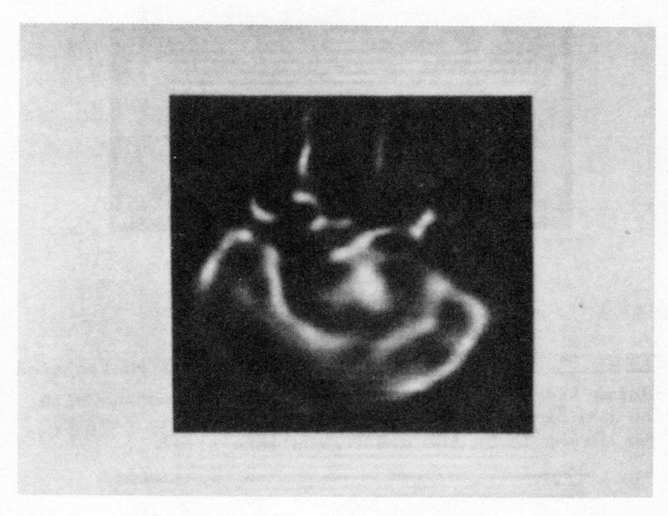

Figure 6 The grey scale complement of the cross-polarized $|S_{21}|$ image of Fig. 5a.

The grey scale complement further processed for contour enhancement by a 3x3 Laplacian operator is shown in Fig. 7.

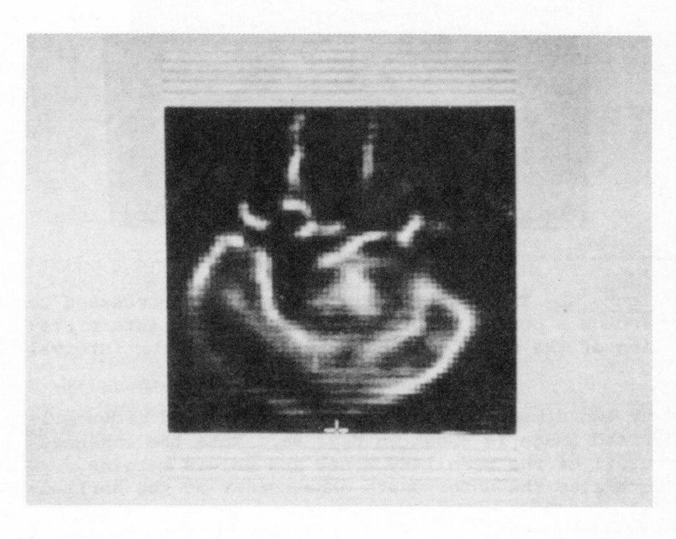

Figure 7 The image of Fig. 6 after processing with a 3x3 Laplacian operator.

The two dimensional band pass filtered and co-polarized $|S_{21}|$ image is shown in Fig. 8a. A contour map corresponding to Fig. 8a is shown in Fig. 8b. The contour inteval was 10, thereby dividing the total range into 25 zones. Note that the cortical lobulations as well as the medullary inner and outer zones are now more easily visualized.

Figure 8a The co-polarized $|S_{21}|$ image of an isolated canine kidney after frequency domain processing with the two dimensional digital filter shown in Fig. 2. The cursor is on the outer medullary stripe.

Figure 8b The image of Fig. 8a has been processed to produce a contour map to aid quantitative interpretation of the monochrome display. The contour interval is 10.

The two dimensional band pass filtered and cross-polarized image is shown in Fig. 9a. Note the enhanced detail of the medullary zones and pelvis margins. Note also the unfortunate enhancement of the horizontal rastering. The contour map at an interval of 50, thereby dividing the range into 5 zones is shown in Fig. 9b. Note the greater dynamic range of the cross-polarized image.

Discussion

The conclusions with respect to a comparison of co-polarized $|S_{21}|$ and cross-polarized $|S_{21}|$ imagery must be conditioned on both physical and biological grounds. The pertinent physical considerations are two-fold: the rectangular aperture produces a non-symmetrical

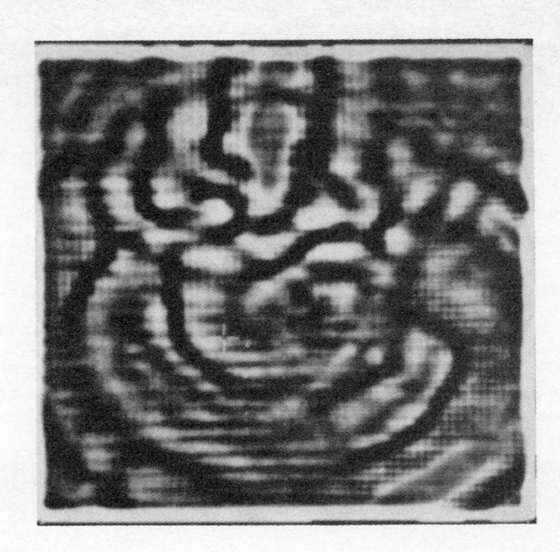

Figure 9a The cross-polarized $|S_{21}|$ image of an isolated canine kidney after frequency domain processing with the two dimensional digital filter shown in Fig. 2. The cursor is in the renal pelvis.

Figure 9b The image of Fig. 9a presented as a contour map with an interval of 50.

modulation transfer function (MTF) since in the near field case resolution is inversely related to aperture area. The narrow (vertical) dimension of the aperture is approximately one-half of the broad (horizontal) dimension. This implies that spatial frequencies near the $\pm f_{max}$ limit will be less attenuated in the plane of elevation than in the plane of azimuth. In addition, the double ridged waveguide weights the electric field within the aperture to further enhance resolution in both elevation and azimuth. The effect of aperture asymmetry is best illustrated by reference to the psd of the brain phantom as shown in Fig. 3b. The increased energy at $\pm f_{max}$ in elevation is easily observed on the vertical axis of the psd display. This asymmetry is evident in the psd (cf., Fig. 5b) of the cross-polarized kidney when the psd is compared to that of the cross-polarized case (cf., Fig. 4b). Thus, the horizontal rastering is more severe in the cross-polarized case. It is not apparent why the MTF asymmetry is greater in the cross-polarized case, but no obvious solution exists with respect to antenna design except to use circular rather than rectangular waveguide to assure a symmetrical aperture. Unfortunately, this is not an attractive solution due to the reduced bandwidth of circular waveguide and its well known propensity for multi-modal propagation.

A second physical consideration is the fact that received signal intensities for the cross-polarized case were lower than for the co-polarized case, a result which is not in itself surprising. This is evidenced, for example, by the fact that details outside of the capsule are virtually lost for the co-polarized case (cf. Fig. 4a), whereas the support rod and its related artifact are obviously well retained in the cross-polarized image (cf. Fig. 5a). The corollary to this observation is some reduction of the advantage afforded by additional attenuation for propagation at the water/target interface for suppression of multipath. This would imply that some additional means of multipath suppression, such as differential propagation delay, would be especially useful in the cross-polarized case (7, 14).

The use of polarization rotation as a means to characterize a heterogeneous biological dielectric could be more expeditiously implemented by a single receiving antenna (water loaded, of course) with dual mode propagation. One alternative would be a square aperture with feeds in each of two orthogonal walls. In this way, the co-polarized and cross-polarized images could be collected simultaneously rather than sequentially as was the case in this experiment. Since data colleciton times are presently in the order of 4.5 hours, this would be a significant advantage. Similarly, both images would represent the same target. The use of linear combinations of the two images for enhancement purposes would then become feasible.

Biological considerations are conveniently presented in the context of image interpretation. Reference to the graphic representation of canine renal anatomy and nephron as shown in F.g (10 and 11) will assist the following discussion.

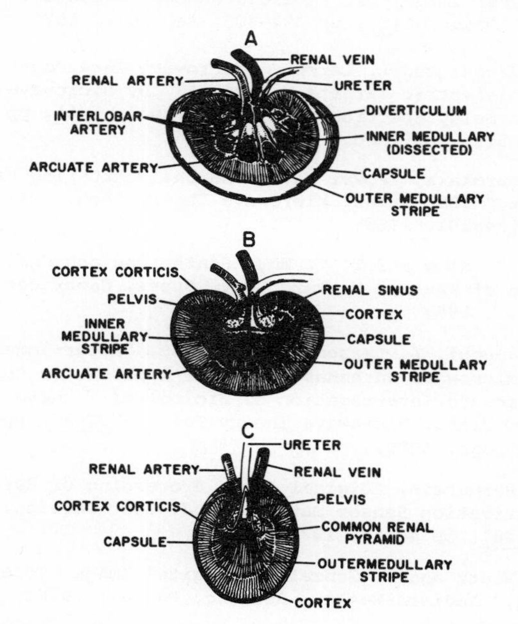

Figure 10 A graphic representation of canine renal anatomy. Note that drawing B represents a dissected specimen.

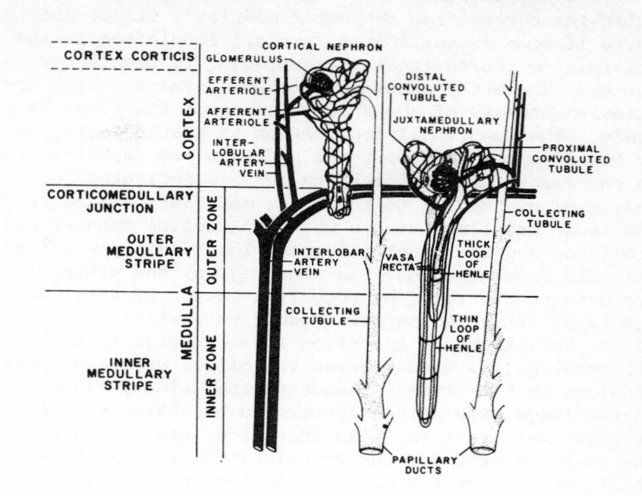

Figure 11 A graphic representation of the canine nephron.

The co-polarized image of Fig. 8 illustrates the major regions of renal anatomy. At the perimeter are the renal capsule and the cortex corticis. The capsule is a thin, fibrous layer which serves to encapsulate the organ. The cortex corticis is a region relatively devoid of glomeruli, but well populated with proximal and distal convoluted tubules. In terms of function at the microscopic level, the proximal and distal tubules are regions of electrolyte, carbohydrate and amino acid transport. The cortex corticis apparently presents lower insertion losses than the deeper cortex as judged by its brighter appearance. Medial to this area is the renal cortex. This is a region where glomeruli predominate with some associated tubular function. Thus, the functional addition is that of blood ultrafiltration. Vascular patterns within the organ serve to lobulate the cortex and provide a demarcation at the cortico-medullary junction of cortex from the outer medullary stripe as shown in the angiogram of Fig. 12.

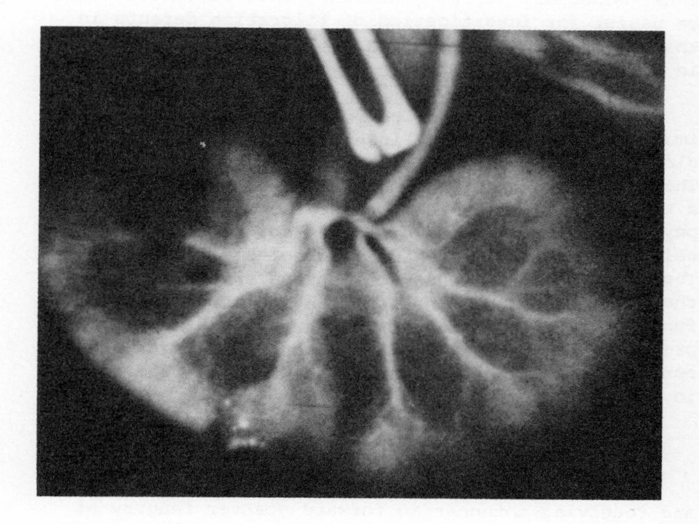

Figure 12 An x-ray angiogram of a canine kidney. Note the cortical lobulations.

The co-polarized image appears to nicely distinguish the cortex from the outer medullary stripe and there is some suggestion of cortical lobulation in the multiple contoured areas corresponding to the deeper cortex. The outer medullary stripe contains a high population density of the thick regions of the loops of Henle. The functional implication is that this region is one of water transport and hydrogen ion secretion. In the image, higher insertion loss is indicated for this region. Medial to the outer medullary stripe is the inner medullary stripe where collecting tubules are prominent and the counter-current loops multiply ultrafiltrate concentration. In this region, the ultrafiltrate is concentrated by roughly a factor of 4-6. In the image, this region corresponds to spatially contoured variations of insertion loss. Medial to this is the renal pelvis which serves to collect urine and move it along to the ureter, thence to the urinary bladder. In the image, the pelvis appears to be characterized by relatively uniform, high insertion loss. Note that the specimen used for the co-polarized scan did have a quasisaculated pelvis with a superior and inferior cistern.

The cross-polarized $|S_{21}|$ image in Fig. 9 illustrates the same general features, but many qualitative distinctions exist in comparison to the co-polarized image. The cross-polarized image is dominated by a perimeter that would appear to be defined by the capsule and cortex at the distal (_in situ_) margin and the inner medullary stripe at the proximal (_in situ_) margin. Neither cortex corticis nor cortical lobulations are well represented. Also, the inferior pole (_in situ_) at the upper right hand margin of the organ in Fig. 9 is represented by a distinct area of low insertion loss.

In general terms, it would appear that the cross-polarized transmission image is more affected by linear or quasilinear (with respect to dimensions of ca. 4mm) dielectric interfaces than is the case with the co-polarized transmission image. This could be an advantage in lamellar or piece-wise linear lamellar areas such as grey matter/white matter interfaces in the brain, facial planes in the liver and muscle/fat, muscle/bone or bone/tendon interfaces in the extremities. In the case of the kidney, cross-polarization would appear to enhance the cortico-medullary junction and the pelvis/inner-medullary stripe junction. This attribute would be useful for investigation of filtration defects in the case of the former and urine concentration defects in the case of the latter.

In either case of co-polarized or cross-polarized images, the biological interpretations are based on regional correspondence with known microscopic function. These interpretations must remain equivocal until such a time as experimental verification by induced pathophysiology can be performed. Also, it is useful to recall that the canine kidney is not papillary (13). Thus, the canine kidney collection system differs from the human kidney. The accompanying structural and functional specialization of the human kidney is not present in the canine kidney, thus the images presented herein and in earleir papers should be viewed with this in mind.

Conclusions

The images produced by co-polarized transmitting and receiving antennas in forward scatter imagery at microwave frequencies has been shown to represent different features of the isolated canine kidney than is the case with cross-polarized antennas. In general terms, the cross-polarized case appears to enhance linear and quasilinear interfaces with respect to distances of ca. 4mm. The co-polarized case appears to enhance features such as cortical lobulations and cortex corticis. In either case, the imagery appears to be relatable to known regional specialization within the renal organ in terms of water electrolyte processing.

Acknowledgements

The authors would like to acknowledge the efforts of Mary Lou Shoo in the preparation and composition of this manuscript, Leon S. Butler for his excellent work in removing and preparing kidney specimens, Maurice E. T. Swinnen for his expertise in interfacing instruments to computers, the U.S. Army Engineer Topographic Laboratory for their aid in processing the image data, and Ted Brinkmann of the Division of Medical Audio-Visual Services, Walter Reed Army Institute of Research for his expert photography.

References

1. E. H. Grant, R. J. Shepard, and G. P. South, "Dielectric Behavior of Biological Molecules in Solution," Oxford Univ. Press, 1978.

2. P. O. Vogelhut, "The Dielectric Properties of Water and Their Role in Enzyme Substrate Interactions," Contract No. NONR-222(92),1962.

3. B. Enander and G. Larson, "Microwave Radiometric Measurements of the Temperature Inside a Body," Electron. Lett., 10, p 317, July 1974.

4. A.H. Barrett and P.C. Meyers, "Subcutaneous Temperatures: A Method of Non-Invasive Sensing," Sci., 190, pp 669-671, 1975.

5. L. Larsen and J. Jacobi, "Microwave Interrogation of Dielectric Targets. Part I: By Scattering Parameters," Med. Phys., 5(6), pp 500-508, Nov./Dec. 1978.

6. L. Larsen and J. Jacobi, "Microwave Scattering Parameter Imagery of an Isolated Canine Kidney," Med. Phys., 6(5), pp 394-403, Sep./Oct. 1979.

7. J. Jacobi and L. Larsen, "Microwave Interrogation of Dielectric Targets. Part II: By Microwave Time Delay Spectroscopy," Med. Phys., 5(6), pp 509-513, Nov./Dec., 1978.

8. K. Kurokawa, "Power Waves and the Scattering Matrix," IEEE Trans. Microwave Theory Tech., MTT-13, pp 194-202, 1965.

9. R.W.P. King and T. T. Wu, "Scattering and Diffraction of Waves," Harvard Univ. Press, Cambridge, Mass., 1959.

10. J. Jacobi, L. Larsen, and C. Hast, "Water-Immersed Microwave Antennas and Their Application to Microwave Interrogation of Biological Targets," IEEE Trans. Microwave Theory Tech., MTT-27, pp 70-78, Jan. 1979.

11. R. Bernstein, "Digital Image Processing of Earth Observation Sensor Data," IBM J. Res. Develop., 20(46), pp 40-57, 1976.

12. P. Wintz and J. Gonzales, "Digital Image Processing," Addison-Wesley, Reading, Penna., 1978.

13. J. G. Baer, "Comparative Anatomy of Vertebrates," Butterworth, London, pp 156-157, 1964.

14. J.H. Jacobi and L.E. Larsen, "Microwave Time Delay Spectroscopic Imagery of Isolated Canine Kidney," Med. Phys., 7(1), pp 1-7, 1980.

Microwave Thermography: Principles, Methods and Clinical Applications*

P.C. Myers†, N.L. Sadowsky††, and A.H. Barrett†

ABSTRACT

We review the physical principles, method of operation, measurement limitations, and potential medical applications of microwave thermography. We present detailed results of a study of breast cancer detection at 1.3 and 3.3 GHz, including the dependence of detection rates on microwave frequency, time, tumor depth, and tumor size. At 1.3 GHz, microwave thermography detects breast cancer as well as infrared thermography (true-positive rate = 0.76 when true-negative rate = 0.63). When the two methods are combined, the true-positive rate increases by about 0.1 over that of either method alone.

INTRODUCTION

Several recent investigations of microwave emission from human tissue show that: the emission spectrum of live human tissue appears thermal over the frequency range 10^4 — 10^{10} Hz [1,2]; at centimeter wavelengths microwave radiometry is capable of detecting variations in subsurface temperature [3-8] and this technique may have diagnostic value in detection of breast cancer and other pathological conditions with associated thermal structure [9,10].

We have called this measurement technique microwave thermography in analogy with the well-known infrared method. In this report we describe the relevant physical principles and instrumentation. We present results of a clinical study of breast cancer detection at 1 and 3 GHz, and we present initial results of several other studies.

PHYSICAL PRINCIPLES

Microwave thermography is possible because (1) human tissue emits thermal radiation, whose intensity at microwave frequencies is proportional to tissue temperature; (2) several medical problems, including cancer and vascular occlusions, are accompanied by local changes in temperature of order 1° C; (3) microwave radiation can penetrate human tissue for e^{-1} distances of several cm; and (4) present-day radiometers can easily detect intensity changes corresponding to emitter temperature changes of less than 1° C. The underlying physics is, therefore, the physics of thermal radiation, and its transfer in a lossy, inhomogeneous medium.

Thermal Radiation

All media with absolute temperature T > 0° K emit electromagnetic radiation toward, and absorb electromagnetic radiation from their surroundings. This thermal emission is a result of the accelerations experienced by individual charges in the course of their thermal motions. Figure 1 shows the intensity spectrum I of black-body emission for T = 100° K, and T = 1000° K. At human body temperature T = 300° K, maximum intensity occurs at a wavelength λ = 10 μm. This has led to the operation of infrared thermographs at wavelengths near 10μm. At a typical microwave frequency of 3 GHz (λ = 10 cm), I is reduced by a factor of $\cong 10^8$ from its maximum; but microwave radiometers, developed primarily for radio astronomy, can easily detect radiation with this intensity.

Radiative Transfer

In order to relate the emergent brightness temperature at the skin surface to the underlying distribution of tissue temperature, it is necessary to make a specific tissue model and to make several simplifying assumptions. These give a highly idealized picture which should be considered a first approximation only. We model a tumor in breast fat, in accordance with our current clinical work. We assume that the tissue is composed of planar layers of fat, skin, and mus-

---

* Manuscript received November 17, 1978
Presented at the Workshop on Diagnosis and Therapy Using Microwaves, 8th European Microwave Conference, Paris, France, September 8, 1978.
† Department of Physics and Research Laboratory of Electronics, Massachusetts Institute of Technology, Cambridge, MA 02139, U.S.A.
††Department of Radiology, Faulkner Hospital, Boston, MA 02130, U.S.A.

Reprinted with permission from *J. of Microwave Power*, vol. 14, no. 2, pp. 105-115, June 1979.

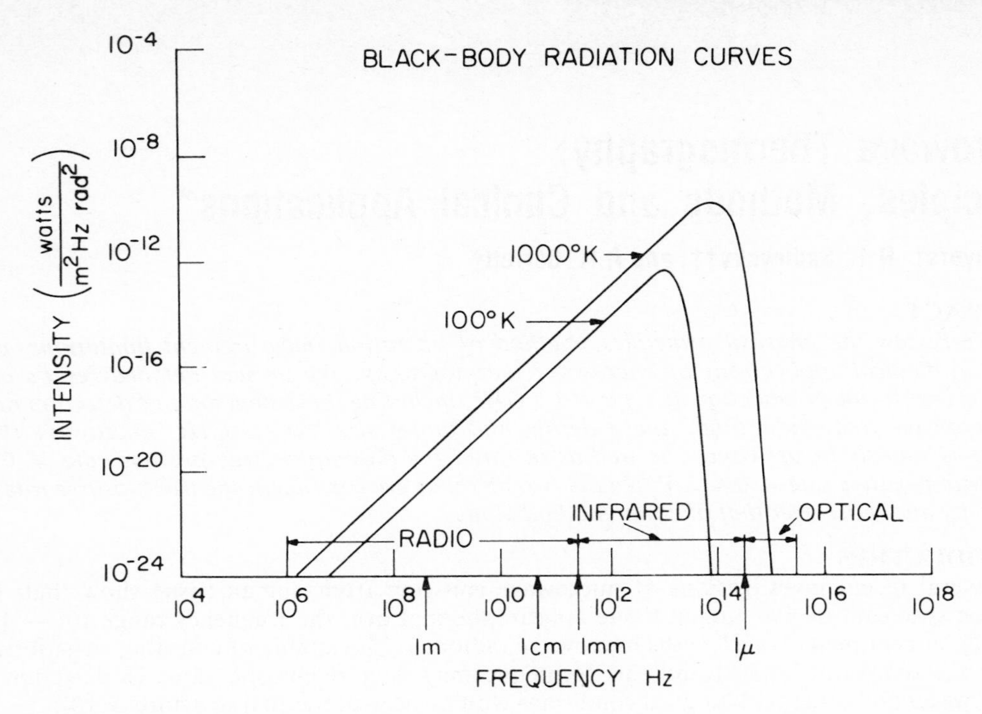

Figure 1 Specific intensity of black-body thermal radiation with frequency, for absolute temperature T = 100 K and T = 1000 K.

cle, with a single inhomogeneity in the fat layer. We ignore all other inhomogeneities such as blood vessels. We ignore all scattering except that which occurs at layer interfaces. We assume that the tissue is isothermal at temperature T, except for the source in the fat layer, which is hotter by $\triangle T$ °C. The background radiation incident on the source has brightness temperature $T_B = T$. The dielectric tissue properties are assumed to be the bulk properties summarized by Johnson and Guy[11].

The radiation transfer depends strongly on the absorption properties of high- and low-water tissue. These properties are shown in Figure 2 in terms of the e^{-1} penetration depth for power. At 3 GHz (free-space wavelength 10 cm) the penetration depth is $\cong$5 cm ($\cong$1 tissue wavelength) in fat and $\cong$0.8 cm ($\cong$0.5 tissue wavelengths) in muscle or skin. The penetration depth decreases with increasing frequency. At frequencies above 10 GHz, escaping thermal emission originates in a skin layer <1 mm deep. This suggests that thermography at millimeter wavelengths may be expected to give results very similar to those of infrared thermography.

The presence of parallel tissue layers makes it necessary to consider possible interference effects due to multiple reflections. Standing waves in the fat layer have been shown to be significant at 2450 MHz in the case of microwave diathermy, where one transmits power from the skin through the fat into the deep muscle tissue [12,13]. In the present case, a tumor imbedded in the fat layer radiates from the fat through the skin into a receiving antenna at the skin surface. Standing waves may, therefore, arise in the skin layer. We estimate the magnitude of this effect by considering the case of maximum interference, which occurs if the layer is $\lambda/4$ in thickness. This condition is met at 3.3 GHz for a skin thickness of 3.4 mm. If the antenna is matched to the skin-fat combination, the transmission-line equations indicate that the twice-reflected wave entering the antenna interferes constructively with the zero-reflected wave, and adds approximately 9% to its amplitude. Since this is the maximum contribution, we neglect this effect in the following discussion.

Neglect of interference effects permits the discussion of radiative transfer in terms of power, and, therefore, at microwave frequencies in terms of temperature. The excess brightness temperature due to the tumor, ΔT_B, incident on the skin-antenna interface is given by the plane-parallel solution to the equation of radiative transfer [14].

$$\Delta T_B = \Delta T(1 - e^{-K_t \Delta z_t})\, e^{-K_f \Delta z_f}\, t_{fs}\, e^{-K_s \Delta z_s}. \quad (1)$$

Here ΔT is the mean excess temperature of the tumor above ambient, K_t, K_f and K_s are the power absorption coefficients (inverse penetration depths) of the tumor, fat, and skin tissue; Δz_t, Δz_f, and Δz_s are the sizes of the tumor, overlying fat layer, and skin layer; t_{fs} is the power transmission coefficient from fat to skin. For illustration, we evaluate (1) for tissue properties at 3 GHz: $K_t = 0.65\ cm^{-1}$[15], $K_f = 0.21\ cm^{-1}$, $K_s = 1.2\ cm^{-1}$, $t_{fs} = 0.75$[11], and for representative dimensions $\Delta z_t = 2$ cm, $\Delta z_f = 2$ cm, $\Delta z_s = 2$ mm. We obtain $\Delta T_B = 0.28\ \Delta T$. Thus for a typical temperature elevation associated with breast carcinoma, $\Delta T = 1.5°$ C[16], one may expect $\Delta T_B \cong 0.4$ °C.

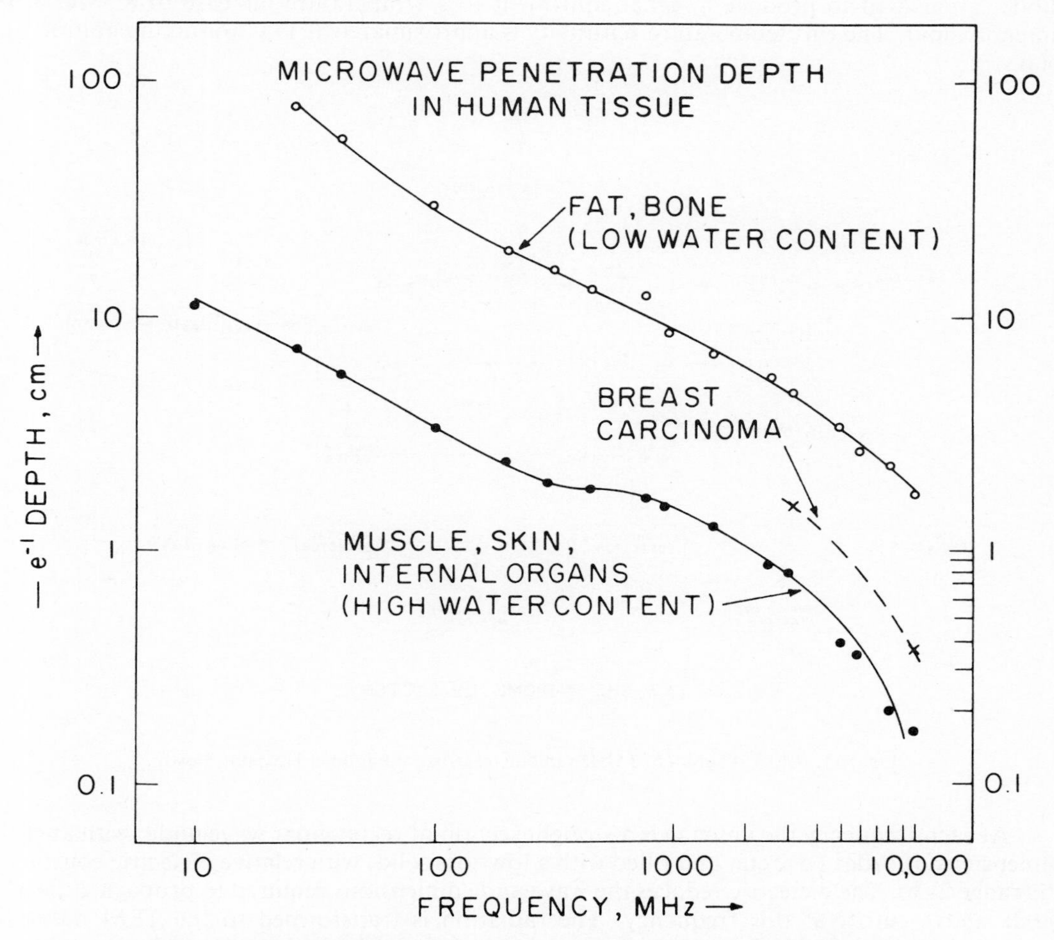

Figure 2 Penetration depth of microwave power in human tissue with frequency.

Resolution

The resolution properties of any antenna depend upon the frequency, the antenna aperture dimensions, the source-aperture distance, and the dielectric properties of the medium being viewed. In general, the limiting factors are diffraction and inhomogeneities in the medium. We discuss here only the rectangular waveguide antenna in direct contact with the skin surface. The antenna response pattern is then primarily the near-field pattern, since for antenna dimensions of 2-3 cm the far-field distance is comparable to the e^{-1} depth. The antenna resolution is then to a first approximation defined by the aperture dimensions.

INSTRUMENTATION

We have built three radiometers for use at 1.3, 3.3, and 6.0 GHz. The 1.3 and 3.3 GHz instruments are in clinical use; the 6.0 GHz unit is largely ready for hospital use, but no clinical data have been taken at 6.0 GHz. All have design similar to the block diagram in Figure 3, which illustrates the 3.3 GHz radiometer. It is a conventional Dicke-switched, or comparison, superheterodyne radiometer, with 100 MHz intermediate frequency bandwidth, centered at 60 MHz. The input to the first stage tunnel-diode amplifier is switched at 8 Hz between the antenna, and a matched load maintained at 22.0 ± 0.1° C by a thermoelectric refrigerator. The resulting square-wave signal is detected at the intermediate frequency by a square-law crystal detector, whose output is synchronously demodulated. The output signal is fed in parallel to an analog strip-chart recorder, and to a digital processor which converts the incoming voltage to a L.E.D. display of temperature in degrees C. Gain calibration is provided by a solid-state noise diode, attenuated to produce a signal equivalent to a temperature increase of 8.5° C at the antenna input. The rms temperature sensitivity is approximately 0.1° C for an integration time of 6 s.

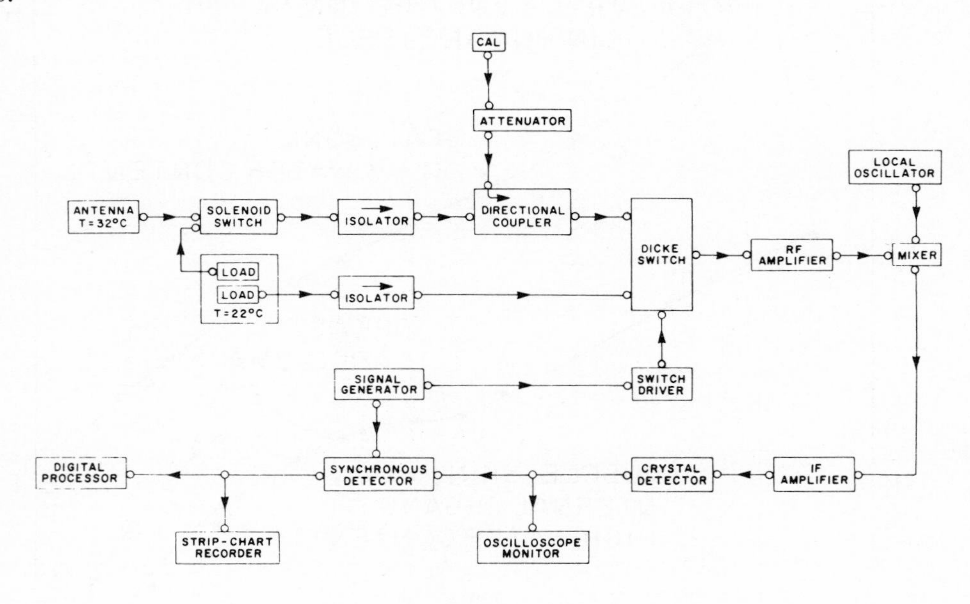

Figure 3 Block diagram of 3 GHz radiometer in use at Faulkner Hospital, Boston.

At each frequency the antenna is a straight section of rectangular waveguide, with aperture dimensions of order 1 x 2 cm, and filled with a low-loss solid, with relative dielectric constant in the range 2–30. The dielectric reduces the waveguide dimensions required to propagate the TE_{10} mode above cutoff at this frequency. The radiation is transformed to the TEM mode for transmission through coaxial cable in the usual way, by a probe in the waveguide. During data-taking, the antenna aperture is placed flush against the skin, in order to eliminate reflective loss at the tissue-air interface, and in order to permit minimum spatial resolution for the given antenna size. Most of the detected power thus originates in the near field of the antenna. Impedance matching between the tissue and the antenna input is achieved by placing a thin slab of higher-dielectric-constant material adjacent to and behind the aperture plane. Matching between the antenna and the coaxial line is achieved by varying the probe length and backshort distance. Network analyzer measurements indicate return loss $\gtrsim$15 dB across the full receiver bandwidth. Thermal conduction effects between the antenna and the tissue have been reduced by heating the antenna, or by placing a thin sheet of insulator at the antenna aperture.

DIAGNOSTIC APPLICATIONS

There are many potential medical applications of microwave radiometry. By analogy with infrared thermography, we may expect these to include detection of subsurface thermal anomalies such as malignant tumors, especially in the female breast; localized inflammations, such as appendicitis; and vascular insufficiency in the limbs and in the brain.

Breast Cancer Detection

For the initial clinical investigation, we have sought to use microwave radiometry to detect breast cancer. The need for a reliable means of early detection is indicated by the fact that approximately one woman in 15 in the United States will get breast cancer in her lifetime. The potential danger of x-rays as a diagnostic method has been underscored by recent claims [17] that each examination involving a dose of 1 rad may increase a woman's lifetime breast cancer incidence risk by approximately 1%, for example, from 7% to 7.07%. The value of infrared thermography in breast cancer diagnosis has also been questioned recently by Moskowitz, *et al.* [18].

A 3.3 GHz radiometer was installed in Faulkner Hospital, Boston, in November 1974. A 1.3 GHz radiometer of higher sensitivity was installed in 1976. Our goal was to examine patients who are examined as part of the Sagoff Breast Cancer Detection Clinic. Most women examined are outpatients having family history of breast cancer, a lump in the breast, or other complaint. Each patient receives a clinical examination, xeromammography and infrared thermography. The xeromammography and infrared thermography films are judged by staff radiologists. A medical technician operates the microwave equipment full-time at the hospital, and examines approximately 30 patients per week. To date we have examined over 4000 women at 3.3 GHz and over 1000 women at both frequencies.

At each frequency, the examination lasts about 10 min from start to finish. The patient opens her robe above the waist, and lies supine on an examining table. A technician places the antenna on the first position to be measured, usually the right nipple, and holds it in place for about 15 s (several integration times). The measured temperature is indicated on a digital L.E.D. display, and recorded on a mapping form. The antenna is placed on the symmetrically opposite position on the other breast for another 15 s; the process is continued until two nine-point grids have been mapped out, one on each breast. The spacing between adjacent grid points depends on the breast size, but is typically 3 cm. The mapping form is completed by adding information about the patient's medical history, and the results of the xeromammography and infrared thermography.

The goal of the data analysis has been to find a quantitative test which can be applied to the measured temperatures, in order to produce high true-positive (TP) and true-negative (TN) detection rates. [The TP rate, or sensitivity, is the fraction of cancers which the criterion identifies correctly; the TN rate, or specificity, is the fraction of normals which the criterion identifies correctly.] The test should be quantitative and objective, to permit rapid assessment of the patient's thermal pattern; and to avoid the training, time and variability in judgment which accompany the requirement of a skilled "reader". The microwave data are well suited to construction of such a test, because they are in numerical form. This contrasts with the usual pictorial data format for mammography or infrared thermography.

We have tested numerous mathematical combinations of the measured temperatures in order to find those which best discriminate between known cancers (confirmed by biopsy) and "normals" (all other cases) [19, 20]. At both 1 GHz and 3GHz, the most significant discriminators rely on right-left asymmetry, as has also been found in infrared thermography. A less significant detection criterion is the presence of a "hot spot" or temperature which departs from the mean of its neighbors by more than a specified threshhold (typically 1.3 °C). The 3 GHz data and detection criteria have been discussed previously [21].

Since the 3 GHz radiometer has been in clinical use, with minor modifications, for over 3 years, we have begun to examine the consistency of its cancer detection performance over time. By dividing approximately 3000 examinations into 6 groups, we find that at TN $\cong$ 0.65, the TP rate has decreased from an average of 0.62 for the first four groups to an average of 0.42 for the last two groups. This decrease appears to be statistically significant, but its origin is not known and is still under investigation.

For the most recent 1000 cases, we have taken data at both 1 and 3 GHz, so it is possible to directly compare the performance of the two frequencies. Figure 4 shows histograms of the 1 GHz temperature asymmetries most effective in distinguishing between cancer and "normal"; the 3 GHz histograms are similar. Figure 4a is a histogram of the corrected temperature difference $\Delta T_{RL} - \langle \Delta T_{RL} \rangle$. The difference ΔT_{RL} is measured between symmetrically opposite points on the chest; there are 9 such values measured for each patient. Each of the 9 mean

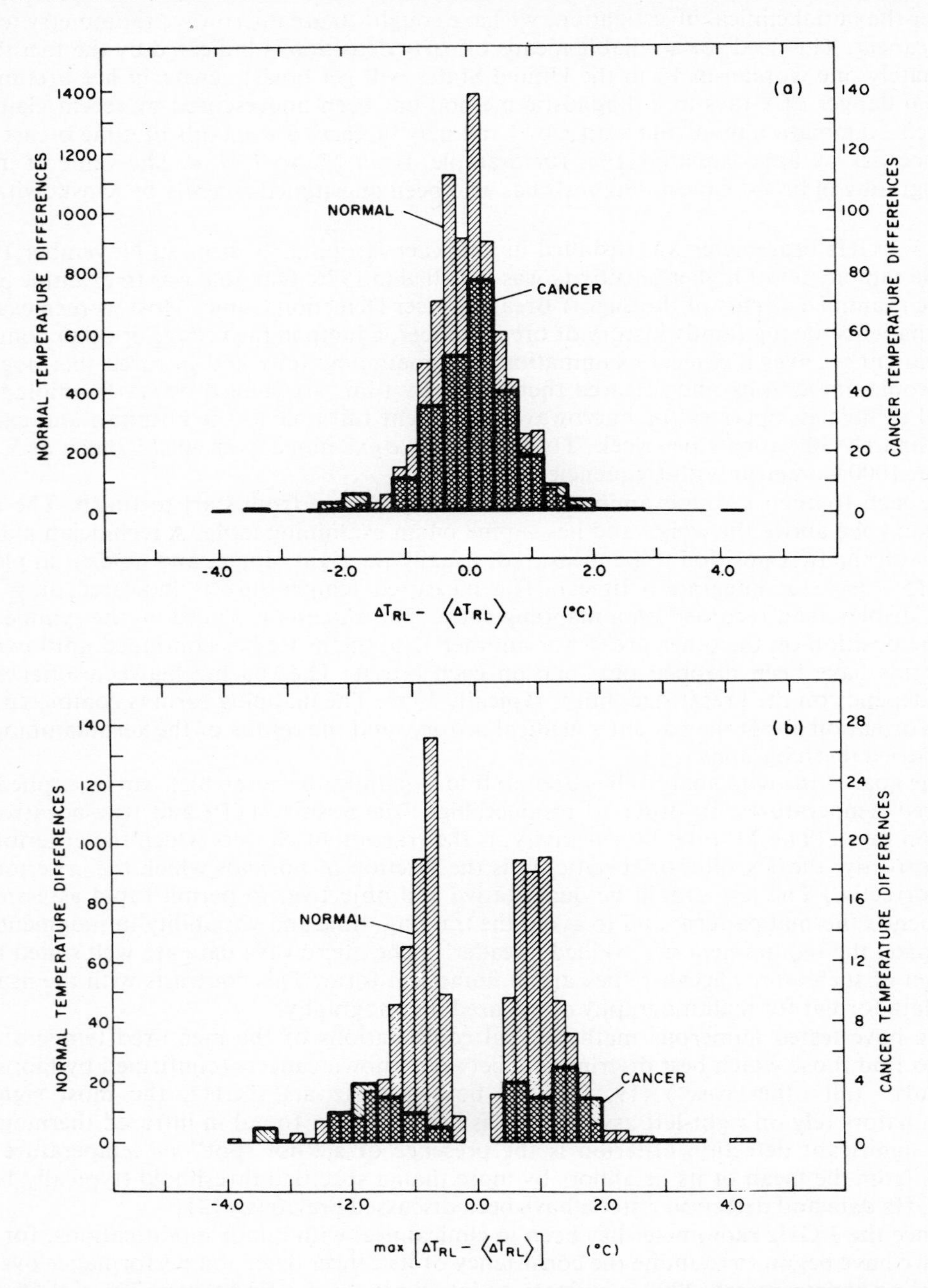

Figure 4 Histograms of 1 GHz corrected right-left temperature differences, used to distinguish between women with breast cancer (confirmed by biopsy) and "normals" (all others). (a), histogram of corrected temperature differences between members of the nine pairs of symmetrically opposite positions on the chest where data are taken for each patient; (b), histogram of the maximum value, for each patient, of the temperature difference in (a).

values $<\triangle T_{RL}>$ is the average over the approximately 1000 normal patients in the sample for one of the 9 pairs of positions where data is taken. These $<\triangle T_{RL}>$ are relatively small ($\sim\pm 0.1$ °C), but distinctly different from each other. Figure 4b is a histogram of the maximum value (regardless of sign) of $\triangle T_{RL} - <\triangle T_{RL}>$ for each patient. To use one of these temperature differences as a detection criterion, one chooses a threshhold value T_0. If the difference exceeds T_0, the criterion indicates "cancer"; otherwise it indicates "normal". The performance of the criterion in terms of TP and TN rates depends on T_0; as T_0 increases, TP decreases and TN increases.

Figure 5 shows the TP and TN performance for $\sim$1000 normal patients and 29 cancers of microwave thermography (M), infrared thermography (I), xeromammography (X), and several combinations, as the corresponding detection threshholds are varied. The 3 GHz microwave curve, labelled M(3), relies primarily on max $[\triangle T_{RL} - <\triangle T_{RL}>]$, as T_0 increases from 0.6 °C (upper left of curve) to 1.4 °C (lower right of curve). At each value of TN, it is clear the TP rate at 1 GHz exceeds the TP rate at 3 GHz by about 10%, i.e. that radiometry at 1 GHz detects more cancers than at 3 GHz. Although the difference appears significant, it should be remembered that the 3 GHz TP rate has decreased by a comparable amount since the start of the 3 GHz measurements.

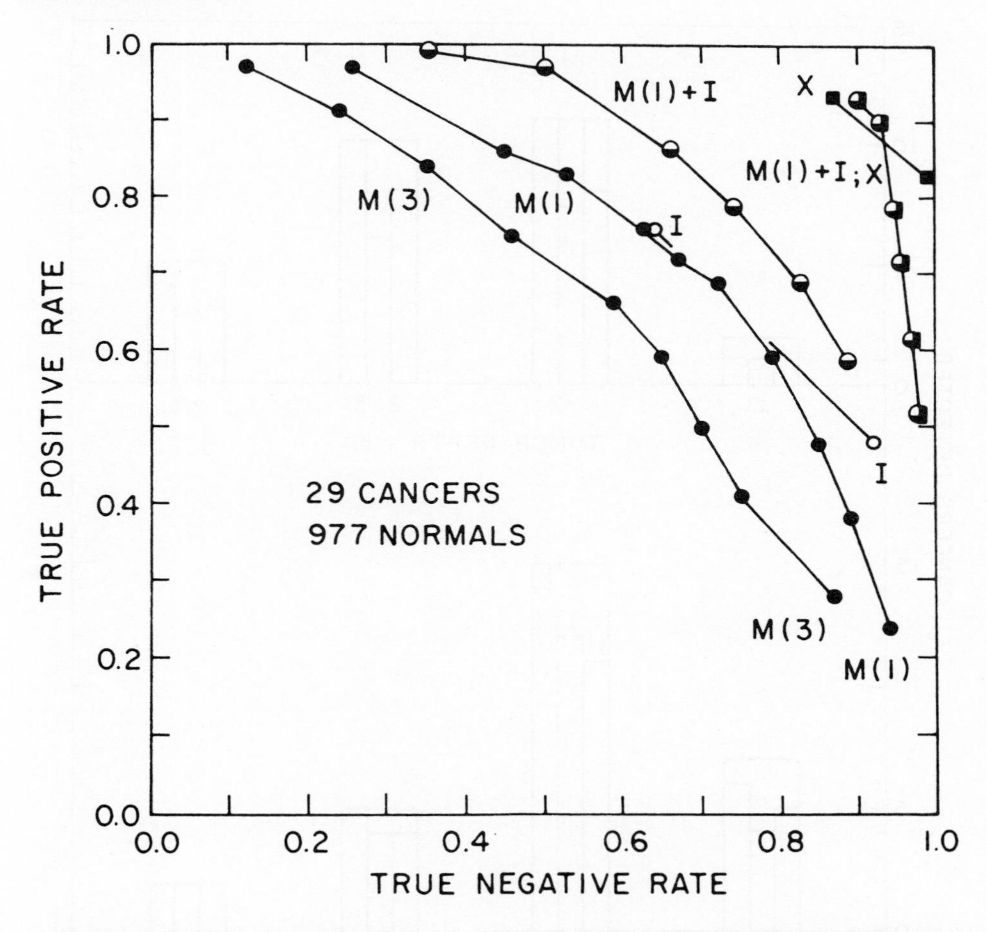

Figure 5 Breast cancer detection performance of various combinations of microwave thermography at 1 GHz [M(1)], and at 3 GHz [M(3)]; of infrared thermography [I], and of xeromammography [X]. True positive rate = fraction of cancers diagnosed correctly; true negative rate = fraction of normals diagnosed correctly. For each method, the family of points depends on the choice of detection threshhold. Curve labelled M(1) + I shows performance of criterion in which cancer is indicated if either 1 GHz or infrared test is positive. Curve labelled M(1) + I; X shows result of two-step examination in which 1 GHz and infrared tests are used in first step; x-ray is then used for follow-up only if first step test is positive.

While the apparent superiority of detection at 1 GHz to detection at 3 GHz must be considered with caution, comparison of the 1 GHz and infrared results shows two definite characteristics. At TN ≅ 0.65, both methods have TP ≅ 0.75; thus their statistical performance is extremely similar. However, this does not mean that the two methods extract the same thermal information. Among the 29 cancer cases examined, the microwave and infrared methods disagree in their diagnoses in 12 cases (41%). This is a higher rate of disagreement than that of the microwave and x-ray method —(24%), or of the infrared and x-ray methods (28%). This lack of correlation is evident when one combines the microwave and infrared criteria: if either one indicates cancer then the combined criterion indicates cancer. The corresponding curve in Figure 5, labelled M(1) + I, has about 10% greater TP than either the microwave or the infrared curve by itself.

The combination of microwave and infrared thermography is appealing because each method sends no radiation into the body, and is therefore completely safe. The curve labelled M(1) + I; X shows how these methods could be used for screening. Here the x-ray is used only to follow up those cases indicated as positive by the combined microwave and infrared examinations. By proper choice of the microwave threshhold, the two-step procedure can detect as

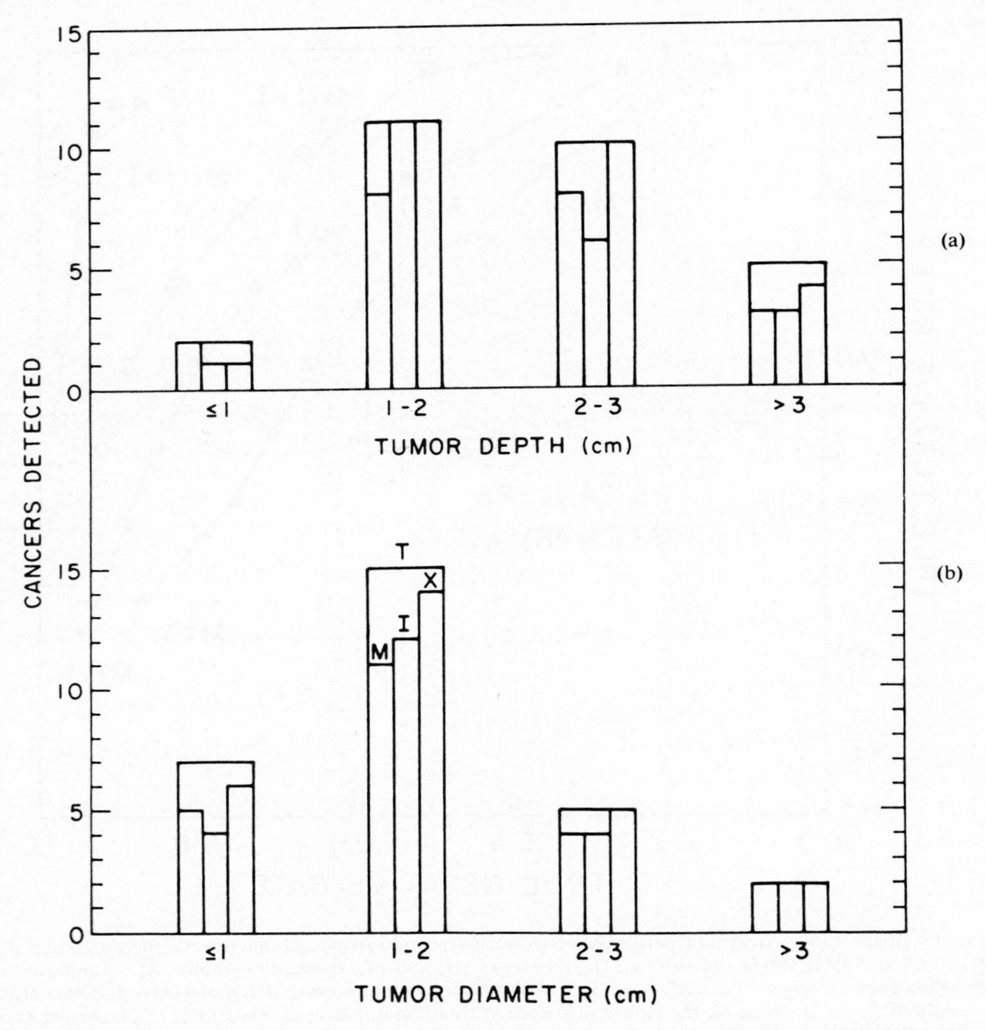

Figure 6 Histograms of the number of cancers detected by 1 GHz microwave thermography (M), infrared thermography (I), and xeromammography (X), in terms of (a) mean tumor diameter and (b) tumor depth; the horizontal line at the top of each group (T) indicates the total number of cancers in that group.

many cancers as can the x-ray alone, with about the same TN rate, but with significant reduction in the number of women exposed to x-ray examination.

We have examined the temperature differences associated with the known cancer cases to see whether any significant correlation exists between temperature difference and size, or between temperature difference and depth. The size and depth data are based on biopsy reports. Their values are uncertain by as much as 1 cm. We find no significant correlation in either case. However, larger tumors are easier to detect for each method, as is evident from the histograms in Figure 6. The lack of correlation between temperature difference and size, and the relative ineffectiveness of the "hotspot" criterion suggest that the change in temperature due to most breast tumors cannot be realistically modeled by a simple local maximum at the tumor site. Although thermistor probe measurements indicate that such maxima are present [16], both the microwave and infrared methods appear to rely on the presence of a more diffuse distribution of elevated temperature for their success in detection.

Other Studies

We have begun three new investigations. (1) a correlation study between forehead temperatures at 1 GHz and 3 GHz, and as measured with a contact thermistor thermometer; and oral temperature, (2) mapping abdominal temperatures of patients having abdominal pain, and (3) mapping the calves of patients suspected of having deep venous thrombosis. To date the results are too few to draw any definite conclusions. However, preliminary results indicate that except possibly for the 1 GHz and 3 GHz data, the forehead and oral temperature correlations are not significant. For the calf temperature maps, normal legs have similar patterns on the posterior side consisting of a rise from knee to calf, and a decrease from calf to ankle.

ACKNOWLEDGEMENTS

We thank J.W. Barrett, D.C. Papa and P.V. Wright for engineering assistance; V. Taylor for data-taking, and H.M. Lindsay and B. Rosen for data analysis. This work was supported by Grant 2 R01 GM20370–05 from the U.S. National Institute of General Medical Sciences.

REFERENCES

1 J. Edrich and P.C. Hardee, *Proc. IEEE*, **62**, 1391, (1974).
2 B. Enander and G. Larson, Trita-Tet-7701 (Royal Institute of Technology, Stockholm) (1977),
3 A.H. Barrett and P.C. Myers, M.I.T. Research Laboratory of Electronics Quarterly Progress Report *No. 107*, 14 (15 October 1972).
4 A.H. Barrett and P.C. Myers, M.I.T. Research Laboratory of Electronics Quarterly Progress Report *No. 109*, 1 (15 April 1973).
5 A.H. Barrett and P.C. Myers, M.I.T. Research Laboratory of Electronics Quarterly Progress Report *No. 112*, 39 (15 January 1974).
6 A.H. Barrett and P.C. Myers, *Bibl. Radiol.*, **6**, 45, (1975).
7 A.H. Barrett and P.C. Myers, *Science*, **190**, 669, (1975).
8 B. Enander and G. Larson, *Electronics Lett.* **10**, 317 (1974).
9 A.H. Barrett, P.C. Myers and N.L. Sadowsky, in Proceedings, Third International Symposium on Detection and Prevention of Cancer (in press; Marcel Dekker, New York) (1978).
10 A.H. Barrett, P.C. Myers and N.L. Sadowsky, *Radio Science*, **12**, (Supplement), 167 (1977).
11 C.C. Johnson and A.W. Guy, *Proc. IEEE*, **60**, 692 (1972).
12 A.W. Guy and J.F. Lehmann, *Trans. IEEE*, **BME-13**, 76 (1966).
13 A.W. Guy, *Trans. IEEE*, **MTT-19**, 214 (1971)
14 S. Chandrasekhar, Radiative Transfer (Oxford University Press) (1950).
15 T.F. England, *Nature*, **166**, 480 (1950).
16 M. Gautherie, Y. Quenneville and Ch. Gros, Functional Explorations in Senology, **93** (1975).
17 B.J. Culliton, *Science*, **193**, 555 (1976).
18 M. Moskowitz, J. Milbrath, P. Gartside, A. Zermeno and D. Mandel, *New England Jour. Med.*, **295**, 249 (1976).
19 C. Jung, unpublished BS thesis, M.I.T. Department of Electrical Engineering (1976).
20 H.M. Lindsay, unpublished BS thesis, M.I.T. Department of Physics (1978).
21 P.C. Myers, and A.H. Barrett, in *The Physical Basis of Electromagnetic Interactions with Biological Systems*, L.S. Taylor and A.Y. Cheung, Eds. (Washington: U.S. Government Printing Office), 309 (1978).

DISCUSSION

Question: Dr. Schmitt, Germany

How do you account for changes in heat exchange using contact probes? (i.e. radiation + convection).

Answer: Dr. Myers, U.S.A.

We have found that controlling the temperature of the antenna to about 32° C gives the minimum variation of detected microwave temperature with time, for most patients. More recently, we have used a layer of "heat-shrink" rubberized plastic about 1 mm thick on the antenna aperture as a thermal insulator. Network analyzer measurements indicate that no significant change in microwave power transfer is introduced when this layer is added.

Question: Dr. Priou, France

Following microwave thermography, microwave holography, cross section of cancer cells versus frequency, better knowledge of the heated pattern of cancer cells in the body, detection, knowledge of the size, depth and view of the cancer cells, what can we expect in the next 5 or 10 years?

Answer: Dr. Myers, U.S.A.

In five to ten years I would expect that microwave thermography will have been explored in an imaging mode, as has been done for infrared and millimeter-wave thermography. The ability to scan and make two-dimensional maps will greatly facilitate the application or microwave thermography to new disease areas such as detection of appendicitis, and vascular insufficiency in the limbs and brain. It may be possible to extract new diagnostic information by comparing maps at several frequencies. We will probably see the development of many combinations of active and passive thermal techniques, such as the use of microwave thermography as a temperature monitor for hyperthermia in cancer treatment, and the use of microwave thermography to measure tissue warming or cooling rates. I do not see any connection with microwave holography, since the thermal radiation sensed by microwave thermography is incoherent.

Question: Dr. Gautherie, France

Could you make additional comments on your diagram showing distribution of cases according to tumor depth?

Answer: Dr. Myers, U.S.A.

It appears that the microwave detection rate does not decrease significantly as tumor depth increases, as is also true for the infrared and x-ray methods. This suggests that we are not viewing a simple "hot-spot" thermal distribution, but rather a complex thermal pattern with significant heating near the surface.

Question: Dr. Schmitt, Germany

A frequently heard argument against microwave thermography is that it is only capable of detecting rather large tumors. What is the clinical experience on early detection?

Answer: Dr. Myers, U.S.A.

Our clinical experience is still very limited, but a 1 GHz we have the results shown in Figure 6. These data indicate that all three methods (microwave, infrared and x-ray) perform about equally well in detecting tumors with diameter = 1 cm as determined from biopsy reports. Our earlier data at 3 GHz was not as good as the present 1 GHz data, suggesting that at 1 GHz, the improved depth sensitivity is a more important factor than the degraded resolution.

Question: Dr. Gautherie, France

Did you carry out comparative measurements on breast patients with inflammatory diseases? This could be interesting since discrimination between carcinomas and abcesses is generally difficult with infrared thermography.

Answer: Dr. Myers, U.S.A.

To date we have not found any significant way to distinguish between benign and cancerous breast tumors, using microwave thermography.

Question: Mr. Maoveni, Iran

Has microwave thermography been used for the diagnosis of skin diseases or disorders (I.E. epithehoma)? If yes, please cite some references.

Answer: Dr. Myers, U.S.A.

We have not tried to use microwave thermography for diagnosis of skin diseases.

Question: Dr. Gautherie, France

Considering the strong changes of penetration depth depending on type of tissues, what is your general viewpoint as to a future development of multifrequency radiometer?

Answer: Dr. Myers, U.S.A.

It is difficult to predict the diagnostic utility of comparative multifrequency radiometry, because of a) the complexity of the thermal distribution to be sensed, b) the possibility of frequency-dependent standing waves between tissue layers, and c) the difficulty of achieving the same resolution pattern at each frequency.

Question: Professor Bianco, Italy

On which basis have the 3 frequencies (1.3, 3.3 and 6.0 GHz) been chosen?

Answer: Dr. Myers, U.S.A.

The curve of penetration depth with frequency, in Figure 2, indicates that one should stay below $\sim$7 GHz in order to have penetration depth greater than typical skin thickness $\sim$3 mm. For frequencies below 1 GHz, poor resolution is a severe limitation. The particular values 1.3, 3.3 and 6.0 GHz were chosen as a matter of convenience.

Question: Dr. Foulds, U.K.

Do you think that the results you obtain using your radiometer may need to be re-interpreted bearing in mind the results of K.M. Ludeke?

Answer: Dr. Myers, U.S.A.

Our diagnostic results are essentially true positive and true negative rates. These are based on the correlation between our measured signals and biopsy findings, and do not depend on the interpretation of these signals in terms of tissue temperatures and emissivity. Of course, it is desirable to develop more precise methods, as indicated by the results of Dr. Ludecke. The methods of the Hamburg group, under Dr. Schilz, and of the Lille group, under Dr. Leroy, are especially interesting because they permit measurement of two independent quantities — tissue temperature and emissivity — instead of the one quantity which we have been measuring, which is essentially the product of the two. It will be interesting to see whether the clinical application of these methods results in an improvement in diagnostic accuracy.

Question: Dr. Douple, U.S.A.

Have you considered the construction of an array of receivers in order to simplify the data collection? Is this feasible (i.e. to produce a single hard copy)? Have you evaluated the cost benefit of your procedure and compared it to mammography (x-rays)? What do you estimate would be the cost of a final product for mass screening?

Answer: Dr. Myers, U.S.A.

An array of N receivers, operating as parallel radiometers, would cost about N times as much as a single receiver, and would be prohibitively expensive, at today's costs, for desirable values of $N \gtrsim 10$. The benefit of reduced measurement time would not be worth the cost.

A comparison of cost vs benefit between microwave thermography and mammography is extremely difficult because the risk of a mammography examination is uncertain, and probably dependent upon dose, patient age, and many other factors. If one assumes zero risk for mammography, then our clinical data indicate that microwave thermography is better used as a complement to mammography than as a replacement for it.

It costs us about $15,000 to buy the parts for each of the radiometers we have built in our laboratory.

Microwave Fixation of Brain Tissue as a Neurochemical Technique—A Review*

J. H. Merritt and J. W. Frazer†

ABSTRACT

Microwave devices have been developed for rapidly inactivating brain enzymes by focusing the power output into the heads of small laboratory animals. The rapid inactivation achieved prevents postmortem changes and permits the measurement of neurochemicals such as acetylcholine at concentrations close to those obtained in vivo. *The technique promises the assay of neurochemical parameters not possible before.*

Microwave energy can be used as a fixative in neurochemical investigations of brain material. Reports in recent years indicate that this technique is gaining acceptance. The purpose of our review is to summarize some key studies and to provide references to the relevant literature.

The specific biochemical events associated with nerve transmission involve the chemical facilitation of a nerve impulse across the space between neurons, the synaptic gap. When a neuron is depolarized, a chemical transmitter is released from the axonal terminal. The transmitter then moves across the synaptic "cleft" and attaches to a specific receptor on the postsynaptic membrane of the next neuron and produces a depolarization which, if of sufficient intensity, will initiate a propagated potential. Repolarization of the presynaptic membrane halts the release of the chemical transmitter, its action being terminated by such processes as uptake into cellular particles and enzymic catabolism.

The whole sequence of events takes place so rapidly that any assay by which one would measure the concentration of the various neurotransmitters as they obtain *in vivo* must include some provision for rapidly halting enzymic processes. Conventionally, enzymic activity is terminated by freezing the brain in liquid nitrogen or cooled Freon. Unfortunately, immersion of the head of an experimental animal in, for example, Freon at −150° C, does not result in immediate freezing of the entire brain. Swaab [1] reported that it required more than 1.5 min for the hypothalamus of a 130 g rat to reach 0° C after being plunged into Freon and that it took 35 s to freeze the cortex. Unfortunately, a few seconds of ischemia will produce large changes in brain levels of, inter alia, ATP, inorganic phosphate, substrates of the glycolytic pathway [2], and cyclic AMP [3].

To overcome slowness of freezing, Veech *et al.* [4] introduced the freeze-blowing technique. Two hollow stainless steel probes are driven into the cranial cavity of an immobilized rat. Air pressure enters one probe and forces the supratentorial portion of the brain out the other and into a hollow aluminum disc precooled to liquid nitrogen temperature. Veech *et al.* concluded that the levels of labile metabolite

* Manuscript received January 3, 1977; in revised form April 15, 1977.

† Radiation Physics Branch, Radiation Sciences Division, USAF School of Aerospace Medicine, Brooks AFB, Texas 78235, U.S.A. Dr. Frazer's present address is: University of Texas Health Sciences Center, San Antonio, Texas 78284 U.S.A.

Reprinted with permission from *J. of Microwave Power*, vol. 12, no. 2, pp. 133–139, June 1977.

such as α-oxoglutarate, pyruvate, glucose-6-phosphate, and adenine nucleotides temperatures that arrest enzymic activity, it has two distinct drawbacks: The *vivo*.

While the freeze-blowing technique quickly cools the brain tissue of rats to temperatures that arrest enzymic activity, it has two distinct drawbacks: The architecture of the brain is destroyed, making dissection impossible; enzymic activity is restored if the tissue thaws.

The rapidity of acetylcholinesterase -catalyzed hydrolysis of acetylcholine made it seem reasonable that a postmortem decrease in levels of this neurotransmitter would occur; indeed, in the rat, whole brain acetylcholine decreases at a rate of 2 nmol/g/min after death [5]. To circumvent this decrease and to establish true *in vivo* brain concentrations of acetylcholine, Stavinoha *et al.* [6] proposed using microwave irradiation as a fixative. The original application of the technique required 11 s of 2450 MHz radiation to inactivate the brain cholinesterase of rats. The investigators later were able to obtain complete and irreversible inactivation of enzymic activity within four seconds [7]. In the meantime, Schmidt *et al.* [8] confirmed the utility of the method as being more accurate and convenient in determination of brain levels of acetylcholine.

Some early applications of microwave fixation involved whole-body radiation of animals in commercial ovens. Since the radiation was diffuse and unfocused, the time for inactivation of brain enzymes was relatively slow (10-20 s) and postmortem changes inevitably occured [8, 9, 10]. Several devices have been described recently that are designed to focus the microwave power on the heads of small animals through some form of waveguide [6, 7, 11, 12, 13, 14]. Brain temperatures of 90° C can be obtained in as little as 300 ms with these instruments [15]. Enzyme systems such as adenyl cyclase, phosphodiesterase, cholinesterase, choline acetyltransferase and NADH-diaphorase [7, 11, 13, 14, 16] are completely and irreversibly inactivated at this temperature.

A further advantage of microwave tissue-fixation is the ease with which the fixed brain can be dissected into discrete parts [7, 8]. The brain tends to separate easily at anatomical boundaries, which results in a more consistent dissection as shown in the standard deviations of weights when the areas from fixed brains are compared with those of dissected fresh brains [8]. See Figure 1.

Stavinoha *et al.* [7] reported little loss of water from rat brain tissue after as long as 10 s of microwave irradiation. Brain tissue temperature after a 10 s exposure was above 90° C. This is reflected in the small ($< 1\%$) increase in the amount of brain protein/g of net-weight. Medina *et al.* [17] observed no significant decrease in water content or protein in the brains of mice after exposure to 400 ms of microwave radiation at a power density that elevated brain temperature to 80° C.

It has now been well established that the acetylcholine content in whole-brain fixed by microwave radiation is higher than in the brain of an animal killed by conventional methods, e.g. by decapitation or by cervical dislocation [5, 6, 7, 8, 11]. The synthetic and degradative enzymes, choline acetyltransferase (EC 2.3.1.6) and cholinesterase (EC 3.1.1.8) respectively, are irreversibly inactivated by sufficient amounts of microwave energy [6, 7, 8, 18]. While whole brain levels of acetylcholine do decline rapidly after death, a recent study has shown that in some brain areas there is actually a postmortem increase in the levels of this neurotransmitter [19], presumably because of geographical differences in the relative activities of the synthetic and degradative enzymes. This last study emphasizes the importance of rapidly terminating enzymes activity in examining the complexities of neurochemical systems.

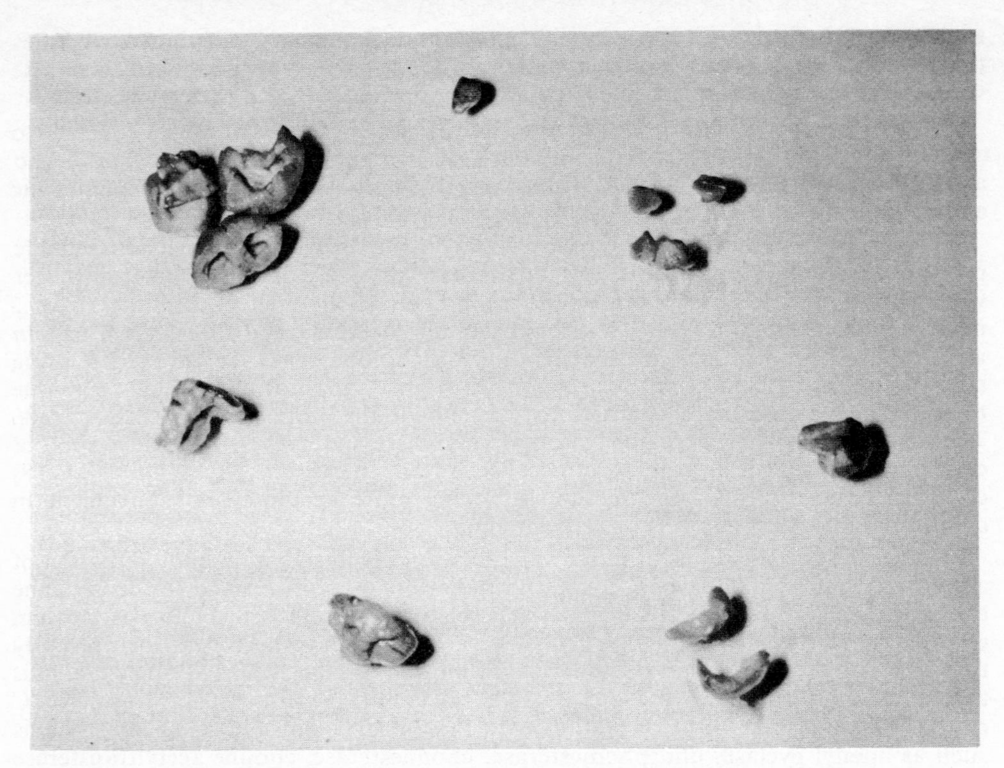

Figure 1 A microwave-fixed brain dissected according to the guidelines of Glowinski and Iversen (J. Neurochem. *13*, 655, 1966). Clockwise from the top: hypothalamus, corpus striatum, midbrain, hippocampus, cerebellum, medulla, and cortex. The microwave-fixed brain is easily dissected since it tends to separate along anatomical lines.

Another set of circumstances is illuminated by recent work on acetylcholine. Total brain content of choline, after a 6-s microwave-inactivation, was 25 nmol/g. If inactivation time of brain was extended beyond death, the choline content rose at about 24 nmol/g/min. Choline and acetylcholine were present in equimolar amounts in the microwave-inactivated brains. The amounts of choline generated after death was more than could have arisen from hydrolysis of acetylcholine [5]. Apparently its source is phosphotidyl choline, a neuronal membrane constituent with a high rate of turnover, and which may be a direct participant in the cholinergic pathway.

In contrast to acetylcholine, the rates for synthesis and metabolism of norepinephrine, dopamine, and serotonin are thought to be relatively slow, so rapid inactivation of brain enzymes does not seem to be a prerequisite for the accurate measurement of *in vivo* levels of these putative neurotransmitters. Merritt *et al.* [20], however, have shown that norepinephrine, dopamine and its metabolic homovanillic acid, and serotonin and its metabolite 5-hydroxindole acetic acid are present in higher concentrations in the brains of mice after sacrifice by microwave irradiation. Although disputed, these findings have now been replicated [22]. Guidotti *et al.* [13] found that the immediate precusor of dopamine, dihydroxyphenylalanine, was nearly two-fold greater in brains of rats killed by microwave irradiation than in brains of decapitated rats. This observation has been confirmed; in addition it was shown that the precusor of serotonin, 5-hydroxytrpytophan, was higher in microwave irradiated rats [23]. These data indicate that there may be a pool of adrenergic and serotonergic neurotransmitters with a very rapid turnover. It appears, then, that the technique of microwave inactivation will permit the study of turnover rates of the putative neurotransmitters in the brains of small laboratory animals [18].

Postmortem increases in the brain levels of the inhibitory transmitter gamma-aminobutyric acid (GABA) have been widely reported. Several recent investigations have been made in which microwave irradiation was used prior to assay of GABA [24, 25, 26]. In these studies GABA was seen to rise rapidly after death if enzymic activity was not terminated rapidly. Balcom *et al.* [24] found an 18% increase in GABA levels in rats sacrificed by decapitation as compared to microwave fixation. The GABA levels observed in irradiated animals in this study were much lower than had previously been reported for animals sacrificed by conventional methods. Balcom *et al.* attributed the lower levels to enzymic activity being terminated by the microwave exposure, thus eliminating a postmortem increase in GABA levels. These same authors have also studied the concentration of brain glutamate, the immediate precusor of GABA, after microwave inactivation [27]. The glutamate levels were the same in animals sacrificed by microwave fixation or by decapitation at room temperature. They concluded that glutamate synthesis continued during the postmortem period since the major increases in GABA in decapitated animals were not reflected in decreases in glutamate.

Cyclic 3′5′-adenosine monophosphate (cyclic AMP) is thought to play an important part in nerve transmission and possibly in nerve function through intracellular mediation of synaptic messages and through the intermediary role of cyclic AMP-dependent protein kinases. It has been known for some time that cyclic AMP accumulates very rapidly in the brain after death [28, 29]. Investigations of the distribution of cyclic AMP in the brain must take this accumulation into account and methods of sacrifice must be used which rapidly inactivate the synthetic and degradative enzymes (adenyl cyclase and phosphodiesterase respectively). As mentioned above, the methods of choice for rapid inactivation of brain enzymes are the freeze-blowing technique and microwave irradiation. For instance, with decapitated rats, the cyclic AMP in brains frozen in liquid nitrogen within 60 s of death is 7-10 times higher than in brains of animals sacrificed by microwave irradiation [13, 16] and is 2 times higher than in freeze-blown brains [30]. In brain areas of the mouse, cyclic AMP is twice as high in animals irradiated at 1.5 kW and requiring 4 s for inactivation of enzymes as in those irradiated at 6 kW and requiring 300 ms for inactivation [31].

One objection raised to using microwave irradiation in assessing *in vivo* concentrations of cyclic AMP was that heat of the irradiation favored the dissociation of bound- to free-nucleotide, with subsequent degradation of phosphodiesterase [30]. This would produce low levels of cyclic AMP as an artifact. On the other hand, the data of Jones *et al.* [16] suggest a correlation between increasing speed of inactivation and decreasing levels of cyclic AMP (see Figure 2). The very rapid inactivation achieved by Jones and Stavinoha [31], in which both adenyl cyclase and phosphodiesterase are completely inactivated within 300 ms, probably permits measurement of cyclic AMP levels close to those that occur *in vivo*, although steady-state levels may not as yet have been achieved. Extremely rapid inactivation with very high power microwave irradiation, in the order of 1 ms, may reveal vanishingly small brain concentrations of cyclic AMP. Unfortunately, devices of this type will be constructed only with some difficulty principally because of cost. While with the present very efficient inactivation systems more than 90% of the output power is absorbed by the head, less than 10% is deposited in the brain

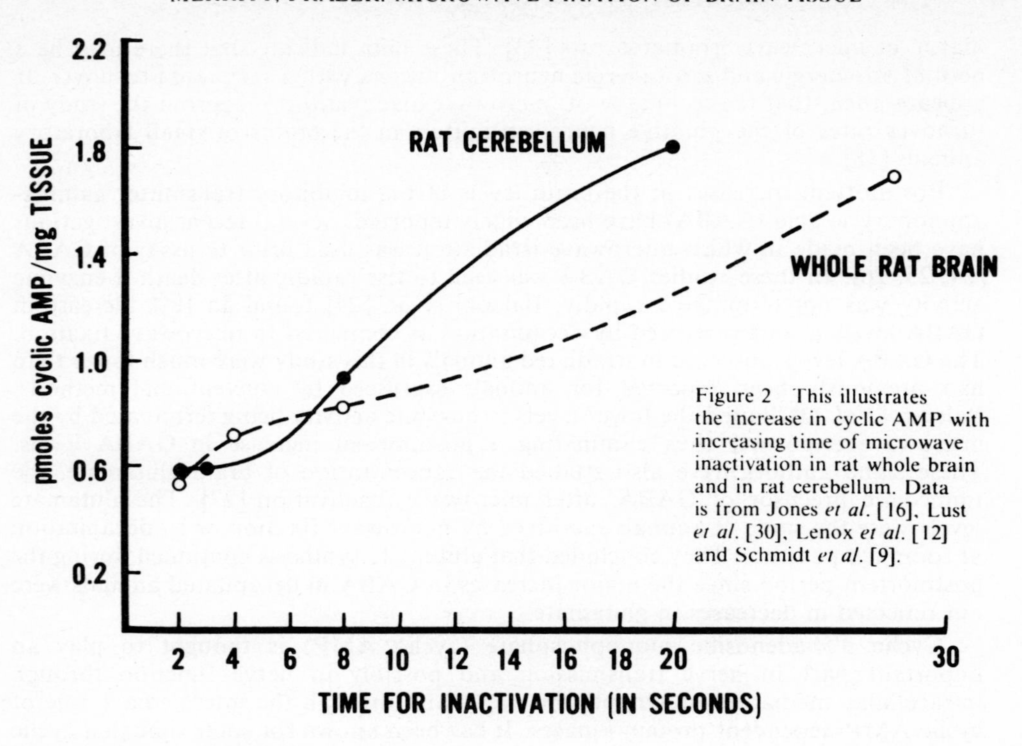

Figure 2 This illustrates the increase in cyclic AMP with increasing time of microwave inactivation in rat whole brain and in rat cerebellum. Data is from Jones *et al.* [16], Lust *et al.* [30], Lenox *et al.* [12] and Schmidt *et al.* [9].

tissue itself. The ultrashort inactivation systems must await improved cavity design.

Lust *et al.* [30] have pointed out that the extremely high metabolic rate of the brain and its limited energy reserves mean that substantial postmortem changes will take place if fixation is not fast enough. The ischemia that occurs as a result of decapitation results in a 4 to 7-fold increase in the glycolytic rate and in changes in substrate and cofactor levels of the glycolytic pathway in brain [2].Even in mice immersed in liquid nitrogen, regional studies of labile metabolites have shown that ischemic changes occur in deeper structures such as the medulla, brain stem, and hypothalamus [1, 32].

In comparing the freeze-blowing technique and microwave irradiation in the study of labile metabolites, Veech *et al.* [4] and Lust *et al.* [30] concluded that freeze-blowing appeared to approximate *in vivo* levels most closely. Phosphocreatine, which is rapidly utilized during ischemia, was much lower in the freeze-blown brains than in microwave-fixed brains, indicating that the former were subjected to less anoxia than the latter [30]. Both studies showed that adenosine triphosphate (ATP), also utilized during ischemia, was higher in the freeze-blown animals, and lactate, which accumulates during anoxia, was lower. Again, the conclusion was that the freeze-blowing technique results in a truer *in vivo* picture of metabolism. The above-mentioned investigators sought to explain the differences in the two techniques (freeze-blowing versus microwave) by such mechanisms as an increase in the rate of brain metabolism due to heat prior to inactivation of the enzymes [4] and the disruption of cell membranes by the intense heat generated by the microwave which would permit extracellular diffusion of substances [30].

In the comparative study made by Veech *et al.* [4], the rats were exposed to 14 s of radiation from a somewhat less powerful microwave transmitter (1.5 kW). The relatively lengthy time of exposure probably produced considerable ischmia before inactivation of the enzymes and may account for the variance between their data and those reported by Medina *et al.* [17]. Using a more powerful microwave apparatus (6 kW) these investigators compared several labile intermediary metabolites in the brains of mice sacrificed by 400 ms of irradiation with those reported for freeze-blowing. Little difference was found between the two techniques.

An interesting application of microwave fixation to neurochemistry has been recently published [33]. In this study, the brain fatty acid levels of rats sacrificed by decapitation were compared with levels in rats whose brains were fixed with microwave heating. Although the free fatty acid in brains of irradiated rats was only insignificantly lower than in decapitated animals whose brains were frozen in liquid nitrogen within 60 s, levels in unfrozen brains in decapitated animals rose dramatically, reaching 100% of initial values in 5 min. Not only that, there was a disproportionate increase of arachidonic acid in the free fatty acids; this postmortem increase, not observed in rats sacrificed by microwave fixation, supports the hypothesis that the rapid production of brain free fatty acids reflects phospholipase activity. In view of the known relationship of arachidonate, prostaglandins, and cyclic AMP activity, the results are consistent and indicate a most fruitful area for research.

Mention should be made of the reports of pharmacologic studies beginning to appear in which microwave fixation is used in the experimental protocol. Longoni *et al.* [34] reported the effect of leptazol, hyoscine, and pentobarbital on the concentration of acetylcholine in discrete areas of brains. Trabucchi *et al.* [35] assessed the effect of dopa, apomorphine, and amphetamine on the turnover rate of acetylcholine. Modak *et al.* [15] studied the effects of pentobarbital on acetylcholine concentration in mouse brain areas. Naik *et al.* [36] reported the effects of muscimol and baclofen on GABA levels in rat brains.

The suitability of the microwave fixation technique to prevent postmortem changes in heat-stable, rapidly changing biochemicals appears well established. The advantages of its application to neurochemistry are manifold and, given higher levels of output power, fixation systems will be able to achieve a rapidity of inactivation never before possible. The technique will permit the measurement of neurochemical parameters, heretofore unattainable, in such situations as behavioral studies and during neurophysiologic events.

REFERENCES

1 Swaab, D., "Pitfalls in the use of rapid freezing for stopping brain and spinal cord metabolism in rat and mouse," J. Neurochem., *18*, 2085-2092 (1971).

2 Lowry, O., Passonneau, J., Hasselberger, F. and Schultz, D., "Effect of ischemia on known substrates and cofactors of glycolytic pathway in brain." J. Biol. Chem., *239*, 18-30 (1964).

3 Breckenridge, B., "Cyclic AMP and drug action." Ann. Rev. Pharmac., *10*, 19-34 (1970).

4 Veech, R., Harris, R., Veloso, D. and Veech, E., "Freeze-blowing: A new technique for the study of brain *in vivo*." J. Neurochem., *20*, 183-188 (1973).

5 Stavinoha, W. and Weintraub, S., "Choline content of rat brain." Science, *183*, 964-965 (1974).

6 Stavinoha, W., Pepelko, B. and Smith, P., "Microwave irradiation to inactivate cholinesterase in the rat brain prior to analysis for acetylcholine." Pharmacologist, *12*, 257 (1970).

7 Stavinoha, W., Weintraub, S. and Modak, A., "The use of microwave heating to inactivate cholinesterase in the rat brain prior to analysis for acetylcholine." J. Neurochem., *20*, 361-371 (1973).

8 Schmidt, D., Speth, R., Welsch, F. and Schmidt, M., "The use of microwave radiation in the determination of acetylcholine in the rat brain." Brain Res., *38*, 377-389 (1972).

9 Schmidt, M., Schmidt, D. and Robison, G., "Cyclic AMP in the rat brain: Microwave irradiation as a means of tissue fixation." Adv. Cyclic Nucleotide Res., *1*, 425-434 (1972).

10 Alderman, J. and Schellenberger, M., "ϒ-Aminobutyric acid (GABA) in the rat brain: Re-evaluation of sampling procedures and the postmortem increase." J. Neurochem., *22*, 937-940 (1974).

11 Butcher, S., Butcher, L., Harms, M. and Jenden, D., "Fast fixation of brain *in situ* by high intensity microwave irradiation: Application to neurochemical studies." J. Microwave Power, *11*(1), 61-65 (1976).

12 Lenox, R., Gandhi, O., Meyerhoff, J. and Grove, M., "A microwave applicator for *in vivo* rapid inactivation of enzymes in the central nervous system." IEEE Trans. Microwave Theor. Tech., *24*, 58-61 (1976).
13 Guidotti, A., Cheney, D., Trabucchi, M., Wang, C. and Hawkins, R., "Focussed microwave radiation: A technique to minimize postmortem changes of cyclic nucleotides, dopa and choline and to preserve brain morphology." Neuropharmac., *13*, 1115-1122 (1974).
14 Butcher, L. and Butcher, S., "Brain temperature and enzyme histochemistry after high intensity microwave irrasiation." Life Sciences, *19*, 1079-1088 (1976).
15 Modak, A., Weintraub, S., McCoy, T. and Stavinoha, W., "Use of 300-msec microwave irradiation for enzyme inactivation: A study of effects of sodium pentobarbital on acetylcholine concentration in mouse brain regions." J. Pharmac. Exptl. Therap., *197*, 245-252 (1976).
16 Jones, D., Medina, M., Ross, D. and Stavinoha, W., "Rates of inactivation of adenylcyclase and phosphodiesterase: Determinants of brain cyclic AMP." Life Sciences, *14*, 1577-1585 (1974).
17 Medina, M., Jones, D., Stavinoha, W. and Ross, D., "The levels of labile intermediary metabolites in mouse brain following rapid tissue fixation with microwave irradiation." J. Neurochem., *24*, 223-227 (1975).
18 Wang, C., "Microwave radiation for the assay of cyclic nucleotides and putative neurotransmitter concentrations in brains of mice and rats." Fed. Proc., *33*, 1591 (1974).
19 Weintraub, S., Modak, A. and Stavinoha, W., "Acetylcholine: Postmortem increases in rat brain regions." Brain Res., *105*, 170-183 (1976).
20 Merritt, J., Medina, M. and Frazer, J., "Neurotransmitter content of mouse brain after inactivation by microwave heating." Res. Comm. Chem. Path. Pharmac., *10*, 751-754 (1975).
21 Weintraub, S., Stavinoha, W., Pike, R., Morgan, W., Modak, A., Koslow, S. and Blank, L., "Evaluation of the necessity for rapid inactivation of brain enzymes prior to analysis of norepinephrine, dopamine and serotonin in the mouse." Life Sciences, *17*, 1423-1427 (1976).
22 Catravas, G., Personal communication (1976).
23 Lindqvist, M., Kehr, W. and Carlsson, A., "Attempts to measure endogenous levels of dopa and 5-hydroxytryptophan in rat brain." J. Neural Trans., *36*, 161-176 (1975).
24 Balcom, G., Lenox, R. and Meyerhoff, J., "Regional τ-aminobutyric acid levels in rat determined after microwave fixation." J. Neurochem., *24*, 609-613 (1975).
25 Richardson, D. and Scudder, C., "Microwave irradiation and brain τ-aminobutyric acid levels in mice." Life Sciences, *19*, 1431-1440 (1976).
26 Knieriem, K. and Medina, M., "Levels of gamma aminobutyric acid in mouse brain following tissue fixation by microwave irradiation." Trans. Am. Soc. Neurochem., *6*, 170 (1975).
27 Balcom, G., Lenox, R. and Meyerhoff, J., "Regional glutamate levels in mouse brain determined after microwave fixation." J. Neurochem., *26*, 423-425 (1976).
28 Ebadi, M., Weiss, B. and Costa, E., "Microassay of adenosine 3'5' monophosphate (cyclic AMP) in brain and other tissues by the luciferin-luciferase system." J. Neurochem., *18*, 183-192 (1974).
29 Uzunov, P. and Weiss, B., "Effect of phenothiazine tranquilizers on the cyclic 3'5'-adenosine monophosphate system of rat brain," Neuropharmac., *10*, 697-708 (1971).
30 Lust, W., Passonneau, J. and Veech, R., "Cyclic adenosine monophosphate, metabolites, and phosphorylase in neural tissue: A comparison of methods of fixation." Science, *181*, 280-282 (1973).
31 Jones, D. and Stavinoha, W., "Levels of cyclic nucleotides in mouse regional brain following 300 ms microwave inactivation." J. Neurochem. (In press).
32 Ferrendelli, J., Gay, M., Sedgwick, W. and Chang, M., "Quick freezing of the murine CNS: Comparison of regional cooling rates and metabolite levels when using liquid nitrogen or Freon-12." J. Neurochem., *19*, 979-987 (1972).
33 Cenedalla, R., Galli, C. and Paoletti, R., "Brain free fatty acid levels in rats sacrificed by decapitation versus focused microwave irradiation." Lipids, *10*, 290-293 (1975).
34 Longoni, A., Mulas, A. and Pepu, G., "Drug effect on acetylcholine levels in discrete brain regions of rats killed by microwave irradiation." Brit. J. Pharmac., *52*, 429P (1974).
35 Trabucchi, M., Cheney, D., Racagni, G. and Costa, E., "*In vivo* inhibition of striatal acetylcholine by L-dopa, apomorphine and (+)-amphetamine." Brain Res., *85*, 130-134 (1975).
36 Naik, S., Guidotti, A. and Costa, E., "Central GABA receptor agonists: Comparison of muscimol and baclofen." Neuropharmac., *15*, 479-484 (1976).

A Microwave Applicator for *In Vivo* Rapid Inactivation of Enzymes in the Central Nervous System

ROBERT H. LENOX, O. P. GANDHI, SENIOR MEMBER, IEEE, JAMES L. MEYERHOFF, AND H. MARK GROVE, SENIOR MEMBER, IEEE

***Abstract*—The paper describes modifications of microwave techniques for *in vivo* rapid inactivation of brain enzymes. These modified techniques offer greater rapidity and homogeneity of inactivation. The microwave-treated brain remains suitable for regional dissection.**

INTRODUCTION

Cyclic adenosine monophosphate (cyclic AMP) has been established as an intracellular mediator of the action of a number of hormones [1]–[3], and may play an important role in the function of the central nervous system [4], [5]. Determination of the concentration of cyclic AMP in various regions of the brain therefore provides an important tool in neurochemical research, e.g., in evaluation of the effects of hormones or drugs in the central nervous system. In order to reliably determine the distribution of cyclic AMP, it is necessary, in the process of sacrificing the animal, to rapidly denature (inactivate) the enzymes that both produce and degrade cyclic AMP, i.e., adenylate cyclase (AC) and phosphodiesterase (PDE), respectively. These enzymes, if left active for even a few seconds, will produce artifactual increases in the levels of cyclic AMP that do not reflect actual endogenous concentrations prior to sacrifice [6]. This rapid anoxic activity of enzymes seen with cyclic AMP during the postmortem period also occurs with many other important metabolites in the brain [7], [8].

Conventional methods of sacrifice use liquid nitrogen to freeze the tissue and thereby stop enzymatic activity. Most freezing methods using immersion into liquid nitrogen inactivate enzymes nonuniformly throughout the brain, and may require up to 90 s for total inactivation [9]. In addition, freezing methods do not permit dissection at room temperature because any postmortem thawing of tissue allows enzymatic activity to resume. The more rapid freezing methods such as freeze-blowing [10] have decreased the time of inactivation considerably, but do not permit regional dissection since there is a loss of anatomical features.

Microwave heating provides a promising approach that is based upon the principle that cyclic AMP is a relatively heat-stable substance [11], while the enzymes (AC and PDE) involved in its metabolism are heat labile and denature irreversibly at temperatures in the range of 65–90°C. Using sufficiently high microwave power, inactivation of enzymes can be achieved with exposure times on the order of 1–2 s or less, permitting subsequent regional dissection of the brain at room temperature. An important requirement in the design of a microwave inactivator, however, is the need for uniform heating of brain regions. This paper addresses one such attempt to design an applicator that would both couple the animal to the field more efficiently and improve upon the uniformity of the heating pattern.

METHODS

Early microwave inactivation procedures for the rat exposed the whole body within an oven cavity at a frequency of 2450 MHz. In our laboratory, the exposure time required for adequate inactivation of enzymes in carefully measured high-intensity regions of the oven ranged from 35–50 s at 2000-W forward power and 250 W reflected. The time required was dependent upon the size of the animal and varied with the field intensity obtained in various regions within the oven cavity. Subsequent reports in the literature demonstrated persistent anoxic artifacts at these times of exposure and the necessity for increasing the rapidity of the inactivation process [12]. In order to significantly increase the efficiency of the microwave inactivation of the brain, microwave exposure was restricted to the animal's head.

The first design, as shown in Fig. 1, consisted of inserting the animal's head through a hole in the broad face of a WR 430 waveguide. In an effort to concentrate the field at the rat head, we placed, on both narrow walls of the waveguide, aluminum shims which tapered to a maximum height in the region of the rat head, as noted in Fig. 2. Although inactivation times required decreased by 86 percent, to 5 s, there was variability of heating portions of the brain because of both the nonuniformity of the field being generated in the waveguide chamber and the mobility of the rat while in the field. The applicator design was modified to restrain the animal's head with a clamping device, eliminating any motion within the field. Studies demonstrated an improved reproducibility of results from animal to animal, but the heating pattern of the brain revealed a significant gradient.

At this point it became quite evident that not only was rapid inactivation required, but as one became increasingly interested in multiple discrete regions of the brain, a requirement of homogeneity of the inactivation process throughout the brain became most essential. In addition, the stress of vigorous immobilization of the animal was considered to be a possible confounding variable. Our laboratory has attempted to address both of these factors in the design and development of the microwave apparatus described as follows.

The arrangement used was a WR 430 waveguide test cell driven by a Varian PPS 2.5 generator and terminated in a short-circuiting

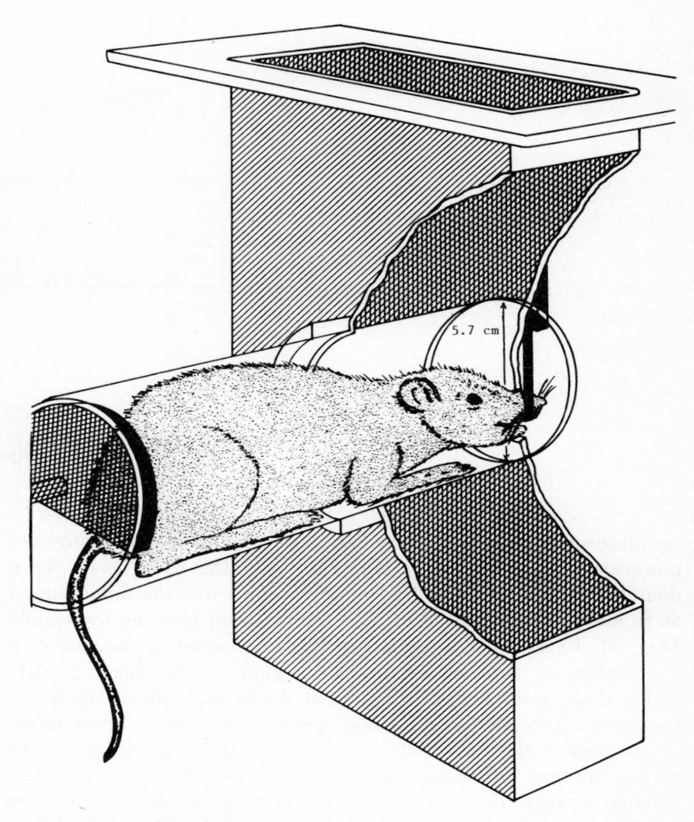

Fig. 1. Broad-face microwave applicator. Animal is placed in a Plexiglas tube which is inserted into the waveguide chamber.

Manuscript received December 10, 1974; revised June 23, 1975.

R. H. Lenox and J. L. Meyerhoff are with the Department of Neuroendocrinology and the Department of Microwave Research, Division of Neuropsychiatry, Walter Reed Army Institute of Research, Washington, DC 20012.

O. P. Gandhi is with the Department of Electrical Engineering, University of Utah, Salt Lake City, UT 84112.

H. M. Grove is with the System Avionics Division, Air Force Avionics Laboratory, Wright-Patterson Air Force Base, OH 45433.

Reprinted from *IEEE Trans. Microwave Theory Tech.*, vol. MTT-24, pp. 58–61, Jan. 1976.

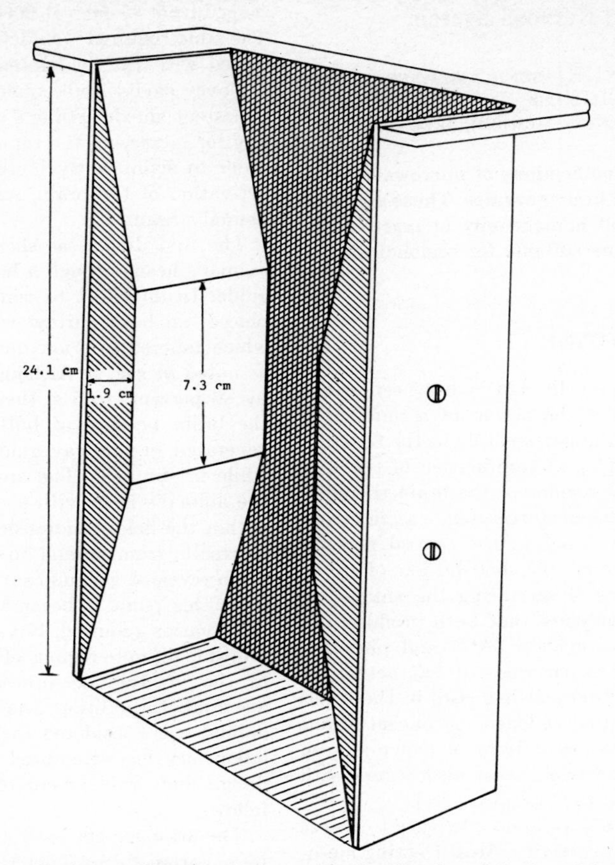

Fig. 2. Design modification of the waveguide chamber of the broad-face microwave applicator.

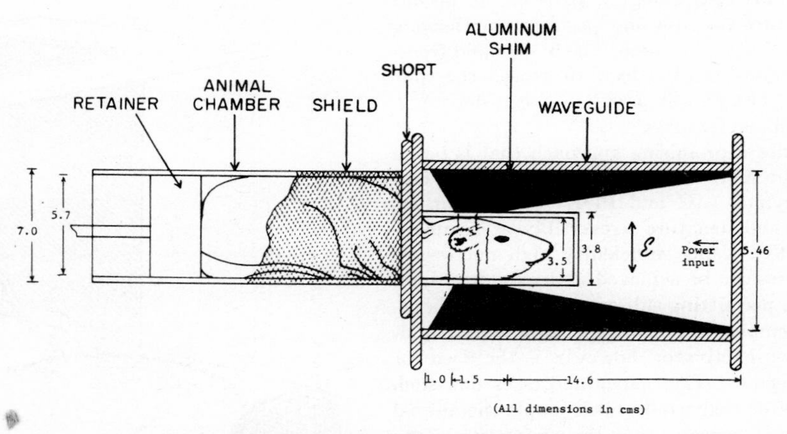

Fig. 3. Shorting endplate microwave applicator.

endplate with a central 3.8-cm-diam hole. The 3.5 kW of microwave power was matched to the complex load (the rat's head) by a double screw tuner. The hole diameter used was the minimum so as to allow the protrusion of only the rat head into the waveguide (Fig. 3). In addition to reducing the hole diameter, the choice of its location in the low-electric-field region at the shorting plate rather than at the high-field region at the broad walls of the waveguide helped to maintain the mode pattern and, hence, the symmetry of fields in the waveguide. Also, the leakage of power out of the waveguide was minimized because of the new hole location. The residual leakage from the hole in the shorting endplate was further reduced by wrapping copper screening around the 7.0-cm-OD Plexiglas tube used to hold the rat body. While allowing the visualization of the rat positioning, the screen acts as a cylindrical waveguide with the lowest TE_{11} mode cutoff frequency of approximately 2.5 GHz. The efficacy of shielding was confirmed by observing a power density of no more than 5 mW/cm² at a distance of 10 cm from the termination during 3500-W power exposures. The rat brain, of dimensions on the order of $2.5 \times 1.8 \times 1.0$ cm, occupied a location centered in the maximum field region at a distance of $\lambda_g/4$ away from the short at the microwave frequency of 2450 MHz. The rat body was located in a 5.7-cm-ID tube external to the waveguide with the rat head in an offset 3.5-cm-ID tube protruding into the waveguide.

To concentrate the microwave power into the rat head and to obtain the uniformity of field distribution over the 2.5-cm longitudinal extent of the rat brain, the shorter dimension of the waveguide was gradually reduced (Fig. 4) to 3.8 cm by using two identical, tapered, aluminum plates attached to the broad faces of the waveguide. The plates tapered down over a length of 14.6 cm (to

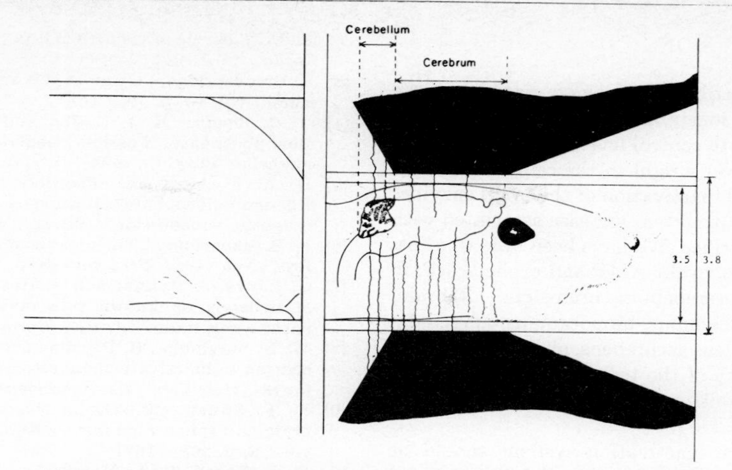

Fig. 4. Design modification of the waveguide chamber of the shorting endplate microwave applicator.

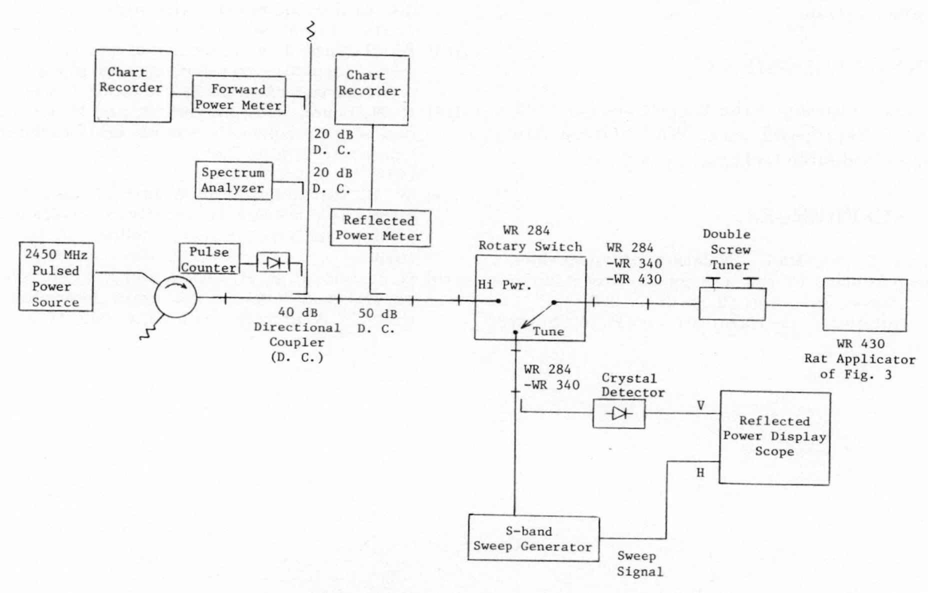

Fig. 5. Block diagram of the enzyme inactivator.

minimize incident power reflection) to a flat region of 1.5-cm length, and then abruptly returned to full waveguide height over an axial length of 1.0 cm. This produced an increase of fields away from the shorting endplate to compensate for the observed dielectric load of the rat head, which had otherwise resulted in a marked concentration of fields close to the short of the waveguide. Without this modification there was disproportionately greater heating of the cerebellum of the rat brain.

The block diagram of the microwave enzyme inactivator is shown in Fig. 5. The power is matched to the rat head by a double screw tuner, giving a fairly broad-band match as shown in Fig. 6. A low-Q match is essential since heating after a few RF pulses does result in altered dielectric properties of the load which, for narrow band matches, produce considerable reflection for subsequent pulses. Precise monitoring of the time duration of applied power is done by means of a counter that counts the number of microwave pulses. In order to accurately determine the applied microwave energy, a recording of the incident and reflected powers is obtained for each of the animals inactivated by the system.

Inactivation of brain regions was evaluated [13] and was shown to be more uniform using the construction outlined. The exposure time necessary for a 325-g rat was 2.8 s, with the animal becoming unconscious within fractions of a second upon application of microwave power [14]. The uniformity of fields along the width

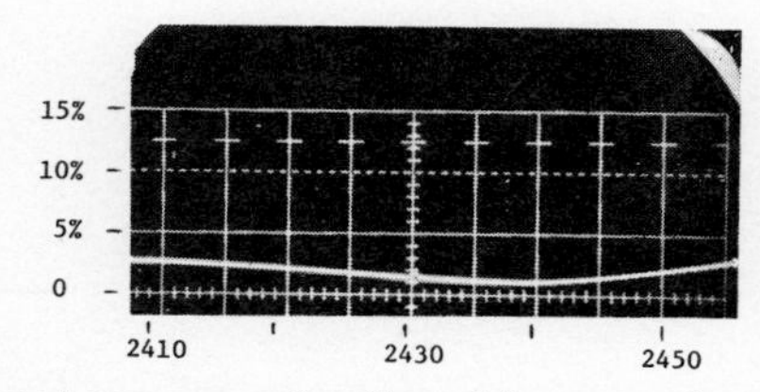

Fig. 6. Reflected power as a function of frequency for a double-screw-tuned rat applicator.

of the waveguide (from ear to ear of the animal) is obtained because of the symmetrical dielectric loading due to the rat head in that dimension, and the consequent concentration of power therein. Improved uniformity of heating of the brain was achieved in the rostral to caudal dimension (cerebrum to cerebellum) due to the tapered aluminum shims. In the vertical plane, however, a slight gradient (ventral greater than dorsal) was observed. This is currently ascribed to the overheating of the abundant muscle (because of its higher electrical conductivity) on the ventral side of the brain. The Plexiglas cylinder used to hold the rat in this study did not prevent rotation of the head in unanesthetized animals. Rotation of greater than 30° caused additional perturbation of the field and contributed to limited reproducibility from animal to animal. Further work is currently in progress to alleviate this difficulty.

CONCLUSIONS

The microwave applicators described herein have resulted in the simultaneous sacrifice and rapid inactivation of brain enzymes in the rat. Present results demonstrate control levels of cyclic AMP to be approximately 0.6-pmole/mg wet weight in the cerebellum. This is indicative of both the very rapid inactivation of the brain enzymes involved and the prevention of postmortem increase associated with more conventional methods of sacrifice. We have been able to measure levels of two cyclic nucleotides, cyclic AMP and cyclic GMP, in 13 distinct regions of the brain: cerebellum, brainstem, midbrain, substantia nigra, thalamus, hypothalamus, hippocampus, amygdala-pyriform cortex, septal nuclei, nucleus accumbens, olfactory tubercle, striatum, and cortex. Applicability of the technique to many putative central nervous system transmitters is being investigated in our laboratory [15].

Users of high-power microwave inactivation systems should be aware of the factors affecting the uniformity and reliability outlined here. We are continuing attempts to modify our microwave parameters to accomodate an increased mobility of the rat with improved uniformity and speed of inactivation.

ACKNOWLEDGMENT

The authors wish to thank P. Brown of the Department of Microwave Research, Division of Neuropsychiatry, Walter Reed Army Institute of Research, for his valuable technical assistance.

REFERENCES

[1] E. W. Sutherland and T. W. Rall, "Relation of adenosine-3′, 5′-phosphate and phosphorylase to the actions of catecholamines and other hormones," *Pharm. Rev.*, vol. 12, p. 265, 1960.

[2] R. W. Butcher, G. A. Robison, J. G. Hardman, and E. W. Sutherland, "The role of cyclic AMP in hormone actions," *Advan. Enzyme Regul.*, vol. 6, p. 357, 1968.

[3] B. Breckenridge, "Cyclic AMP and drug action," *Ann. Rev. Pharmacol.*, vol. 10, p. 19, 1970.

[4] G. R. Siggins, R. J. Hoffer, and F. E. Bloom, "Cyclic adenosine monophosphate: Possible mediator for norepinephrine effects on cerebellar Purkinje cells," *Science*, vol. 165, p. 1018, 1969.

[5] D. A. McAfee, M. Schorderet, and P. Greengard, "Adenosine 3′,5′-monophosphate in nervous tissue: Increase associated with synaptic transmission," *Science*, vol. 171, p. 1156, 1971.

[6] B. Breckenridge, "The measurement of cyclic adenylate in tissues," *Proc. Nat. Acad. Sci.*, vol. 52, p. 1580, 1964.

[7] O. Lowry, J. Passonneau, F. Hasselberger, and D. Schultz, "Effect of ischemia on known substrates and cofactors of the glycolytic pathway in brain," *J. Biol. Chem.*, vol. 239, p. 18, 1964.

[8] W. B. Stavinoha, B. Pepelko, and P. Smith, "The use of microwave heating to inactivate cholinesterase in the rat brain prior to analysis for acetylcholine," *Pharmacologist*, vol. 12, p. 257, 1970.

[9] D. F. Swaab, "Pitfalls in the use of rapid freezing for stopping brain and spinal cord metabolism in rat and mouse," *J. Neurochem.*, vol. 18, p. 2085, 1971.

[10] R. L. Veech, R. L. Harris, D. Veloso, and E. H. Veech, "Freeze-blowing: A new technique of the study of brain *in vivo*," *J. Neurochem.*, vol. 20, pp. 183–188, 1973.

[11] E. W. Sutherland and T. W. Rall, "Fractionation and characterization of a cyclic adenine ribonucleotide formed by tissue particles," *J. Biol. Chem.*, vol. 232, p. 1077, 1958.

[12] W. D. Lust, J. V. Passonneau, and R. L. Veech, "Cyclic adenosine monophosphate, metabolites, and phosphorylase in neural tissue: A comparison of methods of fixation," *Science*, vol. 181, p. 280, 1973.

[13] R. H. Lenox, J. L. Meyerhoff, and H. L. Wray, "Regional distribution of cyclic nucleotides in rat brain as determined after microwave fixation technique," *Proc. Soc. Neurosci.* (Abstract), vol. 4, p. 303, 1974.

[14] W. B. Stavinoha, S. T. Weintraub, and A. T. Modak, "The use of microwave heating to inactivate cholinesterase in the rat brain prior to analysis for acetylcholine," *J. Neurochem.*, vol. 20, p. 361, 1973.

[15] G. J. Balcom, R. H. Lenox, and J. L. Meyerhoff, "Regional λ-aminobutyric acid levels in rat brain determined after microwave fixation," *J. Neurochem.*, vol. 24, p. 609, 1975.

Applications of Microwave Thawing to the Recovery of Deep Frozen Cells and Organs: A Review*

W. A. G. Voss†, R. V. Rajotte† and J. B. Dossetor‡

ABSTRACT

Large microwave insults, at a frequency of 2450 MHz, have been applied to deep-frozen adult canine kidneys, fetal mouse hearts and tissue culture cells, causing controlled temperature changes at rates up to 300°C/min, from −196°C to 23 ± 12°C. This paper describes the two microwave systems which have been used for heating a number of different biological samples, ranging in volume from 2 to 10 ml, and presents some of the results obtained, together with the method used.

The electrical activity of fetal mouse hearts, recovered from −196°C by microwave heating, has survived in a high percentage of cases. The hearts were taken from 17 to 19 day old embryos, frozen in 5 ml samples of Minimum Essential Medium with 25 mm Hepes buffer, 10% dimethyl sulfoxide and 10% fetal calf serum. Subsequent subcutaneous implantation in the ear of syngeneic adult mice was used; in this way electrical activity has been studied for periods up to 35 days. Resumption of electrical activity is obtained by microwave and water bath thawing. Tissue culture cells will also withstand rapid thawing in a resonant microwave system operating at very high (near breakdown) field strengths. Uniform microwave thawing of adult frozen canine kidneys has been obtained at rates between 100 and 300°C/min from −79°C. It is now possible to control the heating to an end point of 23°C with a variation across the organ of < ± 12°C, providing prior perfusion of the organ was complete. Although subsequent functional success has not yet been achieved with frozen thawed canine kidneys, certain areas of viable tissue, with vascular integrity, have been observed.

---

* Joint research undertaken by members of the Medical Research Council of Canada Transplantation Group and the Microwave Laboratories at the University of Alberta, Canada. Based on a paper presented at the International Symposium on Biologic Effects and Health Hazards of Microwave Radiation, Warsaw, October 1973.
Manuscript received November 29, 1973, in revised form May 6, 1974.

† The Surgical Medical Research Institute, The University of Alberta, Edmonton, Alberta, Canada.
‡ MRC Transplantation Group, The University of Alberta Hospital, Edmonton, Alberta, Canada.
Reprints may be obtained from the authors.

Reprinted with permission from *J. of Microwave Power*, vol. 9, no. 3, pp. 181–194, Sept. 1974.

Introduction

The use of energy in the microwave spectrum provides a method of controlling the rate and uniformity of heating of deep-frozen materials. If deep frozen human organs are ever to be recovered from low temperature storage banks, microwave thawing techniques have to be considered. A study of the biological effects of microwave heating on nucleated cells in the range from −196°C to +35°C, is an essential step in the search for methods of viable organ preservation.

High survival rates for several cell types in tissue culture protected with 5% (v/v) dimethyl sulfoxide (DMSO) or 5% (v/v) glycerol thawed by short wave diathermy (27 MHz) have been reported by Silver et al [27]. Microwave and short wave (2450 and 27 MHz) thawing of canine kidneys have been briefly reported by Lehr et al [18], but, so far, there have been no reports of the recovery of viability and function of deep frozen (−196°C) kidneys, by any of the freezing and thawing regimens used. A recent report of success with canine kidneys recovered from −22°C is encouraging [7].

Thawing of biological materials is conventionally achieved in a water bath. A high degree of success has been obtained with cells of many types and certain multicellular structures [22]. Skin and cornea are preserved clinically in the deep frozen state [31, 25]; in fact, an encouraging number of advances have recently been made in the preservation of other multicellular structures using water bath thawing. The cryobiological methods used have been described in a number of books [37, 31] and an excellent review of freezing phenomena has been given recently by Mazur [21]. Chick embryo heart anlage has survived freezing for short periods in liquid nitrogen [9] using ethylene glycol (EG) as a cryoprotective agent. Whittingham, Leibo and Mazur [36] froze 2 - 8 cell mouse embryos to −196°C and −269°C at slow cooling rates (0.3 to 20°C/min) and then thawed them slowly at rates of 4° to 25°C/min in a water bath, with subsequent development of 50 - 70% of 2500 such embryos into blastocysts on culture. When these were placed in pseudo-pregnant mothers, 65% became implanted as pregnancies, and of these over 40% became fetuses or went to term. The cryoprotective agent dimethyl sulfoxide (DMSO), at 1 m concentration, was about twice as effective as an equal concentration of glycerol. These particular compounds penetrate all membranes, reduce the free water content and thereby the damage from high concentrations of intra and extracellular solutes that would otherwise be present at that temperature. Optimum cooling rates have been established in many cases.

Supercooled adult hearts have resumed beating [32, 2, 17] but prior to the work of Offerijns and Krijen in 1972 [24] and Rapatz in 1970 [28], attempts to freeze adult mammalian hearts had met with little success. A detailed review of the subject has been given by Luyet [19]. Rapatz [28] obtained partial resumption of activity in all parts of adult frog hearts after freezing to below −55°C, using EG as a cryoprotective agent. Offerijns and Krijen [24] added DMSO to perfusates of isolated adult rat hearts. With super-cooling to temperatures not below −18°C all hearts survived in 2.1 M DMSO. With freezing to −30°C, young rat hearts (10 - 16 days old) also recovered, but older hearts did not, possibly because of a lower tolerance of adult hearts to high concentrations of extracellular NaCl. Other organs have withstood supercooling.

The small, fully differentiated heart also provides a convenient model for studying the effects of microwave thawing. Not only is it possible to determine the effects of the insult at different power levels, the method of controlled thawing may have immediate applications in, for example, the preservation of adult heart valves, the selective preservation of certain cells, improved preservation techniques for embryos in animal husbandry, muscle in general, and eventually, help to determine whether or not the kidney, the organ in greatest transplant need, can be preserved indefinitely. The kidney is probably the greatest challenge to preservation engineering today. In this paper, our methods and experiences in these laboratories are briefly described, for work at the cell level, with the fetal mouse heart, and with canine kidneys. The histological, microfil, oxygen and para-amino-hippurate studies on frozen-thawed canine kidneys have been described elsewhere recently [26]. All data and photographic information are available [38].

Microwave Heating System Design

In microwave biology studies, waveguide systems are preferable as the fields are known. Multimode cavities are unsatisfactory in this respect but in thawing applications, where the geometry of the material is often irregular, their use is indicated in many cases. The results described in this paper were obtained in two multimode cavities, both designed in an attempt to optimize the uniformity of heating in organs, as well as to control the rate of heating. However, we regard the waveguide applicator, with an optical control system on the sample, and possibly fed by two magnetrons through circulators, as the most promising for further studies. Fairly complex systems, either resonant or non-resonant, appear to be required to achieve uniformity and predictability of heating. There are many theoretical and practical aspects of microwave heating systems which have not been studied in this regard.

The two resonant microwave systems which have been used were cubic structures, the basic design following that of the conventional microwave oven. The smaller unit, a 10″ cube, coupled to a 1KW, 2450 MHz magnetron, was used for thawing tissue culture cells. A larger volume oven was required for thawing organs by our method: a 17″ cube coupled to a 2 KW, 2.45 GHz magnetron, as shown in Figure 1.

Both systems were equipped with rotating turntables on which the samples were placed. In the larger cavity, the turntable was oscillated as well as rotated, through a drive shaft powered from two independent external electric motors. Rotation speeds were 20 rpm and the oscillation rate was selected between 20 and 30 cycles per min. The turntables and drive shafts were made of teflon, a durable material which does not absorb microwave energy. Heating times were determined experimentally in advance; samples that were not recovered in the range 0 to 40°C were rejected. Heating rates were selected to be similar to those obtained in a water bath with 2 to 5 ml samples but lower rates are easily obtained in practice by reducing the amount of microwave power, and the limits of these systems for samples in the volume range 2 to 100 ml were found to be 50 to 500°C/min in initial experiments. Perfused, frozen kidneys were immersed in a 1 litre teflon container filled with fluorocarbon, the material which was also used for perfusing the organ in the initial freezing stage. A groove was made in the turntable to hold the unit. Organs frozen by the method shown diagramatically in Figure 2 were placed in the fluorocarbon, the initial temperature of which was +4°C. Fluorocarbon does not absorb any significant amount of microwave energy.

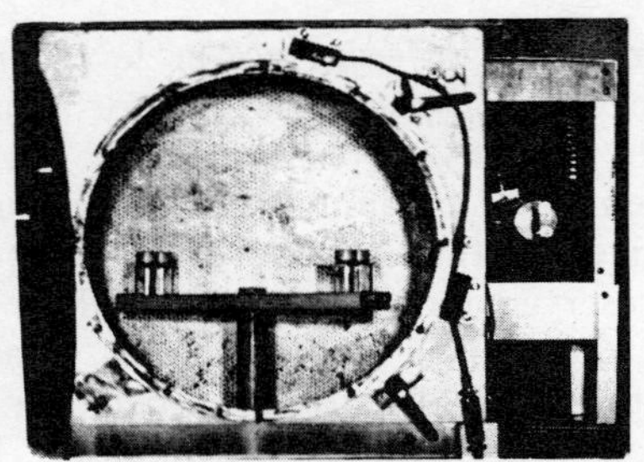

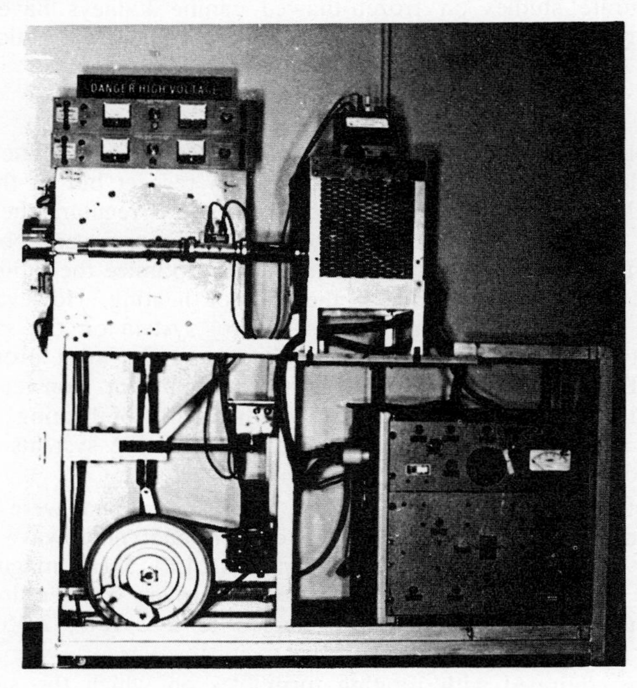

Figure 1 Microwave Thawing System. The two turntables are shown at the top through the open cavity door. The drive mechanism, power supplies and magnetron are shown in the lower part of the illustration.

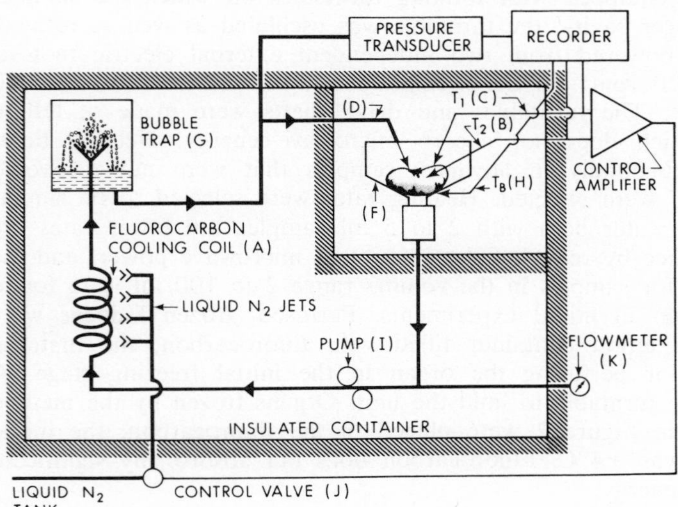

Figure 2 Schematic diagram of the cooling system.

Cubic structures have the greatest number of modes of resonance and appear a logical first choice; the first appreciable mode density, around 2450 MHz, occurs with a 10″ cube, and increases to an effective mean value of one mode per MHz with a 17″ empty cubic structure [12]. In practice the magnetrons used have a bandwidth of about 15 MHz so that an appreciable number of modes are coupled by a distributed radiator. A folded stripline antenna was used as the radiator [13], the size and position of the actual design being determined experimentally. The choice of designs was arbitrary; considerable theoretical work along the lines described by Ketterer et al [16] is essential. As these authors point out, it may be necessary to combine 915 and 2450 MHz heating in order to obtain adequate uniformity of heating. In order to achieve a reasonable degree of uniformity of thawing with 2450 MHz source we have found it essential, so far, to rotate the material in two planes and to accept the fact that the mean reflected power back to the magnetron is about 50%. Thus, we have made a practice of using magnetrons at about half their rated output power. We have also increased the wall loss by using perforated metal linings, as seen in Figure 1. This decreases the energy conversion efficiency of the cavity but allows for a wider range of loads without causing excessive amounts of power reflected back to the magnetron. The average microwave power density within an organ can be determined from the specific heat (a mean value of about 0.8 has been determined), the temperature rise and the time.

The 17″ cubic cavity was used for thawing canine kidneys and fetal hearts, conditions which represent very different degrees of loading. For thawing tissue culture cells, we were interested in applying very high field strengths, to determine the effect on their survival. The smaller cavity was used for this; in the absence of a load, corona discharge was detected around the teflon shaft at 1 KW input power. Rotating two 2ml samples in the oven, with a 50 ml water load in one corner of the cavity gave rise to a significant preferential heating of these samples, at the same time reducing the reflected power back to the magnetron to about 50%. The preferential heating mechanism is not understood but is being investigated. The electric field strengths in the cavity have not been measured but it is assumed that peak values are in the range 1000 to 10,000 volts/cm.

The systems used for the initial work lack refinement and the approach is entirely experimental. For our initial purposes the systems have worked to the extent of providing rapid, reasonably uniform thawing and some basic data. A travelling wave heating system, incorporating forward and reflected power meters to control power density in the samples, together with an optical sensing device to detect the thawing point, has been recently built. Initial measurements on this system indicate that the microwave heating rate of the materials used increases very rapidly just below 0°C. Our objective here is to study the effect of electric field strength on tissue viability. The fetal heart, in both the frozen and unfrozen state, may provide a useful method of studying this critical parameter. In the work reported here using cavities we have operated at a mean heating rate which approximately corresponds to that obtained with 2ml samples thawed in a 35°C water bath. However, the initial and final rates are quite different due to the changing dielectric properties of the liquids and tissue, which are very dependent on the perfusate, the cryoprotective agents (pronounced in the case of DMSO) and the structure of the ice present. The mean absorbed power densities were of the order of 10 watts/cc in the majority of cases.

Microwave Thawing of Tissue Culture Cells

Methods

(a) *Tissue Cultures.* Chinese hamster cells (Puck Strain A) were used to assess the relative viability of microwave thawing in comparison with a 35°C water bath. The techniques of cell growth, handling, and culturing are exactly those described in previous work by our colleagues [2] and involved scoring colony formation 8 - 9 days after the plating of single cells.

(b) *Freezing Regimen.* Cell suspensions (2 ml) in ampoules with either DMSO (5% and 10%, v/v) or DMSO (5%) with hydroxyethyl starch (5%), were frozen in solid CO_2 with alcohol for at least 30 min. Seeding was not used. The cooling rate was approximately 20°C/min as defined by the time taken for the sample to reach −70°C from 23°C when the temperature was measured, in a duplicate, with a copper-constantan thermocouple.

(c) *Microwave Thawing Apparatus.* The 2ml frozen samples, in vitro, were placed in an insulating styrofoam container (transparant to the microwave field) and rotated at 15 rpm by a teflon rod connected to an external electric motor. Certain procedures were taken to protect the microwave generator, as the sample absorbed only about 1% of the available power: in the majority of tests a 50 ml sample of water at 20°C was placed in the oven. A thermal sensor was also used to protect the microwave generator, which is capable of withstanding very high reflected power for short periods.

The transfer time of a sample from the ice bath to the heater was less than 10 sec. After heating, the temperature of the 2ml sample was recorded within 5 sec by a sterilized thermocouple. About 50% of the samples were heated to a useful final temperature, between 0 and 40°C. Others were rejected, if the measured final temperatures were higher than 40°C. Multiple samples and unfrozen controls were assayed. Microwave thawing rates between 100 and 250°C/min (from −79 to +10°C) were used and compared to the colony survival rates with 115°C/min water bath thawing.

Results

The results, which have been presented and discussed in detail previously [35], demonstrate very clearly that no disadvantages to microwave thawing procedures, with a cell system, are apparent if the final temperature is below about 10°C. Thus, the rapid absorption of microwave energy appears acceptable. Survival figures as good as, or better than, those realized with standard thawing procedures can be achieved for Chinese hamster cells. Twice the conventional thawing rate is acceptable and it is to be noted that, theoretically, this is relatively independent of sample shape and size. However, this statement must be qualified as the heating rates in different temperature ranges are quite different, and yet to be measured.

Microwave Thawing of Fetal Mouse Hearts

The fully differentiated fetal heart of the mouse can be reimplanted in the ear of an adult syngeneic mouse and studied electrically over a long period of time. Immunologically, this is equivalent to transplanation between identical twins in man. Microwave energy (at 2450 MHz) was used as one method of thawing and compared to thawing in a water bath.

Materials and Methods

Heart Transplant. Hearts were removed from Balb-c fetuses obtained at 16 - 18 days gestation. The hearts at this time measure approximately 1 mm in diameter and beat rhythmically. Following thawing, the embryonic hearts were implanted directly into the ear of adult syngeneic mice anesthetized by nembutal. The anterior aspect of the ear was injected subcutaneously with 0.1 ml of saline. The injection causes the two layers of skin to separate and thus form a pocket for the fetal heart. A small incision was made to introduce the fetal heart into the subcutaneous space. Fetal hearts implanted in this manner are nourished by the surrounding tissue fluid until abundant capillary supply is formed some days after grafting [8, 5, 14, 15]. DMSO was not removed before implantation in the ear, as it was felt that the concentration used (10% v/v) would not harm the tissue, and would slowly diffuse into the host.

Freezing. The fetal hearts were placed in various prechilled solutions, some of which contained the cryoprotective agent DMSO (10% v/v). The final composition of the solutions containing DMSO was obtained by slowly adding the DMSO over a twenty minute period, after the hearts had first been placed in the DMSO deficient solutions. 5 ml glass vials of the various solutions, each containing one heart, were then immediately placed in a prechilled freezing unit (Linde B-4-1). The freezing rate was maintained between 0.5 and 0.7°C/min, by a thermocouple control system, down to −100°C. The samples were then placed in liquid nitrogen vapor and cooled at 5 - 10°C/min down to −196°C. They were stored for 72 - 216 hrs. at −196°C before being rewarmed.

Thawing Process. Frozen hearts were rewarmed either by placing the 5 ml vials in a 35°C water bath, thereby raising the temperature at a rate of ~150°C/min or in the microwave system described, and shown in Figure 1. Two or four samples were thawed at a time. Additional stationary water samples were included in the cavity to protect the magnetron, as previously described. In these experiments, mean heating rates of 200°C/min were used to thaw the hearts from −196°C to 10°C ± 10°C.

Assay of Graft Function. It is not possible to assess the pump action of the graft; however, the electrical property was evaluated by electrocardiography and contractility could be seen with a magnifying glass through the skin of the external ear. The leads of the ECG machine were attached to the anesthetized test animal by means of small clips. A small clip was also attached to the mouse ear containing the fetal heart and connected proximally to the V-lead or search electrode of the machine. As seen in Figure 3, the electrical activity of both the adult heart (b) and that of the fetal heart (a) can be recorded simultaneously as two distinct sets of rhythmic electrical activities. (For a detailed description of the technique, see Jirsch, Kraft and Diener [14]).

Results

As controls, Balb-c mice received direct syngeneic ear transplants of unfrozen fetal hearts. 95% of these control transplants had electrical activity by the 5th day and continued to function for periods in excess of 90 days. In previous experiments [14] electrical activity of similar unfrozen transplants had existed for periods in excess of one year.

Activity of the frozen-thawed hearts was monitored at intervals over the period of 6 to 30 days after implantation. As in the control groups, those hearts

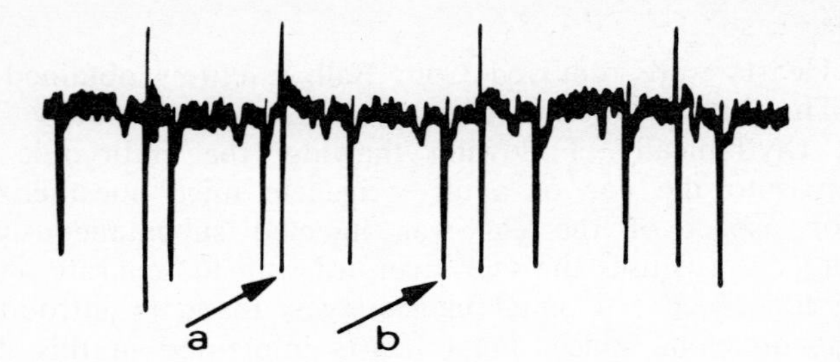

Figure 3 A typical ECG showing normal mouse complexes (arrow b) at approximately 250 beat/min, with the slower superimposed fetal heart rhytym (arrow a).

which had survived the initial period after grafting continued to beat indefinitely, i.e. up to 100 days. There were no differences detectable in electrical activity between control implants and those which had survived freeze-thawing; however, the electrical complex recorded is only a simple biphasic deflection in either instance.

The results [27] showed that suspending solutions containing Eagle's MEM or McCoy's 5a, with 10% fetal calf serum, 10% dimethyl sulfoxide, and Hepes buffer, gave rise to significant survival rates. In contrast, solutions not containing either Hepes buffer, fetal calf serum or DMSO did not give rise to survival of the fetal hearts initially[1]. Pooling the results for the first of the above solutions, the overall survival between 6 - 15 days was 37 of 55 hearts frozen, thawed and implanted, or 67%, and at more than 30 days was 32 of 54, or 59%. It was found (a) that DMSO and fetal calf serum are useful components for survival of fetal mouse hearts exposed to freeze-thaw injury, (b) that microwave thawing is as effective as thawing in a water bath, and encourages the hope that it may be effective with larger tissues where water bath thawing would certainly be ineffective, and (c) that this model can be used, in future work, to assess different cryoprotective agents and freeze-thawing techniques on survival of a multicellular organ which is at or near the upper limit of size for nutritional survival by diffusion and neo-capillary ingrowth, i.e. at or near the upper limit of size for survival without the need for direct vascular anastomosis at time of graft implantation.

The electric field strength experienced by the hearts during thawing in our case would be expected to vary from a peak value of 0 to a few thousand volts/cm as the 1 mm organ in the 5 ml sample is rotated randomly through the resonant electromagnetic fields of the cavity. Only a very small fraction of the stored energy in the cavity is absorbed in the heart and the solution in which

[1] We have subsequently found that fetal calf serum may not be essential. Glycerol has also been used successfully as a cryoprotective agent, providing that a longer diffusion time is allowed.

it is frozen. The absorption of energy increases very rapidly above −10°C; in a microwave cavity it is not possible to monitor the rate between, for example, −60 and −15°C (possibly the most critical range in freezing and thawing). Consequently, we established a mean rate (from −196°C to +10°C) slightly above that possible in a water bath in order to make a comparison: the rates for the water bath and the microwave thawed heart should be about the same in the −60 to −15°C range. Statistically, with a limited number of hearts, there is no significant difference between the two methods of thawing. A histological analysis of the multiple insults of freezing, thawing and implantation is being made; no gross abnormalities have been observed; we feel that the procedure may be useful for studying other forms of microwave insult, in particular those associated with low power pulsed microwave signals.

Microwave Thawing of Kidneys and Kidney Slices

So far we have evidence that the microwave insult is acceptable to one type of nucleated cell and mouse late-fetal hearts at temperatures at below and just above zero. The difficulty of extending these freezing and thawing techniques to large masses or organized tissue is very apparent [1]. Different cells may require different freezing and thawing rates and may be susceptible to different concentrations of cryoprotective agents [21]. Complete perfusion of an organ also poses problems. All these factors mitigate against the eventual recovery of whole organs from the deep frozen state: results prior to last year have not been encouraging [4, 10, 11, 20], but the situation is changing [7, 26].

In attempting to design a complete freeze-thaw system, our objective has been to use the most promising perfusates in a system which allows for the controlled variation of the basic parameters, uniform freezing and thawing rates. If controlled, uniform freezing and thawing cannot be obtained, there seems little purpose in other experiments on whole organ preservation.

The freezing system, shown in Figure 2, is rate controlled by the thermocouples. The complete procedures is shown in Figure 4. The choice of DMSO is a logical start as it is the most penetrating cryoprotective agent known. Fluorocarbon was chosen for subsequent perfusion at temperatures below +4°C as it is biologically and physicochemically acceptable [34, 23, 30, 33], and will perfuse through part of the critical tissue freezing range (−15 to −40°C) without requiring abnormally high perfusion pressures. A typical flow rate of 10cc/min was obtained at −35°C and 200 mmHg. As DMSO and intracellular electrolytes are insoluble in fluorocarbon (FC47, 3 M.Coli) it is assumed that these substances are not removed. Cooling only proceeded if the temperature differential between the thermocouple probes in the kidney was $\leqq 5$°C; this resulted in an average rate of 1°C/min from +4 to −40°C. At about −40°C, environmental cooling was then used down to −79°C at the same rate. Thermocouples were removed at this temperature and the organs stored for periods between 3 hrs and several days in dry ice at −79°C. A similar freezing regimen was used for a series of tests on thin, unperfused, kidney slices, with storage at −79 or −196°C.

For thawing, the frozen kidneys were placed individually in a teflon holder, and clamped into position gently by teflon bolts. This assembly, immersed in fluorocarbon, was then heated in the system shown in Figure 1. The input microwave power was raised to 1.5 KW over a period of 3 seconds and then

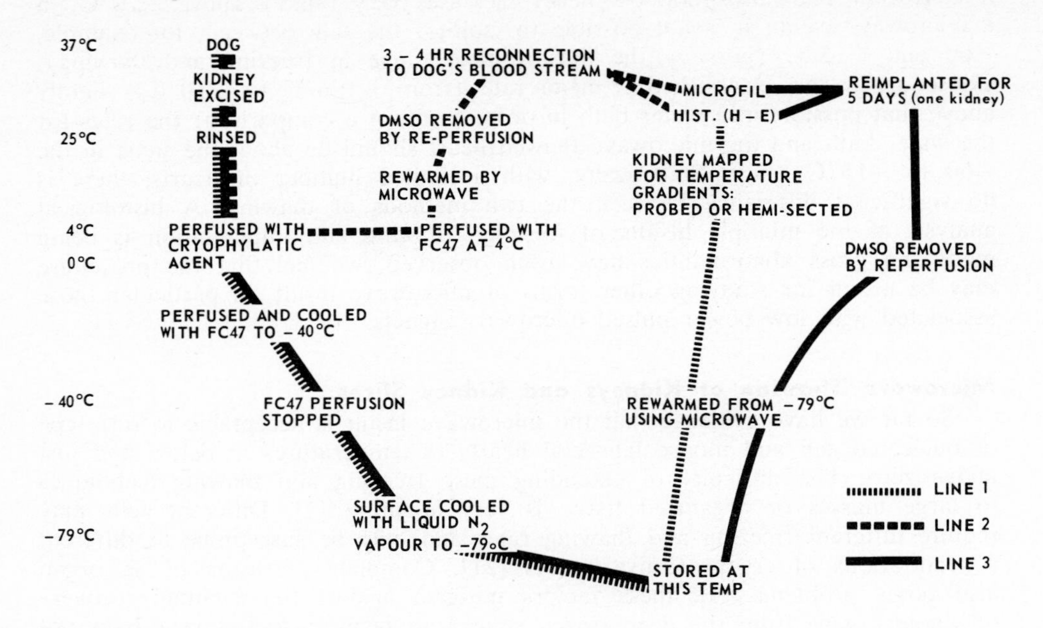

Figure 4 Experimental procedures.

in the experiments reported here, that power maintained for the required time. This incident power level, with 80g to 100g canine kidneys resulted in a thawing rate between 100 and 200°C/min. Other thawing regimens have been used but we have so far no reason to reject this one, although we would like to improve the uniformity of the final temperature to within ± 2°C. After thawing, kidneys were perfused with Ringer's lactate containing either 10% DMSO or 10% Mannitol initially and then Ringer's lactate by itself. This perfusion step was calculated to displace the hyperosmolar fluorocarbon from the vascular tree and remove DMSO from the hyperosmolar interstitial and intracellular fluid by diffusion without subjecting the cells to too great an osmotic gradient[2]. Some experiments were terminated at this point and the thawed organs studied for temperature gradient. In experiments where DMSO was removed, Ringer's lactate containing 10% DMSO was first perfused for 10 minutes followed by a 10 minute perfusion with Ringer's alone. Mannitol in decreasing concentrations was also used to remove the DMSO while at the same time minimizing the osmotic gradient across the cell wall. The Transfer time from the frozen container to the microwave system was typically 2 minutes; the thawing time was of the order of 1 minute. The total time that the kidney had been maintained without normal circulation in this procedure above 4°C was about 1 hr before subsequent manipulations for microfil injection histology or reimplantation.

Uniformity of heating within reasonable limits has been achieved if, and only if, the original perfusion of the kidney was complete. Areas of poor perfusion can be easily seen as pink areas after slicing after thawing and are critically

2 We have subsequently revised our procedure, following the techniques described by Dietzman (7) and our own observations on the importance of slowly removing DMSO. (Added in press).

dependent upon the handling of the organ prior to freezing. A vasodilator (dibenzylene) was injected into the dog prior to nephrectomy. After removal each kidney was rinsed for 5 minutes with lactated Ringer's (Baxter) at 25°C. When whole kidneys were to be frozen, dimeythl sulfoxide was added to the Ringer's in a two-step perfusion process. 10% (v/v) DMSO in Ringer's was perfused for 10 minutes followed by 20% (v/v) DMSO in Ringer's for a further 10 minutes. During this period when the kidney was taking up DMSO from perfusates at either 4 or 25°C, it was immersed in a bath of lactated Ringer's at 45°C. If the procedure described above was not in any way satisfactory, the organ was rejected. A number of experiments were performed to determine the required heating time for a given weight and then, providing the perfusion had been complete, final temperatures in the range 23 ± 12°C were obtained. In a number of cases, this range mas maintained except for small areas which had failed to perfuse completely. It was decided that an average final temperature of 20°C would allow for variations of ± 15°C with a factor of safety. We feel that this can be achieved now, if the perfusion is complete but may represent too great a temperature variation. Temperatures were recorded by hemi-secting the kidneys and probing with thermocouples beneath the surface, looking particularly for areas that were frozen. In all cases these were determined in areas which showed signs of blood and an absence of perfusion. Some typical temperature maps are shown in Figure 5.

Unfrozen kidneys. To study the microwave insult and the effects of the cryoprotective agents, some kidneys were not frozen. They were cooled and perfused to +4°C by the method described and then immediately rewarmed in the microwave system to an acceptable final temperature.

Microfil Injection. This injection technique was used to outline the microcirculation of kidneys frozen-thawed by the methods described. Microfil is a catylisal silastic rubber, which will pass through the glomeruli and post-glomerular capillary circulation into the venous system. The kidney was sliced after the microfil had set, dehydrated with increasing concentrations of alcohol and then made transparent with methyl salicylate. Photographic records were taken at a magnification of 40 to 65X. The histological and microfil studies have been reported [26].

Reimplantation. One kidney, which was deemed to have been frozen and thawed to the best of the above criteria, was reimplanted for 5 days and then removed for histologic and microfil examination. Other kidneys were reconnected externally to the femoral artery and the vein of the same dog and allowed to perfuse for 3 - 4 hrs. Assessment of this tissue, after blood recirculation, was in accord with the usual pathologic criteria. As might be expected, after recirculation, evidence of tissue damage was much greater than could be detected by histology of thawed tissue prior to reperfusion with normal blood.

Results

A group of kidneys which were not frozen but had been permeated with the protective agent, followed by perfusion at 4°C with FC47 and warmed in the microwave system, showed what was judged to be near normal histology and assessed as 'viable' after being anastomosed via the femoral artery in the vein of the same dog for a number of hours. The vascularity of the kidney after microfil injection was judged to be normal.

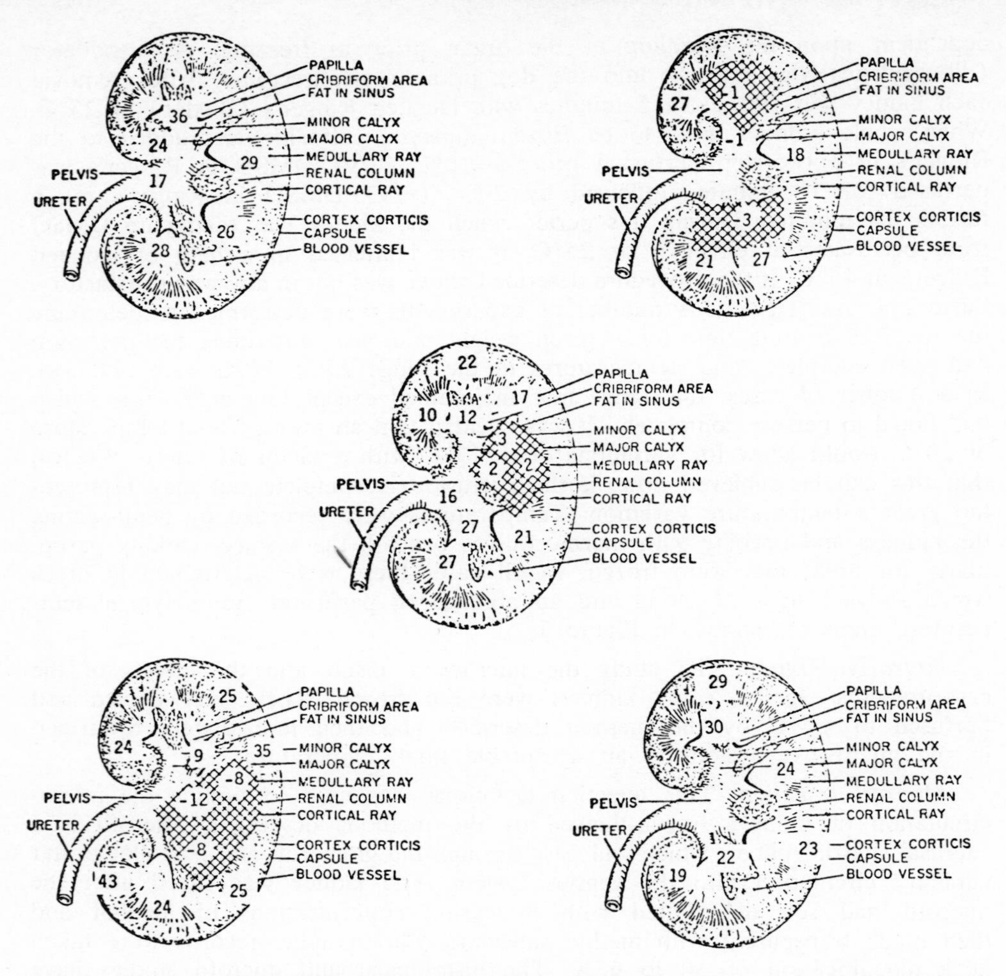

Figure 5 Typical temperature maps obtained by probing beneath the surface of hemisected kidneys immediately after thawing from —79°C. The worst (most extreme) values are shown in °C. Complete perfusion is essential: non-perfused areas are shown hatched. The ureter and cortex were externally filled with fluorocarbon before freezing commenced.

Kidneys thawed from −79°C at rates between 100 and 200°C/min showed possible satisfactory temperature profiles if the perfusion was complete, Figure 5. Overheating of the external ureter was seen on several kidneys. Complete perfusion of all the areas of the kidney with fluorocarbon presents a continuing problem.

The kidney reimplanted in the groin of the same dog after a complete freeze-thaw cycle, as indicated in Figure 4, was examined after 5 days. Although the external appearance was completely normal and no inflammation had occurred, damage to the microcirculation was subsequently apparent. In isolated areas glomeruli and capillaries were well preserved indicating that the blood was being perfused through the organ. Other parts of the kidney showed gross damage, and, in fact, could not be properly injected with microfil. At no time was there evidence that the kidney was capable of forming urine.

An extensive series of tests on kidney slices has indicated damage other than that seen in the whole kidney tests on the microvascular system. Unfrozen slices were used to compare the effects of Cross solution, 10 and 15% DMSO in Cross solution and Culture Medium 199 (Bio-Cult Labs, Glasgow) with and without 10% fetal calf serum and Hepes buffer. It has been found that 10 and 15% DMSO in Cross solution prior to Warburg incubation had little effect on PAH and O_2 uptake. PAH uptake was significantly lower in the other solutions. Slices which were frozen in these solutions and warmed by water bath or microwaves at different heating rates also had good uptake of O_2 but PAH uptake was poor. Abbott [1] has reported similar findings with water bath thawing. Histological examination of all slices, frozen or normal, has proved difficult and unreliable. PAH is the most sensitive index of tubular cell viability and these tests have failed to show preservation with frozen-thawed slices.[3] Our histological studies, with microfil sections and thawing profiles, have been published elsewhere for a series of kidney and kidney slice experiments [26]. We have clearly demonstrated the difficulty in assaying damage due to the freeze-thaw insults, in line with the experiences of Abbott [1]. So far we have no definite reason to reject microwave thawing at the organ level, and we have established a system which should allow the study of the many phenomena associated with freeze-thaw insults at the cell and organ level. The fact that some nucleated cells and one very small multicellular organ will survive a series of gross insults of this type is encouraging.

Acknowledgments

We wish to thank all the members of the MRC Transplantation Group and the Surgical Medical Research Institute at the University of Alberta for their contributions to this continuing work. In particular we would like to thank Dr. M. J. Ashwood-Smith and Mrs. Carole Warby of the University of Victoria for their contributions and cooperation in our joint researches in cryobiology and preservation engineering in Western Canada. We express our gratitude also to the Medical Research Council of Canada and the National Research Council of Canada (Grants A2272 and E2373) for their financial support of this work.

References

1 Abbott, W. M., Cryobiology, 5(6), p. 454 (1969).

2 Ashwood-Smith, M. J., Warby, C., Connor, K. W. and Becker, G., Cryobiology, 9, p. 441 (1972).

3 Barner, H., "Perfusion and Freezing of Rat Heart," *in* Organ Perfusion and Preservation. Ed. Norman, J. C. Appleton Century Crofts, New York, 1968, p. 717*f*.

4 Beltron, J. C. and Blumenthal, H. J., J. Urology, 82(4), p. 424 (1959).

5 Conway, H., Griffith, B. H., Shannon, J. E. and Findley, A., Surgical Forum, p. 596 (1957).

6 Cross, R. J. and Taggart, J. V., Amer. J. Physiology, 161, p. 181 (1949).

7 Dietzman, R. H., Rebelo, A. E., Graham, E. F., Crabo, B. and Lillehei, R. C., Surgery, 74, p. 181 (1973).

3 We now believe this to be caused by osmotic stress from the removal of additives. A revised procedure has given encouraging initial results.

8 Fulmer, R. I., Cramer, A. T., Liebelt, R. A. and Liebelt, A. J., Amer. J. Anatomy, 113, p. 273 (1963).
9 Gonzales, F. and Luyet, B., Biodynamica, 7, p. 1 (1950).
10 Guttman, F. M., Khalessi, A. and Berdorikoff, G., Cryobiology, 6, p. 339 (1970).
11 Halazs, N. A., Miller, S. H. and Devin J., Cryobiology, 7, p. 163 (1970).
12 James, C. R., Tinga, W. R. and Voss, W. A. G. "Energy Conversion in Closed Microwave Cavities" *in* Microwave Power Engineering. Ed.: Okress, E. C., Academic Press, New York, 1968, p. 28 *f*.
13 Johnston, D. A., "Travelling Wave and Resonant Microwave Heating Applicator Design." Ph.D. Thesis, University of Alberta, Canada, 1972, p. 89*f*.
14 Jirsch, D. W., Kraft, N. and Diener, E., Cardiovas. Res. (in press), 1973.
15 Judd, K. P., Allen, C. R., Guibertean, M. J. and Trentin, J. J., Transplantation 1(1), p. 470 (1969).
16 Ketterer, F., Holst, H., Lehr, H. and Kritikos, H., Cryobiology, 10, p. 515 (1973).
17 Karow A. M. Cryobiology, 5, p. 429 (1966).
18 Lehr, H. B., Transplant. Proc., 3, p. 1565 (1971).
19 Luyet, B., Cryobiology, 8, p. 190 (1971).
20 Manax, W. G., Bloch, J. B., Longerbean, J. K., Eyal, Z. and Lillehei, R. C., Cryobiology, 1(2), p. 157 (1964).
21 Mazur, P., Science, 168, p. 939 (1970).
22 Meryman, H. T., Cryobiology, Academic Press, New York, 1966.
23 Nose, T., Federation Proc., 29(5), p. 1789 (1970).
24 Offerijns, F. G. J. and Krijen, H. W., Cryobiology, 9, p. 289 (1972).
25 Pegg, D. E., Biomedical Engineering (London), p. 290 (1970).
26 Rajotte, R. V., "Microwave Thawing of Mammalian Cells, Tissues and Organs." M.Sc. Thesis, University of Alberta, Canada, 1973; Rajotte, R. V., Dossetor, J. B., Voss, W. A. G. and Stiller, C. R., Proc. IEEE, 62, p. 76 (1974).
27 Rajotte, R. V., Jirsch, D. W., Dossetor, J. B., Diener, E. and Voss, W. A. G., Cryobiology, 11, p. 28 (1974).
28 Rapatz, G., Biodynamica, 11, p. 1, 1970.
29 Silver, R. K., Lehr, H. B., Summers, A., Green, A. E. and Coriell, L., Proc. Soc. Exp. Bio. Med., 115, p. 453 (1964).
30 Sloviter, H. A., Yamada, H. and Ogoshi, S., Federation Proc., 29(5), p. 1755 (1970).
31 Smith, A. U., (ed) "Current Trends in Cryobiology", Plenum Press, New York, 1970.
32 Smith, A. U., Federation Proc. Suppl. 15, p. 196 (1965).
33 Spitzer, H. L., Sachs, G. and Clark, L. C., Federation Proc., 29(5), p. 1746 (1970).
34 Triner, L., Berosky, M., Habif, D. V. and Nahas, G. G., Federation Proc., 29(5), p. 1778 (1970).
35 Voss, W. A. G., Warby, C., Rajotte, R. V. and Ashwood-Smith, M. J., Cryobiology, 9, p. 562 (1972).
36 Whittingham, D. G., Liebow, S. P. and Mazur, P., Science, 178, p. 411 (1972).
37 Wolstenholme, G. E. W. and O'Connor, M. (ed.) "The Frozen Cell." Ciba (London), 1970.
38 The authors, at the Surgical Medical Research Institute, The University of Alberta, Edmonton, Canada.

Electromagnetic Techniques for Medical Diagnosis: A Review

MAGDY F. ISKANDER, MEMBER, IEEE, AND CARL H. DURNEY, MEMBER, IEEE

Abstract—In this paper, potential electromagnetic (EM) methods for medical diagnosis are reviewed. These include impedance plethysmography, microwave methods for lung water measurements, EM flowmeters, and microwave radiometry diagnostic techniques. Other techniques that are in preliminary research stages, such as EM imaging and use of microwave Doppler radar to monitor arterial wall movements, are briefly discussed. The principles underlying the operation of each method are described, along with comments about adequacy for medical diagnosis. The important experimental results that identify the advantages and the limitations of each method are presented. In most cases, it is clear that while the electromagnetic diagnostic techniques are attractive and promising, much more research is still needed before these methods are ready for full clinical use. Suggestions for future development and/or possible extensions are discussed.

I. Introduction

Historically, the use of electromagnetic (EM) techniques for medical diagnosis goes back to 1926 when Lambert and Gremels first used measurements of electrical resistance across the lung to monitor the development of pulmonary edema [1]. In 1936, the basic idea of EM flowmeters was first introduced, and since then has been extensively applied to blood-flow measurement [2]-[4]. It is only recently (1971), however, that microwave radiometry was first applied to problems in biology and medicine [5]-[9]. The first published suggestion that microwave techniques might be used in the management of lung disease was made by Süsskind in 1973 [10]. The idea has been extensively examined and experimentally and theoretically evaluated by the group at the University of Utah [11]-[14]. Diagnostic techniques such as EM imaging of the thorax and use of microwave Doppler measurements to monitor arterial wall movement are at preliminary stages of development and are briefly discussed [15]-[18]. It should be noted that other EM methods have been used in a variety of ways to monitor parameters of medical relevance. These include microwave measurements of respiration [19], microwave interrogation of biological targets [20], and measurement of heart dynamics using microwaves [21]. In this review, we focus attention on the basic techniques of EM diagnoses, their advantages and limitations, rather than restricting the discussion to historical development. Because of space limitations, no attempt has been made to be all inclusive, but only the more important and significant EM diagnostic techniques are included.

II. Impedance Plethysmography

The basic idea behind the use of low-frequency EM methods (20-100 kHz) for measuring changes in the fluid accumulation in the lungs is that most of the current flow through the tissue at these frequencies is due to movement of ions in extracellular water. Because air constitutes the major component of the lung's volume (70 to 80 percent), the resistance of the lung is expected to be high and to be sensitive to changes in blood and extracellular fluid volumes [22]. Attempts to monitor the fluid accumulation in the lung by measuring the electrical impedance (EI) directly across the lungs of cats and dogs were first reported by Lambert and Gremels [1]. The impedance cardiograph transmits a low-frequency, low-current (2-4 mA) electrical signal from two outer electrodes. The drop in voltage between two electrodes placed directly on the lung is then measured by using a high-impedance voltmeter. The method was found to be most sensitive in the earliest stages, which is a decided advantage in measuring pulmonary edema.

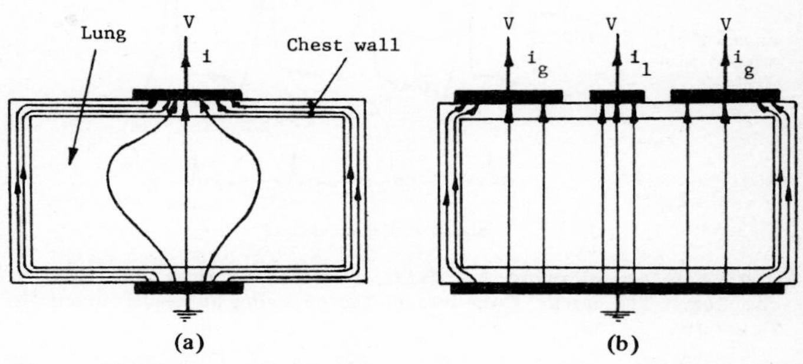

Fig. 1. Impedance measurements across the thorax. (a) With small unguarded electrodes, the major part of the measured current i passes through the low impedance chest wall section. (b) With the guard electrode, most of the chest wall currents flow through the guard ring i_g while the current flowing through the central electrode i_1 is largely directed through the underlying lung.

The development of EI plethysmography [22] in the late 1950's and early 1960's was soon applied to lungs in the thorax [23]. These efforts, however, have been frustrated by the short-circuiting effect of the more conductive surrounding tissues such as the mediastinum and the chest wall. Because of this shunting effect, the method grossly underestimates true lung impedance, with changes of the order of only a few percent attributed to changes in lung water [24], [25].

The use of guarded electrodes [26], [27], and the focusing electrode bridge have greatly improved the sensitivity of the *in vivo* measurements to changes in lung tissue impedance. Graham first suggested surrounding the detecting electrodes with a guard ring, driven at the same potential as the active electrode [26]. With the guard-ring electrode placed over the lateral chest wall, most of the chest wall current should flow through the guard ring, while the current flowing through the central electrode should be largely directed through the underlying lung. The principle of operation of the guarded electrode system is illustrated in Fig. 1.

Manuscript received April 16, 1979; revised August 13, 1979.
The authors are with the Department of Electrical Engineering, University of Utah, Salt Lake City, UT 84112.

Reprinted from *Proc. IEEE*, vol. 68, pp. 126-132, Jan. 1980.

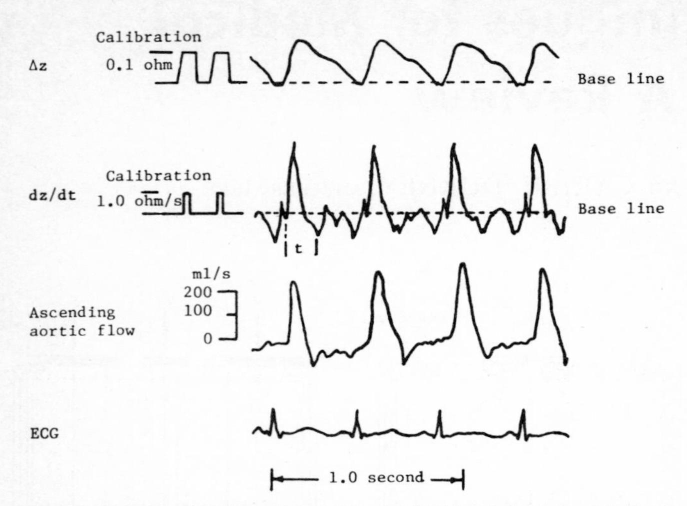

Fig. 2. A record showing Δz, dz/dt, aortic blood flow, and ECG in the dog. The aortic flow was measured using an electromagnetic flowmeter. (After L. W. Baker [30].)

Severinghaus *et al.* [27], further developed the guard-ring principle by adding "focusing" electrodes to obtain a better measurement of lung impedance independently of chest wall and mediastinal structures. They basically introduced a small electrode to detect the skin potential gradient lateral to the detector electrode, which is then used to drive the guard ring to inject sufficient current into the chest virtually to eliminate this lateral gradient. This procedure not only minimizes the lateral (chest wall) current flow from the active electrode, but also avoids problems arising from skin contact resistance in the guard ring. Furthermore, Severinghaus *et al.* [27], divided the guard ring and the skin potential detecting electrode into four quadrants to allow the guard current to match the chest wall current in its quadrants. The sensitivity of this focusing electrode bridge was evaluated by conducting several experiments on dogs and comparing the results with those obtained using the double indicator-dilution technique. The experimental results indicated that the impedance method had about 20 percent the sensitivity of the indicator-dilution method with a reproducibility of ±5 percent. Although the good reproducibility of the experimental measurements is an encouraging advance in the development of the EI method, its poor sensitivity may frustrate widespread clinical use. The possibility of improving the method's sensitivity still exists, but as of now it is not a quantitatively sensitive method.

The use of the EI method as a means of measuring other physiological activities has also been investigated for many years. For example, the technique has been used as a noninvasive measure of stroke volume, myocardial contractility, peripheral vascular disease, and tidal volume [28]. Of particular interest is the use of the EI in measuring the tidal volume and ventilation in critically ill patients and newborn infants [29]. The use of direct spirometry for these subjects has many disadvantages, including the obstruction of the airways and the alteration of the breathing pattern. A comparison of tidal volume signals obtained from the impedance monitor (with the electrodes being properly located high on the midaxillary lines) and from a spirometer clearly demonstrated that the impedance monitor provides a linear response with changes in the lung volume as measured by a spirometer. The linearity of the monitor output continues through tidal breathing volumes up to at least 35 ml [29].

Other interesting results were obtained when the EI method was used to measure cardiac output [30]. A typical record showing the impedance change Δz as a function of time is shown in Fig. 2.

The cardiac output was calculated from the maximum rate of change of the Δz waveform, i.e., $|dz/dt|_{max}$ [30] and the results obtained were compared with the values obtained from other techniques. The comparison clearly showed that the EI method does provide a noninvasive technique to monitor the cardiac output continuously during surgery and also to monitor changes in cardiac output postoperatively [30].

A very recent application of the EI method is to determine the urinary bladder fullness in patients with spinal cord injuries. The exploratory study conducted in the College of Nursing at the University of Utah demonstrated variable changes in impedance in human subjects as the bladder filled [31]. *In vitro* experiments on dogs clearly showed a consistent drop in impedance upon filling the bladders with different solutions.

In summary, many potential applications of the EI method are being explored in a parallel effort with the engineering struggle to improve the sensitivity of the methods.

III. Microwave Methods of Lung Water Measurement

At microwave frequencies, changes in the dielectric properties of tissue are closely related to the amount of water present. The microwave method of lung water measurements basically utilizes these changes in dielectric properties of lung tissue to detect changes in its water content. The method is based on a continuous monitoring of the reflection and/or the transmission coefficient to indicate changes in the permittivity of the lung tissue. This method has the advantage of using highly penetrating microwave signals rather than ultrasonic signals that are both highly attenuated and dispersed in the lung. It should be noted that the microwave radiation method, although based on the complex permittivity of the lung, is fundamentally different from the EI method. In the EI method the signal transmitted through the thorax is primarily in the form of conduction current, in which the net current passing through any cross section is a constant value, independent of the location of the cross section. In the microwave measurements, on the other hand, the transmission consists of electromagnetic waves that are attenuated as they travel through the body. Since the microwave transmission is not dominated by conduction currents, as is the EI method, the microwave method does not suffer from the short-circuiting effect of the highly conducting tissue surrounding the lung, which is characteristic of the EI method. The development of the microwave method, on the other hand, critically depends on the development of a compact applicator that provides maximum coupling to tissue and minimum radiation leakage. The minimum external leakage is a critical requirement, since any leakage radiation reaching the microwave receiver could obscure the information carried by the highly attenuated signal traveling through the thorax.

The sensitivity of the microwave method was examined theoretically with a three-layer planar model representing the front chest wall, lung, and back chest wall [11]. The results clearly indicated that superior sensitivity to changes in lung water content can be achieved by utilizing the transmission coefficient in both magnitude and phase rather than the reflection coefficient. Typically, a 50-fold increase in sensitivity was achieved at 915 MHz by utilizing the transmission coefficient (particularly the phase) instead of any other parameter in the scattering matrix [12]. It was also shown that the best compromise between resolution and attenuation was found to be 915 MHz [13]. These theoretical results were confirmed by conducting experiments on a three-layer phantom, as described elsewhere [13].

The development of a suitable microwave applicator is an important factor in determining the feasibility of using the microwave method in clinical applications. Attempts to use radiation-type applicators such as dielectric loaded waveguides [32] has not been successful because of the excessive leakage radiation. It should be noted that the use of transmission rather than reflection measurements complicates further the design problems by requiring consistent relative positions of the applicators during the course of the test. Iskander and Durney recently developed a new and more efficient method of coupling the microwave energy into the thorax [33]. The method utilizes a surface transmission line (coplanar waveguide) applicator that has proven to couple the energy efficiently and with minimum leakage radiation [14], and can also be used as a receiver.

The feasibility of using the microwave transmission method to measure the fluid accumulation in the lungs was evaluated by conducting several experiments on dogs, using a printed-circuit version of the applicator. The experimental procedure consisted of inducing pulmonary edema in dog's lungs (usually using the high-pressure edema model) and monitoring *in vivo* the changes in the transmitted 915-MHz microwave signal. A continuous recording of the phase of the microwave transmission coefficient showed a waveform with a periodic shape that is related directly to the cycles of inspiration and expiration, with superimposed smaller changes related to the cardiac cycle. The baseline of this waveform shifted as increasing total lung water content developed, and reversed as the condition was reversed. Experimental results from a typical dog experiment are shown in Fig. 3. From these results it is clear that changes in the phase of the microwave transmitted signal agree very well with changes in the pulmonary edema as indicated by the changes in the mean pulmonary arterial pressure. The EM coupler recently developed by the group at the University of Utah has played an important role in the success of this and other similar experiments, since excessive leakage radiation from other applicators produced unsuccessful attempts by others to make similar measurements.

Because of the encouraging initial experience and the promising results obtained from the microwave transmission method, efforts were made to evaluate the accuracy of the method, particularly in detecting early stages of interstitial edema. Several experiments on isolated lungs were conducted in which changes in the microwave signal transmitted through a lobe of an isolated dog's lung were compared with the lung weight as the edema developed. Isolated lung experiments provide the important advantage of eliminating the interference from the extrapulmonary structures which are thought to limit the accuracy of the *in vivo* experiments. Furthermore, the lung weight is known to be the most precise method of determining the total lung fluid content. The results from a typical experiment are shown in Fig. 4. These results clearly show the immediate and direct change in the phase of the microwave transmitted signal as the weight of the isolated lung was changed. These results are, therefore, a strong indication of the sensitivity of the microwave method for the early detection of pulmonary edema.

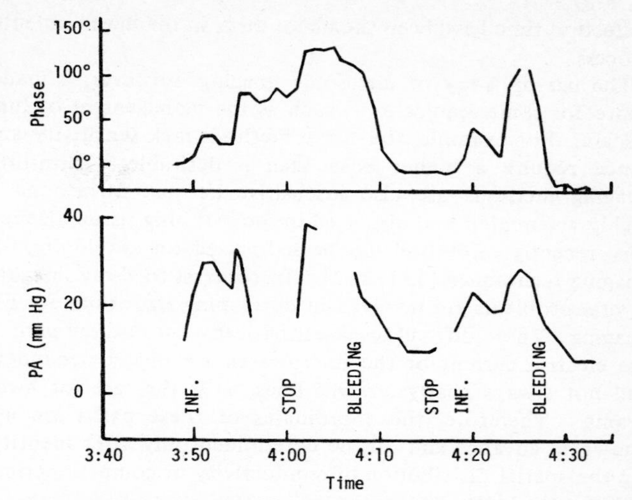

Fig. 3. Experimental results of a dog experiment illustrating the correlation between the phase of the microwave transmitted signal and the PA pressure, with the transfusion of blood and the subsequent bleeding of the dog. (After D. J. Shoff *et al.* [12].)

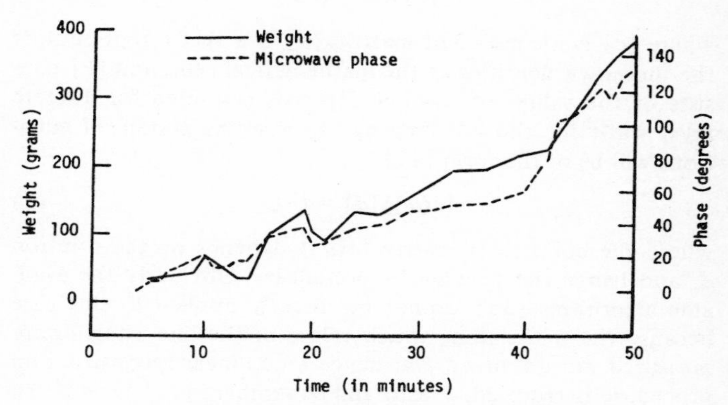

Fig. 4. Results from an isolated lung experiment.

The microwave method was initially used only to detect changes in the lung water content. Recently attempts have been made to calculate the absolute value of the water content of the lung [34]. The method simply involves obtaining a transverse cross-section image of a human thorax, using the computerized axial tomographic X-ray scanner with the microwave applicators in position. This information is used to construct a model of the cross section of the thorax, which is then used in solving for the internal electric fields numerically by the method of moments. The procedure is certainly valuable for quantitatively evaluating the effects of organ size, shape, and relative position on the microwave transmitted signal, but it is expensive because it requires a great deal of computer time [34]. Further studies using this model are certainly required.

IV. Electromagnetic Imaging

The ability to view a body section or layer clearly, without interference from other regions, has long been a goal of medical radiology. In recent years there has been increasing interest in developing X-ray and ultrasound imaging techniques. Improving the resolution of the image and reducing the data collection time have been the major tasks in the developmental process.

The use of X-ray or ultrasonic imaging, however, is inadequate for some applications, such as the management of lung disease. For example, the X-ray methods lack sensitivity and hence require a higher dose than is desirable. Ultrasonic imaging methods are also insensitive because ultrasound is highly attenuated and dispersed in the soft lung tissue. Hence, very recently, attention has been focused on developing EM imaging techniques [15], [17]. In contrast to X-ray imaging, several problems are involved in developing algorithms for EM imaging. These difficulties occur basically because the path of the electric current or the microwaves are object dependent and not always nearly straight lines, as is the case for X-ray beams. Therefore, the coordinates of these paths are not known in advance and can be determined only after identifying the spatial distribution of conductivity or complex permittivity of the body.

As a result, the X-ray imaging problem is usually reduced to finding a solution for a linear system of equations in the form

$$[c][d] = [P] \tag{1}$$

where $[c]$ is a coefficient matrix, $[d]$ is a vector representing the unknown densities at the mathematical cells, and $[P]$ consists of the values of the line integrals measured for discrete rays, while for the EM imaging the resulting system of equations will be of the form [17]

$$[c(d)][d] = [P] \tag{2}$$

where the coefficient matrix $[c(d)]$ depends on the solution d, and hence the problem is nonlinear. Obviously the available algorithms [35] cannot be directly applied in this case because the paths along which values of the line integrals are measured are unknown and hence a nonlinear reconstruction procedure is required. Tasto and Schomberg [17] proposed a simple extension of the well-known algebraic reconstruction technique (ART) [35] to take into account nonlinearities. For dc or low-frequency conductivity (σ) imaging, the proposed extension involves determining the electric potential V interior to the object by solving Laplace's equation $\nabla \cdot (\sigma \nabla V) = 0$ numerically. An initial conductivity profile is first assumed and the potential computation, subject to the given set of boundary conditions, is then performed using the finite difference method [15]. The current entering or leaving each electrode will then follow the gradient of the potential field. Once these stream lines are known, a line integral similar to the line integrals known from X-ray reconstruction can be determined and the reconstruction procedure continued in a manner similar to the X-ray case. If an iterative procedure is used to reconstruct the image, the calculation of the electric potential must be performed after each iteration and the current stream lines should be modified accordingly.

Several experiments with the nonlinear reconstruction technique were carried out using simulated data [15], [17]. The images obtained are shown in Fig. 5 where it is clear that they are generally satisfactory, although there is some blurring and local overshoot [15]. Experiments with measured rather than simulated data revealed that the resulting images of complex objects are not acceptable [17]. A common limitation of the conductivity imaging procedures is the preferred current path through the high-conductivity tissue. For example, it was found very difficult to image a low-conductivity region surrounded by a high-conductivity one because the current generally takes the path of least resistance around the low-conductivity region [15]. This problem is particularly important in imaging the human thorax because of the high-conductance current path through the chest wall. To improve the sensitivity of the measurement to high resistance tissue (e.g., the lungs), Swanson proposed an impedance camera that utilizes an array of guarded electrodes similar to those used in the impedance measurements as described in the second section. Recently, an isoadmittance contour map was obtained by applying the guarded-electrode array to the subject's back [16]. No attempts have been made to use reconstruction algorithms, and the map obtained, although promising, clearly indicated that a great deal of work must be done before the impedance camera can be of clinical use. Furthermore, more work should be done toward the development of a complex permittivity camera based on microwave measurements. This camera will have the advantage of providing both the real and

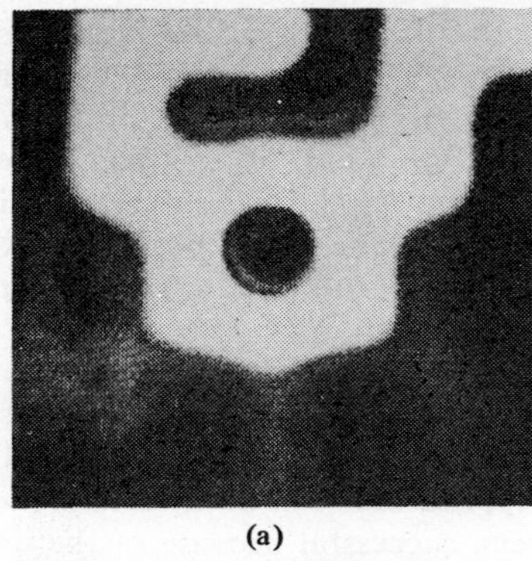

(a)

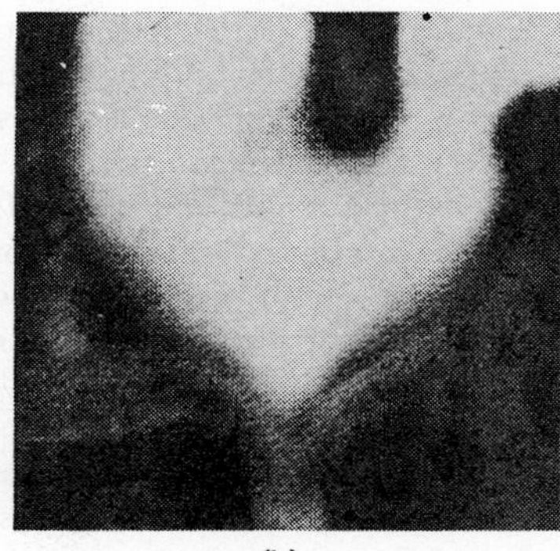

(b)

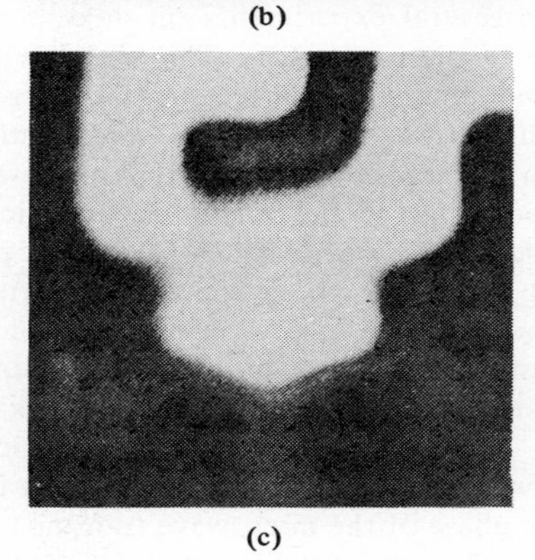

(c)

Fig. 5. (a) A conductivity profile with the white and the black colors representing conductivities of 0.015 S/m and 0.005 S/m, respectively. (b) The conductivity image obtained after one iteration. (c) The image obtained after ten iterations. (After R. J. Lytle *et al.* [15].)

imaginary parts of the complex permittivity, which is a step towards the unique identification of the images. On the basis of the microwave method of lung water measurement (see Section III), it is clear that much improved sensitivity, particularly to high resistance tissue such as the lungs, can be achieved by utilizing the transmission of microwave signals instead of low-frequency currents.

V. Microwave Radiometry and Its Potential Application in Medicine

Radiometry originates from the fact that all bodies above absolute zero temperature emit energy in the form of electromagnetic radiation. Curves of brightness of a perfectly absorbing body (blackbody radiator) as a function of frequency show that the radiation is spread over a very wide frequency band [36]. The frequency at which the maximum radiation occurs depends on the body temperature, and for temperatures of practical interest ($T < 1000$ K), the maximum radiation occurs in the infrared region. This is why this emitted part of the spectrum has been detected and used for more than twenty years to obtain thermographic images.

It is generally difficult to meet the ideal properties of a black body (perfect radiation and perfect absorption) and hence such a body does not exist. Because the radiative properties of objects vary according to their nature, shape, etc., the concept of gray bodies was introduced. The radiative properties of a gray body are different from those of a black body, although the physical laws governing the emission processes are the same in both cases. The energy radiated by an infinite gray body e_g at absolute temperature T(K) and frequency f(Hz) is related to the energy radiated from a black body, e_b, at the same temperature by

$$e_g = \epsilon e_b \text{ W}/(m^2 \text{ Hz}) \tag{3}$$

where the emissivity factor ϵ is between zero and unity and is, in general, a function of temperature, frequency, orientation, composition, and surface conditions of the body. In the frequency region for which $f \ll KT/h$ (microwave region), the brightness is given by the Rayleigh–Jeans expression

$$B = 2f^2 KT\epsilon/c^2 \tag{4}$$

From (4) it is clear that any gray body at an absolute temperature $T > 0$ K is characterized by a continuous RF spectrum of emission whose frequency distribution is proportional to the square of the frequency. This radiation depends on the emissivity ϵ and, therefore, by measuring changes in the emitted energy, variations in the dielectric properties, relative dimensions, and temperature can be detected. While microwave radiometry has been applied to a variety of problems, including radio astronomy, communications, and atmospheric sensing for many years, it has been only recently applied to medical diagnosis. Calculations based on the gray-body theory show that biological systems emit microwave radiation detectable by presently available microwave radiometric techniques [6]. The major advantage of using the emitted radiation at the microwave frequencies is simply the greater penetration depth compared to that obtained when using the infrared radiation. This permits measurements of microwave radiation emitted from structures within the body. Microwave radiometry, on the other hand, has coarser spatial resolution than infrared thermography. The tradeoff between the desired depth of penetration and the necessary resolution is the major consideration in determining the operating frequency range in a radiometer experiment. Therefore, depending on the applications, measurements have been made in the frequency band between 0.6 and 1.0 GHz by Enander *et al.*, to monitor changes in the temperature inside the body [7], at 1.3 GHz and 3.3 GHz by Barrett *et al.* [5], [37], and at 1.7 GHz by Porter [9], both to detect breast cancer, and at 45 GHz to obtain thermographic images of a human subject [8]. Bigu del Blanco *et al.*, have also used radiometric measurements at 9.2 GHz to explore possible means of communications in biological systems and for medical diagnostics [6].

The detection and measurement of thermal radiation at radio frequencies require different techniques from those commonly employed for the monitoring of CW signals of relatively high power levels. For example, the power level of the signal is usually of the order of 10^{-15} to 10^{-20} W, and therefore receivers of high sensitivity and high stability are required. In all the medical applications described earlier, the radiometer used is a broadband receiver of the Dicke type which has a temperature resolution of 0.1 K [5]. In this system the limitations in sensitivity due to receiver instability were reduced by continuously switching the receiver input between the antenna and a comparison noise source at a frequency high enough (30 times/s) so that the gain had no time to change during the one cycle. In most cases, however, the antenna used was a straight section of a rectangular waveguide filled with dielectric [5], [9], with the exception of Enander and Larson who used a small loop antenna designed to give a good match when placed on a human body [7], and Edrich who used an elliptical dish antenna [38].

The feasibility of using the microwave radiometry in detecting breast cancer tumors has been demonstrated by conducting several experiments on phantoms [9]. Results from clinical trials, however, were marginal. For example, based on the true-positive (TP) and true-negative (TN) rates of data analysis [5], the X-ray criterion was clearly found to be superior in both TP and TN to that of either microwave or infrared techniques. In numerical terms, the microwave, infrared, and X-ray TP rates were 0.73, 0.77, and 0.89, respectively, while the TN rates were 0.73, 0.68, and 0.92, respectively [5]. The conclusions are, therefore, tentative and should await the performance of more refined experiments and the development of adequate instrumentation. Of particular interest is improvement of resolution and simultaneous use of multifrequency observation with radiometers of high sensitivity [38]. The former is certainly important at relatively deep distances of 1.5 cm to 5.0 cm, where areas of resolution of 10–50 cm^2 were reported. It should be noted that these limitations on resolution not only degrade the scan quality, but also impose severe restrictions on the system sensitivity. Recent attempts to overcome these difficulties by combining microwave radiometry with infrared thermography, or artificially enhancing the image (heating pattern) by microwave or ultrasound heating have given marginally positive results [37], [39]. It should be emphasized that more effort is required to improve the response patterns of the antennas. Instead of the commonly used openended waveguides with a half-power beamwidth of approximately 50°, much more directive antennas are required. For example, a pencil-beam antenna would considerably improve the resolution and would therefore increase the usefulness of radiometry in diagnostics. Efforts to develop such a technique for clinical use are certainly commendable because microwave thermography is not only noninvasive, but is also passive, and hence completely safe.

VI. Blood-Flow Measurements

Cardiovascular surgery is usually aimed at either the improvement of the heart as a pump or the improvement of the arteries in conducting blood. During such operations, blood-flow measurements play an important role in evaluating the effects of different procedures. Several techniques are available for making *in vivo* blood-flow measurements, but none is in common use [3]. They are either too complicated or inaccurate, or they may involve a considerable risk to the patient. In this section, we describe two electromagnetic techniques for blood-flow measurements.

A. The Electromagnetic Flowmeter.

The general principle of an electromagnetic flowmeter was first introduced by Kolin [2] and since then has reached a high standard of performance and reliability, and can be easily and safely administered during surgery. The principle of operation is based on the familiar Faraday law of EM induction shown in Fig. 6. If blood is flowing in a tube or blood vessel oriented at right angles to the magnetic field, the voltage generated across a diameter perpendicular to the magnetic field and to the direction of flow is given by [3]

$$E = Blu \times 10^{-8} \tag{5}$$

where E is the potential difference in volts, B is the magnetic field strength in gauss, l is the length of the conductor in cm within the field, and u is the speed of the blood flow in centimeter per second. To avoid electrode polarization problems, it is desirable to use an alternating magnetic field as shown in Fig. 6. In this case, an extra term must be added to the right-hand side of (5) to account for the electrodes and their leads acting as a one-turn transformer [3]. From (5) it is clear that by measuring the potential difference across the diameter of the blood vessel, the mean velocity of blood flow can be calculated. The induced voltage may be picked up from electrodes in contact with the outer surface of the vessel wall without opening the artery. Studies performed on flow models have demonstrated a linear relation between the blood flow and the electrical signal as well as excellent reproducibility [3].

In spite of the wide use of the electromagnetic flowmeter in surgical applications, technical problems are sometimes encountered. These include ECG pickup, particularly when working with vessels very close to the heart, zero-point drift and instability, and some other minor artifacts due to vessel drying and varying vessel diameter. Zero-point instability can easily be recognized by clamping the vessel and readjusting the instrument to zero while the problem of ECG signal interference may be recognized, and hence accounted for. Efforts are presently being focused on utilizing the basic idea of the EM flowmeter to noninvasively monitor the blood flow in humans [40].

B. Monitoring of Arterial Wall Movement by Microwave Doppler Radar

Recently, Stuchly *et al.* [18] described a technique for monitoring the movements of arterial walls by using a low-power, X-band Doppler radar. The experimental procedure utilizes the measured doppler shift, which is the difference

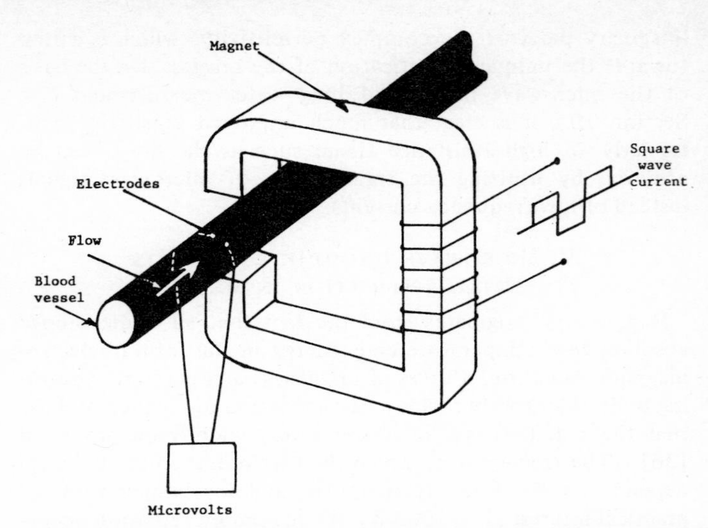

Fig. 6. A schematic diagram illustrating the basic principle of the electromagnetic flowmeter. (After C. Cappelen, Jr. *et al.* [3]).

between the phase of the transmitted and returned signals, to indicate the arterial displacement. The arterial wall velocity towards and away from the radar, on the other hand, is related to the measured doppler frequency, which is the difference between the transmitted and returned signals. To investigate the feasibility of the method, several experiments were performed on human subjects by using a doppler transceiver operating at a frequency of 10.525 GHz [18]. The experimental results clearly illustrated the feasibility of the technique. Many technical and medical questions, however, are still to be answered before this method can be of any clinical value. Special efforts are particularly required to measure absolute displacement of the arterial wall rather than relative, uncalibrated movements. A numerical technique similar to that developed by Iskander *et al.* [34], might be suitable to use in this calibration procedure since it takes into account the scattering properties of the artery and the attenuation between the antenna and the target.

VII. Conclusions

In this paper recent advances in the electromagnetic diagnostic techniques are described. From the preceding discussion, one might assume that these techniques represent a large and rapidly growing area of application. Unfortunately, this has been far from the case so far. Apart from the EM flowmeter, other potential techniques are still in research stages, with their clinical applications being restricted to specialized centers. For example, the sensitivity and the potential use of the electrical impedance method, although it was first used in 1926 to measure changes in lung water content, is still being evaluated by researchers.

Among many factors contributing to slow development in this area of research is the complexity and lack of understanding of the nature of the interaction between electromagnetic waves and the human body. For example, attempts to estimate the sensitivity and the accuracy of a given procedure have frequently been frustrated by the complexity of the body structure and limitations on computer storage and expenses when numerical calculations were performed. This forced workers in this area either to base their conclusions on simple and idealized models, with questionable adequacy and accuracy, or to conduct pilot experiments and ignore the ever necessary basic understanding. The situation is further complicated by the variability of the human body from person to person, not only in structure, but also in reaction to a given experimental procedure.

The interdisciplinary nature of the research in this area is certainly one of the major contributing factors to the past slow progress in this area. Additional difficulties in research are sometimes complicated by difficulties encountered in communication among members of interdisciplinary research teams. It is, however, our hope that the encouraging initial experiences with potential electromagnetic diagnostic techniques and the promising results obtained will attract talented engineers, scientists, and physicians to cooperate in developing this important area.

References

[1] R. K. Lambert and H. Gremels, "On the factors concerned in the production of pulmonary edema," *J. Physiol.* (London), vol. 61, pp. 98-112, 1926.

[2] A. Kolin, "Electromagnetic flowmeter: principle of method and its application to blood flow measurements," *Proc. Soc. Experiment. Biol. Med.*, vol. 35, pp. 53-56, 1936.

[3] C. Cappelen, Jr., and K. V. Hall, "Electromagnetic blood flowmetry in clinical surgery," *Acta. Chir. Scand.*, Suppl. 368, pp. 3-37, 1967.

[4] B. T. Williams, S. Sancho-Fornos, D. B. Clarke, L. D. Abrams, and W. G. Schenk, Jr., "Continuous long-term measurement of cardiac output after open-heart surgery," *Ann. Surg.*, vol. 174, pp. 357-363, 1971.

[5] A. E. Barrett, P. C. Myers, and N. L. Sadowsky, "Detection of breast cancer by microwave radiometry," *Radio Sci.*, vol. 12, no. 6S, pp. 167S-171S, 1977.

[6] J. Bigu del Blanco and C. Romero-Sierra, "Microwave radiometric techniques: A means to explore the possibility of communication in biological systems," in *6th Annual Meet., Neuro-Electric Soc. Symp. Effects of Low-Frequency Magentic and Electric Fields on Biological Communication Processes* (Snowmass-at-Aspen, CO, Feb. 1973) vol. 6; Llaurado, Sances, and Battocletti, Eds. Springfield, IL: C. C. Thomas, 1974.

[7] B. Enander and G. Larson, "Microwave radiometric measurements of the temperature inside the body," *Electron. Lett.*, vol. 10, p. 317, July 1974.

[8] J. Edrich and P. C. Hardee, "Thermography at millimeter wavelengths," *Proc. IEEE* (Lett.), vol. 62, pp. 1391-1392, 1974.

[9] R. A. Porter and H. H. Miller, "Microwave radiometric detection and location of breast cancer," in *Proc. IEEE Electro/78*, Session 30 (Boston, MA, May 23-25), 1978.

[10] C. Süsskind, "Possible use of microwaves in the management of lung disease," *Proc. IEEE* (Lett.), vol. 61, p. 673, 1973.

[11] P. C. Pedersen, C. C. Johnson, C. H. Durney, and D. G. Bragg, "Microwave reflection and transmission measurements for pulmonary diagnosis and monitoring," *IEEE Trans. Biomed. Eng.*, vol. BME-25, pp. 40-48, 1978.

[12] D. J. Shoff, M. F. Iskander, C. H. Durney, and D. G. Bragg, "Noninvasive microwave methods for measuring tissue volume in normal dogs after whole blood infusion," *Med. Res. Eng.*, 1979, in press.

[13] M. F. Iskander, C. H. Durney, D. J. Shoff, and D. G. Bragg, "Pulmonary diagnostics using noninvasive microwave methods," presented at the Int. Symp. Biological Effects of Electromagnetic Waves, Airlie, VA, Oct. 30-Nov. 4, 1977; also accepted for publication in *Radio Sci.*, Dec. 1979.

[14] C. H. Durney, M. F. Iskander, and D. G. Bragg, "Noninvasive microwave methods for measuring changes in lung water content," *Proc. IEEE Electro/78*, Session 30 (Boston, MA, May 23-25), 1978.

[15] R. J. Lytle and K. A. Dines, "An impedance camera: a system for determining the spatial variation of electrical conductivity," Lawrence Livermore Lab., Livermore, CA, UCRL-52413, 1978.

[16] R. P. Henderson, J. G. Webster, "An impedance camera for specific measurements of the thorax," *IEEE Trans. Biomed. Eng.*, vol. BME-25, pp. 250-254, 1978.

[17] M. Tasto and H. Schomberg, "Object reconstruction from projections and some nonlinear extensions," presented at NATO Advanced Study Institute on Pattern Recognition and Signal Processing, June 25-July 4, 1978, Paris, France; also to appear in NATO Advanced Study Institute series.

[18] S. S. Stuchly, M. Goldberg, A. Thansandote, and B. Carraro, "Monitoring of arterial wall movement by microwave Doppler radar," presented at the Symp. Electromagnetic Fields in Biological Systems, Ottawa, Ont., Canada, June 27-30, 1978.

[19] J. C. Lin, "Noninvasive microwave measurement of respiration," *Proc. IEEE* (Lett), vol. 63, p. 1530, 1975.

[20] J. H. Jacobi, L. E. Larson, and C. T. Hast, "Water-immersed microwave antennas and their application to microwave interrogation of biological tergets," *IEEE Trans. Microwave Theory Tech.*, vol. MIT-27, pp. 70-78, 1979.

[21] I. Yamaura, "Measurement of heart dynamics using microwaves (Fourth Rep.)—microwave stethoscope," *Inst. Electron. Commun. Eng. Japan*, vol. TG-EMCJ78-15, pp. 9-14, July 1978.

[22] N. C. Staub, "Pulmonary edema," *Physiol. Rev.*, vol. 54, pp. 678-811, 1974.

[23] R. D. Allison, E. L. Holmes, and J. Nyboer, "Volumetric dynamics of respiration as measured by electrical impedance plethysmography," *J. Appl. Physiol.*, vol. 19, pp. 166-173, 1964.

[24] M. Pomerantz, F. Delgado, and B. Eiseman, "Clinical evaluation of transthoracic electrical impedance as a guide to intrathoracic fluid volume," *Ann. Sur.*, vol. 171, p. 686, 1970.

[25] J. V. Van de Water, I. T. Miller, E. N. C. Milne, E. L. Hanson, G. F. Sheldon, and K. S. Kagey, "Impedance plethysmography: a noninvasive means of monitoring the thoracic surgery patient," *J. Thorac. Cardiov. Sur.*, vol. 60, pp. 641-647, 1970.

[26] M. Graham, "Guard ring use in physiological measurements," *IEEE Trans. Biomed. Electron.*, vol. 12, pp. 197-198, 1965.

[27] J. W. Severinghaus, C. Catron, and W. Noble, "A focusing electrode bridge for unilateral lung resistance," *J. Appl. Physiol.*, vol. 32, pp. 526-530, 1972.

[28] J. C. Denniston and L. E. Baker, "Measurement of pleural effusion by electrical impedance," *J. Appl. Physiol.*, vol. 38, pp. 851-857, 1975.

[29] L. H. Hamilton and W. T. Bruns, "Impedance-measuring ventilation monitor for infants," *Ann. Biomed. Eng.*, vol. 1, pp. 324-332, 1973.

[30] L. E. Baker, W. V. Judy, L. A. Geddes, F. M. Langley, and D. W. Hill, "The measurement of cardiac output by means of electrical impedance," *Cardio. Res. Cent. Bull.*, vol. 9, pp. 135-145, 1971.

[31] J. C. Abbey, "Impedance measurement of urinary bladder fullness," project at the University of Utah College of Nursing, funded by the Division of Nursing of the U.S. Department of Health, Education, and Welfare, 1978.

[32] I. Yamaura, "Measurements of 1.8-2.7 GHz microwave attenuation in human torso," *IEEE Trans. Microwave Theory Tech.*, vol. MTT-25, pp. 707-710, 1977.

[33] M. F. Iskander and C. H. Durney, "An electromagnetic energy coupler for medical applications," *Proc. IEEE* (Lett.), vol. 67, pp. 1463-1465, Oct. 1979.

[34] M. F. Iskander, C. H. Durney, D. G. Bragg, and B. H. Ovard, "A microwave method of estimating absolute value of average lung water content," presented at the Open Symposium on the Biological Effects of Electromagnetic Waves, Helsinki, Finland, August 1-8, 1978; also submitted to *Radio Sci.*

[35] R. A. Brooks and G. Dichiro, "Principles of computer assisted tomography (CAT) in radiographic and radioisotopic imaging," *Phys. Med. Biol.*, vol. 21, pp. 689-732, 1976.

[36] J. D. Krauss, *Radioastronomy*. New York: McGraw-Hill, 1966.

[37] A. H. Barrett and P. C. Myers, "Subcutaneous temperature: a method of noninvasive sensing," *Science*, vol. 190, pp. 669-671, 1975.

[38] J. Edrich *et al.*, "Imaging thermograms at centimeter and millimeter wavelengths," *Ann. NY Acad. Sci.*, 1979.

[39] P. C. Myers, N. L. Sadowsky, and A. H. Barrett, "Microwave thermography: Principles, methods, and clinical applications," presented at the 8th European Microwave Conference, Paris, France, September 8, 1978. Proceedings published in the *J. Microwave Power*, vol. 14, no. 2, 1979.

[40] H. Boccalon, B. Candelon, J. J. Tillie, A. Graulle, H. G. Doll, P. Puel, and A. Enjalbert, "New noninvasive device for pulsatile blood flow measurement," in *Dig. 11th Int. Conf. Med. Biol. Eng.* (Ottawa, Ont., Canada), pp. 428-429, August 1978.

Part VI
Safety Standards

The microwave frequency range has been generally defined as extending from a lower limit of 100-300 MHz to an upper limit of 100-300 GHz. The free-space wavelengths in this range correspond to macroscopic dimensions comparable to man (1-3 meters (m)) on one end and to the easily-visualized minimum dimensions (1-3 millimeters (mm)) on the other end. Historically, microwave safety standards have been developed with varying conformity to this definition, leading to some inadequacy of the standards at the extreme limits.

A more meaningful definition of "microwaves" can be derived from the observation that most dramatic "microwave" biological effects occur in the frequency range in which there is maximum penetration of internal field on the test animal. Such a resonance frequency band of considerable breadth is scaled to the size and shape of the animal and varies with polarization of the incident radiation. It is reasonable to include the resonance frequency band for man in the "microwave" definition, that is, the decade of 30-300 MHz. Thus, the frequency range associated with the ANSI C95 standard, 10 MHz—100 GHz, is in reasonable correspondence with this definition of "microwaves."

It is important to distinguish between *exposure* and *emission* standards. *Exposure* standards specify the maximum permissible values of some field parameter, measured in the absence of the subject to which personnel could be exposed. This field parameter may be that of electric field (V/m), magnetic field (A/m), power density (mW/cm^2), or energy density (mW-h/cm^2). Other alternatives include the squares of the electric and magnetic fields, that is, $|E|^2$ and $|H|^2$, as well as related energy densities, $\frac{1}{4}\epsilon_0 |E|^2$ or $\frac{1}{4}\mu_0 |H|^2$, expressed in joules/cm^3. Note that this energy density expresses the localized volume density of energy at any instant. This is to be distinguished from the earlier-cited energy density (mW-h/cm^2) which expresses the total energy flux through the unit area at some point in given or arbitrary time. Most existing exposure standards apply specifically to the far field, in which case equivalent values of any unspecified field parameter are derived from the well-known far-field equations for a radiating plane wave. The maximum field limit is specified as an average value with an averaging time of seconds to minutes. The safety factor intended in existing standards varies from 10 to over 100 in some cases.

It should be emphasized that, in general, exposure limits have a dependence on time or exposure duration. In general, exposure to higher fields is permitted for short periods of time; there is considerable diversity in handling the question of upper limits for very short periods. In addition, exposure standards will show dependence on frequency with some relaxation allowed for partial vs whole-body exposure.

Emission standards apply to devices or equipment, not to personnel, and specify a maximum value of some field parameter, usually power density (mW/cm^2), of microwave leakage at a specified distance from the external surface of the device. Given a knowledge of the radiation pattern of such leakage, one can compute potential exposure (field, exposure area, and duration) of a person with a given spatial and temporal variation of location with respect to the leakage source. The emission value is not to be confused with exposure, which is generally much lower than emission in the case of small leakage sources.

In recent years, a complicating factor in a logical picture of safety standards was recognized—that of potential hazardous interference to electronic equipment by microwave or RF radiation, particularly to medical devices like the cardiac pacemaker. Hazardous interference has occurred at power density levels as low as 5 orders of magnitude below levels hazardous to the human body. It has been generally recognized that the principal area of deficiency in such problems is the *susceptibility* of the pertinent electronic devices and that in most cases such susceptibility is suppressed by conventional techniques like filtering or shielding.

The standards evolving in this category of hazardous interference are either those specifying *minimum safe distances* from radiation sources (as in the case of electro-explosive devices) to be obeyed by device users, or *maximum susceptibility* (as that being developed for pacemakers) of devices. The former is the responsibility of device *users* while the latter is the responsibility of device manufacturers.

In recent years, there have been studies on a susceptibility standard for pacemakers by the Association for the Advancement of Medical Instrumentation (AAMI) and the Bureau of Medical Devices (Food and Drug Administration; United States Dept. Health, Education, and Welfare). Following the passage of the Medical Device Safety Act of 1976, the Bureau has contracted the development of a susceptibility standard for all medical devices. This standard was in its second draft in early 1977 with the expectation that it would be issued as a "guide" in the near future.

The published articles on safety standards reprinted in this chapter have been selected to authoritatively cover the history and present state of the most important microwave safety standards. The first reprint, by Mumford, presents a definitive history of exposure standards prior to 1961 and techniques for computing potential exposure-field levels in the radiation patterns of various transmitter antennas, particularly those of radar systems. The second reprint, by Michaelson, presents an up-to-date comparison of the principal exposure standards

Manuscript received July 15, 1981.

around the world. The next reprint, by Marha, presents one view of the development of microwave exposure standards in Eastern Europe.

The next two reprints are recent authoritative treatments of the bases for Western exposure standards and their comparison to those in Eastern Europe. The paper by Guy describes the evolution of the present ANSI C95 standard as well as those of Eastern Europe. He points out the now universal acceptance of frequency dependence based on the specific absorption rate (SAR) concept which relates internal fields to external fields and shows how this scales from animal to animal. Lastly, the paper by Tell and Harlen makes a connection between Western exposure standards for microwave/RF and those for other types of heat exposure. The last reprint is a general report on environmental levels in the United States which must be compatible with any environmental standard.

John M. Osepchuk
Editor

Bibliography

[1] ANSI C95.3–1973, "Techniques and instrumentation for the measurement of potentially hazardous electromagnetic radiation at microwave frequencies," IEEE, 1973.

[2] United States Depts. of Army and Air Force, "Control of hazards to health from microwave radiation," *Army Tech. Bull. TB Med. 270, AF Manual AFM 161-7*, Dec. 1965.

[3] United States Dept. of Navy, "Technical manual for RF radiation hazards," Naval Ships Systems Command, NAVSHIPS 0900-005-8000, July 1966.

[4] USSR Ministry of Hygiene, "Temporary safety regulations for personnel in the presence of microwave generators," publication no. 273-58, Nov. 26, 1958.

[5] Czechoslovakia Ministry of Hygiene, "Supplement to the information bulletin for the discipline of industrial hygiene and occupational diseases and for radiation hygiene," Prague, June 1968.

[6] United States Dept. of Labor, "National consensus standards and established federal standards," Occupational Safety and Health Administration, *Federal Register*, part II, p. 10522, May 29, 1971.

[7] United States Dept. of Health, Education, and Welfare—FDA/BRH, "Regulations for administration and enforcement of the Radiation Control for Health and Safety Act of 1968," HHS publication no. FDA 80-8035, July 1980.

[8] Canadian Ministry of Health and Public Welfare, "Radiation emitting device regulations," *Can. Gazette*, part II, vol. 106, no. 5, pp. 266–271, Feb. 15, 1972; also *Can. Gazette*, part I, pp. 2673–2676, July 13, 1974.

[9] H. P. Schwan, "Microwave radiation, biophysical considerations and standards criteria," *IEEE Trans. Bio-Med. Eng.*, vol. BME-19, pp. 304–312, July 1972.

[10] W. W. Mumford, "Heat stress due to RF radiation," *Proc. IEEE*, vol. 57, pp. 171–178, 1969.

[11] I. R. Petrov, Ed., *Influence of Microwave Radiation on the Organism of Men and Animals*, NASA Technical Translation TTF-208, Feb. 1972. Originally published by Meditsina Press, Leningrad, 200 pp., 1970.

[12] W. A. Palmisano and A. Peczenik, "Some considerations of microwave hazards exposure criteria," *Military Med.*, pp. 611–618, July 1966.

[13] R. A. Tell, "Reference data for RF emission hazard analysis," United States Environ. Prot. Agency, report ORP/51D 72-3, June 1972.

[14] R. A. Tell and J. C. Nelson, "Microwave hazard measurements near Varians aircraft radars," United States Environ. Prot. Agency, report RDDRA4 15 (4), pp. 161–226, April 1974.

[15] J. Damelin, "VHF-UHF radiation hazards and safety guidelines," United States Fed. Commun. Comm., Office of the Chief Engineer, report no. 7104, July 19, 1971.

[16] J. M. Osepchuk, R. A. Foerstner, and D. R. McConnell, "Computation of personnel exposure in microwave leakage fields and comparison with personnel exposure standards," *Digest 1973 Symp. on Microwave Power*, Edmonton, Alberta: IMPI, 1973.

[17] C. C. Johnson, "Research needed for establishing a radiofrequency electromagnetic radiation safety standard," *J. Microwave Power*, vol. 8, pp. 367–388, Nov. 1973.

[18] S. Caine, "Status report on activities of the American National Standards C95 Committee on Radiofrequency Radiation Hazards," *Digest 1972 IEEE Symp. on Electromag. Compat.*, pp. 117–124, 1972.

[19] J. M. Osepchuk, "Comparison of potential device interference and biological exposure hazards in microwave leakage fields," *1971 IEEE Int. EMC Symp. Record*, pp. 155–161, 1971.

[20] J. C. Mitchell, W. D. Hunt, W. H. Walter, III, and J. K. Miller, "Empirical studies of cardiac pacemaker interference," *Aerosp. Med.*, pp. 189–195, Feb. 1974.

[21] S. Baranski and P. Czerski, *Biological Effects of Microwaves*, Stroudsburg, PA: Dowden, Hutchinson, and Ross, Inc., 1977 (chapter 6 is on standards).

[22] Draft standard on "Susceptibility of medical devices," United States Dept. of Health, Education, and Welfare, FDA, Bureau of Medical Devices, 1977.

[23] *IMPI Performance Standard on Leakage from Industrial Microwave Systems*, IMPI publication IS-1, New York: IMPI, Aug. 1973.

[24] ANSI C95.4, "Safety guide for the prevention of radio-frequency hazards in the use of electric blasting caps," 1971.

[25] Environmental Health Directorate, Health Protection Branch, Safety-Code-6, "Recommended safe procedures for the installation and the use of radio frequency and microwave devices in the frequency range 10 MHz–300 GHz," 79-EHD-30, Canada: Ministry of National Health and Welfare, Feb. 1979.

[26] S. Gerber, *Standards for General Population Exposure to Microwave and Radiofrequency Emission*, City of Portland, OR, Bureau of Planning, Jan. 8, 1980.

[27] M. G. Shandala, "Biologic effects of electromagnetic waves (EMW) on the basis of standards setting in the USSR," *Proc. 1978 URSI Meeting on Biol. Effects of Electromagnetic Waves*, Helsinki, Finland, 1978.

[28] Hygienic Standard USSR (12.1.006-76): *Occupational Safety Standards System, Electromagnetic Fields of Radiofrequency, General Safety Requirements*, Moscow: Ministry of Health, Jan. 22, 1976.

[29] L. R. Solon, "A public health approach to microwave and radiofrequency radiation," *Bull. of Atomic Scientists*, pp. 51–55, Oct. 1979.

[30] "Criteria for a recommended standard," *Occupational Exposure to Hot Environments*, NIOSH (HSM 72-10269), 1972.

[31] N. A. Leidel, K. A. Busch, and W. E. Crouse, *Exposure Measurement Action Level and Occupational Environmental Variability*, NIOSH Technical Information Report, United States Dept. of Health, Education, and Welfare, publication (NIOSH) 76-131, Cincinnati, OH, Dec. 1975.

Some Technical Aspects of Microwave Radiation Hazards*

W. W. MUMFORD†, FELLOW, IRE

Summary—**Man's ability to generate microwave power has been increasing at the rate of about 15 db per decade. Experiments performed by subjecting animals to high power indicate that hazards to personnel could exist if appropriate safety measures are not adopted and observed.**

This paper reviews the history of the recognition of this potential hazard and the safety measures adopted by the Bell System and others to protect personnel.

Some typical and pertinent research work is discussed, and it is shown how these results have influenced the establishment of criteria for safe and potentially hazardous environments for human beings.

The currently adopted safety limits of the Bell System and others are reviewed in some detail, and a recommended method of calculating power densities is derived, pointing out the limitations of the approximations used.

Some of the commercially available power density meters are mentioned, and their method of operation is described. Their use in surveying a site is discussed, and the shielding effect of wire mesh fences is presented in a nomograph.

Definition of Microwave Radiation

The microwave region is considered to extend from the highest radio frequencies down to the ultra-high frequency band between 300 and 3000 Mc, but radiation hazards may exist at any radio frequency capable of being absorbed by the body. Referring to the electromagnetic spectrum chart of Fig. 1, which shows the wavelength from $(10)^7$ to $(10)^{-12}$ cm, corresponding to frequencies from 3000 cps to 3×10^{16} cps, the four main classifications of radiation are indicated at the top.

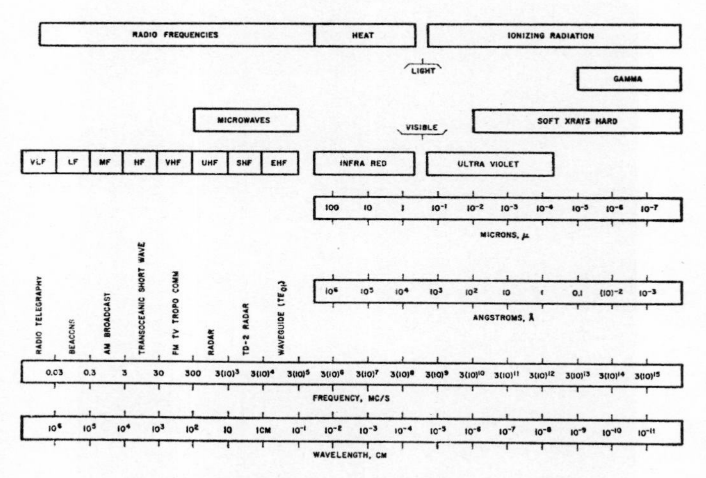

Fig. 1—The electromagnetic spectrum.

* Received by the IRE, May 5, 1960; revised manuscript received, November 25, 1960.

† Bell Telephone Labs., Whippany, N. J.

These are:

1) Radio frequency waves from 10 kc to $(10)^{12}$ cps,
2) Heat waves or infrared rays from $(10)^{12}$ cps to about $4(10)^{14}$ cps, the latter corresponding to a wavelength of about 0.72 $(10)^{-6}$ meters (0.72 micron),
3) The visible spectrum from a wavelength of 0.72 micron (7200 A) to about 3800 A,
4) Ionizing radiation (such as ultra-violet, X-ray and gamma radiation) at wavelengths less than about 3800 A.

The radio frequencies include eight frequency regions corresponding to the eight decades of wavelength they occupy. These eight bands have been called:

1) Very-low frequencies $(10)^7$ to $(10)^6$ cm
2) Low frequencies $(10)^6$ to $(10)^5$ cm
3) Medium frequencies $(10)^5$ to $(10)^4$ cm
4) High frequencies $(10)^4$ to $(10)^3$ cm
5) Very-high frequencies $(10)^3$ to $(10)^2$ cm
6) Ultra-high frequencies $(10)^2$ to (10) cm
7) Super-high frequencies 10 cm to 1 cm
8) Extra-high frequencies 1 cm to $(10)^{-1}$ cm.

Some of the uses of this portion of the electromagnetic spectrum include radio telegraphy, radio broadcasting, radio communication, radar, and millimeter wave waveguide transmission systems. Diathermy, radio frequency furnaces, and other industrial apparatus based upon the heating effect also use this portion of the spectrum. This heating effect can be hazardous.

Some Uses of Microwave Radiation

As indicated on the chart, the "microwave" band includes the UHF, SHF and EHF bands, and the chief uses of microwaves are radar, tropospheric scatter propagation, and relay links. Satellite communication links will use microwaves too, and these systems will probably be more powerful than the present radar and scatter propagation transmitters.

Some typical microwave antennas are shown in Figs. 2–5.

Fig. 2 shows U. S. Army field site with its associated acquisition and tracking radars. The high power acquisition radars are normally used only when the antenna is scanning, and, hence, the average power absorbed by an object at a fixed point is reduced by the fact that the direct beam is pointed in that direction only a fraction of the time it takes for the antenna to

Reprinted from *Proc. IRE*, vol. 49, pp. 427–447, Feb. 1961.

complete one whole revolution. Some of these radars lay down such a strong field that if they were not rotating, the power density might be hazardous to a distance of 500 feet or more. In such cases, interlocks may be used to ensure that the transmitter be idle if the antenna does not scan.

The tracking radars usually do not scan, but rather point toward the target or the missile wherever it may be. Hence, they are potentially more hazardous than the acquisition radars even though their power may be less. Safety considerations may dictate that interlocks be provided to prevent a target tracking radar from pointing in certain critical directions.

Fig. 3 shows a TD-2 relay link installation, which is such a familiar sight across the continent. In this system, the total radiated power is so low anywhere in the beam of the antenna that conditions are safe.

Fig. 4 shows the antennas of a White Alice installation. Here the antennas are fixed in space, and the power radiated is high enough so that the power density reaches a potentially hazardous value. The men in the foreground are posing with a portable field strength measuring device which was used to determine where safety fences should be erected. Tests were made at reduced power to be safe.

Fig. 2—A model of an Army field site.

Fig. 3—A TD-2 microwave relay link tower (courtesy of Mountain States Telephone and Telegraph Company).

Fig. 5 shows an antenna for a tropospheric scatter propagation communication link. Two of these links are now in operation in the Bell System, and more are being planned. The power of this transmitter is high enough to warrant a field strength survey in front of and around the antenna, and here again two men are shown with the simple laboratory model of a field intensity meter. In this system calculations indicated that a protective fence was needed. After the fence was erected, measurements (at half power) indicated that the power density outside the fence would be less than 1/10 mw/cm^2 at full power, a value which is 1/10 that which is considered safe for indefinite exposure.

Fig. 4—Antennas for White Alice.

Fig. 5—Tropospheric scatter antennas

Trend of Power Available

Some of the modern equipments are potentially hazardous, but with the trend toward higher and higher powers, our awareness now of the hazards of the future becomes increasingly important. Let us see where we stand today regarding the amounts of microwave power which man has been capable of producing at 3000 Mc, for example. Fig. 6 shows this trend over the past couple of decades, and the dashed line extending into the future represents one estimate. In 1940, the famous British magnetron [14] was capable of delivering 10 kw of peak-pulsed power into a properly matched antenna. This would be an average power of 10 watts if the duty cycle were one in a thousand.[1] By about 1945 [14], the inprovements made on this type of magnetron resulted in the capability of delivering 1100 kw of peak-pulsed power, or, with the same duty cycle, 1.1 kw of average power. After World War II, there was, naturally, a slowdown in the advance of this art, but by 1957 [43] a klystron was available which was capable of delivering 8.0 kw of average power. At the present time, it appears that the development of higher powered devices has again been stimulated so that by 1965 [58] at least 1000 kw of average power has been predicted. This trend from 10 watts in 1940 to 1000 kw in 1965 represents a rising capability of 50 db in less than 30 years, or a rate of over 15 db per decade.

Thus, it appears appropriate to examine the situation regarding microwave radiation hazards today, lest we be faced with an intolerable situation in the future.

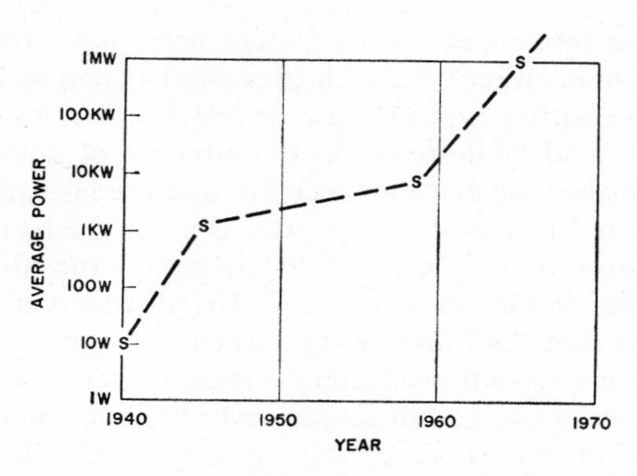

Fig. 6—Trend of power available at 3000 Mc.

Effects of Microwave Radiation

Most of the experimental work to date supports the belief that the chief effect of microwave energy on living tissue is to produce heating. Consequently, exposure to microwave radiation should probably represent no hazard unless overheating is a possibility. Within carefully prescribed limits, the heating effect of radio waves may actually be beneficial; in fact, this is the basis of diathermy, which has long been employed therapeutically. The use of diathermy is so widespread and so well accepted a part of medical procedure that the FCC has assigned seven frequencies, three in the HF or VHF[2] and four in the microwave region,[3] for the operation of medical diathermy equipment.

Heating is a function of the strength of the microwave field, that is, the average power flow per unit area (usually expressed in milliwatts per square centimeter). It is also a function of time. The heating may take place near the surface or deep within the body, the depth of penetration being related to frequency. Frequencies in the region 200 to 900 Mc penetrate deeply, whereas *S*-band (1500–5200 Mc) and *X*-band (5200–11,000 Mc) frequencies used by radars produce heating at or near the surface.

Heating effects, depending on frequency, are 1) a general rise in body temperature, similar to fever, or 2) something more localized, akin to the cooking process in a radar oven, where a steak can be cooked from the inside out. The human body can compensate for a certain amount of heating of the first type through perspiration, if the temperature rise is not too sudden. Consequently, the hazard may be somewhat less in cool weather than on an extremely hot day when the body's cooling mechanism is already working at full capacity. Compensating mechanisms for coping with the second type of heating are less adequate.

Circulating blood acts as a coolant, so that localized heating is least serious in parts such as muscle tissue, which are well equipped with blood vessels. Heating is more of a danger to the brain, the testes and the hollow viscera. The most widespread publicity has related to the effect of microwave radiation on the eyes. The viscous material within the eyeball is affected by heat in much the same manner as the white of an egg, which is transparent at room temperature but becomes opaque white when warmed slightly. In the eyes, as in the egg white, the process is irreversible.

As the surface of the human body is more generously supplied with sensory nerves than the interior, a feeling of warmth may give a warning in case of over-exposure of frequencies which produce surface heating. If the frequencies are such as to cause a general rise in body temperature, the resulting sensation of discomfort may or may not be perceived in time to provide adequate warning. In the case of localized microwave heating deep within the body, it is still less likely that any warning sensations would be noted before damage was done. Hence, it is important to establish limits and to delineate the areas in which a potential health hazard could exist.

[1] Duty cycle is the fraction of the time that the radar is transmitting.

[2] 13.56 Mc
27.12
40.68

[3] 915 Mc
2,450
5,850
18,000

Review of Meetings, Data and Recommendations

Let us review some of the typical, significant and pertinent written publications and conferences on biological effects of microwaves and see how these data and their interpretation have influenced our thinking regarding the adoption of exposure limits and safety regulations. This examination had best be done chronologically in order to interpret the recommendations in their proper perspective.

Much of the material which is about to be presented is published in writing in the proceedings of conferences which have been held recently (*i.e.*, since 1955). Three well documented conferences include:

1) "Symposium on Physiologic and Pathologic Effects of Microwaves," held at the Mayo Foundation House, Rochester, Minn., September 23 and 24, 1955. Holding this meeting at the Mayo Clinic was suggested by Dr. F. G. Hirsch, who was then medical director of Sandia Corporation, Sandia Base, Albuquerque, N. M. It was a particularly appropriate meeting place because microwave diathermy was introduced to the medical profession by the Mayo Clinic. The papers presented at this symposium were subsequently published in the IRE Transactions on Medical Electronics (vol. ME-4; February, 1956). The attendance at this meeting represented the medical profession, governmental agencies, institutions of higher learning, and industry. The data presented yielded sufficient evidence for establishing tolerance levels for experimental animals as a function of exposure time; however, the extrapolation to human beings was still subject to considerable uncertainty. The consensus at that time and the seemingly proper conservative approach was to assume that human beings were just as susceptible as other animals.
2) "The First Annual Tri-Service Conference on Biological Hazards of Microwave Radiation" held at Griffiss Air Force Base, Rome, N. Y., on July 15–16, 1957. This meeting was sponsored by the Air Research and Development Command, Headquarters. It was held "to assemble the A.R.D.C. advisory panel, tri-service representatives and various contractors working in the radiation hazard area to effect an understanding of activities and accomplishments to date." Several interested industrial representatives were also invited. The work to date was described and the program for the future discussed [46], [47]. It was planned to hold periodic meetings to maintain close liaison among the investigators in the field, to expand research programs to include a sampling of frequencies from 200 through 35,000 Mc, to investigate the effects of microwaves combined with X rays, to study methods of microwave field measurement, and to study testicular and ocular damage as a function of frequency and cumulative dosages. The very capable and energetic host was Colonel G. M. Knauf, who presided at the sessions and contributed to the success of the meeting by delivering two papers in addition to his appropriate commentaries and introductions. The Proceedings of this conference were compiled and edited by Dr. E. G. Pattishall, University of Virginia, under project No. 57-13, contract AF 18(600)-1180 with the George Washington University, Washington, D. C.
3) "The Second Annual Tri-Service Conference on Biological Effects of Microwave Energy," held at Griffiss Air Force Base, Rome, N. Y., on July 8, 9, and 10, 1958, sponsored by the Air Research and Development Command, Headquarters, to "bring together key researchers in the bioeffects area so that each department within the Armed Services could discuss on-going and needed research." This meeting was patterned after the first Tri-Service meeting and the Proceedings were compiled and edited by Pattishall and Banghart, Project Directors, University of Virginia, Charlottesville, Va., under contract AF 18(600)-1792, Division of Educational Research, University of Virginia. This document is identified as ARDC-TR-58-54 and ASTIA Document No. AD 131477. Colonel Knauf was again the genial host and chairman in addition to presenting three significant papers.

The three documents which record the presentations at these three meetings contain several comprehensive bibliographies on the subject [41], [42]. Only a few selected references will be quoted here. Each reference was chosen either because it presented a good summary, or represented typical data, or because it appeared to mark a real "milestone" in the advance of knowledge.

Probably the first demonstration of the heating effect of radio frequency energy was that of d'Arsonval in 1890, according to Krusen [30], who gave the address of welcome at the Mayo meeting. He (d'Arsonval) "demstrated that 'high-frequency' electric currents of 10,000 cycles per second produced no muscular contraction in the human being, but causes only heating. Since then physicians have been employing increasingly higher frequencies for heating of living tissues. By 1900, high-frequency currents of 1,000,000 to 3,000,000 cycles per second (long wave diathermy), were in use. . . ."

Krusen continued, " . . . and by 1935, electric currents of still higher frequency, 10,000,000 cycles per second at a wavelength of 30 meters (short-wave diathermy), were being employed." One type of hazard of these machines arose from the fact that the original models were inadequately shielded and improperly operated on communication channels, causing disruption of our overseas short wave circuits.

Brody [28], Medical Liaison Officer, Bureau of Aeronautics, Navy Department, Washington, D. C., speaking at the Mayo meeting, summarized the work conducted between 1935 and 1950 thus: "One heard tales concerning the sterilizing effects of radar beams or other equally serious manifestations. . . . To determine whether or not these suppositions might be based on factual grounds, a clinical study was conducted in 1943 by Daily [6] and in 1945, by Lidman and Cohn [7]. Certainly (in view of our present knowledge), the lack of clinical changes reported resulting from exposures to radar emanations may be attributable to the low power outputs then available. Meanwhile, basic research utilizing animals continued, notably by Clark, Hines, Salisbury and Randall [12], [15]. . . . These studies amply demonstrated the insidious effects of high-power microwaves on certain tissues and pointed out the most significant changes with the use of discrete frequencies."

In 1949, Clark *et al.* [12] at Collins Radio Company, supported by RAND Corporation and the U. S. Air Force developed cataracts in the eyes of rabbits upon exposing them to 3000 mw/cm² for ten minutes at a frequency of 2500 Mc. This corroborated the earlier work of Richardson, Duane and Hines [10] who, in 1947, reported lenticular opacities appeared at about 50°C with intentional overdoses. There still remained, however, the task of gathering enough data to establish a reasonable limit. Meanwhile, in 1948, Imig, Thomson and Hines [11] reported that "testicular degeneration" may occur from microwave heating at a lower temperature than from infrared heating.

Olendorf [13] reported, in 1949, that he had used 2400 Mc radiation to produce lesions in the brains of rabbits.

Thus it was established more than ten years ago that microwaves could be hazardous.

In 1952, Hirsch [19], [23] of Sandia Corporation reported with Parker that lenticular opacities occurred in the eyes of a technician operating a microwave generator. Their report [23] states ". . . In the fall of 1950 he [the technician] set up a microwave test bench and used it from November, 1950, to October, 1951. His equipment included an experimental microwave generator tunable from about 9 to 18 cm wavelength (1660 to 3320 megacycles per second) and having an average power output of 100 watts on a 50 per cent duty cycle. His test line was terminated by a horn antenna which dissipated the power into a room. . . . It has been calculated that the intensity of radiation at a plane coinciding with the rim of the dissipating antenna was about 100 milliwatts per square centimeter.

"A habitual practice of this operator is worthy of mention, since it may well have some bearing on the case. In order to determine whether or not the equipment was generating energy, he made a regular practice of placing his hand in the dissipating antenna . . . and noting the heating effect on his hand. In these circumstances it was necessary for him to look into the antenna in order to place his hand properly. . . .

"On October 11, 1951 he presented himself to one of us [Hirsch] with the chief complaint of an inability to see clearly. . . . This visual disturbance had developed, he said, over a period of about a week or 10 days. He was not aware of any loss of visual acuity prior to that time."

Hirsch continued, as a result of examination, "A diagnosis of bilateral nuclear cataracts with acute chorioretinitis was made" and commented that, "it will be well, therefore, to use this case as a means of recalling the attention of ophthalmologists, industrial physicians, and microwave operators to the potentialities of microwave radiations in order that the use of this form of energy will be accompanied by appropriate respect and precautions."

Hirsch's well documented report attracted widespread attention both in industry and in the Armed Services. On the basis of Hirsch's observation that an estimated 100 mw/cm² was hazardous, organizations began adopting rules and regulations concerning microwave hazards. Some rules specified a hazardous level and others specified a safe level. Naturally, because of the paucity of adequate quantitative data, there was a large spread between the specified hazardous levels and the specified safe levels at that time.

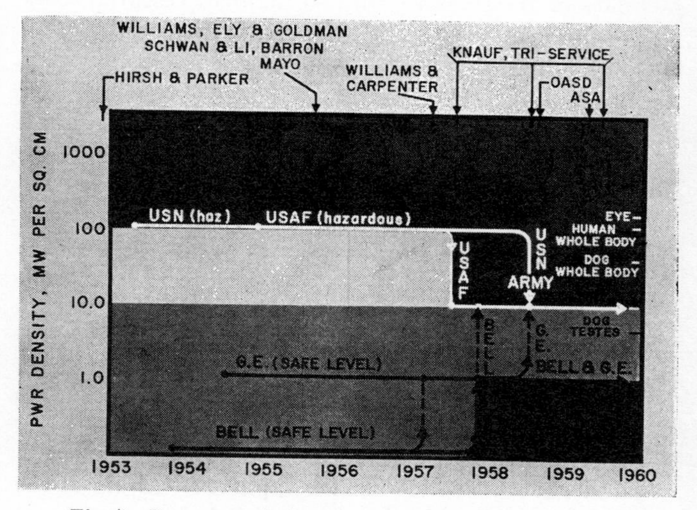

Fig. 7—Recommendations for safe and hazardous levels.

Fig. 7 shows graphically how the various recommendations have progressed from 1953 to date. Power density in mw/cm² is plotted logarithmically on the ordinate, with time in years linearly on the abscissa. At the extreme right the power density levels are indicated at which it has been reported that various effects occur. For example, reports [33], [36], indicated that 155 mw/cm² was the threshold for cataract formation by single exposures to the eye; 100 mw/cm² was estimated to be tolerable under favorable conditions for total immersion of the human body (except for the testes), 40 mw/cm² was lethal to dogs [29], [36], and 5 mw/cm² was considered to be the maximum exposure for no observable change in the testes of dogs [36]. These data will be discussed in more detail later.

With these reported critical levels in mind, let us turn our attention then to the safe and hazardous levels adopted by various organizations through the years following Hirsch's report, and see how the conferences and reports affected these recommendations.

Brody [28] reported at the Mayo meeting that "A density level based on the figure 100 mw/cm² was agreed upon as a damage risk criterion by those attending a symposium at the Naval Medical Research Institute in April, 1953."

The Central Safety Committee of the Bell Telephone Laboratories reviewed the then published material, and on November 16, 1953, one department of the Laboratories issued a bulletin based upon the Safety Committee considerations. Excerpts from this bulletin are quoted. In view of the fact that "about all that has been shown at present is that, at frequencies used in radar and microwave communication, wave fronts impinging on the eye with intensities corresponding to an energy flow of 100 milliwatts per square centimeter may cause injury . . . [therefore] it seems likely that 0.1 milliwatt would be safe." At that time it was believed that a safety factor of 30 db was incorporated in this recommendation. We shall see that subsequent research work reported on dog testes would indicate a safety factor of considerably less than 30 db.

The General Electric Company, Schenectady, N. Y., too, was early in establishing a safe level. Vosburgh [32], consultant, G.E. Health Services, submitted at the Mayo meeting the recorded minutes of a meeting held on June 1, 1954, at Schenectady dealing with "Microwave Radiation; Hazards and Safety Measures" in which recommended safety measures were established. These recommended measures included, among other excellent suggestions, "Appropriate procedures shall be applied to limit the direct and *reflected* intensity to 1 milliwatt per square cm average in all locations to which people require access: a) by the use of shielding or absorbing material, or b) by remote viewing."

The U. S. Air Force was also apparently aware of the potential hazards of microwaves, at least as early as 1954. Major D. B. Williams and Colonel R. S. Fixott [49], in presenting a summary of the SAMUSAF program for research on biomedical aspects of microwave radiation at the first Tri-Service Conference in 1957, stated " . . . only three years ago [1954] the sole data available on microwave tolerance was that an estimated 3000 mw/cm² should be regarded as hazardous for personnel exposure." Apparently, the Hirsch report (1952) had been overlooked.

By November 1, 1954 the U. S. Air Force [24] recognized that " . . . continuous exposure to microwave radiation intensities of 100 milliwatts/cm² can cause damage to human tissues, particularly the eyes." The Hirsch report was apparently accepted.

Colonel Knauf of the U. S. Air Force, [40], [52] speaking at the 106th annual meeting of the American Medical Association in New York, N. Y., on June 7, 1957, said, "About two years ago [1955] . . . a power density of approximately 200 milliwatts per square centimeter seemed to be the point at which a division between effect and no effect could be predicted." Apparently the Hirsch report was being discounted. He continued, "Hence, we published a maximum safe exposure level of 10 milliwatts per square centimeter."

In the meantime, the Mayo meeting had been held September 23 and 24, 1955. It is evident that the excellent papers presented at this meeting had quite an impact on the thinking concerning safe and hazardous levels of microwave radiation. The solid work of Williams *et al.* [33] of the U. S. Air Force in collaboration with Monahan of St. Johns Hospital, London, Eng., on time and power thresholds for the production of opacities was a big contribution. Schwan and Li [31] concluded that 10 mw/cm² should not be applied for more than one hour if total body absorption of radiation is assumed.

While many of the Mayo papers dealt with the production of opacities at levels in the neighborhood of 100 mw and up, one notable contribution, that of Ely and Goldman *et al.* [29] of the Naval Medical Research Institute, National Medical Center, Bethesda, Md., reported *death of animals at less than 50* mw/cm² at a wavelength of 10 cm. In this work they had established that only about 40 per cent of the incident energy had been absorbed. If all the energy had been absorbed, the lethal dose would have been about 20 mw/cm². The estimate of 40 per cent power absorbed seems a reasonable one in light of the work reported at Mayo by Schwan and Li [31] of the Electromedical Group, Moore School of Electrical Engineering and Department of Physical Medicine and Rehabilitation, Schools of Medicine, University of Pennsylvania, Philadelphia, Pa. They had studied the absorption of microwave radiation by the human body as a function of frequency for various skin thicknesses, taking into account the propagation constants of skin, fat and muscle. At frequencies from 150 Mc to 400 Mc, the body could absorb from 30 per cent to 50 per cent of the radiation. Above 900 Mc, they reported that the percentage absorbed fluctuations between 20 and 100 per cent depending on frequency, and thickness of skin and subcutaneous fat.

Also at the Mayo meeting, Barron [26], [27] of Lockheed Aircraft Corporation, Burbank, Calif., reported no serious changes in personnel exposed to fields as high as 13 mw/cm². They had initiated a "comprehensive physical examination program of radar personnel in an effort to determine whether prolonged (years) and short (months) duration exposure to microwave emanations had resulted in transient or permanent biologic damage." Included in the program were 226 personnel with histories of radar contact varying from occasional beam exposure to 4 hours a day and up to 13 years overall. Areas near the transmitter had been zoned, and all personnel were prohibited from entering zone A, where the density exceeded 13 mw/cm², and 88 control subjects were prohibited from entering zones where the density exceeded 1.6 mw/cm^2. Barron reported, "the original examination showed a significant decrease in the polymorphonuclear cells in 25 per cent of the radar personnel as compared with 12 per cent in the control group. An interesting and disproportionate increase in monocytes above 6 per cent and eosinophiles over 4 per cent was also noted in the radar group." (These were not considered to be serious.) "There was no indication of increased significant pathology among the radar group. Ocular examinations revealed a high incidence of pathology for the radar group. However, in all but a single case of retinal hemorrhage, the etiology was known and determined to be entirely unrelated to radar exposure. . . .

"Re-examination was accomplished on 175 subjects following 6 to 9 months of incidental contact with both 3-cm and 10-cm pulsed radar (power densities less than 13 milliwatts per square centimeter). . . . Preliminary observations revealed a decrease of red blood cells in excess of 10 per cent from the original in 30 per cent of subjects; an increase of white blood cells in 50 per cent; and an increase in lymphocytes in 39 per cent. The blood platelets showed no significant unidirectional change." The observed changes were not considered to be serious. *No* changes in the eyes were observed.

After the Mayo meeting (1955), it was apparent that some estimates of hazardous levels were too high and that the original 1953 Bell Laboratories estimate of safe level (0.1 mw/cm^2) was too conservative.

In an urgent action technical order [40] dated June 17, 1957, and also a Rome Air Development Center Regulation [39] dated May 31, 1957, there was established a "hazardous microwave radiation level of 10 milliwatts per square centimeter or greater over the entire microwave spectrum" [50].

Within the Bell Telephone Laboratories, several engineers initiated memoranda, sponsored conferences, and wrote letters to have the safe level limit revised upward. These efforts, in liaison with the engineers at the American Telephone and Telegraph Company, coordinated by Smith, led to the adoption of a potentially hazardous level of 10 mw/cm^2 on October 24, 1957.

The Navy, the Army, and General Electric Company also reported at the second Tri-Service meeting in July, 1958, that 10 mw/cm^2 was being considered the limit. Roman, consultant, Bureau of Ships, reported [65], "The Bureau of Medicine has tentatively established a working level of ten milliwatts per square centimeter as the tolerable dosage for constant exposure to microwave radiation."

Lieutenant Colonel L. C. MacMurray [63], U. S. Army Environmental Health Laboratory, reported, "The U. S. Army has adopted the tentative criteria of 10 milliwatts per square centimeter as being the upper limit of safe exposure to microwave radiation." U. S. Army Regulation No. 40-583 [74] states, "Personnel will not be permitted to work in the field of radiation . . . where the . . . power density exceeds 10 milliwatts per square centimeter."

Vosburgh reported [68] that ". . . the health and hygiene service of the G.E. Co. does recommend that the G.E. Co. strike an agreement with the Armed Services on this value . . . and . . . conclude (concur) with the net recommended ceiling tolerance level of 10 milliwatts per square centimeter."

Thus, there appears to be general agreement between all three Armed Services and at least two industrial organizations on the upper limit of 10 mw/cm^2.

A review of the factors which influenced the establishment of exposure limits would not be complete without mentioning the "California Incident" [37], which received widespread publicity in May, 1957, just prior to the first Tri-Service meeting in Rome. This was the report in which a worker was allegedly killed by radar. Fricker commented at the first Tri-Service meeting "Further investigation revealed that this person died of other causes."[4] Nevertheless, the widespread publicity of this incident did add stimulus to the research work on radiation hazards of microwaves.

The Tri-Service meetings have been helpful indeed in bringing us up to date with the recent advances. In 1957, the reports of Knauf [46], [47], the commentaries of Fricker [45], the analysis of Schwan [48], and the practical engineering report of Dondero [44] were especially worthwhile. At the second Tri-Service meeting in July, 1958, the contributions of Knauf [62], Schwan [66], [67], Hartman [59], Roman [65], MacMurray [63], Herrick [60], Carpenter [57], Michaelson *et al.* [64], Keplinger [61], Susskind and Jacobson [55] are notable. We shall refer to some of these later. The third annual Tri-Service Conference was held at the University of California, Berkeley, August 25, 26 and 27, 1959, and the Proceedings are available [76]. It should be noted that the data presented at this meeting substantiated the previous choice of 10 mw/cm^2 for the upper limit.

Other meetings have been organized to sponsor the dissemination of information. On August 15, 1958, the office of the Assistant Secretary of Defense, Research and Engineering, held a classified meeting at the Pentagon to review electromagnetic radiation effects on materiel and personnel. In attendance were representatives of the three Armed Services, the OASD, a Bureau of Ships contractor, and one man from Bell Telephone Laboratories.

This meeting paved the way for the organization of a "General Conference on Standardization in the Field of Radio-Frequency Radiation Hazards," held May 4, 1959, at the Headquarters of the American Standards Association in New York, N. Y. At this meeting, it was voted that the American Standards Association be requested to establish a project in regard to hazards aris-

[4] See [45], p. 85.

ing from radio-frequency electromagnetic radiation and that the AIEE and the Department of the Navy, Bureau of Ships, be recommended as cosponsors for the proposed project. The future recommendations of this committee will, no doubt, be very helpful to all.

Detailed Examination of Typical, Significant and Pertinent Data

Let us now look at some of the typical, significant and pertinent data in more detail. Clark [12], [15], Richardson [10], and Hirsch [19], [23] had demonstrated cataract formation prior to the work reported by Williams [33] at the Mayo Clinic, and Williams showed how his more recent results tied in satisfactorily with the results of the previous workers. Not only that, but he presented the data in such a way that one could interpret them in terms of a threshold level vs time of exposure for a 50–50 chance of cataract formation in the eyes of rabbits by single exposures of 2400 Mc radiation. Table I summarizes these data, which are plotted on Fig. 8.

TABLE I
Power-Time Threshold for Production of Opacities in Rabbit Eyes (Single Exposure)

Exposure Time	Threshold Power Density
5 min.	0.6 w/cm²
20 min.	0.4 w/cm²
90 min.	0.3 w/cm²
270 min.	<0.22 w/cm²
270 min.	>0.12 w/cm²

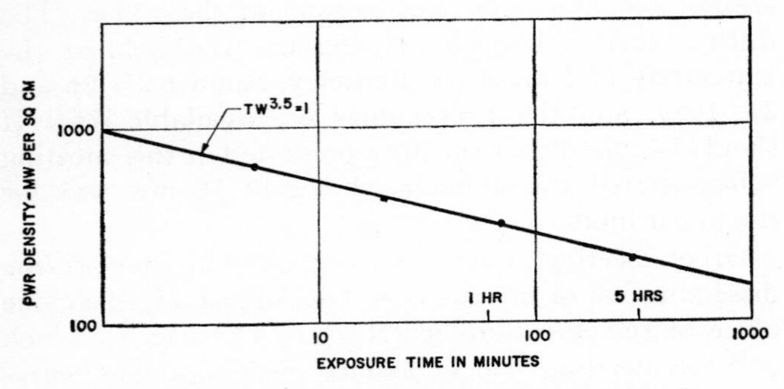

Fig. 8—Threshold for the formation of cataracts in the eyes of rabbits [33].

These observations indicated that the time of exposure for a 50–50 chance of cataract formation varied inversely as the seven-halves power of the power density. Extrapolation of this relationship to an exposure time of one minute would predict an even chance for the formation of cataracts at a power density level of about 1000 mw/cm². Extrapolation to an exposure time of one working day would give about 170 mw/cm².

Conclusions based upon extrapolation both as illustrated above and from animals to man could prove to be misleading. However, until further evidence became available, the more conservative use of the available data seemed to be warranted, and the threshold for producing cataracts in animals has been considered dangerous for human eyes.

Williams also pointed out that the multiple exposures reported by Daily [21] produced opacities at density levels well below the threshold requirements for single exposures. Specifically, ten exposures of 30 minutes each at a level of 150 mw/cm² produced opacities in two of six rabbits. This indicated a possibility that opacities might be produced as a cumulative effect. This cumulative effect was substantiated later by Carpenter [57] at Tufts University, Medford, Mass., and reported at the second Tri-Service meeting in 1958. This is of considerable importance with respect to its connotations in human exposure to low-power levels of radiation over long periods of time.

Since much of the published work dealt with the formation of cataracts, a false impression has been created that the eye is the organ most susceptible to damage by localized exposure to microwave radiation. That the testes were more susceptible than the eye was reported by Ely, Goldman, Hearon, Williams, and Carpenter [36] of the Naval Medical Research Institute in 1957. This may not have been of as much concern since the scrotum is sensitive to the sensation of heat, whereas the eye is not, and the subject would have the warmth as a warning in the scrotum but not in the eye. However, in the exposure of the eye to a fairly uniform wavefront, the sensation of heat in the skin near the eye should give a similar warning. This was observed by Daily *et al.* [21] who reported that "human [eye] exposures of 240 milliwatts per square centimeter were discontinued because of pain or discomfort, yet no other ill effects were observed."

The pioneering work of Ely and Goldman [29], [36] on effects of total immersion of animals in microwave fields had shown that, with a maximum available power density of no more than 100 mw/cm² fatal fevers were induced in rats, rabbits and tranquilized dogs. No animal survived in a treatment which raised the body temperature to 44°C (111.2°F). Continued exposure to 25 mw/cm² maintained a body temperature rise of 1°C, and 50 mw/cm² was lethal to rabbits and dogs.

Subsequently Michaelson, Dondero and Howland [64] of University of Rochester, Rochester, N. Y., who *avoided the use of premedication with tranquilizers*, reported their dogs gave no response to 45 mw/cm² and to 100 mw/cm², but a marked response at 165 mw/cm². A comparison of these data with those of Ely seems to indicate a significant difference. One might think that the use of medication by Ely could be an assignable cause. However, Susskind *et al.* at the University of California [55], present data to show that tranquilized mice withstood more than untreated ones. It was their opinion that the controlling factor was a critical effect of reducing the initial body temperature; considerably more energy was required for the tranquilized mice to reach the critical body temperature. Furthermore,

Keplinger [61], [75] at University of Miami, Coral Gables, Fla., working at 24,000 Mc on unmedicated rats, determined that a level of 28 mw/cm^2 for 139 minutes was lethal and that 24 mw/cm^2 for 450 minutes was not lethal. At this frequency Schwan and Li have calculated that it is possible to have 100 per cent absorption in the skin, fat and muscle. If, indeed, 100 per cent was absorbed, then Keplinger's data are quite compatible with those of Ely and Goldman; namely, that 20 or 25 mw/cm^2 absorbed by the fur-bearing animal may be lethal. Tests conducted at M.I.T. Lincoln Laboratories, Lexington, Mass., are in substantial agreement with this conclusion [34], [35], [56].

These data are for fur-bearing animals. Ely, Goldman, Hearon, Williams and Carpenter [36], in their very thorough manner, point out, however, that man has a much better heat dissipating mechanism than fur-bearing animals. They estimate that the human being is capable of dissipating 1 kw of absorbed power. (This is not as much power as would be incident on a man lying prone in the noonday sun at the equator.) Their discussion as to how they arrived at this figure is substantiated by references to earlier work, and their conclusion seems reasonable if the energy is dissipated in the skin, fat and muscle near the surface where heat exchange with the cooling mechanism is good. However, if the energy is dissipated deeper in the body, in the hollow viscera, for example, as it might be for the longer radio waves, the heat exchange mechanism is no longer as effective and, as in the case of the eye, the internal temperature rise might become intolerable. Evidence of this effect was reported by Hines and Randall [22], who observed temperature rises of more than 40°C in the ileum of anesthetized rabbits after 30 minutes of localized irradiation at 12-cm wavelength, during which time the rectal temperature rose less than 1°C.

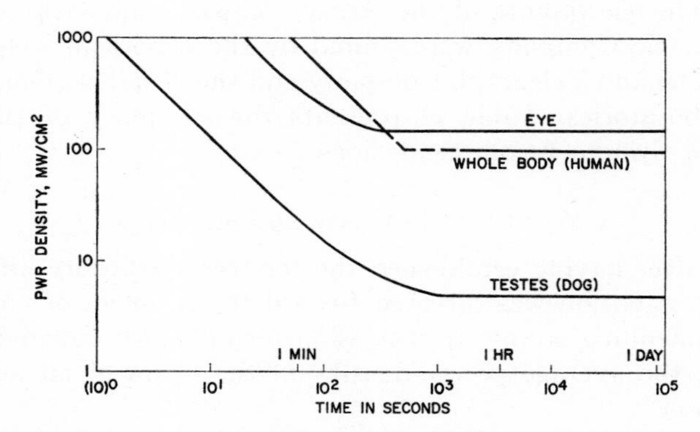

Fig. 9—Threshold levels vs time for three sensitive structures [36].

With the foregoing discussion in mind, let us look at the conclusions of Ely *et al.* [36] in more detail. Fig. 9 shows their preliminary estimates of threshold power densities according to time of exposure for three sensitive structures: the eye, the whole body, and the testes. Power density, in mw/cm^2 is plotted logarithmically on the ordinate, and time of exposure is plotted also logarithmically on the abscissa. The downward slope at short exposure time is based on their estimates of the thermal time constant of the structure. The terminal power density for long exposure is determined by the heat exchange characteristics of the structure and an estimated tolerable terminal temperature which was assumed to be 102.2°F for the whole body, 113.0°F for the eye, and 98.6°F for the testes. The power densities required to maintain these "tolerable" temperatures are 100 mw/cm^2 for the whole body, 155 mw/cm^2 for the eye, and 5 mw/cm^2 for the testes.

Note that the only threshold power density level which is lower than the currently adopted potentially hazardous level of 10 mw/cm^2 is that for the testes, 5 mw/cm^2. The question naturally arises, just how tolerable or how hazardous is a temperature of 98.6°F in the testes? Ely discussed this point briefly at a panel meeting held in New York, N. Y., on October 7, 1959, which was cosponsored by the New York Section of the IRE, the local chapter of the IRE Professional Group on Medical Electronics and the New York Section of the Communications Division of the AIEE. At this meeting Ely presented the graph which indicated that his estimated threshold for testicular exposure was 5 mw/cm^2. In answer to a question from the floor regarding the criterion for this limit, he went on to say that it is well known that undescended testes are sterile; that is, when testes are kept at a temperature of 98.6°F continuously, they become sterile. Furthermore, that in the experiments with dogs this temperature (98.6°F) could be maintained in the testes with a power density of 5 mw/cm^2.

Ely *et al.*[5] comment further on this point thus, "Although the criterion of hazard used in this study [of testicular changes] has been the least demonstrable damage, other factors should be considered in the overall viewpoint. The minimal testicular damage is almost certainly completely reversible. Even considerably more severe testicular insult will probably be reversible, with the only finding being a temporary sterility. An even greater injury can result in permanent sterility, which result would evoke varying reaction."

In regard to the varying reaction, there is the story of a radar station which had chronic failure every Friday night. Investigation disclosed that a technician had cut a large hole in the side wall of the transmitting waveguide and closed it up again with a flap. When this flap was kept closed the radar operated normally.

Each Friday night, however, before his buddies left for weekend leaves, he opened the flap and allowed the microwave power to escape into the room. He charged a fee for an exposure to this microwave power which, he claimed, would render the subject temporarily sterile for the weekend without jeopardizing any other capa-

[5] See [36], p. 91.

bility. Whether or not a temperature of 98.6°F in the testes is tolerable, the technician did a "land office" business for a while.

The possibility that nonthermal effects may exist was suggested by Schwan [48], [66] at both Tri-Service Conferences. Herrick [60] also suggests the possible existence of nonthermal effects in her report on pearl-chain formation. Schowalter of Western Electric Company points out, however, the similarity of this to the Reauleaux effect, which is induced by heat, the difference being only that in Herrick's observations, the pearl-chain formations were aligned (perhaps parallel to the electric vector), whereas in the Reauleaux effect the direction of chain formation is random.

More recently other investigators have reported effects which may be nonthermal. Heller and Teixeira-Pinto at the New England Institute for Medical Research, Ridgefield, Conn. [72] report on a new physical method of creating chromosomal aberrations. They also observed that the application of radio frequency energy to a medium in which paramecia were swimming in random directions caused these micro-organisms to swim parallel to the electric field. A change in the frequency caused them to turn and swim at right angles to the electric field. This work was done with pulsed power in the frequency range of 27 Mc. The power density at which these presumably nonthermal effects were observed was not given in the report.

Heller described his observations and showed some most amazing moving pictures (photo micrographs) of swimming paramecia at the twelfth Annual Conference on Electrical Techniques in Medicine and Biology held in Philadelphia, Pa., November 10, 11, 12, 1959. This meeting was sponsored by the IRE, the AIEE, and the Instrument Society of America. In answer to a question from the floor, he stated that the estimated electric field strength of the radio frequency pulses required to produce the observed nonthermal effects was of the order of 100 to 1000 volts/cm. If we knew the impedance of his medium, the corresponding peak power density could be calculated. For example, if the medium were water, with a relative dielectric constant of 80, the peak power density would be of the order of 3 to 300 watts/cm^2. In a typical radar with a duty cycle of one in a thousand, these peak, pulse power densities could be achieved in free space with an average power density of 3 to 300 mw/cm^2. The lower limit of the estimate is in the range of the currently adopted potentially hazardous level.

Carpenter [57] at Tufts University, has interpreted some of his data in terms of a possible nonthermal effect. Specifically, he found that rabbits' eyes exposed to 140 mw/cm^2 with *continuous waves* for 20 minutes *did not* develop opacities, but that rabbits' eyes exposed to 140 mw/cm^2 average *pulsed power* for 20 minutes *did* develop opacities. If the thermal time constant of the lens were long compared with the pulse length, then the cataractogenic nature of pulsed power densities could indeed be explained in terms of a nonthermal effect.

With the foregoing examples in mind, the implication is inescapable that we must not ignore completely the possibility of harmful nonthermal effects. We must hence keep our minds open for consideration of the adoption in the future of a limit for peak pulse power in addition to the limit already adopted for average power. Perhaps, too, someday we may have enough information on the effects at various frequencies to specify a limit that varies with frequency. More quantitative data are needed, however, before these effects may be taken into account properly.

In the meantime, pending further information, the limit has been based upon average power density. It is understandable that there was some question regarding the safety factor of the adopted limit of 10 mw/cm^2. Some organizations felt that it would be desirable to recommend that the power density in living quarters be kept somewhat below the upper limit. For example, the U. S. Army [63] had made an exhaustive study of the published literature and had made measurements at Army training installations. As a result of this study, they had arrived at some generalized criteria that they believed should be applied to training situations. One of these criteria was that rest areas should be provided for trainees at distances where the power densities were less than 1 mw/cm^2. This conclusion was compatible with Schwan's opinion [31] that 10 mw/cm^2 should not be allowed for more than one hour.

The General Electric Company [68] expressed the opinion that it would become necessary generally to monitor at a 1 mw-mean value in order to make the necessary allowance for harmonics and spurious waves.

Bell System Recommendations in Detail

The viewpoints of the Army, Schwan, and General Electric Company were shared by the American Telephone and Telegraph Company and the Bell Telephone Laboratories. Table II presents the summary of the Bell System recommendations.

Calculation of Power Densities

After having established the tentative exposure limits, attention was directed toward the problem of recommending simple transmission formulas for calculating the average power density in the beam of an antenna.

The field in front of the usual parabolic antenna can be characterized by referring to two separate regions:

1) The "near-field," or Fresnel region, where the radiation is substantially confined within a cylindrical pattern.
2) The "far-field," or Fraunhofer region, beyond the Fresnel region in free space, where the radiation is essentially confined to a conical pattern and

the power density along the beam axis falls off inversely with the square of the distance.

To compute the value of the power density in the near-field, assuming a circular "dish" antenna, use

$$W = \frac{16P}{\pi D^2} = \frac{4P}{A}, \quad (1)$$

where

P = average power output, *not* peak power,
D = diameter of antenna,
A = area of antenna.

If this computation reveals a power density which is less than the limit, then there is no need to proceed with further calculations, since (1) gives the *maximum* power density that can exist on the axis of the beam of a properly focussed antenna. (A defocussed antenna could give more, but that condition is not usual.)

If the computation from (1) reveals a power density greater than the limit, then one assumes that this value may exist any place in the near-field region, and attention is directed to the far-field region.

In the far-field region, the free-space power density on the beam axis may be computed from

$$W = \frac{GP}{4\pi r^2} = \frac{AP}{\lambda^2 r^2}, \quad (2)$$

where λ = wavelength.

The distance from the antenna to the intersection of the near-field (1) with the far-field (2), is given by

$$r_1 = \frac{\pi D^2}{8\lambda} = \frac{A}{2\lambda}. \quad (3)$$

These formulas do not include the effect of ground reflection which could cause a value of power density which is four times the free-space value.

Setting the power density equal to the potentially hazardous level of 10 mw/cm², one may calculate the distance to the boundary of the potentially hazardous zone. This has been done for some of the radars which have round antenna apertures. More involved calculations are necessary when the antenna shape and illumination are more complex. Other calculations have been reported for several types of radars [53], and these are included in Table III.

TABLE II

SUMMARY OF BELL SYSTEM RECOMMENDATIONS

1. For the time being, microwave exposure limits may be classified as follows:

Average Power Density mw/cm²	Classification
Above 10	Potentially hazardous
Between 1 and 10	Safe for incidental or occasional exposure
Below 1	Safe for indefinitely prolonged exposure or permanent assignment

2. Employees are cautioned to abide by the following rules:

a) Never enter an area posted for microwave radiation hazard without verifying that all transmitters have been turned off and will not be turned on again without ample notice.
b) Never look into an open waveguide which is connected to energize transmitters.
c) Never climb poles, towers or other structures into a region of possible high radar field without verifying that all transmitters have been turned off.

TABLE III

NIKE AND OTHER COMMON RADARS—DISTANCE IN FEET FROM RADAR ANTENNA TO BOUNDARY OF POTENTIALLY HAZARDOUS ZONE*

(Arranged in descending order of distances)

Radar Type	Distance in Feet for 0.01 watt/cm²
AN/FPS-16	
Sig C Mod.	1020
Standard Mod.	590
AN/FPS-6	560
Herc. Imp. Acq. HIPAR (Fixed)	550†,‡
AN/MPS-23	530
AN/MPS-14	472
Herc. Imp. TTR	400
AN/TPQ-5	350
AN/FPS-20	338
AN/MPQ-21 (10′)	300
Herc. MTR (Ajax)	270
Ajax Acq. (Fixed)	260†
AN/CPS-9	260
AN/MPQ-21 (7′)	210
AN/MPS-4	205
AN/FPS-8 (40′×14′)	205
Ajax MTR	205
AN/CPS-6B	200
AN/FPS-10	200
AN/MPS-22	185
AN/FPS-18	178
AN/MPS-12	175
AN/MPQ-18	175
AN/FPS-3	172
AN/MPS-7	172
AN/MPQ-21	165
AN/TPS-1G (40′×11′)	150
AN/FPS-36	150
Ajax TTR	132
Herc. Imp. Acq. (Fixed)	130†,‡
Herc. Acq. (Fixed)	130†,‡
AN/FPS-14	109
AN/FPS-4 (narrow pulse)	106
AN/MPS-8 (narrow pulse)	106
AN/TPS-10D (narrow pulse)	106
AN/MPS-10 (C)	105
AN/FPS-8	101
AN/MPS-11	101
Scr 584	70
AN/MPQ-10 (S)	50
AN/TPS-1-D	50
AN/FPS-25	40
AN/FPS-31	27.5
Herc. Imp. Acq. HIPAR (Rot.)	25
Ajax Acq. (Rot.)	8
AN/PPS-4	2.5

* Based on the following assumptions:

1) Free space transmission.
2) No ground reflections. These could double the distances shown.
3) Calculations apply to the axis of the beam, *i.e.*, where the power density is greatest.
4) The beam is considered to be fixed in space, *i.e.*, not scanning.

† Not normally used with fixed antenna.
‡ Interlocks provide assurance that transmitter is idle unless antenna is rotating.

Let us see how those approximate formulas, (1), (2), and (3), were derived and just how accurate they are.

Assume that we have a plane electromagnetic wave impinging on a totally absorbing body. If the projected area of the body is A, then the power absorbed will be the power density times the area or

$$P = WA. \tag{4}$$

If the field strength of the incident wave is E volts per meter in free space, then the power density in watts per square meter will be

$$W = \frac{E^2 \text{ volts per meter}}{377 \text{ ohms}}. \tag{5}$$

Ordinarily it is not necessary to specify the field strength in volts per meter, but only to specify the power density in watts per unit area if the impedance of the medium is known.

For an isotropic radiator in free space, radiating a total average power P in all directions equally, the power density on the surface of a concentric sphere of radius r will be simply the total radiated power divided by the area of that sphere, or

$$W = \frac{P}{A} = \frac{P}{4\pi r^2}. \tag{6}$$

Now if the radiator is not isotropic and radiates with a directivity gain G in a given direction, then the power density at a distance r would be, in the far-field region,

$$W = \frac{GP}{4\pi r^2}. \tag{7}$$

Allowing for 100 per cent ground reflection, which doubles the electric field strength (and hence quadruples the power density), we have

$$W = \frac{GP}{\pi r^2}. \tag{8}$$

If we express the gain G in terms of the antenna area thus

$$G = \frac{4\pi A}{\lambda^2} \tag{9}$$

and insert this in (7) and (8), we have the alternative expressions in terms of antenna area (3).

$$W = \frac{AP}{\lambda^2 r^2} \tag{10}$$

for the free-space power density, and

$$W = \frac{4AP}{\lambda^2 r^2}, \tag{11}$$

which assumes 100 per cent reflection from the ground.

It will be convenient to express the far-field free-space power density (10) in terms of the power density at the antenna aperture, $W_0 = P/A$, and hence $P = W_0 A$. Substituting this expression for P in (10) and dividing both sides by W_0, we have,

$$\frac{W}{W_0} = \left(\frac{A}{\lambda r}\right)^2. \tag{12}$$

For convenience, this may be rewritten thus

$$\frac{W}{W_0} = 4\left(\frac{A}{2\lambda r}\right)^2. \tag{13}$$

This expression applies in the far-field region, but a simple modification makes it applicable to the near-field region for a uniformly illuminated round aperture, thus

$$\frac{W}{W_0} = 4 \sin^2\left(\frac{A}{2\lambda r}\right). \tag{14}$$

This more exact expression is plotted in Fig. 10 along with the lines representing the approximate formulas, (1) and (2).

In this graph and in some subsequent graphs, the relative power density is plotted in decibels on the ordinate and the distance, or more specifically $\lambda r/A$, is plotted logarithmically on the abscissa.

Note the alternate maxima and minima in the near field. The maxima all are 6 db above (4 times) the power density at the aperture. To be realistic, in the near-field region we must assume that a man could be standing at such a distance that he could be exposed to a maximum. Hence the near-field approximation

$$W = 4W_0 = \frac{4P}{A}. \tag{15}$$

This agrees with (1).

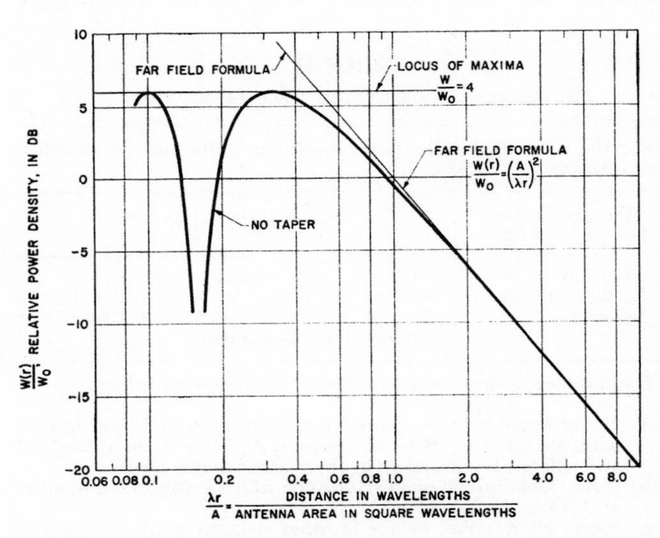

Fig. 10—Power density vs distance for uniformly illuminated round aperture.

Note that the near-field approximation intersects the far-field approximation when

$$r_1 = \frac{A}{2\lambda} = \frac{\pi D^2}{8\lambda} \tag{16}$$

as given in (3).

Also notice that the largest deviation of the approximations from the more exact expression occurs at this distance and this deviation amounts to only 1.5 db which, in many cases, is of little consequence. For all other distances the error will be less than this 1.5 db for the conditions assumed above.

Thus, it appears that these approximations are fairly reliable. They do, however, neglect the loss of power due to "spill over" at the antenna and the effectiveness of the antenna area and hence, in general, will give conservative estimates. Likewise, transmitter powers are often rated as power available *at the generator* so that any transmission line losses between the generator and the antenna would make the estimates even more conservative.

So far, we have assumed uniform illumination of a round aperture, whereas most antennas have the illumination tapered so as to reduce the sidelobes. Often a square-law taper to 10 db down at the edge of the aperture is used. Let us see what this does to *the power density when the total radiated power is kept the same for different tapers.*

Fig. 11 is a plot of the relative power density at the aperture of the antenna as a function of the radial distance from the center of the aperture for different tapers. For uniform illumination the relative power density is, of course, unity all over the aperture. However, for a linear electric-field taper, the power density is more than three times this value at the center, tapering to about 0.3 at the edge. For a square-law taper of the electric field, the power density at the center of the antenna is over twice that for the uniform case and tapers to about 0.2 at the edge.

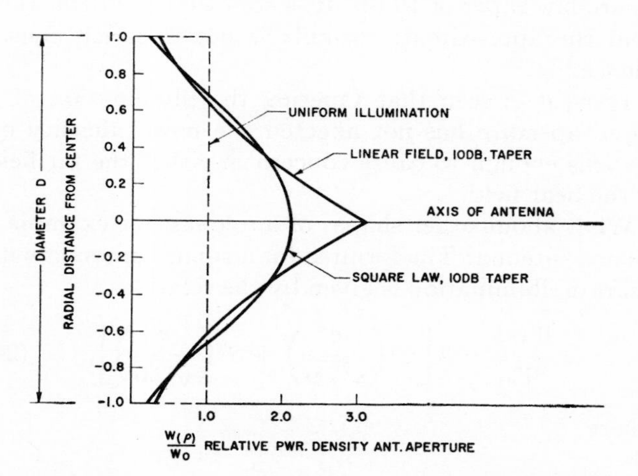

Fig. 11—Comparison of power densities at the aperture for different tapers of illumination.

Judging from those curves, one *might* think that the power density maxima along the axis in the near field and the power density in the far field might be quite different for these three tapers. Such, however, is not the case as will be shown in the following analysis.

We assume a circular aperture of diameter $2a$ with an illumination taper function $A(\rho)$. The field strength at a distance r along the axis of the antenna is given by the well-known relation

$$E_{(r)} = \frac{i\beta E_0}{r} \exp(-i\beta r) \int_0^a A(\rho) \exp(-i\beta(r' - r))\rho d\rho \tag{17}$$

where E_0 is the electric field at the center of the aperture,

$$\beta = \frac{2\pi}{\lambda}$$
$$A(\rho) = \text{normalized amplitude taper}$$
$$\exp(x) = e^x.$$

If we now neglect the angular dependence of the Huygens' sources, then, for distances more than a few diameters away, the following approximation may be used:

$$(r' - r) = \frac{\rho^2}{2r}$$
$$\left(\text{for } \frac{a}{r} \ll 1\right). \tag{18}$$

Thus, when the distance is much greater than the radius, a good approximation is given by

$$E_{(r)} = \frac{i\beta E_0}{r} \exp(-i\beta r) \int_0^a A(\rho) \exp\left(\frac{-i\beta\rho^2}{2r}\right)\rho d\rho. \tag{19}$$

For uniform illumination, $A(\rho) = 1$ and the formula quoted previously is derived.

For a linear taper

$$A(\rho) = 1 - (1-\alpha)\frac{\rho}{a}, \tag{20}$$

where

$$\alpha = \frac{E(\rho = a)}{E(\rho = 0)}.$$

Performing the quadrature results in the following expression:

$$\frac{W(r)}{W_0} = \frac{6}{1 + 2\alpha + 3\alpha^2}\left|1 - \alpha \cos\frac{\pi}{2}\mu_0^2 - \frac{1-\alpha}{\mu_0}C(\mu_0) + i\left(\alpha \sin\frac{\pi}{2}\mu_0^2 + \frac{1-\alpha}{\mu_0}S(\mu_0)\right)\right|^2 \tag{21}$$

where

$$W_0 = \frac{P}{\pi a^2}$$

$$\mu_0^2 = \frac{2a^2}{\lambda r}.$$

C and S are the Fresnel integrals

$$C(\mu) = \int_0^\mu \cos \frac{\pi}{2} v^2 dv$$

$$S(\mu) = \int_0^\mu \sin \frac{\pi}{2} v^2 dv$$

as tabulated by Jahnke and Emde.[6]

This expression is plotted in Fig. 12 where it is seen that the major effect of the taper has been to fill in the deep nulls which existed in the near field for the uniformly illuminated case. The difference between this calculated curve and the approximate far-field formula (2) is less than $\frac{1}{2}$ db in the far field and again about 1.5 db at the intersection where $r = A/2\lambda$.

In the near field, the maxima are within about $\frac{1}{2}$ db of the $W = 4W_0$ line.

For the case of a square-law taper of the electric field, the amplitude function becomes

$$A(\rho) = 1 - (1 - \alpha)\left(\frac{\rho}{a}\right)^2. \tag{22}$$

Performing the indicated quadrature results in the following expression

$$\frac{W(r)}{W_0} = \frac{3}{1 + \alpha + \alpha^2}\left|1 - \alpha \cos \frac{\pi}{2}\mu_0^2 - (1 - \alpha)\frac{\sin \frac{\pi}{2}\mu_0^2}{\frac{\pi}{2}\mu_0^2} + i\left(\alpha \sin \frac{\pi}{2}\mu_0^2 + (1 - \alpha)\frac{1 - \cos \frac{\pi}{2}\mu_0^2}{\frac{\pi}{2}\mu_0^2}\right)\right|^2 \tag{23}$$

where again

$$W_0 = \frac{P}{A},$$

and

$$\mu_0^2 = \frac{2a^2}{\lambda r}.$$

A plot of this expression is given in Fig. 13. One curve and two straight lines are plotted. The straight lines are the approximations of (1) and (12). The curve is for a square-law taper of 10 db. It is seen that the departure from the approximate formula is not of much consequence.

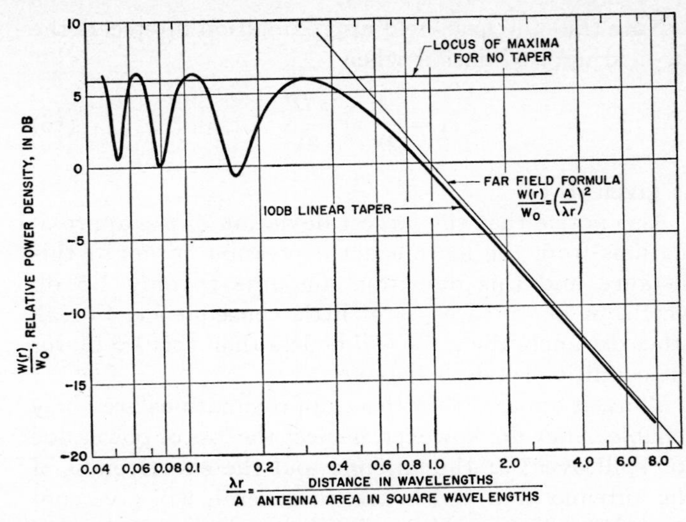

Fig. 12—Power density vs distance for round aperture with linear taper (10 db).

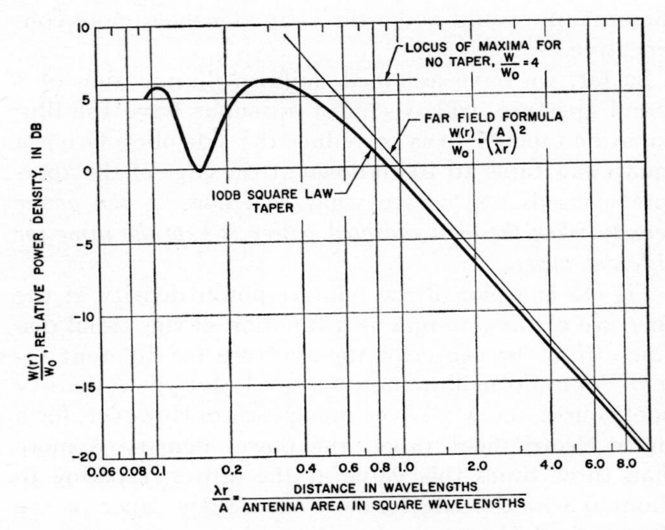

Fig. 13—Power density vs distance for round aperture with square-law taper (10 db).

Thus, it is seen that tapering the illumination of a round aperture has not affected the power density on the axis enough to cause concern in either the far field or the near field.

What about other shapes of antennas, for example a square antenna? The formula for a square antenna with uniform illumination is given by the relation

$$\frac{W(r)}{W_0} = 4\left[C^2\left(\frac{d}{\sqrt{2\lambda r}}\right) + S^2\left(\frac{d}{\sqrt{2\lambda r}}\right)\right], \tag{24}$$

where

$$W_0 = \frac{P}{A} = \frac{P}{d^2}.$$

[6] E. Jahnke and F. Emde, "Table of Functions," Dover Publications, Inc., New York, N. Y., p. 34; 1943.

This is plotted in Fig. 14, together with the approximations based upon *equal area antennas*. For the square antenna the near-field maxima are lower than the approximation, but the one at the greatest distance is only about one db low. In the far field, the asymptotic agreement is excellent.

To summarize these observations, Fig. 15 is presented. This shows the near- and far-field approximations along with the maxima for the round antennas with linear and square-law 10-db tapers and for the square antenna with uniform illumination.

The abscissa is plotted on $\lambda r/A$ which can be applied to either round or square antennas. This shows that the approximate formulas (1), (2), and (3) can be applied to square antennas by choosing a diameter which makes the areas equal.

Thus, the approximate formulas are seen to apply quite well to round and square antennas. For more exact results at the region of crossover between near and far fields, the expression containing $\sin^2$ may be used (14).

For long rectangular antennas, an approximate formula derived by Engelbrecht is

$$\frac{W}{W_0} = \left(\frac{A}{\lambda r}\right)^2 \text{ beyond the distance } r \geq \frac{d_1^2}{2\lambda}, \tag{25}$$

and

$$\frac{W}{W_0} = \frac{2d_2}{\lambda r} \text{ within the distance } r < \frac{d_1^2}{2\lambda} \tag{26}$$

d_1 = wide dimension

d_2 = narrow dimension.

This near-field formula was derived from a more complete analysis for distance such that

$$\frac{d_1}{r} < 1$$

and

$$\frac{d_2^2}{\lambda r} < 1.$$

Reviewing the foregoing discussion, it is seen that a simple calculation of the maximum power density in the near field according to (1) reveals whether or not a hazard is involved in the beam of the antenna. If this power density exceeds 10 mw/cm², then a simple calculation based upon the far field formula (2) reveals the potentially hazardous distance in free space. These formulas apply to uniform illumination of square, round or rectangular apertures or to illumination which is tapered in amplitude for round apertures. For other shapes and tapers a more complicated analysis would be necessary.

To calculate accurately the power density off the axis of the main beam requires the solution of a more difficult mathematical problem. One approach [53] reveals

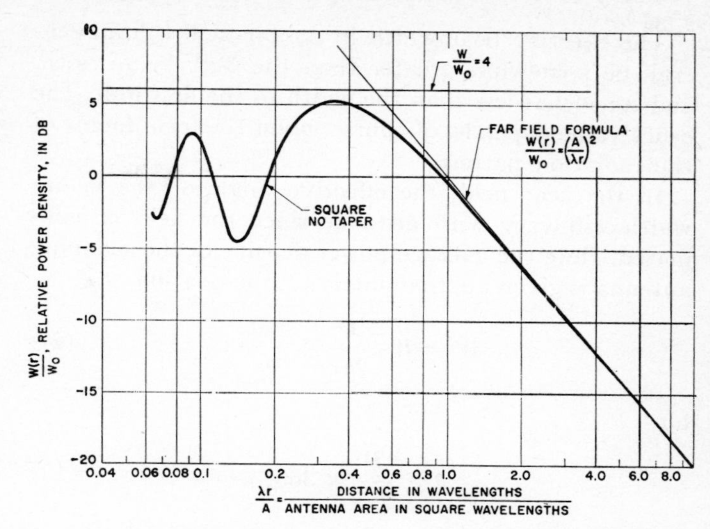

Fig. 14—Power density vs distance for square aperture uniformly illuminated.

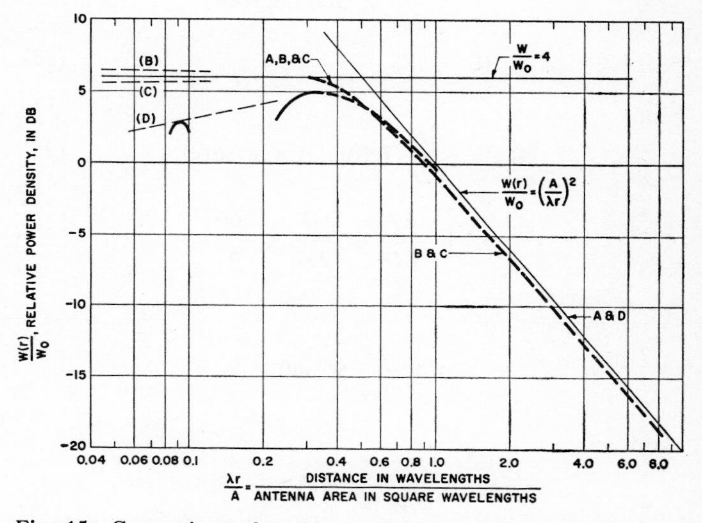

Fig. 15—Comparison of power densities for different shapes and tapers: A. Round—no taper; B. Round—10-db linear taper; C. Round—10-db square-law taper; D. Square—no taper.

that the collimated beam in the near field falls off approximately 12 db per radius.

Many of our radars do not have the simple shapes nor simple illumination tapers that were treated above. In such cases the approximate formulas will not apply directly and a more complete analysis is indicated.

Scanning Antennas

The specified limits for safe and potentially hazardous power densities have been based upon average power. In the case of the scanning antenna, the average power absorbed by a fixed subject will be reduced by the ratio of the effective beamwidth to the scanned angle, if the thermal time constant of the part of the exposed individual is long compared with the scanning period. Accordingly, the potentially hazardous distance is reduced by the square root of this ratio.

The effective beamwidth in the far field will, in general, be somewhat greater than the 3-db beamwidth, and somewhat less than the width to the first null. The exact value depends, of course, upon the form factor of the radiation pattern.

In the near field, the effective angle of the beamwidth will vary with distance since the field is collimated. Here the average power density of the scanning antenna is given approximately by the relation

$$W \doteq W_0 \frac{D}{2\pi r} \times \frac{360}{\theta}, \tag{27}$$

for

$$\theta \geq \frac{D}{2\pi r} \times 360,$$

where θ is the scanned angle in degrees. For

$$\theta < \frac{D}{2\pi r} \times 360 \text{ degrees}$$

$$W \doteq W_0. \tag{28}$$

Since

$$W_0 = \frac{4P}{\pi D^2} \text{ (circular aperture)} \tag{29}$$

$$W \doteq \frac{4P}{\pi D^2} \times \frac{D}{2\pi r} \times \frac{360}{\theta},$$

for

$$\theta \geq \frac{D}{2\pi r} \times 360.$$

And

$$W = \frac{4P}{\pi D^2}, \tag{30}$$

for

$$\theta < \frac{D}{2\pi r} \times 360.$$

Setting $W = 10$ mw/cm², the potentially hazardous distance in the near-field is

$$r_{\text{hazardous}} \doteq \frac{1}{5\pi^2} \cdot \frac{P}{D} \times \frac{360}{\theta} \text{ cm} \tag{31}$$

for

$$\theta \geq \frac{D}{2\pi r} \times 360 \quad \text{and} \quad r \leq \frac{D^2}{4\lambda},$$

where P is in mw, D is in cm.

For

$$\theta < \frac{D}{2\pi r} \times 360,$$

the previous discussion is applicable.

The potentially hazardous level of 10 mw/cm² was based on average power measurements. In fact, many of the experiments on animals were made with continuous-wave oscillators and the deleterious effects appeared to be due solely to the production of heat and the accompanying rise in temperature. If the duty cycle of the radar is one in a thousand, the peak-power density associated with this level would be 10 w/cm². It may be reasonable to assume that this peak-power density may not cause damage by breaking down of tissues. However, if the radar duty cycle were say $(10)^{-7}$ the peak-power density would be 100 kw/cm², approaching the level which causes breakdown in air. At this level, it would be reasonable to assume that there might be a possibility of breakdown of tissue. Somewhere between the two examples just cited, danger due to breakdown of tissue from effects other than heating may be possible. A similar situation exists for very narrow-beamed sweeping antennas, whose exposure cycle is very small. The observation of Heller and Pinto [72] suggests effects other than thermal and if, in fact, these nonthermal effects are due to peak power, rather than average power, a more elaborate specification of safe and hazardous levels will be necessary.

Measurement of Microwave Power Density

The measurement of the power density depends basically upon the determination of the power absorbed in a given area. Knowing the power absorbed by the given area, one computes the power density by dividing the absorbed power by the area.

Devices for measuring microwave power have been available for many years and are described in standard text books. A brief description is given by Southworth,[7] and a comparison of power-measuring techniques is discussed by Sucher of the Microwave Research Institute, Polytechnic Institute of Brooklyn [25]. General details are given by Green, Fisher, and Ferguson [8]. Montgomery devotes one whole chapter to the discussion of microwave power measurements [9].

There are four basic methods of measuring power. These are: calorimetry, bolometry, voltage (and resistance) measuring systems, and measurement in terms of radiation pressure on a reflecting surface [20]. Calorimetric methods are based upon the transfer of the electromagnetic energy into heat, and the power is determined solely by the measurement of temperature, mass, and time. With a knowledge of thermal capacity, only temperature and time need be measured.

Bolometric measurements are based upon the absorption of the power in a temperature-sensitive resistive element. The change in resistance is then used to indicate the power. This method is the one most widely used in commercially available power meters, and the most popular sensing element is the thermistor, de-

[7] See [16], p. 655.

scribed by Southworth: [16]

"Certain mixtures of semiconductors have especially high temperature coefficients of resistivity. Resistor units made up of such materials are therefore thermally sensitive . . . [and] the measurement of power is based upon this resistance change. . . . Power measurements are based on calibration made at [dc or] low frequencies.

"To measure the change of resistance as power is dissipated, the thermistor is often placed in one arm of a dc bridge. By noting the dc power necessary for balance with and without RF power applied to the thermistor, the difference is determined. In a modification of this method, the bridge is balanced before RF power is applied, and the measurement consists merely of noting the meter deflection (due to the bridge unbalance) when RF power is added. It is possible to make such a power meter direct reading. . . ."

There are many commercially available thermistor bridges on the market. However, the electronic circuits of some of these are not shielded adequately to be unaffected by the high-power microwave-pulsed energy of a radar. The peak pulse power may be 1000 or more times the average power, and since in the application to radio-frequency hazards we are currently interested chiefly in average powers from 1 to 100 mw; the peak pulse power may be from 1 to 100 watts. Without adequate shielding, the vacuum tubes associated with some of these power meters may be susceptible to the microwave power interference and readings may be obtained even when the pick-up antenna is disconnected.

There are, however, some commercially available thermistor bridges designed specifically to work in an environment of pulsed microwave radiation. These, when used in conjunction with an appropriately calibrated antenna and a matched thermistor in an RF head, are suitable for power density measurements.

If the radiated power is continuous, rather than pulsed, a much simpler relative power meter using a readily available diode rectifier and dc meter in conjunction with a calibrated antenna may be used. After a calibration in the laboratory, this type of unit is convenient, because of its light weight and small size, to carry around in the vicinity of the antenna for survey purposes or for monitoring. These are not, however, suitable for pulsed-power applications since the high-peak-pulse power may burn out or damage the delicate point contact of the microwave diode.

All of these power measuring devices are used with a suitable antenna whose effective area is known. "Standard" waveguide horn antennas, loops and dipoles are available commercially whose effective areas are known. For example, the effective area of a lossless loop or small dipole (doublet) is $\frac{3}{8}$ of the wavelength squared divided by π [16], [18], and has a gain, over an isotropic antenna (whose effective area is the square of the wavelength divided by 4π) of 1.76 db. A half-wavelength antenna has an absolute gain (over an isotropic radiator) of 2.15 db and an effective area of 0.131 times the square of the wavelength.

Waveguide horns, either circular or rectangular, are amenable to calculation [18], and their effective areas range from about one half the physical area for optimum horns, to about 80 per cent for very long horns. An optimum horn is one whose open-end area is scaled to give the maximum gain for that particular length of the flare in the waveguide. Specifically, optimum circular horns have an effective area equal to 51 per cent of the actual area, while optimum rectangular horns have 49 per cent effectiveness. An infinitely long circular horn has a relative effectiveness of 84 per cent while infinitely long rectangular horns have 81 per cent effectiveness. Such "standard" antennas are used to pick up the energy which is then transmitted by a well-shielded transmission line to the thermistor-sensing element whose resistance has been adjusted so as to match the impedance of the transmission line. All of the power is dissipated in the thermistor, causing its temperature to rise, its resistance to change, which unbalances the dc bridge and the unbalance current indicates the power, from which the power density is calculated.

Recently there have appeared on the market at least four equipments designed specifically for microwave power density measurement. There are:

1) "Broadband Power Density Meter," Model NF-157, Empire Devices, Amsterdam, N. Y., covering the frequency range from 200 Mc to 10,000 Mc with three different RF pick-up probes. The power density for midscale reading is 1 mw/cm² to 1 w/cm². The claimed accuracy is ± 1 db at midscale. The weight is 11 to 13 pounds.
2) "Electromagnetic Radiation Detector," Microline Model 646, Sperry Microwave Electronics Company, Clearwater, Fla. Three units are available to cover three bands (2700 to 3300 Mc, 5400 to 5900 Mc, and 8200 to 12,000 Mc). The range of power density is 1 to 20 mw/cm² with 10 mw/cm² at midscale. The weight is about 8 pounds.
3) "Densiometer," Model 1200, Radar Measurements Corporation, Hicksville, N. Y., covers 5 bands with 4 different antennas: VHF (200 Mc–225 Mc); UHF (400 Mc–450 Mc); S band (2600 Mc–3300 Mc); C band (5000 Mc–5900 Mc), and X band (8500 Mc–10,000 Mc). The power density, for midscale reading is 10 mw/cm². The weight is about 2 pounds. It comes with a handy "exposure meter" type of carrying case.
4) Radiation Monitor, Model B86B1, Sperry Microwave Electronics Company, Clearwater, Fla., covers from 200 to 10,000 Mc with one antenna. The power density for midscale reading is 10 mw/cm². It receives all polarizations simultaneously. The claimed accuracy is ± 2 db at any frequency within the operating range. The weight is 2 pounds.

The devices which rely upon antenna pick-up probes all have two limitations in common: they are frequency sensitive, and they are directive. These limitations are no handicap if the power density meter is being used to survey a particular transmitter site where the frequency, direction and polarization of the radiation are all known. However, for use as a detection device by the telephone serviceman who approaches an area which is merely posted as being hazardous without specifying the frequency and direction of the microwave radiation, they are of rather limited usefulness.

What is needed here is something that responds to all radio frequencies, regardless of polarization and direction of arrival. Perhaps some day we shall have such a device. It will probably be based either on the calorimetric method, using an absorber which acts as a black body, or on the radiation pressure principle. This problem has been and is being given considerable thought in various research centers.

Some people have suggested using the glow discharge in a neon lamp as an indicator. These lamps are known to respond to the peak-pulse power, rather than the average power, since "the ionization and deionization times of neon are presumably much less than the width of the pulse modulation envelope."[8] These lamps are therefore unsuited for indicating average power density.

Further information on the determination of power density at microwave frequencies has been given [24], [44].

Regarding the procedure in making a survey of a particular site in the field, if such a survey seems to be warranted on the basis of a calculated estimate, Dondero [44] points out that, after selecting the appropriate measuring equipment, "it is always good practice to perform initial measurements at some practical (safe) distance from the radiator and gradually work in toward the source of radiation." This provides some safeguard for personnel performing the measurement. This procedure differs from that which has been used in making field strength surveys patterned after the style set in the early 1920's by Bown [1], [2], who measured the coverage of a radio broadcasting station, in that the survey is always made by *approaching* the transmitter. Also, it may be advisable to perform the survey at a known reduction of power and interpret the results in terms of power density for full transmitter power.

Reflection may complicate the field strength pattern. It has been known for many years [3] that obstacles such as trees, buildings, telephone lines, etc., reflect radio waves to a certain extent and produce standing waves in the field strength pattern. It is essential to keep this in mind while making the survey and to move the pick-up antenna around, this way and that, to establish whether the reading is being affected by reflections. If it is, the maximum reading should be recorded.

If a directive antenna is used, the horn should be directed not only toward the transmitting antenna but also around in other directions toward any object nearby which might act as a reflector. If a significant reading is obtained from such a reflecting object, then the total power should be figured, not by adding the two powers, but by figuring the square of the sum of the square roots of the powers. In other words, since the direct and reflected waves are coherent, coming from the same source, their electric field strengths will add vectorially to give a maximum when they are in phase. The corresponding maximum power is proportional to the square of the sum of the field strengths. Since the field strengths are proportional to the square roots of the powers, we have, therefore, the square of the sum of the square roots of the powers.

However, in a complex situation where there are no reflections but several different transmitters, for example, as would exist at the operational field-radar site, then the power densities from the radars on different frequencies should be added directly and not the field strengths, since, in general the radar frequencies will not be identical.

Having once established the boundary of the potentially hazardous zone, where power densities of 10 mw/cm^2 may occur, appropriate measures must be adopted to keep personnel from entering that zone. Barricades or shielding fences may be installed. If gates must be provided for occasional access to the area by personnel, interlocks should be provided to ensure that the transmitter be idle whenever personnel open the gate and enter the area.

In some cases it may be necessary that the transmitter be kept operative while personnel enter into an area where the field is potentially hazardous. In this event, some kind of a shielding arrangement is indicated. Whether this be of the fence or shielded room type or the mobile-portable shielding-garment type, some knowledge of the effectiveness of the shield is necessary in order to be sure that the power density within the shielded area is low enough to be safe. Of course, completely enclosing the area in a water-tight copper container would most surely be adequate, but neither economical nor practical. A wire mesh might also be adequate and much cheaper and more practical from the standpoint of the circulation of air.

Protective Mesh—Transmission Through a Grid of Wires

How effective is a wire screen? It is apparent that the more wide open the mesh, the less the shielding. In order to answer this question, some early published work [5] was reviewed. Laboratory tests were made both at the Bell Telephone Laboratories in Whippany,

[8] See [9], p. 220.

N. J., and at Wheeler Laboratories in Great Neck, N. Y., in an attempt to check the earlier work. It was found that *good* agreement was obtained at the *same frequency* as that of the earlier experiment, but that quite *poor* agreement was obtained at *three times that frequency.*

Looking further into the published literature, the formulas of Schelkunoff and Sharpless [4] and that of Marcuvitz [17] were tried. A somewhat better agreement with the data was obtained, but still neither formula was satisfactory. However, by combining the two formulas properly, since one neglected one thing and the other neglected something else, an empirical formula was derived which checked all of the available experimental data quite well. This empirical formula was then revised so as to make it suitable for use in nomographic form and the nomograph was constructed, as seen in Fig. 16. It applies to normal incidence on a grid of parallel wires of diameter $2r$ having a spacing a between centers. The electric vector is assumed to be parallel to the wires. To use the nomograph, a straight edge is aligned with one point at the left corresponding to the spacing in wavelengths and another point at the right corresponding to the ratio of the spacing to the radius of the wire. At the point where the straight edge crosses the line in the middle labeled "Transmitted Power (db)," the shielding effectiveness of the grid of wires is expressed in decibels.

The modified empirical formula upon which this nomograph was based is the following

$$\frac{P_0}{P_L} = \frac{B^2}{4}, \qquad (32)$$

where

$$|B| = \frac{\lambda}{a} \frac{1}{\ln\left[\frac{\left(.83 \exp \frac{2\pi r}{a}\right)}{\exp\left(\frac{2\pi r}{a}\right) - 1}\right]}$$

P_0 = Incident power,

P_L = Transmitted power.

The accuracy of the nomograph appears to be slightly better than ± 1 db, judging from a comparison with the available measured data.

The results are applicable equally well to a screen of perpendicular wires by ignoring one or the other set of parallel wires forming the mesh.

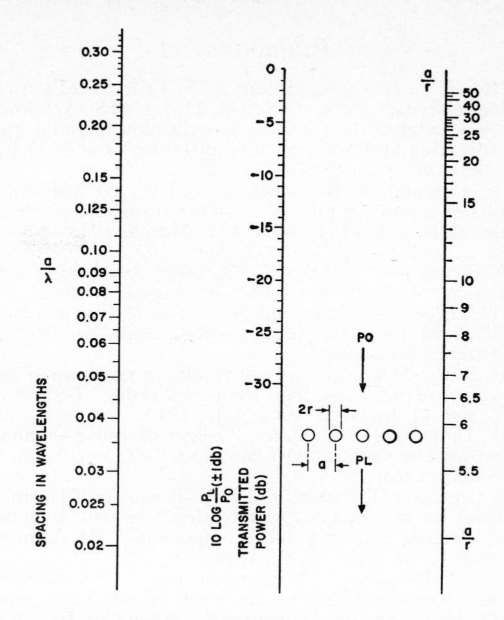

Fig. 16—Transmission through a grid of wires of radius r and spacing a.

Acknowledgment

The author is grateful indeed for the many helpful suggestions of his colleagues within the Bell System and for their cogent observation and comments acquired through frank discussions and internal memoranda. In addition to those whose names appear in the text of this paper, the following contributors have participated:

L. G. Abraham
R. B. Bagby
J. S. Baynard
J. B. Bishop
D. C. Brown
C. E. Becraft
N. H. Brown
G. D. Calvert
H. E. Curtis
A. B. Crawford
R. D. Campbell
R. H. Card
M. R. Dungan
H. W. Evans
D. W. Eitner
R. S. Engelbrecht
G. R. Frantz
R. R. Garreau
E. P. Gelvin, M.D.
E. Halbach
R. R. Hough
W. C. Jakes
D. Jarett
S. F. Knakkergaard
E. W. Kuhlmann, Jr.
J. F. MacMaster
C. E. Martin, M.D.
T. W. Madigan
J. P. Molnar
R. J. Phillips
J. R. Ranucci
R. J. Salhany
G. M. Smith
P. H. Smith
N. F. Schlaack
E. G. Schowalter, M.D.
R. L. Shanahan
W. C. Tinus
W. H. Tidd
W. H. Thatcher
J. B. Vreeland
C. A. Warren
F. E. Willson
H. A. Wenk
L. H. Whitney, M. D.
M. M. Weiss
M. G. Williams.

The motivation for this review arose originally in connection with the research and development program of the NIKE guided missile systems. This phase of the study was sponsored by the U. S. Army Ordnance Corps under contract with Western Electric Company (#DA-30-069-ORD-1955). The liaison personnel, Colonel L. G. Jones, Colonel R. C. Miles, Major R. D. Fuller, and J. J. Turner, stimulated this effort with many fruitful discussions. The advice and comments of Colonel L. C. MacMurray of the U. S. Army Hygiene Agency were helpful indeed.

Bibliography

[1] R. Bown, C. R. Englund, and H. T. Friis, "Radio transmission measurements," Proc. IRE, vol. 11, pp. 115–152; April, 1923.

[2] R. Bown and G. D. Gillett, "Distribution of radio waves from broadcasting stations over city districts," Proc. IRE, vol. 12, pp. 395–409; August, 1924.

[3] C. R. Englund, A. B. Crawford, and W. W. Mumford, "Some results of a study of ultra-short-wave transmission phenomena," Proc. IRE, vol. 21, pp. 464–492; March, 1933. Also *Bell Sys. Tech. J.*, vol. 12, pp. 197–227; April, 1933.

[4] S. A. Schelkunoff and W. M. Sharpless, "Reflecting Systems for Antennas," U. S. Patent 2,292,342; August 4, 1942.

[5] W. D. Hayes, "Gratings and Screens as Microwave Reflectors," M.I.T. Rad. Lab., Cambridge, Mass., Rept. No. 54-20; April 1, 1943 (now declassified).

[6] L. E. Daily, "Clinical study of results of exposure of laboratory personnel to radar and high frequency radio," *U. S. Naval Med. Bull.*, vol. 41, pp. 1052–1065; July, 1943.

[7] B. I. Lidman and C. Cohn, "Effect of radar emanations on hematopoietic system," *Air Surgeons Bull.*, vol. 2, pp. 448–449; December, 1945.

[8] E. I. Green, H. J. Fisher, and J. G. Ferguson, "Techniques and facilities for microwave radar testing," in Bell Telephone Lab. Staff, "Radar Systems and Components," D. Van Nostrand Co., Inc., New York, N. Y.; 1949.

[9] R. N. Griesheimer, "Microwave power measurements," in "Techniques of Power Measurements," C. G. Montgomery, Ed., M.I.T. Rad. Lab. Ser., McGraw-Hill Book Co., Inc., New York, N. Y., vol. 11, ch. 3, pp. 79–218; 1947.

[10] A. M. Richardson, T. D. Duane, and H. M. Hines, "Experimental lenticular opacities produced by microwave irradiations," *Arch. Phys. Med.*, vol. 29, pp. 765–769; December, 1948.

[11] E. J. Imig, J. D. Thomson, and H. M. Hines, "Testicular degeneration as a result of microwave irradiation," *Proc. Soc. Exp. Biol. and Med.*, vol. 69, pp. 383–386; November, 1948.

[12] J. W. Clark, H. M. Hines, and W. W. Salisbury, "Exposure to microwaves," *Electronics*, vol. 22, pp. 66–67; May, 1949.

[13] W. H. Olendorf, "Focal neurological lesions produced by microwave irradiation," *Proc. Soc. Exp. Biol. and Med.*, vol. 72, pp. 432–434; November, 1949.

[14] The Bell Telephone Lab. Staff, "Radar System and Components," D. Van Nostrand Co., Inc., New York, N. Y.; 1949.

[15] J. W. Clark, "Effects of intense microwave radiation on living organisms," Proc. IRE, vol. 38, pp. 1028–1032; September, 1950.

[16] G. C. Southworth, "Principles and Applications of Waveguide Transmission," D. Van Nostrand Co., Inc., New York, N. Y.; 1950.

[17] N. Marcuvitz, "Waveguide Handbook," M.I.T. Rad. Lab. Ser., McGraw-Hill Book Co., Inc., New York, N. Y., vol. 10; 1951.

[18] S. A. Schelkunoff and H. T. Friis, "Antennas Theory and Practice," John Wiley and Sons, Inc., New York, N. Y.; 1952.

[19] F. G. Hirsch, "Microwave Cataracts," presented at NSAE Conf. on Industrial Health, Cincinnati, Ohio; April 25, 1952.

[20] A. L. Cullen, "A general method for the absolute measurement of microwave power," IEE Monog. No. 23; February 15, 1952. Also, *Proc. IEE*, pt. IV, vol. 99, pp. 112–120; April, 1952.

[21] L. Daily, Jr., K. G. Wakim, J. F. Herrick, E. M. Parkhill, and W. L. Benedict, "The effects of microwave diathermy on the eye of the rabbit," *Am. J. Ophthalmol.*, vol. 35, pp. 1001–1017; July, 1952.

[22] H. M. Hines and J. E. Randall, "Possible industrial hazards in the use of microwave radiation," *Electrical Engrg.*, vol. 71, pp. 979–881; October, 1952

[23] F. G. Hirsch and J. T. Parker, "Bilateral lenticular opacities occurring in a technician operating a microwave generator," *AMA Arch. Ind. Health*, vol. 6, pp. 512–517; December, 1952.

[24] USAF "Air Force Radio Frequency Radiation Hazards Handbook," Rome A. F. Depot, Griffiss AFB, N. Y., Tech. Order 31P-1-15; November 1, 1954.

[25] M. Sucher, " A comparison of microwave power measurement techniques," *Proc. Symp. on Modern Advances in Microwave Techniques*, Polytechnic Inst. of Brooklyn, Brooklyn, N. Y., pp. 309–323; November 8–10, 1954.

[26] C. I. Barron, A. A. Love, and A. A. Baraff, "Physical evaluation of personnel exposed to microwave emanations," *J. Aviation Med.*, vol. 26, pp. 442–452; December, 1955.

[27] C. I. Barron, A. A. Love, and A. A. Baraff, "Physical evaluation of personnel exposed to microwave emanations," IRE Trans. on Medical Electronics, vol. ME-4, p. 44; February, 1956.

[28] S. I. Brody, "Military aspects of biological effects of microwave radiation," IRE Trans. on Mecidal Electronics, vol. ME-3, pp. 8–9; February, 1956.

[29] T. S. Ely and D. E. Goldman, "Heat exchange characteristics of animals exposed to 10 cm microwaves," IRE Trans. on Medical Electronics, vol. ME-4, pp. 38–43; February, 1956.

[30] F. H. Krusen, "Address of welcome," IRE Trans. on Medical Electronics, vol. ME-4, pp. 3–4; February, 1956.

[31] H. P. Schwan and K. Li, "The mechanism of absorption of ultrahigh frequency electromagnetic energy in tissues, as related to the problem of tolerance dosage," IRE Trans. on Medical Electronics, vol. ME-4, pp. 45–49; February, 1956. Also in, Proc. IRE, vol. 44, pp. 1572–1581; November, 1956.

[32] B. L. Vosburgh, "Problems which are challenging investigators in industry," IRE Trans. on Medical Electronics, vol. ME-4, pp. 5–7; February, 1956.

[33] D. B. Williams, J. P. Monahan, W. J. Nicholson, and J. J. Aldrich, "Biologic effects studies on microwave radiation, time and power thresholds for the production of 12.3 cm microwaves," IRE Trans. on Medical Electronics, vol. ME-4, pp. 17–22, February, 1956.

[34] M.I.T. Lincoln Lab., Lexington, Mass., Div. 3, Quart. Prog. Rept.; July 15, 1956.

[35] M.I.T. Lincoln Lab., Lexington, Mass., Div. 3, Quart. Prog. Rept.; July 15, 1957.

[36] T. S. Ely, D. E. Goldman, J. Z. Hearon, R. B. Williams, and H. M. Carpenter (USN), "Heating characteristics of laboratory animals exposed to ten-centimeter microwaves," Naval Med. Res. Inst., Bethesda, Md., Res. Rept. Project NM001 056.13.02; March 21, 1957.

[37] J. T. McLaughlin, "Tissue destruction and death from microwave radiation (radar)," *Calif. Med.*, vol. 86, pp. 336–339; May, 1957.

[38] "Industrial Hazards," Hdq., RADC, Griffiss AFB, N. Y., RADC Regulation NR 160-1; May 31, 1957.

[39] USAF "Microwave Radiation Hazards," Rome A. F. Depot, Griffiss AFB, N. Y., Urgent Action Tech. Order 31-1-511; June 17, 1957.

[40] G. M. Knauf, "The Biological Effects of Microwave Radiation on Air Force Personnel," presented at 106th Annual AMA Meeting, New York, N. Y., June 7, 1957.

[41] "Biologic effects of radio frequency energies 1940–1957," *Proc. 1st Annual Tri-Service Conf. on Biological Hazards of Microwave Radiation*, RADC, Grifiss AFB, N. Y.; July 15–16, 1957. Bibliography, Appendix B, pp. 94–103.

[42] "Microwaves and their biological effects," *Proc. 1st Annual Tri-Service Conf. on Biological Hazards of Microwave Radiation*, RADC, Griffiss AFB, N. Y.; July 15–16, 1957. Bibliography, Appendix E, pp. 111–114.

[43] J. S. Burgess, "High power microwave facilities," *Proc. 1st Annual Tri-Service Conf. on Biological Hazards of Microwave Radiation*, RADC, Griffiss AFB, N. Y.; July 15–16, 1957.

[44] R. L. Dondero, "Determination of power densities at microwave frequencies," *Proc. 1st Annual Tri-Service Conf. on Biological Hazards of Microwave Radiation*; RADC, Griffiss AFB, N. Y.; July 15–16, 1957. Appendix F, pp. 115–118.

[45] "Microwave exposure discussion," *Proc. 1st Annual Tri-Service Conf. on Biological Hazards of Microwave Radiation*, RADC, Griffiss AFB, N. Y., pp. 79–88, July 15–16, 1957.

[46] G. M. Knauf, "Program for the investigation of the biological effects of electromagnetic radiation at the Rome Air Development Center," *Proc. 1st Annual Tri-Service Conf. on Biological Hazards of Microwave Radiation*, RADC, Griffiss AFB, N. Y., pp. 35–46; July 15–16, 1957.

[47] G. M. Knauf, "Investigation of the biological effects of electromagnetic radiation," *Proc. 1st Annual Tri-Service Conf. on Biological Hazards of Microwave Radiation*, RADC, Griffiss AFB, N. Y., pp. 89–93; July 15–16, 1957.

[48] H. P. Schwann, "The physiological basis of injury," *Proc. 1st Annual Tri-Service Conf. on Biological Hazards of Microwave Radiation*, RADC, Griffiss AFB, N. Y., pp. 60–63; July 15–16, 1957.

[49] D. B. Williams and R. S. Fixott, "A summary of the SAMUSAF program for research on biomedical aspects of microwave radiation," *Proc. 1st Annual Tri-State Conf. on Biological Hazards of Microwave Radiation*, RADC, Griffiss AFB, N. Y., pp. 6–19; July 15–16, 1957.

[50] R. H. Card, "The Hazards of Radio Transmitters and Their Correction," presented at Natl. Safety Congr., Chicago, Ill.; October 24, 1957.

[51] H. B. Zackison, Sr. (USA), "Information for Protection against Microwave Radiation in Connection with DA Circular 40-2-Army ENGEU 57-155," U. S. Army; December 31, 1957.

[52] G. M. Knauf, "The biological effects of microwave radiation on Air Force personnel," *AMA Arch. Ind. Health*, vol. 17, pp. 48–52; January, 1958.

[53] "Radio Frequency Hazards," Handbook Bu Aer or USAF Tech. Order 31-1-80, April 15, 1958.

[54] "Radio Frequency Radiation Hazards," Handbook U. S. Air

Force Tech. Order 31-1-80, April 15, 1958. Revised May 15, 1958.
[55] C. Susskind and Staff, "Biological Effects of Microwave Radiation," Annual Scientific Rept., University of California, Inst. of Engrg. Res., Ser. 60, No. 205, Contract AF41 (657)-114; June 30, 1958.
[56] Lincoln Lab. M.I.T., Div. 3, Quart. Prog. Rept.; July 15, 1958.
[57] R. L. Carpenter, "Review of work conducted at Tufts University," *Proc. 2nd Annual Tri-Service Conj. on Biological Effects of Microwave Energy*, RADC, Griffiss AFB, N. Y., pp. 146–168; July 8–10, 1958.
[58] H. Davis, "Discussion of long range development plans in the Air Force," *Proc. 2nd Annual Tri-Service Conj. on Biological Effects of Microwave Energy*, RADC, Griffiss AFB, N. Y., pp. pp. 19–32; July 8–10, 1958.
[59] F. W. Hartman, "The pathology of hyperpyrexia," *Proc. 2nd Annual Tri-Service Conf. on Biological Effects on Microwave Energy*, RADC, Griffiss AFB, N. Y., pp. 54–69; July 8–10, 1958.
[60] J. F. Herrick, "Pearl-chain formation," *Proc. 2nd Annual Tri-Service Conf. on Biological Effects of Microwave Energy*, RADC, Griffiss AFB, N. Y., pp. 88–96; July 8–10, 1958.
[61] M. L. Keplinger, "Review of work conducted at University of Miami," *Proc. 2nd Annual Tri-Service Conf. on Biological Effects of Microwave Energy*, RADC, Griffiss AFB, N. Y., pp. 215–233; July 8–10, 1958.
[62] G. M. Knauf, "New concepts in personnel protection," *Proc. 2nd Annual Tri-Service Conf. on Biological Effects of Microwave Energy*, RADC, Griffiss AFB, N. Y., pp. 49–55; July 8–10, 1958.
[63] L. C. MacMurray, "Microwave radiation hazard problems in the U. S. Army," *Proc. 2nd Annual Tri-Service Conf. on Biological Effects of Microwave Energy*, RADC, Griffiss AFB, N. Y., pp. 79–87; July 8–10, 1958.
[64] S. Michaelson, R. Dundero, and J. W. Howland, "Review of work conducted at the University of Rochester," *Proc. 2nd Annual Tri-Service Conf. on Biological Effects of Microwave Energy*, RADC, Griffiss AFB, N. Y., pp. 175–185; July 8–10, 1958.
[65] J. Roman, "Radio frequency hazards aboard Naval ships," *Proc. 2nd Annual Tri-Service Conf. on Biological Effects of Microwave Energy*, RADC, Griffiss AFB, N. Y., pp. 70–87; July 8–10, 1958.
[66] H. P. Schwan, "Molecular response characteristics to ultrahigh frequency fields," *Proc. 2nd Annual Tri-Service Conf. on Biological Effects of Microwave Energy*, RADC, Griffiss AFB, N. Y., pp. 33–48; July 8–10, 1958.
[67] H. P. Schwan, "Survey of microwave absorption characteristics of body tissues," *Proc. 2nd Annual Tri-Service Conf. on Biological Effects of Microwave Energy*, RADC, Griffiss AFB, N. Y., pp. 126–145; July 8–10, 1958.
[68] B. L. Vosburgh, "Recommended tolerance levels of M-W energy; current views of the General Electric Company's health and hygiene service," *Proc. 2nd Annual Tri-Service Conf. on Biological Effects of Microwave Energy*, RADC, Griffiss AFB, N. Y., pp. 118–125; July 8–10, 1958.
[69] W. H. Tidd, "White Alice," *Bell Labs. Rec.*, vol. 36, pp. 278–283; August, 1958.
[70] "Hazards to Health from Microwave Energy " Hdq., Dept. of the Army, U. S. Army Regulation No. 40-583; September 9, 1958.
[71] C. I. Barron and A. A. Baraff, "Medical considerations of exposure to microwaves (radar)," *J. A MA*, vol. 168, pp. 1194–1199; November, 1958.
[72] J. H. Heller and A. A. Teixeira-Pinto, "A new physical method of creating chromosomal aberrations," *Nature*, vol. 183, pp. 905–906; March, 1959.
[73] P. Bailey, "High intensity radiation produces convulsions death in monkey," *Aviation Week*, vol. 70, pp. 29–30; May, 1959.
[74] W. W. Mumford and M. M. Weiss, "Microwave Radiation Hazards," presented at the 4th Annual Meeting of the Health Physics Soc., Gatlinburg, Tenn.; June 18, 1959.
[75] W. B. Deichmann, F. H. Stephens, Jr., M. Keplinger, and K. F. Lampe, "Acute effects of microwave radiation on experimental animals (24,000 mc)," *J. Occ. Med.*, vol. 1, pp. 369–381; July, 1959.
[76] C. Susskind, Ed., "Proceedings of Third Annual Tri-Service Conference on Biological Effects of Microwave Radiating Equipments," RADC, Griffiss AFB, N. Y., Tech. Rept. RADC-TR-59-140; August, 1959.
[77] R. W. Bickmore and R. C. Hansen, "Antenna power densities in the Fresnel region," PROC. IRE, vol. 47, pp. 2119–2120; December, 1959.
[78] J. Hannock and M. M. Weiss, Eds., "Radiation Hazard Management," D. Van Nostrand Co., Inc., Princeton, N. J.; 1961.

Microwave/Radiofrequency Protection Standards: Concepts, Criteria and Applications*

Sol M. Michaelson

Department of Radiation Biology and Biophysics, University of Rochester, School of Medicine and Dentistry, Rochester, NY 14625 U.S.A.

Electromagnetic energies i.e. 300 KHz to 300 MHz (Radiofrequency) and 300 MHz to 300 GHz (Microwaves) can produce biological effects or injury depending on power levels and exposure durations. There is thus a need to set limits on the amount of exposure to radiant energies individuals can accept with safety. Protection standards should be based on scientific evidence but quite often are the result of empirical approaches to various problems reflecting current qualitative and quantitative knowledge.

In considering standards, it is necessary to keep in mind the essential differences between a "personnel exposure" standard and a "performance" standard for a piece of equipment. An exposure standard refers to the maximum safe (incorporating a safety factor) level of power density and exposure time for the whole body or any of its parts. This standard is a guide to people on how to limit exposure for safety. An emission standard (or performance standard) refers not to people but to equipment and specifies the maximum limit of emission close to a device which ensures that likely human exposure will be at levels considerably below personnel exposure limits.

Basic Considerations Illumination of biological systems with MW/RF energy leads to temperature elevation when the rate of energy absorption exceeds the rate of energy dissipation. Whether the resultant temperature elevation is diffuse or confined to specific anatomical sites, depends on: the electromagnetic field characteristics and distributions within the body as well as the passive and active thermoregulatory mechanisms available to the particular biological entity.

Analysis of Literature Elucidation of the biologic effects of microwave exposure requires a careful review and critical analysis of the available literature. Such review requires differentiation of the established effects and mechanisms from speculative and unsubstantiated reports.

Although there is considerable agreement among scientists concerning the biological effects and potential hazards of microwaves, there are areas of disagreement. There also is a philosophical question about the definition of hazard. All effects are not necessarily hazards. In fact, some effects may have beneficial applications under appropriately controlled conditions. MW/RF induced changes must be understood sufficiently so that their clinical significance can be determined, their hazard potential assessed and the appropriate benefit/risk analyses applied. It is important to determine whether an observed effect is irreparable, transient or reversible, disappearing when the electromagnetic field is removed or after some interval of time. Of course, even reversible effects are unacceptable if they transiently impair the ability of the individual to function properly or to perform a required task.

In an analysis of scientific literature to determine the probability of a biological response from exposure to a noxious agent, we must consider the consistency of experimental results claimed, both the nature of the response and the biological system involved, the ability to replicate the results of studies with consistency and whether the results claimed and observations reported can be explained by accepted biological principles.

Experiments with small animals, such as mice and rats, to evaluate the potential effects of MW/RF energy, must be carefully designed and performed. The responses may be the result of another, unrelated agent inadvertently introduced into the experimental design rather than the factor intended to be studied. The fact that a living organism responds to many stimuli is a part of the process of living; such responses are examples of biological "effects". Since biological organisms have considerable tolerance to change, these "effects" may be well within the capability of the organism to maintain a normal equilibrium or condition of homeostasis. If, on the other ha d, an effect is of such an intense nature that it compromises the individual's ability to function properly or overcomes the recovery capability of the individual, then the "effect" should be considered a "hazard". In any discussion of the potential for biological "effects" from exposure to electromagnetic energies we must first determine whether any "effect" can be demonstrated; and then determine whether such an observed "effect" is "hazardous".

When assessing the results of research on biological effects of MW/RF exposure it is important to note whether the techniques used are such that possible effects of intervening factors i.e. noise, vibration, chemicals, variation in temperature, humidity, air flow are avoided and care is taken to avoid population densities that perturb the field to the extent that measurements become meaningless. The sensitivity of the experiment should be adequate to ensure a reasonable probability that an effect would be detected if indeed any exists. The experiment and observational techniques should be objective. Data should be subjected to acceptable analytical methods with no relevant data deleted from consideration. If an effect is claimed, it should be demonstrated at an acceptable level of statistical significance by application of appropriate tests. A given experiment, should be internally consistent with respect to the effect of interest. Finally, the results should be quantifiable and susceptible to confirmation by other investigators.

Principles of Biologic Experiments and Interpretation Proper investigation of the biologic effects of electromagnetic fields requires an understanding and appreciation of biophysical principles and "comparative medicine". Such studies require interspecies "scaling", the selection of biomedical parameters which consider basic

Reprinted with permission from *Book of Papers, vol. II, 5th Int. Congress of the Int. Rad. Prot. Assoc.*, 1980, pp. 407–414.

physiological functions, identification of specific and nonspecific reactions, and differentiation of adaptational or compensatory changes from pathological manifestations.

Scaling Much of the research on biological effects of microwaves has been done with small rodents that have coefficients of heat absorption, field concentration effects, body surface areas, and thermal regulatory mechanisms significantly different from man. Adverse reaction in animals does not prove adverse effect in man, and lack of reaction in animals does not prove that man will not be affected. Even closely related species can differ widely in their response. The literature is replete with "anomalous" reactions. Thus, results of exposure of common laboratory animals cannot be readily extrapolated to man unless some form of "scaling" among different animal species, and from animal to man, can be invoked in an accurate way to obtain a quantitatively valid extrapolation from the actual data observed.

The physical factors that must be considered include: frequency of radiation, intensity, animal orientation with respect to the source, size of animal with respect to the wavelength, portion of the body irradiated, exposure time-tensity factors, environmental conditions (temperature, humidity, air flow), and absorbed heat distribution in the body. In addition, variables such as restraint, metabolic rate, body ratio of volume/surface area, and thermoregulatory mechanisms will affect the biological response to microwaves.

The need for proper dosimetry in experimental procedures and the importance of realistic scaling factors required for extrapolation of data obtained with small laboratory animals to man are clearly required. Maximum absorption in man occurs at 80 MHz and falls off at higher frequencies. Formulas for scaling factors among species are available (1). Five milliwatts per centimeter square, 2450MHz, 5 mW/cm^2 exposure of a small animal such as a mouse or a rat, can result in a thermal effect that could influence the central nervous system and elicit behavioral and other physiologic responses in that animal, but not necessarily in a larger animal or man.

Epidemiology A number of retrospective studies have been done on human populations exposed or believed to have been exposed to MW/RF energies. Those performed in the U.S. (2,3) and Poland (4), have not revealed any relationship of altered morbidity or mortality to MW/RF exposure. Nervous system and cardiovascular alterations in humans exposed to microwaves has been reported in Eastern European literature (5,6,7). Most of the reported effects are subjective, consisting of fatigability, headache, sleepiness, irritability, loss of appetite, and memory difficulties. Psychic changes that include unstable mood, hypochondriasis, and anxiety have been reported. The symptoms are reversible, and pathological damage to neural structures is insignificant. There is considerable difficulty in establishing the presence of, and quantifying the frequency and severity of "subjective" complaints. Individuals suffering from a variety of chronic diseases may exhibit the same dysfunctions of the central nervous and cardiovascular systems as those reported to be a result of exposure to microwaves; thus, it is extremely difficult, if not impossible, to rule out other factors in attempting to relate microwave exposure to clinical conditions.

Protection Guides and Standards: Exposure Standards The first standards for controlling exposure to MW/RF were introduced in the 1950's, in the USA and the USSR. The maximum permissible exposure levels proposed then have remained substantially unchanged, i.e., for continuous exposure these are respectively 10 mW/cm^2 and 10 μW/cm^2. Most countries that developed national standards based them on either the US (8) or the Soviet (9) values. Subsequently, however, some countries have proposed standards intermediate between these extremes.

Current U.S. Government Standards Include: Occupational Standard (general) - OSHA Standard, adopted in 1972, applies to employees in the private sector. An addendum, adopted in 1975, applies to work conditions particularly in the telecommunications industry. OSHA Standards are mandatory for federal employees including the military. Maximum permissible exposure limit is 10 mW/cm^2, for durations greater than 6 min, over the frequency range 10 MHz-100 GHz.

Product Emission Standard - "Radiation Control for Health and Safety Act of 1968" (PL 90-602), administered by HEW/FDA (BRH), provides authority for controlling radiation from electronic devices. BRH microwave oven standard, effective October, 1971: Ovens may not emit (leak) more than 1 mW/cm^2 at time of manufacture and 5 mW/cm^2 subsequently, for the life of the product--measured at a distance of 5 cm and under conditions specified in the standard.

Additionally, there are non-goverment organizations which develop recommended standards and safety criteria, e.g.: American National Standards Institute (ANSI) - A voluntary body with members from government, industry, various associations and the academic community which develops consensus standards (guides) in various areas. ANSI issued a nonionizing radiation safety standard in 1966 with maximum permissible exposures of 10 mW/cm^2, as averaged over any 6 minute period, for frequencies from 10 MHz to 100 GHz which was essentially adopted by OSHA. This standard was reviewed and reissued with minor modifications in 1975. ANSI must review and withdraw, revise or reissue ANSI Standards every five years. Presently Subcommittee 4 of ANSI Committee C-95, which deals with hazards to personnel is reevaluating ANSI's radiofrequency exposure standard for adoption in 1980. The recommendations, based on frequency dependence and specific absorption rates (SAR) state: For human exposure to electromagnetic energy of radiofrequencies from 300 kHz to 100 GHz, the radiofrequency protection guides, in terms of equivalent plane wave free space power density, and in terms of the mean squared electric (E^2) and magnetic (H^2) field strengths as a function of frequency, are:

Frequency (MHz)	Power Density (mW/cm^2)	E^2 (V^2/m^2)	H^2 (A^2/m^2)
0.3 - 3	100	400,000	2.5
3 - 30	$900/f^2$	4,000 $(900/f^2)$	0.025 $(900/f^2)$
30 - 300	1.0	4,000	0.025
300 - 1500	f/300	4,000 (f/300)	0.025 (f/300)
1500 - 100,000	5	20,000	0.125

Note: f is the frequency, in megahertz (MHz)

For near field exposure, the only applicable radiofrequency protection guides are the mean squared electric and magnetic field strengths given in columns 3 and 4. For convenience, these guides may be expressed in equivalent plane wave power density.

For both pulsed and non-pulsed fields, the power density and the mean squares of the field strengths, as applicable, are averaged over any 0.1 hour period and should not exceed the values given in the Table. For situations involving exposure of the whole body, the radiofrequency protection guide is believed to result in energy deposition averaged over the entire body mass for any 0.1 hour period of about 144 joules per kilogram (J/kg) or less. This is equivalent to a specific absorption rate (SAR) of about 0.40 watts per kilogram (W/kg) spatially and temporally averaged over the entire body mass. This recommendation will no doubt be adopted with only minor modifications if any.

The National Institute for Occupational Safety and Health (NIOSH) is developing a criteria document with recommended standards for occupational MW/RF exposures, which is, except for certain modifications, comparable to that recommended by ANSI.

The Radiation Protection Bureau of Health and Welfare Canada is considering "Emission and Exposure Standards for Microwave Radiation". The maximum permissible levels (MPL'S) are 1 $mW\text{-}hr/cm^2$ average energy flux for whole body exposure as averaged over an hour and a maximum exposure during any one minute of 25 mW/cm^2 for occupational settings. The MPL's would apply for the frequency range of 10 MHz-300 GHz. No distinction is made between CW and pulsed waveforms. There is no lower MPL for the general population.

The State Committee on Standards of the Council of Ministers of the USSR has promu gated "Occupational Safety Standards for Electromagnetic Fields of Radiofrequency (GOST 12.1.006-76)," effective January 1, 1977. It specifies the maximum permissible magnitudes of voltage and current density of an EM field in the workplace. It does not apply to personnel of the Ministry of Defense. Maximum permissible RF fields in the workplace must not, during the course of the workday, exceed;

P (mW/cm^2) Stationary Source (See Note 1)	P (mW/cm^2) Rotating, Scanning	E (V/m)	H (A/m)	Frequency Range
			5	60 KHz-1.5 MHz
		50		1.5 MHz-3.0 MHz
		20		3.0 MHz-30 MHz
		10	0.3	30 MHz-50 MHz
		5		50 MHz-300 MHz
0.01	0.1	(entire workday)		300 MHz-300 MHz
0.10	1.0	(2 hr. period during workday)		
1.00	(20 min. period during workday)			

Note 1: Also applies in environments with ambient temperatures above 28°C and/or in the presence of X-ray radiation, except, under these conditions, the maximum during a 20 minute period is restricted to 0.1 mW/cm^2.

There is some indication that the USSR Ministry of Health has endorsed guidelines for maximum exposure limits for the general population which stipulates the maximum allowable levels of electromagnetic energy in human dwellings or in areas of human dwellings, as follows:

P ($\mu W/cm^2$)	E (V/m)	Frequency Range
	20	30-300 KHz
	10	300 KHz-3.0 MHz
	4	3.0-30 MHz
	2	30-300 MHz
5		300 MHz-300 GHz

In 1977, the Polish Ministries of work, Wages and Social Affairs and of Health and Social Welfare promulgated a change in the Polish Standard for occupational exposure. The change extends the frequency range down from 300 to 0.1 MHz;

Hazardous Zone II		Hazardous Zone I		Intermediate Zone		Safe Zone	Frequency Range
Tp		Tp		Tp		Tp	
0	250 A/m 1000 V/m	40/H 150/E	10 A/m 70 V/m	Entire 8 hr workday	2 A/m 20 V/m	No limit	0.1-10 MHz
	300 V/m	3200/E2	20 V/m		7 V/m		10-300 MHz

Tp = Permissible time of exposure/workday (minutes).
E = Electric field (volts/meter).
H = Magnetic field (amps/meter).

In 1976, the Swedish National Board for Industrial Safety, promulgated a nonionizing radiofrequency standard (Worker Protection Authority Instruction No. 111) effective January 1, 1977. This regulation applies to all work which may involve exposure to radiofrequencies between 10 MHz and 300 GHz. The instruction specifically excludes applications involving the treatment of patients. Maximum permissible exposures (as averaged over a six minute period) are:

Power Density	Frequency Range
5 mW/cm^2	10 MHz to 300 MHz
1 mW/cm^2	300 MHz to 300 GHz

The maximum permissible momentary exposure is 25 mW/cm^2.

Emission Standards The best known emission standards concern the maximum permissible leakage from microwave ovens. The Canadian standard (10,11) restricts the maximum leakage to 1 mW/cm^2 at 5 cm from the oven (consumer, commercial and industrial). The U.S. standard (12) specifies a maximum emission level at 5 cm of 1 mW/cm^2 before purchase and 5 mW/cm^2 thereafter which is consistent with standards for the general population in the USSR and Poland (13). The standard applies to domestic and commercial ovens, but not to industrial equipment. This has been adopted in Japan and most of Western Europe.

Conclusion

International cooperation in the development of compatible standards should be encouraged. Towards this end the International Radiation Protection Association (IRPA) charter was broadened in April 1977, to include nonionizing radiation. IRPA has cooperated with the World Health Organization (WHO) in the preparation of a criteria document which is scheduled for 1980. The European Regional Office (ERO) of the World Health Organization is presently preparing a manual on health aspects of exposure to nonionizing radiation. The manual is intended to provide guidance in nonionizing radiation protection and to summarize international experience in the field. Among the topic areas to be included are health aspects of ultraviolet, optical, infrared and laser; microwave RF and ELF fields; ultrasound; licensing, legislation and regulations.

ACKNOWLEDGEMENT

*This paper is based on work performed under contract with The U.S. Department of Energy at the University of Rochester Department of Radiation Biology and Biophysics and has been assigned Report No. UR-3490-1622.

REFERENCES

1. Durney, C.H., C.C. Jacobson, P.W. Barber, H. Massoudi, M.F. Iskander, J.L. Lords, D.K. Ryser, S.J. Allen and J.C. Mitchell (1978). Radiofrequency Radiation Dosimetry Handbook (2nd edition). USAF Report SAM-TR-78-22, Brooks Air Force Base, TX.

2. Robinette, C.D. and C. Silverman (1977). Cause of Death Following Occupational Exposure to Microwave Radiation (Radar) 1950-1974. Pages 337-344, in: Symposium on Biological Effects and Measurement of Radio Frequency/Microwaves. HEW Publication (FDA) 77-8026.

3. Lilienfeld, A.M., J. Tonascia, S. Tonascia, C.H. Libauer, G.M. Canthen, J.A. Markowitz and S. Weida (1978). Foreign Service Health Status Study: Evaluation of Health Status of Foreign Service and other Employees from Selected Eastern European Posts-Final Report July 31, 1978 Contract No. 6025-619073 Dept. of Epidemiol. Johns Hopkins Univ. Baltimore, NTIS PB-288, 1963.

4. Czerski, P. and M. Piotrowski (1972). Proposals for specification of allowable levels of microwave radiation, Medycyna Lotnicza (Polish), No. 39, 127-139.

5. Marha, K., J. Musil and H. Tuha (1968). Electromagnetic Fields and the Living Environment, State Health Publishing House Prague. San Francisco Press, 1971).

6. Petrov, I.R., (ed.) (1970). Influence of Microwave Radiation on the Organism of Man and Animals, Meditsina Press Leningrad.

7. Gordon, Z.V. (1970). Occupational health aspects of radio-frequency electromagnetic radiation. In: Ergonomics and Physical Environmental Factors. Occupational Safety and Health Series, No. 21, International Labour Office, p. 159, Geneva.

8. ANSI (1966). Safety Level of Electromagnetic Radiation with Respect to Personnel, American National Standards Institute, C95.1-1966, C95.1 - 1974, New York.

9. USSR (1958). Temporary Sanitary Rules for Working with Centimeter Waves. Ministry of Health Protection of the USSR.

10. DHEW Canada (1974). Radiation Emitting Devices Regulations. SOR/74-601 23 October 1974. Part III Microwave Ovens Canada Gazette Part V 108, pp. 2822-2825.

11. Repacholi, M.H. (1978). Proposed exposure limits for microwave and radiofrequency radiations in Canada. J. Microwave Power 13, 199-277.

12. USDHEW (1974). Regulations for Administration and Enforcement of the Radiation Control of Health and Safety Act of 1968 paragraph 1030.10 Microwave Ovens DHEW Publ. No. (FDA) 75-8003 pp. 36-37, July, 1974.

13. Baranski, S. and P. Czerski (1976). Biological Effects of Microwaves. Dowden, Hutchinson and Ross, Stroudsburg, PA, 234 p.

Microwave Radiation Safety Standards in Eastern Europe

KAREL MARHA

Abstract—**Research in Eastern Europe on biological effects of microwaves is briefly reviewed and a basic viewpoint involving nonthermal and cumulative effects is presented. Safety standards expressed in terms of dose or irradiation are described based on this viewpoint. It is suggested that differences between these standards and those in the West may become smaller with further study and closer collaboration between researchers in this field.**

Introduction

THE development of microwaves in Socialist countries, particularly in radar applications, has aroused interest in their possible biological effects. It is necessary to consider measures not only for the protection of the health of persons producing, operating, and servicing this equipment, but also for the protection of the population working or living near powerful microwave generators, even if they are not occupationally connected with their operation.

Pioneer efforts in this field began in the Soviet Union in 1953, when a laboratory for the study of electromagnetic waves of radio frequencies for the purpose of systematically studying the biological effects of microwaves and the need for hygiene conditions when working with microwave generators was established at the Institute of Industrial Hygiene and Occupational Diseases, Academy of Medical Sciences of the USSR, Moscow. A similar laboratory was established in Czechoslovakia in 1960 at the Institute of Industrial Hygiene and Occupational Diseases, Prague. Besides these two special laboratories, a considerable number of other laboratories exist, especially in the USSR and in Czechoslovakia, but also in Poland (for example the Institute of Industrial Hygiene in Lodz). At these places experimental research is conducted on the effects of physical factors upon organisms, on the mechanisms of microwave biological effects, and on the study of clinical and hygienic aspects of work in electromagnetic fields. In these laboratories, the effects not only of microwaves, but also of electromagnetic waves of lower frequencies (practically from 0.1 kHz up) are being studied. In the USSR laboratories even exist (for example, at the Institute of Work Safety, Leningrad) which study the biological effects in fields at the power frequency (i.e., 50 Hz) in connection with the production and distribution of electric energy over high-tension networks of 400 kV.

Manuscript received August 13, 1970; revised October 14, 1970.
The author is with the Institute of Industrial Hygiene and Occupational Diseases, Prague, Czechoslovakia.

All this research of considerably complex character is carried out in three basic directions: 1) problems of the mechanism of biological effects of electromagnetic waves; 2) obtaining data for elaboration of standards of maximum admissible fields for the workers and the population; 3) securing protection of the people against adverse effects of electromagnetic fields.

Mechanism of Effects

The basic approach to the solution of questions about the effect of electromagnetic waves upon organisms consists of the admission of the possibility of nonthermal effects (often described in literature, not quite aptly, as specific) along with the heating or thermal effects. This premise strongly influences the actual method of irradiation in experiments. In principle, long-term irradiation of comparatively small intensities is applied. The length of one exposure is usually indirectly proportional to the intensity of the field in order to avoid acute manifestation of the heating effect. This experimental approach already admits, without saying, the possibility of a cumulative effect. Its existence is verified in some experiments. Results of one of them are demonstrated in Fig. 1.

A variety of methods are applied in testing the response of the organism to irradiation. The most frequently used methods are the follow-up of body weight, blood pressure [1], [2], [30], and temperature; the study of the physiology of the higher nervous activity [15], [16], [24] (especially the study of the effect upon conditioned and unconditioned reflexes [2], [6], [14], [26]); the observation of electrophysiology, namely electroencephalography [11], [13]; the follow-up of fertility and estral cycle [9], [10], [19], [22], [28]; and the study of the histochemistry of the nervous system, neurosecretion, and biochemical examinations (especially vitamin C [5], cholinesterase [20], proteins [10], [28], [32], ribonucleic and deoxyribonucleic acid [32], amino acids, residual nitrogen; change in cholinesterase seems to be most sensitive). One of the most sensitive methods already applied for a number of years is the affecting of the irritability of animals [15], [21] (especially mice and rats), as they are especially sensitive to sound stimuli. These audiogenic animals, selected by breeding, undergo an attack similar to epilepsy after a certain sound stimulus; its formation, i.e., length and intensity of the effect of the stimulus, as well as the actual course of the subsequent reaction, may be influenced by irradiation by electromagnetic waves in

Reprinted from *IEEE Trans. Microwave Theory Tech.*, vol. MTT-19, pp. 165–168, Feb. 1971.

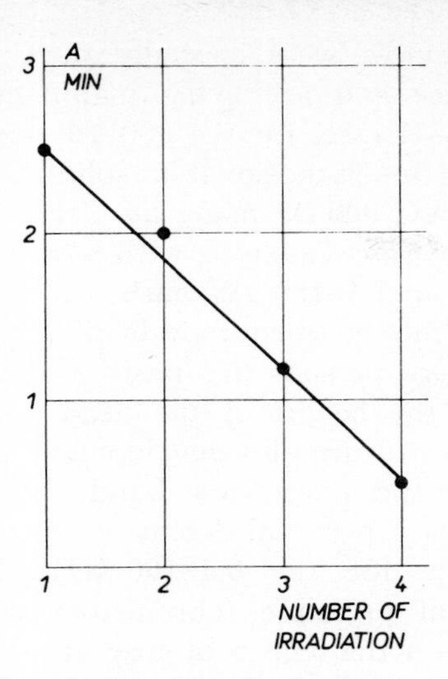

Fig. 1. Shortening of time since the beginning of irradiation (interval A) at which there is the first attempt to escape as a function of the number of exposures. Rats (males, 300 g); exposure, 3 min; time of rest, 10 min; wavelength, 10 cm; pulse operation; repetition frequency, 600 Hz; pulsewidth, 2 μs; power density, 65 mW/cm² [17, pp. 238–241].

the whole width of the frequency spectrum. This method seems to be therefore most suitable for studies of nonthermal effects with relation to physical parameters of the signal, such as frequency, modulation, and so forth.

Recently, research about the effects of the combination of electromagnetic waves with further hazards, be they chemical or physical, has been more widely carried out [1], [12], [24], [25], [31]. The experiments are interesting not only from the theoretical point of view but also from the practical viewpoint, because the generators of electromagnetic waves are increasingly used in various branches of industry as the source of heat energy when processing conducting, semiconducting, or nonconducting substances in induction and dielectric heating. Along with the electromagnetic field there may occur various other factors which in themselves have a noxious effect upon organisms. It is therefore necessary to solve the problem of whether or not the effects of these factors are mutually reinforced. Thus various drugs such as strychnine, nicotine, and cresol are being studied. The effect of the electromagnetic field on rats with lungs dusted with silicotic powder is also being studied. For some time the simultaneous effect of microwaves and soft and hard X-radiation has been studied.

It may be said in general that in practically all the studied effects, it is possible under certain circumstances to observe a so-called double phase [17], i.e. that first stimulation and then inhibition of the studied phenomena occurs. This implies considerable difficulties for proposed clinical methods of examination, because in the presence of a double phase not even the sensitive change of cholinesterase is suitable for clinical studies of the damage of organisms by electromagnetic waves. The higher the frequency, the more significant the double phase seems to be.

Maximum Admissible Values

For the statement of the maximum admissible values of intensity of irradiation in the microwave band, it is necessary to carry out long-duration studies of power density under working conditions near microwave generators in order to evaluate the work place from the point of view of hygiene (e.g., temperature, air moisture, noise), to make clinical–physiological studies of persons working under these conditions, and to carry out experimental research of the biological effects of microwave, mainly in animals.

The basic criterion for the determination of the maximum admissible values of microwave irradiation is the reaction of organisms to small intensities of the field and the accumulation of the biological effect if it is chronic. The idea is that the selected maximum admissible values must not only guarantee protection against possible damage to organisms, but must also exclude adverse subjective complaints such as excessive fatigue, irritation, headache, and so forth under long-duration exposure in microwave fields.

In the USSR such complex follow-up of the health conditions in workers for approximately eight years disclosed that, in many persons, a comparatively small intensity of irradiation (hundreds of microwatts per square centimeter) for a long duration caused some functional changes to occur in the organism, manifested especially in electroencephalographic examination [4], [15], [29]. Experimental research in animals exposed to the chronic effect of the VHF field also disclosed that at an intensity of the field up to 1 mW/cm² functional and morphologic changes [7] (changes in the central nervous system and hypotension) occur. On the basis of the mentioned data and by application of a certain safety coefficient respecting the individual differences among people, a value of 10 μW/cm² per working day was established in the USSR as the maximum admissible value of microwave irradiation. If essential work has to be carried out in environments with higher power density, the period of exposure is limited to 2 h/day at values of 0.01–0.1 mW/cm² and to a maximum of 15–20 min/day at values up to 1 mW/cm². In the latter case the wearing of protective glasses covered by a metal mesh or by a transparent metal film is obligatory.

These values, generally valid in the USSR, were accepted without changes for the military personnel of the Warsaw pact. The maximum value 10 μW/cm² (medium power density) per working day was temporarily accepted in Czechoslovakia for the civil population also, but only for pulsed generators (radars). On the basis of experiments with irradiation of laboratory animals by pulsed and continuous waves in the 10-cm band, we found that with equal thermal parameters (i.e., equal frequency, equal average power density and, therefore, equal heating of the irradiated organism) the pulsed microwaves are biologically much more effective than CW. This is why for CW sources such as microwave ovens, a maximum value of 25 μW/cm² is permitted in

TABLE I

Maximum Admissible Mean Values per Shift or Day of Irradiation by Electromagnetic Waves, Calculated from the Determined Irradiation During 1 Working or Calendar Week, Valid in Czechoslovakia[a]

Frequency Band	Maximum Admissible Mean Values	
	For Workers at HF and UHF Generators	For Other Workers and for the General Population
From 30 kHz to 30 MHz	400 (50 V/m)	120 (5 V/m)
Above 30 MHz to 300 MHz	80 (10 V/m)	24 (1 V/m)
Above 300 MHz to 300 GHz (CW operation)	200 (25 μW/cm²)	60 (2.5 μW/cm²)
Above 300 MHz to 300 GHz (pulse operation)	80 (10 μW/cm²)	24 (1 μW/cm²)

[a] At each value of maximum admissible irradiation, the corresponding maximum mean intensity of the field (volts per meter) in the HF band or the power density (microwatts per square centimeter) in the UHF band is given in parentheses.

Czechoslovakia for 8 h of uninterrupted irradiation. This value must be understood as somewhat arbitrary since it was chosen to agree with the temporary maximum admissible value of the electric field (10 V/m) at frequencies below 300 MHz, using their relation in the far field through the free-space wave impedance.

According to experience gained by us to date in studying the health state of persons working in VHF and UHF environment, it seems probable that it will be possible to increase this value for CW. We believe a tolerable medium admissible value for CW microwaves should be 0.1 mW/cm² for uninterrupted irradiation of 8 h per day. It must be stressed that this is the value measured in the place where a person actually is.

Because we expect the development and manufacture of personal dosimeters in the future, no maximum admissible intensity of the field or power density has been standardized in Czechoslovakia. A dose of integrated power density (or integrated field strength for HF and lower frequencies), with the period in hours during which a person is actually irradiated, has been standardized. This is called the maximum admissible value of irradiation. Table I demonstrates the values presently valid in Czechoslovakia for persons occupationally exposed (these persons are liable to undergo periodic medical controls), as well as for the general population living in the vicinity of power generators of electromagnetic waves.

Health Protection

An essential condition for successful application of standards on maximum admissible intensities of the field, successful clinical evaluation of possible changes in organisms, and finally successful experimental work is a satisfactory field measurement device. It must afford comparatively rapid and accurate measurement of the power density at the place of the irradiated object. Such devices were specially developed for the hygiene service and are being manufactured in the USSR (PO-1 MEDIK for the 300–16 700-MHz band) [7], as well as in Czechoslovakia (solid-state measuring equipment QXC 900 05 made by TESLA Pardubice) [18] for all microwave bands with a basic range without attenuation of 1–150 μW/cm².

Obligatory measurements are in principle conducted at places where persons are positioned during work, especially at the height of the head and the sexual organs. In the future the development of a personal dosimeter for the microwave band also is expected. A prototype of a personal dosimeter was developed in Czechoslovakia for the 0.1–100-MHz band on an electrochemical principle (chronistor) which permits one to ascertain the length of stay at a work place as well as the total irradiation at the height of the head during this period [27]. A similar device is being developed in the USSR [7].

Furthermore, legislative measures and controls in the production plants insure that no installations with industrial generators which might irradiate above tolerance limits at locations where operating personnel are situated are being produced or imported into Czechoslovakia. Each such new installation must be approved by the Chief Hygienist of the Republic before production in series is started. Even after the generators are installed, the system of hygiene establishments controls all generators.

In cases of emergency or other necessary cases, protective clothing with good reflecting properties in the entire microwave band are available for use.

Legislation also provides for increased protection of adolescents and women, as they are most sensitive to the effects of microwaves [3], [9], [22], [23].

Conclusion

In this paper it is possible to give only a brief survey of the problems of the possible microwave hazards in Socialist countries. Some methods of approach mentioned herein differ somewhat from the approaches in other countries. It seems, however, that a more detailed study would disclose that the differences are not as great as they seem on first sight when comparing the maximum admissible values. This would require, however, a deeper analysis and closer contact between the people dealing with these problems.

References

[1] S. B. Aronova, "K voprosu mechanizma dejstvija električeskogo impulsnogo polja na davlenije krovi," *Vop. Kurortol. Fizioterap. Leč h. Kul't.*, vol. 3, 1961, pp. 243–246.

[2] V. A. Baronenko and K. F. Timofejeva, "Vlijanije električeskich polej VČ i UVČ na uslovnoreflektornuju dejatelnost i nekotoryje bezuslovnyje funkcii životnych i čeloveka," *Fiziol. Ž. SSSR im I. M. Sechenova*, vol. 45, no. 2, 1959, pp. 203–207.

[3] G. Cocozza, A. Blasio, and B. Nunziata, "Rilievi sulle embriopatie da onde corte," *La Pediatria Rivista D'igiene Med. Chir. Dell'infazia*, vol. LXVIII, no. 1, 1960, pp. 7–23.

[4] J. Formánek, R. Fischer, and D. Frantíková, "Zdravotnické problémy práce ve vf poli, zejména na vysílacích stanicích," *Zpráva VÚS*, 1961.

[5] P. J. Ganeev: *Trudy VMA im Kirova*, vol. 73, 1957.

[6] Z. V. Gordon, "Voprosy gigieny truda pri rabote s generatorami santimetrovych voln"; *Ž. Gigieny, Epidemol, Mikrobiol. i Imunologii*, vol. 1, 1957, pp. 399–404.

[7] ——, "Voprosy gigeny truda i biologičeskogo dejstvija elektromagnitnych polej sverchvysokich častot," *Medicina*, 1966, pp. 25–125.

[8] ——, personal communication.

[9] S. F. Gorodeckaja, "K charakteristike biologičeskogo dejstvija trechsantimetrovych radiovoln na životnyj organizm," *Voprosy Biofiziki i Mechanizma Dejstvija Ionizirujuščej Radiacii*, Kiev, 1964, pp. 70–74.

[10] J. Grzesik, and F. Kumasza, "Wlyw pole elektromagnetycznego średniej częstotliwosci na organy miesowe i białka krwi biełych myszy," *Med. Pracy*, vol. 11, no. 5, 1960, pp. 323–330.

[11] J. A. Cholodov, "Vlijanije elektromagnitnych polej na centralnuju nervnuju sistemz," *Nauka*, 1966.

[12] D. I. Jakimenko, Lečenije nekotorych nejrotrofičeskich kožnych zabolevanij ultrafioletovym oblučenijem i tokami ultravysokoj častoty v malych dozirovkach," *Vestn Dermatov. i Venerol.*, vol. 3, no. 5, 1961, pp. 33–36.

[13] E. Klimková-Deutschová, *Základy průmyslové neurologie*, Praha, 1956.

[14] N. N. Livšic, "Uslovnoreflektornaja dejatelnost sobak pri lokalnych vozdejstvijach polem UVČ na nekotoryje zony kory bolšich polušarij," *Biofizika*, vol. 2, no. 2, 1957, pp. 197–208.

[15] N. N. Livšic: Rol nervnoj sistemy v reakcijach organizma na dejstvije elektromagnitnogo polja ultravysokoj častoty; *Biofizika*, vol. 2, no. 3, 1957, pp. 378–389.

[16] N. N. Livšic, "Dejstvije polja ultravysokoj častoty na funkciji nervnoj sistemy," *Biofizika*, vol. 3, no. 4, 1958, pp. 426–437.

[17] K. Marha, "Biologické účinky elektromagnetických vln o vysoké frekvenci," *Pract. Lék.*, vol. XV, no. 9, 1963, pp. 238–241, 387.

[18] K. Marha, J. Musil, and H. Tuhá, "Elektromagnetické pole a životní prostredí," *Praha*, 1968.

[19] L. Minecki, "Działanie pól elektromagnetycznych wielkiej częstotliwości na rozwój embrionalny," *Med. Pracy*, vol. XV, no. 6, 1964, pp. 391–396.

[20] S. V. Nikogosjan, "Vlijanije SVČ na aktivnost cholinesterazy v syvorotke krovi i organach u životnych," *O Biol. Dejstviji Sverchvysokich Častot*, Moskva, 1960, pp. 81–85.

[21] J. Novák and V. Černy, "Vliv impulsního elektromagnetického pole na lidský organismus," *ČLČ*, vol. CII, no. 18, 1963, pp. 496–497.

[22] J. A. Osipov, "Gigijena truda i vlijanije na rabotajuščich elektromagnitnych polej radiočastot," *Medicna*, 1965.

[23] A. M. Palladin and F. M. Spaskaja, "K voprosu vlijanija uvf polej na specifičeskije funkciji ženščin rabotajuščich s generatory uvf," *Akuš. i Ginekol.*, vol. 4, 1962, pp. 69–74.

[24] A. S. Presman, "Biologičeskoje dejstvij e mikrovoln," *Usp. So Vrem. Biol*, vol. 51, no. 1, 1961, pp. 82–103.

[25] A. S. Presman and N. A. Levitina, "Vlijanije neteplovogo mikrovolnogo oblučenija na rezistentnost životnych k gamaoblučeniju," *Radiobiol*, vol. 2, no. 1, 1962, p. 170.

[26] T. N. Promtova, "Vlijanije nepreryvnogo električeskogo polja UVČ navysšuju nervnuju dejatelnost sobak v norme i patologii, *Žh. Vysšh. Ner. Dejatel.*, vol. 6, no. 6, 1956, pp. 846–854.

[27] M. Režný, "Prototyp širokopásmového integrátoru elektromagnetického pole," *VÚST A. S. Popova*, Praha, 1969.

[28] M. N. Sadčikova, "Sostojanije nervnoj sistemx pri vozdejstviji polej vysokich i sverchvysokich častot," *Fiz. faktory Vnešnej Sredy*, Moskva, 1960, pp. 177–183.

[29] M. Šercl, "Zur Wirkung der elektromagnetischen Zentimeterwellen auf das Nervensystem des Menschen (Radar)," *Z. Gesamte Hygiene ihre Grenzgebiete*, vol. 7, no. 12, 1961, pp. 897–907.

[30] L. Ulrich and J. Ferlin, "Vliv práce ve vysokovýkonných vysílacích stanicích na nêkteré funkce organismu," *Pract. lék.*, vol. XI, 1959, p. 500.

[31] P. H. Volfovskaja and J. A. Osipov, "K voprosu o kombinirovanom vozdejstviji polja vysokoj častoty i rentgenovskogo izlučenija v proizvodstvennych uslovijach," *Gig Sanit*, vol. 26, no. 5 1961, pp. 18–23.

[32] *O Biol. Vozdejstviji Sverchvysokich Častot*, Moskva, 1960.

NON-IONIZING RADIATION
Dosimetry and Interaction

Arthur W. Guy

A. INTRODUCTION

In the past decade, with the increased involvement of engineers and physicists in research on the biological effects of electromagnetic fields, significant progress has been made toward a better quantitative understanding of the relationship between observed effects and the magnitude of the electromagnetic fields producing the effects. It has been my observation that most researchers feel that the greatest scientific progress has been made in the area of dosimetry, where both the exposure fields and the fields coupled to the body tissues are now being quantified as precisely as in the non-biological areas of radio science. This new technology has been particularly valuable in allowing researchers to improve their experimental methodology and safety regulators to develop a better rationale for exposure standards.

Unfortunately, a few researchers and their well-meaning disciples have resisted the adoption of these dosimetric techniques under the belief that such techniques can be applied only to the interaction mechanisms involving heat. Therefore, they assert, the dosimetric quantities are not suitable for low-level or the so-called non-thermal effects. Such notions are faulty, in fact, somewhat naive, since even to define an effect as low-level or non-thermal requires the quantitative information concerning absorbed energy which these same critics believe has no relevance. The major point of contention appears to be the now widespread and accepted use of the specific absorption rate (SAR) in exposed tissue, rather than the previously used power density, as the important dosimetric quantity. The argument against the use of the SAR appears to be based on the fact that it is a measure of absorbed energy in the tissue which is interpreted by the critics as being exactly equal to the heat generated in the exposed tissue. The critics, therefore, feel that the only physical significance of the SAR is that it is a measure of heating in exposed tissues. What they do not seem to understand is that, regardless of the SAR level or the temperature rise in the exposed tissues, the SAR is a measure of absorbed energy which may or may not all be dissipated as heat. If other mechanisms of interaction are at work in the exposed tissues, some infinitesimal fraction of the SAR must be attributed to this interaction, so

$$SAR = W + N \tag{1}$$

where

$$W = 10^{-3}\rho^{-1}\sigma E^2 \tag{2}$$

is the heating, and

$$N = f(E) + g(H) \tag{3}$$

is the energy absorbed due to possible non-thermal interactions, E is the elelctric field strength, H is the magnetic field strength, f and g are unknown functons of W and H and σ is the electrical conductivity. Virtually all of the seriously proposed and verified non-thermal interactions are related to the electric field strength in the medium since the magnetic permeability of tissue is the same as

Reprinted with permission from *Proc. of the Non-Ionizing Rad. Symp. ACGIH*, Washington, DC, Nov. 26-28, 1979.

free space. It has also been verified by countless numbers of measurements that SAR $\approx$ W in exposed biological tissues since $NW^{-1} << 1$ is infinitesimal. Therefore

$$E = 10^{3/2}\rho^{1/2}\sigma^{-1/2}(SAR)^{1/2} \quad (5)$$

so

$$N = f(10^{3/2}\rho^{1/2}\sigma^{-1/2}(SAR)^{1/2}) \quad (6)$$

Thus, the use of the quantity SAR does not imply that the only interaction mechanism is heat, SAR can be used as a meaningful quantity for relating any biological effects to electromagnetic fields with the possible exception of direct magnetic effects. In general, however, a given magnitude of SAR or electric field strength E implies a certain range of values for H and therefore, may also be useful to a certain extent in quantifying possible H type interactions.

B. QUANTITIES AND UNITS

Figure 1 illustrates the quantities and units used for guides and standards pertaining to the exposure of humans to electromagnetic fields. The guides are specified in terms of $\bar{E}$, expressed in units of V/m; the magnetic field strength, $\bar{H}$, expressed in units of A/m or power density expressed units of mW/cm^2. The latter may be obtained for a radiation field through a measurement of either E or H and relating to power density by the equations shown in Figure 1. These equations have also been used for near zone or nonradiation fields with the understanding that the calculated result is an "apparent" power density (APD) rather than a true power density.

Figure 2 illustrates that even though one does measure the incident external fields or power density, I, with a survey meter, the quantity, by itself, is not useful for determining what is or may be happening inside the body of a test animal or man exposed to the fields. If one is to use the biological results from animal experiments for assessing the biological consequences of human exposure in order to establish safe level of exposure, simply relating the incident energy measured with a survey meter, is not enough. The characteristics of energy absorption are considerably different in animals and man exposed to the same I. The scattered energy, S, transmitted energy, T, and reflected energy between tissue boundaries, R, vary markedly as a function of the exposure wave length, exposed body size and shape, and the exposed body orientation. In addition, any laboratory instrumentation in contact with the tissues of the animal will also perturb and often may considerably enhance the absorbed energy or SAR in the tissues of the exposed subject. In order to relate the biological effects observed in the exposed laboratory animal to potential effects in exposed man, one has to quantify the fields in the tissues. Since the SAR has been the easiest quantity to measure and is an index of the internal electric field strength, its use has been widely accepted. The quantity has been defined by the National Council on Radiation Protection and Measurements (NCRP) as an important quantity for use in nonionizing radiation dosimetry (1). One should understand the difference, however, between the spatially dependent term, SAR, which is defined for a point in the tissue and the term average SAR, which is a whole body average of the SAR.

C. DOSIMETRY

Figure 3 illustrates how the average SAR varies in bodies of various sizes, exposed to 10 mW/cm^2 radiation fields at various frequencies based on the work of Gandhi and his colleagues (2). The curve at the left of the figure illustrates the absorption characteristics for a prolate spheroidal model representing man, 175 cm long, oriented parallel to an incident electric field. The curve shows that the average SAR, for the man model, exposed to radiation at frequencies from 10 through 10,000 MHz, first increases as the square of the frequency, reaching a maximum of 2 W/Kg at about 70 or 80 MHz then decreases gradually, approaching an asymtote of approximately 0.16 W/Kg.

When a a smaller body simulating a rat, 20 cm long is exposed, the maximum absorption occurs at a higher frequency, 650 MHz, and when even a smaller 7.5 cm long body simulating a mouse is exposed, we see that the maximum absorption is within the microwave frequency range. Each body, therefore, has a characteristic resonant frequency depending on the length of the long axis. From the curves we see that the average SAR would be very high for a 7.5 cm long animal exposed to 10 mW/cm^2 at its resonant frequency, but for a man, however, exposed to the same frequency the decrease in coupling would result in an order of magnitude or more decrease in average SAR. This is the kind of problem that we must consider when using the data from animal experiments for the setting of safe radiation exposure guides for man. The star symbols above the curves indicate the maximum SAR that is expected for the three examples based on laboratory experiments and theoretical studies on models (3) (4) (5). For reference, the horizontal dashed lines indicate the level basal metabolic rate for man and the level of SAR that is associated with vigorous heating in clinical diathermy treatments. The vertical dashed line denotes the important Industrial, Scientific, and Medical (ISM) frequency of 2450 MHz.

Figure 4 illustrates the kind of problem one can encounter if one does not take the scaling of dosimetry quantities into account. The small table in the figure compares the average SAR in the body of the mouse to that in the body of a man when each is exposed to 10 and 2000 MHz at various magnetic field strengths corresponding to some existing human exposure standards. The allowed magnetic field strength is used as the specified quantity in this illustration since its coupling characteristics to the body are important in the scaling of dosimetric quantities. In the first case, a mouse exposed to 10 MHz, 10 mW/cm^2 power density will be subjected to a magnetic field strength of 0.5 A/m which would produce circulating eddy currents in the tissue resulting in an average SAR that is proportional to the cross sectional area of the body. Since the cross sectional area of a mouse is small, the SAR is extremely small, only 9.2 W/kg. When a man is exposed to the same fields, we see that the SAR is three orders of magnitude greater as a result of increased body cross section. The value of 4.7 mW/kg, however is still biologically insignificant.

If the magnetic exposure field was increased to 250 A/m (equivalent to an East European safety guide for exposure periods up to 20 minutes), the SAR in the mouse would still be relatively small, only 2.3 W/kg, which is significantly less than the basal metabolic rate of the animal. The average SAR in man, however, exposed to his level would be very high, indeed, reaching a value of 1.2 kW/kg which could prove to be fatal as a result of overheating in a matter of seconds. In another example where the mouse and man are exposed to a 2000 MHz, 0.5 A/m field is illustrated in Figure 4, the mouse would experience an average SAR more than 32 times greater than that experienced by the man. The mouse could suffer biological effects that would not occur in the man in this case. The above examples, illustrate the kind of traps that researchers, uninformed of electromagnetic coupling characteristics, may fall into. In the latter case, if an uninformed researcher observed biological effects in the mouse exposed to 2000 MHz, he would err on the conservative side in assuming that the same exposure was dangerous to a human, when in fact, it could be 32 times less dangerous. On the other hand, if he failed to account for dosimetry scaling, his conclusions pertaining to the former case could lead to disasterous results. I have heard Soviet and East European scientists report negative effects in work involving the exposure of mice and other small animals to magnetic fields that are orders of magnitude above those allowed by the exposure guides of Western countries. One wonders if the lack of dosimetry scaling in their work accounts for the fact that their magnetic field exposure guides are equal to or above those allowed in the Western countries?

D. MEASUREMENT OF AVERAGE SAR

I will now discuss a method for quantifying the average SAR in laboratory animals exposed to electromagnetic fields. One of the techniques that has been used for some years, developed originally by Philips and Hunt at the Battelle Pacific Northwest Research Laboratories (6), involves the use of a specially designed twin-well calorimeter. The operation of of this calorimeter may best be described by noting its construction details as illustrated in Figures 5 through 8. Figure 5 illustrates the inner components of the calorimeter consisting of two brass cylinders each large enough to hold a particular animal of interest. Each cylinder is surrounded by an array of thermocouples attached to the outer walls connected in series so that the individual voltages are additive. The array on the right cylinder, however, is connected to the array on the left cylinder so that voltages are subtractive. Thus, if both cylinders are at the same temperature, the total output voltage from the pair of cylinders would be zero. Any temperature difference between the cylinders, however, would result in a net voltage output from the thermocouple arrays. The twin cylindrical wells are surrounded by an outer oval shaped cylinder designed to be held at constant temperatures by a circulating fluid as shown in Figure 6. The figure illustrates two of such outer jackets, each designed for a different sized twin-well calorimeter. In the final step of construction, the air gaps in between the cylinders illustrated in Figure 7 are filled with a foam material. Two different size calorimeters are shown in Figure 8 with inlet and outlet ports for the circulating fluid. After animals are placed in the wells, a styrofoam cover is attached as shown at the left of the figure.

Figure 9 illustrates the time characteristics of the output voltage from the twin-well calorimeter. If one puts a freshly killed animal in each well at a time -2.5 hours, any temperature difference will result in first a rapid rise in output voltage followed by an exponential decay as the body temperature of the animals approach each other and that of the outer jacket of the twin-well. If heat is added to one animal at time zero hours, there will be a sharp rise followed by a decay in temperature. The cross hatched area between the two curves in Figure 9 is proportional to the added energy. If the output from the calorimeter is allowed to become negligibly small before the heat is added to one of the animals, the time integrated voltage may be used to calculate the total added heat. This system can be calibrated with a known amount of ice at 0 C or by adding a known amount of heat to one of two identical phantom bodies contained in the twin-well calorimeter. After the system is calibrated, any microwave energy added to an exposed animal body may be determined and the average SAR calculated.

The use of a microprocessor system shown in Figure 10 greatly improves the accuracy of the integration over the large number of hours required to quantify the added heat. The data and calculations may then be stored on a floppy disk for later reference. Though twin-well calorimetry offers a very powerful method for quantifying the absorbed energy or the average SAR in the body of an exposed animal, it will not provide the researcher a clear picture of the actual distribution of deposited energy. This is probably one of the main arguments of those who criticize the use of the SAR in the quantification of biological effects.

E. MEASUREMENT OF SAR DISTRIBUTION

Figure 11 illustrates some of the methods that we have used in the past at the University of Washington for quantifying the SAR or the electric field strength at a particular location in exposed (live or dead) tissue. The electric field strength in the tissue may be directly measured with a diode and voltmeter combination. The leads of the dipole serve as a dipole antenna that impresses a radio frequency (RF) voltage across the diode that is proportional to the electric field strength. The RF voltage is rectified by the diode and applied to a voltmeter through high resistance leads that are transparent to the incident fields. The dc voltage may be calibrated to provide a measure of the electric field strength either at the surface of the tissue or in the tissue, depending on where the diode is placed. The measured electric field strength and known electrical conductivity of the tissue may be used to calculate the SAR by equation 5. Another method for point by point measurement of SAR is the use of a standard miniature thermocouple and a glass micro pipette shown at the bottom of Figure 11. The glass pipette is first placed in the tissue with the tip in the region where one wishes to measure the SAR. The thermocouple is then placed in the micro pipette with its tip at the point. Figure 12 illustrates how the thermocouple may be connected to a voltmeter and recorder for recording temperature. The temperature is first recorded prior to exposure of the tissue. The thermocouple is then quickly withdrawn and the microwave fields are applied at a sufficiently high level to produce a rapid but limited temperature rise. The microwaves are then turned off after a brief exposure (appreciably less than a minute in duration). The thermocouple is reinserted and the temperature recorded for a period of time to characterize a temperature versus time curve. The curve is then extrapolated to the end of the exposure period and the net temperature change before and after the microwave exposure is used to calculate the SAR. One should keep in mind that even though the measurement is based on a thermal method, it does not mean that the information obtained cannot be used in determining dosimetric quantities for low-level exposure. The high level exposure provide the same correlation factors relating the incident field to the SAR and the tissue, regardless of exposure levels as long as the temperature changes during the dosimetry measurement are limited to a few degrees.

This above method has been used successfully in cataract research on rabbits. In this work, the SAR was measured as a function of position in the exposed rabbit eye, including the anterior chamber, the lens, the vitreous humor, and the tissue behind the eye by moving the probe along an axis as shown in Figure 13. Based on data obtained from five rabbits in these experiments, the SAR distribution as shown in Figure 14 is obtained. The results are normalized to an incident power density of 1 mW/cm^2 on the scale at the left of the curve and for an applicator input power of 1 W on the scale at the right of the curve. It should be noted for this case, that the measured exposure fields are too close to the applicator (5cm) to be classified as true radiation fields. This 2450 MHz C-director diathermy applicator has been used considerably in the past for cataract research, so the dosimetry corresponding to its use is important in order to determine what is actually occuring in the eye. In the figure it can be seen that the maximum SAR actually occurred in back of the lens, a short distance from the posterior pole. This information was used in a computer program designed to determine the temperature distribution in the exposed rabbit eye using a model as shown in Figure 15. The mathematical model allowed the formulation of a numerical technique for calculating temperature. The formulation included various mechanisms of cooling in addition to the energy deposited by the microwaves. A typical computer printout based on the model is shown in Figure 16 for an incident power density of 150 mW/cm^2. We see that the maximum temperature is not necessarily at the location of maximum SAR since the various cooling mechanisms come into play. The hot spot of 42.8 degrees occuring in the lens near te posterior pole is very close to the location where microwave induced cataracts were observed in test animals exposed to the 150 mW/cm^2 incident power density.

Figure 17 shows a newer commercially available instrument which is more convenient for measuring temperatures in an electromagnetic field environment. The instrument developed by Bowman (7) does not have to be withdrawn from the tissue during microwave exposure, since the leads to the sensor have the same electrical conductivity as typical tissue, and therefore are truely transparent to the microwaves within a tissue medium.

Another powerful method for measuring SAR, developed at the University of Washington by the author some years ago, involves the use of thermography and phantom models (3) (8) (9). Perhaps the most dramatic results obtained with this method for determining SAR distribution patterns were derived from an experiment at the University of Arkansas conducted by Korbel (10) (11). The experiment was one of a few in the USA resulting in behavioral changes in rats exposed to low-level fields which supported results from the Soviet and East European laboratories. In Korbel's experiments, effects were being observed due to exposures as low as 0.1 to 1.0 mW/cm^2. The author felt that it would be interesting to quantify the SAR distribution in the bodies of the exposed rats using the new thermographic technique.

Figure 18 illustrates the original exposure facility at the University of Arkansas where a population of 15 rats were exposed in a resonant 500 MHz cavity (top cover opened to show the animals). The same system was mocked up at the University of Washington as shown in Figure 19. The mock exposure system was identical in most respects except that phantom animals were used, consisting of bottles of synthetic tissue that had the same mass and electrical properties as that of the exposed rats. The water supply bottles and the metal screens at the bottom of each rat cage were carefully positioned to correspond to the conditions of the real experiment.

One of the phantom rats consisted of a more complex model simulating a rat standing on the metal screen in its cage while drinking out of a water supply bottle. Three phantom legs were added to the model for simulating the contact of the three limbs with the metal screen and a crude conical shaped head was added to make contact with the water spout for simulating a rat in a drinking position with his mouth in contact with the water spout. The model was bisected along a sagittal plane so that after exposure it could be separated, as shown on the right in Figure 20, for thermographic observation. The model was assembled and placed with the other phantom rats (type shown at the left in Figure 20) in the cavity for exposure to a very high input power. After the exposure, the model was quickly disassembled and the temperature distribution of the exposed sagittal plane on one half of the model was recorded by a thermograph camera.

The results shown in Figure 21 illustrate a typical thermographic recording. The thermogram on the left of the figure displays the temperature (also SAR) distribution by brightness or intensity proportional to temperature. Clearly, most of the energy was absorbed in the model at the legs that were in contact with the metal screen and at the point of contact between the conical head of the model and the water bottle. Single "B" scan thermograph recordings were also taken through the simulated legs before and after exposure so that the separation of the scans was proportional to the temperature or SAR. The results of the measurements were normalized to correspond to the lowest 0.1 mW/cm^2 APD used in the Korbell experiment. It was surprising to find that the SAR was over 18 mW/cm^3 (18 W/kg) in the simulated legs when the ADP in the cavity was only 0.1 mW/cm^2. For the maximum 1.0 mW/cm^2 ADP level used in the Korbell experiment the maximum SAR was 180 W/kg, which exceeds the SAR normally associated with diathermy treatments in the clinic. Of course the real rats were not subjected to these high levels of SAR all of the time. The maximum absorption would apparently occur only when the animal was in a drinking position. Clearly, what was happening, was the rats body was short-circuiting most of the RF energy by nearly completing a circuit between the upper and the lower plates of the resonant cavity, which were capacitive coupling to the water bottle and the metal screen at the bottom of the cage. This is an example where low-level exposure fields do not necessarily imply a low-level SAR in the body of an exposed animal.

The thermograph technique may also be used to quantify the two-dimensional SAR distribution in the bodies of actual animals (5)(4). After they are sacrificed, they may be frozen with dry ice, cast in a styrofoam block, bisected, and returned to room temperature while each half of the animal's body is covered with a thin plastic film to prevent loss of moisture. The reassembled body may then be exposed and thermographically analyzed in the same manner as used for the phantom models. Figure 22 illustrates results obtained by exposing the head of a cat in connection with investigations of the effects of microwaves on the central nervous system. The thermograms at the top left are based on the 915 MHz 2.5 mW/cm^3 radiation coming from the direction denoted by arrow A. We see from the intensity scan a fairly widespread absorption in the brain of the animal, with a maximum SAR of about 2.27 mW/cm^3 (W/kg) as determined from the double line scans taken vertically through B. When a metal electrode was placed in the head of the animal, however, the SAR became more localized and intense at the tip of the electrode as illustrated by the various thermograph intensity scans taken at different thresholds. The line scans indicate a more than order of magnitude increase in SAR with the presence of the electrode. This type of electrode has been used in this country for measuring CNS effects, due to low-level microwave exposures. Another example of thermographically obtained SAR patterns is that for a rat exposed to 915 MHz 1mW/cm^2 (normalized) radiation as shown in Figure 23. Figure 24 illustrates a thermographic study of the surface of a pigeon cadaver (feathers removed) oriented in a flying position with radiation directed dorsally. The $\bar{E}$, $\bar{H}$, and $\bar{K}$ vectors are illustrated in the figure. These results are representative of those expected for a flying bird exposed to the transmitted microwave beam from a solar power satellite. Note the high SAR in the wings and neck.

F. MEASUREMENT OF SAR IN MAN

In addition to quantifying the SAR in exposed animals, one must also do the same for man if biological effects are to be extropolated from the one species to the other. In essence, we must determine what exposure fields will induce SAR levels in man sufficient to produce biological effects assuming that the effects may occur for the same SAR levels as observed in test animals. Again, it is necessary to resort to phantom models.

Figure 25 illustrates one half of a bisected full-size model of a child composed of synthetic tissue with the same dieletctric properties as muscle tissue. The model consists of a synthetic tissue filled plastic foam shell with internal contours identical in shape as the outer contours of the body of a child. When the half section of the model is joined with the mating half at the sagittal plane, the model may be used to analyze the SAR patterns in a child exposed to various electromagnetic sources. In the case illustrated, the model is used to analyze the SAR patterns when the head of the child is exposed to leakage from a 915 MHz microwave oven.

The oven was modified in order to increase the leakage to the level that would produce sufficient heating to allow thermographic analysis. Again, the reader should keep in mind that even though the relationship between SAR and exposure level is quantified for a high level exposure, the relationship also holds for the case of low-level leakage. After the intact model is exposed to the leaky microwave oven, it is placed in a position, as shown in Figure 26, for recording a thermograph picture. With this orientation, the thermograph patterns shown in Figure 27 were measured. The data on thermograms are normalized for an exposure level of 1 mW/cm^2 measured 5 cm from the oven door. The first vertical column of thermograms correspond to the maximum coupling case where the child is exposed with his face at a distance of 4.8 cm from the oven door; the next is for an exposure distance of 9.9 cm; and the final for an exposure distance of 35.3 cm. The line scans at the bottom of the figures show that the maximum absorption occurs in the vicinity of the eyes and nose of the child model. The highest level of 455 mW/kg SAR was observed for an exposure distance of 4.8 cm. One must keep in mind that the child model is composed of a homogeneous synthetic tissue. For a more exact SAR analysis, one should take into account the bone, the subcutaneous fat, and the complex inner geometry of the body. This would require the development of much more sophisticated models.

A powerful and more convenient way to quantify the SAR patterns in man exposed to various sources is implemented through the use of scale models. With these models, far less power is required for the dosimetric studies and the cost is substantially reduced over that for full scale model studies.

One can facilitate the model fabrication through the use of a pantograph engraving machine, as shown in Figure 28. Two metal or plastic blocks, each with an impressed molded concave surface conforming to the shape of the body of a bisected man (along frontal plane with one half of the model corresponding to the front section and one half corresponding to the back section) are used as master patterns. Each master can be used as a guide for engraving a duplicate mold in styrofoam of suitable size to correspond to a particular model for analyzing the SAR patterns for a desired exposure frequency. The size of the styrofoam models can be controlled by the pantograph adjustments. With proper choice of parameters, the SAR patterns in the model exposed to a model frequency will be identical to those in a full scale model exposed at a full scale frequency. For example, a 1/5 scale model with electrical conductivity of 5 times the actual full scale tissue conductivity, exposed to a frequency 5 times higher than the full scale frequency will be subjected to the same SAR absorption patterns that would occur for the full scale conditions. For the scale model, however, the magnitude of the SAR would be 5 times larger. Figure 29 illustrates the number of engraved styrofoam molds of various sizes and positions. Figure 30 illustrates the completed models (half sections separated) filled with synthetic muscle tissue. A model (half section removed to show orientation) in the exposure position in an anechoic chamber is shown in Figure 31 and the thermographic recording operation after exposure is illustrated in Figure 32.

Figure 33 illustrates the thermographic data for exposure of the figure (exposure of left side of figure, E field parallel to the long axis of the body as denoted by the vectors $\bar{K}$ and $\bar{E}$) to 915 MHz radiation. The results, normalized for 1 mW/cm^2, for the 11.55 scale model are identical to those that would be obtained for a full size man model exposed to 79 MHz radiation. These data correspond to the case of maximum energy absorption when the body is exposed to its characteristic resonant frequency (condition when the long axis of the body is 4/10 of the exposure wavelength). The thermographic recording shows that the maximum SAR level in the exposed figure is approximately 0.96 W/kg or about two to five times higher than the whole body average SAR of 0.2 to 0.4 W/kg reported by Gandhi (C12) for these exposure conditions.

When the model is exposed at frequencies far above resonance, for example, 450 MHz, we see from the thermographic data in Figure 34 that the energy absorption is superficial in the torso of the body and is more concentrated in the limbs. The frontal exposure is denoted by $\bar{E}$' $\bar{H}$, and $\bar{K}$ vector for this case. One of the absorption characteristics that we notice at increasing frequencies above resonance is that even though the average values of SAR decrease significantly, as seen in Figure 3, the maximum values observed thermographically, are reduced only slightly. Though the maximum values of SAR are only slightly reduced, the fact that the absorption occurs over a smaller and smaller region of the body with increasing frequency, implies a necessary reduction of the average SAR.

The results for exposures at frequencies significantly below the resonant frequency are illustrated in Figures 35 and 36. At these frequencies, the SAR patterns in exposed bodies consist of two components, one due to the electric field coupling and the other due to magnetic field coupling (eddy currents). In order to illustrate the characteristics of each, the models were exposed to each field component separately. Figure 35 illustrates the thermographic results for the model exposed to a magnetic field perpendicular to the frontal plane of the body, simulating 31 MHz magnetic field exposure. The rather bizarre SAR pattern can be explained from simple physical principles. If the exposed subject were spherical in shape, the SAR pattern would be toroidal shaped as a result of circulating eddy currents. The magnitude of the SAR would vary as the square of the radial distance starting at zero at the center and reaching maximum at the periphery. In the human body, however, the circulating eddy currents are disturbed by the tissue discontinuities, especially where the currents must flow around sharp corners where the limbs join the body forming wedge shaped regions of air surrounded by tissue. These discontinuities force the currents to assume complex flow patterns resulting in a much higher than average current density or electric field in the axillary and perineal regions. The results indicate that for the 31 MHz exposure frequency a magnetic field strength of 1 A/m will result in a maximum SAR of 2.52 W/kg in the axillary regions of the body.

Figure 36 illustrates the thermographic data for the same man exposed to an electric field parallel to the long axis of the body. The hot spots or maximum SAR occur in the legs and the neck. This pattern results from the induced longitudal body currents (which would be uniform for an exposed spherical model) being constricted to a smaller cross section as they pass from the torso region to the legs and neck..

For exposures of the body to frequencies lower than 31 MHz, the distribution patterns of the SAR would remain the same but with a magnitude that would decrease as the square of the frequency. The patterns would also remain the same in different size bodies of the same shape. For this case, however, the magnitude of the magnetically induced SAR would vary directly as the cross sectional area of the body, while the magnitude of the electrically induced SAR would remain the same. Thus with the model consisting of tissue with homogeneous electrical properties, the measured results for one frequency could be extrapolated all the way down to a frequency of 60 Hz providing an estimate for body currents induced in humans exposed to high voltage transmission lines. The results of such calculations (converting SAR to current density) agree very closely to those measured in the tissues of actual man exposed to high strength 60 Hz electric fields. Thus, the thermographic and scale modelling technique is not only useful in the microwave and RF frequency range, but has applications in the area of low frequency fields. One must keep in mind, however, that when a body is exposed to a combination of electric and magnetic fields, the magnetically and electrically induced fields in the tissue will be superposed resulting in a complex field distribution and SAR pattern dependent on the relative magnitudes and phase of the two exposure field components.

G. APPLICATION TO SAFETY GUIDES

In completing the discussion of dosimetry and scaling, I would like to now direct my attention to the problem of relating these physical aspects to the problem of the setting of guides for safe human exposure to electromagnetic fields. A good example is the approach used by the American National Standards Institute Subcommittee C95.4 (ANSI C95.4 SC) in recommending new safety guides. For purposes of this paper we will discuss only the whole body exposure aspects of the proposed ANSI safety guide.

The 1966 and 1974 ANSI guides were based on the best experimental evidence available at the time they were promulgated as judged by a cross-section of scientists, many of whom are recognized as international authorities on the subject. It should be noted, however, that due to the lack of suitable experimental and theoretical data in the literature, this guide had many weaknesses. There was no specification on how the guides should be modified as a result of different environmental conditions or health conditions of exposed individuals. The guide does not take into account the variation of energy coupling with frequency, modulation, effects, and possible specific frequency effects, nor does it take into account the effects due to low level exposure reported so frequently in Soviet and East European literature.

A number of working groups were set up in the ANSI C95 SC to review both the old and the new literature and to make recommendations for the improvement for the 1974 guide as a basis for the promulgation of a new Radio Frequency Protection Guide (RFPG) in 1980.

Approximately 1700 references (titles, abstracts, or full papers) were reviewed by the ANSI C95.4 SC. Those most pertinent or applicable to developing safety setting criteria were specified. In its new work, the subcommittee especially wanted to make use of the engineering advances made over the last several years that could be used to more accurately extrapolate laboratory data from test animals for recommending maximum incident power density or electromagnetic field strength for human exposure. In most of the world, non-ionizing radiation safety standards today do not take into account this extrapolation process and as a result, many are too conservative over some portions of the spectrum and not conservative enough over other portions as described earlier in this paper. In order to properly extrapolate the laboratory results, it is necessary to convert electromagnetic field exposure level information used for laboratory experiments into the rate of energy absorption in the bodies of the exposed test animals. In addition, it is also necessary to relate human exposure under various levels and conditions to the rate of energy absorption or SAR in the tissues, as the quantity most directly related to the biological effects. Engineering work over the past several years, allows easy calculation and prediction of the average SAR in man and animals exposed under various conditions, and, in many cases, the actual SAR distribution over the entire body may be quantifiable.

The subcommittee members agreed to the following approach in formulating a new RFPG:

1) Select most pertinent literature reporting biological effects in terms of quantitative information.

2) Determine from literature directly or through calculation the average SAR in the exposed subjects where effects were noted.

3) Based on the selected literature and corresponding average SAR, select a maximum allowable SAR for human exposure.

4) Determine maximum allowable whole-body human exposure levels of power density and field strengths to limit SAR to the selected maximum as a function of frequency.

5) Develop criteria for partial-body exposures.

The Subcommittee members consisted of two groups, one heavily weighted by background in engineering and the other heavily weighted with expertise in the life sciences. The engineering group identified the sources that could be best used for extrapolating exposure levels to the average and peak SAR produced in exposed animals as well as man. The life sciences group evaluated the literature at hand and identified that most pertinent for developing a criteria for setting the RFPG.

The life sciences group, through a careful selection process, identified some 28 representative papers covering various biological effects that contained sufficient information to quantify the average SAR and sometimes the peak SAR measured in the test animals used in the reported work. These papers are listed in the Appendix. The engineering group then, through Dosimetry references (2) (12) (13) and the information in each paper, tabulated the exposure parameters and calculated the average SAR in the exposed animals discussed in each reference. Figure 37 is a plot of the SAR values for each of the 28 references, spanning a range of values from .002-20 W/kg.

The entire group then discussed and evaluated the work based on reliability and importance in terms of hazards. The committee analyzed the papers based on thresholds for effects or hazards in terms of agreement with others, replication by other researchers, and confounding factors. It was noted that there was general agreement among many of the papers reporting effects at 0.5 W/kg of average SAR and above. On the other hand, the thresholds for effects occurring below 0.5 W/kg varied over a wide range with no general agreement between papers concerning a level that would constitute a hazard. The majority of the ANSI C95.4 SC felt that the reported effects that were hazardous occurred at whole-body average SAR levels levels of 4 W/kg and above. After considerable deliberation, the majority of the subcommittee agreed that a decrement in working ability by test animals began to occur when test were subjected to average SAR levels of 4-10 W/kg. The majority of the subcommittee membership also agreed that the average SAR allowed for human exposure should be at least an order of magnitude below the threshold level of what they considered the threshold for hazardous effects.

The engineering group especially noted the incident power density (in mW/cm^2) required to limit human body average SAR to a constant level, in order to provide a specification as to the frequency dependence of the standard to account for changes in coupling. Based on the relationship between incident power density and the total body average SAR which is provided in the curves similar to those shown in Figure 38, the engineers formulated a frequency dependence criteria for the standard in terms of a fixed maximum allowed average SAR. The engineering group felt that the RFPG should be specified in terms of a constant power density over the region where maximum coupling (body resonance) would occur for a wide range of different body sizes from an exposed infant in free-space (highest frequency resonance condition) to a full-sized man standing on a ground plane (lowest frequency resonance condition). The group also felt that the RFPG should be increased with the inverse square of frequency below the resonance range and directly with frequency above the range. It was felt, however, that in the upper range the RFPG should be no higher than five times that allowed for the resonance range in order not to exceed the SAR allowed at body resonance.

A proposal by the engineers to set allowable exposure levels to values that would correspond to a constant average value of SAR in an exposed human was adopted. This resulted in the exposure criteria based on limiting the SAR to 0.4 W/kg as shown in Figure 39. The proposed RFPG would limit body exposure to of 1 mW/cm^2 over the frequency range of 30-300 MHz. At frequencies below 30 MHz, the exposure level allowed by the proposed RFPG rises as the square of frequency by the formula shown on the chart up to a maximum of 100 mW/cm^2 at a frequency of 3 MHz. From this frequency down to 300 kHz the proposed RFPG would limit the exposure level to 100 mW/cm^2. At frequencies above 300 MHz, the proposed RFPG would allow exposure to rise with frequency according to the formula given on the chart until a level of 5 mW/cm^2 is reached at 1500 MHz. At frequencies above 1500 MHz, the proposed RFPG specifies a maximum exposure level of 5 mW/cm^2 up to a frequency of 100 GHz.

It is interesting to compare the newly proposed ANSI whole body exposure guide with the old guide, the Soviet Occupational Standard, the standard being considered by the National Institutes of Occupational Safety and Health (NIOSH) and the Soviet General Population Standard as illustrated in Figure 40. For the purpose of this comparison, all of the standards have been characterized in terms of the APD. Since USA standards limit electric and magnetic field exposures to levels equivalent to the allowable plane-wave power density levels, the equivalent APD is the same for each. Soviet standards, on the other hand, are specified only in terms of electric or magnetic field strength at frequencies below 300 MHz with no implied relationship with plane-wave fields. For purposes of comparison, however, the author has denoted the APD for each with a label (E) for that derived from the allowable E field and a (H) for that derived from the allowable H field. The standard being considered by NIOSH is similar to the ANSI RFPG except the 1 mW/cm^2 minimum level covers a broader range down to 10 MHz and the frequency maximum level is limited to 25 mW/cm^2 instead of the 100 mW/cm^2 allowed by the ANSI standard. Also the averaging time is one second for the NIOSH standard instead of the six minutes recommended by the ANSI C95.4 SC. Though the Soviet standards, based on power density at frequencies above 300 MHz and electric field strength below 300 MHz, are three to four orders of magnitude below the newly proposed US standards, their frequency dependence is strikingly similar to that of the proposed US guides. Soviet standards appear to account for the variation of absorbed energy with frequency by step-wise adjustments of levels in contrast with the ramp adjustments for the proposed US standards. It is important to note that the APD for the Soviet magnetic field occupational standards are more than three orders of magnitude greater than that for the electric standards, actually exceeding the level recommended in the proposed US standards. This is exactly what one would expect if dosimetry quantities from magnetic field experiments on small animals are scaled for equivalence to human exposure as previously discussed and illustrated in Figure 4.

Perhaps the biological potency of the fields allowed by the standards may be better illustrated if they are converted into the equivalent average SAR for an average man exposed to the fields as illustrated in Figure 41. Due to the body resonance, we see maximum energy would be absorbed in the vicinity of 80 MHz for all of the standards, becoming less at higher frequencies and significantly less at lower frequencies. It is surprising to note that the magnetic field strength allowed by the Soviet occupational standard would result in an average SAR exceeding the maximum allowed at any frequency by the proposed US standards. One should keep in mind, however, that it may not be possible from a practical standpoint to produce magnetic field strengths allowed by the Soviet standards without producing an associated electric field strength that could exceed the standard.

It is also of interest to compare the standards in terms of the maximum SAR or "hot spot" level that may occur based on the thermographic information discussed previously. This comparison is shown in Figure 42. Note that the maximum SAR levels allowed by the Soviet magnetic field standards are equal to, but do not exceed those allowed by the proposed US standards.

Finally, if the maximum SAR is converted into the maximum equivalent field strength in the exposed tissue, the results in Figure 43 are obtained. Note that the level of this quantity, 30 to 35 V/m, allowed by the proposed ANSI RFPG, varies less with frequency than any of the other standards plotted in the figure.

I would like to conclude this discussion by indicating that the dosimetry and scaling techniques I have discussed have allowed for a significant improvement in the proposed US standards in terms of allowing for differences of energy coupling with exposure frequency. The overall problem is far from resolved, however. We have uncovered only the tip of the iceberg. Considerably more dosimetric and biological data are needed not only for quantifying whole-body exposures, but also for helping to resolve the "partial-body" exposure problem which has not been discussed here. We have hardly begun to resolve the profound differences between the results from Soviet and East European laboratories and our own. The most significant changes in the US standards may occur in the distant future when more data is available to assess the biological effects of long-term exposures to weak, as well as modulated and pulsed electromagnetic fields. We also must carefully determine for a given SAR if there are any "specific frequency effects" that would require additional adjustments of the frequency dependence of the exposure guides discussed in this paper.

COLONEL WANGEMANN: Let's take one quick question or comment before we go on.

MR. ALAN ANDERSON (Bureau of Radiological Health): A couple of perspectives on this idea of SAR from a biologist's viewpoint specifically.

I think it is time to stop attacking it as no advance at all and, as some would indicate, a setback. To me, this is a very big step forward that has been made.

We have taken a step which is somewhat akin to going from the old ranking in from ionizing radiation to having some feel for absorbed dose that you get from the rat.

The problem is that there is one more step to take, and I use the ionizing radiation analogy to keep reminding myself that there is one more step to take, and I would like you to do the same thing.

There is one more unit that we have to get to, and that is a unit which factors in the relative biological effectiveness as a function of frequency.

We have now obtained a value and a technique for measurement which allows us to get at the absorbed dose. That was a necessary step, and it has been a giant step.

We have to continue working to find out what the relative biological effectiveness is of each of these absorbed doses to better refine the safety standards.

Now, that could get us into the nonthermal mechanisms, but let us not depart from thermal too quickly. Temperature has different levels of importance in biological systems.

What we have heard talked about most here is the temperature at the gross level of the body, if you will, base metabolism rate. We have also heard that refined a little bit, and necessarily so, from the standpoint of localized rises in temperature; that kind of energy distribution in the body was critical to begin to determine.

But there is another level of temperature that is important for biology that is not yet considered by this concept of specific absorption rate because it cannot measure small enough volumes, and that concept of temperature is the temperature at the level of molecular motion.

In fact, for biology and for most of the biological processes that are important at, let us say, the recognition of an antibody and an antigen in the immune response. That process is a process that is aided and abetted by brownie in motion, by KT, energy at that low a level in a small volume inside a cell.

That level of temperature we have not yet gotten a handle on, and it is certainly feasible that perturbations in temperature at that level certainly are not going to raise the rectal temperature of even a gnat, may not even show up in the volume sensed by even the miniature probes that are now being developed, and it is an area that we are going to have to look at in the future in terms of defining just what we mean by thermal.

There are also some very small changes in temperature in biology that we are going to have to face up to as important. Thank you.

DR. GUY: I agree with everything you said except where you indicated the SAR had no relevance to this particular case. One still must know what the average fields are within a region of cells to know how they couple to a particular cell, so the application of the dosimetric techniques that I discussed is a necessary first step toward the understanding of the interaction of electromagnetic fields with the individual cells.

References

1 NCRP (1980), Radiofrequency electromagnetic fields: Properties, quantities and units, biophysical interaction and measurement. Publ. No. 67, National Council on Radiation Protection and Measurement: Washington, D.C.

2 Deposition of electromagnetic energy in animals and in models of man with and without grounding and reflector effects.

3 Guy, A. W., "Quantitation of induced electromagnetic field patterns and associated biological effects," in Biologic Effects and Health Hazards of Microwave Radiation Proc. of an Int'l Symp., Warsaw, October, 1973, (Eds. Microwave Radiation, Annals New York Academy of Sciences, Paul E. Tyler (Eds. P. Czerski, et al), pp. 203-216, Polish Medical Publishers, Warsaw 1974.

4 Johnson C. C. and A. W. Guy, "Nonionizing electromagnetic wave effects in biological materials and systems," Proc. IEEE, 60(6):692-718, June, 1972.

5 Guy, A. W., M. D. Webb, and C. C. Sorensen, "Determination of power absorption in man exposed to high frequency electromagnetic fields by thermographic measurements on scale models," IEEE Trans. Biomed. Eng. 23(5):361-371, September, 1976.

6 Phillips, R.D, E. L. Hunt, and N.W. King: Field measurements, absorbed dose, and biologic dosimetry of microwaves. Ann. N.Y. Acad. of Sci. 247:499-509, 1975.

7 Bowman, R.R.: A probe for measuring temperature in radio-frequency-heated material. IEEE Trans. MTT-24:43-45, 1976.

8 Guy, A. W., J.F. Lehmann, J. A. McDougall, and C. C. Sorensen, "Studies on therapeutic heating by electromagnetic energy," in Thermal problems in Biotechnology, ASME, United Engrg. Ctrs., New York, pp. 26-45, 1968.

9 Guy, A. W., "Analyses of electromagnetic fields induced in biological tissues by thermographic studies on equivalent phantom models," IEEE Trans. Microwave Theory Tech. 19(2):205-214, February, 1971.

10 Korbel, S. F., "Behavioral effects of low intensity UHF radiation", Biological Effects and Health Implications of Microwave Radiation Symposium Proceedings- BRH/DBE 70-2. September, 1969.

11 Guy, A. W. , Korbel, S. F., "Dosimetry studies on a UHF cavity exposure chamber for rodents", IMPI Microwave Power Symposium Summaries of Presented Papers, paper 10.4, May 1972.

12 Om P. Gandhi, "Dosimetry-the absorption properties of man and

experimental animals", Bulletin of the New York Academy of Medicine, Volume 55:11, December 1979.

13 Durney, C.H., C.C. Johnson, P.W. Barber, H. Massoudi, M.F. Iskander, J.L. Lords, D.K. Ryser, S.F. Allen, and J.C. Mitchell: Radiofrequency Radiation Dosimetry Handbook, Second Ed., Report No. SAM-TR78-22, prepared by the University of Utah for USAF School of Aerospace Medicine, Brooks AFB, TX, 1978.

APPENDIX

Experimental Reports from the Literature
on Biological Effects of Radio Frequency Electromagnetic Fields
Selected by ANSI C95.4 Subcommittee

1. Environmental Factors

a. Shandala, M.G., M.I. Rudnev, and M.A. Navakatian, Patterns of change in behavioral reactions to low power densities of microwaves. In Abstracts of 1977 International Symposium on the Biological Effects of Electromagnetic Waves, Airlie, VA, Oct. 30-Nov. 4, p. 88.

b. Johnson, R.B., S. Mizumori, R.H. Lovely, and A.W. Guy, Adaptations to microwave exposure as a function of power density and ambient temperature in the rat. In Abstracts of 1978 Symposium on Electromagnetic Fields in Biological Systems, Ottawa, Canada, June 27-30, p. 30.

c. Monahan, J.C., and H.S. Ho, Microwave induced avoidance behavior in the mouse. In Biological Effects of Electromagnetic Waves, C.C. Johnson and M.L. Shore, eds., HEW Publication (FDA) 77-8010, Vol. I, 274-283, 1976.

2. Behavior and Physiology

a. D'Andrea, L.A., O.P. Gandhi, and J.L. Lords, Behavioral and thermal effects of microwave radiation at resonant and nonresonant wavelengths. Radio Science 12(6S):251-256, 1977.

b. Thomas, J.R., E.D. Finch, D.W. Fulk, and L.S. Burch, Effects of low-level microwave radiation on behavioral baselines. Ann. N. Y. Acad. Sci. 247:425-431, 1975.

c. Frey, A.H., Behavioral effects of electromagnetic energy. In Biological Effects and Measurement of Radio Frequency/Microwaves, D.G. Hazzard, ed., HEW Publication (FDA) 77-8026, 11-22, 1977.

d. King, N.W., D.R. Justesen, and R.L. Clarke, Behavioral sensitivity to microwave irradiation. Science 172:398-401, 1971.

e. Frey, A.H., S.R. Feld, and B. Frey, Neural function and behavior: Defining the relationship. Ann. N. Y. Acad. Sci. 247:433-438, 1975.

f. Lin, J.C., A.W. Guy, and L.R. Caldwell, Thermographic and behavioral studies of rats in the near field of 918-MHz radiation. IEEE Trans. on Microwave Theory and Techniques MTT-25:833-836, 1977.

3. Immunology

a. Shandala, M.G., M.I. Rudnev, G.K. Vinogradov, N.G. Belonozhiko, and N.M. Gonchar, Immunological and hematological effects of microwaves at low power densities. In Abstracts of 1977 International Symposium on the Biological Effects of Electromagnetic Waves, Airlie, VA, Oct. 30-Nov. 4, p. 85.

b. Czerski, P., Microwave effects on the blood-forming system with particular reference to the lymphocyte. Ann. N.Y. Acad. Sci. 247:232-241, 1975.

c. Huang, A.T., M.E. Engle, J.A. Elder, J.B. Kinn, and T.R. Ward, The effects of microwave radiation (2450 MHz) on the morphology and chromosomes of lymphocytes. Radio Science 12 (6S):173-177, 1977.

d. Smialowiz, R.J., J.B. Kinn, and J.A. Elder, Exposure of rats _in utero_ through early life to 2450 MHz (CW) microwave radiation: Effects on lymphocytes. Radio Science 14/S, 1979, in press.

e. Smialowiz, R.J., C.M. Weil, J.B. Kinn, and J.A. Elder, Exposure of rats to 425 (CW) microwave radiation: Effects on lymphocytes. J. Microwave Power, in press.

4. Teratology

a. Berman, E., J.B. Kinn, and H.B. Carter, Observations of mouse fetuses after irradiation with 2.45 GHz microwaves. Health Physics 35:791-801, 1978.

5. Central Nervous System/Blood-Brain-Barrier

a. Bawin, S.M., L.K. Kaymarck, and W.R. Adey, Effects of modulated VHF fields on the central nervous system. Ann. N.Y. Acad. Sci. 247:74-80, 1975.

b. Blackman, C.F., J.A. Elder, C.M. Weil, S.G. Boname, D.C. Eichinger, and D.E. House. Induction of calcium in efflux from brain tissue by radio-frequency radiation: Effects of modulation-frequency and field-strength. Radio Science 14/S, 1979, in press.

c. Frey, A.H., S.R. Feld, and B. Frey, Neural function and behavior: Defining the relationship. Ann. N.Y. Acad. Sci. 247:433-438, 1975.

d. Albert, E.N., Light and electron microscope observations on the blood-brain barrier after microwave irradiation. In Biological Effects and Measurement of Radio Frequency/Microwaves, D.G. Hazzard, ed., HEW Publication (FDA) 77-8026, 294-304, 1977.

6. Cataracts: None ≤ 10 mW/cm^2

7. Genetics: None ≤ 10 mW/cm^2

8. Human Studies: None

9. Thermoregulation and Metabolism

a. Lovely, R.H., D.E. Myers, and A.W. Guy, Irradiation of rats by 918 MHz microwaves at 2.5 mW/cm^2: Delineating the dose-response relationship. Radio Science 12(6S):139-146, 1977.

b. Stern, S., L. Margolin, B. Weiss, and S.-T. Lu, Microwaves affect thermoregulatory behavior in rats. Abstract, Bioelectromagnetics Symposium, Seattle, WA, June 18-22, 1979.

c. Adair, E.R., Microwave modification of thermoregulatory behavior: Threshold and suprathreshold effects. Abstract, Bioelectromagnetics Symposium, Seattle, WA, June 18-22, 1979.

d. De Lorge, J.O., Operant behavior and colonic temperature of squirrel monkeys (Saimiri sciureus) during microwave irradiation. NTIS Document No. AD A043706, 33 pp., 1977.

e. Lu, S.-T., N. Lebda, and S.M. Michaelson, Effects of microwave radiation on the rats' pituitary-thyroid axis. Radio Science 14/S, 1979, in press.

10. Biorhythms

a. Lu, S.-T., N. Lebda, and S.M. Michaelson, Effects of microwave radiation on the rats' pituitary-thyroid axis. Radio Science 14/S, 1979, in press.

11. Endocrinology

a. Lovely, R.H., A.W. Guy, R.B. Johnson, and M. Mathews, Alterations of behavioral and biochemical parameters during and consequent to 500 $\mu W/cm^2$ chronic 2450 MHz microwave exposure. In Abstracts of 1978 Symposium on Electromagnetic Fields in Biological Systems, Ottawa, Canada, June 27-30, p. 34.

b. Travers, W.D., Low intensity microwave effects on the synthesis of thyroid hormones and serum proteins, Health Physics 33(6):662, 1978.

12. Development

a. Michaelson, S.M., R. Guillet, and F.W. Heggeness, The influence of microwaves on development of the rat. Radio Science 14/S, 1979, in press.

b. McRee, D.I., and P.E. Hamrick, Exposures of Japanese quail embryos to 2.45 GHz microwave radiation during development. Radiation Research 71(2):355-366, 1977.

c. Johnson, R.B., S. Mizumori, and R.H. Lovely, Adult behavioral deficit in rats exposed prenatally to 918 MHz microwaves. In Developmental Toxicology of Energy Related Pollutants, M. Sikov and D. Malum, eds., 17th Hanford Symposium, DOE, GPO, pp. 281-299, 1979.

13. RF Hearing

14. Hematology

a. Mitchell, D.S., W.G. Switzer, and E.L. Bronaugh, Hyperactivity and disruption of operant behavior in rats after multiple exposures to microwave radiation. Radio Sciences 12(6S):263-271, 1977.

b. Miro, L., R. Loubiere, and A. Pfister, Effects of microwaves on the cell metabolism of the reticulo-historytic system. In Biologic Effects and Health Hazards of Microwave Radiation, P. Czerski, K. Ostrowski, M.L. Shore, C. Silverman, M.J. Suess, and B. Weldeskoz, eds., Polish Medical Publishers, Warsaw, pp. 89-97, 1974.

15. Cardiovascular

a. Reed, J.R., J.L. Lords, and C.H. Durney, Microwave irradiations of the isolated rat heart after treatment with ANS blocking agents. Radio Science 12(6S):161-165, 1977.

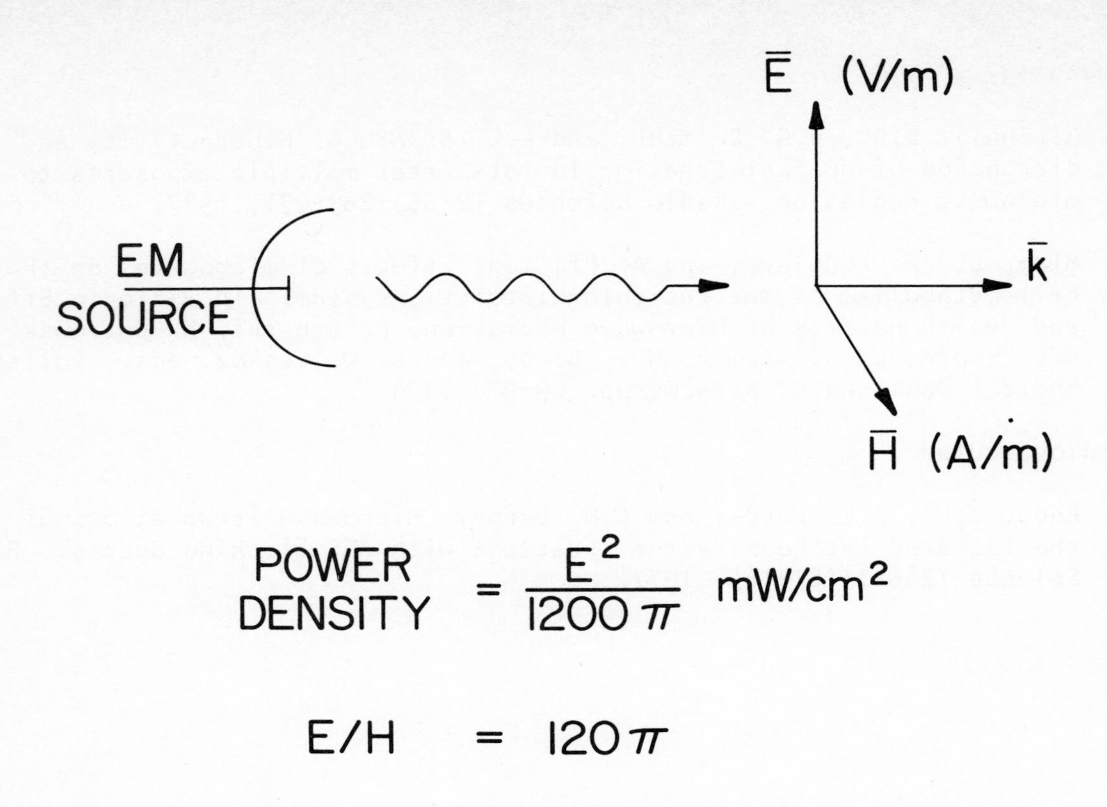

Figure 1 Quantities and units used for specifition of electromagnetic field exposure guides

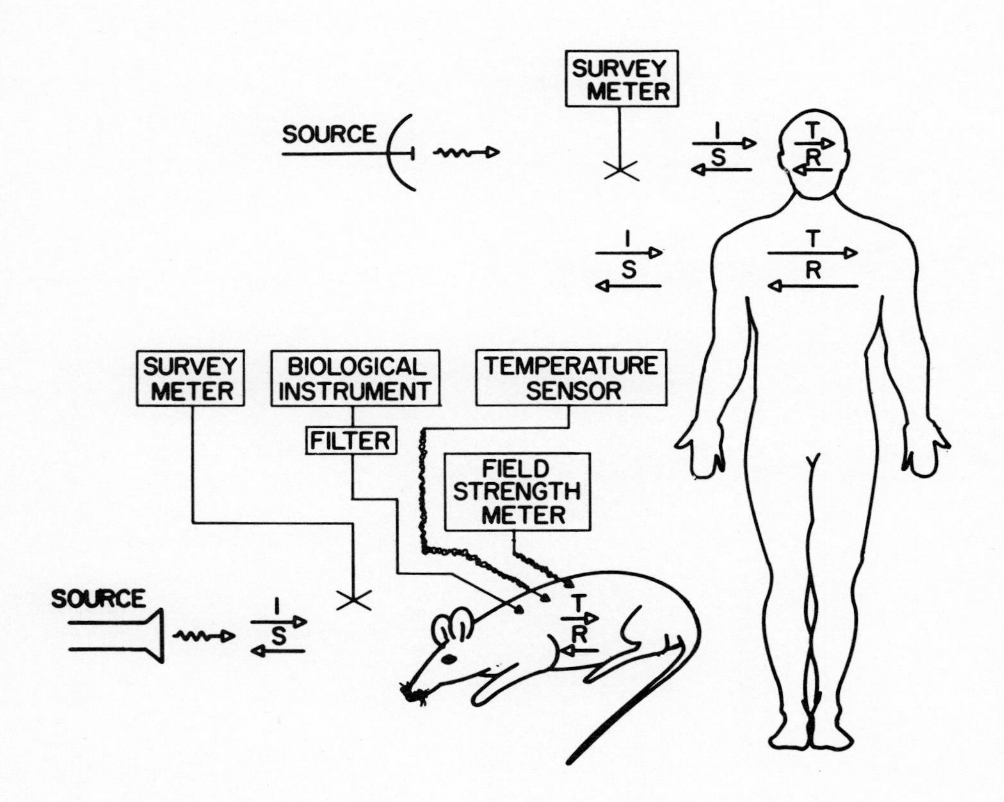

Figure 2 Comparison of experimental animal and human exposure to electromagnetic fields

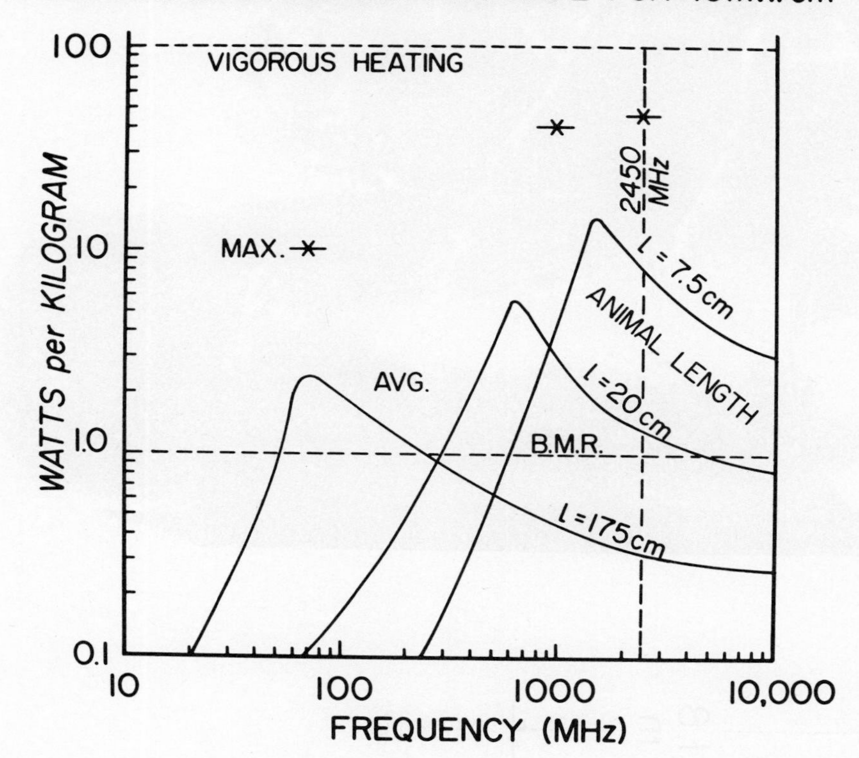

Figure 3 Average SAR in prolate spheroidal models of biological subjects exposed to electromagnetic radiation as a function of frequency [from Gandhi, et al (12)]

SCALING PROBLEM

$\bar{E}$, $\bar{k}$, $\bar{H}$, E_{tr}, E_{tm}, 8 cm, 1.8 m

f (MHz)	H (A/m)	σE_{tr}^2 (W/kg)	σE_{tm}^2 (W/kg)
10	0.5	9.2×10^{-6}	4.7×10^{-3}
10	250	2.3	1.2×10^{3}
2000	0.5	13	0.4

Figure 4 Illustration of the need for scaling dosimetric data from animals to man

Figure 5 Configuration of inner wells and associated thermocouple arrays in construction of twin-well calorimeters

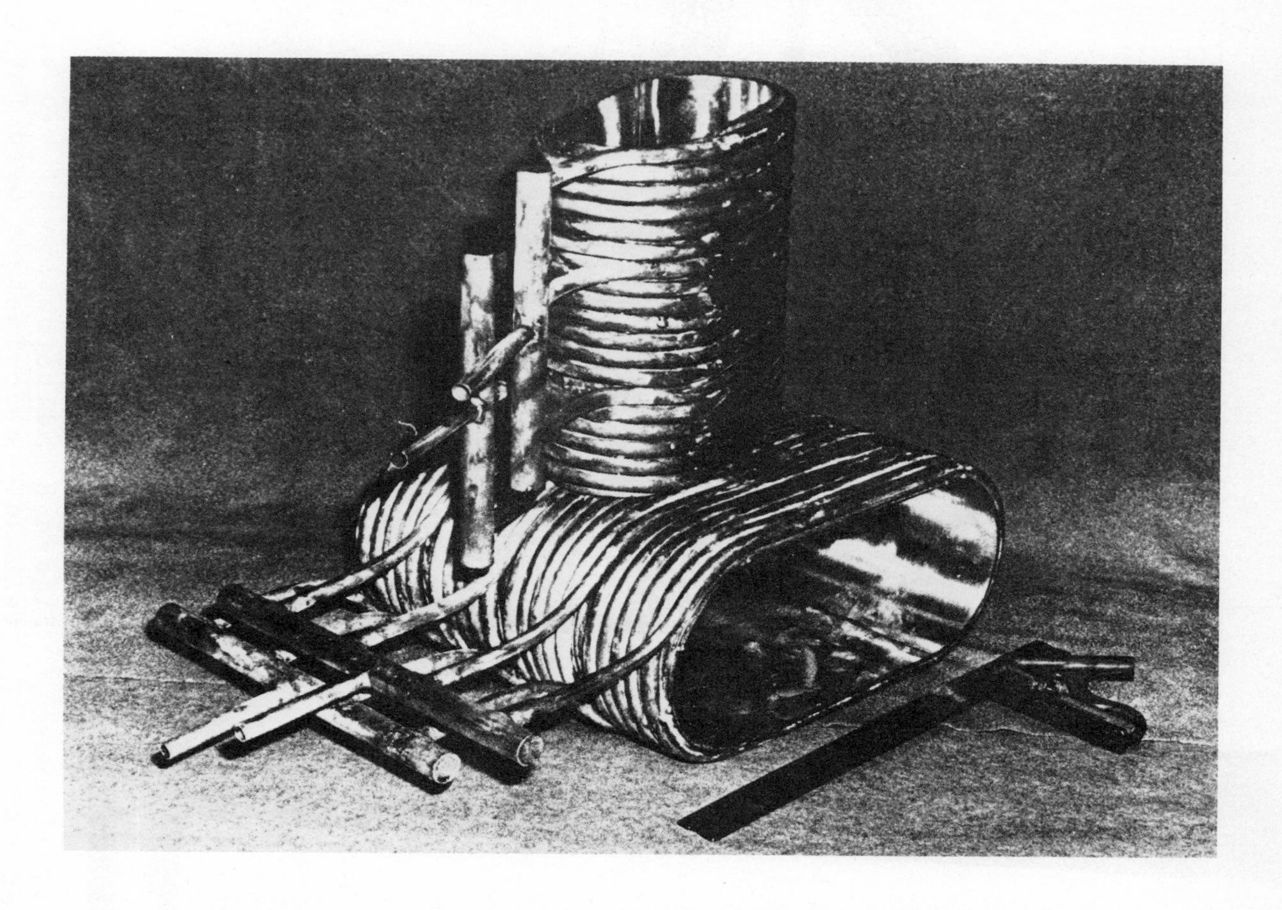

Figure 6 Outer cylinders (temperature stabilized by circulating liquid) used in construction of twin-well calorimeters

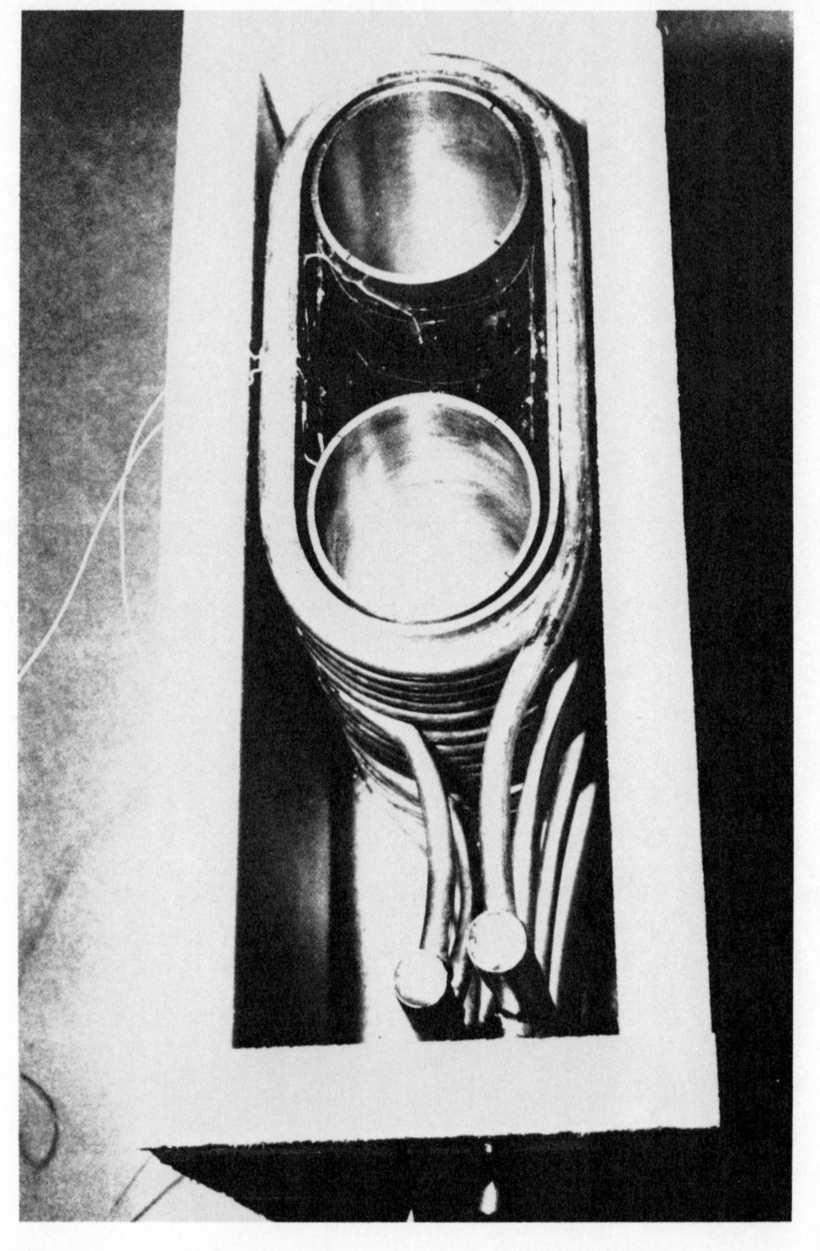

Figure 7 Assembly of twin-well calorimeter before insulation is placed between the inner and outer cylinders

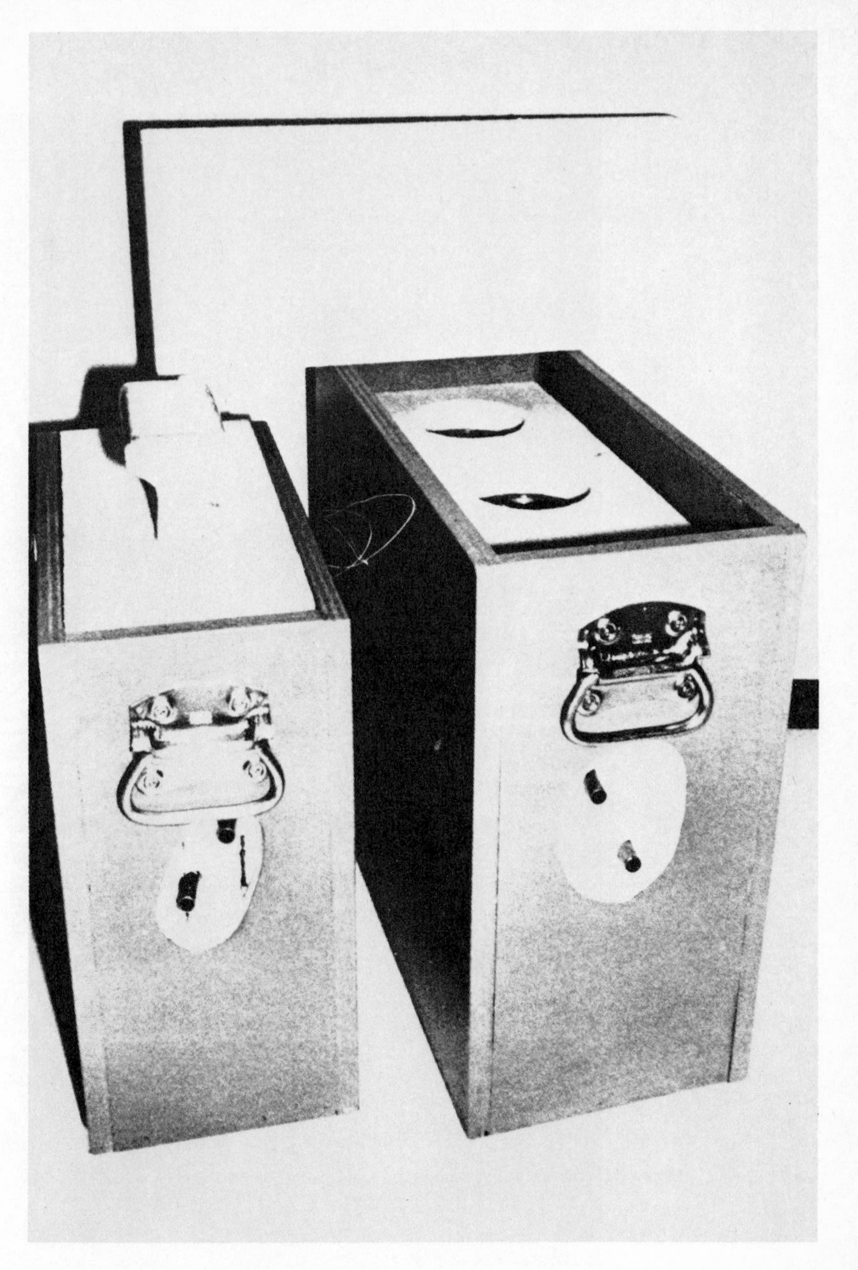

Figure 8 Illustration of two completed twin-well calorimeters used for different sized rats

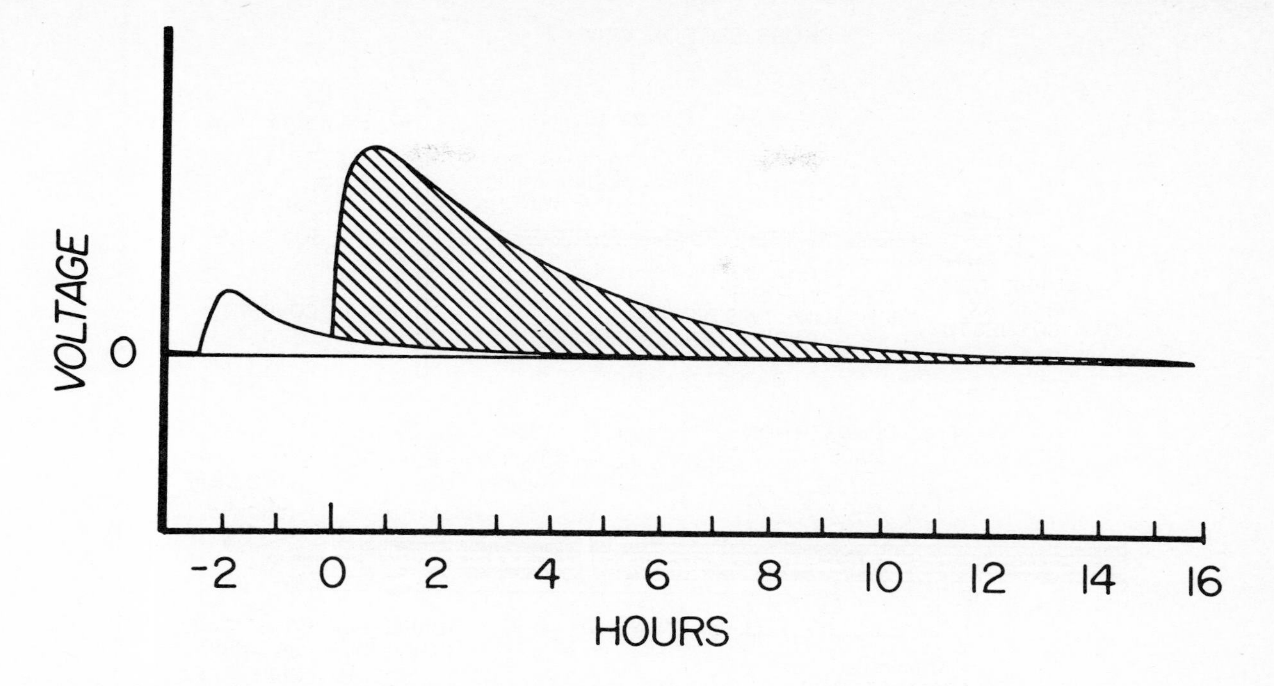

Figure 9 Output voltage from twin-well calorimeter containing rat carcasses

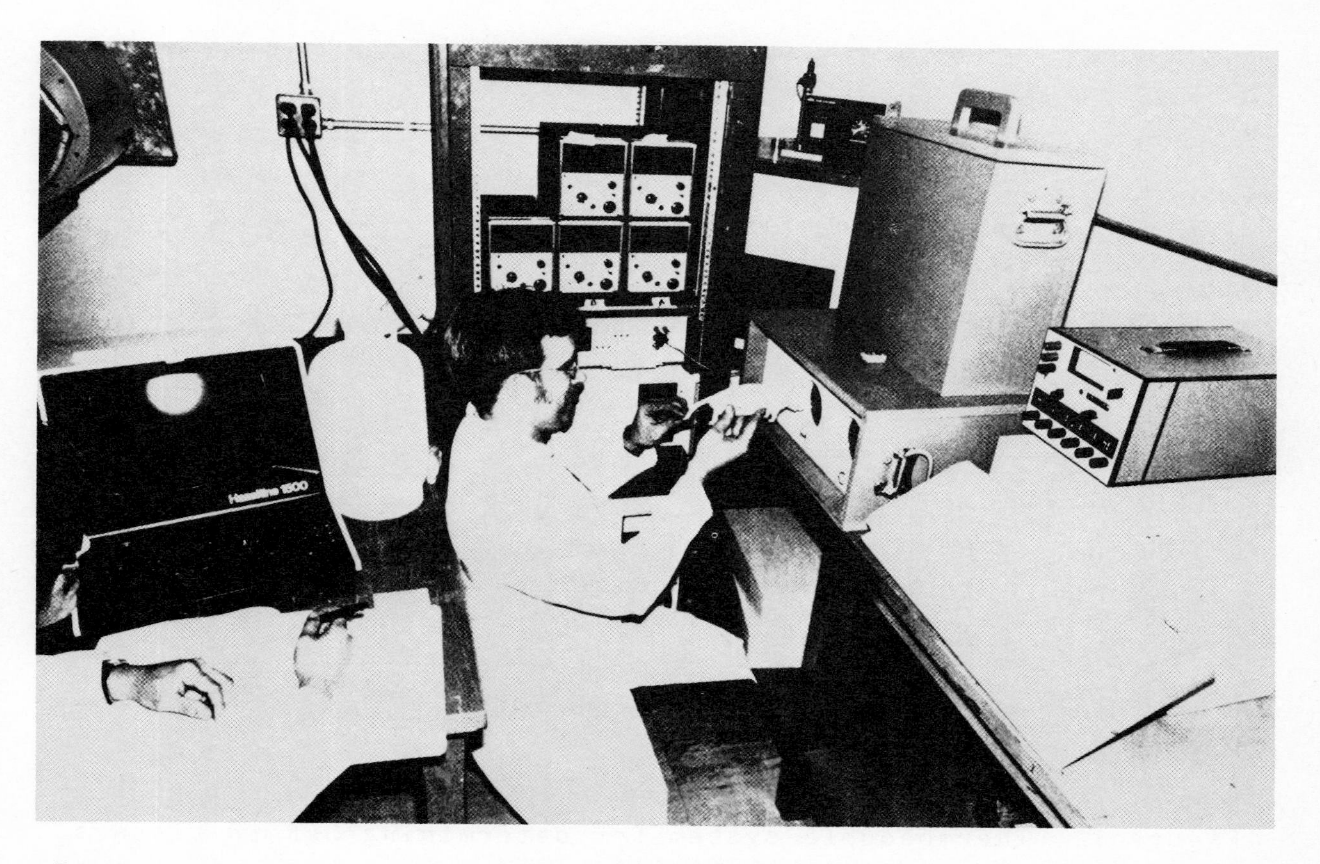

Figure 10 Photograph illustrating micro-processor controlled twin-well calorimeter system

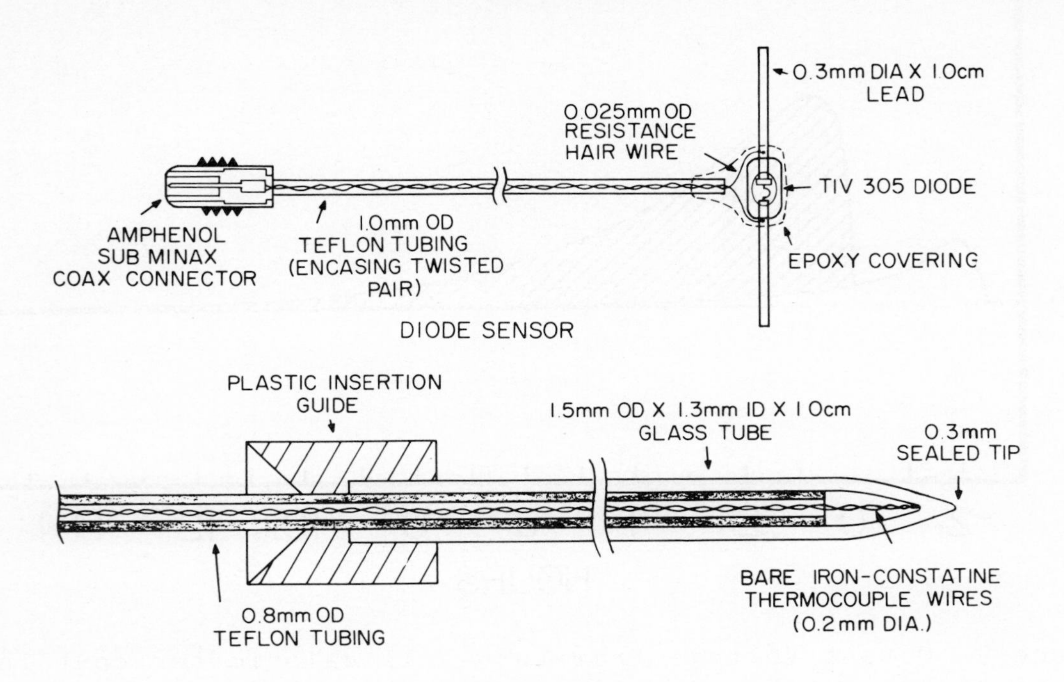

Figure 11 Methods for determining SAR in biological tissues exposed to electromagnetic fields

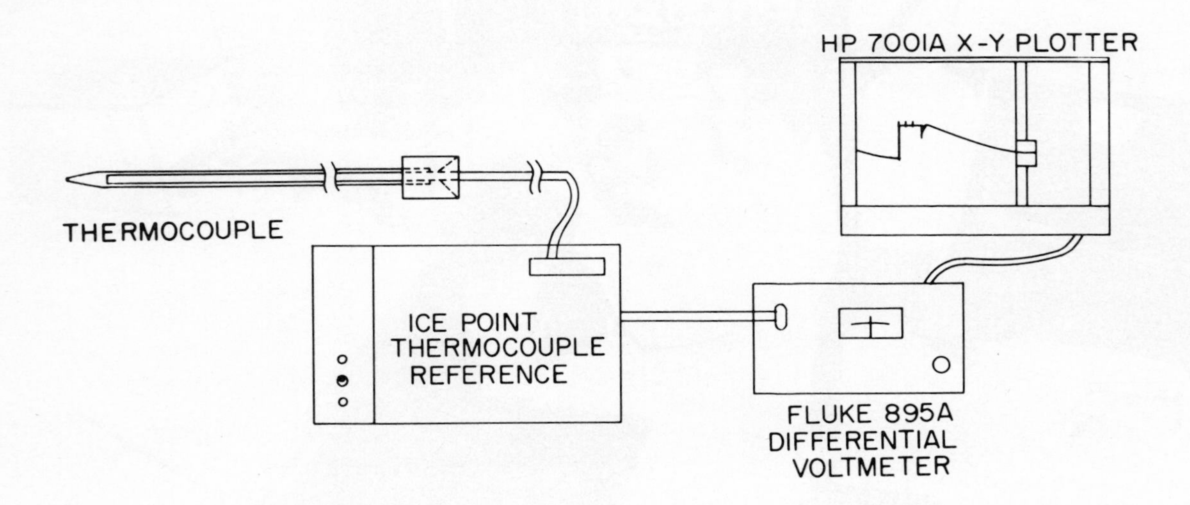

Figure 12 Thermocouple system for determining SAR in tissues exposed to electromagnetic fields

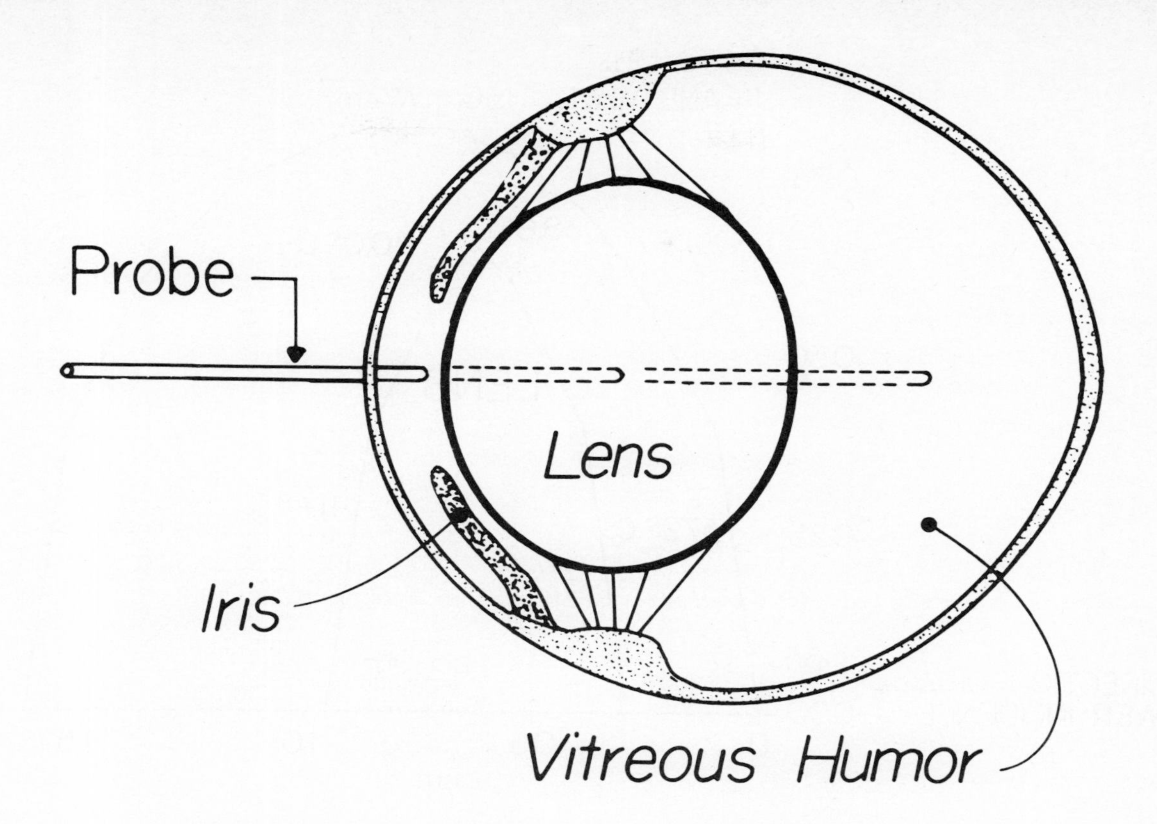

Figure 13 Pathway of probe for measuring SAR in exposed rabbit eye

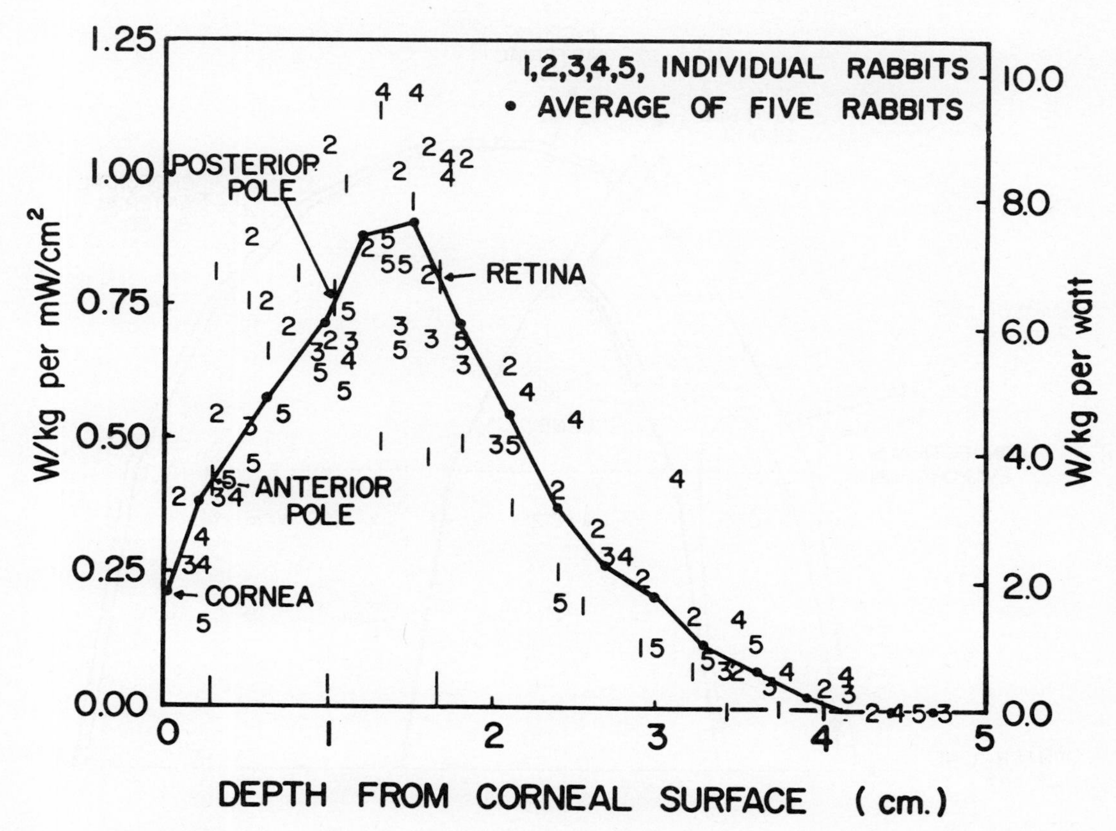

Figure 14 Measured SAR pattern along the axis of an exposed rabbit eye

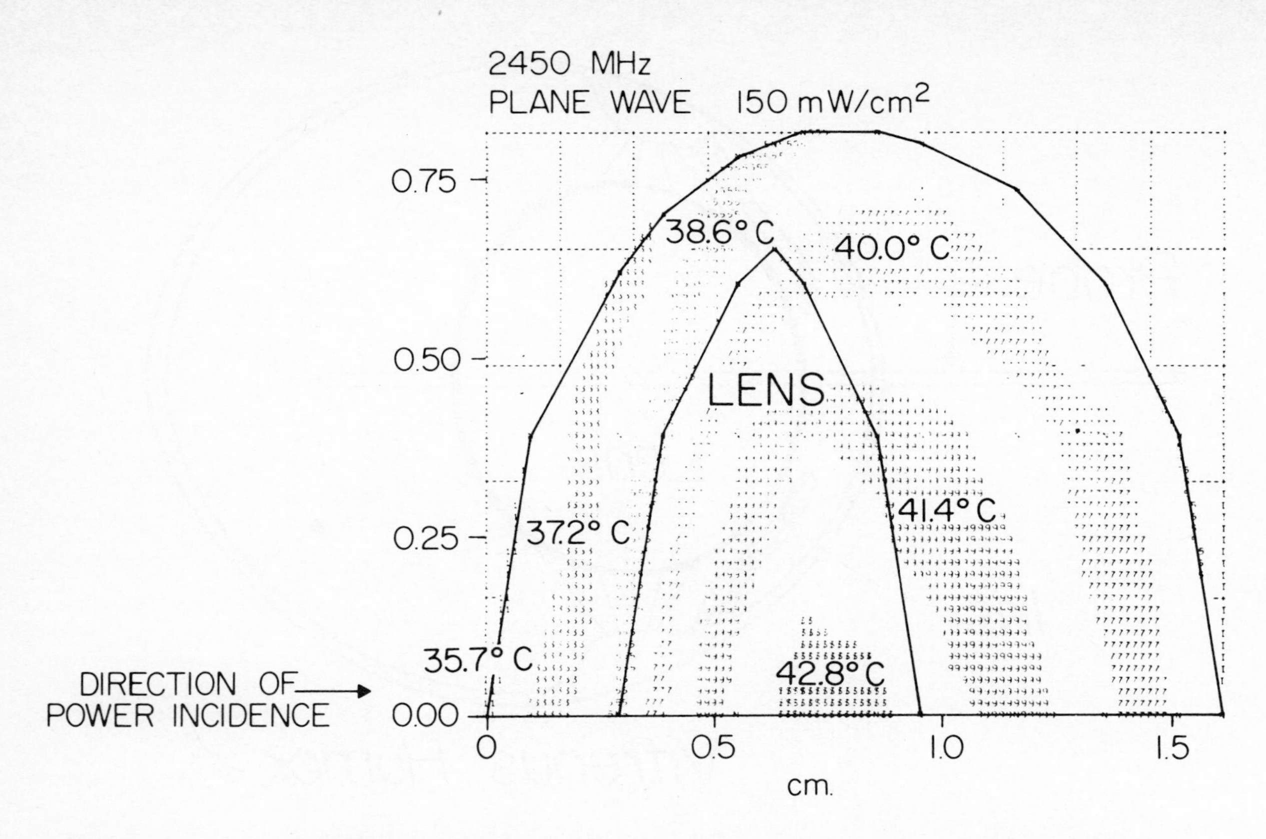

Figure 15 Thermodynamic model for analysing temperature distribution in rabbit eye exposed to microwave radiation

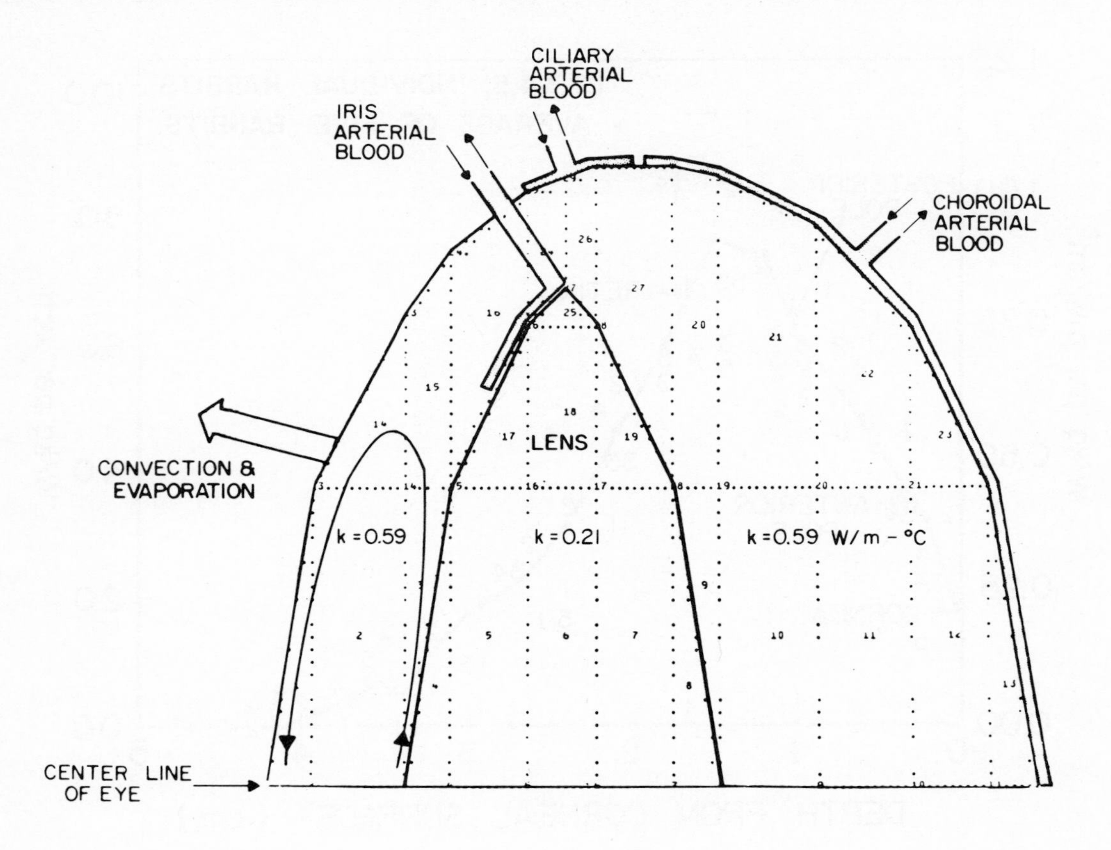

Figure 16 Computer calculation of temperature distribution in rabbit eye exposed to microwave radiation

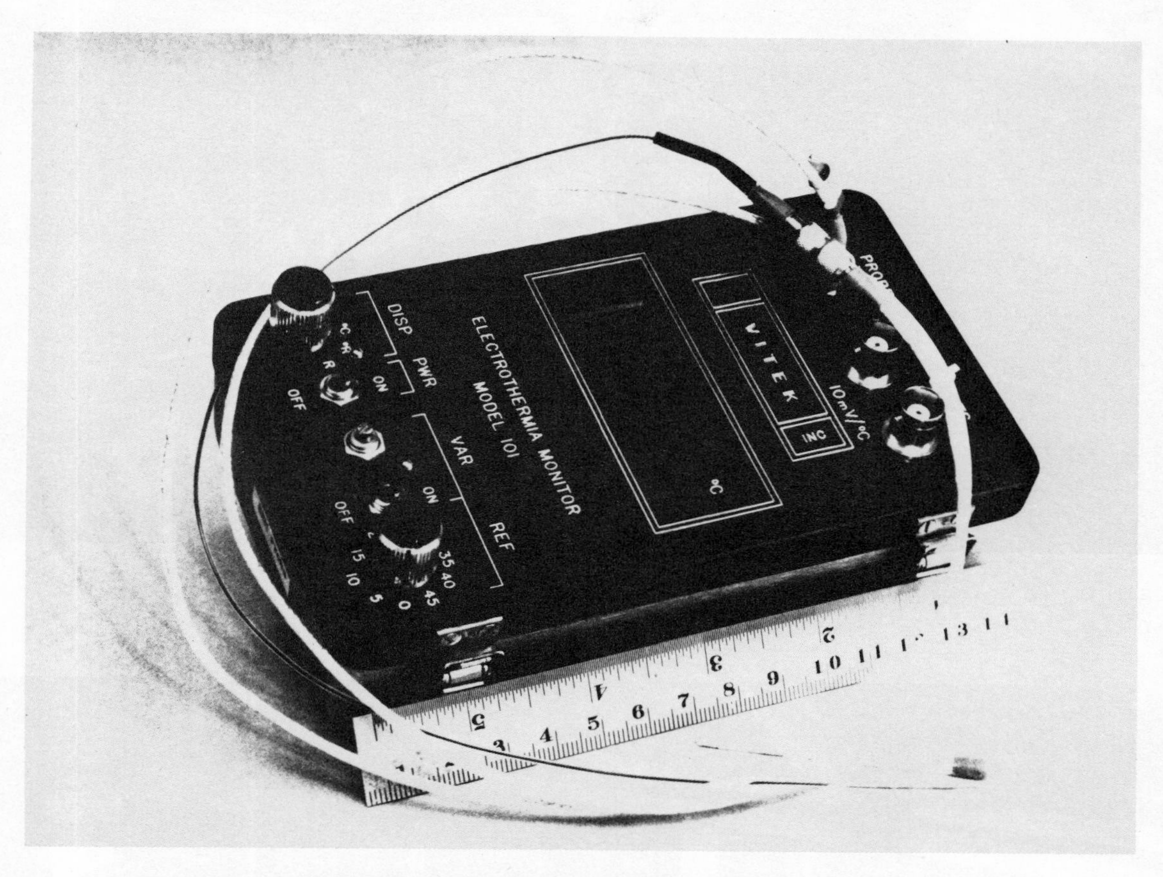

Figure 17 Instrumentation for measuring temperature in tissue exposed to electromagnetic radiaton

Figure 18 Cavity used by Korbel (10)(11) for exposing a population of rats to 500 MHz fields

Figure 19 Exposure of rat phantom models in 500 MHz cavity

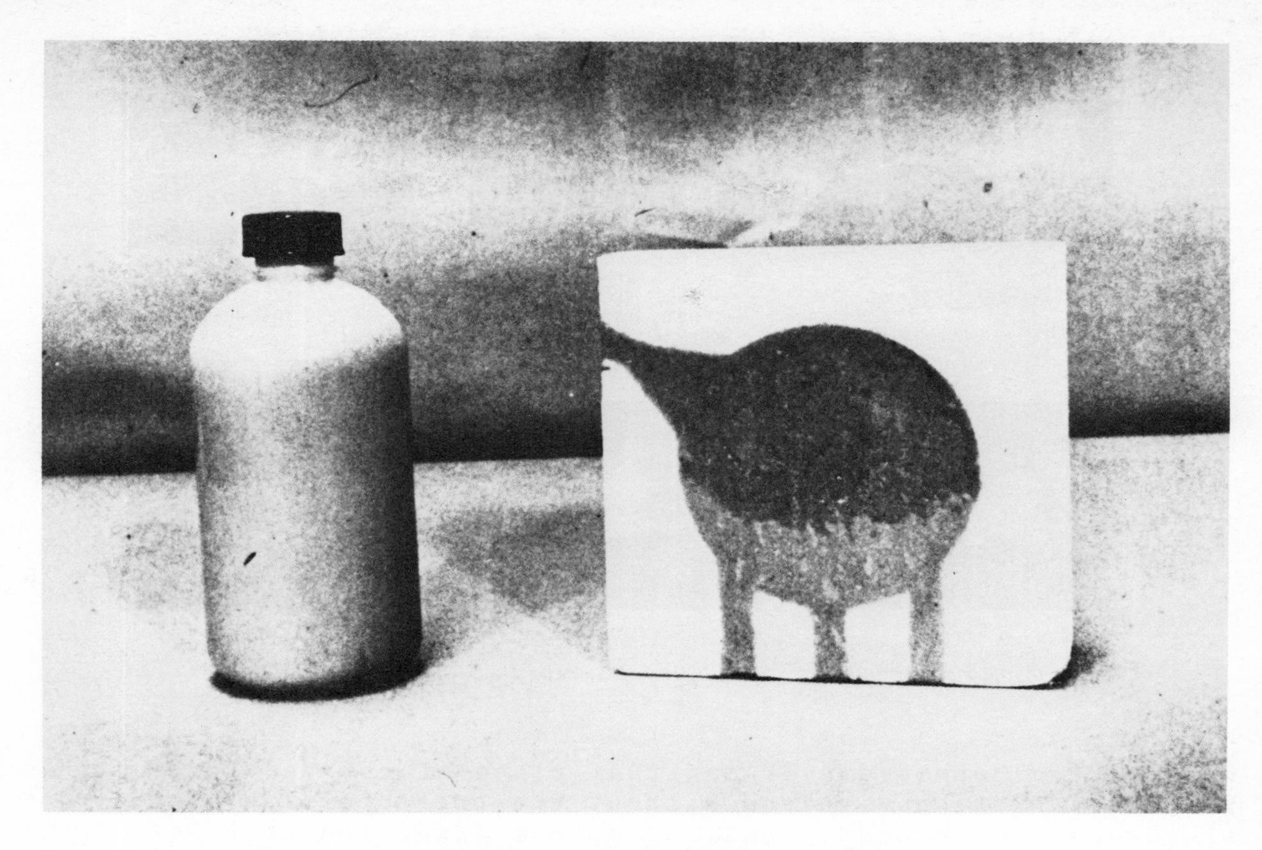

Figure 20 Phantom models consisting of synthetic tissue for simulating rats exposed to electromagnetic fields

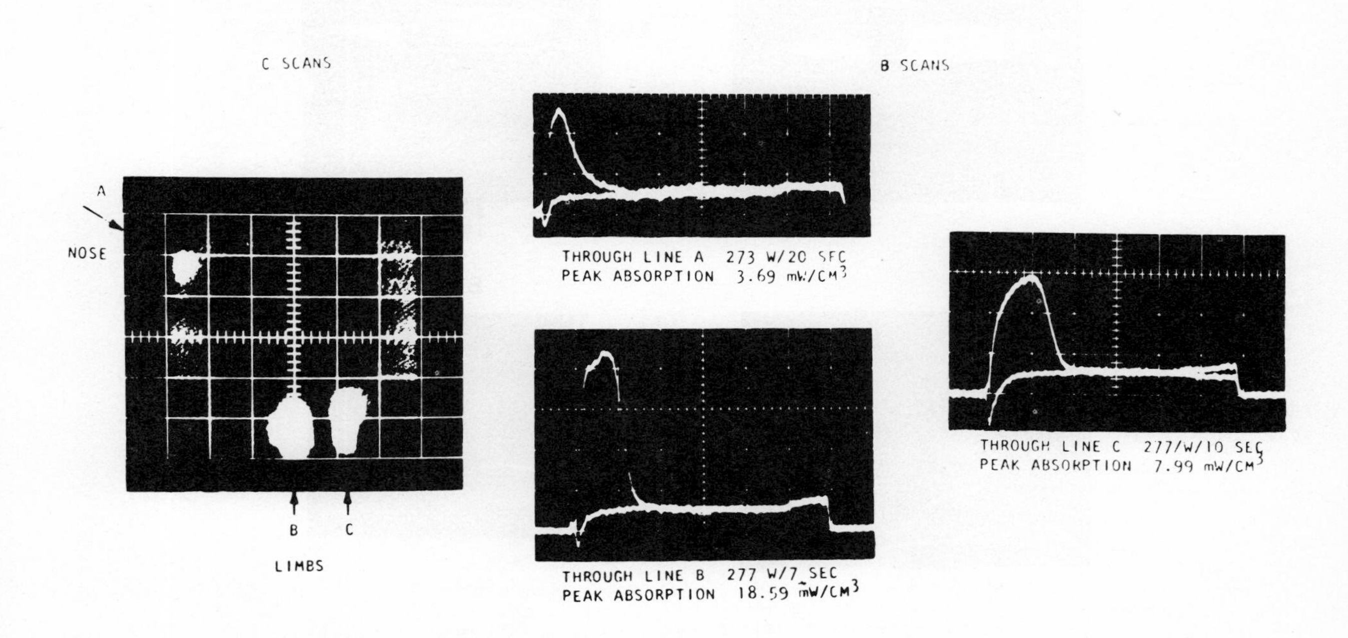

Figure 21 Thermograms of a spherical phantom of rat exposed to 500 MHz cavity fields

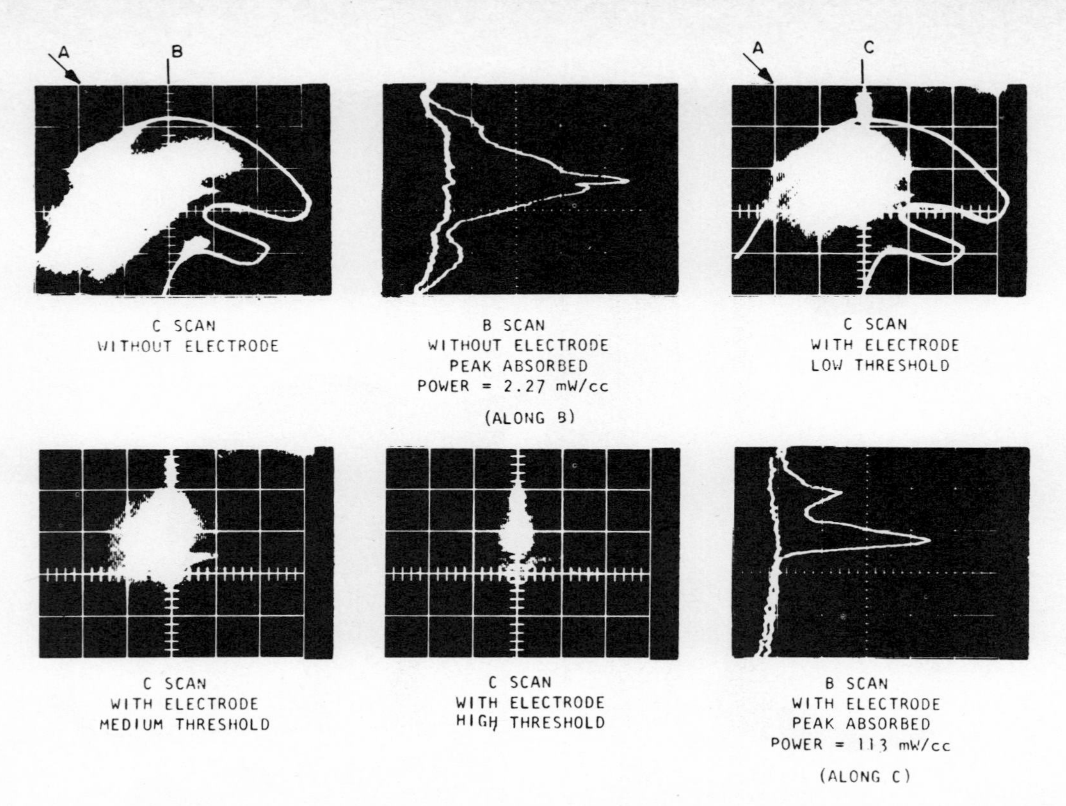

Figure 22 Thermograms of sagittal plane of cat head exposed to 915 MHz radiation with and without the presence of a metal electrode axes B and C (radiation from direction of A with incident power density 2.5 mW/cm^2)

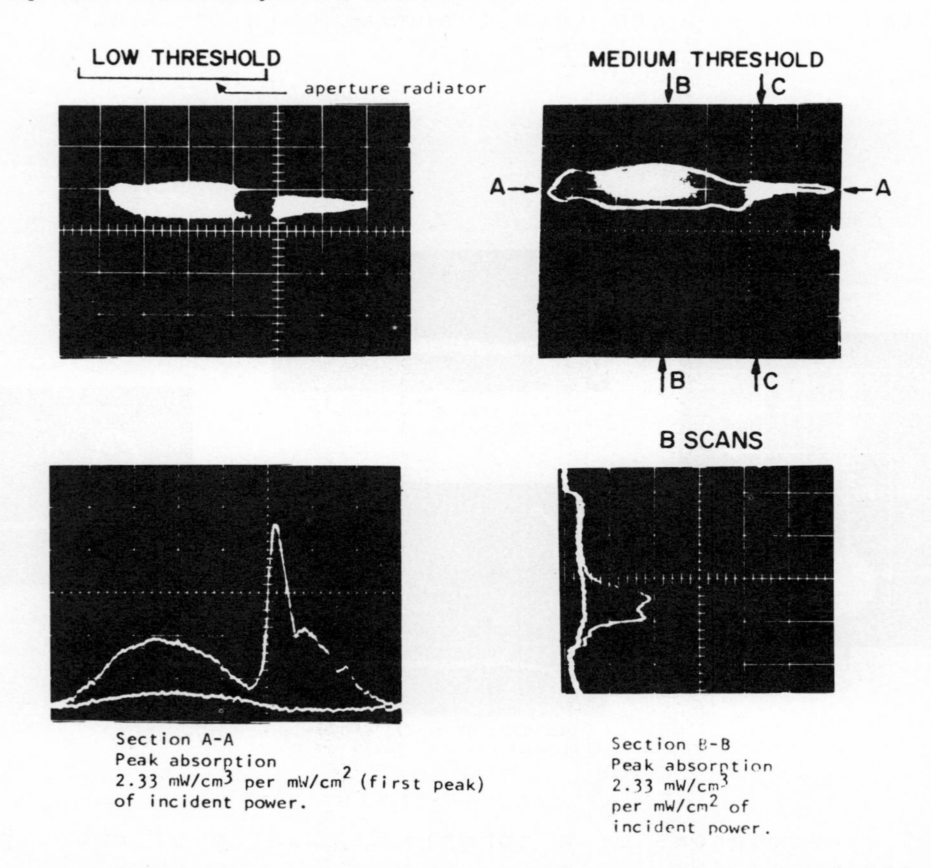

Figure 23 Thermograms of sagittal plane of a rat exposed to 918 MHz radiation

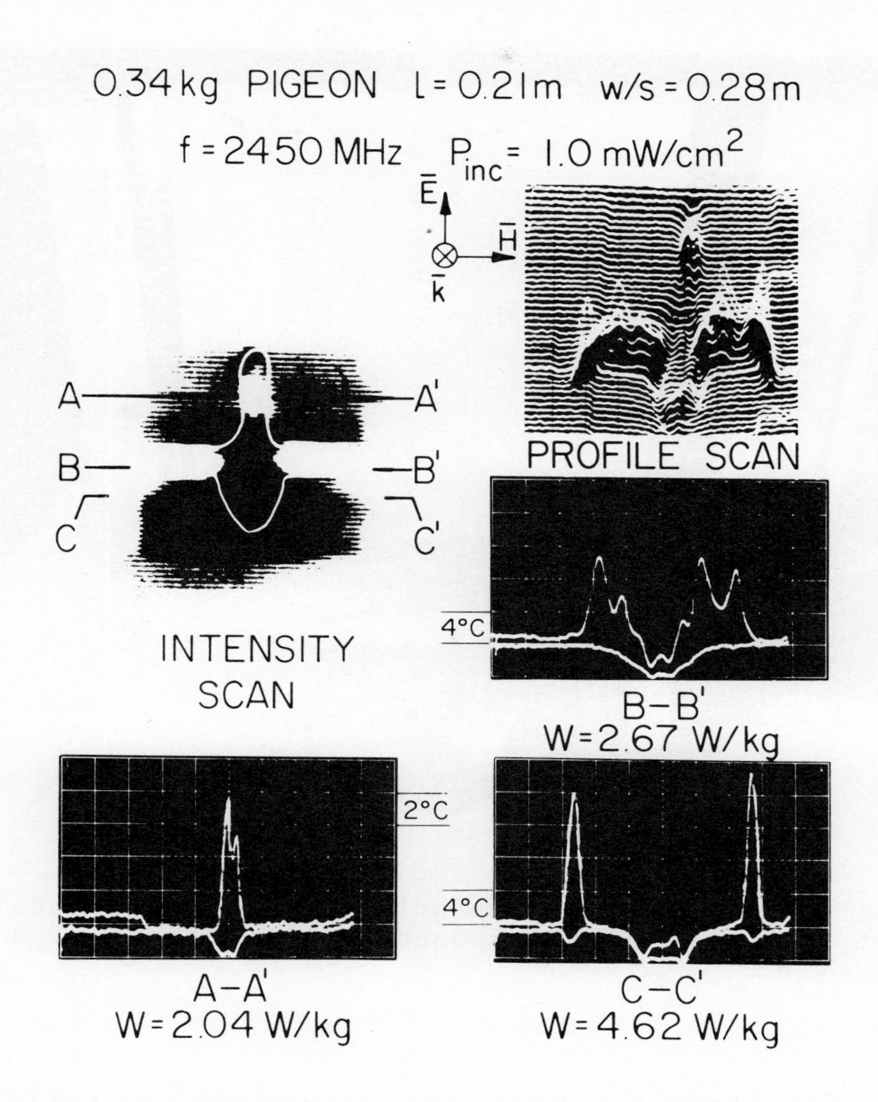

Figure 24 Thermograms of the surface of pigeon exposed to microwave radiation (dorsal exposure)

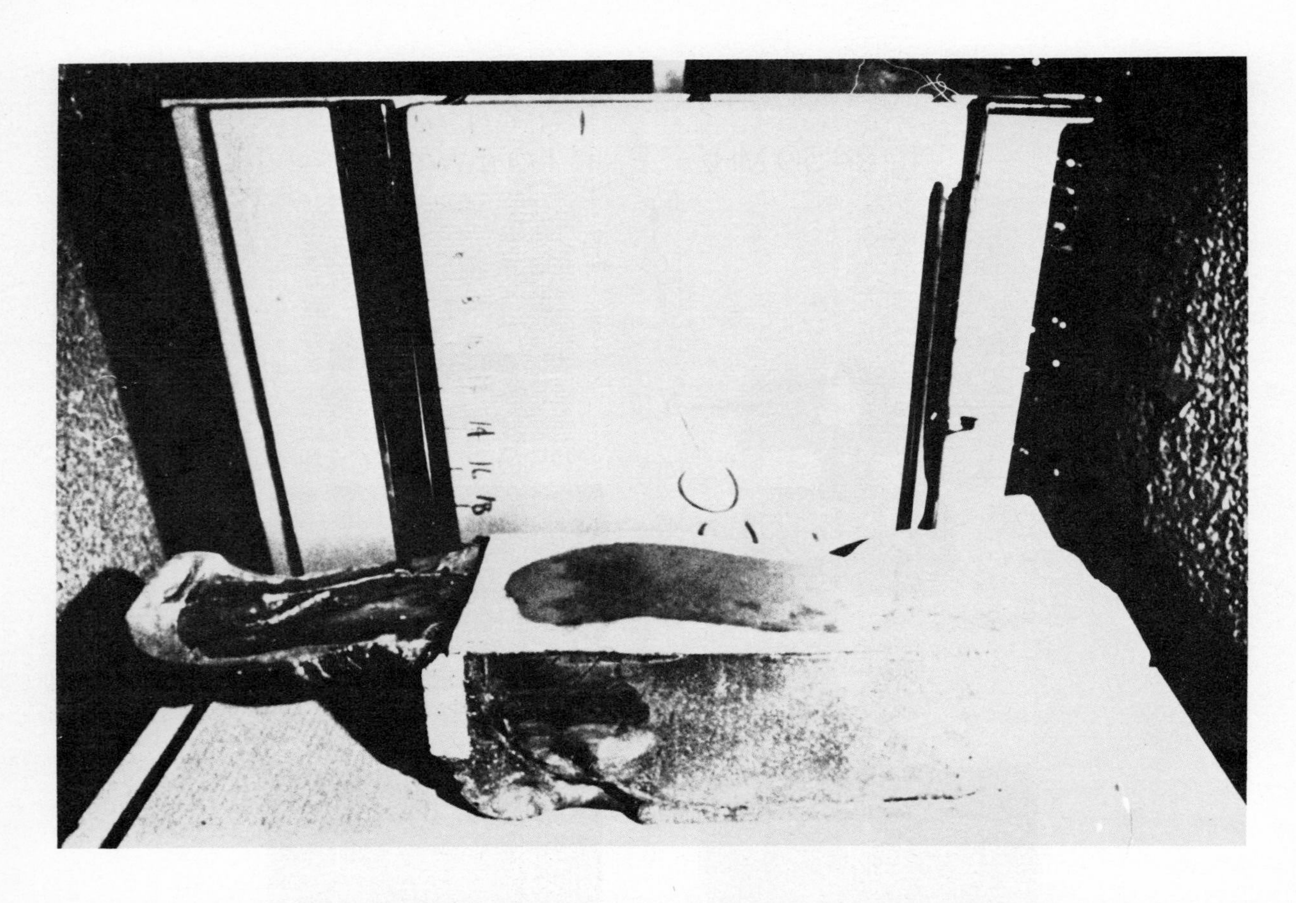

Figure 25 Full-scale tissue phantom of child (one-half of model removed at sagittal plain) exposed to the leakage of a microwave oven

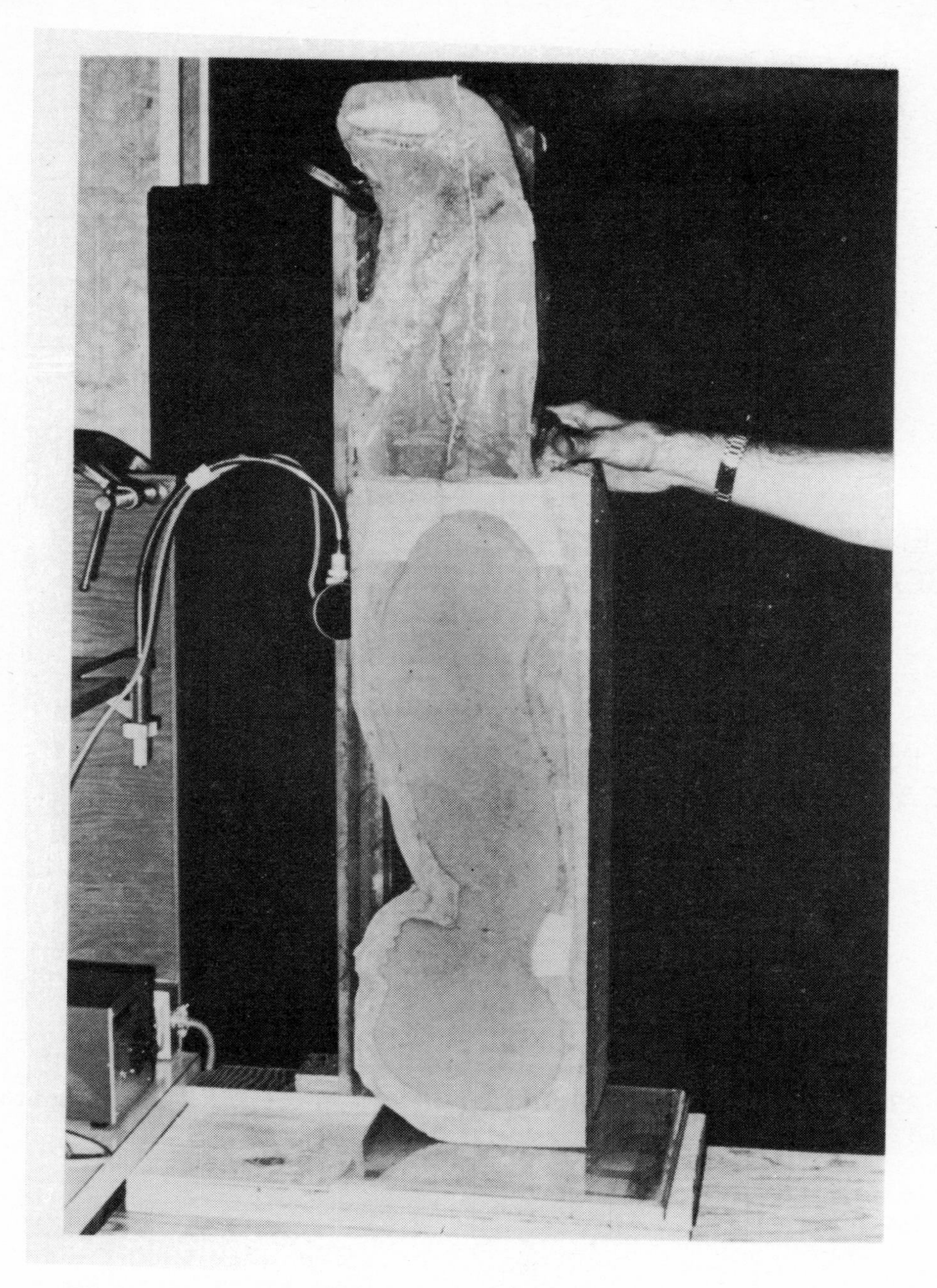

Figure 26 Orientation of a full scale child model for thermographic analysis

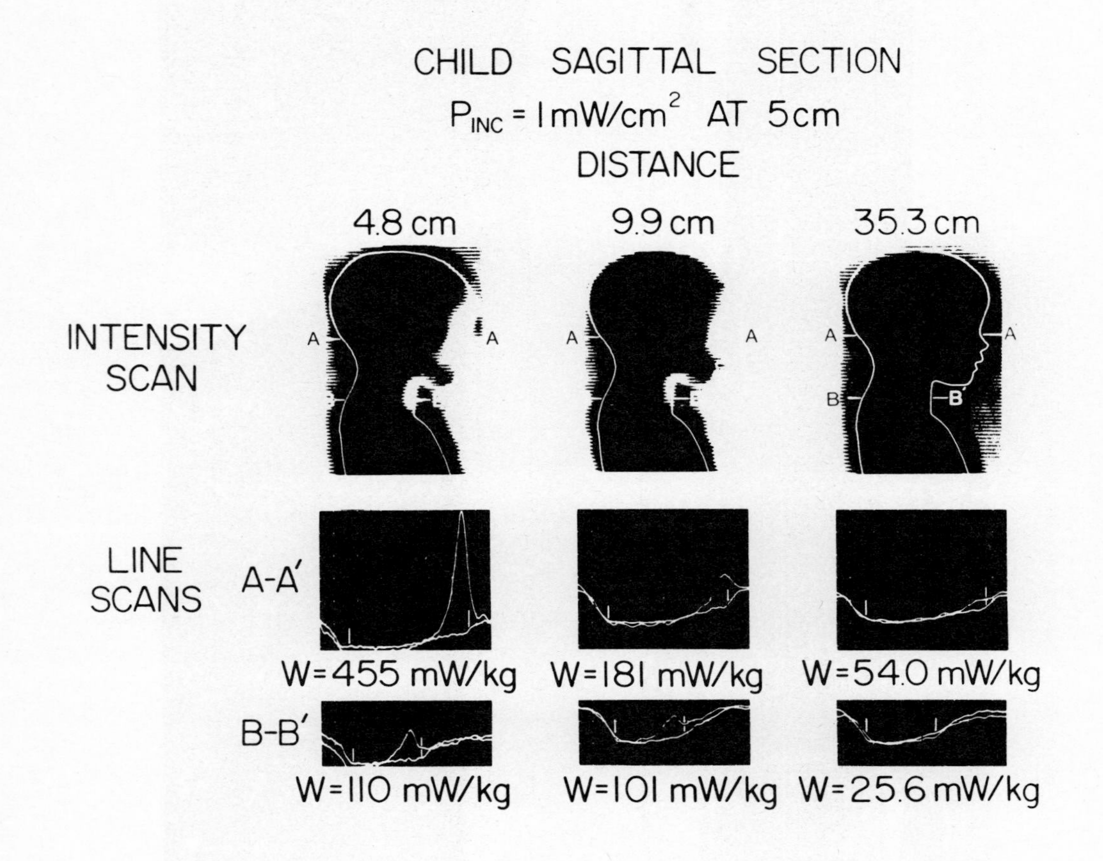

Figure 27 Thermographic study of SAR patterns on sagittal plane of child model exposed to leaky microwave oven

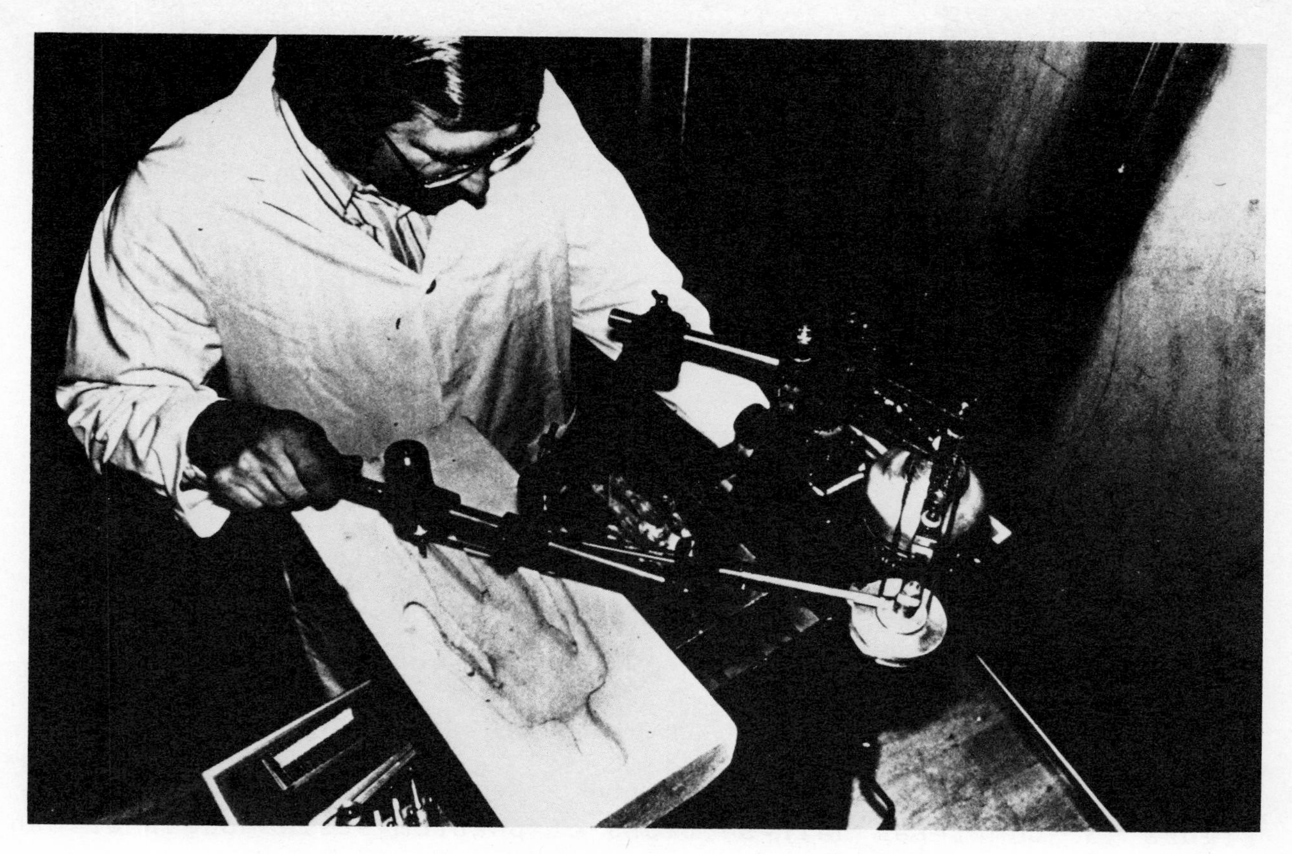

Figure 28 Use of pantograph engraver for fabricating phantom scale models of man

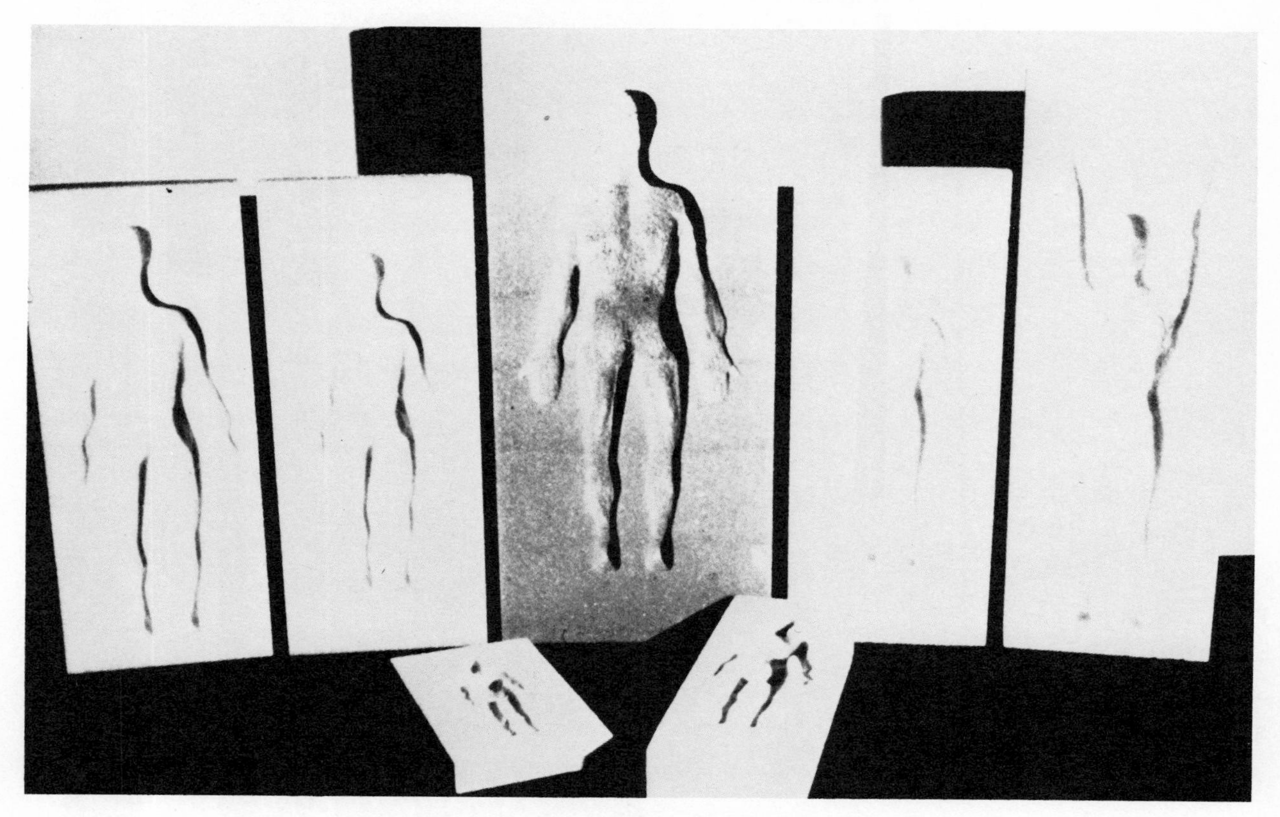

Figure 29 Illustration of half-sections of styrofoam models used for fabricating scale phantom models of man

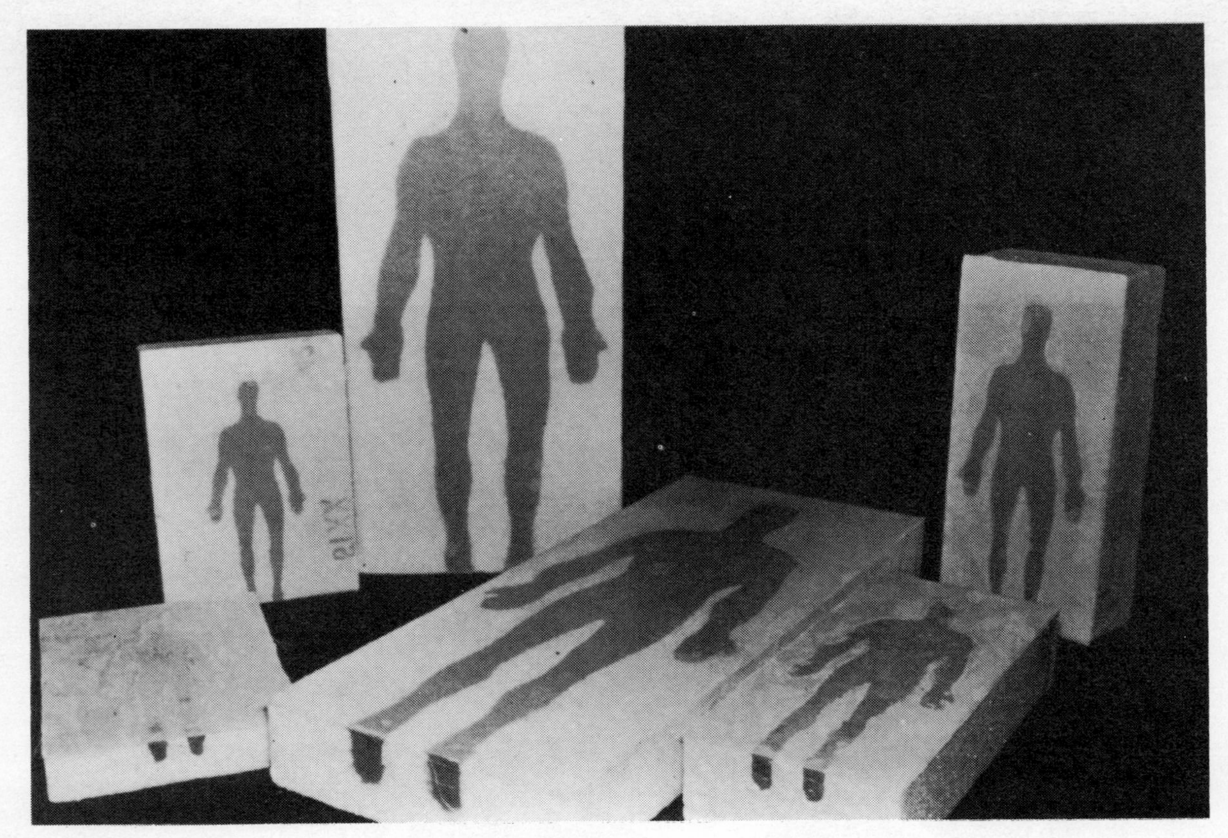

Figure 30 Completed half-sections of phantom scale models of man filled with synthetic tissue

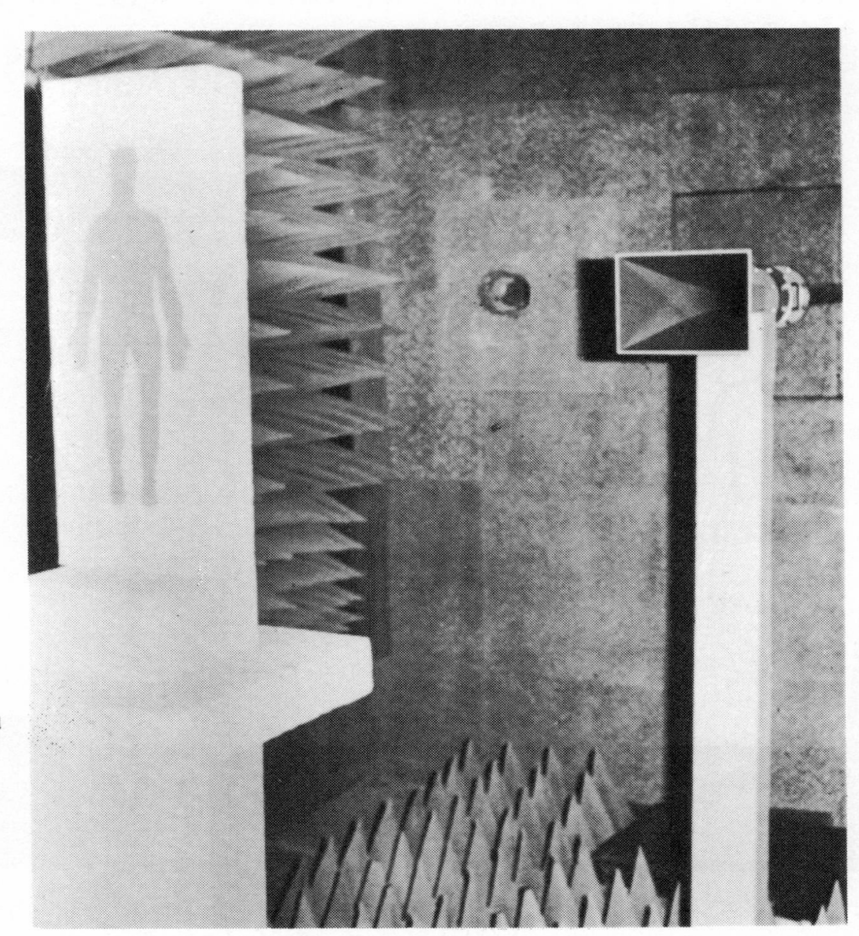

Figure 31
Exposure of phantom scale model (half section removed)

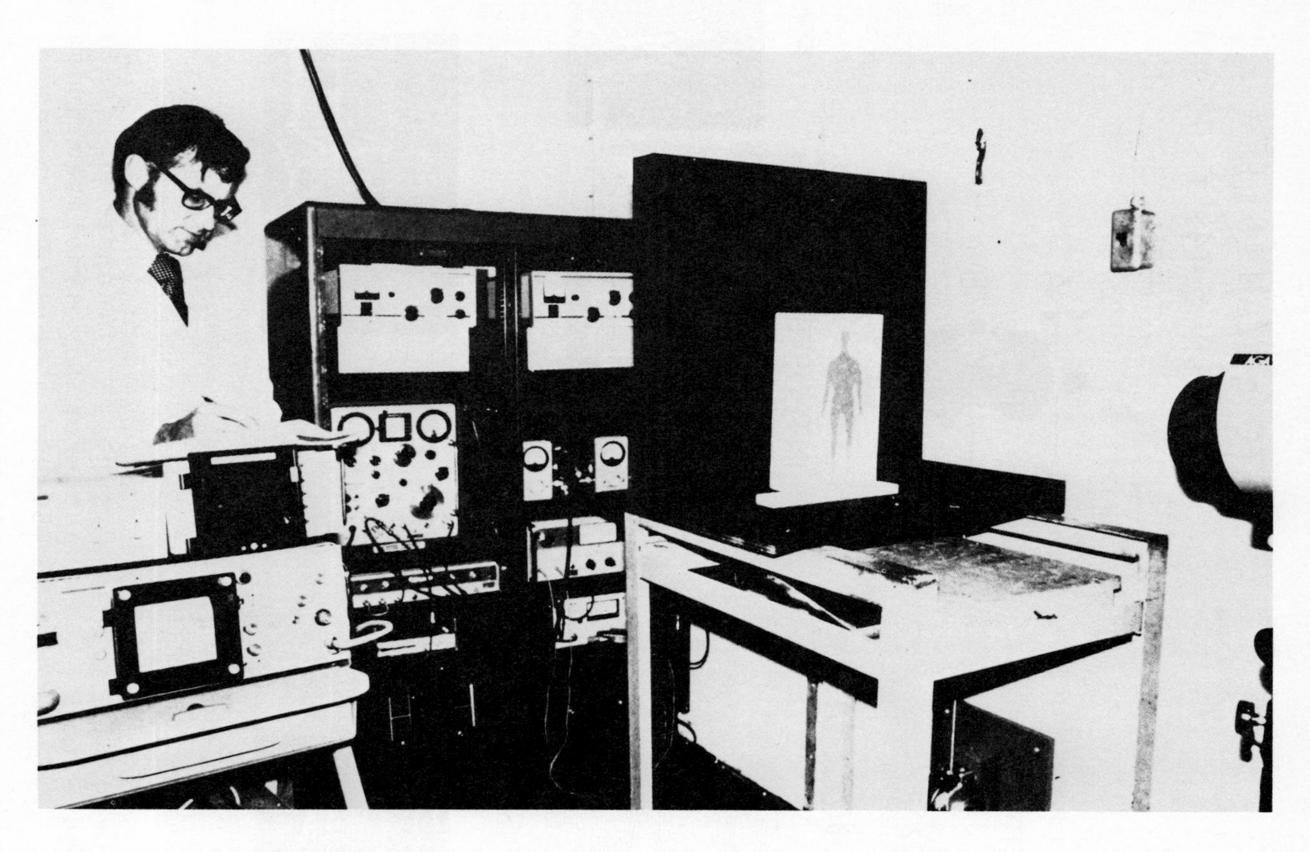

Figure 32 Use of thermograph for measuring SAR patterns in exposed scale models of man

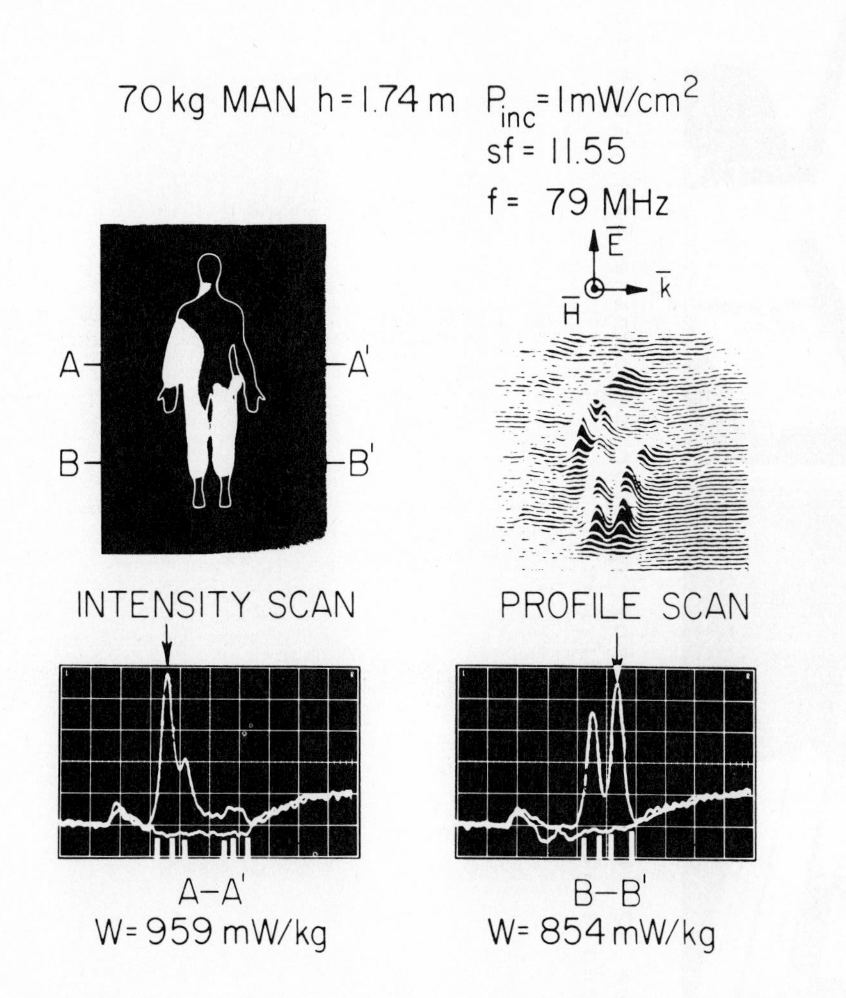

Figure 33 Thermographically measured SAR patterns for simulated human exposure to 79 MHz

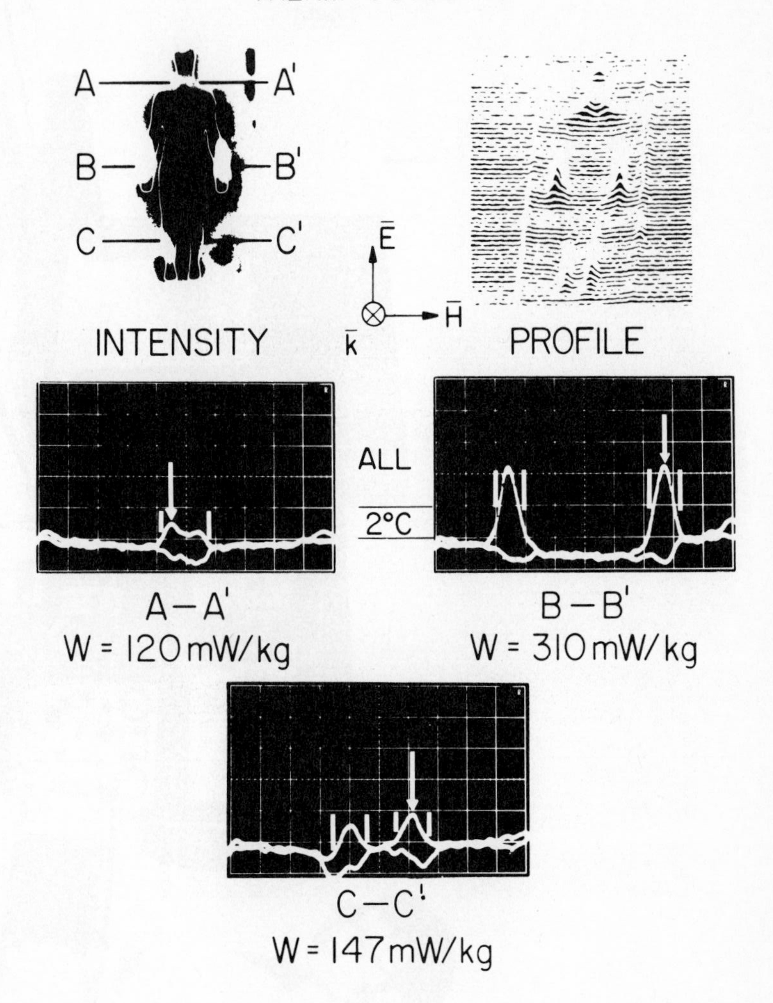

Figure 34 Thermographically measured SAR patterns for simulated human exposure to 450 MHz

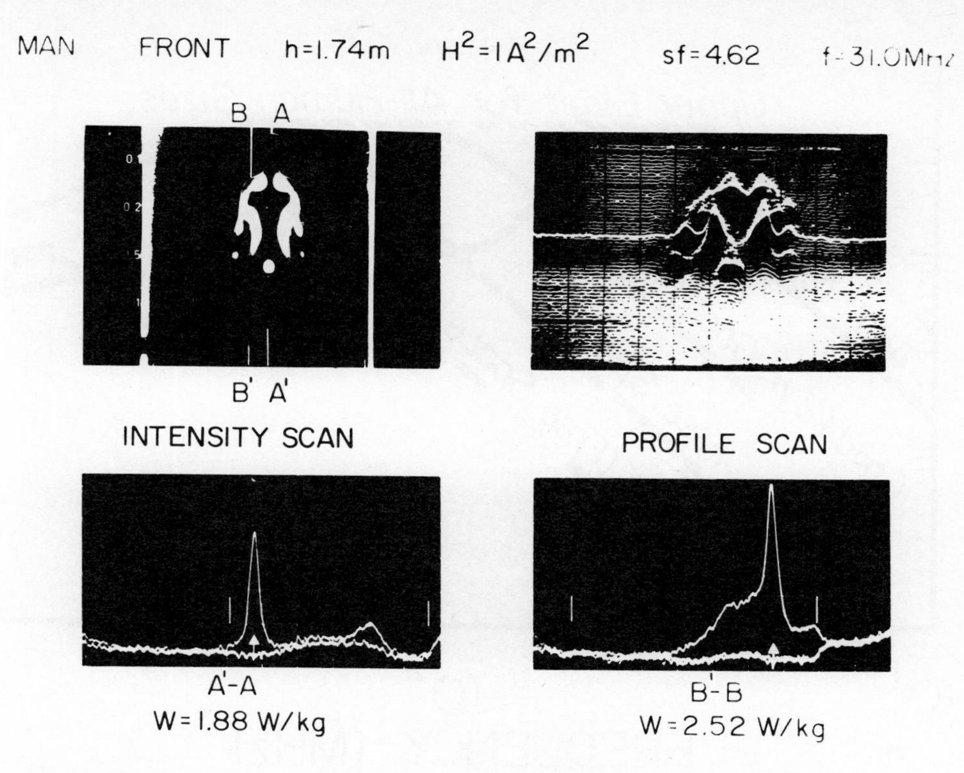

Figure 35 Thermographically measured SAR patterns for simulated human exposure to 31 MHz magnetic field

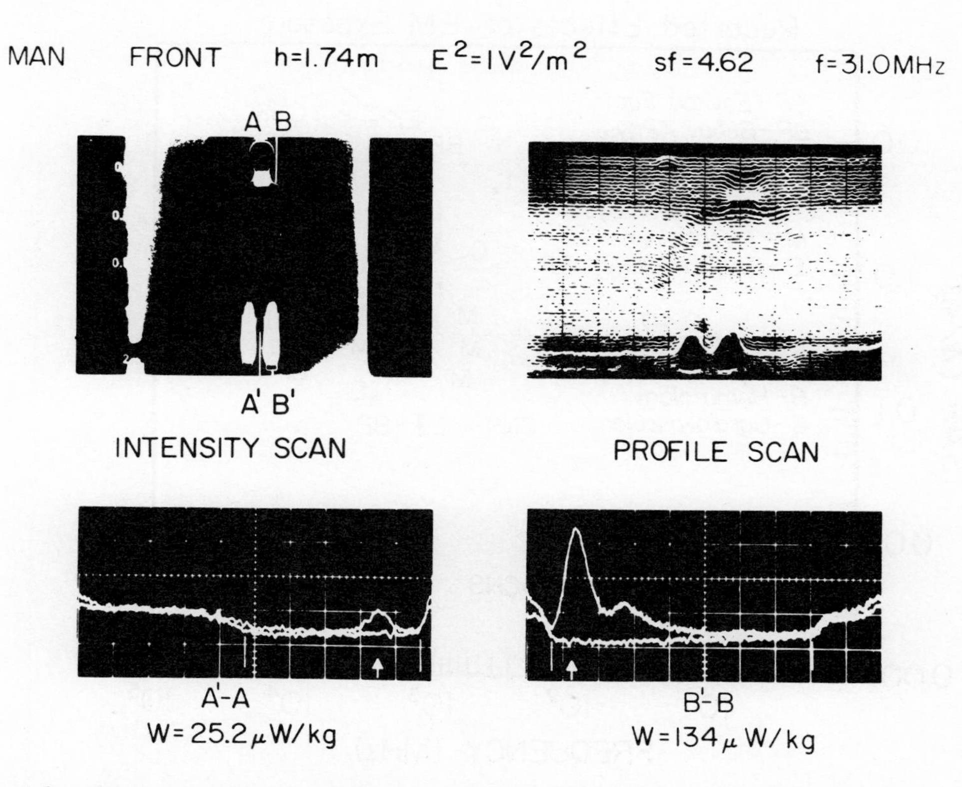

Figure 36 Thermographically measured SAR patterns for simulated human exposure to 31 MHz electric field

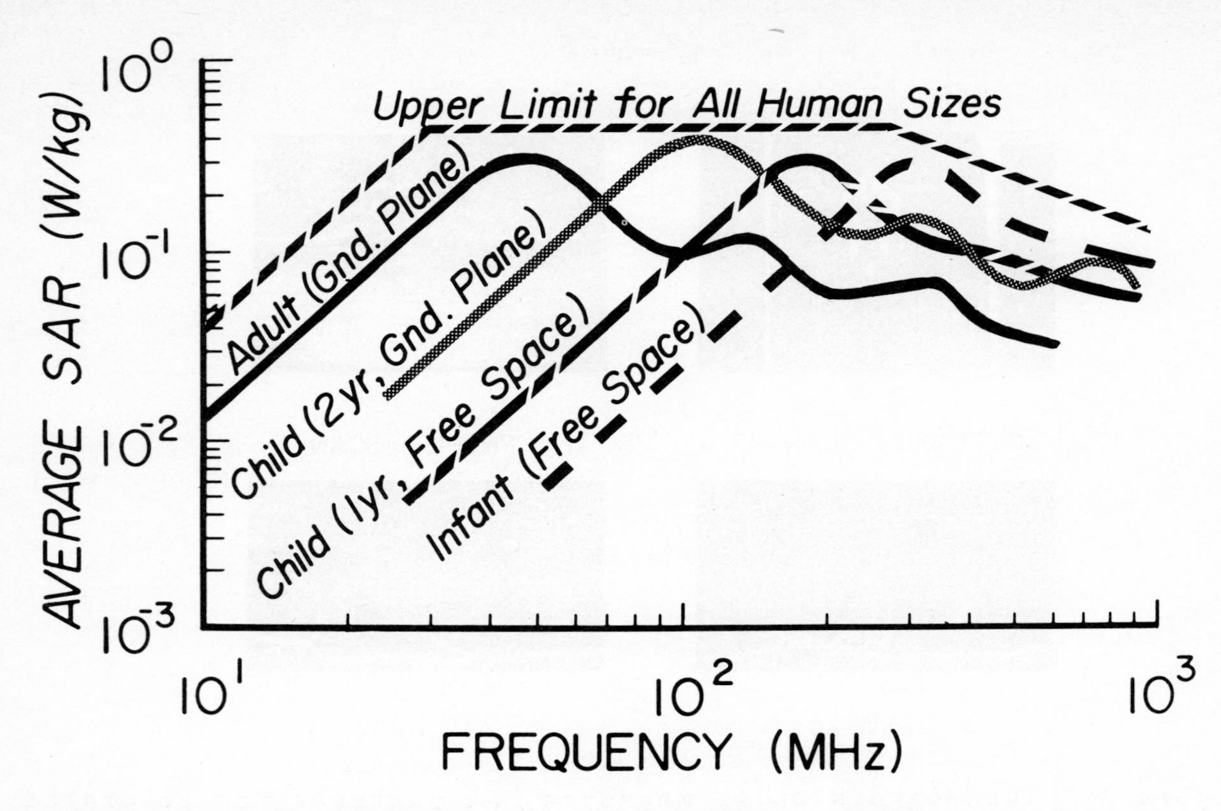

Figure 37 Average SAR vs frequency for human exposure to electromagnetic plane wave radiation [from Gandhi (12)]

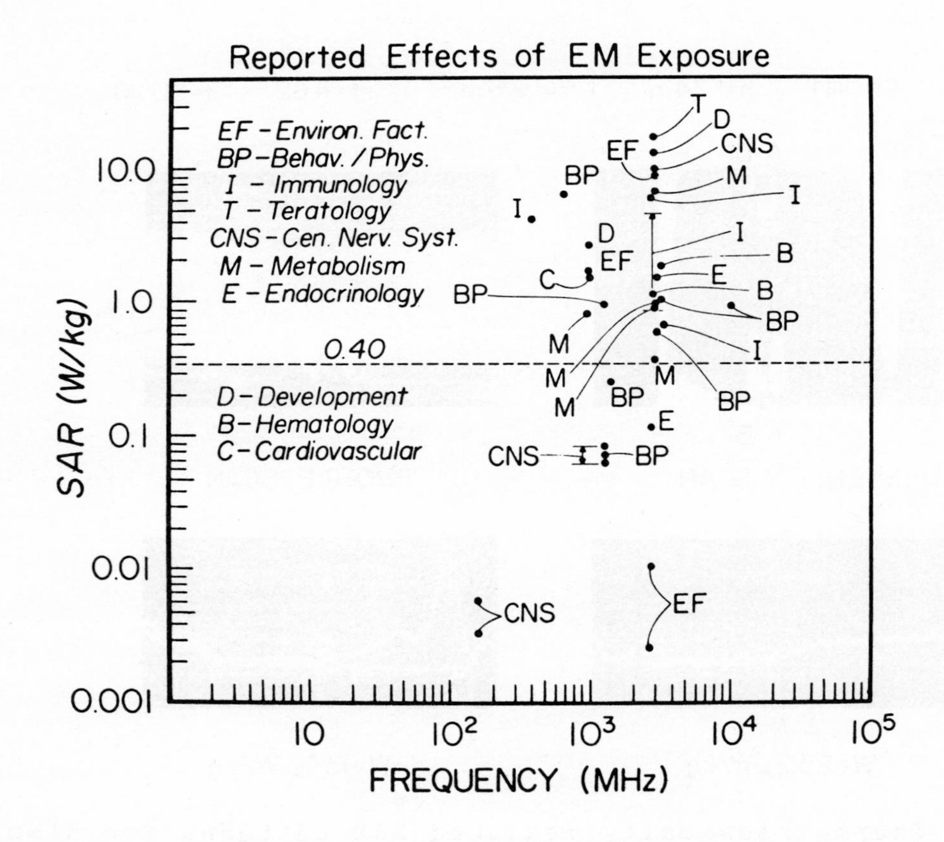

Figure 38 Reported biological effects as a function of average SAR and frequency

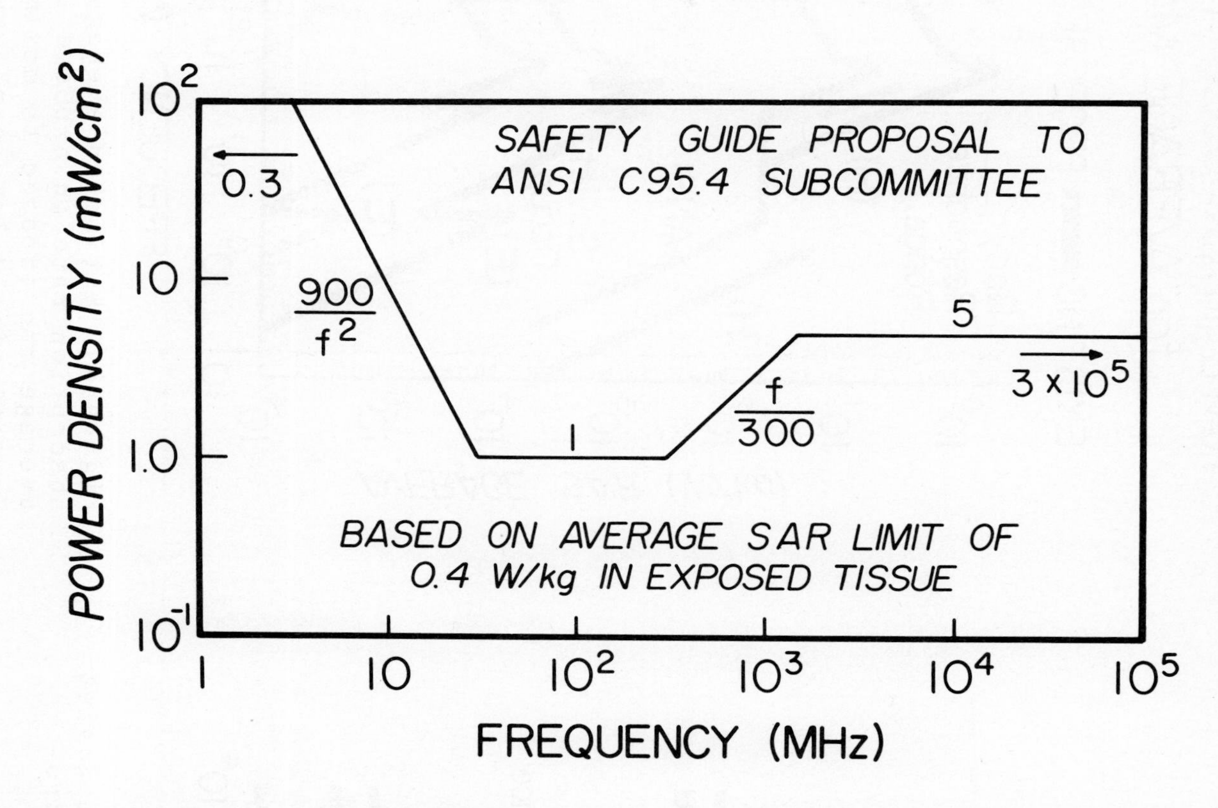

Figure 39 Proposed ANSI C95.4 Subcommittee Human Exposure Guide (based on whole body exposure)

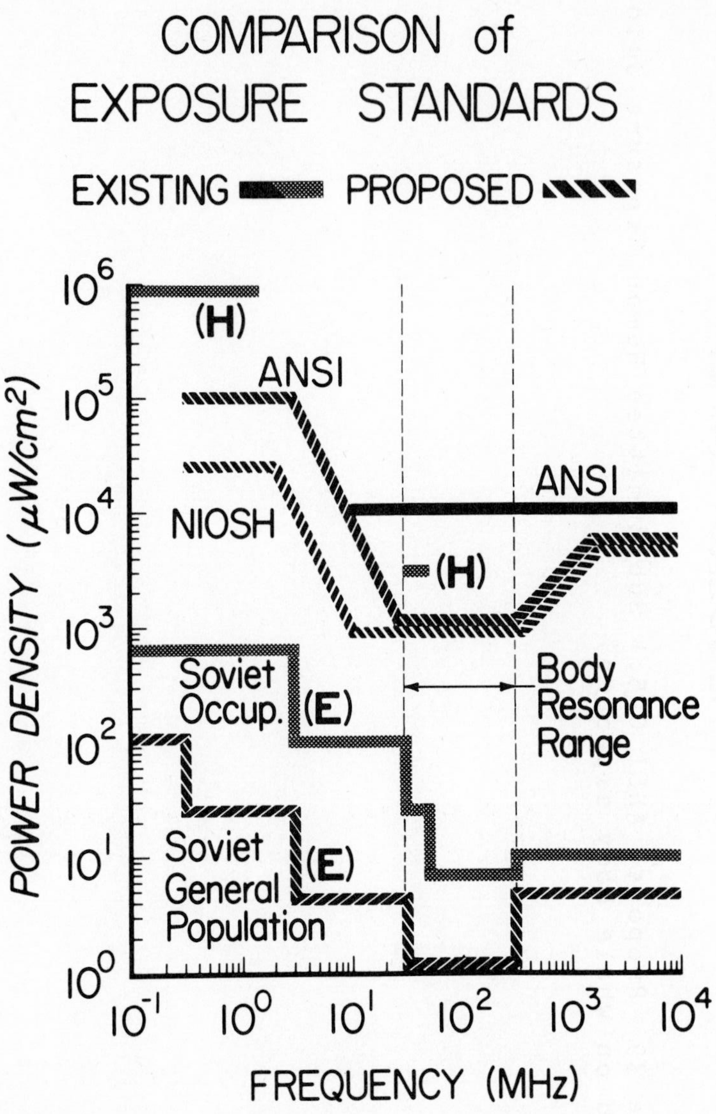

COMPARISON of
EXPOSURE STANDARDS
EXISTING
PROPOSED
POWER DENSITY (μW/cm²)
10⁶
10⁵
10⁴
10³
10²
10¹
10⁰
(H)
ANSI
NIOSH
ANSI
(H)
Soviet Occup.
(E)
Body Resonance Range
Soviet General Population
(E)
10⁻¹
10⁰
10¹
10²
10³
10⁴
FREQUENCY (MHz)

Figure 40 Comparison of American and Soviet standards for human exposure to electromagnetic fields

AVERAGE SPECIFIC ABSORPTION
RATE in EXPOSED TISSUES
for AVERAGE MAN
EXISTING
PROPOSED
AVERAGE SAR (W/kg)
10⁰
10⁻¹
10⁻²
10⁻³
10⁻⁴
10⁻⁵
10⁻⁶
Body Resonance Range
ANSI
(H)
(H)
ANSI
Soviet Occup.
(E)
(E)
Soviet General Pop.
10⁻¹
10⁰
10¹
10²
10³
10⁴
FREQUENCY (MHz)

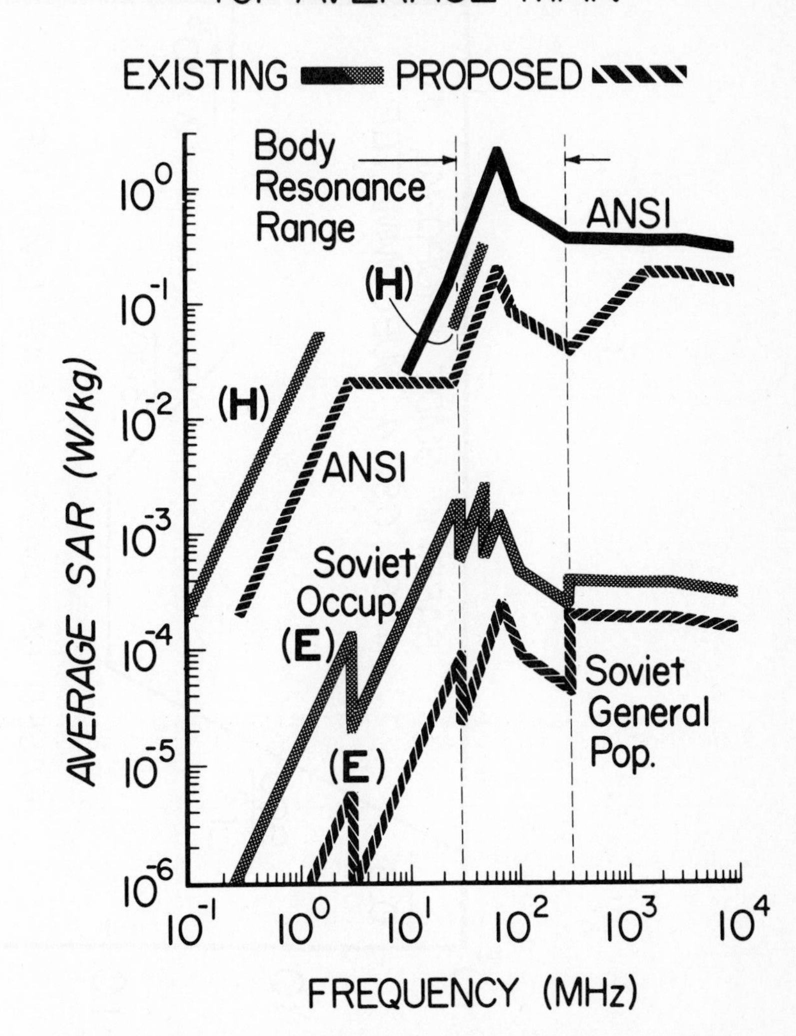

Figure 41 Estimated average specific absorption rate in exposed tissues of average man exposed to maximum electromagnetic field levels allowed by American and Soviet standards

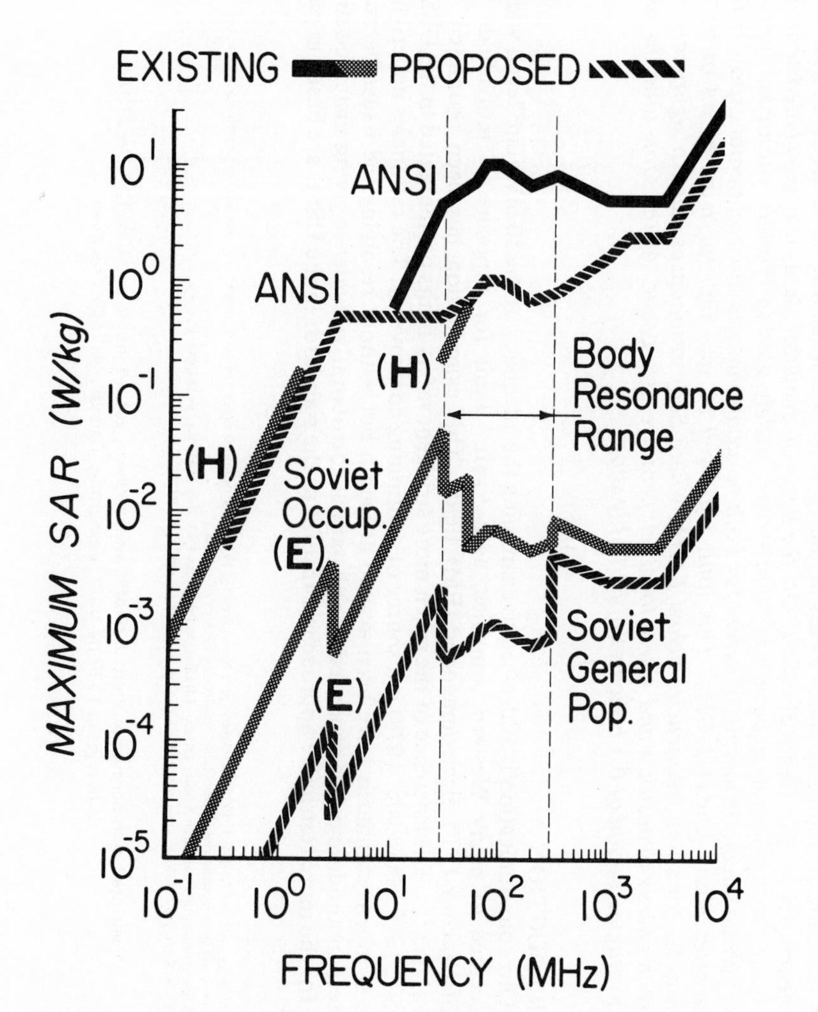

Figure 42 Estimated maximum specific absorption rate in exposed tissues of average man exposed to maximum electro-magnetic field levels allowed by American and Soviet standards

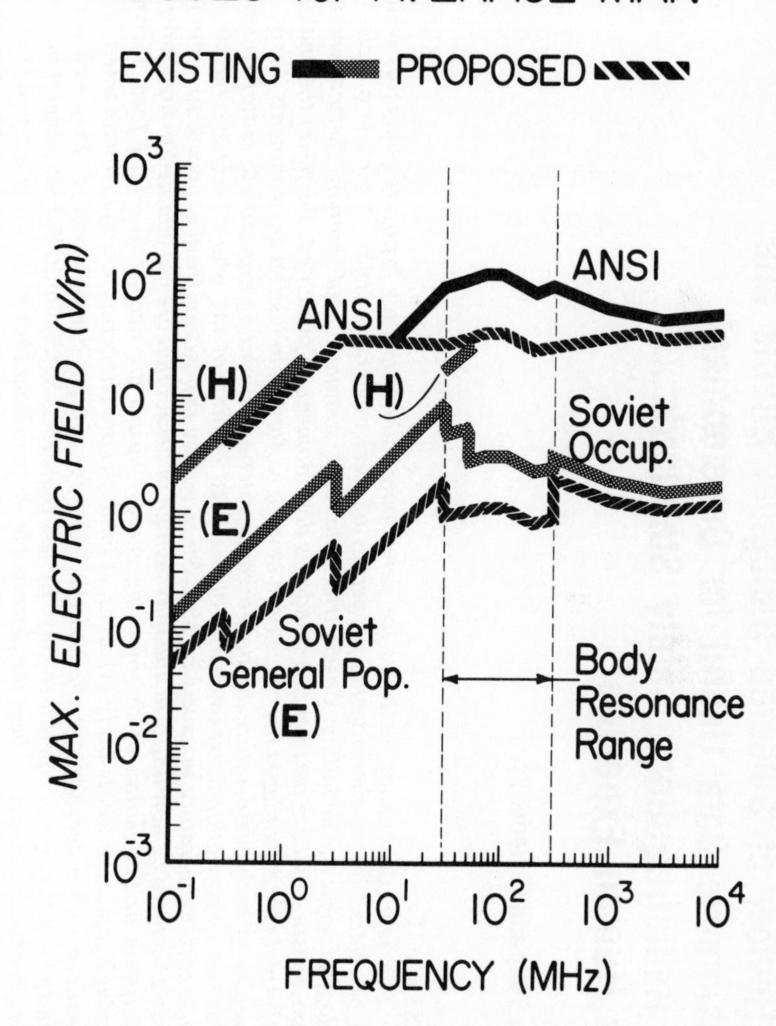

Figure 43 Estimated maximum electric field strengh in exposed tissues of average man exposed to maximum electro-magnetic field levels allowed by American and Soviet standards

A Review of Selected Biological Effects and Dosimetric Data Useful for Development of Radiofrequency Safety Standards for Human Exposure*

R.A. Tell† and F. Harlen††

ABSTRACT

This report examines the bases for developing radiofrequency exposure standards which can be related to the thermogenic properties of electromagnetic fields. A review of selected biological effects, including dosimetric data and simulation of human thermodynamic characteristics that are pertinent to standards development, is presented. Based on the analogy of thermal-stress standards that have been developed for hot industrial environments, limits on increases of body temperature are proposed as criteria for limiting exposure to radiofrequency fields, i.e., occupational exposures involving deep heating of the whole body should not increase core temperature in excess of 1°C. Since energy deposition from exposure to some RF fields is likely to be non-uniform and may be high in tissues that are not adapted to high rates of absorption or dissipation of thermalizing energy, means are needed to adjust focal thermal loading against the whole-body averages. A limit on core temperature is inadequate when focal elevations of temperature are close to the limits for protein denaturation, as may well occur even though the core temperature may rise less than 1°C. Safety limits for the general population are also discussed and here the permissible thermal load should be low enough to cause no more than an insignificant increase in core temperature. Areas needing further research to reduce the uncertainties in developing safe exposure limits for man are delineated. Even in highly adverse environmental conditions the gross thermal load and consequential heat stress from exposure to radiofrequency fields at the 10 mW/cm² level will be small compared with that generated by any physical effort. On the basis of available data, it is concluded that the safe value for continuous exposure to 10 mW/cm², widely used in Western countries, appears to provide an adequate margin of safety for both occupational and environmental exposure for frequencies above about 1 GHz. This limit may well be too high (perhaps by an order of magnitude) for some frequencies below 1 GHz where body resonances cause a significant increase in energy deposition and where local temperature rises occur. At the same time the present averaging period of 0.1 h seems unjustifiably short.

INTRODUCTION

Our communication is addressed to examining the adequacy of the 10 mW/cm² level, widely accepted in many Western countries, as a "safe" limit for exposure of individuals to radiofrequency (RF) electromagnetic (EM) fields. This examination has been prompted by several factors. The existence of the much more stringent safety limits promulgated in the USSR (GOST, 1976; Shandala, 1978) is a source of continuing controversy. The rationales underlying these stringent standards have been argued at length but without resolving the issues because Western methodology has produced no substantial corroborative evidence giving sustained support for the more restrictive approaches. Insofar as the standards in the USSR are based in part

* Manuscript received August 1, 1979; in revised form November 13, 1979. (The opinions expressed in this paper are advanced to promote discussion and are not necessarily representative of the official opinions of EPA or of NRPB.)

† U.S. Environmental Protection Agency Office of Radiation Programs P.O. Box 18416 Las Vegas, Nv. 89114 U.S.A.

†† National Radiological Protection Board Harwell-Didcot Oxfordshire OX11ORQ U.K.

Reprinted with permission from *J. of Microwave Power*, vol. 14, no. 4, pp. 405–424, Dec. 1979.

on effects in small laboratory animals due to RF exposure in the frequency range 1–10 GHz, it is highly pertinent that recent research shows that the energy absorbed in these animals may be as much as 30 times greater than that for man in a field of a given strength. On the other hand, this research also suggests that the "safe" level of radiofrequency exposure, 10 mW/cm^2, used in the United States, United Kingdom and other Western nations, may be too high at lower RF frequencies. Although the American National Standards Institute (ANSI, 1974) voluntary standard for industry and the very similar standards of the UK have been widely upheld as applicable in the work place and even for non-occupational exposures, doubts have been expressed about their applicability to the general population. In view of the absence of any safety standard specifically designated for the population as a whole, several organizations, including the US Environmental Protection Agency and the UK National Radiological Protection Board, have been looking into this situation for some time. These various factors, when taken together, have been effective in creating a sense of uncertainty in the general population and are reflected in the public attitude that where there is no standard, a need for regulation exists (Justesen, 1978). A standard that is 20 years old is seen by many, not as one that has stood the test of time but as one that is obviously out of date (cf Glaser and Dodge, 1975).

Our thesis can be summarized as follows: (1) absorption of RF energy at high field strengths can be dangerously thermogenic to man and animals; (2) Western safety standards were initially established on a thermal basis; (3) we have sufficient new data to warrant re- examination of the adequacy of these thermally based standards; (4) the overwhelming thrust of the biological evidence we have is that thermal reasoning is a sound basis for a safety standard; (5) in terms of hazard the only "specific" effect of RF energy that need be considered is the unique way that the energy may be deposited internally in man, leading to temperature gradients and thermal loads which are difficult, if not impossible, to mimic by other methods.

This communication approaches the problem by viewing man as a biological entity in which various thermoregulatory mechanisms function to maintain a constant body temperature. Over 20 years ago similar arguments were introduced by Schwan and his colleagues (see for example Schwan and Li, 1956; Schwan, 1957) with respect to the development of RF safety standards. In many environmental conditions heat conservation will be more important than heat dissipation, but in viewing absorption of RF energy as a hazard we need to consider only the factor of dissipation. The homeostatic process has to cope with and is stressed by internal heat production via the metabolic functions of the body and external sources of thermal energy, be they high ambient temperature, sunlight or other sources of radiant energy, such as RF waves. The thermal load placed on man by RF fields is considered as only one of many thermal factors present in the environment, and it is proposed that the development of safety standards for man's exposure to such fields should follow a similar path to that proved in developing the thermal stress limits for work in hot environments.

In arriving at our thermally based position, we acknowledge the claims that weak (as averaged) pulsed and sinusoidally modulated fields may have potentially adverse effects. Reference is to reports of alterations of blood-brain barrier or neurocirculatory function, of augmented effux of calcium ions in *in vitro* brain materials, and of altered behavioral response to psychoactive drugs (cf Frey, et al., 1975; Oscar and Hawkins, 1977; Merrit, et al., 1978; Blackman et al., 1979; Bawin, et al., 1975, 1978; Thomas, et al., 1979). Though, in our opinion, the 20- to 30-min exposures that have resulted in the cited weak-field changes of biological function are not implicative of adverse effects, we agree that chronic studies are indicated and must continually be monitored to ensure that any safety guide developed now can be modified to properly reflect new results as they become available. We suggest that a valid scientific and realistic approach to a review of the Western standards is thorough study of the thermogenic properties of RF energy and their physiological effects.

THERMAL SUSCEPTIBILITY OF MAN

The susceptibility of man to thermally stressing environments has been studied extensively for many years. In 1944 Robinson et al. (1945) defined an "index of physiological effect" based on a subject's heart rate, skin temperature, rectal temperature, and rate of sweating during exposure to the environment being evaluated. They developed data that define physiologically equivalent effects associated with a wide range of environmental conditions. Their work suggested that for a given level of work activity, there are combinations of environmental conditions that elicit the same physiological stress; that is, if the environmental conditions are changed and the level of work activity is maintained, a different physiological stress reaction could become evident.

With increased understanding, various limits have been developed, aimed at protecting people who must work in a hot environment. These limits have made use of a number of different parameters that affect man's ability to tolerate thermal loads, the more important of these being air temperature, relative humidity (RH), air velocity, duration of exposure to the hot environment and the work rate during exposure. In developing its own criteria document for a recommended standard for occupational exposure to hot environments (NIOSH 1972), the US National Institute of Occupational Safety and Health reviewed the available proposals for controlling thermal stress in hot environments. NIOSH found that some of the proposals (Yaglou and Minard, 1957; Brief, 1970; Wyndham, 1962) could not be recommended to industrial workers because the margin of safety was too small. The NIOSH recommendations correlate permissible work rate with environmental factors to ensure that any increase in rectal temperature in man should be less than 1°C — they provided evidence that a 1°C rise in rectal temperature should be allowed as a reasonable upper limit for man's thermoregulatory response to heat stress. This view is also endorsed by WHO, but should be recognized as being conservative. Regular diurnal temperature excursions of 1.5°C in any individual are normal, and in directly equivalent circumstances the range of core temperatures of individuals may well exceed 1.0°C. Even after acclimatization there is a difference of 0.5°C ascribable to a tropical as opposed to a temperate climate. In Samson Wright's *Applied Physiology* (1971) it is stated that, after a 3 mile race, rectal temperatures as high as 40–41°C have been recorded in athletes, and that emotional stimuli can induce transitory body temperature excursions in mammals by as much as 2°C (Justesen, et al., 1974).

Figure 1, taken from Lind (1963) and reproduced in the NIOSH report, illustrates the levels of rectal equilibrium temperature of one subject operating at three different work rates in a wide range of climatic conditions. Effective temperature, used in Fig. 1, is an empirical sensory index, combining into a single value the subjective sensation of the thermal effect of

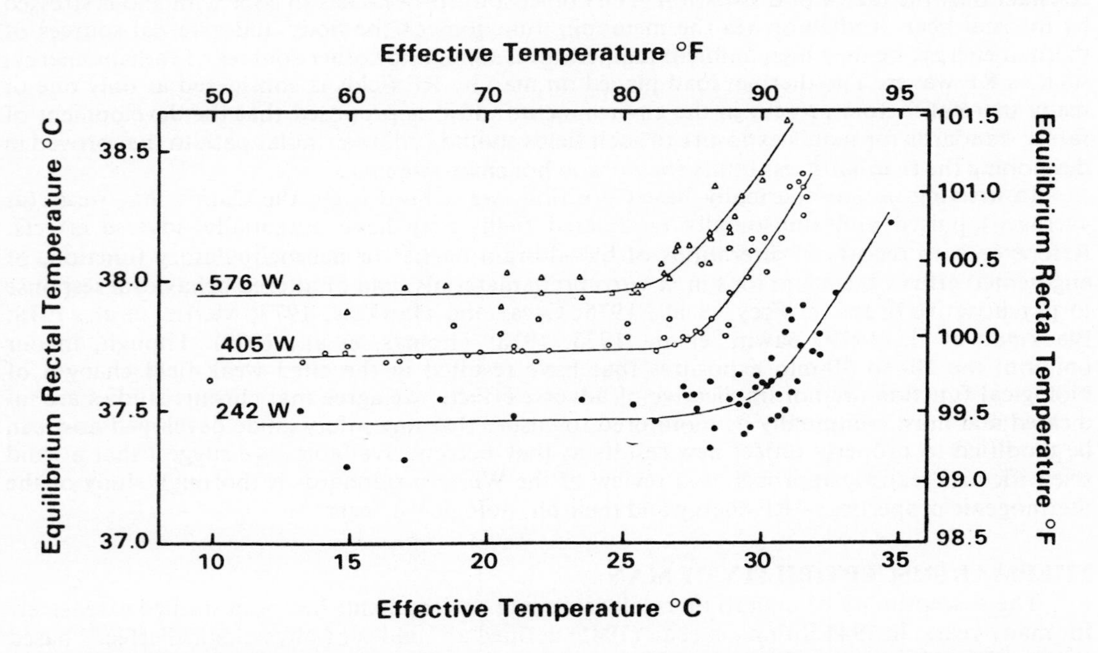

Figure 1 The levels of equilibrium rectal temperature of one subject working at 242, 405, and 567 W in a wide range of climatic conditions. From Lind (1963).

temperature, humidity, and air speed upon the human body. In still air at 100% RH, the effective temperature is the same as the dry bulb temperature, otherwise it is always lower. Lind's results show that the equilibrium rectal temperature in man is essentially independent of the environmental conditions up to a maximum effective temperature; below this temperature it is determined solely by the work rate. The higher the work rate, the higher the equilibrium rectal temperature and the lower the maximum allowable effective temperature.

Each data point in Fig. 1 was developed by an experiment that was continued until the equilibrium rectal temperature was reached, this being a period of 30 to 60 min depending on the combination of heat and work. The original work leading to this finding is credited to Nielson (1938) and subsequently was confirmed to hold only below critical air temperatures by Wyndham (1954). Wyndham and his associates showed that the sweat rate in men was a function of rectal temperature. They found that at rectal temperatures higher than about 38 to 39°C, sweat rate (the principle thermoregulatory process operating at high thermal loads) becomes insensitive to further increases in rectal temperature. At higher rectal temperatures, the thermoregulatory process becomes saturated.

According to Wyndham et al . . . "although the individual may experience difficulty in completing a task when his rectal temperature exceeds 101°F (38.3°C), he is not at the point at which excessive or intolerable conditions occur, with a danger of rectal temperature rising to hyperpyrexial levels, until the thermoregulatory processes are saturated or reach their maximum values. This does not happen until rectal temperatures reach 102–103°F (38.9–39.4°C) . . . It is equally clear that no industrially advanced nation can allow an industry to expose its employees to conditions which would saturate their thermoregulatory, processes, i.e., produce rectal temperatures in excess of 102–103°F (38.9-39.4°C), with a consequent risk of heat stroke."

Subsequent work by Stolwijk et al. (1968) added further data on how rectal temperature varies as a function of heat load and how, within limits, it is essentially independent of ambient air temperature. Figure 2 adapted from their paper demonstrates a linear relationship between rectal temperature and heat produced during exercise for three different ambient air temperatures. The correlation coefficient for the relationship between these two variables was found to be 0.91. When Lind's results are plotted in this way they yield a parallel line displaced to rectal temperatures higher by about 0.3°C.

In the NIOSH guidelines the environmental conditions are expressed in terms of the wet bulb globe temperature (WBGT).[1] Ramsey (1978) has summarized these WBGT guidelines in Table 1 which were arrived at by OSHA's Standard Advisory Committee on Heat Stress (Ramsey, 1975). The values of WBGT shown are intended not as upper limits or tolerance levels, but as levels that indicate the need for instigation of work practices to reduce adverse thermal effects. The values shown in Table 1 assume adult males, normally clothed, acclimatized, physically fit, in good health and nutrition.

It is important to note that in studies of young females (Lofstedt, 1966), it was found that females had higher body temperatures and lower sweat production than did young males for the same heat exposure. In accordance with these findings Dukes-Dubos (1976) has suggested lowering the WBGT threshold limits shown in Table 1 by 0.5°C to compensate for lower heat tolerance and 1°C for lower aerobic capacity in women. In a similar approach,modifications to the threshold values have been suggested for the obese and elderly as supported by the work of Minard (1973) and Henschel (1976). A lowering of the thresholds by 1-2°C is suggested due to these two factors. Minard (1973) and Webb (1964) have pointed out the higher heat storage and physiological stress associated with the unacclimatized state. Wyndham (1973) has discussed similar factors for workers not conditioned to a physical work task; In view of these data yet another modification of the WBGT values in Table 1 are suggested of about 2°C. Other modifications due to higher air velocities and different clothing have been suggested and Table 2 taken from Ramsey (1978) summarizes all of these modifications.

---

[1]The WBGT is a method of quantifying the thermal characteristics of environmental conditions in a single parameter. WBGT is equal to 0.7 wet bulb temperature + 0.3 globe temperature where the globe temperature is obtained via a measurement of air temperature inside a 6 inch diameter black copper sphere (see Walters, 1968; Minard et al., 1964; Taylor et al., 1968). As with "effective temperature," WBGT is always lower than dry bulb temperature unless the measurements are made in still air at 100% RH.

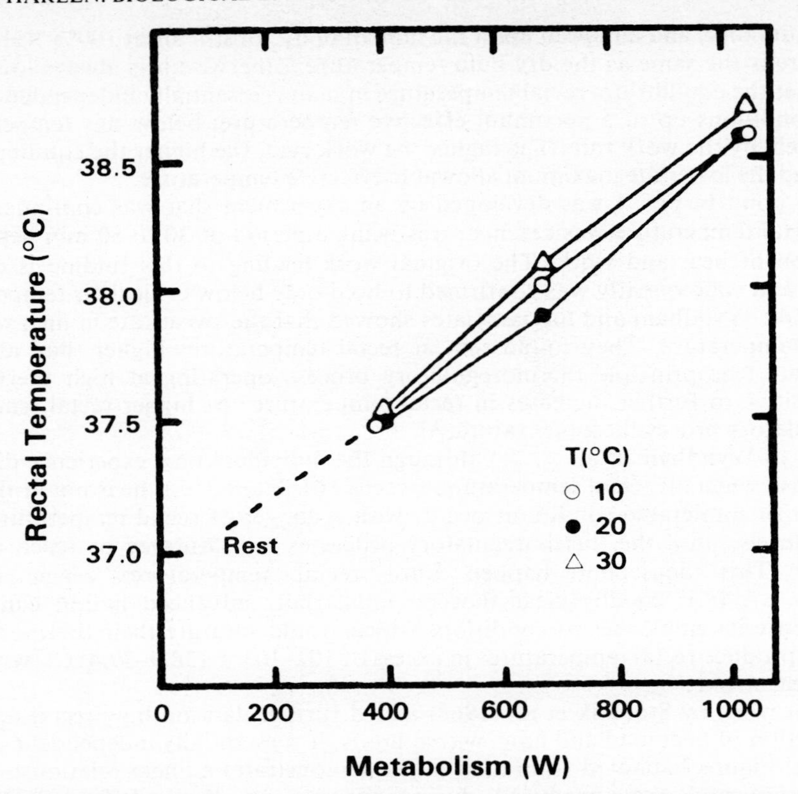

Figure 2 Variation of rectal temperature with metabolic rate at three different environmental temperatures. Adapted from Stolwijk et al. (1968).

Table 1 THRESHOLD WBGT LEVELS (Two hour time weighted average values) (taken from Ramsey, 1978)

Activity Level	Threshold °C	WBGT °F
Light work (less than 233W)	30	86
Moderate work (234–349W)	28	82
Heavy work (350–465W)	26	79
Very heavy work (greater than 465W)	25	77

Table 2 Suggested modifications to threshold WBGT values shown in table 1 (taken from Ramsey, 1978)

Factors	Modification WBGT °C	WBGT °F
1. Unacclimatized, not physically fit	−2	−4
2. Air velocity:		
1.5 m/s (300 fpm) and air temperature 35°C (95°F)	+2	+4
3. Clothing:		
Shorts; semi-nude	+2	+4
Impermeable jacket or body armor	−2	−4
Rain coats, fireman's coat	−4	−7
Completely enclosed suits	−5	−9
4. Obese, elderly	−1 to −2	−2 to −4
5. Female	−1	−2

Givoni and Goldman (1972) developed a series of formulae useful for predicting rectal temperature response to work, environment, and clothing. For our purposes the most interesting aspects of their work are illustrated by Fig. 3. This figure includes predicted and measured rectal temperatures at two work rates, two wind speeds, and three clothing types. The data were obtained at a dry bulb temperature of 35°C and about 50 percent RH. Givoni and Goldman also conclude, "It seems that, from the point of view of the response of rectal temperature to ambient temperature, there are two zones. Below about 30°C and down to the 15°C examined in the present study, rectal temperature at rest or work is independent of the ambient air temperature per se, while above this zone it rises with it."

These findings are relevant to considerations of the thermal load which can be safely imposed on an individual by the absorption of radiofrequency energy. It seems clear that on the basis of classical heat stress concepts, safe work practices dictate maintaining the rectal temperature elevation in man to no more than about 1°C over his normal value. If exposure to radiofrequency fields is added as another component of the environment then corresponding adjustments may be required in work rate or other environmental conditions to maintain the rectal temperature rise within the prescribed limit but this is only likely to be necessary under extreme ambient conditions where heat rejection is difficult. The selection of a 1°C rise included consideration of the obvious fact that strenuous exercise in man results in an increase in average body temperature and that for all activity there is an associated physiologic strain. Maintaining physiologic conditions of heat stress within reasonable limits, according to sound dosimetric and biological research data, should also be acceptable in considering exposure to radiofrequency energy.

On the basis of the foregoing information the following general conclusion is reached. Under most environmental conditions up to about 30°C, the rectal temperature is determined

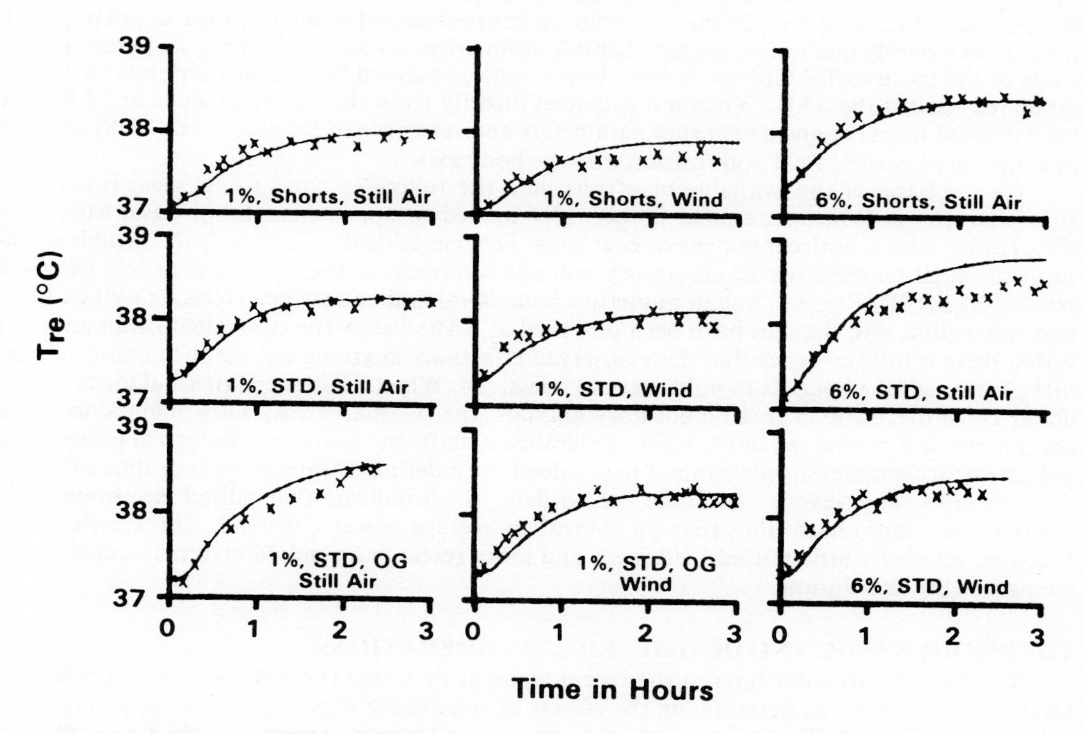

Figure 3 Comparison of predicted (lines) and measured (points) patterns of rectal temperature with two work levels (M = 310W and 470W attained by walking on an inclined treadmill at 1 and 6% grades), two wind speeds (still air = 0.5 m/s, and 5 m/s), and three clothing types (shorts, fatigues, and fatigues and overgarment). Air temperature was 35°C and vapor pressure 21 mm Hg in all experiments. From Givoni and Goldman (1972).

only by the work intensity for heat acclimated, young, physically fit males. In view of the observation by Kraning et al. (1966) that a heat load induced by physical activity is twice as strenuous on the cardiovascular system as an equal load caused by a hot environment, it seems conservative to assume that the heat stress induced by the absorption of RF energy will not produce higher rectal temperatures than an equivalent metabolic rate caused by exercise. In this case, the rectal temperature is used as a measure of the whole-body thermal stress level caused by the RF exposure and is most appropriately used for exposures at frequencies that ensure deep penetration into the body by the electromagnetic fields. An additional consideration, which may become most important under certain conditions of exposure, is the possibility of highly non-uniform deposition of energy within the body of man and subsequent non-uniform increases in localized tissue temperatures, the biological consequences of which may not be properly reflected by deep body average (rectal) temperature. With the proviso that for some conditions it may also be necessary to control local hot spots, it would seem reasonable to define intensity of radiofrequency irradiation as a function of frequencies that could potentially elevate man's rectal temperature by 1°C. Extensive theoretical calculations of body absorption by man have been performed by Durney et al. (1978) and an inversion of these specific absorption rate (SAR) curves from the viewpoint of thermal loads of RF energy in man (Tell, 1978) can be used to outline the limits of the RF fields to which man may reasonably be exposed.

THERMOGENIC BIOEFFECTS OF RADIOFREQUENCY ENERGY

The thermogenic properties of radiofrequency energy have been extensively documented in the literature on biological effects. A significant problem in making use of many of these data, however, is that it is very difficult, if not impossible, to obtain meaningful relationships among the observed effects, the parameters of environment and EM fields, and the potential impact of such exposures on human health. Upon examining the literature we found a surprisingly small number of reports that are directly quantitatively useful for the development of thermally based RF exposure standards. Few of the available reports provide sufficient technical detail to permit other than a purely qualitative insight. Table 3 summarizes a selection of what we perceive to be some of the more useful biological data. In this table, observed biological endpoints have been tabulated against the SAR. When not provided directly from the reports, values of SAR have been derived from the given exposure parameters and reference to Durney et al. (1978) with the assumption of electric field polarization with the body axis.

On the basis of the available bioeffects data the following conclusions seem reasonably supportable: 1. When an animal is thermally loaded to approximately its Basal Metabolic Rate (BMR) value, noticeable signs of heat stress become evident; 2. Clearly discernible elevations of rectal temperature in laboratory animals can occur at thermal loads as low as 20–30 percent of the BMR; 3. Though numerous behavioral and endocrine effects as well as heart and respiration rate changes have been observed at SARs below the equivalent of an animal's BMR, there is little evidence that these observed effects are anything but manifestations of normal physiological responses to mild thermal stress; 4. When RF induced thermal loads exceed about twice the BMR in small laboratory animals, the thermal stress, when maintained on a continuous and protracted basis, leads to significant shifts in various physiological indices that indicate unacceptable chronic strain if maintained for indefinitely long periods of time (order of days, months, and years); 5. There are no data which indicate that pulsed electromagnetic fields are any more hazardous than an equivalent average power CW field. This conclusion is based on relatively little pulsed field work and more research on pulsed effects is called for to strengthen this conclusion.

THERMODYNAMIC AND DOSIMETRIC CONSIDERATIONS

We find of particular interest the recent work of de Lorge (1978a), Guy et al. (1978) and Durney et al. (1978). In determining the effects of microwave exposure on changes in operant behavior in three different species (rats, squirrel and rhesus monkeys), de Lorge (1978a) observed that a noticeable shift occurred within one hour when the rectal temperature was elevated 1°C this time being on the order of the equilibration time for body temperature rise due to a step function increase in thermal load to man (Emery, 1975; Stolwijk and Hardy,

1966). de Lorge observed the interesting fact that the exposure power density required to elevate the animal's rectal temperature 1°C in 60 min appeared linearly proportional to the logarithm of the animal's body mass.

Table 3 Selected thermogenic effects of radiofrequency irradiation according to measured or estimated specific absorption rate (SAR) in W/kg

SAR	Observed Effects	Investigators
190	Produced 45°C in dog thyroid gland, localized thyroid exposure at 2.45 GHz, 2 h, thyroxin release rate 10X control value	Magin et al. (1977)
131	Produced 41°C in dog thyroid gland, localized thyroid exposure at 2.45 GHz, 2 h, thyroxin release rate 3X control value	Magin et al. (1977)
80	Highest incidence of congenitial anomalies in mouse fetuses after 4 min exposure on day 8 and 10 of gestation	Rugh (1976)
58	Produced 39°C in dog thyroid gland, localized thyroid exposure at 2.45 GHz, 2 h, thyroxin release rate 3.5X control value	Magin et al. (1977)
40	Severe heat stress in rats at 2.45 GHz, evidence of blood vessels being sensitive tissue	Polson et al. (1974)
35	In SAR range of 30–40, body temperature was 42.5 — 43.0°C in guinea pig; brain temperature sometimes as much as 2°C lower than rectal temperature, 2.45 GHz	Bruce-Wolfe et al. (1979)
28	5°C increase in brain temperature in rats after 2 h of exposure at 1600 MHz. 3°C increase in body temperature in 10 min, norepinephrine content of hypothalamus reduced below that of hyperthermal controls	Merritt et al. (1977)
25	Reduced oxygen consumption by mice with apparent heat stress at ambient temperature of 24°C, 2.45 GHz	Ho and Edwards (1976)
12	Increased respiration in rabbit 20X greater than heart rate, 2.45 GHz	Birenbaum et al. (1975)
12	Increased heat dissipation rate in mice, 2.45 GHz	Ho and McManaway (1977)
11	Rectal temperature of 42.4°C in rats at 2.45 GHz accompanied by severe heat stress	Phillips et al. (1975)
11	Bradycardia after 1-h exposure in rats at 2.45 GHz, acclimation after repeated exposure	Phillips et al. (1973)
8	Plasma corticosterone levels elevated in rats after 30 min at 2.45 GHz	Lotz and Michaelson (1978)
8	Growth hormone levels lower in rats at 2.45 GHz, 30 or 60 min	Lotz et al. (1977)
7	Change in mitogen stimulated response after 4 h exposure/day *in utero* of rats at 425 MHz	Smialowicz et al. (1977)
6.5	Observed threshold for depression of metabolic rate in rats at 2.45 GHz (lower oxygen consumption)	Phillips et al. (1975)
6.4	No apparent effects in rats in offspring or litter size at 2.45 GHz, 1 h on day 9 and 16 of gestation	Michaelson et al. (1976)
5.6	Upper level of dose rates *in utero* producing some exencephaly in mice, 2.45 GHz, 100 min daily during organogenesis	Berman and Carter (1978)
5.5	No heart rate changes in turtle or tortoise whole-body exposure at 960 MHz	Flanigan et al. (1977)
5.0	Change in mitogen stimulated response after 4 h exposure/day *in utero* of rats at 2450 MHz	Smialowicz et al. (1977)
5.0	Reduced oxygen consumption in rats at 2.45 GHz at ambient temperature of 24°C	Ho and Edwards (1976)
4.9	1°C rise in rectal temperature in rats at 2.45 GHz after about 60 min, ambient temperature of 22.5°C	Houk et al. (1975)
4.8	Noticed transient elevation in plasma corticosterone levels on rats at 2.45 GHz, 30 min	Guillet et al. (1975)
4.8	Ambulatory activity decreased in rats at 2.45 GHz	Sanza and de Lorge (1977)

Table 3 continued

4.3	1°C rise in rectal temperature in rhesus monkey after 60 min at 2.45 GHz	de Lorge (1976)
4	No metabolic effects in rat at 2.45 GHz	Phillips et al. (1975)
3.5	Threshold for elevation in rat rectal temperature after 2 h at 1600 MHz	Merritt et al. (1977)
3.2	Plasma corticosterone levels elevated in rats after 60 min at 2.45 GHz	Lotz and Michaelson (1978)
3.2	Threshold for increase in rectal temperature in rats after 1 h at 2.45 GHz, no changes in adrenal or thyroid gland weights up to 8 h exposure	Lu et al. (1976)
3.1	0.70°C rectal temperature rise in rats at 2.45 GHz at ambient temperature of 22.5°C	Houk et al. (1975)
3.0	"few" W/kg influenced neural output in isolated neural preparations	Seaman et al. (1975)
2.9	Ambulatory activity decreased in rats at 2.45 GHz	Sanza and de Lorge (1977)
2.7	"Not thermally significant" in mice after 100 min at 2.45 GHz	Berman and Carter (1978)
2.5	Rectal temperature elevated 1°C in squirrel monkey in 60 min at 2.45 GHz	de Lorge (1977)
2.5	Lower threshold for latency changes in neuron conduction at a range of different frequencies	Guy et al. (1974)
2.4	Obvious signs of thermal stress in rats after 1 h, for 5 days 2450 MHz	Diachenko and Milroy (1975)
2.1	Bradycardia in isolated rat heart, 960 MHz	Olsen et al. (1977)
2.0	Colonic temperature significantly increased in rats after 60 min at 2.45 GHz	Lotz and Michaelson (1978)
2.0	0.7°C rise in rectal equilibrium temperature after 1.5 h at 26 MHz	Frazer et al. (1976)
1.6	Observable changes in rectal temperature in rats at an ambient temperature of 27.5°C and 2.45 GHz	Houk et al. (1975)
1.6	Circulatory thyrotropin levels depressed in rats after 1 or 2 h at 2.45 GHz	Lu et al. (1977)
1.6	Threshold for some measured parameters in rats exposed for 10 days, 2 h/day at 2400 MHz	Dordevic (1975)
1.6	No difference in body mass of mice exposed 2 min/h daily, for 36 days, 2.45 GHz	McAfee et al. (1973)

In Fig. 4 we plot de Lorge's power density-temperature threshold data and extrapolate the curve to man's body mass of 70 kg and then find that the expected exposure intensity at 2.45 GHz producing a 1°C rectal temperature rise in man would presumably be about 92 mW/cm², assuming that man's thermoregulatory system would function as well as that of the lower animal. Michaelson[1] observed a similar relation when he correlated power density with surface area of the body. It must be recognized that the projected 92 mW/cm² exposure for man does not consider the fact that the principal heating will be surface heating and although the body core temperature may in fact not rise appreciably, localized tissues near the body surface may approach a dangerous level, i.e., that of irreversible denaturation. Thus, great caution and judgement is necessary when trying to evaluate the thermal load placed on an individual from a RF field. In proposing limits for RF exposure it is necessary, therefore, to establish that the thermal load and temperature excursion of local tissues is not sufficient to cause injury or unacceptable stress. Even so, it seems rather obvious that the present safety standard of 10 mW/cm² in use in the US and UK provides an adequate margin of safety both for deep body core and surface temperature elevations in occupational environments at 2.45 GHz.

Our reasoning is predicated on the basis that the depth of penetration for 2.45 GHz radiation is such that about 90 percent of the energy is deposited within the first 4 cm of tissue and that if all of the incident energy were absorbed with no reflection, a maximal rate of energy deposition on the order of 2.5 W/kg would result from a 10 mW/cm² exposure. In a static system, with no thermal diffusion or dissipation mechanisms operative, this rate would lead to no more than a 2-3°C rise in the outer layers of tissue in a 3-h period. In the dynamic system of man with an ability to equilibrate core temperature within 1-2 h (Emery, 1975; Stolwijk and Hardy, 1966) it would be surprising to see a 1°C rise in these tissues. The time required for at-

[1]Personal Communication at 1978 IMPI Meeting, Ottawa, Ontario, Canada, June 27-30, 1978.

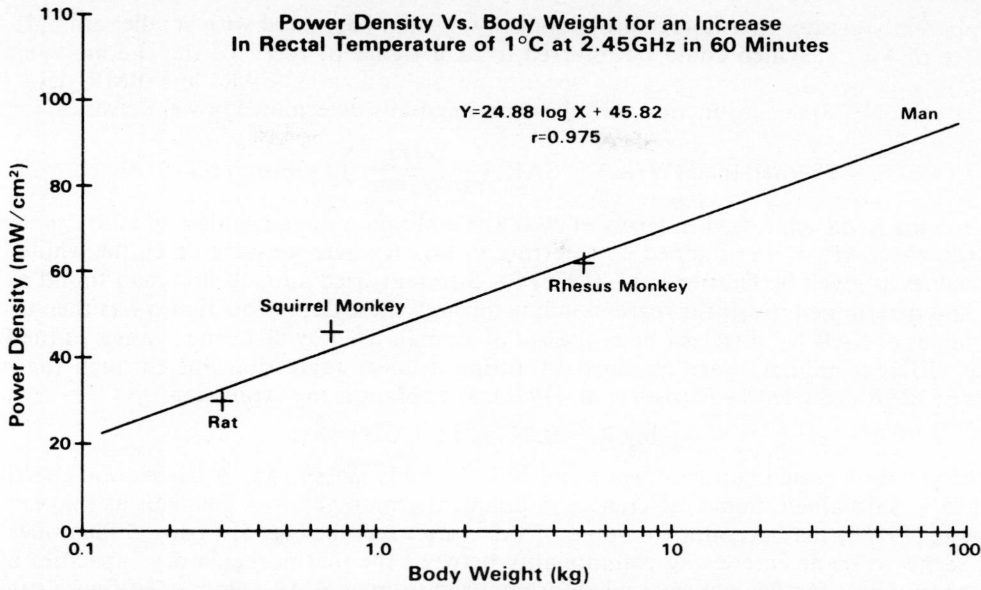

Figure 4 Power density capable of elevating rectal temperature 1°C in 60 min of radiofrequency exposure as a function of body weight using data of de Lorge (1978).

tainment of thermal equilibrium in localized tissues may be quite short and considerably less than 1 h (Kritikos and Schwan, 1979). A pertinent observation by Ely and Goldman (1957) which involved human whole body exposure to high intensity (~ 100 mW/cm^2) microwave radiation at 3 GHz showed a slight net decrease in rectal temperature after sufficient time had elapsed and body sweating was activated.

Durney and his co-workers (1978) have provided us with a convenient means of gleaning additional insight from the data of de Lorge and Michaelson. We found it instructive to form the ratio of the RF thermal load placed on the animal to it's BMR and plotting this quantity vs. body mass. Figure 5 shows the results of this operation and illustrates another way of viewing

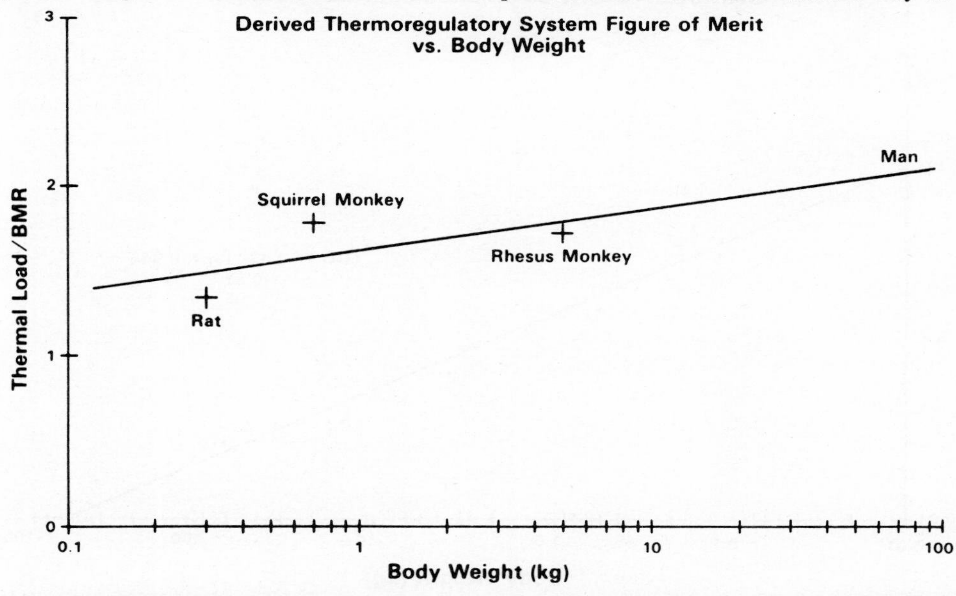

Figure 5 Derived thermoregulatory system figure of merit vs. body mass. The figure of merit is defined as the ratio of the radiofrequency induced thermal load to the basal metabolic rate of the animal resulting in a 1°C rise in rectal temperature in 60 min.

the apparently greater thermoregulatory capacity of man compared with smaller animals. The ordinate of Fig. 5, which could be referred to as a figure of merit of the thermoregulatory capacity, was developed by using the specific absorption rates (SAR) and BMR data from Durney et al. (1978) in conjunction with the experimentally determined power densities, i.e.:

$$\text{Thermal load (W/kg)} = \text{SAR}\left(\frac{\text{W/kg}}{\text{mW/cm}^2}\right) \times \text{S(mW/cm}^2\text{)}$$

By using the BMR expressed in terms of W/kg, we obtain a dimensionless quantity plotted as the ordinale. SARs were obtained by referring to Fig. 6 where we have taken the whole body SAR values as given by Durney et al. (1978) for different sized animals and man found at 2.45 GHz and determined the linear regression line through these data. This figure was then used to read values of SAR for different body masses of animals used by de Lorge. Values of the BMR of the different animals were obtained by fitting a linear regression line through the many values of BMR provided by Durney et al. (1978). We obtained the expression

$$\log R = 0.74 \log M + 0.59 \text{ where}$$

R is the metabolic rate in units of watts and M is the body mass in kg. A correlation coefficient of 0.996 was obtained. Some differences in unique thermoregulatory behavior as suggested by de Lorge (1978b) may account for some of the scatter seen in Fig. 5. Figure 5 illustrates that there seems to be an interesting commonality between the thermoregulatory capacities of the different animals for RF exposure when normalized to their BMR values. The figure suggests that for man, as well as for the smaller animals, when the thermal load developed by RF energy alone is some 1.5–2.0 times the BMR, a 1°C rise in rectal temperature would be expected, presuming that the thermal load is generally distributed in a similar way as it is in the smaller animals. We have projected, via a least-squares fit of the small animal data, a value for the so called thermoregulatory figure of merit for man of some 2 x BMR. Obviously, at 2.45 GHz this thermal distribution will not be similar but at lower frequencies where the depth-of-penetration to body-size ratio is proportionately similar in respect to that in the rat at 2.45 GHz, the figure of merit should hold. By using the methods of Durney et al., we estimate this lower equivalent

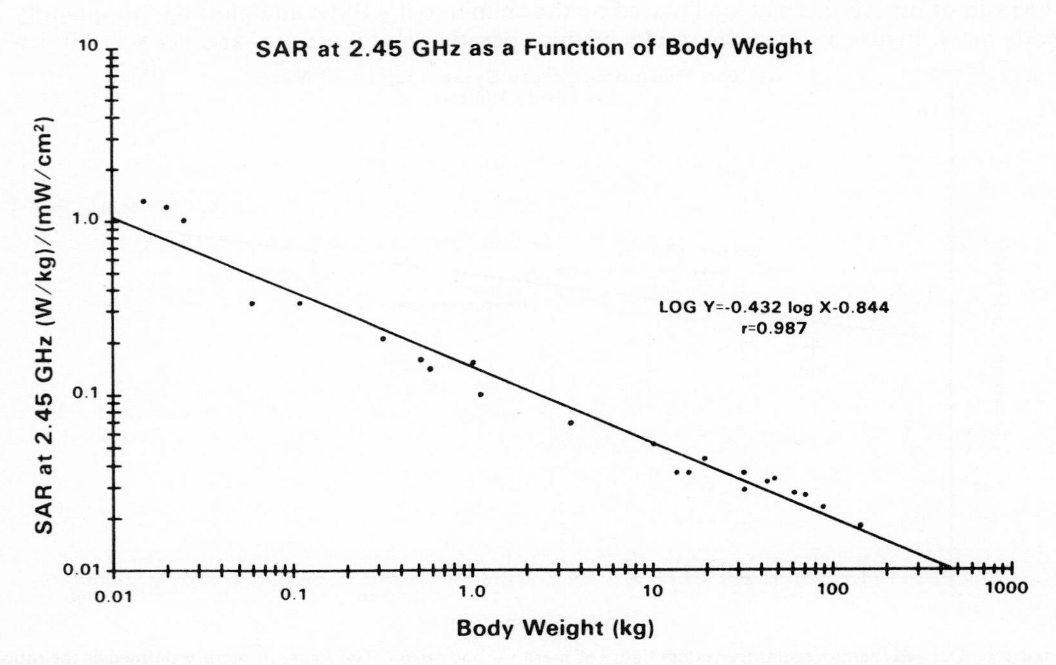

Figure 6 Specific absorption rate (SAR) as a function of body mass at 2.45 GHz with body axis aligned with the electric field using data of Durney et al. (1978).

frequency to be around 350 MHz. Thus, the extension of the above reasoning to lower frequencies than 2.45 GHz seems justified and we suggest that as a convenient boundary, occupational exposures at 10 mW/cm^2 time averaged power density above 1 GHz would appear unquestionably thermally safe.

Due to a lack of pertinent experimental data on elevation of animal's body temperatures at lower frequencies, it is not clear how low in frequency this approach can be taken. The problem of the thermodynamic modelling of the human body has received considerable attention (Fan et al., 1971), but the first attempt to consider the addition of RF energy as a thermal load in a model of man was performed by Guy et al. (1973). Subsequently, Guy et al. (1978) made use of Emery's modeling technique (Emery, 1975) to determine the increase of body temperature in man exposed at two different frequencies. Figure 7 taken from their work shows the results of exposing man at the whole-body resonant frequency of about 80 MHz at 10 mW/cm^2. Thermographic determinations of the distribution of energy at 80 MHz were used to provide input to the thermodynamic model. Results show temperatures at various anatomical locations of the body. Attention is drawn to the hypothalamic temperature, which starts near 36.9°C and reaches an equilibrium value of 37.9°C, thus showing a 1°C rise. If we assume that the hypothalamic temperature is essentially equal to the rectal temperature or no more than about 0.2°C lower, then absorption of RF energy at the observed rate of 170 W leads to a rise of rectal temperature near 1°C in a period of about 3 h. Figure 8 (from Guy et al., 1978) illustrates predicted body temperatures as a function of time for a metabolically generated value of 275 W (170 W due to exercise + 105 W BMR), which could be accomplished by work activity. Good agreement is observed between the predicted 1°C rise of hypothalamic temperature obtained by Guy et al. (1978) and that of Givoni and Goldman (1972), who used an external work level of almost the same value (205 W vs. 170 W) (compare Fig. 8 with Fig. 3). These results indicate a general agreement with the assumption that the irradiation-induced thermal load can be equated to an equal level of work activity (exercise) in terms of the expected increase in general, deep-body temperature.

We have employed the highly useful formulae developed by Givoni and Goldman (1972), which relate body heat production, environmental conditions described by air temperature, vapor pressure, and wind speed, the thermophysical properties of clothing, and the resulting equilibrium rectal temperature in man. These relationships were developed from extensive tests with men and represent an empirical approach to the solution of the heat balance equation for man. They found that:

$$T_{reeq} = 36.75 + 0.004\,M + (0.0218/clo)\,(T_{DB}\text{-}36) + 0.8\exp 0.0047(E_{req}\text{-}E_{max}) \qquad (3)$$

where

T_{reeq}	=	the equilibrium rectal temperature
E_{req}	=	required evaporative cooling = $M + (11.6/clo)(T_{DB}-36)$
E_{max}	=	maximum possible evaporative cooling = 25.5 $(im/clo)(44-\phi Pa)$
M	=	total metabolic rate (W)
T_{DB}	=	dry bulb air temperature
clo	=	thermal resistance of clothing system
im/clo	=	permeability index of clothing
ϕPa	=	vapor pressure of air (mmHg)

The clo and im/clo values are in turn functions of the effective air velocity given by

$$Veff = Vair + 0.004\,(M-105);\ m/s$$

and for a man clothed only in shorts with no shirt, Givoni and Goldman (1978) state values for clo and im/clo as

$$\begin{aligned} clo &= 0.57\,Veff^{-0.30} \\ im/clo &= 1.20\,Veff^{+0.30} \end{aligned} \qquad (4)$$

For a standard military uniform consisting of fatigues they provide

$$\begin{aligned} \text{clo} &= 0.99\, \text{Veff}^{-0.25} \\ \text{im/clo} &= 0.75\, \text{Veff}^{+0.25}. \end{aligned} \tag{5}$$

We have made use of equation 3 to examine the anticipated rise of core temperature under an applied intensity of RF radiation where M includes the heat load provided by RF energy. We assumed a comfortable environment as defined by ASHRAE (1967) consisting of $T_{DB} = 25.0°C$, $\phi Pa = 12$ mmHg or relative humidity of 50%, and an air velocity of 0.13 m/s. Figure 9 presents the results where the equilibrium rectal temperature is plotted as a function of the incident power density *at body resonance*. It is apparent that above approximately 30 mW/cm² the human thermoregulatory system is potentially overloaded and uncontrolled excursion of rectal temperature may occur for sustained exposures. We point out that these calculations assume that the individual is at rest and thus any additional physical activity will reduce the duration of the induction period before overheating occurs.

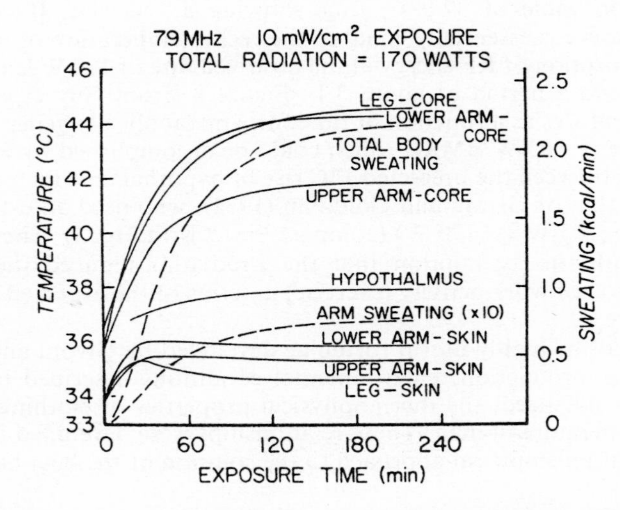

Figure 7 Change in temperature and sweating of man exposed to 79 MHz electromagnetic fields at 10 mW/cm². From Guy et al. (1978).

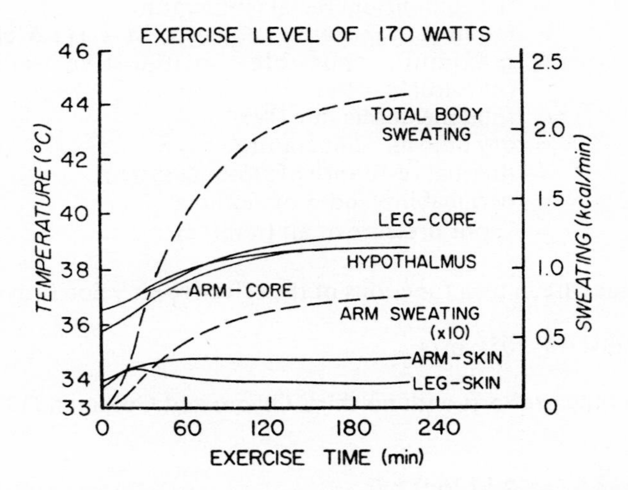

Figure 8 Effect of exercise on tissue temperature and sweating. A total metabolic rate of 275W (170W due to exercise plus 105 basal metabolic rate) is assumed. From Guy et al. (1978).

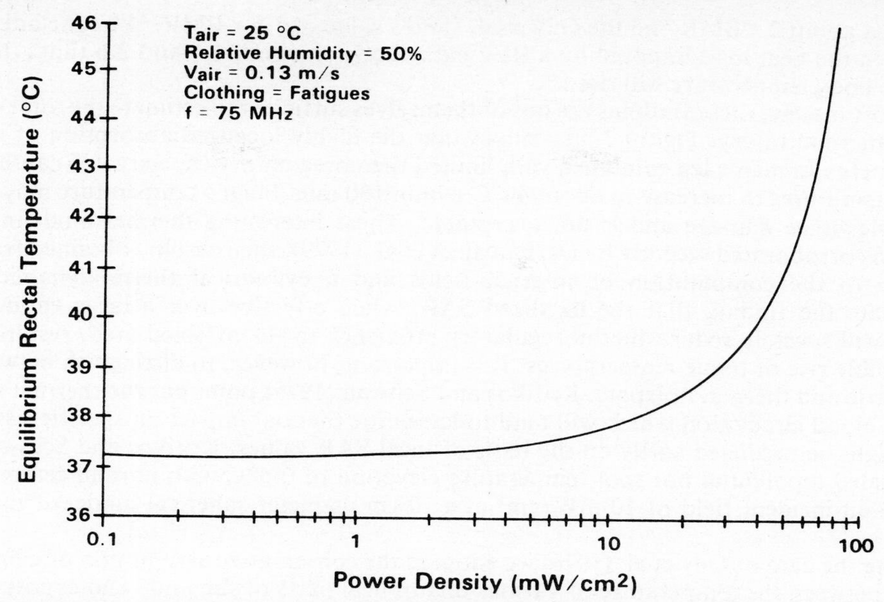

Figure 9 Equilibrium rectal temperature in man dressed in fatigues, at rest in a thermally comfortable environment as a function of incident power density of an electromagnetic field at 75 MHz.

It is informative to also examine the effect of the climatic environment on man's tolerance to applied RF heat stress and we have developed in Fig. 10 a plot of the relative increase in equilibrium rectal temperature in a resting man associated with a fixed RF exposure level of 10 mW/cm² *at body resonance* for various values of the WBGT. Mumford (1969) first questioned this influence of the thermal environment on permissible RF induced heat stress. The comfort zone is normally taken approximately around 20°C WBGT. It should be noted that WBGT's of more than 30°C represent extreme environmental conditions such that only light work should be carried out and the duration of the work should be closely controlled. Again we emphasize the importance of considering the combined action of the applied RF heat load and the ongoing work activity. Finally, using the methods of Givoni and Goldman (1978) we have determined for man wearing shorts, and no other clothing, the applied RF-induced heat load required to elevate his resting rectal temperature by 1°C as 268W or approximately 2.6 x BMR. We observe again the encouraging convergence of values obtained for increasing man's body temperature by approximately 1°C by noting the figure of merit projected for man from small-animal data

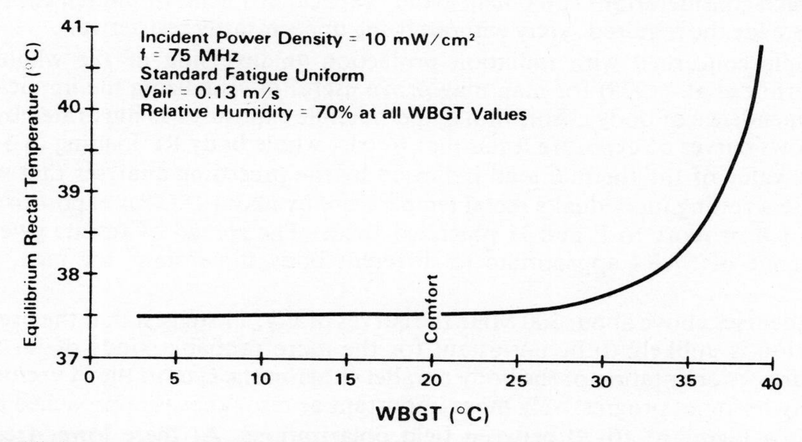

Figure 10 Equilibrium rectal temperature as a function of the WBGT for a constant exposure to a 75 MHz, 10 mW/cm² electromagnetic field.

in Fig. 6 of about 2 x BMR and the Guy et al. (1978) value of 1.6 x BMR. We conclude that for man, when the heat load imposed by a RF field is between about 1.6 and 2.6 times the BMR, deep core body temperature will rise 1 °C.

Unfortunately, these findings are not of themselves sufficient to estimate the total extent of possible thermal injury. Figure 7 also shows that the highly localized absorption of radiofrequency energy in man's leg combined with limited themoregulatory capacity will cause internal thigh temperatures to increase to about 44°C within 180 min. Such a temperature may result in irreversible tissue damage and is not acceptable. These interesting thermal modeling results have been corroborated recently by Deffenbaugh et al. (1979); their results, obtained from a cell approach to the computation of internal fields and a cylindrical thermodynamic model, substantiate the finding that the localized SAR, when effective over a large enough tissue volume, will severely reduce thermoregulatory efficiency to the affected area, resulting in an unacceptable rise of tissue temperatures. It is important, however, to distinguish between electromagnetic and thermal hotspots. Kritikos and Schwan (1979) point out the thermal smearing effect of blood circulation which will tend to lessen the thermal impact on specific tissue areas which might be predicted solely on the basis of local SAR values. Kritikos and Schwan (1979) also revealed a potential hot spot temperature elevation of 0.5°C with normal cerebral blood flow and an incident field of 10 mW/cm^2 in a 10 cm diameter spherical model of the human brain.

Using the data of Guy et al. (1978) we adopted the conservative assumption of a linear relationship between the temperature of various anatomical parts of the body and exposure power density to estimate the intensity that would be associated with a maximum tissue temperature rise of about 1°C, which near body resonance in free space means areas particularly in the region of the thigh. Under this conservative assumption, a limiting exposure of about 1.5 mW/cm^2 would be suggested. A rise of core temperature of 1°C as a limiting criterion for RF exposure is obviously not sufficient to completely protect the body from thermal injury and it is clear that as frequencies approach body resonance, the present 10 mW/cm^2 standard is too permissive for sustained exposure. The standard should in fact be modified to become more restrictive through the body-resonance range.

At frequencies below body resonance the rate of energy absorption falls as the square of frequency and it seems to us that considerable progressive relaxation of the safe limit is warranted, with the proviso that local temperatures are not excessive. Tell et al. (1979) have recently provided experimental confirmation of highly localized current densities in man immersed in low-frequency fields, which indicates that local tissue SARs in small cross-sectional areas may exceed whole body averages by two orders of magnitude. Significant enhancements of SARs, particularly at lower frequencies, are likely to take place within a wavelength of the source where reactive fields will predominate. The selection of a frequency below which permissible levels are allowed to rise might reasonably embody consideration of operational characteristics of sources. Such considerations could help avoid practical problems of implementing protective guides and provide the required safety without being unnecessarily restrictive.

For people concerned with radiation protection an inversion of the whole-body SAR curves of Durney et al. (1978) for man may prove useful in visualizing the implications of the frequency dependence of body absorption on permissible exposures as illustrated by Tell(1978). Figure 11 shows curves of exposure levels that restrict whole body RF loading to 3 W/kg. This is an average value of the thermal load indicated by the preceding analyses that would be expected to raise a resting individual's rectal temperature by about 1°C for exposure durations of the order of 1 h or more to E and H polarized fields. The spread of results given in Fig. 11 reflect the range of SARs appropriate to different body dimensions for men, women and children.

For frequencies above about 500 MHz the curves of Fig. 11 suggest that the precise polarity of the radiation is unlikely to be important for the more probable kinds of far-field (plane-wave) exposures — orientation of the body parallel either to the E or to the H vector. Below 500 MHz, polarity becomes progressively more important as resonance is approached and at 70–80 MHz there is a factor of 10–20 between field polarizations. At these lower frequencies, of course, localized tissue heating as opposed to global rise of body temperature may be the critical

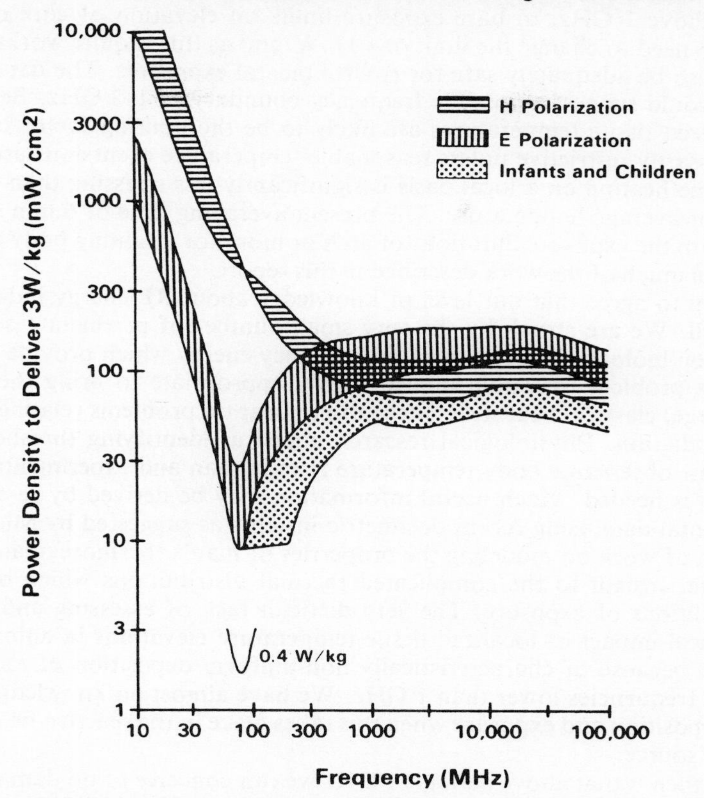

Figure 11 Power density as a function of frequency that will produce a thermal load of 3 W/kg in men, women and children for electric (E) and magnetic (H) field polarization.

factor, and Fig. 11 shows also the values of power density around the region of resonance that would restrict the whole body SAR to less than 0.4 W/kg and temperature rise in the thigh below about 1°C. However, these frequencies correspond to wavelengths in air of about 4 m and it is likely that the more significant exposures will be at distances of less than a wavelength from the source. At distances of less than a wavelength power density is a rather meaningless concept. The E and H fields are no longer in phase. There will be components of various polarizations and the reactive components, which vary much more rapidly with distance than the resistive "radiated" components, will tend to predominate. Further research is needed on these near-field exposures, on partial-body exposures and on the consequences of non-uniform energy deposition in man.

CONCLUSIONS

In the final analysis, it is extremely difficult to relate changes in health to small changes in body temperature. In keeping with the philosophy that if we are to err, it is better to err on the safe side, we find attractive the concept of limiting exposures such that there is no increase in body core temperature, if the cost is acceptable. When an observable increase in body temperature occurs, this must be interpreted as an indication that the physiological system is under some stress and for the general population no significant increase in body core temperature from exposure should be allowed. In the context of diurnal temperature changes in individuals of about 1.5°C, any core temperature increase estimated for exposure to 10 mW/cm² of 2.45 GHz radiation can surely be regarded as insignificant. In the case of extremely high frequencies, where only surface heating occurs, tolerable levels could well be established on a basis similar to that used in setting light and infrared radiation safety guides.

In occupational exposures, it seems reasonable to accept some degree of thermal stress and, for exposures above 1 GHz, to base exposure limits on elevation of core temperature. There is, however, no need to change the limit of 10 mW/cm^2 as this is quite workable occupationally and it may also be adequately safe for environmental exposures. The data of Kritikos and Schwan (1979) would suggest a possible frequency boundary of 1–3 GHz. Below approximately 1 GHz localized tissue temperatures are likely to be the limiting factor but exposure limits will be unnecessarily restrictive unless reasonable temperature excursions are allowed — relatively small volume heating on a local basis is significantly less stressing than a similar increase in body core or average temperature. The present averaging time of 6 min seems rather short as compared with the exposure durations of an h or more for attaining body core thermal equilibrium implied in much of the work described in this report.

We would hasten to agree that our level of knowledge about RF-energy induced thermal stress in man is small. We are struck by the very small number of pertinent research papers within the literature on biological effects of radiofrequency energy which provide much meaningful insight to this problem. It would appear highly appropriate to bring the experiences reported on in the large, classical heat stress literature to bear on problems relating to exposure to radiofrequency radiation. Physiological research aimed at identifying threshold exposure levels which cause just observable body temperature rises in man and experimental animals at different frequencies is needed. Much useful information may be derived by re-evaluation of old animal experimental data using recent dosimetric insights as suggested by Michaelson and Lu (1979). Extension of work on modeling the properties of man's thermoregulatory system is needed to give greater insight to the complicated thermal distributions which occur in man under different conditions of exposure. The very difficult task of assessing and generalizing about the physiological impact of localized tissue temperature elevations in animals and man should be attempted because of characteristically non-uniform deposition of radiofrequency energy, especially at frequencies lower than 1 GHz. We have almost no knowledge of the correlation of energy deposition and exposure when this takes place in the reactive fields less than 1 wavelength from the source.

Our conclusion then is that above perhaps 1 GHz we can conceive of no damaging thermal stress in man at an intensity of 10 mW/cm^2 — the currently accepted guideline in many Western countries for occupational exposure. At lower frequencies the margin of safety can be much lower in localized tissues and unacceptable temperatures are theoretically possible when the exposure duration encompasses several h. This problem is most critical near body resonance. A lowering of the permissible level for continuous whole body exposures to perhaps 1 mW/cm^2, for E field polarization, seems justified on the basis of present research results. The consideration of both core temperature and localized body temperature seems desirable in establishing frequency-dependent exposure limits to RF waves, but we emphasize the need to investigate more carefully the possibility of adverse reactions in animals that may be due to localized heating in the absence of significant whole-body heating.

REFERENCES

American National Standards Institute (1974). Safety levels of electromagnetic radiation with respect to personnel. Report ANSI-C95.1, 1974.

Bawin, S.M., L.K. Kaczmarek, and W.R. Adey (1975). Effects of modulated vhf fields on the central nervous system. In: *Annals of the New York Academy of Sciences,* Vol. 247, pp. 74–81, February.

Bawin, S.M., W.R. Adey, and I.M. Sabbot (1978). Ionic factors in release of $^{45}Ca^{2+}$ from chicken cerebral tissue by electromagnetic fields. *Proc. Natl. Acad. Sci. USA*, Vol. 75, No. 12, pp. 6314–6318, December.

Berman, A. and H.B. Carter (1978). Observations on mouse fetuses exposed to 2.45 GHz microwave radiation (meeting abstract), *Health Physics*, Vol. 33, p. 661.

Bierenbaum, L., I.T. Kaplan, W. Metlay, S.W. Rosenthal, and M.M. Zaret (1975). Microwave and infrared effects on heart rate, respiration rate and subcutaneous temperature of the rabbit. *J. Microwave Power*, Vol. 10(1), pp. 3–18.

Blackman, C.F., J.A. Elder, C.M. Weil, S.G. Benane, D.C. Eichinger, and D.E. House (1979). Induction of Calcium-ion efflux from brain tissue by radiofrequency radiation: effects of modulation and field strength. *Radio Science*, Vol. 14, Number 6S, pp. 93–98, November- December.

Brief (1970). Heating and cooling for man in industry. Am. Ind. Hygiene Assoc., Southfield.

Bruce-Wolfe, V., M. Mathews, and D.R. Justesen (1977). The visually-evoked electrocortical response of the guinea pig after microwave-induced hyperthermia (meeting abstract). In: Abstracts of 1977 International Symposium on the Biological Effects of Electromagnetic Waves, URSI, Airlie, Virginia, October 30-November 4, p. 97.

de Lorge, J.O. (1976). Behavior and temperature in rhesus monkeys exposed to low level microwave irradiation. Naval Aerospace Medical Research Laboratory report NAMRL-1222, Pensacola, Florida, January 19.

de Lorge, J. (1977). Operant behavior and colonic temperature of squirrel monkeys (saimiri sciureus) during microwave irradiation. Naval Aerospace Medical Research Laboratory report NAMRL-1236, Pensacola, Florida, June 8.

de Lorge, J. (1978a). Disruption of behavior in mammals of three different sizes to microwaves: Extrapolation to larger mammals (1978). In: Abstracts of Scientific Papers of 1978 Symposium on Electromagnetic Fields in Biological Systems, Ottawa, Canada, June 27-30.

de Lorge, J.O. (1978b). The effect of 5.62 GHz microwave radiation on complex operant behavior in rats. In: Abstracts 0S2: Biological Effects of Electromagnetic Waves, and presented at XIX General Assembly of the International Union of Radio Science, August 1-8, 1978, Helsinki, Finland.

Defenbaugh, D.M., R.J. Spiegel, and J.E. Mann (1979). A thermal model of the human body exposed to an electromagnetic field, presented at 1979 Bioelectromagnetics Symposium, June 18-22, Seattle, Washington.

Diachenko, J.A. and W.C. Milroy (1975). The effects of high power pulsed and low level cw microwave radiation on an operant behavior in rats. U.S. Naval Weapons Laboratory report NSWC/OL TR-3230, Dahlgreen, Virginia.

Dordevic, Z. (1975). Study of the biological effects of prolonged exposure of rats to microwave radiation having intensities of 5 to 50 mW/cm^2, *Vojnosanit. Pregl*, Vol 32(1), pp. 51-53.

Dukes-Dobos, F.M. (1976). Rationale and provisions of the work practices standard for work in hot environments as recommended by NIOSH . . . Standards for Occupational Exposures to Hot Environments, DHEW publication (NIOSH)-76-100.

Durney, C.H., et al. (1978). Radiofrequency radiation dosimetry handbook (second edition). Technical report SAM-TR-78-22 prepared for USAF School of Aerospace Medicine, Brooks Air Force Base, Texas 78235.

Ely, T.S. and D.E. Goldman (1957). Heating characteristics of laboratory animals exposed to ten-centimeter microwaves. Research report project NM 001 056.13.OZ, Naval Medical Research Institute, Bethesda, Maryland, March 21.

Emery, A.F., R.E. Short, A.W. Guy, K.K. Kraning, and J.C. Lin (1975). The numerical thermal simulation of the human body when absorbing non-ionizing microwave irradiation-with emphasis on the effect of different sweat models. In: Biological Effects of Electromagnetic Waves, selected papers of the USNC/URSI Annual Meeting, Boulder, Colorado, October 20-23, 1975. USDHEW publication (FDA) 77-8011, December 1976.

Fan, L-T, F-T Hsu, and C-L Hwang (1971). A review on mathematical models of the human thermal system. *IEEE Trans. Bio-Medical Engr.*, Vol. BME-18(3), pp. 218-234, May.

Flanigan, W.F., W.R. Lowell, and R.L. Seeley (1977). Absence of ECG and respiratory changes in turtles and tortoises exposed to microwave fields of low density (meeting abstract). In: Abstracts of 1977 International Symposium on the Biological Effects of Electromagnetic Waves, URSI, Airlie, Virginia, October 30-November 4, 1977, p. 31.

Frazer, J.W., J.H. Merritt, S.J. Allen, R.H. Hartzell, J.A. Ratliff, A.F. Chamness, R.W. Detinkler, and T. McLellan (1976). Thermal responses to high-frequency electromagnetic radiation fields, available through National Technical Information Services, Springfield, Virginia 22161, Document AD A032179.

Frey, A.H., S.R. Feld, and B. Frey (1975). Neural function and behavior: defining the relationship. *Anals New York Academy of Sciences*, Vol. 247, pp. 433-439.

Givoni, B., and R.F. Goldman (1972). Predicting rectal temperature response to work, environment, and clothing. *J. Applied Physiology*, Vol. 32, June, pp. 812-822.

Glaser, Z.R. and C.H. Dodge (1975). Biomedical aspects of radiofrequency and microwave radiation: a review of selected Soviet, East European, and Western references. In: *Biological Effects of Electromagnetic Waves, selected papers of the USNC/URSI Annual Meeting*, Boulder, Colorado, October 20-23, 1975. USDHEW publication (FDA) 77-8011, December 1976.

Guillet, R., W.G. Lotz, and S. M. Michaelson (1975). Time-course of adrenal response in microwave-exposed rats (proceedings abstract). In: *Proceedings of the 1975 Annual Meeting of the International Union of Radio Science*, Boulder, Colorado, October 20-23, 1975, p. 316.

Guy, A.W., C.C. Johnson, J.C. Lin, A.F. Emery, and K.K. Kraning (1973). Electromagnetic power deposition in man exposed to high-frequency fields and the associated thermal and physiologic consequences. Technical report SAM-TR-73-13 prepared for USAF School of Aerospace Medicine, Brooks Air Force Base, Texas. NTIS document AD-776821, December.

Guy, A.W., J.C. Lin, and C.K. Chou (1974). Electrophysiological effects of electromagnetic fields on animals. In: *Fundamental and Applied Aspects of Nonionizing Radiation*, Michaelson, S.M., M.W. Miller, R. Magin, and E.L. Carstensen, eds. New York: Plenum Press, pp. 167-211.

Guy, A.W., M.D. Webb, A.F. Emery, and C.K. Chou (1978). Determination of the average SAR and SAR patterns in man and simplified models of man and animals exposed to radiation fields from 50-2450 MHz and the thermal consequences. In: Abstracts of Scientific Papers Open Symposium on the Biological Effects of Electromagnetic Waves held at the XIX General Assembly of the International Union of Radio Science, Helsinki, Finland, August 1-8, 1978.

Henschel, A. (1976). Effects of age and sex on heat tolerance. Standards for Occupational Exposure to Hot Environments. DHEW publication (NIOSH)-76-100.

Ho, H.S. and M. McManaway (1977). Heat-dissipation rate of mice after microwave irradiation. *J. Microwave Power*, Vol. 12(1), pp. 93-100.

Ho, H.S. and W.P. Edwards (1976). Dose rate related effects on the oxygen consumption of mice during and after microwave irradiation (meeting abstract). In: *Proceedings of the 1976 Annual Meeting of the International Union of Radio Science*, Amherst, Massachusetts, October 11-15, 1976, p. 119.

Houk, W.M., S.M. Michaelson, and D.E. Beischer (1975). The effects of environmental temperature on thermoregulatory, serum lipid, carbohydrate, and growth hormone responses of rats exposed to microwaves (proceedings abstract). In: *Proceedings of the 1975 Annual Meeting of the International Union of Radio Science*, Boulder, Colorado, October 20-23, p. 309.

Justesen, D.R. (1978). Neurological and psychological effects of radiofrequency electromagnetic radiations: toward rapproachement between east and west (meeting abstract). In: Abstracts 0S2: Biological Effects of Electromagnetic Waves, International Union of Radio Science, Helsinki, Finland, p. 7.

Justesen, D.R., D.M. Levinson, and L.R. Justesen (1974). Psychogenic stressors are potent mediators of the thermal response to microwave radiation. In: *Biologic Effects and Health Hazards of Microwave Radiation*, Proceedings of an International Symposium, Warsaw, edited by P. Czerski, October 15-18, 1973, published by Polish Medical Publishers, Warsaw, 1974, pp. 134-140.

Kraning, K.K., H.S. Belding, and B.A. Hertig (1966). Use of sweating rate to predict other physiological responses to heat. *J. Applied Physiology*, Vol. 21, pp. 111-117.

Kritikos and Schwan (1979). Potential temperature rise induced by electromagnetic field in brain tissues. *IEEE Trans. Engineering*, Vol. BME-26(1), pp. 29-33, January.

Krupp, J.H., (1977). Thermal response in *macaca mulatto* exposed to 15 and 20 MHz radiofrequency radiation. Available through National Technical Information Services, Springfield, Virginia 22171, Document AD A045508.

Lind, A.R. (1963). A physiological criterion for setting thermal environmental limits for everyday work. *J. Applied Physiology*, Vol. 18, pp. 51-56.

Lofstedt, B. (1966). Human heat tolerance. Dept. Hyg., University of Lund, Lund.

Lotz, W.G. and S.M. Michaelson (1978). Temperature and corticosterone relationships in microwave-exposed rats. *J. Applied Physiology*, Vol. 44(3), pp. 438-445.

Lotz, W.G., S.M. Michaelson, and N.J. Lebda (1977). Growth hormone levels of rats exposed to 2450 MHz (cw) Microwaves (meeting abstract). In: Abstracts of 1977 International Symposium on the Biological Effects of Electromagnetic Waves, URSI, Airlie, Virginia, October 30-November 4, 1977, p. 39.

Lu, S.T., N. Lebda, and S.M. Michaelson (1977). Effects of microwave radiation on the rat's pituitary-thyroid axis (meeting abstract). In: Abstracts of 1977 International Symposium on the Biological Effects of Electromagnetic Waves, USRI, Airlie, Virginia, October 30-November 4, p. 37.

Lu, S.T., N. Lebda, S.M. Michaelson, S. Pettit, and D. Rirera (1976). Thermal and neuroendocrine effects of long term, low level microwave (2450 MHz, cw) irradiation (meeting abstract). In: *Proceedings of the 1976 Annual Meeting of the International Union of Radio Science,* Amherst, Massachusetts, October 11-15, pp. 90-91.

Magin, R.L., S.T. Lu, and S.M. Michaelson (1977). Stimulation of dog thyroid by local application of high intensity microwaves. *Am. J. Physiology*, Vol. 233(5), p. E363-E368.

McAfee, R.D., R. Braus, and J. Fleming (1973). The effect of 2450 MHz microwave irradiation on the growth of mice. *J. Microwave Power*, Vol. 8(1), pp. 111-116.

Merritt, J.H., A.F. Chamness, and S.J. Allen (1978). Studies on blood-brain permeability after microwave irradiation. *Radiation and Environmental Biophysics*, Vol. 15, pp. 367-377.

Merritt, J.H., A.F. Chamness, R.H. Hartzell, and S.J. Allen (1977). Orientation effects on microwave-induced hyperthermia and neurochemical correlates. *J. Microwave Power*, Vol. 12(2), pp. 167-172.

Michaelson, S.M. and S-T Lu (1979). Metabolic and physical scaling in microwave radiofrequency bioeffects studies. Presented at 1979 Bioelectromagnetics Symposium, June 18-22, Seattle, Washington.

Michaelson, S.M., R. Guillet, M.A. Catallo, J. Small, G. Inamine, and F.W. Heggeness (1976). Influence of 2450 MHz cw microwaves on rats exposed in utero (symposium summary). *J. Microwave Power*, Vol. 11(2), pp. 165-166.

Michaelson, S.M., W.M. Houk, N.J.A. Lebda, S.T. Lu, and R.L. Magin (1975). Biochemical and neuroendocrine aspects of exposure to microwaves. *Annals of New York Academy of Sciences*, Vol. 247, pp. 21-45.

Minard, D. (1973). Physiology of heat stress. The Industrial Environment — Its Evaluation and Control. National Institute for Occupational Safety and Health, p. 399.

Minard, D. and R.L. O'Brien (1964). Heat casualties in the Navy and Marine Corps., 1959-1962, with appendices on the field use of the wet bulb globe thermometer index. Research report No. 7, Contract MR 005.01-0001.01, Naval Medical Research Institute, Bethesda, Maryland.

Monahan, J.C., and H.S. Ho (1976). Temperature dependence of microwave avoidance (meeting abstract). In: *Proceedings of the 1976 Annual Meeting of the International Union of Radio Science*, Amherst, Massachusetts, October 11-15, pp. 4-5.

Mumford, W.W. (1969). Heat stress due to rf radiation. *Proceedings of the IEEE*, Vol. 57, pp. 171-178.

Nielsen, M. (1938). Die regulation der korpertemperatur bei muskelarbeit. *Skand. Arch. Physiol.*, Vol. 89, pp. 193-230.

NIOSH (1972). Criteria for a recommended standard . . . Occupational exposure to hot environments. Technical report HSM 72-10269, U.S. Department of Health, Education, and Welfare, National Institute for Occupational Safety and Health.

Olsen, R.G., J.L. Lords, and C.H. Durney (1977). Microwave —induced chronotropic effects in the isolated rat heart. *Annals of Biomed. Eng.*, Vol. 5(4), pp. 395-409.

Oscar, K.J. and T.D. Hawkins (1977). Microwave alteration of the blood-brain barrier system of rats. *Brain Research*, Vol. 126, pp. 281-293.

Phillips, R.D., E.L. Hunt, R.D. Castro, and N.W. King (1975), Thermoregulatory, metabolic, and cardiovascular response of rats to microwaves. *J. Applied Physiology*, Vol. 38(4), pp. 630-635.

Phillips, R.D., N.W. King, and E.L. Hunt (1973). Thermoregulatory, cardiovascular, and metabolic response of rats to single or repeated exposures to 3450 MHz microwaves. *Proc. Microwave Power Symposium* (Inst. Microwave Power, Canada), September, pp. 3B5/1-3B5/4.

Polson, P., D.C.L. Jones, A. Karp, and J.S. Krebs (1974). Mortality in rats exposed to cw microwave radiation at 0.95, 2.45, 4.54, and 7.44 GHz. Stanford Research Institute technical report DAAK02-73-C-0453, Menlo Park, California.

Ramsey, J.D. (1978). Abbreviated guidelines for heat stress exposure. *American Industr. Hyg. Assoc. J.*, Vol. 39(6), pp. 491-495.

Robinson, S., E.S. Turrell, and S.D. Gerking (1945). Physiologically equivalent conditions of air temperature and humidity. *American J. Physiology*, Vol. 143, pp. 21-32.

Rugh, R. and M. McManaway (1976). Comparison of ionizing and microwave radiations with respect to their effects on the rodent embryo and fetus (abstract). *Teratology*, Vol. 14(2), p. 251.

Sanza, J.N. and J..de Lorge (1977). Fixed internal behavior of rats exposed to microwaves at low power densities. *Radio Science*, Vol. 12(6S), pp. 273-277.

Schwan, H.P. (1957). The physiological basis of RF injury. In: *Proceedings of Tri-Service Conference on Biological Hazards of Microwave Radiation*, ed. E.G. Pattishall, held at USAF Rome Air Development Center, New York, 15-16 July 1957.

Schwan, H.P. and K. Li (1956). Hazards due to total body irradiation by radar. *Proceedings of the I.R.E.*, Vol. 44, pp. 1572-1581, November.

Seaman, R.L., H. Wachtel, and W.T. Jones (1975). Stripline techniques in the study of microwave biological effects on isolated neural preparations. In: *Microwave Power Symposium Proceedings*, Waterloo. Ontario, Canada, May 28-30, pp. 99-102.

Shandala, M.G. (1978). Biologic effects of electromagnetic waves (EMW) as the basis of standard setting for the general population in the USSR (meeting abstract). In: Abstracts 0S2: Biological Effects of Electromagnetic Waves, International Union of Radio Science, Helsinki, Finland, p. 6.

Shaposknikov, Yu. G. and I.F. Yares'ko (1974). Influence of low-intensity microwaves on the acid base balance in experimental animals. *Eksp. Khir. Anesteziol.* , Vol. 6, pp. 60-61.

Smialowicz, R.J., J.B. Kinn, C.M. Weil, and T.R. Ward (1977). Chronic exposure of rats to 425 or 2450 MHz microwave radiation effects on lymphocytes (meeting abstract). In: Abstracts of 1977 International Symposium on the Biological Effects of Electromagnetic Waves, URSI, Airlie, Virginia, October 30-November 4, 1977, p. 140.

State committee on Standards of the Council of Ministers of the USSR (1976). Occupational safety standards — electromagnetic fields of radiofrequency-general safety requirements — GOST 12.1.006-76, Moscow.

Stolwijk, J.A., B. Saltin, and A.O. Gagge (1968). Physiological factors associated with sweating during exercise. *J. Aerospace Medicine*, Vol. 39, pp. 1101-1105.

Stolwijk, J.A. and J.D. Hardy (1966). Partitional colorimetric studies of responses of man to thermal transients. *J. Applied Physiol.*, Vol. 21(3), pp. 967-977.

Taylor, N.B.G., L.A. Kuehn, and M.R. Howst (1969). A direct-reading mecury thermometer for the wet bulb globe thermometer Index. *Can. J. of Physiol. and Pharmacol.*, Vol. 47(3).

Tell, R.A. (1978). An analysis of radiofrequency and microwave absorption data with consideration of thermal safety standards. U.S. Environmental Protection Agency technical report ORP/EAD 78-2, P.O. Box 18416, Las Vegas, Nevada 89114.

Tell, R.A., E.D. Mantiply, C.H. Durney, and H. Massoudi (1979). Electric and magnetic field intensities and associated induced body currents in man in close proximity to a 50 kw AM standard broadcast station. Presented at 1979 Bioelectromagnetics Symposium, June 18-22, Seattle, Washington.

Thomas, J.R., L.S. Burch, and S.S. Yeandle (1979). Microwave radiation and chlordiazepoxide:synergistic effects on fixed-interval behavior. *Science* 203 (4387): pp. 1357-1358.

Walters, J.D. (1968). A field assessment of a prototype meter for measuring the wet bulb globe thermometer index. *Brit. J. Indust. Med.*, Vol. 25, pp. 235-240.

Webb, P. (ed) (1964). *Bioastronautics Data Book*, National Aeronautics and Space Administration, NASA, SD-3006.

Wright, S. (1971). *Applied Physiology*, 12th edition, revised by Keele and Nest, Oxford University Press.

Wyndham, C.H. (1962). Tolerable limits of air conditions for men at work in hot mines. *Ergonomics*, Vol. 5, p. 115.

Wyndham, C.H. (1973). The physiology of exercise under heat stress. *Am. Rev. Phys.*, Vol. 35, p. 193.

Wyndham, C.H., N.B. Strydom, J.F. Morrison, C.G. Williams, G.A.G. Bredell, J.S. Maritz, and A. Munro (1965). Criteria for physiological limits for work in heat. *J. Applied Physiology*, Vol. 20(1), pp. 37-45.

Wyndham, C.H., N.B. Strydom, J.F. Morrison, F.D. Dutoit, and J.G. Kraan (1954). Responses of unacclimatized men under stress of heat and work. *J. Applied Physiology*, Vol. 6, pp. 681-686.

Yaglou, C.O. and D. Minard (1957). Control of heat casualties at military training centers. *AMA Arch. Industr. Health*, Vol. 16, pp. 302-316.

Zhuravlev, V.A. (1973). The combined effect of a superhigh-frequency field and an unfavorable micro-climate on the body. *Voen Med Zh*, Vol. 3, pp. 64-67.

RADIOFREQUENCY ENVIRONMENTS IN THE UNITED STATES

David E. Janes, Jr.

U.S. Environmental Protection Agency

ABSTRACT

As part of a program to determine the need for guidelines to control environmental levels of radiofrequency radiation (3 kHz-300 GHz), the Environmental Protection Agency began measuring levels in urban areas in October 1975. Data on environmental levels in the frequency range 0.5-900 MHz have been obtained for 15 urban areas. This paper summarizes the results of these measurements in the *general environment* and gives comparative data from other studies on levels near *specific sources* such as broadcast antennas, radars, walkie-talkies, medical diathermy units, and radiofrequency heat sealers.

INTRODUCTION

The entire population is exposed to radio waves, including microwaves, from a variety of sources. Examples include: radio and television broadcast systems, radars, radio telephones, citizen band radio, microwave relay links, medical diathermy units, heat sealers, and microwave ovens.

It is convenient to define two kinds of exposure. One occurs at distances far from individual sources, and the exposure is due to the superposition of the fields from many sources operating at different frequencies; we will call this the *general radiofrequency environment*. The other kind of exposure occurs so close to a source that the radiofrequency environment is dominated by the source (or sources) at that location; we will call this the *specific source environment*.

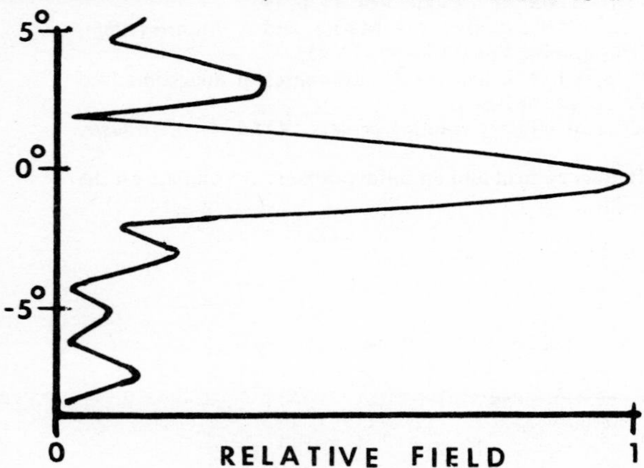

Fig. 1. Vertical Radiation Pattern for a UHF-TV Transmitting Antenna (Degrees above and below the horizontal plane)

The commonly used quantity for specifying exposure to radio waves is power density, i.e., watts per square meter (W/m^2). For historical reasons exposure is often expressed in terms of milliwatts (mW) or microwatts (μW) per square centimeter (cm^2), i.e., 1 W/m^2 = 0.1 mW/cm^2 = 100 $\mu W/cm^2$.

GENERAL RADIOFREQUENCY ENVIRONMENTS

Broadcast Sources. The *general radiofrequency environment* is dominated by radio and television broadcast transmissions [1-5]. Based on measurements made at 486 different locations in 15 cities with a 1970 population of over 44 million, the median exposure in urban areas of the U.S. is 0.005 $\mu W/cm^2$ i.e., 50 percent of the population is exposed to higher and 50 percent is exposed to lower levels. The results of these studies, shown in Tables 1 and 2, indicate that 95 percent of the population is exposed to levels that are less than 0.1 $\mu W/cm^2$. These estimates do not include contributions from AM radio transmission because the absorption of energy by humans at AM frequencies (0.535-1.605 MHz) is orders of magnitude less than at FM and TV frequencies (54-890 MHz) [6]. These estimates do not include refinements such as accounting for population mobility, for exposures at heights greater than 6 meters (20 ft.), for building attenuation, or for periods of time when sources are not transmitting. The results are based on

Table 1. Cumulative Population Exposure for 15 U.S. Cities (54-900 MHz)

Power Density ($\mu W/cm^2$)	Cumulative Percent* of Population
0.002	19.5
0.005	49.5
0.01	68.7
0.02	82.4
0.05	91.4
0.1	94.7
0.2	97.
0.5	98.8
1.0	99.4

* For example, 19.5% are exposed to levels less than 0.002 $\mu W/cm^2$, 68.7% are exposed to levels less than 0.01 $\mu W/cm^2$, etc.

Reprinted from *15th IEEE Int. Conf. on Commun.*, 1979, Boston, MA, June 10-14, vol. 2 of 4, pp. 31.4.1-31.4.5.

Table 2. Population Exposure in 15 U.S. Cities (54-900 MHz)

City	Median Exposure ($\mu W/cm^2$)	Percent Exposed $\leq 1\ \mu W/cm^2$
Boston	0.018	98.50
Atlanta	0.016	99.20
Miami	0.0070	98.20
Philadelphia	0.0070	99.87
New York	0.0022	99.60
Chicago	0.0020	99.60
Washington	0.009	97.20
Las Vegas	0.012	99.10
San Diego	0.010	99.85
Portland	0.020	99.70
Houston	0.011	99.99
Los Angeles	0.0048	99.90
Denver	0.0074	99.85
Seattle	0.0071	99.81
San Francisco	0.002	97.66
All Cities	0.0048	99.44

the population that resides in areas more than several hundred feet from FM and TV broadcast antennas where an unobstructed measurement 6 meters above ground would result in the indicated values [4].

Other Sources. For a number of reasons both low- and high-power sources that operate outside of the broadcast frequency bands do not contribute very much to the *general radiofrequency environment,* although the contribution of the higher powered devices to *specific source environments* may be large. Examples of low-power devices include: microwave relay links, personal radios such as radio telephones and citizens band, and the traffic radars used by law enforcement agencies for measuring the speed of vehicles. Examples of high-power sources include: satellite communications systems, military acquisition and tracking radars, and civilian air traffic control, air route surveillance, and weather radars. Because all of the above mentioned high-power systems use highly directive antennas, their beams have small cross-sections, and only a small volume of space is irradiated at any instant of time. For many of the systems the antenna is high above ground or is angled 2 to 3 degrees above the horizon thereby making exposure to the main beam improbable. Most radar systems rotate which further reduces the average exposure. The contribution of radars to the *general radiofrequency environment* in one large urban area is summarized in Table 3 [7]. The largest value of the power density, 0.001 $\mu W/cm^2$, is less than the median value for the broadcast frequency bands.

Table 3. Typical Urban Radar Environments in San Francisco, CA

Location	No. of Radars Detected	Average Power Density ($\mu W/cm^2$)
Mt. Diablo	8	2.6×10^{-5}
Palo Alto	10	2.7×10^{-4}
Bernal Heights	10	1.1×10^{-3}

SPECIFIC SOURCE RADIOFREQUENCY ENVIRONMENTS

Broadcast Sources. The antennas used for VHF- and UHF-TV broadcasting are highly directive as illustrated in Fig. 1. With antenna patterns such as this, levels at high elevations close to the source can be considerably greater than those found at ground level [8]. Measurements have been made in tall buildings that either support broadcast antennas or are within a city block or so of another tall building that supports a broadcast antenna and the results are summarized in Table 4 [9-10]. The values range from less than 1 $\mu W/cm^2$ to 97 $\mu W/cm^2$ inside buildings to 230 $\mu W/cm^2$ at an unshielded location on the roof of one building. Note that two things are required to obtain these higher levels, high elevation and close proximity (a few hundred ft. or less) to an antenna of a high-power source. The upper floors of tall buildings located far from broadcast antennas are not exposed to levels that differ significantly (factors of 10) from those found near the ground at equivalent distances.

Power densities near the bases of FM towers are typically 1 to 10 $\mu W/cm^2$ [Fig. 9 in Ref. 4]. However, some FM antennas have a grating lobe that is coaxial with the tower in addition to the main lobe illustrated for a TV antenna in Fig. 1 [11]. Fields of 100 to 350 $\mu W/cm^2$ have

Table 4. Radiofrequency Levels in Tall Buildings That Are Located Close to FM and TV Antennas

Location	Power Density ($\mu W/cm^2$)		
	FM	TV	TOTAL
EMPIRE STATE BUILDING (1)			
86th Floor Observatory	15.2		
102nd Floor Observatory			
Near Window	30.7	1.79	32.5
Near Elevator	1.35		
WORLD TRADE CENTER (1)			
107th Floor Observatory	0.10	1.10	1.20
Roof Observatory	0.15	7.18	7.33
PAN AM BUILDING (1)			
54th Floor	3.76	6.52	10.3
ONE BISCAYNE TOWER (2)			
26th Floor	7		
30th Floor	5		
34th Floor	62		
38th Floor	97		
Roof (shielded area)	134		
Roof	148		
SEARS BUILDING (3)			
50th Floor	32	34	66
Roof	201	29	230
FEDERAL BUILDING (3)			
39th Floor	5.7	.73	6.5
HOME TOWER (4)			
10th Floor	18		
17th Floor	0.2		
Roof	119		
Roof	180		
MILAM BUILDING (5)			
47th Floor	35.8	31.6	67

(1) New York (2) Miami (3) Chicago (4) San Diego (5) Houston

been measured in areas that are accessible to transient foot traffic [12]. These levels fall off rapidly with distance from the antenna tower, but levels near a few residences may range from 50 to 100 μW/cm². In a single unusual case, fields near the base of an FM antenna tower ranged between 1000-7000 μW/cm²; exposures in open areas, i.e., not close to conducting structures did not exceed 2000 μW/cm² [13].

Low-Power Sources. For purposes of discussion it is convenient to distinguish between low- and high-power source contributions to the *specific environment*. The distinction is somewhat arbitrary and contains some implicit assumptions on how devices are used and controlled. The four types of low-power sources that will be discussed here are microwave relay links, low-power radar, mobile communications equipment, i.e., radio telephones, citizens band radios, hand-held walkie-talkies, etc., and microwave ovens for home use. The high-power sources remaining to be discussed are: satellite communications, radar, industrial, and diathermy sources.

Microwave relay links used for long distance communications have transmitter powers that usually are 5 watts or less. The maximum power density is calculated to be about 700 μW/cm²; except for service personnel these fields do not occur in accessible locations. Maximum values at ground level are calculated to be less than 1 μW/cm² [14].

The radars used for measuring the speed of vehicles have transmiter powers of about 100 milliwatts (0.1 watts). These devices are either hand-held or vehicle mounted. They are continuous wave rather than pulse modulated and determine speed from the Doppler frequency shift of the returned signal. The maximum calculated power density for typical devices ranges from 170 to 400 μW/cm² at the face of the device and decreases to less than 24 μW/cm² and 0.2 μW/cm² at distances of 3 meters and 30 meters, respectively [15].

Two other low-power radars that are in common use are weather radar in aircraft and navigational radar on small boats. Normally, aircraft radar are not operated on the ground. When they are, power densities for a number of radar-aircraft combinations were less than 10,000 μW/cm² everywhere except directly on the surface of the radome housing one system [16]. Values were less than 1,000 μW/cm² at distances greater than 3.5 meters (11.5 feet) for all 5 systems studied [16]. For marine radars the computed average power density for any of the 6 units that were studied is less than 50 μW/cm² at the antenna's turning circle radius [17]. One of the units has an option for sector scanning, and when operated in this mode the maximum power density was about 250 μW/cm² [17].

Most of the information for the *specific source environments* produced by personal radio devices is for systems mounted on vehicles or for hand-held walkie-talkies. Interpretation of this data is difficult because most of the measurements are made in the near-field and these fields are not uniform over volumes comparable to the size of humans. The absorption patterns for these complex near-fields may differ appreciably from those produced by far-field whole body exposures; the absorption may be higher or lower and the sites of maximum absorption may be different.

Some values for fields in and around vehicles equipped with radios are presented in Table 5. The values range from a few volts per meter to 1,350 volts per meter [18-21]. Except for Ref. 19, only electric fields have been measured. To obtain power density, the magnetic fields also must be known since the impedance of these fields is not, in general, 120π Ohms, the free space impedance for a plane wave. Some authors have defined an "equivalent" free field power density by assuming the impedance value for free space and calculating the power density, S, according to the equations,

$$S(W/m^2) = E^2(V/m)^2/120\pi(\Omega) \quad (1)$$
$$= 120\pi(\Omega)H^2(A/m)^2 \quad (2)$$

However, when used in this manner "equivalent" does not necessarily mean an equivalent heating power density. If one ignores this complication and performs the calculation, then the 2 to 1,350 V/m range corresponds to an "equivalent" power density range of 1 μW/cm² to 483 mW/cm².

Only minimal data is available for the fields produced by hand-held walkie-talkies. In one study electric fields 12 cm from a 2.5 Watt hand-held unit operating at 27.12 MHz were

Table 5. Electric Field Strength In and Around Radio Equipped Vehicles

Frequency (MHz)	Power (watts)	Veh. Type	Field (V/m)	Dist. (m)	Ref.
27.075	4	Sedan	2-7	1	[18]
27.12	4	Sedan	225-1350	0.05	[19]
27.12	4	Sedan	100-610	0.13	[19]
27.12	4	Sedan	21-60	0.6	[19]
27.61	80[a]	Sedan	21-251[b]	(c)	[20]
40.27	110	Sedan	10-190	(c)	[20]
40.27	110	Sedan	75-368[b]	(c)	[20]
40.27	110	Semi	5-475	(c)	[20]
41.31	100	Comp.	5-106[d]	(c)	[21]
41.31	100	Pickup	7-165[d]	(c)	[21]
162.475	110	Sedan	8-201	(c)	[20]
164.45	60	Sedan	5-52[d]	(c)	[21]
164.45	60	S.Wag.	5-64[d]	(c)	[21]
164.45	60	Van	5-95[d]	(c)	[21]

a. Power level used with special authorization of the Interagency Radio Advisory Committee; legal power is 4 watts.
b. Vehicle was placed on a ground plane.
c. See cited reference for antenna type and location.
d. Calculated from the reported electric energy density, U_E, given in the original report from the equation $U_E(nJ/m^3) = 0.0043E^2(V^2/m^2)$.

measured to be as much as 205 V/m, which has an "equivalent" power density of 11 mW/cm^2 [22]. The measured magnetic field was 0.9 A/m which from equation (2) is "equivalent" to 31 mW/cm^2 [21]. Similar values are given for a 4 Watt transmitter [19]. Maximum fields of 200 nJ/m^3 (212 V/m or 11.9 mW/cm^2 "equivalent") have been measured for a 1.8 Watt hand-held unit operating at 164.45 MHz [21]. The measured exposure diminishes by a factor of 10 at distances of 1 or 2 inches from the site of maximum exposure [21].

In the earlier discussion it was noted that the complex near-fields were not well characterized by power density and predictions of thermal impact could not be directly extrapolated from far-field whole body exposures. One study using dielectric phantom models of human heads predicts temperature increases of less than 0.1°C in the region of the eye for a 6 Watt, 150 MHz hand-held unit whose antenna was positioned 0.5 cm from the eye [23]. These techniques have also been extended to higher frequencies [22-25].

Microwave ovens can be considered as low power devices if they meet the performance standard of the Food and Drug Administration which specifies that under standard test conditions, the oven when new, may not leak more than 1000 $\mu W/cm^2$ at any point 5 cm from the oven [26]. This performance may degrade to 5000 $\mu W/cm^2$ over the lifetime of the oven. Simple inverse square calculations show that an oven leaking 5000 $\mu W/cm^2$ at 5 cm will produce levels of about 4 $\mu W/cm^2$ at six feet and 1 $\mu W/cm^2$ at ten feet.

Satellite Communications and Radar Systems. Satellite communications systems are continuous wave sources and radars are usually pulse modulated. A study done by the Electromagnetic Compatibility Analysis Center in 1972 identified 223 continuous wave sources with effective radiated powers greater than 1 megawatt and 375 pulsed sources with *peak* effective radiated powers of 10 gigawatts or greater [27]. The power density in the main beam of these systems can be greater than 10 mW/cm^2 [28,29]. However, as discussed above, the probability of being irradiated by the *main* beam of one of these sources is quite small. Persons who live or work near high power sources, e.g., near airports or military bases, may be exposed to sidelobe or secondary radiation from systems having stationary or slowly moving antennas as well as from many types of radars with rapidly moving antennas. Calculated exposures fall into the range of 10 to 100 $\mu W/cm^2$ at distances up to one-half mile from some of these systems [30]. The motion of the antennas of acquisition radars reduces the time-averaged power density from these systems. For high-power radars, a combination of mitigating factors, i.e., beam motion and antenna elevation angle, make it unlikely that power densities will exceed 50 $\mu W/cm^2$ at distances greater than one-half mile, at least in locations that are accessible to people [29].

Diathermy Units. Both efficiency and leakage of microwave diathermy applicators have been the subjects of recent studies [31,32]. The leakage results are summarized in Table 6 [32]. The measurements were made 5 cm from the interface between the tissue equivalent phantom and the applicator. Under test conditions, conventional applicators had leakage fields 2 to about 9 times the allowable leakage for microwave ovens. Leakage is considerably lower when using one type of contact applicator under the same test conditions.

Table 6. Leakage from Diathermy Applicators[a]

Applicator Type	Leakage (mW/cm^2)	
	Nominal[b]	Maximum[c]
Burdick "B"	10.4	35.5
Burdick "E"	19.0	44.0
Transco[d]	-	0.2

a. Determined using a planar phantom of muscle equivalent material with a 1 cm simulated fat layer.
b. Determined using net power recommended by manufacturer for lower back treatments, 45 and 70 Watts net power for types "B" and "E", respectively.
c. Extrapolated for "effective treatment" conditions as defined by Lehmann et al. [33], i.e., an absorption rate of 235 W/kg.
d. Contact applicator.

Industrial Radio Frequency Sources. High-powered sources are used extensively in industry for heating and drying. The data presented in Table 7 is taken from a study of two synthetic fiber dryers used in the textile industry, an edge gluer from the lumber industry and seven heat sealers used in the plastics industry. The results in Table 7 are presented in terms of field strength and may be expressed in "equivalent" power density using equations 1 or 2 above as appropriate. The cautions in interpreting this "equivalent" power density that were discussed earlier must also be observed here.

Table 7. Electric and Magnetic Fields at Operator Positions near Industrial Radio Frequency Sources. (Selected maximum values from Reference 34).

Source	Power (kw)	Freq. (MHz)	Field Strength[a]	
			Electric (V/m)	Magnetic (A/m)
Fiber Dryer	20	41	319	13.2
Glue Dryer	20	27	221	1.0
Heat Sealer	10	15	831	.5
Heat Sealer	2	22	493	12.1
Heat Sealer	4	30	973	.4

a. Average of two values given in [34].

SUMMARY AND CONCLUSIONS

The low-levels in the *general radiofrequency environment* are dominated by radio and television transmissions. Microwave relay links, low-power radars, mobile communications systems, microwave ovens, etc. make almost negligible contributions. Most of the population (99%) is exposed to levels less than 1 μW/cm². The largest radar field measured in the *general environment* of one urban area was less than the median value for the broadcast band.

The *specific source environment* for most broadcast sources is well below 100 μW/cm². The only fields in excess of this value occur in the immediate vicinity of a few FM antenna towers and on the roofs of tall buildings that are located within a city block or so of FM and TV broadcast antennas. When considering other high-power sources such as radars and satellite communications systems, the potential for exposure of persons to obviously high levels of radiation is small.

Leakage from medical diathermy units can exceed 40 mW/cm² and fields in the vicinity of some industrial sources can exceed 10 mW/cm² in "equivalent" power density. Of the low-power sources, only portable communications equipment produces fields that, at least on cursory examination, appear high, but tissue temperature elevation should not exceed 0.1°C. The interpretation of these "equivalent" power densities is difficult and the difficulty is confounded by partial body exposure and intermittent operation.

REFERENCES

[1] Tell, R., J. Nelson, & N. Hankin (1974). HF Spectral Activity in the Washington, DC Area, Rad. Data & Repts. 15: 549.
[2] Janes, D., R. Tell, T. Athey, & N. Hankin (1977). Radio-frequency Radiation Levels in Urban Areas, Radio Science 12(65): 49-56.
[3] ditto (1977). Nonionizing Radiation Exposure in Urban Areas of the U.S., P. IVth Int. Cong. Int. Rad. Prot. Assoc. 2: 329.
[4] Athey, T., R. Tell, N. Hankin, D. Lambdin, E. Mantiply, & D. Janes (1978). RF Radiation Levels and Population Exposure in Urban Areas of the Eastern U.S., EPA Pub. 520/2-77-008.
[5] Tell, R. & E. Mantiply (1978) Population Exposure to VHF and UHF Broadcast Radiation in the United States, EPA Pub. ORP/EAD 78-5.
[6] Tell, R. (1978). An Analysis of RF and Microwave Absorp. Data With Consideration of Thermal Safety Standards, EPA Pub. ORP/EAD 78-2.
[7] Tell, R. (1977) An Analysis of Radar Exposure in the San Francisco Area, EPA Pub. ORP/EAD-77-3.
[8] Tell, R. & J. Nelson (1974). Calculated Field Intensities Near a High Power UHF Broadcast Installation, Rad. Data & Repts. 15: 401.
[9] Ruggera, P. (1975). Changes in RF E-Field Strengths within a Hospital During a 16-Month Period, DHEW Pub. (FDA) 75-8032.
[10] Tell, R. & N. Hankin (1978). Meas. of RF Field Intensity in Buildings with Close Proximity to Broadcast Stations, EPA Pub. ORP/EAD 78-3.
[11] Tell, R. (1978). Near-Field Radiation Properties of Simple Linear Antennas With Applications to Radiofrequency Hazards and Broadcasting. EPA Pub. ORP/EAD 78-4.
[12] Tell, R. (1978). Unpublished results.
[13] Tell, R. & P. O'Brien (1977). An Investigation of Broadcast Radiation Intensities at Mt. Wilson, California, EPA Pub. ORP/EAD-77-2.
[14] Hankin, N. (1978). Unpublished results.
[15] ditto (1976). Radiation Characteristics of Traffic Radar Systems, EPA Pub. ORP/EAD-76-1.
[16] Tell, R., N. Hankin, & D. Janes (1976). Aircraft Radar Meas. in the Near Field, P. 9th Midyear Topic. Symp. Health Phys. Soc. p. 239.
[17] Peak, D., D. Conover, W. Herman, & R. Shuping (1975). Measurement of Power Density from Marine Radar, HEW Pub. (FDA) 76-8004.
[18] Bronaugh, E., D. Kerns, W. McGinnis (1977). Electromagnetic Emissions from Typical Citizens' Band Mobile Radio Installations in Three Sizes of Vehicles, IEEE Pub. 77CH1231-0 EMC.
[19] Ruggera, P. (1979). Measurements of Electromagnetic Fields in the Close Proximity of CB Antennas, HEW Pub. (FDA) 79-8080.
[20] Adams, J., M. Kanda, J. Shafter, & Y. Wu (1977). Near Field Electric Field Strength Levels of EM Environments Applicable to Automotive Systems, unpublished results, National Bureau of Standards, Boulder, CO.
[21] Lambdin, D. (1979). An Investigation of Energy Densities in the Vicinity of Vehicles with Mobile Communications Equipment and Near a Hand-Held Walkie Talkie, EPA Pub. ORP/EAD 79-2.
[22] Ruggera, P. (1977). Near-Field Measurements of RF Fields, in HEW Pub. 77-8026, p.104.
[23] Balzano, Q., O. Garay, & R. Steel (1977). Energy Deposition in Biological Tissue Near Portable Radio Transmitters at VHF and UHF, in IEEE Pub 77CH1176-7VT, pp 25-39.
[24] ditto (1978). A Comparison Between the Energy Deposition in Portable Radio Operators at 900 MHz and 450 MHz, preprint.
[25] ditto (1978). Energy Deposition in Simulated Human Oper. of 800 MHz Port. Trans., preprint.
[26] Title, 21, Code of Fed. Regs., Part 1030.10.
[27] Tell, R. (1973). Environmental Nonionizing Radiation Exposure, in EPA Pub. EPA/ORP 73-2.
[28] Hankin, N. (1974). An Eval. of Satellite Com. Systems as Sources of Environmental Microwave Radiation, EPA Pub 520/2-74-008.
[29] Larsen, E. & J. Shafer (1977). Surveys of Electromagnetic Field Intensities Near Representative Higher-Power FAA Transmitting Antennas, DOT Pub FAA-RD-77-179.
[30] Hankin, N. (1978). Unpublished results.
[31] Witters, D. & G. Kantor (1978). Free-Space Electric Field Mapping of Microwave Diathermy Applicators, HEW Pub. (FDA) 79-8074.
[32] Bassen, H., G. Kantor, P. Ruggera, & D. Witters (1978). Leakage in the Proximity of Microwave Diathermy Applicators Used on Humans or Phantom Models, HEW Pub 79-8073.
[33] Lehmann, J., A. Guy, J. Stonebridge, & B. deLateur (1978). Evaluation of a Therapeutic Direct-Contact 915 MHz Microwave Applicator for Effective Deep-Tissue Heating in Humans, IEEE Trans. Microwave Theory & Tech. MTT-26(8): 556.
[34] Conover, D., W. Parr, E. Sensintaffer, & W. Murray (1976). Measurement of Electric and Magnetic Field Strengths from Industrial Radiofrequency (15-40.68 MHz) Power Sources, in HEW Pub. (FDA) 77-8011, Vol. II, pp 356-362.

PART VII
Interference Effects: Electromagnetic Compatibility of Cardiac Pacemakers

Electromagnetic interference (EMI) of medical prosthetic devices such as the cardiac pacemaker is a unique biological effect of nonionizing radiofrequency radiation (RFR). The potential hazard for such interactions was well established soon after the development of the demand pacemaker and, in the last ten years, interference tests have been conducted using almost every conceivable RFR source. The bibliography which follows this introduction documents many such tests. During these ten years, manufacturers have continuously improved the EMI characteristics of their products and significantly reduced the potential hazard.

SAM-TR-76-4, on the biological significance of RFR emission on pacemakers, provides a summary of the relative EMI thresholds (susceptibility levels) for the vast majority of cardiac pacemakers in use around January 1976. It includes laboratory data for three frequencies and a wide range of pulse widths, and summarizes, from actual field tests, typical effects one might expect around high-powered RFR emitters prevalent throughout the world. This paper also describes appropriate instrumentation and test procedures for conducting pacemaker EMI tests, and details the significance of interference characteristics in assessing the extent of possible biological effects. The principal conclusions are: (1) the functional design of the demand-type pacemaker of necessity provides a window (heart-signal sensing circuit) by which certain forms of extraneous RF energy, which mimic heart activity, can interact and disrupt normal pacemaker operation; (2) EMI test results (Jan.-Aug. 1975), when compared with data obtained two to three years previously and earlier case histories of adverse interactions, provide remarkable evidence of the overall improvement in EMI thresholds of the newer model pacemakers; and (3) acceptance of a pacemaker electromagnetic compatibility test standard, such as that described in the *Association for the Advancement of Medical Instrumentation Pacemaker Standard* (August 1975), should provide reasonable assurance that pacemakers are compatible with the unrestricted RF environment. The following table provides more recent results of tests conducted at the United States Air Force School of Aerospace Medicine in 1977–1978 and serves as an addendum to Table I of SAM-TR-76-4. These more recent test results are published in SAM-TR-79-20, dated June 1979.

The paper by Simon et al. analyzes the effect of a variety of dental appliances on pacemaker function, presenting data from electromagnetic field intensity measurements in typical dental examining rooms and laboratories. Fourteen patients with permanently implanted pacemakers were evaluated while the dental instruments and laboratory equipment were operating in a manner which simulated a realistic clinical setting. This paper documents the fact that temporary but repetitive pacemaker inhibition is possible where dental equipment is frequently energized in a start-stop sequence. Thus, pacemaker users present potential problems as dental patients. The study concludes that the dental environment is a source of moderate electromagnetic interference, and several basic guidelines are presented for dealing with this situation.

TABLE I
CARDIAC PACEMAKER EMI THRESHOLDS IN VOLTS/METER
450 MHz, 5 pps—Simulated Implant
(Addendum to Table I of SAM-TR-76-4)
(Reference SAM-TR-79-20, June 1979)

Pacemaker	20 ms	1 ms	10 μs
American			
1613	>360	>360	>360
American Optical			
281143	>330	>330	>330
Biotronik			
IDP-44	110	125	330
Cardiac Pacemaker, Inc.			
501 UD	>330	>330	>330
Coratomic			
L-500	>360	>360	>360
Cordis			
Stanicor K (No. D)	>330	>330	>330
Edwards			
8116	>330	>330	>330
20S	>330	>330	>330
General Electric			
A10759/2075A	165	260	235
Medcor			
C (Equiv 3-70C)	>330	>330	>330
Medtronic			
5972	>330	>330	>330
5973	>330	>330	>330
Pacesetter			
BD-101	>330	>330	>330
Stimtech			
3821	30	105	>330
Vitatron			
MIP 42 RT	30	90	200

Note: (1) Maximum E-field available was 330 V/m and 360 V/m.

A clinical study was conducted several years ago in which 53 patients with implanted pacemakers were exposed to magnetic fields (0.5-1.35 G) generated by fixed and portable weapons detectors like those used in most airports today (*J. Amer. Med. Ass.*, vol. 217, pp. 1033-1034, 1971). The patients were continuously monitored during the tests. A few devices demonstrated some minimal effects, which were judged clinically insignificant. It was concluded that persons with any of the permanent pacemakers tested should suffer no ill effects from the airport weapons detectors.

Manuscript received July 15, 1981.

Kahn and Schlentz (1973) describe the mechanisms responsible for cardiac pacemaker electrical interference and the basic approach to achieve electromagnetic compatibility. They point out the effects of field strength, frequency, and modulation parameters (as illustrated in SAM-TR-76-4) on pacemaker EMI susceptibility. The extent to which EM energy is coupled to leads, components, and interconnections is described in terms of the pacemaker system as an antenna. Important considerations are discussed, such as the reduction of incident EM radiation wavelength in traversing from air to biological media and the difference in the unipolar and bipolar antenna aperture. This paper makes it clear that the technical feasibility of protecting cardiac pacemaker systems from most electromagnetic environments has been recognized for several years. Recent EMI test results, as described in SAM-TR-76-4 and SAM-TR-79-20 (Table 1), confirm the manufacturers' success in achieving a high level of electromagnetic compatibility and, in most instances, improving other characteristics as well.

Although some pacemakers marketed today will still encounter brief periods of RF interference, such interactions concerning the general populace are not likely to be clinically significant.

John C. Mitchell
Associate Editor

BIBLIOGRAPHY

Cardiac Pacemaker Electromagnetic Compatibility

[1] H. Brandaleone, "Motor vehicle driving and cardiac pacemakers," *Ann. Intern. Med.*, vol. 81, pp. 548–550, 1974.

[2] J. E. Bridges and E. E. Brueschke, "Hazardous electromagnetic interaction with medical electronics," Chicago, IL: VIIT Research Institute, EMC Symposium, 1970.

[3] R. G. Chrystal, J. A. Kastor, and R. W. Desanctis, "Inhibition of discharge of an external demand pacemaker by an electric razor," *Amer. J. Cardiol.*, vol. 27, pp. 695–697, 1971.

[4] G. F. D'Cunha et al., "Syncopal attacks arising from erratic demand pacemaker function in the vicinity of a television transmitter," *Amer. J. Cardiol.*, vol. 31, pp. 789–791, 1973.

[5] S. Furman et al., "The influence of electromagnetic environment on the performance of artificial cardiac pacemakers," *Ann. Thorac. Surg.*, vol. 6, pp. 90–95, 1968.

[6] F. Giori et al., "A method for testing the radiofrequency tolerance of implantable cardiac pacemakers," Electro Sciences for Medicine, Inc., Oct. 1971.

[7] D. H. Gobeli, "Electromagnetic interference in cardiac pacemakers," presented at 1971 IEEE Int. Symp. on Electromag. Compat., Medtronic, Inc., 1971.

[8] K. A. Hardy, "Measured effects of 450-, 350-, and 250-MHz pulsed and 26-MHz cw radiofrequency fields on cardiac pacemakers," SAM-TR-79-20, USAF School of Aerospace Medicine, Brooks AFB, Texas, June 1979.

[9] O. C. Hood et al., "Anti-hijacking efforts and cardiac pacemakers—report of a clinical study," *Aerosp. Med.*, P. 43, no. 3, pp. 314–322, 1972.

[10] S. N. Hunyor et al., "Interference hazards with Australian noncompetitive ("demand") pacemakers," *Med. J. Aust.*, vol. 2, no. 13, pp. 635–655, 1971.

[11] W. D. Hurt, "Cardiac pacemaker electromagnetic interference (3050 MHz)," SAM-TR-72-36, USAF School of Aerospace Medicine, Brooks AFB, Texas, Dec. 1972.

[12] W. D. Hurt, "Effects of electromagnetic interference (2450 MHz) on cardiac pacemakers," SAM-TR-73-40, USAF School of Aerospace Medicine, Brooks AFB, Texas, Dec. 1973.

[13] W. D. Hurt, J. C. Mitchell, and T. O. Steiner, "Measured effects of square-wave modulated FR fields (450 and 3100 MHz) on cardiac pacemakers," SAM-TR-74-51, USAF School of Aerospace Medicine, Brooks AFB, Texas, Dec. 1974.

[14] D. L. Johnson, "Effect on pacemakers of airport weapon detector," *Can. Med. Assoc. J.*, vol. 110, no. 7, pp. 778–780, 1974.

[15] D. L. Johnson, "Pacemaker inhibition by magnetic stirrers," *Amer. Heart J.*, vol. 85, p. 433, 1973.

[16] A. R. Kahn and R. J. Schlentz, "Design and construction methods for protecting implanted cardiac pacemakers from electromagnetic interference," *Proc. IVth Int. Symp. on Cardiac Pacing*, Groningen, The Netherlands, April 17–19, 1973.

[17] G. R. King et al., "Effect of microwave oven on implanted cardiac pacemaker," *J. Amer. Med. Ass.*, vol. 212, p. 1213, 1970.

[18] D. R. Koerner, "The employee wearing a cardiac pacemaker," *J. Occup. Med.*, vol. 16, pp. 392–394, 1974.

[19] R. J. Merrow, "Microwave interference with pacemakers," *Amer. Soc. Safety Eng. J.*, vol. 1, pp. 35–37, 1973.

[20] J. C. Mitchell et al., "Electromagnetic compatibility of cardiac pacemakers," presented at 1972 IEEE Int. Symp. on Electromag. Compat., Arlington Heights, IL, July 18, 1972, *Symp. Record*, pp. 5–9.

[21] J. C. Mitchell et al., "Empirical studies of cardiac pacemaker interference," *Aerosp. Med.*, vol. 45, no. 2, pp. 189–195, 1974.

[22] J. C. Mitchell, W. D. Hurt, and T. O. Steiner, "EMC design effectiveness in electronic medical prosthetic devices," presented at Seventh Rochester Int. Conf. on Environmental Toxicity—Fundamental and Applied Aspects of Nonionizing Radiation, University of Rochester, Rochester, NY, June 6, 1974.

[23] J. C. Mitchell, "Electromagnetic interference of cardiac pacemakers," AGARD Lecture Series no. 78 on Radiation Hazards, Sept. 1975.

[24] J. C. Mitchell and W. D. Hurt, "The biological significance of radiofrequency radiation emission on cardiac pacemaker performance," SAM-TR-76-4, USAF School of Aerospace Medicine, Brooks AFB, Texas, Jan. 1976.

[25] B. D. McLees and E. D. Finch, "Bibliography on the hazards of artificial cardiac pacemaker exposure to radiofrequency fields and electric shock," Naval Med. Res. Inst., MF12.524.015-0001B, Report no. 2, April 1971.

[26] J. K. O'Donoghue, "Inhibition of a demand pacemaker by electrosurgery," *Chest*, vol. 64, pp. 664–666, 1973.

[27] J. M. Osepchuk, "A simple microwave test for assessing interference susceptibility of pacemakers," *Raytheon Res. Div. Publication T-943*, 1970.

[28] B. Parker, S. Furman, and D. J. W. Escher, "Input signals to pacemakers in a hospital environment," *Ann. N.Y. Acad. Sci.* vol. 167, pp. 823–834, 1969.

[29] B. A. Pickers and M. J. Goldberg, "Inhibition of a demand pacemaker and interference with monitoring equipment by radiofrequency transmissions," *Brit. Med. J.*, vol. 2, pp. 504–506, 1969.

[30] P. S. Ruggera and R. L. Elder, "Electromagnetic radiation interference with cardiac pacemakers," United States Dept. Health, Education, and Welfare, BRH/DEP Pub. 71-5, April 1971.

[31] P. L. Rustan, W. D. Hurt, and J. C. Mitchell, "Microwave oven interference with cardiac pacemakers," *Med. Instrum.*, vol. 7, pp. 185–188, 1973.

[32] S. A. Sanchez, "Danger! When you transmit you can turn off a pacemaker," *QST*, pp. 58–60, Newington, CT:, Am. Radio Relay League, Inc., March, 1973.

[33] R. J. Schlentz, "How to achieve electromagnetic compatibility," Medtronics Inc. report, 1971.

[34] R. J. Schlentz, "*In vitro* measurements of the electromagnetic susceptibility of cardiac pacemakers," presented at 7th Annual AAMI Convention, April 1972.

[35] R. J. Schlentz, "Electromagnetic compatibility of implanted cardiac pacemakers," presented at EMC Meeting, Montreaux, Switzerland, May 20–22, 1975.

[36] A. B. Simon et al., "The individual with a pacemaker in the dental environment," *J. Amer. Dent. Ass.*, vol. 91, pp. 1224–1229, 1975.

[37] N. P. D. Smyth et al., "Effects of active magnetic fields on permanently implanted triggered pacemakers," *Med. Instrum.* vol.7, pp. 189–195, 1973.

[38] N. P. D. Smyth et al., "The pacemaker patient and the electromagnetic environment," *J. Amer. Med. Ass.*, vol. 227, p. 1412, 1974.

[39] E. Sowton, K. Gray, and T. Preston, "Electrical interference in noncompetitive pacemakers," *Brit. Heart J.*, vol. 32, pp. 626–632, 1970.

[40] T. O. Steiner, "Development of a pacemaker monitor with cardiac simulator," SAM-TR-75-7, USAF School of Aerospace Medicine, Brooks AFB, Texas, Feb. 1975.

[41] J. C. Toler, "Energy density: a proposed parameter for assessing electromagnetic performance of demand pacemakers," presented at EMC Meeting, Montreaux, Switzerland, May 20–22, 1975.

[42] A. R. Valentino, D. A. Miller, and J. E. Bridges, "Susceptibility of cardiac pacemakers to 60 Hz magnetic fields," presented at 1972 IEEE Int. Symp. Electromag. Compat., Arlington Heights, IL, July 1972.

[43] R. P. Van Wijk van Brievingh et al., "Measurement techniques for assessing the influence of electromagnetic fields on implanted pacemakers," *Med. Biol. Eng.*, vol. 12, no. 1, pp. 42–48, 1974.

[44] W. H. Walter, III, et al., "Cardiac pulse generators and electromagnetic interference," *J. Amer. Med. Ass.*, vol. 224, pp. 1628–1631, 1973.

[45] B. J. Walz et al., "Cardiac pacemakers—does radiation therapy affect performance?" *J. Amer. Med. Ass.*, vol. 234, pp. 72–73, 1975.

[46] L. Woolley, J. Woodworth, and J. L. Dobbs, "A preliminary evaluation of the effects of electrical pulp testers on dogs with artificial pacemakers," *J. Amer. Dent. Ass.*, vol. 89, pp. 1099–1101, 1974.

[47] R. F. Yatteau, "Radar-induced failure of a demand pacemaker," *New Eng. J. Med.*, vol. 283, pp. 1447–1448, 1972.

[48] "Possible electromagnetic interference with cardiac pacemakers from dental induction casting machines and electrosurgical devices," Reports of Councils and Bureaus, *J. Amer. Dent. Ass.*, vol. 86, p. 426, 1973.

[49] "Pacemaker users face no danger from gun detector," *J. Amer. Med. Ass.*, vol. 217, pp. 1033–1034, 1971.

[50] AAMI Pacemaker Standard, August 1975, prepared by Ass. for the Advancement of Med. Instrumen. under FDA Contract no. 223-74-5083, 1975.

[51] H. J. Bisping, H. Stockberg, J. Meyer, and W. Frik (Rhein Westfal Th, Fak. Med. Radiol. Abt., D-5100 Aachen, Fed. Rep. of Germany), "Radiation therapy in patients with electronic cardiac pacemakers—interferences with pacemakers function by ionizing radiation and other sources of disturbance," *Strahlentherapie*, vol. 153, pp. 456–461, July 1977.

[52] J. R. Marback, R. T. Meoz-Mendex, J. K. Huffman, P. T. Hudgins, and P. R. Almond (Vet. Adm. Hosp., Radiotherapy Serv., Houston, TX 77221), "The effects on cardiac pacemakers of ionizing radiation and electromagnetic interference from radiotherapy machines," *Int. J. Radiat. Oncol. Biol. Phys.*, vol. 4, November-December 1978.

THE BIOLOGICAL SIGNIFICANCE OF RADIOFREQUENCY RADIATION EMISSION ON CARDIAC PACEMAKER PERFORMANCE

J. C. Mitchell and W. D. Hurt

INTRODUCTION

During the past 10 years, the electronic cardiac pacemaker has been developed into a sophisticated prosthetic device. It is applied in medical facilities throughout the world to correct malfunctions (atrioventricular heart block) of the body's electrical conduction system and thus to restore the rhythmic pumping action of the heart.

Pacemakers may be classed as fixed rate (asynchronous) and demand (synchronous or R-wave inhibited). Fixed-rate pacemakers provide a fixed, preset rate of electrical stimuli to the ventricles, which is independent of the electrical and/or mechanical activity of the heart. Demand pacemakers sense the depolarizations of the heart muscle activity and produce their own depolarization signals (electrical stimulus) only if the normal heart depolarizations are not present. The atrial synchronous pacemakers sense the depolarization of the atria, delay the signal to simulate natural conduction time, and then provide the electrical stimulus to the ventricles. The R-wave inhibited demand pacemaker inhibits its output when it senses depolarization of the ventricles if it occurs naturally; i. e., the pacemaker functions only when the AV heart block occurs (11).

Most pacemakers implanted today are the R-wave inhibited type. They contain an electronic timing circuit which is reset by normal depolarization or the pacemaker stimulus. Their sensing circuit is programmed to respond to electrical signals normally generated by the heart. Energy pulses induced externally via the pacemaker leads or circuitry can erroneously cause the pacemaker to inhibit its needed output (4, 8). Thus, the interaction of radiofrequency (RF) electromagnetic radiation fields with cardiac pacemakers represents a unique indirect biological effect. This results primarily from the fact that current pacemaker interference thresholds begin as low as 10 V/m, while peak E-field levels of several hundred volts per meter can be associated with pulsed fields having average power densities well below the acceptable, nonrestricted 10 mW/cm^2 personnel exposure level; i. e., the E-field level of a continuous wave (CW) 10 mW/cm^2 field is about 200 V/m, but it can be much higher for a low duty cycle pulsed source, having the same 10 mW/cm^2 average power density.

Case histories of pacemaker electromagnetic interference (EMI) reported in the open literature (1, 5, 6, 9, 10, 12, 15), combined with

Reprinted from *The United States Air Force School of Aerospace Medicine—Technical Report-76-4*, Jan. 1976.

newer findings such as those discussed at the June 1975 FDA open public hearing on pacemaker interference by antitheft devices, substantiate the potential problem. Also, many cases of pacemaker EMI probably go unreported due to the nature of the interference phenomena. For instance, upon sensing intense 60-Hz external EM radiation, many pacemakers revert to a fixed rate so close to their demand rate that the user would not normally detect the change. Radiofrequency radiation emitters such as air route surveillance radar can cause many pacemakers to miss single beats as the radar beam scans past, an effect most likely unnoticed (8). Even more serious interference may not be identified because, most often, interaction times are short; i.e., either the source of EMI is moved or turned off or the user moves from the particular area of the effect. Additionally, little postmortem followup is available to identify any possible causal relationship to EMI.

Although the potential hazard to individual users of the more sensitive pacemakers is generally acknowledged, controversy will continue as to the clinical significance of this effect of RF radiation on the pacemaker populace (2, 14).

Results of tests conducted in 1974 and 1975 by the USAF School of Aerospace Medicine (USAFSAM), both in the USAFSAM RF laboratory and in close proximity to a series of Air Force RF emitters, demonstrate the significance of EM emission characteristics in the overall assessment of the EMI of cardiac pacemakers.

TEST PROCEDURES

Implant Simulation

Realistic assessment of the effects of radiofrequency electromagnetic radiation (EMR) on cardiac pacemakers requires actual implant conditions or accurate simulation of implantation. Initial EMI studies by USAFSAM were conducted by implanting pacemakers in 18-to 20-kg dogs and effecting a complete atrioventricular heart block (18). This procedure is costly and has obvious disadvantages in having to handle the animals under a variety of test conditions in the laboratory and at remote test sites. Thus, alternate techniques were developed to simulate the pacemaker implant (4, 7, 8).

More recently the Association for the Advancement of Medical Instrumentation (AAMI), working under a contract with the U.S. Food and Drug Administration (FDA), has developed a draft protocol for testing cardiac pacemaker EMI characteristics. They recommend using a 80- x 40- x 20-cm container made of 5-cm-thick, low-dielectric plastic foam (density of 0.035 g/cm^3). The container is filled with 0.03 molar saline solution, and the pacemaker leads are located to place 1 cm of solution between the pacemaker and its lead(s) and the wall of the container. The pacemaker lead is stretched out horizontally, and the pacemaker response is picked up via 2- x 2-cm copper-mesh screen electrodes placed in the solution in each end of the container. The USAFSAM pacemaker test container used in the laboratory studies reported herein was designed to AAMI specifications and is shown in Figure 1. For purposes of this illustration, the front view of the container shows the pacemaker placed outside the test container. During testing, the pacemaker and horizontal lead arrangement is placed inside the test container as described above, with 1 cm of solution between the pacemaker and front wall of the container. A similar arrangement used in all previous USAFSAM tests provided good correlation between this method of simulated implant and the implanted dogs (4, 18). With the many variables (body size, location, orientation, depth of implant) in actual human implants, this procedure for implant simulation is believed sufficient for EMI testing. Additionally, it is recommended that both "free-field" and "simulated-implant" data be reported to bracket actual implant representation.

Instrumentation

Many different types of instrumentation techniques have been used in cardiac pacemaker EMI testing (4, 8, 15-18). The principal requirement is that the instrumentation system be immune to the EM fields encountered in the tests and that it presents to the pacemaker a load and signal simulating those encountered in an actual implant situation, so that the results obtained apply to a human implant. The system should also provide simultaneous real-time recording of the incident EMR signal and the pacemaker response.

Several pacemaker models have interference rates identical to their demand rates, so it is often difficult to determine susceptibility thresholds for pacemakers in EMR fields having pulse repetition rates (PRR) sufficient to cause the pacemaker to revert to its fixed rate. The minimum PRR values range from 3 to 60 Hz depending on the specific pacemaker. Thus, a device to simulate normal heart activity at the pacemaker leads is required so that an R-wave inhibited pacemaker would be inhibited by this simulated activity and would not produce a pulse until it detected interference and reverted to its interference mode. An additional requirement is imposed for a synchronous pacemaker to track the simulated activity up to its interference threshold. A system

Figure 1. USAFSAM testing container used in pacemaker EMI tests.
A: Front view. B: Back view.

of this type has been developed and incorporates a light-emitting diode (LED) fiber optics monitoring system (16). This system was used in essentially all of the field tests at RF emitter sites. Results of such tests are in good agreement with USAFSAM laboratory studies using the AAMI-type test arrangement as described above.

TEST RESULTS

Laboratory Studies

Eighty pacemakers (23 different models) were evaluated to establish their relative electromagnetic interference (EMI) susceptibilities as a function of the radiation frequency, pulse width, and E-field intensity. Figure 2 shows most of the pacemaker types tested. The tests were

Figure 2. Pacemaker models representative of those tested.

performed in the USAFSAM RF anechoic chamber using three different frequencies (450, 1600, and 3200 MHz); a constant pulse repetition rate of 10 pulses per second (pps); pulse widths (duration) of 0.01, 0.02, 0.5, 1, 2, 5, 10, and 20 ms; and E-field levels up to 1200 V/m. The total ranges of pulse widths (PW) and E-field levels were somewhat different for each frequency.

The 450-MHz radiation source was provided by Microwave Cavity Laboratory (MCL) model 15022 power generator, amplified by an MCL model 10110 power amplifier (up to 1000 W), and fed via an air dielectric "Heliax" transmission line to an EMCO model 3101 conical logspiral antenna.

The 1600- and 3200-MHz radiation sources were provided by a Cober model 1831 high-power microwave generator (peak power up to 1800 W). The 1600-MHz signal was fed via a 7/8" air dielectric "Heliax" transmission line to an AEL model H-5001 horn in the anechoic chamber. The 3200-MHz signal was fed via flexible waveguide to a Struthers model 1109 antenna horn.

A multiple-pulse generator designed and built at USAFSAM to supply external modulation for both the MCL and Cober generators was used to provide the special pulse modulation used in these tests.

Figure 3 shows the antenna and test container configuration in the anechoic chamber for the 450-MHz tests. Similar arrangements were used for the 1600- and 3200-MHz tests. Figure 4 illustrates the test arrangement and monitoring system used. Many of the pacemakers were tested on both sides; i.e., after a set of data was collected, the pacemaker was turned over with its other side facing the incident radiation field. This can change some pacemaker thresholds by a factor of 2-3.

The RF field at the test location was measured (mapped) with a National Bureau of Standards (NBS) electric field intensity meter (model EDM-1B) and/or a Narda electromagnetic radiation monitor (model 8316). Less than 3-dB variation in the field was measured across the front of the pacemaker test container. The solution was then added to the container and the transmitter output power adjusted to equal that used for the field mapping measurements. A corresponding field measurement was made using a Singer field intensity analyzer (model EMA-910) or a Fairchild interference analyzer (model EMC-25) to establish a correlation factor between the field analyzer readings and the power density at the test location. The field analyzer was connected to a monitoring antenna which was located above and behind the pacemaker test container.

Figure 3. Pacemaker test configuration in the USAFSAM anechoic chamber.

The copper-mesh electrodes in the solution were connected to an ECG amplifier by approximately 10 ft of lossy line, and the signal conductor was routed via a shielded path through the anechoic chamber wall. The amplifier output was fed into a computing counter so that the pacemaker rate could be displayed, and also to one channel of a dual-channel strip-chart recorder. The field intensity analyzer output was recorded on the second channel.

The EMI threshold data obtained under simulated-implant conditions are summarized in Table 1 for each of the pacemaker types (models) tested. Where more than one of a particular model was evaluated, the lowest threshold value is presented because it is believed such values are sufficiently representative of the threshold value of these devices

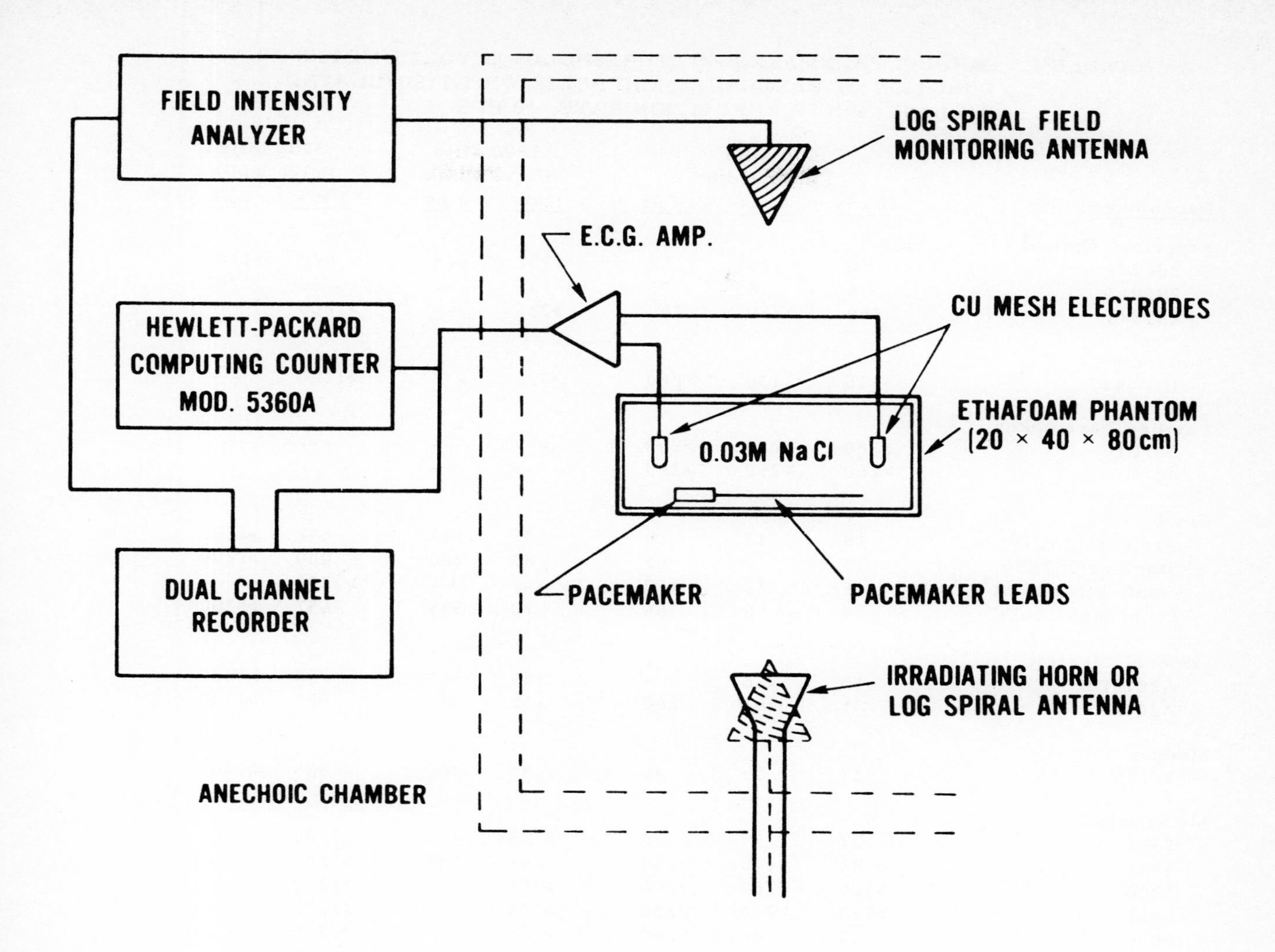

Figure 4. Schematic of pacemaker test configuration and monitoring instrumentation.

in use. The EMI threshold value in rms volts per meter represents the E-field strength at which the pacemaker rate falls below 50 beats per minute (bpm) or exceeds 120 bpm--arbitrarily defined here as a clinically significant adverse effect. Although essentially all these data were obtained for a constant pulse repetition rate of 10 pps, the PRR was occasionally decreased to confirm the interference threshold. This is necessary for this test configuration for pacemakers that remain in their fixed interference-rejection mode at a PRR of 10 pps, but that cut off when the PRR is lowered further.

TABLE 1. CARDIAC PACEMAKER EMI THRESHOLDS IN VOLTS/METER AS A FUNCTION OF FREQUENCY AND PULSE WIDTH (SIMULATED-IMPLANT; PULSE REPETITION RATE, 10 PPS)

Pacemaker	450 MHz Pulse width			1600 MHz Pulse width		3200 MHz Pulse width	
	20 ms	1 ms	10 μs	1 ms	2 μs	1 ms	2 μs
American Optical							
281003	6	13	62	132	623	620	>1200
281013	6	15	73	---	---	790	>1200
281143	>260	>260	>260	>725	---	>1200	---
Biotronik							
IDP-44	116	130	>260	>725	---	>1200	---
Cardiac Pacemaker							
301UD	>260	>260	>260	---	---	---	---
401BD	>260	>260	>260	---	---	---	---
Cordis							
Atricor 133C7[a]	18	19	28	35	148	642	>1200
Stanicor 143E7	15	13	65	83	380	400	>1200
Omni Atri. 164A[a]	9	12	110	---	---	>1200	---
Omni Stan. 162C	4	6	65	70	536	657	>1200
General Electric							
A2072D[b]	16	28	---	188	363	790	>1200
A2075A[b]	31	82	260	130	543	983	>1200
Medcor							
3-70A	23	23	46	34	209	325	>1200
Medtronic							
5942	37	39	260	>725	---	>1200	---
5944	82	110	>260	251	758	>1200	---
5950	>260	>260	>260	>725	---	>1200	---
5951	>260	>260	>260	>725	---	>1200	---
9000	22	26	>260	>725	---	>1200	---
Pacesetter							
BD-101	>260	>260	>260	>725	---	>1200	---
Starr-Edwards							
8114	14	21	---	56	139	357	>1200
8116	>260	>260	>260	>725	---	>1200	---
Stimtech							
3821	26	78	>260	199	536	>1200	---
Vitatron							
MIP-40-RT	37	116	207	226	623	579	>1200

NOTE: The maximum E-field levels available were 260, 725, and 1200 V/m for the 450, 1600, and 3200 MHz frequencies, respectively.

[a] These EMI thresholds represent the E-field level at which these cardiac pacemakers begin to synchronize with the incident RF signal; however, under worst-case conditions of these tests, these pacemakers did not exceed the manufacturer's design limits to ~150 bpm.

[b] These EMI thresholds were observed at 5 pps; at 10 pps the General Electric pacemakers generally revert to fixed rate.

These test results demonstrate that cardiac pacemaker EMI is strongly dependent on the frequency and pulse width of the incident radiation, and there remains a wide range of EMI thresholds among the different manufacturers and pacemaker models currently marketed. Of the three frequencies used, the 450-MHz source resulted in the lowest EMI thresholds. Data were obtained for about eight different pulse widths, but only the range is presented here to illustrate the general trend--that shorter pulse widths will result in higher EMI thresholds. Although the longest duration pulse used in these tests was 20 ms at the 450-MHz frequency, there is some indication that pulses wider than 20-30 ms may reverse the trend of EMI threshold values in some models of pacemakers (13).

In these tests, the antenna horns were placed with the E-field vector parallel to the horizontal lead arrangement for maximum coupling of RF to the pacemaker. Some pacemakers have better case attenuation and lead RF filtering, causing their EMI thresholds to shift more than others as the incident RF frequency is increased.

Five of the 11 manufacturers have essentially resolved the potential EMI problem, demonstrating the technical feasibility of developing cardiac pacemakers to be compatible with the overall RF environment. Progress in this respect has been remarkable compared with the EMI state of technology 3 years ago.

Field Test Studies

Tests were conducted in close proximity to a variety of RF radiation sources to assess potential EMI effects as a function of E-field level and corresponding distance from the respective emitters.

The emitter characteristics ranged in operating frequencies from 35 kHz to 9 GHz, pulse widths (PW) from 1 to 2000 μs, pulse repetition rates from 20 to 400 pps, and peak output powers from 0.02 to 32 MW.

Figure 5 illustrates the instrumentation system used in these field-test studies. In each instance, a vehicle was instrumented to accommodate the equipment necessary to monitor and record the real-time E-field level and cardiac pacemaker response on a dual-channel strip-chart recorder. For each test location, the pacemaker test container and field monitoring antenna were located 20-25 ft from the test vehicle.

The specific test equipment used to measure E-field exposure levels varied, depending on the electromagnetic frequency which was being

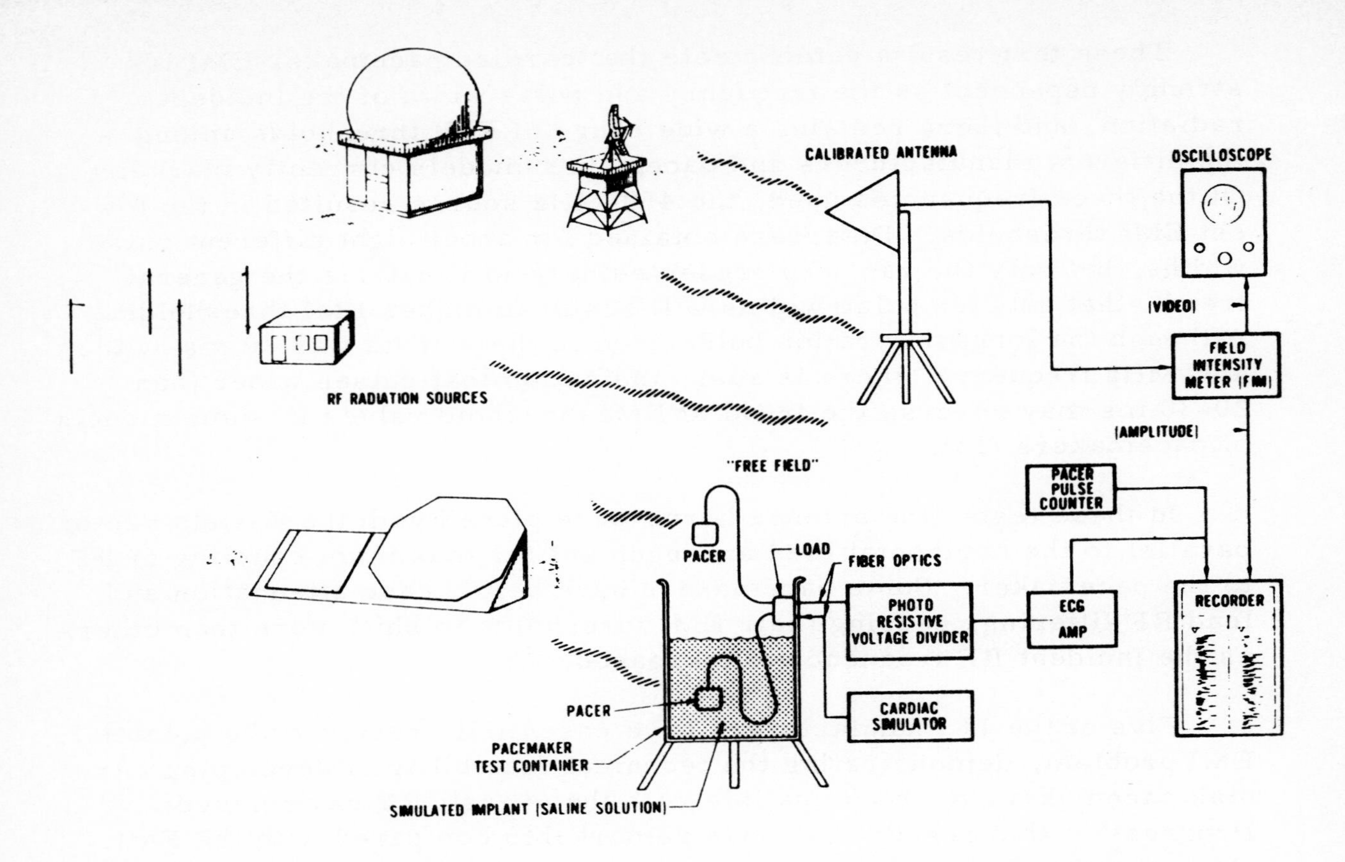

Figure 5. Instrumentation system for cardiac pacemaker EMI field tests.

measured. In general, this equipment consisted of a field intensity meter (FIM), associated calibrated antennas, and a strip-chart recorder. The FIM is an internally calibrated receiver capable of measuring amplitudes of electromagnetic energy at the FIM input. Applying known calibration factors for the respective antenna enables an operator to measure the field intensity at the antenna. The field intensity in this study is referred to as the E-field level. An oscilloscope was intermittently used to observe the pulse rate and width of the incident signal (RF radiation).

To assure maximum accuracy, pulse and signal generators were used before each field test, to calibrate the field measurement equipment, and after each field test, to validate the proper functioning of the field measurement equipment. For such calibration tests, known amplitudes of electromagnetic energy were inserted into the input of the FIM. This signal substitution was at the operating frequency, pulse rate, and pulse width of the radiation source being tested. All test equipment was certified and was within the normal calibration cycle.

The fiber optics telemetry system used to monitor the pacemaker is designed to present the pacemaker with a load comparable to that produced by the heart, and to prevent EMI pickup in the telemetry link (16). The load network was mounted on a subminiature audio jack to which the pacemaker leads were attached with "flea-clips." A light-emitting diode was mounted in a mating plug so that when the pacemaker fired, approximately one-half of the pacemaker output current passed through the LED, causing it to flash during the pulse. The LED was optically coupled to a length of light pipe which was terminated by a photoresistor in a voltage divider circuit. The output of the voltage divider was amplified by an ECG amplifier and fed to the strip-chart recorder for a permanent record. An electromechanical digital counter in parallel with the recorder was used to numerically determine the pacemaker pulse rate in beats per minute.

To determine whether demand-type pacemakers were operating in their interference-rejection fixed-rate mode or in the normal demand-rate mode, a cardiac simulator system was used to simulate cardiac activity to the pacemaker. This device consists of a variable-frequency pulsating-light source optically coupled via a light pipe and photocell to the pacemaker input leads. If the pacemaker were in its demand mode, the signals from the simulator would, in effect, turn off the pacemaker; if the pacemaker were in the interference-rejection fixed-rate mode, the simulator would have no effect.

The pacemakers were tested in both "free-field" and "simulated-implant" configurations as illustrated in Figure 5. The following paragraphs summarize typical simulated-implant test results by emitter type/function, and frequency. Systems which operate at frequencies greater than 5 GHz are omitted since few, if any, EMI effects were found for such systems.

Height-Finder Radar--Frequency ~2700-2900 MHz; pulse width (duration) ~2 μs; pulse repetition rate ~300-400 pps; peak power ~3-5 MW; scans a -2° to +32° elevation sector with a 3-s-cycle period, often changing its azimuth setting. Dependent on the tower height and terrain elevations, these systems can produce E-field levels of 1000-3000 V/m at distances of 200-2000 ft on the ground in the main beam. However, due to the relatively high operating frequency, short pulse duration, narrow beam width, and the fact that the intense portion of the beam traverses a fixed location rather rapidly, pacemaker interference from these systems is minimal. Characteristically, the more sensitive devices will skip a few beats (5-10 bpm) in close proximity to these systems in the sector scanned by the main beam. Tests conducted using four different height-finder radar over a 2-year period revealed no significant pacemaker interference under simulated-implant conditions.

Search Radar--Frequency ~200-500 MHz; pulse width (duration) ~15-30 μs; pulse repetition rate ~300-400 pps; peak power ~3-5 MW; scans 360° (azimuth) at a fixed elevation angle at 5 rpm. Such systems are usually located on a 50-75 ft tower. They produce a peak E-field level (designated E1) of 150-200 V/m at distances of 500-2500 ft on the ground as the main beam passes overhead. Secondary (E-field level greater than 15° on either side of main beam center) peak E-field levels of 30-80 V/m are typical on the ground at distances of 200-500 ft. The maximum secondary E-field level is designated E2. Many of the pacemakers tested skipped single beats coincident with the passage overhead of the main beam about every 12 seconds, out to distances of about a mile. Closer to these systems where E2 exceeds the threshold of some pacemakers and the moving antenna lobe structure mimics a low PRR, potentially serious pacemaker interference can occur. Most of the adverse effects (significant interference) recorded at distances greater than ~600 ft from these emitters were from pacemaker models that are being rapidly phased out of service and replaced with less sensitive devices.

Search Radar--Frequency ~1250-1350 MHz; pulse width (duration) ~6 μs; repetition rate ~200-400 pps; peak power ~2.5-10 MW; scans 360° (azimuth) at a fixed elevation angle at 5 rpm. These systems are usually located on a 50-75 ft tower. They produce a peak E-field (E1) of 150-400 V/m on the ground at distances of 200-2000 ft as the main beam passes overhead. The E2 levels ranged from 50 to 100 V/m at 150-400 ft distances. These systems affect pacemakers in the same manner as the 200-500 MHz search systems, but the extent of effects is less because of the higher operating frequencies. There was a significant difference in the free-field and simulated-implant data.

Search Radar--Frequency 2400-2900 MHz; operating parameters of these systems are analogous to those of the other search systems except the peak output power may vary over a wider range extending to 15 MW. Peak E-field levels (E1 and E2) of ~150 and 100 V/m, respectively, were recorded at 275 ft. These systems did not produce any significant pacemaker interference in any of these tests, primarily because such systems were located on a 50-75 ft tower and the operating frequency is significantly attenuated by the implant. The same was true for the search systems operating at frequencies greater than 5 GHz.

Ground to Air Telemetry Transmitter--Frequency 250-300 MHz. These systems are capable of 15-20 kW operation, using a fixed omnidirectional antenna (located on a 70 ft pole) to propagate a 33 pps square wave modulated signal. The periods of transmission vary from 2 to 5 s. The pulse periods are separated by 1.5-s intervals during which one or two pulses of ~35-ms pulse width occur. Although the measured E-field levels did not exceed ~10 V/m, the square wave pulsed signal causes many of the pacemakers to miss up to ~10 bpm within 500 ft of this system. Except for the older more sensitive pacemakers, only one pacemaker, the General Electric model A2075, was significantly affected under the simulated-implant tests.

LORAN-C Transmitter--This system operates at a frequency of 100 kHz, propagating a 9-pulse group repetition rate of 10/s with 1 ms between pulses, and a 250-μs pulse width. An output power of 1 MW is fed to antennas suspended from four 600 ft towers. No significant pacemaker interference was observed from this system, even for tests conducted within 50 ft of the main antenna feed line.

EMP Simulators--Electromagnetic pulse (EMP) simulators are unique sources of RF emission which produce intense pulses (up to 100,000 V/m) in 0.5 μs with ~90% of the frequency components below 10 MHz. Tests were conducted using single pulses at 5, 25, and 50 kV/m. On the basis of pacemaker response recordings made before and after exposure, it was determined the pacemakers were not seriously disrupted. Tests conducted using a repetitively pulsed (2-100 pps) source and peak E-field levels from 300 to 6000 V/m established an EMI threshold of 500 V/m under simulated implant conditions. The effective short pulse probably accounts for the relatively high EMI threshold, although the effect of the EMP frequency spectrum is not well established.

High-Powered Phased Array Surveillance and Track Radar--Frequency 400-450 MHz; pulse widths (duration) up to 60 μs; pulse repetition rates up to 200 pps; peak power 32 MW; about 5000 transmitting elements; RF beam electronically formed and scanned. The RF emission from this type of emitter, producing a series of rapidly varying pulses at essentially all locations on the ground in the primary scan sector, is much different from a normal search radar. Significant pacemaker interference can be expected out to a distance of about 2500 ft.

Multiple-Frequency Emitters--Environmental sites where more than one source of RF emission produces complex E-field patterns are not uncommon. One example is the air route surveillance radar that usually operate in conjunction with one or more height finder radar. Tests conducted under these conditions are difficult to assess, but generally do not appear to represent a significant threat. More complex situations, where perhaps 2-10 different sources are being propagated in the same direction, generally require an empirical evaluation to assess the potential risk to pacemaker users.

Field Tests Summary--The results of these field test studies are summarized as follows:

(1) Many of the pacemakers tested demonstrated some form of periodic interference resulting in reductions in pacemaker rate by 5-6 bpm. From a clinical viewpoint, this type of interference is generally judged insignificant; i.e., the loss of not more than one beat in any 10-s period (13).

(2) EMI thresholds observed in the field tests agree quite well with the data from the laboratory tests.

(3) With the exception of several of the older more sensitive pacemakers, and perhaps one or two current models, there were few examples of serious pacemaker disruption under the simulated-implant conditions in any of these tests.

(4) The potential hazard to pacemaker users in the vicinity of radar complexes has been significantly reduced in the last 2 years with the phaseout of several popular but sensitive model pacemakers, and more importantly, their replacement with new state-of-the-art RF resistant devices.

(5) The results of these field test studies are directly applicable to 90% of the RF emitters in this country today.

DISCUSSION

The laboratory data summarized in Table 1 illustrate the extent to which cardiac pacemaker interference depends on the emission characteristics of the radiation source. Such findings were further substantiated in essentially all the field test studies at RF emitter sites. Additionally, comparison of free-field to simulated-implant test data readily demonstrates that shielding (RF signal attenuation) of both the pacemaker and lead(s) is a major protective factor, particularly at the higher frequencies. This points out that in many cases of pacemaker EMI, the user could effectively eliminate the potential hazard by rotating or relocating his body to increase the shielding between the pacemaker and the source of interference. This may inadvertently happen in many potential interference situations.

Included in these tests were 23 pacemaker models manufactured by 11 different companies. Based on the laboratory tests at 450 MHz, using a 1-ms pulse duration at 10 pps: (1) 7 of the 23 models tested have EMI thresholds greater than 200 V/m, (2) 10 of 23 have EMI thresholds greater than 100 V/m, (3) 12 of 23 have EMI thresholds greater than 50 V/m, and (4) 8 of the 11 that have EMI thresholds below 50 V/m are older models which are essentially phased out now or will be in the coming months. These test data substantiate the success most manufacturers have had in developing prosthetic devices that are compatible with the electromagnetic environment.

Extrapolation of these field test data to the vast majority of operational RF emitters or emitter complexes in the CONUS reveals few that produce E-field levels (E2) greater than 100 V/m for sufficient time periods to significantly disrupt normal pacemaker function. Thus, if pacemakers were designed and tested to be compatible with the minimum E-field level, viz 200 V/m, associated with the unrestricted 10 mW/cm^2 personnel exposure level, potential EMI situations would be substantially reduced or effectively eliminated. Such actions will eventually achieve overall RF environmental compatibility.

CONCLUSIONS

Cardiac pacemaker EMI is strongly dependent on the frequency, pulse width, real-time E-field level, and effective pulse repetition rate of the incident radiation signal. To a lesser extent but also important is polarization of the field with respect to pacemaker and lead orientation.

Based on the many variables encountered in RF field measurement, implantation simulation, and pacemaker parameters, the reported EMI threshold values can vary by a factor of about 6 dB. Notwithstanding this fact, the values listed in Table 1 are believed most representative of the current state of technology. They can be used as a guide in selecting pacemakers with the better EMI characteristics and/or for assessing potential EMI interactions with a wide range of RF emitters.

Probably the largest variable in assessing the biological impact of pacemaker EMI on man is the heart condition and general state of health of the individual user. A person totally dependent on the pacemaker would obviously suffer greater consequences than one who required only periodic assistance from the pacemaker to regulate heart rate.

Many of the pacemakers included in the field tests, periodically skipped single beats in close proximity to the emitters. However, this is not surprising when one considers this type of pacemaker is designed to sense the relatively low-level heart signal. These pacemakers undoubtedly respond to many other types of pulsed RF emission in much the same manner. Clinically, this type of EMI is judged insignificant.

On the other hand, pacemakers having EMI thresholds of 100 V/m or less may be expected to encounter serious disruption of normal function in close proximity to certain high-power pulsed emitters.

In general, these current test results, when compared with such studies conducted 2-3 years ago, provide remarkable evidence of the overall improvement in the EMI characteristics of currently marketed cardiac pacemakers. These findings also demonstrate the technical feasibility of manufacturing high-quality RF-resistant pacemakers.

Such improvements will likely continue through the mutual efforts of physicians and manufacturers, recognizing the potential EMI threat in an environment where pulsed-RF sources are commonplace and increasing in number. Also, acceptance of a pacemaker EMC test standard would provide reasonable assurance that all new devices would be compatible with the unrestricted RF environment.

BIBLIOGRAPHY

1. D'Cunha, G. F., et al. Syncopal attacks arising from erratic demand pacemaker function in the vicinity of a television transmitter. Am J Cardiol 31:789-791 (1973).

2. Editorial. Microwaves and pacemakers - just how well do they go together? JAMA 221:957-959 (1972).

3. Hurt, W. D. Effects of electromagnetic interference (2450 MHz) on cardiac pacemakers. SAM-TR-73-40, Dec 1973.

4. Hurt, W. D., et al. Measured effects of square-wave modulated RF fields (450 and 3100 MHz) on cardiac pacemakers. SAM-TR-74-51, Dec 1974.

5. King, G. R., et al. Effect of microwave oven on implanted cardiac pacemakers. JAMA 212:1213 (1970).

6. Koerner, D. R. The employee wearing a cardiac pacemaker. J Occup Med 16:392-394 (1974).

7. Mitchell, J. C., et al. EMC design effectiveness in electronic medical prosthetic devices. Proceedings of the Seventh Rochester International Conference on Environmental Toxicity, Fundamental and Applied Aspects of Non-Ionizing Radiation, University of Rochester, N.Y., 5-7 June 1974.

8. Mitchell, J. C., et al. Empirical studies of cardiac pacemaker interference. Aerospace Med 45(2):189-195 (1974).

9. O'Donoghue, J. K. Inhibition of a demand pacemaker by electrosurgery. Chest 64:664-666 (1973).

10. Pickers, B. A., and M. J. Goldberg. Inhibition of a demand pacemaker and interference with monitoring equipment by radiofrequency transmissions. Br Med J 2:504-506 (1969).

11. Ruggera, P. S., and R. L. Elder. Electromagnetic radiation interference with cardiac pacemakers. U.S. Department of Health, Education and Welfare, PHS, BRH/DEP 71-5, Apr 1971.

12. Sanchez, S. A. When you transmit you can turn off a pacemaker. QST for March 1973.

13. Schlentz, R. J. Electromagnetic compatibility of implanted cardiac pacemakers. Paper presented 20-22 May 1975, EMC Symposium Montreux, Switzerland.

14. Smyth, N. P. D., et al. The pacemaker patient and the electromagnetic environment. JAMA 227:1412 (1974).

15. Sowton, E., K. Gray, and T. Preston. Electrical interference in non-competitive pacemakers. Br Heart J 32:626-632 (1970).

16. Steiner, T. O. Development of a pacemaker monitor with cardiac simulator. SAM-TR-75-7, Feb 1975.

17. Van Wijk Van Brievingh, R. P., et al. Measurement techniques for assessing the influence of electromagnetic fields on implanted pacemakers. Med Biol Eng 12(1):42-48 (1974).

18. Walter, W. H., III, et al. Cardiac pulse generators and electromagnetic interference. JAMA 224:1628-1631 (1973).

ABBREVIATIONS AND ACRONYMS

AV	atrioventricular
bpm	beats per minute
cm	centimeter(s)
CW	continuous wave
dB	decibel(s)
E-field	electric field strength (V/m)
EMI	electromagnetic radiation interference
EMR	electromagnetic radiation
Hz	hertz
kHz	kilohertz
LED	light-emitting diode
MHz	megahertz
μs	microsecond(s)
ms	millisecond(s)
mW	milliwatt(s)
MW	megawatt(s)
pps	pulse(s) per second
PRR	pulse repetition rate
PW	pulse width
RF	radiofrequency
rms	root mean square
s	second
V/m	volt(s) per meter

Although the dental environment is a source of light to moderate electromagnetic interference, there is only a small risk that the operation of pacemakers may be affected.

The individual with a pacemaker in the dental environment

Arthur B. Simon, MD
Beverly Linde, MS, RN
Gerald H. Bonnette, DDS, Ann Arbor
Robert J. Schlentz, MS, Minneapolis

A systematic examination of the potential hazards of electromagnetic interference with pacemakers in patients in the dental environment was carried out with the use of several pulse generators of three manufacturers. Only one model was affected by the dental operatory equipment tested. In two of three patients 2-second periods of asystole were noted; this was accompanied by symptoms in one.

Permanent pacemakers implanted in three dogs were unaffected by short or sustained stimulation with pulp vitality testers. A permanent pacemaker implanted in one dog was unaffected by short or sustained stimulation with an ultrasonic cleaner.

Reliance on severe symptoms alone is a poor method of assessing interference in the clinical situation. Guidelines for protection of the patient wearing a pacemaker in the dental environment are given.

Permanent implantable cardiac pacemakers have been used extensively during the past decade for the treatment of a variety of cardiac rhythm disturbances, principally major atrioventricular conduction defects and symptomatic sinus node disease.[1,2] Soon after the development of noncompetitive (demand) pacemakers, it became widely recognized that environmental sources of electromagnetic radiation also would inhibit the normal function of demand pacemakers.[3] The patient can be affected by the fields that radiate from an interfering source or he can become part of a current path by contact with the interference source. Several case reports of interference to commercially available pacing units by electrical appliances,[3] medical diathermy,[3] electrocautery,[4] microwave ovens,[5] radar,[6] and television transmitters[7] are available, although the magnitude of the risk to the patient is controversial.[8] The degree of pacemaker susceptibility varies from manufacturer to manu-

Reprinted with permission from *J. American Dental Assoc.*, vol. 91, no. 6, pp. 1224–1229, Dec. 1975.

facturer and various models from the same manufacturer will react differently.[3,9] In 1973, the ADA Council on Dental Materials and Devices reported on the warning of the Food and Drug Administration that electrocautery (conducted interference) and dental induction casting machines (radiated interference) could interfere with normal pacemaker function.[10]

This report of a systematic examination of the potential danger of the effect of dental devices on pacemaker function is intended to provide a realistic and more complete appraisal of the extent of the problem; an overall view of other potential problems for the patient with a pacemaker in the dental environment is presented also.

Methods

Inasmuch as it would be impossible to establish field intensity levels for all available dental office environments and to measure their effects on all currently available implantable pacemakers, it was elected to limit the scope of the examination to one environment and one patient population.

The study was divided into three sections. The first assessed the potential risk of the dental office environment by systematically examining the electromagnetic field intensity in several dental examining rooms and laboratories of the dental service of the University of Michigan Medical Center. Electromagnetic radiation intensity was measured with an electromagnetic field intensity meter. Electric field intensities were measured for the equipment found in several dental examining rooms, the dental radiology section, a dental utility workroom, and in the dental laboratory and induction casting room of the University of Michigan School of Dentistry (Table 1).

The second section of the study involved exposure of patients to these fields. After they had given their informed consent, 14 patients with properly operating, permanently implanted pacemaker systems from three manufacturers were requested to enter the dental clinic and dental laboratories while dental instruments and the laboratory equipment were operated in a fashion simulating the clinical situation, although no dental treatment was being performed. Patients were selected from the pacemaker clinic of the section of cardiology. Patients who were likely to be dependent on their pacemaker during the examination, as well as patients who were intermittently in sinus rhythm, were selected to provide a cross-section of possible cardiac conditions.

Patients were exposed to the electromagnetic radiation from a motorized dental chair, by sitting in it while it was moved in its four directions. Simulated on-off runs of belt-driven and air-driven handpieces were made at a distance of 1 m. A hand-held dental engine was held 1 to 6 cm from the pacemaker. The patients also were brought near a polisher powered by ¼ hp motor, a suction device, and radiographic equipment including a panoramic rotator. In the dental laboratory, three patients were sequentially ex-

Table 1 ▪ Equipment used in tests of pacemakers in the dental environment.

Equipment	Name or model	Manufacturer
Electromagnetic field intensity meter. (The device measures the electromagnetic environment up to 200 mhz. Thus, both discrete frequency and impulsive signals such as those coming from arcing motors can be accurately measured.)	E-Field Sensor, Model EFS-1/LMT	Instruments for Industry, Inc. Farmingdale, NY
Dental chair	Dynadjust	Ritter Co., Rochester, NY 14603
Belt-driven dental engine	Model C	Ritter Co., Rochester, NY 14603
Turbine handpiece		Air Drive MD Co.
Electrotorque handpiece		Kerr Mfg. Co., Romulus, Mich 48174
Polisher	Ticonium polisher and grinder	Ticonium Co., Albany, NY 12207
Suction apparatus		Nelson-Harvard Co.
Panoramic chair	Panorex model PAN 3s	S. S. White Co., Philadelphia, 19102
Induction casting machine	Ticomatic (270 khz, 6 w)	Ticonium Co., Albany, NY 12207
Power mixer	Vac-U-Vestor & Power-Mixer	Whip-Mix Corp., Louisville, 40217
Model trimmer		Handler Mfg. Co.
Dental laboratory lathe	Model 26	Handler Mfg. Co.
Dental pulp vitality tester. (This unit has a simulation rate of 435 pulses/sec, a stimulation duration of 140 μsec, and an output of up to 165 v through a biphasic waveform across 100,000 ohm resistance.)	Vitapulp	Pelton & Crane Co., Charlotte, NC 28203
Ultrasonic tooth cleaner	Model 1010M	Cavitron Corp., Long Island City, NY 11101

posed to an induction casting machine, a power mixer, a model trimmer, and a dental lathe. Selected patients also had several teeth stimulated with a 110-v, line-powered or battery-operated pulp tester.

All patients were continuously monitored electrocardiographically during each exposure. The examination with a specific instrument was discontinued when an effect on the pulse generator was seen or after 5 to 12 repetitive simulations with no effect were performed. All measurements with the electric field sensor were made 5 to 10 cm from the pulse generator. Throughout the clinical tests, simultaneous measurements of the electromagnetic fields and the pacemaker system performance were made. If no response was noted in the usual dental patient position, the patient was purposely moved closer to the instrument being evaluated (often only a few centimeters away) in order to create a "worse case" situation, and the tests were repeated.

These pacemaker models were tested: Medtronic models* 5842, 5942 (two patients), 5943, 5944, 5945, and 9000; Cordis models† Ectocor 144C and Omni-Stanicor 162C; and General Electric‡ A2072D (now discontinued), A2075A, A2075D (two patients), and A2072AA (now discontinued). Eight patients were paced intravenously. Their pulse generators were located just below the clavicle on the right or left side. Six patients were paced with epicardial leads. Their pulse generators were implanted in the abdominal wall near the belt line.

The third part of the project was an analysis of the potential risk from one line-powered dental pulp vitality tester. Because of the inability to stimulate the teeth of patients repeatedly at high stimulation levels, three dogs with surgically induced atrioventricular block whose cardiac rhythms were maintained with a permanently implanted demand pacemaker were used. All were anesthetized with pentobarbital before the test. The hearts of all dogs were paced from an endocardial lead system of the same manufacturer with the pacemaker pulse generator implanted subcutaneously. All had been paced from one to ten months. These models of pacemakers were evaluated: Medtronic models 5944 and 5945; and GE model A2075A. The dogs' jaws were extended with standard retractors, and several teeth were stimulated with the pulp tester at various voltage outputs. The test was conducted with use of both the head and a more distal location (usually the hind leg) as an anode.

Table 2 ▪ Field intensity measurements.

	Distance		Intensity (v/m)	
Instrument or equipment	(m)	Min	Mean	Max
Induction casting machine	2.0	3.2	3.5	4.3
Panoramic radiography rotator and chair	1.0	<1.0	2.6	6.8
Belt-driven handpiece (sample 1)	1.0	<1.0	1.1	2.3
Electrotorque handpiece	1-10 cm	<1.0	1.3	2.7
Belt-driven handpiece (sample 2)	1.0	1.0	2.5	9.0
Belt-driven handpiece	0.5	4.0	5.0	6.5
Chair	1.0	<1.0	1.8	2.7
Air-turbine handpiece	1.0	<1.0	<1.0	1.0
Suction apparatus	0.2	0	0	0
Laboratory lathe	0.2	0	0	0
Ultrasonic tooth cleaner	1 cm	1.0	1.0	1.0

Thus, the entire pacing system was part of the current path. An additional dog with surgically induced atrioventricular block was exposed to radiation from an ultrasonic tooth cleaner. His heart was paced with a Medtronic model 5942 and an endocardial lead system.

Results

The mean value and range of the field intensity measurements for each instrument evaluated are shown in Table 2. The range of intensity of electromagnetic radiation from each instrument was broad, and results would vary from day to day or on repetitive measurements during the same day. The reasons for this are not known, but variables such as the position of the engine brushes on the motor commutator and the time with respect to the power line cycle when the equipment was turned on probably are the causes.

No dental instrument tested had any influence on these pacemakers tested: Medtronic models 5842, 5942, 5943, 5944, 5945, and 9000; Cordis models Ectocor 144C and Omni-Stanicor 162C; or the GE A2072D and A2072AA.

Two of three GE A2075 pulse generators were repetitively inhibited by the dental chair, panoramic chair, and two belt-driven drills with average interference levels of 1 v/m. An example of pacemaker inhibition caused by these instruments is shown in the illustration panel A. Periods of asystole of 2.1 to 2.3 seconds were repetitively produced with a single on-off cycle. Characteristically, inhibition occurred when the device in question was starting or stopping. This inhibition was prevented by conversion of the unit to asynchronous mode. Only one of the two patients who demonstrated inhibition was symptomatic, and he only mildly so; therefore, reliance on severe symptoms alone is a poor method of assessing interference with pacemaker function

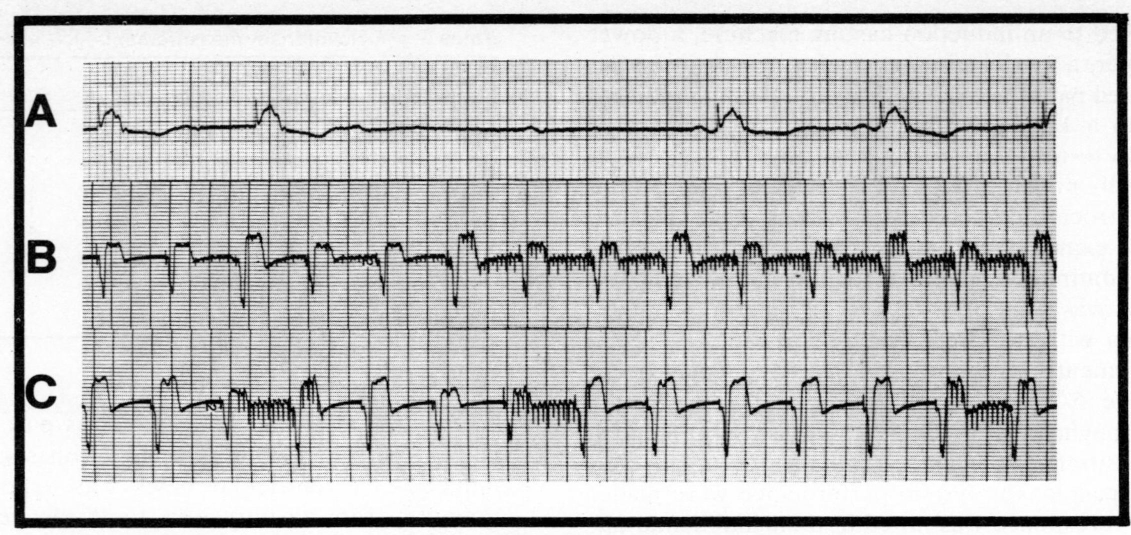

Electrocardiographic rhythm strips. Panel A: 2.3-sec pause caused by operation of belt-driven drill. Each heavy vertical line equals 0.2 sec. Panel B: example of artifacts caused by pulp vitality tester used in continuous fashion. No inhibition noted. Panel C: example of artifacts caused by the pulp vitality tester used in short intermittent bursts. Normal pacemaker function noted.

in these clinical circumstances. No instances of reversion to fixed rate were encountered with any device at any distance.

The field strength also was measured during all tests. While handpieces were running at high speeds or during continuous motor rotation, no electromagnetic radiation levels were detectable and no inhibition or reversion was seen. The induction casting machine, however, emitted continuously elevated levels throughout testing. No effect on any pacemaker was noted during the use of the pulp testing instrument in either the patient or dog studies, at any voltage in either the continuous or intermittent stimulation mode. An example of the electrocardiogram artifact produced by this instrument is presented in the illustration (panels B and C).

Discussion

In many situations, the patient wearing a pacemaker is exposed to electromagnetic radiation that is potentially strong enough to have an influence on his pacemaker. Improvements in pacemaker circuitry and design have lessened the risk, but isolated incidents of pacemaker inhibition continue to occur.[4,6,7]

Currently available pacemakers operate in either the fixed rate (asynchronous) mode, in which the pacemaker paces the heart continuously at a predetermined rate, or the demand (synchronous or noncompetitive) mode, in which the heart is paced only in the absence of spontaneous cardiac rhythm. The latter type of unit "senses" intrinsic cardiac rhythm and is inhibited from pacing the heart if an electrical signal from the heart, indicating depolarization, is detected by the electronic circuitry. The hearts of most patients with implanted pacemakers are paced with the latter type of device.

External electromagnetic energy can interfere with the sensing function of the pacemaker. These signals may mimic ventricular depolarization of the heart and therefore inhibit the pacemaker. If these signals are strong enough, the pacemaker can operate asynchronously (reversion to fixed rate operation). The implanted pacemaker is not damaged by the electromagnetic radiation and will resume normal operation when the patient moves from the electromagnetic field.

In dental equipment in which intermittent start-stop use is frequent, temporary but repetitive inhibition is possible. This may result in long asystolic intervals accompanied by dizziness or seizure (Stokes-Adams attacks) or it may cause other serious cardiac arrhythmias.

The person wearing a pacemaker in the dental environment presents a potential problem as a dental patient, as a dentist or as a dental auxiliary, or as a bystander in the dental office. Because of the large number of dental equipment manufacturers and 40 or more pacemaker producers, it is impossible to test all possible combinations of clinical situations.

This study has shown that the dental environment is a source of moderate electromagnetic interference; that this radiation is sufficient to temporarily inhibit a currently available demand pacemaker of one manufacturer, and that the patient with a pacemaker can be adversely affected by certain dental equipment in normal use. This latter risk appears to be small.

Because of the limited combinations of dental equipment and pacing systems tested, the full extent of the risk remains incompletely explored. However, several basic guidelines and recommendations can be derived from the results of this study, taken together with reports of others.

—The presence of a permanently implanted pacemaker should be detected by the routine medical history questionnaire given to any dental patient. Since all patients with a pacemaker have some underlying heart disease as the genesis of their rhythm disturbance, this is an extension of the cardiac portion of the medical history.

—The type of pacemaker (asynchronous or demand) and the indication for pacing should be recorded for each patient. If possible, the level of pacemaker dependency (what arrhythmias and symptoms result during times of temporary cessation of pacing for the patient) also should be obtained.

—The patient's personal physician or cardiologist should always be contacted for specific guidance. There is a wide variation in the severity of underlying heart disease of each patient with a pacemaker, and since there are a number of possible indications for pacemaker implantation, no recommendations that can apply to all patients can be made to the dentist.

—Isolated cases of endocarditis with permanently implanted pacemakers have been reported; since the pulse generator and electrodes are usually covered by endothelium shortly after implantation, routine endocarditis prophylaxis is not indicated because of the presence of the pacemaker itself. However, some patients with permanent pacemakers have coexistent congenital or acquired valvular disease, and prophylaxis may be required for these patients. This again emphasizes the need for inquiries into each patient's medical history for information as to the origin of the underlying heart disease.

—Because electrosurgery has previously been shown to have adverse effects on normal pacemaker function in operative situations other than dental work, repetition of these experiments here was not justified. However, electrocautery can be performed in the patient wearing a demand pacemaker by using short bursts (1 second or less in duration) spaced 10 or more seconds apart. This procedure allows the heart to maintain sufficient output if interference occurs. Application of a strong magnet over the pulse generator will convert it to asynchronous mode. Monitoring of mutual signs during electrosurgical tool use is essential.

—Since asynchronous pacemakers are influenced only by very intense electromagnetic fields, no special precautions in the dental office (other than those for electrosurgery) are necessary for the patients with these fixed rate pacemakers.

—Selected units of the GE A2075 series of pacemakers have shown inhibition in the presence of some dental equipment. These patients, if dependent on the pacemaker, should have it temporarily converted to the asynchronous mode by the application of a strong magnet over the pulse generator during the dental procedure. The pacemaker then operates as a fixed rate pacemaker.

—If doubt exists concerning the susceptibility of a specific pacemaker system, if the amount of radiation from a specific dental apparatus is unknown, or if the degree of pacemaker dependency is uncertain, and continuous electrocardiographic monitoring during the procedure is not feasible, conversion of the pacing unit to asynchronous mode would seem to be indicated.

—The situation for persons who wear a pacemaker and who work with or in the vicinity of induction casting equipment should be evaluated on an individual basis.

—Within the framework of this study, pulp testers caused no interference with pacemaker function. However, this finding is at variance with another report.[11] Therefore, caution should be exercised with any such device.

—The ultrasonic dental cleaning apparatus caused no interference with the pacemaker function, as tested here, under a limited number of observations.

—Although not assessed in this series, pectoralis major myopotentials capable of inhibiting unipolar pulse generators have been previously reported.[12] A patient undergoing dental treatment may consciously or unconsciously use his arms to perform isometric work during oral manipulation. Such activity may inhibit normal pacemaker function. This problem may be avoided by having the patient relax and consciously avoid muscle tension of the upper extremities or, again, by converting the unit to fixed rate mode.

We thank these persons for their help in performing these experiments: Lupe Cardenas; James D'Innocenzo; Edward Dootz; Kenneth Exworthy; Lynn Hansen; David Himley; Larry Johnson; Amy Klaenhammer; Kristina Larson, DVM; Peter Morawetz, PhD; Gary Robinson; Robert Scott; Chuck Scharte; and Mark Taube.

Dr. Simon is assistant professor of internal medicine at the Heart Station of the University Hospital, University of Michigan Medical Center, Ann Arbor, 48104. Miss Linde is a clinical specialist nurse in cardiology at the University Hospital. Dr. Bonnette is professor and chairman of the hospital dental service of the University of Michigan Medical Center. Mr. Schlentz is manager of the Electromagnetic Compatibility Department, Medtronic, Inc., Minneapolis. Address requests for reprints to Dr. Simon.

*Medtronic, Inc., Minneapolis, 55418.
†Cordis, Inc., Miami, Fla 33137.
‡General Electric Co., Inc., Milwaukee, 53201.

1. Furman, S., and Esher, D.J.W. Principles and techniques of cardiac pacing. New York, Harper & Row, 1970.
2. Chokshi, D.S., and others. Treatment of sinoatrial rhythm disturbances with permanent cardiac pacing. Am J Cardiol 32: 215 Aug 1973.
3. Sowton, E.; Gray, K.; Preston, T. Electrical interference in non-competitive pacemakers. Br Heart J 32:626 Sept 1970.
4. Green, L., and Meredith, J. Transurethral operations employing high frequency electrical currents in patients with demand pacemakers. J Urology 103:446 Sept 1972.
5. Bonney, C.H.; Ruston, P.L.; and Ford, G.E. Evaluation of effects of the microwave oven and radar electromagnetic radiation on noncompetitive cardiac pacemakers. IEEE Trans Biomed Eng 20:357 Sept 1973.
6. Yatteau, R.F. Radar-induced failure of a demand pacemaker. N Engl J Med 283:1447 Dec 24, 1970.
7. D'Cunha, G.F., and others. Syncopal attacks arising from erratic demand pacemaker function in the vicinity of a television transmitter. Am J Cardiol 31:789 June 1973.
8. Smyth, N.P.D., and others. The pacemaker patient and the electromagnetic environment. JAMA 227:1412 March 25, 1974.
9. Smyth, N.P., and others. Effect of an active magnetometer on permanently implanted pacemakers. JAMA 221:162 July 10, 1972.
10. Council on Dental Materials and Devices. Possible electromagnetic interference with cardiac pacemakers from dental induction casting machines and electrosurgical devices. JADA 86:426 Feb 1973.
11. Woolley, L.H.; Woodworth, J.; and Dobbs, J.L. A preliminary evaluation of the effects of electrical pulp testers on dogs with artificial pacemakers. JADA 89:1099 Nov 1974.
12. Ohm, O.I., and others. Interference effect of myopotentials on function of unipolar demand pacemakers. Br Heart J 36:77 Jan 1974.

Author Index

Subject Index

Editor's Biography

John M. Osepchuk is Consulting Scientist in the Research Division of Raytheon Company, Lexington, Massachusetts. He received B.A. and M.A. degrees in Engineering Science and Applied Physics and a Ph.D. in Applied Physics, all from Harvard University.

Upon joining Raytheon in 1950, Dr. Osepchuk conducted research on ridge-waveguide and magnetrons, and helped develop the first high-power backward-wave oscillator in the United States. During 1956 and 1957, he was technical liaison for Raytheon at the microwave tube research laboratories of Compagnie Generale de Telegraphie san Fils in Paris, France, From 1957 to 1962, Dr. Osepchuk was head of several research projects on crossed-field devices.

From 1962 to 1964, Dr. Osepchuk was chief microwave engineer for Sage Laboratories in Natick, MA. In 1964, he rejoined the Raytheon Research Division in Waltham, MA. He has directed various projects in the fields of microwaves, image tubes, and physical electronics. In recent years, he has consulted for Amana and other Raytheon Divisions on radiation hazards, and investigated various aspects of Radarange technology, especially those involving leakage and safety. He was appointed consulting scientist in December, 1974.

He has published and presented many papers in the fields of microwaves (tubes, ferrites, plasmas, pacemaker interference) and radiation hazards. He was guest editor for the special issue (February 1971) on Biological Effects of Microwaves of the *IEEE Transactions on Microwave Theory and Techniques.* He was editor of the *Journal of Microwave Power* (1970–1971). He has contributed to one book and holds over ten patents.

Dr. Osepchuk was national lecturer (1977–1978) of the IEEE MTT Society on "Microwave Radiation Hazards in Perspective." In addition, he was general chairman of the 1978 Symposium on Electromagnetic Fields in Biological Systems, which was cosponsored by IEEE-MTT-S and IMPI. He was on the Program Committee and a session chairman for a symposium on "Health Aspects of Nonionizing Radiation," held April 9–10, 1979 under the sponsorship of the New York Academy of Medicine.

Dr. Osepchuk is a Fellow of the IEEE and the International Microwave Power Institute, and a member of Phi Beta Kappa, Sigma Xi, and the Bioelectromagnetics Society. He is a past chairman of the Boston Section of the IRE Professional Group on Electron Devices, and a past member of the National Administrative Committee of the IEEE Group on Microwave Theory and Techniques and the Board of Governors of the International Microwave Power Institute. He is also a member of various committees of ANSI C95, the Association of Home Appliance Manufacturers (AHAM), and is presently a member of the IEEE Committee on Man and Radiation, and the National Administrative Committee of the IEEE Society on Social Implications of Technology.